TABLE OF CONTENTS

PART 1

TABLE OF CONTENTS

PART 2

Proceedings of Symposia in Pure Mathematics
Volume **40** (1983), Part 2

WEAK SIMULTANEOUS RESOLUTION FOR DEFORMATIONS OF GORENSTEIN SURFACE SINGULARITIES

HENRY B. LAUFER[1]

I. Introduction. Let $\lambda\colon \mathcal{V} \to T$ be a (flat) deformation of the two-dimensional isolated hypersurface singularity (V, p). We assume that T is reduced. This paper contains a partial answer to the two well-known questions [**49**, p. 115]:

(a) Does $\mu_t := \mu^{(3)}(V_t)$ constant imply weak simultaneous resolution?

(b) Does $\mu^*(V_t)$ constant imply strong, or at least weak, simultaneous resolution?

The converses to (a) and (b) are known to be true [**31**, **49**, **7**]. In this paper we shall prove

(a) (Theorem 6.4) If each V_t has a singularity p_t such that (V_t, p_t) is homeomorphic to (V, p), then $\lambda\colon \mathcal{V} \to T$ has a weak simultaneous resolution. It is known [**35**], except for surface singularities, that μ_t constant implies that (V_t, p_t) has constant topological type.

(b) If $\mu^*(V_t)$ is constant, then $\lambda\colon \mathcal{V} \to T$ has a weak simultaneous resolution. This follows immediately from Theorem 6.4, the fact that constant $\mu^*(V_t)$ implies the Whitney conditions [**48**] and the Thom-Mather Theorem (see also [**50**]). We shall discuss strong simultaneous resolution in another paper.

Let us start by only requiring that (V, p) be a purely two-dimensional singularity. Let $\pi\colon M \to V$ be a resolution of V. Let K be the canonical divisor on M. Let $\mathcal{S}_m = \pi_*\mathcal{O}(mK)$, a coherent sheaf of modules on X. Then [**41**, **19**] we may blow-up V at $\mathcal{S}_m$, $\phi\colon X \to V$. Then (Theorem 3.3), for $m \geqslant 3$, $\phi\colon X \to V$ is the RDP resolution of V, i.e., X is obtained from the minimal resolution of V by blowing down the rational -2 curves. In case (V, p) is Gorenstein, $\mathcal{S}_m$ is a sheaf of ideals

1980 *Mathematics Subject Classification*. Primary 32G11; Secondary 14B07.
[1] Research partially supported by NSF Grant MCS-8102621.

0082-0717/81/0000-0644/$08.50

and this result was shown by Shepherd-Barron [46, Corollary 6, p. 7] with $m \geqslant 4$. The canonical map $\mathcal{S}_m \otimes \mathcal{S}_n \to \mathcal{S}_{m+n}$ is surjective for $m \geqslant 2$ and $n \geqslant 3$ (Theorems 3.2 and 3.5). This includes the $m = n = 3$ case, as conjectured by Reid (see [46, Corollary 5, p. 5]).

Now suppose that (V, p) is normal and Gorenstein. Let $\pi: M \to V$ be the minimal resolution of V. The Gorenstein condition implies that we may take K to be supported on A, the exceptional set in M. Then $K \cdot K$ is defined, and depends on choosing the minimal resolution. Then $\lambda: \mathcal{V} \to T$ has a very weak simultaneous resolution (Definition 4.1) after a finite base change if and only if $K_t \cdot K_t$ is constant (Theorem 5.7). For multiple singularities, sum the individual $K_t \cdot K_t$. Some important aspects of the proof of Theorem 5.7 were done by Shepherd-Barron [46] under the hypothesis that K has no base points. Theorem 5.7 is proved as follows. We first take T to be 1-dimensional and λ to be algebraic. Then the $\mathcal{S}_{m,t}$ fit together to form a coherent sheaf of ideals $\mathcal{I}_m$ on $\mathcal{V}$. The $K \cdot K$ condition implies that $\mathcal{I}_m$ is normally flat. A construction similar to blowing up $\mathcal{V}$ at $\mathcal{I}_m$ gives the simultaneous RDP resolution. This resolves to a very weak simultaneous resolution after a finite base change [8]. The result for arbitrary reduced T then follows from general considerations [10, 40, 28].

$-K_t \cdot K_t$ is upper semicontinuous (Theorem 5.2). Let $h_t = \dim H^1(M_t, \mathcal{O})$. Then $K_t \cdot K_t$ constant implies that h_t is constant (Theorem 5.3).

If V_t has a singularity p_t such that (V_t, p_t) is homeomorphic to (V, p), then λ has a weak simultaneous resolution (Theorem 6.4). No base change is required via Wahl's result [52]. The proof of Theorem 6.4 makes essential use of Neumann's theorem [38].

The author would like to thank Professors Hironaka, Lipman, Schlessinger, Teissier and Wahl for useful and stimulating conversations.

II. Known preliminaries. Throughout this paper, dim denotes dimension as a complex vector space.

Recall [24] how the Riemann-Roch theorem may be extended to the one-dimensional singular case. Let C be a compact one-dimensional (reduced) analytic space all of whose singularities are plane curve singularities, i.e. C can be locally embedded in two-dimensional manifolds. Let $\pi: X \to C$ be a resolution, i.e. a normalization, of C. Let $\mathcal{O}_C$ be the structure sheaf on C. Let $\mathcal{O}_X$ be the structure sheaf on X. $\pi_*(\mathcal{O}_X)$ is then the sheaf of germs of weakly holomorphic functions on C. Let $\mathcal{Q} := \pi_*(\mathcal{O}_X)/\mathcal{O}_C$. Then $\mathcal{Q}$ is supported precisely at the singular points of C. Let $\delta := \sum \dim \mathcal{Q}_x$, where the sum is over all singular points x of C. Let $\mathfrak{c}$ be the conductor [24, p. 116]. Then there is a natural inclusion $\pi_*(\mathcal{O}_X(-\mathfrak{c})) \subset \mathcal{O}_C$. Let $\mathcal{R} := \mathcal{O}_C/\pi_*(\mathcal{O}_X(-\mathfrak{c}))$. Then [24, Theorem 1.1, p. 115] also $\delta = \sum \dim \mathcal{R}_x$, where the sum is again over all singular points of C. Observe that the degree of the divisor $\mathfrak{c}$ is 2δ.

Let L be a line bundle over C. Let $\mathcal{O}(L)$ be the sheaf of germs of sections of L. Let L_X be the pull-back of L of X. Let $\mathcal{O}(L_X)$ be the sheaf of germs of sections of

L_X. Since π^{-1} maps Stein covers to Stein covers, there is an exact sequence

$$(2.1)\qquad \begin{aligned} 0 &\to \Gamma(X, \mathcal{O}(L_X - \mathfrak{c})) \to \Gamma(C, \mathcal{O}(L)) \to \Gamma(C, \mathfrak{R}) \\ &\to H^1(X, \mathcal{O}(L_X - \mathfrak{c})) \to H^1(C, \mathcal{O}(L)) \to 0. \end{aligned}$$

(2.1) corresponds to [**24**, (1.11), p. 116].

Now suppose that C is an irreducible curve. Let g' be the arithmetic genus of C, i.e., $g' = \dim H^1(C, \mathcal{O}_C)$. Let g be the geometric genus of C, i.e. the genus of X. Then $g' = g + \delta$. Let $c(L)$ be the Chern class of L. Then $c(L_X) = c(L)$. From (2.1), it follows that for $c(L) > 2g' - 2$,

$$(2.2)\qquad \dim \Gamma(C, \mathcal{O}(L)) = c(L) - 1 - g' \quad \text{and} \quad H'(C, \mathcal{O}(L)) = 0.$$

Recall [**45**, Proposition IV, p. 75] that if C is an irreducible subvariety of the two-manifold M and if K is the canonical bundle on M,

$$(2.3)\qquad g' = 1 + \tfrac{1}{2}(C \cdot C + C \cdot K).$$

We now wish to re-examine the proof of [**27**, Theorem 3.2, p. 603; **23**, p. 246] for the nongood resolution case. Let M be a strictly pseudoconvex two-manifold with connected exceptional set A. Let $A = \bigcup A_i$ be the decomposition of A into irreducible components. Let L be a line bundle over M with $\mathcal{O}(L)$ its sheaf of germs of sections. Recall that $L \cdot A_i$ is just the Chern class of L restricted to A_i. Suppose that $L \cdot A_i \geqslant K \cdot A_i$ for all i. Then [**23**, p. 246], $H^1(M, \mathcal{O}(L)) = 0$.

In proving this, we may proceed as follows. Let $Z_0 = 0$, $Z_1 = A_{i_1}, \ldots, Z_k = A_{i_k} + Z_{k-1}, \ldots, Z_l = Z = A_{i_l} + Z_{l-1}$ be a computation sequence for Z, the fundamental cycle [**27**, Proposition 4.1, p. 607; **29**, p. 1259]. $A_i \cdot Z \leqslant 0$, all i; $A_{i_k} \cdot Z_{k-1} > 0$ for $1 < k \leqslant l$; $A_{i_1} \cdot Z_0 = 0$, of course.

Now consider the computation sequence in reverse order: Let $A_{j_1} = A_{i_l}, \ldots, A_{j_k} = A_{i_r}$ with $r = l - k + 1, \ldots, A_{j_l} = A_{i_1}$, $1 \leqslant k \leqslant l$. Let $Y_0 = 0 = Z - Z_l$, $Y_1 = A_{j_1} = Z - Z_{l-1}, \ldots, Y_k = Y_{k-1} + A_{j_k} = Z - Z_{l-k}, \ldots, Y_l = Z - Z_0 = Z$. Look at $(L - Y_{k-1}) \cdot A_{j_k}$. Let g_i' denote the arithmetic genus of A_i. Let $r = l - k + 1$. Recall [**45**, Proposition IV. 5, p. 75] that $A_i \cdot K = 2g_i' - 2 - A_i \cdot A_i$. Then

$$(2.4)\qquad \begin{aligned} (L - Y_{k-1}) \cdot A_{j_k} &= (L - Z + Z_r) \cdot A_{i_r} = \big(L - Z + Z_{r-1} + A_{i_r}\big) \cdot A_{i_r} \\ &\geqslant 2g_i' - 2 - A_{i_r} \cdot Z + A_{i_r} \cdot Z_{r-1}. \end{aligned}$$

For $r > 1$, which is equivalent to $k < l$, we have $A_{i_r} \cdot Z_{r-1} > 0$. So $(L - Y_{k-1}) \cdot A_{j_k} > 2g_{i_k}' - 2$. For $r = 1$, or $k = l$, we choose A_{i_1} so that $A_{i_1} \cdot Z < 0$. Such an A_{i_1} exists since $Z \cdot Z < 0$. Then also $(L - Y_{l-1}) \cdot A_{j_l} > 2g_{j_l}' - 2$. Recall that for a divisor $D \geqslant 0$, $\mathcal{O}(L - D)$ denotes those germs of sections of L which vanish to the order given by D. By (2.2), for all $1 \leqslant k \leqslant l$, the line bundle $L - Y_{k-1}$ has trivial cohomology on A_{j_k}. Upon reaching $L - Z = L - Y_l$, we may replace L by $L - Z$ and repeat the argument. So, as in the proof of [**27**, Theorem 3.2, p. 603], which is essentially [**14**, §4 Satz 1, p. 355] we have proved

PROPOSITION 2.1. *Let M be a strictly pseudoconvex two-manifold with connected exceptional set $A = \bigcup A_i$. Let L be a line bundle on M such that $L \cdot A_i \geqslant K \cdot A_i$ for* *all i. Let* $0 = Y_0, \ldots, Y_k, \ldots, Y_l = Z$ *be defined as above, with* A_{i_1}, *the first element of the computation sequence, satisfying* $A_{i_1} \cdot Z < 0$. *Then for all* $n \geq 0$ *and all* $1 \leq k \leq l$, $H^1(M, \mathcal{O}(L - nZ - Y_{k-1})) = 0$. *Also, all sections of* $L - nZ - Y_{k-1}$ *restricted to* A_{j_k} *lift to sections of L on M.*

III. Tricanonical resolution. In this section, we let V be a purely two-dimensional (reduced) space. V may have nonisolated singularities. Let $\xi: M' \to V$ be a resolution of V [**20**]. Let $\mathcal{O}(K)$ be the canonical sheaf on M', i.e. the sheaf of germs of holomorphic 2-forms (M' is of dimension two). Let $\mathcal{O}(mK)$ denote $\mathcal{O}(K) \otimes \cdots \otimes \mathcal{O}(K)$ with m factors; $\otimes$ is taken over the structure sheaf of M'. So $\mathcal{O}(mK)$ is the sheaf of m-pluricanonical forms.

Let L be a line bundle on a space M. Recall that $x \in M$ is a *base point* for L if all elements of $\Gamma(M, \mathcal{O}(L))$ vanish at x.

The next theorem is due to Shepherd-Barron [**46**, Theorem 2, p. 2] in the important case that $L = 2K$.

THEOREM 3.1. *Let V be a Stein two-dimensional space. Let* $\pi: M \to V$ *be the minimal resolution of V. Suppose that L is a line bundle over M with* $L \cdot A_i \geq 2K \cdot A_i$ *for all irreducible exceptional curves* A_i *on M. Then L has no base points on M.*

PROOF. The normalization $\tilde{V}$ of V is also Stein; use [**13**] to see this. $\tilde{V}$ has only isolated singularities. M is also the minimal resolution of $\tilde{V}$, $\tilde{\pi}: M \to \tilde{V}$. $\tilde{\pi}_*(\mathcal{O}(L))$ is a coherent sheaf on $\tilde{V}$ [**13**] and is isomorphic to $\mathcal{O}(L)$ on the regular points of $\tilde{V}$. So by Cartan's Theorem A, we need only prove the theorem under the restrictions that M is strictly pseudoconvex with a connected exceptional set A. Moreover, we need only show that L has no base points on A.

Let $A = \bigcup A_i$ be the decomposition of A into irreducible components. Recall that on a minimal resolution, $K \cdot A_i \geq 0$ for all i. Moreover, $K \cdot A_i = 0$ if and only if $g_i' = 0$ and $A_i \cdot A_i = -2$.

Recall the Y_k used in Proposition 2.1. We shall prove, by induction on k, that L has no base points on supp(Y_k). We first consider the case $k = 1$. $L \cdot A_{j_1} \geq 2K \cdot A_{j_1} \geq 2g_{j_1}'$. Let L_1 be the restriction to A_{j_1} of the line bundle L. By (2.2), L_1 has no base point at any regular point of A_{j_1}. We may also see, as follows, using only $c(L_1) \geq 2g_{j_1}' - 1$, that L_1 has no base point at any singular point of A_{j_1}. We use (2.1). Let L_X denote the pull-back of L_1 to X, a normalization of $A_{j_1} = C$. Let $g = g_{j_1}$.

$$c(L_x - \mathfrak{c}) = c(L_1) - 2\delta \geq 2g - 1.$$

So $H^1(X, \mathcal{O}(L_x - \mathfrak{c})) = 0$. Then $\Gamma(C, \mathcal{O}(L_1)) \to \Gamma(C, \mathfrak{R})$ is onto. At a singular point x of C, $\mathfrak{R}_x \neq 0$ and so $\Gamma(C, \mathcal{O}(L_1))$ has an element which is nonzero at x. By Proposition 2.1, elements of $\Gamma(A_{j_1}, \mathcal{O}(L_1))$ lift to $\Gamma(M, \mathcal{O}(L))$.

Now suppose that the induction step is true for $k - 1$, i.e. L has no base point on supp(Y_{k-1}). The computation of (2.4) shows that for $K \cdot A_{j_k} \geq 1$, $(L - Y_{k-1}) \cdot A_{j_k} \geq 2g_{j_k}'$. So, as in the previous paragraph, L has no base points on $A_{j_k} - \text{supp}(Y_{k-1})$. This completes the induction step for $K \cdot A_{j_k} \geq 1$.

Now consider the remaining case, $K \cdot A_{j_k} = 0$. Then $(L - Y_{k-1}) \cdot A_{j_k} \geq -1$. If $(L - Y_{k-1}) \cdot A_{j_k} \geq 0$, then we are done, as above. For the remaining subcase, $(L - Y_{k-1}) \cdot A_{j_k} = -1$, we have that $L \cdot A_{j_k} = K \cdot A_{j_k} = 0$. Then L restricted to A_{j_k}, denoted by L_k, is the trivial bundle. By the induction hypothesis, since $k \geq 2$, L has a section which is nonzero at a point in $\operatorname{supp}(Y_{k-1}) \cap A_{j_k}$. Since L_k is the trivial bundle on A_{j_k}, this section is nonzero at all points of A_{j_k}. This completes the proof of Theorem 3.1.

Theorem 3.2 below is somewhat stronger than [**46**, Corollary 5(i), p. 5].

Retain the notation of the proof of Theorem 3.1. Let $\mathfrak{F}$ be a sheaf on M. Recall that $\Gamma(A, \mathfrak{F}) := \operatorname{dir\,lim} \Gamma(U, \mathfrak{F})$, where U varies over all open neighborhoods of A.

If L_1 and L_2 are line bundles, we let $L_1 + L_2$ denote $L_1 \otimes L_2$.

THEOREM 3.2. *Let M be a strictly pseudoconvex two-manifold with connected exceptional set A. Suppose that M is the minimal resolution of its blow-down, i.e. $A_i \cdot K > 0$ for all irreducible components of A. Let L_1 and L_2 be line bundles on M such that $L_1 \cdot A_i \geq 2K \cdot A_i$ and $L_2 \cdot A_i \geq 3K \cdot A_i$ for all i. Then the canonical map*

$$\Gamma(A, \mathcal{O}(L_1)) \otimes_{\mathbf{C}} \Gamma(A, \mathcal{O}(L_2)) \to \Gamma(A, \mathcal{O}(L_1 + L_2))$$

is onto.

PROOF. We first do a formal computation, using essentially infinitesimal neighborhoods of A. We then show that the formal computation suffices.

Recall the Y_k of (2.4). Let $C_s = nZ + Y_k$ for $n \geq 0$, $0 \leq k < l$; $s = nl + k$. We shall show that, for all s, the map

$$\tau\colon \Gamma(A, \mathcal{O}(L_1)) \otimes_{\mathbf{C}} \Gamma(A, \mathcal{O}(L_2)) \to \Gamma(A, \mathcal{O}(L_1 + L_2))/\Gamma(A, \mathcal{O}(L_1 + L_2 - C_s))$$

is onto. By taking successive quotients

$$\begin{gathered} \Gamma(A, \mathcal{O}(L_1 + L_2))/\Gamma(A, \mathcal{O}(L_1 + L_2 - C_1)) \\ \Gamma(A, \mathcal{O}(L_1 + L_2 - C_1))/\Gamma(A, \mathcal{O}(L_1 + L_2 - C_2)) \\ \vdots \\ \Gamma(A, \mathcal{O}(L_1 + L_2 - C_{s-1}))/\Gamma(A, \mathcal{O}(L_1 + L_2 - C_s)), \end{gathered}$$

it suffices to show that

$$\text{(3.1)} \qquad \operatorname{im} \tau \supset \Gamma(A, \mathcal{O}(L_1 + L_2 - C_{s-1}))/\Gamma(A, \mathcal{O}(L_1 + L_2 - C_s)).$$

By Proposition 2.1, $\Gamma(A, \mathcal{O}(L_1 + L_2 - C_{s-1}))/\Gamma(A, \mathcal{O}(L_1 + L_2 - C_s))$ may be identified with $\Gamma(A, \mathcal{O}(L_1 + L_2 - C_{s-1})/\mathcal{O}(L_1 + L_2 - C_s))$.

Let A_t be such that $C_s = C_{s-1} + A_t$. By Theorem 3.1, there exists $f \in \Gamma(A, \mathcal{O}(L_1))$ such that f is nonzero at all singular points of A. Multiplication by f gives an injective map

$$\text{(3.2)} \qquad \begin{aligned} \tau_f\colon \Gamma(A_t, \mathcal{O}(L_2 - C_{s-1})/\mathcal{O}(L_2 - C_s)) \\ \to \Gamma(A_t, \mathcal{O}(L_1 + L_2 - C_{s-1})/\mathcal{O}(L_1 + L_2 - C_s)). \end{aligned}$$

By (2.2) and (2.4), $\operatorname{im}\tau_f$ has dimension $(L_2 - C_{s-1})\cdot A_t + 1 - g_t'$ and codimension $L_1 \cdot A_t$. So $\operatorname{im}\tau_f$ is precisely the subspace of

$$\Gamma(A_t, \mathcal{O}(L_1 + L_2 - C_{s-1})/\mathcal{O}(L_1 + L_2 - C_s))$$

of elements which vanish on A_t to the order given by (f), the divisor of f. There are thus $L_1 \cdot A_t$ vanishing conditions on $\operatorname{im}\tau_f$ as a subspace of

$$\Gamma(A_t, \mathcal{O}(L_1 + L_2 - C_{s-1})/\mathcal{O}(L_1 + L_2 - C_s)).$$

By Proposition 2.1, $\operatorname{im}\tau \supset \operatorname{im}\tau_f$.

Now consider the first C_i such that $A_t \subset \operatorname{supp}(C_i)$, i.e. $A_t \not\subset \operatorname{supp}(C_{i-1})$ and $C_i = C_{i-1} + A_t$. Let L be a line bundle on M. Then, on A_t, sections of $L - C_{i-1}$ may be identified with those sections of L on A_t which vanish at $\operatorname{supp}(C_{i-1}) \cap A_t$ in prescribed ways. By construction, $\operatorname{supp}(C_{i-1}) \cap A_t \cap \operatorname{supp}(f) = \varnothing$.

Consider

(3.3)

$$\begin{aligned}&\Gamma(A_t, \mathcal{O}(L_1 - C_{i-1})/\mathcal{O}(L_1 - C_i)) \otimes_{\mathbf{C}} \Gamma(A_t, \mathcal{O}(L_2 - C_{s-1})/\mathcal{O}(L_2 - C_s))\\ &\quad \overset{\alpha}{\to} \Gamma(A_t, \mathcal{O}(L_1 + L_2 - C_{i-1} - C_{s-1})/\mathcal{O}(L_1 + L_2 - C_{i-1} - C_{s-1} - A_t))\\ &\quad \overset{\iota}{\to} \Gamma(A_t, \mathcal{O}(L_1 + L_2 - C_{s-1})/\mathcal{O}(L_1 + L_2 - C_s)),\end{aligned}$$

whjere α multiplies sections and ι is the inclusion map described just above. To prove (3.1), it suffices that $\operatorname{im}(\iota \circ \alpha)$ contain elements with any given prescribed zeros strictly less than (f). All of the prescribed zeros are in $\operatorname{supp}(f)$. The total number of prescribed zeros, counting multiplicities, is at most $L_1 \cdot A_t - 1$. This is essentially a vanishing condition on $\operatorname{im}\alpha$.

By (2.2), in order for a section to have prescribed zeros of total order r in a line bundle L, it suffices that $c(L) \geqslant 2g' + r$. The condition $c(L) \geqslant 2g' + r$ also allows the section to be nonzero at any finite number of prescribed points. So multiplying prescribed sections will add prescribed zeros. In α of (3.3), we may allocate, as follows, the $L_1 \cdot A_t - 1$ vanishing conditions between

$$\Gamma(A_t, \mathcal{O}(L_1 - C_{i-1})/\mathcal{O}(L_1 - C_i)) \quad \text{and} \quad \Gamma(A_t, \mathcal{O}(L_2 - C_{s-1})/\mathcal{O}(L_2 - C_s)).$$

In case $L_1 \cdot A_t > 0$ and $L_2 \cdot A_t > 0$, by (2.4), $(L_1 - C_{i-1})\cdot A_t \geqslant 2g_t'$ and $(L_2 - C_{s-1})\cdot A_t \geqslant 2g_t'$. So it suffices for the allocation that

$$(3.4) \qquad (L_1 - C_{i-1})\cdot A_t + (L_2 - C_{s-1})\cdot A_t \geqslant 4g_t' + L_1 \cdot A_t - 1.$$

Since $L_2 \cdot A_t \geqslant 3K \cdot A_t$ with strict inequality in the case $K \cdot A_t = 0$, $L_2 \cdot A_t \geqslant 2K \cdot A_t + 1$. Then (3.4) follows from (2.4).

The remaining case is that $L_1 \cdot A_t = 0$ or $L_2 \cdot A_t = 0$. For $L_1 \cdot A_t = 0$, f is nonzero on A_t and (f) gives no vanishing conditions. The map τ_f of (3.2) is already onto. For $L_2 \cdot A_t = 0$, reverse the roles of L_1 and L_2. Again (f) gives no vanishing conditions and τ_f of the new (3.2) is onto. So (3.1) is true for all C_s and so τ is surjective.

We now continue as in [**26**, Theorem 7.5, p. 137]. Let $f_1, f_2 \in \Gamma(A, \mathcal{O}(L_1))$ have no common zeros. Then the sheaf map $\mathcal{O}(L_2) \oplus \mathcal{O}(L_2) \to \mathcal{O}(L_1 + L_2)$ given by $(g_1, g_2) \to (f_1 \otimes g_1 + f_2 \otimes g_2)$ is surjective near A. Let $\mathcal{K}$ be ker γ, so that

$$0 \to \mathcal{K} \to \mathcal{O}(L_2) \oplus \mathcal{O}(L_2) \to \mathcal{O}(L_1 + L_2) \to 0$$

is exact. For all $n \geqslant 0$, multiplying by the ideal sheaf $\mathcal{O}(-nZ)$ gives the exact sequence

$$0 \to \mathcal{K} \cdot \mathcal{O}(-nZ) \to \mathcal{O}(L_2 - nZ) \oplus \mathcal{O}(L_2 - nZ) \to \mathcal{O}(L_1 + L_2 - nZ) \to 0.$$

$$\begin{array}{ccccccc}
\Gamma(A, \mathcal{O}(L_2 - nZ)) & \oplus & \Gamma(A, \mathcal{O}(L_2 - nZ)) & \to & \Gamma(A, \mathcal{O}(L_1 + L_2 - nZ)) & \to & H^1(A, \mathcal{K} \cdot \mathcal{O}(-nZ)) \\
 & \downarrow & & & \downarrow & & \downarrow \sigma_* \\
\Gamma(A, \mathcal{O}(L_2)) & \oplus & \Gamma(A, \mathcal{O}(L_2)) & \xrightarrow{\gamma_*} & \Gamma(A, \mathcal{O}(L_1 + L_2)) & \to & H^1(A, \mathcal{K})
\end{array}$$

is commutative with exact rows. By [**14**, §4 Satz 1, p. 355], σ_* is the zero map for n sufficiently large. Choose such an n. Then any $h \in \Gamma(A, \mathcal{O}(L_1 + L_2 - nZ))$ is in the image of γ_*. This concludes the proof of Theorem 3.2.

Return to V, our given purely two-dimensional analytic space. Let $\xi\colon M' \to V$ be a resolution of V [**20**]. Let $\mathcal{S}_m = \xi_*(\mathcal{O}(mK))$. Since all resolutions of V may be obtained from each other via quadratic transformations, $\mathcal{S}_m$ is, in fact, independent of the choice of the resolution ξ. So when convenient, we may take $\xi\colon M' \to V$ to be $\pi\colon M \to V$, the minimal resolution.

Recall that M' is also a resolution of the normalization of V. Observe also that in case V is normal and has only Gorenstein singularities, $\mathcal{S}_m$ is a sheaf of ideals.

Let $\mathcal{S}$ be a coherent analytic sheaf on V. $\mathcal{S}$ is necessarily locally free off a nowhere dense subvariety S of V. Suppose that $\mathcal{S}$ has constant rank on $V - S$. Then [**41**, pp. 268–272] there is a unique proper modification mapping $\phi\colon X \to V$ which is the monoidal transformation of V with respect to $\mathcal{S}$. In [**41**], it is assumed that V is irreducible. But irreducibility is used only to conclude that $\mathcal{S}$ has constant rank on $V - S$. In case $\mathcal{S}$ should be a sheaf of ideals, ϕ coincides with the usual monoidal transformation at an ideal [**41**, p. 271]. We are especially interested in the case that $\mathcal{S} = \mathcal{S}_m$. We may choose S to be the singular locus of V. Then $\mathcal{S}_m$ has rank 1. ϕ_m is then given locally as follows. Let $y \in V$. Let $\lambda_1, \ldots, \lambda_p$ be generators of $\mathcal{S}_m$ near y. Near y, let $\mu_m\colon V - S \to V \times \mathbf{P}^{p-1}$ be given by $\mu_m(z) = (z, (\lambda_1(z)\colon \cdots \colon \lambda_p(z)))$. Then μ_m is an embedding. X_m is the closure in $V \times \mathbf{P}^{p-1}$ of im μ_m. $\phi_m\colon X_m \to V$ is induced by projection onto V in $V \times \mathbf{P}^{p-1}$. Then $\phi_m\colon X_m \to V$ is (locally) the monoidal transformation of V with respect to $\mathcal{S}$.

Recall the RDP (rational double point) resolution of V [**36**], as follows: Let $\pi\colon M \to V$ be the minimal resolution of V. Blow down in M all nonsingular irreducible curves A_i such that $g_i = 0$, $A_i \cdot A_i = -2$ and $\pi(A_i)$ is just a point. Call the blown-down space N. All of the singularities of N are rational double points. π induces a proper modification mapping $\psi\colon N \to V$, called the RDP resolution of V.

Theorem 3.3 below is similar to [**46**, Corollary 6, p. 7]. See also [**36**, Theorem, p. 38].

THEOREM 3.3. *Let V and $\mathcal{S}_m$ be defined as above. Let $\phi: X \to V$ be the monoidal transformation with respect to $\mathcal{S}_m$. Let $\tilde{X}$ denote the normalization of X. Let $\alpha: \tilde{X} \to V$ be the induced map. For $m \geq 2$, $\alpha: \tilde{X} \to V$ is the RDP resolution of V. For $m \geq 3$, X is already normal and $\phi: X \to V$ is the RDP resolution of V.*

Using the same example as in [**46**, p. 8] we can see that the conclusions of Theorem 3.3 do not always hold for $m = 1$. Also, for $m = 2$, X need not be normal. Thus, consider the singularity $p = (0,0,0)$ in $V = \{(x, y, z) \mid z^2 = x^3 + y^6\}$. p is a simple elliptic singularity [**44**]. Let $\pi: M \to V$ be the minimal resolution. Let $A = \pi^{-1}(p)$ be the exceptional set. Then A is an elliptic curve with $A \cdot A = -1$. $\mathcal{S}_1$ is isomorphic to the sheaf of ideals generated by (x, y, z). Blowing up at $\mathcal{S}_1$ and then normalizing yields [**31**, Lemma 2.2, p. 317] a space with a simple elliptic singularity equivalent to $(0,0,0)$ in $\{(x, y, z) \mid z^2 = y(x^3 + y^3)\}$. $\mathcal{S}_2$ is isomorphic to the sheaf of ideals generated by (x, y^2, z). Blowing up at $\mathcal{S}_2$ yields a nonnormal space.

Theorem 3.3 is really a local theorem, as may be seen as follows. Let $\omega: \tilde{V} \to V$ be the normalization of V, a local construction in all dimensions. Since $\tilde{V}$ is of dimension two, it has only isolated singularities. Hence resolving $\tilde{V}$ (and thereby resolving V) is a local construction. Hence $\mathcal{S}_m$ is defined locally, and so $\phi: X \to V$ is defined locally. Thus in proving Theorem 3.3, we may assume that V is Stein.

Suppose that V is reducible at $y \in V$. Let $V = \bigcup V_j$ be the decomposition of V into irreducible components. Then in resolving V, we may separately resolve each V_j, $\pi_j: M_j \to V_j$. Let $\mathcal{S}_{m,j}$ denote $\pi_j(\mathcal{O}(mK))$. Then $\mathcal{S}_{m,y}$, the stalk of $\mathcal{S}_m$ at y, has a natural direct sum decomposition $\mathcal{S}_{m,y} \approx \oplus \mathcal{S}_{m,j,y}$. Let $\phi_j: X_j \to V_j$ be the monoidal transformation of V_j with respect to $\mathcal{S}_j$. Then X is isomorphic to the disjoint union of the X_j. So in proving Theorem 3.3, we may assume that V is irreducible at y.

Let $\omega: \tilde{V} \to V$ be the normalization of V. Then $\omega^{-1}(y)$ is just one point, $\tilde{y}$, since V may be assumed to be irreducible at y. Suppose that ω resolves V, i.e. $\tilde{V}$ is smooth at $\tilde{y}$. $\mathcal{O}(mK)$ is isomorphic to $\mathcal{O}_{\tilde{V}}$ near $\tilde{y}$. Then $\mathcal{S}_m$ is isomorphic to $\tilde{\mathcal{O}}_V$, the sheaf of germs of weakly holomorphic functions. Let z_1 and z_2 be local coordinates near $\tilde{y}$. Then z_1 and z_2 are also weakly holomorphic functions near y. So generators for $\mathcal{S}_m$ include 1, z_1 and z_2. Then $X \approx \tilde{V}$.

Recall [**41**] the definitions of the various pull-backs of a sheaf. Let $\psi: N \to V$ be a proper modification. Let $\mathcal{S}$ be a coherent sheaf on V. $\psi^*\mathcal{S}$ has stalk at $x \in N$ given by $(\psi^*\mathcal{S})_x := \mathcal{S}_{\psi(x)} \otimes \mathcal{O}_{N,x}$, where $\mathcal{O}_{N,x}$ is given the structure of an $\mathcal{O}_{V,\psi(x)}$ module via ψ and $\otimes$ is taken over $\mathcal{O}_{V,\psi(x)}$. Let $T(\psi^*\mathcal{S})$ be the torsion subsheaf of $\psi^*\mathcal{S}$. Then $\mathcal{S} \circ \psi := \psi^*\mathcal{S}/T(\psi^*\mathcal{S})$.

PROPOSITION 3.4. *Let $\psi: N \to V$ be a proper modification of the two-dimensional space V. Suppose that N is normal. Let $\tau: M \to N$ be a resolution of N. Let $\pi = \psi \circ \tau$. Suppose that $\pi: M \to V$ is the minimal resolution. Recall $\mathcal{S}_m = \pi_*\mathcal{O}(mK)$ from above. Let $m \geq 2$. Then $\mathcal{S}_m \circ \pi = \mathcal{O}(mK)$. If $\tau_*(\mathcal{O}(mK))$ is locally free, then $\mathcal{S}_m \circ \psi = \tau_*(\mathcal{O}(mK))$.*

PROOF. Suppose that we have already shown that $\tau_*(\mathcal{O}(mK))$ locally free implies that $\mathcal{S}_m \circ \psi = \tau_*(\mathcal{O}(mK))$, for all τ. Letting $\tau\colon M \to N$ be the identity map then implies that $\mathcal{S}_m \circ \pi = \mathcal{O}(mK)$. So it suffices to prove the last statement of the proposition.

Observe that $\psi_*(\tau_*(\mathcal{O}(mK))) = \pi_*(\mathcal{O}(mK)) = \mathcal{S}_m$. Also, since the constructions of the various sheaves are local with respect to V, we may assume that V is Stein.

Let $x \in N$. Let $\lambda_1,\ldots,\lambda_p$ generate $\mathcal{S}_{m,\psi(x)}$. Each λ_i is the direct image of a λ'_i defined in a neighborhood of $\pi^{-1}(\psi(x))$ in M. There is an exact sequence near $\psi(x)$

$$\mathcal{O}_V^q \to \mathcal{O}_V^p \to \mathcal{S}_m \to 0 \tag{3.5}$$

with λ_i the image of $(0,\ldots,0,1,0,\ldots,0)$. Tensoring (3.5) over $\mathcal{O}_V$ with $\mathcal{O}_N$ gives

$$\mathcal{O}_N^q \to \mathcal{O}_N^p \to \psi^*\mathcal{S}_m \to 0$$

which is exact at each point near $x \in N$.

There is a map of coherent sheaves $\alpha\colon \mathcal{O}_N^p \to \pi_*(\mathcal{O}(mK))$ given by $\alpha(0,\ldots,0,1,0,\ldots,0) = \pi_*(\lambda'_i)$. α induces a map $\beta\colon \psi^*\mathcal{S}_m \to \tau_*(\mathcal{O}(mK))$. $\operatorname{im}\alpha = \operatorname{im}\beta$. β is an isomorphism off a nowhere dense subvariety N' of N. $\ker\beta$, since it is supported on N', is a torsion sheaf. $\tau_*(\mathcal{O}(mK))$ is assumed locally free (necessarily of rank 1), so $\ker\beta = T(\psi^*\mathcal{S}_m)$ and $\operatorname{im}\beta = \mathcal{S}_m \circ \psi$. Then also $\operatorname{im}\alpha = \mathcal{S}_m \circ \psi$.

Let $\alpha'\colon \mathcal{O}_M^p \to \mathcal{O}(mK)$ be given by $\alpha'(0,\ldots,0,1,0,\ldots,0) = \lambda'_i$. By Theorem 3.1, α' is onto. By hypothesis, $\tau_*(\mathcal{O}(mK))$ is locally free, necessarily of rank one. So $\Gamma(\tau^{-1}(x), \mathcal{O}(mK))$ has a single generator (over $\mathcal{O}_{N,x}$). By Theorem 3.1, the single generator for $\Gamma(\tau^{-1}(x), \mathcal{O}(mK))$ is nowhere zero on $\tau^{-1}(x)$. Then mK restricts to the trivial line bundle on $\tau^{-1}(x)$. By Theorem 3.1, there is a λ'_i which is nowhere zero on $\tau^{-1}(x)$. So $\mathcal{O}(mK) \approx \mathcal{O}_M$ near $\tau^{-1}(x)$. Since N is normal, τ_* is an isomorphism on holomorphic functions. So α is onto. This concludes the proof of Proposition 3.4.

We shall now complete the proof of Theorem 3.3. We do the $m \geqslant 2$ part first. Let $\psi\colon N \to V$ be the RDP resolution of V. Recall that N is obtained from the minimal resolution $\pi\colon M \to V$ by blowing down the curves A_i on M such that $A_i \cdot K = 0$. mK restricts to the trivial line bundle near A_i such that $A_i \cdot K = 0$. Let $\tau\colon M \to N$. Then $\tau_*(\mathcal{O}(mK))$ is locally free. By Proposition 3.4, $\mathcal{S}_m \circ \psi = \tau_*(\mathcal{O}(mK))$ is locally free. So by the universal property of $\phi\colon X \to V$, the monoidal transformation with respect to $\mathcal{S}_m$, there is a proper modification $\lambda\colon N \to X$. Recall that $X \approx \tilde{V}$, the normalization of V, wherever $\tilde{V}$ is a manifold. N is also the RDP resolution of $\tilde{V}$. So λ is an isomorphism except possibly on the $\tau(A_i)$, where A_i is an exceptional curve on M, which blows down via π. Each point of V is the image of only finitely many A_i. So each point of X is also the image of only finitely many $\tau(A_i)$. If $A_i \cdot K = 0$, then $\tau(A_i)$ is a point. If $A_i \cdot K \neq 0$, $\lambda(\tau(A_i))$ cannot be a point, for then $\tau_*(\mathcal{O}(mK))$ would be the trivial sheaf in a neighborhood of $\tau(A_i)$, which it is not. So λ maps $\tau(A_i)$ to a 1-dimensional set, i.e. λ is a finite map on $\tau(A_i)$. So $\lambda^{-1}(x)$ is a finite set for all $x \in X$. Since λ is a proper modification and N is normal, λ is precisely the normalization map and $N \approx \tilde{X}$. This concludes the $m \geqslant 2$ part of Theorem 3.3.

Recall that S denotes the singular locus of V. Let $y \in V$. Every element of $\mathcal{S}_{m,y}$ is determined by its restriction (as a germ) to $V - S$. On $V - S$, tensor product (of tensors) gives a natural isomorphism $\mathcal{S}_m \otimes \mathcal{S}_n \approx \mathcal{S}_{m+n}$. For all $y \in V$, there is an induced natural map $\mathcal{S}_m \otimes_{\mathbf{C}} \mathcal{S}_n \to \mathcal{S}_{m+n}$. Theorem 3.2 gives immediately

THEOREM 3.5. *The natural map* $\mathcal{S}_m \otimes \mathcal{S}_n \to \mathcal{S}_{m+n}$ *is surjective for* $m \geqslant 2$ *and* $n \geqslant 3$.

Let us observe that for the $m \geqslant 3$ part of Theorem 3.3, we can in fact choose m to be arbitrarily large. Let $\phi_m: X_m \to V$ temporarily denote the monoidal transformation with respect to $\mathcal{S}_m$. Let $\lambda_1, \ldots, \lambda_p$ be generators of $\mathcal{S}_{m,y}$, $y \in V$. X_m is the closure in $V \times \mathbf{P}^{p-1}$ of im μ_m, with $\mu_m: V - S \to V \times \mathbf{P}^{p-1}$ given by $\mu_m(z) = (z, (\lambda_1(z): \cdots : \lambda_p(z)))$. Let r be a positive integer. By Theorem 3.5, $\mathcal{S}_{rm,y}$ is generated by tensor products of the λ_i having r factors. There are in fact $n := \binom{r+p-1}{p-1}$ such products. X_{rm} is the closure in $V \times \mathbf{P}^{p-1}$ of im μ_{rm}, defined analogously. Embed $\mathbf{P}^{p-1}$ in $\mathbf{P}^{n-1}$ via the Veronese embedding. This induces the desired isomorphism between X_m and X_{rm}.

We may reorder the λ_i as is convenient.

With m large, $mK \cdot A_i$ can be made as large as desired off of the A_i with $K \cdot A_i = 0$. So all of the line bundles to follow can be chosen to satisfy the hypotheses of Proposition 2.1 and Theorem 3.1.

As observed above, we may assume that V is Stein and irreducible at y. $\pi: M \to V$ is, again, the minimal resolution. Upon restricting to a smaller neighborhood of y, we may assume that M is strictly pseudoconvex with connected exceptional set $A = \cup A_i = \pi^{-1}(y)$. As in the paragraph following the proof of Proposition 3.4, there is an induced map $\gamma: M \to X$. Let $\lambda'_i \in \Gamma(A, \mathcal{O}(mK))$ give λ_i as its direct image. Then γ may be realized by extending to A the map $\gamma_{M-A}: M - A \to V \times \mathbf{P}^{p-1}$ given by $\gamma_{M-A}(z) = (\pi(z), (\lambda'_1(z): \cdots : \lambda'_p(z)))$. So $\gamma(z) = (\pi(z), (\lambda'_1(z): \cdots : \lambda'_p(z)))$.

$A' := \cup A_j$ with $A_j \cdot K = 0$. Let A'' be a connected component of A'. mK is a trivial line bundle in some neighborhood of A''. So, as above, $\gamma(A'')$ is a point. Let us verify that otherwise γ separates points. Consider $x_1, x_2 \in A$, $x_1 \neq x_2$ and x_1 and x_2 not in the same A''. By Theorem 3.1, mK has a selection which is nonzero at both x_1 and x_2. So for γ to separate x_1 and x_2, it suffices to find a section of mK which vanishes at one of x_1 and x_2, but not at both. There are two cases. The first is that x_1 (or x_2) is in A''. Then let Z_x be the fundamental cycle of A''. Then for all i (and large m), $(mK - Z_x) \cdot A_i \geqslant 2K \cdot A_i$. So by Theorem 3.1 applied to $mK - Z_x$, mK has a section which vanishes at x_1 but not at x_2.

The other case is that neither x_1 nor x_2 is in A''. Suppose that $x_1 \in A_1$. Let Z' be the least cycle on A such that $Z' \geqslant A_1$ and $(mK - Z') \cdot A_i \geqslant 0$ for all i. By choosing m large, we can insure that A_1 appears with coefficient 1 in Z' and that the only other A_i having nonzero coefficients in Z' satisfy $A_i \cdot K = 0$. For m still larger, $(mK - Z') \cdot A_i \geq 2K \cdot A_i$, all i. So by Theorem 3.1, for $x_2 \notin A_1$, mK has a section vanishing at x_1 and not at x_2. The remaining subcase is that x_1 and x_2 are not in distinct A_i, say $x_1, x_2 \in A_1 \cap A_2 \cap \cdots \cap A_j$. Let Y_s from Proposition 2.1 be the first Y_k in which an A_i, $1 \leq i \leq j$, appears. Then $mK - Y_{s-1}$ has a large Chern class on A_1. So sections of $\mathcal{O}(mK - Y_{s-1})/\mathcal{O}(mK - Y_s)$ separate x_1 and x_2. By Proposition 2.1, these sections lift to sections of mK.

$\tau: M \to N$, with N the RDP resolution of V. So we have shown, so far, that $\lambda: N \to X$ is one-to-one (and onto).

Let us now show normality at $\gamma(x)$, $x \in A''$. $\gamma(x) = \gamma(A'')$. Let Z_x be the fundamental cycle of A''. For m large, $(mK - 2Z_x) \cdot A_i \geq K \cdot A_i$, all i. Since mK is trivial near A'', $\mathcal{O}(mK - Z_x)/\mathcal{O}(mK - 2Z_x) \approx \mathcal{O}(-Z_x)/\mathcal{O}(-2Z_x)$. Let $\mathfrak{m}$ be the maximal ideal at $\tau(x) = \tau(A'')$. $\mathfrak{m}/\mathfrak{m}^2$ is isomorphic to $\Gamma(A'', \mathcal{O}(-Z_x)/\mathcal{O}(-2Z_x))$ [**1**, Theorem 4, pp. 132–133].

Finally for $x \notin A'$, we must show that γ is biholomorphic near x. Let $x \in A_1$. Let Z' as above be the least cycle on A such that $Z' \geq A_1$ and $(mK - Z') \cdot A_i \geq 0$ for all i. Let $Z'' = Z' - A_1$. Then $\mathcal{O}(mK - Z'')/\mathcal{O}(mK - Z')$ is the sheaf of germs of sections of a line bundle L over the curve A_1. By making m large, L can be chosen to have arbitrarily large Chern class. Then, by (2.1), for m large, L has no base points and generators over $\mathbf{C}$ of $\Gamma(A_1, \mathcal{O}(L))$ embed A_1 in projective space. For m large, by Proposition 2.1,

$$\Gamma(m, \mathcal{O}(mK - Z'')) \to \Gamma(A_1, \mathcal{O}(mK - Z'')/\mathcal{O}(mK - Z'))$$

is surjective. Let $\lambda'_1, \ldots, \lambda'_t \in \Gamma(M, \mathcal{O}(mK - Z''))$ project onto a basis of $\Gamma(A_1, \mathcal{O}(L))$. Let $\lambda'_1(x) \neq 0$. There are two cases. If x is a singular point of A_1, the embedding of A_1 in projective space via the λ'_i necessarily extends to show that γ is biholomorphic near x. For x a regular point of A_1, choose λ'_2 so that in $\Gamma(A_1, \mathcal{O}(L))$, λ'_2 maps to a section of L with a first order zero at x. By Theorem 3.1 applied to $mK - Z'$, mK has a section λ'_3 which vanishes on A_1 to first order near x. Then $(\lambda'_1 : \lambda'_2 : \lambda'_3)$ embeds M in $\mathbf{P}^2$ near x. Then γ is again biholomorphic near x.

This concludes the proof of Theorem 3.3.

IV. Very weak simultaneous resolution—special case. We now specialize to the case where V has only normal Gorenstein singularities. Let $\pi: M \to V$ be the *minimal* resolution of V. Look at V near a singularity p. Since p is assumed to be Gorenstein, near p there is a meromorphic two-form ω which is holomorphic and nonzero on $V - p$ [**5**, Proposition 5.1, p. 16 and **16**, Satz 3.1, p. 278]. The divisor of $\pi^*(\omega)$ is then supported on $A = \pi^{-1}(p)$ and is a representative for the canonical divisor K. We also let K denote the divisor of $\pi^*(\omega)$. Then $K \cdot K$ is a well-defined integer, which is an invariant of the singularity p. Observe again that the definition of $K \cdot K$ depends on choosing the minimal resolution for V. Nonminimal resolutions will give more negative values for the self-intersection of the canonical divisor. Observe that by Theorem 3.1, $-m^2 K \cdot K$ is the multiplicity of the ideal $\mathcal{S}_{m,p}$ for $m \geq 2$ [**51**, p. 420]. See also (4.3) below. In case V has multiple singularities, we define $K \cdot K$ for V as $K \cdot K := \Sigma(K \cdot K)_p$, with the sum over all singular points of V.

Let (V, p) be the germ of V at p. Let $\lambda: \mathcal{V} \to T$ be the germ of a (flat) deformation of (V, p). Let 0 denote the initial point in T. Arbitrarily small representatives for λ, $\mathcal{V}$ and T will also be denoted by λ, $\mathcal{V}$ and T. We assume that T is reduced. $V_t := \lambda^{-1}(t)$. Flatness implies that V_t is two-dimensional. We allow V_t to have multiple singularities. We can see as follows that each singularity of V_t is normal and Gorenstein: Replacing T by a resolution of T [**18**], we may assume without loss of generality that T is smooth. The special fiber $V = V_0$ of $\mathcal{V}$ is normal, so $\mathcal{V}$ is normal. Then V_t is normal. So V_t is also Cohen-Macauley since it is of dimension two. By [**17**, 1.40(a), p. 13], $\mathcal{V}$ is Gorenstein. So also V_t is Gorenstein and $K_t \cdot K_t$ is defined as $K \cdot K$ for V_t.

Let $\Pi: \mathfrak{M} \to \mathcal{V}$ be the germ of a holomorphic map. $M_t := (\lambda \circ \Pi)^{-1}(t)$. Let Π_t denote the restriction of Π to M_t. Recall [**49**, Definition 2, pp. 72–73]

DEFINITION 4.1. The map germ $\Pi: \mathfrak{M} \to \mathcal{V}$ is a *very weak simultaneous resolution* of the germ of the (flat) deformation $\lambda: \mathcal{V} \to T$, T reduced, if for all sufficiently small representatives of λ, the germ Π has a representative such that:

(0) Ψ is a proper modification map,

(i) $\lambda \circ \Pi: \mathfrak{M} \to T$ is a flat map,

(ii) $\Psi_t: M_t \to V_t$ is a resolution of V_t, all t.

Recall [**15**, Satz 2.3, p. 244] in Definition 4.1 that with M_0 assumed to be smooth in (ii), flatness in (i) just means that $\lambda \circ \Pi$ is a locally trivial deformation of M_0.

DEFINITION 4.2. The map germ $\Psi: \mathfrak{N} \to V$ is a *simultaneous RDP resolution* of $\lambda: \mathcal{V} \to T$, T reduced, if for all sufficiently small representatives of λ, the germ Ψ has a representative such that:

(0) Ψ is a proper modification map,

(i) $\lambda \circ \Psi: \mathfrak{N} \to T$ is a flat map,

(ii) $\Psi_t: N_t \to V_t$ is the RDP resolution of V_t.

Given a very weak simultaneous resolution $\Pi: \mathfrak{M} \to V$ we may simultaneously blow down all the exceptional curves of the first kind in the M_t [**25**, Theorem 3, p. 85; **43**, Satz 2, p. 548; **4**, Lemma 2.1, p. 334]. After finitely many such simultaneous blow-downs, we may assume that $\Pi: \mathfrak{M} \to V$ is a very weak simultaneous resolution with each $\Pi_t: M_t \to V_t$ the minimal resolution. Let $A_{t,i}$ be an irreducible component of A_t, the exceptional set in M_t, such that $A_{t,i} \cdot K_t = 0$, i.e., $A_{t,i}$ is a nonsingular rational curve with $A_{t,i} \cdot A_{t,i} = -2$. By [**32**, Proposition 2.3, p. 4], $A_{t,i}$ is homologous in $\mathfrak{M}$ to a (unique) cycle D_i on $A_0 := A$ such that $D_i \cdot K = 0$. Since $\Pi_0: M_0 \to V_0$ is the minimal resolution, all components A_j of A with $\operatorname{supp}(A_j) \subset \operatorname{supp}(D_i)$ satisfy $A_j \cdot K = 0$. So blowing down a neighborhood in $\mathfrak{M}$ of all the A_j in A such that $A_j \cdot K = 0$ [**52**, **42**] yields a simultaneous RDP resolution $\Psi: \mathfrak{N} \to V$ of $\lambda: \mathcal{V} \to T$.

By a *finite base change* for $\lambda\colon \mathcal{V} \to T$, we mean a germ of a deformation $\lambda'\colon \mathcal{V}' \to T'$ which is induced from λ via a finite proper surjective map $\alpha\colon T' \to T$, $\alpha^{-1}(0) = 0'$. α expresses T' as a finite branched cover of T. By **[8]** if $\lambda\colon \mathcal{V} \to T$ has a simultaneous RDP resolution, then after a finite base change, λ has a very weak simultaneous resolution.

Recall the definition for X to be a Moišezon space [**37**, **3**, p. 126; **41**]. X is then the reduced analytic space associated to an algebraic space. By [**3**], for every point $p \in X$ there exist a projective algebraic space Y, and a holomorphic map $\sigma\colon Y \to X$ such that σ has a local holomorphic inverse near p. So in proving Theorem 4.3 below, we may without loss of generality assume that $\mathcal{V}_t$ is projective algebraic.

THEOREM 4.3. *Let $\lambda\colon \mathcal{V} \to T$ be the germ of a (flat) deformation of the germ (V, p) of a normal Gorenstein two-dimensional singularity. Suppose that $K_t \cdot K_t$ is constant. Suppose that the map germ $\lambda\colon \mathcal{V} \to T$ has a representative $\lambda_r\colon \mathcal{V}_t \to \mathbf{P}^1$ such that $\mathcal{V}_r$ is a Moišezon space and λ_r is a meromorphic function on $\mathcal{V}_r$. Then λ has a simultaneous RDP resolution.*

(V, p) may split into more than one singularity. For example, let $p = (0,0,0)$, $V = \{z^2 = x^3 + y^{12}\}$ and $\mathcal{V} = \{z^2 = x^3 + (y^2 + t)^6\}$. For $V = V_0$, on the minimal resolution, $A = A_1 \cup A_2$ with A_1 and A_2 nonsingular and intersecting transversely. $g_1 = 1$, $g_2 = 0$; $A_1 \cdot A_1 = -1$, $A_2 \cdot A_2 = -2$; $K = -2A_1 - A_2$; $K \cdot K = -2$. For $t \neq 0$, V_t has two singularities, at $x = z = 0$, $y^2 = t$. Each singularity is equivalent to $z^2 = x^3 + y^6$ at $(0,0,0)$ with $K \cdot K = -1$. So $K_t \cdot K_t = -2$, all t. So $\lambda\colon \mathcal{V} \to T = \{t\}$ has a simultaneous RDP resolution. In fact, following the proof of Theorem 4.3 below, we blow up $\mathcal{V}$ at the ideal

$$\mathcal{I}_3 = \left(z, (y^2 + t)^3, x(y^2 + t), x^2\right).$$

We then see that a 2:1 base change is required for a very weak simultaneous resolution.

An outline of the proof of Theorem 4.3 is as follows. Since V is assumed to be Gorenstein, there is a nonzero holomorphic 2-form ω on $V - p$. Division by $\omega^{\otimes m}$ exhibits $\mathcal{S}_m$ as a sheaf of ideals. Take $m \geqslant 3$. We wish to blow up $\mathcal{V}$ simultaneously along the ideal sheaves $\mathcal{S}_{m,t}$, at least in case each V_t has a single singularity. To do so, we need that all of the $\mathcal{S}_{m,t}$ have the same Hilbert function. We also need a coherent ideal sheaf $\mathcal{I}_m$ on $\mathcal{V}$ such that $\mathcal{I}_m$ restricts to $\mathcal{S}_{m,t}$ on V_t, for all $t \in T$ sufficiently near to 0. By Theorem 3.3, the blow-up of V_t at $\mathcal{S}_{m,t}$ is the RDP resolution. A slightly different argument is actually used so as to include the case of V_t having multiple singularities.

Let $\pi\colon M \to V$ be the minimal resolution of V. Let $D \geqslant 0$ be a cycle on $A = \pi^{-1}(p)$, $A = \bigcup A_i$. Recall that $\mathcal{O}(-D)$ is the ideal sheaf of germs of functions on M which vanish on A_i to that order given by D, $\mathcal{O}_D := \mathcal{O}/\mathcal{O}(-D)$. Let $h^i(\mathcal{O}_D)$ denote dim $H^i(M, \mathcal{O}_D)$. Recall the Riemann-Roch Theorem [**45**, Proposition IV, 4, p. 75]

$$(4.1) \qquad \chi(D) := h^0(\mathcal{O}_D) - h^1(\mathcal{O}_D) = \tfrac{1}{2}(D \cdot D + D \cdot K).$$

For any coherent sheaf $\mathcal{F}$ on M, $H^2(M, \mathcal{F}) = 0$ since M is noncompact of dimension two [**47**]. Assuming V Stein, let $h = \dim H^1(M, \mathcal{O})$. h is often denoted by p_g.

$$0 \to \mathcal{O}(-D) \to \mathcal{O} \to \mathcal{O}_D \to 0$$

is exact. So if $H^1(M, \mathcal{O}(-D)) = 0$, then $h^1(\mathcal{O}_D) = h$. So by (4.1),

$$h^0(\mathcal{O}_D) = h - \tfrac{1}{2}(D \cdot D + D \cdot K). \tag{4.2}$$

By Proposition 2.1,

$$h^0(\mathcal{O}_{-mK}) = h - \frac{K \cdot K}{2}(m^2 - m). \tag{4.3}$$

Since $H^1(M, \mathcal{O}(mK)) = 0$, all of the obstructions to lifting will vanish. So, as ideals, with $n \geqslant m \geqslant 1$,

$$\begin{aligned} \dim \mathcal{S}_m/\mathcal{S}_n &= h^0(\mathcal{O}_{-nk}) - h^0(\mathcal{O}_{-mk}) \\ &= -\frac{K \cdot K}{2}\left[(n^2 - m^2) - (n - m)\right]. \end{aligned}$$

Now look at the Hilbert function $H(\nu) := \dim (\mathcal{S}_m)^\nu/(\mathcal{S}_m)^{\nu+1}$ of $\mathcal{S}_m$. By Theorem 3.5, $(\mathcal{S}_m)^\nu = \mathcal{S}_{m\nu}$. So

$$H(\nu) = -\frac{K \cdot K}{2}\left[m^2(2\nu + 1) - m\right], \qquad \nu \geqslant 1,$$

$$H(0) = h - \frac{K \cdot K}{2}(m^2 - m), \quad \text{from (4.3).}$$

We shall now construct the ideal sheaf $\mathcal{I}_m$.

We are given the map germ $\lambda: \mathcal{V} \to T$ of Theorem 4.3. In this proof, V will denote the germ of an analytic space. Recall that $V = V_0$. In this proof, all representatives of germs will have a subscript r. By blowing up $\mathcal{V}_r$, away from p, we may assume that λ_r is a (holomorphic) map, i.e., the zero locus of λ_r is disjoint from the pole locus of λ_r. Let Π_r: $\mathfrak{M}_r \to \mathcal{V}_r$ be a projective algebraic resolution of $\mathcal{V}_r$ [**18**]. Let $\mathfrak{M}$ be the germ of $\mathfrak{M}_r$ at $\Pi_r^{-1}(p)$. Let Π: $\mathfrak{M} \to \mathcal{V}$ be the germ of Π_r at $\mathfrak{M}$. $\mathfrak{M}_r$ is nonsingular of dimension 3, but Π is not necessarily a simultaneous resolution. We choose $\mathfrak{M}_r$ so that the exceptional set $\mathcal{E}$ in $\mathfrak{M}$ has normal crossings. Let $\rho = \lambda \circ \Pi$: $\mathfrak{M} \to T$. We take $0 \in T$ to correspond to 0 in $\mathbf{P}^1$. Let $M_t = \rho^{-1}(t)$. Observe that M_0 need not be reduced. Let $E_t = M_t \cap \mathcal{E}$. Let $M' \subset M_0$ be the proper transform under Π of V. Let ξ denote the restriction of Π to M'. Then we may assume that ξ: $M' \to V$ is a resolution (which is not necessarily the minimal resolution π: $M \to V$) and that M' meets $\mathcal{E}$ transversely. $M' \cap \mathcal{E} = A'$ is the exceptional set in M'. By performing additional blow-ups, we may assume that A' does not intersect the closure in $\mathfrak{M}$ of $\mathcal{E} - E_0 = \cup E_t$, $t \neq 0$. Π: $\mathfrak{M} - M_0 \to \mathcal{V} - V_0$ is a weak simultaneous resolution (see Definition 5.1).

Let $K_{\mathfrak{M}}$ be the canonical bundle on $\mathfrak{M}_r$. Let t be a coordinate for T. On M_{t_0}, define $\mathcal{O}(K_{t_0}) := \mathcal{O}(K_{\mathfrak{M}})/(t - t_0)\mathcal{O}(K_{\mathfrak{M}})$. The vector field $\partial/\partial t$ induces a nowhere zero section of the normal bundle of M_{t_0} in $\mathfrak{M}$, i.e. M_{t_0} has a trivial normal bundle. So K_{t_0} coincides with the usual canonical bundle on M_{t_0} at the regular points of M_{t_0}. To restrict a 3-form on $\mathcal{M}$ to a 2-form on M_{t_0}, contract with $\partial/\partial t$ and then restrict the coefficients to M_{t_0}. There is the same construction for mK_{t_0}, i.e. $\mathcal{O}(mK_{t_0}) := \mathcal{O}(mK_{\mathcal{M}})/(t-t_0)\mathcal{O}(mK_{\mathcal{M}})$.

Since V is Gorenstein, so is $\mathcal{V}$. There exists Ω on $\mathcal{V}$, a holomorphic nonzero 3-form on $\mathcal{V} - \mathrm{Sing}(\mathcal{V})$. The restriction of Ω to V_t, i.e. the contraction with $\partial/\partial t$ followed by the restriction of the coefficients, gives ω_t, a holomorphic nonzero 2-form on $V_t - \mathrm{Sing}(V_t)$. On $\mathcal{V}$, $\mathcal{O}\cdot\Omega^{\otimes m} \approx \mathcal{O}$ can be thought of as meromorphic sections of $mK_{\mathcal{V}}$ which are holomorphic on $\mathcal{V} - \mathrm{Sing}(\mathcal{V})$.

We now define $\mathcal{I}_m$. Our definition has been chosen to be independent of any resolution of $\mathcal{V}$. Let $\mathcal{I}_m$ be the sheaf of ideals in $\mathcal{O}\cdot\Omega^{\otimes m}$ given by those sections of $\mathcal{O}\cdot\Omega^{\otimes m}$ which for $t \neq 0$ restrict to elements of $\mathcal{S}_{m,t}$ on V_t. Thus the restriction of elements of $\mathcal{I}_{m,p}$ to $V - p$ consists of (the germs at p of) those meromorphic sections of mK_V which are holomorphic on $V - p$ and have extensions to $\mathcal{O}\cdot\Omega^{\otimes m}$ such that on each V_t, $t \neq 0$, the extension extends holomorphically to M_t. We first need that $\mathcal{I}_m$ is coherent. Recall $\Pi\colon \mathcal{M} \to \mathcal{V}$. Since we may choose small representatives for Π which are proper maps, $\Pi_*(\mathcal{O}(mK_{\mathcal{M}}))$ is coherent. $\Pi_*(\mathcal{O}(mK_{\mathcal{M}})) \subset \mathcal{I}_m$. In fact, on $\mathcal{V} - V_0$, where Π is a weak simultaneous resolution, $\Pi_*(\mathcal{O}(mK_{\mathcal{M}})) = \mathcal{I}_m$ because: Look at α, an element of $\mathcal{O}\cdot\Omega^{\otimes m}$, and suppose $\alpha \notin \Pi_*(\mathcal{O}(mK_{\mathcal{M}}))$. We need that $\alpha \notin \mathcal{I}_m$. $\alpha \notin \Pi_*\mathcal{O}(mK_{\mathcal{M}})$ means that $\Pi^*(\alpha)$ is not holomorphic on $\mathcal{M}$, i.e. that $\Pi(\alpha)$ has poles P on $\mathcal{E}$, the exceptional set in $\mathcal{M}$. P has dimension 2. Let Z be the zero set of $\Pi^*(\alpha)$. Then $P \cap Z$ has dimension 1. So $E_t := \mathcal{E} \cap M_t$, the exceptional set in M_t, can be contained in $P \cap Z$ for only a discrete subset S of $T - \{0\}$. Then $E_t \cap (P - (P \cap Z)) \neq \varnothing$ for $t \in T - \{0\} - S$. Then the restriction of α to M_t is not in $\mathcal{S}_{m,t}$ for $t \in T - \{0\} - S$.

Since $\mathcal{O}_{\mathcal{V},p}$ is Noetherian, $\mathcal{I}_{m,p}$, the stalk of $\mathcal{I}_m$ at the initial singular point p, is finitely generated over $\mathcal{O}_{V,p}$. $\Pi_*(\mathcal{O}(mK_{\mathcal{M}}))$, which is contained in $\mathcal{I}_m$ and which equals $\mathcal{I}_m$ on $\mathcal{V} - V_0$, is finitely generated near p. Hence $\mathcal{I}_m$ is finitely generated near p and hence is coherent.

We define $\mathcal{I}_{m,t_0} := \mathcal{I}_m/(t-t_0)\mathcal{I}_m$. Temporarily replace Π over $T - \{0\}$ by the simultaneous minimal resolution. Then $H^1(M_t, \mathcal{O}(mK_t)) = 0$ for $t \neq 0$. Then by [**42**, Bemerkung, pp. 94–95], $\mathcal{I}_{m,t} = \mathcal{S}_{m,t}$, $t \neq 0$.

Observe that the natural map $\mathcal{I}_{m,0} \to \mathcal{O}_V\cdot\omega^{\otimes m}$ is an injection. Indeed, if α is an element of $\mathcal{I}_m$ which vanishes upon restriction to $V - p$, then α/t has removable singularities on $V - p$ and so defines an element β of $\mathcal{I}_m$. Then $\alpha = t\beta$, i.e. $\alpha \in t\mathcal{I}_m$.

Let $\mathcal{Q}_m := \mathcal{O}\cdot\Omega^{\otimes m}/\mathcal{I}_m$. Then $\mathcal{Q}_m$ is supported on $\mathrm{Sing}(\mathcal{V})$. $\lambda_*(\mathcal{Q}_m)$ can have no torsion. For if an element α of $\mathcal{O}\cdot\Omega^{\otimes m}$ maps to 0 in $\mathcal{Q}_m$ above $t \neq 0$, then α restricts to an element of $\mathcal{S}_{m,t}$ for $t \neq 0$ and so α is an element of $\mathcal{I}_m$, by the definition of $\mathcal{I}_m$. So $\lambda_*(\mathcal{Q}_m)$ is a locally free sheaf on the one-dimensional space T, including $t = 0$, of rank $\dim \mathcal{O}_{V_t}\cdot\omega_t^{\otimes m}/\mathcal{S}_{m,t}$, $t \neq 0$. By (4.3),

$$\text{(4.4)} \qquad \operatorname{Rank} \lambda_*(\mathcal{Q}_m) = h_t - \frac{K_t \cdot K_t}{2}(m^2 - m), \qquad t \neq 0.$$

Of course, in (4.4) $K_t \cdot K_t$ is the same for all $t \neq 0$ because $K_t \cdot K_t$ is defined topologically and Π is a weak simultaneous resolution. h_t is also independent of t for $t \neq 0$; let $m = 0$ in (4.4).

The proofs of Lemmas 4.4 and 4.5 below are postponed until later in this section. Observe that the hypotheses of Lemmas 4.4 and 4.5 do not assume that $K_t \cdot K_t$ is constant. $\mathcal{S}_m := \mathcal{S}_{m,0}$.

LEMMA 4.4. *Let $\lambda: \mathcal{V} \to T$ be a germ of a flat deformation of the germ (V, p) of a normal Gorenstein two-dimensional singularity with T smooth of dimensiion 1. If $\mathcal{I}_{m,0} \not\subset \mathcal{S}_m$, then there exists $c > 0$ such that for all large ν,*

$$\dim \mathcal{I}_{m\nu,0} / \left(\mathcal{S}_{m\nu} \cap \mathcal{I}_{m\nu,0}\right) \geqslant c\nu^2.$$

LEMMA 4.5. *Let $\lambda: \mathcal{V} \to T$ be a germ of a flat deformation of the germ (V, p) of a normal Gorenstein two-dimensional singularity. Suppose that the map germ $\lambda: \mathcal{V} \to T$ has a representative $\lambda_r: \mathcal{V}_r \to \mathbf{P}^1$ such that $\mathcal{V}_r$ is a projective algebraic variety and λ_r is a meromorphic function on $\mathcal{V}_r$. Then, as a function of m, $\dim \mathcal{S}_m/(\mathcal{S}_m \cap \mathcal{I}_{m,0})$ is bounded by a linear function of m.*

Consider the inclusions given in (4.5). Think of each inclusion as an injection. (4.5) is commutative. All sheaves are over V.

$$\text{(4.5)} \qquad \begin{array}{ccc} \mathcal{O} \cdot \omega^{\otimes m\nu} & \overset{\alpha}{\supset} & \mathcal{I}_{m\nu,0} \\ \beta \cup & & \cup \gamma \\ \mathcal{S}_{m\nu} & \underset{\delta}{\supset} & \left(\mathcal{S}_{m\nu} \cap \mathcal{I}_{m\nu,0}\right) \end{array}$$

The codimension of $\operatorname{im} \alpha$ is given by (4.4). The codimension of $\operatorname{im} \beta$ is given by (4.3) with $h = h_0$ and $K \cdot K = K_0 \cdot K_0$. Lemma 4.5 estimates the codimension of $\operatorname{im} \delta$. Lemma 4.4 is about the codimension of $\operatorname{im} \gamma$.

Fix m. In (4.5), look at the rate of quadratic growth of the codimensions of the inclusions as $\nu \to \infty$. Codim$(\beta \circ \delta)$ is asymptotic to $(K_0 \cdot K_0)(m^2\nu^2)$. $\operatorname{im}(\beta \circ \delta) = \operatorname{im}(\alpha \circ \gamma)$. Since we have not yet used that $K_t \cdot K_t$ is constant, we have proved the

PROPOSITION 4.6. *Let $\lambda: \mathcal{V} \to T$ be as in Lemma 4.5. Then $-K_0 \cdot K_0 \geqslant -K_t \cdot K_t$,*

We now use the hypothesis that $K_0 \cdot K_0 = K_t \cdot K_t$. Then (4.5) and Lemma 4.4 imply that $\mathcal{I}_{m,0} \subset \mathcal{S}_m$. With $\nu = 1$, (4.5) becomes:

$$\text{(4.6)} \qquad \begin{array}{ccc} \mathcal{O} \cdot \omega^{\otimes m} & \overset{\alpha}{\supset} & \mathcal{I}_{m,0} \\ \beta \cup & & \| \\ \mathcal{S}_{m} & \underset{\delta}{\supset} & \left(\mathcal{S}_{m} \cap \mathcal{I}_{m,0}\right) \end{array}$$

Then $h_0 \leqslant h_t$. But $h_0 \geqslant h_t$ [**11**, Théorème 1, p. 144]. So,

PROPOSITION 4.7. *Let $\lambda: \mathcal{V} \to T$ be as in Theorem 4.3. Then h_t is constant.*

Then from (4.6), $\mathcal{S}_m = \mathcal{S}_m \cap \mathcal{I}_{m,0} = \mathcal{I}_{m,0}$.

We now show that blowing up a suitable $\mathcal{W}$ at essentially the ideal sheaf $\mathcal{I}_m$, $m \geqslant 3$, yields the desired simultaneous RDP resolution. We wish to apply [**18**, Definition 8, p. 226], where we need a smooth center for the monoidal transform.

$\lambda_*(\mathcal{I}_m/\mathcal{I}_m^2)$ is a locally free sheaf on T. Let $f_1,\ldots,f_l$ be elements of $\mathcal{I}_m$ whose images in $\lambda_*(\mathcal{I}_m/\mathcal{I}_m^2)$ generate $\lambda_*(\mathcal{I}_m/\mathcal{I}_m^2)$ over $\mathcal{O}_T$. Then the map $f = (f_1,\ldots,f_l, t)$: $\mathcal{V} \to \mathbf{C}^{l+1}$ is proper. Let $\mathcal{W} := f(\mathcal{V})$. f: $\mathcal{V} \to \mathcal{W}$ is the normalization map. Let $(x_1,\ldots,x_l, t)$ be local coordinates for $\mathbf{C}^{l+1}$. Let $\mathcal{J}$ be the ideal sheaf on $\mathcal{W}$ generated by $(x_1,\ldots,x_l)$. Then $f_*(\mathcal{I}_m) = \mathcal{J}$. Let $(z_1,\ldots,z_n)$ be local coordinates for an ambient neighborhood of V. Then since all products $z_i f_j$, $1 \leqslant i \leqslant n$, $1 \leqslant j \leqslant l$, are in $\mathcal{I}_m$, $f_*(z_i f_j)$ is holomorphic on $\mathcal{W}$. Then f: $\mathcal{V} - \operatorname{loc}(\mathcal{I}_m) \to \mathcal{W} - \operatorname{loc}(\mathcal{J})$ is a biholomorphic map.

By construction, we may identify $\mathcal{J}_{t_0} := \mathcal{J}/(t - t_0)\mathcal{J}$ with $\mathcal{S}_{m,t_0}$. Then $\dim \mathcal{J}_t^\nu/\mathcal{J}_t^{\nu+1} = \dim \mathcal{S}_{m,t}^\nu/\mathcal{S}_{m,t}^{\nu+1}$, $\nu \geqslant 1$, and we have shown that this value is independent of t. $\dim \mathcal{J}_t^0/\mathcal{J}_t := \dim \mathcal{O}_{W,t}/\mathcal{J}_t$ equals 1 for all t. So $\mathcal{W}$ is normally flat along $\operatorname{loc}(\mathcal{J})$. By [**18**, Definition 8, p. 226 and Lemma 6, p. 216] and Theorem 3.3, the monoidal transformation Φ: $\mathcal{M} \to \mathcal{W}$ of $\mathcal{W}$ at $\mathcal{J}$ gives a simultaneous RDP resolution of $\mathcal{W}$. By Proposition 4.7 and [**43**, Satz 2, p. 548], we may simultaneously blow down the exceptional sets in Ψ, $\mathcal{J}$: $\mathcal{M} \to \tilde{\mathcal{W}}$. Since $\mathcal{M}$ is normal, $\tilde{\mathcal{W}}$ is normal. Then the induced map ω: $\tilde{\mathcal{W}} \to \mathcal{W}$ is the normalization map. But f: $\mathcal{V} \to \mathcal{W}$ is also the normalization map. Hence $\tilde{\mathcal{W}} \approx \mathcal{V}$.

This concludes the proof of Theorem 4.3 except for the proofs of Lemmas 4.4 and 4.5.

PROOF OF LEMMA 4.4. Suppose that $\sigma \in \mathcal{I}_{m,0}$ is not an element of $\mathcal{S}_m$. Then on the minimal resolution M of V, σ has a pole of order $r > 0$ on some component A_i of A. $\tau := \sigma \otimes \cdots \otimes \sigma$, with ν factors, has a pole of order $r\nu$ on A_i. Since all holomorphic functions on V extend to holomorphic functions on $\mathcal{V}$, multiplying τ by a holomorphic function f with a zero on A_i of order less than $r\nu$ gives an element of $\mathcal{I}_{m\nu,0} - \mathcal{S}_{m\nu}$.

We see, as follows, that there are many such functions f. Let D be a cycle on A such that for all A_j, $D \cdot A_j \leqslant 0$ and $(D + A_i) \cdot A_j \leqslant -K \cdot A_j$. Then, Proposition 2.1, $H^1(M, \mathcal{O}(-sD - A_i)) = 0$ for all positive integers s. Then

$$\Gamma(M, \mathcal{O}(-sD))/\Gamma(M, \mathcal{O}(-sD - A_i)) \approx \Gamma\big(M, \mathcal{O}_{A_i}(-sD)\big).$$

Let g be the genus of A_i. Then, by the Riemann-Roch Theorem,

$$\dim \Gamma\big(M, \mathcal{O}_{A_i}(-sD)\big) \geqslant sD \cdot A_i + 1 - g.$$

Let d_i be the coefficient of A_i in D. Then

$$\dim \mathcal{I}_{m\nu,0}/\big(\mathcal{S}_{m\nu} \cap \mathcal{I}_{m\nu,0}\big) \geqslant \Sigma(-sD \cdot A_i + 1 - g),$$

for $sd_i < r\nu$. Take $0 < c < ((-A_i \cdot D)/2)(r/d_i)^2$ to complete the proof.

PROOF OF LEMMA 4.5. Recall Π_r: $\mathcal{M}_r \to \mathcal{V}_r$, the above projective algebraic resolution of $\mathcal{V}_r$ such that $\rho_r := \lambda_r \circ \Pi_r$: $\mathcal{M}_r \to \mathbf{P}^1$ is a holomorphic map. $\mathcal{M}$ is the germ of $\mathcal{M}_r$ at $E_0 := \Pi_r^{-1}(p)$. Π: $\mathcal{M} \to \mathcal{V}$ is the germ of Π_r. $V_r := \lambda_r^{-1}(0)$. M_r' is the proper transform of V_r under Π_r. We may assume that ξ_r, the restriction of

Π_r to M'_r, is a resolution of V_r and that M'_r meets the exceptional set in $\mathfrak{M}_r$ transversely. M' is the germ of M'_r at E_0. $M_{0,r} := \rho_r^{-1}(0)$. M_0 is the germ of $M_{0,r}$ at E_0.

We recall [**11**, p. 143] the canonical extension by 0 from $\Gamma(M', \mathcal{O}(mK))$ to $\Gamma(M_0, \mathcal{O}(mK))$, where $\mathcal{O}(K_0)$ is the canonical sheaf on $M_{0,r}$. A' is the exceptional set in M'. Let (x, y, z) be local coordinates in $\mathfrak{M}$ near $(0,0,0) \in A'$. $M' = \{z = 0\}$. $\mathcal{E}$, the exceptional set in $\mathfrak{M}$, has reduced equation either $xy = 0$ or $x = 0$. For convenience, we have chosen $\mathfrak{M}$ so that $A' \cap \overline{(\mathcal{E} - E_0)} = \varnothing$. So $\rho: \mathfrak{M} \to T$ is given respectively by $t = x^a y^b z u(x, y, z)$ or $t = x^a z u(x, y, z)$ with $a, b \geq 1$ and $u(0,0,0) \neq 0$. The second case is essentially a special case of the first case, so we shall take $t = x^a y^b z u := x^a y^b z u(x, y, z)$. Recall that $M_0 = \{t = 0\}$. Let $\tau' \in \Gamma(M', \mathcal{O}(mK))$. τ' may be given locally by

$$\tau' = f(x, y)(dx \wedge dy)^m. \tag{4.7}$$

Extend τ' to τ on M_0 by

$$\tau = (x^a y^b u)^m f(x, y)(dx \wedge dy \wedge dz)^m. \tag{4.8}$$

Off of E_0, the vector field $\theta = (1/x^a y^b u)\partial/\partial z$ satisfies $\theta(t) = 1$ on M' and so demonstrates the triviality of the normal bundle to $M' - A'$. Contracting τ with θ, m times, gives τ', as needed. On M_0, the computed equation for τ is independent of the choices which we have made.

Recall the resolution $\xi: M' \to V$, obtained by taking the germ of ξ_r above at A'. Let $\pi: M \to V$ be the minimal resolution. There is a blowing down map $\alpha: M' \to M$. Let J be the divisor of the Jacobian of the map α. Recall that $\mathcal{O}(K)$ denotes the canonical sheaf on both M and M'. Then

$$\alpha^*: \Gamma(M, \mathcal{O}(mK)) \to \Gamma(M', \mathcal{O}(mK - mJ)) \text{ is an isomorphism.}$$

Let D be a positive divisor on A, the exceptional set in M. Let D' be the total transform of D via α'. Then $\alpha^*: \Gamma(M, \mathcal{O}(mK - D)) \to \Gamma(M', \mathcal{O}(mK - mJ - D'))$ is also an isomorphism.

Let D be a cycle on A such that $D \cdot A_i \leq 0$ for all irreducible components A_i of A and such that $K - D$ has no base points on M. By Theorem 3.1, it suffices to take $D = -K$. Recall that $K_{\mathfrak{M}}$ denotes the canonical sheaf on $\mathfrak{M}_r$, and also on $\mathfrak{M}$. We shall show below that, using the above identification α^*, (4.7) and (4.8), every element of $\Gamma(M, \mathcal{O}(mK - D))$ lifts to an element of $\Gamma(M, \mathcal{O}(mK_{\mathfrak{M}}))$. (This was shown for $D = 0$ by Shepherd-Barron [**46**, Theorem 12, pp. 14–15].) Since

$$\dim \Gamma(M, \mathcal{O}(mK))/\Gamma(M, \mathcal{O}(mK - D)) = \chi(-mK + D) - \chi(-mK),$$

this will prove the lemma.

For $\tau^* \in \Gamma(M, \mathcal{O}(mK - D))$, let $\tau' := \alpha^*(\tau^*)$. Let τ denote the canonical extension of τ' by 0 to $\Gamma(M_0, \mathcal{O}(K_0))$. By [**13**], $\Gamma(M, \mathcal{O}(mK - D))$ is finitely generated over $\Gamma(V, \mathcal{O}_V)$. By induction on m, $m \geq 1$, we shall find $\{\tau_j^*\}$, $1 \leq j \leq l$, and effective divisors F_m on $\mathcal{V}_r$, $p \notin \operatorname{supp} F_m$, such that, letting F_m^* denote the total transform of F_m under Π_r:

(a) The τ_j^* generate $\Gamma(M, \mathcal{O}(mK - D))$ over $\Gamma(V, \mathcal{O}_V)$.

(b) The τ_j^* have no common zeros.

(c) The map $(\tau_1^* : \cdots : \tau_l^*)$: $M \to \mathbf{P}^{l-1}$ is (somewhere) nondegenerate.

(d) Each τ_j has a representative $\tau_{j,r} \in \Gamma(M_0, \mathcal{O}(mK_0) \otimes \mathcal{O}(F_m^*))$ which lifts to $\tilde{\tau}_{j,r} \in \Gamma(M_r, \mathcal{O}(mK + F_m^*))$.

This will prove the lemma.

We start the induction at $m = 1$. Since $\mathcal{O}(K - D)$ has no base points on M, $\mathcal{O}(K - J - D')$ has no base points on M'. Then we may choose meromorphic sections $\tau'_{j,r}$ of $K - J - D'$ on M'_r such that (a), (b), and (c) are satisfied for the germs τ_j^* at A of the $(\alpha^*)^{-1}(\tau'_{j,r})$. $\xi_r(\tau'_{j,r})$ is a meromorphic canonical form on V_r which is not necessarily holomorphic at p and at some other subvarieties of V_r. So there is an effective divisor G_1 on V_r, which we also choose to be ample, such that all of the τ_j have representatives $\tau_{j,r} \in \Gamma(M_0, \mathcal{O}(K_0) \otimes \mathcal{O}(G_1^*))$. $\mathcal{O}(G_1^*)$ is a quasi-positive sheaf [**16**, p. 226]. Let $M_\infty := \rho_r^{-1}(\infty)$ be the polar divisor of ρ_r. Then $M_\infty - M_{0,r}$ is equivalent to the trivial divisor. Then [**16**, Satz 2.2, p. 273], $H^1(\mathfrak{M}_r, \mathcal{O}(K + G_1^* + M_\infty - M_{0,r})) = 0$. Let $F_1 = G_1 + \lambda^{-1}(\infty)$. Then

$$(4.9)\quad 0 \to \mathcal{O}(K_{\mathfrak{M}} + F_1^* - M_{0,r}) \to \mathcal{O}(K_{\mathfrak{M}} + F_1^*) \to \mathcal{O}(K_0) \otimes \mathcal{O}(G_1^*) \to 0$$

is exact since $\operatorname{supp}(M_{0,r}) \cap \operatorname{supp}(M_\infty) = \varnothing$ by construction. The long exact cohomology sequence for (4.9) proves the $m = 1$ case.

We now assume the existence of the τ_j^* for m as in (a)–(d). We must construct F_{m+1} and $\sigma_k^* \in \Gamma(M, \mathcal{O}((m+1)K - D))$ satisfying (a)–(d) for $(m+1)$. We first want a quasipositive sheaf $\mathfrak{T}$ on $\mathfrak{M}_r$ such that on $\mathfrak{M}$ (i.e. near E_0), $\mathfrak{T}$ is isomorphic to the subsheaf of $\mathcal{O}(mK_{\mathfrak{M}})$ which is generated by the $\tilde{\tau}_j$. $\mathfrak{T}$ will be locally free of rank 1. To assure quasipositivity, it suffices that $\mathfrak{T}$ have no base points and that the (holomorphic) map into projective space given by a basis of $\Gamma(M_r, \mathfrak{T})$ be nondegenerate.

We have chosen the $\{\tau_j^*\}$ so that $(\tau_1^*, \ldots, \tau_l^*)$: $M \to \mathbf{P}^{l-1}$ is nondegenerate. Think of ρ_r: $\mathfrak{M}_r \to \mathbf{P}^1$ as a nonconstant meromorphic function on $\mathfrak{M}_r$. Then $(\tilde{\tau}_1, \ldots, \tilde{\tau}_l, \rho_r\tilde{\tau}_1, \ldots, \rho_r\tilde{\tau}_l)$: $\mathfrak{M} \to \mathbf{P}^{2l-1}$ is nondegenerate. Let $\mathfrak{T}$ be the subsheaf of $\mathcal{O}(mK_{\mathfrak{M}} + F_m^* + M_\infty)$ generated by the $\tilde{\tau}_{j,r}$ and the $\rho_r \cdot \tilde{\tau}_{j,r}$. We now look at $\mathfrak{T}$ near M'. Let τ' in (4.7) be a local generator for $\mathcal{O}(mK - mJ - D')$ as a subsheaf of $\mathcal{O}(mK)$. In (4.7), τ' has a divisor equal to $mJ + D'$. Let c and d be such that $mJ + D' = (x^c y^d)$. Then

$$(4.10)\qquad \tau' = x^c y^d v(x, y)(dx \wedge dy)^m, \quad \text{with } v(x, y) \neq 0.$$

τ' extends via (4.8) to

$$(4.11)\qquad \tau = x^{ma+c} y^{mb+d} u^m(x, y) v(x, y)(dx \wedge dy \wedge dz)^m$$

with $\tau \in \Gamma(M_0, \mathcal{O}(mK_0))$. Then a local lifting of τ to $\tilde{\tau} \in \mathcal{O}(mK_{\mathfrak{M}})$ will be such that the divisor of $\tilde{\tau}$ equals $(x^{ma+c} y^{mb+d})$. That is, there is a divisor H_m, supported on E_0 near M' such that near M', on $\mathfrak{M}$,

$$\mathcal{T} = \mathcal{O}(mK_{\mathfrak{M}} + F_m^* + M_\infty - H_m) \tag{4.12}$$

and such that $H_m \cap M' = mJ + D'$.

So $\mathcal{T}$ is locally free near M'. By [**41**, §2, pp. 268–272 or **19**, §1, pp. 314–320], we may modify $\mathfrak{M}_r$ so that $\mathcal{T}$ becomes locally free on $\mathfrak{M}_r$. This modification does not change $\mathfrak{M}_r$ near M' since $\mathcal{T}$ is already locally free near M'. We may extend our definition of H_m above and say that there is a divisor H_m on $\mathfrak{M}_r$ such that $\mathcal{T} = \mathcal{O}(mK_{\mathfrak{M}} + F_m^* + M_\infty - H_m)$ on $\mathfrak{M}_r$. Near M', H_m is supported on E_0 and $H_m \cap M' = mJ + D'$. Then observe that p is an isolated point of $\Pi_r(M_r' \cap \operatorname{supp} H_m)$.

Since $\mathcal{O}((m+1)K - D)$ has no base points on M, $\mathcal{O}((m+1)K - (m+1)J - D')$ has no base points on M'. Then we may choose initial meromorphic sections $\sigma_{k,r}'$ of $\mathcal{O}((m+1)K - (m+1)J - D')$ on M_r' such that (a), (b) and (c) are satisfied for the germs $(\alpha^*)^{-1}(\sigma_k')$ of the $\sigma_{k,r}'$ at A'. $\xi_r(\sigma_{k,r}')$ is a meromorphic pluricanonical form on V_r which is not necessarily holomorphic at p and at some other subvarieties of V_r. By multiplying $\xi_r(\sigma_{k,r}')$ by a meromorphic function, we may assume that $\xi_r(\sigma_{k,r}')$ has a zero of any desired order at $\Pi_r(M_r' \cap \operatorname{supp} H_m) - p$. Recall that σ_k is the extension via (4.7) and (4.8) of σ_k' from M' to M_0. So there is an effective divisor G_{m+1} on $\mathcal{V}_r$, which we also choose to be ample, such that all of the σ_k have representatives $\sigma_{k,r}$ on $M_{0,r}$ such that, letting $S := \operatorname{supp}(M_{0,r}) - \operatorname{supp}(E_0)$,

$$\sigma_{k,r|S} \in \Gamma(S, \mathcal{O}((m+1)K_0) \otimes \mathcal{O}(G_{m+1}^* - H_m)). \tag{4.13}$$

$\mathcal{O}(G_{m+1}^*)$ is a quasipositive sheaf. So also $\mathcal{T} \otimes \mathcal{O}(G_{m+1}^*)$ is quasipositive. Then [**16**, Satz 2.2, p. 273], $H^1(M_r, \mathcal{O}(K_{\mathfrak{M}} + G_{m+1}^* + M_\infty - M_{0,r}) \otimes T) = 0$.

Let $F_{m+1} = G_{m+1} + F_m + 2\lambda^{-1}(\infty)$. Using (4.12), we see that

$$\begin{aligned} 0 &\to \Gamma(M_r, \mathcal{O}((m+1)K_{\mathfrak{M}} + F_{m+1}^* - M_{0,r} - H_m)) \\ &\to \Gamma(M_r, \mathcal{O}((m+1)K_{\mathfrak{M}} + F_{m+1}^* - H_m)) \\ &\to \Gamma(M_{0,r}, \mathcal{O}((m+1)K_0) \otimes \mathcal{O}(F_{m+1}^* - H_m)) \\ &\to 0 \end{aligned}$$

is an exact sequence.

In order to complete the proof of the lemma, we need that

$$\sigma_{k,r} \in \Gamma(M_{0,r}, \mathcal{O}((m+1)K_0) \otimes \mathcal{O}(F_{m+1}^* - H_m)).$$

(4.13) says that this is true on $\operatorname{supp}(M_{0,r}) - \operatorname{supp}(E_0)$. So we need to examine σ_k near E_0. Recall that the divisor of σ_k' is at least $(m+1)J + D' \geqslant mJ + D'$. Then, as in (4.10) and (4.11), any local lifting $\tilde{\sigma}_k$ of σ_k has a divisor $C \geqslant (x^{(m+1)a+c}y^{(m+1)b+d})$. Locally, $H = (x^{ma+c}y^{mb+d})$ and $E_0 = (x^a y^b)$. So $\tilde{\sigma}_k \in \mathcal{O}((m+1)K_{\mathfrak{M}} - H_m)$ and in fact

$$\tilde{\sigma}_k \in \mathcal{O}((m+1)K_{\mathfrak{M}} - H_m - M_0)$$

on $\operatorname{supp}(E_0) - M'$. So $\sigma_k \equiv 0$ in $\mathcal{O}((m+1)K_0) \otimes \mathcal{O}(-H_m)$ on $\operatorname{supp}(E_0) - M'$, near M'. So σ_k may be further extended by 0 (regardless of H_m on the rest of E_0) to $\sigma_k \in \Gamma(M_0, \mathcal{O}((m+1)K_0) \otimes \mathcal{O}(-H_m))$. That is, $\sigma_{k,r} \in \Gamma(M_{0,r}, \mathcal{O}((m+1)K_0) \otimes \mathcal{O}(F^*_{m+1} - H_m))$. This completes the proof of the lemma.

V. Very weak simultaneous resolution–general case. Weak simultaneous resolutions will appear in our proofs, so let us recall the definition [**49**, Définition 3.1.1, p. 105]:

DEFINITION 5.1. The map germ $\Pi: \mathfrak{M} \to V$ is a *weak simultaneous resolution* of the germ of the (flat) deformation $\lambda: \mathcal{V} \to T$ along the image T_1 of a section $\sigma: T \to V$, T reduced, if for all sufficiently small representatives of λ, the germ Π has a representative such that:

(0)–(ii) of Definition 4.1 are satisfied and

(iii) the map induced by restriction $\tilde{\rho} := \widetilde{\lambda \circ \Pi}: (\Pi^{-1}(T_1))_{\text{red}} \to T$ is simple, i.e. a locally trivial deformation.

We are interested, of course, in the case where $\lambda: \mathcal{V} \to T$ is the germ of a deformation of the normal two-dimensional singularity (V, p). By Definition 5.1(iii) and [**32**, Proposition 2.5, p. 5], if $\Pi: \mathfrak{M} \to \mathcal{V}$ is a weak simultaneous resolution along T_1, then $\Pi^{-1}(T_1)$ is the full exceptional set in $\mathfrak{M}$ and each V_t has just one singularity. Conversely, if $\Pi: \mathfrak{M} \to \mathcal{V}$ is a very weak simultaneous resolution from Definition 4.1, let $\mathcal{E}$ be the exceptional set in $\mathfrak{M}$ and $\bar{\rho}: \mathcal{E}_{\text{red}} \to T$ be the induced map. If $\bar{\rho}$ is simple, then each V_t has just one singularity, p_t. $\sigma: T \to \mathcal{V}$ given by $\sigma(t) = p_t$ is a section, i.e. is holomorphic, seen as follows. Let $\sigma_1: T \to \mathfrak{M}$ be a section with $\sigma_1(T) \subset \mathcal{E}_{\text{red}}$. σ_1 exists since $\bar{\rho}$ is assumed to be simple. Let $\sigma = \Pi \circ \sigma_1$. So we may define weak simultaneous resolution without introducing a section σ.

For each t, $\mathcal{E}_{\text{red},t}$ has only plane curve singularities. Plane curve singularities have a well-understood equisingularity theory [**55**, **49**]. So $\lambda: \mathcal{V} \to T$ has a weak simultaneous resolution if and only if λ has a very weak simultaneous resolution $\Pi: \mathfrak{M} \to \mathcal{V}$ such that $\bar{\rho}: \mathcal{E}_{\text{red}} \to T$ is an equisingular deformation of $A := \mathcal{E}_{\text{red},0}$.

(V, p) is our given germ of a normal Gorenstein singularity. by [**2**, Theorem 3.8, p. 32], we may assume that V has a projective algebraic representative. Let $\tau': \mathcal{X}' \to U'$ be the versal deformation of (V, p). τ' is often called the miniversal or semiuniversal deformation of (V, p). τ' is the germ of a deformation. By [**10**], we may assume that τ' has a projective algebraic representative $\tau'_r: \mathcal{X}'_r \to U'_r$. Let $\tau: \mathcal{X} \to U$ and $\tau_r: \mathcal{X}_r \to U_r$ be the reductions of $\tau': \mathcal{X}' \to U'$ and $\tau'_r: \mathcal{X}'_r \to U'_r$ respectively.

THEOREM 5.2. *Let $\lambda: \mathcal{V} \to T$ be the germ of a (flat) deformation of the normal Gorenstein two-dimensional singularity (V, p) with T a reduced analytic space. Then $-K_0 \cdot K_0 \geqslant -K_t \cdot K_t$ for all $t \in T$ sufficiently near to 0. Also, $S := \{t \in T \mid K_t \cdot K_t = K_0 \cdot K_0\}$ is a subvariety of T.*

PROOF. To prove this theorem, it suffices to prove it for the special case that $\lambda: \mathcal{V} \to T$ is $\tau: \mathcal{X} \to U$, the reduction of the versal deformation of (V, p). Let

$\Pi: \mathfrak{M} \to \mathfrak{X}$ be a (not necessarily simultaneous) resolution of τ. Π has a projective algebraic representative $\Pi_r: \mathfrak{M}_r \to \mathfrak{X}_r$. Then, off a nowhere dense proper subvariety U' of U, Π is a weak simultaneous resolution. Then $K_u \cdot K_u$ is constant on each connected component of $U - U'$. Let U_i be an (analytically) irreducible component of U. Let $\gamma: \mathbf{P}^1 \to U_r$ be a holomorphic map such that $\gamma(0) = 0$ and $\gamma(x) \in U_i - U'$ for x near 0 but $x \neq 0$. γ induces a deformation of (V, p) over $\mathbf{P}^1$. By Proposition 4.6, $-K_0 \cdot K_0 \geqslant -K_u \cdot K_u$ for $u \in U_i - U'$. Hence $-K_0 \cdot K_0 \geqslant -K_u \cdot K_u$ for $u \notin U'$.

Now restrict τ to U', a projective algebraic subvariety of U. Then, arguing as before, there exists U'', a nowhere dense subvariety of U', such that $-K_0 \cdot K_0 \geqslant -K_u \cdot K_u$ for $u \in U' - U''$. After a finite number of steps, we conclude that $-K_0 \cdot K_0 \geqslant -K_u \cdot K_u$ for all $u \in U$, u sufficiently near to 0.

The above argument also shows that S is a subvariety.

THEOREM 5.3. *Let $\lambda: \mathfrak{V} \to T$ be the germ of a (flat) deformation of the normal Gorenstein two-dimensional singularity (V, p) with T a reduced analytic space. If $K_t \cdot K_t$ is constant, then also h_t is constant.*

PROOF. It suffices to prove the theorem for the special case that $\lambda: \mathfrak{V} \to T$ is $\tau: \mathfrak{X} \to U$, the reduction of the versal deformation of (V, p). Where τ has a very weak simultaneous resolution, h_t is locally constant by [**52**, **42**]. Using Proposition 4.7, the result then follows as in the proof of Theorem 5.2.

PROPOSITION 5.4. *Let $\lambda: \mathfrak{V} \to T$ be the germ of a (flat) deformation of the normal Gorenstein two-dimensional singularity (V, p) with T a reduced analytic space. Suppose that λ has a very weak simultaneous resolution. Then $K_t \cdot K_t$ is constant.*

PROOF. This proposition is known; see [**46**, Proposition 10, p. 10]. Another proof is as follows. Resolving T, we may assume that $\mathfrak{V}$ is Gorenstein. Then there is a nonzero holomorphic canonical form Ω on $\mathfrak{V} - \operatorname{Sing}(\mathfrak{V})$. Let $\Pi: \mathfrak{M} \to \mathfrak{V}$ be the very weak simultaneous resolution of $\mathfrak{V}$. Then $\Pi^*(\Omega)$ has a continuously varying divisor K_t on each M_t. Since $K_t \cdot K_t$ is integer valued, $K_t \cdot K_t$ is constant.

Observe the following. Let $\omega: \mathfrak{M} \to T$ be a deformation of a resolution of M with T reduced. Then [**52**, **42**], ω simultaneously blows down if and only if $h_t := \dim H^1(M_t, \mathcal{O})$ is constant. So in the case that (V, p) is Gorenstein, for ω, h_t constant implies that $K_t \cdot K_t$ is constant. For λ of Theorem 5.3, $K_t \cdot K_t$ constant implies that h_t is constant.

What about converses to these implications? There are many known examples (see [**9**, **21**, **54**]) of deformations $\lambda: \mathfrak{V} \to T$ where $h_t = 1$ for all t and $K_t \cdot K_t$ changes.

Now look at deformations $\omega: \mathfrak{M} \to T$ with (V, p) Gorenstein. With (V, p) other than rational or minimally elliptic, i.e. with $h := h_0 \geqslant 2$, there are deformations ω which induce topologically trivial deformations of the exceptional sets in M_t such that the blow down of M_t is not Gorenstein [**29**, Theorem 4.3, p. 1282]. Since deformations of a Gorenstein singularity are also Gorenstein, ω cannot simultaneously blow down. $K_t \cdot K_t$ may be defined topologically, even for $t \neq 0$, using rational cycles. $K_t \cdot K_t$ is constant.

Here is another example. Let A be a Riemann surface of genus $g \geqslant 3$ with a hyperelliptic Weierstrass point q. Let N be the line bundle $-q$ on A. Let M be the total space of N. Then A, thought of as the 0-section of N, blows down in M. Now deform $A_0 := A$ in a family such that A_t is not hyperelliptic for $t \neq 0$. Let K_t denote the canonical bundle on A_t. Deform $N_0 := N$ so that $-(2g-2)N_t = K_t$ for all t. Then blowing down A_t in M_t gives a Gorenstein singularity for all t. But for $t \neq 0$, $h_t < h_0$.

LEMMA 5.5. *Let* $\lambda: \mathcal{V} \to T$ *be the germ of a (flat) deformation of* (V, p) *with* T *the germ of a smooth* 1-*dimensional space. Let* t *be a local coordinate for* T. *Then either* λ *is the trivial deformation, or else there is a smallest integer* $n \geqslant 1$ *such that for any map* $g: T \to U$ *which induces* λ *from* $\tau: \mathcal{X} \to U$,

(i) g *is uniquely determined and nonzero on the nonreduced subspace* $\{t^{n+1} = 0\}$ *and*

(ii) g *is the zero map on the nonreduced subspace* $\{t^n = 0\}$.

PROOF. Suppose that λ is not the trivial deformation. Let n be the greatest integer such that λ induces the trivial deformation on $\{t^n = 0\}$. We need to show that an induced map $g: T \to U$ is uniquely determined on $\{t^{n+1} = 0\}$ (see [**6**, p. 102]). It suffices that for $2 \leqslant m \leqslant n$, every automorphism of λ over $\{t^m = 0\}$ lift to an automorphism over $\{t^{m+1} = 0\}$. Since λ is the trivial deformation over $\{t^m = 0\}$, an automorphism of λ is a family of automorphisms of (V, p) over $\{t^m = 0\}$. The automorphisms of λ over $\{t^2 = 0\}$ are $\Gamma(V, T_V)$, vector fields on V. Integrating a vector field extends the infinitesimal automorphism to T.

For general $m \geqslant 2$, we prove by induction that all automorphisms of (V, p) over $\{t^m = 0\}$ extend to $\{t^{n+1} = 0\}$. By the induction hypothesis, the family of automorphisms of λ over $\{t^{m-1} = 0\}$ extends to $\{t^{n+1} = 0\}$. Composing λ over $\{t^m = 0\}$ with the inverse of this extension, we may assume that λ is the trivial family of automorphisms over $\{t^{m-1} = 0\}$. Let T' be 1-dimensional and smooth with local coordinate ε. Via $\varepsilon = t^m$, λ over $\{t^m = 0\}$ may be induced from λ', a family of automorphisms of (V, p) over $\{\varepsilon^2 = 0\}$. Then, as for $m = 2$, the automorphisms of λ' extend to automorphisms over T'. This induces an extension of λ over $\{t^m = 0\}$ to $\{t^{n+1} = 0\}$.

Recall [**34**, Corollary 2.12, p. 190] as follows. Our notation in this paper differs slightly from that in [**34**]. Let $\pi: M \to V$ be the minimal resolution of (V, p), a normal two-dimensional Gorenstein singularity. Let $\omega: \mathfrak{M} \to Q$ be a 1-convex map which is the versal deformation of M. Q is smooth. Let $T_a = \{q \in Q \mid h_q := \dim H^1(M_q, \mathcal{O}) = h_0\}$ be the simultaneous blow-down subspace of Q. T_a is reduced. Let $\omega_a: \mathfrak{M}_a \to T_a$ be the restriction of ω to $\omega^{-1}(T_a)$. Let $\pi_a: \mathcal{V}_a \to T_a$ be the simultaneous blow-down of $\mathfrak{M}_a$. Let $f: T_a \to U$ induce π_a from $\tau: \mathcal{X} \to U$, the reduction of the versal deformation of (V, p). Then [**34**, Corollary 2.12, p. 140], f is finite and proper. See also [**4** and **53**] for related constructions.

THEOREM 5.6. *Let* $f\colon T_a \to U$ *be as above. Let* $S = \{u \in U \mid K_u \cdot K_u = K_0 \cdot K_0\}$. *Then* $f(T_a) = S$.

PROOF. $f(T_a)$ is a subvariety of S by [**13**] and Proposition 5.4. We shall assume that $f(T_a) \neq S$ and reach a contradiction. Let S_1 be an irreducible component of S such that $S_1 \not\subset f(T_a)$. Then $\dim[S_1 \cap f(T_a)] < \dim S_1$. Temporarily blow up S at 0, $\phi\colon S' \to S$. Let $E = \phi^{-1}(0)$. Let primes$'$ denote proper transforms via ϕ. For any irreducible subvariety Z of S, $\dim(E \cap Z') + 1 = \dim Z' = \dim Z$. So there exists $y \in E \cap S_1'$ such that $y \notin E \cap (S_1 \cap f(T_a))'$.

By Theorem 5.2, S is a subvariety of U. Let $S = \bigcup S_i$ be the decomposition of S into irreducible components. Observe as follows that for each S_i, there is a (projective algebraic) subvariety $\bar{S}_{i,r}$ of U_r such that S_i is an (analytically) irreducible component of the germ of $\bar{S}_{i,r}$ at 0: Let $\Pi_r\colon \mathfrak{M}_r \to \mathfrak{X}_r$ be a not necessarily simultaneous resolution of $\mathfrak{X}_r$. Then off a nowhere dense subvariety of U_r' of U_r, Π_r is a weak simultaneous resolution. Take $U_r' \supset \mathrm{Sing}(U_r)$. So $K_u \cdot K_u$ is constant on each connected component of $U_r - U_r'$. If S_i is an irreducible component of U, then let $\bar{S}_{i,r} = U_r$, and we are done. If not, S_i is contained in the germ of U_r' at 0. Now repeat the argument using U_r'.

Then there is a holomorphic map $\gamma_r'\colon \mathbf{P}^1 \to \bar{S}_{1,r}'$ such that $\gamma_r'(0) = y$, $\gamma_r := \phi_r \circ \gamma_r'$ is not constant, and for t near 0 in $\mathbf{P}^1$, $t \neq 0$, $\gamma_r(t) \in S_1 - f(T_a) - \bigcup S_i$, $i \neq 1$. γ_r induces a deformation $\lambda\colon V \to T$. By Theorem 4.3 and [**8**], there is a finite base change $\alpha\colon T_1 \to T$ such that the induced deformation $\lambda_1\colon \mathcal{V}_1 \to T_1$ has a very weak simultaneous resolution, $\Pi\colon \mathfrak{M}_1 \to \mathcal{V}_1$. $\lambda_1\colon \mathcal{V}_1 \to T_1$ is induced from $\tau\colon \mathfrak{X} \to U$ via $\gamma \circ \alpha\colon T_1 \to U$.

Let $\rho_1 := \lambda_1 \circ \Pi\colon \mathfrak{M}_1 \to T_1$. Then the deformation ρ_1 may be induced from $\omega\colon \mathfrak{M} \to Q$ [**33**, Theorem 5, p. 515]. Let us recall part of the proof of [**33**, Theorem 5, p. 515]: $B := Q \times T_1$. We form a deformation $\tau\colon \mathcal{Y} \to B$ such that $\tau_{|O\times T_1} \approx \rho_1$ and $\tau_{|Q\times O} \approx \omega$. There is a submersion $\sigma\colon B \to Q$ which induces τ from ω. We may take $\tau_{|Q\times O}\colon Q \to Q$ to be the identity map. $R := \sigma(O \times T_1) \subset T_a$.

Since ρ_1 is not the trivial deformation and T_1 is one-dimensional, $\sigma\colon O \times T_1 \to T_a$ is proper and R is (the germ of) an irreducible subvariety. Let $B_1 := \sigma^{-1}(R)$, an irreducible subvariety of B. In B_1, R may be identified with $R \times O$, which is transverse to $O \times T_1$. $K_b \cdot K_b$ is constant on Y_b for $b \in B_1$. Since h_t is constant on $M_{1,t}$, $t \in T_1$, h_b is also constant on Y_b for $b \in B_1$. Hence we may simultaneously blow down $\tau\colon \mathcal{Y} \to B$ over B_1, yielding $\tau_1\colon \mathcal{W} \to B_1$, a (flat) deformation of (V, p). On $O \times T_1 \subset B_1$, τ_1 restricts to a deformation isomorphic to $\lambda_1\colon \mathcal{V}_1 \to T_1$. On $R \times O \subset B_1$, τ_1 restricts to a deformation isomorphic to π_a restricted to R. Let $g\colon (O \times T_1) \cup (R \times O) \to U$ be given by $g_{|O\times T_1} = \gamma \circ \alpha$ and $g_{|R\times O}$ be the restriction of $f\colon T_a \to U$ to R. Since $O \times T_1$ and $R \times O$ are transverse in B_1, g is holomorphic and induces τ_1 restricted to $(O \times T_1)$ and $(R \times O)$. By [**40**, Théorème (b), pp. 162–163] g may be extended to a holomorphic map $g_1\colon B_1 \to U$. Necessarily, $g(B_1) \subset S$.

Since B_1 is irreducible, $g_1(B_1)$ is contained in an irreducible component S_j of S. Since $g_1(B_1) \supset \gamma \circ \alpha(T_1)$ by construction, in fact $g_1(B_1) \subset S_1$. Then $g_1(R \times O) \subset S_1 \cap f(T_a)$. But the deformation $\lambda_1: \mathcal{V}_1 \to T_1$ may be induced, by construction, from $\tau: \mathcal{X} \to U$ restricted to $\tau^{-1}(g(R)) = \tau^{-1}(g_1(R \times O))$. This contradicts Lemma 5.5 and our selection above, in the first paragraph of the proof, of $y \in (E \cap S_1') - [E \cap (S_1 \cap f(T_a))']$.

THEOREM 5.7. *Let $\lambda: \mathcal{V} \to T$ be the germ of a (flat) deformation of the normal Gorenstein two-dimensional singularity (V, p) with T a reduced analytic space. Then λ has a very weak simultaneous resolution, possibly after a finite base change, if and only if $K_t \cdot K_t$ is constant.*

PROOF. If λ has a very weak simultaneous resolution, then $K_t \cdot K_t$ is constant by Proposition 5.4.

Conversely, assume that $K_t \cdot K_t$ is constant. $\lambda: \mathcal{V} \to T$ is induced from $\tau: \mathcal{X} \to U$ via a holomorphic map $g: T \to U$ with $g(T) \subset S$ of Theorem 5.6. $f: T_a \to U$ has $f(T_a) = S$ by Theorem 5.6. Let $T' = T \times_U T_a$ be the fiber product [**12**, Corollary 0.32, p. 29]. Then the map $\alpha: T' \to T$ is finite and proper since f is finite and proper. α is surjective since $g(T) \subset f(T_a)$. Then α is a finite base change. Via the map $\beta: T' \to T_a$ we see that the induced deformation $\lambda': \mathcal{V}' \to T'$ has a very weak simultaneous resolution.

VI. Weak simultaneous resolution. Recall Neumann's theorem [**38**]. Let (V, p) be the germ of a normal two-dimensional singularity. Let $\pi': M' \to V$ be the minimal good resolution. Let A' be the exceptional set in M'. Then the oriented homotopy type of $V - p$ determines the topology of the pair (M', A'). Hence also if $\pi: M \to V$ is the minimal resolution, the oriented homotopy type of $V - p$ determines the topology of the pair (M, A). In most cases, (M, A) is already determined by the fundamental group of $V - p$. Observe that Neumann's theorem implies that the oriented homotopy type of $V - p$ determines the topology of (V, p).

Let $\lambda: \mathcal{V} \to T$ be the germ of a deformation of the normal Gorenstein two-dimensional singularity (V, p) with T reduced. If one of the singularities p_t of V_t is topologically equivalent to p in V, then in fact p_t is the only singularity in V_t. Indeed, since $K_t \cdot K_t$ is the sum of $K \cdot K$ over all singularities in V_t, Theorem 5.2 implies that $t \in S$. By Theorem 5.7, we may resolve λ over S, after a finite base change. Choose the minimal very weak simultaneous resolution. Then by [**32**, Proposition 2.5, p. 5] and Neumann's theorem, p_t is the only a singularity in V_t.

LEMMA 6.1. *Let $\lambda: \mathcal{V} \to T$ be the germ of a (flat) deformation of the normal Gorenstein two-dimensional singularity (V, p) with T a reduced analytic space. Then $W = \{t \in T \mid V_t$ has a singularity p_t such that (V_t, p_t) is homeomorphic to $(V, p)\}$ is a subvariety of T.*

PROOF. As observed above, $W \subset S$, S from Theorem 5.2. As above, after a finite base change, resolve λ above S with the minimal very weak simultaneous resolution, $\Pi: \mathcal{M} \to \mathcal{V} \to S$. We may ignore base changes. By Neumann's theorem, for $w \in W$, A_w has the same intersection matrix as $A_0 = A$. Then, as in the proof of [**34**, Theorem 2.11, p. 189], each irreducible component is homologous in $\mathcal{M}$ to an irreducible component $A_{w,i}$ of A_w. Let S' be the maximal subvariety of S above which all of the A_i weakly lift [**32**, Proposition 2.7, p. 6]. Then $W \subset S'$. Let $\alpha: S'' \to S'$ be a resolution of S', a proper map. It suffices to prove the lemma near each point of S'' for the induced deformation over S''. Since S'' is smooth, each A_i locally lifts to above S''. Let $\mu(A_s)$ be the sum of the Milnor numbers for the (plane curve) singularities of A_s, $s \in S''$. Then $\mu(A_w) = \mu(A)$ for $w \in W$. Then W is the subvariety of S'' where $\mu(A_s)$ is constant.

LEMMA 6.2. *Let* $\lambda: \mathcal{V} \to T$ *be as in Theorem* 4.3. *Suppose additionally that each* V_t *has a singularity* p_t *such that* (V_t, p_t) *is homeomorphic to* (V, p). *Then* λ *has a weak simultaneous resolution such that each fiber is a minimal good resolution.*

PROOF. Let $\Psi: \mathcal{N} \to \mathcal{V}$ be the RDP resolution of Theorem 4.3. By Neumann's theorem, each fiber N_t has an isomorphic set of rational double points. Since μ is upper semicontinuous, the deformation which Ψ induces on the rational double points is trivial. So there is a minimal weak simultaneous resolution of the singularities of $\mathcal{N}$. This gives a very weak simultaneous resolution of λ, $\Pi: \mathcal{M} \to V$. The fibers M_t are all homeomorphic near the exceptional sets A_t. As in [**34**, p. 189], Π induces an equisingular deformation of $A = A_0$. Then additional simultaneous blow-ups of the M_t along suitable singularities of the A_t give a weak simultaneous resolution. The fibers can all be chosen to be minimal good resolutions.

Let (V, p) be as above. Let $\pi': M' \to V$ be the minimal good resolution of V, the MGR in the resolution of [**52**]. Let $\omega': \mathcal{M}' \to Q'$ be the versal deformation of M' [**33**]. Let A' be the (reduced) exceptional set in M'. As in [**52**, p. 333], let $\mathcal{S}$ be the sheaf of vector fields on M' which map Id A' to Id A'. Let θ' be the tangent sheaf on M'. Let $\mathcal{N} := \theta'/\mathcal{S}$. Then $\mathcal{N} \approx \oplus \mathcal{N}_i$, where $\mathcal{N}_i$ is the normal sheaf to A'_i, an irreducible component of A'. As in [**32**, (2.2), p. 7],

$$0 \to H^1(M', \mathcal{S}) \to H^1(M', \theta) \to H^1(M', \mathcal{N}) \to 0$$

is an exact sequence. By the construction in [**33**, p. 313], there is a submanifold P of Q', $0 \in P$, with tangent space $T_{P,0} \approx H^1(M', \mathcal{S})$ such that all of the A'_i lift to above P. By [**34**, Theorem 3.5, p. 194], P is the maximal subspace of Q' above which all of the A'_i lift weakly.

Let $T_w = \{w \in P \mid h_w = h_0\}$ be the simultaneous blow-down-subspace of P. Let $\lambda_w: \mathcal{V}_w \to T_w$ be the induced deformation. Since the tangent space to T_w is contained in $H^1(M', \mathcal{S})$, the induced map from T_w to ES of Wahl [**52**, especially p. 335] is an embedding. Recall [**52**, Theorem 4.6(b), p. 341], the induced map $ES \to U'$ is injective. So any $g: T_w \to U$ which induces λ_w from the reduced versal deformation $\tau: \mathcal{X} \to U$ is an embedding (see [**22**, pp. 205—18] for a treatment in the analytic category).

THEOREM 6.3. *Let* $g\colon T_w \to U$ *be as above. Let* $W = \{u \in U \mid X_u$ *has a singularity* p_u *such that* (X_u, p_u) *is homeomorphic to* $(V, p)\}$. *Then* $g(T_w) = W$.

PROOF. $g(T_w) \subset W$ be Definition 5.1. The rest of the proof of Theorem 6.3 is the same as the proof of Theorem 5.6 via the changes:

"W" replaces "S" and the condition $\{X_u$ has a singularity p_u such that (X_u, p_u) is homeomorphic to $(V, p)\}$ replaces the condition $\{K_u \cdot K_u = K_0 \cdot K_0\}$.

"T_w" replaces "T_a".

Lemma 6.1 replaces Theorem 5.2.

Lemma 6.2 replaces Theorem 4.3 and [**8**]. The base change $\alpha\colon T_1 \to T$ is not needed.

$\omega\colon \mathfrak{M}' \to Q'$ replaces $\omega\colon \mathfrak{M} \to Q$.

THEOREM 6.4. *Let* $\lambda\colon \mathcal{V} \to T$ *be the germ of a (flat) deformation of the normal Gorenstein two-dimensional singularity* (V, p) *with* T *a reduced analytic space. Then* λ *has a weak simultaneous resolution if and only if each* V_t *has a singularity* p_t *such that* (V_t, p_t) *is homeomorphic to* (V, p).

PROOF. This theorem follows immediately from Definition 5.1, Theorem 6.3, the versality of $\tau\colon \mathfrak{X} \to U$ and the fact [**52**, Theorem 4.6(b), p. 341] that $g\colon T_w \to W$ is biholomorphic.

REFERENCES

1. M. Artin, *On isolated rational singularities of surfaces*, Amer. J. Math. **88** (1966), 129–136.

2. ______, *Algebraic approximation of structures over complete local rings*, Inst. Hautes Etudes Sci. Publ. Math. **36** (1969), 23–58.

3. ______, *Algebraization of formal modules*. II. *Existence of modifications*, Ann. of Math. (2) **91** (1970), 88–135.

4. ______, *Algebraic construction of Brieskorn's resolutions*, J. Algebra **29** (1974), 330–348.

5. H. Bass, *On the ubiquity of Gorenstein rings*, Math. Z. **82** (1963), 8–28.

6. A. Beauville, *Foncteurs sur les anneaux artiniens. Application aux déformations verselles*, Astérisque **16** (1974), 82–104.

7. J. Briancon and J. Speder, *Les conditions de Whitney impliquent "μ(*) constant"*, Ann. Inst. Fourier (Grenoble) **26** (1976), 153–164.

8. E. Brieskorn, *Singular elements of semi-simple algebraic groups*, Proc. Internat. Congr. Math., Nice 1970, Vol. 2, Gauthier-Villars, Paris, 1971, pp. 279–284.

9. ______, *Die Hierarchie der* 1-*Modularen Singularitäten*, Manuscripta Math. **27** (1979), 183–219.

10. R. Elkik, *Algébrisation du module formel d'une singularité isolée*, Astérisque **16** (1974), 133–144.

11. ______, *Singularités rationnelles et deformations*, Invent. Math. **47** (1978), 139–147.

12. G. Fischer, *Complex analytic geometry*, Lecture Notes in Math., vol. 538, Springer-Verlag, Berlin and New York, 1976.

13. H. Grauert, *Ein Theorem der analytischen Garbentheorie und die Modulräume komplexer Strukturen*, Inst. Hautes Etudes Sci. Publ. Math. **5** (1960).

14. ______, *Uber Modifikationen und exzeptionelle analytische Mengen*, Math. Ann. **146** (1962), 331–368.

15. H. Grauert and H. Kerner, *Deformationen von Singularitäten komplexer Räume*, Math. Ann. **153** (1964), 236–260.

16. H. Grauert and O. Riemenschneider, *Verschwindungssätze für analytische Kohomologiegruppen auf komplexen Räumen*, Invent. Math. **11** (1970), 263–292.

17. J. Herzog and E. Kunz, *Der kanonische Modul eines Cohen-Macaulay-Rings*, Lecture Notes in Math., vol. 238, Springer-Verlag, Berlin and New York, 1971.

18. H. Hironaka, *Resolution of singularities of an algebraic variety over a field of characteristic zero*. I, II, Ann. of Math. (2) **79** (1964), 109–326.

19. H. Hironaka and H. Rossi, *On the equivalence of embeddings of exceptional complex spaces*, Math. Ann. **156** (1964), 313–333.

20. F. Hirzebruch, *Über vierdimensionale Riemannsche Flächen mehrdeutiger analytischer Funktionen von zwei komplexen Veränderlichen*, Math. Ann. **126** (1953), 1–22.

21. U. Karras, *Deformations of cusp singularities*, Proc. Sympos. Pure Math., vol. 30, Amer. Math. Soc., Providence, R. I., 1977, pp. 37–44.

22. ______, *Methoden zur Berechnung von Algebraischen Invarianten und zur Konstruktion von Deformationen normaler Flächensingularitäten*, Habilitationsschrift, Dortmund, 1981.

23. M. Kato, *Riemann-Roch theorem for strongly pseudoconvex manifolds of dimension* 2, Math Ann. **222** (1976), 243–250.

24. K. Kodaira, *On compact complex analytic surfaces*. I, Ann. of Math. (2) **71** (1960), 111–152.

25. ______, *On stability of compact submanifolds of complex manifolds*, Amer. J. Math. **85** (1963), 79–94.

26. H. Laufer, *Normal two-dimensional singularities*, Ann. of Math. Studies, no. 71, Princeton Univ. Press, Princeton, N. J., 1971.

27. ______, *On rational singularities*, Amer. J. Math. **94** (1972), 597–608.

28. ______, *Deformations of resolutions of two-dimensional singularities*, Rice Univ. Studies **59** (1973), 53–96.

29. ______, *On minimally elliptic singularities*, Amer. J. Math. **99** (1977), 1257–1295.

30. ______, *On μ for surface singularities*, Proc. Sympos. Pure Math., vol. 30, Amer. Math. Soc., Providence, R. I., 1977, pp. 45–49.

31. ______, *On normal two-dimensional double point singularities*, Israel J. Math. **31** (1978), 315–334.

32. ______, *Ambient deformations for exceptional sets in two-manifolds*, Invent. Math. **55** (1979), 1–36.

33. ______, *Versal deformations for two-dimensional pseudoconvex manifolds*, Ann. Scuola Norm. Sup Pisa Cl. Sci. **7** (1980), 511–521.

34. ______, *Lifting cycles to deformations of two-dimensional pseudoconvex manifolds*, Trans. Amer. Math. Soc. **266** (1981), 183–202.

35. D. Lê and C. Ramanujan, *The invariance of Milnor's number implies the invariance of the topological type*, Amer. J. Math. **98** (1976), 67–78.

36. J. Lipman, *Double point resolutions of deformations of rational singularities*, Compositio Math. **38** (1979), 37–42.

37. B. Moišezon, *On n-dimensional compact varieties with n algebraically independent meromorphic functions*, Amer. Math. Soc. Transl. **63** (1967), 51–177.

38. W. Neumann, *A calculus for plumbing applied to the topology of complex surface singularities and degenerating complex curves*, Trans. Amer. Math. Soc. **268** (1981), 299–344.

39. P. Orlik and P. Wagreich, *Algebraic surfaces with k^* action*, Acta Math. **138** (1977), 43–81.

40. G. Pourcin, *Déformation de singularités isolées*, Astérisque **16** (1974), 161–173.

41. O. Riemenschneider, *Characterizing Moisezon spaces by almost positive coherent analytic sheaves*, Math. Z. (1971), 263–284.

42. ______, *Bermerkungen zur Deformationstheorie nicht-rationaler Singuläritaten*, Manuscripta Math. **14** (1974), 91–100.

43. ______, *Familien komplexer Räume mit streng pseudokonvexer spezieller Faser*, Comment. Math. Helv. **51** (1976), 547–565.

44. K. Saito, *Einfach-elliptische Singuläritaten*, Invent. Math. **23** (1974), 289–325.

45. J. P. Serre, *Groupes algébriques et corps de classes*, Actualités Sci. Indust., no. 1264, Hermann, Paris, 1959.

46. N. Shepherd-Barron, *Some questions on singularities in* 2 *and* 3 *dimensions*, Thesis, Mathematics Institute, University of Warwick, 1980.

47. Y.-T. Siu, *Analytic sheaf cohomology groups of dimension n of n-dimensional complex spaces*, Trans. Amer. Math. Soc. **143** (1969), 77–94.

48. B. Teissier, *Cycles évanescents, sections planes et conditions de Whitney*, Astérisque **7-8** (1973), 285–362.

49. ______, *Résolution simultanée*. I, II, Sém. sur les Singularités des Surfaces (Demazure et al., Editors), Lecture Notes in Math., vol. 777, Springer-Verlag, Berlin and New York, 1980.

50. J.-L. Verdier, *Stratifications de Whitney et théorème de Bertini-Sard*, Invent. Math. **36** (1970), 295–312.

51. P. Wagreich, *Elliptic singularities of surfaces*, Amer. J. Math. **92** (1970), 419–454.

52. J. Wahl, *Equisingular deformations of normal surface singularities*, Ann. of Math. (2) **104** (1976), 325–356.

53. ______, *Simultaneous resolution and discriminantal loci*, Duke Math. J. **46** (1979), 341–375.

54. ______, *Elliptic deformations of minimally elliptic singularities*, Math. Ann. **253** (1980), 241–262.

55. O. Zariski, *Studies in equisingularity*. I, II, Amer. J. Math. **87** (1965), 507–536; 972–1006.

STATE UNIVERSITY OF NEW YORK AT STONY BROOK

Proceedings of Symposia in Pure Mathematics
Volume **40** (1983), Part 2

INTRODUCTION TO LINEAR DIFFERENTIAL SYSTEMS

LÊ DŨNG TRÁNG AND ZOGHMAN MEBKHOUT

CONTENTS

Introduction. In this exposition we would like to introduce the reader to the theory of linear differential systems for studying "Singularities". This theory is sometimes called the theory of $\mathcal{D}_X$-modules, where $(X, \mathcal{O}_X)$ is a complex analytic manifold of dimension n. Because of recent applications of this theory, it seems to us important to make an introduction to the subject and write down a survey of some basic results of this theory. In this paper, as in the prior lectures **[56, 57, 59]**,

---

1980 *Mathematics Subject Classification.* Primary 14, 32, 34, 35.

we try to explain the theory of $\mathcal{D}_X$-modules in a simple and natural manner. Thus we shall omit the proofs. If we state or use somebody's result, we shall say it explicitly and give a reference for it.

After the work of *B. Malgrange* [**45**] in the algebraic aspects of partial differential equations, the systematic use of *Grothendieck* theory was carried out by the *M. Sato* school in the early seventies. In [**69**] we find a microlocal classification of the linear systems at a generic point of the characteristic variety. A little later, *B. Malgrange* discovered the relation between the *Bernstein-Sato* polynomial of the singularity of a hypersurface and the monodromy of this singularity. He proved [**47**] that the roots of the *Bernstein-Sato* polynomial are rational numbers in the case of an isolated singularity. Using *Hironaka*'s resolution of the singularities, *M. Kashiwara* proved the rationality of the roots of this polynomial in the general case [**33**]. Recently *B. Malgrange* found another nice proof of this result [**49**]. From this moment it became clear that the theory of linear partial differential systems was a powerful tool to study the topology of singular complex analytic spaces embedded in X. This was the main purpose of the second author as it is pointed out in the introduction of the lecture [**57**]. After proving the singular *Poincaré* lemma in this context [**53**] based on *Grothendieck*'s comparison theorem [**20**] and the theory of *holonomic systems*, he proved the "duality theorem" [**54**] which is a common generalization of *Poincaré* duality and *Serre* duality. To understand the *Poincaré-Verdier* [**75**] duality in this context he was led to prove that the derived category of regular holonomic complexes is equivalent to the derived category of constructible complexes on X. We would like to describe below this result which generalizes *Deligne*'s result [**12**] on local systems solving the so-called *Riemann-Hilbert* problem. If X is the variety of the Borel subgroups of a complex linear semisimple group, this result is the missing link in [**16**] which relates the *Verma* modules (algebra) to *Schubert* cells (topology) in order to prove the *Kazhdan-Lusztig* multiplicity formulas. This was done in [**3** and **6**]. In relation with the notion of perversity introduced by *M. Goresky* and *R. MacPherson* in [**17**], *P. Deligne* proved that the category of holonomic regular systems is equivalent to the category of perverse sheaves. It is also expected [**8**] that this result will be used to put a pure mixed *Hodge* structure on the intersection homology groups of a singular projective variety.

We shall restrict ourselves to the differential operator point of view since our main interest is the topology of singularities. We shall not discuss the microlocal point of view: the reader aiming to know it is referred to [**69, 34, 36**]. In this context one may find similar notions and results concerning microdifferential systems; in particular there is a notion of regular microdifferential systems (cf. [**36**]). We shall use freely the derived categories, since the most important results are stated in the "language" of derived categories. We refer the reader to [**25** and **74**] for an introduction to the derived categories.

We heartily thank *Marie-Jo Lécuyer* for the excellent job of typing she did for us.

2. The one-dimensional case. In this section we try to motivate and help the reader by listing classical results about linear differential equations in the one-dimensional case. To accustom the reader and to make the whole presentation comprehensive and coherent with what follows we have presented these results in the language used afterwards.

(2.1) *Solutions of a linear differential equation.* Let: $P(x, d/dx) = a_0(x)d^m/dx^m + \cdots + a_m(x)$ be an mth order complex analytic differential operator defined on the open disc D centered at 0 in $\mathbf{C}$, i.e. the functions a_i are complex analytic functions on D ($i = 0, \ldots, m$). We denote $D^* := D - \{0\}$. Let $\mathcal{O}_U$ be the sheaf of holomorphic functions on the open set U of $\mathbf{C}$. The well-known Cauchy theorem on the existence and unicity of the solutions of linear differential equations (cf. [**30**, §12.1]) can be stated as follows:

THEOREM (2.1.1) (CAUCHY). *Suppose that, for any $x \in D^*$, $a_0(x) \neq 0$, then the kernel* $(\operatorname{Ker} P)|_{D^*}$ *of P: $\mathcal{O}_{D^*} \to \mathcal{O}_{D^*}$ is a local system on D^* and its cokernel* Coker P *is zero.*

(2.1.2) Recall that a local system on a topological space X is a sheaf of $\mathbf{C}$-vector spaces which is locally constant and locally isomorphic to a constant sheaf $\mathbf{C}^p$, with $p \in \mathbf{N}$.

Let $x_0 \in D^*$. One may consider any $\varphi \in (\operatorname{Ker} P)_{x_0}$ as a solution at x_0 of the linear differential equation $Pf = 0$.

(2.1.3) Let γ be a loop of D^* at x_0. The analytic continuation of a solution $\varphi \in (\operatorname{Ker} P)_{x_0}$ of $Pf = 0$ along γ defines $\gamma_* \varphi \in \mathcal{O}_{D,x_0}$ after a turn along γ. Of course, $\gamma_* \varphi$ is another solution of $Pf = 0$ and one knows that it depends only on the homotopy class of γ in $\pi_1(D^*, x_0)$. We call $\gamma_* \varphi$ the *monodromy transform* of φ along γ. Thus we have a representation for the fundamental group of $\pi_1(D^*, x_0)$ on $(\operatorname{Ker} P)_{x_0}$. As $\gamma_* \varphi$ is another determination of the multivalued function Φ on D^* obtained from φ, the same representation of $\pi_1(D^*, x_0)$ induces a representation on $\mathcal{V}_{\Phi,x_0}$, the $\mathbf{C}$-vector space generated in $\mathcal{O}_{D,x_0}$ by the determinations of Φ at x_0. As $\mathcal{V}_{\Phi,x_0} \subset (\operatorname{Ker} P)_{x_0}$, the $\mathbf{C}$-vector $\mathcal{V}_{\Phi,x_0}$ is finite dimensional and Φ is a complex analytic multivalued function of *finite determination*.

Notice that this representation of $\pi_1(D^*, x_0)$ on $\mathcal{V}_{\Phi,x_0}$ defines a local system $\mathcal{V}_\Phi^*$ on D^* which is a subsheaf of $(\operatorname{Ker} P)_{|D^*}$.

(2.1.4) One may consider P: $\mathcal{O}_D \to \mathcal{O}_D$ and wonder what may happen at 0. This is given by

THEOREM (2.1.5) (CAUCHY-MALGRANGE [**46**]). *Suppose that $a_0(x) \neq 0$ for any $x \in D^*$; then the kernel* Ker P *and the cokernel* Coker P *of P: $\mathcal{O}_D \to \mathcal{O}_D$ are constructible sheaves which are local systems on D^* and the support of* Coker P *is* $\{0\}$.

(2.1.6) Recall that a sheaf of $\mathbf{C}$-vector spaces on a complex analytic space X is a *constructible sheaf* on X if there is a complex analytic stratification $(X_\alpha)_{\alpha \in A}$ of X such that the restriction of the sheaf on each X_α is a local system.

In view of proving (2.1.5) it is enough to prove that P induces a finite index operator of $\mathcal{O}_{D,0}$ i.e. $(\operatorname{Ker} P)_0$ and $(\operatorname{Coker} P)_0$ are finite-dimensional spaces on **C**. In [**46**] this fact is proved by *B. Malgrange* who gives altogether the value of the index $\chi(P,0)$ of P at 0,

$$\chi(P,0) := \dim(\operatorname{Ker} P)_0 - \dim(\operatorname{Coker} P)_0 = m - v(a_0)$$

where v is the valuation of an analytic function of $\mathcal{O}_{D,0}$ at 0.

(2.2) *$\mathcal{D}$-modules in dimension one.*

(2.2.1) Instead of considering the differential equation $Pf = 0$ of (2.1), one may consider the quotient $\mathcal{D}/\mathcal{D}P$ of the sheaf $\mathcal{D}$ of differential operators on D by the left ideal generated by P. Then $\mathcal{D}/\mathcal{D}P$ is a coherent left $\mathcal{D}$-module with the presentation

$$0 \to \mathcal{D} \xrightarrow{P} \mathcal{D} \to \mathcal{D}/\mathcal{D}P \to 0$$

where $\mathcal{D} \xrightarrow{P} \mathcal{D}$ is the right multiplication by P. Applying the functor $\mathcal{H}om_{\mathcal{D}}(\cdot\,, \mathcal{O}_D)$, where $\mathcal{O}_D$ has the obvious structure of left $\mathcal{D}$-module, one has the exact sequence of **C**-sheaf

$$0 \to \mathcal{H}om_{\mathcal{D}}(\mathcal{D}/\mathcal{D}P, \mathcal{O}_D) \to \mathcal{O}_D \xrightarrow{P} \mathcal{O}_D \to \mathcal{E}xt^1_{\mathcal{D}}(\mathcal{D}/\mathcal{D}P, \mathcal{O}_D) \to 0$$

where $\mathcal{O}_D \xrightarrow{P} \mathcal{O}_D$ is the **C**-morphism defined by the differential operator P itself, because, as one may check easily by oneself, there is a canonical isomorphism $\mathcal{H}om_{\mathcal{D}}(\mathcal{D}, \mathcal{O}_D) \xrightarrow{\sim} \mathcal{O}_D$.

Thus one finds another interpretation of $\operatorname{Ker} P$ and $\operatorname{Coker} P$ in (2.1.5)

$$\operatorname{Ker} P \cong \mathcal{H}om_{\mathcal{D}}(\mathcal{D}/\mathcal{D}P, \mathcal{O}_D), \qquad \operatorname{Coker} P \cong \mathcal{E}xt^1_{\mathcal{D}}(\mathcal{D}/\mathcal{D}P, \mathcal{O}_D),$$

i.e. these sheaves are respectively isomorphic to the 0th and 1st cohomologies of the complex $0 \to \mathcal{O}_D \xrightarrow{P} \mathcal{O}_D \to 0$.

Notice that Theorem (2.1.5) says that these 0th and 1st cohomology sheaves have supports of respective codimension 0 and 1 in D; this will be a general fact (cf. (4.6)).

(2.2.2) More generally one considers a coherent left $\mathcal{D}$-module $\mathcal{M}$ on D. On a sufficiently small disc D one has a resolution (cf. [**31**]):

$$0 \to \mathcal{D}^r \xrightarrow{\psi} \mathcal{D}^q \xrightarrow{\varphi} \mathcal{D}^p \to \mathcal{M} \to 0.$$

Applying the functor $\mathcal{H}om_{\mathcal{D}}(\cdot\,, \mathcal{O}_D)$ one obtains the complex **C**-vector spaces

$$0 \to \mathcal{O}_D^p \xrightarrow{\varphi^*} \mathcal{O}_D^q \xrightarrow{\psi^*} \mathcal{O}^r \to 0.$$

One may check that the cohomology of this complex does not depend on the chosen presentation and is zero in dimension different from 0 and 1. This complex defines *the holomorphic solutions* of the coherent left $\mathcal{D}$-module $\mathcal{M}$; more precisely the holomorphic solutions of the coherent left module $\mathcal{M}$ are the object of some derived category represented by this complex (cf. (3.1.6)). A coherent left $\mathcal{D}$-module will be called a *linear differential system on D*.

(2.3) *Regular singular equations.*

(2.3.1) Let us consider again the situation of (2.1) defined by an mth order differential operator. Let $x_0 \in D^*$. As we saw in (2.1.3) a solution $\varphi \in (\operatorname{Ker} P)_{x_0}$ of $Pf = 0$ defines a complex analytic multivalued function Φ on D^* which has finite determination, i.e. $\mathcal{V}_{\Phi, x_0}$ is a finite-dimensional **C**-vector space. Moreover if $\gamma \in \pi_1(D^*, x_0)$, it defines T_γ: $\mathcal{V}_{\Phi, x_0} \to \mathcal{V}_{\Phi, x_0}$ by $T_\gamma(\varphi) = \gamma_* \varphi$. We call T_γ the *monodromy of* Φ *along* γ. If γ is the generator of $\pi_1(D^*, x_0)$ we shall call $T = T_\gamma$ the *monodromy of* Φ *around* 0.

In [**19**, §§410–411] one may find the following statement.

THEOREM (2.3.2). *Let Φ be a complex analytic multivalued function on D^*. Let T: $\mathcal{V}_{\Phi, x_0} \to \mathcal{V}_{\Phi, x_0}$ be the monodromy of Φ around* 0. *For any determination φ of Φ at x_0 there are a finite number of pairs $(\alpha_i, \nu_i) \in \mathbf{C} \times \mathbf{N}$ $(i = 1, \ldots, r)$ and of (uniform) analytic functions c_{α_i, ν_i} on D^* $(i = 1, \ldots, r)$ such that*

(1) $\beta_j = \exp(2\pi \alpha_j \sqrt{-1})$ *is an eigenvalue of T $(j = 1, \ldots, r)$.*

(2) $\nu_j \leqslant l$ *where $l + 1$ is the size of the Jordan block of T associated to β_j.*

(3) *We have*

$$\varphi(x) = \sum_{i=1}^{r} c_{\alpha_i, \nu_i}(x) x^{\alpha_i} (\operatorname{Log} x)^{\nu_i}$$

where x^{α_i} and $(\operatorname{Log} x)^{\nu_i}$ are determinations at x_0 of the corresponding multivalued functions.

(2.3.3) The functions c_{α_i, ν_i} above may have a singularity at 0. If all c_{α_i, ν_i} are meromorphic at 0, one says that Φ is *regular at* 0. If one nonzero c_{α_i, ν_i} has an essential singularity at 0, one says Φ is *irregular at* 0.

(2.3.4) One says that the differential equation $Pf = 0$ is *regular singular at* 0, if on a sufficiently small punctured disc D^* centered at 0 all its solutions at $x_0 \in D^*$ define complex analytic regular multivalued functions on D^*.

Let us recall the classical result of *Fuchs*:

THEOREM (2.3.5) (FUCHS, CF. [**30**, §15.3]). *The differential equation $Pf = 0$ is regular singular at 0 if and only if the order of the pole of $a_k(x)/a_0(x)$ at 0 is at most k $(k = 0, \ldots, m)$.*

There is another way to understand regularity due to *B. Malgrange* (cf. [**46**]) and which will be generalized in higher dimension (cf. (4.1.4)).

Let $\hat{P}$: $\hat{\mathcal{O}}_{D,0} \to \hat{\mathcal{O}}_{D,0}$ be the operator defined by P on the $\mathfrak{m}$-completion $\hat{\mathcal{O}}_{D,0}$ of $\mathcal{O}_{D,0}$ at 0.

THEOREM (2.3.6) (MALGRANGE [**46**]). *For any analytic differential operator P defined on a neighborhood of* 0 *in D, the operator $\hat{P}$ has finite index and*

$$\chi(\hat{P}, 0) = \dim(\operatorname{Ker} \hat{P}) - \dim(\operatorname{Coker} \hat{P}) = \sup_{0 \leqslant p \leqslant m} \left(p - v(a_{m-p})\right).$$

COROLLARY (2.3.7). *The following assertions are equivalent.*

(i) *The differential equation $Pf = 0$ has a regular singularity at* 0.

(ii) $\chi(P, 0) = \chi(\hat{P}, 0)$.

(iii) *The complexes* $0 \to \mathcal{O}_{D,0} \xrightarrow{P} \mathcal{O}_{D,0} \to 0$ *and* $0 \to \hat{\mathcal{O}}_{D,0} \xrightarrow{\hat{P}} \hat{\mathcal{O}}_{D,0} \to 0$ *are quasi-isomorphic.*

(2.3.8) Let us recall that two complexes $C_1^{\cdot}$ and $C_2^{\cdot}$ are quasi-isomorphic if there is a morphism between them which induces an isomorphism of the cohomology.

(2.3.9) Let us denote by $\mathbf{R}\,\mathcal{H}\mathrm{om}_{\mathcal{D}}(\mathcal{M}, \mathcal{O}_D)$ the complex $0 \to \mathcal{O}_D \xrightarrow{P} \mathcal{O}_D \to 0$, with $\mathcal{M} = \mathcal{D}/\mathcal{D}P$. Then the complex $0 \to \mathcal{O}_{D,0} \xrightarrow{P} \mathcal{O}_{D,0} \to 0$ can be denoted by $\mathbf{R}\,\mathcal{H}\mathrm{om}_{\mathcal{D}}(\mathcal{M}, \mathcal{O}_D) \otimes_{\mathbf{C}_D} \mathbf{C}_Y$ where $\mathbf{C}_Y$ is the sheaf with fiber $\mathbf{C}$ on $Y := \{0\}$ and fiber 0 elsewhere.

Beside of it $\mathbf{R}\,\mathcal{H}\mathrm{om}_{\mathcal{D}}(\mathcal{M}, \hat{\mathcal{O}}_{D,0})$ can denote $0 \to \hat{\mathcal{O}}_{D,0} \xrightarrow{\hat{P}} \hat{\mathcal{O}}_{D,0} \to 0$, as it is obtained from the presentation $0 \to \mathcal{D} \to \mathcal{D} \to \mathcal{D}/\mathcal{D}P \to 0$ of $\mathcal{M} := \mathcal{D}/\mathcal{D}P$ by applying the functor $\mathcal{H}\mathrm{om}_{\mathcal{D}}(\cdot, \hat{\mathcal{O}}_{D,0})$, as $\hat{\mathcal{O}}_{D,0}$ is naturally a left $\mathcal{D}$-module (here we have identified $\hat{\mathcal{O}}_{D,0}$ and the sheaf $\varprojlim_n \mathcal{O}_D/\mathfrak{m}_0^n$ the fiber of which over 0 is $\hat{\mathcal{O}}_{D,0}$ and is 0 on D^*). Our statement (2.3.7) says that $Pf = 0$ is regular singular at 0 if and only if $\mathbf{R}\,\mathcal{H}\mathrm{om}_{\mathcal{D}}(\mathcal{M}, \mathcal{O}_D) \otimes_{\mathbf{C}_D} \mathbf{C}_Y$ and $\mathbf{R}\,\mathcal{H}\mathrm{om}_{\mathcal{D}}(\mathcal{M}, \hat{\mathcal{O}}_{D,0})$ are quasi-isomorphic. This allows us to define a linear differential system as in (2.2.2) with regular singularity at 0.

(2.4) *The classical Riemann-Hilbert problem.*

(2.4.1) In [Abh. Braunschweig. Wiss. Gesellsch. **7** (1857)] *Riemann* considered the following problem (cf. [**30**, §15.92]).

Find a multivalued analytic function F on $\mathbf{P}^1 - \{a_1, a_2, a_3\}$ (where a_1, a_2, a_3 are three distinct points of the complex projective line $\mathbf{P}^1$) such that

(i) the local system $\mathcal{V}_F$ defined by F on $\mathbf{P}^1 - \{a_1, a_2, a_3\}$ is at most of rank two,

(ii) in a neighbourhood U_i of the point a_i outside of a_i the determinations of F generate a local system generated by $(z - a_i)^{r_i} f_i(z)$ and $(z - a_i)^{s_i} g_i(z)$ where f_i and g_i are analytic on U_i and nonzero at a_i.

In the case where $a_1 = 0$, $a_2 = 1$, $a_3 = +\infty$ and $s_1 = s_2 = 0$, *Riemann* proved that such a multivalued function F is the solution of the hypergeometric equation

$$P_0 F = z(1 - z)\frac{d^2F}{dz^2} + ((r_1 + r_2 - 2)z + 1 - r_1)\frac{dF}{dz} - r_3 s_3 F = 0.$$

In the case considered above, *Papperitz* (in Math. Ann. **25** (1885)) proved that the multivalued function F is a solution of the generalized hypergeometric equation (cf. [**30**, §15.92]).

(2.4.2) One may view the problem of *Riemann* in the following way. Let $\{a_1, \ldots, a_k\}$ be distinct points of $\mathbf{P}^1$ and $\alpha \in \mathbf{P}^1 - \{a_1, \ldots, a_k\}$. Consider $\mathcal{V}$ a local system on $\mathbf{P}^1 - \{a_1, \ldots, a_k\} = U$ or equivalently a representation of $\pi_1(U, \alpha)$ in a finite-dimensional vector space $\mathcal{V}_\alpha$. Is it possible to find a differential system with regular solutions on U the solutions of which define the local system $\mathcal{V}$ on U? (Cf. *D. Hilbert*, Actes Congr. Internat. Math. Paris, 1900; 21st Hilbert problem.)

Notice that, in general, one does not obtain a linear differential equation of order m but a linear differential system. In this form the *Riemann-Hilbert* problem was solved by *G. Birkhoff* in (Proc. Nat. Acad. Sci. U.S.A. **49** (1913)). Actually a similar problem can be considered on a curve of positive genus outside a finite number of points. Namely let $U = X - \{a_1, \ldots, a_k\}$ where X is a curve of genus g and $\mathcal{V}$ be a local system on U. One may generalize the *Riemann-Hilbert* problem to this case by looking for integrable connections with regular singular points instead of differential systems (cf. [**39**] for instance). In this new formulation, *P. Deligne* has solved the problem in higher dimensions (cf. [**12**]). But one may reformulate the problem by looking for a differential system in the sense of (2.2.2) the holomorphic solutions of which define the local system $\mathcal{V}$ on U: but one must add a condition of "regularity" on the differential system to make the problem nontrivial, as $\mathcal{O}_X \otimes_{\mathbf{C}_X} j_! \mathcal{V}$, where $j_! \mathcal{V}$ is the constructible sheaf extension of $\mathcal{V}$ by adding a zero fiber on $a_1, \ldots, a_k$, is a $\mathcal{D}_X$-module the holomorphic solutions of which define $\mathcal{V}$ by restriction on U, but is in no way a "good" candidate to solve this generalized *Hilbert-Riemann* problem (cf. §4, Regular holonomic systems). This later point of view can be generalized in higher dimension, as we try to describe in §4, but as we shall see the situation is much more complicated, as we have to consider complexes of sheaves with constructible cohomology which become objects of derived categories (cf. Examples (4.4)).

(2.5) *Relation with algebraic geometry.* We are going to give only one relation with algebraic singularities. This relation is a source of the strong interest in the category of $\mathcal{D}$-modules.

(2.5.1) Let f: $(\mathbf{C}^{n+1}, 0) \to (\mathbf{C}, 0)$ be the germ of a complex analytic function on $(\mathbf{C}^{n+1}, 0)$. Assume that 0 *is an isolated critical point of f*. Then f induces a mapping φ: $X \to D$ where $X = \mathring{B}_\varepsilon \cap f^{-1}(D)$, $\mathring{B}_\varepsilon$ being an open ball of sufficiently small radius $\varepsilon > 0$ centered at 0 in $\mathbf{C}^{n+1}$ and D an open disc of radius $0 < \eta \ll \varepsilon$ centered at 0 in $\mathbf{C}$. According to [**62**], φ defines a fibration $X - f^{-1}(0) \to D^*$ which does not depend on ε, η with $1 \gg \varepsilon > 0$ and $\varepsilon \gg \eta > 0$. Actually the quoted result of [**62**] states that $\mathbf{R}^p\varphi_*\mathbf{C}_X$ is a constructible sheaf on D ($p \in \mathbf{N}$).

Recall that $\mathbf{R}^p\varphi_*\mathbf{C}_X$ is the $\mathbf{C}$-sheaf on D associated to the presheaf which makes the $\mathbf{C}$-vector space $H^p(\varphi^{-1}(U), \mathbf{C})$ correspond to the open set U of D.

As 0 is an isolated critical point of f at 0, $\mathbf{R}^p\varphi_*\mathbf{C}_X = 0$ except if $p = 0$, n and $\mathbf{R}^n\varphi_*\mathbf{C}_X$ is a constructible sheaf which has a zero fiber at 0 and which is a local system of rank μ on D^* with

$$\mu = \dim_{\mathbf{C}} \mathbf{C}\{z_0, \ldots, z_n\} / (\partial f/\partial z_0, \ldots, \partial f/\partial z_n)$$

(this number μ is called the *Milnor number of f at* 0), when $n \geqslant 1$.

(2.5.2) As *E. Brieskorn* in [**9**], consider the following period integrals: let ω be an n-form on $(\mathbf{C}^{n+1}, 0)$; suppose X is small enough such that ω is defined on X; consider a continuous family of n-cycles $\gamma(t)$ of $\varphi^{-1}(t)$ for $t \in D^*$; by a theorem of Cauchy the integral $\int_{\gamma(t)} \omega = g(t)$ defines a complex analytic multivalued function on D^*. According to a theorem due to *M. Nilsson* (cf. [**63**]) this multivalued function $g(t)$ is regular at 0. One may prove that all the functions obtained in this way are the solutions of a differential system which is regular in the sense of (2.3.9). In [**9**] *E. Brieskorn* described explicitly this differential system which is called the *Gauss-Manin connection*. In the following, we shall see that this system is essentially the unique regular one which has the above period integrals as solutions (cf. (4.6.2)).

(2.5.3) Let $g(t) = \int_{\gamma(t)} \omega$ be a period integral as described in (2.5.2). According to (2.3.2) one can find $\alpha_1, \ldots, \alpha_r$ such that

$$g(t) = \sum_{\substack{\alpha \in A_1 \cup \cdots \cup A_r \\ \nu \leqslant l}} c_{\alpha,\nu} t^{\alpha} (\operatorname{Log} t)^{\nu}$$

where $c_{\alpha,\nu} \in \mathbf{C}$, $A_i := \alpha_i + \mathbf{N}$ and $c_{\alpha_i,r} \neq 0$ for some ν, $l \geqslant \sup(l_i)$ where $l_i + 1$ is the size of the Jordan block associated to the eigenvalue $\beta_i = \exp(2\pi\sqrt{-1}\,\alpha_i)$ of the monodromy of g around 0.

The monodromy of g around 0 being given by the monodromy of the fibration $X - f^{-1}(0) \to D^*$ defined by φ, the local version of the monodromy theorem (cf. [**11, 40, 43**]) shows that one has $\alpha_i \in \mathbf{Q}$ and $l_i \leqslant n$ $(i = 1, \ldots, r)$. The numbers α_i defined above are called *the exponents of* g. The smallest exponent among those of all possible period integrals will be called *the Arnold exponent of f at* 0 (cf. [**1**]).

Lately *A. N. Varchenco* [**71, 72, 73**] proved that in an analytic family f_t with Milnor number at 0 constant, the Arnold exponent is constant.

It would be a good thing to compute these exponents using the theory of $\mathcal{D}$-modules as it develops (compare with [**78**]).

3. Linear differential systems.

(3.1) *Generalities.* Let $(X, \mathcal{O}_X)$ be a smooth complex analytic manifold of dimension n. Let $\mathcal{D}_X = \mathcal{D}$ be the sheaf of differential operators of finite order with holomorphic coefficients. The sheaf $\mathcal{D}$ is a coherent sheaf of noncommutative rings with units. On local coordinates $x = (x_1, \ldots, x_n)$ a section of $\mathcal{D}$ is an ordinary differential operator

$$P = P\left(x, \frac{\partial}{\partial x}\right) = \sum_{|\alpha| \leqslant m} a_\alpha(x) \frac{\partial^\alpha}{\partial x^\alpha}$$

where the $a_\alpha(x)$ are holomorphic functions and $a_\alpha(x) \neq 0$ for some α, $|\alpha| = m$, and m is the order of P and the $\alpha = (\alpha_1, \ldots, \alpha_n)$ are multi-indices of length $|\alpha| = \alpha_1 + \cdots + \alpha_n$.

DEFINITION (3.1.1). A *linear differential system* is a left coherent $\mathcal{D}$-module $\mathcal{M}$ (compare with (2.2.2)).

Since $\mathcal{D}$ is coherent, a coherent $\mathcal{D}$-module $\mathcal{M}$ is a $\mathcal{D}$-module having locally on X a presentation

$$\mathcal{D}^p \xrightarrow{A} \mathcal{D}^q \to \mathcal{M} \to 0,$$

where A is a matrix of differential operators.

(3.1.2) For example if I is a left ideal of $\mathcal{D}$ generated by a finite number of differential operators the cyclic $\mathcal{D}$-module $\mathcal{D}/I$ is a linear differential system. The sheaf $\mathcal{O}_X$ of holomorphic functions is a left $\mathcal{D}$-module. In local coordinates $x = (x_1, \ldots, x_n)$ we have the isomorphism

$$\mathcal{D}/\left(\frac{\partial}{\partial x_1}, \ldots, \frac{\partial}{\partial x_n}\right) \xrightarrow{\sim} \mathcal{O}_X.$$

The $\mathcal{D}$-module $\mathcal{O}_X$ is coherent and called the *de Rham* system.

(3.1.3) Remember that the *characteristic variety* of a differential operator $P = \sum_{|\alpha| \leq m} \cdot a_\alpha(x)\partial^\alpha/\partial x^\alpha$ of order m is the hypersurface $S\check{S}(P)$ of the cotangent bundle T^*X of X defined by the *principal symbol* $\sigma(P) = \sum_{|\alpha| = m} a_\alpha(x)\xi^\alpha$ which is a homogeneous function in the cotangent coordinates $\xi = (\xi_1, \ldots, \xi_n)$. More generally the characteristic variety of the cyclic module $\mathcal{D}/I$, where I is a left ideal of finite type, is the variety defined by the ideal generated by all the principal symbols of the operators in I. In fact it is possible to choose the generators $P_1, \ldots, P_p$ of I such that the principal symbols $\sigma(P_1), \ldots, \sigma(P_p)$ define the characteristic variety $S\check{S}(I)$ of the system $\mathcal{D}/I$ [**65**].

To define the characteristic variety $S\check{S}(\mathfrak{M})$ of a general system $\mathfrak{M}$ let us recall what is a good filtration of $\mathfrak{M}$. An increasing filtration $(\mathfrak{M}_k)_{k \in \mathbf{Z}}$ by coherent $\mathcal{O}_X$-submodules of $\mathfrak{M}$ is a *good filtration* if

(i) $\bigcup_k \mathfrak{M}_k = \mathfrak{M}$ and $\mathcal{D}(m)\mathfrak{M}_k \subset \mathfrak{M}_{k+m}$ where $\mathcal{D}(m)$ denotes the sheaf of differential operators of order $\leq m$,

(ii) locally there exists some integer j_0 such that $\mathfrak{M}_{j+m} = \mathcal{D}(m)\mathfrak{M}_j$ if $m \geq 0$, $j > j_0$.

Locally any system $\mathfrak{M}$ has a good filtration, e.g. the quotient filtration induced a local presentation by the natural filtration of $\mathcal{D}$. If $\mathcal{D}^p \xrightarrow{A} \mathcal{D}^q \xrightarrow{\varphi} \mathfrak{M}$ is a local presentation and $\mathfrak{M}_k = \varphi(\mathcal{D}(k))$, then $(\mathfrak{M}_k)_{k \in \mathbf{Z}}$ is a good local filtration of $\mathfrak{M}$ [**65**]. But globally good filtrations may not exist. The sheaf $\mathcal{D}$ has a global good filtration $\mathcal{D}(m)_{m \in \mathbf{Z}}$. The space Specan $\bigoplus_{m \geq 0} \mathcal{D}(m)/\mathcal{D}(m-1)$ is the cotangent bundle T^*X of X. If $(\mathfrak{M}_k)_{k \in \mathbf{Z}}$ is a good filtration of a system $\mathfrak{M}$ the characteristic variety $S\check{S}(\mathfrak{M})$ of $\mathfrak{M}$ is the support of the grad($\mathcal{D}$)-module $\bigoplus_k \mathfrak{M}_k/\mathfrak{M}_{k-1}$ where $\operatorname{grad}(\mathcal{D}) = \bigoplus_{m \geq 0} \mathcal{D}(m)/\mathcal{D}(m-1)$. The characteristic variety does not depend on the good filtration chosen locally thus the above construction patches globally. If we use the sheaf $\check{\mathcal{E}}_X$ of microdifferential operators [**69**] then an intrinsic definition is

$$S\check{S}(\mathfrak{M}) = \operatorname{support}\left(\check{\mathcal{E}}_X \underset{\pi^{-1}\mathcal{D}}{\otimes} \pi^{-1}\mathfrak{M}\right),$$

where $\pi\colon T^*X \to X$ is the natural projection. Since we will not use the sheaf $\check{\mathcal{E}}_X$ we refer to [**33, 69**] for its definition. We have the basic theorem [**69**]:

THEOREM (3.1.4). *The characteristic variety of a system $\mathfrak{M}$ is an involutive subspace of T^*X.*

Remember that T^*X has a symplectic structure. The *Poisson* bracket $\{f, g\}$ of two homogeneous functions on T^*X is defined: if $x = (x_1,\dots,x_n)$ is a system of local coordinates and $\xi = (\xi_1,\dots,\xi_n)$ is the associate cotangent system of coordinates, then

$$\{f, g\} = H_f(g) = \sum_{i=1}^{n}\left(\frac{\partial f}{\partial \xi_i}\frac{\partial g}{\partial x_i} - \frac{\partial f}{\partial x_i}\frac{\partial g}{\partial \xi_i}\right).$$

An analytic subspace of T^*X is *involutive* if it is a homogeneous subspace of T^*X, whose reduced ideal I is stable by the *Poisson* bracket. So if an involutive subspace is not empty, its dimension is at least equal to $\dim X = n$. We refer to [**15**] for another algebraic proof of involutivity and comments.

In the following we shall need the notion of

(3.1.5) *Derived categories.* For a complete treatment of derived categories and derived functors we refer the reader to [**25, 74**]. Here we shall only give a few essential results needed in our exposition. If $\mathcal{Q}$ is a sheaf of rings with units on X, we denote $\mathrm{mod}(\mathcal{Q})$ the category of left $\mathcal{Q}$-modules. The category $\mathrm{mod}(\mathcal{Q})$ is embedded in its derived category $D(\mathcal{Q})$. The objects of $D(\mathcal{Q})$ are complexes of $\mathrm{mod}(\mathcal{Q})$ such that a morphism of complexes is an isomorphism in $D(\mathcal{Q})$ if and only if it induces isomorphisms in the cohomology. The category is not abelian but it has the structure of a triangulated category. Moreover if F is a left (resp. a right) exact functor from $\mathrm{mod}(\mathcal{Q})$ into an abelian category $\mathcal{B}$, under some assumptions it has a derived right (resp. left) functor $\mathbf{R}F$ (resp. $\mathbf{L}F$) which is a functor from $D(\mathcal{Q})$ to the derived category $D(\mathcal{B})$ of $\mathcal{B}$. In this paper we use basically the left exact functor $\mathcal{H}om_{\mathcal{Q}}(*, *)$ and the right exact functor $* \otimes_{\mathcal{Q}} *$. We denote by $\mathbf{R}\,\mathcal{H}om_{\mathcal{Q}}(*, *)$ and $* \otimes^{\mathbf{L}}_{\mathcal{Q}} *$ their derived functors. *All complexes are bounded.*

(3.1.6) Just as in the one dimension case we have a functor from the category $\mathrm{mod}(\mathcal{D})$ of left $\mathcal{D}$-modules to the category $\mathrm{mod}(\mathbf{C}_X)$ of sheaves of complex vector spaces

$$\mathcal{M} \to \mathcal{H}om_{\mathcal{D}}(\mathcal{M}, \mathcal{O}_X)$$

which associate to a $\mathcal{D}$-module $\mathcal{M}$ the sheaf of its holomorphic solutions. Its derived functors are the functors

$$\mathcal{M} \to \mathcal{E}xt^i_{\mathcal{D}}(\mathcal{M}, \mathcal{O}_X)$$

and only a finite number of these functors are not zero since the homological dimension of $\mathcal{D}$ is equal to $n = \dim X$ (cf. [**31, 5**]). The functor $\mathcal{H}om_{\mathcal{D}}(\cdot, \mathcal{O}_X)$ extends into a functor from the derived category $D(\mathcal{D})$ of $\mathrm{mod}(\mathcal{D})$ to the derived category $D(\mathbf{C}_X)$ of $\mathrm{mod}(\mathbf{C}_X)$

$$\mathcal{M} \to \mathbf{R}\,\mathcal{H}om_{\mathcal{D}}(\mathcal{M}, \mathcal{O}_X)$$

which associates to a complex $\mathcal{M}$ of $\mathcal{D}$-modules its holomorphic solutions complex. For a general system $\mathcal{M}$ the sheaves $\mathcal{E}xt^i_{\mathcal{D}}(\mathcal{M}, \mathcal{O}_X)$ could be quite complicated. For geometrical purposes we must restrict ourselves to a smaller class of systems, namely *the maximally overdetermined systems* now called *holonomic systems*.

(3.2) *Holonomic systems*: *examples*.

DEFINITION (3.2.1). *A linear system* $\mathfrak{M}$ *is called holonomic if* $\dim S\check{S}(\mathfrak{M}) = n$.

This notion is so important that we give several examples.

EXAMPLES (3.2.2). Let $\mathfrak{M} = \mathfrak{D}/\mathfrak{D}P$ where P is a nonzero differential operator. One can check that the characteristic variety of $\mathfrak{M}$ is defined by principal symbol $\sigma(P)$ of P, thus $\mathfrak{M}$ is holonomic if $\dim X = 1$ and not holonomic if $\dim X \geqslant 2$. This makes the theory quite complicated in higher dimension.

EXAMPLES (3.2.3). *Remember that an integrable connection is a left* $\mathfrak{D}$*-module which is free of finite type as* $\mathcal{O}_X$*-module*. For instance the *de Rham* system $\mathcal{O}_X$ is an integrable connection and, because of the isomorphism of (3.1.2), locally its holomorphic solution is the *de Rham* complex of X which is quasi-isomorphic to the constant sheaf $\mathbf{C}_X$ by the *Poincaré* lemma. The characteristic variety of an integrable connection is the zero section of T^*X. So an integrable connection is an holonomic module. In fact *any holonomic system with a characteristic variety equal to the zero section of* T^*X *is an integrable connection* (this fact is an easy consequence of the theorem of *Cauchy-Kowalewskaïa* [**31**]). Since the characteristic variety of a general holonomic system could have "singularities" this shows that the holonomic system is the right generalization of integrable connections.

The next example is important.

EXAMPLE (3.2.4). Let $Y \subset X$ be a reduced divisor of X with arbitrary singularities. The sheaf $\mathcal{O}[*Y]$ of meromorphic functions on X with poles in Y is a left $\mathfrak{D}$-module. In fact a deep result says that $\mathcal{O}[*Y]$ is holonomic [**33**].

Lately one found out the characteristic variety of $\mathcal{O}[*Y]$ in terms of the "geometry" of Y. The description of this characteristic variety is rather complicated in the general case to be done shortly here (cf. [**44**]). In the particular case Y has only isolated singularities $S\check{S}(\mathcal{O}[*Y])$ is the union of the zero section $T^*_X X$ of the cotangent bundle of X, the closure of the cotangent bundle in X of the nonsingular part of Y and the fibers of the cotangent bundle of X over the singular points of Y (cf. [**44**]). This system $\mathcal{O}_x[*Y]$ which one may call the localization of the *de Rham system along* Y plays an important role in this theory. This is closely related to the *Bernstein-Sato* polynomial theory [**4**].

EXAMPLE (3.2.5). Let f: $(\mathbf{C}^n, 0) \to (\mathbf{C}, 0)$ be a germ of an analytic function. The *Bernstein-Sato polynomial of* f called also the *b-function* is the unitary polynomial $b(s)$ of one variable of minimal degree such that there exists a differential operator $P(s, x, d/dx) \in \mathfrak{D} \otimes_{\mathbf{C}} \mathbf{C}[s] = \mathfrak{D}[s]$ with polynomial coefficients in s and the relation

$$P(s, x, d/dx) f^{s+1} = b(s) f^s.$$

See [**4** and **5**]. We are going to sketch briefly *Kashiwara's* proof of the existence of the *Bernstein-Sato* polynomial $b(s)$ of f [**33**]. According to *Bernstein* [**4**] and *Kashiwara* [**33**] we associate a holonomic system to f in the following way.

Let us consider the sheaf $\mathcal{O}[s, f^{-1}] = \mathcal{O}[f^{-1}] \otimes_{\mathbf{C}} \mathbf{C}[s]$ where $\mathcal{O}[f^{-1}]$ is the sheaf of meromorphic functions with poles on the hypersurface $f^{-1}(0)$ and the free module $\mathcal{O}[s, f^{-1}]\mathbf{f}^s$ generated by the symbol $\mathbf{f}^s$ over $\mathcal{O}[s, f^{-1}]$. The sheaf $\mathcal{O}[s, f^{-1}]\mathbf{f}^s$ has a left $\mathcal{D}[s]$ structure, namely the action of s and the action of the holomorphic functions are the obvious ones; if $\mathfrak{X}$ is a vector field and g a section of $\mathcal{O}[s, f^{-1}]$ the action of $\mathcal{D}$ is defined by

$$\mathfrak{X}(g\mathbf{f}^s) = \mathfrak{X}(g)\mathbf{f}^s + sg\mathfrak{X}(f)f^{-1}\mathbf{f}^s.$$

Now let us consider the submodule $\mathcal{D}[s]\mathbf{f}^s$ generated by $\mathbf{f}^s$ over $\mathcal{D}[s]$ and its submodule $\mathcal{D}[s]f\mathbf{f}^s$. Then a basic result [**33**] states that the $\mathcal{D}$-module

$$\mathcal{M}_f \stackrel{\text{def}}{=} \mathcal{D}[s]\mathbf{f}^s/\mathcal{D}[s]f\mathbf{f}^s$$

is holonomic. The action of s on $\mathcal{M}_f$ is $\mathcal{D}$-linear and belongs to $\mathcal{E}\mathrm{nd}_{\mathcal{D}}(\mathcal{M}_f)_0$. If we know that $\mathcal{E}\mathrm{nd}_{\mathcal{D}}(\mathcal{M}_f)_0$ is a complex vector space of finite dimension then $b(s)$ is just the minimal polynomial of s. This means that

$$b(s)\mathbf{f}^s \equiv 0 \quad \text{in } \mathcal{M}_f \quad \text{or} \quad b(s)\mathbf{f}^s \in \mathcal{D}[s]f\cdot\mathbf{f}^s.$$

Thus

$$b(s)\mathbf{f}^s = P(s, x, \partial/\partial x)f\cdot\mathbf{f}^s.$$

The most important property of holonomic systems is that their holomorphic solution complexes are constructible. Recall that a (bounded) complex $\mathcal{F}$ of sheaves of vector complex spaces on X is called constructible if its cohomology sheaves are constructible on X (cf. (2.1.6)). Thus the assertion is

THEOREM (3.2.6) [**32**]. *If $\mathcal{M}$ is a holonomic system on X, the sheaves $\mathcal{E}\mathrm{xt}^i_{\mathcal{D}}(\mathcal{M}, \mathcal{O}_X)$ are constructible.*

For example if $\mathcal{M}$ is an integrable connection the *Cauchy* existence theorem (compare to (2.1)) states that $\mathcal{H}\mathrm{om}_{\mathcal{D}}(\mathcal{M}, \mathcal{O}_X)$ is a local system and the *Poincaré* lemma states that $\mathcal{E}\mathrm{xt}^i_{\mathcal{D}}(\mathcal{M}, \mathcal{O}_X) = 0$ if $i \geqslant 1$.

We note $D(\mathcal{D})_h$ (resp. $D(\mathbf{C}_X)_c$) the derived category of bounded complexes of $\mathcal{D}$-modules (resp. of $\mathbf{C}_X$-modules) with holonomic cohomology (resp. with constructible cohomology). It follows from Theorem (3.2.6) that the functor $\mathbf{R}\,\mathcal{H}\mathrm{om}_{\mathcal{D}}(*, \mathcal{O}_X)$ sends the category $D(\mathcal{D})_h$ into the category $D(\mathbf{C}_X)_c$. We will denote this functor $\mathbf{S}$,

$$\mathbf{S}(\mathcal{M}) = \mathbf{R}\,\mathcal{H}\mathrm{om}_{\mathcal{D}}(\mathcal{M}, \mathcal{O}_X).$$

(3.3) *Operations on holonomic systems.* In this section we define three important operations on holonomic systems which give holonomic systems. The definitions and the use of these operations are rather technical but they are needed in (3.5.4) and §4 below. More comments about definitions and the proofs of the theorems can be found in [**5**, **33** or **65**].

(3.3.1) *Algebraic local cohomology.* Let $Y \subset X$ be an analytic subspace defined by an ideal $\mathcal{J}_Y$. Let j: $U = X - Y \hookrightarrow X$ be the natural inclusion. If $\mathcal{M}$ is a

$\mathfrak{D}$-module, the sheaf $\underline{\Gamma}_Y(\mathfrak{M})$ of sections supported by Y is also a $\mathfrak{D}$-module since the operation of $\mathfrak{D}$ preserves supports. As in Grothendieck [20] we defined the functor of algebraic local cohomology by

$$\underline{\Gamma}_{[Y]}(\mathfrak{M}) \overset{\text{def}}{=} \varinjlim_k \mathcal{H}om_{\mathcal{O}_X}\left(\mathcal{O}_X/\mathfrak{F}_Y^k, \mathfrak{M}\right).$$

Then $\underline{\Gamma}_{[Y]}(\mathfrak{M})$ is the subspace of $\underline{\Gamma}_Y(\mathfrak{M})$ of sections annihilated by a power of the ideal $\mathfrak{F}_Y$ [25]. But $\underline{\Gamma}_{[Y]}(\mathfrak{M})$ is also a $\mathfrak{D}$-module because the action of a vector field on $\underline{\Gamma}_Y(\mathfrak{M})$ preserves $\underline{\Gamma}_{[Y]}(\mathfrak{M})$ [52, p. 98]. The functor $\underline{\Gamma}_{[Y]}(*)$ is a functor from mod($\mathfrak{D}$) to mod($\mathfrak{D}$). It has a derived functor from $D(\mathfrak{D})$ into $D(\mathfrak{D})$,

$$\underline{\mathbf{R}\Gamma}_{[Y]}(\mathfrak{M}) \overset{\text{def}}{:=} \mathbf{R} \varinjlim_k \mathcal{H}om_{\mathcal{O}_X}\left(\mathcal{O}_X/\mathfrak{F}_Y^k, \mathfrak{M}\right).$$

For the same reason we defined the functor from $D(\mathfrak{D})$ to $D(\mathfrak{D})$ by

$$\mathbf{R}\left[j_*\right]\mathfrak{M} \overset{\text{def}}{:=} \mathbf{R} \varinjlim_k \mathcal{H}om_{\mathcal{O}_X}\left(\mathfrak{F}_Y^k, \mathfrak{M}\right).$$

The complex $\underline{\mathbf{R}\Gamma}_{[Y]}(\mathfrak{M})$ is a subcomplex of $\underline{\mathbf{R}\Gamma}_Y(\mathfrak{M})$ and the complex $\mathbf{R}[j_*]\mathfrak{M}$ is a subcomplex of $\mathbf{R}\, j_* j^{-1}\mathfrak{M}$. We have a triangle in $D(\mathfrak{D})$

$$(*)\qquad \begin{array}{ccc} & \mathbf{R}\left[j_*\right]\mathfrak{M} & \\ {}^{+1}\swarrow & & \nwarrow \\ \underline{\mathbf{R}\Gamma}_{[Y]}(\mathfrak{M}) & \longrightarrow & \mathfrak{M} \end{array}$$

In the case $\mathfrak{M}$ is the *de Rham* system $\mathcal{O}_X$ and Y is a hypersurface in X, one may check that $\mathbf{R}[j_*]\mathcal{O}_X = \mathcal{O}[*Y]$, the localization of the *de Rham* system along Y (cf. (3.2.4)).

THEOREM (3.3.2). *Then the complex* $\underline{\mathbf{R}\Gamma}_{[Y]}(\mathfrak{M})$ *belongs to* $D(\mathfrak{D})_h$ *if* $\mathfrak{M}$ *belongs to* $D(\mathfrak{D})_h$.

This theorem means that the triangle $(*)$ is a triangle in $D(\mathfrak{D})_h$ if $\mathfrak{M}$ is in $D(\mathfrak{D})_h$.

In the case $\mathcal{O}_X = \mathfrak{M}$ this theorem is proved in [53] by reducing it to the case where Y is a hypersurface by a suitable *Mayer-Vietoris* sequence. This later case is precisely the example (3.2.4) and we saw it closely related to the *Bernstein-Sato* polynomial of Y. The generalization is due to *Kashiwara* [35]. Theorem (3.3.2) gives many examples of holonomic systems.

(3.3.3) *Integration along the fibers*. This section is devoted to what corresponds to "direct images" of $\mathfrak{D}$-modules and involves rather technical definitions.

Let $\pi\colon \tilde{X} \to X$ be a morphism of smooth complex varieties. We defined a sheaf on $\tilde{X}$ [68, 69, 33] by

$$\mathfrak{D}_{X \leftarrow \tilde{X}} \overset{\text{def}}{:=} p_* \varinjlim_k \mathcal{E}xt^n_{\mathcal{O}_{\tilde{X}\times X}}\left(\mathcal{O}_{\tilde{X}\times X}/\mathfrak{F}_\pi^k, p^*\Omega_{\tilde{X}}\right)$$

where

$\mathfrak{I}_\pi$: is the ideal of the graph of π in $\tilde{X} \times X$;

$\Omega_{\tilde{X}}$: is the sheaf of the holomorphic forms in $\tilde{X}$ of degree dim $\tilde{X}$;

p: is the projection from $\tilde{X} \times X$ onto $\tilde{X}$;

q: is the projection from $\tilde{X} \times X$ onto X.

The sheaf $\mathcal{D}_{X \leftarrow \tilde{X}}$ has the structure of a right $\mathcal{D}_{\tilde{X}}$-module coming from the natural structure of a right $\mathcal{D}_{\tilde{X}}$-module of $\Omega_{\tilde{X}}$ (see [**65**, p. 124]) and a structure of a left $\pi^{-1}\mathcal{D}_X$-module. Let $\tilde{\mathcal{M}}$ be a left $\mathcal{D}_{\tilde{X}}$-module. Then the complex

$$\int_\pi \tilde{\mathcal{M}} \overset{\text{def}}{=} \mathbf{R}\pi_*\left(\mathcal{D}_{X \leftarrow \tilde{X}} \otimes^{\mathbf{L}}_{\mathcal{D}_{\tilde{X}}} \tilde{\mathcal{M}}\right)$$

is a complex of left $\mathcal{D}_X$-modules. See [**65**, p. 136] for comprehensive comments of these structures.

THEOREM (3.3.4). *Suppose that π is projective and $\tilde{\mathcal{M}}$ has a global good filtration. Then the complex $\int_\pi \tilde{\mathcal{M}}$ has a $\mathcal{D}_X$ holonomic cohomology if $\tilde{\mathcal{M}}$ is holonomic.*

This theorem is proved by *Kashiwara* [**33**]. The condition that $\tilde{\mathcal{M}}$ has global good filtration reduces this theorem to the *Grauert-Remmert* theorem on the coherence of the direct images of a coherent $\mathcal{O}_{\tilde{X}}$-module.

(3.3.5) *Inverse images*. For this section a comprehensive reference is [**65**, p. 110] where the structures are explained.

Let $\pi: \tilde{X} \to X$ be a morphism of analytic smooth complex varieties. We use the same notations. We defined the sheaf on $\tilde{X}$ by

$$\mathcal{D}_{\tilde{X} \to X} \overset{\text{def}}{:=} p_* \varinjlim_k \mathcal{E}xt^n_{\mathcal{O}_{\tilde{X} \times X}}\left(\mathcal{O}_{\tilde{X} \times X}/\mathfrak{I}^k_\pi, q^*\Omega_X\right),$$

where Ω_X is the sheaf of n holomorphic forms in X with its natural structure of a right $\mathcal{D}_X$-module (cf. [**65**, p. 129]). This sheaf is a right $\pi^{-1}\mathcal{D}_X$-module and a left $\mathcal{D}_{\tilde{X}}$-module [**65**]. Let $\mathcal{M}$ be a left $\mathcal{D}_X$-module. Then the complex

$$\mathbf{L}\pi^*\mathcal{M} \overset{\text{def}}{=} \mathcal{D}_{\tilde{X} \to X} \otimes^{\mathbf{L}}_{\pi^{-1}} \pi^{-1}\mathcal{M}$$

is a complex of left $\mathcal{D}_{\tilde{X}}$-modules.

THEOREM (3.3.6). *If $\mathcal{M}$ is holonomic, the complex $\mathbf{L}\pi^*\mathcal{M}$ has a $\mathcal{D}_{\tilde{X}}$-holonomic cohomology.*

This theorem is proved by *Kashiwara*, answering a question from *Malgrange* in a manuscript entitled *B-functions and restriction of holonomic systems* (1977) (cf. [**35**]).

(3.4) *Local duality theorem*. Let $\mathcal{M}$ be a left-holonomic system on X. Then the sheaf $\mathcal{E}xt^i_{\mathcal{D}}(\mathcal{M}, \mathcal{D})$ vanishes if $i \neq n$ [**31**]. The sheaf $\mathcal{E}xt^n_{\mathcal{D}}(\mathcal{M}, \mathcal{D})$ has a right structure of a $\mathcal{D}$-module. The dual $\mathcal{M}^*$ of $\mathcal{M}$ is defined by

$$\mathcal{M}^* \overset{\text{def}}{=} \mathcal{H}om_{\mathcal{O}_X}(\Omega_X, \mathcal{E}xt^n_{\mathcal{D}}(\mathcal{M}, \mathcal{D})).$$

The sheaf $\mathcal{M}^*$ has a canonical structure of a left $\mathcal{D}$-module (see [**65**, p. 129]). In fact $\mathcal{M}^*$ is holonomic [**32**]. The functor $\mathcal{M} \to \mathcal{M}^*$ is an exact contravariant functor such that $\mathcal{M}^{**} \rightleftarrows \mathcal{M}$ [**32**]. Let $\mathcal{M}$ be a complex of $D(\mathcal{D})_h$; then $\mathcal{M}^*$ is also a complex of $D(\mathcal{D})_h$. One can prove $\mathcal{O}_X^* \simeq \mathcal{O}_X$ (cf. [**31** and **52**]) which implies

$$\mathbf{S}(\mathcal{M}^*) \overset{\text{def}}{:=} \mathbf{R}\,\mathcal{H}om_{\mathcal{D}}(\mathcal{M}^*, \mathcal{O}_X) \overset{\sim}{\to} \mathbf{R}\,\mathcal{H}om_{\mathcal{D}}(\mathcal{O}_X, \mathcal{M}).$$

The complex $\mathbf{R}\,\mathcal{H}om_{\mathcal{D}}(\mathcal{O}_X, \mathcal{M})$ is called the *de Rham* complex of $\mathcal{M}$ and noted $DR(\mathcal{M})$. It is proved (see [**52**] for example) that if $\mathcal{M}$ is a left $\mathcal{D}$-module, the complex $DR(\mathcal{M})$ is isomorphic in $D(\mathbf{C}_X)$ to the complex

$$0 \to \mathcal{M} \to \mathcal{M} \otimes_{\mathcal{O}_X} \Omega_X^1 \to \cdots \to \mathcal{M} \otimes_{\mathcal{O}_X} \Omega_X^n \to 0.$$

Let $\mathcal{F}$ be a complex of $D(\mathbf{C}_X)$, we denote by $\check{\mathcal{F}}$ the *Verdier dual* of $\mathcal{F}$:

$$\check{\mathcal{F}} \overset{\text{def}}{:=} \mathbf{R}\,\mathcal{H}om_{\mathbf{C}_X}(\mathcal{F}, \mathbf{C}_X).$$

THEOREM (3.4.1). *Let $\mathcal{M}$ be a complex of $D(\mathcal{D})_h$. Then we have canonical isomorphisms in $D(\mathbf{C}_X)_c$,*

$$\mathcal{S}(\mathcal{M}^*) \simeq DR(\mathcal{M}) \overset{\sim}{\to} \mathbf{S}(\mathcal{M})\check{}.$$

This theorem is proved in ([**55**, Théorème 1.1, Chapitre III], or *Théorèmes de bidualité locales pour les $\mathcal{D}$-modules holonomes*, Ark. Mat. **20** (1982), 111–122). It uses *Poincaré-Serre-Verdier* dualities. Remember that the composition functor $\mathcal{M} \to \mathcal{M}^*$ with itself is isomorphic to the identity functor of $D(\mathcal{D})_h$ and the biduality theorem [**77**] states that the composition of the functor $\mathcal{F} \to \check{\mathcal{F}}$ with itself is isomorphic to the identity functor of $D(\mathbf{C}_X)_c$. *Verdier*'s biduality [**77**] requires the use of the resolution of singularities and thus it is more difficult to prove it than the biduality of holonomic systems which follows easily from the coherence as $\mathcal{D}$-modules. This is the first example of the dictionary between analytic results concerning differential systems (= holonomic $\mathcal{D}$-modules) and geometric results concerning constructible complexes. We can state Theorem (3.4.1) by saying that the following diagram is commutative:

$$\begin{array}{ccc} D(\mathcal{D})_h & \overset{\mathbf{S}}{\to} & D(\mathbf{C}_X)_c \\ *\downarrow & & \downarrow\check{} \\ D(\mathcal{D})_h & \overset{\mathbf{S}}{\to} & D(\mathbf{C}_x)_c \end{array}$$

Theorem (3.4.1) is a local one. The next theorem is a global one.

(3.5) *Global duality theorem*. In this part we shall show that the duality theorem for $\mathcal{D}$-modules is a common generalization of the *Poincaré* duality and the *Serre* duality.

Let $\mathcal{M}$ be a bounded complex of left $\mathcal{D}$-modules. The complex of global holomorphic solutions is the complex of complex vector spaces

$$\mathbf{R}\Gamma(X; \mathbf{S}(\mathcal{M})) \stackrel{\text{def}}{=} \mathbf{R}\,\mathbf{Hom}_{\mathcal{D}}(\mathcal{M}, \mathcal{O}_X);$$

we denote

$$\mathbf{E}^i(\mathcal{M}) \stackrel{\text{def}}{=} \mathbf{Ext}^i_{\mathcal{D}}(X; \mathcal{M}, \mathcal{O}_X)$$

its ith cohomology space. The global *de Rham* complex with compact support of $\mathcal{M}$ is the complex of complex vector spaces

$$\mathbf{R}\Gamma_c(X; DR(\mathcal{M})) \stackrel{\text{def}}{=} \mathbf{R}\,\mathbf{Hom}_{\mathcal{D},c}(\mathcal{O}_X, \mathcal{M}),$$

where c is the family of the compact sets of X. We note

$$\mathbf{E}^c_i(\mathcal{M}) \stackrel{\text{def}}{=} \mathbf{Ext}^{2n-i}_{\mathcal{D},c}(X; \mathcal{O}_X, \mathcal{M})$$

its $(2n-i)$th cohomology space. Then we have the *Yoneda* pairing

$$\mathbf{E}^c_i(\mathcal{M}) \times \mathbf{E}^i(\mathcal{M}) \to \mathbf{E}^c_0(\mathcal{O}_X) = H^{2n}_c(X; \mathbf{C}).$$

By composition with the *trace map*

$$H^{2n}_c(X; \mathbf{C}) \to \mathbf{C}$$

we get a pairing

$$A(\mathcal{M}): \mathbf{E}^c_i(\mathcal{M}) \times \mathbf{E}^i(\mathcal{M}) \to \mathbf{C}.$$

THEOREM (3.5.1) [**54**]. *Let $\mathcal{M}$ be a bounded complex of $\mathcal{D}$-modules with coherent cohomology. Then there is a unique pair of topologies Q.F.S-Q.D.F.S. on $\mathbf{E}^i(\mathcal{M})$ and $\mathbf{E}^c_i(\mathcal{M})$ such that $A(\mathcal{M})$ induces a perfect pairing between the separated spaces associated to them. Moreover if $\mathcal{M}$ belongs to $D(\mathcal{D})_h$ then $A(\mathcal{M})$ induces an isomorphism between $\mathbf{E}^i(\mathcal{M})$ and the algebraic dual of $\mathbf{E}^c_i(\mathcal{M})$.*

We recall that Q.F.S. means quotient of *Fréchet-Schwartz*, and Q.D.F.S. means quotient of dual of *Fréchet-Schwartz*.

EXAMPLE (3.5.2) (POINCARÉ DUALITY). Let $\mathcal{M}$ be the *de Rham* system $\mathcal{O}_X$. We know that the characteristic variety $S\check{S}(\mathcal{O}_X)$ of $\mathcal{O}_X$ is the zero section of T^*X (cf. (3.2.3)). This means that $\mathcal{O}_X$ is holonomic. By the *Poincaré* lemma we have

$$\mathbf{E}^i(\mathcal{O}_X) = H^i(X; \mathbf{C}), \qquad \mathbf{E}^c_i(\mathcal{O}_X) = H^{2n-i}_c(X; \mathbf{C}).$$

If we apply the "duality theorem" to $\mathcal{O}_X$ we get an analytic proof of the *Poincaré* duality for X.

EXAMPLE (3.5.3) (SERRE DUALITY). Let $\mathcal{M}$ be the system $\mathcal{D}$. The characteristic variety $S\check{S}(\mathcal{D})$ of $\mathcal{D}$ is the whole cotangent bundle T^*X. It is easy to see [**52**] that

$$\mathbf{E}^i(\mathcal{D}) = H^i(X; \mathcal{O}_X) \quad \text{and} \quad \mathbf{E}^c_i(\mathcal{D}) = H^{n-i}_c(X; \Omega_X).$$

If we apply the "duality theorem" to $\mathcal{M} = \mathcal{D}$ we find *Serre* duality.

It appears that *Poincaré* duality and *Serre* duality are limit cases of the duality theorem as the dimensions of the characteristic varieties of the systems $\mathcal{O}_X$ and $\mathcal{D}$ are respectively n and $2n$.

In these two examples the characteristic varieties have no singularities. The next example deals with singularities and has been an important step to our solution of the *Riemann-Hilbert* problem. But we need first two important results. Let Y be an analytic subspace of X. The complex $\underline{\mathbf{R}\Gamma}_{[Y]}(\mathcal{O}_X)$ is a subcomplex of $\underline{\mathbf{R}\Gamma}_Y(\mathcal{O}_X)$ (cf. (3.3.1)) and lives in $D(\mathcal{D})_h$ (cf. Theorem (3.3.2)).

THEOREM (3.5.4) (GROTHENDIECK'S COMPARISON THEOREM) [53]. *The natural morphism* $DR(\underline{\mathbf{R}\Gamma}_{[Y]}(\mathcal{O}_X)) \to DR(\underline{\mathbf{R}\Gamma}_Y(\mathcal{O}_X))$ *of* $D(\mathbf{C}_X)_c$ *is an isomorphism.*

This theorem is proved in [53] by reducing to *Grothendieck*'s theorem [20]. But since

$$DR\big(\underline{\mathbf{R}\Gamma}_Y(\mathcal{O}_X)\big) = \underline{\mathbf{R}\Gamma}_Y\big(DR(\mathcal{O}_X)\big) = \underline{\mathbf{R}\Gamma}_Y(\mathbf{C}_X)$$

by the *Poincaré* lemma, Theorem (3.5.4) computes the *Borel-Moore* homology of Y

$$\mathbf{E}_i^c\big(\underline{\mathbf{R}\Gamma}_{[Y]}(\mathcal{O}_X)\big) = H_i^c(Y;\mathbf{C}).$$

To compute the cohomology of Y we have

THEOREM (3.5.5) (SINGULAR POINCARÉ LEMMA [53]). *We have a natural isomorphism in* $D(\mathbf{C}_X)_c$

$$\mathbf{C}_Y \xrightarrow{\sim} \mathbf{S}\big(\mathbf{R}\Gamma_{[Y]}(\mathcal{O}_X)\big).$$

This theorem is proved in [53] using the theory of holonomic systems and the local duality (cf. (3.4)). By Theorem (3.5.5) we have

$$\mathbf{E}^i\big(\underline{\mathbf{R}\Gamma}_{[Y]}(\mathcal{O}_X)\big) \simeq H^i(Y;\mathbf{C}).$$

EXAMPLE (3.5.6). If we apply the duality theorem to $\mathcal{M} = \mathbf{R}\Gamma_{[Y]}(\mathcal{O}_X)$ we find the Poincaré duality for a *singular* space Y embedded in X from a differential system point of view. The complex $\mathbf{R}\Gamma_{[Y]}(\mathcal{O}_X)$ as we are going to see in §4 is "regular", and it solves the *Riemann-Hilbert* problem for the constructible sheaf $\mathbf{C}_Y$.

EXAMPLE (3.5.7) (VERDIER-DUALITY). Let $\mathcal{F}$ be a complex of $D(\mathbf{C}_X)_c$. Theorem (4.3.1) (see §4) says that there is a unique regular holonomic complex $\mathcal{M}$ such that $S(\mathcal{M}) \simeq \mathcal{F}$.

So by this result we get

$$\mathbf{E}^i(\mathcal{M}) = \mathbf{H}^i(X;\mathcal{F}), \qquad \mathbf{E}_i^c(\mathcal{M}) = \mathrm{Ext}^{2n-i}_{\mathbf{C}_{X,c}}(X;\mathcal{F},\mathbf{C}_X).$$

If we apply the duality theorem to this $\mathcal{M}$ we get the *Verdier* duality [75] for the constructible complex $\mathcal{F}$ from the point of view of differential systems. This is a progress in comparison with [57] where holonomic systems of infinite order were needed. In fact, as it was already pointed out in [57], this example was the motivation of the results stated in §4.

The proof of the duality theorem uses *Deligne*'s "méthode de la déscente cohomologique" [13] and a theorem of *Verdier* [76], cf. [54].

(3.6) *Holonomic system of infinite order*. To study specific problems in analytic geometry, we must introduce the sheaf $\mathcal{D}_X^\infty = \mathcal{D}^\infty$ of *differential operators of infinite order* because of the existence of essential singularities which lead to the irregularity of differential systems. This sheaf was introduced by *M. Sato* [68] using local cohomology in developing the theory of hyperfunctions. The sheaf $\mathcal{D}^\infty$ is a sheaf of noncommutative rings with unit faithfully flat over its subsheaf $\mathcal{D}$ [69]. In local coordinates $x = (x_1, \ldots, x_n)$, a local section of $\mathcal{D}^\infty$ looks like a series

$$\sum_{|\alpha|=0}^{\infty} \frac{a_\alpha(x)}{\alpha!} \frac{\partial^\alpha}{\partial x^\alpha}$$

where $a_\alpha(x)$ are holomorphic functions such that

$$\lim_{|\alpha| \to +\infty} |a_\alpha(x)|^{1/|\alpha|} = 0$$

uniformly on compact sets. The *de Rham* system $\mathcal{O}_X$ is a left $\mathcal{D}^\infty$-module. More generally we define a *holonomic system of infinite order* to be a left $\mathcal{D}^\infty$-module $\mathcal{M}^\infty$ such that locally there exist an holonomic $\mathcal{D}$-module $\mathcal{M}$ and an isomorphism

$$\mathcal{D}^\infty \otimes_{\mathcal{D}} \mathcal{M} \xrightarrow{\sim} \mathcal{M}^\infty.$$

In general it is fundamental and difficult to find out if a given $\mathcal{D}^\infty$-module is holonomic or not. Here is a fundamental example. Let $Y \subset X$ be an analytic subspace of X and consider the complex of the local cohomology $\underline{\mathbf{R}\Gamma}_Y(\mathcal{O}_X)$. We have a natural morphism

$$(*) \qquad \mathcal{D}^\infty \otimes_{\mathcal{D}} \mathbf{R}\Gamma_{[Y]}(\mathcal{O}_X) \to \mathbf{R}\Gamma_Y(\mathcal{O}_X).$$

THEOREM (3.6.1) [53] (LOCAL ALGEBRAIC-ANALYTIC COMPARISON THEOREM). *The morphism* $(*)$ *is an isomorphism.*

Theorem (3.6.1) says that the local cohomology sheaves $\underline{H}_Y^k(\mathcal{O}_X)$ are holonomic systems of infinite order. Some special cases were treated in [50, 51]. The proof of Theorem (3.6.1) uses Theorem (3.5.5) and the following.

THEOREM (3.6.2) (BIDUALITY THEOREM). *Let* $\mathcal{M}^\infty$ *be an holonomic system of infinite order. Then there is a natural isomorphism*

$$\mathcal{M}^\infty \to \mathbf{R}\,\mathcal{H}om_{\mathbf{C}_X}(\mathbf{R}\,\mathcal{H}om_{\mathcal{D}^\infty}(\mathcal{M}^\infty, \mathcal{O}_X), \mathcal{O}_X).$$

This theorem is proved in [55, Théorème 2.1, Chapitre III] and called (maybe improperly) the biduality theorem for $\mathcal{D}^\infty$-modules. The proofs use the local duality theorem and some *nuclear topological spaces*. This theorem means that a $\mathcal{D}^\infty$-holonomic module is completely determined by its holomorphic solution complex. We understand it as a generalization of the Frobenius principle which determines an integrable connection from its horizontal sections. We denote by $D(\mathcal{D}^\infty)_h$ the category of bounded complexes of left $\mathcal{D}^\infty$-modules with holonomic cohomology. *It is not* clear that this category is triangulated. But this is a consequence of Theorem (4.3.2) of the next section. From Theorem (3.2.6) it results that the functor $\mathbf{R}\,\mathcal{H}om_{\mathcal{D}^\infty}(*, \mathcal{O}_X)$ sends the category $D(\mathcal{D}^\infty)_h$ to the category $D(\mathbf{C}_X)_c$. We denote by $\mathbf{F}$ the functor

$$\mathbf{F}\colon D(\mathcal{D}^\infty)_h \to D(\mathbf{C}_X)_c, \qquad \mathbf{F}(\mathcal{M}^\infty) = \mathbf{R}\,\mathcal{H}om_{\mathcal{D}^\infty}(\mathcal{M}^\infty, \mathcal{O}_X).$$

If $\mathcal{M}$ is a complex of $D(\mathcal{D})_h$, *we shall denote by* $\mathcal{M}^\infty$ *the complex* of $D(\mathcal{D}^\infty)_h$ defined by $\mathcal{M}^\infty = \mathcal{D}^\infty \otimes_{\mathcal{D}} \mathcal{M}$.

By Theorem (3.6.2) if $\mathcal{M}^\infty$ is a complex of $D(\mathcal{D}^\infty)_h$ then its support is the same as of the support of the constructible complex $\mathbf{F}(\mathcal{M}^\infty)$.

4. Regular holonomic systems. In this section we shall define the notion of regular holonomic systems and state the resolution of a problem of *Riemann-Hilbert* type for constructible sheaves.

(4.1) DEFINITION. We use the notations of §§2 and 3. We have the following proposition (cf. [**55**, Chapitre III, Proposition 3.3] or *Théorèmes de bidualité pour les $\mathcal{D}$-modules holonomes*, Ark. Mat. **20** (1982), 111–124).

PROPOSITION (4.1.1). *Let $\mathcal{M}$ be a complex of $D(\mathcal{D})_h$ and Y be an analytic subspace of X. Then the following conditions are equivalent*:

(a) *the natural homomorphism of $D(\mathbf{C}_X)$ is an isomorphism*

$$DR\big(\underline{\mathbf{R}\Gamma}_{[Y]}(\mathcal{M})\big) \to DR\big(\underline{\mathbf{R}\Gamma}_Y(\mathcal{M})\big);$$

(b) *the natural homomorphism of $D(\mathbf{C}_X)$ is an isomorphism*

$$\mathbf{S}(\mathcal{M}) \otimes_{\mathbf{C}_X} \mathbf{C}_Y \to \mathbf{S}\big(\underline{\mathbf{R}\Gamma}_{[Y]}(\mathcal{M})\big);$$

(c) *the natural homomorphism of $D(\mathcal{D}^\infty)$ is an isomorphism*

$$\mathcal{D}^\infty \otimes_{\mathcal{D}} \underline{\mathbf{R}\Gamma}_{[Y]}(\mathcal{M}) \to \underline{\mathbf{R}\Gamma}_Y(\mathcal{M}^\infty).$$

It is proved in [**53**] that, if $\mathcal{M}$ is the *de Rham* system $\mathcal{O}_X$, then (a) $\Rightarrow$ (b) $\Rightarrow$ (c). For a general $\mathcal{M}$ the proof of the equivalence of the conditions (a), (b) and (c) is similar. The reader may find in [**59**, §5] a report on these conditions. The condition (a) characterizes a regular meromorphic connection in the sense of *Deligne* (cf. [**12**, Proposition 6.8]).

DEFINITION (4.1.2). *A complex $\mathcal{M}$ is called regular along Y, if it satisfies the equivalent conditions* (a), (b) *and* (c) *of Proposition* (4.1.1).

DEFINITION (4.1.3). *A complex $\mathcal{M}$ is called regular, if it is regular along any subspace of X.*

Let Y be defined by an ideal $\mathfrak{I}_Y$. The sheaf of *Zariski* holomorphic function

$$\mathcal{O}_{(X\hat{|}Y)} \overset{\text{def}}{:=} \varprojlim_k \mathcal{O}_X/\mathfrak{I}_Y^k$$

is a left $\mathcal{D}$-module.

PROPOSITION (4.1.4). *If $\mathcal{M}$ is holonomic complex, there is a natural isomorphism in $D(\mathbf{C}_X)$*

$$\mathbf{R}\mathcal{H}om_{\mathcal{D}}\left(\underline{\mathbf{R}\Gamma}_{[Y]}(\mathcal{M}), \mathcal{O}_X\right) \xrightarrow{\sim} \mathbf{R}\mathcal{H}om_{\mathcal{D}}\left(\mathcal{M}, \mathcal{O}_{(X\hat{|}Y)}\right).$$

By Proposition (4.1.4) if $\mathcal{M}$ is regular along Y, in the derived category its "formal" solutions are "convergent" by the condition (b) of Proposition (4.1.1). This is, in the one-dimensional case, the *Malgrange* characterization of a differential equation with a regular singularity [**46**, Theorem 1.4]. (Compare with (2.3.9).) Proposition (4.1.4) is proved in [**55**, Chapter II, Proposition 6.1].

We denote $D(\mathcal{D})_{hr}$ the derived category of regular complexes. It is a full and triangulated subcategory of $D(\mathcal{D})_h$ stable by local cohomology. Moreover, to test the regularity of a complex using the *Mayer-Vietoris* sequence it is enough to test it along a hypersurface [**53**]. But it is a difficult problem to decide whether a complex $\mathcal{M}$ is regular or not. In higher dimensions the main tool is the resolution of singularities (see [**24**]). The result of [**53**] can be stated as follows:

THEOREM (4.1.5). *The de Rham system $\mathcal{O}_X$ is regular.*

This means that if Y is an analytic subspace of X we have the isomorphism in $D(\mathbf{C}_X)_c$:

$$\mathbf{C}_Y \simeq \mathbf{S}(\mathcal{O}_X) \otimes_{\mathbf{C}} \mathbf{C}_Y \xrightarrow{\sim} \mathbf{S}\left(\mathbf{R}\Gamma_{[Y]}(\mathcal{O}_X)\right) \simeq DR\left(\mathcal{O}_{(X\hat{|}Y)}\right) \simeq \Omega^{\cdot}_{(X\hat{|}Y)}.$$

So the formal completion of the *de Rham* complex along Y is a resolution of the constant sheaf $\mathbf{C}_Y$ on Y. This is a result of *Deligne* (1969) and is the first important theorem in *crystalline cohomology* over $\mathbf{C}$. See [**26** and **27**] for other proofs. *Deligne* proves from this result that if X is an algebraic variety over $\mathbf{C}$ then the category of constructible sheaves on X is equivalent to the category of "pro-module crystals" [**14**]. The relation with Theorem (4.3.1) is not clear for the moment.

Let $\mathcal{M}$ be a complex of $D(\mathcal{D})_h$. If the cohomology modules of $\mathcal{M}$ are regular then of course the complex $\mathcal{M}$ is regular. The converse is also true but we do not need it in §(4.3).

PROPOSITION (4.1.6). *A holonomic complex $\mathcal{M}$ is regular if and only if its cohomology modules are regular.*

Let Y be an analytic subspace of X and j be the natural inclusion of $U = X - Y$ in X. Then condition (c) of Proposition (4.1.1) gives us an isomorphism in $D(\mathcal{D}^\infty)_h$ if $\mathcal{M}$ is regular.

$$(*) \qquad \mathcal{D}^\infty \otimes_{\mathcal{D}} \mathbf{R}\left[j_*\right]\mathcal{M} \xrightarrow{\sim} \mathbf{R}j_*j^{-1}\mathcal{M}^\infty.$$

To prove Proposition (4.1.6) it is enough to show that, if $\mathcal{M}_i$ is a cohomology module of the complex $\mathcal{M}$ that the homomorphism of $D(\mathcal{D}^\infty)_h$,

$$(**) \qquad \mathcal{D}^\infty \otimes_{\mathcal{D}} \mathbf{R}\left[j_*\right]\mathcal{M}_i \to \mathbf{R}j_*j^{-1}\mathcal{M}_i^\infty,$$

is an isomorphism for any Y. But we can suppose that Y is an hypersurface *by Mayer-Vietoris*. The isomorphism (**) follows from the isomorphism (*) if we know that the functors $[j_*]$ and j_*j^{-1} are exact. It is trivial for the first functor since Y is a hypersurface. We can prove that j_*j^{-1} is also exact by reducing it to the functor $[j_*]$ using results of the next section or using the biduality Theorem (3.6.2) and a "dévissage" as in [7, Chapitre VIII] (see Theorem (4.5.2)).

(4.2) *Operations on regular complexes*. As in (3.3) we shall state here what happens to the property of regularity under some operations on the $\mathcal{D}$-modules.

(4.2.1) *Direct images*. Let $\pi\colon \tilde{X} \to X$ be a projective morphism of smooth complex varieties and a complex $\tilde{\mathfrak{M}}$ of $D(\mathcal{D}_{\tilde{X}})_h$ such that the cohomology modules of $\tilde{\mathfrak{M}}$ have a *global good filtration*. We have the following theorem [**55**, Chapitre V, Théorème 4.3.1].

THEOREM (4.2.2). *The complex* $\mathfrak{M} = \int_\pi \tilde{\mathfrak{M}}$ *is regular if* $\tilde{\mathfrak{M}}$ *is regular and we have the isomorphism in* $D(\mathbf{C}_X)_c$,

$$\mathbf{R}\pi_*\mathbf{S}(\tilde{\mathfrak{M}})[\dim \tilde{X}] \xrightarrow{\sim} \mathbf{S}(\mathfrak{M})[\dim X].$$

The proof of the first assertion uses a result of Grauert-Remmert as in the original proof of *Grothendieck* [**20**]. The second assertion is proved in [**61**] and the *relative duality* theorem for holonomic systems.

(4.2.3) *Inverse images*. Let $\pi\colon \tilde{X} \to X$ be a morphism of smooth complex varieties and $\mathfrak{M}$ a complex of $D(\mathcal{D}_X)_h$. By Theorem (3.3.6) the complex $\tilde{\mathfrak{M}} = \mathbf{L}\pi^*\mathfrak{M}$ is a complex in $D(\mathcal{D}_{\tilde{X}})_h$.

THEOREM (4.2.4). *If* $\mathfrak{M}$ *is a regular complex then* $\tilde{\mathfrak{M}}$ *is a regular complex and we have the isomorphism in* $D(\mathbf{C}_{\tilde{X}})_c$,

$$DR(\tilde{\mathfrak{M}})[2\dim \tilde{X}] \xrightarrow{\sim} \pi^! DR(\mathfrak{M})[2\dim X].$$

Let $\pi^!$ denote the extraordinary inverse image functor (cf. [**77**]). By *Verdier* duality [**75**] and Theorem (4.2.4) we have the isomorphism in $D(\mathbf{C}_{\tilde{X}})_c$,

$$\mathbf{S}(\tilde{\mathfrak{M}}) \xleftarrow{\sim} \pi^{-1}\mathbf{S}(\mathfrak{M}).$$

Theorem (4.2.4) is proved in [**61**]. To get the duality formula of Theorem (4.2.2) the regularity of $\tilde{\mathfrak{M}}$ *is not necessary*, but in the duality formula of Theorem (4.2.4) the regularity of $\mathfrak{M}$ is *necessary*.

(4.3) *The Riemann-Hilbert problem*. In this section we show how the *Riemann-Hilbert* problem extended to differential systems is solved (cf. (2.4)). In the context of regular integrable connections this has been solved by *P. Deligne* [**12**] (cf. [**39**] for a survey of this case). We would like to give here the precise results and the structures of the proofs will be given in (4.5).

We denote by $\mathbf{S}_r$ *the restriction* to the subcategory $D(\mathcal{D})_{hr}$ of the functor $\mathbf{S}$ which associates to a regular holonomic complex $\mathcal{M}$ its complex of holomorphic solutions,

$$\mathbf{S}_r : D(\mathcal{D})_{hr} \to D(\mathbf{C}_X)_c, \qquad \mathcal{M} \mapsto \mathbf{S}_r(\mathcal{M}) = \mathbf{R}\,\mathcal{H}om_{\mathcal{D}}(\mathcal{M}, \mathcal{O}_X).$$

THEOREM (4.3.1) [**60**]. *The functor* $\mathbf{S}_r$ *is an equivalence of triangulated categories.*

In some sense this theorem solves a generalized *Riemann-Hilbert* problem where one seeks differential systems with regular singularities in the sense of (4.1.2) which has a given complex of $\mathbf{C}_X$-sheaves with constructible cohomology. Actually the answer to the problem of this context shows that one does not avoid the language of derived categories as the given complex of $\mathbf{C}_X$-sheaves with constructible cohomology is not the solution of a regular differential system but the solution of a complex of differential systems with regular cohomology which is unique (up to isomorphism) in the adequate derived category. For a more detailed survey of this point of view we refer the reader to [**59**].

The functor $\mathbf{S}_r$ has no obvious inverse (cf. [**37**]). That is why we shall consider the following functors:

$$\mathbf{F} : D(\mathcal{D}^\infty)_h \to D(\mathbf{C}_X)_c \text{ defined on the object } \mathcal{M}^\infty \text{ of } D(\mathcal{D}^\infty)_h \text{ by}$$

$$\mathcal{M}^\infty \mapsto \mathbf{F}(\mathcal{M}^\infty) = \mathbf{R}\,\mathcal{H}om_{\mathcal{D}^\infty}(\mathcal{M}^\infty, \mathcal{O}_X)$$

and

$$\mathbf{G} : D(\mathbf{C}_X)_c \to D(\mathcal{D}^\infty) \text{ defined on the object } \mathcal{F} \text{ of } D(\mathbf{C}_X)_c \text{ by}$$

$$\mathcal{F} \mapsto \mathbf{G}(\mathcal{F}) = \mathbf{R}\,\mathcal{H}om_{\mathbf{C}_X}(\mathcal{F}, \mathcal{O}_X).$$

Of course the complex $\mathbf{G}(\mathcal{F})$ is a complex of $D(\mathcal{D}^\infty)$ because $\mathcal{O}_X$ is a left $\mathcal{D}^\infty$-module.

THEOREM (4.3.2) [**55**]. *If* $\mathcal{F} \in D(\mathbf{C}_X)_c$, *the complex* $\mathbf{G}(\mathcal{F})$ *belongs to* $D(\mathcal{D}^\infty)_h$ *and the functors* $\mathbf{F}$ *and* $\mathbf{G}$ *are inverse to each other.*

This theorem answers question $(\mathrm{HR})'_1$ of [**59**]. Somehow this is the answer to another generalized *Riemann-Hilbert* type problem for $\mathcal{D}^\infty$-modules. As it is pointed out in [**59**, §2] the proofs of Theorems (4.3.1) and (4.3.2) are closely related. A local proof of Theorem (4.3.1) is necessary in the proof of Theorem (4.3.2). The proof of Theorem (4.3.1) requires a global version of *Hironaka* resolution of singularities [**29**] (see Remark (4.6.4)). In §(4.5) we shall explain the structure of the proofs.

By Theorems (4.3.1) and (4.3.2) the three categories $D(\mathbf{C}_X)_c$, $D(\mathcal{D})_{hr}$ and $D(\mathcal{D}^\infty)$ are equivalent by the functors $\mathbf{S}_r$, $\mathbf{F}$ and $\mathbf{T}$

$$\begin{array}{ccc} D(\mathcal{D})_{hr} & \xrightarrow{\mathbf{S}_r} & D(\mathbf{C}_X)_c \\ \mathbf{T}\downarrow & & \| \\ D(\mathcal{D}^\infty)_h & \xrightarrow{\mathbf{F}} & D(\mathbf{C}_X)_c \end{array}$$

where $\mathbf{T}$ is the functor

$$\mathcal{M} \to \mathbf{T}(\mathcal{M}) = \mathcal{D}^\infty \otimes_{\mathcal{D}} \mathcal{M}.$$

Any result in each of these categories has potentially its analogue in the two others. The theorems of §(4.2) are examples of this fact. But these theorems are only results of *existence* and *unicity*. Given a situation in one category it is in general a difficult problem to find *explicitly* the analogue situations in the others. To illustrate this and to help the reader to understand, we shall give several examples. These examples are already in **[60]**. These examples show why in spite of the simplicity of Theorems (4.3.1) and (4.3.2), it took a long time to find a good formulation.

In the Examples (4.4.5), (4.4.6), a single sheaf $\mathcal{F}$ gives rise to complexes $\mathcal{M}$ and $\mathcal{M}^\infty$. In the Examples (4.4.8), (4.4.9), a *complex* $\mathcal{F}$ gives rise to a single module $\mathcal{M}$.

(4.5) *Structures of the proofs of Theorems* (4.3.1) *and* (4.3.2).

First step (*local resolution*). To prove Theorems (4.3.1) and (4.3.2) we make a "dévissage" of a constructible complex and use the resolution of the singularities. Then we come to prove the basic following Theorem (4.5.1). Let $Z \subset Y$ be two analytic subspaces of X such that Z contains the singular part of Y. Let $\mathcal{F}$ be a constructible complex such that its cohomology sheaves are locally constant on $W = Y \backslash Z$ and zero outside W, i.e. zero on Z and on $X \backslash Y$. Let $p = \operatorname{codim} Y$.

THEOREM (4.5.1). *Up to an isomorphism of* $D(\mathcal{D})_{hr}$ *there exists a unique regular complex* $\mathcal{M}$ *such that*

(i) $\mathbf{R}\,\mathcal{H}om_{\mathbf{C}_X}(\mathcal{F}, \mathcal{O}_X) \xrightarrow{\sim} \mathcal{D}^\infty \otimes_{\mathcal{D}} \mathcal{M}$.

(ii) $\mathbf{S}_r(\mathcal{M}) \simeq \mathbf{F}(\mathbf{R}\,\mathcal{H}om_{\mathbf{C}_X}(\mathcal{F}, \mathcal{O}_X)) \simeq \mathcal{F}$.

Let $(\tilde{Y}, \tilde{Z})$ be an embedded resolution of the pair (Y, Z), that is to say a projectif morphism of smooth complex varieties π

$$\begin{array}{ccccc} \tilde{Z} & \subset & \tilde{Y} & \subset & \tilde{X} \\ \downarrow & & \downarrow \pi_Y i & & \downarrow \pi \\ Z & \subset & Y & \subset & X \end{array}$$

such that $\tilde{Y}$, the strict transform of Y by π, is smooth and $\tilde{Z} = \pi_Y^{-1}(Z)$ is a normal divisor crossing in $\tilde{Y}$. The complex $\mathcal{M}$ of Theorem (4.5.1) is the integration along the fiber $\int_{\pi \circ i} \tilde{\mathcal{M}}_D$ of the *Deligne* extension (Example (4.4.4) above) associated to $\tilde{\mathcal{F}} = \pi_Y^{-1}\mathcal{F}$ on $\tilde{Y}$. This proves the *existence* of $\mathcal{M}$. Formula (ii) follows from condition (c) of Proposition (4.1.1) or from the relative duality (Theorem (4.2.2)) which avoids the use of the sheaf $\mathcal{D}^\infty$. To prove the *unicity* of $\mathcal{M}$ we use the duality Theorem (4.2.4). Let $\mathcal{N}$ be a regular complex such that $\mathbf{S}_r(\mathcal{N}) \xrightarrow{\sim} \mathcal{F}$. Let π: $\tilde{X} \to X$ be a resolution of the pair (Y, Z). We prove the formula

$$\mathbf{R}\pi_* \mathbf{R}\,\mathcal{H}om_{\mathcal{D}_{\tilde{X}}}\left(\mathbf{L}\pi^*\mathcal{N}, \int_i \tilde{\mathcal{M}}_D\right) \xrightarrow{\sim} \mathbf{R}\,\mathcal{H}om_{\mathcal{D}_X}\left(\mathcal{N}, \int_{\pi \circ i} \tilde{\mathcal{M}}_D\right)$$

(4.4) *Examples.*

TABLE I

$\mathcal{F} \in D(\mathbf{C}_X)_c$	$\mathcal{M} \in D(\mathcal{D}_X)_{hr}$ such that $\mathcal{F} = \mathbf{S}_r(\mathcal{M})$	$\mathcal{M}^\infty = \mathbf{G}\mathcal{F} = \mathbf{R}\,\mathcal{H}om_{\mathbf{C}_X}(\mathcal{F}, \mathcal{O}_X) = \mathcal{D}_X^\infty \otimes_{\mathcal{D}_X} \mathcal{M}$; $\mathcal{F} = \mathbf{F}(\mathcal{M}^\infty)$
EXAMPLE (4.4.1). $\mathcal{F} = \mathbf{C}_X$	$\mathcal{M} = \mathcal{O}_X =$ the *de Rham* system (*Poincaré* lemma)	$\mathcal{M}^\infty = \mathcal{D}_X^\infty \otimes_{\mathcal{D}_X} \mathcal{O}_X = \mathcal{O}_X$
EXAMPLE (4.4.2). Let $Y \subset X$ be a divisor with normal crossings of X; $U = X \setminus Y \hookrightarrow_j X$; $\mathcal{F} = j_!\underline{\mathbf{C}}_U$.	$\mathcal{M} = \mathcal{O}[*Y] =$ the sheaf of the meromorphic functions in X with poles in Y. (*Atiyah-Hodge*, 1954) [**2**].	$\mathcal{M}^\infty = \mathcal{D}_X^\infty \otimes_{\mathcal{D}_X} \mathcal{O}[*Y] = j_*\mathcal{O}_U$ ([**50, 51**]; 1974).
EXAMPLE (4.4.3). Let $Y \subset X$ be a hypersurface with arbitrary singularity of X; $U = X \setminus Y \hookrightarrow_j X$, $\mathcal{F} = j_!\mathbf{C}_U$.	$\mathcal{M} = \mathcal{O}[*Y] = \varinjlim_k \mathcal{H}om_{\mathcal{O}_X}(\mathcal{J}_Y^k; \mathcal{O}_X)$ (*Grothendieck* [**20**]; 1963).	$\mathcal{M}^\infty = \mathcal{D}_X^\infty \otimes_{\mathcal{D}_X} \mathcal{O}[*Y] = j_*\mathcal{O}_U$ ([**52**; 1975).
EXAMPLE (4.4.4). Let $Y \subset X$ be a divisor with normal crossings of X. $U = X \setminus Y \hookrightarrow_j X$. Let $\mathcal{L}$ be a local system on U and $\mathcal{F} = j_!\mathcal{L}$.	$\mathcal{M} = \mathcal{M}_D =$ *Deligne* extension (*Deligne* [**12**]; 1969).	$\mathcal{M}^\infty = \mathcal{D}_X^\infty \otimes_{\mathcal{D}_X} \mathcal{M}_D = j_*(\mathcal{O}_U \otimes_{\mathbf{C}} \mathcal{L})$. ([**55**]; 1978)

In the first four examples, there are no complexes. To a single constructible sheaf corresponds a single module $\mathcal{M}$ and a single module $\mathcal{M}^\infty$. The next five examples show that it is not the general situation.

TABLE II

EXAMPLE (4.4.5). Let $Y \subset X$ be an analytic subspace defined by an Ideal $\mathfrak{I}_Y$. Let $\mathcal{F} = \mathbf{C}_Y$.	$\mathcal{M} = \underline{\mathbf{R}\Gamma_{[Y]}}(\mathcal{O}_X)$ = the local algebraic cohomology of $\mathcal{O}_X$ (singular *Poincaré* lemma-Theorem 1.1 of [**53**]; 1976).	$\mathcal{M}^\infty = \mathcal{D}_X^\infty \otimes_{\mathcal{D}_X} \underline{\mathbf{R}\Gamma_{[Y]}}(\mathcal{O}_X) = \underline{\mathbf{R}}\Gamma_Y(\mathcal{O}_X)$ (Theorem 1.2 of [**53**]; 1976).
EXAMPLE (4.4.6). Let $Z \subset Y \subset X$ be closed analytic subspaces of X such that Z contains the singular part of Y; $X \setminus Z \hookrightarrow_j X$. Let $\mathcal{L}$ be a monodromy on $Y \setminus Z$ and $\mathcal{F} = j_!\mathcal{L}$.	$\mathcal{M} = \int_\pi \mathcal{M}_D$ = integration along the fibers of the *Deligne* extension for a suitable resolution of the singularities. ([**55**]; Chapitre V; 1978). See [**33**] for the meaning of $\int_\pi$ or 3.3.2.	$\mathcal{M}^\infty = \mathbf{R}\,\mathcal{H}om_{\mathbf{C}_X}(\mathcal{F}, \mathcal{O}_X) = \mathcal{D}_X^\infty \otimes_{\mathcal{D}_X} \mathcal{M}$ ([**55**, Chapitre V]; 1978).
EXAMPLE (4.4.7). Let $\tilde{X} \xrightarrow{\pi} X$ be a proper morphism of smooth varieties and $\tilde{\mathcal{F}}$ a complex in $D(\mathbf{C}_{\tilde{X}})_c$. $\mathcal{F} = \mathbf{R}\pi_*\tilde{\mathcal{F}}$.	$\mathcal{M}_{\mathrm{G.M}}$ = the *Gauss-Manin* system associated to $(\pi, \tilde{\mathcal{F}})$. ([**59**, §7]; 1979).	Cf. Example (4.4.6)
EXAMPLE (4.4.8). Let $Y \subset X$ be a closed analytic subspace of X and π_Y the complex of *Deligne-MacPherson* of intersection homology sheaves [**18**].	$\mathcal{M}$ = single and simple module ([**3**, **6**]; 1980).	Cf. Example (4.4.6).
EXAMPLE (4.4.9). Let $f\colon (\mathbf{C}^n, 0) \to (\mathbf{C}, 0)$ be a germ of an analytic function and $\check{\mathcal{F}} = \mathbf{R}\psi_f$ be the complex of vanishing cycles [**23**].	In [**49**] *B. Malgrange* has built the corresponding regular $\mathcal{D}$-module. He quotes that independently *M. Kashiwara* and *Beilinson* and *I. Bernstein* have obtained similar results.	Cf. Example (4.4.6).

where $\tilde{\mathcal{M}}_D$ is the *Deligne* extension associated to $\tilde{\mathcal{F}} = \pi_Y^{-1}\mathcal{F}$. But over $\tilde{Y} = \pi_Y^{-1}(Y)$ the functor $\mathbf{S}_r$ is fully faithful since $\tilde{Z} = \pi_Y^{-1}(Z)$ is a normal crossing divisor in $\tilde{Y}$ and the cohomology sheaves of $\tilde{\mathcal{F}}$ are locally constant in $\tilde{W} = \tilde{Y} - \tilde{Z}$ [**12**]. By Theorem (4.2.4) the complexes $\mathbf{L}\pi^*\mathcal{N}$ and $\int_i \tilde{\mathcal{M}}_D[-P]$ have the same holomorphic solution complex $\tilde{\mathcal{F}}$. Thus by the faithfulness of $\mathbf{S}_r$ in this case, we get a morphism $\mathbf{L}\pi^*\mathcal{N} \to \int_i \tilde{\mathcal{M}}_D[-P]$ and by the previous isomorphism a morphism

$$(*) \qquad \mathcal{N} \to \int_{\pi \circ i} \tilde{\mathcal{M}}_D$$

which corresponds to the identity morphism $\mathcal{F} \to \mathcal{F}$. Applying the functor $\mathbf{T}$, $\mathcal{M} \to \mathcal{D}^\infty \otimes_{\mathcal{D}} \mathcal{M}$ we get a morphism

$$(**) \qquad \mathcal{D}^\infty \otimes_{\mathcal{D}} \mathcal{N} \to \mathcal{D}^\infty \otimes_{\mathcal{D}} \int_{\pi \circ i} \tilde{\mathcal{M}}_D[-P]$$

which is an isomorphism by the biduality Theorem (3.6.2) since $\mathbf{S}_r(\mathcal{N}) = \mathbf{S}_r(\int_{\pi \circ i} \tilde{\mathcal{M}}_D) = \mathcal{F}$. But $\mathcal{D}^\infty$ is faithfully flat over $\mathcal{D}$, the morphism $(*)$ is already an isomorphism and the unicity follows and Theorem (4.5.1) is proved.

Second step ("*dévissage*"). To prove some local *property* of a regular complex $\mathcal{M}$ we proceed by induction on the dimension of the support Y of $\mathcal{M}$. With the Mayer-Vietoris sequence and by induction, we can suppose that Y is irreducible. In this case the singular support Z of $\mathcal{M}$, which is an analytic subspace of Y containing the singular locus of Y such that the cohomology sheaves of the complex $\mathbf{S}_r(\mathcal{M})$ are locally constant outside Z, is a proper subspace on Y and $\dim Z < \dim Y$. We have a triangle of $D(\mathcal{D})_{hr}$

$$\begin{array}{ccc} & \mathbf{R}\left[j_*\right]\mathcal{M} & \\ \swarrow & & \nwarrow \\ \underline{\mathbf{R}\Gamma}_{[Z]}(\mathcal{M}) & \to & \mathcal{M} \end{array}$$

if j: $W = Y \backslash Z \hookrightarrow X$ is the natural injection. But $\dim Z < \dim Y$ and by the induction hypothesis the *property* holds for $\underline{\mathbf{R}\Gamma}_{[Z]}(\mathcal{M})$. Thus, because of the above triangle, if the *property* holds for $\mathbf{R}[j_*]\mathcal{M}$, then it holds for $\mathcal{M}$. But $\mathcal{M}$ is regular along Z and condition (b) of Proposition (4.1.1) tells that the cohomology sheaves of the holomorphic solution complex $\mathbf{S}_r(\mathbf{R}[j_*]\mathcal{M})$ are locally constant on $W = Y - Z$ and zero outside W. Then Theorem (4.5.1) tells us that $\mathbf{R}[j_*]\mathcal{M}$ is isomorphic with the integration $\int_{\pi \circ i} \tilde{\mathcal{M}}_D$ of the *Deligne* extension $\tilde{\mathcal{M}}_D$ associated to $\tilde{\mathcal{F}} := \pi_Y^{-1}\mathbf{S}_r(\mathbf{R}[j_*]\mathcal{M})$ where π is an embedded resolution of the pair (Y, Z). We are reduced to prove the *property* holds for *Deligne* extensions. Using this "dévissage" we prove [**61**, Theorem (2.4.1)] the following theorem.

THEOREM (4.5.2). *The dual complex $\mathcal{M}^*$ of a regular complex $\mathcal{M}$ is regular.*

Let $\mathcal{M}_i$ $(i = 1, 2)$ be two regular complexes. Using Theorem (4.5.2) and the same "dévissage" we prove that the homomorphism in $D(\mathbf{C}_X)_c$,

$$\mathbf{R}\,\mathcal{H}om_{\mathcal{D}}(\mathcal{M}_1, \mathcal{M}_2) \overset{\mathbf{S}_r}{\to} \mathbf{R}\,\mathcal{H}om_{\mathbf{C}_X}(\mathcal{F}_2, \mathcal{F}_1),$$

is an isomorphism if $\mathcal{F}_i = \mathbf{S}_r(\mathcal{M}_i)$ $(i = 1, 2)$. This means that the functor $\mathbf{S}_r$ is fully faithful [**61**].

Let $\mathcal{F}$ be a complex of $D(\mathbf{C}_X)_c$, then we prove that the complex $\mathbf{R}\,\mathcal{H}om_{\mathbf{C}_X}(\mathcal{F}, \mathcal{O}_X)$ is a complex of $D(\mathcal{D}^\infty)_h$. To avoid using the fact that the category $D(\mathcal{D}^\infty)_h$ is triangulated we use the faithfulness of $\mathbf{S}_r$ and reduce to the case where $\mathcal{F}$ is locally constant outside a subspace Z of its support Y containing the singular locus of Y and zero outside $W = Y \setminus Z$. This case is treated by formula (i) of Theorem (4.5.1). The biduality theorem shows that $\mathbf{G}$ is an inverse of $\mathbf{F}$ and Theorem (4.3.2) follows [**59**].

Third step (*global resolution*). (See Remark (4.6.4).) To prove Theorem (4.3.1) it remains to show that the functor $\mathbf{S}_r$ is essentially surjective. We want to show that any complex $\mathcal{F}$ of $D(\mathbf{C})_c$ looks like $\mathbf{S}_r(\mathcal{M})$. The functor $\mathbf{S}_r$ is fully faithful so we can suppose that $\mathcal{F}$ is a constructible sheaf. By induction on the dimension of the support of $\mathcal{F}$ we can suppose that $\mathcal{F}$ is locally constant on a *Zariski* open smooth set of its support which is not empty and zero outside it. Using *global* [**29**] resolution of the singularities we are reduced to Theorem (4.5.1).

For details we refer the reader to [**55** and **61**].

REMARK (4.5.3). Let $\mathcal{M}$ be of $D(\mathcal{D})_h$ which is not necessarily regular. Then by Theorem (4.3.1) and Theorem (4.3.2) there exists a unique regular complex $\mathcal{M}_r$ such that

$$\mathcal{F} = \mathbf{S}(\mathcal{M}) = \mathbf{S}_r(\mathcal{M}_r) = \mathbf{F}(\mathcal{D}^\infty \otimes_{\mathcal{D}} \mathcal{M}) = \mathbf{F}(\mathcal{D}^\infty \otimes_{\mathcal{D}} \mathcal{M}_r).$$

It follows that globally in X we have the isomorphism

$$(*) \qquad \mathcal{D}^\infty \otimes_{\mathcal{D}} \mathcal{M} \xrightarrow{\sim} \mathcal{D}^\infty \otimes_{\mathcal{D}} \mathcal{M}_r$$

see [**59**, §7] where the result is stated locally in X.

This formula helps to prove results on irregular complexes. For example let Y be a hypersurface and $j\colon U = X \setminus Y \hookrightarrow X$ the natural inclusion. We want to prove that the functor $j_* j^{-1}$ is exact on the category of holonomic systems. Let $\mathcal{M}$ be a holonomic $\mathcal{D}$-module; then by this previous isomorphism we have

$$\mathbf{R}\, j_* j^{-1} \mathcal{D}^\infty \otimes_{\mathcal{D}} \mathcal{M} \simeq \mathbf{R}\, j_* j^{-1} \mathcal{D}^\infty \otimes_{\mathcal{D}} \mathcal{M}_r \simeq \mathcal{D}^\infty \otimes_{\mathcal{D}} \mathbf{R}[j_*] \mathcal{M}_r$$

because of condition (c) of Proposition (4.1.1) since $\mathcal{M}_r$ is regular. But the functor $[j_*]$ is exact and this shows that the functor $j_* j^{-1}$ is exact on the category of holonomic systems.

REMARK (4.5.4). The functor $\mathbf{S}\colon D(\mathcal{D})_h \to D(\mathbf{C}_X)_c$ is not an equivalence of categories. We must add some new structure to the category $D(\mathbf{C}_X)_c$ to classify the irregular complexes. If $\dim X = 1$, *B. Malgrange* told us that he completely solved the problem (cf. [**48**]).

(4.6) *Perverse complexes*. The functors $\mathbf{S}_r$ and $\mathbf{F}$ are equivalences of derived categories. As we have seen in the examples of (4.4), a single module of $D(\mathcal{D})_{hr}$ can have a whole complex of $D(\mathbf{C}_X)_c$ as holomorphic solutions just as a constructible sheaf can be the solution of a whole complex of $D(\mathcal{D})_{hr}$. But the complexes of $D(\mathbf{C}_X)_c$ which are the holomorphic solution complexes of a single $\mathcal{D}$-module have remarkable properties and are called the *perverse complexes* (for the middle perversity) [**18**].

DEFINITION (4.6.1). *A complex $\mathcal{F}$ of $D(\mathbf{C}_X)_c$ is called perverse if*

(i) $h^i(\mathcal{F}) = 0$ *for* $i < 0$, *and the codimension of the support of the sheaf* $h^i(\mathcal{F})$ *is* $\geqslant i$.

(ii) $h^i(\mathcal{F}^\vee) = 0$ *for* $i < 0$, *and the codimension of the support of the sheaf* $h^i(\mathcal{F}^\vee)$ *is* $\geqslant i$.

Remember that $\mathcal{F}^\vee = \mathbf{R}\,\mathcal{H}om_{\mathbf{C}_X}(\mathcal{F}, \mathbf{C}_X)$ and $h^i(\mathcal{F})$ is the i-cohomology sheaf of $\mathcal{F}$. We have the following lemma.

LEMMA (4.6.2). *Let $\mathcal{M}$ be a holonomic system on X. Then the complex $\mathcal{F} = \mathbf{S}(\mathcal{M})$ of holomorphic solutions of $\mathcal{M}$ is perverse.*

The property (i) of Definition (4.6.1) is obtained by *M. Kashiwara* in [**32**, Theorem 4.1]. By the local duality Theorem (3.4.1) we have $\mathcal{F}^\vee = \mathbf{S}(\mathcal{M}^*)$; thus the complex $\mathcal{F}$ satisfies the condition (ii) of (4.6.1).

Combining this lemma with Theorem (4.3.1) we are led to

THEOREM (4.6.3) (DELIGNE [**7**]). *The restriction of the functor* $\mathbf{S}_r$ *induces an equivalence between the category* $\mathrm{mod}(\mathcal{D})_{hr}$ *of holonomic and regular $\mathcal{D}$-modules and the category of perverse complexes.*

Since the category $\mathrm{mod}(\mathcal{D})_{hr}$ is abelian and local the category of perverse complexes is *abelian* and *local*.

Let $\mathcal{M}$ be a holonomic regular system, Lemma (4.6.2) above shows $\mathbf{S}_r(\mathcal{M})$ is perverse. Conversely let $\mathcal{F}$ be a perverse complex, then Theorems (4.3.1) and (4.3.2) give a holonomic and regular complex $\mathcal{M}$ of $D(\mathcal{D})_{hr}$ such that

$$\mathcal{F} = \mathbf{S}_r(\mathcal{M}) = \mathbf{F}\big(\mathbf{R}\,\mathcal{H}om_{\mathbf{C}_X}(\mathcal{F}, \mathcal{O}_X)\big) = \mathbf{F}(\mathcal{D}^\infty \otimes_{\mathcal{D}} \mathcal{M}).$$

But to see that $\mathcal{M}$ is a single $\mathcal{D}$-module, because of the faithful flatness of $\mathcal{D}^\infty$ over $\mathcal{D}$, it is enough to see that $\mathbf{R}\,\mathcal{H}om_{\mathbf{C}_X}(\mathcal{F}, \mathcal{O}_X)$ is a single $\mathcal{D}^\infty$-module or that

$$\mathcal{E}xt^i_{\mathbf{C}_X}(\mathcal{F}, \mathcal{O}_X) = 0 \quad \text{if } i \neq 0.$$

This can be done by "dévissage" on $\mathcal{F}$ and triangulation [**7**]. This proves Theorem (4.6.3).

The characteristic sheaf $\mathbf{C}_Y$ of a complex subspace of X is not perverse if Y is not a complete intersection since Theorem (3.5.5) says that

$$\mathbf{C}_Y \xrightarrow{\sim} \mathbf{S}_r\big(\underline{\mathbf{R}\Gamma_{[Y]}}(\mathcal{O}_X)\big)$$

and $\underline{\mathbf{R}\Gamma_{[Y]}}(\mathcal{O}_X)$ is a complex in general.

Let us give an example of a perverse sheaf. Consider the Example (4.4.6) of §(4.4) where $Y = X$ and Z is a hypersurface. Let $\mathcal{L}$ be a local system on $U = X \backslash Z \hookrightarrow_j X$. We have (cf. (4.4.6))

$$\mathcal{F} = j_!\mathcal{L} = \mathbf{S}_r(\mathcal{M}) = \mathbf{F}\big(\mathbf{R}\,\mathcal{H}om_{\mathbf{C}_X}(j_!\mathcal{L}, \mathcal{O}_X)\big).$$

But on another hand,

$$\mathbf{R}\,\mathcal{H}om_{\mathbf{C}_X}(j_!\mathcal{L}, \mathcal{O}_X) \cong \mathbf{R}j_*(\mathcal{O}_U \otimes_{\mathbf{C}_U} \mathcal{L}^\vee) \cong j_*(\mathcal{O}_U \otimes_{\mathbf{C}_U} \mathcal{L}^\vee) \cong \mathcal{D}^\infty \otimes_{\mathcal{D}} \mathcal{M}.$$

By the faithful flatness of $\mathcal{D}^\infty$ over $\mathcal{D}$, the complex $\mathcal{M}$ is a regular holonomic system: it implies that $j_!\mathcal{L}$ is perverse. In fact this result can be proved directly by topological methods. Theorem (4.3.1) shows that the system $\mathcal{M}$ such that $\mathbf{S}_r(\mathcal{M}) = j_!\mathcal{L}$ is unique up to isomorphism: one calls it the *Deligne system* associated to the local system $\mathcal{L}$. The "solutions" $j_!\mathcal{L}$ can be given by *Nilsson* class functions [**5**, **36**] (compare to [**12**, Chapitre III, Théorème 1.8]).

An important perverse complex is the intersection homology a complex of *Deligne*, *Goresky* and *MacPherson* [**18**] of a singular space $Y \subset X$ of pure dimension m. Let

$$Y = Y_m \supset Y_{m-1} \supset \cdots Y_0 \supset Y_{-1} = \varnothing$$

be a *Whitney* filtration of Y where Y_i is a closed analytic subspace, $Y_i - Y_{i+1}$ is nonsingular of pure dimension i. Let $U_i = Y_m - Y_{m-i}$ and $j_i\colon U_i \to U_{i+1}$ be the natural inclusion. By definition the intersection homology complex is the complex [**18**]

$$\pi_Y = \mathcal{C}_{\leqslant m-1}\mathbf{R}\, j_{m*} \cdots \mathcal{C}_{\leqslant 0}\mathbf{R}\, j_{1*}(\mathbf{C}_{U_i})$$

where, for a complex $\mathcal{F}$ of $D(\mathbf{C}_X)$ and a natural number p, $\mathcal{C}_{\leqslant p}\mathcal{F}$ is the truncated complex which has the following property:

$$h^i(\mathcal{C}_{\leqslant p}\mathcal{F}) = \begin{cases} h^i(\mathcal{F}) & \text{if } i \leqslant p; \\ 0 & \text{if } i > p. \end{cases}$$

The complex $\pi_Y[m-n]$ is perverse and Theorem (4.6.3) shows there is a unique regular holonomic $\mathcal{D}$-module $\mathcal{L}(Y, X)$ such that

$$\mathbf{S}_r(\mathcal{L}(Y, X)) \xrightarrow{\sim} \pi_Y[m-n].$$

There is an explicit description of $\mathcal{L}(Y, X)$ in [**7**] in terms of the local cohomology functors of Y and $Z = \operatorname{sing} Y$. It is proved in [**18**] that $\pi_Y[m-n] \xrightarrow{\sim} (\pi_Y[m-n])^\vee$ which is a local result. From the *Verdier* global duality we get the *Poincaré* duality for the middle intersection groups of Y, if Y is compact. These results are topological. But according to our dictionary between analytic results and topological results, we can also prove these results using linear differential systems. In fact it is enough to check that $\mathcal{L}(Y, X) \xrightarrow{\sim} \mathcal{L}(Y, X)^*$ and use the local duality theorem for the $\mathcal{D}$-modules (Theorem (3.4.1)). Then use the global duality for the $\mathcal{D}$-module (Theorem (3.5.1)) to find the *Poincaré* duality for the middle intersection groups of Y.

REMARK (4.6.4). Following a remark of *B. Malgrange* one finds that, since the category of perverse complexes is a *local* category, it is possible to avoid the use of the *global* resolution of singularities in the proof of Theorem (4.3.1) if instead we use Theorem (4.6.3) which requires only the *local* resolution of singularities. In fact by Theorem (4.3.2) to prove that the functor $\mathbf{S}_r$ is essentially surjective by "dévissage" we are reduced to the case of perverse complexes and need only the *local* resolution of singularities.

(4.7) *Hodge structure on the middle homology groups.* It is conjectured in **[10]** that the middle homology groups of a complex projective variety Y carry a *pure Hodge* structure. We would like to describe *Brylinski*'s conjecture to solve this problem.

Let $\mathfrak{M}$ be a holonomic module. We have seen that locally $\mathfrak{M}$ has many good filtrations but globally good filtrations may not exist. But if $\mathfrak{M}$ is regular it is expected **[7]** that $\mathfrak{M}$ has a canonical global good filtration $(\mathfrak{M})_{k\in\mathbf{Z}}$. This filtration induces a filtration $F^{\cdot}$ on $DR(\mathfrak{M})$,

$$F^p(DR(\mathfrak{M})): 0 \to \mathfrak{M}_{-p} \to \Omega^1_X \otimes_{\mathcal{O}_X} \mathfrak{M}_{-p+1} \to \cdots \to \Omega^n_X \otimes_{\mathcal{O}_X} \mathfrak{M}_{-p+n} \to 0.$$

If X is the projective space, *Brylinski* conjecture states that the filtration $F^{\cdot}(DR(\mathfrak{L}(X, Y)))$ induces a pure *Hodge* structure on

$$\mathbf{H}^k(X, DR(\mathfrak{L}(Y, X))) = \mathbf{H}^k(Y, \pi_Y[m-n])$$

of weight $k+n-m$.

REMARK (4.7.1). *Deligne* proved that, if $\mathfrak{M}$ is a coherent algebraic $\mathfrak{D}$-module on the projective space $\mathbf{P}^n$, then it has a global good filtration.

REFERENCES

1. V. I. Arnold, *Points critiques des fonctions différentiables et leurs formes normales*, Uspehi Mat. Nauk **30** (1975), 3–65. (Russian)

2. M. Atiyah and W. V. Hodge, *Integrals of second kind on algebraic varieties*, Ann. of Math. (2) **62** (1955), 56–91.

3. A. Beilinson and I. N. Bernstein, *Localisation de g-modules*, Note aux C. R. Acad. Sci. Paris Sér. A-B **292** (1981).

4. I. N. Bernstein, *The analytic continuations of generalized functions with respect to a parameter*, Functional Anal. Appl. **6** (1972), 26–40.

5. J. E. Björk, *Rings of differential operators*, North-Holland, Amsterdam, 1979.

6. J. L. Brylinski and M. Kashiwara, *Démonstration de la conjecture de Kazhdan-Lusztig sur les modules de Verma*, Note aux C. R. Acad. Sci. Paris Sér. A (1980), 373–376.

7. J. L. Brylinkski, *Contributions à la théorie des groupes*, Thèse de Doctorat d'État, Université d'Orsay, Juin 1981.

8. ______, *Systèmes holonomes à singularités régulières et filtrations de Hodge*. II, Comptes Rendus de la Conférence, Analyse et Topologie sur les Espaces Singuliers (Luminy, Juillet 1981), Astérisque (to appear).

9. E. Brieskorn, *Die Monodromie der isolierten Singularitäten von Hyperflächen*, Manuscripta Math. **2** (1970), 103–161.

10. J. Cheeger, M. Goresky and R. MacPherson, *L^2-cohomology and intersection homology for singular algebraic varieties*, Differential Geometry Sem. (I.A.S.S. Yau, Editor), Ann. of Math. Studies, Princeton Univ. Press, Princeton, N. J., 1981.

11. C. H. Clemens, *Picard Lefschetz theorem for families of non singular algebraic varieties acquiring ordinary singularities*, Trans. Amer. Math. Soc. **136** (1969), 93–108.

12. P. Deligne, *Equations différentielles à points singuliers réguliers*, Lecture Notes in Math., vol. 163, Springer-Verlag, Berlin-Heidelberg-New York, 1970.

13. ______, *Théorie de Hodge*. II, III, Inst. Hautes Études Sci. Publ. Math. **40** (1972), 5–57; **44** (1974), 1–77.

14. ______, Exposés à l'Inst. Hautes Études Sci., Mars 1970 (unpublished).

15. O. Gabber, *The integrability of the characteristic variety*, Amer. J. Math. **103** (1981), 445–468.

16. S. Gel'fand and R. MacPherson, *Verma modules and Schubert cells: a dictionary*, Inst. Hautes Études Sci., no. M/80/45 (Nov. 1980) (preprint).

17. M. Goresky and R. MacPherson, *Intersection homology theory*. I, Topology **19** (1980), 135–162.

18. ______, *Intersection homology*. II, Inst. Hautes Études Sci. (1981) (preprint).

19. E. Goursat, *Cours d'analyse mathématique*. II, Gauthier-Villars, Paris, 1929.

20. A. Grothendieck, *On the de Rham cohomology of algebraic varieties*, Inst. Hautes Études Sci. Publ. Math. **29** (1966), 93–103.

21. A. Grothendieck et al., *Dix esposés sur la cohomologie des schémas*, North-Holland, Amsterdam, 1968.

22. A. Grothendieck, *Cohomologie l-adique et fonctions L*, SGA V (le Bois-Marie, 1965-66), Lecture Notes in Math., vol. 589, Springer-Verlag, Berlin-Heidelberg-New York, 1977.

23. A. Grothendieck (with P. Deligne et N. Katz), *Groupes de monodromie en géométrie algébrique*, SGA VII (1967-689), Lecture Notes in Math., vols. 288 et 340, Springer-Verlag, Berlin-Heidelberg-New York, 1973.

24. A. Grothendieck, *Travaux de Heisuke Hironaka sur la résolution des singularités*, Actes C.I.M. (Nice, 1970), vol. I, Gauthier-Villars, Paris, 1971, p. 79.

25. R. Hartshorne, *Residues and duality*, Lecture Notes in Math., vol. 20, Springer-Verlag, Berlin-Heidelberg-New York, 1966.

26. ______, *On the de Rham cohomology of an algebraic variety*, Inst. Hautes Études Sci. Publ. Math. **45** (1975), 6–99.

27. M. Herrera and D. Liebermann, *Duality and the de Rham cohomology of infinitesimal neighborhoods*, Invent. Math. **13** (1971), 97–126.

28. H. Hironaka, *Resolution of singularities of an algebraic variety over a field of characteristic zero*. I, II, Ann. of Math. (2) **79** (1964).

29. ______, *Desingularization of analytic varieties*, Actes C.I.M. (Nice, 1970), vol. 2, Gauthier-Villars, Paris, 1971, pp. 627–631.

30. E. L. Ince, *Ordinary differential equations*, Dover, New York, 1956.

31. M. Kashiwara, *Algebraic study of systems of partial differential equations*, Master thesis, University of Kyoto (1971). (Japanese)

32. ______, *On the maximally overdetermined systems of linear differential equations*. I, Pub. Res. Inst. Math. Sci., Kyoto Univ. **10** (1975), 563–579.

33. ______, *B-functions and holonomic systems*, Invent. Math. **38** (1976), 33–53.

34. M. Kashiwara and I. Oshima, *Systems of differential equations with regular singularities and their boundary value problem*, Ann. of Math. (2) **106** (1977), 145–200.

35. M. Kashiwara, *On the holonomic systems of linear differential equations*. II, Invent. Math. **49** (1978), 121–135.

36. M. Kashiwara and T. Kawai, *On holonomic systems of micro-differential equations*. III. *Systems with regular singularities*, Publ. Res. Inst. Math. Sci. Kyoto Univ. **17** (1981), 813–979.

37. M. Kashiwara, *Systèmes holonomes et distributions tempérées*, Sém. Goualouic-Schwartz (1979-80), exposé No. XIX, École Polytechnique, Palaiseau.

38. N. Katz, *The regularity theorem in algebraic geometry*, Actes C.I.M. (Nice, 1970), vol 1, Gauthier-Villars, Paris, 1971, pp. 436–449.

39. ______, *An overview of Deligne's work in Hilbert's twenty-first problem*, Proc. Sympos. Pure Math., vol. 28, Amer. Math. Soc., Providence, R. I., 1976, pp. 537–557.

40. A. Landman, *On Picard-Lefchetz transformation for algebraic manifolds acquiring general singularities*, Trans. Amer. Math. Soc. **181** (1973), 89–126.

41. Lê D. T., *Some remarks in relative monodromy*, Nordic Summer School Sympos. in Math. (Oslo, August 1976), Sijthoff Noordhoff, 1977.

42. ______, *The geometry of the monodromy theorem*, C. P. Ramanujam, A Tribute, Studies in Math., no. 8, Tata Institute, Bombay, 1978, pp. 157–173.

43. ______, *Le théorème de la monodromie singulière*, Note aux C. R. Acad. Sci. Paris Sér. A-B **288** (1978), 985–988.

44. Lê D. T. and Z. Mebkhout, *Variétés caractéristiques et variétés polaires* C. R. Acad. Sci. Paris. (to appear).

45. B. Malgrange, *Systèmes différentiels à coefficients constants*, Sém. Bourbaki, No. 246, 1962-63.

46. ______, *Sur les points singuliers réguliers des équations différentielles*, Enseign. Math. **20** (1976), 147–176.

47. ______, *Le polynôme de I. N. Bernstein d'une singularité isolée*, Lecture Notes in Math., vol. 459, Springer-Verlag, Berlin-Heidelberg-New York, 1976.

48. ______, *Rapport sur le théorème de l'indice de Boutet de Monvel et de Kashiwara*, Comptes Rendus de la Conf., Analyse et Topologie sur les Espaces Singuliers (Luminy, Juillet 1981), Astérisque (to appear).

49. ______, *Le polynôme de I. N. Bernstein-Sato et cycles évanescents*, Comptes Rendus de la Conf., Analyse et Topologie sur les Espaces Singuliers (Luminy, Juillet 1981), Astérisque (to appear).

50. Z. Mebkhout, *La valeur principale des fonctions à singularités essentielles*, Note aux C. R. Acad. Sci. Paris Sér. A-B **280** (1975), 205–207.

51. ______, *La valeur principale et le résidu simple des formes à singularités essentielles*, Fonctions de Plusieurs Variables Complexes. II, Lecture Notes in Math., vol. 482, Springer-Verlag, Berlin-Heidelberg-New York, 1975, pp. 190–215.

52. ______, *La cohomologie locale d'une hypersurface*, Fonctions de Plusieurs Variables. III, Lecture Notes in Math., vol. 670, Springer-Verlag, Berlin-Heidelberg-New York, 1977, pp. 89–119.

53. ______, *Local cohomology of analytic spaces*, Publ. Res. Inst. Math. Sci. Kyoto Univ. **12** (1977), 247–256.

54. ______, *Théorèmes de dualité pour les $\mathcal{D}_X$-modules cohérents*, Note aux C. R. Acad. Sci. Sér. A-B **287** (1977), 785–787; Math. Scand. **50** (1982), 25–43.

55. ______, *Cohomologie locale des espaces analytiques complexes*, Thèse de Doctorat d'État, Université Paris VII (1979), 126 pp.

56. ______, *Dualité de Poincaré*, Sém. sur les Singularités (Paris VII), Publ. de l'Université Paris VII, no. 7, 1980, 139–182.

57. ______, *The Poincaré-Serre-Verdier duality*, Proc. Conf. Algebraic Geom. (Copenhagen, 1978), Lecture Notes in Math., vol. 732, Springer-Verlag, Berlin-Heidelberg-New York, 1979.

58. ______, *Sur le problème de Hilbert-Riemann*, Note aux C. R. Acad. Sci. Paris Sér. A-B **290** (1980), 415–417.

59. ______, *Sur le problème de Hilbert-Riemann*, Proc. Les Houches (1979), Lecture Notes in Phys., vol. 126, Springer-Verlag, Berlin-Heidelberg-New York, 1980, pp. 90–110.

60. ______, *The Hilbert-Riemann problem in higher dimension*, Proc. Conf. Generalized Functions Appl. in Math. Phys. (Moscow, November 1980), Steklov Inst., 1981, pp. 334–341.

61. ______, *Une autre équivalence de catégories* Compositio Math. (1982), (to appear).

62. J. Milnor, *Singular points of complex hypersurfaces*, Ann. of Math. Studies, No. 61, Princeton Univ. Press, Princeton, N. J., 1968.

63. N. Nilsson, *Some growth and ramification properties of certain integrals on algebraic manifolds*, Ark. Mat. **5** (1965), 463–476.

64. ______, *Monodromy and asymptotic properties of certain multiple integrals*, Ark. Mat. **18** (1980), 181–198.

65. F. Pham, *Singularités des systèmes de Gauss-Manin*, Progress in Math., Birkhäuser, Boston, 1979.

66. ______, *Structures de Hodge mixtes associées à un germe de fonctions à point critique isolé*, Comptes Rendus de la Conf., Analyse et Topologie sur les Espaces Singuliers (Luminy, Juillet 1981), Astérisque (to appear).

67. M. Sato, *Hyperfunctions theory*. I, II, J. Fac. Sci. Univ. Tokyo Sect. IA Math. **8** (1960), 387–437.

68. ______, *Hyperfunctions and partial differential equations*, Proc. Internat. Conf. Functional Analysis and Related Topics, Univ. of Tokyo Press, 1969.

69. M. Sato, T. Kawai and M. Kashiwara, *Microfunctions and pseudo-differential equations*, Lecture Notes in Math., vol. 287, Springer-Verlag, Berlin-Heidelberg-New York, 1973, pp. 265–529.

70. J. Sherk and J. Steenbrink, unpublished preprint.

71. A. N. Varchencko, *The asymptotics of holomorphic forms determine a mixed Hodge structure*, Soviet Math. Dokl. **22** (1980), no. 3.

72. ______, *Asymptotic mixed Hodge structure on vanishing cohomology*, Izv. Akad. Nauk SSSR Ser. Mat. **45** (1981), 540–591.

73. ______, *Hodge properties of Gauss-Manin connections*, Funkcional. Anal. i Priložen **14 (1)** (1980), 36–37. (Russian)

74. J. L. Verdier, *Catégories dérivées état* 0, SGA 4 1/2, Lecture Notes in Math., vol. 569, Springer-Verlag, Berlin-Heidelberg-New York, 1977, pp. 262–311.
75. ______, *Dualité dans les espaces localement compacts*, Sém. Bourbaki, exposé n° 300 (1965-66).
76. ______, *Topologies sur les éspaces de cohomologie d'un complexe de faisceaux analytiques à cohomologie cohérente*, Bull. Soc. Math. France **99** (1972), 337–342.
77. ______, *Classe d'homologie associée à un cycle*, Sém. de l'École Norm. Sup., Astérisque **36-37** (1976), 101–151.
78. T. Yano, these PROCEEDINGS.

CENTRE DE MATHEMATIQUES DE L'ECOLE POLYTECHNIQUE, PALAISEAU, FRANCE

UNIVERSITE PARIS VII, FRANCE

Proceedings of Symposia in Pure Mathematics
Volume 40 (1983), Part 2

CYCLES EVANESCENTS, SECTIONS PLANES ET CONDITIONS DE WHITNEY. II

LÊ D. T. ET B. TEISSIER

ABSTRACT. Given a partition $X = \bigcup X_\alpha$ of a complex analytic space X into nonsingular constructible subspaces, we state several local conditions of algebraic, topological and combinatorial nature which are equivalent to the fact that this partition is a Whitney stratification of X. In particular this result contains a converse to a suitable improvement of the Thom-Mather topological triviality theorem (in the complex-analytic case).

0. Introduction. Dans son article [**W1**], H. Whitney introduit la notion de stratification régulière comme outil dans l'étude des espaces analytiques complexes singuliers. En même temps, R. Thom (voir [**Th 1, 2, 3, 4**]) étend ces idées au cas différentiable, en introduisant la notion d'ensemble stratifié, et au cas relatif, introduisant en particulier la notion de morphisme sans éclatement. Il donne les théorèmes de trivialité topologique locale le long des strates d'une stratification ou d'un morphisme (premier et deuxième théorèmes d'isotopie). Le programme de R. Thom a été précisé et développé par J. Mather dans [**Ma1, Ma2**]. Dans [**Ło**] S. Łojasiewicz montre que tout ensemble semianalytique a une stratification régulière. Dans [**Sch**], M. H. Schwartz donne une définition des classes de Chern pour un espace analytique complexe muni d'une stratification régulière (voir aussi [**B-S**]).

En relation avec sa théorie de l'équisingularité O. Zariski (cf. [**Z1**]) a été amené à donner dans le cas d'hypersurfaces dont le lieu singulier est non singulier et de codimension un des conditions algébriques pour que la partie non singulière et la partie singulière de l'hypersurface forment une stratification régulière. Dans [**H6**] H. Hironaka montre que la condition de régularité de Whitney entraine l'équimultiplicité. Dans [**Te2**] B. Teissier donne un critère numérique pour la régularité de Whitney d'une hypersurface le long de son lieu singulier supposé non singulier. Dans [**Te5**], il énonce que ce critère est nécessaire et suffisant pour obtenir la condition de régularité dans cette situation géométrique. En fait la réciproque du résultat de [**Te2**] a été démontrée par J. Briançon et J. P. Spéder dans [**Br-Sp2**].

1980 *Mathematics Subject Classification.* Primary 32C40, 32B30, 32C42.

0082-0717/82/0000-0560/$08.25

Du point de vue topologique, dans la situation de O. Zariski ci-dessus, i.e. une hypersurface à lieu singulier non singulier de codimension un, O. Zarsiki dans [**Z.1**] et Lê D. T. montrent que la trivialité topologique de l'hypersurface le long de son lieu singulier donne la condition de régularité (cf. [**Lê2, L-R**]). Cette situation particulière a laissé espérer que la trivialité topologique le long des strates d'une stratification implique la condition de régularité de Whitney (cf. [**Te2, Te4**]). Le résultat de B. Teissier de [**Te2**] cité ci-dessus s'interprète de la façon suivante: dans le cas d'une hypersurface à lieu singulier non singulier, on a la condition de Whitney si l'on a la trivialité topologique locale quand on coupe l'hypersurface par un "drapeau générique" de sous-espaces non singuliers contenant le lieu singulier. C'est en fait ce résultat que nous avons en particulier généralisé ici et qui contient la "bonne" réciproque du théorème de Thom-Mather (cf. §5). J. Briançon et J. P. Spéder ont en effet donné dans [**Br-Sp1**] un exemple d'une hypersurface dans $\mathbf{C}^4$ topologiquement triviale le long de son lieu singulier qui est une courbe non singuliére, et ne satisfaisant pas la condition de régularité de Whitney.

L'étude de la condition de régularité amène naturellement à celle des limites d'espaces tangents et des limites de sécantes (cf. [**W2**]). C'est ainsi que H. Hironaka dans [**H6** et **H3**] (et dans des conférences à l'Inst. Hautes Etudes Sci. en 1968 non publiées) a donné des conditions algébro-géométriques impliquant la condition de régularité de Whitney. Ce type de résultat a été étendu par J. P. Spéder dans [**Sp**] et G. Canuto et J. P. Spéder [**C-S**]. L'étude géométrique des limites d'espaces tangents a été commencée dans [**He-Lê**] par J. P. G. Henry et Lê D. T. En utilisant ces concepts, V. Navarro dans [**N**] démontre dans le cas particulier où la petite strate est de dimension 1 une conjecture de B. Teissier selon laquelle la condition de régularité de Whitney se conserve par intersection par un sous-espace non singulier général contenant la petite strate, ce qui donne en particulier une autre démonstration du théorème d'Hironaka déjà cité sur l'équimultiplicité le long d'une strate d'une stratification régulière (voir aussi [**Na-Tr**]).

Par ailleurs de divers points de vue a été introduit le concept de courbe polaire locale et de variété polaire locale (cf. [**Lê3, Te3**]). Dans le cas de singularités d'intersections complètes la relation entre la multiplicité de ces variétés polaires et la topologie locale de la singularité et de ses sections linéaires génériques (cf. [**Lê3, Te3, Lê5, Lê6, Te2**]) a donné les idées nécessaires pour comprendre d'une part que les multiplicités des variétés polaires locales s'interprètent comme des invariants combinatoires de la topologie locale et d'autre part de préciser la relation entre l'obstruction d'Euler locale que R. MacPherson (cf. [**MP**]) associe à un point singulier et la géométrie locale de la singularité (cf. [**Lê-Te**]). En fait les variétés polaires locales fournissent le lien entre la structure des limites d'espaces tangents en un point singulier et la topologie locale des singularités et de leurs sections linéaires. Ce lien permet d'une part de définir numériquement une stratification naturelle qui donne une construction explicite du cycle de MacPherson-Schwartz (cf. [**Lê-Te**]) d'un espace singulier et d'autre part un critère numérique pour la condition de régularité de Whitney qui implique que cette stratification est régulière. Ce critère numérique a été énoncé par B. Teissier dans [**Te6**], mais sa démonstration, correcte dans le cas des hypersurfaces, contenait une erreur réparée par J. P. G. Henry et M. Merle dans [**He-M2**] et par B. Teissier dans [**Te1**] en utilisant l'idée de J. P. G. Henry et M. Merle.

Dans cet article nous utilisons la plupart des idées ci-dessus pour obtenir une réciproque raisonnable au théorème de Thom-Mather (cf. §5) dans le cadre analytique complexe. Cette réciproque énonce des critères algébriques, combinatoires, homotopiques et topologiques (cf. Théorème (5.3.1)) qui impliquent la condition de régularité de Whitney.

Les caractéristiques d'Euler-Poincaré locales utilisées dans la réciproque du théorème de Thom-Mather généralisent les $\mu^{(i)}$ de [**Te2**] (cf. [**Te5**, Chapter VI, §4]), et ont été introduites par M. Kashiwara [**K**] et par A. Dubson dans [**D**] pour calculer l'obstruction d'Euler locale de R. MacPherson. Ils ont a observé (cf. [loc.cit.]) que ces caractéristiques étaient constantes le long des strates d'une stratification régulière. J. P. Brasselet et M. H. Schwartz [**B-S**] ont montré la constance de l'obstruction d'Euler locale le long des strates d'une stratification régulière. Signalons d'autres travaux significatifs sur les stratifications régulières. Dans le cas réel (différentiable, sous-analytique) par D. Trotman [**Tr**], V. Navarro et D. Trotman [**Na-Tr**], J. L. Verdier [**V**], H. Hironaka [**H4**], M. Goresky [**G**] ainsi que des travaux particuliers dans le cadre analytique complexe dans le cas d'espaces analytiques dont le lieu singulier est non singulier de codimension un [**St, D-F, B-G-G, Bu-G**].

Pour pouvoir énoncer notre théorème principal (5.3.1), nous sommes amenés à des préliminaires sur l'étude de la topologie locale des singularités, en particulier nous introduisons la notion de système fondamental à un paramètre de bons voisinages, la notion de morphisme descriptible et d'équivalence d'homotopie descriptible, et nous démontrons à l'instar de [**Lê1**] une généralisation du théorème de fibration de Milnor [**Mi**]. Nous démontrons aussi que les caractéristiques d'Euler-Poincaré locales de Kasiwara et Dubson sont constantes le long des strates d'une stratification régulière.

Dans tout le texte on dira stratification de Whitney au lieu de stratification régulière et condition de Whitney au lieu de condition de régularité de Whitney.

Nous pensons qu'un travail analogue à celui que nous faisons ici devrait être fait pour la condition de Thom (comparer à [**Sa**]).

Nous remercions Claudine Harmide pour son excellent travail de frappe.

1. Rappels.

(1.1) *Variétés polaires locales* (cf. [**LêTe**, §2; **Te1**, Chapitre IV]).

(1.1.1) Soient X un espace analytique complexe réduit purement de dimension d, $x \in X$ et $(X, x) \subset (\mathbf{C}^{N+1}, 0)$ un plongement local défini au voisinage de x.

Pour tout entier k, $0 \leqslant k \leqslant d-1$, notons G_k la Grassmannienne des noyaux des projections linéaires $p\colon \mathbf{C}^{N+1} \to \mathbf{C}^{d-k+1}$. Soit $\mathring{P}_k(X, p)$ l'ensemble des points de la partie non singulière X^0 de X qui sont critiques pour la restriction de p à X^0. L'adhérence $P_k(X, p)$ de $\mathring{P}_k(X, p)$ dans un représentant assez petit de (X, x) est un sous-espace analytique fermé de ce représentant.

(1.1.2) Il existe un ouvert de Zariski dense U_k de la Grassmannienne G_k tel que, pour toute projection $p\colon \mathbf{C}^{N+1} \to \mathbf{C}^{d-k+1}$ dont le noyau appartient à U_k, le germe $(P_k(X, p), x)$ est soit vide, soit réduit, de codimension k et sa classe d'équisingularité (donc sa multiplicité) ne dépend pas du choix de p (cf. [**Te1**, Chapitre IV, §3]). Un tel germe est appelé variété polaire locale générale de codimension k de (X, x) et est noté $(P_k(X), x)$. On a $(P_0(X), x) = (X, x)$. On note $M^*_{X,x}$ la suite $(m_x(P_0(X)), \ldots, m_x(P_{d-1}(X)))$ des multiplicités en x des variétés polaires locales générales de (X, x). Le premier terme de cette suite est la multiplicité de X en x et $(P_k(X), x) = \varnothing$ si et seulement si sa multiplicité en x est nulle. Avec les hypothèses faites, le point x de X est non singulier si et seulement si $M^*_{X,x} = (1, 0, \ldots, 0, 0)$.

(1.1.3) Dans [**Te1**, Chapitre IV, §3] on démontre

THÉORÈME (B. TEISSIER). *La suite $M^*_{X,x}$ ne dépend que de l'algèbre analytique $\mathcal{O}_{X,x}$ de (X, x).*

La définition des variétés polaires locales et les résultats généraux sur la semi-continuité de la multiplicité impliquent

(1.1.4) PROPOSITION. *Pour tout entier ν, l'ensemble des points $x \in X$ tels que $m_x(P_k(X)) \geqslant \nu$ est un sous-espace analytique fermé de X.*

(1.2) *Stratifications et condition de Whitney.* Dans ce paragraphe et la suite, nous utiliserons la notion d'ensemble sous-analytique (réel) et ses propriétés (cf. [**H1, H2, Ha**]).

(1.2.1) DÉFINITION. Soit X un ensemble sous-analytique. Soit $(X_\alpha)_{\alpha \in A}$ une famille localement finie de sous-ensembles sous-analytiques non singuliers connexes de X. On dit que la famille $(X_\alpha)_{\alpha \in A}$ est une *stratification sous-analytique* de X si:

(1) la famille $(X_\alpha)_{\alpha \in A}$ est une partition de X;

(2) la fermeture $\overline{X}_\alpha$ de X_α dans X et $\overline{X}_\alpha - X_\alpha$ sont des sous-ensembles sous-analytiques de X, pour chaque $\alpha \in A$. (Cette condition est ici pour mémoire, car en fait elle est toujours vérifiée (cf. [**H2**]).)

Les sous-ensembles X_α sont appelés *strates* de la stratification.

(1.2.2) DÉFINITION. Dans le cas d'un espace analytique complexe réduit X, si la stratification sous-analytique $(X_\alpha)_{\alpha \in A}$ est telle que $\overline{X}_\alpha$ et $\overline{X}_\alpha - X_\alpha$, pour tout $\alpha \in A$, soient des sous-espaces analytiques complexes fermés de X, on dit que la stratification $(X_\alpha)_{\alpha \in A}$ de X est une *stratification analytique complexe* de X.

(1.2.3) DÉFINITION (CF. [**W1, H1, H2**]). Soit X un ensemble sous-analytique (resp. un espace analytique complexe réduit) muni d'une stratification $(X_\alpha)_{\alpha \in A}$ sous-analytique (resp. analytique complexe). On dit que cette stratification satisfait *la propriété de frontière* si, pour tout $(\alpha, \beta) \in A \times A$ tel que $X_\alpha \cap \overline{X}_\beta \neq \varnothing$, on a $X_\alpha \subset \overline{X}_\beta$.

Dans ce cas $\overline{X}_\alpha$ et $\overline{X}_\alpha - X_\alpha$, pour tout $\alpha \in A$, sont union de strates de la stratification donnée de X.

(1.2.4) DÉFINITION (CF. [**W1, H1**]). Soient X un ensemble sous-analytique, M un sous-ensemble sous-analytique non singulier de X, Y un sous-ensemble sous-analytique non singulier de la fermeture $\overline{M}$ de M dans X (qui est sous-analytique). On dit que *le couple* (M, Y) *satisfait la condition de Whitney en un point* $y \in Y$ s'il existe un plongement local de (X, y) dans $(\mathbf{R}^{N+1}, 0)$ tel que, pour toute suite $(x_n, y_n)_{n \in \mathbf{N}}$ de couples de points dans $M \times Y$ qui tend vers (y, y) et pour laquelle la direction limite T des espaces tangents $T_{x_n}M$ et la direction limite $l_{\mathbf{R}}$ des directions réelles de sécantes $\overline{x_n y_n}$ dans $\mathbf{R}^{N+1}$ existent, on a l'inclusion $l_{\mathbf{R}} \subset T$.

On vérifie que, si la condition est vérifié par un plongement local, elle est vérifiée par tous les plongements locaux.

On dit que le couple (M, Y) satisfait la condition de Whitney, si (M, Y) satisfait la condition de Whitney en tout point $y \in Y$.

(1.2.5) REMARQUE. Dans le cas où X est un espace analytique complexe réduit, on considère un sous-espace analytique complexe M non singulier dont la fermeture $\overline{M}$ dans X est aussi un sous-espace analytique complexe et un sous-espace analytique complexe non singulier Y de $\overline{M}$ contenu dans $\overline{M} - M$. On définit, comme dans (1.2.4), que le couple (M, Y) satisfait la condition de Whitney en un point $y \in Y$: on remarque que les espaces tangents $T_{x_n}M$ sont complexes, que par conséquent T est aussi un espace complexe et que, si la direction réelle $l_{\mathbf{R}}$ est contenue dans T, l'unique direction complexe l qu'elle définit est aussi contenue dans T.

(1.2.6) DÉFINITION. Soit X un ensemble sous-analytique (resp. un espace analytique complexe réduit) muni d'une stratification $(X_\alpha)_{\alpha \in A}$ sous analytique (resp. analytique complexe). On dit que la stratification $(X_\alpha)_{\alpha \in A}$ est une *stratification de Whitney* si:

(1) la stratification $(X_\alpha)_{\alpha \in A}$ satisfait la propriété de frontière;

(2) pour tout $(\alpha, \beta) \in A \times A$ tel que $X_\alpha \subset \overline{X}_\beta$ le couple (X_β, X_α) satisfait la condition de Whitney.

Dans [**W1**], on démontre (cf. [**H1**] pour le cas sous-analytique)

(1.2.7) THÉORÈME (H. WHITNEY-H. HIRONAKA). *Soit X un ensemble sous-analytique* (*resp. un espace analytique complexe réduit*). *Soit* $(\Phi_i)_{i \in I}$ *une famille localement finie de sous-ensembles fermés de X sous-analytiques* (*resp. analytiques complexes*). *Il existe une stratification* $(X_\alpha)_{\alpha \in A}$ *sous-analytique* (*resp. analytique complexe*) *de X qui est une stratification de Whitney et telle que chaque* Φ_i, $i \in I$, *soit union de strates.*

En utilisant le premier théorème d'isotopie de Thom et Mather (cf. [**Th 1–4** et **Ma2**]), on démontre

(1.2.8) THÉORÈME (R. THOM-J. MATHER). *Soient X un ensemble sous-analytique* (*resp. un espace analytique complexe réduit*) *et* $(X_\alpha)_{\alpha \in A}$ *une stratification de Whitney de X sous-analytique* (*resp. analytique complexe*). *Soit* $x \in X$ *et* X_α *la strate qui contient x. Pour tout plongement analytique local* $(X, x) \subset (\mathbf{R}^{N+1}, 0)$ *(resp.* $(X, x) \subset (\mathbf{C}^{N+1}, 0)$*), il existe un système fondamental de voisinages* $(U_j)_{j \in J}$ *de* 0 *dans* $\mathbf{R}^{N+1}$ *(resp.* $\mathbf{C}^{N+1}$*) et un système compatible d'homéomorphismes*

$$\varphi_j \colon U_j \rightleftarrows (X_\alpha \cap U_j) \times (H \cap U_j),$$

où H est un sous-espace analytique réel (resp. complexe) non singulier de $\mathbf{R}^{N+1}$ *(resp.* $\mathbf{C}^{N+1}$*) qui coupe transversalement* X_α *en x avec* $X_\alpha \cap U_j \cap H = \{x\}$*, et au demeurant quelconque, homéomorphismes qui induisent des homéomorphismes* $\varphi_{j,\beta}$: $U_j \cap \overline{X} \rightleftarrows (X_\alpha \cap U_j) \times (H \cap U_j \cap \overline{X}_\beta)$ *pour tout* $\beta \in A$ *tel que* $x \in \overline{X}_\beta$.

(1.2.9) REMARQUE. Dans l'énoncé précédent on peut choisir des ouverts U_j qui sont des bons voisinages au sens de Prill [**P**] de x dans chacun des $\overline{X}_\beta$ tels que $x \in \overline{X}_\beta$.

(1.3) *Caractérisation algébrique de la condition de Whitney.* Dans [**Te1** et **He-M2**] on démontre

(1.3.1) THÉORÈME (B. TEISSIER). *Soient X un espace analytique complexe réduit purement de dimension d, Y un sous-espace analytique complexe non singulier de X et* $x \in Y$*. Les conditions suivantes sont équivalentes*:

(i) *la suite* $M^*_{X,y}$ *(cf.* (1.1.2)*) est indépendante de* $y \in Y$ *au voisinage de x*;

(ii) *le couple* (X^0, Y)*, où* X^0 *désigne la partie non singulière de X, satisfait la condition de Whitney en x.*

On a aussitôt

(1.3.2) COROLLAIRE. *L'ensemble des points* $y \in Y$ *en lesquels le couple* (X^0, Y) *ne satisfait pas les conditions de Whitney est un fermé analytique strict de Y.*

Rappelons enfin comment l'on utilise ce résultat pour construire une stratification de Whitney canonique d'un espace analytique complexe réduit (cf. [**Lê-Te3**, 6.1; **Te1**, VI, §3]).

Posons $F_0 = X$, $F_1 = \operatorname{Sing} X$ (lieu singulier de X).

Supposons avoir construit $F_0, F_1, \ldots, F_{l-1}$. Soit $F_{l-1} = \bigcup_{j \in J_{l-1}} F_{l-1,j}$ la décomposition de F_{l-1} en composantes irréductibles. Pour chaque $j \in J_{l-1}$, notons B_j l'ensemble des points $x \in F_{l-1,j}$ où l'une des suites $M^*_{F_{k,j},x}$ (où $0 \leqslant k \leqslant l-2$ et $j \in J_k$) ne prend pas la valeur qu'elle prend en un point général de $F_{l-1,j}$. On définit F_l comme le sous-espace analytique fermé de F_{l-1} qui est réunion du lieu singulier de F_{l-1} et des fermés analytiques B_j ($j \in J_{l-1}$) qui forment une famille localement finie comme on le vérifie immédiatement.

Le sous-espace analytique fermé F_l est rare dans F_{l-1}, et l'on obtient ainsi une filtration

$$X = F_0 \supset F_1 \supset \cdots \supset F_{l-1} \supset F_l \supset \cdots$$

où, pour tout $x \in X$, $\dim_x F_l < \dim_x F_{l-1}$ et par conséquent, pour tout $x \in X$, il existe un voisinage ouvert U et un entier l_0 tel que $F_l \cap U = \varnothing$ quand $l \geqslant l_0$.

Il résulte du Théorème (1.3.1)

(1.3.3) COROLLAIRE. *La famille* $(X_\alpha)_{\alpha\in A}$ *des composantes connexes des différences* $F_k - F_{k+1}$ *est une stratification de Whitney analytique complexe de* X *et pour toute stratification de Whitney analytique complexe* $(Z_\gamma)_{\gamma\in C}$ *de* X, *chaque strate* Z_γ *est contenue dans une strate* X_α.

(1.3.4) REMARQUE. Etant donné une famille $(\Phi_i)_{i\in I}$ localement finie de sous-espaces analytiques complexes fermés de X (comme dans (1.2.7)), une modification de la construction précédente donne aussi une stratification de Whitney analytique complexe de X minimale parmi celles où chaque Φ_i est réunion de strates.

De (1.1.3) et (1.3.3), on obtient

(1.3.5) COROLLAIRE. *Tout espace analytique complexe réduit* X *possède une stratification de Whitney analytique complexe minimale qui ne dépend que de la structure analytique de* X; *en particulier, si* $i: U \to X$ *est une immersion ouverte, les composantes connexes des images réciproques par* i *des strates de la stratification de Whitney analytique complexe minimale de* X *donnent les strates de la stratification de Whitney analytique complexe minimale de* U.

(1.4) *Stratifications de Thom.*

(1.4.1) Soit $f: X \to Y$ un morphisme sous-analytique (resp. analytique complexe) d'ensembles sous-analytiques (resp. espaces analytiques complexes réduits). Nous dirons que f est *stratifiable* s'il existe des stratifications de Whitney $(X_\alpha)_{\alpha\in A}$ et $(Y_\beta)_{\beta\in B}$ de X et Y respectivement telles que, pour tout $\alpha \in A$, il existe $\beta(\alpha) \in B$ pour lequel f induise une submersion analytique (resp. analytique complexe) $f_\alpha: X_\alpha \to Y_{\beta(\alpha)}$.

(1.4.2) EXEMPLE. Un morphisme sous-analytique (resp. analytique complexe) qui est propre est stratifiable (cf. [**H3**, §3, Theorem]).

(1.4.3) Dans le cas où $f: X \to Y$ est propre, surjectif, sous-analytique (resp. analytique complexe) et stratifié, le premier théorème d'isotopie de Thom et Mather déjà invoqué dans (1.2.8) implique que, pour toute strate Y_β, $\beta \in B$, le morphimse f induit une fibration topologique localement triviale de $f^{-1}(Y_\beta)$ sur Y_β.

(1.4.4) Soit $f: X \to Y$ un morphisme sous-analytique (resp. analytique complexe) stratifiable. Soient $(X_\alpha)_{\alpha\in A}$ et $(Y_\beta)_{\beta\in B}$ des stratifications de Whitney de X et Y respectivement telles que, pour tout $\alpha \in A$, il existe $\beta(\alpha) \in B$ pour lequel f induise une submersion analytique (resp. analytique complexe) $f_\alpha: X_\alpha \to Y_{\beta(\alpha)}$. On dit que *le couple* (X_β, X_α) *tel que* $X_\alpha \subset \overline{X_\beta}$ *satisfait la condition de Thom en* $x \in X_\alpha$ *relativement à* f s'il existe un plongement analytique (resp. analytique complexe) de (X, x) dans $(\mathbf{R}^{N+1}, 0)$ (resp. de (X, x) dans $(\mathbf{C}^{N+1}, 0)$) et une extension analytique F de f à $(\mathbf{R}^{N+1}, 0)$ (resp. $(\mathbf{C}^{N+1}, 0)$) telle que la restriction de F à un voisinage de x dans X coincide avec la restriction de f à ce voisinage, et que, pour toute suite $(x_n)_{n\in\mathbf{N}}$ de points de X_β qui tend vers x et pour laquelle la suite des espaces tangents $T_{x_n}(F^{-1}(F(x_n)) \cap X_\beta)$ a une limite T, on ait

$$T \supset T_x\big(F^{-1}(F(x)) \cap X_\alpha\big).$$

On dit que la stratification de Whitney $(X_\alpha)_{\alpha \in A}$ satisfait *la condition de Thom* relativement à f si, pour tout couple (X_β, X_α) de strates telles que $X_\alpha \subset \overline{X}_\beta$, le couple (X_β, X_α) satisfait la condition de Thom relativement à f en tout point $x \in X_\alpha$. Dans ce cas on dit aussi que la stratification de Whitney $(X_\alpha)_{\alpha \in A}$ satisfait *la condition a_f de Thom* (cf. [**Th1**; **H3**, §5]).

(1.4.5) Nous dirons qu'un morphisme sous-analytique (resp. analytique complexe) f: $X \to Y$ est *un morphisme de Thom* sous-analytique (resp. analytique complexe) s'il est stratifiable et s'il existe une stratification de Whitney $(X_\alpha)_{\alpha \in A}$ de X qui satisfait la condition de Thom relativement à f.

2. Morphismes de Thom locaux.

(2.1) *Morphismes descriptibles.*

(2.1.1) Définition. On dit qu'un morphisme analytique complexe f: $X \to Y$ d'espaces analytiques complexes réduits est un *morphisme descriptible* s'il existe une stratification $(Y_\beta)_{\beta \in B}$ *analytique complexe* de Y telle que, pour tout $\beta \in B$, f induise une fibration topologique localement triviale f_β: $f^{-1}(Y_\beta) \to Y_\beta$.

Si Y est irréductible, on appellera *fibre générale d'un morphisme descriptible*, la fibre en un point de la strate dense dans Y.

(2.1.2) D'après (1.4.3), si f: $X \to Y$ est un morphisme analytique complexe propre d'espaces analytiques comples réduits, le morphisme f est descriptible (il n'est pas nécessaire de supposer que f est surjectif!).

(2.1.3) Proposition. *Soit f: $X \to Y$ un morphisme descriptible tel que*:

(1) *la stratification $(Y_\beta)_{\beta \in B}$ de Y telle que f induise une fibration topologique localement triviale f_β: $f^{-1}(Y_\beta) \to Y_\beta$ est finie, i.e.* Card $B < +\infty$;

(2) *les caractéristiques d'Euler-Poincaré $\chi(X)$ de X et $\chi(Y_\beta)$ des strates Y_β, $\beta \in B$, sont finies.*

Alors on a l'égalité:

$$\chi(X) = \sum_{\beta \in B} \chi(F_\beta)\chi(Y_\beta)$$

où F_β est une fibre de f_β.

Preuve. Comme dans [**S**], on remarque qu'un espace analytique complexe réduit est triangulable d'après [**H4**] et que le bord de l'étoile, i.e. l'entrelacement, d'une réunion de strates, est de caractéristique d'Euler-Poincaré nulle. Ceci implique que le complémentaire d'une telle réunion dans son "voisinage tubulaire" est de caractéristique d'Euler-Poincaré nulle. La suite de Mayer-Vietoris donne alors immédiatement la relation cherchée.

(2.1.4) Remarque. Dans [**MP**] le même argument que dans la preuve de (2.1.3) donne que la correspondance qui, à une variété algébrique compacte complexe X fait correspondre le groupe abélien $\mathbf{F}(X)$ des fonctions constructibles sur X et à un morphisme de tels variétés f: $X \to Y$ fait correspondre l'homomorphisme

$\mathbf{F}(f)$: $\mathbf{F}(X) \to \mathbf{F}(Y)$ qui, à la fonction caractéristique $\mathbf{1}_w$ d'une sous-variété W de X, fait correspondre la fonction constructible dans $\mathbf{F}(Y)$ définie par $y \mapsto \chi(f^{-1}(y) \cap W)$, est un foncteur de la catégorie des variétés algébriques compactes complexes dans la catégorie des groupes abéliens. En fait cet argument montre que l'on peut étendre ce foncteur à la catégorie des espaces analytiques réduits stratifiables par des stratifications dont les strates sont de caractéristiques d'Euler-Poincaré finies et dont les morphismes sont descriptibles avec de telles stratifications.

(2.1.5) REMARQUE. Dans le cas sous-analytique une notion similaire à celle des morphismes descriptibles ne conduit pas à une formule aussi simple que celle de (2.1.3).

(2.1.6) DÉFINITION. Soient f: $X \to Y$ et f': $X' \to Y'$ deux morphismes descriptibles. Nous dirons que l'on a un *homéomorphisme ou une équivalence topologique* (resp. *une équivalence d'homotopie*) *descriptible de f sur f'* s'il existe:

(i) des stratifications analytiques complexes $(Y_\beta)_{\beta \in B}$ et $(Y'_\beta)_{\beta \in B}$ de Y et Y' respectivement telles que f et f' induisent respectivement pour tout $\beta \in B$ des fibrations f_β: $f^{-1}(Y_\beta) \to Y_\beta$ et f'_β: $f'^{-1}(Y'_\beta) \to Y'_\beta$,

(ii) des homéomorphismes (resp. équivalence d'homotopie) g: $X \to X'$, h: $Y \to Y'$ tels que $f' \circ g = h \circ f$,

tels que, pour tout $\beta \in B$, g et h induisent un isomorphisme (resp. une équivalence d'homotopie fibrée) entre les fibrations topologiques f_β et f'_β.

(2.2) *Systèmes fondamentaux à un paramètre de bons voisinages.*

(2.2.1) DÉFINITION. Soient $a > 0$ et I l'intervalle $]0, a]$. On dit que la famille $(U_t)_{t \in I}$ est *un système fondamental à un paramètre de bons voisinages de* 0 *dans* $\mathbf{R}^{N+1}$ si:

(i) pour tout $t \in I$, U_t est un voisinage ouvert relativement compact sous-analytique de 0 dans $\mathbf{R}^{N+1}$;

(ii) le couple $(\overline{U}_t, \overline{U}_t - U_t)$ est homéomorphe à $(\mathbf{B}^{N+1}, \mathbf{S}^N)$;

(iii) le sous-ensemble $\mathfrak{U} = \bigcup_{t \in I}(U_t \times \{t\})$ est un sous-ensemble sous-analytique de $\mathbf{R}^{N+1} \times \mathbf{R}$;

(iv) le sous-ensemble $\mathfrak{U}_1 = \bigcup_{t \in I}(\partial U_t \times \{t\})$ (où $\partial U_t = \overline{U}_t - U_t$) de $\mathbf{R}^{N+1} \times \mathbf{R}$ a une stratification de Whitney $(\Sigma_\beta)_{\beta \in B}$ sous-analytique et la restriction π_β à Σ_β de la projection de $\mathfrak{U}_1$ sur I est lisse pour tout $\beta \in B$;

(v) pour tout t, $t' \in I$, $t > t'$, $\overline{U}_t$ contient strictement $\overline{U}_{t'}$, $\partial U_t \cap \partial U_{t'} = \varnothing$ et $(U_t)_{t \in I}$ est un système fondamental de voisinages de 0 dans $\mathbf{R}^{N+1}$.

(2.2.2) REMARKQES ET EXEMPLES.

(a) D'après le (iv) de (2.2.1), pour tout $t \in I$, on a l'égalité $\pi_\beta^{-1}(t) = \Sigma_{\beta,t} \times \{t\}$ et $(\Sigma_{\beta,t})_{\beta \in B}$ est une stratification de Whitney sous-analytique de ∂U_t: en fait d'après **[Ch]**, il suffit de vérifier la propriété de frontière, qui provient de ce que $\Sigma_{\beta,t}$ est dense dans $\overline{\Sigma}_\beta \cap \pi^{-1}(t)$.

(b) Pour tout $x \in \overline{U}_a - \{0\}$, il existe un $t_x \in I$ et un seul tel que $x \in \partial U_{t_x}$.

En effet une application immédiate du premier théorème d'isotopie de Thom-Mather **[Th1; Ma**, § 8] montre que $\mathfrak{U}_1$ est homéomorphe à $\partial U_a \times I$ en utilisant la condition (iv) de (2.2.1).

En fait le même argument donne un homéomorphisme de $\mathfrak{U} \cup \mathfrak{U}_1$ avec $\bar{U}_a \times I$.

La projection sur le premier facteur donne une application continue $\mathfrak{U}_1 \to \bar{U}_a$ qui induit un homéomorphisme φ de $\mathfrak{U}_1 \cup \{0,0\}$ dans $\bar{U}_a$. Comme le bord ∂U_t, pour tout $t \in I$, est homéomorphe à $\mathbf{S}^N$, l'homéomorphisme φ est nécessairement surjectif. En effet si φ n'est pas surjectif, soit $\xi \notin \operatorname{Im} \varphi$. Evidemment $\xi \neq 0$ et dans $\bar{U}_a - \{0, \xi\}$ les images par φ de $\partial U_a \times \{a\}$ et $\partial U_t \times \{t\}$ pour t assez petit ne peuvent pas être homologues dans $\bar{U}_\alpha - \{0, \xi\}$ ce qui contredit le fait qu'ils le sont dans $\mathfrak{U}_1$. Ceci montre donc que pour tout $x \in \bar{U}_a - \{0\}$ il existe $t_x \in I$ tel que $(x, t_x) \in \mathfrak{U}_1$: si $x = 0$, on peut définir $t_0 = 0$. L'application τ: $\bar{U}_a \to \bar{I}$ définie par $\tau(x) = t_x$ est évidemment continue puisqu'elle est la composée de φ^{-1} et de la projection sur $\bar{I}$.

(c) Pour tout $a > 0$, $(B_t)_{t\in I}$, où B_t est la boule ouverte de $\mathbf{R}^{N+1}$ centrée en 0 et de rayon $t \in]0, a]$, est un système fondamental à un paramètre de bons voisinages de 0 dans $\mathbf{R}^{N+1}$.

(d) Soit p: $\mathbf{R}^{N+1} \to \mathbf{R}^p$ une projection linéaire et soit B'_u la boule ouverte de $\mathbf{R}^p$ de centre 0 et de rayon u. Soit r: $]0, a] \to \mathbf{R}$ une fonction sous-analytique strictement croissante telle que $r(t)$ soit majoré par $t/2$, pour tout $t \in]0, a] = I$. La famille $(B_t \cap p^{-1}(B'_{r(t)}))_{t\in I}$ est un système fondamental à un paramètre de bons voisinages de 0 dans $\mathbf{R}^{N+1}$.

(e) Soient ε_i: $]0, a] \to \mathbf{R}$ $(i = 0, \ldots, N)$ une famille finie de fonctions positives strictement croissantes sous-analytiques telles que $\lim_{t\to 0} \varepsilon_i(t) = 0$, pour tout $i = 0, \ldots, N$. La famille $(\prod_{i=0}^N (]-\varepsilon_i(t), +\varepsilon_i(t)[))_{t\in I}$ est un système fondamental à un paramètre de bons voisinages de 0 dans $\mathbf{R}^{N+1}$.

(2.2.3) Définition. Soient $(X, 0)$ un germe d'espace analytique complexe réduit et X un représentant de ce germe qui est fermé dans un voisinage ouvert U de 0 dans $\mathbf{C}^{N+1}$ et pour lequel on a une stratification de Whitney $(X_\alpha)_{\alpha\in A}$ analytique complexe telle que $0 \in \bar{X}_\alpha$, pour tout $\alpha \in A$. Soit $(U_t)_{t\in I}$ un système fondamental à un paramètre de bons voisinages de 0 dans $\mathbf{C}^{N+1}$ contenus dans U et dont $(Z_\beta)_{\beta\in B}$ est la stratification de Whitney sous-analytique de

$$\mathfrak{U}_1 = \bigcup_{t\in I} (\partial U_t \times \{t\}).$$

On dit que $(U_t)_{t\in I}$ est *un système fondamental à un paramètre de bons voisinages relativement à la stratification* $(X_\alpha)_{\alpha\in A}$ si:

(i) pour tout $\alpha \in A$ et tout $\beta \in B$, la strate $X_\alpha \times I$ de $X \times I$ coupe Z_β transversalement dans $\mathbf{C}^{N+1} \times I$;

(ii) la restriction à l'intersection de $X_\alpha \times I$ et de Z_β de la projection sur I est lisse;

(iii) les composantes connexes $(S_\gamma)_{\gamma\in C}$ des intersections $(X_\alpha \times I) \cap Z_\beta$ qui sont non vides forment une stratification de Whitney sous-analytique de $(X \times I) \cap \mathfrak{U}_1$.

(2.2.4) Remarque. D'après [**Ch**], la condition (i) n'assure pas que l'ensemble des composantes connexes des intersections $(X_\alpha \times I) \cap Z_\beta$ qui sont non vides forme une stratification de $(X \times I) \times \mathcal{U}_1$ qui satisfait la propriété de frontière (1.2.3).

Il est donc raisonnable de demander dans (iii) cette propriété.

(2.2.5) LEMME. *Soient X et sa stratification de Whitney $(X_\alpha)_{\alpha \in A}$ analytique complexe comme dans* (2.2.3). *On suppose en outre que l'on a une famille finie $Y_1, \ldots, Y_k$ de sous-espaces analytiques complexes fermés de X qui contiennent* 0 *et qui sont unions de strates. Si $(U_t)_{t \in I}$ est un système fondamental à un paramètre de bons voisinages relativement à la stratification $(X_\alpha)_{\alpha \in A}$, alors $(U_t \cap X)_{t \in I}$ est un système de bons voisinages de* 0 *dans $\mathbf{C}^{N+1}$ au sens de Prill* (*cf.* [**P**]) *relativement à $Y_1, \ldots, Y_k$.*

PREUVE. Nous allons montrer, pour tout t, $t' \in I$, $t \geqslant t'$, que $\overline{U}_{t'} \cap X$ est un rétracte par déformation de $\overline{U}_t \cap X$ par des rétractions qui préservent les Y_j $(1 \leqslant j \leqslant k)$, ce qui montrera que $\overline{U}_{t'} \cap Y_j$ $(1 \leqslant j \leqslant k)$ et $\overline{U}_{t'} \cap (X - Y_j)$ $(1 \leqslant j \leqslant k)$ sont aussi des rétractes par déformation de $\overline{U}_t \cap Y_j$ $(1 \leqslant j \leqslant k)$ et $\overline{U}_t \cap (X - Y_j)$ $(1 \leqslant j \leqslant k)$ respectivement.

Notons $\overline{\mathcal{U}} = \mathcal{U} \cup \mathcal{U}_1$, où $\mathcal{U} = \bigcup_{t \in I}(U_t \times \{t\})$, $\mathcal{U}_1 = \bigcup_{t \in I}((\overline{U}_t - U_t) \times \{t\})$. Les hypothèses sur la famille $(U_t)_{t \in I}$ donnent une stratification de Whitney sous-analytique de $\overline{\mathcal{U}}$:

$$\overline{\mathcal{U}} = \bigcup_{\beta \in B} \Sigma_\beta \cup \mathcal{U}.$$

Nous avons aussi la stratification de Whitney triviale de $X \times I$:

$$X \times I = \bigcup_{\alpha \in A} (X_\alpha \times I).$$

Par hypothèse les composantes connexes $(S_\gamma)_{\gamma \in C}$ des intersections $(X_\alpha \times I) \cap Z_\beta$ qui sont non vides forment une stratification de Whitney sous-analytique de $(X \times I) \cap \mathcal{U}_1 = \mathcal{X}_1$. On notera $\mathcal{X} = (X \times I) \cap \overline{\mathcal{U}}$.

La restriction π à $\mathcal{X}_1$ de la projection sur I est propre et l'hypothèse faite sur le système $(U_t)_{t \in I}$ dit que la restriction π_γ de π à S_γ, pour tout $\gamma \in C$, est lisse sur I.

Le premier théorème d'isotopie de Thom-Mather [**Th1**–**4**; **Ma2**, §8] nous donne un champ de vecteurs v continu et intégrable sur $\mathcal{X}_1 \cap \pi^{-1}([t', t])$, qui est tangent aux strates S_γ et dont la projection sur $[t', t]$ est le champ de vecteurs unité. A l'aide de ce champ de vecteurs v nous allons construire dans $\mathcal{X} \cap \pi^{-1}(]t', t])$ un champ de vecteurs qui va réaliser la déformation par rétraction cherchée.

Tout d'abord dans $\mathcal{X} \cap (\overline{U}_{t'} \times]t', t[)$ nous considérons le champ de vecteurs qui relève trivialement le champ de vecteurs unité de $]t', t]$.

Soit (x, t_1) un point de $\mathcal{X} \cap \pi^{-1}(]t', t]) - \overline{U}_{t'} \times]t', t]$. Le champ de vecteurs V est défini par

$$V_{(x, t_1)} = \left(V_x, V_{t_1}\right)$$

où

$$\begin{cases} V_x = \dfrac{(t_x - t')v_x}{(t_1 - t')} & \text{où } (v_x, 1) \text{ est la valeur de } v \text{ en } (x, t_x). \\ V_{t_1} = 1, \end{cases}$$

Ce champ est bien continu et intégrable, puisque v l'est et que t_x est continu.

La projection de V sur $]t', t]$ est donc le champ de vecteurs unité. Sa restriction à $\mathcal{X}_1 \cap \pi^{-1}(]t', t])$ coincide avec v, tandis que sa restriction à

$$(\partial U_{t'} \times]t', t]) \cap (X \times I)$$

est le relèvement trivial du champ de vecteurs unités de $]t', t]$.

L'intégration de la projection du champ de vecteurs sur $\overline{U}_t$ donne pour tout $t_1 \in]t', t]$ un homéomorphisme ρ_{t,t_1} de $\overline{U}_t$ sur $\overline{U}_{t_1}$ dont la restriction à $\overline{U}_{t'}$ induit l'identité de $\overline{U}_{t'}$. L'application

$$r_{t,t'}: \overline{U}_t \to \overline{U}_{t'}$$

définie par $r_{t,t'}(x, t) = \lim_{t_1 \to t'} \rho_{t,t_1}(x, t)$ est la rétraction cherchée.

(2.2.6) Exemple. Soit $(X, 0)$ un germe d'espace analytique complexe. Soit X un représentant de $(X, 0)$ fermé dans le voisinage ouvert U de 0 de $\mathbf{C}^{N+1}$. Pour toute stratification de Whitney $(X_\alpha)_{\alpha \in A}$ analytique complexe, il existe ε_0 tel que $(B_\varepsilon)_{\varepsilon \in I}$, avec $I =]0, \varepsilon_0]$, soit un système fondamental à un paramètre de bons voisinages de 0 dans $\mathbf{C}^{N+1}$ relativement à $(X_\alpha)_{\alpha \in A}$ (cf. [**Lê1**]).

(2.2.7) Remarque. Nous pouvons caractériser un système fondamental à un paramètre de bons voisinages de 0 dans $\mathbf{R}^{N+1}$ de la façon suivante:

(2.2.7.1) Proposition. *Soit $(U_t)_{t \in I}$ un système fondamental de voisinages sous-analytiques de 0 dans $\mathbf{R}^{N+1}$ tel que, pour tout $t \in I$, le couple $(\overline{U}_t, \overline{U}_t - U_t)$ soit homéomorphe à $(\mathbf{B}^{N+1}, \mathbf{S}^N)$. Le système $(U_t)_{t \in I}$ est un système fondamental à un paramètre de bons voisinages de 0 dans $\mathbf{R}^{N+1}$ si et seulement s'il existe une stratification de Whitney sous-analytique $(\Sigma'_\beta)_{\beta \in B}$ d'un voisinage U de $\overline{U}_a$ telle que $0 \in \overline{\Sigma}'_\beta$ et une fonction sous-analytique propre $\tau: U \to \mathbf{R}_+$ telle que $\tau^{-1}(0) = 0$, la restriction de τ à Σ'_β soit de rang un sauf éventuellement en 0 et $U_t = \{x \in \mathbf{R}^{N+1}, \tau(x) < t\}$.*

Nous laissons les détails de la preuve au lecteur, mais nous faisons remarquer que la fonction τ construite dans (2.2.2)(b) est sous-analytique. Par ailleurs $\mathcal{U}_1$ et Σ_β sont les graphes dans $\mathbf{R}^{N+1} \times \mathbf{R}$ de la restriction de τ à $\overline{U}_a - \{0\}$ et à $\Sigma'_\beta \cap \overline{U}_a - \{0\}$. De plus $\Sigma'_\beta \cap \tau^{-1}(t)$ est le $\Sigma_{\beta,t}$ défini dans (2.2.2)(a).

Il n'est peut-être pas nécessaire de supposer à l'avance que $(\overline{U}_t, \overline{U}_t - U_t)$ soit homéomorphe à $(\mathbf{B}^{N+1}, \mathbf{S}^N)$, mais nous ne savons pas comment nous passer de cette hypothèse.

(2.2.7.2) Ce point de vue permet également de caractériser les systèmes fondamentaux à un paramètre de bons voisinages de 0 dans $\mathbf{C}^{N+1}$ relativement à une stratification de Whitney $(X_\alpha)_{\alpha \in A}$ analytique complexe d'un espace analytique complexe réduit X qui contient 0.

Soit $(U_t)_{t\in I}$ un système fondamental à un paramètre de bons voisinages de 0 dans $\mathbf{C}^{N+1}$. Reprenons les notations de (2.2.7.1) et appelons $(\Sigma'_\beta)_{\beta\in B}$ la stratification de Whitney sous-analytique du voisinage U de 0 de $\mathbf{C}^{N+1}$ qui lui est associée et $\tau\colon U \to \mathbf{R}_+$ la fonction sous-analytique qui lui correspond.

Soit X un espace analytique complexe fermé de U et $(X_\alpha)_{\alpha\in A}$ une stratification de Whitney analytique complexe de X telle que $0 \in \overline{X}_\alpha$. La condition (i) de (2.2.3) est équivalente à la transversalité de X_α et Σ'_β, pour tout $\alpha \in A$ et pour tout $\beta \in B$. La condition (ii) signifie que la restriction de τ aux intersections non vides $X_\alpha \cap \Sigma'_\beta$ est de rang un. Si (i) est vérifié la condition (iii) est équivalente à demander que les composantes connexes $(S'_\gamma)_{\gamma\in C}$ des intersections non vides des $X_\alpha \cap \Sigma'_\beta$ forment une stratification avec la propriété de frontière (d'après [**Ch**]). Comme dans (2.2.2)(a), on peut alors montrer que pour tout $t \in I$, $(S'_\gamma \cap \tau^{-1}(t))_{\gamma\in C} = (S'_{\gamma,t})_{\gamma\in C}$ est une stratification de Whitney de $X \cap \tau^{-1}(t) = X \cap \partial U_t$.

(2.2.8) Exemple. La remarque (2.2.7) permet de donner une classe importante d'exemples de systèmes fondamentaux à un paramètre de bons voisinages.

Soit $p\colon U \to \mathbf{R}_+$ une fonction analytique *réelle* définie sur un voisinage de 0 dans $\mathbf{C}^{N+1}$ et telle que $p^{-1}(0) = \{0\}$. Le théorème de Bertini-Sard (cf. [**H2**, **V**]) dans le cadre analytique réel nous permet de supposer, quitte à choisir un voisinage ouvert U de 0 dans $\mathbf{C}^{N+1}$, que 0 est le seul point critique de p dans U. Soit X un sous-espace analytique complexe réduit fermé de U qui contient 0. Quitte à supposer U assez petit, X a une stratification de Whitney analytique complexe $(X_\alpha)_{\alpha\in A}$ telle que $0 \in \overline{X}_\alpha$. Le théorème de Bertini-Sard appliqué à la restriction de p à X_α et le fait que A est un ensemble fini, car une stratification de Whitney est localement finie, impliquent qu'il existe $\varepsilon_0 > 0$ tel que, pour tout ε, $\varepsilon_0 \geqslant \varepsilon > 0$, $\partial U_\varepsilon = \{x \in U,\ p(x) = \varepsilon\}$ est transverse à $(X_\alpha)_{\alpha\in A}$.

Le système $U_\varepsilon = \{x \in U,\ p(x) < \varepsilon\}$, $\varepsilon \in\,]0, \varepsilon_0]$, est donc un système fondamental à un paramètre de bons voisinages de 0 dans $\mathbf{C}^{N+1}$ relativement à $(X_\alpha)_{\alpha\in A}$ puisque la propriété de frontière exigée implicitement dans le (iii) de (2.2.3) est automatiquement satisfaite dans ce cas où ∂U_ε est non singulière et composé d'une seule strate fermée (cf. [**Ch**]).

L'exemple (2.2.6) correspond donc au cas où p est la fonction "carré de la distance à 0", i.e. $p(x) = \|x\|^2$.

(2.2.9) Définition. Soit S un ensemble sous-analytique non singulier. On dit que $(U_t)_{t\in I}$, avec $I =]0, a]$, est *un système fondamental à un paramètre de bons voisinages de* $\{0\} \times S$ dans $\mathbf{R}^{N+1} \times S$ s'il existe une stratification de Whitney $(\Sigma'_\beta)_{\beta\in B}$ sous-analytique d'un voisinage ouvert sous-analytique $U(S)$ de $\{0\} \times S$ dans $\mathbf{R}^{N+1} \times S$ dans laquelle $\{0\} \times S$ est une strate notée Σ'_0 et $\{0\} \times S \subset \overline{\Sigma}'_\beta$, et une fonction $\tau\colon U(S) \to \mathbf{R}_+$ sous-analytique telle que:

(i) $\tau^{-1}(0) = \{0\} \times S$;

(ii) la restriction τ_β de τ à Σ'_β est de rang un, pour tout $\beta \in B$;

(iii) $U_t = \{(x, s) \in U(S) \times S,\ \tau(x, s) < t\}$ et $(\Sigma'_\beta \cap \tau^{-1}(t))_{\beta\in B}$ est une stratification de Whitney sous-analytique de $\partial U_t = \overline{U_t} - U_t$;

(iv) la projection p sur S est de rang maximum sur chaque Σ'_β et pour tout $s \in S$, $(U_t \cap p^{-1}(s))_{t\in I}$ est un système fondamental à un paramètre de bons voisinages de $(0, s)$ dans $\mathbf{R}^{N+1} \times \{s\}$.

(2.2.10) EXEMPLE. Soit $p\colon V \to \mathbf{R}_+$ une fonction analytique réelle telle que $p^{-1}(0) = \{0\}$. On a une fonction $\tau\colon V \times S \to \mathbf{R}_+$ définie par $\tau(x, s) = p(x)$. En posant $U(S) = V \times S$ et en stratifiant trivialement par $U(S) - \{0\} \times S$ et $\{0\} \times S$, pour $a > 0$ assez petit, le système $(U_t = \{(x, s) \in U(S), \tau(x, s) < t\})_{t\in]0,a]}$ est un système fondamental à un paramètre de bons voisinages de $\{0\} \times S$ dans $\mathbf{R}^{N+1} \times S$.

(2.2.11) DÉFINITION. Soit S un espace analytique complexe réduit qui est sous-analytiques. Soit X un sous-espace analytique complexe réduit de $\mathbf{C}^{N+1} \times S$ qui est sous-analytique et qui contient $\{0\} \times S$. On suppose que X a une stratification de Whitney $(X_\alpha)_{\alpha\in A}$ analytique complexe qui contient la strate $\{0\} \times S$ notée X_0 et pour laquelle $\{0\} \times S \subset \overline{X}_\alpha$ pour $\alpha \in A$. On dit qu'un système fondamental à un paramètre de bons voisinages $(U_t)_{t\in I}$ de $\{0\} \times S$ dans $\mathbf{C}^{N+1} \times S$ est *un système de bons voisinages relativement à la stratification* $(X_\alpha)_{\alpha\in A}$ si:

(i) les composantes connexes $(S'_\gamma)_{\gamma\in C}$ des intersections non vides de X_α et Z'_β pour $\alpha \in A$ et $\beta \in B$ forment une stratification de Whitney;

(ii) la restriction aux intersections non vides des X_α et Z'_β de la projection p sur S est de rang maximum;

(iii) le système $(U_t \cap p^{-1}(s))_{t\in I}$ est un système fondamental à un paramètre de bons voisinages de $(0, s)$ dans $\mathbf{C}^{N+1}$ relativement à $(X_\alpha \cap p^{-1}(s))_{\alpha\in A}$ pour tout $s \in S$.

(2.3) *Morphismes de Thom locaux et cycles évanescents*. En (2.1.2) nous avons vu qu'un morphisme propre est descriptible. Dans le cas d'un germe de morphisme analytique complexe $f\colon (X, 0) \to (Y, 0)$, en général il n'existe pas de représentants $f\colon X \to Y$ avec X et Y assez petits pour que f induise un morphisme descriptible. Comme R. Thom le remarque dans [**Th1**], c'est le cas quand $Z \xrightarrow{\pi} \mathbf{C}^2$ est l'éclatement de 0 dans $\mathbf{C}^2$ et que l'on considère le germe de $\pi\colon (Z, x) \to (\mathbf{C}^2, 0)$ avec $x \in \pi^{-1}(0)$.

Dans le cas où $f\colon (\mathbf{C}^{n+1}, 0) \to (\mathbf{C}, 0)$, J. Milnor a remarqué que f induisait un morphisme descriptible localement en choisissant convenablement un représentant de f. Dans [**Lê1**], on trouve que ce résultat s'étend à tout germe de fonction f: $(X, 0) \to (\mathbf{C}, 0)$ défini sur un germe d'espace analytique complexe réduit quelconque. La raison invoquée dans [**Lê1**] pour qu'une telle situation apparaisse est que l'on peut stratifier un représentant X de $(X, 0)$ de telle sorte que la stratification $(X_\alpha)_{\alpha\in A}$ de Y soit analytique complexe et satisfasse simultanément la condition de Whitney et la condition de Thom (cf. (1.4.4)). Nous allons voir dans la suite que ce type de résultat se généralise.

Enonçons dans notre cadre le résultat (Theorem 1.1) de [**Lê1**]:

(2.3.1) THÉORÈME. *Soit* $f_0\colon (X, 0) \to (\mathbf{C}, 0)$ *un germe de fonction analytique complexe sur le germe d'espace analytique complexe réduit* $(X, 0)$. *Soit* $f\colon X \to D$ *un représentant de* f_0 *tel que* X *soit un sous-espace analytique complexe de* $\mathbf{C}^{N+1}$ *et ait une stratification de Whitney* $(X_\alpha)_{\alpha\in A}$ *analytique complexe qui satisfait la condition* a_f *de Thom. Soit* $(U_t)_{t\in I}$ *un système fondamental à un paramètre de bons voisinages de* 0 *dans* $\mathbf{C}^{N+1}$ *relativement à* $(X_\alpha)_{\alpha\in A}$. *Alors pour tout* $t \in I$, *il existe* η_t, *tel que, pour tout* η, $0<\eta\leqslant\eta_t$, f *induise un morphisme descriptible*:

$$\varphi_{t,\eta}\colon X\cap U_t\cap f^{-1}(D_\eta)\to D_\eta$$

où D_η *est le disque ouvert de* $\mathbf{C}$ *de centre* 0 *et de rayon* η. *De plus pour tout* t, $t'\in I$ *et tout* η, η', $0<\eta\leqslant\eta_t$, $0<\eta'\leqslant\eta_{t'}$, *il existe une équivalence d'homotopie descriptible de* $\varphi_{t,\eta}$ *sur* $\varphi_{t',\eta'}$.

DÉMONSTRATION. La démonstration de ce théorème est essentiellement la même que celle de [**Lê1**].

Le point crucial est qu'il existe un représentant X de $(X,0)$ qui a une stratification de Whitney $(X_\alpha)_{\alpha\in A}$ analytique complexe qui satisfait la condition a_f de Thom et ceci est assuré par un résultat de H. Hironaka [**H3**, §5].

Quitte à choisir un représentant X de $(X,0)$ plus petit, on peut supposer que $0\in\overline{X}_\alpha$, pour tout $\alpha\in A$.

On note comme précédemment $\mathfrak{U}=\bigcup_{t\in I}(U_t\times\{t\})$, $\mathfrak{U}_1=\bigcup_{t\in I}(\partial U_t\times\{t\})$, $\overline{\mathfrak{U}}=\mathfrak{U}\cup\mathfrak{U}_1$, $(\Sigma_\beta)_{\beta\in B}$ la stratification de Whitney de $\mathfrak{U}$, $(\Sigma'_\beta)_{\beta\in B}$ la stratification d'un voisinage de $\overline{U}_a$ associée (cf. (2.2.7.1)), etc.

D'après (2.2.2)(a), on a une stratification de Whitney $(\Sigma_{\beta,t})_{\beta\in B}$ sous-analytique de $\partial U_t=\overline{U}_t-U_t$: en fait $\Sigma_{\beta,t}=\Sigma'_\beta\cap\tau^{-1}(t)$ (cf. (2.2.7.1)).

D'après l'hypothèse faite sur le système $(U_t)_{t\in I}$, pour tout $\beta\in B$ et tout $\alpha\in A$, Σ'_β coupe X_α transversalement, les composantes connexes $(S'_\gamma)_{\gamma\in C}$ des intersections non vides $X_\alpha\cap\Sigma'_\beta$ forment une stratification de Whitney sous-analytique de X, et pour tout $t\in I$, $(S'_{\gamma,t})_{\gamma\in C}$, avec $S'_{\gamma,t}=S'_\gamma\cap\tau^{-1}(t)$, est une stratification de Whitney sous-analytique de $X\cap\partial U_t$.

D'après le théorème de Bertini-Sard, il existe $\bar\eta$ tel que, pour tout η, $0<\eta\leqslant\bar\eta$, la restriction de f à X_α, pour tout $\alpha\in A$, n'a aucune valeur critique dans $\overline{D}_\eta-\{0\}$.

Soit $t\in I$. Supposons que, pour tout $\eta>0$, il existe $x_\eta\neq0$ tel que:

(i) $|f(x_\eta)|<\eta$;

(ii) $x_\eta\in X\cap\partial U_t$ et X_{α_η} et $\Sigma_{\beta_\eta,t}$ contiennent x_η;

(iii) $f^{-1}(f(x_\eta))\cap X_{\alpha_\eta}$ ne coupe pas $\Sigma_{\beta_\eta,t}$ transversalement dans $\mathbf{C}^{N+1}$.

On peut alors trouver une suite de points $x_n\in X\cap\partial U_t$ telle que:

(i) $\lim_{n\to\infty}x_n=x\in f^{-1}(0)\cap X\cap\partial U_t$ et $x\in X_\alpha\cap\Sigma_{\beta,t}$;

(ii) pour tout $n\in\mathbf{N}$, $x_n\in X_{\alpha'}\cap\Sigma_{\beta',t}$;

(iii) $T_{x_n}(X_{\alpha'}\cap f^{-1}(f(x_n)))$ et $T_{x_n}(\Sigma_{\beta',t})$ ne se coupent pas transversalement dans $\mathbf{C}^{N+1}$;

(iv) $\lim_{n\to\infty}T_{x_n}(X_{\alpha'}\cap f^{-1}(f(x_n)))=T$, $\lim_{n\to\infty}T_{x_n}(\Sigma_{\beta',t})=T_1$.

Or $T\supset T_x\,(X_\alpha\cap f^{-1}(f(x)))$ d'après la condition a_f de Thom satisfaite par $(X_\alpha)_{\alpha\in A}$ et $T_1\supset T_x\,(\Sigma_{\beta,t})$ car $(\Sigma_{\beta,t})_{\beta\in B}$ est une stratification de Whitney sous-analytique de ∂U_t.

L'hypothèse faite sur le système $(U_t)_{t\in I}$ implique la transversalité de $T_x(\Sigma_{\beta,t})$ et de $T_x(X_\alpha \cap f^{-1}(f(x)))$ ce qui contredit le (iii) ci-dessus. Par conséquent il existe $\eta_t > 0$ tel que, pour tout η, $0 < \eta \leqslant \eta_t$, la restriction de f aux strates $S'_{\gamma,t}$ soit de rang maximum.

Ainsi, si l'on a choisi $\eta_t \leqslant \bar{\eta}$, on est assuré que la restriction de f à X_α n'a pas de valeurs critiques dans $\bar{D}_{\eta_t} - \{0\}$.

Le premier théorème d'isotopie de Thom-Mather [**Th1**, **Ma2**] implique alors que f induit une fibration topologique de $X \cap \bar{U}_t \cap f^{-1}(D_{\eta_t} - \{0\})$ sur $D_{\eta_t} - \{0\}$. Ceci montre que $\varphi_{t,\eta}$ est un morphisme descriptible pour tout η, $0 < \eta \leqslant \eta_t$. Bien évidemment pour tout η, η', $0 < \eta' \leqslant \eta \leqslant \eta_t$, on a un homéomorphisme descriptible de $\varphi_{t,\eta}$ sur $\varphi_{t,\eta'}$.

Soient t, $t' \in I$. Pour obtenir la dernière assertion de (2.3.1) il suffit de montrer que, pour η assez petit et non nul, les fibres $f^{-1}(\xi) \cap U_t$ et $f^{-1}(\xi) \cap U_{t'}$ sont homéomorphes pour $0 < |\xi| < \eta$ (cf. [**Do**]).

Pour cela on procède de façon similaire à [**Lê1**, Theorem 1.1]. Nous venons de voir que la restriction de f aux intersections non vides $S'_{\gamma,t} \cap f^{-1}(\bar{D}_{\eta_t} - \{0\})$ est de rang deux. Montrons que ceci reste vrai pour tout t_1 assez proche de t. Sinon comme ci-dessus on construit une suite de points (x_n, t_n) de $X \times I$ tels que:

(i) $\lim_{n\to\infty} x_n = x \in X \cap \partial U_t$ et $x \in X_\alpha \cap \Sigma'_\beta$;

(ii) pour tout $n \in \mathbf{N}$, $x_n \in X_{\alpha'} \cap \Sigma_{\beta',t_n}$;

(iii) $T_{x_n}(X_{\alpha'} \cap f^{-1}(f(x_n)))$ et $T_{x_n}(\Sigma_{\beta',t_n})$ ne se coupent pas transversalement dans $\mathbf{C}^{N+1}$;

(iv) $\lim_{n\to\infty} T_{x_n}(X_{\alpha'} \cap f^{-1}(f(x_n))) = T$, $\lim_{n\to\infty} T_{x_n}(\Sigma_{\beta',t_n}) = T_1$.

Or $T \supset T_x(X_\alpha \cap f^{-1}(f(x)))$ puisque $(X_\alpha)_{\alpha\in A}$ satisfait la condition a_f de Thom et $T_1 \supset T_x(\Sigma_{\beta,t})$ car $(\Sigma_\beta)_{\beta\in B}$ est une stratification de Whitney sous-analytique de $\mathfrak{U}$ et ceci contredit la transversalité de $T_x(\Sigma_{\beta,t})$ et $T_x(X_\alpha \cap f^{-1}(f(x)))$ dans $\mathbf{C}^{N+1}$.

Un argument de compacité montre alors qu'il existe $\tilde{\eta}$ tel que, pour tout $t_1 \in [t', t]$, la restriction de f aux intersections non vides $S'_{\gamma,t_1} \cap f^{-1}(D_{\tilde{\eta}} - \{0\})$ est de rang deux.

Soit ξ, $0 < |\xi| < \tilde{\eta}$. Les composantes connexes des intersections des S'_γ avec $f^{-1}(\xi)$ forment une stratification de Whitney sous-analytique $(S''_\delta)_{\delta\in D}$ de $f^{-1}(\xi)$.

En appliquant à nouveau le premier théorème d'isotopie de Thom-Mather [**Th1**, **Ma2**] on peut construire un champs de vecteurs v différentiable sur les intersections non vides $f^{-1}(\xi) \cap S'_\gamma \cap (\bar{U}_t - U_{t'})$ qui est continu sur $f^{-1}(\xi) \cap (\bar{U}_t - U_{t'})$, intégrable et tel que en tout point $x \in f^{-1}(\xi) \cap S'_\gamma \cap (\bar{U}_t - U_{t'})$ on ait $(d\tau_\gamma)_x(v_x) < 0$. L'intégration de ce champs de vecteurs donne l'homéomorphisme cherché. (En fait en raffinant un peu cette démonstration on obtiendrait un homéomorphisme descriptible de $\varphi_{t,\eta}$ sur $\varphi_{t',\eta'}$, mais nous n'aurons pas besoin de ce résultat dans cet article.)

(2.3.2) COROLLAIRE. *Soit* f_0: $(X, 0) \to (\mathbf{C}, 0)$ *un germe de fonction analytique complexe sur le germe d'espace analytique complexe réduit* $(X, 0)$. *Soit* f: $X \to D$ *un représentant de* f_0 *tel que* X *soit un sous-espace analytique complexe de* $\mathbf{C}^{N+1}$ *et ait une stratification de Whitney* $(X_\alpha)_{\alpha\in A}$ *analytique complexe qui satisfait la condition*

a_f de Thom. Soient $(U_t)_{t\in I}$ et $(V_t)_{t\in I}$ deux systèmes fondamentaux à un paramètre de bons voisinages de 0 dans $\mathbf{C}^{N+1}$ relativement à $(X_\alpha)_{\alpha\in A}$. Alors on a une équivalence d'homotopie descriptible entre les morphismes $\varphi_{t,\eta}$ et $\psi_{t,\eta}$ qui leur sont associés par le Théorème (2.3.1), *pour tout $t \in I$ et quand $0 < \eta \ll t$.*

PREUVE. Soient t_1, t_2, t_3, t_4 tels que

$$U_{t_1} \supset V_{t_2} \supset U_{t_3} \supset V_{t_4}.$$

En choisissant $\xi \neq 0$ assez petit, on a

$$\varphi^{-1}_{t_1,\eta}(\xi) \supset \psi^{-1}_{t_2,\eta}(\xi) \supset \varphi^{-1}_{t_3,\eta}(\xi) \supset \psi^{-1}_{t_4,\eta}(\xi)$$

et la dernière assertion de (2.3.1) implique que ces inclusions sont des équivalences d'homotopie ce qui donne le résultat cherché d'après [**Do**].

(2.3.3) REMARQUE. La fonction $t \mapsto \eta_t$ du Théorème peut-être choisie sous-analytique. L'ensemble $C = \{(t, \eta)$ tels que $t \in I$ et $0 < \eta \leqslant \eta_t\}$ est alors sous-analytique de dimension deux et pour tout $(t, \eta) \in C$ la classe d'équivalence d'homotopie descriptible du morphisme descriptible associé à $f_0\colon (X, 0) \to (\mathbf{C}, 0)$ et au système $(U_t)_{t\in I}$ est en fait un invariant analytique (et en fait topologique) du germe f_0.

Nous avons vu que le point crucial de l'existence de morphismes descriptibles associés localement à un germe $f_0\colon (X, 0) \to (\mathbf{C}, 0)$ provient de la possibilité de trouver une stratification de Whitney *analytique complexe* $(X_\alpha)_{\alpha\in A}$ d'un représentant X de $(X, 0)$ qui satisfait la condition a_f de Thom. Cette possibilité est en fait rare, mais dans cet article, nous allons nous restreindre à la situation des projections linéaires assez générales:

(2.3.4) LEMME. *Soit $(X, 0)$ un germe d'espace analytique complexe réduit. Soit $X \subset \mathbf{C}^{N+1}$ un plongement d'un représentant de ce germe dans $\mathbf{C}^{N+1}$. On supposera que X est fermé dans un voisinage ouvert U de 0 et que $(X_\alpha)_{\alpha\in A}$ est une stratification de Whitney analytique complexe de X telle que $0 \in \overline{X_\alpha}$. Il existe un ouvert dense $\Omega_{N+1,i}((X_\alpha)_{\alpha\in A})$ de l'espace des projections linéaires de $\mathbf{C}^{N+1}$ sur $\mathbf{C}^i$ tel que, pour tout $p \in \Omega_{N+1,i}((X_\alpha)_{\alpha\in A})$ il existe un voisinage ouvert U_p de 0 dans $\mathbf{C}^{N+1}$ tel que la stratification de Whitney analytique complexe induite par $(X_\alpha)_{\alpha\in A}$ dans U_p satisfait la condition a_p de Thom en tout point de $(p^{-1}(0) \cap X - \{0\}) \cap U_p$.*

PREUVE. Comme $(X_\alpha)_{\alpha\in A}$ est une partition localement finie et que $0 \in \overline{X_\alpha}$, l'ensemble A est fini. D'après [**Lê-Te**, (4.1.8)], pour tout $\alpha \in A$, il existe un ouvert de Zariski dense $\Omega_{i,\alpha}$ de l'espace des projections $\mathbf{C}^{N+1} \to \mathbf{C}^i$ tel que, pour tout $p \in \Omega_{i,\alpha}$, il existe un voisinage ouvert $U_{p,\alpha}$ de 0 dans $\mathbf{C}^{N+1}$ tel que le lieu critique $C(p_\alpha)$ de $p|_{X_\alpha \cap U_{p,\alpha}}$ soit vide ou de dimension $i - 1$, que $\overline{C(p_\alpha)}$ soit la variété polaire locale de $\overline{X_\alpha}$ en 0 et que la restriction de p à $\overline{C(p_\alpha)}$ soit finie.

Notons $\Omega_{N+1,i}((X_\alpha)_{\alpha\in A}) = \bigcap_{\alpha\in A}\Omega_{i,\alpha}$ et avec $p \in \Omega_{N+1,i}((X_\alpha)_{\alpha\in A})$, $U_p = \bigcap_{\alpha\in A} U_{p,\alpha}$. Choisissons donc $p \in \Omega_{N+1,i}$ et $x \in (p^{-1}(0) \cap X - \{0\}) \cap U_p$. Soit $x \in X_\alpha$. Comme la restriction de p à $\overline{C(p_\alpha)}$ est finie et que $x \neq 0$, l'espace tangent $T_x X_\alpha$ est transverse à $T_x(p^{-1}(0))$. Soit $(x_n)_{n\in\mathbf{N}}$ une suite de points de X qui tend vers x. On peut supposer $x_n \in X_\beta$ et $\lim_{n\to\infty} T_{x_n}(X_\beta) = T$. Comme

$(X_\alpha)_{\alpha\in A}$ est une stratification de Whitney, on a $T \supset T_x(X_\alpha)$. Comme $\lim_{n\to\infty} T_{x_n}(p^{-1}(p(x_n)))$ est égal à $T_x(p^{-1}(0))$, on obtient aussitôt

$$\lim_{n\to\infty} T_{x_n}\big(p^{-1}(p(x_n)) \cap X_\beta\big) = \lim_{n\to\infty} T_{x_n}\big(p^{-1}(p(x_n))\big) \cap \lim_{n\to\infty} T_{x_n}(X_\beta)$$
$$= T_x\big(p^{-1}(p(x))\big) \cap T \supset T_x\big(p^{-1}(0)\big) \cap T_x(X_\alpha).$$

Ce qui démontre notre lemme.

(2.3.5) COROLLAIRE. *Considérons la situation de* (2.3.4). *Soient* $p \in \Omega_{N+1,i}((X_\alpha)_{\alpha\in A})$ *et* $(U_t)_{t\in I}$ *un système fondamental à un paramètre de bons voisinages de* 0 *dans* $\mathbf{C}^{N+1}$ *relativement à* $(X_\alpha)_{\alpha\in A}$ *contenus dans* U_p (*cf.* (2.3.4)). *Il existe* $\eta_{i,t} > 0$, *tel que pour tout* η, $0 < \eta \leqslant \eta_{i,t}$, *le morphisme*

$$\varphi_{t,\eta}\colon U_t \cap p^{-1}(B_\eta) \cap X \to B_\eta,$$

où B_η *est la boule ouverte de* $\mathbf{C}^i$ *de centre* 0 *et de rayon* η, *soit descriptible et que, pour tout* t, $t' \in I$ *et tout* $\eta > 0$, $0 < \eta \leqslant \inf(\eta_{i,t}, \eta_{i,t'})$, *on ait une équivalence d'homotopie descriptible de* $\varphi_{t,\eta}$ *sur* $\varphi_{t',\eta}$.

PREUVE. Nous procédons de façon analogue à celle de [**Lê6**]. Soit donc $p \in \Omega_{N+1,i}((X_\alpha)_{\alpha\in A})$. D'après le Lemme (2.3.4), en raisonnant comme dans la démonstration du Théorème (2.3.1), on obtient qu'il existe un voisinage ouvert V de 0 dans $\mathbf{C}^i$ tel que, pour tout $\xi \in V$, la partie non singulière de $p^{-1}(\xi) \cap X_\alpha$ coupe transversalement les strates $\Sigma_{\beta,t}$ de ∂U_t dans $\mathbf{C}^{N+1}$. Soit Δ_α l'image de $\overline{C(p_\alpha)}$ par p. Le premier théorème d'isotopie de Thom-Mather déjà cité implique immédiatement que p induit une fibration C^∞ localement triviale

$$\varphi^*\colon \overline{U}_t \cap p^{-1}\Big(V - \bigcup_{\alpha\in A} \Delta_\alpha\Big) \cap X \to V - \bigcup_{\alpha\in A} \Delta_\alpha.$$

Pour obtenir que p induit un morphisme descriptible $\varphi\colon \overline{U}_t \cap p^{-1}(V) \cap X \to V$, il faut modifier la stratification $(X_\alpha)_{\alpha\in A}$ de la façon suivante: on considère la stratification $(X'_{\alpha'})_{\alpha'\in A'}$ de $p^{-1}(V) \cap X - \bigcup_{\alpha\in A} \overline{C(p_\alpha)}$ induite par $(X_\alpha)_{\alpha\in A}$ (cf. (1.3.5)) et on stratifie $p^{-1}(V) \cap \bigcup_{\alpha\in A} \overline{C(p_\alpha)}$ par $(X''_{\alpha''})_{\alpha''\in A''}$ de telle sorte que $(X'_{\alpha'}, X''_{\alpha''})_{\alpha'\in A', \alpha''\in A''}$ soit une stratification de Whitney analytique complexe et qu'il existe une stratification de Whitney $(Y_\delta)_{\delta\in D}$ analytique complexe de $(\bigcup_{\alpha\in A} \Delta_\alpha) \cap V$ telle que l'image d'une strate $X''_{\alpha''}$ soit une strate $Y_{\delta(\alpha'')}$ et que p induise un morphisme lisse $p_{\alpha''}$ de $X''_{\alpha''}$ sur $Y_{\delta(\alpha'')}$. Ce morphisme $p_{\alpha''}$ est en fait fini, puisque la restriction de p à $\bigcup_{\alpha\in A} \overline{C(p_\alpha)}$ est finie. En appliquant le premier théorème d'isotopie de Thom-Mather au morphisme induit par p de $\overline{U}_t \cap p^{-1}(Y_\delta) \cap X$ sur Y_δ, on obtient que le morphisme φ ci-dessus est descriptible et en choisissant $\eta_{i,t}$ pour que $\overline{B}_{\eta_{i,t}} \subset V$, on obtient la première conclusion de (2.3.5). Pour prouver la seconde conclusion de (2.3.5), il faut cependant choisir $\eta_{i,t}$ de telle sorte que $\overline{B}_{\eta_{i,t}} \cap Y_\delta \neq \varnothing$ si et seulement si $0 \in \overline{Y}_\delta$ et que ∂B_η soit transverse à Y_δ pour tout η, $0 < \eta \leqslant \eta_{i,t}$, ce qui est possible d'après le théorème de Bertini-Sard. On montre alors de la même façon que dans le Théorème (2.3.1) que, pour tout $\delta \in D$, tel que $B_\eta \cap Y_\delta \neq \varnothing$ les fibres $p^{-1}(\xi) \cap \overline{U}_t \cap X$ et $p^{-1}(\xi) \cap \overline{U}_{t'} \cap X$ sont homéomorphes.

Ceci démontre la deuxième assertion de (2.3.5) et achève sa preuve.

(2.3.6) REMARQUE. Dans (2.3.5) on aurait pu formuler l'équivalence d'homotopie descriptible entre $\varphi_{t,\eta}$ et $\varphi_{t',\eta'}$ si $0 < \eta \leqslant \eta_{i,t}$ et $0 < \eta' \leqslant \eta_{i,t'}$. Pour cela il suffit de montrer que l'on a un homéomorphisme stratifié de B_η sur $B_{\eta'}$ tel que l'image de $B_\eta \cap Y_\delta$ soit $B_{\eta'} \cap Y_\delta$. Ceci se fait comme dans [**B-V**], en utilisant encore le premier théorème d'isotopie de Thom-Mather appliqué à cette stratification $(Y_\delta)_{\delta \in D}$ de $(\bigcup_{\alpha \in A} \Delta_\alpha) \cap V$ et la fonction "carré de la distance à 0" dans $\mathbf{C}^i$.

(2.3.7) REMARQUE. On a obtenu en particulier que, pour tout $t \in I$, pour tout $\xi \in B_\eta - \bigcup_{\alpha \in A} \Delta_\alpha$, $0 < \eta \leqslant \eta_{i,t}$, les fibres générales $p^{-1}(\xi) \cap \overline{U_t} \cap X$ des morphismes descriptibles $\varphi_{t,\eta}$ sont homéomorphes.

En particulier, les morphismes $X \to \mathbf{C}^i$ induits par des projections linéaires générales d'un espace $\mathbf{C}^N$ contenant X admettent en $0 \in X$ une théorie locale des cycles évanescents, alors que ce n'est pas le cas pour un morphisme quelconque $X \to \mathbf{C}^i$ dès que $i > 1$.

Par un raisonnement analogue à celui qui nous a donné (2.3.2), on démontre

(2.3.8) COROLLAIRE. *Considérons la situation de* (2.3.4). *Soient* $p \in \Omega_{N+1,i}((X_\alpha)_{\alpha \in A})$, $(U_t)_{t \in I}$ *et* $(V_t)_{t \in I}$ *de systèmes fondamentaux à un paramètre de bons voisinages de* 0 *dans* $\mathbf{C}^{N+1}$ *relativement à* $(X_\alpha)_{\alpha \in A}$. *Soient* $\varphi_{t,\eta}$ *et* $\psi_{t,\eta}$ *les morphismes descriptibles associés à ces systèmes par* (2.3.5) *avec* $0 < \eta \leqslant \eta_{i,t}$ *et* $0 < \eta \leqslant \eta_{i,t}$. *Alors on a une équivalence d'homotopie descriptible de* $\varphi_{t,\eta}$ *sur* $\psi_{t,\eta}$, *et leurs fibres générales sont homéomorphes.*

Nous aurons besoin dans la suite du corollaire pratique (2.3.10) suivant:

Considérons la situation de (2.3.4) et le système fondamental à un paramètre de voisinages $(B_t)_{t \in I}$ des boules de centre 0 et de rayon t de $\mathbf{C}^{N+1}$, avec $I =]0, a]$ et le nombre $a > 0$ assez petit pour que ce système soit un système fondamental à un paramètre de bons voisinages de 0 dans $\mathbf{C}^{N+1}$ relativement à $(X_\alpha)_{\alpha \in A}$ (cf. (2.2.6)).

Notons encore $(Y_\delta)_{\delta \in D}$ la stratification de $\bigcup_{\alpha \in A} \Delta_\alpha$ construite dans la preuve de (2.3.5). Pour tout $t \in I$, on peut choisir $a'_t > 0$ tel que $a'_t \leqslant \eta_{i,t}$ et que la fonction $t \mapsto a'_t$ soit sous-analytique strictement croissante et $a'_t \leqslant t/2$ et que $(B'_{t'})_{t' \in I'_t}$, avec $I'_t =]0, a'_t]$, soit un système fondamental à un paramètre de bons voisinages de 0 dans $\mathbf{C}^i$ relativement à $(Y_\delta)_{\delta \in D}$.

On a vu dans (2.2.2)(d) que le système $(B_t \cap p^{-1}(B'_{a'_t}))_{t \in I}$ est un système fondamental à un paramètre de bons voisinages de 0 dans $\mathbf{C}^{N+1}$. En choisissant convenablement la fonction $t \mapsto a'_t$ on peut supposer que ce système est aussi un système fondamental à un paramètre de bons voisinages de 0 dans $\mathbf{C}^{N+1}$ relativement à $(X_\alpha)_{\alpha \in A}$.

On a le lemme suivant avec les notations de (2.3.4)

(2.3.9) LEMME. *Il existe un ouvert de Zariski dense* $\Omega'_{N+1,i} \subset \Omega_{N+1,i}((X_\alpha)_{\alpha \in A})$ *de l'espace des projections linéaires de* $\mathbf{C}^{N+1}$ *sur* $\mathbf{C}^i$ *et un ouvert dense* $\Omega''_{i,j}$ *de l'espace des projections linéaires de* $\mathbf{C}^i$ *sur* $\mathbf{C}^j$ $(j \leqslant i)$ *tels que pour tout* $p \in \Omega'_i$ *et* $q \in \Omega''_{i,j}$ *on ait* $q \circ p \in \Omega_{N+1,j}((X_\alpha)_{\alpha \in A})$.

Si l'on choisit $p \in \Omega_i'$ et $q \in \Omega_{i,j}'' \cap \Omega_{i,j}((Y_\delta)_{\delta \in D})$ la projection $\pi = q \circ p$ induit des morphismes constructibles:

$$\theta_{t,\eta}: \pi^{-1}(B_\eta'') \cap B_t \cap X \to B_\eta'',$$
$$\tilde{\theta}_{t,\eta}: p^{-1}\big(q^{-1}(B_\eta'') \cap B_{a_t'}'\big) \cap B_t \cap X \to B_\eta''$$

dont les fibres générales sont $\pi^{-1}(\xi) \cap B_t \cap X$ et $\pi^{-1}(\xi) \cap B_{a_t'}' \cap B_t$, si ξ est un point assez général de B_η''. Le Corollaire (2.3.8) donne le résultat suivant dont nous aurons besoin

(2.3.10) COROLLAIRE. *Les fibres générales de $\theta_{t,\eta}$ et $\tilde{\theta}_{t,\eta}$ ont le même type d'homotopie.*

3. Caractéristiques d'Euler-Poincaré évanescentes.

(3.1) *Résumé des résultats obtenus.*

(3.1.1) Soit $(X,0)$ un germe d'espace analytique complexe réduit équidimensionnel de dimension d. Soient X un représentant de $(X,0)$ fermé dans un voisinage ouvert U de 0 dans $\mathbf{C}^{N+1}$ et $(X_\alpha)_{\alpha \in A}$ une stratification de Whitney analytique complexe de X. Soient B_t et $B_{t'}'$ les boules ouvertes de $\mathbf{C}^{N+1}$ et $\mathbf{C}^i$ respectivement ($i \leqslant N+1$) centrées en 0 et de rayons respectifs t et t'. D'après (2.3.5) et (2.3.7), si $p \in \Omega_{N+1,i}((X_\alpha)_{\alpha \in A})$, pour t assez petit non nul et $0 < t' \ll t$, le type d'homotopie de la fibre générale

$$p^{-1}(\xi) \cap B_t \cap X = F_{N+1,i}(X,0)$$

où $\xi \in B_{t'}'$ est un point assez général, ne dépend pas de t. D'après (2.3.8) ce type d'homotopie est le même en remplaçant B_t par U_t où $(U_t)_{t \in I}$ est un système fondamental à un paramètre de bons voisinages de 0 dans $\mathbf{C}^{N+1}$ relativement à $(X_\alpha)_{\alpha \in A}$.

(3.1.2) Bien évidemment $F_{N+1,i}(X,0) = \varnothing$ si $i \geqslant d+1$ et, quel que soit le plongement local de $(X,0)$ dans $(\mathbf{C}^{N+1},0)$, $F_{N+1,d}(X,0)$ est un ensemble fini de $m_0(X)$ points, où $m_0(X)$ est la multiplicité de X en 0. De façon générale $\dim F_{N+1,i}(X) = d - i$ si $i \leqslant d$. Le but de ce paragraphe est de montrer que le type d'homotopie de $F_{N+1,i}(X,0)$ ne dépend ni du plongement local de $(X,0)$ dans $(\mathbf{C}^{N+1},0)$, ni du choix de la stratification de Whitney analytique complexe $(X_\alpha)_{\alpha \in A}$ de X choisie, ni du choix de la projection de p dans $\Omega_{N+1,i}((X_\alpha)_{\alpha \in A})$ $(i \leqslant d)$.

Il en résultera que la famille des types d'homotopie de $F_{N+1,0}, \ldots, F_{N+1,d}$ est un invariant analytique du germe $(X,0)$. On appelle la famille de ces types d'homotopie le *type d'homotopie évanescent* de $(X,0)$. En particulier les caractéristiques d'Euler-Poincaré $\chi(F_{N+1,0}(X,0)), \ldots, \chi(F_{N+1,d}(X,0))$ (comparer à [**D**]) seront des invariants analytiques de $(X,0)$. On appellera ces caractéristiques d'Euler-Poincaré *les caractéristiques d'Euler-Poincaré évanescentes* de $(X,0)$, on notera $\chi_i(X,0)$ pour $\chi(F_{N+1,i}(X,0))$; on remarquera que l'on a : $\chi_d(X,0) = m_0(X)$, et $\chi_0(X,0) = 1$ puisque $F_{N+1,0} = X \cap B_t$ qui est contractile (cf. [**B-V**]).

On montrera aussi que, si $(X_\alpha)_{\alpha \in A}$ est une stratification de Whitney analytique complexe de X, la famille $(\chi_0(X, x), \ldots, \chi_d(X, x))$ est indépendante de $x \in X_\alpha$, pour tout $\alpha \in A$.

Rappelons aussi que dans [**D**] A. Dubson utilise ces caractéristiques d'Euler-Poincaré évanescentes pour calculer le nombre d'Euler de $(X, 0)$ (cf. [**MP**]). Nous donnerons dans le paragraphe suivant la relation entre les caractéristiques d'Euler-Poincaré évanescentes associées à une stratification de Whitney analytique complexe $(X_\alpha)_{\alpha \in A}$ de X et les multiplicités des variétés polaires locales en 0 des $\overline{X}_\alpha$, avec $\alpha \in A$: cette relation redonne grâce aux résultats de [**Lê-Te**, (§5)] une démonstration du résultat de A. Dubson [**D**].

(3.2) *Cas de familles paramétrées*. Nous allons considérer la situation suivante:

(3.2.1) Soient X un sous-espace analytique complexe réduit fermé d'un ouvert U de $\mathbf{C}^{N+1}$, $(X_\alpha)_{\alpha \in A}$ une stratification de Whitney analytique complexe de X, $x \in X$ et X_{α_0} la strate qui contient x. On suppose que dans l'ouvert U la strate X_{α_0} est définie par l'idéal engendré par les fonctions analytiques $\zeta_1, \ldots, \zeta_{k_0}$ définies sur l'ouvert U où le système $\{\zeta_1, \ldots, \zeta_{k_0}\}$ peut s'étendre en un système $\{\zeta_1, \ldots, \zeta_{N+1}\}$ de coordonnées de U qui s'annulent en x. On supposera pour simplifier que $x = 0$.

Soit S un espace analytique complexe non singulier connexe. On suppose que l'on a une famille analytique complexe paramétrée par S de projections linéaires p_s: $\mathbf{C}^{N+1} \to \mathbf{C}^i$, i.e. définie par P_0: $\mathbf{C}^{N+1} \times S \to \mathbf{C}^i \times S$, telle que, pour tout $s \in S$, la projection p_s soit dans $\Omega_{N+1,i}((X_\alpha)_{\alpha \in A})$ (cf. (2.3.4)), que p_s soit transverse à X_{α_0} en tout point de X_{α_0} et que son noyau soit transverse aux cônes tangents en tout point de X_{α_0} des variétés polaires associées à p_s des adhérences des strates X_α de la stratification considérée de X. On note P la restriction de P_0 à $X \times S$.

Dans la suite on munira $X \times S$ de la stratification de Whitney analytique complexe définie par $(X_\alpha \times S)_{\alpha \in A}$.

D'après Chapitre V, §2 de [**Te1**], la variété polaire locale relative de la restriction P_α de P à $\overline{X}_\alpha \times S$ est la réunion des variétés polaires de $\overline{X}_\alpha \times \{s\}$ associées à p_s, pour tout $\alpha \in A$.

(3.2.2) Pour $\varepsilon > 0$, on notera T_ε, l'ensemble des points (z, s) de $\mathbf{C}^{N+1} \times S$ tels que $(\sum_{i=1}^{k_0} |\zeta_i(z)|^2)^{1/2} \leqslant \varepsilon$. Le bord de T_ε est noté ∂T_ε.

On a vu dans (2.2.10) que, pour $\varepsilon > 0$ assez petit, le système des intérieurs $\mathring{T}_\varepsilon$ est un système fondamental à un paramètre de bons voisinages de $X_{\alpha_0} \times S$ dans $\mathbf{C}^{N+1} \times S$ si de plus U et S sont sous-analytiques. A l'aide de (2.2.8) qui assure la condition (iii) de (2.2.11), on obtient que, pour tout ouvert $S' \subset S$ qui est sous-analytique, connexe, relativement compact et tel que $\overline{S'} \subset S$, i.e. $S' \subset\subset S$, le système $(\mathring{T}_\varepsilon \cap (\mathbf{C}^{N+1} \times S'))_{\varepsilon \in I}$, avec $I =]0, \varepsilon_0]$ et $\varepsilon_0 > 0$ assez petit, est un système fondamental à un paramètre de bons voisinages de $X_{\alpha_0} \times S'$ dans $(\mathbf{C}^{N+1} \times S)$ relativement à la stratification de Whitney analytique complexe $(X_\alpha \times S')_{\alpha \in A}$ de $X \times S'$.

(3.2.3) LEMME. *Si le voisinage ouvert U de x dans $\mathbf{C}^{N+1}$ est assez petit, la stratification $(X_\alpha \times S')_{\alpha \in A}$ de $X \times S'$ satisfait la condition de Thom relativement à la projection P en tout point de $P^{-1}(P(X_{\alpha_0} \times S')) - X_{\alpha_0} \times S'$.*

PREUVE. La preuve de ce lemme est tout à fait analogue à celle de (2.3.4), car les hypothèses faites sur P montrent que, si le voisinage U est assez petit, la restriction de P aux variétés polaires relatives C_α de P_α est un morphisme fini. Nous laissons donc au lecteur le soin de faire précisément cette preuve.

Par ailleurs:

(3.2.4) LEMME. *Il existe $\varepsilon_0 > 0$ tel que pour tout ε, $0 < \varepsilon \leq \varepsilon_0$, ∂T_ε coupe transversalement $X_\alpha \times S'$ dans $\mathbf{C}^{N+1} \times S$ pour tout $\alpha \in A$.*

PREUVE. L'ensemble des points de $X_\alpha \times S$ où $X_\alpha \times S$ n'est pas transverse à ∂T_ε est un sous-ensemble semianalytique A_α de $X_\alpha \times S$. D'autre part pour chaque $s \in S$ le théorème de Bertini-Sard montre que la restriction à $X_\alpha \times \{s\}$ de la fonction $x \mapsto (\sum_{i=1}^{k_0} |\zeta_i(x)|^2)^{1/2}$ n'a que des valeurs critiques isolées à l'intérieur de $T_\varepsilon \cap (\mathbf{C}^{N+1} \times \{s\})$.

Comme $S' \subset\subset S$, $\mathring{T}_\varepsilon \cap (\mathbf{C}^{N+1} \times S')$ ne rencontre qu'un nombre fini de strates $X_\alpha \times S'$ et il existe $\varepsilon_0 > 0$ tel que $A_\alpha \cap T_\varepsilon = \varnothing$, pour tout $0 < \varepsilon \leq \varepsilon_0$ et $\alpha \in A$.

Nous avons alors une version "à paramètre" de (2.3.5)

(3.2.5) THÉORÈME. *Il existe $\varepsilon_0 > 0$ tel que pour tout ε, $0 < \varepsilon \leq \varepsilon_0(x)$, il existe $\eta_\varepsilon > 0$, tel que pour tout η, $0 < \eta \leq \eta_\varepsilon$, P induit un morphisme descriptible*

$$\varphi_{\varepsilon,\eta}\colon \mathring{T}_\varepsilon \cap P^{-1}(B'_\eta \times S') \cap (X \times S') \to B'_\eta \times S'.$$

DÉMONSTRATION. En utilisant le fait énoncé dans (3.2.3) que l'on a la condition de Thom en tout point de $P^{-1}(P(X_{\alpha_0} \times S')) - X_{\alpha_0} \times S'$ et la transversalité dans $\mathbf{C}^{N+1} \times S'$ de $X_\alpha \times S'$ et ∂T_ε pour ε assez petit, on montre, comme dans (2.3.5), en utilisant la compacité de $\bar{S}'$, et le fait que $X_{\alpha_0} \times S'$ n'est adhérent qu'à un nombre fini de $X_\alpha \times S'$, qu'il existe $\varepsilon_0 > 0$ tel que pour tout ε, $0 < \varepsilon \leq \varepsilon_0$, il existe η_ε tel que, pour tout η, $0 < \eta \leq \eta_\varepsilon$, $(X_\alpha \times S') \cap P^{-1}(\xi, s')$ coupe ∂T_ε transversalement dans $\mathbf{C}^{N+1} \times S'$, pout tour $\xi \in B'_\eta$ et $s' \in S'$. Le premier théorème d'isotopie de Thom-Mather déjà cité montre alors que P induit une fibration topologique localement triviale φ^* de $T_\varepsilon \cap P^{-1}(B'_\eta \times S' - \Delta) \cap (X \times S')$ sur $B'_\eta \times S' - \Delta$, où $\Delta = \bigcup_{\alpha \in A} P(C_\alpha)$ qui est vide ou une hypersurface de $\mathbf{C}^i \times S'$, puisque la restriction de P à C_α est finie et que $C_\alpha = \varnothing$ ou est de dimension $i - 1$. Afin d'obtenir le morphisme descriptible cherché, il suffit de stratifier convenablement la restriction de P à $\bigcup_{\alpha \in A} C_\alpha$ de façon analogue à ce que nous avons fait dans (2.3.5).

(3.2.6) LEMME. *Pour tout ε, $0 < \varepsilon \leq \varepsilon_0$, il existe r_ε, tel que*

$$\left(\mathring{T}_\varepsilon \cap P^{-1}(B'_{r_\varepsilon} \times S')\right)_{0 < \varepsilon \leq \varepsilon_0(x)}$$

soit un système fondamental à un paramètre de bons voisinages de $\{x\} \times S'$ dans $\mathbf{C}^{N+1} \times S'$ relativement à $(X_\alpha \times S')_{\alpha \in A}$.

PREUVE. Il suffit de montrer que les "coins" définis par $\partial T_\varepsilon \cap P^{-1}(\partial B'_{r_\varepsilon} \times S')$ et les "côtés" $\mathring{T}_\varepsilon \cap P^{-1}(\partial B'_{r_\varepsilon} \times S')$ sont bien transverses à $X_\alpha \times S'$, pour tout $\alpha \in A$. La transversalité des "coins" est assurée dès que $r_\varepsilon > 0$ est assez petit, car la restriction P_α de P à la variété polaire relative C_α de $\overline{X}_\alpha \times S'$ est finie et que les fibres de P_0 au-dessus de $\{0\} \times S'$ sont transverses à ∂T_ε d'après (3.2.3). La transversalité des "côtés" est assurée par la compacité de $\overline{S}'$ et le fait que, pour tout $s \in S'$, $X_\alpha \times \{S'\}$ coupe transversalement $P^{-1}(\partial B'_r)$ dès que $r > 0$ est assez petit d'après le théorème de Bertini-Sard.

(3.2.7) COROLLAIRE. *Pour tout $s \in S'$, la fibre générale du morphisme descriptible de* (2.3.5) *défini sur le germe* $(X \times \{s\}, (x, s))$ *par* $p_s \times \{s\}$ *a le même type d'homotopie.*

PREUVE. En fait toutes les fibres générales du morphisme descriptible $\varphi_{\varepsilon,\eta}$ sont homéomorphes. Il suffit de constater que $\varphi_{\varepsilon,\eta}$ induit au-dessus de $\mathbf{C}^i \times \{s\}$ le morphisme descriptible de (2.3.5) défini sur $X \times \{s\}$ par p_s et que la fibre générale de ce morphisme est une fibre générale de $\varphi_{\varepsilon,\eta}$.

(3.2.8) COROLLAIRE. *Si U est assez petit, pour tout $y \in X_{\alpha_0}$, et pour tout $s \in S'$, la fibre générale du morphisme descriptible de* (2.3.5) *défini sur le germe* $(X \times \{s\}, (y, s))$ *par $p_s \times \{s\}$ a le même type d'homotopie.*

PREUVE. Soient $y \in X_{\alpha_0}$ et $0 < \varepsilon \leqslant \varepsilon_0$. Soit Γ un sous-ensemble compact et connexe de X_{α_0} qui contient x et y. En chaque point $z \in \Gamma$ on a un morphisme descriptible

$$\varphi_{\varepsilon,\eta}(z)\colon \mathring{T}_\varepsilon \cap P^{-1}\big(B'_\eta(z) \times S'\big) \cap X \times S' \to B'_\eta(z) \times S'$$

où $\varphi_{\varepsilon,\eta}(z)$ est défini par P comme dans (3.2.5) et $B'_\eta(z)$ une boule ouverte de $\mathbf{C}^i$ centrée en $P(z)$ et de rayon η assez petit. On trouve un nombre fini $z_1, \ldots, z_l$ de points de Γ avec $z_1 = x$ et $z_l = y$ tels que les boules $B'_{\eta_i}(z_i)$ correspondantes recouvrent $P(\Gamma)$. On trouve alors que les fibres générales des $\varphi_{\varepsilon,\eta_i}(z_i)$ sont homéomorphes, ce qui, combiné avec le Corollaire (3.2.7), donne (3.2.8).

(3.3) Nous sommes maintenant en mesure de montrer que les fibres générales des morphismes constructibles considérés dans (2.3.5) ont un type d'homotopie qui est un invariant analytique de $(X, 0)$ et qu'il est constant le long des strates d'une stratification de Whitney analytique complexe. On reprend les notations de (3.2).

(3.3.1) On se ramène immédiatement à la situation suivante

$$X \subset \mathbf{C}^{N+1} \subset \mathbf{C}^{N+M+2}$$

où X est fermé dans l'ouvert U où le système de coordonnées $\{\zeta_1, \ldots, \zeta_{N+1}\}$ s'étend en un système de coordonnées $\{\zeta_1, \ldots, \zeta_{N+1}, \xi_1, \ldots, \xi_{M+1}\}$ d'un ouvert $U \times V$ de $\mathbf{C}^{N+M+2}$. On plonge $X \times \mathbf{C}$ dans $\mathbf{C}^{N+M+2} \times \mathbf{C}$ au moyen des restrictions à X des fonctions $\zeta_1, \ldots, \zeta_{N+1}, t\xi_1, \ldots, t\xi_{M+1}, t$ où t est une coordonnée sur $\mathbf{C}$. La stratification donnée par $(X_\alpha \times \mathbf{C})_{\alpha \in A}$ est une stratification de Whitney analytique complexe de $X \times \mathbf{C}$. On peut supposer que $\{x\} \times \mathbf{C}$ est une strate.

Nous allons nous placer dans une situation analogue à celle décrite par (3.2.1): On appelle T_ε l'ensemble des points de $\mathbf{C}^{N+M+2} \times \mathbf{C}$ tels que

$$\sum_{i=1}^{N+1} |\zeta_i(z)|^2 + \sum_{j=1}^{M+1} |t\xi_j(z)|^2 \leqslant \varepsilon^2.$$

On supposera U assez petit pour qu'il existe q: $\mathbf{C}^{N+1} \to \mathbf{C}^r$ une projection telle que q soit transverse à X_{α_0} en tout point de X_{α_0} et que le noyau de q soit transverse aux cônes tangents en tout point de X_{α_0} des variétés polaires associées à q des adhérences strates X_α de la stratification considérée de X. Ceci est assuré si $q \in \Omega_{N+1,i}((X_\alpha)_{\alpha\in A})$ et U assez petit, parce que ces variétés polaires sont équimultiples le long de X_{α_0} d'après [**Te1**, Chapitre V, Théorème 1.2], puisque $(X_\alpha)_{\alpha\in A}$ est une stratification de Whitney analytique complexe de X.

On note p_0 la composée de q et de la projection $\mathbf{C}^{N+M+2} \to \mathbf{C}^{N+1}$ sur les $N+1$ premiers facteurs.

Soit p_1: $\mathbf{C}^{N+M+2} \to \mathbf{C}^i$ une projection linéaire transverse à X_{α_0} en tout point de X_{α_0} et dont le noyau Ker p_1 est transverse aux cônes tangents en tout point de X_{α_0} des variétés polaires associées à p_1 des adhérences des strates X_α de la stratification $(X_\alpha)_{\alpha\in A}$ de X.

On notera $p_t = (1-t)p_0 + tp_1$. Soit P_0: $\mathbf{C}^{N+M+2} \times \mathbf{C} \to \mathbf{C}^i \times \mathbf{C}$ l'application définie par $P_0(z,t) = (p_t(z), t)$. On appellera P la restriction de P_0 à $X \times \mathbf{C}$.

Puisque P_0 induit pour $t=0$ et $t=1$ des projections transverses p_0 et p_1 à X_{α_0} en tout point de X_{α_0} et dont les noyaux respectifs sont transverses aux cônes tangents en tout point de X_{α_0} des variétés polaires, associées à p_0 et p_1 respectivement, des adhérences $\overline{X}_\alpha$ pour tout $\alpha \in A$, pour tout t sauf un nombre fini, p_t est transverse à X_{α_0} en tout point de X_{α_0} et son noyau est transverse aux cônes tangents en tout point de X_{α_0} des variétés polaires associées à p_t des espaces $\overline{X}_\alpha$. Soit S' un ouvert relativement compact sous-analytique de $\mathbf{C}$ qui contient 0 et 1 et dont l'adhérence S' ne contient que des points t pour lesquels les projections p_t vérifient les conditions de transversalité ci-dessus.

Nous sommes maintenant dans les conditions décrites dans (3.2.1) et nous pouvons appliquer les résultats de (3.2.7) et (3.2.8) qui nous permettent de conclure que le type d'homotopie des fibres générales des morphismes descriptibles considérés est indépendant des plongements locaux $(X, 0) \subset (\mathbf{C}^{N+1}, 0)$ et des coordonnées choisis et reste constant le long d'une strate de Whitney d'une stratification de Whitney analytique complexe.

(3.3.2) *Conclusion.* Soit X un représentant d'un germe $(X, 0)$ d'espace analytique complexe réduit et soit $X = \bigcup_{\alpha\in A} X_\alpha$ une stratification de Whitney analytique complexe de X. Comme nous l'avons dit dans (3.1.2) le type d'homotopie des fibres $F_{N+1,i}(\overline{X}_\alpha, x)$ définies à l'aide d'un plongement local $(X, x) \subset (\mathbf{C}^{N+1}, 0)$ est indépendant de ce plongement et du point x sur sa strate dans $(X_\alpha)_{\alpha\in A}$. En fait notre démonstration nous a donné un morphisme descriptible $X \cap p^{-1}(B'_\eta) \cap B_\varepsilon \to B'_\eta$ avec p dans un ouvert de Zariski dense Ω_i de l'espace des projections de $\mathbf{C}^{N+1}$ sur $\mathbf{C}^i$ et $1 \gg \varepsilon > 0$, $\varepsilon \gg \eta > 0$. La classe d'équivalence d'homotopie descriptible de ce morphisme descriptible ne dépend ni du plongement, ni de x sur sa strate.

Ainsi ces morphismes descriptibles pour $0 \leqslant i \leqslant d$ (resp. leurs fibres générales) définissent des classes d'équivalence d'homotopie descriptible (resp. type d'homotopie) qui sont des invariants analytiques du germe (X, x), constants sur les strates d'une stratification de Whitney analytique complexe, comme on le voit en choisissant la stratification de Whitney analytique complexe canonique (cf. (1.3.3)).

La famille des types d'homotopie des $F_{N+1,i}(\overline{X}_\alpha, x)$ pour les $\overline{X}_\alpha$ contenant x est appelé le *type d'homotopie évanescent total du germe* (X, x) quand $(X_\alpha)_{\alpha \in A}$ est la stratification de Whitney analytique complexe canonique.

Soit $x \in X$ avec $x \in X_\alpha \subset \overline{X}_\beta$ et soit $X \subset \mathbf{C}^{N+1}$ un plongement local de X en x dans $\mathbf{C}^{N+1}$. On notera $\chi_i(\overline{X}_\beta, X_\alpha)$ la caractéristique d'Euler-Poincaré de $F_{N+1,i}(\overline{X}_\beta, x)$ qui est constante le long de X_α. On pose $\dim X_\alpha = d_\alpha$.

Remarquons que si $i \leqslant d_\alpha$, $F_{N+1,i}(\overline{X}_\beta, x)$ est contractible et que $\chi_i(\overline{X}_\beta, X_\alpha) = 1$. On convient que $\chi_i(\overline{X}_\beta, X_\alpha) = 0$ pour $i > d_\beta$.

On notera $\chi^*(X, x) = (\chi(F_{N+1,0}(X, x)), \ldots, \chi(F_{N+1,d}(X, x)))$.

4. Interprétation topologique des multiplicités polaires.

(4.1) Dans ce paragraphe nous donnons une formule qui permet de calculer les multiplicités en chaque point des variétés polaires des adhérences des strates d'une stratification de Whitney analytique complexe en fonction des caractéristiques d'Euler-Poincaré évanescentes en ce point introduites au paragraphe précédent. Nous prenons les notations de (3.3.2).

(4.1.1) THÉORÈME. *Soit* $X = \bigcup_{\alpha \in A} X_\alpha$ *un espace analytique complexe réduit muni d'une stratification de Whitney analytique complexe* $(X_\alpha)_{\alpha \in A}$. *Soit* $x \in X_{\alpha_0}$. *On a l'égalité*:

$$\chi_{d_{\alpha_0}+1}(X, X_{\alpha_0}) - \chi_{d_{\alpha_0}+2}(X, X_{\alpha_0})$$
$$= \sum_{\alpha \neq \alpha_0} (-1)^{d_\alpha - d_{\alpha_0} - 1} m_x\left(P_{d_\alpha - d_{\alpha_0} - 1}(\overline{X}_\alpha, x)\right)\left(1 - \chi_{d_\alpha + 1}(X, X_\alpha)\right).$$

DÉMONSTRATION. Démontrons d'abord le théorème dans le cas où $X_{\alpha_0} = \{x\}$. Le résultat cherché est en fait local, car si $x \notin \overline{X}_\alpha$, on a $P_k(\overline{X}_\alpha, x) = \varnothing$ et $m_x(P_k(\overline{X}_\alpha, x)) = 0$. On peut donc supposer que X est fermé dans un voisinage ouvert U de 0 dans $\mathbf{C}^{N+1}$ assez petit tel que $x \in \overline{X}_\alpha$. Nous avons besoin pour notre démonstration d'adapter à notre situation les résultats (2.3.5) et (2.3.10), précisément dans le cas où $i = 2$ et $j = 1$.

(4.1.2) LEMME. *Il existe un ouvert de Zariski dense* Ω *contenu dans* $\Omega_{N+1,2}((X_\alpha)_{\alpha \in A})$ *tel que, pour* $p \in \Omega$:

(i) *il existe* $a > 0$ *tel que* $(B_t)_{t \in I}$, *avec* $I =]0, a]$, *soit un système fondamental à un paramètre de bons voisinages de* 0 *dans* $\mathbf{C}^{N+1}$ *relativement à* $(X_\alpha)_{\alpha \in A}$ *contenus dans l'ouvert* U_p *défini dans* (2.3.4);

(ii) *il existe* $(\eta_t)_{t \in I}$ *tel que le morphisme*

$$\varphi_{t,\eta}\colon B_t \cap p^{-1}(B'_\eta) \cap X \to B'_\eta$$

induit par p sur la boule ouverte B'_η *de* $\mathbf{C}^2$ *centrée en* 0 *et de rayon* η, $0 < \eta \leqslant \eta_t$, *soit descriptible*;

(iii) *la courbe polaire* Γ_α *définie par p sur* $\overline{X}_\alpha \cap B_a \cap p^{-1}(B'_{\eta_\alpha})$ *en x est réduite et sa multiplicité en x est la multiplicité d'une courbe polaire générale de* $\overline{X}_\alpha$ *en x*;

(iv) *aucune direction limite en* 0 *de bisécantes à* $\Gamma_\alpha - \{x\}$ *n'est contenue dans* Ker p. *En particulier* Ker p *est transverse à* Γ_α *en x, le morphisme analytique p restreint à* Γ_α *induit un homéomorphisme de* Γ_α *sur son image dans* $\mathbf{C}^2$.

Preuve. Les conditions (i) et (ii) sont réalisées si $p \in \Omega_{N+1,2}((X_\alpha)_{\alpha \in A})$ d'après le Corollaire (2.3.5).

Pour obtenir la propriété (iii) on procède comme dans **[Lê-Te**, (2.2.1.2)]. Considérons la modification de Nash $\nu_\alpha\colon N(\overline{X}_\alpha) \to \overline{X}_\alpha$ de $\overline{X}_\alpha$ munie du plongement naturel $N(\overline{X}_\alpha) \subset U \times G_\alpha$, où G_α est la Grassmannienne des plans de dimension $d_\alpha = \dim X_\alpha$ dans $\mathbf{C}^{N+1}$. On choisit une stratification de Whitney analytique complexe $(Z_{\alpha\beta})_{\beta \in B_\alpha}$ de $N(\overline{X}_\alpha)$ telle que $\nu_\alpha^{-1}(x)$ soit une réunion de strates. Soit Ω_α l'ouvert de Zariski dense de l'espace des projections linéaires de $\mathbf{C}^N$ sur $\mathbf{C}^2$ formé des projections p telles que la variété de Schubert $c_{d_\alpha - 1}(\operatorname{Ker} p)$ soit transverse dans G_α à la stratification de $\nu^{-1}(x)$ dans le sens de **[Lê-Te]**. D'après l'argument de **[Lê-Te]** l'espace $U \times c_{d_\alpha - 1}(\operatorname{Ker} p)$ coupe transversalement dans $U \times G_\alpha$ l'espace $N(\overline{X}_\alpha) - \nu^{-1}(\operatorname{Sing} \overline{X}_\alpha)$ selon un sous-ensemble vide ou non singulier et de dimension un si U est assez petit. Dans ce dernier cas l'adhérence de ce sous-ensemble est égale à $N(\overline{X}_\alpha) \cap (U \times c_{d_\alpha - 1}(\operatorname{Ker} p))$. Quitte à se restreindre à un ouvert $\Omega'_\alpha \subset \Omega_\alpha$ on peut supposer que la multiplicité en x de Γ_α est celle d'une courbe polaire générale de $\overline{X}_\alpha$ en x et que l'on a (iv) d'après **[Te1**, Chapitre V, lemme-clé].

On choisit alors $\Omega = \bigcap_{\alpha \in A} \Omega'_\alpha \cap \Omega_{N+1,2}((X_\alpha)_{\alpha \in A})$ qui est un ouvert de Zariski dense de $\Omega_{N+1,2}((X_\alpha)_{\alpha \in A})$ car l'ensemble A est fini.

Si $p \in \Omega$ et $t \in \,]0, a]$ comme dans (4.1.2), la caractéristique d'Euler-Poincaré de la fibre générale du morphisme descriptible $\varphi_{t,\eta}$ (avec $0 < \eta \leqslant \eta_t$) de (4.1.2)(ii) est égale à $\chi_2(X, \{x\})$ avec les notations de (3.3.2). Pour calculer la différence $\chi_1(X, \{x\}) - \chi_2(X, \{x\})$ cherchée dans (3.1) avec $\{x\} = X_{\alpha_0}$, on a besoin du résultat énoncé dans 2.3.9 qui donne dans ce cas où $i = 2$ et $j = 1$:

(4.1.3) Lemme. *Il existe un ouvert de Zariski dense* Ω' *de l'espace des projections linéaires de* $\mathbf{C}^{N+1}$ *sur* $\mathbf{C}^2$ *contenu dans l'ouvert* Ω *de* (4.1.2) *et un ouvert de Zariski dense* Ω'' *de l'espace des projections linéaires de* $\mathbf{C}^2$ *sur* $\mathbf{C}$ *tels que, pour* $p \in \Omega'$ *et* $q \in \Omega''$:

(i) *il existe* $a > 0$, *tel que, pour tout* $t \in \,]0, a]$, *il existe* η_t, *tel que, avec* $0 < \eta \leqslant \eta_t$, *p induit un morphisme descriptible*

$$\varphi_{t,\eta}\colon p^{-1}(B'_\eta) \cap B_t \cap X \to B'_\eta;$$

(ii) *soit* $\Delta = \bigcup_{\alpha \in A} p(\Gamma_\alpha)$, *où* Γ_α, *défini dans* (4.1.2)(iii), *est la courbe polaire de p sur* $\overline{X}_\alpha$, *alors* Ker q *est transverse à* Δ *en* 0 *et il existe* η'_t, $0 < \eta'_t \ll \eta_t$, *tel que, pour tout* $u \in \mathbf{C}$, $0 < |u| \leqslant \eta'_t$, $\varphi_{t,\eta}$ *induit un morphisme descriptible*

$$\tilde{\varphi}_{t,\eta}\colon p^{-1}\big(B'_\eta \cap q^{-1}(u)\big) \cap B_t \cap X \to B'_\eta \cap q^{-1}(u).$$

PREUVE. Il suffit de bien interpréter (2.3.9), en utilisant (2.3.4).

(4.1.4) REMARQUE. D'après (2.3.10) et (3.3.2), $p^{-1}(B'_\eta \cap q^{-1}(u)) \cap B_t \cap X$ a sa caractéristique d'Euler-Poincaré égale à $\chi_1(X, \{x\})$.

(4.2) Fin de la démonstration de (4.1.1) quand $X_{\alpha_0} = \{x\}$.

En utilisant (4.1.3), comme les fibres générales de $\varphi_{t,\eta}$ et $\tilde{\varphi}_{t,\eta}$ sont les mêmes, nous allons utiliser la Proposition (2.1.3) pour évaluer $\chi_1(X, \{x\})$.

(4.2.1) Les fibres de $\tilde{\varphi}_{t,\eta}$ qui ne sont pas générales sont précisément les fibres au-dessus des points de $B'_\eta \cap q^{-1}(u) \cap \Delta$. Posons $\Delta_\alpha = p(\Gamma_\alpha)$ et notons $\Delta_\alpha = \bigcup_{i \in A_\alpha} \Delta_{\alpha,i}$ la décomposition de Δ_α en composantes irréductibles. Les fibres de $\tilde{\varphi}_{t,\eta}$ sur les points de $B'_\eta \cap q^{-1}(u) \cap \Delta_{\alpha,i}$ sont homéomorphes, car $q \in \Omega''$ (cf. (2.3.9)) et les sous-espaces $\Delta_{\alpha,i} - \{0\}$ sont parmi les strates Y_δ introduites dans (2.3.5) car nous sommes dans le cas où $i = 2$. Par ailleurs le cardinal de l'ensemble fini $B'_\eta \cap q^{-1}(u) \cap \Delta_{\alpha,i}$ est la multiplicité de $\Delta_{\alpha,i}$ en 0 qui n'est autre que celle de la composante $\Gamma_{\alpha,i}$ de Γ_α au-dessus de $\Delta_{\alpha,i}$ d'après (4.1.2)(iv). Donc la Proposition (2.1.3) donne dans ce cas

(4.2.1.1)
$$\chi_1(X, \{x\}) = \chi_2(X, \{x\})\Big(1 - \sum_{\alpha \in A} m_x(\Gamma_\alpha)\Big) + \sum_{\alpha \in A} \sum_{i \in A_\alpha} \chi_{\alpha,i} \cdot m_x(\Gamma_{\alpha,i})$$

où $\chi_{\alpha,i}$ est la caractéristique d'Euler-Poincaré de la fibre "singulière" au-dessus d'un point de $B'_\eta \cap q^{-1}(u) \cap \Delta_{\alpha,i}$. En effet on a d'après (2.3.10)

$$\chi_1(X, \{x\}) = \chi\big(p^{-1}\big(B'_\eta \cap q^{-1}(u)\big) \cap B_t \cap X\big),$$
$$\chi_2(X, \{x\}) = \chi\ \ \big(\text{fibre générale de } \tilde{\varphi}_{t,\eta}\big),$$

et $1 - \sum_{\alpha \in A} m_x(\Gamma_\alpha) = \chi(B'_\eta \cap q^{-1}(u) - \Delta)$. Le calcul de $\chi_{\alpha,i}$ est assuré par

(4.2.2) LEMME. *On a l'égalité*:

(4.2.2.1)
$$\chi_{\alpha,i} = \chi_2(X, \{x\}) + (-1)^{d_\alpha}\chi_{d_\alpha+1}(X, X_\alpha) + (-1)^{d_\alpha - 1}.$$

PREUVE. Soit $\xi_{\alpha,i}$ un point de $D \cap \Delta_{\alpha,i}$ où $D = B'_\eta \cap q^{-1}(u)$. On note $x_{\alpha,i}$ l'unique point de $\Gamma_{\alpha,i}$ au-dessus de $\xi_{\alpha,i}$ (d'après (4.1.2)(iv)). Soit $B_{\alpha,i}$ une boule de $\mathbf{C}^{N+1}$ centrée en $x_{\alpha,i}$ et de rayon assez petit. Si ξ est dans D assez proche de $\xi_{\alpha,i}$, $\tilde{\varphi}^{-1}_{t,\eta}(\xi) - B_{\alpha,i}$ et $\tilde{\varphi}^{-1}_{t,\eta}(\xi_{\alpha,i}) - B_{\alpha,i}$ sont homéomorphes, par application du premier théorème d'isotopie de Thom et Mather puisque les points où $\tilde{\varphi}_{t,\eta}$ n'est pas "transverse" aux strates sont intérieurs à $B_{\alpha,i}$. La suite exacte de Mayer-Vietoris et le fait déjà cité (cf. [S]) que $\chi(\tilde{\varphi}_{t,\eta}^{-1}(\xi) \cap \partial B_{\alpha,i}) = 0$ donnent l'égalité

$$(4.2.2.2) \qquad \chi_{\alpha,i} = \chi\big(\tilde{\varphi}_{t,\eta}^{-1}(\xi)\big) - \chi\big(\tilde{\varphi}_{t,\eta}^{-1}(\xi) \cap B_{\alpha,i}\big) + 1$$

puisque $\chi(\tilde{\varphi}_{t,\eta}^{-1}(\xi_{\alpha,i}) \cap B_{\alpha,i})$ vaut 1 car d'après [B-V], $\tilde{\varphi}_{t,\eta}^{-1}(\xi_{\alpha,i}) \cap B_{\alpha,i}$ est contractile si $B_{\alpha,i}$ est de rayon assez petit.

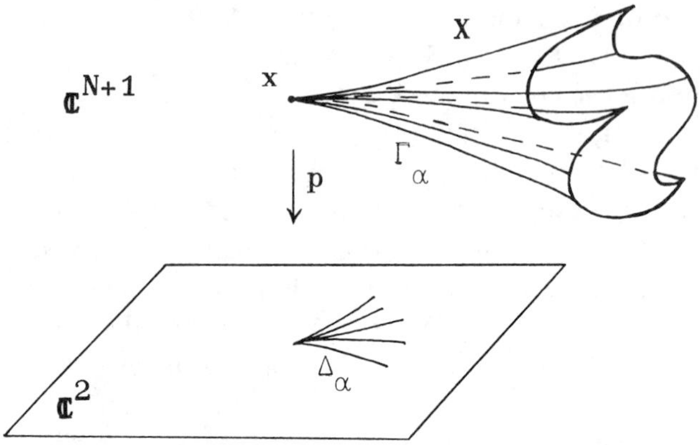

La preuve de (4.2.2) est terminée, si nous pouvons estimer $\chi(\tilde{\varphi}_{t,\eta}^{-1}(\xi) \cap B_{\alpha,i}) = \chi'_{\alpha,i}$.

Pour cela on utilise le Lemme (2.3.9) et on considère une projection linéaire $p_{\alpha,i}\colon \mathbf{C}^{N+1} \to \mathbf{C}^{d_\alpha+1}$ assez générale parmi celles dont le noyau $\operatorname{Ker} p_{\alpha,i}$ contient $\operatorname{Ker} p$. Soit $p'_{\alpha,i}\colon \mathbf{C}^{d_\alpha+1} \to \mathbf{C}^2$ la projection linéaire telle que $p'_{\alpha,i} \circ p_{\alpha,i} = p$. Si les choix faits sont assez généraux, comme dans (2.3.9) on peut trouver des morphismes descriptibles

$$\Psi_{\alpha,i}\colon p_{\alpha,i}^{-1}\big(B'_{\alpha,i} \cap p'^{-1}_{\alpha,i}(B''_{\alpha,i})\big) \cap B_{\alpha,i} \cap X \to B'_{\alpha,i} \cap p'^{-1}_{\alpha,i}(B''_{\alpha,i})$$

induit par $p_{\alpha,i}$ avec $B'_{\alpha,i}$ et $B''_{\alpha,i}$ de rayon assez petit.

Si ξ est un point assez général de $B''_{\alpha,i}$ le Corollaire (2.3.10) nous dit précisément que la caractéristique d'Euler-Poincaré de $U_{\alpha,i} = p_{\alpha,i}^{-1}(B'_{\alpha,i} \cap p'^{-1}_{\alpha,i}(\xi)) \cap B_{\alpha,i} \cap X$ est égale à $\chi'_{\alpha,i}$. Le morphisme descriptible $\Psi_{\alpha,i}$ induit un morphisme descriptible $\tilde{\Psi}_{\alpha,i}$ de $U_{\alpha,i}$ sur $U'_{\alpha,i}$, où $U'_{\alpha,i} = p'^{-1}_{\alpha,i}(\xi) \cap B'_{\alpha,i}$.

Par ailleurs l'image Y_α de X_α par $p_{\alpha,i}$ est une hypersurface de $\mathbf{C}^{d_\alpha+1}$ qui est non singulière en $y_{\alpha,i} = p_{\alpha,i}(x_{\alpha,i})$. Puisque p est choisi de telle sorte que $\Gamma_{\alpha,i}$ est non singulière en $x_{\alpha,i}$ la restriction de p à X_α a un point critique quadratique ordinaire en $x_{\alpha,i}$ et par conséquent, $y_{\alpha,i}$ est un point quadratique ordinaire de la restriction de $p'_{\alpha,i}$ à Y_α. Les points généraux de $\tilde{\Psi}_{\alpha,i}\colon U_{\alpha,i} \to U'_{\alpha,i}$ sont ceux de $p'^{-1}_{\alpha,i}(\xi) \cap B'_{\alpha,i} - Y_\alpha$ et les fibres générales de $\tilde{\Psi}_{\alpha,i}$ ont pour caractéristique d'Euler-Poincaré $\chi_{d_\alpha+1}(X, X_\alpha)$, tandis que les fibres de $\tilde{\Psi}_{\alpha,i}$ au-dessus de $p'^{-1}_{\alpha,i}(\xi) \cap B'_{\alpha,i} \cap Y_\alpha$ sont homéomorphes et contractiles (cf. (3.3.2)) et la Proposition (2.1.3) donne dans ce cas

$$\chi(U_{\alpha,i}) = 1 + (-1)^{d_\alpha-2} + (-1)^{d_\alpha-1}\chi_{d_\alpha+1}(X, X_\alpha)$$

puisque $p'^{-1}_{\alpha,i}(\xi) \cap B'_{\alpha,i} \cap Y_\alpha$ est homéomorphe à la fibre de Milnor d'une singularité quadratique ordinaire de dimension $d_\alpha - 2$. On a donc

$$\chi'_{\alpha,i} = 1 + (-1)^{d_\alpha - 2} + (-1)^{d_\alpha - 1}\chi_{d_\alpha+1}(X, X_\alpha)$$

puisque $\chi'_{\alpha,i} = \chi(U_{\alpha,i})$ comme nous l'avons vu plus haut.

Finalement la formule de (4.2.2.1) en résulte, puisque (4.2.2.2) donne

$$\begin{aligned}\chi_{\alpha,i} &= \chi_2(X, \{x\}) - \chi'_{\alpha,i} + 1\\ &= \chi_2(X, \{x\}) + (-1)^{d_\alpha}\chi_{d_\alpha+1}(X, X_\alpha) + (-1)^{d_\alpha - 1}.\end{aligned}$$

Ceci et (4.2.1.1) donnent la formule de (4.1.1) cherchée quand $\{x\} = X_{\alpha_0}$.

(4.3) Dans le cas général, soit $I_{\alpha_0,i} = \{(H_{\alpha_0}, L_i) \in G_{d_{\alpha_0}} \times G_i \,|\, H_{\alpha_0} \supset L_i\}$ la variété d'incidence. C'est une variété algébrique irréductible, munie des deux projections $\pi_{\alpha_0}\colon I_{\alpha_0,i} \to G_{\alpha_0}$ et $\pi_i\colon I_{\alpha_0,i} \to G_i$. D'après les résultats généraux sur les variétés polaires [**Lê-Te**, 4.1.8; **Te1**, Chapitre IV], il existe un ouvert de Zariski dense $U \subset G_{d_{\alpha_0}}$ tel que, pour $H_{\alpha_0} \in U$, on ait $H_{\alpha_0} \cap X_{\alpha_0} = \{x\}$, et pour tout $\alpha \in A$

$$m_x\Big(P_{d_\alpha - d_{\alpha_0} - 1}\big(\overline{X}_\alpha \cap H_{\alpha_0}, x\big)\Big) = m_x\Big(P_{d_\alpha - d_{\alpha_0} - 1}\big(\overline{X}_\alpha, x\big)\Big).$$

D'autre part d'après (2.3.9), il existe un ouvert de Zariski dense $V \subset G_i$ tel que, pour $L \in V$, on ait l'égalité: $\chi(X \cap B_t \cap (L + u)) = \chi_i(X, X_{\alpha_0})$, pour $u \in \mathbf{C}^{N+1} - L$ assez petit. Comme $I_{\alpha_0,i}$ est irréductible, l'ouvert $\pi_{\alpha_0}^{-1}(U) \cap \pi_i^{-1}(V)$ est dense dans $I_{\alpha_0,i}$ et l'application du théorème dans le cas particulier où $X_{\alpha_0} = \{x\}$ à l'intersection $X \cap H_{\alpha_0}$ fournit l'égalité cherchée dans le cas général.

5. Les réciproques du théorème de Thom-Mather.

(5.1) Soient X un espace analytique complexe réduit et $(X_\alpha)_{\alpha \in A}$ une stratification analytique complexe de X.

Rappelons les diverses conditions d'incidence utilisées:

(5.1.1) (**Whitney en x**) Le couple de strates (X_β, X_α) satisfait la condition de Whitney (cf. (1.2.4)) en un point x de X_α.

(5.1.2) (**Whitney au voisinage de x**) Le couple de strates (X_β, X_α) satisfait la condition de Whitney en tout point d'un voisinage de x dans X_α. Dans ce cas on écrira que (X_β, X_α) vérifie $W_x(X_\beta, X_\alpha)$.

(5.1.3) (**Whitney au voisinage de x**)* Pour tout plongement local de (X, x) dans $(\mathbf{C}^{N+1}, 0)$ et, pour tout entier $0 \leqslant i \leqslant N - d_\alpha$, il existe un ouvert dense U_i de la Grassmannienne des plans de codimension i de $\mathbf{C}^N$ contenant $T_x X_\alpha$ tel que, pour tout espace non singulier H de $\mathbf{C}^{N+1}$ contenant X_α et tel que $T_x H \in U_i$, on ait $W_x(X_\beta \cap H, X_\alpha)$. Dans ce cas on dira que (X_β, X_α) satisfait $W^*_x(X_\beta, X_\alpha)$.

(5.1.4) (**Equimultiplicité polaire**) L'application de X_α dans $\mathbf{N}^{d_\beta}$, qui, à $x \in X_\alpha$, associe la suite $M^*_{\overline{X}_\beta, x}$ des multiplicités en x des variétés polaires locales générales de $\overline{X}_\beta$, est constante au voisinage de x dans X_α. Dans ce cas on dira que (X_β, X_α) satisfait $(M^* \text{ constant})_x(X_\beta, X_\alpha)$.

(5.1.5) **(Equiévanescence)** Pour tout plongement local de (X, x) dans $(\mathbf{C}^{N+1}, 0)$ et, pour tout entier $1 \leqslant j \leqslant d_\alpha$, il existe un ouvert Ω_j de l'espace des projections linéaires de $\mathbf{C}^{N+1}$ sur $\mathbf{C}^j$ tel que, pour tout $p \in \Omega_j$, il existe $\varepsilon_0 > 0$, tel que, pour tout ε, $0 < \varepsilon \leqslant \varepsilon_0$, il existe η_ε, tel que pour tout η, $0 < \eta \leqslant \eta_\varepsilon$, le morphisme $\varphi_{\varepsilon,\eta,\beta}\colon \overline{X}_\beta \cap B_\varepsilon \cap p^{-1}(B'_\eta) \to B'_\eta$, induit par p, est une fibration topologique localement triviale dont les fibres sont contractiles. Dans ce cas on dira que (X_β, X_α) satisfait la condition $EV_x(X_\beta, X_\alpha)$.

(5.1.6) **(Equiévanescence)*** Pour tout plongement local de (X, x) dans $(\mathbf{C}^{N+1}, 0)$ et, pour tout entier $1 \leqslant j \leqslant d_\alpha$ et pour tout entier $0 \leqslant i \leqslant N - d_\alpha$, il existe un ouvert dense $V_{i,j}$ de l'espace produit de celui des projections linéaires de $\mathbf{C}^{N+1}$ sur $\mathbf{C}^j$ et de la Grassmannienne des plans de codimension i de $\mathbf{C}^{N+1}$ contenant $T_x X_\alpha$ tel que, pour toute projection p et tout espace non singulier H de $\mathbf{C}^{N+1}$ contenant X_α et tels que le couple $(p, T_x H) \in V_{i,j}$, on ait la condition d'équiévanescence de (5.1.5) de $X_\beta \cap H$ le long de X_α en x. Dans ce cas on dira que (X_β, X_α) satisfait la condition $EV_x^*(X_\beta, X_\alpha)$.

(5.1.7) **(Equiévanescence numérique)*** Pour tout plongement local (X, x) dans $(\mathbf{C}^{N+1}, 0)$, pour tout entier $1 \leqslant j \leqslant d_\alpha$ et pour tout entier $0 \leqslant i \leqslant N - d_\alpha$, il existe un ouvert dense $W_{i,j}$ de l'espace produit de celui des projections linéaires de $\mathbf{C}^{N+1}$ sur $\mathbf{C}^j$ et de la Grassmannienne des plans de codimension i de $\mathbf{C}^N$ contenant $T_x X_\alpha$ tel que, pour toute projection p et tout espace non singulier H de $\mathbf{C}^{N+1}$ contenant X_α et tels que le couple $(p, T_x H) \in W_{i,j}$, il existe $\varepsilon_0 > 0$, tel que, pour tout ε, $0 < \varepsilon \leqslant \varepsilon_0$, il existe η_ε, tel que, pour tout η, $0 < \eta \leqslant \eta_\varepsilon$, les fibres du morphisme $\varphi_{\varepsilon,\eta,\beta,H}\colon \overline{X}_\beta \cap H \cap B_\varepsilon \cap p^{-1}(B'_\eta) \to B'_\eta$ induit par p, aient une caractéristique d'Euler-Poincaré égale à un. Dans ce cas on dira que le couple (X_β, X_α) satisfait la condition $EVN_x^*(X_\beta, X_\alpha)$.

(5.1.8) **(χ^* constant)** Soient (X_β, X_α) un couple de strates, et $x \in X_\alpha$. Pour tout plongement local $(X, x) \subset (\mathbf{C}^{N+1}, 0)$, l'application $\chi^*_{\overline{X}_\beta}$ qui à tout point $z \in X_\alpha$ au voisinage de x associe la suite $\chi^*(\overline{X}_{\beta,z}) = (\chi_1(\overline{X}_\beta, \{z\}), \ldots, \chi_{N-d_\beta}(\overline{X}_\beta, \{z\})) \in \mathbf{Z}^{N-d_\beta}$ (cf. (3.3.2)), est constante sur X_α au voisinage de x. On dira que X_β satisfait la condition "χ^* constant" le long de X_α en x, ou que (X_β, X_α) satisfait la condition $(\chi^* \text{ constant})_x\,(X_\beta, X_\alpha)$.

(5.1.9) **(Equisingularité topologique)** Pour tout couple de strates (X_β, X_α), soit $x \in X_\alpha$; pour tout plongement local $(X, x) \subset (\mathbf{C}^{N+1}, 0)$, il existe une rétraction locale $r\colon (\mathbf{C}^{N+1}, 0) \to (X_\alpha, x)$ et $\varepsilon_0 > 0$, tels que, pour tout ε, $0 < \varepsilon \leqslant \varepsilon_0$, il existe η_ε tel que, pour tout η, $0 < \eta \leqslant \eta_\varepsilon$, on ait un homéomorphisme de $B_\varepsilon \cap r^{-1}(B_\eta \cap X_\alpha)$ sur $(r^{-1}(x) \cap B_\varepsilon) \times (X_\alpha \cap B_\eta)$ compatible avec la rétraction r et qui induit un homéomorphisme de $\overline{X}_\beta \cap B_\varepsilon \cap r^{-1}(B_\eta \cap X_\alpha)$ sur $(\overline{X}_\beta \cap r^{-1}(x) \cap B_\varepsilon) \times (X_\alpha \cap B_\eta)$. On dira que l'espace $\overline{X}_\beta$ est topologiquement équisingulier le long de X_α en x: dans ce cas on notera $TT_x(X_\beta, X_\alpha)$.

(5.1.10) **(Equisingularité topologique)*** Pour tout couple de strates (X_β, X_α), soit $x \in X_\alpha$; pour tout plongement local $(X, x) \subset (\mathbf{C}^{N+1}, 0)$, pour tout j, $1 \leqslant j \leqslant N - d_\alpha + 1$, il existe un ouvert dense $\mathfrak{U}_j$ de la Grassmannienne des sous-espaces de codimension j de $\mathbf{C}^{N+1}$ contenant $T_x X_\alpha$ tel que, pour tout sous-espace non singulier H, contenant X_α et tel que $T_x H \in \mathfrak{U}_j$, on a que $\overline{X}_\beta \cap H$ est topologiquement équisingulier le long de X_α en x. Dans ce cas l'on dira que (X_β, X_α) satisfait la condition $TT_x^*(X_\beta, X_\alpha)$.

(5.1.11) *Notations.* Etant donnée une condition d'incidence $C_x(X_\beta, X_\alpha)$ où $C = W$ ou W^* ou EV etc., nous dirons qu'une stratification $X = \bigcup_{\alpha \in A} X_\alpha$ satisfait la condition (C) si $C_x(X_\beta, X_\alpha)$ est réalisée pour tous les triplets (X_β, X_α, x) tels que $x \in X_\alpha$.

(5.2) *Le théorème de Thom-Mather* (cf. [**Th**, **Ma1**, **Ma2**]). On a

(5.2.1) THÉORÈME (THOM-MATHER). *Avec les notations de* (5.1), *si tout couple de strates* (X_β, X_α) *satisfait la condition* $W_x(X_\beta, X_\alpha)$ *de* (5.1.2) *ci-dessus, alors tout couple de strates* (X_β, X_α) *satisfait la condition* $TT_x(X_\beta, X_\alpha)$ *de* (5.1.9) *en tout point* x *de* X_α.

(5.2.2) REMARQUE. Si $\overline{X}_\beta$ est topologiquement équisingulier le long de X_α en x, le couple (X_β, X_α) satisfait la condition $EV_x(X_\beta, X_\alpha)$ de (5.1.5).

En résumé on peut noter

$$(W) \Rightarrow (TT) \Rightarrow (EV)$$

où la propriété (P) signifie que l'on a $P_x(X_\beta, X_\alpha)$ pour tout couple de strates (X_β, X_α) d'une stratification donnée $(X_\alpha)_{\alpha \in A}$ de X en tout point x de X_α.

(5.2.3) D'après [**Te1**, Chapitre V, Remarque 1.3], soit (X_β, X_α) un couple de strates qui satisfait $W_x(X_\beta, X_\alpha)$ en $x \in X_\alpha$ (cf. (5.1.2)), alors (X_β, X_α) satisfait $W_x^*(X_\beta, X_\alpha)$. On a donc en fait:

$$\begin{array}{ccccc} (W) \Rightarrow (W^*) \Rightarrow & (TT^*) & \Rightarrow & (EV^*) \\ & \Downarrow & & \Downarrow \\ (W) \searrow\!\!\searrow & (TT) & \Rightarrow & (EV) \end{array}$$

De plus d'après [**Te1**, Chapitre V, Théorème 1.2] on a: $(W) \Leftrightarrow (M^* \text{ constant})$.

(5.3) *Les réciproques du théorème de Thom-Mather.* Il est faux que (TT) implique (W) (cf. [**B-S**]). Par conséquent le théorème de Thom-Mather (5.2.1) n'admet pas de réciproque. Ce que nous appellerons réciproque du théorème de Thom-Mather est l'implication $(TT^*) \Rightarrow (W)$ que nous allons maintenant démontrer. Nous allons en fait démontrer plus:

(5.3.1) THÉORÈME. *Soient X un espace analytique complexe réduit et $(X_\alpha)_{\alpha \in A}$ une stratification analytique complexe de X. Les conditions suivantes sont équivalentes*:

(i) *la stratification $(X_\alpha)_{\alpha \in A}$ de X satisfait la propriété de frontière et pour tout couple de strates (X_β, X_α) et tout point $x \in X_\alpha$, le couple (X_β, X_α) satisfait la condition de Whitney en x*;

(ii) *on a (W^*) pour la stratification $(X_\alpha)_{\alpha \in A}$*;

(iii) *on a (TT^*) pour la stratification $(X_\alpha)_{\alpha \in A}$*;

(iv) *on a (EV^*) et la propriété de frontière pour la stratification $(X_\alpha)_{\alpha \in A}$*;

(v) *on a (EVN^*) et la propriété de frontière pour la stratification $(X_\alpha)_{\alpha \in A}$*;

(vi) *on a $(\chi^* \text{ constant})$ pour la stratification $(X_\alpha)_{\alpha \in A}$*;

(vii) *on a $(M^* \text{ constant})$ pour la stratification $(X_\alpha)_{\alpha \in A}$.*

DÉMONSTRATION. Comme nous avons vu plus haut que (i) $\Leftrightarrow$ (vii) d'après [**Te1**, Chaptire V, Théorème (1.2)] et que (i) $\Leftrightarrow$ (ii) $\Rightarrow$ (iii) $\Rightarrow$ (iv). il nous reste à montrer que (iv) $\Rightarrow$ (v) $\Rightarrow$ (vi) $\Rightarrow$ (vii).

Or, (iv) $\Rightarrow$ (v) est évident.

Montrons l'implication (v) $\Rightarrow$ (vii).

Soit (X_β, X_α) un couple de strates telles que $X_\alpha \subset \overline{X}_\beta$ et soit $x \in X_\alpha$. Nous allons montrer que les variétés polaires en x de l'adhérence $\overline{X}_\beta$ sont équimultiples le long de X_α au voisinage de x. Ceci donnera (vii) d'après [**Te1**, Chapitre V, (1.1.2)].

(5.3.1.1) Soit $(X, x) \subset (\mathbf{C}^{N+1}, 0)$ un plongement local de X en x. En utilisant (2.3.9), il existe un ouvert de Zariski dense Ω de l'espace des projections linéaires de $\mathbf{C}^{N+1}$ sur $\mathbf{C}^{i+1}$ $(1 \leqslant i \leqslant d_\beta)$ tel que, pour $p \in \Omega$, il existe un ouvert dense Ω' de l'espace des projections linéaires de $\mathbf{C}^{i+1}$ sur $\mathbf{C}^i$ tel que, pour un choix convenable de $(\eta', \eta, \varepsilon)$, et $q \in \Omega'$ on ait:

(1) le morphisme $\varphi_{\varepsilon,\eta,\beta}$: $B_\varepsilon \cap p^{-1}(B'_\eta) \cap \overline{X}_\beta \to B'_\eta$ induit par p est descriptible. Soit Δ l'image par p de l'union des variétés polaires relativement à p des fermetures de strates $\overline{X}_{\beta'}$ telles que $X_\alpha \subset \overline{X}_{\beta'} \subset \overline{X}_\beta$; Nous appellons Δ le *discriminant* de $\varphi_{\varepsilon,\eta,\beta}$;

(2) la projection linéaire q restreinte à $\Delta \cap B'_\eta \cap q^{-1}(B'_\eta)$, induit un morphisme descriptible $\Psi_{\eta,\eta',\beta}$ au-dessus de $B''_{\eta'} \subset \mathbf{C}^i$;

(3) le morphisme $\theta_{\varepsilon,\eta,\eta',\beta}$: $B_\varepsilon \cap p^{-1}(B'_\eta \cap q^{-1}(B''_{\eta'})) \cap \overline{X}_\beta \to B''_{\eta'}$ induit par $q \circ p$ est descriptible.

Considérons d'abord le cas où $1 \leqslant i \leqslant d_\alpha$. Dans ce cas on choisit $(\eta', \eta, \varepsilon)$ de telle sorte que:

(4) pour un point ξ de B''_η assez général pour $\theta_{\varepsilon,\eta,\eta',\beta}$, la fibre $q^{-1}(\xi) \cap B'_\eta$ coupe transversalement Δ et la fibre de $\varphi_{\varepsilon,\eta,\beta}$ au-dessus de tout point de $\varphi_{\varepsilon,\eta,\beta}(X_\alpha) \cap B'_\eta$ ainsi que $\varphi^{-1}_{\varepsilon,\eta,\beta}(q^{-1}(\xi) \cap B'_\eta)$ sont de caractéristique d'Euler-Poincaré égale à 1 d'après l'hypothèse (v) de (5.3.1).

Fixons $(\eta', \eta, \varepsilon)$.

Posons $\varphi = \varphi_{\varepsilon,\eta,\beta}$, $\Psi = \Psi_{\eta',\eta,\beta}$, $\theta = \theta_{\varepsilon,\eta,\eta',\beta}$.

Notons $Z = \varphi^{-1}(q^{-1}(\xi) \cap B'_\eta)$ et $\chi(Z) = 1$, $S = q^{-1}(\xi) \cap B'_\eta$ et φ_S: $Z \to S$ le morphisme descriptible induit par φ. Soit $\Delta_S = \Delta \cap S$.

Montrons par récurrence sur $\dim X_\beta - \dim X_\alpha$ que $\Delta = \varnothing$. Ceci est évident si $\dim X_\beta = \dim X_\alpha$, puisque X_α est non singulier en x. Supposons donc que, $\dim X_\beta - \dim X_\alpha = k + 1$ et pour tout β', tels que $\dim X_{\beta'} - \dim X_\alpha \leqslant k$ avec $k \geqslant 0$, le discriminant (défini comme dans (5.3.1.1)(1)) de la restriction à $\overline{X}_{\beta'}$ d'une projection sur $\mathbf{C}^{i+1}$ assez générale, telle que p, soit vide. Le discriminant Δ est alors l'image par φ de la variété polaire de dimension i de $\overline{X}_\beta \cap B_\varepsilon \cap p^{-1}(B'_\eta)$ relativement à la projection linéaire p.

Si $1 \leqslant i \leqslant d_\alpha - 1$, la fibre générale de φ_S est de caractéristique d'Euler-Poincaré égale à 1 et on a, pour tout $s \in \Delta_S$

$$\chi(\varphi_S^{-1}(s)) = 1 + (-1)^{d_\beta - i}$$

car au-dessus de s, il n'y a qu'un point critique de φ_S quand p est assez général (cf. (4.1.2)(iv)) et ce point critique est un point critique quadratique ordinaire de la restriction de φ_S à $Z \cap X_\beta$. La Proposition (2.1.3) donne, si Δ est non vide, avec $\mathrm{Card}(S \cap \Delta) = m_0 \neq 0$;

$$\chi(Z) = 1 - m_0 + m_0\left(1 + (-1)^{d_\beta - i}\right) = 1 + m_0(-1)^{d_\beta - i}$$

ce qui est contradictoire avec $\chi(Z) = 1$. Il faut donc que $m_0 = 0$, d'où le résultat cherché pour les variétés polaires de $\overline{X}_\beta$ en x qui sont de dimension $i < d_\alpha$, puisqu'elles sont vides!

Si $i = d_\alpha$, la fibre générale de φ_s est de caractéristique d'Euler-Poincaré égale à un certain entier χ_0 et on a, pour tout $s \in \Delta_s$

$$\chi\left(\varphi^{-1}(s)\right) = \chi_0 + (-1)^{d_\beta - d_\alpha}$$

pour les mêmes raisons que ci-dessus. La fibre au-dessus d'un point de $S \cap p(X_\alpha)$ est de caractéristique d'Euler-Poincaré égale à 1. La Proposition (2.1.3) donne dans ce cas

$$\chi(Z) = (1 - m_0 - 1)\chi_0 + m_0\left(\chi_0 + (-1)^{d_\beta - d_\alpha}\right) + 1 = 1 + m_0(-1)^{d_\beta - d_\alpha}$$

et on conclut, comme ci-dessus, que la variété polaire de $\overline{X}_\beta$ en x de dimension d_α est vide.

Si $i \geqslant d_\alpha + 1$, d'après [**Te1**, Chapitre I, 5.5], pour montrer que la variété polaire de dimension i de $\overline{X}_\beta$ en x est équimultiple le long de X_α au voisinage de x, il suffit de montrer qu'un espace non singulier H de codimension $i - d_\alpha$ qui contient X_α et dont l'espace tangent $T_x H$ est assez général, intersecte cette variété polaire exactement selon X_α au voisinage de x, si cette variété polaire n'est pas vide.

Puisque la projection p est assez générale, l'image inverse par p d'un sous-espace H' non singulier de $\mathbf{C}^{i+1}$ et de dimension $d_\alpha + 1$ qui contient $p(X_\alpha)$ et dont l'espace tangent $T_{p(x)}\,H'$ est assez général est un espace non singulier H de codimension $i - d_\alpha$ qui contient X_α et dont le plan tangent $T_x H$ est assez général. Puisque $(W) \Rightarrow (W^*)$, la situation du couple $(X_\beta \cap H, X_\alpha)$ est analogue à la situation précédente où $i = d_\alpha$. On en déduit que l'on a $(P_{d_\beta - i}(\overline{X}_\beta, x) - X_\alpha) \cap H = \varnothing$ d'où l'équimultiplicité cherchée.

Ceci achève la démonstration de (v) $\Rightarrow$ (vii).

Il nous reste à montrer (vi) $\Leftrightarrow$ (vii).

L'implication (vii) $\Rightarrow$ (vi) provient de ce que (vii) et (i) sont équivalents d'après [**Te1**, Chapitre V, Théorème (1.2)] et que (i) implique (vi) d'après (3.3.2).

Pour établir que (vi) implique (vii), considérons un couple de strates (X_β, X_α) tel que $X_\alpha \subset \overline{X}_\beta$ et soit $\overline{X}_\beta = \bigcup_{\beta' \in B} X_{\beta'}$ la stratification de $\overline{X}_\beta$ induite par celle de X. Il suffit de montrer que cette stratification est une stratification de Whitney. Pour cela nous procédons par récurrence sur $\dim X_\beta - \dim X_\alpha$.

Si $\dim X_\beta - \dim X_\alpha = 0$, il n'y a rien à démontrer. Supposons l'implication (vi) $\Rightarrow$ (vii) démontrée pour les couples $(X_{\beta'}, X_{\alpha'})$ avec $X_{\alpha'} \subset \overline{X}_{\beta'}$ tels que $\dim X_{\beta'} - \dim X_{\alpha'} < \dim X_\beta - \dim X_\alpha$. Ceci implique d'une part que $\bigcup_{\beta' \in B, \beta' \neq \beta} \overline{X}_{\beta'}$ est une stratification de Whitney analytique complexe et d'autre part que $\bigcup_{\beta' \in B, \beta' \neq \alpha} X_{\beta'}$ est aussi une stratification de Whitney. Il nous reste à démontrer que X_β satisfait la condition de Whitney le long de X_α en tout point de X_α. D'après [**Te1**, Chapitre V, Théorème (1.2)] ceci revient à démontrer que l'on a $(M^* \text{ constant})_x (X_\beta, X_\alpha)$ en tout point $x \in X_\alpha$.

Soit $Z \subset X_\alpha$ le sous-ensemble analytique formé des points de X_α où (X_β, X_α) ne satisfait pas la condition de Whitney, i.e. où $(M^* \text{ constant})_x (X_\beta, X_\alpha)$ n'est pas satisfaite (cf. [**Te1**, Chapitre VI, (2.1)] et comparer à [**Lê-Te**, (6.1.5)]). Nous pouvons stratifier $Z = \bigcup_{\gamma \in C} Z_\gamma$ de telle façon que la stratification

$$\overline{X}_\beta = \bigcup_{\substack{\beta' \in B \\ \beta' \neq \alpha}} X_{\beta'} \cup (X_\alpha - Z) \bigcup_{\gamma \in C} Z_\gamma$$

soit une stratification de Whitney analytique complexe. Soit $x \in Z$ et soit Z_γ une strate de dimension maximum parmi celles auxquelles x est adhérent. Nous allons montrer que $Z_\gamma = \varnothing$, ce qui impliquera que $Z = \varnothing$.

Supposons $Z_\gamma \neq \varnothing$ et soit $y \in Z_\gamma$. Posons pour simplifier $Z_\gamma = X_{\alpha'}$. Considérons un voisinage U de y dans $\overline{X}_\beta$ tel que $U \cap Z_{\gamma'} = \varnothing$, pour tout $\gamma' \in C, \gamma' \neq \gamma$. Nous allons montrer que, si U est assez petit,

$$(M^* \text{ constant})_z (X_\beta, X_\alpha)$$

est satisfaite en tout point $z \in U$, ce qui impliquera $U \cap Z_\gamma = \varnothing$ et donc $Z = \varnothing$. Puisque le couple $(X_\beta \cap U, X_{\alpha'} \cap U)$ satisfait la condition de Whitney, on a $P_k(\overline{X}_\beta, z) = \varnothing$ pour tout $k \geqslant d_\beta - d_\alpha$, et tout $z \in X_{\alpha'} \cap U$ et donc en tout point $z \in X_\alpha \cap U$ puisque $d_\alpha > d_{\alpha'}$ (cf. [**Te1**, Chapitre V, Théorème (1.2)]).

Pour établir la constance de $m_z(P_k(\overline{X}_\beta, z))$ pour tout $z \in X_\alpha \cap U$, nous utilisons la formule du Théorème (4.1.1) appliquée à la stratification de Whitney analytique complexe

$$\overline{X}_\beta \cap U = \bigcup_{\substack{\beta' \neq \alpha \\ X_{\beta'} \cap U \neq \varnothing}} (X_{\beta'} \cap U) \cup ((X_\alpha - X_{\alpha'}) \cap U) \cup (X_{\alpha'} \cap U)$$

quand U est assez petit:

(5.3.1.2)

$$\begin{cases} \chi_{d_{\alpha'}+1}\left(\overline{X}_\beta \cap U, X_{\alpha'} \cap U\right) - \chi_{d_{\alpha'}+2}\left(\overline{X}_\beta \cap U, X_{\alpha'} \cap U\right) \\ \quad = \displaystyle\sum_{\substack{\beta' \neq \alpha' \\ X_{\beta'} \cap U \neq \varnothing}} (-1)^{d_{\beta'} - d_{\alpha'} - 1} m_z\left(P_{d_{\beta'} - d_{\alpha'} - 1}\left(\overline{X}_{\beta'}, z\right)\right)\left(1 - \chi_{d_{\beta'}+1}\left(\overline{X}_\beta, X_{\beta'}\right)\right) \end{cases}$$

pour tout $z \in X_{\alpha'} \cap U$. D'après l'hypothèse (vi)

$$\chi_{d_{\alpha'}+i}\left(\overline{X}_\beta \cap U, X_{\alpha'} \cap U\right) = \chi_{d_{\alpha'}+i}\left(\overline{X}_\beta, X_\alpha\right) \quad \text{pour tout } i.$$

Donc si $d_{\alpha'} + i \leqslant d_\alpha$ on a

$$\chi_{d_{\alpha'}+i}\left(\overline{X}_\beta \cap U, X_{\alpha'} \cap U\right) = 1$$

puisque

$$\chi_k\big(\overline{X}_\beta \cap U, X_\alpha \cap U\big) = 1,$$

pour $0 \leqslant k \leqslant d_\alpha$. Par ailleurs $m_z(P_{d_{\beta'}-d_{\alpha'}-1}(\overline{X}_{\beta'}, z))$ est constant sur $U \cap X_\alpha$ pour tout $\beta' \neq \beta$.

Choisissons $z \in U - X_{\alpha'}$ et soit $(X_{\alpha''}, z)$ un germe en z de sous-espace non singulier contenu dans X_α et de dimension $d_{\alpha'}$. Dans un ouvert V de z dans $\mathbf{C}^{N+1}$ assez petit, la stratification analytique complexe

$$\overline{X}_\beta \cap V = \bigcup_{\substack{\beta' \in B - \{\alpha\} \\ X_{\beta'} \cap V \neq \varnothing}} (X_{\beta'} \cap V) \cup ((X_\alpha - X_{\alpha''}) \cap V) \cup (X_{\alpha''} \cap V)$$

est de Whitney et le Théorème (4.1.1) donne

(5.3.1.3)

$$\begin{cases} \chi_{d_{\alpha'}+1}\big(\overline{X}_\beta \cap V, X_{\alpha''} \cap V\big) - \chi_{d_{\alpha'}+2}\big(\overline{X}_\beta \cap V, X_{\alpha''} \cap V\big) \\ = \displaystyle\sum_{\substack{\beta' \neq \alpha' \\ X_{\beta'} \cap V \neq \varnothing}} (-1)^{d_{\beta'}-d_{\alpha'}-1} m_{z'}\big(P_{d_{\beta'}-d_{\alpha'}-1}\big(\overline{X}_{\beta'}, z'\big)\big)\big(1 - \chi_{d_{\beta'+1}}\big(\overline{X}_\beta, X_{\beta'}\big)\big) \end{cases}$$

pour tout $z' \in X_{\alpha''} \cap V$.

L'hypothèse (vi) donne $\chi_{d_{\alpha'}+i}(\overline{X}_\beta \cap V, X_{\alpha''} \cap V) = \chi_{d_{\alpha'}+i}(\overline{X}_\beta, X_\alpha)$.

Si $d_{\alpha'} + 1 < d_\alpha$, $m_{z'}(P_{d_\alpha - d_{\alpha'}-1}(X_\alpha, z')) = 0$ puisque X_α est non singulier en z' et de plus on a

$$\chi_{d_{\alpha'}+1}\big(\overline{X}_\beta, X_\alpha\big) = \chi_{d_{\alpha'}+2}\big(\overline{X}_\beta, X_\alpha\big) = 1.$$

Les formules (5.3.1.2) et (5.3.1.3) donnent respectivement:

$$(-1)^{d_\beta - d_{\alpha'}} m_z\big(P_{d_\beta - d_{\alpha'}-1}\big(\overline{X}_\beta, z\big)\big) = \sum_{\substack{\beta' \neq \beta \\ \beta' \in B - \{\alpha\}}} (-1)^{d_{\beta'}-d_{\alpha'}-1} m_z\big(P_{d_{\beta'}-d_{\alpha'}-1}\big(\overline{X}_{\beta'}, z\big)\big)\big(1 - \chi_{d_{\beta'}+1}\big(\overline{X}_\beta, X_{\beta'}\big)\big)$$

où $B = \{\beta', X_{\beta'} \cap U \neq \varnothing\}$, pour tout $z \in X_{\alpha'} \cap U$, et

$$(-1)^{d_\beta - d_{\alpha'}} m_z\big(P_{d_\beta - d_{\alpha'}-1}\big(\overline{X}_\beta z'\big)\big) = \sum_{\substack{\beta' \neq \beta \\ \beta' \in B' - \{\alpha\}}} (-1)^{d_{\beta'}-d_{\alpha'}-1} m_{z'}\big(P_{d_{\beta'}-d_{\alpha'}-1}\big(\overline{X}_{\beta'}, z'\big)\big)\big(1 - \chi_{d_{\beta'}+1}\big(\overline{X}_{\beta'}, X_{\beta'}\big)\big)$$

où $B' = \{\beta', X_{\beta'} \cap V \neq \varnothing\}$, pour tout $z' \in X_{\alpha''} \cap V$.

Pour un choix convenable de U on a $B = B'$ et la comparaison entre les deux formules donne

$$m_z\big(P_{d_\beta - d_{\alpha'}-1}\big(\overline{X}_\beta, z\big)\big) = m_{z'}\big(P_{d_\beta - d_{\alpha'}-1}\big(\overline{X}_\beta, z'\big)\big)$$

pour tout $z \in X_{\alpha'} \cap U$ et tout $z' \in X_{\alpha''} \cap V$, parce que l'hypothèse de récurrence a donné $m_z(P_{d_{\beta'}-d_{\alpha'}-1}(\overline{X}_{\beta'}, z))$ égal à $m_{z'}(P_{d_{\beta'}-d_{\alpha'}-1}(\overline{X}_{\beta'}, z'))$ quand $\beta' \neq \beta$ et

$\beta' \in B - \{\alpha\}$. On en déduit que la valeur de $m_z(P_{d_\beta - d_{\alpha'} - 1}(\overline{X}_\beta, z))$ est constante sur $X_\alpha \cap U$, puisque z' est un point général de X_α.

Si $d_{\alpha'} + 1 = d_\alpha$, $m_{z'}(P_{d_\alpha - d_{\alpha'} - 1}(X_\alpha, z')) = 1$ pour $z' \in X_\alpha \cap V$ puisque X_α est non singulier en z'. La formule (5.3.1.2) donne

$$
\begin{aligned}
1 - \chi_{d_\alpha + 1}(\overline{X}_\beta, X_\alpha) &= (-1)^{d_\beta - d_\alpha} m_z\big(P_{d_\beta - d_\alpha}(\overline{X}_\beta, z)\big) \\
&+ \sum_{\substack{\beta' \neq \beta \\ \beta' \in B - \{\alpha\}}} (-1)^{d_{\beta'} - d_\alpha} m_z\big(P_{d_{\beta'} - d_\alpha}(\overline{X}_{\beta'}, z)\big)\big(1 - \chi_{d_{\beta'} + 1}(\overline{X}_\beta, X_{\beta'})\big) \\
&+ m_z(P_0(X_\alpha, z))\big(1 - \chi_{d_\alpha + 1}(\overline{X}_\beta, X_\alpha)\big)
\end{aligned}
$$

pour tout $z \in X_{\alpha'} \cap U$.

La formule (5.3.1.3) donne

$$
\begin{aligned}
1 - \chi_{d_\alpha + 1}(\overline{X}_\beta, X_\alpha) &= (-1)^{d_\beta - d_\alpha} m_{z'}\big(P_{d_\beta - d_\alpha}(\overline{X}_\beta, z')\big) \\
&+ \sum_{\substack{\beta' \neq \beta \\ \beta' \in B - \{\alpha\}}} (-1)^{d_\beta - d_\alpha} m_{z'}\big(P_{d_{\beta'} - d_\alpha}(\overline{X}_{\beta'}, z')\big)\big(1 - \chi_{d_{\beta'} + 1}(\overline{X}_\beta, X_{\beta'})\big) \\
&+ m_{z'}(P_0(X_\alpha, z'))\big(1 - \chi_{d_\alpha + 1}(\overline{X}_\beta, X_\alpha)\big)
\end{aligned}
$$

pour tout $z' \in X_{\alpha''} \cap V$.

En utilisant l'hypothèse de récurrence comme ci-dessus, on conclut que $m_z(P_{d_\beta - d_\alpha}(\overline{X}_\beta, z))$ est constant sur $X_\alpha \cap U$.

Ceci montre que, pour tout $k \geqslant d_\beta - d_{\alpha'} - 1$ les multiplicités $m_x(P_k(\overline{X}_\beta, x))$ sont constantes sur $X_\alpha \cap U$ (en fait on a remarqué qu'elles sont nulles si $k \geqslant d_\beta - d_{\alpha'}$).

Soient $x \in X_{\alpha'} \cap U$ et un plongement local de (X, x) dans $(\mathbf{C}^{N+1}m, 0)$. Quitte à rétrécir U on peut choisir un sous-espace non singulier Y_α de dimension $d_{\alpha'} + 1$ de X_α contenant $X_{\alpha'}$. La stratification

$$
\begin{aligned}
(5.3.1.3) \qquad \overline{X}_\beta \cap U = &\bigcup_{\beta' \in B - \{\alpha\}} (X_{\beta'} \cap U) \cup ((X_\alpha - Y_\alpha) \cap U) \\
&\cup ((Y_\alpha - X'_\alpha) \cap U) \cup (X'_\alpha \cap U)
\end{aligned}
$$

est une stratification de Whitney analytique complexe.

Soit H un sous-espace non singulier de U de dimension l, $d_{\alpha'} + 1 \leqslant l \leqslant N$ assez général qui contient Y_α.

Quitte à rétrécir encore U, la stratification induite sur $\overline{X}_\beta \cap H \cap U$ par la stratification (5.3.1.3) satisfait encore la condition (χ^* constant) de (vi). D'après ce qui précède et le fait déjà utilisé dans (5.2.3) que $(W) \Rightarrow (W^*)$, tous les couples de strates de la stratification précédente sauf peut-être $(\overline{X}_\beta \cap H \cap U, X_{\alpha'} \cap U)$ satisfont les conditions de Whitney. En remplaçant U par un voisinage d'un point général de $X_{\alpha'}$, d'après le théorème de Whitney **[W1]**, on peut enfin supposer que cette stratification est de Whitney. Le raisonnement ci-dessus montre alors que la multiplicité $m_z(P_{d_\beta - d_{\alpha'} - 1 - N + l}(\overline{X}_\beta \cap H, z))$ est constante lorsque z parcourt

$Y_\alpha \cap U$: la multiplicité $m_z(P_{d_\beta - d_{\alpha'} - 1 - N + l}(\overline{X}_\beta, z))$ est constante sur $X_\alpha \cap U$, puisque, d'après [**Lê-Te**, (4.1.8)] elle est égale à la multiplicité précédente en tout point de $\overline{X}_\beta \cap H \cap U$ et est constante sur $(X_\alpha - X'_\alpha) \cap U$ qui rencontre $Y_\alpha \cap U$. Comme on peut choisir $d_{\alpha'} + 1 \leqslant l \leqslant N$, on obtient l'équimultiplicité de toutes les variétés polaires de $\overline{X}_\beta$ le long de $X_\alpha \cap U$, ce qui contredit l'hypothèse $X_{\alpha'} \neq \varnothing$.

Ceci montre que (vi) entraine (vii) et achéve la démonstration du Théorème (5.3.1).

(5.3.2) COROLLAIRE. *Soit $(X_\alpha)_{\alpha \in D}$ une stratification analytique complexe d'un espace analytique complexe réduit X. C'est une stratification de Whitney si et seulement si pour chaque $\alpha \in A$, let type d'homotopie évanescent total* (*cf.* 3.3.2) *du germe (X, x) est constant pour $x \in X_\alpha$.*

BIBLIOGRAPHIE

[**B-S**] J. P. Brasselet et M. H. Schwartz, *Sur les classes de Chern d'un ensemble analytique complexe*, Astérisque **82–83** (1981), 93–147.

[**B-G-G**] J. Briançon, A. Galligo et J. M. Granger, *Déformations équisingulières des germes de courbes gauches réduites*, Mém. Soc. Math. France (N.S.) **1** (1980).

[**B-H-S**] J. Briançon, J. P. G. Henry et J. P. Speder, *Les conditions de Whitney en un point sont analytiques*, Note aux C. R. Acad. Sci. Paris Sér. A-B **282** (1976), 279.

[**Br-Sp1**] J. Briançon, et J. P. Speder, *La trivialité topologique n'implique pas les conditions de Whitney*, Note aux C. R. Acad. Sci. Paris Sér. A **280** (1975), 365.

[**Br-Sp2**] ______, *Les conditions de Whitney impliquent μ^* constant*, Ann. Inst. Fourier (Grenoble) **26** (1976), 153–163.

[**Bu**] R. O. Buchweitz, *On Zariski's criterion for equisingularity and non-smoothable monomial curves*, Thèse d'Etat, Paris VII (1981), voir aussi ce volume.

[**Bu-G**] R. O. Buchweitz and G. M. Greuel, *The Milnor number and deformations of complex curve singularities*, Invent. Math. **58** (1980), 241.

[**B-V**] D. Burghelea and A. Verona, *Local homological properties of analytic sets*, Manuscript Math. **7** (1972), 55–66.

[**Ch**] D. Cheniot, *Sur les sections transversales d'un ensemble stratifié*, Note aux C. R. Acad. Sci. Paris Sér. A-B **275** (1972), 915–916.

[**C-S**] G. Canuto et J. P. Speder, *Un critère d'éclatement pour les conditions de Whitney*, Ann. Scuola Norm. Sup. Pisa Cl. Sci. (3) **27** (1973).

[**D-F**] R. Draper and K. Fischer, *Derivations into the integral closure*, Preprint, George Mason University, 1981.

[**D**] A. Dubson, *Classes caractéristiques des variétés singulières*, Note aux C. R. Acad. Sci. Paris Sér. A **287** (1978), 237.

[**Do**] A. Dold, *Partitions of unity in the theory of fibrations*, Ann. of Math. (2) **78** (1963), 223–255.

[**G**] M. Goresky, *Triangulation of stratified objects*, Proc. Amer. Math. Soc. **72** (1978).

[**Ha**] R. M. Hardt, *Stratification of real analytic mappings and images*, Invent. Math. **28** (1975).

[**H1**] H. Hironaka, *Subanalytic sets*, Volume in Honour of K. Akizuki, Kinokuniya, Tokyo, 1973.

[**H2**] ______, *Introduciton to real-analytic sets and real analytic maps*, Quaderni del gruppo... Instituto L. Tonelli Via Buonarrotti, Pisa.

[**H3**] ______, *Stratification and flatness in "Real and complex singularities"*, Nordic Summer School, Oslo 1976, Sijthoff and Noordhoff, Alphen aan den Rijn, Netherlands 1977.

[**H4**] ______, *Triangulation of algebraic sets*, Proc. Amer. Math. Soc. Sympos. on Algebraic Geometry, Arcata, 1974, Proc. Sympos. Pure Math., Vol. 29, Amer. Math. Soc., Providence, R.I, 1975 pp. 165–185.

[**H5**] ______, *Equivalence and deformations of isolated singularities*, Woods Hole Seminar on Algebraic Geometry, 1964. (multigraphié)

[H6] ______, *Normal cones in analytic Whitney stratifications*, Inst. Hautes Etudes Sci. Publ. Math. **36** (1970), (volume dédié à Zariski).

[He-Lê] J. P. G. Henry et Lê D. T., *Limites d'espaces tangents*, Sém. Norguet, Lecture Notes in Math., vol. 482, Springer-Verlag, Berlin-Heidelberg-New York, 1975.

[He-M1] J. P. G. Henry et M. Merle, *Limites d'espaces tangents et transversalité de variétés polaires*, Actes Conf. sur les Singularités de La Rabida, Lecture Notes in Math., Springer-Verlag, Berlin and New York (à paraitre).

[He-M2] ______, *Limites de normales, conditions de Whitney et éclatement d'Hironaka*, Prépublication, Centre de Math. Ecole Polytechnique, Avril 1982.

[K] M. Kashiwara, Proc. Japan Acad. **49** (1973), 803–804. Voir aussi: Cours de Kashiwara à Paris-Nord, 1976–77.

[Lê1] Lê D. T., *Some remarks on relative monodromy*, In "Real and Complex Singularities", Nordic Summer School, Oslo 1976, Sijthoff and Noordhoff, 1977.

[Lê2] ______, *Sur un critère d'équisingularité*, Note C. R. Acad. Sci. Paris Sér. A-B **272** (1971), 138–140.

[Lê3] ______, *Calcul du nombre de cycles évanouissants d'une hypersurface complexe*, Ann. Inst. Fourier (Grenoble) **23** (1973), 261–270.

[Lê4] ______, *Calcul du nombre de Milnor d'une singularité isolée d'intersection complète*, Funkcional Anal. i Priložen. **8** (1974), 45–52.

[Lê5] ______, *Topological use of polar curves*, Proc. Sympos. Pure Math., Vol. 29, Amer. Math. Soc., Providence, R.I., 1974, pp. 507–512.

[Lê6] ______, *Vanishing cycles on analytic sets*, Proc. Conf. on Algebraic Analysis, Res. Inst. Math. Sci. Kyoto, July 1975.

[L-R] Lê D. T. and C. P. Ramanujam, *The invariance of Milnor's number implies the invariance of the topological type*, Amer. J. Math. **98** (1976), 67–78.

[Lê-Te] Lê D. T. et B. Teissier, *Variétés polaires locales et classes de Chern des variétés singulières*, Ann. of Math. (2) **114** (1981), 457–491.

[Ło] S. Łojasiewicz, *Ensembles semi-analytiques* (Mimeographie Inst. Hautes Etudes Sci. 1965).

[Ma1] J. Mather, *Notes on topological stability*, Harvard Univ. Preprint, July 1970.

[Ma2] ______, *Stratifications and mappings*, Dynamical Systems, Academic Press, New York, 1973.

[MP] R. MacPherson, *Chern classes for singular algebraic varieties*, Ann. of Math. (2) **100** (1974), 423–432.

[Mi] J. Milnor, *Singular points of complex hypersurfaces*, Ann. of Math. Studies, Princeton Univ. Press, Princeton, N.J., 1968.

[Na] V. Navarro, *Conditions de Whitney et sections planes*, Invent. Math. **61** (1980), 199–266.

[Na-Tr] V. Navarro and D. Trotman, *Whitney regularity and generic wings*, Ann. Inst. Fourier (Grenoble) **31** (1981), 87–111.

[P] D. Prill, *Local classification of quotients of complex manifolds by discontinuous groups*, Duke Math. J. **34** (1967), 375–386.

[Sa] C. Sabbah, *Morphismes analytiques stratifiés sans éclatement et cycles évanescents*, Conf. de Luminy sur Algèbre et Analyse sur les Espaces Singuliers, Astérisque (a paraitre).

[Sch] M. H. Schwartz, *Classes caractéristiques définies par une stratification d'une variété analytique complexe*, Note aux C. R. Acad. Sci. Paris Sér. A-B **260**, 3262–3264 et 3535–3537.

[S] D. Sullivan, *Combinatorial invariants of analytic spaces*, Proc. Liverpool Singularities Sympos. I, Lecture Notes in Math., vol. 192, Springer-Verlag, Berlin-Heidelberg-New York, 1971, pp. 165–168.

[Sp] J. P. Speder, *Equisingularité et conditions de Whitney*, Amer. J. Math. **97** (1975), 571–588.

[St] J. Stutz, *Analytic sets as branched coverings*, Trans. Amer. Math. Soc. **166** (1972), 241–259.

[Te1] B. Teissier, *Multiplicités polaires, sections planes, et conditions de Whitney*, Actes de la Conf. de La Rabida, Lecture Notes in Math., Springer-Verlag, Berlin and New York (a paraitre).

[Te2] ______, *Cycles évanescents, sections planes et conditions de Whitney*, Singularités à Cargèse, 1972, Astérisque **7-8** (1973).

[Te3] ______, *Variétés polaires*. I, *Invariants polaires des singularités d'hypersurfaces*, Invent. Math. **40** (1977), 267–292.

[Te4] ______, *Introduction to equisingularity problems*, Proc. Sympos. Pure Math., vol. 29, Arcata 1974, Amer. Math. Soc., Providence, R.I., 1974.

[Te5] ______, *Cycles évanouissants et conditions de Whitney*, Note C. R. Acad. Sci. Paris Sér. A **276** (1973), 1051–1054.

[Te6] ______, *Variétés polaires locales et conditions de Whitney*, Note aux C. R. Acad. Sci. Paris Sér. A-B **290** (1980), 799.

[Th1] R. Thom, *Ensembles et morphismes stratifiés*, Bull. Amer. Math. Soc. **75** (1969), 240–284.

[Th2] ______, *Local differential properties of analytic varieties*, Bombay Colloquium, Oxford Univ. Press, 1967.

[Th3] ______, *Proprietés différentielles locales des ensembles analytiques*, Sem. Bourbaki **281** (1964–65).

[Th4] ______, *La stabilité topologique des applications polynomiales*, L'enseignement Math. **8** (1962).

[Tr] D. Trotman, Thèse, Orsay, 1980.

[V] J. L. Verdier, *Stratifications de Whitney et Théorème de Bertini Sard*, Invent. Math. **36** (1976), 295–312.

[W1] H. Whitney, *Tangents to an analytic variety*, Ann. of Math. (2) **81** (1964), 496–549.

[W2] ______, *Local properties of analytic sets*, Differential and Combinatorial Topology, Sympos. Honour of M. Morse (S. S. Cairns, Editor), Princeton Univ. Press, Princeton, N.J., 1965, p. 205.

[Z1] O. Zariski, *Studies in equisingularity*. I, II, III, Amer. J. Math. **87** (1965); **90** (1968).

[Z2] ______, *Some open questions in the theory of singularities*, Bull. Amer. Math. Soc. **77** (1971), 481–491.

[Z3] ______, *Collected papers*. vol. IV. *Equisingularity on algebraic varieties*, MIT Press, Cambridge, Mass., 1979.

Centre de Mathématiques de l'École Polytechnique, Palaiseau, France

Proceedings of Symposia in Pure Mathematics
Volume **40** (1983), Part 2

REPORT ON THE PROBLEM SESSION

LÊ D. T. AND B. TEISSIER

The questions that follow were either asked during the problem session or sent to the organizers afterwards. We tried to gather them into groups of related interest and to add references when we could. Quite a few of the questions seem not to be new, and may have been published elsewhere. We have taken the liberty to change the order in which the questions have been asked, and to add comments to some of them.

Question 1 (*David Liebermann*). Given an isolated singularity of hypersurface: $f(z_1,\ldots,z_n) = 0$, with $f \in \mathbf{C}\{z_1,\ldots,z_n\} = \mathcal{O}_n$, let $j(f)$ be the ideal generated by $\partial f/\partial z_1,\ldots,\partial f/\partial z_n$ in $\mathcal{O}_n$. Is there an homogeneous space X, the cohomology ring $H^*(X,\mathbf{C})$ of which is isomorphic to the Artin algebra $\mathcal{O}_n/j(f)$ (or $\mathcal{O}_n/(f, j(f))$? For example, if $f = z_1^{a_1} + \cdots + z_n^{a_n}$, take $X = \mathbf{P}_{\mathbf{C}}^{a_1-1} \times \cdots \times \mathbf{P}_{\mathbf{C}}^{a_n-1}$.

Comment. One may allow singular spaces X, and consider their intersection homology ring in the sense of Goresky-MacPherson (cf. [**G-MI** and **G-MII**]).

The next questions concern isolated singularities of hypersurfaces and especially the singularities that can be obtained by deforming a given one.

Question 2 (*Joseph Steenbrink*). In an analytic family of complex analytic hypersurfaces with isolated singularities, what jumps are possible for the Milnor number? For example, what does it mean that the Milnor number changes by 1 exactly?

Comment. This, and other questions in the same direction, can be deemed to be questions on the geometry of the versal deformations of an isolated singularity of hypersurface (see [**Tj, Te1**]). More precisely, let F: $X \to S$ be such a versal deformation, and let C be its critical locus and $D \subset S$ be its discriminant hypersurface where $D = F(C)$. Let $D = \bigcup D_\nu$ be the equimultiplicity partition of

---

1980 *Mathematics Subject Classification.* Primary 32C40; Secondary 32C05, 14D05.

0082-0717/82/0000-0975/$03.25

D: $D_\nu = \{s \in D / m_s(D) = \nu\}$. According to [**Te5** and **Lê1**], the map $F \mid C = n$: $C \to D$ induces an étale covering $(n^{-1}(D_\nu))_{\text{red}} \to D_\nu$, so for each connected component of D_ν, the number of critical points of F lying over a point $s \in D_\nu$ is constant, their Milnor number is well defined (the sum of these Milnor numbers is ν, see [**Te1**]) and according to the theorems of Lê-Ramanujam [**L-R**] for $\dim F^{-1}(s) \neq 2$, the topological type of these points as singular points in $F^{-1}(s)$ is well defined. The problem is therefore to study what possibilities for ν and these topological types actually occur. This can be called the problem of topological adjacencies, to be compared with the problem of adjacencies stated by Arnol'd [**A**] where one asks whether a given analytic type of singularity occurs in an arbitrary small deformation of another.

A simpler problem of this type, stated by Teissier in [**Te4, Te2**] is the following: What is the maximum ν such that there is a component of D_ν above which there are only ordinary quadratic singularities? For plane curves this number is the well-known invariant δ. In higher dimensions, some interesting bounds were given by Yomdin (see [**Te2**, 5.6.4]). The lower bound can probably be improved.

In the same vein, recall that N. A'Campo has shown in [**AC2**] that if $f(z_1, \ldots, z_n)$ has a deformation which has only ordinary quadratic singularities (in a small ball around the singular point 0) and gives at most 2 critical values, then f is of the form $\tilde{f}(z_1, z_2, z_3) + z_4^2 + \cdots + z_n^2$ up to isomorphism, where $\tilde{f}(z_1, z_2, z_3) = 0$ defines a simple singularity of surface. This can be also seen as a theorem on the geometry of the discriminant of simple singularities.

By the way we recall that the problem of the nonsingularity of the strata D_ν (see [**Te5**]) is still open for $n \geqslant 3$. (For $n = 2$ see [**Wa** and **Te6**].)

Question 2*bis* (*C. T. C. Wall*). Study the possible splittings of critical points and critical values in deformations of simple elliptic singularities.

Comments supplied by T. Wall. In the case of simple singularities this was answered by [**Ly**] (see also [**AC2**]) and [**W1**] in the case of $\tilde{E}_6$. Note that here also there are exceptional moduli (as in Pham's bimodal example [**P2**] or [**Te2** or **Be**]) corresponding to the 24 decomposition types with 2 critical values which are yet to be understood.

Question 2*ter* (*C. T. C. Wall*). Let X be a double plane in $\mathbf{C}^2 \times \mathbf{P}^1$ given by $(x, y, (u : v))$ by an equation of the form $f = a(x, y)u^2 + 2b(x, y)uv + c(x, y)v^2 = 0$ (a, b, c complex analytic). Suppose a, b, c all vanish at $(0, 0)$ but at least one of them is regular there (so the line $x = y = 0$ (L) on X is not a singular line). Is the sum of the Milnor numbers of X on L one less than the Milnor number of the singularity of the discriminant $\Delta = \{b^2 - ac = 0\}$ at 0?

Comments supplied by C. T. C. Wall. This problem is linked to [**W2**]; In the two following examples the answer to the question is yes.

EXAMPLE 1.

$$f = x^4u^2 + 2y^2uv + xv^2, \qquad \Delta = y^4 - x^5.$$

EXAMPLE 2.

$$f = y^2u^2 + 2xuv + y^2v^2, \qquad \Delta = x^2 - y^4.$$

Question 3 (*André Galligo*). What is new about the smoothness of the μ-constant stratum?

Comments. Given an isolated singularity of hypersurface X: $f(z_1, \ldots, z_n) = 0$ in $(\mathbf{C}^n, 0)$, one defines the sequence μ^* of Milnor numbers of general plane sections of X: $\mu^*_{X,0} = (\mu^{(n)}(X,0), \mu^{(n-1)}(X,0), \ldots, \mu^{(1)}(X,0), \mu^{(0)}(X,0))$, where $\mu^{(i)}(X,0) = \mu(X \cap H^{(i)}, 0)$, where $H^{(i)}$ is a general plane of dimension i through 0 (see [**Te1**]). For the connection between the constancy of μ and that of μ^* in an analytic family of hypersurfaces with isolated singularities, see [**Te1** and **L-T**]. It is perhaps more natural to consider the subspaces D^*_ν of D_ν where the singularities of the fibres of F have not only given μ's, but given μ^*'s. We ask the following questions:

Question 4 (*Lê-Teissier*). Compute μ^* from the geometry of the discriminant at 0 (it is known that $\mu^{(n)}$ is the multiplicity of the discriminant (cf. [**Te1**, **Te2**])).

Question 5 (*Cf*. [**Te2**, §5]). Are the D^*_ν nonsingular?

Comments. All these questions have analogies for the discriminant of the family of all projective plane curves of a given degree d. The corresponding questions are also open.

Question 6. E. Brieskorn asks what is new concerning the question of O. Zariski [**Z**]: Is the multiplicity of a hypersurface X at $p \in X$ an invariant of the local topological type of X at p (as an embedded subspace of $\mathbf{C}^{n+1}$)?

Comments. D. Trotman [**Tr**] and R. Ephraim [**E**] have proved that multiplicity is a C^1-invariant for hypersurfaces. Gau (these PROCEEDINGS) has proved the corresponding result for a reduced complex-analytic space $X^r \subset \mathbf{C}^N$ of arbitrary embedding codimension. He has also given an example showing that multiplicity is *not* an invariant of the topological type in codimension $\geqslant 2$.

D. Trotman reports that V. Navarro has shown that there cannot be two singular hypersurfaces of the same topological type, having multiplicity 2 and 3 respectively.

Recall also that thanks to results of A'Campo [**AC1**] if the Euler-Poincaré characteristic of the complement in $\mathbf{P}^N$ of the projective variety associated to the tangent cone $C_{X,0}$ at 0 of a hypersurface $X^N \subset \mathbf{C}^{N+1}$ is nonzero, the multiplicity of X at 0 is equal to the least integer m for which the mth power h^m of the local monodromy of X at 0 has a nonzero Lefschetz number. Since the monodromy is an invariant of the topological type, this indicates a natural way to investigate Zariski's question. For instance, in the case of surfaces in $\mathbf{C}^3$, the only case left to understand is the case of surfaces having a tangent cone which is set-theoretically the union of two distinct planes. Can such a surface have the same topological type as a surface not having this property? As an exercise, using this method, one can check that a surface in $\mathbf{C}^3$ having at 0 a singularity of multiplicity 2 cannot have the same local topological type at 0 as another surface of multiplicity different from 2. Another direction is suggested by

Question 7 (*E. Looijenga*). For a normal surface singularity, can one compute the multiplicity from the data given by a minimal resolution? This would allow one to apply W. Neumann's result (cf. [**N**]).

Comment. A question weaker than Zariski's question is the following: In an analytic family of hypersurfaces with isolated singularity, does "μ constant" imply equimultiplicity? Since the condition "μ constant" is equivalent to local topological triviality in dimension $\neq 2$ (cf. [**L-R**]) and in view of Laufer's result that topological triviality for a family of surfaces with isolated singularity implies very weak simultaneous resolution (these PROCEEDINGS) one can ask also whether very weak simultaneous resolution in a family of hypersurfaces implies equimultiplicity.

Question 8 (*E. Brieskorn*). Study the action, defined by Gabrielov, of the braid group on infinite root systems, i.e. on symmetric matrices with -2 on the diagonal which correspond to the infinite root system. This problem comes up in the study of the monodromy in a geometric distinguished basis.

Question 9 (*P. Orlik*). What is new about the old question of Thom: To find a nice polyhedron inside the Milnor fibre which is a deformation retract of the Milnor fibre, and vanishes at the singularity. If a group acts on the situation can you get an invariant polyhedron? Is it linked to curvatures?

Comments. In [**P1** and **Mi3**] one sees that Pham has built such a polyhedron for Pham-Brieskorn singularities: $z_1^{a_1} + \cdots + z_n^{a_n} = 0$. A method due to Oka [**O**] builds such a polyhedron for the Milnor fibre of the Thom-Sebastiani sum $f(z_1, \ldots, z_n) + g(w_1, \ldots, w_n)$ from polyhedra for the Milnor fibres of f and g.

Also using the methods of Lê in [**Le2, Le3**], one can build inductively a vanishing polyhedron which is a deformation retract of the Milnor fibre, of dimension half the real dimension of this fibre but which is not invariant under the monodromy. The relation to the metric properties of the Milnor fibre is connected to the following questions:

Question 10 (*Teissier*). Is there a meaning to the notion of "vanishing cycle of smallest volume in its homology class"? If this is the case, the "rate of vanishing" of such a cycle $\gamma_i(t)$ on $f^{-1}(t) \cap \mathbf{B}_\varepsilon$, i.e. the exponent ε_i in an expansion $\mathrm{Vol}(\gamma_i(t)) = a_i t^{\varepsilon_i} (\mathrm{Log}\, t)^{\beta_i} + \cdots$, $a_i \in \mathbf{C}^*$, should be related to the polar invariants of [**Te3** and **Le4**].

Comments. The volumes in question should be computed by integrals on the $\gamma_i(t)$ of real analytic differential forms, and by Nilsson's regularity theorem [**Ni**] should be functions of t of Nilsson's class. There should be a connection between the ε_i and the exponents of the Gauss-Manin connection; see Varchenko's description in [**V**] of the mixed Hodge structure on the cohomology of the Milnor fibre. See also Lichtin's paper in these PROCEEDINGS.

Question 10 seems to be related to the following:

Question 11 (*Fary*). Can one use the ideas of Meeks-Yau [**M-Y**] to exhibit embedded spheres of minimal area in the Milnor fibres?

Question 12 (*Lê D. T.*). Take the algebraic link obtained by intersecting a plane curve singularity $(X, 0) \subset (\mathbf{C}^2, 0)$ and a sphere $\mathbf{S}_\varepsilon$ centered at 0 and of sufficiently small radius ε. Is a Milnor fibre on $\mathbf{S}_\varepsilon$ isotopic in $\mathbf{S}_\varepsilon$ to a minimal surface in $\mathbf{S}_\varepsilon$ with the metric induced by $\mathbf{C}^2$ having L as boundary? (See [**Le5**].)

Comment. The answer is yes for $xy = 0$ and $x^p - y^q = 0$, with $(p, q) = 1$.

Question 13 (*Kyoji Saito*). (1) To what extent do the exponents of an isolated singularity of a hypersurface determine it?

(2) In an analytic family of isolated singularities where the Milnor number varies, what happens to the exponents? (There is a conjecture of Arnol'd for the case where the Milnor number jumps by 1.)

Comments. There are several ways of defining "exponents" of an isolated singularity of hypersurfaces. Saito refers to his own, described in his lectures. There are others: Steenbrink [**S**] defines exponents which are adequate logarithms of the eigenvalues of the monodromy acting on the space of vanishing cycles endowed with its mixed Hodge structure; Varchenko [**V**] defines exponents using the leading term of the asymptotic expansions of the period integrals on the Milnor fibre, and finally one can also call exponents the roots of the b-function, see Yano [**Y**] and Malgrange [**M**]. All these exponents have in common that the exponential $2\pi i \times$ (exponent) is an eigenvalue of the monodromy.

The smallest exponent appearing in Varchenko's definition is Arnol'd's exponent and is constant in a μ-constant family (see [**V**]). We are led to the following question.

Question 14 (*Lê D. T.*). Describe geometric conditions on a family of hypersurfaces with isolated singularity which imply the constancy of the b-function.

Comment. μ-constant is obviously necessary, but there is an example of Yano namely $x^3 + y^7 + txy^5 = 0$, where μ is constant, but not the b-function. One can observe that the family of generic polar curves is not equisingular. One may wonder whether a condition of equisingularity on the polar varieties [**L-T**] implies the constancy of the b-function.

Question 15 (*R. Smith*). What happens to the ramification divisor of the Gauss mapping and its degree for a hypersurface in an analytic family which acquires singularities?

Comment. Again, this is linked to the polar varieties. See [**Pi**].

Question 16 (*S. Zucker*). Take a family of smooth projective varieties over $\mathbf{D}^* \times \mathbf{D}^*$ (where $\mathbf{D}^*$ is a punctured disk); extend it over $\mathbf{D} \times \mathbf{D}$. Can one control the singularity over 0 of the inverse image of a smooth curve in $\mathbf{D} \times \mathbf{D}$ containing 0?

Comment. Using the theory of $\mathcal{D}$-modules one can get at least some cohomological control.

Question 17 (*R. Langevin*). In 1954 [**Mi1**] J. Milnor constructed an invariant $\bar{\mu}(C_1, C_2, C_3)$ detecting "higher order linking" for three real closed curves (C_1, C_2, C_3) which is defined modulo the g.c.d. of the linking numbers of the pair (C_i, C_j). Give an interpretation of this invariant in the case of the algebraic link coming from a reduced complex curve with three branches.

In 1953 [**Mi2**] J. Milnor studied the following geometric invariant of a knot C in $\mathbf{R}^3$

$$\mathfrak{M} = \inf_{\substack{\text{(isotopy class} \\ \text{of } C)}} \left[\int_C (k + |\tau|)\, ds \right]$$

where k and τ are the curvature and torsion of the curve C in $\mathbf{R}^3$ and s is the arc length parameter. What does the analogous integral

$$\inf_{\substack{\text{(isotopy class} \\ \text{of } C)}} \left[\int_C (k + |\tau| + 1)\, ds \right]$$

give for an algebraic knot in the unit sphere of dimension 3? Has the rescaled integral $\int_{(f=0)\cap S_\varepsilon} (k + |\tau| + 1/\varepsilon)\, ds$ an interesting limit when ε goes to zero?

Question 18 (*J. Fine*). Given $f(z_1, \ldots, z_N)$ nondegenerate with respect to its Newton diagram in the coordinates $z_1, \ldots, z_N$ (see [**K**]). What happens to the Newton diagram when one makes a coordinate change?

Question 19 (*J. Wahl*). Consider the normal surface singularities with resolution dual graph

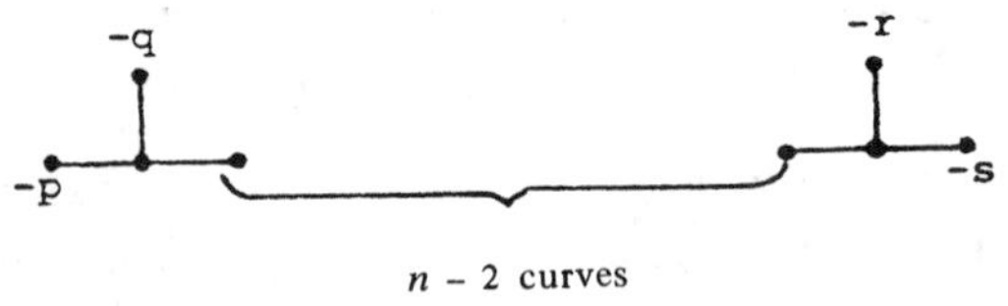

$p, q, r, s \geqslant 2$, not all $= 2$, $n \geqslant 2$. These singularities are minimally elliptic, of "degree" $d = p + q + r + s - 8$. The graph above uniquely determines the analytic type. They are hypersurfaces iff $d \leqslant 3$, in which case they are *bimodular* in the sense of Arnol'd (and, when $d = 4$, bimodular for C. T. C. Wall). Some other points:

(a) If $p = q = r = s = 2$, these are (à la Kodaira) degenerate fibres in an elliptic pencil of type $\tilde{D}_n$ (= affine Dynkin diagram).

(b) After the cusps (= hyperbolic) singularities, they are the simplest ones which admit no $\mathbf{C}^*$-action for the graph type.

Problem. (a) Where do these singularities come from? i.e., give a description of the local rings via automorphic forms, or Fourier series, or the like, as is done for all simpler singularities.

(b) Find natural global (= compact) complex surfaces on which these singularities sit (for cusps, use Inoué surfaces; if you have a $\mathbf{C}^*$-action, use equivariant completion). "Natural" excludes the artificiality of taking an elliptic pencil with $\tilde{D}_n$ degenerate fibre, and blowing up and blowing down. Observe that if you consider the following diagram you get a rational singularity, which is never a complete intersection. The facts are

(a) This singularity is a $\mathbf{Z}/2$-quotient of a cusp singularity.

(b) The fundamental group π_1 of the neighborhood boundary is solvable.

$$1 \to \mathbf{Z} \to \pi_1 \to \mathbf{Z}/2 \to 1,$$

and one can describe exactly how this π_1 acts on $H \times H$ to get $H \times H/\pi_1$ = some punctured neighbourhood of the singularity.

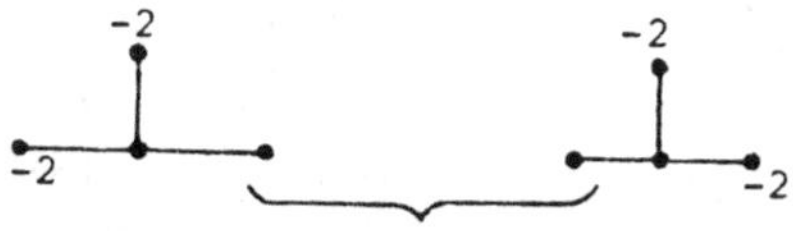

Question 20 (*A. Iarrobino*). Are peelable Artin algebras (cf. [Iarrobino's paper in these PROCEEDINGS]) alignable—do they have a deformation to $C[x]/(x^n)$? In other words, do finite length-n map germs that are smoothable in $n - 1$ steps through multigerms of lengths $(n - 1, 1)$, $(n - 2, 1, 1), \ldots,$ and $(1, \ldots, 1)$, also lie in the closure $\bar{\Sigma}^{1,\ldots,1}$ of the Morin singularities having the same length. This was shown for maps from $\mathbf{C}^2$ by Briançon and Galligo [**B-G**]; recently Gaffney showed that any minimal dimension component of the subspace of complete intersections in Hilb $\mathbf{C}[[x_1, \ldots, x_r]]$ consists of peelable algebras. The question asks if there is only one such minimal-dimension component. (See *Deformations of zero-dimensional complete intersections* in these PROCEEDINGS.)

Question 21 (*old*). Are there rigid zero-dimensional singularities?

Question 22 (*I. R. Porteous*). The genesis of umbilics. Suppose P is a point of a smoothly deformable smooth surface M in $\mathbf{R}^3$, with no rib-line of M passing through P. How should we deform M so that P becomes a generic umbilic of M?

Simply pressing a ball-bearing into M at P is rather too naive! Since there has to be at least one rib-line through any generic umbilic, we must first of all produce a rib-line from somewhere, either by shifting one over from elsewhere or by giving birth to a new rib-line at P. A new rib-line will appear as a point (a nongeneric A_4 point) and this will most simply grow to a simple closed rib-line with two generic A_4 points on it. At some point of it other than the A_4 points a nongeneric umbilic may then be born, the associated cubic form being necessarily orthogonal—that is, having its Hessian directions mutually orthogonal. This nongeneric umbilic on the rib-line will at once become twins, one, the female, say, being a hyperbolic star and the other, the male, being a monster, this being subtly different from the creation myth in Genesis 2, verses 21–22.

What are the possibilities for the future development and adolescence of these two umbilics bearing in mind that a D_5 singularity specialises to an A_4, but a D_4 does not?

Questions (*by Edward Bierstone, Pierre D. Milman, Gerald W. Schwartz*). Let φ: $M \to N$ be a proper real analytic mapping between real analytic manifolds. If $x \in M$, let φ_x^*: $\mathcal{O}_{\varphi(x)} \to \mathcal{O}_x$ denote the induced homomorphism of local rings.

Question 23. Suppose $N = \mathbf{R}^n$. Let $Y = \varphi(M)$. If $y \in Y$, let $\underline{m}_y$ be the maximal ideal of $\mathcal{O}_y$, and $\mathcal{T}_y = \bigcap_{x \in \varphi^{-1}(y)} \operatorname{Ker} \varphi_x^*$. We define the kth *order tangent bundle* T^kY of Y by $T^kY = \bigcup_{y \in Y} T_y^k Y$, where

$$T_y^k Y = \left\{\xi \in \left(\underline{m}_y/\underline{m}_y^{k+1}\right)^*: \xi \perp \left(\mathcal{T}_y + \underline{m}_y^{k+1}\right)/\underline{m}_y^{k+1}\right\}.$$

Is T^kY subanalytic?

Question 24. Let $P = \{x \in M: r(x) < \dim \mathcal{O}_{\varphi(x)}/\operatorname{Ker} \varphi_x^*\}$, where $r(x)$ denotes the general rank of φ near x; P is a closed subset of M. Is P analytic (or semianalytic)?

Question 25. Let $x \in P$, $y = \varphi(x)$, and let Q_y denote the local flattener of φ at x, in the sense of Hironaka (Q_y is a germ of an analytic subspace of N at y). Is $Q_Y \subset \varphi(P)_y$, where $\varphi(P)_y$ denotes the germ of $\varphi(P)$ at y?

Question 26. If $x \in M$, let $\hat{\varphi}_x^*$: $\hat{\mathcal{O}}_{\varphi(x)} \to \hat{\mathcal{O}}_x$ denote the induced homomorphism of completed local rings. Investigate the analogues of (1), (2), (3), with φ_x^* replaced by $\hat{\varphi}_x^*$.

Question 27. Let $B \subset A$ be compact subanalytic subsets of $\mathbf{R}^n$. Assume that $A - B$ is a real analytic manifold of dimension k with $\overline{A - B} = A$ (so that $\dim B < \dim A$). Does there exist a finite collection $\{(V_i, A_i, U_i)\}$ such that

(1) Each V_i is an open subset of $\mathbf{R}^k$, A_i is a compact subanalytic subset of V_i such that $A_i = \overline{\operatorname{int} A_i}$, and u_i: $V_i \to \mathbf{R}^n$ is real analytic.

(2) For each i, $u_i \mid \operatorname{int} A_i$ is an analytic isomorphism onto its image.

(3) The union $\bigcup_i u_i(\operatorname{int} A_i) = A - B$?

Such a collection exists if A is Nash subanalytic.

Question 28 (*Y. Yomdin*). Let f: $M^n \to \mathbf{R}^k$ be a stable mapping. Let $h = f + \varphi$, φ: $M^n \to \mathbf{R}^k$. By definition, if h is not equivalent to f, then φ is "big".

(1) For each type of stable singularities of f what are inequalities on φ to guarantee equivalence of f and $h = f + \varphi$?

(2) Are there weaker conditions on φ under which the structures of singularities of f and $h = f + \varphi$, although they do not necessarily coincide, are "close" one to another?

A more precise version of the last question in a simple special case is the following.

Let $B = \{x \in \mathbf{R}^n \mid |x| \leqslant 1\}$, $\mathring{B} = \{x \mid |x| < 1\}$. Let $f\colon B \to \mathbf{R}$, $f(x) = |x|^2$, and let φ be a C^∞-smooth function, supp $\varphi \subset \mathring{B}$.

Does there exist an integral function $\nu_n\colon \mathbf{Z}_+ \to \mathbf{Z}_+$, $\lim_{k\to\infty} \nu(k) = \infty$, such that if $h = |x|^2 + \varphi$ has more than k critical points in $\mathring{B}$, then at some $x \in \mathring{B}$,

$$|d^{\nu_n(k)}\varphi(x)| \geqslant C_{n,k},$$

where $C_{n,k} > 0$ is a constant, depending only on n and k?

Questions by C. T. C. Wall.

(1) *Sufficiency of jets* for topological (or C^k) equivalence. Most of the results in the literature concern only right-equivalence or V-equivalence.

Question 29. Can one extend them to topological (right-left) equivalence, or to left equivalence? The most general problem asks about C^r-equivalence of C^l realizations of the k-jet z ($l \geqslant \max(r, k)$): when is the number of equivalence classes (a) equal to 1, (b) finite?

(2) One would like to show that ∞-C^0-determined implies C^0-stable in a punctured neighbourhood of 0 and hence that it implies multitransversality results (cf. Conjecture 32). Argument fails since results on C^0 stability always need a compactness hypothesis. How can one get around this? A particular problem is

Question 30. Prove that ∞-C^0-determined $\Rightarrow$ the map germ is topologically cone-like.

Question 31. Does C^1-stable imply C^∞-stable?

(The answer may depend on the dimensions of source and target.)

Conjecture 32. There is a stratification $\mathbb{S}$ of the jet space J^∞ (strictly of a subset of J^∞ with complement of infinite codimension) such that a proper C^∞-map f is C^0-stable $\Leftrightarrow$ f is multitransverse to $\mathbb{S}$.

Note. $\mathbb{S}$ cannot be Whitney regular. It should be (a)-regular. Clearly if $\mathbb{S}$ exists it is unique.

REMARK. One can certainly achieve this on subsets of J^∞ with complement of fairly large but finite codimension. The most difficult part is proving C^0-triviality of versal unfoldings (Damon's paper **[D2]** gives many useful cases here).

Question 33. Let $f\colon (\mathbf{R}^n, 0) \to (\mathbf{R}, 0)$ be an analytic germ whose complexification has an isolated singular point with Milnor number μ. Then C. T. C. Wall has shown the *semicharacteristic* ($= \frac{1}{2}$sum of Betti numbers) of the link $S_\varepsilon^{n-1} \cap f^{-1}(0)$ is $\equiv \mu + 1 \pmod 2$.

Problem. Does a corresponding result hold for complete intersections?

Question 34 (*Tim Poston*). Folding theory? Given a mapping with a particular stable singularity structure, or a family $\mathbf{R}^k \times \mathbf{R}^n \to \mathbf{R}$ with a particular stable bifurcation (catastrophe set, not topologically conical, when can a single more degenerate singularity be found such that the initial picture is just a slice (embedded? immersed?) of its universal unfolding? (This would be a systematic of finding an "organizing centre".)

Global versus Local structure. In many practical cases, an organizing centre found and locally analyzed gives a description of which numerical work proves globally correct. What global bounds on which derivatives can be shown to guarantee this apparent miracle?

Question 35 (*Golubitsky*). *Problem on symmetry breaking.* Let g: $\mathbf{R}^n \to \mathbf{R}^n$ be the germ of a C^∞-mapping commuting with a given linear irreductible action of a compact Lie group Γ. Suppose that g has codimension 1 with respect to contact equivalences commuting with Γ. Let G: $\mathbf{R}^n \times \mathbf{R} \to \mathbf{R}^n$ be the universal unfolding of g.

Problem. Classify the isotropy subgroups corresponding to solutions of $G = 0$. In particular, suppose $G(x, \lambda) = 0$ and Σ_x is the isotropy subgroup of Γ corresponding to $x \in \mathbf{R}^n$. Then, is Σ_x a maximal subgroup in the lattice of isotropy subgroups of Γ (with the given representation)?

Question 36 (*C. T. C. Wall*). There are fairly ample lists of hyperplane and surface singularities, but little else apart from the lists in Mather VI: the nice dimensions. More precisely, map-germs which are 0-modal or 1-modal or 2-modal for R-equivalence were listed by Arnol'd: 0-modal for K-equivalence by Mather (completed by Giusti [**G**] and Damon [**D1**]).

(a) List the 1-modals for K-equivalence (I have done this for $n > p$, Gibson for $n = p = 2$. The cases $n = p = 3$ and $n < p$ look more complicated).

(b) List 0-modal germs for $\mathcal{A}$-equivalence (Bruce and Gaffney have done this for $n = 1$, $p = 2$; there are partial lists for $n = p = 2$ and for $n = 3$, $p = 2$).

Question 37 (*N. A. Baas*). *Stability and singularities of composed mappings.* Consider sequences of mappings

$$X_1 \xrightarrow{f_1} X_2 \xrightarrow{f_2} \cdots \xrightarrow{f_{n-1}} X_n$$

where the X_i's are C^∞-manifolds and the mappings are C^∞ and proper. We consider the compositions as elements of the following space,

$$F = C^\infty(X_1, X_2) \times C^\infty(X_2, X_3) \times \cdots \times C^\infty(X_{n-1}, X_n)$$

where each component space has the Whitney-topology. Let

$$A = \mathrm{Homeo}(X_1) \times \cdots \times \mathrm{Homeo}(X_n)$$

and define α: $A \times F \to F$ by

$$(a_1,\ldots,a_n) \times (f_1,\ldots,f_{n-1}) = (a_2 f_1 a_1^{-1},\ldots,a_n f_{n-1} a_{n-1}^{-1}).$$

DEFINITION 1. $f = (f_1, \ldots, f_{n-1})$ is topologically stable iff it is an interior point of its A-orbit.

Conjecture 1. Topologically stable compositions between compact manifolds are dense (Baas, Institute of Advanced Study, Princeton-notes, 1974).

Problem. To obtain classification of stable germs of compositions in terms of diagrams of associated algebras. Is it so that the most naturally associated algebra-composition

$$Q_p(f_1) \leftarrow Q_p(f_2) \leftarrow \cdots \leftarrow Q_p(f_{n-1})$$

up to isomorphism classifies the orbit of $(f_1, \ldots, f_{n-1})$ for p suitable?

REFERENCES

[A] V. I. Arnol'd, *Normal forms for functions near degenerate critical points*, Functional Anal. Appl. **6** (1972), 254–272.

[AC1] N. A'Campo, *La fonction zeta d'une monodromie*, Comment. Math. Helv. **50** (1975), 233–248.

[AC2] ______, *Le groupe de monodromie du déploiement des singularitiés de courbes planes*. II, Proc. Internat. Congr. Math. (Vancouver, 1974), vol. 1, p. 395.

[B-G] J. Briançon et A. Galligo, *Déformations distinguées d'un point de* $\mathbf{R}^2$ *ou* $\mathbf{C}^2$, Astérisque No. 7–8 (S.M.F., 1973), 129–138.

[Be] P. Berthelot, *Classification topologique universelle des singularités, d'après F. Pham*, Astérique No. 16 (S.M.F., 1973).

[D1] J. Damon, *Topological properties of discrete algebras types*. I: *the Hilbert-Samuel function*, Advances in Math. Suppl. Ser. **5** (1978), 83–118; II: *Real and complex algebras*, Amer. J. Math. **101** (1979), 1219–1248.

[D2] ______, *Finite determinacy and topological triviality*, Invent. Math. **62** (1980), 299–324.

[E] R. Ephraim, *C^1-preservation of multiplicity*, Duke Math. J. **43** (1976), 797–803.

[G] M. Giusti, *Classification des singularités simples d'intersections complètes*, thèse d'état Paris VII, 1981, these PROCEEDINGS.

[G-M1] M. Goresky and R. MacPherson, *Intersection homology theory*, Topology **19** (1980), 135–162.

[G-M2] ______, *Intersection homology*. II (to appear).

[K] A. G. Kushnirenko, *Newton polyhedron and Milnor numbers*, Funkcional Anal. i Priložen. **9** (1975), 74–75. (Russian)

[L-R] Lê D. T. and C. P. Ramanujam, *The invariance of Milnor's number implies the invariance of the topological type*, Amer. J. Math. **98** (1976), 67–78.

[L-T] Lê D. T. et B. Teissier, *Variétés polaires locales et classes de Chern des variétés singulières*, Ann. of Math. (2) **114** (1981), 457–491.

[Le1] Lê D. T., *Une application d'un théorème d'A'Campo à l'équisingularité*, Indag. Math. **76** (1973), 403–409.

[Le2] ______, *The geometry of the monodromy theorem*, C. P. Ramanujam, a Tribute, Tata Inst., Bombay, 1978, pp. 157–173.

[Le3] ______, *Remarks on relative monodromy*, Real and Complex Singularities (Oslo, 1976), Noordhoff-Sijthoff, 1977.

[Le4] ______, *La monodromie n'a pas de point fixe*, J. Fac. Sci. Tokyo **22** (1975), 409–427.

[Le5] ______, *Surfaces minimales et singularités*, Acta Math. Vietnam **4** (1979), 76–79.

[Ly] O. V. Lyashko, *Decomposition of simple singularities of functions*, Functional Anal. Appl. **10** (1976), 122–128.

[M] B. Malgrange, *Sur les polynômes de I. N. Bernstein*, Séminaire Goulaouic-Schwartz, École Polytechnique, 1974.

[Mi1] J. Milnor, *Link groups*, Ann. of Math. (2) **59** (1954).

[Mi2] ______, *On total curvature of closed space curves*, Math. Scand. **1** (1953).

[Mi3] ______, *Singular points of complex hypersurfaces*, Ann. of Math. Studies, Princeton Univ. Press, Princeton, N.J., 1968.

[M-Y] W. Meeks, III and S.-T. Yau, *Topology of 3-dimensional manifolds and the embedding problem in minimal surface theory*, Ann. of Math. (2) **112** (1980), 441–484.

[N] W. Neumann, *A calculus for plumbing applied to the topology of complex surface singularities and degenerating complex curves*, Trans. Amer. Math. Soc. **268** (1981), 299–344.

[Ni] N. Nilsson, *Some growth and ramification properties of certain integrals on algebraic manifolds*, Ark. Math. **5** (1964).

[O] Mutsuo Oka, thèse, Orsay, 1975.

[P1] F. Pham, *Formules de Picard-Lefschetz généralisées et ramification des intégrales*, Bull. Soc. Math. France **93** (1965), 333–367.

[P2] ______, *Remarque sur l'équisingularité universelle*, preprint, Univ. de Nice, 1970.

[Pi] R. Piene, *Polar classes of singular varieties*, Ann. Sci. École Norm. Sup. **11** (1978).

[S] J. Steenbrink, *Mixed Hodge structure on the vanishing cohomology*, Real and Complex Singularities (Oslo, 1976), Noordhoff-Sijthoff, 1977.

[Te1] B. Teissier, *Cycles évanescents, sections planes et conditions de Whitney*, Astérique No. 7–8 (S.M.F., 1973).

[Te2] ______, *The hunting of invariants in the geometry of discriminants*, Real and Complex Singularities (Oslo, 1976), Noordhoff-Sijthoff, 1977.

[Te3] ______, *Variétés polaires*. I, Invent. Math. **40** (1977), 267–292.

[Te4] ______, *Sur la version catastrophique de la règle des phases de Gibbs*, "Singularities à Cargése 1975", Publ. Univ. de Nice, 1975.

[Te5] ______, *Déformations à type topologique constant*. II, Astérisque No. 16 (S.M.F., 1973).

[Te6] *Appendix to the course of O. Zariski*: "*Modules de branches planes*", École Polytechnique, Publ. Centre de Math., 1973.

[Tj] G. N. Tjurina, *Flat locally semi-universal deformations of isolated singularities of complex spaces*, Izv. Akad. Nauk SSSR Ser. Math. **33** (1969), 1026–1058. (Russian)

[Tr] D. Trotman, thése, Université d'Orsay, 1980.

[V] A. N. Varchenko, *Gauss-Manin connection of isolated singular point and Bernstein polynomial*, Bull. Soc. Math. (2) **104** (1980), 205–223. See also *Hodge properties of Gauss-Manin connection*, Functional Anal. Appl. **14** (1980), 36–37.

[Wa] J. Wahl, *Equisingular deformations of plane algebroid curves*, Trans. Amer. Math. Soc. **193** (1974), 143–170.

[W1] C. T. C. Wall, *Affine cubic functions*. IV, Philos. Trans. Roy. Soc. London Ser. A (1981).

[W2] ______, *Singularities of nets of quadrics*, Compositio Math. **42** (1980), 187–212.

[Y] T. Yano, *On the theory of b-functions*, Publ. Res. Inst. Math. Sci. Kyoto Univ. **14** (1978), 111–202. See also, Yano's lecture in these PROCEEDINGS.

[Z] O. Zariski, *Some open questions in the theory of singularities*, Bull. Amer. Math. Soc. **77** (1971), 481–491.

CENTRE DE MATHÉMATIQUES DE L'ÉCOLE POLYTECHNIQUE, PALAISEAU, FRANCE

Proceedings of Symposia in Pure Mathematics
Volume **40** (1983), Part 2

EXTENDED ARTIN GROUPS

HARM VAN DER LEK

1. Introduction. We want to compute a presentation of the fundamental group of the complement of the discriminant of a semiuniversal deformation of *simply elliptic* singularities and *cusp* singularities. There are two steps:

(1) Give an appropriate *description* of this complement (i.e. one which is a good starting point for a topologist).

(2) Determine the fundamental group.

For step (1), due to Looijenga [7], we need the concept of *generalized root system* [**6**] and *extended Coxeter group*. The groups that arise as fundamental groups we propose to call *extended Artin groups*, because they can be described as Artin groups extended with translations (though this is not a group extension in the usual sense).

For *simple singularities* (A_l, D_l, E_6, E_7, E_8) this program is done in the work of Arnold and Brieskorn [**2** step (1), **3** step (2)]. In §2 we recall this, also because we need that notion.

2. Coxeter, Artin and braid groups. We recall some definitions and facts from [**1** and **3**]. A group W is called a *Coxeter group* (and the pair $(W, \{\sigma_1, \ldots, \sigma_l\})$ a *Coxeter system*) if it admits a presentation of the following form:

2.1 generators: $\sigma_1, \ldots, \sigma_l$, relations: $(\sigma_i\sigma_j)^{m_{ij}} = 1$, $i, j \in \{1, \ldots, l\}$ where $\{m_{ij}\}_{i,j\in\{1,\ldots,l\}}$, with $m_{ij} \in \mathbf{N} \cup \{\infty\}$ and $m_{ii} = 1$ is the *Coxeter matrix*. In other words, W is generated by (in the cases we are interested in, only finitely many) elements of order two; and by saying what the orders of the products of two such are, we have given a complete set of the relations. The relations 2.1 are clearly equivalent to the following:

2.2 a. $\sigma_i^2 = 1$, $i = 1, \ldots, l$; b. $\sigma_i\sigma_j\sigma_i \cdots = \sigma_j\sigma_i\sigma_j \cdots$ (each side m_{ij} factors).

The *Artin group* (named this way and studied in [**4**]) belonging to W (or to $\{m_{ij}\}$), denoted by A_W, is the group we get by forgetting the relation 2.2a. So it

---

1980 *Mathematics Subject Classification*. Primary 57R45.

0082-0717/81/0000-0645/$02.00

has a presentation:

2.3 generators: $s_1, \ldots, s_l$, relations: $s_i s_j s_i \cdots = s_j s_i s_j \cdots$ (m_{ij} factors, $i \neq j$).

For each Coxeter group W there exists a faithful representation in a real vectorspace V, such that the generators become reflections [**1**], so we may assume $W \subset \mathrm{Gl}(V)$. There exists an open cone $\mathring{I} \subset V$ such that W acts properly discontinuously on $\mathring{I}$ (also proved in [**1**]). We have: $\mathring{I} = V$ iff W is finite.

The group W acts also on the domain $\Omega := \mathrm{Im}^{-1}(\mathring{I}) = \{x + \sqrt{-1}\,y \mid y \in \mathring{I}\} \subset V_{\mathbf{C}}$. Then $Y := \Omega - \{\text{reflection hyperplanes}\}$ is just the union of regular orbits. We put $X := Y/W$, the *regular orbit space*.

In the case that W is of type A, D or E the pair $(V_{\mathbf{C}}/W, V_{\mathbf{C}}/W - X) \doteq$ ({all orbits}, {singular orbits}) is homeomorphic to the pair (base space, discriminant) of the semiuniversal deformation of a simple singularity.

So X is then the complement of the discriminant of the semiuniversal deformation of a simple singularity. So step (1) is done in this case.

Deligne [**5**] has shown that X is a $K(\pi, 1)$ in this case. Step (2) is done by Brieskorn [**3**]. He proves that the fundamental group $\pi_1(X)$ is the Artin group belonging to W given by 2.3. We can show that for this last result on $\pi_1(X)$ it is not necessary to assume that W is finite.

THEOREM. *Let W be any Coxeter group. The fundamental group $\pi_1(X)$ of the space of regular orbits as described above has a presentation as the Artin group belonging to W given by* 2.3.

Our proof of this theorem (and the theorem of §3) is elementary. It is inspired by Deligne [**5**] and uses the notion 'gallery'.

EXAMPLE. All this is clearly illustrated in the case A_l. W is then S_{l+1}, the permutation group on $l+1$ elements. Then $m_{ij} = 2$ if $|i-j| > 1$ and $m_{ij} = 3$ if $|i-j| = 1$. So the *Coxeter* graph is

$$A_l : \circ\!\!-\!\!-\!\!\circ\!\!-\!\!- \cdots -\!\!-\!\!\circ .$$

$W = S_{l+1}$ acts on $\mathbf{C}^{l+1}$ by permuting coordinates. So we can think of X as the space of *unordered* $l+1$ tuples of *different* complex numbers. Choose $(0, 1, 2, \ldots, l) \in X$ as base point. Then we can imagine an element of $\pi_1(X)$ as a *braid* on $l+1$ *strings* and we can see in the diagram why the relations $\sigma_i \sigma_{i+1} \sigma_i = \sigma_{i+1} \sigma_i \sigma_{i+1}$ holds.

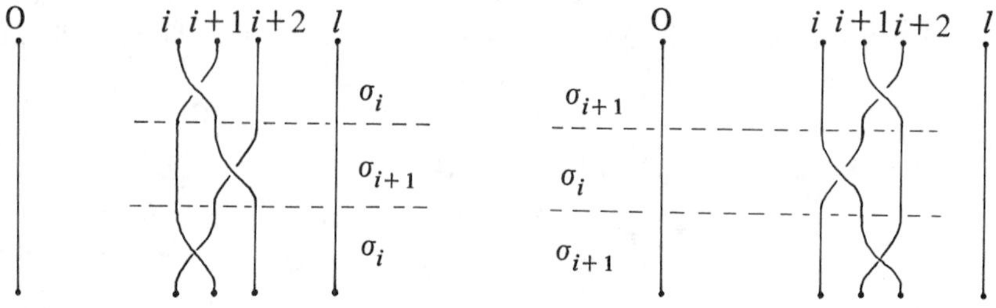

3. Extended Coxeter and Artin groups. There are some other singularities, whose complements of the discriminant of a semiuniversal deformation can be described as a regular orbit space (certain simply elliptic singularities and cusp singularities).

We briefly recall the description of this orbit space, arising from *generalized root systems*, as in [**6**]. In addition to the Coxeter group $W \subset \mathrm{Gl}(V)$ we have now also a *root lattice* $Q \subset V$ as follows:

Let $\{n_{ij}\}$ be a so-called (*generalized*) *Cartan matrix*. This is an $l \times l$ matrix with 2's on the diagonal and nonpositive integers outside the diagonal such that $n_{ij} = 0$ iff $n_{ji} = 0$. $\{\alpha_1, \dots, \alpha_l\} \subset V$ resp. $\{\check{\alpha}_1, \dots, \check{\alpha}_l\} \subset V^*$ are lin. indep. sets (the *root basis resp. dual root basis*) s.t. $\langle \alpha_i, \check{\alpha}_j \rangle = \check{\alpha}_j(\alpha_i) = n_{ij}$. The fundamental reflections are given by $s_i(x) := x - \langle x, \check{\alpha}_i \rangle \alpha_i$. $(W, \{s_1, \dots, s_l\})$ is a Coxeter system with Coxeter matrix $\{m_{ij}\}$ determined by the n_{ij} by $m_{ij} = 2, 3, 4, 6, \infty$ if $n_{ij}n_{ji} = 0, 1, 2, 3$ or $\geqslant 4$ resp. Let Q be $\bigoplus_i^l \mathbf{Z}\alpha_i$. Since $\sigma_i(Q) \subset Q$, W acts on Q. The semidirect product of W and Q is denoted by $\tilde{W}$ and called the *extended Coxeter* (or Weyl) *group*. For later use we remark that this group $\tilde{W}$ can be described as follows:

Add to the free product of W and Q the relations $wt = w(t)w$ for all $w \in W$ and $t \in Q$ or equivalently:

Add to the free product of W and Q the relations

$$s_i t = s_i(t) s_i \qquad (= t t_i^{-\check{\alpha}_i(t)} s_i) \tag{3.1}$$

for all $i \in \{1, \dots, l\}$ and $t \in Q$. In this section we use the letters Ω, Y and X for objects, that are different from but similar to those of §2.

The group $\tilde{W}$ acts on the domain $\Omega := \{x + \sqrt{-1}\,y \mid y \in \mathring{I}\}$. The action of Q is given by $t_i(x + \sqrt{-1}\,y) = x + \alpha_i + \sqrt{-1}\,y$. Again $Y := \Omega - \{$reflection hyperplanes of $\tilde{W}\}$ is the subspace where the action is free. The reflection hyperplanes are of the form $\{z \in \Omega \mid \check{\alpha}(z) = n\}$; $\check{\alpha}$ a dual root, $n \in \mathbf{Z}$.

The *regular orbit space* $X := Y/\tilde{W}$ is now also homeomorphic to the complement of the discriminant of a semiuniversal deformation of some singularities for certain Cartan matrices (see Looijenga [7, Chapter III]).

So step (1) is done. We describe now our results on step (2) in this case. First we describe how a presentation for $\pi_1(X)$ can be obtained by changing a presentation of $\tilde{W}$ in a suitable equivalent presentation and then deleting some relations:

presentation of $\tilde{W}$	equivalent presentation of $\tilde{W}$	presentation of $\pi_1(X) = A_{\tilde{W}}$
generators: $\sigma_1, \dots, \sigma_l, \tau_1, \dots, \tau_l$		generators: $s_1, \dots, s_l, t_1, \dots, t_l$
	1a. $\sigma_i^2 = 1$	
1. $(\sigma_i\sigma_j)^{m_{ij}} = 1$	1b. $\sigma_i\sigma_j\sigma_i \cdots = \sigma_j\sigma_i\sigma_j \cdots$ m_{ij} factors, $i \neq j$	1b. $s_is_js_i \cdots = s_js_is_j \cdots$ m_{ij} factors, $i \neq j$
2. $\tau_i\tau_j = \tau_j\tau_i$	2. $\tau_i\tau_j = \tau_j\tau_i$	2. $t_it_j = t_jt_i$
3. $\sigma_i\tau_j = \tau_j\tau_i^{-n_{ji}}\sigma_i$	3a. $(i = j)\ \sigma_i\tau_j = \tau_i^{-1}\sigma_i$	
	3b. $(-n_{ji} = 2r)\ \sigma_i\tau_j = \tau_j\tau_i^r\sigma_i\tau_i^{-r}$	3b. $s_it_j = t_jt_i^rs_it_i^{-r}$
	3c. $(-n_{ji} = 2r+1)\ \sigma_i\tau_j = \tau_j\tau_i^{r+1}\sigma_i^{-1}\tau_i^{-r}$	3c. $s_it_j = t_jt_i^{r+1}s_i^{-1}t_i^{-r}$

So we claim that we get a presentation for $\pi_1(X)$ which is now by definition the extended Artin group $A_{\tilde{W}}$ by forgetting the relations: 1a. $\sigma_i^2 = 1$, $i = 1,\ldots,l$, and 3b. $\sigma_i \tau_i = \tau_i^{-1}\sigma_i$.

The elements $s_1,\ldots,s_l$, $t_1,\ldots,t_l \in \pi_1(X, *)$ are obtained as follows. Choose as base point $* \in X$ the point $p(\sqrt{-1}\,a)$ where p denotes the projection $Y \to X$. For $i = 1,\ldots,l$ define

$$S_i\colon [0,1] \to Y,$$
$$t \mapsto \phi(t)\alpha_i + \sqrt{-1}\,((1-t)a + t\sigma_i(a))$$

where $\phi\colon [0,1] \to [0,\tfrac{1}{2})$ continuous of the form

$$T_i\colon [0,1] \to Y,$$
$$t \mapsto t\alpha_i + \sqrt{-1}\,a$$

e.g. $\phi(t) = \frac{1}{4}\sin \pi t$. Here a is a point in the so-called fundamental chamber C: $\mathring{I} \supset C := \{x \in V \mid \check{\alpha}_i(x) > 0, i = 1,\ldots,l\}$. Now $p \circ S_i$ and $p \circ T_i$ are closed curves in X and they represent the classes s_i and t_i.

For instance for $\tilde{A}_2$ see the diagram:

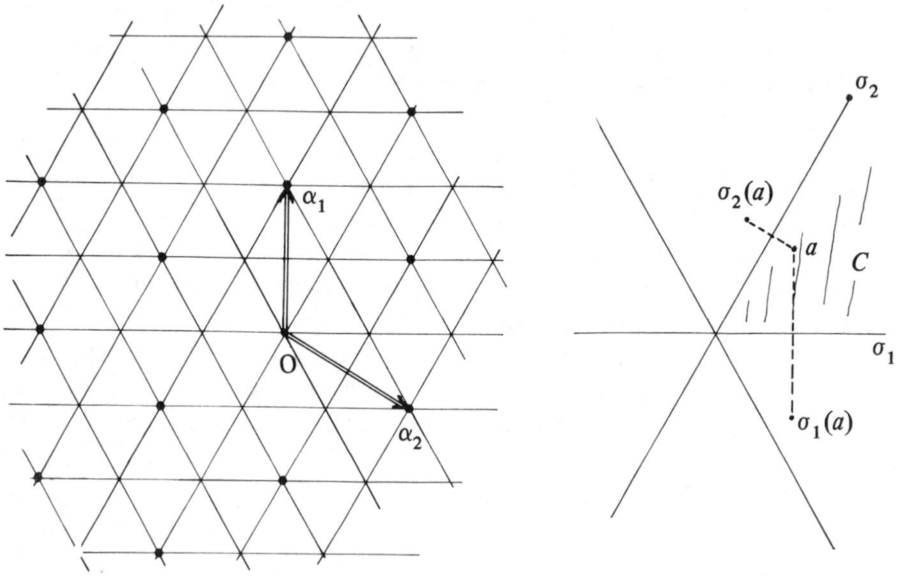

real part imaginary part

The group $A_{\tilde{W}}$ can more conveniently be described in the following way (compare 3.1).

THEOREM. *The fundamental group $\pi_1(X, *)$ can be described as follows: Add to the free product of A_W and Q the following relations:*

$$s_i^{(n)}t = \sigma_i(t)s_i^{(n+\check{\alpha}_i(t))}, \qquad i = 1,\ldots,l, \quad t \in Q, \tag{3.2}$$

where $s_i^{(2m)} := t_i^m s_i t_i^m$ $(m \in \mathbf{Z})$, *and* $s_i^{(2m-1)} := t_i^m s_i^{-1} t_i^m$.

We do not prove the theorem here. Only remark that $s_i^{(n)}$ corresponds to a path in Y which can be described as follows:

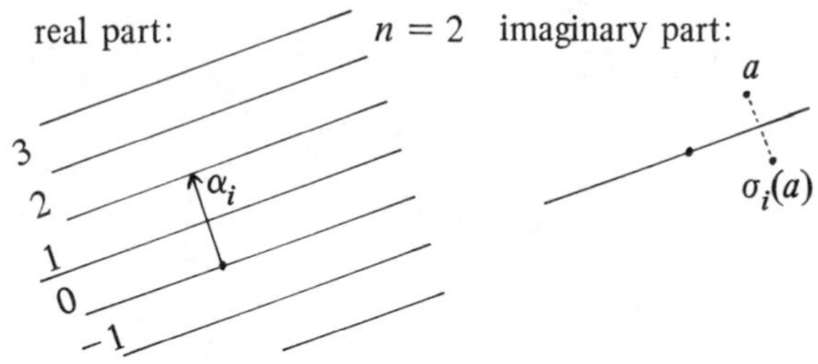

The im. part goes from a to $\sigma_i(a)$. The real part starts and ends in 0, but when the im. part is at level $\check{\alpha}_i = 0$ then the real part has to be in the layer $\{x \in V \mid n < \check{\alpha}_i(x) < n + 1\}$. With this description you can also see in a picture why the relation 3.2 holds.

REFERENCES

1. N. Bourbaki, *Groupes et algèbres de Lie*, Hermann, Paris, 1968, Chapitres 4, 5 et 6.

2. E. Brieskorn, *Singular elements of semi-simple algebraic groups*, Actes Congr. Internat. Math. (Nice, 1970), Gauthier-Villars, Paris, 1971.

3. ______, *Die Fundamentalgruppe des Raumes der regulären Orbits einer endlichen komplexen Spiegelungs gruppe*, Invent. Math. **12** (1971), 57–61.

4. E. Brieskorn and K. Saito, *Artin-Gruppen and Coxeter-Gruppen*, Invent. Math. **17** (1972), 245–271.

5. P. Deligne, *Les immeubles des groupes de tresses généralisés*, Invent. Math. **17** (1972), 273–302.

6. E. Looijenga, *Invariant theory for generalized root systems*, Invent. Math. **61** (1980), 1–32.

7. ______, *Rational surfaces with an anti-canonical cycle* (to appear).

KATHOLIEKE UNIVERSITEIT, NIJMEGEN, THE NETHERLANDS

Proceedings of Symposia in Pure Mathematics
Volume 40 (1983), Part 2

DIFFERENTIABLE STRUCTURES ON COMPLETE INTERSECTIONS. II

ANATOLY S. LIBGOBER AND JOHN W. WOOD[1]

1. Introduction. In this paper we describe some work on the homotopy and diffeomorphism classification of nonsingular complete intersections. This is in part a survey of earlier results (especially [**LW2, LW3**]) and in part an extension of those results to the even-dimensional case. R. Thom observed that the diffeomorphism type of the complete intersection $X_n(\mathbf{d})$ is determined by the dimension n and the multidegree $\mathbf{d} = (d_1,\ldots,d_r)$. In fact these invariants determine the isotopy class of the embedding of X in CP_{n+r}. A basic problem is to describe X up to homotopy or diffeomorphism in terms of a minimal set of invariants computed from n and $\mathbf{d}$. For example when $n = 1$, X is diffeomorphic to a connected sum of tori and the number of summands is determined by the Euler characteristic $e = d\{2 - \Sigma_{j=1}^r(d_j - 1)\}$ where $d = \Pi_{j=1}^r d_j$ is the total degree.

By the Lefschetz theorem on hyperplane sections the inclusion $X_n \hookrightarrow CP_{n+r}$ is an n-equivalence, that is $\pi_i X \to \pi_i CP_{n+r}$ is an isomorphism for $i < n$ and an epimorphism for $i = n$. From this and Poincaré duality it follows that the homology groups of X_n and CP_n are the same except that the middle-dimensional group, $H = H_n(X; Z)$, is free of rank generally much greater than 1. (The rank can be computed from the Euler characteristic.)

Intersection pairing (i.e., Poincaré duality) makes H a unimodular bilinear form space which is symmetric for n even and skew symmetric for n odd. Study of this form leads to connected sum decompositions of X, the first step in our study of differentiable structures. The skew symmetric case gives rise to a Kervaire invariant studied by several authors, see especially [**B3**]. Differentiable structures in this case are studied in part I of this paper [**LW3**]. Connected sum decomposition in the symmetric case is treated in [**LW2**]; we summarize these results in §2.

1980 *Mathematics Subject Classification*. Primary 57R55, 14M10.
[1]The authors were partially supported by the National Science Foundation.

Circumstances in which the homotopy type of X is determined by the integral cohomology ring are presented in §3. For this in §4 we construct degree one maps between nonsingular complete intersections from certain algebraic maps of X to projective space which are one-to-one with singular image, generalizing the notion of cuspidal projection for curves. The application of surgery theory is given in §5 and in §6 results on diffeomorphism, isotopy, and the moduli space of complete intersections are described.

Unless otherwise specified we will always assume $n > 2$; most of our methods do not work when $n = 2$.

The case $n = 3$ illustrates the program we would like to carry out in higher dimensions. A sequence of papers [**W, J, Z**] gives a complete classification of simply connected 6-manifolds up to homotopy or diffeomorphism type. For a complete intersection there is a smooth connected sum decomposition

$$X_3(\mathbf{d}) = M \# S^3 \times S^3 \# \cdots \# S^3 \times S^3$$

where M is simply connected and has the same homology module as CP_3. If $x \in H^2(X; Z)$ is a generator, $x^3 \cap [M] = d$ is the total degree. The homotopy type of M is determined by d, the Stiefel-Whitney class w_2, and when d and w_2 are both even, the Pontryagin class $p_1 \bmod 48$. The diffeomorphism type of M is determined by d, w_2, and p_1; see [**LW3**, §9] for more detail. We have used a computer search to find examples of homotopy equivalence and of diffeomorphism:

$$X_3(15, 14, 3, 3, 2) \simeq X_3(18, 7, 6, 5)$$

and

$$X_3(16, 10, 7, 7, 2, 2, 2) = X_3(14, 14, 5, 4, 4, 4)$$

are the examples in each case with smallest Chern class c_1.

2. Connected sum decompositions. We begin by recalling results which permit us to decompose X up to diffeomorphism as a connected sum $X = M \# N$ where N is a smooth, $(n-1)$-connected, almost parallelizable manifold and where rank $H_n M$ is as small as possible.

First let us consider the case when n is odd. Then the minimal rank of $H_n M$ is either 2 or 0. The former case holds if and only if the following two circumstances happen.

(a) There is no homologically trivial n-sphere embedded in X with nontrivial normal bundle.

In this case one can construct a natural nondegenerate quadratic form on $H_n(X; Z/2)$ associated with the intersection pairing.

(b) The Arf invariant of this quadratic form is nontrivial.

In terms of the multidegree of the complete intersection, (a) holds if and only if the binomial coefficient $\binom{m+l}{m+1}$ is even where $n = 2m + 1 \neq 1$, 3, or 7 and l is the number of even entires in $\mathbf{d} = (d_1, \ldots, d_r)$. Provided (a) holds, the Arf invariant, called the Kervaire invariant $K(X)$, is given as follows:

(i) If d is odd

$$K(x) = \begin{cases} 0 & \text{if } d \equiv \pm 1 \bmod 8, \\ 1 & \text{if } d \equiv \pm 3 \bmod 8. \end{cases}$$

(ii) If d is even $K(X) = 1$ if and only if $n \equiv 1 \bmod 8$, $l = 2$, and $8 \nmid d$.

This was proven by Browder **[B3]** with special cases (and alternate methods) in **[M, W1, W3, L, O]**.

In the piecewise linear or topological category however one always has a decomposition $X \approx K\#N$ where rank $H_n K = 0$. The cohomology structure of K is given by $H^*(K; Z) = Z[x, y]/\{x^m = dy, y^2 = 0\}$, $n = 2m + 1$. We call a simply connected CW-space with this cohomology ring a d-twisted homology CP_n. For $d = 1$ such a space is homotopy equivalent to CP_n. We will call K a *core* and M a *smooth core* of X.

The situation for n even is quite different, rank $H_n K$ can never be 0. The reason is that not all homology classes in $H_n X$ are spherical. More precisely the image of the Hurewicz map $\pi_n X \to H_n X$ is the orthogonal complement in $H_n X$ to the class h Poincaré dual to $x^{n/2}$ where x is the generator of $H^2(X)$. Geometrically h is the class of a section of X by a linear subspace of CP_{n+r} of dimension $n/2 + r$. The following results are proved in **[LW2]**:

(1) There is an orthogonal decomposition $H_n X = A \oplus B$ where $h \in A$ and rank $A \leq 5$.

(2) To any such homology decomposition corresponds a topological connected sum decomposition $X \approx K\#N$ such that $A = H_n K$, $B = H_n N$, and $N - D^{2n}$ is smooth, parallelizable, and $(n-1)$-connected.

But $\partial(N - D^{2n})$ is not necessarily a smooth S^{2n-1}. We call the manifold K corresponding to an A of minimal rank a *core* of X. It depends on choices made in its construction. The precise value of the minimal rank A, in fact the entire structure of A as a unimodular bilinear form space, is determined by d and the type. The type of A is the same as the type of H which is even iff the binomial coefficient $\binom{m+l}{l}$ is even where $n = 2m$ and l is the number of even entries in $\mathbf{d}$, see **[LW2**, §3].

Note that in fact $N = \alpha(S^n \times S^n)\#\beta(V \cup D^{2n})$ where V is the manifold obtained by plumbing of tangent bundles over S^n according to the graph E_8. Here α and β are determined by the Euler characteristic and signature of X and the rank and signature of A. For example if the intersection form on H has even type, then rank $A = 2$, $\alpha = \frac{1}{2}(e(x) - n - 2 - \sigma(x))$, and $\beta = \frac{1}{8}\sigma(X)$.

To construct a smooth connected sum decomposition of X let $\beta = q\beta_n + r$, $0 \leq r < \beta_n$, where β_n is the order of the group of homotopy spheres in dimension $2n - 1$. Since ∂V generates this group, $\beta_n(V \cup D^{2n}) = (V \natural \cdots \natural V) \cup D^{2n}$ is a smooth manifold and we have

(3) $X = M\#N$ where $N = \alpha(S^n \times S^n)\#q\beta_n(V \cup D^{2n})$ and $M = K\#r(V \cup D^{2n})$ are smooth.

M is called a *smooth core* of X; its homotopy type is determined by the homotopy type of the core K and by $\sigma(X)$.

3. Homotopy type. We are interested in the question under what circumstances the homotopy type of $X_n(\mathbf{d})$ or of a core K is determined by simple invariants. For example when n is odd K is a d-twisted homology CP_n. Assuming that d does not have small divisors it is shown in [**LW3**, §2] that K is homotopy equivalent to the $2n$-skeleton of the fibre of the map $CP_n \to K(Z/d, n+1)$. The result is

THEOREM 3.1. *If n is odd and if $p \mid d$ implies $2p \geqslant n+3$, then*:
(i) *Any two d-twisted homology CP_n's are homotopy equivalent.*
(ii) *The homotopy type of $X_n(\mathbf{d})$ is determined by n, d, and the Euler characteristic.*

Note that these invariants are all consequences of the integral cohomology ring. When n is even the core K is more complicated and we do not have a similar, obstruction theoretic characterization. Nevertheless we make the following

Conjecture. If $p \mid d$ implies $2p \geqslant n+3$, then the homotopy type of $X_n(\mathbf{d})$ and its core are determined by the integral cohomology ring.

This is true for n odd and also, for certain multidegrees, when n is even by 3.2 below. On the other hand, 3.4 gives an example where d has a small divisor in which cohomology operations distinguish between homotopy types. Under the assumption of the conjecture, however, by 3.5 cohomology operations are determined by the ring structure. We pose the question when the homotopy type or p-homotopy type is determined by the integral cohomology, for example assuming $2p \geqslant n+3$. A related fact due to Deligne, Griffiths, Morgan, and Sullivan is that any Kähler manifold is formal (see [**D**]) and in particular the rational homotopy type is determined by the rational cohomology. Pete Bousfield pointed out to us that if a finite complex is formal its p-homotopy type is determined for sufficiently large p.

W. Browder has made the appealing conjecture that the homotopy type of a complete intersection is determined by the integral cohomology ring and the type of the tangent sphere bundle as a spherical fibration.

We say a multidegree is pairwise relatively prime if d_i is prime to d_j for $i \neq j$.

THEOREM 3.2. *Let X and X' be two n-dimensional complete intersections with pairwise relatively prime multidegrees. Then they have homotopy equivalent cores if and only if they have the same total degree.*

PROOF. Note that since we assume $n > 2$, the total degree is determined by the cohomology ring of the core so the condition is necessary. Let $p_1^{e_1} \cdots p_k^{e_k}$ be the prime decomposition of the total degree of X. We will show that X has a core homotopy equivalent to a fixed core $\overline{K}$ of $\overline{X} = X_n(p_1^{e_1}, \ldots, p_k^{e_k})$. In the next section we construct a map $\varphi\colon X \to \overline{X}$ such that $\varphi^*\bar{x} = x$ where x and $\bar{x}$ are generators for $H^2(\ ;Z)$. Since X and $\overline{X}$ have the same total degree, φ is a map of degree one.

The theorem then follows from

PROPOSITION 3.3. *Let $\varphi\colon X \to \bar{X}$ be a continuous map of degree one between complete intersections of dimension $n > 2$ which commutes up to homotopy with the inclusion in projective space and let $\bar{K}$ be a given core of $\bar{X}$. Then there is a core K of X which is homotopy equivalent to $\bar{K}$.*

PROOF. Let $\bar{\pi}\colon \bar{X} \to \bar{K}$ be the map collasping $\bar{N}$ to a point. If we split X as $K\#N$ where $\bar{\pi} \circ \varphi$ can be factored as $\bar{\varphi} \circ \pi$,

$$\begin{array}{ccc} X & \xrightarrow{\varphi} & \bar{X} \\ \pi \downarrow & & \downarrow \bar{\pi} \\ K & \xrightarrow{\bar{\varphi}} & \bar{K} \end{array}$$

so that $\bar{\varphi}$ induces an isomorphism on homology, then by Whitehead's theorem $\bar{\varphi}$ will be a homotopy equivalence.

Suppose first that n is even. Since $\bar{\pi} \circ \varphi$ has degree one, Poincaré duality defines a splitting $\beta\colon H_*\bar{K} \to H_*X$ of $\bar{\pi} \circ \varphi$, cf. [**B1**, I.2.5]. It follows from the definition of β that $\beta(h_{\bar{x}}) = h_x$ so we may take $A = \beta H_n\bar{K}$ as the unimodular summand of H_nX containing h_x and apply (2) of §2 to obtain a connected sum decomposition $X \approx K\#N$. Now $N - D^{2n}$ has the homotopy type of a bouquet of n-spheres. Such a sphere, $f\colon S^n \to X$, satisfies $f_*[S^n] \in A^\perp$, so $\varphi_*f_*[S^n] \in (H_n\bar{K})^\perp$. Since $\pi_n\bar{X}$ is a subgroup of $H_n\bar{X}$ [**W2**, Lemma 2] it follows that $\bar{\pi} \circ \varphi \circ f$ is null-homotopic so that $\bar{\pi} \circ \varphi$ can be factored as required above.

Now suppose n is odd. Then $\pi_nX = Z/d \oplus H_nX$ and similarly for $\bar{X}$; further $\varphi_\#$ restricts to an isomorphism on Z/d. Then $\pi_n\bar{K} = Z/d$ and $(\pi \circ \varphi)_\#$ restricts to an isomorphism. Hence the symplectic basis for H_nX given by embedded copies of $S^n \vee S^n$ in X used in the handle removing argument of [**W1**, §1] can be modified so that these spheres lie in $\ker(\bar{\pi} \circ \varphi)_\#$. This basis gives rise to a decomposition $X \approx K\#N$ such that $\bar{\pi} \circ \varphi$ factors through K as required. This completes the proof of Proposition 3.3.

The rest of §3 is not essential for the sequel; it concerns when cohomology operations can be used to distinguish homotopy types of cores and when they are formal consequences of the ring structure.

PROPOSITION 3.4. *If $2p \leqslant n + 2$, $X_n(p, p)$ and $X_n(p^2)$ do not have homotopy equivalent cores.*

PROOF. We will show these cores are distinguished by their mod p characteristic class q_1, see [**MS**, p. 229]. First there is a Wu formula characterizing q_1 in terms of the action of the mod p Steenrod algebra, hence q_1 is a homotopy invariant. Next $\pi^*q_1K = q_1X$ where $\pi\colon X \to K$ is the collapsing map since q_1 is determined by the Pontryagin classes $p_1,\ldots,p_l$ $(l = (p-1)/2)$ and these correspond under π^* for $2(p-1) < 2n$ since N is almost parallelizable. Hence $q_1(K)$ is determined by $q_1(X)$. Finally q_1 belongs to a multiplicative sequence in the Pontryagin classes and so can be computed from its value on line bundles: $q_1(\gamma^{\otimes d}) = (dx)^{p-1}$. It follows tht $q_1(X_n(p, p)) = (n + 3 - 2p^{p-1})x^{p-1} \bmod p$ and $q_1(X_n(p^2)) = (n + 2 - p^{2(p-1)})x^{p-1} \bmod p$. For $2p \leqslant n + 2$, the class x^{p-1} is indivisible, hence these classes are different mod p.

For $p = 2$ replace q_1 by w_2 in the argument. Except when $p = 2$ and $n \equiv 2$ or 4 mod 8, the cores have isomorphic cohomology rings. (In the exceptional case the type of the intersection pairing is different.)

This method of detecting counterexamples cannot work when d does not have small divisors:

THEOREM 3.5. *If the total degree d of $X_n(\mathbf{d})$ satisfies $p \mid d \Rightarrow 2p \geqslant n + 3$, then the action of the Steenrod algebra (for any prime) is determined by the cohomology ring of X.*

PROOF. If n is odd this follows from 3.1 so we may assume n is even. Let E be the subgroup of $H^n(X; Z)$ dual to the vanishing cycles, $E = h^{\perp}$. The subgroup generated by $x^{n/2}$ and E has index d in $H^n(X; Z)$. Fix a prime p. Since $\dim x = 2$, $P^j x^k = \binom{k}{j} x^{k+j(p-1)}$ (see [**SE**, p. 78]) and we need only consider the action of P^j on $H^n(X; Z/p)$. The only primes p for which $P^j(H^n)$ lies in a nonzero group satisfy $2p \leqslant n + 2$ and these primes do not divide d. Therefore $x^{n/2}$ and E generate $H^n(X, Z/p)$. But the Steenrod operations vanish on E by the following, since X is a hyperplane section.

LEMMA 3.6. *Let V_n be a nonsingular hyperplane section and let E be the subgroup of $H^n(V)$ dual to the vanishing cycles. Then the Steenrod operations vanish on E.*

PROOF. E is generated by classes dual to homology classes represented by embedded spheres with normal bundle isomorphic to their tangent bundle [**AF**]. Let $f\colon S^n \hookrightarrow V$ be dual to $\alpha \in H^n(V)$, let N be the normal bundle of f, and N^* the Thom space. Let $p\colon V \to N^*$ denote the map which collapses the complement of N to the base point. Then $p^*(u) = \alpha$ where u generates $H^n(N^*)$. By naturality it is enough to show the Steenrod operations vanish on u. This follows from the fact that N^* is homeomorphic to the product $S^n \times S^n$ modulo the diagonal.

4. Special projections. The existence of the degree one map between complete intersections needed for 3.2 is a consequence of the following

LEMMA 4.1. *If a and b are relatively prime, then there is a degree one map*

$$\varphi\colon X_n(ab, c_1, \ldots, c_k) \to X_n(a, b, c_1, \ldots, c_k).$$

PROOF. By repeated use of Bertini's theorem the hypersurfaces given by the equations

$$z_0^a + \cdots + z_m^a = 0,$$

$$\lambda_0 z_0^b + \cdots + \lambda_m z_m^b = 0, \qquad \lambda_0 \neq 0,$$

and k equations of the form

$$f_i = \mu_1 z_1^{c_i} + \cdots + \mu_m z_m^{c_i} = 0, \qquad i = 1, \ldots, k,$$

have nonsingular intersection for generic choice of the coefficients. Bertini's theorem applies since the only base point of such a system, $[1,0,\dots,0]$, does not satisfy the first equation. Thus we may represent $X_n(a, b, c_1,\dots,c_k)$ up to diffeomorphism by this complete intersection, cf. [**W3**, §2].

The image of this variety under projection to CP_{m-1} from $[1,0,\dots,0]$, that is under the map sending $[z_0,\dots,z_m]$ to $[z_1,\dots,z_m]$, is given by the equations

$$(-z_1^a - \cdots - z_m^a)^b = \left(-\frac{\lambda_1}{\lambda_0} z_1^b - \cdots - \frac{\lambda_m}{\lambda_0} z_m^b\right)^a$$

and $f_i = 0$ for $i = 1,\dots,k$. The image is a singualr variety V. If a given point $[z_1,\dots,z_m] \in V$ is the image of $[z_0, z_1,\dots,z_m]$, then z_0^a and z_0^b are determined by $(z_1,\dots,z_m)$. Since a and b are relatively prime, z_0 is determined. Therefore π: $X_n(a, b, c_1,\dots,c_k) \to V$ is 1-1 and hence a homeomorphism.

Let ψ be the retraction induced by degeneration of a nonsingular complete intersection $X_n(ab, c_1,\dots,c_k)$ onto V. Then $\varphi = \pi^{-1} \circ \psi$. These maps commute with inclusion into CP_m so the generators of $H^2(; Z)$ correspond. Since the total degrees are the same, φ has degree one.

Another consequence of this construction is

COROLLARY 4.2. *If* $\mathbf{d} = (d_1,\dots,d_r)$ *is pairwise relatively prime then there is a* 1-1 *algebraic map of* $X_n(\mathbf{d})$ *onto a singular hypersurface* V *of degree* d. *In particular* $X_n(\mathbf{d})$ *and* V *are homeomorphic.*

Given $X_n(\mathbf{d}) \subset CP_{n+r}$ the question of algebraically embedding X in lower codimension has been much studied. Very recently L. Astey and S. Gitler have obtained nonimmersion results in the smooth category. In the case of curves, 1-1 algebraic maps are called cuspidal projections and have been studied by R. Piene [**P**]. §3 implies necessary conditions for such maps to exist.

PROPOSITION 4.3. *Let* $\overline{X} = X_n(\mathbf{d})$ *be a complete intersection which admits a* 1-1 *algebraic map into* CP_{n+1}. *Then* $\overline{X}$ *and the hypersurface* $X = X_n(d)$ *have homotopy equivalent cores.*

PROOF. The image of $\overline{X}$ in CP_{n+1} is a singular hypersurface V and the map $\overline{X} \to V$ is a homeomorphism. The retraction of X on V provides a continuous, degree one map $X \to \overline{X}$ to which Proposition 3.3 applies. Hence the cores are homotopy equivalent.

As a result applying 3.4 we get

COROLLARY 4.4. *If* $2p \leqslant n + 2$, $X_n(p, p)$ *does not admit a* 1-1 *algebraic map into* CP_{n+1}.

5. Application of surgery theory. In this section we apply the exact sequence of surgery theory to a smooth core of an even-dimensional complete intersection. The treatment here is analogous to that in [**LW3**, §§3 and 4] for the case when M

is a d-twisted homology CP_n for n odd. The main consequence is

THEOREM 5.1. *The number of distinct diffeomorphism classes of complete intersections of even dimension n and total degree d with a core K of a fixed homotopy and with given Euler characteristic and Pontryagin classes is less than a bound depending only on n.*

Let M be a fixed smooth core and denote by $hS(M)$ the set of smooth manifolds homotopy equivalent to M.

LEMMA 5.2. *If the middle Betti number of M is $\leqslant b$, the number of elements of $hS(M)$ with a given set of Pontryagin classes is bounded by a function of n and b.*

PROOF. The exact sequence of surgery theory gives an inclusion of $hS(M)$ in $[M, G/O]$, see [**B1**, II4.10]. There is also an exact sequence

$$\tilde{\pi}_s^0(M) \to [M, G/O] \to \widetilde{KO}^0(M)$$

where the first term is reduced stable cohomotopy and the last is reduced real K-theory, see [**LW3**, §3]. (This sequence is induced from the sequence of fibrations $SG \to G/O \to BSO$.) If the homotopy equivalence $f\colon M_1 \to M$ represents an element of $hS(M)$, its image in $KO^0(M)$ is given by $f^{-1*}\tau M_1 - \tau M$. If M_1 and M have the same Pontryagin classes, then this image lies in the torsion subgroup of $KO^0(M)$, cf. [**B2**, Lemma 2.24]. Also the set of elements in $[M, G/O]$ with the same image in $KO^0(M)$ is bounded by the order of $\tilde{\pi}_s^0(M)$. Thus Lemma 5.2 follows from

LEMMA 5.3. *Let M be a core of a complete intersection of even dimension n with middle Betti number less than or equal to b. Then*:

(i) *$\pi_s^0(M)$ is a finite group of order bounded by a function of b and n (and independent of the degree).*

(ii) *$KO(M)$ has only even torsion of order $\leqslant 2^b$.*

PROOF. The first assertion follows immediately from the Atiyah-Hirzebruch spectral sequence for stable cohomotopy. The bound depends on b and the orders of the stable homotopy groups of spheres in dimensions $\leqslant 2n$, see [**LW3**, Theorem 3.3(ii)].

Since $H^*(M; Z)$ is free and concentrated in even dimensions, the corresponding spectral sequence for KO-theory shows that $KO(M)$ has no odd torsion. Replacing the map $f\colon M \to CP_n$ by an inclusion the cohomology of the pair is given by

$$H^q(CP_n, M; Z) = \begin{cases} \text{free of rank} \leqslant b-1, & q = n+1, \\ Z/d, & q = n+3, n+5, \ldots, 2n+1, \\ 0, & \text{otherwise.} \end{cases}$$

It follows from the spectral sequence for the pair that the even torsion in $KO^1(CP_n, M)$ is at most of order 2^{b-1}. Since the torsion in $KO^0(CP_n)$ has order at most 2, assertion (ii) follows.

For the odd-dimensional case in [**LW3**, §§4 and 5] with additional assumptions more precise bounds are obtained.

PROOF OF 5.1. For any X in a set of complete intersections with fixed invariants as described in the theorem let $X = M \# N$ be a differentiable connected sum decomposition where M is a minimal smooth core. The Pontryagin classes of X determine the signature so by §2 the diffeomorphism type of N is the same for each X as is the homotopy type of M. The middle Betti number of M is bounded by $b = 5 + 8\beta_n$ which depends only on n. The Pontryagin classes of M are determined by those of X for if $\pi: X \to M$ is the collapsing map, $\pi^*: H^*(M; Z) \to H^*(X; Z)$ is injective and $\pi^* p_j(M) = p_j(X)$ for $2j < n$ since the summand N is almost parallelizable. The top Pontryagin class is then determined by the signature, $\sigma(M) = \sigma(X) - \sigma(N)$. Therefore the theorem follows from Lemma 5.2.

6. Diffeomorphic complete intersections, isotopy, and moduli spaces. As a consequence of the results of §§3 and 5 we have the following

THEOREM 6.1. *In any dimension $n \neq 2$ and for any integer k there are k distinct multidegrees for which the corresponding complete intersections are all diffeomorphic.*

PROOF. For n odd see [**LW3**, 6.3]. We give here the proof for n even. It suffices to produce a sufficiently large number of complete intersections with the same invariants as required in Theorem 5.1. The key result is a counting argument due to A. O. L. Atkin, see [**LW3**, §6] for the proof.

PROPOSITION 6.2. *Given integers n and N there are integers r and d such that the number of distinct unordered r-tuples $(d_1, \ldots, d_r)$ with product d and with the same first n symmetric functions is greater than N. Moreover d may be taken to be a product of $2r$ distinct primes.*

It follows from [**LW3**, 6.1] that the Euler characteristic and Pontryagin classes of the corresponding n-dimensional complete intersections are the same. Since each multidegree is pairwise relatively prime, the cores are homotopy equivalent by 3.2. The result follows by taking N greater than k times the bound in 5.1.

For $n = 2$ the recent work of Mike Freedman shows that the homeomorphism type of these manifolds is determined by the homotopy type and hence there are many homeomorphic complete intersections with different multidegrees.

We conclude by quoting some results from [**LW3**] which hold for both even and odd n and which relate equivalence up to diffeomorphism to isotopy and analytic equivalence. First, examples of 5.1 never exist with small codimension because in that case the differential structure on X determines the multidegree. More precisely we have

THEOREM 6.3. *If $r = \operatorname{codim}(X_n \subset CP_{n+r}) \leqslant n/2$, $n > 2$, then d and the Pontryagin classes of X_n determine certain symmetric functions of d from which r and $d_1, \ldots, d_r$ can be recovered. If r is given and $r \leqslant 1 + n/2$, $d_1, \ldots, d_r$ can still be recovered.*

An example shows the conditions on r are sharp: $X_3(12,10)$ and $X_3(15,4,2)$ have the same degree and Pontryagin class, in fact $X_3(12,10) = X_3(15,4,2) \# 13440(S^3 \times S^3)$. By 6.3 this would not be possible if both had codimension 2 or if either had codimension 1.

Using this result and Thom's observation for the low codimension case and Haefliger's work on isotopy classes of embeddings in higher codimension we deduce

THEOREM 6.4. *If X_n and Y_n are complete intersections in CP_{n+r} and if X_n is diffeomorphic to Y_n, then there is a diffeomorphism of CP_{n+r} isotopic to the identity carrying X_n to Y_n.*

On the other hand elementary considerations show

THEOREM 6.5. *If X_n and Y_n are complete intersections of dimension $n \geqslant 2$ (and with $c_1 \neq 0$ if $n = 2$) and if X_n is analytically equivalent to Y_n, then there is a projective linear transformation of CP_{n+r} carrying X_n to Y_n. Further they have the same multidegree.*

As was pointed out by B. Moishezon this implies

COROLLARY 6.6. *The moduli space of complex structures on the smooth manifold underlying a complete intersection can have arbitrarily many irreducible components.*

Indeed if X and Y are complete intersections in the same component they are connected by a sequence of small deformations. Then by a result of Sernesi [S] X would be analytically equivalent to a complete intersection of the same multidegree as Y so by 6.5 X and Y would have the same multidegree. Thus 6.1 implies the corollary.

REFERENCES

[AF] A. Andreotti and T. Frankel, *The second Lefschetz theorem on hyperplane sections*, Global Analysis, Princeton Univ. Press, Princeton, N.J., 1969, pp. 1–20.

[B1] W. Browder, *Surgery on simply connected manifolds*, Springer-Verlag, Berlin and New York, 1972.

[B2] ———, *Surgery and the theory of differentiable transformation groups*, Proc. Tulane Sympos. on Transformation groups, Springer-Verlag, Berlin and New York, 1968, pp. 1–46.

[B3] ———, *Complete intersections and the Kervaire invariant*, Lecture Notes in Math., vol. 763, Springer-Verlag, Berlin and New York, 1979, pp. 88–108.

[D] P. Deligne, P. Griffiths, J. Morgan and D. Sullivan, *Real homotopy theory of Kähler manifolds*, Invent. Math. **29** (1975), 245–274.

[J] P. Jupp, *Classification of certain 6-manifolds*, Proc. Cambridge Philos. Soc. **73** (1973), 293–300.

[L] A. Libgober, *Geometric procedure for killing the middle-dimensional homology groups of algebraic hypersurfaces*, Proc. Amer. Math. Soc. **63** (1977), 198–202.

[LW1] A. Libgober and J. Wood, *Diffeomorphic complete intersections with different multidegrees*, Bull. Amer. Math. Soc. (N. S.) **2** (1980), 459–461.

[LW2] ———, *On the topological structure of even dimensional complete intersections*, Trans. Amer. Math. Soc. **267** (1981), 637–660.

[LW3] ———, *Differentiable structures on complete intersections*. I, Topology **21** (1982), 469–482.

[MS] J. Milnor and T. Stasheff, *Characteristic classes*, Princeton Univ. Press, Princeton, N.J., 1974.

[M] S. Morita, *The Kervaire invariant of hypersurfaces in complex projective space*, Comment. Math. Helv. **50** (1975), 403–419.

[O] S. Ochanine, *Signature modulo* 16, *invariants de Kervaire généralisés et nombres caractéristiques dans la K-théorie réelle*, Mem. Soc. Math. France (N.S.) **5**.

[P] R. Piene, *Cuspidal projections of space curves*, Math. Ann. **256** (1981), 95–119.

[S] E. Sernesi, *Small deformations of global complete intersections*, Boll. Un. Mat. Ital. **12** (1975), 138–146.

[SE] N. Steenrod and D. B. A. Epstein, *Cohomology operations*, Princeton Univ. Press, Princeton, N.J., 1962.

[W] C. T. C. Wall, *Classification problems in differential topology*. V. *On certain* 6-*manifolds*, Invent. Math. **1** (1966), 355–374.

[W1] J. Wood, *Removing handles from nonsingular algebraic hypersurfaces in* CP_{n+1}, Invent. Math. **31** (1975), 1–6.

[W2] ______, *A connected sum decomposition for complete intersections*, Proc. Sympos. Pure Math., vol. 32, 2, Amer. Math. Soc., Providence, R.I, 1978, pp. 191–193.

[W3] ______, *Complete intersections as branched covers and the Kervaire invariant*, Math. Ann. **240** (1979), 223–230.

[Ž] A. V. Žubr, *Classification of simply connected six-dimensional spinor manifolds*, Math. U.S.S.R.-Izv. **9** (1975), 793–812.

UNIVERSITY OF ILLINOIS AT CHICAGO

Proceedings of Symposia in Pure Mathematics
Volume 40 (1983), Part 2

ALEXANDER INVARIANTS OF PLANE ALGEBRAIC CURVES

A. LIBGOBER[1]

1. Introduction. The aim of this paper is to describe our work [**7, 8, 9**] on the fundamental groups of the complement to irreducible plane algebraic curves. The approach is based on the methods used so far for the study of knot groups. We associate with such a curve an invariant of the fundamental group of the complement, which we call the Alexander module of the curve, and which is essentially equivalent to the factor of this fundamental group by its second commutator subgroup. We show how it depends on the types of singularities of the curve and compute the "rationalized" Alexander module in terms of the position of singularities. The key tool is the study of the surfaces associated with any curve C, namely the cyclic coverings of the projective plane with branching set C and possibly the line L in infinity.

The study of the plane singular curves was initiated by O. Zariski in the beautiful series of papers [**17–22**] which appeared between 1929 and 1936. Most of the ideas described below are taken from those works. In particular Zariski observed and exploited the connection between nontriviality of the first Betti number of the cyclic branched covering and noncommutativity of the fundamental group of the complement to the branching locus. Our Alexander modules give the quantitative measure of this relationship.

2. Experimental data: examples. Before explaining our main results on the fundamental groups of the complement to the plane curves we shall describe known examples of computations of those groups. From now on we shall refer to the fundamental groups of the complement to the plane curve as the groups of the curve (cf. the terminology: the knot group is the fundamental group of the complement to the knotted sphere). Note that constructing computable examples is not an easy task.

1980 *Mathematics Subject Classification.* Primary 14F20, 57M05, 14H20, 57M10, 14J17, 57M15.
[1]Supported by an NSF Grant.

0082-0717/81/0000-C192/$03.25

EXAMPLE 2.1. Let C be a nonsingular algebraic curve of degree d in P^2. Then $\pi_1(P^2 - C) = Z/d$ is the cyclic group of order d. If C has as singular points only nodes (i.e. all singularities locally can be given by equation $x^2 - y^2 = 0$) then $\pi_1(P^2 - C) = Z/d$ as well. This was known as Zariski's conjecture and proved by Deligne and Fulton in 1979 [**4, 6**].

EXAMPLE 2.2. Let V be a nonsingular algebraic surface and $\varphi\colon V \to P^2$ be a generic projection. Let $C \subset P^2$ be the branching locus of φ. Then C has nodes and cusps as singularities (the latter locally are given by equation $x^2 = y^3$), and C usually has an interesting fundamental group. The concrete computations were made by B. Moishezon [**12**] in the case where V is a hypersurface in P^3. Let n be the degree of the hypersurface V. Then C is the curve of degree $n(n-1)$ with $n(n-1)(n-2)(n-3)/2$ nodes and $n(n-1)(n-2)$ cusps; $\pi_1(P^2 - C)$ is the Artin's braid group $B(n)$ (see [**1**]), factorized by the center. In the case $n = 3$ this was computed already by Zariski [**17**]: one obtains the famous six-cuspidal sextic with all cusps belonging to a conic, and $\pi_1(P^2 - C)$ is given in [**17**] as the free product $Z/2 * Z/3$.

EXAMPLE 2.3. The following construction goes back to Zariski (cf. [**20**] and also [**5**]). Let V be a nonsingular projective variety over $\mathbf{C}$, L an invertible sheaf on V, and E a subspace of $H^0(V, L)$. Then $P(E)$, the projective space associated to E, can be thought as the set of divisors D_x ($x \in P(E)$) on V belonging to the linear system E. Let $\mathrm{Disc}(E)$ be the subset of $P(E)$ such that D_x is singular.

In many cases $\mathrm{Disc}(E)$ is the singular hypersurface in $P(E)$ with nontrivial $\pi_1(P(E) - \mathrm{Disc}(E))$. For generic plane P^2 in $P(E)$, $\mathrm{Disc}(E) \cap P^2$ is the curve C for which by Zariski's theorem [**21**] $\pi_1(P^2 - C) = \pi_1(P(E) - \mathrm{Disc}(E))$, i.e. one obtains C with a nontrivial fundamental group. The concrete computations were made in the following cases:

(a) (Zariski [**20**]). Let $V = P^1$, $L_1 = O(n)$ and $E = H^0(P^1, O(n))$. In down to earth terms $P(E)$ can be thought of as the set of unordered n-tuples of points on P^1 (which is parametrized by P^n) and $\mathrm{Disc}(E)$ as the set of n-tuples which have coincident points. The curve $C = \mathrm{Disc}(E) \cap P^2$ for generic P^2 has degree $2(n-1)$, $(3n-6)$ cusps and $2(n-2)(n-3)$ nodes (in fact C is dual to a generic rational nodal curve of degree n). $\pi_1(P^2 - C)$ is the braid group of sphere (see [**1**]), i.e. has a presentation with generators $\sigma_1 \cdots \sigma_{n-1}$ and relations $\sigma_i\sigma_j = \sigma_j\sigma_i$ for $|i-j| \geqslant 2$, $\sigma_i\sigma_{i+1}\sigma_i = \sigma_{i+1}\sigma_i\sigma_{i+1}$, and $\sigma_1 \cdots \sigma_{n-2}\sigma_{n-1}^2\sigma_{n-2} \cdots \sigma_1 = 1$.

(b) (cf. [**5**]). Let $V = P^2$ (resp. $V = P^1 \times P^1$), $L = O_{P^2}(3)$, (resp. $L = O_{P^1}(2) \otimes O_{P^1}(2)$) and E be the complete linear system defined by L. In this case $P(E) - \mathrm{Disc}(E)$ can be thought of as the space of nonsingular plane cubic curves (resp. nonsingular sections of a fixed quadric in P^3 by other quadrics). The construction above yields the curve C_3 of degree 12 with 24 cusps and 21 nodes (resp. the curve $C_{2,2}$ of degree 12 with 24 cusps and 22 nodes). The fundamental groups of those curves are given as the following semidirect products:

$$0 \to K_{27} \to \pi_1(P^2 - C_3) \to \mathrm{SL}_2(Z) \to 1,$$
$$0 \to Q_{64} \to \pi_1(P^2 - C_{2,2}) \to \mathrm{SL}_2(Z) \to 1$$

where K_{27} (resp. Q_{64}) is the unique nonabelian group of order 27 (resp. 64) and of exponents 3 (resp. 4).

Note that the partial information on $\pi_1(P(H^0(P^3, O_{P^3}(3)) - \text{Disc}(E)))$ (i.e. on the π_1 for the space of nonsingular cubic surfaces) is obtained in [**10**].

(c) [**22**]. Let V be an elliptic curve, D be any divisor of degree n, and $E = H^0(V, L(D))$. The curve $C = \text{Disc}(E) \cap P^2$ for generic P^2 is the curve of degree $2n$ with $3n$ cusps and $2n(n-3)$ nodes. It can be identified with the dual to generic plane curve of degree n and genus 1. The fundamental group $\pi_1(P^2 - C)$ can be identified with $H_n = \text{Ker}(B_n(T) \xrightarrow{\varphi} H_1(T))$ where $B_n(T)$ is the braid group of torus T [**1**], and φ is the natural projection. In [**22**] Zariski gives an explicit presentation for H_n.

EXAMPLE 2.4 (OKA [**13**]). Let $O_{p,q}$ be given by equation $(X^p + Y^p)^q + (Y^q + Z^q)^p = 0$ where p and q are relatively prime. This curve has pq singularities given locally by equation $X^p + Y^q = 0$. Oka's computation yields $\pi_1(P^2 - O_{p,q}) = Z_p * Z_q$. (Case $p = 2$, $q = 3$ is again the case considered by Zariski in Example 2.2 ($n = 3$).)

3. Definition of the Alexander module. Although in the geometric problems the main role is apparently played by fundamental groups of the complement to projective curves, we will study the groups of curves in an affine plane. Moreover for simplicity we shall assume that the line L in infinity is in a general position relative to C. Most of the results below can be modified to include the case of arbitrary position of L (cf. [**7**]). In the case of generic L the groups of affine and projective curves are related by the central extension (cf. [**19**])

$$0 \to Z \to \pi_1(P^2 - (C \cup L)) \to \pi_1(P^2 - C) \to 0. \tag{3.1}$$

Therefore essentially no information is lost by switching to affine curves.

Let G denote $\pi_1(P^2 - (C \cup L))$, and G' and G'' be the first and the second commutator subgroups of G. We have the extension

$$0 \to G'/G'' \to G/G'' \to G/G' \to 0. \tag{3.2}$$

One can easily see that

$$G/G' = H_1(P^2 - (C \cup L), Z) = Z. \tag{3.3}$$

Hence to describe G/G'' it is enough to describe G'/G'' as the module over the Z-group ring of Z, i.e. over the ring $Z[t, t^{-1}]$ of Laurent polynomials with integral coefficients.

DEFINITION 3.1. The Alexander module $A_1(Z)$ of a plane algebraic curve is the group G'/G'' considered as a $Z[t, t^{-1}]$-module.

One can give a slightly more geometric description of the Alexander modules as follows (cf. [**11**]). The group G'/G'' can be identified with $H_1(\widetilde{P^2 - C \cup L})$, the first homology group of the infinite cyclic covering of $P^2 - C \cup L$. The action of $G/G' = Z$ on G'/G'' from the sequence (3.2) coincides with the action of Z on

G'/G'' by the deck transformation. This motivates the following

DEFINITION 3.2. Let R be a ring. The ith Alexander module of the curve C with coefficients in R is the group $H_i(P^2 - C \cup L, R)$ considered as a $R[t, t^{-1}]$ module. If R is a field then the Alexander modules have a more simple structure.

LEMMA 3.1 (CF. [**11**, 7]). *Assume that R is a field. Then $A_1(R)$ is a $R[t, t^{-1}]$-torsion module. In particular, because $R[t, t^{-1}]$ is a principal ideal domain, we have the cyclic decomposition*

$$A_1(R) = \bigoplus_i R[t, t^{-1}]/(\lambda_i) \tag{3.4}$$

where (λ_i) is the principal ideal generated by the polynomial λ_i.

DEFINITION 3.3. The (global) Alexander polynomial $\Delta_C(R)$ of the curve C is $\prod_i \lambda_i$.

In the cases when the fundamental group of C is known one can easily compute the corresponding Alexander module and polynomial (cf. [7, **8**]). Main results of this paper give information on $A_1(Q)$ and $\Delta_C(R)$ which does not depend on the knowledge of $\pi_1(P^2 - C)$. Here are the computations of Alexander polynomials for examples given in §2.

EXAMPLE 3.1. Let C_n be the branching curve of a generic projection of a nonsingular hypersurface in P^3. Then

$$\Delta_{C_n}(Z) = \begin{cases} t^2 - t + 1, & n = 3, 4, \\ 1, & n \geqslant 5. \end{cases} \tag{3.5}$$

EXAMPLE 3.2. Let C be a 3-cuspidal quartic (Example 2.3(a), $n = 3$). Then $\pi_1(P^2 - C)$ is the metacyclic group of order 12, and

$$\Delta_C(F_p) = \begin{cases} 1, & p \neq 3, \\ t + 1, & p = 3 \end{cases} \tag{3.6}$$

(F_p is the prime field of characteristic p).

DEFINITION 3.4. Let p_i be a singularity of a curve C. The local Alexander polynomial of p_i is the characteristic polynomial of the monodromy operator of the singularity p_i. Equivalently, the Alexander polynomial of p_i is the Alexander polynomial in the sense of the knot theory of the link of singularity p_i. We denote it by $\Delta_{p_i}(R)$.

EXAMPLE 3.3. The local Alexander polynomial of the node is equal to $t - 1$.

The local Alexander polynomial of the cusp $x^2 = y^3$ is equal to $t^2 - t + 1$. More generally for the singularity $x^P = y^q$ with relatively prime p and q one obtains as the Alexander polynomial

$$\Delta_{p,q} = \frac{(t^{pq} - 1)(t - 1)}{(t^p - 1)(t^q - 1)}.$$

REMARK 3.1. The definition of the local Alexander polynomial requires modification in the case when L is not in general position relative to C. Then one should consider the points of tangency of L and C as singular points (cf. [7]).

DEFINITION 3.5. Let $S^3 = \partial T(L)$ be the boundary of the tubular neighbourhood of L. The Alexander polynomial of C in infinity is the Alexander polynomial of the link $C \cap \partial T(L) \subset \partial T(L)$. We denote it as $\Delta_{\infty,C}(R)$.

EXAMPLE 3.4 (CF. [7]). Let C be in general position relative to L, and let the degree of C equal d. Then

$$\Delta_{\infty,C}(R) = (t-1)(t^d-1)^{d-2}.$$

4. Divisibility theorem and applications.

THEOREM 4.1 [7]. *Let R be a field and let C be an irreducible plane algebraic curve. Then*:

(1) $\Delta_C(R)$ *divides the product* $\prod \Delta_{p_i,c}(R)$ *of the local Alexander polynomials of all singularities.*

(2) $\Delta_C(R)$ *divides* $\Delta_{\infty,C}(R)$.

We shall indicate two applications of this theorem by specializing it to the case of curves which have as singularities only cusps and nodes. The first one is

COROLLARY 4.1 (CF. [8]). *Let C be a cuspidal curve.*

(a) *If* $\deg C \equiv \pm 1\ (6)$ *then G' is a perfect group.*

(b) *If* $\deg C \equiv \pm 2\ (6)$ *then G'/G'' has only 3-torsion.*

(c) *If* $\deg C \equiv 3\ (6)$ *then G'/G'' has only 2-torsion.*

Indeed, the Alexander polynomial $\Delta_C(F)$ should divide $\prod \Delta_{p_i,C} = (t^2-t+1)^{\kappa}\cdot(t-1)^{\delta}$ and $\Delta_{\infty,C} = (t-1)(t^d-1)^{d-2}$, where κ is the number of cusps and δ is the number of nodes. If $p=2$ (resp. $p=3$), then those two polynomials do not have common roots besides $t=1$ (i.e., $\Delta_C(F_p)=1$) unless $\deg C \equiv 0, \pm 2\ (6)$ (resp. unless $\deg C \equiv 0, 3\ (6)$).

For other p's $\prod \Delta_{p_i,C}$ and $\Delta_{\infty,C}$ do not have common roots besides $t=1$ if $\deg C \not\equiv 0\ (6)$. This clearly implies Corollary 4.1.

REMARK 4.1. R. Randell explained to me that this corollary can be deduced from the results of [**14**] as well.

REMARK 4.2. G'/G'' can indeed have 3-torsion as shown by Example 3.2.

Another application of Theorem 4.1 is the following

COROLLARY 4.2 [**18, 9**]. *Let $F_k(C)$ be a desingularisation of the k-fold cyclic multiple plane, i.e. the cyclic covering of P^2 branched over C and possibly the line L in infinity. Then the irregularity of $F_k(C)$ is zero unless both k and* $\deg C$ *are divisible by* 6.

The proof of this corollary is based on the following

LEMMA 4.1 (CF. [7]). *Let $F_k(C)$ be as in Corollary* 4.2 *and let λ_i be polynomials defined by the cyclic decomposition* (3.4) *of $A_1(Q)$. Let c_i^k be the number of the common roots of λ_i and t^k-1. Then* $\operatorname{rk} H_1(F_k(C), Q) = \Sigma_i c_i^k$.

Clearly the divisibility theorem and Lemma 4.1 imply Corollary 4.2. Indeed, if $6 \mid \deg C$ then $\Delta_C(Q) = 1$. If $6 \mid \deg C$ then $\Delta_C(Q)$ is equal to $(t^2 - t + 1)^l$ for some l. In both cases $\Delta_C(Q)$ and hence all λ_i are relatively prime to $t^k - 1$, i.e. $c_i^k = 0$.

More detailed analysis of $F_k(C)$ leads to the following result.

THEOREM 4.2 (CF. [**8**]). *Let $F_k(C)$ be as in Corollary 4.2. If all singularities of C are unibranched or nodes and $\exp(2\pi i/k)$ is a root of none of the Alexander polynomials of branches of C, then $F_k(C)$ is simply connected.*

COROLLARY 4.3. *If k is a power of a prime and singularities of C are as in Theorem 4.2 then $F_k(C)$ is simply connected.*

This follows from the standard fact that the Alexander polynomial can be normalized as $\Delta_C(1) = 1$ and the fact that no divisor φ of $t^{p^l} - 1$ has property $\varphi(1) = 1$.

We conclude this section with a result in a different direction. Let $\mathcal{C}$ denote an immersed Riemannian surface in CP^2 with possibly nonlocally flat points. It is interesting to know how the algebraicity of $\mathcal{C}$ affects the topology of the complement of $\mathcal{C}$ in CP^2. One can define in a similar way the Alexander polynomial of $\mathcal{C}$. However the divisibility theorem as above is false for nonalgebraic immersions. In fact, $\Delta_{\mathcal{C}}$ fails to be even cyclotomic. The dichotomy between algebraic and nonalgebraic cases is reflected also in $A_2(Z)$.

THEOREM 4.3 [**8**]. *Let $\mathcal{C}$ be an immersed surface in CP^2 and $A_i(Z)$ be the ith Alexander module of $\mathcal{C}$ (see Definition 3.2). Then:*

(a) $A_i(\mathcal{C}) = 0$ *for* $i > 2$.

(b) *If $\mathcal{C}$ is algebraic then $A_2(Z)$ is a free $Z[t, t^{-1}]$ module of rank* $\operatorname{rk} H_1(\mathcal{C}) + \deg \mathcal{C} - 1$.

REMARK 4.3. Another interpretation of the global Alexander polynomial was given by R. Randell (see [**15**]).

5. Computation of $A_1(Q)$. In this section, the Alexander module $A_1(Q)$ will be computed in terms of dimensions of certain linear systems of curves which have "prescribed behavior" at the singular points of the curve C. To describe these linear systems we need to introduce several notions.

PROPOSITION 5.1. *Let $f(x, y)$ be a germ of an analytic function which has an isolated singularity at the origin and let φ be any polynomial. Let $\psi_\varphi(n)$ denote the minimal k such that $z^k\varphi$ belongs to the adjoint ideal* [**2**] *of the singularity $z^k = f(x, y)$. Then $\psi_\varphi(n)$ is either zero, or has the form $[\kappa_\varphi n]$ where κ_φ is a rational number.* ([ ] *denotes the integral part.*)

The function $\psi_\varphi(n)$ can be computed in terms of an embedded desingularisation of f. In the case when f is generic in the sense of Kouchnirenko one can use the following description of the adjoint ideal given by Tessier and Merle. According to [**16**] in the case when $F(x, y, z)$ has a nondegenerate isolated singularity the adjoint ideal is generated by the monomials $x^{i_0}y^{i_1}z^{i_2}$ such that (i_0+1, i_1+1, i_2+1) is inside the Newton polyhedron of the singularity $F(x, y, z)$.

DEFINITION 5.1. A constant of quasiadjunction of the singularity $f(x, y)$ is one of the numbers κ_φ for various φ. A constant of quasiadjunction of the curve C is one of the constants of quasiadjunction of the singularities of C.

EXAMPLE 5.1. (a) Let $f(x, y) = x^2 + y^2$. This singularity has no constants of quasiadjunction. (b) Let $f(x, y) = x^2 + y^3$. There is only one constant of quasi-adjunction, namely $\kappa_1 = \frac{1}{6}$ corresponding to $\varphi = 1$. (c) Let $f(x, y) = x^2 + y^5$. There are two constants of quasiadjunction: $\kappa_1 = \frac{3}{10}$ and $\kappa_x = \frac{1}{10}$ corresponding to $\varphi = 1$ and $\varphi = x$.

These computations follow from the explicit computation of the Newton polyhedra for corresponding singularities.

DEFINITION 5.2. Let κ be a constant of quasiadjunction of the curve C. Then $\mathcal{A}_x$ is the sheaf of ideals in $\mathcal{O}_{p^2}$ such that $\Gamma(U, \mathcal{A}_\kappa)$ consists of $\varphi \in \Gamma(U, \mathcal{O}_{p^2})$ such that $\kappa_\varphi < \kappa$ at any singular point of C belonging to U.

DEFINITION 5.3. The superabundance $s_\kappa(n)$ of the linear system $\Gamma(P^2, \mathcal{A}_x(n))$ of the curves of degree n defined by sheaf A_κ is $\dim H^1(P^2, \mathcal{A}_\kappa(n))$.

DEFINITION 5.4. The factor of quasiadjunction corresponding to κ is the polynomial $\Delta_\kappa = (t - \exp(2\pi i\kappa))(t - \exp(-2\pi i\kappa))$.

Now we are in position to formulate the main result.

THEOREM 5.1. *Let C be an irreducible plane algebraic curve such that all singularities have semisimple monodromy (e.g., unibranched singularities, nodes, etc.). Then*

$$A_1(\mathbf{R}) = \bigoplus_{\kappa} \bigoplus_{s_\kappa(d-3-\kappa d)} \mathbf{R}[t, t^{-1}]/(\Delta_\kappa) \tag{5.1}$$

where $\bigoplus_{s_\kappa(n)}$ denotes the direct sum of $s_\kappa(n)$ copies of the cyclic module $\mathbf{R}[t, t^{-1}]/(\Delta_\kappa)$. ($\mathbf{R}$ *is the field of reals.*)

COROLLARY 5.1. *Assume C has as singularities only the cusps and nodes. Then*

$$A_1(Q) = \bigoplus_{s} Q[t, t^{-1}]/t^2 - t + 1 \tag{5.2}$$

where s is the superabundance of the system of curves of degree $d - 3 - d/6$ passing through the cusps of C.

Indeed, in the case of cuspidal curves, $\kappa = \frac{1}{6}$, $\Delta_{1/6} = t^2 - t + 1$ and $\Gamma(P^2, \mathcal{A}_{1/6}(n))$ consists of the curves of degree n passing through the cusps.

COROLLARY 5.2. *Assume C has as singularities only the singularities given locally by $x^2 + y^5 = 0$. Then*

$$A_1(Q) = \bigoplus_{s} Q[t, t^{-1}]/(t^4 - t^3 + t^2 - t + 1) \tag{5.3}$$

where s is the superabundance of the system of curves of degree $d - 3 - 3d/10$ passing through the singularities of C.

For the curve with singularities as in Corollary 5.2, one has as constants of quasiadjunction $\frac{3}{10}$ and $\frac{1}{10}$. Because $A_1(\mathbf{R})$ given in Theorem 5.1 in fact can be defined over Q, we obtain $s_{3/10}(d-3-3d/10) = s_{1/10}(d-3-d/10)$. Curves in the linear system $\Gamma(P^2, \mathcal{Q}_{3/10}(n))$ are the curves of degree n just passing through the singularities of C. (Note that $\Gamma(P^2, \mathcal{Q}_{1/10}(n))$ consists of the curves of degree n which are passing through the singularities of C and whose tangents at the singular points of C coincide with the tangents of C.)

EXAMPLE 5.2. Let C_{6k} (resp. C_{10k}) be given in the equation

$$f_{2k}^3 + f_{3k}^3 = 0 \tag{5.4a}$$

$$\left(\text{resp. } f_{2k}^5 + f_{5k}^2 = 0\right) \tag{5.4b}$$

where f_m denotes the defining polynomial of a nonsingular curve of degree m. Then C_{6k} (resp. C_{10k}) has as singularities only cusps (resp. only the singularities given locally by $x^2 + y^5 = 0$).

They are located at the intersection points of the curves $f_{2k} = 0$ and $f_{3k} = 0$ (resp. $f_{2k} = 0$ and $f_{5k} = 0$). By the Cayley-Bacharach theorem [**3**] the superabundance of the system of the curves of degree $5k - 3$ (resp. $7k - 3$) passing through the singularities of C_{6k} (resp. C_{10k}) is equal to 1. Hence the Alexander module is given by

$$A_1(Q) = \begin{cases} \text{for } C_{6k}: & Q[t, t^{-1}]/t^2 - t + 1, \\ \text{for } C_{10k}: & Q[t, t^{-1}]/t^4 - t^3 + t^2 - t + 1. \end{cases}$$

The proof of Theorem 5.1 will appear in [**9**]. Here we shall note that it uses the semisimplicity of the action of t on G'/G'' which follows from the proof of the divisibility theorem, and a combination of Lemma 4.1 with the appropriate generalization of Zariski's [**19**] arguments.

REFERENCES

1. J. Birman, *Braids, links and mapping class groups*, Ann. of Math. Studies, No. 82, Princeton Univ. Press, Princeton, N. J., 1975.

2. P. Blass and J. Lipman, *Remarks on adjoints and arithmetic genus of algebraic varieties*, Amer. J. Math. **101** (1979).

3. P. Griffith and J. Harris, *Principles of algebraic geometry*, Wiley, New York, 1978.

4. P. Deligne, *Le groupe fondamental du complement d'une courbe plane n'ayant que des points doubles ordinaires est abelian*, Sem. Bourbaki no. 543 (Nov. 1979).

5. I. Dolgachev and A. Libgober, *On the fundamental group of the complement to a discriminant variety*, Lecture Notes in Math., vol. 862, Springer-Verlag, Berlin and New York, 1981.

6. W. Fulton, *On the fundamental group of the complement to a node curve*, Ann. of Math. (2) **111** (1980), 407–409.

7. A. Libgober, *Alexander polynomials of plane algebraic curves and cyclic multiple planes*, Duke Math. J. **49** (1982).

8. ______, *Alexander modules of plane algebraic curves*, Preprint (1981).

9. ______, *On G/G″ for the fundamental group of the complement to plane algebraic curves*.

10. ______, *On the fundamental group of the space of cubic surfaces*, Math. Z. **162** (1978), 63–67.

11. J. Milnor, *Infinite cyclic coverings*, Topology of Manifolds (Michigan State Univ.), Prindle, Weber & Schmidt, Boston, 1967.

12. B. Moishezon, *Stable branch curves and braid monodomies*, Lecture Notes in Math., vol. 862, Springer-Verlag, Berlin and New York, 1981.

13 M. Oka, *Some plane curves whose complement have non-abelian fundamental groups*, Math. Ann. **218** (1978), 55–65.

14. R. Randell, *Some topology of Zariski surfaces*, Lecture Notes in Math., vol. 788, Springer-Verlag, Berlin and New York, 1979.

15. ______, *Milnor fibres and Alexander polynomials of plane curves*, these PROCEEDINGS.

16. B. Tessier and M. Merle, *Condition d'adjunction, d'apres DuVal*, Lecture Notes in Math., vol. 777, Springer-Verlag, Berlin and New York, pp. 229–295.

17. O. Zariski, *On the problem of existence of algebraic functions of two variables possessing a given branch curve*, Amer. J. Math. **51** (1929), 305–328.

18. ______, *On the linear connection index of the algebraic surface* $z^n = f(x, y)$, Proc. Nat. Acad. Sci. U.S.A. **15** (1929), 494–501.

19. ______, *On the irregularity of cyclic multiple planes*, Ann. of Math. (2) **32** (1931), 485–511.

20. ______, *On the Poincare group of rational plane curves*, Amer. J. Math. **58** (1930), 607–619.

21. ______, *On the Poincare group of an algebraic hypersurface*, Ann. of Math. (2) **38** (1937), 131–141.

22. ______, *The topological discriminant group of a Riemann surface of genus p*, Amer. J. Math. **59** (1937), 335–358.

UNIVERSITY OF ILLINOIS AT CHICAGO CIRCLE

Proceedings of Symposia in Pure Mathematics
Volume **40** (1983), Part 2

A CONNECTION BETWEEN POLAR INVARIANTS AND ROOTS OF THE BERNSTEIN-SATO POLYNOMIAL

BEN LICHTIN[1]

1. One subject in the study of the b-function as yet untouched is the algebro-geometric understanding of some, if not all, of its roots. One does not expect all of its roots to be equally interesting geometrically because of the "jumping" phenomenon in an equisingular family of hypersurfaces discovered by Yano [**19**]. Nonetheless, a reasonable point of view asks which, if any, of the roots have an interesting algebro-geometric expression.

This paper summarizes the results of [**11**] which addresses this question.

2. We briefly recall the two subjects of basic interest.

1. *Polar invariants*. Given a representative of an analytic germ $f: (\mathbf{C}^n, \bar{0}) \to (\mathbf{C}, 0)$ with isolated singularity at $\bar{0}$ and defined in a neighborhood $\mathfrak{U}$ of $\bar{0}$, let Jf be its Jacobian ideal, generated by the partials $\partial f/\partial z_i$.

Let $\pi: \mathfrak{X} \to \mathfrak{U}$ be the normalized blowing up of Jf in $\mathfrak{U}$. Let $\mathfrak{D} = \bigcup \mathfrak{D}_q$ be the exceptional divisor expressed as a union of its irreducible components.

Denote by m_q the value of the pullback of the maximal ideal $\mathfrak{M}_n$ in $\mathcal{O}_{\mathfrak{U},\bar{0}}$ along $\mathfrak{D}_q$. We write this by

$$m_q = v_{\mathfrak{D}_q}(\pi^* \mathfrak{M}_n).$$

Let $e_q = v_{\mathfrak{D}_q}(\pi^*(f)) - m_q$. Clearly, for $f \in \mathfrak{M}_n^3$, $e_q \geqslant 0$. It turns out (see [**15** and **16**]) that $e_q = v_{\mathfrak{D}_q}(\pi^*(Jf))$.

The ratios $\{e_q/m_q\}$ are the polar invariants of f. They appear in many different geometric contexts associated to f (see [**16**, **17**, . . .]).

1980 *Mathematics Subject Classification*. Primary 32B.

[1]Supported in part by NSF grant MCS 77-18723 AO3 at the Institute for Advanced Study.

Of interest here is that found in Teissier's 1974 Arcata talk (or Lê's 1973 Annales d'Institut Fourier paper). He observed that the polar invariants measure the vanishing rate of the heights of relative vanishing cycles, where relative is with respect to a "generic" hyperplane.

In more detail, let H be a 'generic' hyperplane (to be specified below) and choose coordinates so that $\{z_1 = 0\} = H$. Consider the fibers $X_t = f^{-1}(t)$ in $\mathfrak{U}$ for t near to zero in C.

The exact sequence of relative homology for the pair $(X_t, X_t \cap H)$ reduces to the short exact sequence

$$0 \to H_{n-1}(X_t) \to H_{n-1}(X_t, X_t \cap H) \xrightarrow{\partial} H_{n-2}(X_t \cap H) \to 0.$$

Because H is generic for $\mathfrak{U}$, $\dim_{\mathbf{C}} H_{n-2}(X_t \cap H) = \mu^{(n-1)}$ [**15**]. Thus, there are $\mu + \mu^{(n-1)}$ relative vanishing cycles forming a basis for $H_{n-1}(X_t, X_t \cap H)$.

We find a realization of them by considering the critical points of the height function $|z_1|$ from H. These are the points in X_t whose tangent planes are parallel to H. Thus, for H "generic" in the sense of [**17**], such points must be in the polar curve S of f associated to the direction vector $(1 : 0 : \cdots : 0)$. These critical points generate gradient cells which generate the relative cycles of $H_{n-1}(X_t, X_t \cap H)$.

Let $S = \bigcup \Gamma_q$ be the decomposition into its irreducible components. Then we can parametrize Γ_q as

$$z_1 = v^{m_q}, \quad z_i = \varphi_i(v), \quad \operatorname{ord} \varphi_i \geqslant m_q, \qquad i = 1, \ldots, n \qquad [\mathbf{16}, \mathbf{17}].$$

The critical points p of $|z_1|$ in X_t must satisfy the equations

$$t = v^{e_q + m_q} U(v), \qquad U(0) \neq 0,$$
$$|z_1(p)| = |v|^{m_q}, \quad \text{for some } q.$$

So, the height of $|z_1(p)|$ (or equivalently, the corresponding relative gradient cell) vanishes to order $m_q/(e_q + m_q) = 1/(e_q/m_q + 1)$ in t.

Inspired by this observation, Teissier asked how the polar invariants might also appear in the description of the vanishing rate of the "volumes" of the cycles in the fibers of f, as $t \to 0$.

In the classical study of the Gauss-Manin connection, one observes that periods of the type $\int_{\gamma_t} \omega_t$ solve the differential equation underlying the connection. Here, ω_t is a section of the sheaf of relative differential $(n-1)$ forms along the fibers X_t and γ_t is a section of the homology sheaf $\bigcup_{t \neq 0} H_{n-1}(X_t, \mathbf{C})$ [**10**]. Thinking roughly of these periods as weighted volumes of γ_t is what is intended by use of the quotation marks.

It is well known that these periods have a general expression of the form

$$\text{(2-1)} \qquad \sum C_{p,q} t^{\lambda_q} (\log t)^p \quad \text{as } t \to 0,$$

where the λ_q runs through finitely many arithmetic progressions of positive rational numbers.

For given $\{\gamma_t\}$ and ω_t, the least such λ_q measures in some sense the vanishing volume rate of the γ_t and the e_q/m_q measure the height of the vanishing relative cycles.

So, a natural question is in what way the e_q/m_q appear in the expressions for some of the least λ_q appearing as exponents in expansions (2-1).

The point of this article and [**11**] is to give an indication to the answer of this question, at least for the simplest case where $n = 2$ and the curve defined by f is irreducible.

As discussed in §5, the preliminary conclusions to this question that could be inferred from A'Campo [**1**] and Lê [**8**] are given a more precise form by the analysis summarized here.

2. *Bernstein-Sato polynomial* ("*b-function*"). The key intermediate tool is the local b-function of f, $b_f(s)$ globally defined by Bernstein [**2**] (and adapted to the local situation by Björk [**4**] and Kashiwara [**6**]) as the monic generator of the ideal $\mathcal{I} = \{\beta(s)$: there is a linear partial differential operator $\mathcal{P}(s, x, D_x)$, analytic in x, and polynomial in the indeterminate s such that

$$\mathcal{P}(s, x, D_x) f^{s+1} = \beta(s) f^s\}.$$

$\mathcal{I} \neq (0)$ is Bernstein's existential proof of the nonvacuousness of the concept in the global (polynomial in x coefficients) case.

For any of the three proofs of existence, if f is a real polynomial and $f_{\mathbf{C}}$ its complexification, $b_f(s) = b_{f_{\mathbf{C}}}(s)$.

For the moment, assume f is a real polynomial (cf. Remark below).

We can then use a characterization of the roots of b_f in terms of the poles of the analytic continuation of the generalized functions

$$I_{\pm}(s, \psi) = \int_{\mathbf{R}^n} f_{\pm}^s \psi \, |dx| , \qquad I_{\text{abs}} = I_+ + I_- ,$$

where ψ is a test function in $C_0^\infty(\mathbf{R}^n, \mathbf{C})$, functions with compact support in a sufficiently small neighborhood of the origin in $\mathbf{R}^n$.

It is well known that the poles of the three generalized functions lie on finitely many arithmetic progressions of negative rational numbers [**3**]. The fact that b_f exists and is a polynomial says that only finitely many arithmetic progressions, consisting only of poles, exist and start with the roots of b_f.

One sees this by exploiting the defining relation of b_f. Let $\mathcal{B}(s) = \prod_1^d \Gamma(s - \beta_j)$ where $b_f(s) = \prod_1^d (s - \beta_j)$ and Γ is the standard gamma function.

Then $\mathcal{P} f^{s+1} = b_f(s) f^s$ implies for *complex valued* s with $\operatorname{Re} s \gg 0$

$$\mathcal{P}(s, x, D_x) \frac{f_{\pm}^{s+1}}{\mathcal{B}(s+1)} = \frac{f_{\pm}^s}{\mathcal{B}(s)}. \tag{2-2}$$

Noting that for $\operatorname{Re}(s) \gg 0$ the generalized functions $f_{\pm}^s/\mathcal{B}(s)$ have no poles, we can then use (2-2) to extend analytically to $\mathbf{C}$ and find poles only where $\mathcal{B}$ has poles. But these occur only for the elements in the sequences

$$\{\beta_j, \beta_j - 1, \beta_j - 2, \ldots\}_{j=1}^{d}.$$

So, the starting poles of $f_\pm^s$ are at least within the set of roots of b_f.

One can see how the poles of $I_\pm(s, -)$ relate to the set of exponents $\{\lambda_q\}$ in the fractional power-Nilsson expansions for the periods $\int_{\gamma_t} \omega_t$ by noting that the integral over a sufficiently small ball $\mathbf{B}_n$ about $\bar{0}$ in $\mathbf{R}^n$ containing the support of ψ is

$$\int_{\mathbf{B}_n} f_+^s \psi \, |dx| = \int_0^\varepsilon t^s \left(\int_{f=t} \psi/df \right) dt \quad \text{for an appropriate } \varepsilon > 0. \tag{2-3}$$

The relative form ψ/df is defined by the relation $df \wedge \psi/df = \psi\, dx$ [**9**].

Now, $\int_{f=t} \psi/df = \sum \mu_{p,q} t^{\lambda_q - 1} (\log t)^p$ as $t \to 0^+$. So the right-hand side is a sum of integrals

$$\mu_{p,q} \int_0^\varepsilon t^{s+\lambda_q-1} (\log t)^p \, dt.$$

If we rescale ε to 1, we can use the identity in the region $\operatorname{Re}(s + \mu) > 0$

$$\int_0^1 t^{s+\mu-1} (\log t)^p \, dt = \frac{d^p}{ds^p} \left(\frac{1}{\mu + s} \right) \tag{2-4}$$

to observe that a pole of the analytic continuation of the left-hand side of (2-3) in s must be a pole of the analytic continuation of the right-hand side of (2-3) via (2-4) in s. Thus, the starting poles β_j of $I_\pm(s, -)$ must appear in the Nilsson expansions for certain periods $\int_{\gamma_t} \omega_t$ as $-\beta_j$ and be equal in this form to some λ_q. Note that here the cycle γ_t is in the fiber $f_{\mathbf{C}}^{-1}(t)$, as described in Malgrange [**13**].

Hence, if we relate polar invariants of $f_{\mathbf{C}}$ to the roots of $b_{f_{\mathbf{C}}}(s) = b_f(s)$, implicitly we will have related the former to certain exponents interpretable as describing the vanishing volume rates in the sense already described.

REMARK. If f is a complex polynomial, more care has to be taken with respect to the choice of cycle γ_t, cf. [**12**].

3. Following Varcenko [**18**], we define the array of multiplicities for a resolution $\pi\colon X \to \mathcal{U}$ of the isolated singularity at $\bar{0}$ defined by f in the neighborhood $\mathcal{U}$ of $\bar{0}$ as the set of pairs of integral n-tuples.

$\mathcal{Q} = \{((m_1, \ldots, m_n); (b_1, \ldots, b_n))$: centered at some point $p \in \pi^{-1}(\bar{0})$ there are local coordinates $(y_1, \ldots, y_n)$ so that $f \circ \pi(y_1, \ldots, y_n) = y_1^{m_1} \cdots y_n^{m_n}$ (local unit at p) and $\det d\pi(y_1, \ldots, y_n) = y_1^{b_1} \cdots y_n^{b_n}$ (local unit at p)$\}$. We now form the set of ratios for each $e \in \mathbf{N}$, $\mathcal{R}_e = \{-(e + b_j)/m_j$: and m_j, b_j are jth coordinates of tuples $(m_1, \ldots, m_n)$, $(b_1, \ldots, b_n)$, the pair of which belongs to $\mathcal{Q}\}$ and are interested in the set $\mathcal{R}_1$, the set of starting values for the arithmetic progressions which contain the poles of $f_\pm^s$ (as is well known, see [**3**, **10**, **18**, ...]). We then say that a ratio s' in $\mathcal{R}_e$ is determined by the element $((m_1, \ldots, m_n); (b_1, \ldots, b_n))$ if $s' = -(e + b_j)/m_j$ for some $j = 1, \ldots, n$.

In general, it is possible to formulate a negative criterion, one which assures that an element of $\mathfrak{R}_1$ is *not* a pole of $f_{\pm}^s$ or $|f|^s$ for real polynomial f [10]. On the other hand, Varcenko observed [18] that the largest value in $\mathfrak{R}_1$ satisfies a particular property insuring that it is always a pole of $f_{\pm}^s$. We call this

Condition A. A ratio s_0 satisfies Condition A if whenever $s_0 \in \mathfrak{R}_e$ for some e, that is, $s_0 = -(e + b_j)/m_j$ where $((m_1,\ldots,m_j,\ldots,m_n)); ((b_1,\ldots,b_j,\ldots,b_n))$ is the associated element of $\mathfrak{A}$, then necessarily (i) $e = 1$ and (ii) for those k for which $-(1 + b_k)m_k \neq s_0$ then $s_0 m_k + b_k > -1$.

In words, the ratio s_0 satisfies Condition A if whenever it arises as an element of some $\mathfrak{R}_e$ determined by certain elements of $\mathfrak{A}$, necessarily $e = 1$ and s_0 is at least as large as any other starting value determined by these same elements of $\mathfrak{A}$.

In this case, that s_0 is a pole of $f_{\pm}^s$ now follows from an adaptation of the description of Gel'fand and Shilov of the residue distributions of the generalized functions $x_{\pm}^s$—first to the case of monomials $(x_1)_{\pm}^{s_1} \cdots (x_n)_{\pm}^{s_n}$ and then to the local monomial form for $f \circ \pi$ on the real analytic manifold X. See [10, 18] for the details of this description.

4. We state here the main results of our analysis in [11], concerning the array of multiplicities for the canonical resolution [7] of an irreducible plane curve with characteristic sequence $(n, \beta_1,\ldots,\beta_g)$ [14]. We are able to find a precise ordering structure for the ratios in $\mathfrak{R}_1$ (cf. §3) which informs us exactly what the expression for the residue of $|f|^\lambda$ must be at certain ratios $\lambda = -r_i/R_i$, $i = 1,\ldots,g$, in $\mathfrak{R}_1$, for f real. As seen in §5, the $-r_i/R_i$ are easily related to the polar invariants and characteristic ratios $\beta_1/n,\ldots,\beta_g/n$ of the complexified branch. From this, the main theorem in [11] will follow (cf. (4-5)).

We first state the form of the canonical resolution most suitable for (4-5). Let f be a complex polynomial with singularity at $\bar{0}$ in $\mathbf{C}^2$ with defining branch $(n, \beta_1,\ldots,\beta_g)$.

(4-1) THEOREM. (1) *There are g "blocks" $\mathfrak{B}_0, \mathfrak{B}_1,\ldots,\mathfrak{B}_{g-1}$ of sequences of quadratic transformations $X_{i-1} \leftarrow X_i$ which may be divided into rows for which the length and number of rows are computable directly from the characteristic sequence.*

Within the ith block, write each row of complex manifolds X_i as

$$X_{M_j} \leftarrow X_{M_j+1} \leftarrow \cdots \leftarrow X_{M_j+1} \quad \text{for } j = 1,\ldots,q.$$

The initial manifold X_{M_1} of this block is also denoted as $X_{\mathfrak{B}_i}$.

(2) *For each transformation $X_{M_d+\varepsilon} \leftarrow X_{M_d+\varepsilon+1}$ there appears at least one new "distinguished" element $(m^{(M_d+\varepsilon+1)}; b^{(M_d+\varepsilon+1)})$ of the array of multiplicities $\mathfrak{A}$. However, for $i = 0, 1,\ldots,g-1$ the transformations $X_{\mathfrak{B}_{i+1}-1} \leftarrow X_{\mathfrak{B}_{i+1}}$ produce two elements of the array $\mathfrak{A}$. (Note that for each i, this transformation is considered as the last quadratic transformation in the ith block.)*

(3) *For each row of each block of the resolution sequence, there are linear recurrences which generate the $m^{(j)}$ and $b^{(j)}$ pairs, composing the distinguished element $(m^{(j)}; b^{(j)})$ of $\mathfrak{A}$ (cf. (2)). The initial values of the recurrences are $m^{(M_k+1)}$, $b^{(M_k+1)}$ if we consider the kth row.*

(4) *In each X_t (except for the final manifold X_{res}, in which $f \circ \pi$ is locally normal crossing everywhere along the exceptional divisor) there is exactly one chart denoted $\mathfrak{U}_{1,t}$ in which the strict transform of the curve, defined by f near the origin in $\mathbf{C}^2$, meets components of the exceptional divisor in nonnormal crossing points.*

(5) *At the point at which the strict transform in $X_{\mathfrak{B}_i}$ meets the divisor in $\mathfrak{U}_{1,\mathfrak{B}_i}$, there are local coordinates so that*

$$f \circ \pi(x_{\mathfrak{B}_i}, y_{\mathfrak{B}_i}) = y_{\mathfrak{B}_i}^{R_i} f_{\mathfrak{B}_i}(x_{\mathfrak{B}_i}, y_{\mathfrak{B}_i}) \cdot (\textit{local unit}),$$

$$\det d\pi(x_{\mathfrak{B}_i}, y_{\mathfrak{B}_i}) = y_{\mathfrak{B}_i}^{r_i - 1} \cdot (\textit{local unit}).$$

Here, the number of characteristic numbers for the curve defined by $f_{\mathfrak{B}_i}$ drops by one.

(6) *Let*

$$\lambda_l(M_j + \varepsilon) = \frac{-\left(1 + b_l^{(M_j+\varepsilon)}\right)}{m_l^{(M_j+\varepsilon)}} \quad \textit{for } l = 1, 2 \textit{ and } j = 1, \ldots, q$$

be all the distinct possible starting poles (membership in $\mathfrak{R}_1$) arising from the $X_{M_j+\varepsilon}$ sitting inside the ith block of the resolution sequence. Then $-r_i/R_i \geqslant \lambda_l(M_j + \varepsilon)$ and equality holds iff $r_i = 1 + b_l^{(M_j+\varepsilon)}$ and $R_i = M_l^{(M_j+\varepsilon)}$.

(7) *For all manifolds X_j appearing in all blocks prior to $\mathfrak{B}_i$, $r_i \geqslant b_l^{(j)}$, $R_i \geqslant m_l^{(j)}$, and $R_i = m_l^{(j)}$ implies $r_i = 1 + b_l^{(j)}$.*

(8) *$-r_i/R_i > -r_{i+1}/R_{i+1}$ for $i = 1, 2, \ldots, g-1$.*

(9) *For $i \geqslant 1$, $r_i = (\beta_i + n)/e^{(i)}$, and*

$$R_i = \frac{\beta_i e^{(i-1)} + \beta_{i-1}(e^{(i-2)} - e^{(i-1)}) + \cdots + \beta_1(n - e^{(1)})}{e^{(i)}}$$

where $e^{(i)} =$ g.c.d. $(n, \beta_1, \ldots, \beta_i)$.

PROOF. Cf. [**11**].

Note. If f is a real polynomial, the resolution for f is obtained from the resolution for the complexification $f_{\mathbf{C}}$ in an obvious way. It is the real analytic manifold X_{res} (real part of the complex manifold X_{res} obtained for $f_{\mathbf{C}}$) which is used to study the poles of $|f|^\lambda$. Moreover, the array of multiplicities for f is the same set as that for $f_{\mathbf{C}}$, as can be easily seen.

To arrive at the main Theorem (4-5) it is necessary to study the ordering of the ratios in (6) within one block $\mathfrak{B}_i$ of the resolution sequence.

To this end it is necessary to distinguish between two possibilities that can occur at the *start* of the ith block.

Case (A). ord $f_{\mathfrak{B}_i}(x_{\mathfrak{B}_i}, 0) <$ ord $f_{\mathfrak{B}_i}(0, y_{\mathfrak{B}_i})$.

Case (B). ord $f_{\mathfrak{B}_i}(0, y_{\mathfrak{B}_i}) <$ ord $f_{\mathfrak{B}_i}(x_{\mathfrak{B}_i}, 0)$.

For our purposes here, let us assume that we start the ith block in Case (A). This then forces the following relations between certain of the ratios (with the notation as in (4-1)(1)):

(4-2) THEOREM. (i) $\lambda_1(M_k) = \lambda_2(M_{k+1} + 1)$ *for k even.*
(ii) $\lambda_2(M_k) = \lambda_1(M_{k+1} + 1)$ *for k odd.*
Moreover, one also has that
(iii) $\lambda_2(M_k) < \lambda_2(M_k + 1)$ *for k odd and*
(iv) $\lambda_1(M_k) > \lambda_1(M_k + 1)$ *for k even.*
Finally, we have
(v)

$$\lambda_1(M_j+1) \underset{+}{<} \lambda_1(M_j+2) = \lambda_2(M_j+1) \underset{+}{<} \lambda_1(M_j+3)$$
$$\underset{+}{<} \cdots \underset{+}{<} \lambda_1(M_j+k_{j+1}) < \lambda_2(M_j+k_{j+1}) = \lambda_2(M_{j+1})$$

if k_{j+1} is the length of the jth row and if j is even.
If j is odd the ordering reverses in the sense that
(vi) $\lambda_1(M_j + k_{j+1}) \underset{+}{<} \lambda_1(M_j + k_{j+1} - 1) \underset{+}{<} \cdots \underset{+}{<} \lambda_1(M_j + 1) \underset{+}{<} \lambda_2(M_j + 1)$.

PROOF. Cf. [**11**].

(4-2) is more succinctly expressed in a diagram. Thus, there is alternating decrease then increase across rows with the additional property that the largest ratio in an even numbered row is smaller than the smallest ratio in the following odd numbered row. Note however that this picture is not entirely accurate in that the ratios within each segment $[M_j + 1, M_{j+1}]$ will only be piecewise-linear monotonically increasing/decreasing in general.

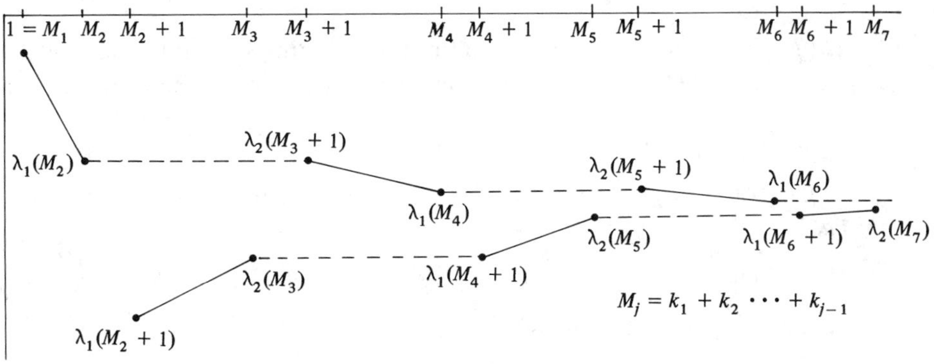

This remains true even for the last (qth) row, if this is an even numbered row. However, now the relevant ratio to compare with $-\lambda_2(M_q + k_{q+1}) = -\lambda_2(M_{q+1})$, is "behind not ahead" of it. Namely we have

$$-\lambda_2(M_{q-1} + 1) < -\lambda_2(M_q + k_{q+1}).$$

Since the assumption that q even implies $-\lambda_2(M_q + k_{q+1}) = r_{i+1}/R_{i+1}$ [**11**] we see that q implies $\lambda_2(M_{q-1} + 1) > -r_{i+1}/R_{i+1}$. Unfortunately, the additional pair that appears in $X_{\mathfrak{B}_{i+1}} = X_{M_q + k_{q+1}}$ (cf. (4-1)(2)) is exactly

$$\left(\left(R_{i+1}, m_2^{(M_{q-1}+1)}\right); \left(r_{i+1} - 1, b_2^{(M_{q-1}+1)}\right)\right),$$

so that there is definitely a chart in $X_{\mathfrak{B}_{i+1}}$ in which $-r_{i+1}/R_{i+1}$ does not satisfy Condition A of §3. An analogous situation holds if q is odd. So we conclude that determining if the (global) residue of $|f|^\lambda$ is zero at $\lambda = -r_{i+1}/R_{i+1}$, for $i = 1, 2, \ldots, g-1$, will be a subtler process than that involved with the residue calculation at $\lambda = -r_1/R_1$. For this latter ratio *always* satisfies Condition A. This follows from Theorems (4-1) and (4-2) but is not a new result. It has been discovered mutually independently by Varcenko [**18**], Igusa (cf. his article in Internat. Colloq. Algebraic Geometry and Complex Analysis 1977), and M. Saito (unpublished).

To proceed further, it is useful to make two observations. Firstly, [**11**]

(4-3) $$\frac{-r_{i+1}}{R_{i+1}} \cdot m_2^{(M_{q-1}+1)} + \left(b_2^{(M_{q-1}+1)} + 1\right) = \frac{r_i e^{(i)} - R_i}{R_{i+1}}.$$

Secondly, calculation of the residue at $-r_{i+1}/R_{i+1}$ is considerably "simplified" by imposing the assumption

(4-4) $$\gcd(r_i, R_i) = 1 \quad \text{for } i = 2, \ldots, g.$$

For this tells us exactly in which charts does $-r_{i+1}/R_{i+1}$ appear as a possible starting pole for $|f|^\lambda$. This is crucial for the evaluation of the residue and it is not yet clear how to obtain information on these smaller poles of $f_\pm^\lambda$ or $|f|^\lambda$ without an assumption like (4-4).

Set

$$p_i = \frac{r_i e^{(i)} - R_i}{R_{i+1}} \quad \text{and} \quad q_i = \frac{R_{i+1} - r_{i+1} e^{(i+1)}}{R_{i+1}}.$$

Let $B(p, q)$ denote the analytic continuation of the standard beta function $\beta(p, q) = \int_0^1 x^{p-1}(1-x)^{q-1}\,dx$, into the region $\mathrm{Re}(p) < -1$, $\mathrm{Re}(q) < -1$, as in [**5**].

The main result of [**11**] is

(4-5) THEOREM. *Up to a constant factor* (*which is nonzero and detailed in* [**11**]), *the value*

$$\mathrm{Res}\left(|f|^\lambda, \psi\right)\big|_{\lambda = -r_{i+1}/R_{i+1}} = \psi(\bar{0})\left[B(p_i, q_i) + B(q_i, -p_i - q_i) + B(p_i, -p_i - q_i)\right].$$

From this, we immediately obtain

(4-6) COROLLARY. *For $\psi \in C_0^\infty(\mathbf{R}^2, \mathbf{C})$ with $\psi(\bar{0}) \neq 0$, the regularization of $\langle |f|^\lambda, \psi\rangle$ possesses a pole at $\lambda = -r_i/R_i$ for each $i = 1, 2, \ldots, g$.*

PROOF. Since $0 < |p_i|, |q_i| < 1$, it is easy to show by using standard identities for $B(p, q)$ that the bracketed sum in (4-5) is *not* zero. See [**11**] for details.

5. We rewrite r_i/R_i to exhibit the polar invariants for the branch, computed by Merle [**14**].

Introduce $\bar{\beta}_i$, $i = 0, 1, \ldots, g$, by way of

$$\bar{\beta}_0 = n, \quad \bar{\beta}_1 = \beta_1, \quad \bar{\beta}_{q+1} = \frac{e^{(q-1)}}{e^{(q)}} \cdot \bar{\beta}_q + \left(\beta_{q+1} - \beta_q\right).$$

It is then straightforward to see that

$$\frac{e^{(q)}}{e^{(q+1)}} \cdot \bar{\beta}_{q+1} = R_{q+1} \quad \text{for } q = 1, 2, \ldots, g.$$

The polar invariants for the branch are the values

$$n-1, \bar{\beta}_1 - 1, \frac{\bar{\beta}_2}{n_1} - 1, \ldots, \frac{\bar{\beta}_q}{n_1 \cdots n_{q-1}} - 1, \ldots, \frac{\bar{\beta}_g}{n_1 \cdots n_{g-1}} - 1,$$

where we use $n_{q+1} = e^{(q)}/e^{(q+1)}$.

In the notation of §1, $e_q/m_q = (\bar{\beta}_q/n_1 \cdots n_{q-1}) - 1$. Thus,

(5-1) $$r_i/R_i = e^{(i)}\left(\frac{\beta_i}{n} + 1\right)\left(\frac{m_i}{e_i + m_i}\right) \quad \text{for } i = 2, 3, \ldots, g.$$

Recalling the interpretation of the rightmost factor from §2, we see that under the conditions of Theorem (4-5), the vanishing rates of the heights of gradient cells ("vanishing relative cycles") appear as factors of certain roots of $b_f(s)$ with the characteristic ratios of the branch as parts of another factor.

Prior to the results discussed here, the only possible source of speculation concerning the relationship between polar invariants and roots of the Bernstein polynomial was found in A'Campo and Lê's work on the characteristic polynomial of monodromy for plane curves.

From their work certain aspects of the denominators of some of the roots of the Bernstein polynomial were evident.

Fom the numbers (recall $e^{(i)} = (n, \beta_1, \ldots, \beta_i)$) for $i = 1, \ldots, g$

$$\lambda_i = \frac{\beta_i e^{(i-1)} + \beta_{i-1}(e^{(i-2)} - e^{(i-1)}) + \cdots + \beta_1(n - e^{(0)})}{e^{(i-1)} \cdot e^{(i)}}.$$

In [**8**], Lê shows

(5-2) PROPOSITION. *There must be roots of the Bernstein polynomial which have* $(\lambda_i \cdot e^{(i-1)})$ *as their denominator, for* $i = 1, 2, \ldots, g$.

It is easy to connect $\lambda_i \cdot e^{(i-1)}$ to polar invariants via Merle's formulae. First, it is clear that $\bar{\beta}_i = \lambda_i \cdot e^{(i)}$.

Now, $n_1 \cdots n_{i-1} = n/e^{(i-1)}$. Thus,

$$\frac{e_i}{m_i} + 1 = \frac{\bar{\beta}_i}{n_1 \cdots n_{i-1}} = \frac{\lambda_i \cdot e^{(i-1)} \cdot e^{(i)}}{n},$$

so that

$$\lambda_i \cdot e^{(i-1)} = \frac{n}{e^{(i)}}\left(\frac{e_i}{m_i} + 1\right).$$

In this way, then, it is evident that we should have expected to find the factors $1/((e_i/m_i) + 1)$ in some of the poles of $f_{\pm}^{s}$. Our analysis, however, was motivated by the desire to make precise in exactly what way these factors do appear in at least some of the poles.

Extensions of these results to multiple planes (locally defined by $z^k - f(x, y)$, f irreducible) are available using formulae for the polar invariants and b-functions for a Thom-Sebastiani polynomial, as computed by Teissier [**17**] and Yano [**19**].

REFERENCES

1. N. A'Campo, *Sur la monodromie des singularités isolées d'hypersurfaces complexes*, Invent. Math. **20** (1973), 147–169.
2. I. N. Bernstein, *Analytic continuation of generalized functions with respect to a parameter*, Functional Anal. Appl. **6** (1972), 273–285.
3. I. N. Bernstein and S. I. Gel'fand, *Meromorphic property of the functions* $p^{\lambda'}$, Functional Anal. Appl. **3** (1969), 68–69.
4. J. E. Björk, *Rings of differential operators*, Van Nostrand, Princeton, N. J., 1980.
5. I. M. Gel'fand and G. Shilov, *Les distributions*, Dunod, Paris, 1972.
6. M. Kashiwara, *B-functions and holonomic systems*, Invent. Math. **38** (1976/77), 33–53.
7. F. Kmety, *Resolution des singularités des courbes*, Lecture Notes in Math., vol. 409, Springer-Verlag, Berlin and New York, 1973.
8. Lê Dung Trang, *Noueds algébriques*, Compositio Math. **25** (1972), 281–321.
9. J. Leray, *Le calcul différential et intégral sur une variété analytique complexe* (*Problème de Cauchy* III), Bull. Soc. Math. France **87** (1959), 81–180.
10. B. Lichtin, *On the differential equations associated to an analytic function near a singular point*, Thesis, Ohio State University, 1978.
11. ______, *A numerical study of the array of multiplicities for an irreducible plane curve* (to appear).
12. ______, *Applications of the* $\mathfrak{D}$*-module notion of integration to a determination of poles of* $|f(x, y)|^s$ *and to the eigenvalues of the Gauss-Manin connection residue* (to appear).
13. B. Malgrange, *Intégrales asymptotiques et monodromie*, Ann. Sci. École Norm. Sup. **7** (1974), 405–430.
14. M. Merle, *Invariants polaires des courbes planes*, Invent. Math. **41** (1977), 103–111.
15. B. Teissier, *Cycles evanescente, sections planes, et conditions de Whitney*, Singularités à Cargese, Asterisque **7**, **8** (1973), 285–362.
16. ______, *Introduction to equisingularity problems*, Proc. AMS Conf. on Algebraic Geometry, 1974.
17. ______, *Variétés polaires*. I. *Invariants polaires des singularités d'hypersurfaces*, Invent. Math. **40** (1977), 267–292.
18. A. Varcenko, *Newton polyhedra and estimation of oscillating integrals*, Functional Anal. Appl. **10** (1977), 175–196.
19. T. Yano, *Theory of b-functions*. I, Publ. Res. Inst. Math. Sci. (1978), 119–201.

HARVARD UNIVERSITY

Proceedings of Symposia in Pure Mathematics
Volume 40 (1983), Part 2

ESTIMATES AND FORMULAE FOR THE C^0 DEGREE OF SUFFICIENCY OF PLANE CURVES

BEN LICHTIN[1]

1. The invariant of an isolated singularity to be computed, in a limited set of cases here, is the "C^0 degree of sufficiency." This is done in different ways for two classes of functions of two complex coordinates. First, we find a class of functions generic (in a stronger sense than usual) with respect to their Newton Polygon for which the C^0 degree is expressible in terms of the polygon. Second, if the function f defines a plane curve, we use a result of Kuo-Lu [**3**] to express the C^0 degree in terms of algebro-geometric invariants associated to the irreducible components of the curve.

2. We start with

(2-1) DEFINITION. If f: $(C^n, \bar{0}) \to (\mathbf{C}, 0)$ is a germ of an analytic function defined in a neighborhood U of $\bar{0}$, we say that the integer ν is the C^0 degree of sufficiency of f at $\bar{0}$ if the set of terms of degree at most ν, f_ν, suffices to determine the germ f up to topological equivalence.

That is, if for any analytic germ g of degree at least $\nu + 1$, there is a germ of a local homeomorphism φ: $(C^n, \bar{0}) \to (C^n, \bar{0})$ so that

$$f_\nu = (f + g) \circ \varphi.$$

(Equality as germs at $\bar{0}$.)

(A weaker version of the idea, v sufficiency, where germs of the zero loci are used instead, is equivalent to C^0 sufficiency when the dimension of the range equals *one*.)

A basic characterization of the C^0 sufficiency property is due to Kuo [**2**] in the real case and Chang-Lu [**1**] in the complex case.

1980 *Mathematics Subject Classification*. Primary 32B.

[1]Supported in part by NSF grant MCS 77-18723 A03.

The νth jet of f at the origin is C^0 sufficient if the Lojasiewicz-type inequality (à la Lojasiewicz's work on the division problem)

$$\left\| \left(\frac{\partial f}{\partial z_1}(z), \ldots, \frac{\partial f}{\partial z_n}(z) \right) \right\| \geqslant \varepsilon |z|^{\nu-\delta} \tag{2-2}$$

holds. Here, this is intended to mean that in a given neighborhood U of $\bar{0}$, there are ε, $\delta > 0$ such that (2-2) is true for all $z \in U$.

Note. The way in which (2-2) is used is to build a flow from f_ν to $f_\nu + g$ by integrating the vector field

$$\begin{Bmatrix} \operatorname{grad}_z(f_\nu + tg)(p) & \text{if } p = (z, t), z \neq 0 \\ (\bar{0}, 1) & \text{if } z = 0 \end{Bmatrix}$$

in $(n + 1)$-space. Then (2-2) is used to show that the flows from points in $U \times \{0\}$ to $U \times \{1\}$ are continuous in z and do not merge en route. This then constructs the local homeomorphism.

Computing the C^0 degree ν_f of f remains to be done.

In an algebro-geometric setting (over **C**), the answer, as given by Teissier [**8**], is

$$\nu_f = [\![\max\{e_q/m_q\}]\!] + 1 \tag{2-3}$$

where $\{e_q/m_q\}$ are the polar invariants of f. $\operatorname{Max}_q\{e_q | m_q\}$ is the Lojasiewicz exponent for f.

Teissier uses the normalized blowing up of the Jacobian ideal $\mathcal{J}f$ to obtain (2-3). On the other hand, it is well known that underlying the Newton polygon nondegeneracy assumption is the ability to construct a resolution for f by blowing up monomial ideals only [7]. The former is more complicated than the latter. Indeed, one can rarely, if ever, make the ideal $\mathcal{J}f$ invertible by use of the toroidal modification of Khovanskii and Varcenko.

What is needed is an alternative characterization of the $\{e_q/m_q\}$. In [**4**], Lejeune-Jalabert and Teissier showed that the best exponent θ for which (2-2) is true can be expressed as

$$\sup_{\gamma \in \Omega} \left[\frac{\min_i(\operatorname{ord}_t(f_{z_i} \circ \gamma))}{\min_j(\operatorname{ord}_t(z_j \circ \gamma))} \right] \tag{2-4}$$

where Ω is the set of all analytic paths through $\bar{0}$ in U.

Evaluation of (2-4) can be adapted to the toroidal modification situation. For by imposing sufficiently strong (but still generic) nondegeneracy conditions, one can without too much difficulty reduce Ω to monomial curves only and then express (2-4) in terms of $\Gamma_+(f)$. This is only possible so far for $n = 2$.

The result is

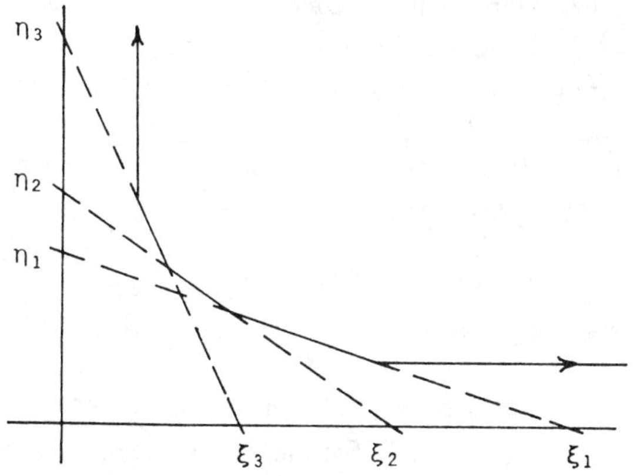

(2-5) THEOREM. *For $n = 2$ and for a germ f satisfying both the standard Kushnirenko nondegeneracy condition and an additional nondegeneracy condition* [**5**], *ν_f is at most one more than the largest axis intercept of any of the support lines first meeting $\Gamma_+(f)$ in a compact facet of maximal dimension* (= 1, *here*).

$$\nu_f \leqslant [\![\max\{\eta_i, \xi_j\}]\!] + 1.$$

For the proof and precise expression of the extra nondegeneracy condition, please see [**5**].

It should be pointed out that, although the extra condition is a generic one, its verification for a particular f cannot be made without recourse to the partials f_{z_i} by way of $\Gamma_+(\mathcal{J}f)$.

One should expect this since the invariant being computed depends both on f and $\mathcal{J}f$. A clearer way to see this is to refer to Teissier's paper [**8**].

There is another Lojasiewicz inequality:

For given U, there are ε, $\alpha > 0$ so that for all $z \in U$,

(2-6) $$|f_{z_1}(z), \ldots, f_{z_n}(z)| \geqslant \varepsilon |f(z)|^\alpha.$$

The best θ of (2-2) and best α of (2-6) are related by

(2-7) $$\theta_{\text{best}} = \frac{\alpha_{\text{best}}}{1 - \alpha_{\text{best}}}.$$

Indeed, for technical reasons, (2-5)'s proof uses (2-6) rather than (2-2) and then expresses the bound for ν_f in terms of a bound for α_{best}.

It follows that we should expect the nondegeneracy conditions, needed to express ν_f in terms of $\Gamma_+(f)$, to incorporate some information about $\Gamma_+(\mathcal{J}f)$. For the proof of (2-5), this certainly is the case.

3. A precise expression for ν_f, for an arbitrary analytic germ f: $(C^2, \bar{0}) \to (C, 0)$, can be given in terms of the characteristic numbers of the analytic branches, composing f, and their intersection numbers. This result was attributed to the author by Teissier [8] and proved by him, using Merle's formula for the polar invariants of a branch. Here, a more concrete proof is given and extends the quoted formula of [8] to the reducible case.

To derive the formula, a result of Kuo and Lu [3] is required and stated here for the irreducible case.

Let $f(x, y) = h(x, y)\prod_{i=1}^{n}(y - A_i(x^{1/n}))$ be the fractional power series decomposition of f. h is a power series of x, y and A_i are "determinations" of the branch.

Define $C_i = \max_{j\neq i} \operatorname{ord}_x(A_j(x^{1/n}) - A_i(x^{1/n}))$, for each $i = 1,\dots,n$.

Let $g_{C_i}(A_i) = A_i(x^{1/n}) - \Sigma_{b_j/n > C_i} a_j x^{b_j/n} + \eta x^{C_i}$ where η is a "generic" parameter value.

Set $l_i = \operatorname{ord}_x f(x, g_{C_i}(A_i))$, $i = 1,\dots,n$, and $\alpha = \max\{l_i - 1\}$.

Then [3] shows that $\nu_f = [\![\alpha]\!] + 1$. If the characteristic sequence is $(n, \beta_1,\dots,\beta_g)$, we show

(3-1) THEOREM. $\nu_f = [\![\mu + \beta_g - 1/n]\!] + 1$.

PROOF. Let $e_i = (n, \beta_1,\dots,\beta_{i-1})$. We show

(1) C_i is always a characteristic exponent β_j/n for some j.

(2) The number of determinations which have a fixed C_j as "order of contact" with a fixed determination is expressible as $e_i - e_{i+1}$ for some $i = 1,\dots,g$. Indeed, if $C_j = \beta_i/n$, this number is $e_i - e_{i+1}$.

PROOF OF (1). Let $A_1(x^{1/n}) = \Sigma_{i=1}^{\infty} a_i x^{b_i/n}$ be any of the n determinations. Then the $n - 1$ other determinations are obtained from A_1 by substitutions t_i: $x^{1/n} \to \omega^i x^{1/n}$, ω a primitive nth root of 1. Label the A_j so that $A_j(x^{1/n}) = A_1(\omega^j x^{1/n})$.

It suffices to find $\varepsilon_i = \operatorname{ord}_x(A_i - A_1)(x^{1/n})$ because one can relabel some other A_j so as to play the role of A_1. Thus, the set of $\{\varepsilon_i\}$ is independent of the choice of A_1.

Now write out the $A_i(x^{1/n})$ as follows:

$$A_1(x^{1/n}) = a_1(x) + a_{\beta_1} x^{\beta_1/n} + \sum_{\beta_1 < b_j < \beta_2} a_j x^{b_j/n} + a_{\beta_2} x^{\beta_2/n} + \cdots,$$

$$A_i(x^{1/n}) = a_1(x) + (\omega^{i\beta_1} a_{\beta_1} x^{\beta_1/n} + \sum \omega^{ib_j} a_j x^{b_j/n} + \omega^{i\beta_2} a_{\beta_2} x^{\beta_2/n} + \cdots.$$

Hence,

$$(A_i - A_1)(x^{1/n}) = (\omega^{i\beta_1} - 1)a_{\beta_1} x^{\beta_1/n} + \sum (\omega^{ib_j} - 1)a_j x^{b_j/n} + (\omega^{i\beta_2} - 1)a_{\beta_2} x^{\beta_2/n} + \cdots.$$

Now filter $\mathcal{C}_n = \langle \omega \rangle$ by the subgroups $\mathcal{C}_{e_1} \supset \mathcal{C}_{e_2} \supset \cdots \supset \{1\}$ where $\mathcal{C}_{e_j} = \langle \omega^{n/e_j} \rangle, j = 1,\dots,g$.

For each $\tau \in \mathcal{C}_n$, there is a unique J such that $\tau \in \mathcal{C}_{e_J} - \mathcal{C}_{e_{J+1}}$. So, $\tau = \omega^i = \omega^{mn/e_J}$ for some $m = 1,\dots,e_J$. $\tau \notin \mathcal{C}_{e_{J+1}}$ implies $mn/e_J \not\equiv 0\ (n/e_{J+1})$, that is $m \not\equiv 0\ (e_J/e_{J+1})$.

We *claim*: $\operatorname{ord}_x[A_i - A_1](x^{1/n}) = \beta_J/n$.

PROOF. For $k \leqslant J-1$, $e_J \mid \beta_k$. Hence, $mne_k/e_J \equiv 0\ (n)$, that is, $\omega^{mn\beta_k/e_J} = 1$.

Secondly, for $k \leqslant J-1$, $\beta_k < b_j < \beta_{k+1}$ implies $mnb_j/e_J \equiv 0\ (n)$. This is because $(n, \beta_1, \ldots, \beta_k) = (n, \beta_1, \ldots, \beta_k, b_1) = \cdots = (n, \beta_1, \ldots, \beta_k, b_l)$ if $b_1/n, \ldots, b_l/n$ are all the numbers appearing as exponents of x between β_k/n and β_{k+1}/n.

Finally, $mn\beta_J/e_J \not\equiv 0\ (n)$. For *if not*, then $m\beta_J \equiv 0\ (e_J)$. But $m \not\equiv 0\ (e_J/e_{J+1})$ implies $me_{J+1} \not\equiv 0\ (e_J)$. Since $(\beta_J, e_J) = e_{J+1}$, necessarily $\beta_J = e_{J+1} \cdot e$ with $(e, e_{J+1}) = 1$. Thus, $m\beta_J = me_{J+1} \cdot e \not\equiv 0\ (e_J)$. But $(e, e_J) = 1$. This implies $me_{J+1} \equiv 0\ (e_J)$, i.e., $m \equiv 0\ (e_J/e_{J+1})$, contrary to a prior conclusion.

Hence, C_i is always a characteristic exponent.

PROOF OF (2). To count the number of determinations A_i with order of contact a given β_j/n with A_1, set $\mathfrak{D}_p = \{A_i(x^{1/n})\colon \operatorname{ord}_x[A_i - A_1](x^{1/n}) \geqslant \beta_p/n\}$. $|\mathfrak{D}_p - \mathfrak{D}_{p+1}|$ may be counted as follows.

To say $A_i \in \mathfrak{D}_p$ is equivalent to saying that $I \equiv \omega^{i\beta_1} \equiv \omega^{i\beta_2} \equiv \cdots \equiv \omega^{i\beta_{p-1}}$, that is, $i\beta_1 \equiv i\beta_2 \equiv \cdots \equiv 0\ (n)$. Hence, i is a solution of these $p-1$ congruences mod (n). It is well known that the number of such solutions in $\{0, 1, \ldots, n-1\}$ is $(n, \beta_1, \ldots, \beta_{p-1}) = e_p$.

So, $|\mathfrak{D}_p - \mathfrak{D}_{p+1}| = e_p - e_{p+1}$.

We see that (1) and (2) allow us to write

$$l = \max\{l_i\} = \frac{\beta_1}{n}(e_0 - e_1) + \cdots + \frac{\beta_g}{n}(e_g - 1) + \beta_g/n$$

where the last β_g/n is needed because we are computing, as well, $\operatorname{ord}_x(A_i - g_{C_i}(A_i))$ with η generic.

Hence, using a formula in [**6**] for μ,

$$l = \frac{\mu}{n} + \frac{\beta_g}{n} + \left(\frac{n-1}{n}\right) = \frac{\beta_1}{n}(e_0 - e_1) + \cdots + \frac{\beta_g}{n} e_g.$$

Setting $\alpha = l - 1$ and solving for α gives the formula as claimed above for $\nu_f = [\![\alpha]\!] + 1$.

When f is analytically reducible at $\bar{0}$, we can extend the analysis in a straightforward way.

Let $\{\Gamma_i\}_1^B$ be the branches composing the factors of f. Let $(n_i, \beta_{1,i}, \ldots, \beta_{g_i})$ be their characteristic sequences. Let $\{A_{j,i}(x^{1/n^i})\}$ be the determinations of the ith branch.

For the ith branch fixed, define the quantities

$$C_{k,j}^{l,i} = \operatorname*{ord}_x\left[A_{j,i}(x^{1/n_i}) - A_{k,l}(x^{1/n_l})\right] \quad \text{for } l \neq i,$$

and $s_{j,i} = \max_{k, l \neq i} C_{k,j}^{l,i}$. Note that $(\Gamma_i, \Gamma_l)_{\bar{0}} = \Sigma_{k,j} C_{k,j}^{l,i}$. $\{\alpha_i = l_i - 1\}_1^B$ are the numbers computed above for each of the branches. Then

$$L_{i,j} = \operatorname*{ord}_x f\left(x, g_{s_{j,i}}\left(A_{j,i}(x^{1/n_i})\right)\right) = \sum_{i \neq q} (\Gamma_i, \Gamma_q)_{\bar{0}} + l_i$$

is a value L_i independent of $j = 1, \ldots, n_i$.

Thus, $\nu_f = [\![\alpha]\!] + 1$ for

$$\alpha = \max\{L_i - 1\} = \max_i \left\{ \sum_{q \neq i} (\Gamma_q, \Gamma_i)_{\bar{0}} + [\![\frac{\mu_i + \beta_{g_i}, -1}{n_i}]\!] \right\}.$$

References

1. S. S. Chang and Y. C. Lu, *On C^0 sufficiency of complex jets*, Canad. J. Math. **25** (1973), 874–880.

2. T. C. Kuo, *On C^0 sufficiency of jets of potential functions*, Topology **8** (1969), 167–171.

3. T. C. Kuo and Y. C. Lu, *On analytic function germs of two complex variables*, Topology **16** (1977), 299–310.

4. M. Lejeune-Jalabert and B. Teissier, Seminar on Integral Closure of Ideals and Equisingularity Theory, Fourier Inst., Fac. Sci. 38402 St. Martin d'Aeres.

5. B. Lichtin, *Estimation of Lojasiewicz exponents and Newton polygons*, Invent. Math. **64** (1981), 417–429.

6. J. Milnor, *Singularities of complex hypersurfaces*, Ann. of Math. Studies, No. 61, Princeton Univ. Press, Princeton, N.J., 1968.

7. D. Mumford, G. Kempf et al., *Toroidal embeddings*. I, Lecture Notes in Math., vol. 339, Springer-Verlag, Berlin and New York, 1973.

8. B. Teissier, *Variétés polaires*. I, *Invariants polaires des singularités d'hypersurfaces*, Invent. Math. **40** (1977), 267–293.

Harvard University

Proceedings of Symposia in Pure Mathematics
Volume **40** (1983), Part 2

QUASI-ORDINARY SINGULARITIES OF SURFACES IN $\mathbf{C}^3$

JOSEPH LIPMAN[1]

Contents

1. Overview. Let me begin by saying what quasi-ordinary singularities are, and what they may be good for.

A singular point P on a surface $F \subset \mathbf{C}^3$ is *quasi-ordinary* if there is a finite map of analytic germs π: $(F, P) \to (\mathbf{C}^2, 0)$ whose discriminant locus Δ (the curve in $\mathbf{C}^2$ over which π ramifies) has a normal crossing at 0 (i.e. 0 is either a smooth point or an ordinary double point of Δ). Given π, we can choose local coordinates x, y, z such that $\pi(x, y, z) = (x, y)$ and such that the germ $(F, P) \subset (\mathbf{C}^3, 0)$ is defined by an equation

$$f(X, Y, Z) = Z^m + g_1(X, Y)Z^{m-1} + \cdots + g_m(X, Y) = 0$$

where the g_i are power series; and "P quasi-ordinary" means that, for some such choice of x, y, z, the *discriminant* $D(X, Y)$ of f (f being considered as a polynomial in Z) is of the form

$$D(X, Y) = X^a Y^b \varepsilon(X, Y), \qquad \varepsilon(0,0) \neq 0.$$

Such singularities arise naturally in the Jungian approach to desingularization, where one begins with a projection of an arbitrary $F \subset \mathbf{C}^3$ into $\mathbf{C}^2$, then applies blowups to the discriminant locus until it has no singularities other than ordinary double points (cf. [$\mathbf{L}_2$, lecture 2] and [$\mathbf{Z}_2$]). In this way, using no more than desingularization of curves, one can modify any surface locally to one with quasi-ordinary singularities. This, first of all, reduces the problem of resolving

1980 *Mathematics Subject Classification*. Primary 14B05, 32C40; Secondary 57Q45.

[1]Partially supported by NSF grant MCS-7903444 at Purdue University.

surface singularities to the quasi-ordinary case; and, secondly, suggests that quasi-ordinary singularities might be useful in analyzing arbitrary surface singularities.

For example, the results on quadratic and monoidal transforms to be given in §3 allow us to describe quite explicitly how to resolve quasi-ordinary singularities by a sequence of transformations in which no singularities other than quasi-ordinary ones appear; and this provides a basis for a proof that a family of germs of surfaces in $\mathbf{C}^3$ which is *equisingular* (in the sense that it has a projection whose discriminant locus is an equisingular family of curves in $\mathbf{C}^2$) admits a *simultaneous embedded resolution* (starting, à la Jung, with a simultaneous resolution of the family of discriminant curves).

Now I will describe the questions to be considered here. (Other questions are not hard to come by: as will soon become apparent, almost any question about plane curve singularities immediately suggests an analogous—and more difficult—one for quasi-ordinary singularities, and not just in the two-dimensional case!) For the most part, proofs, or even complete statements of results, are not given; they can be found in [$\mathbf{L_1}$].

For simplicity we assume unless otherwise indicated that our singularities are locally irreducible.

A quasi-ordinary singularity P lends itself to detailed analysis, because some neighborhood V of P can be parametrized by a *fractional power series* $\zeta = H(X^{1/n}, Y^{1/n})$; in other words, with suitable local coordinates we have $(x, y, z) \in V \Leftrightarrow z = H(s, t)$ where $s^n = x$ and $t^n = y$. What this amounts to is that if $f = f(Z)$ is as above (and irreducible), with discriminant $D = X^a Y^b \varepsilon$, then the roots of f can be represented in the form

$$(1.1) \qquad \zeta_i = H_i(X^{1/n}, Y^{1/n}) = H\big(\omega_{i1} X^{1/n}, \omega_{i2} Y^{1/n}\big), \qquad 1 \leqslant i \leqslant m,$$

with ω_{ij} an n-th root of unity ($j = 1, 2$). This goes back to Jung, the underlying idea being that the local fundamental group of the complement of a plane curve at an ordinary double point is $\mathbf{Z} \times \mathbf{Z}$. (For an algebraic proof cf. [$\mathbf{A_1}$, Theorem 3].) Since

$$X^a Y^b \varepsilon = D = \prod_{i \neq j} \big(H_i(X^{1/n}, Y^{1/n}) - H_j(X^{1/n}, Y^{1/n})\big),$$

unique factorization of (fractional) power series gives

$$(1.2) \quad \zeta_i - \zeta_j = H_i - H_j = X^{u/n} Y^{v/n} \varepsilon_{ij}(X^{1/n}, Y^{1/n}), \qquad \varepsilon_{ij}(0,0) \neq 0.$$

(Here u, v depend on i, j.) The fractional monomials $X^{u/n} Y^{v/n}$ so obtained are the *characteristic monomials* of ζ. The exponents $(u/n, v/n)$ are called the *distinguished pairs* of ζ; they satisfy certain conditions described in the appendix at the end of this section. This parallels exactly the situation for plane curves, where we use fractional (Puiseux) power series and characteristic monomials in one variable X.

In the case of plane curves, two suitably normalized fractional power series in X parametrize *equivalent* singularities if and only if they have the same characteristic monomials [$\mathbf{Z}_4$, Proposition 4.6]. (By "suitably normalized" we mean that the parameter X is chosen to be transversal, so that the denominator n is as small as possible, viz. $n =$ the multiplicity of the singularity.) For a quasi-ordinary singularity P on a surface F, it turns out that quite a lot of the geometry of F near P is determined by the characteristic monomials of a parametrization, for example: the local topology, the tangent cone and the multiplicity at P, the nature of the singular locus, the equivalence class of the singularity of a generic plane curve germ on F at P, or of a generic plane section of F transversal near P to a component of the singular locus. So we may anticipate that the distinguished pairs will provide a good basis for *classifying* quasi-ordinary singularities.

The first problem with this idea is one of uniqueness. Namely, there may be many different possible choices of the projection π, and correspondingly of ζ; and furthermore some projections may have a discriminant with only a normal crossing while the discriminant of the generic projection has a more complicated singularity! (The simplest example of this phenomenon is provided by the surface $Z^3 = XY$.) So we are led to the question which is the central one dealt with in this paper:

(1.3) *For a given quasi-ordinary singularity, are the distinguished pairs of a somehow normalized parametrization ζ as above independent of the possible choices of ζ?*

The answer turns out to be yes. And this is not just a technical matter: approaches to the question which are suggested by the plane curve case lead to algebraic and geometric considerations which add much substance to the idea of classifying quasi-ordinary singularities by distinguished pairs.

For instance, for plane curve singularities one knows, via the classical knot theory of Brauner, Burau, Kähler and Zariski (cf. [**R**]), that *the characteristic monomials of any normalized parametrization are determined by the local topology and vice-versa* (hence the characteristic monomials do not depend on the parametrization). For quasi-ordinary singularities, the distinguished pairs still determine the local topology, but the argument given for this in §2, using saturation, does not give much geometric insight, and does not help much with the converse question. So it would be nice to *find a topological interpretation of the distinguished pairs*. Such an interpretation might well involve an interesting higher-dimensional analogue of the knot theory for curves.

Saturation-theoretic criteria for equivalence of plane curve singularities [$\mathbf{Z}_3$, §2; $\mathbf{Z}_5$, II, §7] motivate one method for getting an affirmative answer to (1.3) for a large class of quasi-ordinary singularities characterized, for example, by the condition $\dim C_4 = 2$, where C_4 is the Whitney cone consisting of "limits of tangents" (cf. beginning of §2). The result for such a singularity P is that *the absolute saturation of the local ring of P determines the distinguished pairs.*

Yet another—and finally successful—approach is suggested by the following well-known result (cf. [$\mathbf{Z}_1$, p. 8] or [$\mathbf{A}_2$, §4]): for an irreducible plane curve singularity Q, the characteristic monomials of any normalized Puiseux parametrization determine and are determined by the sequence of multiplicities of the local rings appearing in the resolution of Q by successive quadratic transformations (where by "resolution..." we mean the sequence of one-dimensional local domains $R_1 < R_2 < \cdots < R_m$ such that R_1 is the local ring of Q, R_{i+1} is obtained from R_i by blowing up the maximal ideal of R_i ($1 \leqslant i < m$), and R_m is regular while R_{m-1} is not). A corresponding result for a quasi-ordinary singularity P should run along the following lines:

There is some natural way of resolving P, say by a sequence of quadratic and "permissible" monoidal transformations (i.e. blowing up points, resp. smooth equimultiple curves), such that certain invariants associated with the local rings appearing in this resolution determine and are determined by the distinguished pairs of any normalized parametrization of P.

In particular, the distinguished pairs are indeed independent of the parametrization.

It *is* possible to establish such a connection between resolutions and distinguished pairs [**L$_1$**, §§5, 6]. The precise statement is too detailed to give here; but let us at least look at some of the problems which arise in carrying through the idea.

First of all, here is what is meant by a "normalized" parametrization of a quasi-ordinary singularity: because of the following lemma, which is a two-dimensional elaboration of the "inversion formula" for plane curve singularities [**Z$_3$**, §3], we can always find a parametrization of the form

$$\zeta = X^{a/n}Y^{b/n}H(X^{1/n}, Y^{1/n}), \qquad H(0,0) \neq 0,$$

where

(i) a and b are not both divisible by n,

(ii) if $a + b < n$, then both $a > 0$ and $b > 0$,

(iii) labelling the distinguished pairs $(\lambda_i, \mu_i)_{1 \leqslant i \leqslant s}$ of ζ so that $\lambda_1 \leqslant \lambda_2 \leqslant \cdots \leqslant \lambda_s$ and $\mu_1 \leqslant \mu_2 \leqslant \cdots \leqslant \mu_s$ (cf. Proposition (1.5)) we have $(\lambda_1, \lambda_2, \ldots, \lambda_s) \geqslant (\mu_1, \mu_2, \ldots, \mu_s)$ (lexicographically).

LEMMA (1.4). *Let $\zeta = X^{a/n}H(X^{1/n}, Y^{1/n})$ be a parametrization with $0 < a < n$ and $H(0,0) \neq 0$. Then, by a definite procedure (given in* [**L$_1$**, §2]) *we can derive from ζ a parametrization of the form $\zeta' = X^{n/a}H'(X^{1/n}, Y^{1/n})$ with $H'(0,0) \neq 0$; and the distinguished pairs of ζ' depend only on those of ζ (cf. Table* (3.4)).

Second, while resolving we want to stay inside the class of quasi-ordinary singularities, so that we can argue by induction on the number of steps in the resolution process. We need then to show that *a quadratic or permissible monoidal transform P' of a quasi-ordinary singularity P is again quasi-ordinary*; and moreover, *from any normalized parametrization ζ for P we can construct one for P' whose distinguished pairs depend only on those of ζ.*

This result is discussed in more detail in §3. For a simple example, blow up the origin on the surface $Z^4 = XY^2$.

Third, which "infinitely near" points shall we consider (where infinitely near points are those which can be obtained from P by a sequence of quadratic and permissible monoidal transformations)? After all, a quadratic transformation produces a whole curve of infinitely near points. And how do we know at any stage of resolution whether to blow up a point or a "permissible" curve (and which curve, if there is more than one)? Roughly speaking, these problems are handled as follows. In the case of quadratic transformations, i.e. blowing up a point Q, it turns out to be sufficient to consider only those finitely many points in the fibre over Q at which the blown-up surface is *not equisingular* along the fibre. As for the choice of what to blow up, what we do is to blow up a curve whenever possible (and a point otherwise), and if more than one curve can be blown up at any stage, we can always choose the one which appeared as the "exceptional curve" (inverse image of the blown-up curve or point) at the previous stage. In this way, we can associate to each quasi-ordinary singularity P a "resolution tree" consisting of finitely many infinitely near points, with branching structure depending only on the distinguished pairs of a normalized parametrization of P. (See below for an illustration of how the singular locus—and in particular the possible curves to be blown up—depends only on the distinguished pairs.)

Finally, which numbers do we attach to the various points in a resolution tree? The numbers we use must be intrinsically associated to the points; and furthermore, given any normalized parametrization for such a point, the numbers in question should be determined by the corresponding distinguished pairs, hence ultimately by the distinguished pairs of some parametrization of the original point P. Conversely, these numbers should give us enough information so that when we look at the entire tree, with its attached numbers, we can reconstruct the original distinguished pairs. Some possibilities which present themselves (as numbers attached to a singularity P) are: the multiplicity, the number of components of the Zariski tangent cone; the number of curves in the singular locus, together with the multiplicities of these curves on the surface, and the multiplicity of P on the curves; the characteristic pairs and intersection numbers of the branches of the plane curve singularities obtained by taking a plane section transversal to the singular locus (away from P), etc. In practice this is more than enough information; and the reconstruction of distinguished pairs from a resolution tree can be carried out, by a very tedious induction based on the transformation formulas of distinguished pairs under quadratic and monoidal transformations (cf. Table (3.4)). Moreover, we do not need the whole tree, but only one specific path through it. Once again, precise statements can be found in [$\mathbf{L_1}$, §§5, 6]. For a simplified treatment, cf. [**Lu**].

We close this section with some examples to illustrate the relation between distinguished pairs and some geometric invariants (tangent cone, singular locus).

(**A**) *Tangent cone*. If a quasi-ordinary P is represented locally, as before, by an equation

$$f(X, Y, Z) = \prod_{i=1}^{m} (Z - \zeta_i) = 0$$

then the Zariski tangent cone (= Whitney's C_3) is the zero set of the leading form f_0 of f (f_0 is the sum of the monomials in f of lowest total degree). The "algebraic" tangent cone is the Spec of the graded ring $\mathbf{C}[X, Y, Z]/f_0$. One finds that this graded ring is determined up to isomorphism by the smallest distinguished pair of any normalized parametrization. The possibilities are typified by equations of the form $Z^n = X^a Y^b$ (with one distinguished pair, viz. $(a/n, b/n)$), for which

$$\begin{aligned} f_0 &= Z^n && \text{if } a+b>n, \\ f_0 &= Z^n - X^aY^b && \text{if } a+b=n, \\ f_0 &= X^aY^b && \text{if } a+b<n. \end{aligned}$$

(We assume that the equation is normalized, so that if $a + b < n$ then both a and b are greater than zero.)

(B) *Singular locus.* (i) Every *normal* quasi-ordinary singularity can be given by an equation of the form $Z^n = XY$.

(ii) There may be one or two smooth curves in the singular locus. For example:

$$\begin{aligned} Z^3 &= X^2Y && (\text{Sing. locus: } [Z = X = 0]). \\ Z^5 &= X^2Y^3 && (\text{Sing. locus: } [Z = X = 0] \cup [Z = Y = 0]). \end{aligned}$$

(iii) The following is typical of the worst that can happen: the singular locus consists of two plane curves, one of which is itself singular. For an example, consider the normalized fractional series

$$\zeta = X^{3/2} + X^{7/4} + X^{9/4}Y^{1/4} + X^{19/8}Y^{11/8}.$$

Multiplying $X^{1/8}$, $Y^{1/8}$ by various 8th roots of unity, one finds that ζ has 32 distinct conjugates ζ_i; $f(X, Y, Z) = \Pi_i(Z - \zeta_i)$ has order 32; and the corresponding singularity has multiplicity 32. The discriminant is

$$\prod_{i \neq j} (\zeta_i - \zeta_j) = X^{1724}Y^{92}\varepsilon(X, Y), \qquad \varepsilon(0,0) \neq 0.$$

Either X or Y vanishes on the singular locus S; and we find that S consists of two curves:

$$\begin{aligned} &C_1\colon X = Z = 0 \qquad (32\text{-fold line}), \\ &C_2\colon Y = 0,\ Z = X^{3/2} + X^{7/4} \end{aligned}$$

(2-fold curve, having itself a 4-fold point at the origin).

Exercise. What happens to the multiplicity of C_2 if we replace the term $X^{9/4}Y^{1/4}$ in ζ by $X^{9/4}Y^{3/4}$? by $X^{9/4}Y^{5/4}$? (Answers (cf. [$\mathbf{L}_1$, p. 39]): 6, 8.)

Appendix. *Characterization of distinguished pairs.* The following description of distinguished pairs turns out to be quite useful. If (λ, μ), (σ, τ) are ordered pairs of rational numbers, we write $(\lambda, \mu) \leqslant (\sigma, \tau)$ to signify $\lambda \leqslant \sigma$ and $\mu \leqslant \tau$. We write $(\lambda, \mu) < (\sigma, \tau)$ if $(\lambda, \mu) \leqslant (\sigma, \tau)$ and $(\lambda, \mu) \neq (\sigma, \tau)$. Also, we let Γ_n be the set of non-negative rational numbers α such that $n\alpha$ is an integer. As usual, $\mathbf{Z}$ denotes the set of all integers.

PROPOSITION (1.5). *Let* $\zeta = \sum_{(\alpha,\beta)\in\Gamma_n\times\Gamma_n} c_{\alpha\beta}X^\alpha Y^\beta$ *be a fractional power series. Then* ζ *parametrizes some quasi-ordinary singularity if and only if there exist pairs* $(\lambda_i, \mu_i) \in \Gamma_n \times \Gamma_n$ $(1 \leq i \leq s)$ *such that*

(1) $(0,0) < (\lambda_1, \mu_1) < (\lambda_2, \mu_2) < \cdots < (\lambda_s, \mu_s)$,

(2) $c_{\lambda_i\mu_i} \neq 0$ *for* $1 \leq i \leq s$,

(3) *if* $c_{\lambda\mu} \neq 0$ *then* $(\lambda, \mu) \in \mathbf{Z} \times \mathbf{Z} + \sum_{(\lambda_i,\mu_i)\leq(\lambda,\mu)} \mathbf{Z}(\lambda_i, \mu_i)$,

(4) $(\lambda_j, \mu_j) \notin \mathbf{Z} \times \mathbf{Z} + \sum_{(\lambda_i,\mu_i)<(\lambda_j,\mu_j)} \mathbf{Z}(\lambda_i, \mu_i)$ $(1 \leq j \leq s)$.

If such pairs exist, they are uniquely determined by ζ; *in fact they are the distinguished pairs of* ζ.

2. Saturation and local topology. In the sequel, $\mathbf{C}\langle \cdots \rangle$ denotes a ring of convergent power series with complex coefficients. We say that

$$\zeta = H(X^{1/n}, Y^{1/n}) \in \mathbf{C}\langle X^{1/n}, Y^{1/n} \rangle$$

is a *quasi-ordinary branch* if ζ parametrizes a quasi-ordinary singularity, i.e. if for any two distinct conjugates $\zeta_i \neq \zeta_j$ of ζ (cf. (1.1)) we have (as in (1.2))

$$\zeta_i - \zeta_j = X^{u/n}Y^{v/n}\varepsilon_{ij}(X^{1/n}, Y^{1/n}), \qquad \varepsilon_{ij}(0,0) \neq 0.$$

The distinguished pairs (λ_i, μ_i) *of a quasi-ordinary branch* ζ *determine the* $\mathbf{C}\langle X, Y\rangle$*-saturation of* $\mathbf{C}\langle X, Y, \zeta\rangle$, *and hence* [$\mathbf{Z}_3$, *Theorem* 6.1] *determine the local topology of the corresponding analytic germ together with its embedding in* $\mathbf{C}^3$.

Indeed, as is readily checked, the saturation in question is the $\mathbf{C}\langle X, Y\rangle$-algebra generated by all the monomials $X^\lambda Y^\mu$ for which (λ, μ) satisfies

$$(2.1) \qquad (0,0) \leq (\lambda, \mu) \in \mathbf{Z} \times \mathbf{Z} + \sum_{(\lambda_i,\mu_i)\leq(\lambda,\mu)} \mathbf{Z}(\lambda_i, \mu_i)$$

(cf. Proposition (1.5) for notation).

In case $(1, 1) \leq (\lambda_1, \mu_1)$, the above saturation is the *absolute saturation* $\tilde{A}$ of $A = \mathbf{C}\langle X, Y, \zeta\rangle$ (cf. [$\mathbf{Z}_5$, III, paragraph preceding Proposition 3.6]). Now $\tilde{A}$ depends only on A (a nontrivial fact!); and, I claim, *the distinguished pairs of* ζ *can be recovered from* $\tilde{A}$, hence do not depend on the choice of ζ.

To justify this claim, note first that the condition $(1, 1) \leq (\lambda_1, \mu_1)$ is intrinsic to A: it is in fact equivalent to the condition that the singular locus of Spec(A) consists of two nonsingular m-fold curves intersecting transversally, where m is the multiplicity of A (cf. example B(ii) in §1). (It is also equivalent to the Whitney cone C_4 for the corresponding singularity P having dimension 2; or to *every* transversal projection of a neighborhood of P to $(\mathbf{C}^2, 0)$ having a normal-crossing discriminant, where "transversal" means "with local degree equal to the multiplicity of P".) Now since saturation commutes with localization [$\mathbf{Z}_5$, III, Proposition 1.2], and since any saturation of a one-dimensional local domain is a local domain with the same multiplicity [$\mathbf{Z}_5$, I, Proposition 2.5], we see that the singular locus of Spec$(\tilde{A})$ consists of two m-fold curves (m is also the multiplicity of $\tilde{A}$, cf. [$\mathbf{Z}_5$, III, Theorem 4.1]). Let ν_1, ν_2 be the discrete valuations belonging to the integral closures of the local rings of these curves. (The integral closures are discrete valuation rings because

$$\mathbf{C}\langle X, Y\rangle \subset \tilde{A} \subset \mathbf{C}\langle X^{1/n}, Y^{1/n}\rangle$$

and because XY vanishes on the singular locus.) Then, with suitable labelling,

$$\nu_1(X) = \nu_2(Y) = m, \quad \nu_2(X) = \nu_1(Y) = 0;$$

and from Proposition (1.5) and (2.1) we find that if

$$S(\tilde{A}) = \{m^{-1}(\nu_1(\alpha), \nu_2(\alpha)) \mid \alpha \in \tilde{A}\}$$

then (λ_1, μ_1) is the smallest pair in $S(\tilde{A}) - (\mathbf{Z} \times \mathbf{Z})$, (λ_2, μ_2) is the smallest pair in $S(\tilde{A}) - (\mathbf{Z} \times \mathbf{Z}) - \mathbf{Z}(\lambda_1, \mu_1)$ and, in general, for $1 \leq j \leq s$, (λ_j, μ_j) is the smallest pair in $S(\tilde{A}) - (\mathbf{Z} \times \mathbf{Z}) - \Sigma_{i<j}\mathbf{Z}(\lambda_i, \mu_i)$. The claim is thereby proved.

In general, when there is no condition on (λ_1, μ_1), we know for other reasons (indicated in §1) that the distinguished pairs of a normalized ζ depend only on $A = \mathbf{C}\langle X, Y, \zeta\rangle$, so that the $\mathbf{C}\langle X, Y\rangle$-saturation A^* described in (2.1) is still determined up to isomorphism by A (independently of the choice of X, Y, ζ). But A^* may be strictly larger than $\tilde{A}$. It would be interesting to have a more direct intrinsic description of A^*. This could lead again to an affirmative answer to (1.3).

Of course one would like a more explicit description of the local topology, as one has via knots for plane curve singularities (cf. [**R**]).

Conversely, it would be nice to know *whether or not the local topology determines the distinguished pairs.*

(I do not know whether the local topology determines even the multiplicity.)

In attempting to understand the local topology of a surface $F \subset \mathbf{C}^3$ having a quasi-ordinary singularity at the origin 0, one thinks first of the intersection of F with a small 5-sphere S^5 centered at 0. Since the singularity at 0 is usually not isolated (example (B)(i), §1), $F \cap S^5$ may not be a manifold. It might then be better to look at the normalization $\bar{F} \to F$, induced say by a linear projection $\bar{F} \subseteq \mathbf{C}^r \xrightarrow{\lambda} \mathbf{C}^3$, and the resulting map

$$L = \bar{F} \cap \lambda^{-1}(S^5) \cap B \to S^5$$

where B is a suitable neighborhood of 0 in $\mathbf{C}^r$. The link L of $\bar{F}$ is a manifold obtained from a 3-sphere by factoring out a finite cyclic group action, i.e. L is a "lens space". (This corresponds to the fact that the origin on $\bar{F}$ is a quotient singularity: for suitable n the fraction field of $\mathbf{C}\langle X^{1/n}, Y^{1/n}\rangle$ is a cyclic Galois extension of that of $\mathbf{C}\langle X, Y, \zeta\rangle$.)

This raises the question, which seems to me worthwhile pursuing, and perhaps not too difficult for knowledgeable topologists:

Can one relate the distinguished pairs of ζ to some homotopy invariants of the above map $L \to S^5$?

3. Quadratic and monoidal transforms. We say for convenience that a local ring A is quasi-ordinary if $A \cong \mathbf{C}\langle X, Y, \zeta\rangle$ with ζ a quasi-ordinary branch (cf. §2), i.e. A is the analytic local ring of some quasi-ordinary singularity. (We could just as well work with *formal* power series over any algebraically closed field of characteristic zero.)

Let A be a quasi-ordinary local ring, and let t: $T \to \operatorname{Spec}(A)$ be the quadratic transformation, i.e. the map obtained by blowing up the maximal ideal $\mathfrak{m}$ of A. After choosing three generators X, Y, ζ of $\mathfrak{m}$, we have an embedding of the closed fibre $t^{-1}(\mathfrak{m})$ into the projective plane $\mathbf{P}^2_{\mathbf{C}}$. By the *quadratic transform of A in the direction* $(\alpha:\beta:\gamma)$ we mean the analytic local ring of the closed point on T corresponding to $(\alpha, \beta, \gamma) \in t^{-1}(\mathfrak{m}) \subseteq \mathbf{P}^2_{\mathbf{C}}$. (The reader may prefer to rephrase all this entirely in terms of local analytic geometry.)

A *monoidal transform* of A is a local ring A' obtained by blowing up a *permissible curve*, i.e. a height one prime ideal p in A such that A and A_p have the same multiplicity. If $A = \mathbf{C}\langle X, Y, \zeta\rangle$ with a normalized quasi-ordinary branch ζ having distinguished pairs $(\lambda_1, \mu_1) < (\lambda_2, \mu_2) < \cdots$, we find that if $\lambda_1 \geqslant 1$ then p may be the ideal (X, ζ), if $\mu_1 \geqslant 1$ then p may be the ideal (Y, ζ), and there are no other possibilities for p. We see then that A' is a finite A-module, uniquely determined by p.

DEFINITION (3.1). *Let A be a quasi-ordinary local ring. We say that A' is a* special transform *of A if A' is not a regular local ring and if one of the following conditions holds*:

(1) A has a permissible curve p, and A' is the corresponding monoidal transform of A.

(2) A has no permissible curve, A' is a quadratic transform of A, and there is a curve q in the singular locus of A whose proper transform passes through A' (i.e. some prime ideal in A' contracts to q).

(3) A has no permissible curve, and A' is a quadratic transform of A in a direction corresponding to a singular point of the (reduced) closed fibre of the quadratic transformation of $\operatorname{Spec}(A)$.

When A is represented by a normalized quasi-ordinary branch, then it is easily verified that any special quadratic transform of A must occur in one of the directions (1:0:0), (0:1:0), (0:0:1).

REMARK (3.2). It can be shown that a quadratic transform A' of A is nonspecial if and only if the blow-up T is *equisingular* along the closed fibre at the point whose analytic local ring is A'.

THEOREM (3.3). *Let A be a quasi-ordinary local ring. Any special transform A' of A is again a quasi-ordinary local ring. If ζ is a normalized branch representing A, i.e. $A \cong \mathbf{C}\langle X, Y, \zeta\rangle$, then, by one of the processes given in* [$\mathbf{L}_1$, §3], *we can find a "standard" quasi-ordinary branch ζ' (not necessarily normalized) which represents A', and whose distinguished pairs depend only on those of ζ and on the process employed, the exact nature of the dependence being as in Table* 3.4.

TABLE 3.4

Transformation	*Distinguished Pairs of Resulting Branch* (omit $i = 1$ if the corresponding pair consists of integers)
LEMMA (1.4)	$(\lambda_i + 1 - \lambda_1)/\lambda_1, \mu_i$
MONOIDAL TRANSFORMATION	
Center (X, ζ)	$\lambda_i - 1, \mu_i$
Center (Y, ζ)	$\lambda_i, \mu_i - 1$
QUADRATIC TRANSFORMATION	
"Transversal Case" $(\lambda_1 + \mu_1 \geqslant 1)$	
Direction (1:0:0)	$\lambda_i + \mu_i - 1, \mu_i$
Direction (0:1:0)	$\lambda_i, \lambda_i + \mu_i - 1$
"Non-Transversal Case" $(\lambda_1 + \mu_1 < 1)$	
Direction (1:0:0)	$\lambda_i + [(1 + \mu_i)(1 - \lambda_1)/\mu_1] - 2, [(1 + \mu_i)/\mu_1] - 1$
Direction (0:1:0)	$\mu_i + [(1 + \lambda_i)(1 - \mu_1)/\lambda_1] - 2, [(1 + \lambda_i)/\lambda_1] - 1$
Direction (0:0:1)	$(\lambda_i(1 - \mu_1) + \mu_i\lambda_1)/1 - \lambda_1 - \mu_1, (\lambda_i\mu_1 + \mu_i(1 - \lambda_1))/1 - \lambda_1 - \mu_1$

We have given the distinguished pairs only for "special" directions in the case of quadratic transformations. This information can always be used to give us the distinguished pairs for non-special directions, cf. [$\mathbf{L_1}$, §4].

To illustrate what is involved in Theorem (3.3), let us examine more closely the quadratic transform A' in a direction $(1{:}0{:}\gamma)$, in case A is represented by a normalized quasi-ordinary branch

$$\zeta = X^{u/n}Y^{v/n}H(X^{1/n}, Y^{1/n}), \qquad H(0,0) \neq 0,$$

where u, v, n are integers with $u + v < n$ ("non-transversal case"). Let $f(X, Y, Z)$ be the minimum polynomial of ζ over $\mathbf{C}\langle X, Y\rangle$, of degree, say, m. Then $A' = \mathbf{C}\langle X, Y, Z\rangle/f'$ where

$$f'(X, Y, Z) = [f(X, XY, X(Z + \gamma))]/X^{(mu+mv)/n}.$$

It is clear that $f'(0, Y, 0) = Y^{mv/n}\cdot$(unit in $\mathbf{C}\langle Y\rangle$). Thus, by the Weierstrass preparation theorem, there is a *unique* power series $g'(X, Y, Z)$ such that $g' \in \mathbf{C}\langle X, Z\rangle[Y]$ is a monic polynomial of degree mv/n in Y, and such that $g' = f'\cdot$(unit in $\mathbf{C}\langle X, Y, Z\rangle$); g' is called the distinguished polynomial in Y associated with f'. Since $\mathbf{C}\langle X, Y, Z\rangle/(f') = \mathbf{C}\langle X, Y, Z\rangle/(g')$, it will be sufficient for our purpose to study the roots of g'.

Let $G(Z^{1/v})$, $G_i(X^{1/n}, Y^{1/n})$ be such that

$$[G(Z^{1/v})]^v = Z + \gamma, \qquad [G_i(X^{1/n}, Y^{1/n})]^v = H_i(X^{1/n}, Y^{1/n}).$$

Let $\xi = G(Z^{1/v})$, $\xi_i = G_i(X^{1/n}, X^{1/n}Y^{1/n})$. Then

$$f'(X, Y, Z) = \prod_{i=1}^{m} \left\{[X^{(n-u-v)/nv}\xi]^v - [Y^{1/n}\xi_i]^v\right\}$$

$$= \pm \prod_{i=1}^{m} \prod_{j=1}^{v} \left\{\omega_j X^{(n-u-v)/nv}\xi - Y^{1/n}\xi_i\right\},$$

where ω_j runs through the vth roots of unity.

Let W be an indeterminate and let $E_i(X,Y,W)$ be such that $E_i(0,0,0) \neq 0$, and

$$E_i(X,Y,W)(W - YG_i(X,XY)) = Y - W\bar{G}_i(X,W), \qquad \bar{G}_i(0,0) \neq 0.$$

Since $G_i(0,0) \neq 0$, the existence of E_i is guaranteed by the preparation theorem. Setting

$$\varepsilon_i = E_i\left(X^{1/n}, Y^{1/n}, \omega_j X^{(n-u-v)/nv}\xi\right)$$

and

$$\bar{\xi}_i = \bar{G}_i\left(X^{1/n}, \omega_j X^{(n-u-v)/nv}\xi\right)$$

we have

$$\varepsilon_i\left(\omega_j X^{(n-u-v)/nv}\xi - Y^{1/n}\xi_i\right) = Y^{1/n} - \omega_j X^{(n-u-v)/nv}\xi\bar{\xi}_i.$$

Hence for some unit ε in $\mathbf{C}\langle X^{1/nv}, Y^{1/n}, Z^{1/v}\rangle$ we have

$$f'(X,Y,Z) = \varepsilon \prod_{i=1}^{m} \prod_{j=1}^{v} \left\{Y^{1/n} - \omega_j X^{(n-u-v)/nv}\xi\bar{\xi}_i\right\}.$$

The double product on the right is clearly the distinguished polynomial in $Y^{1/n}$ associated with $f'(X,Y,Z)$ when $f'(X,Y,Z)$ is thought of as an element of $\mathbf{C}\langle X^{1/nv}, Y^{1/n}, Z^{1/v}\rangle$; but so also is $g'(X,Y,Z)$. By uniqueness we must have

$$g'(X,Y,Z) = \prod_{i=1}^{m} \prod_{j=1}^{v} \left\{Y^{1/n} - \omega_j x^{(n-u-v)/nv}\xi\bar{\xi}_i\right\}.$$

It follows that the roots of g' (considered as a polynomial in Y over $\mathbf{C}\langle X, Z\rangle$) are the n-th powers of the fractional power series $\omega_j X^{(n-u-v)/nv}\xi\bar{\xi}_i$. Thus the roots of g' are fractional power series which are non-units.

Now we have to show that the roots of g' are quasi-ordinary branches, with distinguished pairs depending only on those of ζ, and on whether or not $\gamma = 0$. (For uniformity of notation, we can interchange Y and Z.) For this we study the behavior of the semigroup generated by the pairs of exponents of the non-zero terms in a power series under the operations used above (extraction of v-th roots, passage to associated distinguished polynomial...); and then we apply Proposition (1.5). After obtaining Table 3.4, we can calculate the number of conjugates of ζ' (a root of g'), and compare this number with the degree mv/n of g', to conclude that g' is irreducible when $\gamma = 0$ (g' need not be irreducible if $\gamma \neq 0$). Details are in [$\mathbf{L_1}$].

References

[$\mathbf{A_1}$] S. S. Abhyankar, *On the ramification of algebraic functions*, Amer. J. Math. **77** (1955), 575–592.

[$\mathbf{A_2}$] ———, *Singularities of algebraic curves*, Analytic Methods in Mathematical Physics, Conf. Proc., Gordon & Breach, New York, 1970, pp. 3–14.

[$\mathbf{L_1}$] J. Lipman, *Quasi-ordinary singularities of embedded surfaces*, Thesis, Harvard University, 1965.

[$\mathbf{L_2}$] ———, *Introduction to resolution of singularities*, Algebraic Geometry, Proc. Sympos. Pure Math., vol. 29, Amer. Math. Soc., Providence, R.I., 1975, pp. 187–230.

[**Lu**] I. Luengo, *On the structure of embedded algebroid surfaces*, these PROCEEDINGS.

[**R**] J. E. Reeve, *A summary of results in the topological classification of plane algebroid singularities*, Rend. Sem. Mat. Univ. e Politec. Torino **14** (1954–55), 159–187.

[Z_1] O. Zariski, *Algebraic surfaces*, second ed., Ergebnisse der Math., vol. 61, Springer-Verlag, Berlin, Heidelberg and New York, 1971.

[Z_2] ______, *Exceptional singularities of an algebroid surface and their reduction*, Rend. Accad. Naz. Lincei Cl. Sci. Fis. Mat. Nat. Ser. VIII, **43** (1967), 135–146; reprinted in Oscar Zariski, *Collected papers*, vol. 1, MIT Press, Cambridge, pp. 532–543.

[Z_3] ______, *Studies in equisingularity*. III. *Saturation of local rings and equisingularity*, Amer. J. Math. **90** (1968), 961–1023; reprinted in *Collected papers*, vol. 4, pp. 96–158.

[Z_4] ______, *Contributions to the problem of equisingularity*, Centro Internazionale Matematico Estivo (C.I.M.E.), Questions on Algebraic Varieties. III ciclo, Varenna, 7–17 Settembre 1969, Edizioni Cremonese, Roma, 1970, pp. 261–343; reprinted in *Collected papers*, vol. 4, pp. 159–237.

[Z_5] ______, *General theory of saturation and of saturated local rings*. I, Amer. J. Math. **93** (1971), 573–648; II, Amer. J. Math. **93** (1971), 872–964; III, Amer. J. Math. **97** (1975), 415–502; all reprinted in *Collected papers*, vol. 4.

PURDUE UNIVERSITY

Proceedings of Symposia in Pure Mathematics
Volume **40** (1983), Part 2

THE SMOOTHING COMPONENTS OF A TRIANGLE SINGULARITY. I

EDUARD LOOIJENGA

Introduction. Results of a general nature pertaining to the topology of the semiuniversal deformation of a normal surface singularity are scarce. For example, nothing in the way of a general result is known about the possible singularities on the fibres of such a semiuniversal deformation, let alone the more delicate question of possible combinations of singularities on a single fibre. In fact, there is not even a recipe for finding out whether the singularity is smoothable (i.e. whether there is a deformation possessing nonsingular fibres). Singularities for which such information in available are few. They include the rational double points, the simply-elliptic singularities and the cusp singularities of degree $\leqslant 5$. In these cases our understanding of the semiuniversal deformation comes from an identification of its base space with an explicitly given orbit space, so that the singularities lying over a given point are described by the corresponding isotropy group (which is actually the local monodromy group).

We propose to carry out a similar program for the class of normal surface singularities which follow the cusp singularities in complexity: namely the triangle singularities introduced by Dolgachev (and whose definition will be recalled in §1). Their deformation theory has already been the subject of several investigations: Pinkham [**11**, **12**] has in two Comptes Rendus notes analyzed the nonsingular fibres of the semiuniversal deformation and connected them with the period mapping for $K3$ surfaces. Brieskorn [**1**] listed all the nonrational singularities to which a given triangle hypersurface singularity may deform, while Wahl [**16**] extended much of this to the nonhypersurface case. For the three simplest cases ($D_{2,3,7}$, $D_{2,3,8}$ and $D_{2,3,9}$ in our notation), Brieskorn [**2**] related the base space of the semiuniversal deformation to the Baily-Borel compactification.

We have four main results, three of which concern the so-called negatively graded part S of the base space of a triangle singularity. (This is of codimension

1980 *Mathematics Subject Classification.* Primary 14J17, 14B12, 14D22, 32J05.

0082-0717/82/0000-0646/$03.50

one in the base and is a slice to the equisingular direction. In the complete intersection case, the semiuniversal deformation is topologically trivial in the equisingular direction and hence restricting our attention to the slice is harmless from a topological point of view.) The first result gives a workable algorithm to compute the number of smoothing components of S (and shows, by the way, that this is a rather subtle invariant). The second result gives a description of the set S_f of $s \in S$ over which the fibre has at most rational double points as an orbit space. (This is a stronger and more general form of an earlier announcement of [**8**].) These two results make up Proposition 4. Our third result, Proposition 5, describes a certain partial compactification in a rather general context. The immediate application is in Proposition 6, but this construction is also useful for identifying moduli spaces of $K3$ surfaces coming from geometric invariant theory (like those of Shah). We found it worthwhile to discuss this briefly at the end of §4.

The first draft of this paper was written during the conference and thus I had the opportunity to benefit from useful conversations with Henry Pinkham, Michael Schlessinger and Jonathan Wahl.

The original lecture was intended to be partly expository. I have tried to preserve this a little in the notes: the first section recalls Dolgachev's construction of the triangle singularities and the second section is largely a review of work of Pinkham. The proofs of the propositions in the later sections will appear in sequels to this paper.

1. Construction of the triangle singularities. We describe Dolgachev's [**5**, **6**] construction of the triangle singularities. Suppose we are given positive integers $p \leqslant q \leqslant r$ with $\frac{1}{p} + \frac{1}{q} + \frac{1}{r} < 1$. Then there is triangle in the upper half-plane $\mathcal{H}_+$ with angles π/p, π/q, π/r, which is unique up to isometry. The hyperbolic reflections in the sides of this triangle generate a group G of isometries of $\mathcal{H}_+$ which has the full closed triangle Δ as a fundamental domain. The subgroup $G_+ \subset G$ (of index two) of orientation preserving elements acts as a group of fractional linear transformations on $\mathcal{H}_+$ whose fundamental domain is the union of Δ and a reflected copy Δ' of Δ. By making the appropriate identification on $\Delta \cup \Delta'$ we see that $\mathcal{H}_+/G_+$ is homeomorphic to the two-sphere.

The G_+-action on $\mathcal{H}_+$ naturally extends to the total space of its tangent bundle $T\mathcal{H}_+$ and the inclusion $\mathcal{H}_+ \subset T\mathcal{H}_+$ as zero section determines an inclusion $\mathcal{H}_+/G_+ \subset T\mathcal{H}_+/G_+$ of orbit spaces. Collapsing $\mathcal{H}_+/G_+$ to a point $\{x_0\}$ yields a space Y. It is not hard to see that any G_+-automorphic form of degree $n \geqslant 0$ on $\mathcal{H}_+$ (that is a G_+-invariant expression $f(z)\,dz^n$, $z \in \mathcal{H}_+$, f holomorphic on $\mathcal{H}_+$) determines a continuous function on Y. There are enough of these to give Y the structure of a normal affine surface. With this structure x_0 is the only singular point of Y. The germ (Y, x_0) (as well as any germ isomorphic to it) will be called a *triangle singularity* (with *Dolgachev numbers* p, q, r), here abbreviated as $D_{p,q,r}$-*singularity*. The natural graded structure $A = \bigoplus_{n=0}^{\infty} A_n$, A_n the space of G_+-automorphic forms of degree n, of the coordinate ring of Y is geometrically reflected by the good $\mathbf{C}^*$-action on Y coming from scalar multiplication in the fibres of $T\mathcal{H}_+$. We further observe that the 2-form $t^{-2}\,dz\,dt$ on $T\mathcal{H}_+ - \mathcal{H}_+$ (where t is the fibre coordinate on $T\mathcal{H}_+$ corresponding to dz) is $\mathrm{SL}_2(\mathbf{R})$-invariant and hence determines a holomorphic nowhere zero 2-form ω_0 on $Y - \{x_0\}$. Clearly $\lambda \in \mathbf{C}^*$ transforms ω_0 into $\lambda^{-1}\omega_0$.

2. The fibres of the semiuniversal deformation. As mentioned in the Introduction this is to a large extent a report of work of Pinkham. Our aim is to investigate the deformation theory of the singularity (Y, x_0). As it is easier to treat the deformation theory of a variety (a compact object) than a germ, we first compactify Y by adding a curve to it and then deform this compactification in a way which preserves the added curve. Algebra provides a simple way to compactify Y: just take $Y = \mathrm{Spec}(A) \subset \mathrm{Proj}\, A[t] =: \overline{Y}$, where the grading on $A[t]$ is gotten by putting $\deg t = 1$ and retaining the grading of A. Topologically, $\overline{Y}$ is obtained from Y by adding a point to Y for each $\mathbf{C}^*$-orbit $\mathbf{C}^* \cdot y$ in $Y - \{x_0\}$, so that $\lambda \cdot y$ converges to this point as $|\lambda| \to \infty$. Hence $\overline{Y} - Y$ can be identified with $\mathcal{H}_+/G_+$. We are not satisfied yet, as it appears that $\overline{Y}$ has three singular points on $\overline{Y} - Y$, corresponding to the (irregular) orbits of the vertices of Δ. They are rational double points of type A_{p-1}, A_{q-1} and A_{r-1}. Minimal resolution of these gives a projective compactification X_0 of Y with $X_0 - \{x_0\}$ nonsingular. The curve at infinity, $D_0 := X_0 - Y$, has as its irreducible components *nodal* curves (by this I mean a nonsingular rational curve with self-intersection -2), which meet each other transversally and in at most one point. Its dual intersection graph is the famous $T_{p,q,r}$-diagram

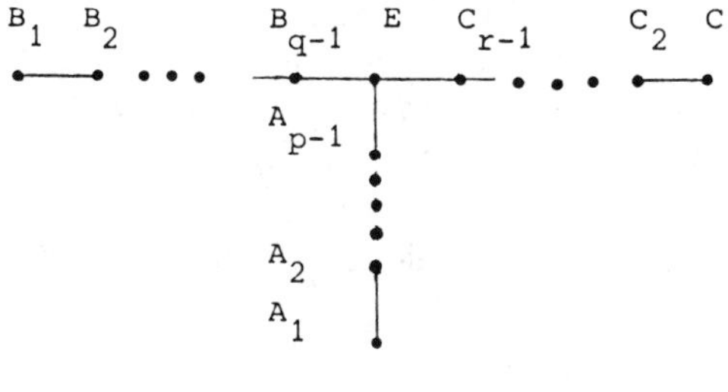

(vertices represent irreducible components of D_0, edges intersection points of these). A curve with the above labeling of its irreducible components will be called a $T_{p,q,r}$-curve. It is not hard to verify that the $\mathbf{C}^*$-action extends to X_0 and that the 2-form ω_0 extends without zeros to $X_0 - \{x_0\}$.

Now let

$$\begin{array}{ccc} (X_0, D_0) & \overset{\iota}{\subset} & (\mathcal{X}, \mathcal{D}) \\ \downarrow & & \downarrow \pi \\ \{0\} & \subset & S \end{array}$$

be a deformation of the pair (X_0, D_0) which preserves D_0. This means that besides the usual condition ($(S,0)$ a pointed germ of an analytic space, π proper and flat, ι an isomorphism of X_0 onto the closed fibre of π), we are further given an extension $\mathcal{D}$ of D which meets every fibre of π in a $T_{p,q,r}$-curve. Now assume that π is semiuniversal for this class of deformations. Then by a general result of Pinkham [**10**] the $\mathbf{C}^*$-action on X_0 extends (at least formally) to a $\mathbf{C}^*$-action on $\mathcal{X}$ and S which makes π equivariant. As the formal $\mathbf{C}^*$-action on $(S,0)$ is good, there is no difficulty in lifting this to an actual $\mathbf{C}^*$-action (which is then also good). So we may take for S an affine space. The germ of π at $\iota(x_0)$ is a deformation of the singularity (X_0, x_0). This deformation is not semiuniversal, although it is almost so: it just corresponds to the negatively graded part of the semiuniversal deformation of (X_0, x_0). In case (X_0, x_0) is a complete intersection, π is locally versal outside x_0. Using this, first Wirthmüller in the hypersurface case and later Damon [**4**] in the complete intersection case, have shown that π is in an appropriate sense 'topologically semiuniversal' at x_0. This is likely to hold for all the triangle singularities, so restricting our attention to this part is probably fairly harmless (compare the discussion at the end of §4).

The first thing we want to do is to give an *a priori* description of all the fibres of π. For this, we observe that $H^1(\mathcal{O}_{X_0}) = 0$ (which is easy to check) and hence (by the semicontinuity of $s \mapsto \dim H^1(\mathcal{O}_{X_s})$), $H^1(\mathcal{O}_{X_s}) = 0$ for all $s \in S$. It can further be shown that on the nonsingular part of any fibre X_s, $s \neq 0$, there lives a unique holomorphic 2-form ω_s such that

(i) ω_s has no zeros,
(ii) $\lambda^*\omega_{\lambda s} = \lambda^{-1}\omega_s$ for all $s \in S$, $\lambda \in \mathbf{C}^*$ and
(iii) $\{\omega_s\}_{s \in S}$ is a holomorphic family. Thus we find

PROPOSITION 1. *Every fibre X_s of π is a normal regular surface endowed with a $T_{p,q,r}$-curve D_s and a holomorphic nowhere zero 2-form ω_s, both defined on the nonsingular part of X_s. Moreover $X_s - D_s$ is affine.*

From work of Pinkham [**13**] it follows that this proposition has the following converse:

PROPOSITION 2. *Let X be a normal regular compact surface endowed with a $T_{p,q,r}$-curve D and a holomorphic nowhere zero 2-form ω, both defined on the nonsingular part of X. Assume that $X - D$ is affine. Then there is a unique $s \in S$ such that there is an isomorphism $X \to X_s$ which maps any irreducible component of D onto the corresponding irreducible component of D_s and pulls ω_s back to ω. Such an isomorphism is also unique.*

For the nonsingular fibres, this is stated in Pinkham [**12**]. The two propositions imply that π defines a moduli space for the triples (X, D, ω). What then are these surfaces? To answer this, let (X, D, ω) be such a triple and let $p\colon \tilde{X} \to X$ minimally resolve the singularities of X. The pull-back $\tilde{\omega}$ of ω will be a meromorphic 2-form on X. We distinguish two cases:

(a) X has only rational double points. Then $\tilde{\omega}$ is holomorphic and without zeros so $\tilde{X}$ is a $K3$ surface. (We recall that a nonsingular compact surface is said to be a $K3$ surface if it is regular and admits a nowhere zero holomorphic 2-form.)

(b) X has at least one singularity x which is not a rational double point. Then x is the only such point. If $\tilde{C}$ denotes the reduced curve $p^{-1}(x)$, then the divisor of $\tilde{\omega}$ is $-\tilde{C}$. Furthermore, $\tilde{X}$ is obtained by successively blowing up points on the regular part of a cubic curve C in $\mathbf{P}^2$, such that $\tilde{C}$ is the strict transform of C. We distinguish three cases: If $b_1(C)$ is the first Betti number of C, then

$b_1(C) = 2$ (C is nonsingular): (X, x) is a simply-elliptic singularity,
$b_1(C) = 1$: (X, x) is a cusp singularity,
$b_1(C) = 0$: (X, x) is a triangle singularity.

We put $S_f = \{s \in S\colon X_s$ has at most rational double points$\}$, to which we will refer as the *rational double point part* of S. It is a Zariski open subset of S. Any irreducible component of S meeting S_f is called a *negative smoothing component*. The study of S_f is the subject of the next section.

3. The rational double point part. The main goal of this section is to characterize the $K3$ surfaces which contain a $T_{p,q,r}$-curve and to describe their moduli space. Let us begin by recalling a few facts concerning $K3$ surfaces. For a $K3$ surface X, we have $H_2(X, \mathbf{Z}) \cong \mathbf{Z}^{22}$ and the intersection form on $H_2(X, \mathbf{Z})$ is even unimodular and of signature $(3, 19)$. These properties characterize $H_2(X, \mathbf{Z})$: it is then isomorphic to $(-E_8)^{\oplus 2} \oplus H^{\oplus 3}$ where E_8 denotes the root lattice generated by the root system E_8 and H stands for the hyperbolic lattice $\mathbf{Z}^2$ with form $\left(\begin{smallmatrix}0&1\\1&0\end{smallmatrix}\right)$. We will refer to the latter as the $K3$ *lattice* and denote it by L. If ω is a nonzero holomorphic 2-form on X (which is unique up to scalar multiple), then integration of ω over 2-cycles in X defines a homomorphism $\omega\colon H_2(X, \mathbf{Z}) \to \mathbf{C}$. The kernel of ω is just the Picard group $\operatorname{Pic}(X)$ of X. Relative to the dual form on $\operatorname{Hom}(H_2(X, \mathbf{Z}), \mathbf{C})$ we have $\omega \cdot \bar{\omega} > 0$, by elementary Hodge theory. If $x \in \operatorname{Pic}(X)$ is such that $x \cdot x \geqslant -2$, then x or $-x$ is represented by an effective divisor.

It will further be convenient to have at our disposal some concepts and notation related to the $T_{p,q,r}$-graph. We let $Q_{p,q,r}$ stand for the free $\mathbf{Z}$-module on the $p + q + r - 2$ symbols $a_1, \ldots, a_{p-1}, b_1, \ldots, b_{q-1}, c_1, \ldots, c_{r-1}, e$. We denote its canonical basis by B and endow $Q_{p,q,r}$ with inner product whose matrix on B is the intersection matrix of a $T_{p,q,r}$-curve. The associated quadratic form is nondegenerate and of signature $(1, p + q + r - 3)$. Observe that in the situation of §2, any fibre X_s of π is in a natural way endowed with an embedding $Q_{p,q,r} \to \operatorname{Pic}(X_s)$. We define the *fundamental isotropic element* $n \in Q_{p,q,r}$ as the unique nonzero $\mathbf{Z}_+$-linear combination of the basis elements with $n \cdot n = 0$ and $n \cdot \alpha \in \{0, 1\}$ for all $\alpha \in B$. If we write $n = \Sigma_{\alpha \in B} n_\alpha \alpha$ then n is given by

(1) 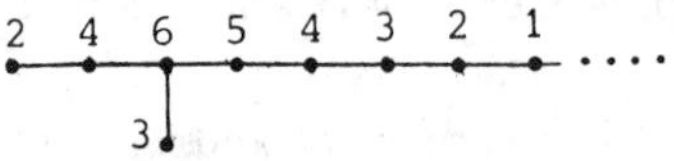 if $p = 2$, $q = 3$,

(2) 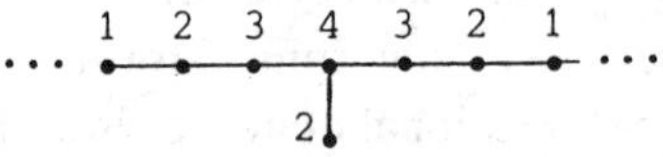 if $p = 2$, $4 \leqslant q$,

(3) 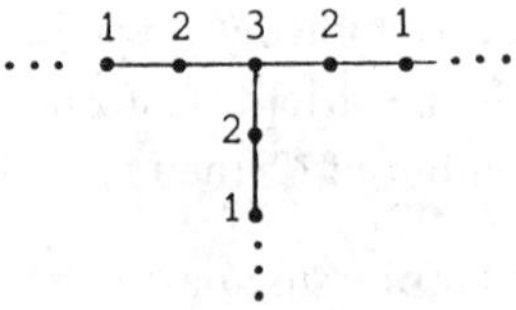 if $p \geqslant 3$,

where the numbers are the nonzero coefficients n_α. We let B_0 denote the set of $\alpha \in B$ with $n_\alpha \neq 0$. Notice that in case (j) B_0 forms an affine root basis of type $\hat{E}_{9-j}$.

Now suppose we are given a lattice M and an embedding j: $Q_{p,q,r} \to M$. We say that an element $e \in M$ is *critical* (*relative* j) if $e \cdot e = -2$, $\mathbf{Z}e + \mathbf{Z} \cdot B_0$ a negative semidefinite lattice and there is no $e' \in \mathbf{Z} \cdot B_0$ such that $e - e'$ is orthogonal to B_0. A more conceptual way to phrase this is that $B_0 \cup \{e\}$ generates an affine root system which strictly contains the one generated by B_0. If the latter is of type $\hat{E}_{9-j}$ then the former (here we invoke the classification of affine root systems) must be of type $\hat{E}_{9-k}$ for some $k < j$. In particular, there will not be any critical vectors in case (1). We say the embedding j is *good* if no critical vector in M is proportional to a vector in the image of j. It is not hard to show that the identity map is good. Hence every primitive embedding is good. The reason for introducing all these notions is

PROPOSITION 3. *Let X be a $K3$ surface endowed with an embedding j: $Q_{p,q,r} \to$ $\mathrm{Pic}(X)$, such that $j(n)$ is the class of an effective divisor (we know that either $j(n)$ or $-j(n)$ is such). Then there exists a $w \in \mathrm{Aut}(H_2(X, \mathbf{Z}))$ which leaves the orthogonal complement of $\mathrm{Pic}(X)$ pointwise fixed and maps each $j(\alpha)$, $\alpha \in B$, on the class of a nodal curve (so that w maps $\Sigma_{\alpha \in B}\, \alpha$ onto the class of a $T_{p,q,r}$-curve) if and only if there are no critical vectors in $\mathrm{Pic}(X)$ relative j.*

The proof actually shows that w can be taken to be a product of reflections with respect to the vectors of norm -2 in $\mathrm{Pic}(X)$. We explain how this proposition, combined with the Torelli theorem for $K3$ surfaces, leads to a description of the rational double point part S_f of S as an orbit space. If j: $Q_{p,q,r} \subset L$ is an embedding of $Q_{p,q,r}$ in the $K3$ lattice L, then its orthogonal complement has signature $(3, 19) - (1, p + q + r - 3) = (2, 22 - p - q - r)$, from which it follows that the set

$$\Omega_j = \{\omega \in \mathrm{Hom}(L, \mathbf{C}) \colon j^*\omega = 0,\ \omega \cdot \omega = 0,\ \omega \cdot \bar{\omega} > 0\}$$

has two connected components. We choose one of them and denote it by Ω. We further let Γ be the group of $\gamma \in \operatorname{Aut}(L)$ with $\gamma \circ j = j$ and $\gamma(\Omega) = \Omega$. Then Γ acts properly discontinuously on Ω.

Going back to the situation of §2, we let $S(j, \Omega)$ be the set of $s \in S_f$ for which there exists an isomorphism ϕ of L onto $H_2(\tilde{X}_s, \mathbf{Z})$ such that $\phi \circ j$ is just the canonical embedding of $Q_{p,q,r}$ in $\operatorname{Pic}(\tilde{X}_s)$ and $\phi^*\tilde{\omega}_s \in \Omega$, where $\tilde{\omega}_s$ is the pull-back of ω_s under the minimal resolution $\tilde{X}_s \to X_s$. As two such ϕ differ by an element of Γ, we get a well-defined map Φ: $S(j, \Omega) \to \Omega/\Gamma$, $s \mapsto \Gamma$-orbit of $\phi^*\tilde{\omega}_s$. Proposition 2 combined with the Torelli theorem for $K3$ surfaces implies that Φ is injective. But a consequence of Proposition 3 is that Φ need not be surjective. To see this, set for any critical vector $e \in L$, $\Omega(e) = \{\omega \in \Omega \colon \omega(e) = 0\}$ and put $\Omega' = \Omega - \bigcup\{\Omega(e)\colon e \text{ critical}\}$. It is easy to see that $\Omega' \neq \varnothing$ if and only if j is good. From Proposition 3 it follows that Φ maps into Ω'/Γ. On the other hand, the surjectivity of the period map for $K3$ surfaces implies that Φ maps onto Ω'/Γ. So we have

PROPOSITION 4. *The map* Φ: $S(j, \Omega) \to \Omega'/\Gamma$ *is an isomorphism. This also sets up a bijective correspondence between the negative smoothing components of S and the* $\operatorname{Aut}(L)$*-equivalence classes of pairs* (j, Ω), *where* $j\colon Q_{p,q,r} \subset L$ *is a good embedding and* Ω *is a connected component of* Ω_j.

The interest of the last part of this proposition is that it reduces the determination of the number of negative smoothing components to a question which is of a lattice theoretic nature and for which effective tools are available, cf. Nikulin [**9**]. Let me illustrate this with two examples.

EXAMPLE 1. For $r \geqslant 7$, $Q_{2,3,r}$ is isomorphic to $(-E_8) \oplus H \oplus (-A_{r-7})$ (here E_8, A_s denote the root lattices of the corresponding root systems). For $s \leqslant 10$ it is easy to embed $(-A_s)$ primitively in $(-E_8) \oplus H \oplus H$ and so there is a primitive embedding $Q_{2,3,r} \to L$ for $7 \leqslant r \leqslant 17$. For $s = 7$, 8 we can embed A_s imprimitively in E_s and hence $Q_{2,3,7+s}$ imprimitively in L. Since all embeddings of $Q_{2,3,r}$ are good, it follows that there exists a smoothing component for all r, $7 \leqslant r \leqslant 17$, and that there are at least two for $r = 14$, 15. For $r = 17$ we also have at least two smoothing components (each a single $\mathbf{C}^*$-orbit), but for a different reason: if j: $Q_{2,3,17} \hookrightarrow L$ is the primitive embedding above, then the choice of a component of Ω_j amounts to the choice of an orientation of the rank two lattice orthogonal to $j(Q_{2,3,17})$. As there is no automorphism of L which is the identity on the image of j, and reverses the orientation of the orthogonal complement, the two components of Ω_j determine distinct smoothing components of S.

EXAMPLE 2. When (X_0, x_0) is a complete intersection, S is nonsingular and so there is only one smoothing component. This implies that there is in this case only one $\operatorname{Aut}(L)$-equivalence class of good embeddings of $Q_{p,q,r}$ in L. For instance, $D_{4,4,4}$ is a hypersurface singularity and there are two $\operatorname{Aut}(L)$ equivalence classes of embeddings $Q_{4,4,4} \hookrightarrow L$; one is primitive (hence good), the other is imprimitive (and bad).

An interesting question is: for which triples (p, q, r) do we have at least one smoothing component? For reasons of signature, a necessary condition is that $p + q + r \leqslant 22$ (J. Wahl [**17**] has shown that this is already implied by the existence of an arbitrary smoothing of the singularity (X_0, x_0)). From Nikulin [**9**, Corollary 1.12.3], it follows that we can embed $Q_{p,q,r}$ in L primitively if $p + q + r \leqslant 21$. Pinkham has just made the computations when $p + q + r = 22$, see the appendix. The result is that we have good embeddings in all cases except for $(p, q, r) = (2, 10, 10)$. He also proves that $D_{2,10,10}$ has no smoothings at all.

Let me close this section with interpreting the torsion of the cokernel of a good embedding $j: Q_{p,q,r} \hookrightarrow L$: if X_s is a nonsingular fibre of π over the corresponding smoothing component, then this torsion is naturally isomorphic to $H_1(X_s - D_s, \mathbf{Z})$.

4. Structure of the smoothing components. We outline how to obtain a Satake-like description of the Stein completion of Ω'/Γ (where we exclude a few triples (p, q, r), see below). This Stein extension must be the normalization of $\bar{S}(j, \Omega)$ and thus we get a good picture of the strata in $\bar{S}(j, \Omega)$. As our construction is meaningful in a somewhat more general set-up, we start out with the following data:

(i) A $\mathbf{Q}$-vector space V of dimension $n \geqslant 2$ endowed with an inner product $(\cdot)$ of signature $(2, n-2)$. This induces an inner product on its dual, which we also denote by $(\cdot)$.

(ii) An arithmetic group $\Gamma \subset \mathrm{Aut}(V)$ which leaves each of the two components of $\{\omega \in \mathrm{Hom}(V, \mathbf{C}): (\omega \cdot \omega) = 0, (\omega \cdot \bar{\omega}) > 0\}$ invariant.

(iii) A collection $\mathcal{E}$ of negative definite $\mathbf{Q}$-lines in V which is a finite union of Γ-orbits in $\mathbf{P}(V)$.

Now let Ω be a component of $\{\omega \in \mathrm{Hom}(V, \mathbf{C}): (\omega \cdot \omega) = 0, (\omega \cdot \bar{\omega}) > 0\}$. For $l \in \mathcal{E}$ we put $\Omega(l) = \{\omega \in \Omega: \omega \mid l = 0\}$. It is not hard to show that $\Omega(l) \neq \varnothing$ and that the collection $\{\Omega(l): l \in \mathcal{E}\}$ is locally finite on Ω. We put $\Omega' := \Omega - \bigcup\{\Omega(l): l \in \mathcal{E}\}$. My purpose is to describe the Stein completion of Ω'/Γ à la Satake, that is, as an orbit space $\hat{\Omega}'/\Gamma$. It will simplify matters if we make the following hypothesis:

(iv) There is no $(n-2)$-dimensional negative (semidefinite) subspace of V spanned by elements of $\mathcal{E}$.

If F is any subspace of V we denote by Ω'_F the image of Ω' under the restriction map $\mathrm{Hom}(V, \mathbf{C}) \to \mathrm{Hom}(F, \mathbf{C})$. Now we define (for the moment just) a set $\hat{\Omega}'$ as the disjoint union of Ω' and the Ω'_F, where F is of type 0, 1, 2 in the sense below.

Type 0. A subspace $F \subset V$ is of Type 0 if F is negative definite and spanned by elements of $\mathcal{E}$ (condition (iv) says that then $\dim F \leqslant n - 3$). This includes the case $F = \{0\}$.

Type 1. Let $F_0 \subset V$ be an isotropic line. Then $F_0^{\perp}/F_0$ has signature $(1, n-3)$. So the set of $x \in F_0/F_0$ with $(x \cdot x) > 0$ determines an open (Lobatchevski) disc D in the associated real projective space of $F_0^{\perp}/F_0$. The orthogonal complement of any $l \in \mathcal{E}$ with $l \subset F_0$ determines a hyperplane in D. The collection of these hyperplanes is locally finite and determines a partition $\mathcal{P}_{F_0}$ of D into polyhedra.

These polyhedra may have infinitely many faces. By condition (iv) there are no vertices and so none of these polyhedra is bounded. For each such polyhedron P we let $F(P)$ denote the subspace of V spanned by F_0 and the $l \in \mathfrak{S}$ with $l \subset F_0^{\perp}$ and l orthogonal to P. By definition, $F(P)$ is of Type 1.

Type 2. Let $F_0 \subset V$ be an isotropic subspace of dim 2. Then the subspace F of V spanned by F_0 and the $l \in \mathfrak{S}$ orthogonal to F_0 is by definition of Type 2. Condition (iv) implies that $\dim F \leqslant n - 3$.

In the definition of $\hat{\Omega}'$ the term "disjoint union" is meant to be relative to the way we indexed the subspace F: if P and P' are distinct elements of $\mathcal{P}_{F_0}$ with $F(P) = F(P')$, then $\Omega_{F(P)}$ and $\Omega_{F(P')}$ are nevertheless disjoint in $\hat{\Omega}'$.

PROPOSITION 5. *We can endow $\hat{\Omega}'$ with a Γ-invariant topology which induces the given topology on each stratum Ω'_F and is such that $\Omega'_F \leqslant \Omega'_G$ implies $F \subset G$. Moreover, $\hat{\Omega}'/\Gamma$ is locally compact Hausdorff and the analytic structure on Ω'/Γ extends to $\hat{\Omega}'/\Gamma$, identifying $\hat{\Omega}'/\Gamma$ with the Stein completion of Ω'/Γ. The natural $\mathbf{C}^*$-action on V induces one on Ω (defined by $\lambda \cdot \omega = \omega \circ \lambda^{-1}$) and hence on Ω/Γ. This $\mathbf{C}^*$-action extends to a good $\mathbf{C}^*$-action on $\hat{\Omega}'/\Gamma$, the (unique) fixed point being the punctual stratum $\Omega'_{\{0\}}$. This gives $\hat{\Omega}'/\Gamma$ the structure of a normal affine variety.*

In case $\mathfrak{S} = \varnothing$, $\hat{\Omega}'/\Gamma$ is a weighted homogeneous cone associated to the Baily-Borel compactification of $\Omega/\Gamma \times \mathbf{C}^*$.

Returning to the case at hand, we take for V the orthogonal complement of $j(Q_{p,q,r})$ in $L_{\mathbf{Q}}$ and for $\mathfrak{S}$ the orthogonal projection in V of the lines spanned by the critical vectors in L. Then (i), (ii) and (iii) are obviously satisfied. We check when (iv) is satisfied. Recall that a critical vector enlarges the affine root system generated by B_0 to an affine root system of higher rank. So if $F \subset V$ is a negative (semidefinite) subspace spanned by elements of $\mathfrak{S}$, then $\tilde{F} := F + \mathbf{Q} \cdot B_0$ is a negative subspace of $L_{\mathbf{Q}}$ spanned by its integral elements of norm -2 and these elements project in $\tilde{F}/\tilde{F} \cap \tilde{F}$ onto an irreducible root system. If B_0 is of type $\hat{E}_{9-j}$, then this root system will be of type E_{9-i} for some $i < j$. Since $\dim \tilde{F} \cap \tilde{F}^{\perp} \leqslant 2$, it then follows that $\dim F \leqslant j + 1$. This is smaller than $\dim V - 2 = 22 - p - q - r$ if $(p, q, r) = (2, 3, r)$ and $r \leqslant 14$ $(j = 1)$ or $(2, q, r)$, $q, r \geqslant 4$ and $q + r \leqslant 16$ $(j = 2)$ or $p, q, r \geqslant 3$ and $p + q + r \leqslant 17$ $(j = 3)$. (This includes all triples for which the corresponding triangle singularity is a complete intersection.) So in these cases condition (iv) is fulfilled and we have the Satake-description $\hat{\Omega}'/\Gamma$ of the Stein completion of Ω'/Γ. Since $\bar{S}(j, \Omega)$ is also Stein (it is an affine variety) the isomorphism $\Phi^{-1}\colon \Omega'/\Gamma \to \bar{S}(j, \Omega)$ of Proposition 4 extends to a morphism $\hat{\Phi}^{-1}\colon \hat{\Omega}'/\Gamma \to \bar{S}(j, \Omega)$. This is a normalization of $\bar{S}(j, \Omega)$.

PROPOSITION 6. *With (p, q, r) and $\hat{\Phi}^{-1}\colon \hat{\Omega}'/\Gamma \to \bar{S}(j, \Omega)$ as above, the composite*

$$\hat{\Omega}' \to \hat{\Omega}'/\Gamma \xrightarrow{\hat{\Phi}^{-1}} \bar{S}(j, \Omega)$$

maps any stratum Ω'_F *of* $\hat{\Omega}'$ *of type resp.* 0, 1, 2 *onto a stratum of* $\bar{S}(j,\Omega)$ *which parametrizes fibres with resp. a triangle singularity, a cusp, a simply-elliptic singularity. The monodromy group of that singularity in the appropriate smoothing component can be identified with the group of* $\gamma \in \Gamma$ *which leave* F *pointwise fixed and in case* F *is of type* 1, $F = F(P)$, *leave the associated polyhedron* P *invariant.*

In the cases $(2,3,r)$, $r \leqslant 14$, this proposition tells us that the inclusion of $S(j,\Omega)$ in the normalization of $\bar{S}(j,\Omega)$ can be identified with the weighted homogeneous cone associated to the Baily-Borel compactification. In the hypersurface case ($r = 7,8,9$) this is due to Brieskorn [**2**].

It also follows that in the cases that Proposition 6 applies, $\hat{\Phi}^{-1}$ induces a bijective correspondence between the Γ-orbits of two-dimensional isotropic subspaces of V and connection strata in the normalization of $\bar{S}(j,\Omega)$ which parametrize fibres with a simply-elliptic singularity. This checks with recent results of Brieskorn [**3**] in the hypersurface case.

I want to conclude with two variants on the construction above—the first pertains to the base space of a semiuniversal deformation of a triangle singularity (and not just the negatively graded piece of it) while the second has to do with geometric invariant theory of polarized $K3$ surfaces.

Let the hypotheses (i)–(iv) be in force again and let U be the connected component of $\{\omega \in \operatorname{Hom}(V,\mathbf{C}): |(\omega\cdot\omega)| < (\omega\cdot\bar{\omega})\}$ which contains Ω. The inner product $\omega \mapsto (\omega\cdot\omega)$ defines a function $q: U \to \mathbf{C}$ which is $\Gamma \times \mathbf{C}^*$ $\mathbf{C}^*$-equivariant if we let Γ act trivially on $\mathbf{C}$ and $\mathbf{C}^*$ act on $\mathbf{C}$ by $\lambda\cdot t = \lambda^{-2}t$. Let $U' \subset U$ be obtained by removing from U the $\omega \in U$ which 'vanish' on some $l \in \mathfrak{S}$. For a subspace $F \subset V$ of Type 0, 1 or 2 the image of U' under the natural map $\operatorname{Hom}(V,\mathbf{C}) \to \operatorname{Hom}(F,\mathbf{C})$ is precisely Ω'_F. We let $\hat{U}'$ be the disjoint union of U' and the $\Omega'_F \times \mathbf{C}$. Then the analogue of Proposition 5 holds for $\hat{U}'/\Gamma$ except for the last part: $\hat{U}'/\Gamma$ has a $\mathbf{C}^*$-action, but this action fails to be good. Notice that we have natural $\mathbf{C}^*$-equivariant projection $\hat{U}'/\Gamma \to \mathbf{C}$ whose fibre over $0 \in \mathbf{C}$ may be identified with $\hat{\Omega}'/\Gamma$. I conjecture that this construction stands in the same relation to the base space of a semiuniversal deformation of a triangle singularity as $\hat{\Omega}'/\Gamma$ to the negatively graded piece of it.

The second variant concerns a partial compactification of $\Omega/(\Gamma \times \mathbf{C}^*)$ related to $\hat{\Omega}'/\Gamma$. If F is of Type 1 or 2 we let Ω_F denote the image of Ω in $\operatorname{Hom}(F,\mathbf{C})$ and we denote by $\hat{\Omega}_F$ the disjoint union of Ω and these Ω_F. Then $\hat{\Omega}$ can be given a topology such that $\hat{\Omega}/\Gamma$ is in a natural way a quasiprojective variety with $\mathbf{C}^*$-action. The orbit space $\hat{\Omega}/(\Gamma \times \mathbf{C}^*)$ will be projective. It can be obtained from the Satake-Baily-Bore! compactification $\Omega^*/(\Gamma \times \mathbf{C}^*)$ of $\Omega/(\Gamma \times \mathbf{C}^*)$ by blowing up the closure of $\cup\{\Omega(l): l \in \mathfrak{S}\}/(\Gamma \times \mathbf{C}^*)$ in $\Omega^*/(\Gamma \times \mathbf{C}^*)$ (this is a Weil divisor).

The application we have in mind is the following: Let L, as before, denote the $K3$ lattice and choose for any even positive integer d an indivisible element $h \in L$ with $(h\cdot h) = d$. Let V denote the orthogonal complement of h in $L_{\mathbf{Q}}$ and choose a connected component Ω of $\{\omega \in \operatorname{Hom}(V,\mathbf{C}): (\omega\cdot\omega) = 0, (\omega\cdot\bar{\omega}) > 0\}$. We take for Γ the group of automorphisms of L which leave h and Ω invariant. The Torelli theorem for $K3$ surfaces asserts that $\Omega/(\Gamma \times \mathbf{C}^*)$ is the moduli space of primitively polarized $K3$ surfaces (possibly with rational double points) of degree d. We let, for $i = 1, 2$, $\mathfrak{S}_i$ denote the set of lines in V which are the orthogonal projection of lines in $L_{\mathbf{Q}}$ spanned by $\varepsilon \in L$ with $\varepsilon \cdot \varepsilon = 0$, $\varepsilon \cdot \overset{*}{h} = i$. It is not hard to show that $\mathfrak{S}_i$ is a single Γ-orbit. The image of $\Omega(l)$, $l \in \mathfrak{S}_i$ in $\Omega/(\Gamma \times \mathbf{C}^*)$ parametrizes polarized $K3$ surfaces X which admit an elliptic fibration such that the polarization is of degree i when restricted to the fibres. In that case the map $X \to \mathbf{P}^{(d+1)/2}$, defined by the polarization, maps each fibre to a $\mathbf{P}^{i-1}$. According to Saint Donat there is converse to this. Using his result it follows that we set $\mathfrak{S} = \mathfrak{S}_1$ for $d = 2$ and $\mathfrak{S} = \mathfrak{S}_1 \cup \mathfrak{S}_2$ for $d \geqslant 4$, then $\Omega'/(\Gamma \times \mathbf{C}^*)$ is a moduli space for primitively polarized $K3$ surfaces (possibly with rational double points) for which the polarization defines a double cover of $\mathbf{P}^2$ for $d = 2$ and is an embedding in $\mathbf{P}^{(d+1)/2}$ for $d \geqslant 4$. The condition (iv) is satisfied if $d \neq 4$ and so $\hat{\Omega}/(\Gamma \times \mathbf{C}^*)$ is then defined. It is not unreasonable to expect a relationship between $\hat{\Omega}/(\Gamma \times \mathbf{C}^*)$ and compactifications of Ω/Γ obtained by doing invariant theory in $\mathbf{P}^{(d+1)/2}$. For $d = 2$ this is the case: $\hat{\Omega}/(\Gamma \times \mathbf{C}^*)$ is precisely the moduli space constructed by Shah [**14**]. If we drop condition (iv) it is still possible to define $\hat{\Omega}/\Gamma$. We shall not do this here, but only remark that in case $d = 4$ (when (iv) is not met) the corresponding $\hat{\Omega}/(\Gamma \times \mathbf{C}^*)$ parametrizes the (degenerate) $K3$ surfaces found by Shah [**15**]. The proof uses a hyperbolic polyhedron in 18-dimensional hyperbolic space which was recently discovered by Vinberg and Kaplanskaja.

REFERENCES

1. E. Brieskorn, *Die Hierarchie der* 1-*modularen Singularitäten*, Manuscripta Math. **27** (1979), 183–219.

2. ______, *The unfolding of the exceptional singularities*, Nova Acta Leopoldina (N.F.) 52 **240** (1980).

3. ______, *Die Milnorgitter derpexzeptionellen unimodularen Singularitäten*, Preprint, 1981.

4. J. Damon, *Topological versality in versal unfoldings* these PROCEEDINGS.

5. I. V. Dolgachev, *Quotient-conic singularities of complex spaces*, Funct. Anal. **8** (1974), 75–76.

6. ______, *Automorphic forms and quasi-homogeneous singularities*, Funct. Anal. Appl. **9** (1975), 67–68.

7. U. Karras, *On pencils of curves and deformations of minimally elliptic singularities*, Math. Ann. **297** (1980), 43–65.

8. E. Looijenga, *Homogeneous spaces associated to certain semi-universal deformations*, Proc. Internat. Congr. Math. (Helsinki, 1978), vol. 2, Acad. Sci. Fenn., Helsinki, 1980, pp. 529–536.

9. V. V. Nikulin, *Integral symmetric bilinear forms and some of their applications*, Math. USSR-Izv. **14** (1980), 103–166.

10. H. Pinkham, *Deformations of algebraic varieties with* $\mathbf{C}^*$-*action*, Asterisque **20** (1974).

11. ______, C. R. Acad. Sci. Paris Sér. A-B **284** (1977), 615–618.

12. ______, C. R. Acad. Sci. Paris. Sér. A-B **284** (1977), 1515–1518.

13. ______, *Deformations of normal surface singularities with* $\mathbf{C}^*$-*action*, Math. Ann. **232** (1978), 65–84.

14. J. Shah, *A complete moduli space for* $K3$ *surfaces of degree* 2, Ann. of Math. (2) **112** (1980), 485–510.

15. ______, *Degenerations of* $K3$ *surfaces of degree* 4, Trans. Amer. Math. Soc. **263** (1981), 271–308.

16. J. Wahl, *Elliptic deformations of minimally elliptic singularities*, Math. Ann. **253** (1980), 241–262.

17. ______, *Derivations of negative weight and non smoothability of certain singularities* (preprint).

KATHOLIEKE UNIVERSITEIT, NIJMEGEN, THE NETHERLANDS

Proceedings of Symposia in Pure Mathematics
Volume **40** (1983), Part 2

ON THE STRUCTURE OF EMBEDDED ALGEBROID SURFACES

IGNACIO LUENGO

Introduction. It is well known that one can resolve a singular point P of a plane algebraic curve by means of successive quadratic transformations and that, at least in the irreducible case, one can use the sequence of multiplicities of the successive points to classify satisfactorily such singularities.

For a singular point Q of a locally embedded algebraic surface S, one can resolve a neighbourhood of Q in S by means of quadratic transformations and monoidal transformations with nonsingular center (see [$\mathbf{L}_2$ and $\mathbf{Z}_3$]). What we do here is to represent all the singular points that we obtain from Q by means of this type of transformation by a tree (in the sense of graph theory) $A(S, Q)$ in such a way that any given sequence

$$\cdots \to S_i \to S_{i-1} \to \cdots \to S_1 \to S$$

of formal quadratic or monoidal transformations (see §1) will be represented by a branch in $A(S, Q)$. Our aim is to compare singularities by means of this tree. We expect that such an approach will shed some light on the general problem of classifying singularities of embedded algebraic surfaces.

This work is essentially devoted to the construction of $A(S, Q)$, but we will also see that in two cases in which a satisfactory classification is known (normal double points and irreducible quasiordinary singularities) we can obtain the same classification using $A(S, Q)$.

The two main aspects of the construction are: (a) Given any singular point P of a surface, if we make a quadratic transformation with center P, the new obtained points form the exceptional divisor D, which is isomorphic to a plane curve. For all the points of D except for a finite number of them, the surface we obtain is equisingular along D in the sense of Zariski. So in order to describe all the

1980 *Mathematics Subject Classification*. Primary 14J17, 14B05.

Key words and phrases. Singularities of surfaces, resolution, equisingularity, quasiordinary singularities.

obtained points it is enough to consider only those in which the surface is not equisingular along D plus a generic point in each component of D. (b) There is a difference with the curve singularity case as a point P in a surface S may have many essentially different desingularization processes. We will see in §3 an example of two surfaces having a process of desingularization in common (this means that they have a part of the tree isomorphic) but nevertheless the whole trees are not isomorphic.

In the case of reducible plane curve singularities, it is known that the equality of the trees is a weaker condition than Zariski equisingularity, for instance the curves of equations

$$(Y^3 + X^5) \cdot (Y^3 + X^7) \quad \text{and} \quad (Y^2 + X^5) \cdot (Y^4 + X^7)$$

have singularities at the origin whose tree is

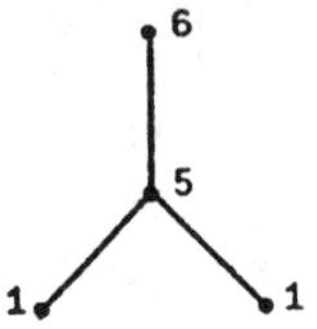

and they are not equisingular. However this kind of example does not occur in the surface case. Consider, for instance, the two above equations as defining surfaces S_1 and S_2 (in the variables X, Y, Z). Then $A(S_1, 0) \neq A(S_2, 0)$.

This paper is a summary of my work **[Lu]**, which contains an extensive treatment of these questions, and the proofs can be found there.

1. The tree of a singularity. For the rest of the paper the word *surface* will stand for an embedded algebroid surface, i.e. an affine scheme $S = \operatorname{Spec}(\mathfrak{O})$, where $\mathfrak{O}$ is a two dimensional complete local ring of embedding dimension 3 defined over a fixed algebraically closed field $k \subset \mathfrak{O}$, which will be assumed to be of characteristic zero.

We will consider *formal quadratic transformations*, that is morphisms $T\colon S' \to S$, where $S' = \operatorname{Spec}(\mathfrak{O}')$ and $\mathfrak{O}'$ is the completion of the local ring in $\operatorname{Bl}_m(\mathfrak{O}) = \operatorname{Proj}\left(\bigoplus_{i \geq 0} m^i\right)$ of a closed point of the exceptional divisor $D = \operatorname{Proj}\left(\bigoplus_{i \geq 0} m^i/m^{i+1}\right)$ and T is the morphism induced by the natural map $\operatorname{Bl}_m(\mathfrak{O}) \to S$. We will also consider *formal monoidal transformations with center a permissible curve* $W \subset S$, i.e., $W = \operatorname{Spec}(\mathfrak{O}/p)$ and p is a height one regular ideal of $\mathfrak{O}$ such that $\mathfrak{O}$ and $\mathfrak{O}_p$ have the same multiplicity and T is defined in the same way as above. Note that S has only a finite number of formal monoidal transforms $S_1, \ldots, S_r$ with center W [$\mathbf{Z}_2$, Proposition 7.2]. We denote by $TM_W(S)$ the set $\{S_1, \ldots, S_r\}$.

Let us define what we understand by a tree.

DEFINITION 1.1. A tree E is a system $E = \{X_i, \pi_i\colon X_i \to X_{i-1}\}_{i \in I}$ consisting of finite sets X_i and maps π_i where $I = \{0, 1, \ldots, n\}$, $n \in \mathbf{N}$, or $I = \mathbf{N}$, X_0 has only one element and the X_i's are pairwise disjoint.

Given a tree E, we associate to it a graph $G(E) = (V(E), L(E))$ whose vertices are $V(E) = \bigcup_{i\in I} X_i$ and the oriented edges are $L(E) = \{(\pi_i(H), H) \mid i \in I - \{0\}, H \in X_i\}$. $G(E)$ is a tree in the usual sense of graph theory.

Later on we will associate to any surface S a tree $E(S)$, whose edges will represent either quadratic or monoidal transformations. In order to distinguish them we will associate a weight to each vertex and edge. This justifies the following

DEFINITION 1.2. A *weighted tree* is a triple (E, α, β) where E is a tree and α: $V(E) \to \mathbf{N}$, β: $L(E) \to \{0, 1\}$ are maps.

An isomorphism of weighted trees is an isomorphism of the corresponding graphs which preserves the weights.

Let S be as above and D the exceptional divisor of the quadratic transformation and $D = D_1 \cup \cdots \cup D_s$ its decomposition in irreducible components. For any point $P \in D$, we denote by S_P the corresponding formal quadratic transform.

PROPOSITION 1.3. *The set $\Sigma(S) = \{P \in D \mid S_P$ is singular and S_P is not Zariski equisingular along $D\}$ is finite, and for each index i, if $P, P' \in D_i - \Sigma(S)$, then either S_P and $S_{P'}$ are both nonsingular surfaces or the corresponding transversal sections to D are equisingular.*

Let us assume $m(S) > 1$ ($m(S)$ = multiplicity of S). Let $W_1, \ldots, W_r$ be the permissible curves of S, $\Sigma(S) = \{P_1, \ldots, P_t\}$ and $\bar{P}_i$ a generic point of D_i.

DEFINITION 1.4. $T(S) = TM_{W_1}(S) \cup \cdots \cup TM_{W_r}(S) \cup \{S_{P_1}, \ldots, S_{P_t}\} \cup \{S_{\bar{P}_1}, \ldots, S_{\bar{P}_s}\}$.

For each i, $1 \leqslant i \leqslant s$, if $m(S_{\bar{P}_i}) > 1$, $S_{\bar{P}_i}$ is equisingular along D, and according to Theorem 7.4 of [$\mathbf{Z}_2$], $S_{\bar{P}_i}$ can be desingularized by means of a sequence of monoidal transformations with centers in D_i and its proper transforms. By this reason only formal monoidal transforms of $S_{\bar{P}_i}$ appear in the tree of S. These $S_{\bar{P}_i}$ and the surfaces obtained by formal monoidal transformations from them will be called *generic surfaces*. For a generic surface H we define:

$$\bar{T}(H) = TM_{\mathrm{Sing}(H)}(H).$$

Now we construct $E(S)$ as follows: First, we define I and the sets X_i. Set $X_0 = \{S\}$. If S is nonsingular ($m(S) = 1$), $I = \{0\}$ and $E(S) = \{X_0\}$. If $m(S) > 1$, set $X_1 = T(S)$. If $m(H) = 1$ for all $H \in X_1$ then $I = \{0, 1\}$. Otherwise we define

$$X_2 = \bigcup_{H\in B} T(H) \cup \bigcup_{H\in B'} \bar{T}(H),$$

where B is the set of surfaces $H \in X_1$ such that $m(H) > 1$ and B' is the set of generic surfaces H in X_1 such that $m(H) > 1$. We construct $X_3, X_4, \ldots$ in the same way. If there is an integer n for which $m(H) = 1$ for all $H \in X_n$ we set $I = \{0, 1, \ldots, n\}$ and in the other case we set $I = \mathbf{N}$.

For every $H \in X_i$, there is only one H' such that $H \in T(H')$ or $H \in \bar{T}(H')$. We define the map π_i: $X_i \to X_{i-1}$ by $\pi_i(H) = H'$.

Finally, let us define the weights on $E(S)$. For $H \in V(E(S))$, we put $\alpha(H) = m(H)$; and if $(H', H) \in L(E(S))$, we define

$\beta(H', H) = 0$ if H is a formal quadratic transform of H',

$\beta(H', H) = 1$ if H is a formal monoidal transform of H'.

DEFINITION 1.5. The weighted tree $A(S) = (E(S), \alpha, \beta)$ constructed above will be called the *tree of infinitely near singularities of S*.

EXAMPLE 1.6. Let S be the surface of equation $Z^2 + Y^3 + X^5$. Then $A(S)$ is isomorphic to the following graph:

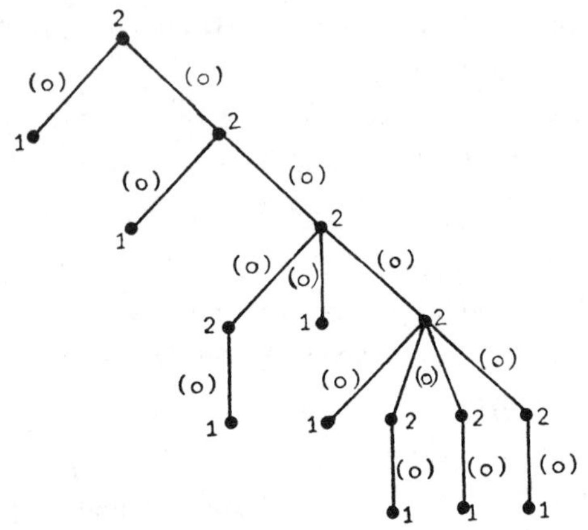

In this case permissible curves do not appear in the process because S is absolutely isolated, i.e., it has an isolated singularity at its closed point and by means of successive quadratic transformations we get only a finite number of singular points. Using the results of [$\mathbf{Z}_3$] one can prove the following

PROPOSITION 1.7. *S is absolutely isolated if and only if $A(S)$ is finite.*

2. Tree of quasiordinary singularities of S. A surface singularity is *quasiordinary or jungian* if it has a finite projection on a plane such that its discriminant has an ordinary double point at the origin. These kinds of singularities play an important role in the resolution of singularities of surfaces (see [$\mathbf{Z}_3$ and $\mathbf{L}_2$]). Given a general surface S, all the quasiordinary singularities which one can obtain from S by quadratic or monoidal tansformations are contained in a subtree which can be deduced from $E(S)$ as follows:

Define $X'_0 = X_0$. If S is quasiordinary, then $E'(S) = \{X'_0\}$. Otherwise we put for any i

$$X'_i = X_i - \{H \in T(S) \mid \text{the branch of } E(S) \text{ joining } S \text{ and } H \text{ contains a quasiordinary surface } H'' \neq H, \text{ not generic}\}.$$

Let $I' = \{i \in I \mid X'_i \neq \varnothing\}$ and $E'(S)$ be the subtree of $E(S)$ consisting of the sets X'_i, $i \in I'$, and the restrictions of π_i to X'_i. Its main property is

THEOREM 2.1. *$E'(S)$ is a finite tree.*

Since each X_i' is finite, $E'(S)$ finite means that each branch of $E'(S)$ is finite. This is equivalent to:

"For every infinite sequence

$$\cdots \to S_i \overset{T_j}{\to} S_{i-1} \overset{T_{i-1}}{\to} \cdots \to S_1 \overset{T_1}{\to} S_0 = S$$

where T_i is a formal quadratic or monoidal transformation and $m(S_i) > 1$, there exists an i such that S_i is quasiordinary".

The method of the proof is similar to Proposition 1.2 of Zariski [$\mathbf{Z_3}$]. The essential difference is that we have to consider both formal quadratic and monoidal transformations instead of only quadratic ones.

In this way we obtain a tree $A'(S) = (E'(S), \alpha, \beta)$. But since we stop the construction of $E'(S)$ each time we find a nongeneric quasiordinary singularity, it is clear that $A'(S)$ provides less information than $A(S)$. However, for the quasiordinary singularities its geometric structure can be determined in another way, namely by means of its characteristic pairs.

Let S be a quasiordinary surface, and $F(Z, X, Y)$ an equation for it such that the discriminant of F with respect to Z is $X^a Y^b \varepsilon(X, Y)$ with $\varepsilon(0,0) \neq 0$. Then the roots of F as a polynomial in Z are power series in $X^{1/d}$, $Y^{1/d}$, that is, there are $\xi_1, \ldots, \xi_n \in k[[X^{1/d}, Y^{1/d}]]$ such that

$$F(Z, X, Y) = \prod_{i=1}^{n} \left(Z - \xi_i(X^{1/d}, Y^{1/d})\right).$$

Then if $i \neq j$, $\xi_i - \xi_j$ divides to $X^a Y^b$ in $k[[X^{1/d}, Y^{1/d}]]$, so $\xi_i - \xi_j = X^{\lambda_{ij}} Y^{\mu_{ij}} \varepsilon_{ij}(X^{1/d}, Y^{1/d})$ with $\varepsilon_{ij}(0,0) \neq 0$ and $(\lambda_{ij}, \mu_{ij}) \in (1/d)(\mathbf{N} \times \mathbf{N})$. The pairs of rational numbers (λ_{ij}, μ_{ij}) are called *characteristic pairs of F*.

If S is an *irreducible* quasiordinary surface, then the set of characteristic pairs of F is totally ordered, and it determines most of the geometry of S, as for instance the tangent cone, the structure of the singular locus, etc.... This can be found in [$\mathbf{L_1}$], where it is also proved that if one chooses a suitable normalized equation F for S, its characteristic pairs only depend on S, and these "normalized" characteristic pairs, NCP(S), allow a classification of such singularities. If S is reducible, each irreducible component S_i of S is again quasiordinary, so it has normalized characteristic pairs NCP(S_i), and for each two different components S_i and S_j one can define intersection multiplicities in a natural way.

Thus we can modify the weights in the tree $A'(S)$ replacing the multiplicity in a nongeneric quasiordianry surface of $A'(S)$ by the normalized characteristic pairs of their irreducible components and the intersection multiplicities. The new "weighted tree" obtained in this way will be denoted by $AQ(S)$ and will be called the *tree of quasiordinary singularities of S*.

In order to analyze the difference of information contained in both trees $A(S)$ and $AQ(S)$, the first question arising naturally is

Question 2.2. Is it true that for any quasiordinary surfaces S and S' one has $A(S) \simeq A(S')$ if and only if $AQ(S) \simeq AQ(S')$?

We will see in the next section that the answer is yes if S and S' are irreducible.

3. Irreducible quasiordinary singularities. Let S be an irreducible quasiordinary surface. From [**L**$_1$] it follows that all the information contained in $T(S)$ (i.e., $\Sigma(S)$, number of components of the tangent cone, etc...) can be obtained from the set NCP(S). Morevoer, all the nongeneric surfaces of $T(S)$ are again irreducible quasiordinary and its normalized characteristic pairs are obtained from those of S by means of the explicit formulae of [**L**$_1$, p. 78].

From this, it is not hard to see

PROPOSITION 3.1. *If S and S' are irreducible quasiordinary surfaces* NCP(S) = NCP(S') *then the weighted trees $A(S)$ and $A(S')$ are isomorphic.*

To see the converse, it is enough to compute the NCP(S) from $A(S)$. Recall that we have formulae giving the variation of the characteristic pairs between any two consecutive points in $A(S)$. In the case of plane curves, the fact that the tree determines the characteristic pairs is showed by induction on the length of the tree using variation formulae as above. In our case, $A(S)$ is not finite, so we will replace it by an appropriate subtree.

Thus, for any general surface S with $m(S) > 1$ put $T^*(S) = T(S) - \{S_{P_1}, \dots, S_{P_r}\}$ where $P_1, \dots, P_r$ are the points of D corresponding to the tangent directions to the permissible curves of S (note that these points actually lie in $\Sigma(S)$). Making the same construction of §1 with $T^*(S)$ instead of $T(S)$ one gets a subtree of $A(S)$ which will be denoted by $A^*(S)$.

THEOREM 3.2. *For any surface S, $A^*(S)$ is finite.*

This result corresponds to the geometric fact that for any sequence of quadratic transformations in directions not tangent to permissible curves the multiplicity drops.

As for irreducible curve singularities one can give an algorithm to compute NCP(S) of a quasiordinary surface S from $A^*(S)$. From this fact we obtain the following

THEOREM 3.3. *If S and S' are irreducible quasiordinary surfaces and $A^*(S) \simeq A^*(S')$ then* NCP(S) = NCP(S').

Let us remark that $A^*(S)$ can be determined (by an algorithm) from the abstract tree $A(S)$, that is $A(S) \simeq A(S')$ implies $A^*(S) \simeq A^*(S')$. So we get

COROLLARY 3.4. *If S and S' are as above, the following statements*

(i) NCP(S) = NCP(S'),

(ii) $A(S) \simeq A(S')$,

(iii) $A^*(S) \simeq A^*(S')$,

are equivalent.

This gives us a new proof of one of the main results of [**L**$_1$], namely, that the normalized characteristic pairs are uniquely determined by S.

Looking at the equivalence of (ii) and (iii) in 3.4, one can ask whether it is true for any surface. The following example gives an answer to this question in the negative sense.

EXAMPLE 3.5. Let S and S' be the surfaces of respective equations $Z^4 + X^5Y$ and $Z^4 + X^2YZ^2 + X^5Y$. Then one can see that $A^*(S) = A^*(S')$. Both of them have a permissible curve with ideal $p = (X, Z)$ and the equation of the quadratic transforms in the corresponding directions, say S_P and $S'_{P'}$ are respectively $Z'^4 + X'^5Y'^2$ and $Z'^4 + X'^2Y'Z'^2 + X'^5Y'^2$. Continuing in this way, one can easily see that $A(S_P) \neq A(S'_{P'})$, whence $A(S) \neq A(S')$.

4. Normal double points. Let S be a normal surface of multiplicity 2, and $\{x, y, z\}$ a coordinate system of S such that the equation of S in this system is given by $Z^2 + F(X, Y)$. The curve $C(S)$ of equation $F(X, Y)$ is reduced and its isomorphism class depends only on S (not on $\{x, y, z\}$) (see Theorems 2 and 2a of [**K**]).

In this case the equisingularity class of $C(S)$ is a good invariant for S. In fact we have

THEOREM 4.1. *Let S and S' be two normal surfaces of multiplicity* 2. *Then the following statements*:

(i) *the plane curves $C(S)$ and $C(S')$ are equisingular*,

(ii) $A(S) \simeq A(S')$,

(iii) $AQ(S) \simeq AQ(S')$,

(iv) $A^*(S) \simeq A^*(S')$,

are equivalent.

The equivalence of (ii) and (iii) follows from the fact that for surfaces of multiplicity 2 it is not difficult to give an affirmative answser to Question 2.2. As in Theorem 3.2, the main point is to find an algorithm to describe $C(S)$ from $AQ(S)$ (or $A^*(S)$).

If now we consider the case of a complex analytic surface X and a normal double point $P \in X$, then the local ring $\mathfrak{O}_{X,P}$ is Cohen-Macaulay and so emb.dim.$(\mathfrak{O}_{X,P}) = \text{mult}(X, P) + 1 = 3$; therefore, X is locally embedded at P and we can associate to P the germ of algebraic curve $C(X, P)$ as before.

On the other hand, we can consider the resolution of a neighbourhood U of P in X by the canonical process of Zariski (i.e., quadratic transformations plus normalizations as in [$\mathbf{Z_1}$]). Let $\pi\colon \tilde{U} \to U$ be such a resolution and $N = \pi^{-1}(P)$. Then $N = \bigcup_{i=1}^n E_i$ where E_i are compact irreducible curves. We can consider now either the topological or differentiable type of the embedding $N \hookrightarrow U$, which is determined by the topological type of the singularities of the E_i's, the genus of E_i's and the intersection numbers (E_i, E_j). Laufer has proved in [**La**] that if one classifies the normal double points by the above topological type, one gets the same classification as the one obtained by the equisingularity class of $C(X, P)$, and so by the tree $A(X, P)$. So from many points of view this is a satisfactory classification of such singularites.

REFERENCES

[**K**] D. Kirby, *The structure of an isolated multiple point of a surface*. I, Proc. London Math. Soc. **6** (1956), 597–609.

[**La**] H. Laufer, *On normal two-dimensional double point singularities*, Israel J. Math. **31** (1978), 315–335.

[$\mathbf{L_1}$] J. Lipman, *Quasi-ordinary singularities of embedded surfaces*, Thesis, Harvard University (1965).

[$\mathbf{L_2}$] ______, *Introduction to resolution of singularities*, Proc. Sympos. Pure. Math., vol. 29, Amer. Math. Soc., Providence, R.I., 1975, pp. 187–230.

[$\mathbf{L_3}$] ______, *Quasi-ordinary singularities of surfaces in* C^3, these PROCEEDINGS.

[**Lu**] I. Luengo, *Sobre la estructura de las singularidades de superficies algebroides sumergidas*, Monograf. Mat. Instituto "Jorge Juan" no. XXV, Madrid, 1981.

[$\mathbf{Z_1}$] O. Zariski, *The reduction of the singularities of an algebraic surface*, Ann. of Math. (2) **40** (1939), 639–689.

[$\mathbf{Z_2}$] ______, *Studies in equisingularity*. II, Amer. J. Math. **87** (1965), 972–1006.

[$\mathbf{Z_3}$] ______, *Exceptional singularities of an algebroid surface and their reduction*, Atti Accad. Naz. Lincei Rend. Cl. Sci. Fis. Mat. Natur. (8) **43** (1967), 135–146.

UNIVERSIDAD COMPLUTENSE DE MADRID, SPAIN

Proceedings of Symposia in Pure Mathematics
Volume **40** (1983), Part 2

SOME NEW SURFACES OF GENERAL TYPE

RICHARD MANDELBAUM

1. Introduction. Suppose V is a minimal surface of general type. Let $\chi = \chi(V) = 1 - q + p_g$, where $q = q(V)$ is the irregularity of V and $p_g = p_g(V)$ is the geometric genus. Let $c_1 = c_1(V)$, $c_2 = c_2(V)$ be its first and second Chern class respectively. Then as a result of work of Yau [**Y**] and Miyaoka [**My**] (see also [**Per**]) it can be shown that

$$2\chi - 6 \leqslant c_1^2 \leqslant 9\chi.$$

If $\sigma = \sigma(V)$ is the signature of V then $\sigma(V) \geqslant 0$ is equivalent to $8\chi \leqslant C_1^2 \leqslant 9\chi$, or equivalently $2c_2 \leqslant c_1^2 \leqslant 3c_2$. The sector $2c_2 \leqslant c_1^2 \leqslant 3c_2$ is probably the least understood area of surfaces of general type. Zappa [**Z**] conjectured that algebraic surfaces with arbitrarily large signature do not exist. This was disproved by Borel [**B**] (see also [**H1**]) who gave examples of actions on the unit ball in $\mathbf{C}^2$ which give rise to surfaces with arbitrarily large signatures. All these Borel surfaces lie on the line $C_1^2 = 9\chi$.

Kodaira [**K**], Hirzebruch [**H2**], and Atiyah [**A**] all have given examples of classes of surfaces of minimal type which lie in the sector $2c_2 < C_1^2 \leqslant 3c_2$. All these examples were in fact Riemann surface bundles $F_{g_1} \to B \to F_{g_2}$ with base space a Riemann surface, having the property that $\pi_1(F_{g_2})$ acted nontrivially on $\pi_*(B)$. (If the action is trivial then by the work of Chern, Hirzebruch and Serre [**A, CHS**] we must have $\sigma(B) = 0$.) In particular Hirzebruch [**H2**] showed that for any $g \geqslant 1$ and $n \geqslant 2$ there exists a minimal surface of general type $\chi(n, g)$ such that

1. $X(n, g)$ is a F_{g_1}-bundle over F_{g_2} with

$$g_2 = 2n^{4g-2}(g-1) + 1, \qquad g_1 = n(2g-1).$$

2.

$$\sigma(X) = 8/3(n^2-1)(g-1)n^{4g-3},$$
$$c_2(X) = 8[n(2g-1) - 1](g-1)n^{4g-2},$$
$$c_1^2(X) = 3\sigma + 2c_2 = 8n^{4g-3}(g-1)[n^2(4g-1) - (2n+1)].$$

1980 *Mathematics Subject Classification*. Primary 57R45.

3.

$$\frac{c_1^2}{c_2} - 2 = 3\frac{\sigma}{c_2} = \frac{n^2-1}{n[n(2g-1)-1]}$$

so

$$\lim_{g\to\infty} c_1^2/c_2 - 2 = 0, \qquad \lim_{n\to\infty} c_1^2/c_2 - 2 = 1/(2g-1).$$

4. The 'smallest' example of a positive signature surface is $X(2,2)$ which is a genus $g_1 = 2\cdot 3 = 6$ bundle over a genus $g_2 = 2\cdot 2^6 + 1 = 129$ base with characteristic numbers $\sigma = 256$, $c_1^2 = 23\cdot 256$, $c_2 = 10\cdot 256$.

We note that *all* of the examples heretofore produced are $K(\pi, 1)$'s, being curve bundles over curves.

We now give a modification of the Hirzebruch-procedure of **[H2]** which produces new examples of minimal surfaces of general type which are *not* $K(\pi,1)$'s and which fill out the sector $2 \leqslant c_1^2/c_2 \leqslant 2\frac{1}{3}$ somewhat more densely. Our 'smallest' example $X(3,1,1)$ satisfies $\sigma(X) = 90$, $c_2(X) = 2106$, $c_1^2 = 4582$ and is a genus 6 fibration over a genus $g_3 = 3^4 + 1 = 82$ base.

2. Method 1. Let R be a compact Riemann surface and suppose $\mathfrak{T}$ is an involution of R with $b = 2c$ fixed points such that $R/\mathfrak{T} = S$ is a Riemann surface of genus g. Let $T \xrightarrow{f} R$ be the universal $H_1(R, \mathbf{Z}_n)$ cover of R and set $d = \deg f$. Note that $g_R = g(R) = 2g - 1 + c$ and, $d = n^{2g_R}$, and $g(T) = n^{2g_R}(g_R - 1) + 1$.

Let $\Gamma_{f'}$, $\Gamma_{\mathfrak{T}f}$ be the graphs of f, $\mathfrak{T}f$ respectively in $T \times R$ and note that Γ_f intersects $\Gamma_{\mathfrak{T}f}$ transversely in $2cd$ points and that as a homology class $\Gamma_f - \Gamma_{\mathfrak{T}f}$ is divisible by n. Blow $T\times R$ up at the $2cd$ points in $\Gamma_n \cap \Gamma_{\mathfrak{T}f}$ to get $V = \overline{T\times R}$ and let D_f, $D_{\mathfrak{T}f}$ be the strict images of Γ_f, $\Gamma_{\mathfrak{T}f}$. Let E be the homology class of $D_f - D_{\mathfrak{T}f}$.

Then again as a homology class we have that E is divisible by n and has $D_f \cup (-D_{\mathfrak{T}f})$ as a nonsingular representative. Let $|E| = D_f \cup (-D_{\mathfrak{T}f})$.

Let $X(n, g, c)$ be the n-fold branched cover of V ramified over $|E|$ with projection map ρ: $X \to V$. Let $V \to T\times R \to T$ be the obvious projection and let π: $X \to T$ be the induced projection. It is easily computed that the genus of a generic fiber is $g_1 = ng_R = n(2g-1) + nc$. By somewhat more complicated computations involving the G-signature theorem (see **[H2]**), we get

$$\begin{aligned}
\sigma(X(n,g,c)) &= \tfrac{4}{3}n^{2g_R-1}(n^2-1)(g_R-1) - \tfrac{2}{3}n^{2g_R-1}c(n^2+2)\\
&= n^{2c}\sigma(X(n,g)) + \tfrac{2}{3}cn^{4g-3+2c}(n^2-4),\\
c_2(X(n,g,c)) &= 4n^{2g_R}(g_R-1)(ng_R-1) + 2n^{2g_R+1}c\\
&= n^{2c}c_2(X(n,g)) + 2cn^{4g-2+2c}[8n(g-1) + 2nc + 3n - 2],\\
c_1^2(X(n,g,c)) &= n^{2c}c_1^2(X(n,g))\\
&\quad + 2cn^{4g-3+2c}[16n^2(g-1) + 4n^2c + 7n^2 - 4n - 4].
\end{aligned}$$

Therefore

$$\frac{c_1^2}{c_2}(n, g, c) = 2 + \frac{2(n^2 - 1)(g_R - 1) - c(n^2 + 2)}{2n(ng_R - 1)(g_R - 1) + cn^2}$$

and thus

$$L = \lim_{n \to \infty} \frac{c_1^2}{c_2} - 2 = \frac{4(g - 1) + c}{4(2g - 1)(g - 1) + c(8g - 5 + 2c)},$$

$$\lim_{g \to \infty} \frac{c_1^2}{c_2} - 2 = \lim_{c \to \infty} \frac{c_1^2}{c_2} - 2 = 0.$$

In particular we have for L:

$g \backslash c$	0	1	2	3	4	5	c
0	—	—	—	—	—	1/29	$\frac{c-4}{2c^2-5c+4}$
1		1/5	1/7	1/9	1/11	...	$\frac{1}{2c+3}$
2	1/3	1/5	3/20	8/63	...	...	$\frac{c+4}{2c^2+11c+12}$
3	1/5	9/61	10/86	...	...	...	$\frac{c+8}{2c^2+19c+40}$
4	1/7	13/113	...				$\frac{c+12}{2c^2+27c+84}$
g	$\frac{1}{2g-1}$	$\frac{4g-3}{8g^2-4g+1}$	$\frac{4g-2}{8g^2-4g+2}$				

As we noted before the 'smallest' example is $X(3, 1, 1)$.

It is easy to see that wherever $c > 0$ we have a surface which is not a $K(\pi, 1)$.

3. Method 2. We again suppose R is a Riemann surface and γ, σ are commuting automorphisms of R of orders q and p respectively.

Let $T \xrightarrow{f} R$ be the universal $H_1(R, \mathbf{Z}_n)$ cover of R and let $\tilde{\sigma}$ be a lift of σ to an automorphism of T. Set $d = \deg f = n^{2g}R$, where $g_R = \text{genus}(R)$.

Now for any self-map T of a space X, let $\text{Fix}\, T = \{x \in X \mid T(x) = x\}$ and $\nu(T) = \text{card}(\text{Fix}\, T)$.

We suppose that $\text{Fix}\, \sigma \neq \varnothing$ and pick $y_0 \in \text{Fix}\, \sigma$, $x_0 \in f^{-1}(y)$. We can now characterize $\tilde{\sigma}$ uniquely by demanding that $\tilde{\sigma}(x_0) = x_0$.

Lastly we shall assume that

$$f(\text{Fix}\, \tilde{\sigma}) \subset \text{Fix}\, \gamma.$$

Now let Γ_f be the graph of f in $T \times R$ and setting $f' = \gamma \circ f$ let $\Gamma_{f'}$ be the graph of f' in $T \times R$.

Now $\tilde{\sigma} \times \sigma$ acts on $T \times R$ by $\tilde{\sigma} \times \sigma(x, y) = (\tilde{\sigma}(x), \sigma(y))$ and this is a proper discontinuous action the complement of the $\nu(\tilde{\sigma})\nu(\sigma)$ fixed points. We set $V' = T \times R/\tilde{\sigma} \times \sigma$ with projection map w: $T \times R \to V'$. V' is a normal analytic surface with cyclic quotient singularities at no more than $\nu(\tilde{\sigma})\nu(\sigma)$ points. Let $V \xrightarrow{\pi} V'$ be its minimal resolution. Let D_f, $D_{f'}$ be proper transforms of $w(\Gamma_f)$, $w(\Gamma_{f'})$ in V. Lastly blow V up along $D_f \cap D_{f'}$ to get a new surface $\tilde{V}$ and proper transforms $\tilde{D}_f$, $\tilde{D}_{f'}$. Set $E = \tilde{D}_f - \tilde{D}_{f'}$. It can then be verified that E is a nonsingular curve in $\tilde{V}$ with two disjoint components and as a divisor that E is divisible by n. We can thus form the n-sheeted branched cover $X \to \tilde{V}$ totally ramified over E.

We can now calculate the invariants of X by making use of the G-signature theorem (see **[H2]**) applied to $\tilde{V}$ and calculating the invariants of $\tilde{V}$. We illustrate these calculations by means of a few examples.

Let R be a Riemann surface of genus g and let σ be an involution on R. We again suppose $\sigma\gamma = \gamma\sigma$ and $f(\operatorname{Fix}\tilde{\sigma}) \subset \operatorname{Fix}\sigma$. Now blow $T \times R$ up at the $\nu(\sigma)\nu(\tilde{\sigma}) + dV(\sigma) - \nu(\tilde{\sigma})$ distinct points in $\operatorname{Fix}(\tilde{\sigma} \times \sigma) \cup (\Gamma_f \cap \Gamma_{f'})$ and let $\widetilde{T \times R}$, $\tilde{\Gamma}_f$, $\tilde{\Gamma}_{f'}$ be the appropriate objects. We note that $\tilde{\Gamma}_f \cdot \tilde{\Gamma}_f = \Gamma_f \cdot \Gamma_f - d\nu(\sigma)$. It is also easy to verify that w can be extended to a map $\tilde{w}\colon \widetilde{T \times R} \to \tilde{V}$ with $\tilde{w}(\tilde{\Gamma}) = \tilde{D}_f$. Then

$$E \circ E = 2\tilde{D}_f \circ \tilde{D}_f = \tilde{\Gamma}_f \circ \tilde{\Gamma}_f = \Gamma_f^2 - d\nu(\sigma).$$

But $\Gamma_f^2 = \chi(T) = d\chi(R)$ so that $E \circ E = d[\chi(R) - \gamma(\gamma)]$. Using the G-signature theorem we obtain $\sigma(X) = n\sigma(\tilde{V}) - [(n^2-1)/3n]E \circ E$, $\sigma(\widetilde{T \times R}) = 2\sigma(\tilde{y}) - \frac{1}{2}R_{\tilde{w}}^2$ where $R_{\tilde{w}}$ is the ramification locus of $\tilde{w}$ in $\tilde{V}$. Now

$$\sigma(\widetilde{T \times R}) = -[\nu(\sigma)\nu(\tilde{\sigma}) + d\nu(\gamma) - \nu(\tilde{\sigma})] \quad \text{while } R_{\tilde{w}}^2 = -2(\sigma)V(\tilde{\sigma}).$$

Thus

$$\sigma(V) = \tfrac{1}{2}[\sigma(T \times R) - \nu(\sigma)\nu(\tilde{\sigma})] = -\tfrac{1}{2}(d\nu(\gamma) - \nu(\tilde{\sigma})) - \nu(\sigma)\nu(\tilde{\sigma}).$$

So

$$\sigma(X) = \frac{n}{2}(\nu(\tilde{\sigma}) - d\nu(\gamma) - n\nu(\sigma)\nu(\tilde{\sigma})) + \frac{n^2-1}{3n}d(2g_R - 2 + \nu(\gamma)).$$

Thus

$$\sigma(X) = \frac{1}{6}\left[4(g_R - 1)(n^2-1) - \nu(\gamma)(n^2+2)\right]n^{2g_R-1} - \frac{n}{2}\nu(\tilde{\sigma})(2\nu(\sigma) - 1).$$

Now let us suppose that R is hyperelliptic and σ is the hyperelliptic involution on R. Thus $\nu(\sigma) = 2g + 2$.

Case I. n is even. In this case an explicit calculation shows that all the fixed points of $\tilde{\sigma}$ lie over y_0 and $\nu(\tilde{\sigma}) = 2^{2g_R}$. As in **[FK**, p. 247] if $g \geq 3$ there exists a γ commuting with σ which has as fixed points y_0 and exactly one other point y_∞ so that $\nu(\gamma) = 2$ and of course $f(\operatorname{Fix}\tilde{\sigma}) = \{y_0\} \subset \operatorname{Fix}\gamma$.

We now compute

$$\sigma(X) = \frac{1}{6}\left[4(g-1)(n^2-1) - 2(n^2+2)\right]n^{2g-1} - \frac{n}{2} \cdot 2^{2g}(4g+3).$$

The 'minimal' examples occur for $n = 4$, $g = 3$ when we compute that $\sigma(X) = 97 \cdot 2^7$ and has fibers of genus 6 and a base of genus $32 \cdot 255$.

Case II. n is odd. In this case $\tilde{\sigma}$ has exactly one fixed point over every fixed point of σ. Thus $\nu(\sigma) = \nu(\tilde{\sigma}) = 2g + 2$ and $f(\text{Fix } \tilde{\sigma}) = \text{Fix } \sigma \subset \text{Fix } \gamma$ so we must have $\gamma = \sigma$. Thus

$$\begin{aligned}\sigma(X) &= \frac{1}{6}\left[4(g-1)(n^2-1) - (2g+2)(n^2+2)\right]n^{2g-1} - \frac{n}{2}(2g+2)(4g+3)\\ &= \frac{1}{3}g(n^2-4)n^{2g-1} - n(g+1)(4g+3).\end{aligned}$$

So, if $g = 2$

$$\sigma(X) = \tfrac{2}{3}(n^2-4)n^3 - 33n$$

with the smallest example of positive signature occurring for $n = 5$, where $\sigma(X) = 5 \cdot 317$ and X has fibers of genus 10 and a base of genus 622. If $g = 3$ then $\sigma(X) = (n^2-4)n^5 - 60n$ so that $n = 3 \Rightarrow \sigma = 1025$ and X has fibers of genus 9 and a base of genus 1454.

We note that in all of the above examples X has $\nu(\sigma)\nu(\tilde{\sigma})$ simply-connected fibers and so $\pi_1(X) = \pi_1(\text{Base})$. Again it is easy to see that none of the X's are $K(\pi, 1)$'s. We leave the calculation of $c_1^2(X)$ as an exercise for the interested reader. In a future paper **[Man]** we intend to explore the above type of construction in greater detail, with special emphasis on more complicated singularities.

References

[A] M. F. Atiyah, *The signature of fibre-bundles*, Global Analysis, Princeton Univ. Press, Princeton, N. J., 1969, pp. 73–84.

[B] A. Borel, *Compact Clifford-Klein forms of symmetric spaces*, Topology **2** (1963), 111–122.

[CHS] S. S. Chern, F. Hirzebruch and J. P. Serre, *The index of a fibered manifold*, Proc. Amer. Math. Soc. **8** (1957), 587–596.

[FK] H. Farkas and I. Kra, *Riemann surfaces*, Springer-Verlag, Berlin and New York, 1980.

[H1] F. Hirzebruch, *Topological methods in algebraic geometry*, 3rd ed., Springer-Verlag, Berlin and New York, 1960, §22.3.

[H2] ______, *The signature of ramified coverings*, Global Analysis, Princeton Univ. Press, Princeton, N. J., 1969, pp. 253–265.

[K] K. Kodaira, *A certain type of irregular algebraic surface*, J. Analyse Math. **19** (1967), 207–215.

[Man] R. Mandelbaum (to appear).

[My] Y. Miyaoka, *On the Chern numbers of surfaces of general type*, Invent. Math. **42** (1977), 225–237.

[Per] Ulf Persson, *On Chern invariants of surfaces of general type*, preprint.

[Y] S. T. Yau, *Calabi's conjecture and some new results in algebraic geometry*, Proc. Nat. Acad. Sci. U.S.A. **74** (1977), 1798–1799.

[Z] G. Zappa, *Sopra une probabile diseguaglianza tra i cautteri invariantivi di una superficie algebrica*, Rend. Mat. **14** (1955), 455–464.

University of Rochester

Proceedings of Symposia in Pure Mathematics
Volume **40** (1983), Part 2

DISTANCE FROM A SUBMANIFOLD IN EUCLIDEAN SPACE

JOHN N. MATHER

Let M be a smooth (i.e. C^∞) compact submanifold without boundary of Euclidean space. Let ρ_M be the function on $\mathbf{R}^r$ which measures the distance to M. Thus

$$\rho_M(x) = \min_{y \in M} \|y - x\|,$$

where $\| \ \|$ denotes the Euclidean norm. Generally, ρ_M will not be C^∞ at all points. We will call points where ρ_M is not C^∞ *singularities* of ρ_M.

F. Almgren asked me what can be said about ρ_M at its singularities. As it turns out, known methods in the theory of singularities of mappings give a great deal of information for generic M, at least when the dimension of the ambient space is small (i.e. $\leqslant 7$). However, the precise results which are needed to answer Almgren's question do not seem to be in the literature, although closely related results are in articles of Buchner [**3**, **4**], Looijenga [**7**], Milman [**10**, **11**], Milman and Waksman [**12**], Yomdin [**19**], and Wall [**18**]. The idea of applying singularities of mappings to this sort of problem is due to Thom [**14**, **15**] who also indicated how it might be applied to many other geometric problems, as well. After the basic results in singularities of mappings were obtained, Wall wrote an exposition [**18**] explaining how Thom's idea of applying singularities of mappings to geometric problems could be carried through in many cases. One of the cases which Wall considers is the distance function. However, Wall's comments on the distance function are terse, and they do not directly address Almgren's question. Another reference is Looijenga's thesis [**7**], which contains a transversality lemma which we will use, as well as many other results related to the study of the distance function.

Buchner [**3**, **4**] applied methods very similar to those which we use here to the study of the cut-locus. Milman [**10**, **11**], Milman and Waksman [**12**], and Yomdin

1980 *Mathematics Subject Classification.* Primary 57R45.

[19] have applied these methods to the study of the structure of the central set of a manifold, which is a problem very similar to the one we study in this exposition. (The central set of M is the set of singularities of ρ_M.) In several brilliant survey articles, Arnold [1] has shown how ideas similar to those discussed in this article come up in many contexts.

Since most of what I will do in this exposition is a slight variation on known results, most of it will be obvious to experts. I hope to make this material known to a wider audience. In contrast to Wall's and Arnold's expositions, this article focuses on a single problem. My original intention was to give a detailed explanation of how to apply the method of singularities of mappings to it. But I have not found time for that, so I will just give an outline.

At this summer school, I learned from Yomdin of a new method [5], based on the theory of Lipschitz functions, which gives very interesting information concerning ρ_M, independent of the ambient dimension r. In particular, Yomdin proved that for generic M, the distance function is a topological Morse function.

1. Normal forms. Let u be the germ at the origin of a continuous (but not necessarily differentiable) function on $\mathbf{R}^r$. Let $x \in \mathbf{R}^r$. We will say that u is a *normal form* for ρ_M at x if there exists a germ ϕ: $(\mathbf{R}^r, x) \to (\mathbf{R}^r, 0)$ of a smooth diffeomorphism such that

$$\rho_M = u\phi + c \tag{1}$$

in a neighborhood of x, where c is a constant.

This article gives a short list of germs at the origin of continuous functions on $\mathbf{R}^r$, for $r \leqslant 7$. The main theorem concerns a generic compact submanifold without boundary M of $\mathbf{R}^r$, for $r \leqslant 7$. To such a submanifold we will associate a finite set E_M in $\mathbf{R}^r \setminus M$. The main result is that for $x \in \mathbf{R}^r \setminus M \setminus E_M$, the germ of ρ_M at x has one of the normal forms in the list. This sort of result is not quite covered by the results of Yomdin [19], Milman [10, 11] and Milman and Waksman [12], who study the structure of the central set, but do not find normal forms for ρ_M.

In the next section, we will explain what we mean by a generic submanifold.

2. Generic submanifolds. We will let $\mathcal{S}_{rk}$ denote the set of smooth compact k-dimensional submanifolds, without boundary, of $\mathbf{R}^r$. We will provide $\mathcal{S}_{rk}$ with the C^∞ topology.

The C^∞ topology on $\mathcal{S}_{rk}$ may be defined in terms of the more familiar C^∞ topology on mappings, as follows. Let M be a smooth compact manifold without boundary. Let $\mathcal{E}(M, \mathbf{R}^r)$ denote the space of smooth embeddings of M in $\mathbf{R}^r$, provided with the C^∞ topology. Let $\mathfrak{M}$ denote a collection of k-dimensional smooth compact manifolds, without boundary, such that every element of $\mathcal{S}_{rk}$ is diffeomorphic to a member of $\mathfrak{M}$. We write $\amalg_{\mathfrak{M}} \mathcal{E}(M, \mathbf{R})$ for the topological direct sum of the spaces $\mathcal{E}(M, \mathbf{R})$. We let

$$I\colon \coprod_{\mathfrak{M}} \mathcal{E}(M, \mathbf{R}) \to \mathcal{S}_{rk}$$

be defined by $I(\phi) = \text{image } \phi$. This is obviously a surjective mapping. By the C^∞ topology on $\mathcal{S}_{rk}$, we mean the quotient topology for this mapping.

It can be shown that there is a complete metric on $\mathcal{S}_{rk}$ such that the C^∞ topology is the underlying topology of that metric.

Consequently, by a well-known theorem of Baire, a countable intersection of open dense subsets of $\mathcal{S}_{rk}$ is dense. A subset of $\mathcal{S}_{rk}$ which contains a countable intersection of open dense subsets is said to be *residual*.

The usual interpretation of "generic M" (for topologists) is the following: One says, "A generic k-dimensional submanifold of $\mathbf{R}^r$ has property $\mathcal{P}$" to mean that the set of $M \in \mathcal{S}_{rk}$ which has property $\mathcal{P}$ is a residual subset of $\mathcal{S}_{rk}$. Thus, one does not define the notion of a "generic submanifold", only what it means for a generic submanifold to have a given property.

As it turns out, the basic theorem of this paper is valid for an open dense set of $M \in \mathcal{S}_{rk}$ (when $r \leq 7$), not just a residual set.

3. Genericity in the sense of measure? From some points of view, the notion of genericity defined in terms of residual sets is not very satisfactory. Measure theorists (and others) may prefer to define genericity in terms of measure. From such a point of view, defining genericity as we did in the previous section can be misleading: a subset of the real line can be residual, and yet be of measure zero [**13**]. Even an open dense set can have arbitrarily small measure.

The difficulty with the measure theoretic viewpoint in our context is that there seems to be no appropriate measure for $\mathcal{S}_{rk}$. There are several ways around this difficulty, all of which seem to involve some compromise which is not entirely satisfactory.

Here is one way, suggested by Arnold, in a similar context. The idea is to study families of k-dimensional submanifolds of $\mathbf{R}^r$, parameterized by compact manifolds.

To be explicit, we consider a smooth compact manifold P, possibly with boundary. A mapping $\phi\colon P \to \mathcal{S}_{rk}$ will be said to be *smooth* if for each $p \in P$, there is an open neighborhood U of p in P and a smooth mapping

$$\Phi\colon U \times M \to \mathbf{R}^r,$$

where $M = \phi(p)$, such that $\Phi(p' \times M) = \phi(p')$, for all $p' \in U$.

Let $\mathcal{C}(P, \mathcal{S}_{rk})$ denote the set of smooth mappings of P into $\mathcal{S}_{rk}$. We may define the C^∞ topology on this set, as follows.

Let M denote a smooth compact k-dimensional manifold without boundary. We let $\mathcal{PE}(P \times M, \mathbf{R}^r)$ denote the smooth embeddings of M in $\mathbf{R}^r$, parameterized by P. We provide $\mathcal{PE}(P \times M, \mathbf{R}^r)$ with the C^∞ topology. We let

$$I\colon \mathcal{PE}(P \times M, \mathbf{R}^r) \to \mathcal{C}(P, \mathcal{S}_{rk})$$

be defined by

$$I(\Phi)(p) = \Phi(p \times M).$$

If P is contractible, then I is surjective, and we define the C^∞ topology on $\mathcal{C}(P, \mathcal{S}_{rk})$ to be the quotient topology for this mapping. In the general case, P

may be covered by the interiors of a finite family $P_1, \ldots, P_k$ of compact contractible submanifolds of P. These will be codimension zero submanifolds, which may have corners. The restriction mappings

$$\mathcal{C}(P, \mathcal{S}_{rk}) \to \mathcal{C}(P_i, \mathcal{S}_{rk})$$

induce an inclusion

$$\mathcal{C}(P, \mathcal{S}_{rk}) \subset \prod_{i=1}^{k} \mathcal{C}(P_i, \mathcal{S}_{rk}).$$

We provide $\mathcal{C}(P, \mathcal{S}_{rk})$ with the induced topology for this inclusion.

Suppose that $r \leqslant 7$. Let $\mathcal{G}^*_{rk}$ denote the set of $M \in \mathcal{S}_{rk}$ for which the basic theorem of this paper is valid. As we mentioned in the previous section, $\mathcal{G}^*_{rk}$ contains an open and dense subset $\mathcal{G}_{rk}$ of $\mathcal{S}_{rk}$. But, we also have the following stronger theorem. For any compact manifold P, there is an open and dense set $\mathcal{G}_{Prk}$ in $\mathcal{C}(P, \mathcal{S}_{rk})$ such that for any $\phi \in \mathcal{G}_{Prk}$, we have that $\phi^{-1}(\mathcal{G}_{rk})$ has full measure in P. This result is valid for the set $\mathcal{G}_{rk}$ which we define in §12.

4. The Basic Theorem. *Let $r \leqslant 7$. Let $M \subset \mathbf{R}^r$ be a generic compact smooth submanifold, without boundary. There exists a finite set $E_M \subset \mathbf{R}^r$, such that if $x \in \mathbf{R}^r \setminus (M \cup E_M)$, then ρ_M has one of the following normal forms at x:*

$A_0(t) = t_1,$
$A_1(t) = \min(t_1, t_2),$
$A_2(t) = \min(t_1, t_2, t_3),$
$\cdots$
$A_6(t) = \min(t_1, t_2, \ldots, t_7),$
$B_2(t) = B_2(t_1, t_2, t_3) = \min_{\xi \in \mathbf{R}}(\xi^4 + t_1\xi^2 + t_2\xi + t_3),$
$B_3(t) = \min(B_2(t_1, t_2, t_3), t_4),$
$\cdots$
$B_6(t) = \min(B_2(t_1, t_2, t_3), t_4, t_5, t_6, t_7),$
$B_5'(t) = \min(B_2(t_1, t_2, t_3), B_2(t_4, t_5, t_6)),$
$B_6'(t) = \min(B_2(t_1, t_2, t_3), B_2(t_4, t_5, t_6), t_7),$
$C_4(t) = \min_{\xi \in \mathbf{R}}(\xi^6 + t_1\xi^4 + t_2\xi^3 + \cdots + t_5),$
$C_5(t) = \min(C_4(t), t_6),$
$C_6(t) = \min(C_4(t), t_6, t_7),$
$D_6(t) = \min_{\xi \in \mathbf{R}}(\xi^8 + t_1\xi^6 + t_2\xi^5 + \cdots + t_7).$

Here, $t = (t_1, \ldots, t_r) \in \mathbf{R}^r$. The set of points x at which ρ_M has one of these normal forms is a smooth (but not usually closed) submanifold of $\mathbf{R}^r$ having the codimension indicated by the subscript, e.g., the set of $x \in \mathbf{R}^r$ where ρ_M has the normal form A_i is a smooth submanifold of codimension i. A particular normal form occurs only if its codimension is $\leqslant r - 1$.

According to the interpretation of genericity we have given, there is an open, dense set $\mathcal{G}_{rk} \subset \mathcal{S}_{rk}$ such that if $M \in \mathcal{G}_{rk}$, then the conclusion of the theorem holds. Moreover, $\mathcal{G}_{rk}$ has the property which we described in §3.

5. The ellipse. The basic theorem is true for an ellipse in $\mathbf{R}^r$. In this section, we will describe ρ_M, when M is an ellipse.

Let M be an ellipse in $\mathbf{R}^r$. We denote the centers of the circles which osculate the ellipse at the endpoints of its major axis by P_1 and P_2. We let L be the line segment joining P_1 and P_2.

First, we consider the case $r = 2$. Let $x \in \mathbf{R}^2$. The function $\|y - x\|$, defined for $y \in M$, takes its minimum value at a single point of M and is nondegenerate there, except when $x \in L$. Consequently, A_0 is a normal form for ρ_M at all points of the plane which are not in $M \cup L$.

Each of the functions $\|y - P_1\|$ and $\|y - P_2\|$ takes its minimum at exactly one point, and is degenerate there. If $x \in L \setminus \{P_1, P_2\}$, then $\|y - x\|$ takes its minimum at two points $\mu_1, \mu_2 \in M$, and is nondegenerate at both of these points. Let $\mathbf{0}$ denote the center of the ellipse. The two circles which pass through x and are centered at μ_1, μ_2 resp. are tangent if and only if $x = \mathbf{0}$. Hence, for $x \in L \setminus \{\mathbf{0}, P_1, P_2\}$, we have that A_1 is a normal form for ρ_M at x.

It follows that the basic theorem is true for the ellipse in the plane, if we set $E_M = \{\mathbf{0}, P_1, P_2\}$.

The ellipse in three dimensional space presents a more interesting problem. Let E denote the plane which contains the ellipse. Let $\tilde{L}$ denote the strip in three dimensional space which is perpendicular to E and whose intersection with E is L. The strip $\tilde{L}$ is bounded by two lines l_1 and l_2 which are perpendicular to E and intersect E in points P_1 and P_2, resp.

Let x be a point of space. The function $\|y - x\|$, defined for $y \in M$, takes its minimum value at a single point of M and is nondegenerate there, except when $x \in \tilde{L}$. Consequently, A_0 is a normal form for ρ_M at all points of the plane which are not in $M \cup \tilde{L}$.

If $x \in l_1 \cup l_2$, then the function $\|y - x\|$ takes its minimum at exactly one point in M and is degenerate there. If, in addition, $x \notin \{P_1, P_2\}$, then B_2 is a normal form for ρ_M at x. In fact, the local diffeomorphism ϕ, for which (1) holds (with $u = B_2$), can be taken to be real analytic. The reason for this will be explained in §§13 and 14.

If $x \in \tilde{L} \setminus (l_1 \cup l_2)$, then $\|y - x\|$ takes its minimum at two points μ_1, μ_2 in M and is nondegenerate at those points. The two spheres through x which are centered at μ_1 and μ_2 are tangent if and only if $x = \mathbf{0}$. Thus, for $x \in \tilde{L} \setminus (l_1 \cup l_2 \cup \mathbf{0})$, we have that A_1 is a normal form for ρ_M at x.

It follows that the basic theorem is true for the ellipse in space, if we set $E_M = \{\mathbf{0}, P_1, P_2\}$.

More generally, the basic theorem is true for the ellipse in $\mathbf{R}^r$. In this case, we still take E to be the plane in $\mathbf{R}^r$ which contains the ellipse. We take l_1 and l_2 to be the $(n - 2)$-dimensional affine subspaces of $\mathbf{R}^n$ which contain P_1 and P_2, resp. and are perpendicular to E. We let $\tilde{L}$ denote the region bounded by l_1 and l_2 in the affine subspace of $\mathbf{R}^r$ which contains the major axis of the ellipse and is perpendicular to E.

Then A_0 is a normal form for ρ_M at x in $\tilde{L} \cup M$, B_2 is a normal form for ρ_M at x in $l_1 \cup l_2 \setminus \{P_1, P_2\}$ and A_1 is a normal form for ρ_M at $x \in \tilde{L} \setminus (l_1 \cup l_2 \cup \mathbf{0})$. We set $E_M = \{\mathbf{0}, P_1, P_2\}$.

The only case in which there is any difficulty in obtaining these normal forms is when $x \in l_1 \cup l_2 \setminus \{P_1, P_2\}$. How to treat this case will be explained in §14.

6. Method of finding normal forms: Relation with catastrophe theory. In [**19**], Yomdin used results in catastrophe theory to obtain normal forms for the central set of a submanifold. These techniques are discussed in Wall [**18**], and used to discuss the cut-locus in Buchner [**3, 4**].

We will find the normal forms for ρ_M by a slight modification of the usual catastrophe theory technique.

The normal forms which we will find are closely related to the classification theorem in catastrophe theory which is known as "Thom's theorem". This theorem was announced in lectures by Thom in the late 1950s and formed the mathematical foundation of his book *Topology and morphogenesis*, whose publication was delayed until 1972. Whether or not this theorem applies to morphogenesis, it has very wide applicability to geometric, physical, and engineering problems. Many of these applications have been pointed out by Arnold [**1**], who has obtained a number of deep theorems along these lines. Earlier Zeeman [**20**] pointed out some other such applications and made some speculations concerning the applicability of "Thom's theorem" to the social sciences and biology.

Thom was the first to point out the very wide applicability of his theorem. As just one small example of his ideas, I had a conversation with him around 1968 or 1969 in which he discussed the possibilities of applying catastrophe theory to the study of the central set and the distance function. However, he never completed this work.

Thom suggested the main lines of the proof of his theorem. Thom's transversality theorem is a key ingredient in the proof, as is Malgrange's preparation theorem, which Thom conjectured. Another key ingredient is the idea, due to H. Levine, of reducing the equivalence problem to a linear problem by introducing a one-parameter family and differentiating. This appears in Levine's notes of Thom's lectures at Bonn in 1960 (cf. [**17**]).

I was the first to give the complete proof. The methods of this proof are very similar to the method which I used in my papers on stability of mappings [**8**], in particular, in the proof that infinitesimal stability implies stability. When I met Thom at the Battelle Conference in Seattle in the summer of 1967, he asked me whether the methods I had used to prove infinitesimal stability implies stability could also be used to prove "Thom's theorem". I checked that it could.

I was not very interested in writing up the proof, because it was so similar to the methods I had already used very extensively in my papers on stability of mappings. But when I went to Warwick in the summer of 1969, Zeeman persuaded me to write up the proof.

I supplied Zeeman with the handwritten *Notes on right equivalence*. Zeeman and Trotman [**21**] completed the proof and added some material of their own.

Zeeman and Trotman's article is an excellent exposition of this material. In their introduction, they give a brief and accurate account of the development of catastrophe theory.

Another excellent exposition is due to Bröcker and Lander. The result we need is a slight modification of Bröcker and Lander's Theorem 14.8 (which is based on results in my *Notes on right equivalence*). The account given by Bröcker and Lander in Chapter XIV is very readable and we recommend it. In the remainder of this section, we state both Bröcker and Lander's Theorem 14.8 and the modification of it which we will need. A further modification, which will also be needed, will be stated in §7.

Let η: $(\mathbf{R}^n, 0) \to \mathbf{R}$ be the germ at the origin of a C^∞ function. For any integer $r \geq 0$, we think of $\mathbf{R}^n$ as the subspace of $\mathbf{R}^{n+r}$ where the last r coordinates vanish. By an *r-parameter* unfolding of η, one means a germ f: $(\mathbf{R}^{n+r}, 0) \to \mathbf{R}$ of a C^∞ function such that $f \mid \mathbf{R}^n = \eta$. This unfolding will be denoted (r, f).

The *category of unfoldings of the germ* η is defined as follows.

DEFINITION. Let (r, f) and (s, g) be unfoldings of η. A morphism

$$(\phi, \alpha) \colon (r, f) \to (s, g)$$

consists of

(I) a C^∞ germ of a mapping ϕ: $(\mathbf{R}^{n+r}, 0) \to (\mathbf{R}^{n+s}, 0)$ satisfying $\phi \mid \mathbf{R}^n =$ identity,

(II) a C^∞ germ of a mapping Φ: $(\mathbf{R}^r, 0) \to (\mathbf{R}^s, 0)$ such that $\pi_s\phi = \Phi\pi_r$, where π_r: $\mathbf{R}^{n+r} \to \mathbf{R}^r$ denotes the projection on the last r coordinates, and

(III) a C^∞ germ of a function α: $(\mathbf{R}^r, 0) \to \mathbf{R}$ such that

$$f = g \circ \phi + \alpha \circ \pi_r.$$

Note that (II) amounts to saying that ϕ has the form

$$\phi(x, u) = (\phi_1(x, u), \Phi(u)),$$

where $x \in \mathbf{R}^n$, $u \in \mathbf{R}^r$, $\phi_1(x, u) \in \mathbf{R}^n$, $\Phi(u) \in \mathbf{R}^s$. We will say that Φ: $(\mathbf{R}^r, 0) \to (\mathbf{R}^s, 0)$ is the germ *associated* to ϕ.

Composition is defined by

$$(\phi, \alpha)(\psi, \beta) = (\phi \circ \psi, \beta + \alpha \circ \Psi),$$

where Ψ is the germ associated to ψ.

An unfolding (s, g) of η is said to be *versal* if for every unfolding (r, f) of η, there is a morphism of (r, f) into (s, g).

The basic theorem of catastrophe theory characterizes versal unfoldings. In order to state this theorem, we need the following notations. We let $\mathcal{E}_n$ denote the ring of germs at the origin of C^∞ real valued functions on $\mathbf{R}^n$. We let $\mathfrak{m}_n$ denote the unique maximal ideal of $\mathcal{E}_n$. Let $x_1, \dots, x_n$ denote the coordinates of $\mathbf{R}^n$. We may think of the x_i's as functions on $\mathbf{R}^n$, or as germs of functions. If we think of x_i in the latter way, then $x_1, \dots, x_n \in \mathfrak{m}_n$ and, in fact, $x_1, \dots, x_n$ generate $\mathfrak{m}_n$ as an ideal in $\mathcal{E}_n$ (Hadamard's Lemma).

Given a C^∞ germ η: $(\mathbf{R}^n, 0) \to \mathbf{R}$, we define $\langle \partial\eta \rangle$ to be the ideal in $\mathcal{E}_n$ generated by $\partial\eta/\partial x_1, \dots, \partial\eta/\partial x_n$. Given an r-parameter unfolding f of η, we define V_f to be the $\mathbf{R}$-vector subspace of $\mathcal{E}_n$ generated by $\partial f/\partial u_1 \mid \mathbf{R}^n, \dots, \partial f/\partial u_r \mid \mathbf{R}^n$, where

$u_1, \ldots, u_r$ denote the coordinates of $\mathbf{R}^n$. In the standard notation of commutative algebra, we may rewrite these definitions as

$$\langle \partial\eta \rangle = \mathcal{E}_n \cdot \left\{ \frac{\partial n}{\partial x_1}, \ldots, \frac{\partial \eta}{\partial x_n} \right\},$$

$$V_f = \mathbf{R} \cdot \left\{ \frac{\partial f}{\partial u_1} \bigg| \mathbf{R}^n, \ldots, \frac{\partial f}{\partial u_r} \bigg| \mathbf{R}^n \right\}.$$

The following is the basic result of catastrophe theory.

BASIC THEOREM OF CATASTROPHE THEORY. *An unfolding (r, f) of η is versal if and only if*

$$\mathcal{E}_n = \langle \partial\eta \rangle + V_f + \mathbf{R},$$

where $\mathbf{R} \subset \mathcal{E}_n$ is embedded as the constants. Two r-parameter versal unfoldings of η are isomorphic (in the category of unfoldings of η).

This statement is not obviously equivalent to Theorem 14.8 of [**2**]. However, the proof of Theorem 14.8 of [**2**, Chapter XVI] also proves our statement. Theorem 16.2 of [**2**] implies the part of our statement which is not in Theorem 14.8 of [**2**].

For what we do in this paper, we need a modified version of the basic theorem of catastrophe theory. For this purpose, we will introduce what we call the *restricted category of unfoldings of* η. The objects in this new category are still the unfoldings of η. But the morphisms (which we will call restricted morphisms) are a subset of the previously defined morphisms.

DEFINITION. A morphism (ϕ, α) will be said to be *restricted* if $\alpha = 0$. In this case, we will speak of the morphism ϕ (in place of $(\phi, 0)$).

An unfolding (s, g) of η will be said to be *strictly versal* if for every unfolding (r, f) of η, there is a restricted morphism of (r, f) into (s, g).

The proof of the following analogue of the basic theorem of catastrophe theory requires only obvious modifications of the proof in Chapter XVI of [**2**].

BASIC THEOREM OF RESTRICTED CATASTROPHE THEORY. *An unfolding (r, f) of η is strictly versal if and only if*

$$\mathcal{E}_n = \langle \partial\eta \rangle + V_f.$$

Two r-parameter strictly versal unfoldings of η are strictly isomorphic (i.e., isomorphic in the restricted category of unfoldings of η).

Note that from the definition of a general morphism (ϕ, α), it follows that $\alpha(0) = 0$. This follows from (I) and (III) in the definition, together with the fact that $f | \mathbf{R}^n = g | \mathbf{R}^n = \eta$.

7. Method of finding normal forms: multi-germs. The restricted catastrophe theory which we introduced in the last section is not general enough for the applications which we wish to make. We also need a restricted catastrophe theory of unfoldings of multi-germs. By a multi-germ, we mean a germ of a C^∞ function

$\eta\colon (M, S) \to \mathbf{R}$, where M is a (finite dimensional) manifold, and S is a finite set. If $S = \{x_1, \ldots, x_k\}$, with $x_i \neq x_j$ for $i \neq j$, then such a multi-germ is the same was a k-tuple $(\eta_1, \ldots, \eta_k)$, where $\eta_i\colon (M, x_i) \to \mathbf{R}$ is a germ of a smooth function. We do not require that M have the same dimension at the various points of x_i.

We identify M with $M \times 0$ in $M \times \mathbf{R}^r$. By an r-parameter *unfolding* of η, one means a germ $f\colon (M \times \mathbf{R}^r, S \times 0) \to \mathbf{R}$ of a C^∞ function such that $f\,|\,M = \eta$. This unfolding will be denoted (r, f).

Such an unfolding of $\eta = (\eta_1, \ldots, \eta_k)$ is the same as a k-tuple $((r, f_1), \ldots, (r, f_k))$, where (r, f_i) is an r-parameter unfolding of η_i in the sense previously defined.

The category of unfoldings of the multi-germ η is defined as follows.

DEFINITION. Let (r, f) and (s, g) be unfoldings of η. A *morphism*

$$(\phi, \alpha)\colon (r, f) \to (s, g)$$

consists of

(I) a C^∞ germ of a mapping $\phi\colon (M \times \mathbf{R}^r, S \times 0) \to (M \times \mathbf{R}^r, S \times 0)$ satisfying $\phi\,|\,M = \text{identity}$,

(II) a C^∞ germ of a mapping $\Phi\colon (\mathbf{R}^r, 0) \to (\mathbf{R}^s, 0)$ such that $\pi_s\phi = \Phi\pi_r$, where $\pi_r\colon M \times \mathbf{R}^r \to \mathbf{R}^r$ denotes the projection on the second factor, and

(III) a C^∞ germ of a function $\alpha\colon (\mathbf{R}^r, 0) \to \mathbf{R}$ such that

$$f = g \circ \phi + \alpha \circ \pi_r.$$

The germ ϕ is the same as a k-tuple $(\phi_1, \ldots, \phi_k)$, where $\phi_i\colon (M \times \mathbf{R}^r, x_i \times 0) \to (M \times \mathbf{R}^r, x_i \times 0)$ is a germ of a C^∞ mapping. Therefore, we may associate to a morphism $(\phi, \alpha)\colon (r, f) \to (s, g)$ a k-tuple $((\phi_1, \alpha), \ldots, (\phi_k, \alpha))$, where $(\phi_i, \alpha)\colon (r, f_i) \to (r, g_i)$ is a morphism in the category of unfoldings of η_i. Conversely, given a k-tuple $((\phi_1, \alpha_1), \ldots, (\phi_k, \alpha_k))$, where $(\phi_i, \alpha_i)\colon (r, f_i) \to (r, g_i)$ is a morphism in the category of unfoldings of η_i, it defines a morphism $(\phi, \alpha)\colon (r, f) \to (s, g)$ in the category of unfoldings of η if and only if $\alpha_i = \alpha_j$ and $\Phi_i = \Phi_j$, for all $1 \leq i, j \leq k$, where $\Phi_i\colon (\mathbf{R}^r, 0) \to (\mathbf{R}^r, 0)$ is the germ associated to ϕ_i.

We will call $\Phi\colon (\mathbf{R}^r, 0) \to (\mathbf{R}^s, 0)$, which appears in (II), the germ *associated* to ϕ.

Composition is defined by

$$(\phi, \alpha)(\psi, \beta) = (\phi \circ \psi, \beta + \alpha \circ \Psi),$$

where Ψ is the germ associated to ψ.

An unfolding (s, g) of η is said to be *versal* if for every unfolding (r, f) of η, there is a morphism of (r, f) into (s, g).

In order to state the basic theorems of catastrophe theory and restricted catastrophe theory for unfoldings of multi-germs, we need the following notations. We let $\mathcal{E}(M)_S$ denote the ring of germs at S of C^∞ functions. We let $\mathfrak{m}_S$ denote the ideal of $\mathcal{E}(M)_S$ consisting of functions at S. Obviously,

$$\mathcal{E}(M)_S = \mathcal{E}(M)_{x(1)} \times \cdots \times \mathcal{E}(M)_{x(k)},$$
$$\mathfrak{m}_S = \mathfrak{m}_{x(1)} \cdots \mathfrak{m}_{x(k)}.$$

Let $\eta \in \mathcal{E}(M)_S$. Then $\eta = (\eta_1, \ldots, \eta_k)$, where $\eta_i \in \mathcal{E}(M)_{x(i)}$. In the previous section, we defined the ideal $\langle \partial\eta_i \rangle \subset \mathcal{E}(M)_{x(i)}$. (Actually, we gave the definition only when $(M, x_i) = (\mathbf{R}^n, 0)$. We may generalize to arbitrary (M, x_i) by choosing a local system of coordinates for M, centered at x. It is easily checked that the ideal $\langle \partial\eta_i \rangle$ is independent of which local coordinate system is chosen.) We set

$$\langle \partial\eta \rangle = \langle \partial\eta_1 \rangle \times \cdots \times \langle \partial\eta_k \rangle.$$

Then $\langle \partial\eta \rangle$ is an ideal in $\mathcal{E}(M)_S = \mathcal{E}(M)_{x(1)} \times \cdots \times \mathcal{E}(M)_{x(k)}$.

Given an r-parameter unfolding f of η, we define V_f to be the $\mathbf{R}$-vector subspace of $\mathcal{E}(M)_S$ generated by $\partial f/\partial u_1 \mid M, \ldots, \partial f/\partial u_r \mid M$, where $u_1, \ldots, u_r$ denote the coordinates of $\mathbf{R}^r$. Note that $\dim_{\mathbf{R}} V_f \leqslant r$.

BASIC THEOREM OF CATASTROPHE THEORY FOR MULTI-GERMS. *An unfolding (r, f) of η is versal if and only if*

$$\mathcal{E}(M)_S = \langle \partial\eta \rangle + V_f + \mathbf{R},$$

where $\mathbf{R} \subset \mathcal{E}(M)_S$ is embedded as the constants. Two r-parameter versal unfoldings of η are isomorphic (in the category of unfoldings of η).

Again, this may be proved by only trivial modifications of the proof given in Chapter XVI of [**2**]. The same remark applies to the next theorem as well. See also [**9** and **21**], where multi-germs are considered.

Now we state the analogous theorem for the *restricted category* of unfoldings of a multi-germ η. The objects in this category are unfoldings of η. The morphisms are defined as follows.

DEFINITION. A morphism (ϕ, α) in the category of unfoldings of η will be said to be *restricted* if $\alpha = 0$. In this case, we will speak of the morphism ϕ (in place of $(\phi, 0)$).

An unfolding (s, g) of η will be said to be *strictly versal* if for every unfolding (r, f) of η, there is a restricted morphism of (r, f) into (s, g).

BASIC THEOREM OF RESTRICTED CATASTROPHE THEORY FOR MULTI-GERMS. *An unfolding (r, f) of η is strictly versal if and only if*

$$\mathcal{E}(M)_S = \langle \partial\eta \rangle + V_f.$$

Two r-parameter strictly versal unfoldings of η are strictly isomorphic.

8. A catastrophe theory view of the distance function. We suppose M is a smooth, compact submanifold without boundary of Euclidean space $\mathbf{R}^r$. We define

$$\sigma_M \colon M \times \mathbf{R}^r \to \mathbf{R}$$

by $\sigma_M(y, x) = \| y - x \|$, for $y \in M$, $x \in \mathbf{R}^r$. Obviously, $\rho_M(x) = \min_{y \in M} \sigma_M(y, x)$. Moreover, σ_M is smooth on $M \times (\mathbf{R}^r \setminus M)$.

Let $x \in \mathbf{R}^r \setminus M$, and set $S_x = \{y \in M \colon \sigma_M(y, x) = \rho_M(x)\}$. Let $\sigma_{M,x}$ denote the germ of σ_M at $S_x \times x$, and let $\eta_x = \sigma_{M,x} \mid M \times x$, so η_x is a germ of a real valued function on M at S_x.

THEOREM. *Suppose S_x is a finite set, and $\sigma_{M,x}$ is a strictly versal unfolding of η_x. Suppose*

$$\eta\colon (N, S) \to \mathbf{R}$$

is a germ of a smooth mapping at a finite set S in a smooth manifold. Suppose η is right equivalent to η_x, i.e., there exists a germ $\psi\colon (M, S_x) \to (N, S)$ of a smooth diffeomorphism and a constant c, such that $\eta_x = \eta \circ \psi + c$. Suppose

$$f\colon (N \times \mathbf{R}^r, S \times 0) \to \mathbf{R}$$

is a strictly versal unfolding of η. Let U be a neighborhood of S in N, and V a neighborhood of 0 in $\mathbf{R}^r$. Let $\tilde{f}\colon U \times V \to \mathbf{R}$ be a representative of f. For $x' \in V$, let

$$\tilde{G}_U(x') = \min_{y \in U} \tilde{f}(y, x'),$$

and let G_U denote the germ of $\tilde{G}_U$ at 0. Then G_U is independent of U, if U is a sufficiently small enough neighborhood of S in N. Moreover, G_U is a normal form for ρ_M at x.

PROOF. Let id denote the germ at 0 of the identity mapping of $\mathbf{R}^r$. Since f is a strictly versal unfolding of η, we have that $f \circ (\psi \times \text{id}) + c$ is a strictly versal unfolding of $\eta_x = \eta \circ \psi + c$. By the basic theorem of restricted catastrophe theory for multi-germs, it follows that $f \circ (\psi \times \text{id}) + c$ and $\sigma_{M,x}$ are isomorphic objects in the restricted category of unfoldings of η_x.

Let $\phi = (\phi, 0)\colon (r, \sigma_{M,x}) \to (r, f \circ (\psi \times \text{id}) + c)$ be an isomorphism in this category. By definition of "restricted morphism", we have

$$\sigma_{M,x} = f \circ (\psi \times \text{id}) \circ \phi + c \tag{2}$$

in M. By the definition of ρ_M, S_x, and σ_M we have

$$\rho_M(x') = \min_{y \in U'} \sigma_M(y, x'), \tag{3}$$

for any neighborhood U' of S_x in M, provided that x' is in a sufficiently small neighborhood of x in $\mathbf{R}^n$.

Let $\Phi\colon (\mathbf{R}^r, x) \to (\mathbf{R}^r, 0)$ be the germ of a diffeomorphism associated to ϕ. Since $\pi_r \circ (\psi \times \text{id}) \circ \phi = \pi_r \circ \phi = \Phi \circ \pi_r$, it follows from (2) and (3) that the germ G_U is independent of U when U is a small enough neighborhood of S in M, and

$$\rho_{M,x} = G_U \circ \Phi + c. \tag{4}$$

In words, this says that G_U is a normal form for ρ_M at x.

9. Versality and transversality. To prove the basic theorem of this paper, using the theorem of §8, we will need to show that for $r \leqslant 7$ and generic $M \subset \mathbf{R}^r$, there exists a finite set $E_M \subset \mathbf{R}^r$ such that if $x \in \mathbf{R}^r \setminus (M \cup E_M)$, then $\sigma_{M,x}$ is a strictly versal unfolding of η_x.

The method of proof is to express the condition that $\sigma_{M,x}$ be a strictly versal unfolding of η_x as a transversality condition and then prove a result analogous to Thom's transversality theorem. This will give us a residual set of M having the required properties.

To show that $\mathcal{G}_{rk}$ contains an open and dense subset of $\mathcal{S}_{rk}$ and to get the properties of $\mathcal{G}_{rk}$ listed in §3 will then require further arguments.

In this section, we will express the condition that $\sigma_{M,x}$ be a strictly versal unfolding of η_x as a transversality condition. In the next section, we will state the analogue of Thom's transversality theorem. No proof will be given in this paper.

We will assume that the reader is familiar with the language of jets, which is explained in the Thom-Levine notes [**17**], Guillemin and Golubitsky [**6**], the author's articles [**8**] and many other places.

We fix a positive integer α. Let M be a manifold. We let $M^{(l)}$ denote the manifold to l-tuples $(y_1,\ldots,y_l)$ satisfying $y_i \neq y_j$, for $i \neq j$. If $y \in M$, we let $J(M)_y$ denote the $\mathbf{R}$-vector space of α-jets of mappings $(M, y) \to \mathbf{R}$. If $S \in M^{(l)}$, we let $J(M)_S = J(M)_{y(1)} \times \cdots \times J(M)_{y(l)}$, where $S = (y_1,\ldots,y_l)$. The *l-fold multi-jet bundle* ${}_lJ(M)$ has base $M^{(l)}$ and fiber $J(M)_S$ over $S \in M^{(l)}$.

If M is a smooth submanifold of $\mathbf{R}^r$ and $x \in \mathbf{R}^r \setminus M$, we define $\sigma_M|_x$: $M \to \mathbf{R}$ by $\sigma_M|_x(y) = \|y - x\|$. We let

$$j^{(l)}(\sigma_M|_x)\colon M^{(l)} \to {}_lJ(M)$$

denote the l-fold multi α-jet extension of $\sigma_M|_x$. We define

$$j_1^{(l)}\sigma_M\colon M^{(l)} \times (\mathbf{R}^r \setminus M) \to {}_lJ(M)$$

by

$$j_1^{(l)}\sigma_M(S, x) = j^{(l)}(\sigma_M|_x)(S).$$

This is the *l-fold multi α-jet extension of σ_M, with respect to the first factor.* Obviously, $j_1^{(l)}\sigma_M$ is a smooth mapping.

Two elements $z \in J(M)_S$ and $z' \in J(M)_{S'}$ are *right equivalent* if there exist a constant c and a germ ϕ: $(M, S) \to (M, S')$ such that

$$z = z' \circ \phi + c.$$

For any $z \in J_M^{(l)}$, the right equivalence class of z, denoted $\mathcal{R}z$, is defined to be the set of all $z' \in J(M)$ which are right equivalent to z. It is known that $\mathcal{R}z$ is a submanifold of ${}_lJ(M)$.

Let $x \in \mathbf{R}^r \setminus M$. We will use the same notation as in §8, i.e., $S_x = \{y \in M: \sigma_M(y, x) = \rho_M(x)\}$, $\sigma_{M,x} =$ the germ of σ_M at $S_x \times x$, and $\eta_x = \sigma_{M,x} \mid M \times x$. We suppose that S_x has finitely many elements, and let l be the number of elements in S_x. We set $z_x = j_1^{(l)}\sigma_M(S_x, x)$.

THEOREM 9.1. *There exists α_0, depending only on r, such that if $\alpha \geqslant \alpha_0$, then $\sigma_{M,x}$ is a versal unfolding of η_x if and only if $j_1^{(l)}\sigma_M$ is transverse to $\mathcal{R}z_x$ at (S_x, x).*

What we need is a necessary and sufficient condition for $\sigma_{M,x}$ to be a *strictly* versal unfolding of η_x. To state this condition, we observe that for any $z = (z_1,\ldots,z_l) \in \mathcal{R}z_x$, we have that $t(z_1) = \cdots = t(z_l)$, where $t(z_i)$ denotes the target of z_i. This is because z_x has this property, in view of the definition of S_x, and then z has this property also, since z is right equivalent to z_x. We let $t(z) = t(z_1) = \cdots = t(z_l)$. Obviously, t: $\mathcal{R}z_x \to \mathbf{R}$ is a smooth mapping.

THEOREM 9.2. *If* $\alpha \geqslant \alpha_0$, *then* $\sigma_{M,x}$ *is a strictly versal unfolding of* η_x *if and only if* $j_1^{(l)}\sigma_M$ *is transverse to* $\mathcal{R}z_x$ *at* (S_x, x) *and*

$$t \circ j_1^{(l)}\sigma_M\colon \left(j_1^{(l)}\sigma_M\right)^{-1}(\mathcal{R}z_x) \to \mathbf{R}$$

is nondegenerate at (S_x, x).

We may take $\alpha_0 = 8$, if $r \leqslant 7$.

Theorem 9.1 follows from the argument given in Chapter XVI of [**2**]. There is no difficulty modifying this argument to give Theorem 9.2.

10. Genericity conditions. In §4, we asserted that there is an open, dense set $\mathcal{G}_{rk} \subset \mathcal{S}_{rk}$ such that if $M \in \mathcal{G}_{rk}$, then the conclusion of the basic theorem holds. In §12, we define $\mathcal{G}_{rk}$. The proof that it is open and dense and that it has the properties claimed in §3 will be given in a future article. In this section, we state some genericity conditions, as a preliminary to defining $\mathcal{G}_{rk}$.

Let N be a k-dimensional manifold and S' a subset of N having l elements. A subset W of $J(N)_{S'}$ will be said to be *right invariant* if for every germ of a diffeomorphism $\phi\colon (N, S') \to (N, S')$, we have that $\phi^*W = W$, where $\phi^*\colon J(N)_{S'} \to J(N)_{S'}$ denotes the induced mapping. For such a set, we may define a subset W_M of $J(M)$, as follows. Let $S \subset M$ have l elements. Let $z \in J(M)_S$. Let $\phi\colon (N, S') \to (M, S)$ be a germ of a diffeomorphism. We define $W_{M,S} = W_M \cap J(M)_S$ by the condition that $z \in J_{M,S}$ if and only if $\phi^*z \in W$. In view of the right invariance of W, this condition is independent of ϕ. Obviously, $W_M = \amalg_{S \in M^{(l)}} W_{M,S}$ is a bundle over $M^{(l)}$, and it is a smooth submanifold of ${}_lJ(M)$ if and only if W is a smooth submanifold of $J(N)_{S'}$.

DEFINITION. Suppose W is a right invariant smooth submanifold of $J(N)_{S'}$. Let $M \in \mathcal{S}_{rk}$. We will say that M is W-generic if the following conditions are satisfied:

(I) $j_1^{(l)}\sigma_M\colon M^{(l)} \times (\mathbf{R}^r \setminus M) \to {}_lJ(M)$ is transverse to W_M, and

(II) $j_1^{(l)}\Lambda_M\colon M^{(l)} \times \Sigma \to {}_lJ(M)$ is transverse to W_M, where Σ is the unit sphere in the dual space $\mathbf{R}^{r*}$ to $\mathbf{R}^r$, and $\Lambda_M\colon M \times \Sigma \to \mathbf{R}$ is defined by $\Lambda_M(y, L) = L(y)$, $y \in M \subset \mathbf{R}^r$, $L \in \mathbf{R}^{r*}$.

Looijenga's theorem [**7**] asserts that the set of $M \in \mathcal{S}_{rk}$ satisfying (I) forms a residual set. This was generalized by Wall [**18**], who observed that the set of $M \in \mathcal{S}_{rk}$ satisfying (I) and (II) forms a residual set. For general W there is no reason why the set of $M \in \mathcal{S}_{rk}$ satisfying (I) and (II) should contain an open and dense set, although this is the case for most W relevant to singularity theory.

11. The basic collection of submanifolds of the jet bundle. To study submanifolds of $\mathbf{R}^r$, for $r \leqslant 7$, we will apply the results of Looijenga and Wall of the previous section to a collection $\mathcal{W}^l$ of submanifolds of $J(N)_{S'}$. We define $\mathcal{W}^l$ in this section.

Recall that $J(N)_{S'}$ consists of multi α-jets. In this section, we take $\alpha = 8$.

It is known (splitting lemma [**2**]) that any jet with vanishing first derivative is right equivalent to $g(x_1,\ldots,x_{k-r}) \pm x_{k-r+1}^2 \pm \cdots \pm x_n^2$, where g has order $\geqslant 3$.

Moreover, g and $\pm x_{k-r+1}^2 \pm \cdots \pm x_n^2$ are determined up to right equivalence by the original jet. We will call the pair $(k-r, g)$ the *principal part* of the original jet. We will call $\pm x_{k-r+1}^2 \pm \cdots \pm x_n^2$ the *quadratic part* of the original jet. In case $k-r=0$, i.e., the original jet has nondegenerate second order part, we let (0) denote the *principal part*. We write $[z]$ for the principal part of z.

We let P denote the set of $z \in J(\mathbf{R}^k)_0$ whose quadratic part is nonnegative. (Note that P is not a submanifold.) We define the following subsets of $J(\mathbf{R}^k)_0$:

$W_0^A = \{z \in P: [z] = (0)\}$,
$W_2^B = \{z \in P: [z] = (1, x^4)\}$,
$W_4^C = \{z \in P: [z] = (1, x^6)\}$,
$W_6^D = \{z \in P: [z] = (1, x^8)\}$,
$W_8^E = \{z \in P: [z] = (1, 0)\}$,
$W_7^F = \{z \in P: \exists \mu > 0, \mu \neq 1 \text{ such that } [z] = (2, (x_1^2 + x_2^2)(x_1^2 + \mu x_2^2))\}$,
$W_8^G = \{z \in P: [z^{(3)}] = (c, 0), z \notin W_6^F, c \geqslant 2\}$,

where $z^{(3)}$ denotes the 3-jet of z.

We let $\mathfrak{W}_0$ denote the collection $\{W_0^A, \ldots, W_8^G\}$ of subsets of P listed above. This is a partition of P. (To check this, remember that $\alpha = 8$, i.e., we are considering 8-jets.)

Each of the above sets, except W_8^G, is a submanifold of $J(\mathbf{R}^k)_0$ and W_8^G is a finite union of right invariant submanifolds. Let $\mathfrak{W}_1 = \{W_0^A, \ldots, W_7^F\} \cup$ a finite partition of W_8^G into right invariant submanifolds. Then W_1 is a partition of P.

Let $\sigma = \Sigma_{W \in \mathfrak{W}_1} n_W W$ be a finite linear combination of elements of $\mathfrak{W}_1$, whose coefficients n_W are nonnegative integers. We set $|\sigma| = \Sigma_{W \in \mathfrak{W}_1} n_W$ and let W^σ denote the submanifold of $(J(\mathbf{R}^k)_0)^l$ consisting of $(z_1, \ldots, z_l)$ such that exactly n_W of the z_i are in W, for each $W \in \mathfrak{W}_1$, and $t(z_1) = \cdots = t(z_l)$. We let $\mathfrak{W}^l = \{W^\sigma: |\sigma| = l\}$.

If N is a k-dimensional manifold and S' has exactly l elements, then $J(N)_{S'}$ may be identified with $(J(\mathbf{R}^k)_0)^l$. Each $W \in \mathfrak{W}^l$ is right invariant. It follows that if we identify each W^σ such that $|\sigma| = l$ with its image in $J(N)_{S'}$, the result is independent of how we identify $J(N)_{S'}$ with $(J(\mathbf{R}^k)_0)^l$. Thus, we may think of the set of W with $|\sigma| = l$ as a collection of submanifolds of $(J(\mathbf{R}^k)_0)^l$ or as a collection of submanifolds of $J(N)_{S'}$, whichever is more convenient.

Recall that Σ is the unit sphere in the dual space $\mathbf{R}^{r*}$ to $\mathbf{R}^r$, and Λ_M: $M \times \Sigma \to \mathbf{R}$ is defined by $\Lambda_M(y, L) = L(y)$, $y \in M$, $L \in \mathbf{R}^{r*}$. For $L \in \Sigma$, we define $\rho_M^*(L) = \min\{L(y): y \in M\}$. We set $S_L = \{y \in M: L(y) = \rho_M^*(L)\}$, $\Lambda_{M,L}$ = the germ of Λ_M at $S_L \times L$, and $\eta_L = \Lambda_{M,L} | M \times \{L\}$.

DEFINITION. Let $M \in \mathfrak{S}_{rk}$, $r \leqslant 7$. We will say M is *generic* if the following conditions are satisfied:

(I) If $x \in \mathbf{R}^r \setminus M$, then the number $l = |S_x|$ of points in S_x is $\leqslant k+1$. Moreover, if W is the unique member of $\mathfrak{W}^l$ such that $j_1^{(l)}\sigma_M(S_x, x) \in W_M$, then $j_1^{(l)}\sigma_M$ is transverse to W_M at (S_x, x).

(II) If $L \in \Sigma$, then $l = |S_L| \leqslant k+1$. Moreover, if W is the unique member of $\mathfrak{W}^l$ such that $j_1^{(l)}\Lambda_M(S_L, L) \in W_M$, then $j_1^{(l)}\Lambda_M$ is transverse to W_M at (S_L, L).

THEOREM (FOR $r \leqslant 7$). *The set of generic $M \subset \mathbf{R}^r$ forms an open and dense set in $\mathcal{S}_{rk}$. We have a stability result for generic $M \subset \mathbf{R}^r$: Let Z_M denote the central set of M. If M is generic and M' is sufficiently close to M in the C^∞ topology, then there is a C^∞ diffeomorphism $\phi: \mathbf{R}^r \to \mathbf{R}^r$ such that $\phi M = M'$ and $\phi Z_M = Z_{M'}$.*

The proof will be given in a subsequent publication. The "condition at infinity" ((II) above) must be imposed to insure that the set of generic M is open. The idea of such a condition is due to Wall, who studied a very similar condition in [**18**]. The fact that the set of generic M is open and the stability result follow from the arguments which prove that multi-transversality implies stability in [**8**].

Suppose $W = W^\sigma \in \mathcal{W}^l$. Given generic $M \subset \mathbf{R}^r$, we define $W_M^* = \{x \in \mathbf{R}^r \setminus M$: S_x has exactly l elements and $j_1^{(l)}(S_x, x) \in W_M\}$ and $W_M^\# = \{L \in \Sigma$: S_L has exactly l elements and $j_1^{(l)}(S_L, L) \in W_M\}$. The hypothesis that M is generic implies that each of W_M^* and $W_M^\#$ is a smooth submanifold, or empty.

Writing cod$(A; B)$ for "codimension of A in B", we have

$$\operatorname{cod}(W_M^*; \mathbf{R}^r) = \operatorname{cod}\left(W_M^\#; \Sigma\right) = \operatorname{cod}\left(W; \left(J(\mathbf{R}^k)_0\right)^l\right) - k, \tag{5}$$

where cod$(A; B)$ is understood to be meaningless when $A = \varnothing$.

The collection of manifolds $\{W_M^*: W \in \mathcal{W}\}$ forms a partition of $\mathbf{R}^r$ into a finite family of submanifolds. Likewise, the collection $\{W_M^\#: W \in \mathcal{W}\}$ forms a partition of Σ into a finite family of submanifolds. In either case, some of the W_M^* and $W_M^\#$ can be empty. Both of these collections are Whitney stratifications (a notion introduced by Thom—see [**15**] for definitions).

Note that $W_M^* = \varnothing$ if the right side of (5) is $> r$; $W_M^\# = \varnothing$ if the right side of (5) is $> r - 1$.

For each $W \in \mathcal{W}$, we have that $\rho_M \mid W_M^*$ and $\Lambda \mid W_M^\#$ are C^∞.

Let $x \in W_M^*$. The fact that M is generic implies that $\sigma_{M,x}$ is a versal deformation of η_x [**2**, Chapter XVI]. This, in turn, shows that η_x determines the germ Z_x of the central set of M at x up to C^∞ equivalence.

Similarly, the fact that M is generic implies that the necessary and sufficient condition for $\sigma_{M,x}$ to be a strictly versal deformation of η_x is that x not be a critical point of $\rho_M \mid W_M^*$.

Let $x \in \mathbf{R}^r \setminus M$. Let $W \in \mathcal{W}$ be such that $x \in W_M^*$. We will say that x is *M-critical* if $\rho_M \mid W_M^*$ has a critical point at x. Let E_M denote the set of M-critical points. The basic theorem of this paper will be true if E_M is finite; for, at any $x \in \mathbf{R}^r \setminus M \setminus E_M$, $\rho_{M,x}$ will have one of the normal forms in the list in the statement of the basic theorem.

12. Definition of $\mathcal{G}_{rk}$, $r \leqslant 7$. Let $M \subset \mathbf{R}^r$ be generic. Consider $W \in \mathcal{W}^l$ such that W_M^* is nonempty. If $x \in \overline{W_M^*} \setminus W_M^*$, we define the *tangent cone* $C_x\overline{W_M^*}$ to be the set of all $v \in \mathbf{R}^r$ such that there exist a sequence $y_1, y_2, \ldots \in W_M^*$, $y_i \to x$, and a sequence $\lambda_1, \lambda_2, \ldots$ of positive numbers such that $\lambda_i(y_i - x)$ converges to v. It can be shown that if $v \in C_x\overline{W_M^*}$, then the directional derivative $D_v\rho_M$ exists.

DEFINITION. Let $M \subset \mathbf{R}^r$ be generic. We will say that M is *strictly generic* if for each $l \leqslant k+1$ (where $k = \dim M$) and each $W \in \mathcal{W}^l$, we have

(III) $\rho_M \mid W_M^*$ has only nondegenerate critical points,

(IV) if $x, y \in E_M$, then $\rho_M(x) \neq \rho_M(y)$, and

(V) for each $x \in \overline{W_M^*} \setminus W_M$, there exists $v \in C_x W_M^*$ such that $D_v \rho_M \neq 0$.

We let $\mathcal{G}_{rk}$ denote the subset of $\mathcal{S}_{rk}$ consisting of those M which are strictly generic.

THEOREM (FOR $r \leqslant 7$). *$\mathcal{G}_{rk}$ is an open and dense subset of $\mathcal{S}_{rk}$. If $M \in \mathcal{G}_{rk}$ and $x \in \mathbf{R}^r \setminus M \setminus E_M$, then $\rho_{M,x}$ has one of the normal forms listed in the main theorem. Moreover, we have the following stability result. Let M, M' be in the same connected component of $\mathcal{G}_{rk}$. Then $\rho_M(E_M)$ and $\rho_{M'}(E_{M'})$ have the same number of points. Let ψ: $\mathbf{R} \to \mathbf{R}$ be an order-preserving C^∞ diffeomorphism such that $\psi\rho_M(E_M) = \rho_{M'}(E_{M'})$. Then there exists a homeomorphism ϕ: $\mathbf{R}^r \to \mathbf{R}^r$ such that $\phi M = M'$, $\phi(E_M) = E_{M'}$ and $\rho_{M'} \circ \phi = \psi \circ \rho_M$. Moreover, ϕ may be chosen to be C^∞ except on E_M.*

This is a new result. The density does not follow from Looijenga's transversality theorem [7], and openness and stability is related to, but does not follow directly from, the stability results of [8]. Proofs will appear in a subsequent publication.

13. Analytic M. Let $r \leqslant 7$ and let $M \in \mathcal{G}_{rk}$ be analytic. If $x \in \mathbf{R}^r \setminus M \setminus E_M$, then $\rho_{M,x}$ has one of the local normal forms listed in the main theorem and the change of coordinates ϕ which gives (1) may be taken to be analytic. This is because the whole theory of versal and strictly versal unfoldings goes through without change in the analytic case.

14. The ellipse. We will prove the assertion in §5 that B_2 is a normal form for ρ_M at x, when $x \in l_1 \cup l_2 \setminus \{P_1, P_2\}$.

By symmetry, there is no loss of generality in supposing $x \in l_1 \setminus P_1$. Then S_x is reduced to one point, the endpoint of the major axis nearest l_1. The function $\|y - x\|$, $y \in M$, differs from $\|y - P_1\|$, $y \in M$, by a constant. Recall that a circle C is said to have first order (resp., second order) contact with M at a point y, if it is tangent to M at y (resp., osculates M at y). Let C be the circle centered at P_1 through y. It is easily checked by analytic geometry that C has third order but not fourth order contact with M at S_x. Therefore, the germ at S_x of the function $y \mapsto \|y - P_1\|$, $y \in M$, is right equivalent to $\xi \mapsto \xi^4$. Hence, η_x (which is the germ at S_x of the function $y \mapsto \|y - x\| = \sqrt{\|y - P_1\|^2 + \|P_1 - x\|^2}$, $y \in M$) is right equivalent to $\xi \mapsto \xi^4$.

By the theorem in §8, it is enough to check that $\sigma_{M,x}$ is a strictly versal unfolding of η_x, in order to see that B_2 is a normal form for ρ_M at x. To check this, it is enough to check that

$$\mathcal{E}_1 = \langle \partial\eta \rangle + V_\sigma$$

(where $\sigma = \sigma_{M,x}$, $\eta = \eta_x$), by the basic theorem of restricted catastrophe theory.

Since η is right equivalent to $\xi \to \xi^4$, $\langle \partial\eta \rangle$ is generated by ξ^3. Hence, it is enough to check that the $r \times 3$ matrix

$$\begin{bmatrix} \dfrac{\partial\sigma}{\partial t_1}, & \dfrac{\partial^2\sigma}{\partial t_1 \partial\xi}, & \dfrac{\partial^3\sigma}{\partial t_1 \partial\xi^2} \\ \cdots & \cdots & \cdots \\ \dfrac{\partial\sigma}{\partial t_r}, & \dfrac{\partial^2\sigma}{\partial t_r \partial\xi}, & \dfrac{\partial^3\sigma}{\partial t_r \partial\xi^2} \end{bmatrix}$$

has rank 3, when evaluated at $S_x \times x$. Here, $t_1, \ldots, t_r$ denote the coordinates of $\mathbf{R}^r$.

For obvious symmetry reasons, it is enough to consider the case when $r = 3$. By a Euclidean motion of $\mathbf{R}^3$, we may reduce to the case when the ellipse is in the (t_1, t_2) plane, its major axis is in the t_1 axis, and its minor axis is in the t_2 axis.

It is easy to see that $\partial^2(\sigma^2)(S_x, x)/\partial t_3 \partial\xi = \partial^3(\sigma^2)(S_x, x)/\partial t_3 \partial\xi^2 = 0$. Moreover, since x is not in the (t_1, t_2) plane, $\partial(\sigma^2)(S_x, x)/\partial t_3 \neq 0$. Moreover, $\partial^2(\sigma^2)(S_x, x)/\partial t_1 \partial\xi = 0$, $\partial^2(\sigma^2)(S_x, x)/\partial t_1 \partial\xi^2 \neq 0$, since x is in the (t_1, t_3) plane. Finally, $\partial^2(\sigma^2)(S_x, x)/\partial t_2 \neq 0$, since the line joining P_1 to S_x is perpendicular to the tangent to M at S_x.

These facts show that the above matrix, with σ replaced by σ^2, has rank 3. It follows that the original matrix has rank 3.

Therefore $\sigma_{M,x}$ is a strictly versal unfolding of η_x and we obtain that B_2 is a normal form for ρ_M at x.

REFERENCES

1. V. I. Arnold, *Singularity theory: selected papers*, London Math. Soc. Lecture Note Series, vol. 53, Cambridge Univ. Press, London, 1981.

2. Th. Bröcker, *Differentiable germs and catastrophes* (translated from German by L. Lander), London Math. Soc. Lecture Note Series, vol. 17, Cambridge Univ. Press, London, 1975.

3. M. Buchner, *Stability of the cut locus in dimensions less than or equal to six*, Invent. Math. **43** (1977), 199–231.

4. ______, *The structure of the cut locus in dimensions less than or equal to six*, Compositio Math., **37** (1978), 103–119.

5. F. H. Clarke, *On the inverse function theorem*, Pacific J. Math. **64** (1976), 97–102.

6. V. Guillemin and M. Golubitsky, *Stable mappings and their singularities*, Springer-Verlag, New York, 1974.

7. E. J. N. Looijenga, *Structural stability of smooth families of C^∞-functions*, Doctoral thesis, Universiteit van Amsterdam, 1974.

8. J. N. Mather, *Stability of mappings*.

I. *The division theorem*, Ann. of Math. (2) **87**, (1968), 89–104.

II. *Infinitesimal stability implies stability*, Ann. of Math. (2) **89** (1969), 254–291.

III. *Finitely determined map-germs*, Publ. Math. Inst. Hautes Études Sci. **35** (1968), 127–156.

IV. *Classification of stable germs by* **R**-*algebras*, Publ. Math. Inst. Hautes Études Sci. **37** (1969), 223–248.

V. *Transversality*, Adv. in Math. **4** (1970), 301–336.

VI. *The nice dimensions*, Proc. Liverpool Singularities Sympos. I, Lecture Notes in Math., vol. 192 Springer-Verlag, Berlin and New York, 1971, pp. 207–253.

9. ______, *Notes on right equivalence*, Warwick, 1969 (unpublished manuscript).

10. D. Milman, *Eine geometrische Ungleihung und ihre Avendung* (*Die zentrale Menge des Gebites und die Erkennung des Gebites durch sie*), General Inequalities, vol. 2, E. F. Beckenbach, Editor (to appear).

11. ______, *The central function of the boundary of a domain and its differential properties* J. Geom. **14** (1980), 182–202.

12. D. Milman and Z. Waksman, *On topological properties of the set of a bounded domain in* $\mathbf{R}^m$ J. Geom. **15** (1981), 1–7.

13. J. C. Oxtoby, *Measure and category*, Springer-Verlag, New York, 1971.

14. R. Thom, *Les singularités des applications différentiables*, Ann. Inst. Fourier (Greoble) **6** (1955–56), 43–87.

15. ______, *Ensembles et morphismes stratifiés*, Bull. Amer. Math. Soc. **75** (1969), 240–284.

16. ______, *Stabilité structurelle et morphogénèse*, Benjmain, 1972; English transl., *Structural stability and morphogenesis*, Benjamin, Reading, Mass., 1975.

17. R. Thom and H. Levine, *Singularities of differentiable mappings*, Math. Institute, Universitaet, Bonn (1959), Also included in the Bonner Math. Shriften and reprinted in Proc. Liverpool Singularities Sympos. I, Lecture Notes in Math., vol. 192, Springer-Verlag, Berlin and New York, 1971, pp. 1–89.

18. C. T. C. Wall, *Geometric properties of generic differentiable manifolds*, Proc. III Latin American School of Math., 597 (1976), pp. 707–774.

19. Y. Yomdin, *Local structure of a generic central set*, Compositio Math. **43** (1981), 225–238.

20. E. C. Zeeman, *Catastrophe theory: selected papers*, 1972–1977, Advanced Book Program, Addison-Wesley, Reading, Mass., 1977.

21. E. C. Zeeman and D. Trotman, *The classification of elementary catastrophes of codimension* $\leqslant 5$, in [**20**], pp. 496–561.

PRINCETON UNIVERSITY

Proceedings of Symposia in Pure Mathematics
Volume 40 (1983), Part 2

ON THE TOPOLOGY OF DELIGNE'S WEIGHT FILTRATION

CLINT McCRORY[1]

Let X be a complex algebraic variety. Deligne has defined a *weight filtration* of H^iX, the ith rational cohomology group of X. It is an increasing filtration of length $2i+1$:

$$H^iX = W_{2i}H^iX \supset W_{2i-1}H^iX \supset \cdots \supset W_0H^iX \supset W_{-1}H^iX = 0.$$

There is also a weight filtration of H_iX, the ith rational homology group of X. It is a decreasing filtration of length $2i+1$:

$$H_iX = W^0H_iX \supset W^1H_iX \supset \cdots \supset W^{2i}H_iX \supset W^{2i+1}H_iX = 0.$$

The weight filtrations on cohomology and homology are dual to each other; in other words $W^qH_iX = \{z \in H_iX \mid \langle y, z\rangle = 0 \text{ for all } y \in W_{q-1}H^iX\}$. (Throughout this paper, all homology and cohomology groups will be with rational coefficients.)

An introduction to the weight filtration of cohomology is given in [**5**, **6** and **9**]. In this note I will emphasize the weight filtration of homology. If X is compact and smooth, then $H_iX = W^iH_iX$ and $W^{i+1}H_iX = 0$, i.e. every class in H_iX has weight i. If $f: X \to Y$ is an algebraic map, then $f_*(W^qH_iX) = (W^qH_iY) \cap f_*(H_iX)$. In other words, the induced morphism on homology is *strictly* compatible with the weight filtration.

For example, if M is a compact smooth variety, and f: $M \to X$ is an algebraic map, then every i-dimensional homology class in the image of f_* has weight i, i.e. $f_*(H_iM) \subset W^iH_iX$ and $f_*(H_iM) \cap W^{i+1}H_iX = 0$. It follows that if $z \in H_iX$ is represented by an algebraic cycle $Z \subset X$, then z has weight i, for we can take M to be a resolution of Z.

In this note I will consider primarily compact (in fact, projective) but possibly singular varieties. For such a variety X, $W^{i+1}H_iX = 0$. A completely dual

1980 *Mathematics Subject Classification*. Primary 14F45.
[1]Supported in part by NSF grant MCS-8102759.

discussion can be made for smooth, but not necessarily compact varieties. For such a variety X, $H_iX = W^iH_iX$.

Suppose X is the union of two smooth hypersurfaces A and B which intersect transversely in the smooth projective variety M. The weight filtration on H_iX can be computed from the Mayer-Vietoris sequence

$$H_iA \oplus H_iB \xrightarrow{j} H_iX \xrightarrow{\partial} H_{i-1}(A \cap B).$$

The morphisms j and ∂ are strictly compatible with the weight filtration, and A, B, $A \cap B$ are smooth and compact. Therefore $H_iX = W^{i-1}H_iX \supset W^iH_iX \supset W^{i+1}H_iX = 0$, and $W^iH_iX = \mathrm{Kernel}(\partial) = \mathrm{Image}(j)$. This analysis can be generalized to any projective hypersurface X with *normal crossings*, i.e. X is a codimension one subvariety of a smooth projective variety M, and X is a union of smooth hypersurfaces which meet transversely in M (cf. [**10**]). The weight filtration for a variety with normal crossings coincides with the filtration associated with the "Mayer-Vietoris spectral sequence" for the homology of X [**9**, §4; **1**].

By Hironaka's resolution of singularities Theorem [**10**], for any subvariety X of the smooth projective variety M, there exists an algebraic map f: $M' \to M$ such that the total transform $X' = f^{-1}(X)$ is a hypersurface with normal crossings in the smooth projective variety M', and $f|(M' - X')$: $M' - X' \to M - X$ is an isomorphism. Let $f' = f|X'$: $X' \to X$. I will refer to the diagram

$$\begin{array}{ccc} X' & \subset & M' \\ f'\downarrow & & \downarrow f \\ X & \subset & M \end{array}$$

as a *resolution of* $X \subset M$ *to normal crossings*.

Since f: $M' \to M$ is a degree-one map, f_*: $H_iM' \to H_iM$ is surjective for all i. And since $f|(M' - X')$: $M' - X' \to M - X$ is an isomorphism, f_*: $H_i(M', X') \to H_i(M, X)$ is an isomorphism for all i. A diagram chase using the long exact homology sequences of (M', X') and (M, X) gives that f'_*: $H_iX' \to H_iX$ is surjective for all i. Therefore, $W^qH_iX = f'_*(W^qH_iX')$. This observation can be used to construct the mixed Hodge structure of a projective variety X (cf. [**1, 15, 16**]).

MacPherson suggested in 1976 that resolution to normal crossings might be used to prove Verdier's conjecture that the weight filtration on the cohomology of a projective variety X should be related to the topological filtration associated with the "Poincaré duality spectral sequence" for X. The latter filtration was first studied by Zeeman in his thesis (1954, cf. [**14, 11, 12**]).

Zeeman's filtration of homology is a topological invariant, but not a homotopy invariant. It is defined for any coefficient group. For the rational homology of a projective variety X of complex dimension n, the filtration of H_iX has length $2n - i + 1$:

$$H_iX = S^0H_iX \supset S^1H_iX \supset \cdots \supset S^{2n-i}H_iX \supset S^{2n-i+1}H_iX = 0.$$

This filtration is dual to the "support filtration" of cohomology which Zeeman studied originally. If X is a smooth projective variety, then $H_iX = S^{2n-i}H_iX$. The morphism of homology induced by a continuous map is not compatible with the Zeeman filtration, in general.

Zeeman's filtration has the following topological characterization for triangulable spaces [**11, 12**]. S^qH_iX is the set of all classes $z \in H_iX$ such that $z \in \text{Image}[H_i(X - A) \to H_iX]$ for all closed subspaces A of X such that the topological dimension of A is less than q.

Zeeman's filtration can also be described using a stratification of X. Suppose that X is a complex projective variety, and $X = X_n \supset X_{n-1} \supset \cdots \supset X_0$ is a complex stratification of X, i.e. each subspace X_k is a subvariety of X of complex dimension k, and the subspaces $X_k - X_{k-1}$ are smooth varieties which are the strata of a Whitney stratification of X. Then $z \in S^qH_iX$ if and only if z is represented by a (PL) cycle Z such that $|Z| \cap X_k$ has codimension $\geqslant q$ in X_k for all k, i.e. $\dim(|Z| \cap X_k) \leqslant 2k - q$ for all k, where $|Z|$ denotes the support of the cycle Z. It follows that S^qH_iX can be described using the intersection homology groups $IH_i^{\bar{p}}X$ of Goresky and MacPherson [**7, 8**]. If X is a *normal* (i.e. locally irreducible) n-dimensional variety,

$$S^qH_iX = \text{Image}\left[IH_i^{\bar{p}}X \to H_iX\right],$$

where $\bar{p} = (p_2, p_3, \ldots, p_{2n})$ is the largest perversity vector with $p_k \leqslant 2n - i - q$ for all k.

THEOREM 1. *For an n-dimensional hypersurface X with normal crossings in a smooth complex projective variety M, the Deligne weight filtration and the Zeeman filtration coincide*:

$$W^{i-q}H_iX = S^{2n-i-q}H_iX \quad \textit{for all } q \geqslant 0.$$

EXAMPLE 1. Let X be a "complex tetrahedron", the union of four complex projective planes in general position in $\mathbf{C}P^3$. Let X_0 be the set of vertices of the tetrahedron—the four points which lie in three of the planes of X. Let X_1 be the union of the edges of the tretrahedron—the points which lie in at least two of the planes of X. Then $X = X_2 \supset X_1 \supset X_0$ is a complex stratification of X. The Mayer-Vietoris spectral sequence shows that H_2X has a basis $\{y, z\}$ over $\mathbf{Q}$ such that y has weight 2 and z has weight 0. The class y is represented by a complex line Y in X which does not meet X_0. Since Y is in general position with the stratification of X, $y \in S^2H_2X$. The class z is represented by a *real* tetrahedron Z, the union of the four triangles whose vertices are the points of X_0. The support of any cycle homologous to Z must contain X_0, so $z \notin S^1H_2X$. In H_3X, all nonzero classes have weight 2, i.e. $H_3X = W^2H_3X \supset W^3H_3X = 0$. The support of any essential 3-cycle must contain X_0, so $H_3X = S^0H_3X \supset S^1H_3X = 0$.

THEOREM 2. *If X is an n-dimensional hypersurface of a smooth complex projective variety M, then*

$$W^{i-q}H_iX \supset S^{2n-i-q}H_iX \quad \textit{for all } q \geqslant 0.$$

As a corollary, if $i > n$ then $W^{2i-2n}H_iX = H_iX$ since $W^{2i-2n}H_iX \supset S^0H_iX$. This corollary has been proved by Deligne for an arbitrary complex variety [**4**, 8.3.10].

EXAMPLE 2. Let X be the cubic cone in $\mathbf{C}P^3$, $X = \{[w, x, y, z] \mid x^3 + y^3 + z^3 = 0\}$. X has an isolated singularity at $x_0 = [1, 0, 0, 0]$. A resolution of $X \subset \mathbf{C}P^3$ to normal crossings is obtained by blowing up $\mathbf{C}P^3$ at x_0. Then $X' = A \cup B \subset (\mathbf{C}P^3)'$, where $A = f^{-1}(x_0)$ and B is the strict transform of X, i.e. B is the closure of $f^{-1}(X - x_0)$ in $(\mathbf{C}P^3)'$. Since X is homeomorphic to the Thom space of a complex line bundle over the tōrus, H_3X' has dimension two. Every class in H_3X is the image of a class in H_3X' which comes from B, and so has weight 3. Thus $H_3X = W^3H_3X$. On the other hand, the support of any essential 3-cycle of X must contain x_0, so $S^1H_3X = 0$.

I conjecture that Theorem 2 is true for any complex projective variety X of complex dimension n. I announced a proof of this conjecture in 1978 [**13**], but Deligne found a mistake in my proof. Several special cases can be verified. For example, it is not hard to see that if X is normal, then $W^iH_iX \supset S^{2n-i}H_iX$, which implies Deligne's observation that $H_iX = W^iH_iX$ for a compact rational homology manifold X.

Now I turn to the proof of Theorem 1. Both the Zeeman filtration and the Deligne filtration arise by expressing the rational homology of X as the hypercohomology of a filtered complex of sheaves on X. For an introduction to derived categories of sheaves, see [**8**]. By and large, I will use the notation and terminology of [**8**].

Let $\underline{\underline{D}}_X^{\cdot}$ be the (rational) dualizing complex of X:

$$0 \to \underline{\underline{D}}_X^{-2n} \to \underline{\underline{D}}_X^{-2n+1} \to \cdots \to \underline{\underline{D}}_X^{-1} \to \underline{\underline{D}}_X^{0} \to 0,$$

and let S be its canonical filtration [**3**, 1.4.6]. A complex naturally isomorphic to $\underline{\underline{D}}_X^{\cdot}$ is the complex $\underline{\underline{C}}_X^{\cdot}$ of rational singular chains on X:

$$\Gamma(U; \underline{\underline{C}}_X^{-i}) = C_i(X, X - U),$$

where $C_i(X, X - U) = C_i(X)/C_i(X - U)$, and $C_i(X)$ (resp. $C_i(X - U)$) is the group of singular chains of X (resp. $X - U$) with rational coefficients. The hypercohomology $\mathcal{H}$ of the complex of sheaves $\underline{\underline{C}}_X^{\cdot}$ is H_*X, and the filtration induced by S on H_*X is the Zeeman filtration. More precisely,

$$S_{-p}\mathcal{H}^{-i}(\underline{\underline{D}}_X^{\cdot}) = S^{p-i}H_iX.$$

The spectral sequence E of the filtered complex $(\underline{\underline{D}}_X^{\cdot}, S)$ is the Poincaré duality spectral sequence of X (cf. [**12**]).

Let $\underline{\underline{\mathbf{Q}}}_{X(\cdot)}$ be the Mayer-Vietoris simplicial resolution of $\underline{\underline{\mathbf{Q}}}_X$, the constant sheaf on X [**16**, I.1] (see also [**9**, §4]):

$$0 \to \underline{\underline{\mathbf{Q}}}_X \to \Pi_*\underline{\underline{\mathbf{Q}}}_{X(0)} \to \Pi_*\underline{\underline{\mathbf{Q}}}_{X(1)} \to \Pi_*\underline{\underline{\mathbf{Q}}}_{X(2)} \to \cdots.$$

Here $X(q)$ is a variety constructed from X, $\underline{\underline{\mathbf{Q}}}_{X(q)}$ is the constant sheaf on $X(q)$, and $\Pi\colon X(q) \to X$ is an algebraic map. The variety $X(q)$ is defined as follows. Let

$V_1, \ldots, V_s$ be the irreducible components of X. Then $X(q)$ is the disjoint union of the varieties $V_{i_0} \cap \cdots \cap V_{i_q}$ as $\{i_0, \ldots, i_q\}$ runs over all the subsets of cardinality $q+1$ of the set $\{1, \ldots, s\}$. Thus $X(q)$ is a smooth variety of dimension $n - q$. The map Π: $X(q) \to X$ is the disjoint union of the inclusion maps $V_{i_0} \cap \cdots \cap V_{i_q} \subset X$.

The differential δ_q: $\Pi_* \underline{\underline{\mathbf{Q}}}_{X(q)} \to \Pi_* \underline{\underline{\mathbf{Q}}}_{X(q+1)}$ is the simplicial coboundary map. More precisely, if $x \in X$ let $I(x) = \{i \mid 1 \leqslant i \leqslant s, x \in V_i\}$. The stalk of $\Pi_* \underline{\underline{\mathbf{Q}}}_{X(q)}$ over x is the rational vector space with basis all the subsets of cardinality $q+1$ of the set $I(x)$. In other words, this stalk is the qth simplicial cochain group, with rational coefficients, of the simplex $\Delta(x)$ whose vertex set is $I(x)$. On stalks over x, the differential δ_q is just the qth simplicial coboundary map of the simplicial cochain complex of $\Delta(x)$. Let W be the filtration of $\underline{\underline{\mathbf{Q}}}_{X(\cdot)}$ defined by

$$(W_{-q} \underline{\underline{\mathbf{Q}}}_{X(\cdot)})^i = \begin{cases} \underline{\underline{\mathbf{Q}}}_{X(i)} & \text{if } i \geqslant q, \\ 0 & \text{if } i < q. \end{cases}$$

(This is Deligne's "bête" filtration [**3**, 1.4.7].) The hypercohomology of the complex of sheaves $\underline{\underline{\mathbf{Q}}}_{X(\cdot)}$ is H^*X, and the filtration induced by W on H^*X is the weight filtration. More precisely,

$$W_{-q} \mathcal{H}^i(\underline{\underline{\mathbf{Q}}}_{X(\cdot)}) = W_{i-q} H^i X.$$

The spectral sequence E' of the filtered complex $(\underline{\underline{\mathbf{Q}}}_{X(\cdot)}, W)$ is the Mayer-Vietoris spectral sequence of X (cf. [**9**, §4]).

Let α: $X \to M$ be the inclusion of X as a hypersurface of the smooth projective variety M. I will show that the filtered complexes $\alpha_* \underline{\underline{D}}^{\boldsymbol{\cdot}}_X$ and $\alpha_* \underline{\underline{\mathbf{Q}}}_{X(\cdot)}$ are *Verdier dual* [**8**, 1.6]. For a precise statement of this duality, I need some notation on shifting indices of complexes and filtrations.

If $(A^{\boldsymbol{\cdot}}, F)$ is a complex with an increasing filtration F, then $(A^{\boldsymbol{\cdot}}[j], F[q])$ is the filtered complex defined by $A^i[j] = A^{i+j}$ and $F_p[q] A^i[j] = F_{p-q} A^{i+j}$ (cf. [**14**, I.6]).

If $A^{\boldsymbol{\cdot}}$ and $B^{\boldsymbol{\cdot}}$ are complexes, then the complex $\operatorname{Hom}^{\boldsymbol{\cdot}}(A^{\boldsymbol{\cdot}}, B^{\boldsymbol{\cdot}})$ has $\operatorname{Hom}^k(A^{\boldsymbol{\cdot}}, B^{\boldsymbol{\cdot}}) = \bigoplus_{j-i=k} \operatorname{Hom}(A^i, B^j)$. If F is an increasing filtration of $A^{\boldsymbol{\cdot}}$ and G is an increasing filtration of $B^{\boldsymbol{\cdot}}$, an increasing filtration (F, G) of the complex $\operatorname{Hom}^{\boldsymbol{\cdot}}(A^{\boldsymbol{\cdot}}, B^{\boldsymbol{\cdot}})$ is defined by

$$(F, G)_q \operatorname{Hom}(A^i, B^j) = \bigoplus_p \{ f \in \operatorname{Hom}(A^i, B^j) \mid f(F_{p-1} A^i) = 0,$$
$$f(F_p A^i) \subset G_{p-q} B^j \}.$$

For discussions of the sheaf homomorphism functor $\underline{\underline{\operatorname{Hom}}}^{\boldsymbol{\cdot}}$, its derived functor $R\underline{\underline{\operatorname{Hom}}}$ and the derived functor Rf_* of the direct image functor f_*, see [**8**, 1.4–1.5].

A *filtered quasi-isomorphism* of filtered complexes of sheaves, f: $(\underline{\underline{A}}^{\boldsymbol{\cdot}}, F) \to (\underline{\underline{B}}^{\boldsymbol{\cdot}}, G)$, is a filtration-preserving map of complexes such that the induced map of cohomology sheaves

$$\underline{\underline{H}}^i(F_p A^{\boldsymbol{\cdot}} / F_{p-1} A^{\boldsymbol{\cdot}}) \to \underline{\underline{H}}^i(G_p B^{\boldsymbol{\cdot}} / G_{p-1} B^{\boldsymbol{\cdot}})$$

is an isomorphism for all i and all p.

Let $\underline{\underline{D}}^{\cdot}_M$ be the dualizing complex of M, and let S be its canonical filtration.

THEOREM 3. *Let X be an n-dimensional hypersurface with normal crossings in the smooth complex projective variety M, with inclusion $\alpha\colon X \to M$. There is a filtered quasi-isomorphism*

$$\left(R\ \underline{\underline{\mathrm{Hom}}}^{\cdot}_{\mathbf{Q}}\left(\alpha_* \underline{\underline{\mathbf{Q}}}_{X(\cdot)}, \underline{\underline{D}}^{\cdot}_M\right), (W, S)\right) \simeq \left(\alpha_* \underline{\underline{D}}^{\cdot}_X, S[2n]\right).$$

PROOF OF THEOREM 1. Taking hypercohomology of the above equation yields Theorem 1. On the one hand, $\alpha_* = R\alpha_* = R\alpha_!$ and by Verdier duality [**8**, 1.6]

$$(W, S)_q \mathcal{H}^{-i} R\ \underline{\underline{\mathrm{Hom}}}^{\cdot}_{\mathbf{Q}}\left(R\alpha_! \underline{\underline{\mathbf{Q}}}_{X(\cdot)}, \underline{\underline{D}}^{\cdot}_M\right) = W^{i-q} H_i X.$$

(The Verdier duality argument is spelled out below in the proof of Theorem 3.) On the other hand,

$$S_q[2n]\mathcal{H}^{-i}\left(R\alpha_! \underline{\underline{D}}^{\cdot}_X\right) = S_{q-2n}\mathcal{H}^{-i}\left(R\alpha_! \underline{\underline{D}}^{\cdot}_X\right) = S^{2n-i-q} H_i X.$$

REMARK. Theorem 3 is much stronger than Theorem 1. Theorem 3 implies the isomorphism of two spectral sequences which converge to $H_* X$: the Poincaré duality spectral sequence E and the Mayer-Vietoris spectral sequence E', with $E_{r+1} \cong E'_r$ for $r \geqslant 1$. Mixed Hodge theory can be used to show that the Mayer-Vietoris spectral sequence E' collapses, i.e. $E'_2 = E'_\infty$ (cf. [**9**, §4]). So Theorem 3 has the surprising topological consequence that $E_3 = E_\infty$ for a hypersurface X with normal crossings in a smooth projective variety.

The following result will be important in the proof of Theorem 3.

PROPOSITION 1. *Let $Y \subset Z$ be subvarieties of the smooth complex projective variety M, with inclusion maps $\alpha\colon Y \to M$ and $\beta\colon Z \to M$. There is a commutative diagram of maps in the derived category of sheaves on M:*

$$\begin{array}{ccc}
R\ \underline{\underline{\mathrm{Hom}}}^{\cdot}_{\mathbf{Q}}\left(\alpha_* \underline{\underline{\mathbf{Q}}}_Y, \underline{\underline{D}}^{\cdot}_M\right) & \underset{\approx}{\overset{\theta(\alpha)}{\rightarrow}} & \alpha_* \underline{\underline{D}}^{\cdot}_Y \\
\downarrow r & & \downarrow s \\
R\ \underline{\underline{\mathrm{Hom}}}^{\cdot}_{\mathbf{Q}}\left(\beta_* \underline{\underline{\mathbf{Q}}}_Z, \underline{\underline{D}}^{\cdot}_M\right) & \underset{\approx}{\overset{\theta(\beta)}{\rightarrow}} & \beta_* \underline{\underline{D}}^{\cdot}_Z
\end{array}$$

where $\theta(\alpha)$ and $\theta(\beta)$ are quasi-isomorphisms and r is induced by restriction of the constant sheaf from Z to Y.

PROOF. $\theta(\alpha)$ is identified with the Verdier duality isomorphism for the complexes of sheaves $\underline{\underline{\mathbf{Q}}}_Y$, $\underline{\underline{D}}^{\cdot}_M$, and the map α, using that $\alpha_* = R\alpha_* = R\alpha_!$, and that $\alpha^! \underline{\underline{D}}^{\cdot}_M \simeq \underline{\underline{D}}^{\cdot}_Y$ since M is smooth. The same argument applies to $\underline{\underline{\mathbf{Q}}}^{\cdot}_Z$, $\underline{\underline{D}}^{\cdot}_M$, and β, to give $\theta(\beta)$. The diagram commutes by the naturality of Verdier duality.

REMARKS. (1) Taking the ith hypercohomology of this diagram gives the classical Lefschetz duality diagram:

$$\begin{array}{ccc}
H^{2m-i}(M, M - Y) & \underset{\approx}{\rightarrow} & H_i Y \\
\downarrow & & \downarrow \\
H^{2m-i}(M, M - Z) & \underset{\approx}{\rightarrow} & H_i Z
\end{array}$$

(2) The map s can be described by replacing $\underline{\underline{D}}_Y^{\cdot}$ by $\underline{\underline{C}}_Y^{\cdot}$ and $\underline{\underline{D}}_Z^{\cdot}$ by $\underline{\underline{C}}_Z^{\cdot}$. On stalks over $y \in Y$, s is the homomorphism $C_*(Y, Y - \{y\}) \to C_*(Z, Z - \{y\})$ induced by inclusion.

PROOF OF THEOREM 3. Since $\underline{\underline{\mathbf{Q}}}_{X(\cdot)}$ is a resolution of $\underline{\underline{\mathbf{Q}}}_X$, there is a quasi-isomorphism

$$R\ \underline{\underline{\mathrm{Hom}}}^{\cdot}_{\mathbf{Q}}\left(\alpha_*\underline{\underline{\mathbf{Q}}}_{X(\cdot)}, \underline{\underline{D}}_M^{\cdot}\right) \simeq R\ \underline{\underline{\mathrm{Hom}}}^{\cdot}_{\mathbf{Q}}\left(\alpha_*\underline{\underline{\mathbf{Q}}}_X, \underline{\underline{D}}_M^{\cdot}\right).$$

So Proposition 1 applied to $\alpha\colon X \to M$ gives a quasi-isomorphism

$$R\ \underline{\underline{\mathrm{Hom}}}^{\cdot}_{\mathbf{Q}}\left(\alpha_*\underline{\underline{\mathbf{Q}}}_{X(\cdot)}, \underline{\underline{D}}_M^{\cdot}\right) \simeq \alpha_*\underline{\underline{D}}_X^{\cdot},$$

which is automatically a filtered quasi-isomorphism for the *canonical* filtrations. But the given filtration (W, S) on $R\,\underline{\underline{\mathrm{Hom}}}^{\cdot}_{\mathbf{Q}}(\alpha_*\underline{\underline{\mathbf{Q}}}_{X(\cdot)}, \underline{\underline{D}}_M^{\cdot})$ is not compatible with the canonical filtration. The solution to this dilemma is to show that $R\,\underline{\underline{\mathrm{Hom}}}^{\cdot}_{\mathbf{Q}}(\alpha_*\underline{\underline{\mathbf{Q}}}_{X(\cdot)}, \underline{\underline{D}}_M^{\cdot})$ is quasi-isomorphic to a complex of sheaves $\underline{\underline{D}}^{\cdot}$ on which the filtration (W, S) is comparable with the canonical filtration, and such that the identity map of $\underline{\underline{D}}^{\cdot}$ is a filtered quasi-isomorphism with respect to these two filtrations.

Let $\underline{\underline{D}}^{\cdot}$ be the simple complex associated to the following double complex, i.e. $\underline{\underline{D}}^r = \bigoplus_{p-q=r} \alpha_*\Pi_*\underline{\underline{D}}^p_{X(q)}$:

$$\begin{array}{ccccccccccc}
\alpha_*\Pi_*\underline{\underline{D}}^{-2n}_{X(0)} & \to & \alpha_*\Pi_*\underline{\underline{D}}^{-2n+1}_{X(0)} & \to & \alpha_*\Pi_*\underline{\underline{D}}^{-2n+2}_{X(0)} & \to & \alpha_*\Pi_*\underline{\underline{D}}^{-2n+3}_{X(0)} & \to & \alpha_*\Pi_*\underline{\underline{D}}^{-2n+4}_{X(0)} & \to & \cdots \\
\uparrow & & \uparrow & & \uparrow & & \uparrow & & \uparrow & & \\
0 & \to & 0 & \to & \alpha_*\Pi_*\underline{\underline{D}}^{-2n+2}_{X(1)} & \to & \alpha_*\Pi_*\underline{\underline{D}}^{-2n+3}_{X(1)} & \to & \alpha_*\Pi_*\underline{\underline{D}}^{-2n+4}_{X(1)} & \to & \cdots \\
\uparrow & & \uparrow & & \uparrow & & \uparrow & & \uparrow & & \\
0 & \to & 0 & \to & 0 & \to & 0 & \to & \alpha_*\Pi_*\underline{\underline{D}}^{-2n+4}_{X(2)} & \to & \cdots \\
\uparrow & & \uparrow & & \uparrow & & \uparrow & & \uparrow & & \\
\vdots & & \vdots & & \vdots & & \vdots & & \vdots & &
\end{array}$$

The qth row is the complex $\alpha_*\Pi_*\underline{\underline{D}}^{\cdot}_{X(q)}$, where $\underline{\underline{D}}^{\cdot}_{X(q)}$ is the dualizing complex of $X(q)$. The vertical maps are simplicial boundary maps. For $x \in X$, the fiber of $\Pi\colon X(q) \to X$ is identified with the set of q-simplexes of the simplex $\Delta(x)$ with vertex set $I(x) = \{i \mid 0 \leqslant i \leqslant s,\ x \in V_i\}$. Each q-simplex of $\Delta(x)$ corresponds to an intersection $V_{i_0} \cap \cdots \cap V_{i_q}$ of irreducible components of X, and the face operators of $\Delta(x)$ correspond to inclusions $V_{i_0} \cap \cdots \cap V_{i_q} \to V_{i_0} \cap \cdots \cap \hat{V}_{i_p} \cap \cdots \cap V_{i_q}$. For each such inclusion, there is an induced map of complexes

$$s\colon \alpha_*\Pi_*\underline{\underline{D}}^{\cdot}_{V_{i_0}\cap\cdots\cap V_{i_q}} \to \alpha_*\Pi_*\underline{\underline{D}}^{\cdot}_{V_{i_0}\cap\cdots\cap\hat{V}_{i_p}\cap\cdots\cap V_{i_q}}$$

as in Proposition 1. Therefore the stalk of $\alpha_*\Pi_*\underline{\underline{D}}^{\cdot}_{X(q)}$ is a system of coefficients on $\Delta(x)$. The map of complexes $\alpha_*\Pi_*\underline{\underline{D}}^{\cdot}_{X(q)} \to \alpha_*\Pi_*\underline{\underline{D}}^{\cdot}_{X(q-1)}$ on the stalk over x is just the simplicial boundary map of this coefficient system.

By Proposition 1, for each $V = V_{i_0} \cap \cdots \cap V_{i_q}$ there is a quasi-isomorphism

$$R\ \underline{\underline{\mathrm{Hom}}}^{\cdot}_{\mathbf{Q}}\left(\alpha_*\Pi_*\mathbf{Q}_V, \underline{\underline{D}}_M^{\cdot}\right) \simeq \alpha_*\Pi_*\underline{\underline{D}}_V^{\cdot},$$

and these quasi-isomorphisms are compatible with the inclusion maps $V_{i_0}\cap\cdots\cap V_{i_q}\to V_{i_0}\cap\cdots\cap\hat{V}_{i_p}\cap\cdots\cap V_{i_q}$. Therefore $R\,\underline{\underline{\mathrm{Hom}}}^{\cdot}_{\mathbf{Q}}(\alpha_*\mathbf{Q}_{X(\cdot)},\underline{\underline{D}}^{\cdot}_M)$ is quasi-isomorphic to the simple complex $\underline{\underline{D}}^{\cdot}$ associated with the double complex $\alpha_*\Pi_*\underline{\underline{D}}^{\cdot}_{X(\cdot)}$, as desired.

Now it is easy to see that the canonical filtration T on $\underline{\underline{D}}^{\cdot}$ is compatible with the filtration (W,S) induced by the quasi-isomorphism $R\,\underline{\underline{\mathrm{Hom}}}^{\cdot}_{\mathbf{Q}}(\alpha_*\mathbf{Q}_{X(\cdot)},\underline{\underline{D}}^{\cdot}_M)\cong\underline{\underline{D}}^{\cdot}$. By definition, $T_{-2n+k}\underline{\underline{D}}^r=\underline{\underline{D}}^r$ for $r<-2n+k$, $T_{-2n+k}\underline{\underline{D}}^{-2n+k}\subset\underline{\underline{D}}^{-2n+k}$, and $T_{-2n+k}\underline{\underline{D}}^r=0$ for $r>-2n+k$, i.e. $T_{-2n+k}(\alpha_*\Pi_*\underline{\underline{D}}^p_{X(q)})=0$ for $p-q>-2n+k$. On the other hand, $(W,S)_k(\alpha_*\Pi_*\underline{\underline{D}}^p_{X(q)})=\alpha_*\Pi_*\underline{\underline{D}}^p_{X(q)}$ for $q\leqslant r$, and $(W,S)_k(\alpha_*\Pi_*\underline{\underline{D}}^p_{X(q)})=0$ for $q>r$. But $\alpha_*\Pi_*\underline{\underline{D}}^p_{X(q)}=0$ for $p<-2n+2q$. Therefore $T_{-2n+k}(\underline{\underline{D}}^{\cdot})\subset(W,S)_k(\underline{\underline{D}}^{\cdot})$.

Finally, the identity is a filtered quasi-isomorphism $(\underline{\underline{D}}^{\cdot},T[2n])\cong(\underline{\underline{D}}^{\cdot},(W,S))$, since

$$\begin{aligned}\underline{\underline{H}}^{\cdot}(T_{-2n+k}\underline{\underline{D}}^{\cdot}/T_{-2n+k-1}\underline{\underline{D}}^{\cdot})&\cong\alpha_*\Pi_*\mathbf{Q}_{X(k)}[-2n+2k]\\&\cong\underline{\underline{H}}^{\cdot}((W,S)_k\underline{\underline{D}}^{\cdot}/(W,S)_{k-1}\underline{\underline{D}}^{\cdot}).\end{aligned}$$

This completes the proof of Theorem 3.

Theorem 3 is related to F. Elzein's Proposition II.4.1 of [**16**]. I am grateful to him for suggesting that his proof could be adapted to my situation. He proves there is a filtered quasi-isomorphism

$$\left(R\,\underline{\underline{\mathrm{Hom}}}^{\cdot}_{\mathbf{Q}}(\underline{\underline{\mathbf{Q}}}_{(M-X)(\cdot)},\underline{\underline{\mathbf{Q}}}_M),(W,S)\right)\cong\left(R\beta_*\underline{\underline{\mathbf{Q}}}_{M-X},S\right),$$

where $\beta\colon M-X\to X$ is the inclusion map, $\underline{\underline{\mathbf{Q}}}_{(M-X)(\cdot)}$ is a simplicial resolution of $\beta_!\underline{\underline{\mathbf{Q}}}_{M-X}$, W is the weight filtration, and S is the canonical filtration. A corollary of his result is that the weight filtration of $H^*(M-X)$ is a shift of the Leray filtration with respect to the inclusion $\beta\colon M-X\to X$ (cf. [**3**, 3.1.8]). Both Theorem 3 and Elzein's result are implied by the fact that, in the filtered derived category of sheaves on M, the following two distinguished triangles are Verdier dual:

$$\begin{array}{ccc}\alpha_*\underline{\underline{\mathbf{Q}}}_{X(\cdot)} & \xrightarrow{[1]} & \underline{\underline{\mathbf{Q}}}_{(M-X)(\cdot)}\\ \nwarrow & & \swarrow\\ & \underline{\underline{\mathbf{Q}}}_M & \end{array}$$

$$\begin{array}{ccc}R\alpha_*\underline{\underline{D}}^{\cdot}_X & \xleftarrow{[1]} & R\beta_*\underline{\underline{\mathbf{Q}}}_{M-X}[2n+2]\\ \searrow & & \nearrow\\ & \underline{\underline{\mathbf{Q}}}_M[2n+2] & \end{array}$$

Proof of Theorem 2. I will prove the dual statement in rational cohomology:

$$W_{i-q}H^iX\subset S_{2n-i-q}H^iX\quad\text{for all } q\geqslant 0,$$

where S_pH^iX is the set of all classes $y \in H^iX$ such that $\langle y, z \rangle = 0$ for all $z \in S^{p+1}H_iX$. It follows from [**12**] and the triangulability of subanalytic pairs [**17**] that $y \in S_pH^iX$ if and only if there is a closed (real) subanalytic subset A of X such that $\dim_{\mathbf{R}} A \leqslant p$ and $y \mid (X - A) = 0$.

Let

$$\begin{array}{ccc} X' & \subset & M' \\ f' \downarrow & & \downarrow f \\ X & \subset & M \end{array}$$

be a resolution of $X \subset M$ to normal crossings. As I showed above, $(f')_*$: $H_*X' \to H_*X$ is surjective; therefore $(f')^*$: $H^*X \to H^*X'$ is injective. I will need a stronger result.

PROPOSITION 2. *Let f': $X' \to X$ be a resolution to normal crossings, U an open subanalytic subset of X, and $U' = (f')^{-1}U$. Then $(f' \mid U')^*$: $H^*U \to H^*U'$ is injective.*

PROOF. Let V be an open subanalytic subset of the ambient manifold M such that $U = V \cap X$, and let $V' = f^{-1}(V)$, so $U' = V' \cap X'$. Consider the long exact cohomology sequences of the pairs (V, U) and (V', U'):

$$\begin{array}{ccccccccccc} \cdots & \to & H^i(V',U') & \to & H^iV' & \to & H^iU' & \to & H^{i+1}(V',U') & \to & \cdots \\ & & \uparrow & & \uparrow & & \uparrow & & \uparrow & & \\ \cdots & \to & H^i(V,U) & \to & H^iV & \to & H^iU & \to & H^{i+1}(V,U) & \to & \cdots \end{array}$$

The outer vertical arrows are isomorphisms, since f maps $V' - U'$ isomorphically onto $V - U$. The arrow $H^iV \to H^iV'$ is injective, since it is induced by a proper degree one map of smooth manifolds. A diagram chase shows that $H^iU \to H^iU'$ is injective, as desired.

Now I can prove that $W_{i-q}H^iX \subset S_{2n-i-q}H^iX$ for all $q \geqslant 0$. The morphism $(f')^*$ is injective and strictly compatible with the weight filtration, and $W_{i-q}H^iX' = S_{2n-i-q}H^iX'$ by Theorem 1. So I just have to show that, for all $p \leqslant 2n - i$, if $(f')^*y \in S_pH^iX'$, then $y \in S_pH^iX$. Suppose $(f')^*y \in S_pH^iX'$. Then there exists a closed subanalytic subspace A' of X' such that $\dim_{\mathbf{R}} A' \leqslant p$ and $(f')^*y \mid (X' - A') = 0$. Let $A = f'(A')$. Then A is also closed and subanalytic. By Proposition 2, $(f')^* \mid (X - A)$: $H^*(X - A) \to H^*(X' - A')$ is injective, so $y \mid (X - A) = 0$. But $\dim_{\mathbf{R}} A \leqslant \dim_{\mathbf{R}} A' \leqslant p$, so $y \in S_pH^iX$. This completes the proof of Theorem 2.

REMARK. It is crucial in this proof that X is a *hypersurface* of the smooth variety M. Suppose that $\dim X = n$ and $\dim M = n + k + 1$, $k > 0$. Then $\dim X' = n + k$. If $y \in W_{i-q}H^iX$ then $(f')^*y \in W_{i-q}H^iX'$, by the naturality of the weight filtration. Theorem 1 implies that $(f')^*y \in S_{2(n+k)-i-q}H^iX'$. The argument above, using Proposition 2, shows that $y \in S_{2(n+k)-i-q}H^iX$. But $S_{2n-i-q}H^iX$ is strictly smaller than $S_{2(n+k)-i-q}H^iX$, in general.

REFERENCES

1. M. Anderson, *Hodge theory for projective varieties*, mimeographed notes, Princeton Univ., Princeton, N.J., 1973.

2. J. Cheeger, M. Goresky and R. MacPherson, *The L^2-cohomology and intersection homology of singular algebraic varieties* Sem. Differential Geometry (S. T. Yau, editor), Annals of Math. Studies No. 102, pp. 303–340.

3. P. Deligne, *Théorie de Hodge*. II, Inst. Hautes Études Sci. Publ. Math. **40** (1971), 5–58.

4. ______, *Théorie de Hodge*. III, Inst. Hautes Études Sci. Publ. Math. **44** (1975), 5–77.

5. ______, *Poids dans la cohomologie des variétés algébriques*, Proc. Internat. Congr. Math. (Vancouver, 1974), pp. 79–85.

6. A. Durfee, *A naive guide to mixed Hodge theory*, these PROCEEDINGS.

7. M. Goresky and R. MacPherson, *Intersection homology theory*, Topology **19** (1980), 135–162.

8. ______, *Intersection homology theory*. II, Invent. Math. (to appear).

9. P. Griffiths and W. Schmid, *Recent developments in Hodge theory*, Discrete Subgroups of Lie Groups and Applications to Moduli (Tata Institute, 1973), Oxford Univ. Press, London, 1975, pp. 31–127.

10. H. Hironaka, *Resolution of singularities of an algebraic variety over a field of characteristic zero*, Ann. of Math. (2) **79** (1964), 109–326.

11. C. McCrory, *Poincaré duality in spaces with singularities*, Ph. D. Thesis, Brandeis University, May, 1972.

12. ______, *Zeeman's filtration of homology*, Trans. Amer. Math. Soc. **250** (1979), 147–166.

13. ______, *The support filtration of cohomology*, Notices Amer. Math. Soc. **25** (1978), A–649.

14. E. C. Zeeman, *Dihomology*. III. *A generalization of the Poincaré duality for manifolds*, Proc. London Math. Soc. (3) **13** (1963), 155–183.

15. F. Elzein, *Structures de Hodge mixtes*, C. R. Acad. Sci. Paris Sér. A–B **292** (1981), 409–412.

16. ______, *Mixed Hodge structures*, preprint, Univ. Paris VII, 1981.

17. H. Hironaka, *Triangulations of algebraic sets*, Proc. Sympos. Pure Math., vol. 29, Amer. Math. Soc., Providence, R.I., 1975, pp. 165–185.

UNIVERSITY OF GEORGIA

Proceedings of Symposia in Pure Mathematics
Volume 40 (1983), Part 2

TOPOLOGICAL TYPES OF POLYNOMIAL MAP GERMS

ISAO NAKAI

This is a summary of the forthcoming paper [N], in which we will introduce some topological families of polynomial mappings.

The problem of singularities of smooth mappings has been long studied, and these studies have made some topological properties of generic mappings clear. But very little has been known about topological classification of polynomial mappings with bounded degree, hitherto. In this summary and the paper [N], we shall offer some results about this question.

We denote the set of polynomial mappings $f\colon \mathbf{K}^n \to \mathbf{K}^p$, $f(0) = 0$ of degree k with coefficients in $\mathbf{K}$, by $P(n, p, k; \mathbf{K})$ for $\mathbf{K} = \mathbf{R}, \mathbf{C}$, and the set of their germs at $0 \in \mathbf{K}^n$, by $P_0(n, p, k; \mathbf{K})$. We say $f, g \in P(n, p, k; \mathbf{K})$ are *topologically equivalent* (as germs in $P_0(n, p, k; \mathbf{K})$) if there are homeomorphisms $\phi\colon \mathbf{K}^n \to \mathbf{K}^n$, $\psi\colon \mathbf{K}^p \to \mathbf{K}^p$ (germs of homeomorphisms with $\phi(0) = 0$, $\psi(0) = 0$) such that $f \circ \phi = \psi \circ g$. We denote the set of topological equivalence classes of $P(n, p, k; \mathbf{K})$, $P_0(n, p, k; \mathbf{K})$ by $P(n, p, k; \mathbf{K})/\text{top}$, $P_0(n, p, k; \mathbf{K})/\text{top}$, respectively.

Here we recall the results about topological types of polynomial mappings.

(1) R. Thom had shown that the following polynomial mappings $f_t\colon \mathbf{R}^3 \to \mathbf{R}^3$, $f_t(0) = 0$, $t \in \mathbf{R}$, defined by

$$f_t(x, y, z) = \Big(\big(x(x^2 + y^2 - a^2) - 2ayz\big)^2\big((ty - x)(x^2 + y^2 - a^2) - 2az(y - t_x)\big)^2, x^2 + y^2 - a^2, z\Big), \qquad (x, y, z) \in \mathbf{R}^3,$$

are topologically equivalent if and only if $t = \pm t'$ in 1962 [T]. This result tells us that $P(n, p, k; \mathbf{R})/\text{top}$ is an infinite set for $n, p \geqslant 3$ and $k \geqslant 12$.

(2) T. Fukuda proved that the sets of topological equivalence classes of polynomial functions $P(n, 1, k; \mathbf{K})/\text{top}$ and polynomial function germs $P_0(n, 1, k; \mathbf{K})/\text{top}$ are finite sets for any $n, k < \infty$ and $\mathbf{K} = \mathbf{R}, \mathbf{C}$, in 1976 [F].

1980 *Mathematics Subject Classification*. Primary 57R45.

0082-0717/81/0000-0197/$02.25

(3) K. Aoki has proved that $P_0(2,2,k;\mathbf{K})/\mathrm{top}$ is a finite set for any $k<\infty$ and $\mathbf{K}=\mathbf{R}$, $\mathbf{C}$, in 1980 [**A**]. This result is an affirmative answer to the following conjecture of Thom that $P(n,2,k;\mathbf{K})/\mathrm{top}$, $P_0(n,2,k;\mathbf{K})/\mathrm{top}$ are finite sets for any $k<\infty$ and $\mathbf{K}=\mathbf{R},\mathbf{C}$ (see e.g., [**F**]).

(4) Recently, this result has been extended by C. Sabbah in [$\mathbf{S_2}$]. Namely, he has shown that $P(2,p,k;\mathbf{C})/\mathrm{top}$ is a finite set for $p,k<\infty$.

Contrary to the conjecture of Thom, we have

THEOREM. $P(n,p,k;\mathbf{K})/\mathrm{top}$, $P_0(n,p,k;\mathbf{K})/\mathrm{top}$, $\mathbf{K}=\mathbf{R},\mathbf{C}$, *are infinite sets if* $n,p,k\geqslant 3$ *or* $n\geqslant 3$, $p\geqslant 2$, $k\geqslant 4$.

Our simplest examples for the Theorem are as follows:

EXAMPLE A. $f_e\colon(\mathbf{K}^{3+r},0)\to(\mathbf{K}^{2+s},0)$ is defined by

$$f_e(x,y,z,w)=((e_1x-y)(e_2x-y)(e_3x-y),\\ (e_4x-y)(e_5x-y)(e_6x-y)z,0),$$

for $(x,y,z)\in\mathbf{K}^3$, $w\in\mathbf{K}^r$, where $e=(e_1,\dots,e_6)\in\mathbf{K}^6$.

EXAMPLE B. $g_a\colon(\mathbf{K}^{3+r},0)\to(\mathbf{K}^{3+s},0)$ is defined by

$$g_a(x,y,z,w)=\big(a_1x^2+a_2xy+a_3y^2,\\ \big(a_4x^2+a_5xy+a_6y^2\big)z,\big(a_7x^2+a_8xy+a_9y^2\big)z,0\big),$$

for $(x,y,z)\in\mathbf{K}^3$, $w\in\mathbf{K}^r$, where $a=(a_1,\dots,a_9)\in\mathbf{K}^9$.

In the remainder of this paper, we will explain that the families f_e, g_a have infinitely many different topological types (more precisely, f_e contains a two-dimensional family of topological types and g_a contains a one-dimensional family of topological types). We remark here that f_e is flat in the sense of algebraic geometry but does not admit any Thom regular stratification for generic $e\in\mathbf{K}^6$, while g_a is not flat nor does it admit any Thom regular stratification for generic $a\in\mathbf{K}^9$ (a complex map germ $f\colon(\mathbf{C}^n,0)\to(\mathbf{C}^p,0)$ is flat if and only if $f^{-1}(f(x))$ are of the same dimension for any x near $0\in\mathbf{C}^n$, see [**H**]).

First we consider f_e and we assume that $\mathbf{K}=\mathbf{R}$ and $r=s=0$ for simplicity. Here we consider the following stratification (S,S') of f_e such that $f_e^{-1}(0)$ is a union of strata, defined by

$$S=\{\mathbf{R}^3-H_4\cup H_5\cup H_6\cup XY, H_4\cup H_5\cup H_6-XY\cup Z,\\ XY-H_1\cup\cdots\cup H_6, H_i\cap XY-0, i=1,\dots,6, Z-0,0\},$$

$$S'=\{\{(u,v)\in\mathbf{R}^2\mid v\neq 0\},\{(u,v)\in\mathbf{R}^2\mid u\neq 0, v=0\},(0,0)\},$$

where $H_i=\{(x,y,z)\in\mathbf{R}^3\mid e_ix-y=0\}$ for $i=1,\dots,6$, $XY=\{(x,y,z)\in\mathbf{R}^3\mid z=0\}$, $Z=\{(x,y,z)\in\mathbf{R}^3\mid y=z=0\}$ and $e_i\neq e_j$ for $i\neq j$. Now we pay attention to Thom's A_{f_e}-condition of (S,S') for f_e. Then we see the condition is not satisfied at any point $x=(0,0,z)\in Z$, $z\neq 0$.

Let S^1 denote the unit circle centered at $0\in\mathbf{R}^2$ in the target space of f_e, and let L_p, $p\in S^1$, denote the ray starting from 0 through p. Now, with the A_{f_e}-condition in mind, we define a subset $M(f_e)(p)\subset Z$ by the set of points $z\in Z$ such that there is a sequence $x_i \in \mathbf{R}^3$ with the following properties:

(1) x_i converge to z, $f_e(x_i) \notin f_e(\Sigma(f_e))$ and the rays $\widehat{0f_e(x_i)}$ converge to L_p,

(2) $\ker df_e(x_i)$ converge to a subspace $T \subset T_z\mathbf{R}^3$ and $T \not\supset \ker d(f_e | Z)(z) = T_z Z$.

Then $M(f_e)$ defines a mapping $M(f_e)$: $(S^1, (1,0)) \to (Z^\omega, \{0\})$ which assigns $M(f_e)(p)$ for any $p \in S^1$, where Z^ω denotes the set of all compact subsets ($\varnothing$ is compact) of Z with the topology induced from the natural metric of nonempty compact subsets of Z, in $Z^\omega - \{\varnothing\}$. By a direct computation, $M(f_e)$ is of the form

$$M(f_e)(\theta) = \{d_1\theta, \ldots, d_i\theta\}$$

with real numbers $d_1, \ldots, d_i \neq 0$ and the coordinate $\theta = y/x$ on S^1 near $(1,0)$ (see §3 in [**N**]). From this it follows that $M(f_e)$ is continuous in the following sense.

In general, we say a mapping f: $X \to Z^\omega$ from a topological space X to the set Z^ω of compact subsets of a metric space Z, with the natural metric in $Z^\omega - \{\varnothing\}$, is *continuous* if for any open set $U \subset Z^\omega - \{\varnothing\}$, $f^{-1}(U \cup \{\varnothing\})$ is open in X. We say two germs of continuous mappings f, g: $(X, x_0) \to (Z^\omega, \{z_0\})$ are *conjugate* if there are germs of homeomorphisms ϕ: $(X, x_0) \to (X, x_0)$ and ψ: $(Z, z_0) \to (Z, z_0)$ such that $\psi^\omega f = g\phi$, where ϕ^ω: $(Z^\omega, \{z_0\}) \to (Z^\omega, \{z_0\})$ is defined by $\psi^\omega(A) = \psi(A)$ for $A \in Z^\omega$.

Using set-theoretical topology, we can prove the next proposition (for the proof, see [**N**]).

PROPOSITION A. *There is an open subset $U \subset \mathbf{R}^6$ such that, if f_e, $f_{e'}$, e, $e' \in U$, are topologically equivalent then $M(f_e)$, $M(f_{e'})$: $(S^1, (1,0)) \to (Z^\omega, \{0\})$ are conjugate.*

Proposition A is just the key of our construction of topological moduli. The reader may feel strange that $M(f_e)$, defined by using differentiability of f_e, is a topological invariant. The reason is due to the fact that the lack of the A_{f_e}-condition reflects on the topological properties of fibres. Let us explain this more precisely in the following.

We begin by preparation of a notion of $B(X, Y)$. Let $X, Y \subset Z$ be topological spaces. $B(X, Y)$ denotes the set of points $x \in X$ such that for any open neighbourhood U of x in X, there is a point $x' \in U \subset X$ such that two germs (X, Y, x), (X, Y, x') are not topologically equivalent. We apply this notion to our map germs with $X = Z$-axis in $\mathbf{R}^3$, $Y = f_e^{-1}(L_p)$, $Z = \mathbf{R}^3$, where L_p denotes the ray starting from the origin through p. Then we can see that $M(f_e)(p) = B(Z, f_e^{-1}(L_p))$, $p \in S^1$ (Assertions 3.1–3.3 in [**N**]). This relationship holds in general in the complex case but does not hold in general in the real case.

Next, we consider a relation of $M(f_e)$ with topological triviality of f_e. Let us consider the sequence of map germs,

$$(f_e \times \mathrm{id}, \pi): (\mathbf{R}^3 \times \mathbf{R}^6, (0,0)) \xrightarrow{f_e \times \mathrm{id}} (\mathbf{R}^2 \times \mathbf{R}^6, (0,0)) \xrightarrow{\pi} (\mathbf{R}^6, 0),$$

and take a natural stratification (S, S', S'') of the sequence which induces the previous stratification (S, S') to f_e: $(\mathbf{R}^2 \times \{e\}) \to (\mathbf{R}^2 \times \{e\})$ for generic $e \in \mathbf{R}^6$. Roughly speaking, difference of conjugacy classes of $M(f_e)$ along one stratum $U \in S''$ in the parameter space $\mathbf{R}^6$ gives an obstruction to lifting any vector field v on U to an integrable stratified vector field on $\mathbf{R}^3 \times U$ and $\mathbf{R}^3 \times U$ via $(f_e \times \mathrm{id}, \pi)$ (for the terminology, see **[G,M]**). More precisely, v can be lifted to such vector fields if and only if there are diffeomorphisms $\psi_{e,t}$: $(S^1 - (1,0),(1,0)) \to (S^1 - (1,0),(1,0))$ and $\phi_{e,t}$: $(\mathbf{R} - \{0\}, 0) \to (\mathbf{R} - \{0\}, 0)$, $e \in U$, $t \in \mathbf{R}$, with $\psi_{e,0} = 1$ and $\phi_{e,0} = 1$ such that $M(f_e) \circ \psi_{e,t} = (\phi_{e,t})^\omega \circ M(f_{e'})$, $\psi_{e',t'} \circ \psi_{e,t} = \phi_{e,t+t'}$ and $\phi_{e',t'} \circ \phi_{e,t} = \phi_{e,t+t'}$ for $e \in U$ and $t, t' \in \mathbf{R}$, where $e' = \exp(tv)(e)$ and $\psi_{e,t}$, $\phi_{e,t}$ are considered as germs of homeomorphisms on $(S^1,(1,0))$, $(\mathbf{R}, 0)$, respectively, and $(\phi_{e,t})^\omega$: $\mathbf{R}^\omega \to \mathbf{R}^\omega$ is defined by $(\phi_{e,t})^\omega(A) = \phi_{e,t}(A)$ for any compact subset $A \subset \mathbf{R}$.

We return to our problem. Proposition A says that $M(f_e)$ is a topological invariant for f_e, $e \in U$. To see that there are infinitely many different conjugacy classes of $M(f_e)$; it is convenient to define the second invariant, namely convergent ratio, for $M(f_e)$ which is defined in [N], precisely. Recall that $M(f_e)$ is of the form $M(f_e)(\theta) = \{d_1\theta, \dots, d_k\theta\}$ with real numbers d_i, $0 < |d_1| < \cdots < |d_k|$, and the coordinate $\theta = y/x$ on the unit circle $S^1 \subset \mathbf{R}^2$. By Assertions 3.1–3.3 in [N], $k = 4$ for $e \in U \subset \mathbf{R}^6$.

Let $a_{e,i}$, $b_{e,i}$ be convergent sequences of positive numbers defined by $a_{e,0} = a_0$, $b_{e,0} = b_0$, $a_{e,i+1} = |d_1|^2/|d_2|^2 \cdot a_{e,i}$, $b_{e,i+1} = |d_1|^2/|d_3|^2 \cdot b_{e,i}$ with arbitrary a_0, $b_0 > 0$.

For a pair of sequences of positive numbers (a, b) convergent to 0, we define a function $N(a, b)$: $\mathbf{N} \to \mathbf{N}$ which assigns for $n \in \mathbf{N}$ the number $\max\{i \mid a_n < b_i\}$, and we define $K(a, b) = \lim_{n\to\infty} N(a, b)(n)/n$. It is easy to see that $K(a, b)$ is independent of shift of indices of sequences but only on topological behavior of (a, b) near the origin, and we call $K(a, b)$ the convergent ratio of (a, b) (for the details, see [N]). Now we define $K_e = K(a_e, b_e)$. Then K_e is independent of a choice of a_0, b_0.

We assume that $M(f_e)$, $M(f_{e'})$ are conjugate, i.e., there are germs of homeomorphisms ϕ: $(S^1,(1,0)) \to (S^1,(1,0))$, ψ: $(\mathbf{R}, 0) \to (\mathbf{R}, 0)$ such that $M(f_{e'}) \circ \phi = \psi^\omega \circ M(f_e)$. Then, ψ sends the sequences a_e, b_e to the sequences $a_{e'}$, $b_{e'}$ with $a_{e',0} = \psi(a_{e,0})$, $b_{e',0} = \psi(b_{e,0})$. So it follows that $K_e = K_{e'}$. So, combining this with Proposition A, we see that K_e is a topological invariant for f_e, $e \in U$. By a direct computation we can show that k_e takes infinitely many different values for $e \in U$, so it follows that f_e, $e \in U$, takes infinitely many different topological types.

Next, we consider the map germs g_a. It is easy to see that $g_a^{-1}(0)$ is of dimension 1 and nonsingular fibres are of dimension 0 for generic $a \in \mathbf{R}^9$. So, g_a is neither flat nor admits any Thom regular stratification. We put a similar argument to g_a as f_e. First we define $M(g_a)(p)$, $p \in S^2$, by the set of points $z \in Z$-axis such that there is a sequence $x_i \in \mathbf{R}$, $x_i \to z$, $g_a(x_i) \notin g_a(\Sigma(g_a))$, $0\widehat{g_a}(x_i) \to L_p$, where L_p is the ray from the origin in $\mathbf{R}^3$ through $p \in S^2 \subset \mathbf{R}^3$.

Then $M(g_a)$ defines a continuous map germ $M(g_a)$: $(S^2, (1,0,0)) \to (\mathbf{R}^\omega, \{0\})$, and we have the following proposition (for the proof, see [N]).

PROPOSITION B. *There is an open set $U \subset \mathbf{R}^9$ and if g_a, $g_{a'}$, a, $a' \in U$, are topologically equivalent then the associated map germs $M(g_a)$, $M(g_{a'})$ are conjugate.*

To see that there are infinitely many different conjugacy classes of $M(g_a)$ for $a \in U$, we need a more delicate argument using the continuity of $M(g_a)$, which can be found in [N]. By this, the family g_a, $a \in U$, has infinite topological types.

For the complex case, we put a similar argument and prove the same result as the real case. Our theorem follows immediately from our examples f_e, g_a.

We remark here a relationship between our definition of $M(f_e)$, $M(g_a)$ and some concept in algebraic geometry. One who knows an elementary notion of blowing up and strict transformation in algebraic geometry (see [H]) will find that our map germs $M(f_e)$, $M(g_a)$ are closely related to strict transformations of f_e, g_a with respect to the blowing ups at the origin in the target spaces. One approach from this direction can be found in [$\mathbf{S_1}$, $\mathbf{S_2}$].

Last of all, we state open parts of our problem.

First, it is a surprising fact that it is not known at all whether or not $P(n, p, 2; \mathbf{K})/\text{top}$, $P_0(n, p, 2; \mathbf{K})/\text{top}$ are finite sets, nor what topological types polynomial map germs of degree 2 have, except for the case $p = 1$.

Secondly, it is not known whether $P(n, 2, 3; \mathbf{K})/\text{top}$, $P_0(n, 2, 3; \mathbf{K})/\text{top}$ are finite sets or not, for any $n \geqslant 3$.

REFERENCES

[A] K. Aoki and H. Noguchi, *On topological types of map germs of plane to plane*, Mem. School of Science & Engineering, Waseda University, No. 44, 1980, pp. 133–156.

[F] T. Fukuda, *Types topologiques des polynômes*, Inst. Hautes Etudes Sci. Publ. Math. **46** (1976), 87–106.

[G] C. G. Gibson et al., *Topological stability of smooth mappings*, Lecture Notes in Math., vol. 552, Springer-Verlag, Berlin and New York, 1976.

[H] H. Hironaka, *Stratifications and flatness*, Real and Complex Singularities, Oslo, 1976, Noordhoff, Groningen, 1977, pp. 199–265.

[M] J. Mather, *Stratifications and mappings*, Dynamical Systems (M. M. Peixoto, Editor), Academic Press, New York, 1973, pp. 195–232.

[N] I. Nakai, *On topological types of polynomial map germs*, preprint.

[$\mathbf{S_1}$] C. Sabbah, *Morphismes analytiques stratifiés sans éclatement et cycles évanescents*, preprint.

[$\mathbf{S_2}$] ______, *Le types topologique éclate d'une application analytique*, preprint.

[T] R. Thom, *La stabilité topologique des applications polynomiales*, Enseignement Math. **8** (1962), 24–33.

KYOTO UNIVERSITY, JAPAN

Proceedings of Symposia in Pure Mathematics
Volume **40** (1983), Part 2

ABELIAN COVERS OF QUASIHOMOGENEOUS SURFACE SINGULARITIES

WALTER D. NEUMANN[1]

Throughout this article (V, p) will be a quasihomogeneous surface singularity, that is, V is a complex surface with a normal singularity at $p \in V$ with a good $\mathbf{C}^*$-action. "Good" means that p is in the closure of each $\mathbf{C}^*$-orbit (the only (V, p) which admit nongood $\mathbf{C}^*$-actions are cyclic quotient singularities).

Denote $V_0 = V - \{p\}$. The orbit space $V_0/\mathbf{C}^*$ is a complex curve X. In X there are just a finite number of points, $x_1, \ldots, x_n$, say, corresponding to $\mathbf{C}^*$-orbits in V_0 with nontrivial isotropy $\mathbf{Z}/\alpha_1, \ldots, \mathbf{Z}/\alpha_n$, say.

THEOREM 1. *Suppose* (V, p) *is as above with* X *rational (we say then* (V, p) *is quasirational). Let* $(\tilde{V}^{\mathrm{ab}}, p)$ *be the universal abelian cover of* (V, p) *branched at* p. *Then, if* $n \leq 2$, $(\tilde{V}^{\mathrm{ab}}, p) \cong (\mathbf{C}^2, 0)$, *and otherwise* $(\tilde{V}^{\mathrm{ab}}, p) \cong (V_A(\alpha_1, \ldots, \alpha_n), 0)$, *where* $V_A(\alpha_1, \ldots, \alpha_n)$ *is the Brieskorn complete intersection*

$$V_A(\alpha_1, \ldots, \alpha_n) = \{z \in \mathbf{C}^n \mid a_{i1} z_1^{\alpha_1} + \cdots + a_{in} z_n^{\alpha_n} = 0,\ i = 1, \ldots, n-2\}$$

for some coefficient matrix $A = (a_{ij})$. *Such an* A *can be determined as follows: there exist unique* $\lambda_1, \ldots, \lambda_{n-3} \in \mathbf{C} - \{0, 1\}$ *for which*

$$(C \cup \infty; \lambda_1, \ldots, \lambda_{n-3}, 1, 0, \infty) \cong (X, x_1, \ldots, x_n)$$

(analytic isomorphism), and then

$$A = \begin{pmatrix} 1 & \cdots & 0 & 0 & 1 & \lambda_1 \\ \vdots & \ddots & & \vdots & \vdots & \vdots \\ 0 & & 1 & 0 & 1 & \lambda_{n-3} \\ 0 & \cdots & 0 & 1 & 1 & 1 \end{pmatrix}$$

is suitable.

1980 *Mathematics Subject Classification*. Primary 32C40.

[1] Research partially supported by the NSF.

0082-0717/82/0000-0168/$03.50

EXAMPLE. Let $\Gamma \subset \mathrm{PSL}(2, \mathbf{R})$ be a Fuchsian group with $\mathbf{H}/\Gamma$ compact, $\mathbf{H}$ the complex upper half-plane. Then Γ acts also on the complex tangent bundle $T\mathbf{H}$ of $\mathbf{H}$. In the orbit space the zero section $\mathbf{H}/\Gamma \subset T\mathbf{H}/\Gamma$ can be blown down to give a quasihomogeneous singularity (V, p), which is quasirational if and only if $\mathbf{H}/\Gamma$ is rational. In this case the theorem was proved by the author [**8**] and independently by Dolgachev [**2**], by computing an appropriate ring of automorphic forms for Γ. Special cases, as well as the analogous result for finite $\Gamma \subset \mathrm{SU}(2)$, had been done by F. Klein [**3, 4**] and J. Milnor [**6**].

For a given collection of data $(X; x_1, \ldots, x_n; \alpha_1, \ldots, \alpha_n)$, as in the theorem, there is a countable infinity of possible isomorphism types for (V, p). The examples just discussed give exactly one (V, p) for each such collection of data, except for the cases $(\alpha_1, \ldots, \alpha_n) = (2, 3, 6), (2, 4, 4), (3, 3, 3)$, or $(2, 2, 2, 2)$. In [**9**] we prove directly that two (V, p)'s with the same $(X; x_1, \ldots, x_n; \alpha_1, \ldots, \alpha_n)$ have the same universal abelian covers, so the theorem then follows, except for the four cases just mentioned. These cases admit easy direct proofs, since their universal abelian covers can be identified as suitable cones on elliptic curves. One is then done. However, the theorem admits a much more elementary proof, which was obscured by the special nature of the examples described above. It is this proof which we present here, plus some simple applications.

Here is our plan of action. If $n \leqslant 2$ (notation as in the theorem) then (V, p) is a cyclic quotient singularity (see for instance [**12** or **9**]) so the theorem is clear. So assume $n \geqslant 3$. Since (V, p) is determined by $V_0 = V - \{p\}$, we must just show that the universal abelian cover of V_0 is isomorphic to $V_A(\alpha_1, \ldots, \alpha_n)_0$. Denote $H = H_1(V_0; \mathbf{Z})$. The universal abelian cover $\tilde{V}_0^{\mathrm{ab}} \to V_0$ is the unique connected regular covering with covering transformation group H. We must thus just exhibit a free holomorphic H-action on $V_A(\alpha_1, \ldots, \alpha_n)_0$ with orbit space V_0. We do this in §2, and prove it in §3.

In §4 we apply this to a discussion of the coordinate ring of V and its Poincaré series, the main result being a computation of the equivariant Poincaré series for the universal abelian cover of (V, p) (formulae (4.2) and (4.4)), which codes the Poincaré series for all abelian covers of (V, p).

In §5 we mention some deformations of V resulting from our main theorem in special cases.

1. Topology of (V, p). We retain the notation of the introduction and assume $n \geqslant 3$.

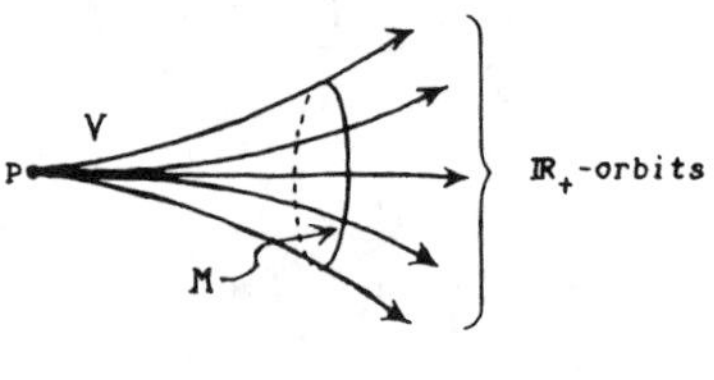

Consider $\mathbf{R}_+ \subset \mathbf{C}^*$ acting on V. The $\mathbf{R}_+$-orbits are rays emanating from p, as in the schematic pictured. We can thus identify $M = V_0/\mathbf{R}_+$ with the link of the singularity. Since $\mathbf{C}^* = \mathbf{R}_+ \times S^1$, we have $M/S^1 = V_0/\mathbf{C}^* = X$, and $\pi\colon M \to X$ is a Seifert fibering, whose fibers are the S^1 orbits in M. To such a Seifert fibering is associated a Seifert invariant

$$((\alpha_1, \beta_1), \ldots, (\alpha_n, \beta_n)), \qquad \gcd(\alpha_i, \beta_i) = 1,$$

which is determined as follows (we are using the unnormalized Seifert invariant of [**7**, **10**]).

Consider the points $x_1, \ldots, x_n \in X$ over which the singular fibers of M lie. Let $D_1, \ldots, D_n$ be disjoint disks about $x_1, \ldots, x_n$ in X. Let $X' = X - \operatorname{int}(D_1 \cup \cdots \cup D_n)$ and $M' = \pi^{-1}(X')$. Then $\pi\colon M' \to X'$ is a genuine circle bundle over a connected 2-manifold with boundary, so it admits a section $s\colon X' \to M'$. Let $Q_i = s(\partial D_i)$ and let H_i be a circle fiber in $\pi^{-1}(\partial D_i)$. Let O_i be $\pi^{-1}(x_i)$. Then we have homology relations $H_i \sim \alpha_i O_i$, $Q_i \sim -\beta_i O_i$ in $H_1(\pi^{-1}(D_i))$, which determine the (α_i, β_i).

By changing the choice of the sections we can change each β_i within its residue class modulo α_i, so long as $\Sigma_i(\beta_i/\alpha_i)$ is held constant. $e(M \to X) = -\Sigma_i(\beta_i/\alpha_i)$ is thus an invariant, called the euler number. (For the reader more familiar with the normalized Seifert invariant we mention that the normalized invariant $(b, (\alpha_1, \beta_1), \ldots, (\alpha_n, \beta_n))$ is equivalent to the unnormalized invariant $((\alpha_1, \beta_1 + b\alpha_1), (\alpha_2, \beta_2), \ldots, (\alpha_n, \beta_n))$.)

We recall from [**1**, **12**, **13**, **9**]

PROPOSITION 1.1. (1) *The quasirational quasihomogeneous singularity* (V, p) *is determined by the data*:

(i) *the analytic isomorphism type of* $(X; x_1, \ldots, x_n)$ *and*

(ii) *the Seifert invariant* $((\alpha_1, \beta_1), \ldots, (\alpha_n, \beta_n))$ (*up to the indeterminancy described above*).

(2) *These data occur for some quasirational quasihomogeneous singularity if and only if* $e = -\Sigma(\beta_i/\alpha_i)$ *is negative.*

Since $V_0 \cong M \times \mathbf{R}$, we have $H_1(M; \mathbf{Z}) = H_1(V_0; \mathbf{Z}) = H$. The standard computation (see e.g. [**14**, **11**, **7**, **10**], etc.) gives

$$\pi_1(M) = \pi_1(V_0) = \left\langle q_1, \ldots, q_n, h \mid q_i^{\alpha_i} h^{\beta_i} = 1, q_1 \cdots q_n = 1 \right\rangle$$

so

$$H = \operatorname{ab}\left\langle q_1, \ldots, q_n, h \mid q_i^{\alpha_i} h^{\beta_i} = 1, q_1 \cdots q_n = 1 \right\rangle,$$

where we are writing H multiplicatively for later convenience, and the notation is chosen to suggest the geometric interpretation of the generators. We shall need the following facts about H.

LEMMAS.

$$|H| = \alpha_1 \cdots \alpha_n |e|. \tag{1.2}$$

$$|\langle h \rangle| = \alpha |e|, \quad \textit{where } \alpha = \operatorname{lcm}(\alpha_1, \ldots, \alpha_n). \tag{1.3}$$

PROOF OF (1.2). The order of H equals the absolute value of the determinant of the relation matrix

$$|H| = \pm\det\begin{bmatrix} 1 & 1 & \cdots & 1 & 0 \\ \alpha_1 & 0 & & 0 & \beta_1 \\ \vdots & \vdots & & & \\ 0 & 0 & \cdots & \alpha_n & \beta_n \end{bmatrix}.$$

A simple induction shows this determinant is $(-1)^{n+1}\alpha_1 \cdots \alpha_n e$.

PROOF OF (1.3). Let $G(\alpha_i)$ denote the cyclic group of order α_i and denote $K = G(\alpha_1) \times \cdots \times G(\alpha_n)$. Let g_i be the generator of $G(\alpha_i)$. Then $g_1 \times \cdots \times g_n$ generates a cyclic subgroup of order α in K. The map $K \to H/\langle h \rangle$ given by $g_i \mapsto q_i$ gives an exact sequence

$$1 \to \langle g_1 \times \cdots \times g_n \rangle \to K \to H/\langle h \rangle \to 1.$$

Thus $|H/\langle h \rangle| = \alpha_1 \cdots \alpha_n/\alpha$, so $|\langle h \rangle| = \alpha |H|/\alpha_1 \cdots \alpha_n = \alpha |e|$.

2. Restatement of the main theorem. If $f = (f_1,\ldots,f_n)\colon \mathbf{C}^n \to \mathbf{C}^{n-2}$ is given by

$$f_i(z_1,\ldots,z_n) = a_{i1}z_1^{\alpha_1} + \cdots + a_{in}z_n^{\alpha_n}$$

then $V_A(\alpha_1,\ldots,\alpha_n) = f^{-1}(0)$. An elementary computation [5] shows $V_A(\alpha_1,\ldots,\alpha_n)$ is a surface with isolated singularity at the origin if and only if every $(n-2)\times(n-2)$ minor of $A = (a_{ij})$ is nonzero. We shall say briefly that A is *good*.

Let $P_A \subset \mathbf{C}P^{n-1}$ be the linear subspace $P_A = \{[w_1,\ldots,w_n] \in \mathbf{C}P^{n-1} \mid a_{i1}w_1 + \cdots + a_{in}w_n = 0,\ i = 1,\ldots,n-2\}$. The condition that A be good is easily seen to mean:

(a) $P_A \cong \mathbf{C}P^1$;

(b) $P_A \cap \{w_i = 0\}$ is a singleton $\{x_i\}$ for each $i = 1,\ldots,n$;

(c) $x_i \neq x_j$ for $i \neq j$.

For example, if

$$A = \begin{pmatrix} 1 & \cdots & 0 & 0 & 1 & \lambda_1 \\ \vdots & \ddots & \vdots & \vdots & \vdots & \\ 0 & & 1 & 0 & 1 & \lambda_{n-3} \\ 0 & \cdots & 0 & 1 & 1 & 1 \end{pmatrix}$$

($\lambda_i \in \mathbf{C} - \{0,1\}$, $\lambda_i \neq \lambda_j$ for $i \neq j$), then $[w_1,\ldots,w_n] \mapsto -w_{n-1}/w_n$ gives an isomorphism $(P_A; x_1,\ldots,x_n) \to (\mathbf{C} \cup \{\infty\}; \lambda_1,\ldots,\lambda_{n-3}, 1, 0, \infty)$. In fact any good A can be put in this "normal form" by a linear coordinate change in $\mathbf{C}^{n-2}$ and a diagonal one in $\mathbf{C}^n$.

Let $\mathrm{Diag}(n) \subset \mathrm{GL}(n,\mathbf{C})$ be the group of diagonal matrices. The homomorphism $\varphi\colon \mathbf{C}^* \to \mathrm{Diag}(n)$, $\varphi(t) = \mathrm{diag}(t^{\alpha/\alpha_1},\ldots,t^{\alpha/\alpha_n})$ defines the standard effective $\mathbf{C}^*$-action on $\mathbf{C}^n$ preserving $V_A(\alpha_1,\ldots,\alpha_n) \subset \mathbf{C}^n$. We shall denote $\varphi(\mathbf{C}^*)$ by G and for $x \in \mathbf{Q}$ we denote $\varphi(\exp(2\pi i x))$ by $\varepsilon(x) \in G$.

The cyclic group $G(\alpha_j)$ acts on $\mathbf{C}^n$ and on $V_A(\alpha_1,\ldots,\alpha_n) \subset \mathbf{C}^n$ by multiplication by α_jth roots of unity in the jth coordinate. This embeds $K = G(\alpha_1) \times \cdots \times G(\alpha_n)$ in $\mathrm{Diag}(n)$, and we identify K with its image. $K \cap G$ is the cyclic group of order α generated by $g_1 \times \cdots \times g_n = \varepsilon(1/\alpha)$, where $g_j \in G(\alpha_j)$ is a generator.

Let $H = \mathrm{ab}\langle q_1,\ldots,q_n, h \mid q_j^{\alpha_j} h^{\beta_j} = 1,\ q_1 \cdots q_n = 1\rangle$ be as in §1, with $\gcd(\alpha_j, \beta_j) = 1$ for $j = 1,\ldots,n$, and with $e = -\Sigma\,\beta_j/\alpha_j < 0$. Define a homomorphism

$$\chi\colon H \to K \cdot G \subset \mathrm{Diag}(n)$$

by $\chi(q_j) = g_j \cdot \varepsilon(\beta_j/e\alpha\alpha_j)$, $\chi(h) = \varepsilon(-1/e\alpha)$. This gives an H-action on $V_A(\alpha_1,\ldots,\alpha_n)$.

In view of our previous discussion, the following is a more explicit version of our main theorem.

THEOREM 2.1. *H acts freely on $V_A(\alpha_1,\ldots,\alpha_n)_0 = V_A(\alpha_1,\ldots,\alpha_n) - \{0\}$. The orbit space $(V, p) = (V_A(\alpha_1,\ldots,\alpha_n)/H, 0)$ is the quasirational quasihomogeneous surface singularity with classifying data, as in Proposition* 1.1, *as follows*:

$$(P_A;\, x_1,\ldots,x_n;\, (\alpha_1, \beta_1),\ldots, (\alpha_n, \beta_n)).$$

Thus $V_A(\alpha_1,\ldots,\alpha_n)_0$ is the universal abelian cover of V_0.

The $\mathbf{C}^*$-action on $V_A(\alpha_1,\ldots,\alpha_n)/H$ is of course the induced action of

$$G/(\chi(H) \cap G) \cong \mathbf{C}^*.$$

As we show during the proof, $\chi(H) \cap G = \langle \varepsilon(1/\alpha e)\rangle$.

3. Proof. We first show $\chi\colon H \to K \cdot G$ is injective. Indeed,

$$\chi(H)/(\chi(H) \cap G) = (\chi(H)\cdot G)/G = (K \cdot G)/G = K/(K \cap G),$$

which has order $\alpha_1 \cdots \alpha_n/\alpha$. On the other hand $\chi(H) \cap G$ contains the element $\chi(h) = \varepsilon(-1/\alpha e)$, so it has order at least αe. Thus $\chi(H)$ has order at least $(\alpha_1 \cdots \alpha_n/\alpha)\alpha e = \alpha_1 \cdots \alpha_n e = |H|$, so χ is injective. It also follows that $\chi(H) \cap G = \langle \chi(h)\rangle$, as mentioned after Theorem 2.1.

To see that H acts freely on $V_A(\alpha_1,\ldots,\alpha_n)_0$, first note that the condition that the coefficient matrix A be good is equivalent to the statement $V_A(\alpha_1,\ldots,\alpha_n)_0 \subset (\mathbf{C}^n)^0$, where $(\mathbf{C}^n)^0 = \{z \in \mathbf{C}^n \mid z_j = 0 \text{ for at most one } j\}$.

LEMMA 3.1. *H acts freely on $(\mathbf{C}^n)^0$.*

PROOF. The H-action on $\mathbf{C}^n$ is given by $\chi = \mathrm{diag}(\chi_1,\ldots,\chi_n)\colon H \to \mathrm{Diag}(n)$, where $\chi_j\colon H \to \mathbf{C}^*$ is the character

$$\chi_j(q_l) = \begin{cases} \exp 2\pi i\big(\beta_l/e\alpha_l\alpha_j\big), & l \neq j, \\ \exp 2\pi i\big(\beta_j/e\alpha_j^2 + 1/\alpha_j\big), & l = j, \end{cases}$$

$$\chi_j(h) = \exp 2\pi i(-1/e\alpha_j).$$

These characters satisfy the relations

$$\text{(3.2)} \qquad \begin{cases} \chi_1^{\beta_1} \cdots \chi_n^{\beta_n} = 1, \\ \chi_i^{\alpha_i} = \chi_j^{\alpha_j}, \qquad i, j = 1,\ldots,n. \end{cases}$$

Now suppose $g \in H$ fixes a point $z \in (\mathbf{C}^n)^0$. If no coordinate of z is zero then $\chi_j(g) = 1$ for all j, that is $\chi(g) = 1$, so $g = 1$. If some coordinate z_j is zero, then $\chi_i(g) = 1$ for $i \neq j$. Choose $a, b \in \mathbf{Z}$ with $a\alpha_j + b\beta_j = 1$. By (3.2) $\chi_j^{\beta_j}(g) = 1$ and $\chi_j^{\alpha_j}(g) = 1$. Thus $\chi_j(g) = (\chi_j^{\alpha_j}(g))^a(\chi_j^{\beta_j}(g))^b = 1$. Thus again $\chi(g) = 1$, so $g = 1$, completing the proof.

Although we shall not need the following lemma for our main theorem, it is worth remarking.

LEMMA 3.3. *The Pontryagin dual* $\hat{H} = \operatorname{Hom}(H, \mathbf{C}^*)$ *has a presentation in terms of the* χ_j (*with* χ_0 *defined as the common value of* $\chi_j^{\alpha_j}$):

$$\hat{H} = \operatorname{ab}\left\langle \chi_0, \chi_1, \ldots, \chi_n \,|\, \chi_1^{\beta_1} \cdots \chi_n^{\beta_n} = 1, \chi_j^{\alpha_j}\chi_0^{-1} = 1 \right\rangle.$$

PROOF. That the stated relations hold is (3.2). That the χ_j generate $\hat{H}$ is the injectivity of χ. Thus we must just show the group presented as above has the same order as does $\hat{H}$, namely $|H|$. But the above group and H have relation matrices which are mutual transposes, so they have equal orders.

From now on we shall identify H with the subgroup $\chi(H) \subset \operatorname{Diag}(n)$.

Denote $V_A(1,\ldots,1) = \{z \in \mathbf{C}^n \,|\, a_{j1}z_1 + \cdots + a_{jn}z_n = 0,\ j = 1,\ldots,n-2\}$. Then $V_A(\alpha_1,\ldots,\alpha_n)/K \cong V_A(1,\ldots,1)$ by the map $(z_1,\ldots,z_n) \mapsto (z_1^{\alpha_1},\ldots,z_n^{\alpha_n})$. The action of $G \cong \mathbf{C}^*$ on $V_A(\alpha_1,\ldots,\alpha_n)$ projects to an ineffective action on $V_A(1,\ldots,1)$, namely $t \in \mathbf{C}^*$ acts by multiplication by t^α. Made effective, it becomes the induced action of $G_K := G/(K \cap G) \cong \mathbf{C}^*$ on $V_A(1,\ldots,1)$. Thus

$$V_A(\alpha_1,\ldots,\alpha_n)_0/KG = V_A(1,\ldots,1)_0/G_K = P_A.$$

Since $KG = HG$, we get a commutative diagram

$$\begin{array}{ccc} V_A(\alpha_1,\ldots,\alpha_n)_0 & \xrightarrow{/K} & V_A(1,\ldots,1)_0 \\ \downarrow /H & & \downarrow /G_K \\ V_0 & \xrightarrow{/G_H} & P_A \end{array}$$

where $G_H = G/(H \cap G) = G/\langle \varepsilon(1/\alpha e)\rangle \cong \mathbf{C}^*$.

Denote the map $V_0 \to P_A$ by π. If $x = [w_1,\ldots,w_n] \in P_A$, then the isotropy of the orbit $\pi^{-1}(x)$ depends only on which coordinates w_i are zero. Thus the only possible orbits with nontrivial isotropy are $\pi^{-1}(x_1),\ldots,\pi^{-1}(x_n)$. We must next show that $\pi^{-1}(x_j)$ has isotropy of order α_j.

For simplicity we take $j = 1$. Let $V_1 = \{z \in V_A(\alpha_1,\ldots,\alpha_n) \,|\, z_1 = 0\}$, so $V_1/H = \pi^{-1}(x_1)$. Let H_1, K_1, G_1 be the images of H, K, and G in the quotient $\operatorname{Diag}(n-1)$ of $\operatorname{Diag}(n)$ which forgets the first component, so H, K, and G act as H_1, K_1, G_1 on V_1. By Lemma 3.1, H_1 is isomorphic to H so we may identify them. Let $p\colon G \to G_1$ be the projection. The isotropy of $G_H = G/(H \cap G) = G/\langle \varepsilon(1/\alpha e)\rangle$ acting on $\pi^{-1}(x_1) = V_1/H$ is $p^{-1}(H_1 \cap G_1)/\langle \varepsilon(1/\alpha e)\rangle$, so we want to show $p^{-1}(H_1 \cap G_1) = \langle \varepsilon(1/\alpha_1\alpha e)\rangle$. Choose $a, b \in \mathbf{Z}$ with $a\alpha_1 + b\beta_1 = 1$. Then $p(\varepsilon(1/\alpha_1\alpha e)) = h^{-a}q_1^b \in H_1$, so $\langle \varepsilon(1/\alpha_1\alpha e)\rangle \subseteq p^{-1}(H_1 \cap G_1)$. To show equality it is enough to show the inclusion $\langle h^{-a}q_1^b\rangle \subseteq H_1 \cap G_1$ is an equality.

Now

$$H_1/(H_1 \cap G_1) = H_1G_1/G_1 = K_1G_1/G_1 = K_1/(K_1 \cap G_1)$$
$$= (G(\alpha_2) \times \cdots \times G(\alpha_n))/\langle g_2 \times \cdots \times g_n \rangle.$$

On the other hand $H_1/\langle h^{-a}q_1^b \rangle = \mathrm{ab}\langle q_2,\ldots,q_n \mid q_j^{\alpha_j} = 1, j = 2,\ldots,n;\ q_2 \cdots q_n = 1\rangle$, since the relations $q_1^{\alpha_1}h^{\beta_1} = q_1^b h^{-a} = 1$ imply $q_1 = h = 1$. Thus $H_1 \cap G_1 = \langle h^{-a}q_1^b \rangle$ as claimed.

It remains only to show that the β_j of the Seifert invariant are the β_j with which we have been working. Recall from §1 that the Seifert invariant is computed in terms of a section to the Seifert fibering on the complement of the singular fibers. We can give such a section explicitly. Let

$$(\mathbf{C}^n)' = \{(z_1,\ldots,z_n) \in \mathbf{C}^n \mid z_j \neq 0 \text{ for all } j\},$$
$$R = \Big\{z = (r_1e^{i\theta_1},\ldots,r_ne^{i\theta_n}) \mid r_j > 0, j = 1,\ldots,n;$$
$$\sum_1^n r_j^2 = 1;\ \sum \beta_j\theta_j = 0 \ (\mathrm{mod}\ 2\pi)\Big\}.$$

An elementary computation shows that R is H-invariant and moreover that R intersects any G-orbit in $(\mathbf{C}^n)'$ in an orbit of $\langle \varepsilon(1/\alpha e)\rangle = G \cap H$. Thus the image $S = R/H \subset (\mathbf{C}^n)'/H$ is a section to the action of $G_H = G/(H \cap G)$ on $(\mathbf{C}^n)'/H$. If we intersect with V_0 we thus get a section in V_0 from which we can compute Seifert invariants. For instance, to find the Seifert pair (α_1, β_1') for the orbit $\pi^{-1}(x_1) \subset V_0$, we choose a small disk $B \subset V_0$ transverse to this orbit, so $G_HB \subset V_0$ is a tubular neighborhood of the orbit. Then, by definition, β_1' is the homology class in $H_1(G_HB; \mathbf{Z}) = \mathbf{Z}$ of $S \cap \partial(G_HB)$ with suitable orientation (in §1 we worked with the corresponding Seifert 3-manifold M, but this makes no difference, since $V_0 = M \times \mathbf{R}_+$).

Now $G_HB = GD'/\langle h^{-a}q_1^b \rangle$, for some disk $D' \subset \mathbf{C}^n$ obtained by lifting B. D' is a disk transverse to the coordinate axis $\{z_1 = 0\}$. For any other small disk $D \subset \mathbf{C}^n$ transverse to $\{z_1 = 0\}$, the homology class of $S \cap \partial(GD/\langle h^{-a}q_1^b \rangle)$ in $H_1(GD/\langle h^{-a}q_1^b\rangle; \mathbf{Z})$ will still be β_1' since any two such discs are isotopic. Choose $D = \{(z, 1,\ldots,1) \mid |z| < 1\}$ so

$$GD = \{(t^{\alpha/\alpha_1}z, t^{\alpha/\alpha_2},\ldots,t^{\alpha/\alpha_n}) \mid |z| < 1, t \in \mathbf{C}^*\}.$$

The homology of $GD/\langle h^{-a}q_1^b\rangle$ is represented by the central orbit $z = 0$, on which $h^{-a}q_1^b$ acts as $\varepsilon(1/\alpha_1\alpha e)$, so a cycle representing this homology can be parametrized as

$$O_1 = \{(0, t^{\alpha/\alpha_2},\ldots,t^{\alpha/\alpha_n}) \mid t = e^{i\theta}, \theta_0 \leqslant \theta \leqslant \theta_0 + |2\pi/\alpha_1\alpha e|\},$$

for any θ_0. Consider the curve in GD:

$$\{(t^{e\alpha/\beta_1 + \alpha/\alpha_1}, t^{\alpha/\alpha_2},\ldots,t^{\alpha/\alpha_n}) \mid t = e^{i\theta}, 0 \leqslant \theta \leqslant |2\pi\beta_1/\alpha_1\alpha e|\}.$$

This curve goes from $(1,\ldots,1)$ to $q_1^{\pm 1}(1,\ldots,1)$ in R. It thus projects to a closed curve in $S \cap (GD/\langle h^{-a}q_1^b\rangle)$ (note that $q_1 \in \langle h^{-a}q_1^b\rangle$, in fact $q_1 = (h^{-a}q_1^b)^{\beta_1}$), and it is not hard to verify that it is a parametrization of $S \cap \partial(GD/\langle h^{-a}q_1^b\rangle)$. Since it is clearly homologous to $\pm\beta_1 O_1$, we have shown $\beta_1' = \pm\beta_1$. A careful check of orientations shows $\beta_1' = +\beta_1$, but this is also clear from the fact that the euler number must be negative and thus agree with e. The proof is thus completed.

4. The Poincaré series. We describe the Poincaré series of the graded affine ring of V. This has been computed by Dolgachev, and also follows directly from Theorem 5.1 of Pinkham [**13**], in both cases also in the nonquasirational case. See also Wagreich [**15**]. Our approach only gives the quasirational case but in a way that codes the corresponding information also for any abelian cover of (V, p).

The coordinate ring R of a variety V with $\mathbf{C}^*$-action is naturally graded by $R_n = \{f \in R \mid f(tz) = t^n f(z)\}$. Recall that the Poincaré series of a graded algebra $R = \oplus R_n$, with $\dim R_n < \infty$ for each n, is $p(R) = \Sigma_n \dim(R_n)x^n$. If R also has an action of a group H preserving the grading of R, then for $g \in H$ we can define an equivariant Poincaré series $p(R, g) = \Sigma_n \operatorname{tr}(g \mid R_n)x^n$. This can also be interpreted as a power series in x with coefficients in the character ring of H. As such we write it $p(R, H)$.

Let $R(V_A)$ be the coordinate ring of $V_A = V_A(\alpha_1,\ldots,\alpha_n)$. It is generated by the coordinate functions $z_1,\ldots,z_n$, of degree $\alpha/\alpha_1,\ldots,\alpha/\alpha_n$ respectively, satisfying the relations $f_1(z_1,\ldots,z_n) = 0,\ldots,f_{n-2}(z_1,\ldots,z_n) = 0$, each of degree α. Since this is a complete intersection, a standard computation shows (see for instance [**15**])

$$p(R(V_A)) = (1 - x^\alpha)^{n-2} \prod_{i=1}^{n} (1 - x^{\alpha/\alpha_i})^{-1}. \tag{4.1}$$

Now on V_A we have the action of H described in §2, with $V_A/H = V$. The coordinate ring $R(V)$ is the ring $R(V_A)^H$ of H-invariant functions on V_A. The equivariant version of (4.1) is

$$p(R(V_A), H) = (1 - \chi_0 \cdot x^\alpha)^{n-2} \prod_{i=1}^{n} \left(1 - \chi_i \cdot x^{\alpha/\alpha_i}\right)^{-1}, \tag{4.2}$$

where the characters χ_i are as in Lemma 3.3. Indeed the factor $(1 - x^{\alpha/\alpha_i})^{-1}$ in $p(R(V_A))$, which represents the contribution of the generator $z_i \in R(V_A)$, must be weighted by the character χ_i by which H acts on z_i, while the factors $(1 - x^\alpha)$, which represent the contributions of the relations $f_i = 0$, are weighted by the character χ_0 by which H acts on f_i.

(4.2) gives an element of $\mathbf{C}[[x]][\hat{H}]$, where $\hat{H} = \operatorname{Hom}(H, \mathbf{C}^*)$ is the group of irreducible characters on H. Clearly, $p(R(V)) = p(R(V_A)^H)$ is the coefficient of the trivial character $\mathbf{1}$ in $p(R(V_A), H) \in \mathbf{C}[[x]][\hat{H}]$. Actually, this is not quite true; the grading on $R(V_A)^H = R(V)$ is the one induced from the $\mathbf{C}^*$-action on V_A, which gives the Nth power of the standard $\mathbf{C}^*$-action on V, where $N = \alpha \mid e \mid$ (see §3), so we have

$$p(R(V))(x^N) = \text{coefficient of } \mathbf{1} \text{ in } p(R(V_A), H), \tag{4.3}$$

as a power series in x^N instead of x.

(4.3) can be simplified as follows, as was shown to me by Don Zagier. First write (4.2) as

$$p(R(V_A), H) = (1 - \chi_0 \cdot x^\alpha)^{-2} \prod_{i=1}^{n} \frac{1 - (\chi_i \cdot x^{\alpha/\alpha_i})^{\alpha_i}}{(1 - \chi_i \cdot x^{\alpha/\alpha_i})} \tag{4.4}$$

$$= \sum_{s_0=0}^{\infty} (1 + s_0) \chi_0^{s_0} x^{\alpha s_0} \prod_{i=1}^{n} \sum_{s_i=0}^{\alpha_i - 1} \chi_i^{s_i} x^{(\alpha/\alpha_i)s_i}.$$

The coefficient of $\mathbf{1}x^s$ in this is

$$\begin{cases} \sum (1 + s_0), \quad \text{sum over all } (s_0, \dots, s_n) \text{ satisfying:} \\ 0 \leqslant s_0;\ 0 \leqslant s_i \leqslant \alpha_i - 1;\ s_0 + \dfrac{s_1}{\alpha_1} + \cdots + \dfrac{s_n}{\alpha_n} = \dfrac{s}{\alpha}; \\ \chi_0^{s_0} \chi_1^{s_1} \cdots \chi_n^{s_n} = \mathbf{1}. \end{cases} \tag{4.5}$$

Now, by Lemma 3.3, the only relations in $\hat{H}$ are of the form

$$\chi_0^{-\mu_1 - \cdots - \mu_n} \chi_1^{l\beta_1 + \mu_1\alpha_1} \cdots \chi_n^{l\beta_n + \mu_n\alpha_n} = \mathbf{1},$$

with $l, \mu_1, \dots, \mu_n \in \mathbf{Z}$. With this notation, in (4.5) we can write $s_0 = -\Sigma \mu_i$, $s_i = l\beta_i + \mu_i\alpha_i$, and we are summing over all $l, \mu_1, \dots, \mu_n$ satisfying $0 \leqslant s_0 = -\Sigma \mu_i$, $0 \leqslant l\beta_i + \mu_i\alpha_i \leqslant \alpha_i - 1$, $l\beta_1/\alpha_1 + \cdots + l\beta_n/\alpha_n = s/\alpha$, or equivalently,

$$0 \leqslant s_0 = -\sum \mu_i; \quad \mu_i = -\left[\frac{l\beta_i}{\alpha_i}\right]; \quad s = -e\alpha l = Nl.$$

Thus there is at most one possible $(s_0, \dots, s_n)$ for given s, and the coefficient of $\mathbf{1} \cdot x^s$ is

$$\begin{cases} 0 & \text{if } s \not\equiv 0 \pmod{N}; \\ \max(0, 1 + s_0) & \text{if } s = lN, \text{ with } s_0 = \sum_{i=1}^{n} \left[\dfrac{l\beta_i}{\alpha_i}\right]. \end{cases}$$

Thus by (4.3) we have finally

$$p(R(V)) = \sum_{l=0}^{\infty} \max\left(0, 1 + \sum \left[\frac{l\beta_i}{\alpha_i}\right]\right) \cdot x^l. \tag{4.6}$$

This agrees with the results of Pinkham and Dolgachev already mentioned.

An elementary calculation shows

$$1 + \sum_{i=1}^{n} \left[\frac{l\beta_i}{\alpha_i}\right] = -el + \frac{\chi}{2} + b_l,$$

where $\chi = 2 - \Sigma_{i=1}^{n}(\alpha_i - 1)/\alpha_i$ and b_l is a periodic function of l with zero average. It follows that

$$p(R(V)) = \frac{-ex}{(1 - x)^2} + \frac{\chi/2}{(1 - x)} + \frac{P}{Q}, \tag{4.7}$$

with $P, Q \in \mathbf{Q}[x]$, and Q cyclotomic, $Q(1) \neq 0$. It is not hard to give an explicit formula for P/Q, but it is not very instructive, so we omit the derivation. The precise result is that $p(R(V))$ is the sum of all but the (finitely many) terms with negative coefficient in the power series

$$\frac{-ex}{(1-x)^2}+\frac{\chi/2}{(1-x)}+\sum_{i=1}^{n}\sum_{\zeta^{\alpha_i}=1,\,\zeta\neq 1}\frac{1}{\alpha_i(1-\zeta)}\cdot\frac{1}{1-\zeta^{\beta_i}x}.$$

In a similar way one can compute the coefficient in $p(R(V_A), H)$ of any character $\chi\in\hat{H}$. Moreover if (V', p) is any abelian cover of (V, p), classified by the subgroup $H'\subset H$ say, then in analogy to (4.3) we have

(4.8) $$p(R(V'))(x^{N'}) = \text{coefficient of } \mathbf{1} \text{ in } p(R(V_A), H'),$$

where $N'=|H'\cap G|$ and $p(R(V_A), H')$ is just $p(R(V_A), H)$ with all the characters restricted to H'. By the orthogonality relations for characters, this can also be written

(4.8)′ $$p(R(V'))(x^{N'})=\frac{1}{|H'|}\sum_{g\in H'}p(R(V_A), g).$$

5. Deformations. Our main theorem leads to interesting examples of deformations of (V, p) in some cases. Namely suppose the character χ_0 is trivial (this is equivalent to requiring that $e=-1/\alpha$). Then, in the notation of §2, $f\colon \mathbf{C}^n\to\mathbf{C}^{n-2}$ induces a map $g\colon \mathbf{C}^n/H\to\mathbf{C}^{n-2}$ which is a deformation of V. In fact $g^{-1}(0)=V$, while a generic fiber $g^{-1}(w)$ has, for each $1\leqslant i<j\leqslant n$, exactly $\alpha/\mathrm{lcm}(\alpha_i,\alpha_j)$ cyclic quotient singularities of order $\gcd(\alpha_i,\alpha_j)$, and has no other singularities. More generally, the same observation yields deformations of the covering of (V, p) corresponding to $\operatorname{Ker}\chi_0$, with only cyclic quotient singularities in the generic fiber.

References

1. P. E. Conner and F. Raymond, *Holomorphic Seifert fibrations*, Proc. Conf. on Transformation Groups, Amherst 1973, Lecture Notes in Math., vol. 299, Springer-Verlag, Berlin and New York, 1974, pp. 124–204.

2. I. V. Dolgachev, Notes in preparation, see also: *Automorphic forms and quasihomogeneous singularities*, Funkcional Anal. i Priložen **9** (2) (1975), 67–68.

3. F. Klein, *Lectures on the icosahedron and the solution of equations of the fifth degree* (1888), reprint, Dover, New York, 1956.

4. F. Klein and R. Fricke, *Vorlesungen über die Theorie der elliptischen Modulfunktionen*, Teubner, Leipzig, 1890.

5. H. Hamm, *Exotische Sphären als Ungebungsränder in speziellen komplexen Räumen*, Math. Ann. **197** (1972), 44–56.

6. J. Milnor, *On the 3-dimensional Brieskorn manifold $M(p, q, r)$*, Papers Dedicated to the Memory of R. H. Fox, Ann. of Math. Studies, No. 84, Princeton Univ. Press, Princeton, N. J., 1975, pp. 175–225.

7. W. D. Neumann, *S^1-actions and the α-invariant of their involutions*, Bonn. Math. Schr. **44** (1971).

8. ______, *Brieskorn complete intersections and automorphic forms*, Invent. Math. **42** (1977), 285–293.

9. ______, *Geometry of quasihomogeneous surface singularities*, these PROCEEDINGS.

10. W. D. Neumann and F. Raymond, *Seifert manifolds, μ-invariant, and orientation reversing maps*, Proc. Alg. and Geom. Topology (Santa Barbara 1977), Lecture Notes in Math., vol. 664, Springer-Verlag, Berlin and New York, 1978, pp. 162–196.

11. P. Orlik, *Seifert manifolds*, Lecture Notes in Math., vol. 291, Springer-Verlag, Berlin and New York, 1972.

12. P. Orlik and P. Wagreich, *Algebraic surfaces with k^*-action*, Acta Math. **138** (1977), 43–81.

13. H. Pinkham, *Normal surface singularities with* $\mathbf{C}^*$*-action*, Math. Ann. **227** (1977), 183–193.

14. H. Seifert, *Topologie dreidimensionaler gefaserter Räume*, Acta Math. **60** (1933), 147–238.

15. P. Wagreich, *The structure of quasi-homogeneous singularities*, these PROCEEDINGS.

UNIVERSITY OF MARYLAND

Proceedings of Symposia in Pure Mathematics
Volume **40** (1983), Part 2

GEOMETRY OF QUASIHOMOGENEOUS SURFACE SINGULARITIES

WALTER D. NEUMANN[1]

This is a slightly expanded version of my talk, the first part of which was purely expository, describing the analytic classification of normal quasihomogeneous surface singularities (from now on QH-singularity for short, our base field is **C**). This classification is well known; the earliest version seems to be Conner and Raymond [**CR**, especially §13], and the most explicit probably Pinkham [**P1**, Theorem 2.1], who bases his version on Orlik and Wagreich's in [**OW**] (see also [**W2**]). Our approach is different from either, and seemed worth reproducing here. It is in terms of "Seifert line bundles" on complex curves, a concept easily generalizable to higher dimensional base spaces and maybe of use in other contexts.

The second part of the talk, and the part to which the title refers, is new material. The main result is a natural "geometric structure" on the link M of any QH-singularity (V, p), such that the geometric structure on M determines the analytic structure on (V, p) and vice versa, in a one-one fashion.

The archetypal example of this is due to Klein [**Kl1**]. Let $I \subset \mathrm{SO}(3)$ be the icosahedral group, that is the group of orientation preserving symmetries of an icosahedron. The universal cover of SO(3) is the group SU(2), also describable as the group S^3 of unit quaternions, and if we lift I to S^3 we get the binary icosahedral group $I' \subset S^3$ of order 120. Klein gave a homeomorphism

$$I' \backslash S^3 \cong \Sigma(2,3,5)$$

where $\Sigma(2,3,5) = V(2,3,5) \cap S^5$ with $V(2,3,5) = \{z \in \mathbf{C}^3 \mid z_1^2 + z_2^3 + z_3^5 = 0\}$. That is, the link $\Sigma(2,3,5)$ of the QH-singularity $(V(2,3,5), 0)$ has the geometric structure $I' \backslash S^3$. Klein gave other examples, including a treatment for $V(2,3,7)$ in [**Kl2**] and [**KF**], see also Brieskorn [**B**]. These results were generalized by Milnor [**M**] to arbitrary $V(a_1, a_2, a_3)$ and by Dolgachev [**D**] and the author [**N1**] to

1980 *Mathematics Subject Classification*. Primary 32C40, 32L05, 53C30.
[1] Partially supported by the National Science Foundation.

complete intersections of $n-2$ copies of the n-dimensional Brieskorn variety $V(a_1,\ldots,a_n)$ in general position. The present result generalizes all of these, without however, giving as explicit a description of the geometric structure as was possible in these special examples. In the general case of rational base curve these more explicit results still hold however, see §7 and [**N4**].

Returning to $V(2,3,5)$, we should say that Klein's formulation is an analytic isomorphism between $I'\backslash\mathbf{C}^2$ and $V(2,3,5)$. Our general correspondence is analogous. Namely, the relevant geometries are certain simply connected 3-dimensional homogeneous spaces $\mathbf{X}=G/K$. To each of them we give a G-invariant complex structure on $\mathbf{X}\times\mathbf{R}_+$. To a geometric structure $M=\Pi\backslash\mathbf{X}$ on a singularity link M is thus associated a complex structure $\Pi\backslash\mathbf{X}\times\mathbf{R}_+$ on $M\times\mathbf{R}_+$, and this is then $V-\{p\}$.

The result is slightly more general than stated in this introduction. Also we give an analogous result for cusp singularities and singularities with dual resolution graphs of the form

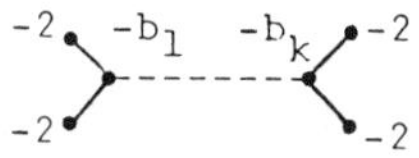

Since in these cases both the geometric structure on the link, and the singularity itself are rigid, this is of more philosophical interest. No other singularity links M^3 admit geometric structures.

The proofs of the geometric results will appear in detail in [**N3**]. Here we just sketch them in a typical case (§6), using the analytic classification of the first part of this paper.

1. Topology of a QH-singularity. By a QH-singularity we mean a normal surface singularity (V,p) with a good $\mathbf{C}^*$-action. "Good" means p is in the closure of every $\mathbf{C}^*$-orbit. Denote $V_0=V-\{p\}$. $V_0/\mathbf{C}^*$ is a complex curve, which we denote X. Consider $\mathbf{R}_+\subset\mathbf{C}^*$ acting on V. After renormalizing by the automorphism $t\mapsto t^{-1}$ of $\mathbf{C}^*$ if necessary, the $\mathbf{R}_+$-orbits are rays emanating from p, as in the schematic picture.

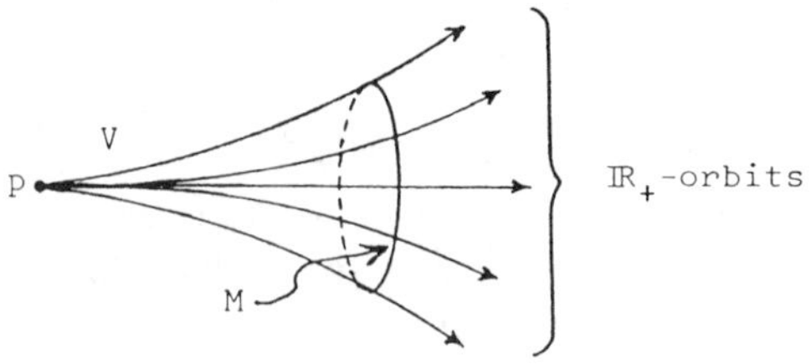

We can thus identify $V_0/\mathbf{R}_+$, with the link M of the singularity. Since $\mathbf{C}^*=\mathbf{R}_+\times S^1$, we have $M/S^1=V_0/\mathbf{C}^*=X$, and the map $M\to X$ is a Seifert fibration of M whose fibers are the S^1-orbits of M.

The topology of V_0, and hence also (V, p), is determined by M, since V_0 is $\mathbf{C}^*$-equivariantly homeomorphic to $M \times \mathbf{R}_+$. M itself is determined (up to S^1-equivariant orientation preserving diffeomorphism) by its Seifert-invariant

$$\{g; b; (\alpha_1, \beta_1), \ldots, (\alpha_n, \beta_n)\}.$$

Here $g = \text{genus}(X)$, and each pair (α_i, β_i) satisfies $0 < \beta_i < \alpha_i$ and $\gcd(\alpha_i, \beta_i) = 1$, and it codes the topology near a singular orbit of the S^1-action on M. The invariant

$$e(M \to X) = -b - \sum \beta_i/\alpha_i$$

generalizes the usual euler number of a nonsingular S^1-bundle.

PROPOSITION ([**NR**, §5; **P1**, §2]). *M occurs as the link of a QH-singularity if and only if $e(M \to X) < 0$.*

A sharper result is in [**N2**, Corollary 6]. This proposition completes the classification of topological types of QH-singularities.

We are using the orientation conventions for Seifert-invariants most prevalent in the literature. A reversal of orientation replaces β_i by $\alpha_i - \beta_i$, b by $-b - r$, and these values are sometimes used instead, particularly in the context of resolution of singularities, where they occur naturally. This is true for instance in [**P1**], quoted above.

2. Analytic classification. With notation as in §1, let x_i denote the point in X over which the (α_i, β_i) orbit lies.

THEOREM ([**CR**, **OW**, **P1**]). *The set of isomorphism types of QH-singularities (V, p) with fixed Seifert-invariant and fixed analytic type of $(X, x_1, \ldots, x_n)$ is isomorphic to* $\mathrm{Jac}(X)$ *modulo an action of* $\mathrm{Aut}(X, x_1, \ldots, x_n)$.

We just sketch our argument; the details are not hard to fill in. Let $\overline{V} = V_0 \times_{\mathbf{C}^*} \mathbf{C}$, that is, $V_0 \times \mathbf{C}$ factored by the $\mathbf{C}^*$-action $t(v, z) = (tv, t^{-1}z)$. Writing $\overline{V} = (V_0 \times_{\mathbf{C}^*} \mathbf{C}^*) \cup (V_0 \times_{\mathbf{C}^*} \{0\}) = V_0 \cup X$, we see that $\overline{V}$ is obtained from V_0 by adding a 0 to each $\mathbf{C}^*$ orbit. We call $X \subset \overline{V}$ the *zero section*. The projection $V_0 \times \mathbf{C} \to V_0$ induces $\overline{V} \to V_0/\mathbf{C}^* = X$ which is a "Seifert line bundle", that is, a complex line bundle except that for each i the fiber over x_i is a singular fiber of the form $\mathbf{C}/\mu_{\alpha_i}$, where μ_{α_i} is the group of α_ith roots of unity.

We shall prove the theorem by classifying Seifert line bundles, so we digress to explain how the Seifert line bundle $E = (\overline{V}, X)$ determines (V, p). We can form $\overline{V} = V_0 \times_{\mathbf{C}^*} \mathbf{C}$ for any Seifert $\mathbf{C}^*$-bundle $V_0 \to X$, not just complements of QH-singular points, so Seifert $\mathbf{C}^*$-bundles and Seifert line bundles are equivalent concepts. We define the Seifert-invariant of such a bundle to be the Seifert-invariant of the associated Seifert fibered $M^3 = V_0/\mathbf{R}_+$, and the euler number $e(V_0 \to X) = e(\overline{V} \to X)$ to be $e(M \to X)$. The latter equals the self-intersection number $X \cdot X$ of the zero section $X \subset \overline{V}$ ($\overline{V}$ is a $\mathbf{Q}$-homology manifold and so intersection numbers can be defined $\mathbf{Q}$-Poincaré dual to cup product). By

Grauert's criterion [G], which can be generalized to this situation, we can blow down $X \subset \overline{V}$ to get a normal singularity (V, p) if and only if $X \cdot X < 0$. $\overline{V}$ is in fact the first stage in a resolution of (V, p). It has only cyclic quotient singularities, at the points $x_i \in X \subset \overline{V}$, and these resolve by linear configurations of exceptional curves to give the familiar star shaped resolution configuration for (V, p).

Let $\mathrm{Pic}(X, (x_1, \alpha_1), \ldots, (x_n, \alpha_n))$ be the set of isomorphism classes of Seifert line bundles $E = (\overline{V} \to X)$ which have singular fibers only over $\{x_1, \ldots, x_n\}$ and which are locally isomorphic, near x_i, to the following:

$$(D \times \mathbf{C})/\mu_{\alpha_i}^{\beta_i} \to D/\mu_{\alpha_i} \cong D,$$

for some β_i with $0 \leqslant \beta_i < \alpha_i$. Here D denotes $\{z \in \mathbf{C} \mid |z| < 1\}$ and μ_α^β denotes the αth roots of unity μ_α acting on $D \times \mathbf{C}$ by $t(z_1, z_2) = (tz_1, t^{-\beta} z_2)$. We do not require $\gcd(\alpha_i, \beta_i) = 1$. Thus the Seifert-invariant will have the form $(g; b, (\alpha'_1, \beta'_1), \ldots, (\alpha'_n, \beta'_n))$, where $\alpha'_i = \alpha_i / \gcd(\alpha_i, \beta_i)$, $\beta'_i = \beta_i / \gcd(\alpha_i, \beta_i)$, and pairs $(\alpha'_i, \beta'_i) = (1, 0)$ refer to nonsingular fibers and may be deleted.

Tensor product $\otimes$ is defined for Seifert line bundles (do it locally in the cyclic cover and then factor by μ_{α_i}, or globally: use Seifert $\mathbf{C}^*$-bundles and form $V_0 \oplus V'_0$ by a pullback, factor $V_0 \oplus V'_0$ by the C^*-action $t(v, v') = (tv, t^{-1}v')$, and then normalize to get $V_0 \otimes V'_0$). This gives a semigroup structure on $\mathrm{Pic}(X, \{(x_i, \alpha_i)\})$ such that β_i: $\mathrm{Pic}(X, \{(x_i, \alpha_i)\}) \to \mathbf{Z}/\alpha_i$ is a homomorphism. The kernel of $\beta = \beta_i \times \cdots \times \beta_n$: $\mathrm{Pic}(X, \{(x_i, \alpha_i)\}) \to \Pi \mathbf{Z}/\alpha_i$ consists of line bundles with no singular fibers, that is, $\mathrm{Ker}(\beta) = \mathrm{Pic}(X)$, which is a group. Hence, $\mathrm{Pic}(X, \{(x_i, \alpha_i)\})$ is a group and moreover, its identity component will be $\mathrm{Pic}_0(X) = \mathrm{Jac}(X)$. Denote $\mathrm{Pic}(X, \{(x_i, \alpha_i)\})/\mathrm{Jac}(X)$ by H. We obtain

$$\begin{array}{ccccccccc}
 & & & & 0 & & 0 & & \\
 & & & & \downarrow & & \downarrow & & \\
0 & \to & \mathrm{Jac}(X) & \to & \mathrm{Pic}(X) & \xrightarrow{d} & \mathbf{Z} & \to & 0 \\
 & & \| & & \downarrow & & \downarrow i & & \\
0 & \to & \mathrm{Jac}(X) & \to & \mathrm{Pic}(X, \{(x_i, \alpha_i)\}) & \xrightarrow{c} & H & \to & 0 \\
 & & & & \downarrow & & \downarrow & & \\
 & & & & \Pi \mathbf{Z}/\alpha_i & = & \Pi \mathbf{Z}/\alpha_i & & \\
 & & & & \downarrow & & \downarrow & & \\
 & & & & 0 & & 0 & &
\end{array}$$

with exact rows and columns, where the top row is classical, with d equal to degree or chern class, and the right-hand column is induced by the rest of the diagram. The projection map c should be thought of as a *chern class for Seifert bundles*. By the following lemma it is an abstract version of the Seifert-invariant.

LEMMA. (i) *$H = \langle g_0, g_1, \ldots, g_n \mid \alpha_i g_i = g_0 \rangle$ with $i\colon \mathbf{Z} \to H$ given by $i(1) = g_0$. In particular any $g \in H$ can be uniquely written in the form $g = bg_0 + \Sigma \beta_i g_i$ with $0 \leqslant \beta_i < \alpha_i$.*

(ii) *If $E \in \mathrm{Pic}(X, \{(x_i, \alpha_i)\})$ has $c(E) = -(bg_0 + \sum \beta_i g_i)$ with $0 \leqslant \beta_i < \alpha_i$ then its Seifert-invariant is $(g; b; (\alpha_1', \beta_1'), \ldots, (\alpha_n', \beta_n'))$ with $\alpha_i' = \alpha_i/\gcd(\alpha_i, \beta_i)$, $\beta_i' = \beta_i/\gcd(\alpha_i, \beta_i)$.*

The content of this lemma is a formula for the Seifert-invariant for $\overline{V} \otimes \overline{V}'$ in terms of the Seifert-invariants of $\overline{V}$ and $\overline{V}'$. It follows easily from the definition of the Seifert-invariant of $\overline{V}$ in terms of a nonzero continuous section to $\overline{V}$ over $X - \{x_i\}$ (see [**NR**]) and the observation that such sections in $\overline{V}$ and $\overline{V}'$ give one in $\overline{V} \otimes \overline{V}'$. The lemma can also be proved using only naturality properties of the euler number for nonsingular bundles to give an independent introduction to the Seifert-invariant (exercise).

The exact sequence

$$0 \to \mathrm{Jac}(X) \to \mathrm{Pic}(X, \{(x_i, \alpha_i)\}) \to H \to 0$$

is the desired classification of Seifert bundles. This is also the form in which it was proved, by very different methods, in [**CR**]. In [**CR**], H arises as the second cohomology group $H^2(\Gamma)$ of a certain Fuchsian or euclidean group Γ (unless (V, p) is a quotient singularity). The connection will become clear in §6.

3. Geometry of 3-manifolds. A *geometric structure* on a manifold M shall mean a complete locally homogeneous riemannian metric of finite volume. Locally homogeneous means any two points have isometric neighborhoods. On the universal cover $\mathbf{X}$ of M such local isometries extend to global ones, so $\mathbf{X}$ is a homogeneous space. We can thus write

$$\mathbf{X} = G/K, \qquad M = \Pi\backslash\mathbf{X} = \Pi\backslash G/K,$$

where $G = G(\mathbf{X})$ is the isometry group of $\mathbf{X}$, $K = G_x$ the isotropy group of some point $x \in \mathbf{X}$ (so K is well defined up to inner automorphisms of G), $\Pi \subset G$ is a discrete subgroup which acts freely on $\mathbf{X} = G/K$ with finite volume quotient. Π is of course isomorphic to $\pi_1(M)$. $G^+ = G^+(\mathbf{X})$ will denote orientation preserving isometries.

To avoid trivial distinctions, we consider two metrics on $\mathbf{X}$ with the same isometry group G to be equivalent, and we only allow maximally symmetric metrics, that is, we assume no metric on $\mathbf{X}$ has strictly larger isometry group than G. The *geometry* $(\mathbf{X}, G)$ will be relevant to 3-manifold theory if $\dim \mathbf{X} = 3$ and $\mathbf{X}$ admits finite volume quotients $\Pi\backslash\mathbf{X}$ with $\Pi \subset G$ discrete. Thurston [**T**] has pointed out that there are exactly 8 such geometries. They are most easily computed in terms of K_0, the identity component of the group $K = G_x$ of isometries fixing a point $x \in \mathbf{X}$.

$K_0 =$	$\mathbf{X} =$			$K =$
SO(3)	$\mathbf{S}^3$	$\mathbf{E}^3$	$\mathbf{H}^3$	O(3)
SO(2)	$\mathbf{S}^2 \times \mathbf{E}^1$		$\mathbf{H}^2 \times \mathbf{E}^1$	$\mathrm{O}(2) \times \mathrm{O}(1)$
SO(2)		N	$\widetilde{\mathrm{PSL}}$	O(2)
$\{1\}$		S		D_8

Here $\mathbf{S}^n$, $\mathbf{E}^n$, and $\mathbf{H}^n$ are spherical, euclidean, and hyperbolic geometry. N, $\widetilde{\mathrm{PSL}}$, and S are certain lie groups with left invariant metrics. Namely $\widetilde{\mathrm{PSL}}$ is the universal cover of $\mathrm{PSL}(2, \mathbf{R}) = G^+(\mathbf{H}^2)$; this geometry can also be described as the universal cover of the unit tangent bundle $T^1\mathbf{H}^2$ of $\mathbf{H}^2$, with natural metric. N is the group of real 3×3 upper triangular unipotent matrices. It is more usefully described for us as the group structure

$$(a, b; c)(a', b'; c') = (a + a', b + b'; c + c' + ab' - a'b) \quad \text{on } \mathbf{R}^2 \times \mathbf{R}.$$

Its center is $\{0\} \times \mathbf{R}$, giving a central extension $0 \to \mathbf{R} \to N \to \mathbf{R}^2 \to 1$. Finally S is a split extension $1 \to \mathbf{R}^2 \to S \to \mathbf{R} \to 1$, and can be described explicitly as the group structure

$$(a, b; c)(a', b'; c') = (a + e^c a', b + e^{-c} b'; c + c') \quad \text{on } \mathbf{R}^2 \times \mathbf{R}.$$

The isometry group for each of $\mathbf{X} = \widetilde{\mathrm{PSL}}$, N, and S is a semidirect product $G(\mathbf{X}) = T \cdot K$, where T is the group $\widetilde{\mathrm{PSL}}$, N, or S respectively, acting by left translations, and K is given in the table and acts on T as follows. For $\mathbf{X} = \widetilde{\mathrm{PSL}}$, $K = \mathrm{O}(2) \subset G(\mathbf{H}^2)$ acts by conjugation on $G^+(\mathbf{H}^2) = \mathrm{PSL}(2, \mathbf{R})$, so lift this action to an action on $T = \widetilde{\mathrm{PSL}}$. For N, with coordinates $\mathbf{R}^2 \times \mathbf{R}$ as above, $K = \mathrm{O}(2)$ acts standardly on $\mathbf{R}^2$ and by determinant on $\mathbf{R}$. For S, K is dihedral of order 8 generated by τ: $(a, b; c) \mapsto (b, a; -c)$ and σ: $(a, b; c) \mapsto (-a, b; c)$.

If $M^3 \to F$ is a Seifert fibration of a closed connected oriented M^3 over a possibly nonorientable surface F, the Seifert-invariant $(g; b; (\alpha_1, \beta_1), \ldots, (\alpha_n, \beta_n))$ is still defined, but we use negative g to indicate nonorientable F. Thus the euler number

$$e(M \to F) = -b - \sum_{i=1}^{n} \frac{\beta_i}{\alpha_i}$$

is still defined. We also define, with $\chi(F)$ equal to the euler characteristic,

$$\chi(M \to F) = \chi(F) - \sum_{i=1}^{n} \frac{\alpha_i - 1}{\alpha_i}.$$

THEOREM. *Let M be a closed connected oriented 3-manifold which admits a geometric structure. Then*

(i) *the geometry $\mathbf{X}$ in question is uniquely determined by M;*

(ii) *If M admits an $\mathbf{H}^3$ structure we shall not discuss it;*

(iii) *M admits an S-structure if and only if either*

(a) *M can be fibered over* S^1 *with fiber* $S^1 \times S^1$ *such that the monodromy* $h \in \mathrm{SL}(2, \mathbf{Z})$ *has* $|\operatorname{trace}(h)| \geqslant 3$, *or*

(b) *M is a twisted double of the orientation* [0, 1]*-bundle over the Klein bottle, but cannot be Seifert fibered*;

(iv) *M admits an* **X***-structure,* $\mathbf{X} \neq \mathbf{H}^3$, *S, if and only if M admits a Seifert fibering* $M \to F$. *The relevant geometry is as follows*:

$e \backslash \chi$	> 0	$= 0$	< 0
$= 0$	$\mathbf{S}^2 \times \mathbf{E}^1$	$\mathbf{E}^3$	$\mathbf{H}^2 \times \mathbf{E}^1$
$\neq 0$	$\mathbf{S}^3$	N	$\widetilde{\mathrm{PSL}}$

$\chi = \chi(M \to F)$
$e = e(M \to F)$

We call the six geometries of part (iv) the *Seifert geometries*.

REMARKS. (1) For $\mathbf{X} = N$ or $\widetilde{\mathrm{PSL}}$, $G(\mathbf{X}) = G^+(\mathbf{X})$, so all manifolds with an **X**-structure are orientable. For $\mathbf{X} = S^3$ the latter is still true but less obvious.

(2) The theorem is valid with minor changes also for nonorientable and/or noncompact M, and with slightly more change even for arbitrary lattices $\Pi \subset G(\mathbf{X})$, $\mathbf{X} \neq \mathbf{H}^3$, see [**N3**].

(3) The invariants χ and e arise naturally for QH-singularities. For instance, Dolgachev has shown, and it also follows easily from Pinkham [**P1**, Theorem 5.1], that the Poincaré series for the graded affine ring of (V, p) is a rational function of the form

$$p(t) = -et/(1-t)^2 + \chi/2(1-t) + P(t)/Q(t),$$

where $Q(t)$ is cyclotomic and not divisible by $(t-1)$. See also [**W2**, §2 and **N4**, §4].

(4) With a natural normalization of the metrics on the geometries, the volume of M is $4\pi^2\chi^2/|e|$ in the Seifert case when $e\chi \neq 0$, and is indeterminate, depending on the geometric structure, in all other cases with $\mathbf{X} \neq \mathbf{H}^3$. For $\mathbf{X} = N$ it is $l^2|e|$, where $l =$ (length of a fiber of M).

If **X** is a Seifert geometry other than $\mathbf{S}^3$ or $\mathbf{E}^3$, then the set of isometries which fix a point $x \in \mathbf{X}$ fixes a tangent direction at that point up to sign, so **X** has a $G(\mathbf{X})$-invariant tangent line field. This line field gives a foliation of **X**, which in fact fibers **X** over $\mathbf{S}^3$, $\mathbf{E}^2$, or $\mathbf{H}^2$ in a way which is obvious for $\mathbf{X} = \mathbf{S}^2 \times \mathbf{E}$ and $\mathbf{H}^2 \times \mathbf{E}$, and which is visible from our description of **X** for $\mathbf{X} = N$ and $\mathbf{X} = \widetilde{\mathrm{PSL}}$. It is this *vertical fibration* of **X** which induces the Seifert fibration of $M = \Pi \backslash \mathbf{X}$ (except for some $(\mathbf{S}^2 \times \mathbf{E}^1)$-structures on $S^2 \times S^1$). Each of the geometries $\mathbf{S}^3$ and $\mathbf{E}^3$ also fibers geometrically over $\mathbf{S}^2$ and $\mathbf{E}^2$, respectively ($\mathbf{S}^3 \to \mathbf{S}^2$ is Hopf fibration), but this fibration is only well defined up to isometries, so the subgroup $G^{\mathrm{fib}} \subset G$ which preserves this fibration is only determined up to conjugation. The different conjugates of G^{fib} correspond to the different "*vertical fibrations*" of **X**; if $M = \Pi \backslash \mathbf{X}$ is a geometric 3-manifold with $\mathbf{X} = \mathbf{S}^3$ or $\mathbf{E}^3$, then Π is in some conjugate of G^{fib} and a Seifert fibration of M can again be induced from the corresponding vertical fibration of $\mathbf{X}$. If $M = T^3 = S^1 \times S^1 \times S^1$ one must choose the conjugate of G^{fib} correctly—otherwise one just gets a foliation of M.

We call the Seifert fibration of $M = \Pi \backslash \mathbf{X}$, induced as above from a vertical fibration of $\mathbf{X}$, a *geometric Seifert fibration*. It is unique for $\mathbf{X} \neq \mathbf{S}^3, \mathbf{E}^3$, but may depend on a choice of conjugate of G^{fib} containing Π for $\mathbf{X} = \mathbf{S}^3$ or $\mathbf{E}^3$.

If $\mathbf{X}$ is a Seifert geometry, let $G_c = G_c(\mathbf{X}) \subset G^+(\mathbf{X})$ be the subgroup preserving a vertical fibration on $\mathbf{X}$ as an oriented fibration. Then $G_c = (G^{\text{fib}})_0$ if $\mathbf{X} = \mathbf{S}^3$ or $\mathbf{E}^3$ and $G_c = G_0$ otherwise. The centre C of G_c is S^1 for $\mathbf{X} = \mathbf{S}^3$ and is $\mathbf{R}$ otherwise.

PROPOSITION. *Let $M = \Pi \backslash \mathbf{X}$ with $\mathbf{X}$ a Seifert geometry. Then $\Pi \subset G_c$ (respectively Π is in some conjugate of G_c if $\mathbf{X} = \mathbf{S}^3$ or $\mathbf{E}^3$) if and only if M can be Seifert fibered with orientable base. If $M \neq S^2 \times S^1$, then $C/C \cap \Pi \cong S^1$ and it acts on M inducing the geometric Seifert fibration $M \to M/S^1$ (for $M = T^3$ this holds only for suitable conjugates of G_c containing Π).*

Not every geometric structure on $S^2 \times S^1$ admits a geometric Seifert fibration, but any Seifert fibration, of any M, is geometric for some geometric structure on M.

4. Geometry of holomorphic Seifert $\mathbf{C}^2$-bundles. For each Seifert geometry $\mathbf{X}$ let $G_{\mathbf{C}}(\mathbf{X}) = G_c(\mathbf{X}) \times \mathbf{R}_+$, acting transitively on $\mathbf{X} \times \mathbf{R}_+$ in the obvious way. We shall describe a complex analytic structure on $\mathbf{X} \times \mathbf{R}_+$ such that $G_{\mathbf{C}} = G_{\mathbf{C}}(\mathbf{X})$ acts by complex analytic maps.

On $\mathbf{E}^1 \times \mathbf{R}_+$ take the complex structure

$$\mathbf{E}^1 \times \mathbf{R}_+ \to \mathbf{C}, \qquad (\theta, r) \mapsto \theta - i \cdot \ln(r).$$

The notation is chosen to suggest that $\mathbf{E}^1 \times \mathbf{R}_+$ is polar coordinates in $\tilde{\mathbf{C}}^*$, the universal cover of $\mathbf{C}^*$. Now the complex structures on $\mathbf{S}^2 \times \mathbf{E}^1 \times \mathbf{R}_+$, $\mathbf{E}^2 \times \mathbf{E}^1 \times \mathbf{R}_+$, and $\mathbf{H}^2 \times \mathbf{E}^1 \times \mathbf{R}_+$ are the obvious ones: $\mathbf{C}P^1 \times \mathbf{C}$, $\mathbf{C} \times \mathbf{C}$, $\mathbf{H} \times \mathbf{C}$ (we use $\mathbf{H}$, as opposed to $\mathbf{H}^2$, to denote the upper half-plane with complex structure, instead of hyperbolic metric).

On $\mathbf{S}^3 \times \mathbf{R}_+$ take the obvious structure as $\mathbf{C}^2 - \{0\}$, by considering $\mathbf{S}^3 \times \mathbf{R}_+$ as polar coordinates, with G_c acting as $U(2)$.

On $N \times \mathbf{R}_+$, coordinatize N as in §3 and take the complex structure

$$N \times \mathbf{R}_+ \to \mathbf{C} \times \mathbf{C}, \qquad ((a, b; c), r) \mapsto \left(a + ib, c - i\left(a^2 + b^2 - 2\ln(r)\right)/2\right).$$

For $\widetilde{\mathrm{PSL}} \times \mathbf{R}_+$ recall first that we can identify $\mathrm{PSL} = \mathbf{Z} \backslash \widetilde{\mathrm{PSL}}$ with the unit tangent bundle $T^1\mathbf{H}^2$ of $\mathbf{H}^2$. Thus $\mathrm{PSL} \times \mathbf{R}_+$ can be taken as polar coordinates in $T_0\mathbf{H}$ (the bundle of nonzero tangent vectors), so $\widetilde{\mathrm{PSL}} \times \mathbf{R}_+$ is identified with $(T_0\mathbf{H})^{\tilde{}}$. Since $T_0\mathbf{H} \cong \mathbf{H} \times \mathbf{C}^*$, we have $\widetilde{\mathrm{PSL}} \times \mathbf{R}_+ \cong (\mathbf{H} \times \mathbf{C}^*)^{\tilde{}} \cong \mathbf{H} \times C$.

In each case denote $\mathbf{X}_{\mathbf{C}} = \mathbf{X} \times \mathbf{R}_+$ with the above complex structure. The center $C \times \mathbf{R}_+$ of $G_{\mathbf{C}}$ can be identified as $\mathbf{C}^*$ acting by multiplication on $\mathbf{X}_{\mathbf{C}} = \mathbf{C}^2 - \{0\}$ when $\mathbf{X}^3 = \mathbf{S}^3$, and as $\mathbf{C}$, acting by translations in the second factor of $\mathbf{X}_{\mathbf{C}} = (-) \times \mathbf{C}$ in each of the other cases.

Let $M = \Pi\backslash\mathbf{X}$ be a geometric Seifert manifold with geometric Seifert fibration, as in the last proposition of §3. Thus $\Pi \subset G_c$ and $C/C \cap \Pi = S^1$. Then

$$M \times \mathbf{R}_+ = \Pi\backslash\mathbf{X} \times \mathbf{R}_+ = \Pi\backslash\mathbf{X}_\mathbf{C}$$

gives a complex structure on $M \times \mathbf{R}_+$ and $S^1 \times \mathbf{R}_+$ acts as $(C/C \cap \Pi) \times \mathbf{R}_+ \cong \mathbf{C}^*$, acting holomorphically. That is, $M \times \mathbf{R}_+$ receives the structure of a holomorphic Seifert $\mathbf{C}^*$-bundle.

THEOREM. *Normalize by fixing* vol(M) *in the* $\chi = 0$ *cases and fixing the length of a fiber of* M *in the* $e = 0$ *cases (see Remark* (4)). *Then if double brackets represent "set of equivalence classes of ..." with an appropriate isomorphism concept, the above construction defines a bijection*:

$$(1)\qquad \left\{\left\{\begin{array}{l}\textit{Geometric structures on } M\\ \textit{with geometric Seifert}\\ \textit{fibration.}\end{array}\right\}\right\} \leftrightarrow \left\{\left\{\begin{array}{l}\textit{Holomorphic structures on}\\ M \times \mathbf{R}_+ \textit{ as a Seifert}\\ \mathbf{C}^*\textit{-bundle.}\end{array}\right\}\right\}.$$

In particular:

$$(2)\qquad \left\{\left\{\begin{array}{l}\textit{Geometric structures on } M\\ \textit{with geometric Seifert}\\ \textit{fibration with negative } e.\end{array}\right\}\right\} \leftrightarrow \left\{\left\{\begin{array}{l}\textit{QH-singularities } (V, p)\\ \textit{with link homeomorphic to}\\ M.\end{array}\right\}\right\}.$$

Unless (V, p) is a cyclic quotient singularity, so M is a lens space, the singularity (V, p) has a unique good $\mathbf{C}^*$-action and M has a unique geometric Seifert fibration. So excluding lens spaces, and assuming M admits a Seifert fibration with negative e, we get

$$(3)\qquad \{\{\textit{Geometric structures on } M\}\} \leftrightarrow \left\{\left\{\begin{array}{l}\textit{Normal singularities}\\ (V, p) \textit{ with link homeo-}\\ \textit{morphic to } M \textit{ which}\\ \textit{admit a good } \mathbf{C}^*\textit{-action}\end{array}\right\}\right\}.$$

For a cyclic quotient singularity (V, p) the geometric structure on its link M is unique and geometric Seifert fibrations of M correspond one-one with $\mathbf{C}^*$-actions on (V, p).

5. Cusps and the geometry S. There is a correspondence, analogous to (3) above, connected with the geometry S. Coordinatize S as $\{(a, b; c) \mid a, b, c \in \mathbf{R}\}$ as in §3. Let $G_c \subset G(S)$ be the subgroup of index 4, generated by $G_0 = S$ (acting on itself by left translations) and the element τ: $(a, b; c) \to (b, a; -c)$ mentioned in §3. Then G_c acts on $\mathbf{H} \times \mathbf{H}$ as follows:

$$(a, b; c)(z_1, z_2) = (e^c z_1 + a, e^{-c} z_2 + b),$$
$$\tau(z_1, z_2) = (z_2, z_1).$$

The map $\mathbf{H} \times \mathbf{H} \to S \times \mathbf{R}_+$, $(z_1, z_2) \mapsto ((\operatorname{Re} z_1, \operatorname{Re} z_2; \ln \operatorname{Im} z_1), \operatorname{Im} z_1 \cdot \operatorname{Im} z_2)$ is a G_c-equivariant homeomorphism. Thus an S-structure $M \cong \Pi\backslash S$, with $\Pi \subset G_c$

on a manifold M leads to a complex structure $M \times \mathbf{R}_+ \cong \Pi \backslash S \times \mathbf{R}_+ \cong \Pi \backslash (\mathbf{H} \times \mathbf{H})$ on $M \times \mathbf{R}_+$.

THEOREM. (1) *A manifold M which admits an S-structure admits a unique one of any chosen volume.*

(2) *The following are equivalent*:

(a) $M \cong \Pi \backslash S$ *with* $\Pi \subset S$;

(b) *M fibers over* S^1 *with fiber* T^2 *and monodromy of trace* $\geqslant 3$;

(c) *M is homeomorphic* (*preserving orientation*) *to the link of a cusp singularity* (*that is a singularity* (V, p) *with cyclic resolution graph*);

(d) *Statement* (c) *is true and* $V - \{p\} = \Pi \backslash (\mathbf{H} \times \mathbf{H})$.

(3) *The following are equivalent*:

(a) $M \cong \Pi \backslash S$ *with* $\Pi \subset G_c$, $\Pi \not\subset S$.

(b) *M is homeomorphic* (*respecting orientation*) *to the link of a singularity* (V, p) *with resolution graph of the form*

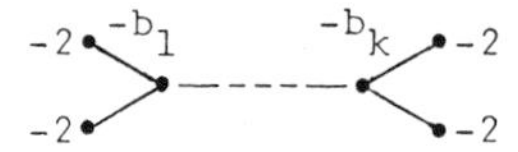

$b_i \geqslant 2$ *for all* i,
$b_i \geqslant 3$ *for some* i.

(c) *Statement* (b) *is true and* $V - \{p\} = \Pi \backslash (\mathbf{H} \times \mathbf{H})$.

REMARKS. 1. The analytic structure on these singularities is unique, by Laufer [**L**]. Their description as "cusps" of discrete quotients of $\mathbf{H} \times \mathbf{H}$ is how they originally arose, in the work Hirzebruch et al. on Hilbert modular surfaces.

2. The double cover of the singularity of (3)(b) above, determined by $\Pi \cap S \subset \Pi$, must be a cusp. It is the cusp with resolution graph

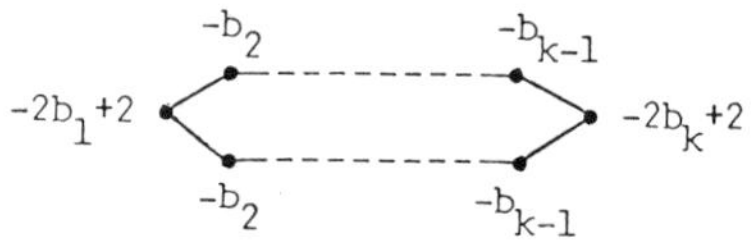

3. In §§4 and 5 we have dealt with every singularity link which admits a geometric structure. Indeed, a singularity link M cannot admit an $\mathbf{H}^3$-structure, since no plumbed manifold admits an $\mathbf{H}^3$-structure. We have dealt with all Seifert manifolds which are singularity links by [**N2**]. Thus the only case remaining was S-structures. One can either compute all orientable S-manifolds and compare with [**N2**]. This is not hard, they turn out to be all 3-manifolds which can be plumbed according to a cyclic plumbing graph or a graph as in (3)(b) above but without the restrictions on b_1 and b_k. It is easier to observe that an S-manifold has solvable fundamental group. We have all singularity links with solvable fundamental group by Wagreich's list [**W1**].

6. Ideas of proof. We illustrate the proof of the theorem of §3 by sketching why a manifold M^3, Seifert fibered as $M \to X$ with orientable base, has a $\widetilde{\mathrm{PSL}}$ structure when $\chi < 0$ and $e \neq 0$. Let Π be $\pi_1(M)$. The center of Π is $\mathbf{Z}$, generated by the class of a nonsingular fiber, and we have an exact sequence

$$0 \to \mathbf{Z} \to \Pi \to \Gamma \to 1,$$

where $\Gamma = \Gamma(g; \alpha_1, \ldots, \alpha_n)$ is a Fuchsian group with signature $(g; \alpha_1, \ldots, \alpha_n)$ (if χ had been $= 0$ or > 0 then Γ would be euclidean or spherical respectively, instead of Fuchsian).

Now the group H of §2 is in fact $H^2(\Gamma; \mathbf{Z})$ and the "chern class" $c \in H$ described there is just the classifying element for this exact sequence. Note that $H^2(\Gamma; \mathbf{R}) = H \otimes \mathbf{R} \cong \mathbf{R}$ by the map $g_0 \mapsto 1$, and if $i\colon \mathbf{Z} \to \mathbf{R}$ is the inclusion, then $i_* c \in H^2(\Gamma; \mathbf{R}) = \mathbf{R}$ is just $e(M \to X)$, by the lemma in §2.

We need to embed Π in $G_0 = G_0(\widetilde{\mathrm{PSL}})$ in the following way:

$$\begin{array}{ccccccccc} 0 & \to & \mathbf{Z} & \to & \Pi & \to & \Gamma & \to & 1 \\ & & \downarrow ? & & \downarrow ? & & \downarrow \alpha & & \\ 0 & \to & \mathbf{R} & \to & G_0 & \to & (G^+(\mathbf{H}^2) = \mathrm{PSL}(2, \mathbf{R})) & \to & 1 \end{array}$$

It is not hard to see that this is just what is necessary to give $\Pi \backslash \widetilde{\mathrm{PSL}}$ the desired Seifert fibered structure. The map α is given to us by choosing a complex structure (or a hyperbolic orbifold structure) on $(X, x_1, \ldots, x_n)$. Let us form the following pushout and pullback extensions $(*)$ and $(**)$:

$$\begin{array}{cccccccccc} & 0 & \to & \mathbf{Z} & \to & \Pi & \to & \Gamma & \to & 1 \\ & & & \downarrow i & & \downarrow & & \| & & \\ (*) & 0 & \to & \mathbf{R} & \to & i_*\Pi & \to & \Gamma & \to & 1 \\ (**) & 0 & \to & \mathbf{R} & \to & \alpha^*\Gamma & \to & \Gamma & \to & 1 \\ & & & \| & & \downarrow & & \downarrow \alpha & & \\ & 0 & \to & \mathbf{R} & \to & G_0 & \to & \mathrm{PSL}(2, \mathbf{R}) & \to & 1 \end{array}$$

It suffices to show that $(*)$ and $(**)$ are isomorphic. But we have already observed that $(*)$ is classified by $i_* c = e = e(M \to X) \in \mathbf{R} = H^2(\Gamma, \mathbf{R})$, and it is not hard to show that $(**)$ is classified by $\chi \in \mathbf{R} = H^2(\Gamma, \mathbf{R})$ (this follows, for instance, from the computation of Seifert-invariants of $\Gamma \backslash \mathrm{PSL}(2, \mathbf{R})$, see [**EHN** or **RV**]). Thus, letting $\beta\colon \mathbf{R} \to \mathbf{R}$ be multiplication by χ/e, the following diagram can be completed, and we are done:

$$\begin{array}{cccccccccc} (*) & 0 & \to & \mathbf{R} & \to & i_*\Pi & \to & \Gamma & \to & 1 \\ & & & \downarrow \beta & & \downarrow \cong & & \| & & \\ (**) & 0 & \to & \mathbf{R} & \to & \alpha^*\Gamma & \to & \Gamma & \to & 1 \end{array}$$

The various different $\widetilde{\mathrm{PSL}}$-structures on M lying over a fixed structure on X are classified by the different homorphisms $\gamma\colon \Pi \to G_0$ for which

$$\begin{array}{ccccccccc} 0 & \to & \mathbf{Z} & \to & \Pi & \xrightarrow{\pi} & \Gamma & \to & 1 \\ & & \downarrow & & \downarrow \gamma & & \downarrow \alpha & & \\ 0 & \to & \mathbf{R} & \to & G_0 & \to & \mathrm{PSL}(2, \mathbf{R}) & \to & 1 \end{array}$$

commutes, up to automorphisms of Π fixing Γ. Given one γ, any other γ' is given by $\gamma'(g) = \gamma(g) \cdot \psi\pi(g)$ for some $\psi \in \mathrm{Hom}(\Gamma, \mathbf{R})$, with $\psi \in \mathrm{Hom}(\Gamma; \mathbf{Z})$ if γ' is related to γ by an automorphism of Π. Thus $\mathrm{Hom}(\Gamma, \mathbf{R})/\mathrm{Hom}(\Gamma, \mathbf{Z})$ classifies the

$\widetilde{\mathrm{PSL}}$-structures on M. The correspondence of §4 sets up a map $\mathrm{Hom}(\Gamma, \mathbf{R})/\mathrm{Hom}(\Gamma, \mathbf{Z}) \to \mathrm{Jac}(X)$ and to prove the theorem of §4 for this case, we must show this is bijective. But it is a map of real tori of equal dimension, so it is enough to show injectivity. A computation shows that this is equivalent to the image of the period mapping $H^0(X, \Omega^1) \to \mathrm{Hom}(\Gamma, \mathbf{C}) = H^1(X, \mathbf{C})$ being tranverse to $\mathrm{Hom}(\Gamma, \mathbf{R}) = H^1(X, \mathbf{R})$, which is true, by classical complex curve theory.

The proofs of the results of §§3 and 4 in the other cases are similar. Details will be given in [**N3**].

7. An application. One interpretation of the results of §4 is that the cocycles which are used in other approaches to classify (Seifert) $\mathbf{C}^*$-bundles over a curve X can be put in a very special form. This implies that one can interpret the affine coordinate ring of a QH-singularity (V, p) as a suitable ring of automorphic forms which transform by characters, rather by some general cocycle "multiplier system." But this is really a nonapplication, in that it would be of interest only if the type of ring of χ-automorphic forms which arises were a type which has independent interest, say to number theorists, which seems not. Nevertheless, as an application of this train of thought we get

THEOREM. *Let (V, p) be a QH-singularity with $V_0/\mathbf{C}^* = (X, x_1, \ldots, x_n)$ and Seifert invariant $(g; b; (\alpha_1, \beta_1), \ldots, (\alpha_n, \beta_n))$. Suppose X is rational, that is, $g = 0$. Let $n \geqslant 3$ (i.e., (V, p) is not a cyclic quotient singularity). Then the universal abelian cover $(\tilde{V}^{\mathrm{ab}}, p)$ of (V, p), branched only at p, is isomorphic to $(V_A(\alpha_1, \ldots, \alpha_n), 0)$, where $V_A(\alpha_1, \ldots, \alpha_n)$ is the Brieskorn complete intersection*

$$V_A(\alpha_1, \ldots, \alpha_n) = \{z \in \mathbf{C}^n \mid a_{i1} z_i^{\alpha_1} + \cdots + a_{in} z_n^{\alpha_n} = 0,\ i = 1, \ldots, n-2\}$$

for suitable coefficient matrix $A = (a_{ij})$. In fact, if one determines $\lambda_1, \ldots, \lambda_{n-3} \in \mathbf{C}$ by the unique analytic isomorphism $(X, x_1, \ldots, x_n) \cong (\mathbf{C} \cup \infty, \lambda_1, \ldots, \lambda_{n-3}, 1, 0, \infty)$, then one can take

$$A = \begin{pmatrix} 1 & & \bigcirc & & 1 & \lambda_1 \\ & \ddots & & & \vdots & \vdots \\ & & 1 & & 1 & \lambda_{n-3} \\ & \bigcirc & & 1 & 1 & 1 \end{pmatrix}.$$

PROOF. We just sketch the proof, for a reason given below. Assume $\chi < 0$. We first note a general fact that if, for $i = 1, 2$,

$$\begin{array}{ccccccccc} 0 & \to & \mathbf{Z} & \to & \Pi_i & \to & \Gamma & \to & 1 \\ & & \gamma_i \mid \mathbf{Z} \downarrow & & \gamma_i \downarrow & & \alpha \downarrow & & \\ 0 & \to & A & \to & G & \to & Q & \to & 1 \end{array}$$

is a map of central extensions, then $\gamma_1([\Pi_1, \Pi_1]) = \gamma_2([\Pi_2, \Pi_2])$. Now our $V_0 = V - p$ is classified by an embedding $\Pi \to G_0 = G_0(\widetilde{\mathrm{PSL}})$ which fits in a diagram

$$\begin{array}{ccccccccc} 0 & \to & \mathbf{Z} & \to & \Pi & \to & \Gamma & \to & 1 \\ & & \downarrow & & \downarrow\gamma & & \downarrow\alpha & & \\ 0 & \to & \mathbf{R} & \to & G_2 & \to & \mathrm{PSL}(2,\mathbf{R}) & \to & 1 \end{array}$$

Also $\gamma \mid [\Pi, \Pi]$ corresponds to the universal abelian cover of V_0, so our initial comment shows that if we prove the theorem for just one (V, p) with given α, then it follows for all such (V, p). But such a proof was given (in fact for $V_0 = \alpha(\Gamma)\diagdown T_0\mathbf{H})$ in [**N1**], see also [**D**].

So this proves the case $\chi < 0$. The analogous proof applies for $\chi > 0$, and would apply too for $\chi = 0$ except that in this case the necessary examples had only been analyzed by ad hoc means in [**M** and **N1**], and this analysis does not give the information we need. This gap could be filled, but this is not necessary, since a much more elementary proof of the theorem will appear elsewhere in these PROCEEDINGS [**N4**]. The above is however, essentially how I found the result, with the observation about commutator subgroups replaced by the corresponding observation about rings of χ-automorphic forms.

REFERENCES

[**B**] E. Brieskorn, *The unfolding of exceptional singularities*, Leopoldina Symposium: Singularities (Thüringen, 1978), Nova Acta Leopoldina, (N.F.) **52** (1981), no. 240, 65–93.

[**CR**] P. E. Conner and F. Raymond, *Holomorphic Seifert fibrations* (Proc. Conf. Transformation Groups, Amherst, 1973), Lecture Notes in Math., vol. 299, Springer-Verlag, Berlin and New York, 1974, pp. 124–204.

[**D**] I. V. Dolgachev, in preparation; see also, *Automorphic forms and quasihomogeneous singularities*, Funkcional-Anal. i Priložen **9** (**2**) (1975), 67–68.

[**EHN**] D. Eisenbud, U. Hirsch and W. D. Neumann, *Transverse foliations of Seifert bundles and self homeomorphisms of the circle*, Comment. Math. Helv. **56** (1981), 638–660.

[**G**] H. Grauert, *Über Modifikation und exceptionelle analytische Mengen*, Math. Ann. **146** (1962), 331–368.

[**K**] U. Karras, *Klassifikation 2-dimensionaler Singularitäten mit auflösbaren lokalen Fundamentalgruppen*, Math. Ann. **213** (1975), 231–255.

[**Kl1**] F. Klein, *Lectures on the icosahedron and the solution of equations of the fifth degree* (1888), reprinted Dover Publications Inc., New York, 1956.

[**Kl2**] ______, *Über die Transformation seibenter Ordnung der elliptischen Funktionen*, Math Ann. **14** (1878), 428–471.

[**KF**] F. Klein and R. Fricke, *Vorlesungen über die Theorie der elliptischen Modulfunktionen*, Teubner, Leipzig, 1890.

[**L**] H. B. Laufer, *Taut two-dimensional singularities*, Math. Ann. **205** (1973), 131–164.

[**M**] J. Milnor, *On the 3-dimensional Brieskorn manifold $M(p, q, r)$* (papers dedicated to the memory of R. H. Fox), Ann. of Math. Studies, no. 84, Princeton Univ. Press, Princeton, N. J., 1975, pp. 175–225.

[**N1**] W. D. Neumann, *Brieskorn complete intersections and automorphic forms*, Invent. Math. **42** (1977), 285–293.

[**N2**] ______, *A calculus for plumbing applied to the topology of complex surface singularities and degenerating complex curves*, Trans. Amer. Math. Soc. **268** (1981), 299–344.

[**N3**] ______, *Geometry of 3-manifolds and normal surfaces singularities* (tentative title), in preparation.

[**N4**] ______, *Abelian covers of quasihomogeneous surface singularities*, these PROCEEDINGS.

[**NR**] W. D. Neumann and F. Raymond, *Seifert manifolds, plumbing, μ-invariant, and orientation reversing maps* (Proc. Alg. and Geom. Topology, Santa Barbara, 1977), Lecture Notes in Math., vol. 644, Springer-Verlag, Berlin and New York.

[**OW**] P. Orlik and P. Wagreich, *Algebraic surfaces with k*-action*, Acta Math. **138** (1977), 43–81.

[**P1**] H. Pinkham, *Normal surface singularities with* **C****-action*, Math. Ann. **227** (1977), 183–193.

[**P2**] ______, *Deformations of normal surface singularities with* **C****-action*, Math. Ann. **232** (1978), 65–84.

[**RV**] F. Raymond and A. T. Vasquez, 3-*manifolds whose universal coverings are Lie groups*, Topology Appl. **12** (1981), 161–179.

[**T**] W. P. Thurston, *Three dimensional manifolds, Kleinian groups and hyperbolic geometry* Bull. Amer. Math. Soc. (N.S) **6** (1982), 357–381.

[**W1**] P. Wagreich, *Singularities of complex surfaces with solvable local fundamental group*, Topology **11** (1971), 51–72.

[**W2**] ______, *The structure of quasi-homogeneous singularities*, these PROCEEDINGS.

UNIVERSITY OF MARYLAND, COLLEGE PARK

Proceedings of Symposia in Pure Mathematics
Volume **40** (1983), Part 2

ON THE STABILITY OF THE NEWTON BOUNDARY

MUTSUO OKA

1. Statement of the results. We consider an analytic function $f(x, y, t)$ defined in a neighborhood of the origin of $C^2 \times C$ which gives a μ-constant deformation of the plane curves. Namely we assume that

(i) $f(0, 0, t) = 0$ and

(ii) f_t (= the restriction of f to $C^2 \times \{t\}$) has an isolated critical point at 0 with the Milnor number $\mu(f_t) = \mu(f_0)$.

We will prove

THEOREM 1.1. *Let f_t be a μ-constant deformation and assume that f_0 is convenient. Then there exists an analytic family of coordinates $\phi_t(x, y) = (x(t), y(t))$ for $|t| < \eta$ where η is a small positive number such that*:

(i) *$\phi_t(0) = 0$ and $\phi_0(x, y) = (x, y)$.*

(ii) *The Newton boundary of f_t in $\phi_t(x, y)$ coincides with $\Gamma(f_0)$.*

For the proof, we use the equivalence of μ-constancy and equisingularity of f_t. As immediate corollaries, we get

COROLLARY 1.2. *Suppose that $\Gamma(f_0)$ is nondegenerate. $\Gamma(f_t; \phi_t)$ is also nondegenerate for $|t| < \eta$.*

COROLLARY 1.3. *$\nu(f_t) \geqslant \nu(f_0)$ and $\alpha(f_t) \leqslant \alpha(f_0)$ for $|t| < \eta$.*

See §2 for the definitions of $\alpha(f_t)$ and $\nu(f_t)$.

It is an open problem whether Corollary 1.2 holds or not for higher dimensional cases. We will strengthen Corollaries 1.2 and 1.3 in §5 (Theorem 5.1 and Corollaries 5.3 and 5.4).

1980 *Mathematics Subject Classification.* Primary 32B30.

2. Definitions. Let $f(z_1,\ldots,z_n)$ be an analytic function defined in a neighborhood of $0 \in C^n$ and we assume $f(0) = 0$. Let $\Sigma a_{\nu_1,\ldots,\nu_n} z_1^{\nu_1} \cdots z_n^{\nu_n}$ be the Taylor expansion of f at 0. Let $\Gamma_+(f; z)$ be the convex hull of the union

$$\left\{(\nu_1,\ldots,\nu_n) + (R^+)^n;\ a_{\nu_1\cdots\nu_n} \neq 0\right\} \ (R^+ = \{x \in R;\ x \geqslant 0\}).$$

Let $\Gamma(f; z)$ be the union of the compact faces of $\Gamma_+(f; z)$ and we call $\Gamma(f; z)$ *the Newton boundary of f in $z = (z_1,\ldots,z_n)$*. Let $\Gamma_-(f; z)$ be the cone of $\Gamma(f; z)$ with the origin [**2**]. In general, a different coordinate gives a different Newton boundary.

For a closed face Δ of $\Gamma(f; z)$, we associated a weighted homogeneous polynomial $f_\Delta(z)$ which is defined by $\Sigma_{(\nu_1\cdots\nu_n)\in\Delta} a_{\nu_1,\ldots,\nu_n} z_1^{\nu_1} \cdots z_n^{\nu_n}$. f is called nondegenerate on Δ if f_Δ: $(C^*)^n \to C$ has no critical points. $\Gamma(f; z)$ is called *nondegenerate* if f is nondegenerate on every face Δ of $\Gamma(f; z)$. f is called *convenient in z* if $\Gamma(f)^i = \{x \in \Gamma(f; z);\ x_j = 0 \text{ for } j \neq i\}$ is nonempty for $i = 1,\ldots,n$.

DEFINITION 2.1. *The Newton number $\nu(f)$ of f* is the supremum of the Newton number $\nu(\Gamma_-(f; z))$ among every possible analytic coordinate $z = (z_1,\ldots,z_n)$. (See [**2**] for the definition of the Newton number $\nu(\Gamma_-(f; z))$.) We call $z = (z_1,\ldots,z_n)$ *a Newton coordinate* for f if $\nu(\Gamma_-(f; z)) = \nu(f)$.

DEFINITION 2.2. Suppose that f has an isolated critical point at 0. We define *the degeneracy index $\alpha(f)$* of f by $\mu(f) - \nu(f)$. This is a nonnegative integer and $\alpha(f) = 0$ if f has a nondegenerate Newton boundary.

EXAMPLE 2.3. Let $f(x, y) = x^k y^k (x + y)^k + \alpha x^n + y^m$. We assume that $n \geqslant m \geqslant 3k + 1$, α is a nonzero complex number and $\alpha \neq (-1)^{n-1}$ if $n = m$. We can see easily that (x, y) is a Newton coordinate for f, $\mu(f) = (2m + n)(k - 1) + 3k + 1$, $\nu(f) = (m + n)(k - 1) + 3k^2 + 1$ and $\alpha(f) = (k - 1)(m - 3k)$.

3. Equisingularity. Let C be a germ of a plane curve at 0 defined by a reduced analytic function $f(x, y)$ and let $C_1,\ldots,C_r$ be its irreducible components. Let $m = m(C)$ (= the multiplicity of C at 0) and let Δ be the face of $\Gamma(f)$ defined by $\{(x, y) \in \Gamma(f);\ x + y = m\}$. We call Δ *the tangential face* of f. Let $f_\Delta(x, y) = \Pi_{\nu=1}^t (a_\nu x + b_\nu y)^{m_\nu}$. Let $C^\nu = \{C_j;\ C_j \text{ is tangent to } a_\nu x + b_\nu y = 0\}$. $C^1,\ldots,C^t$ are called the *tangential components* of C. We denote the proper transform of C_j by C_j'.

Let D be another germ of a plane curve with r irreducible components $D_1,\ldots,D_r$. A bijection ϕ: $\{C_1,\ldots,C_r\} \to \{D_1,\ldots,D_r\}$ is called *an equivalence* if:

(i) ϕ is tangential stable, i.e. $\phi(C^\nu)$ is a tangential component of D, say $D^{\phi(\nu)}$ for $\nu = 1,\ldots,t$.

(ii) $m(C_i) = m(\phi(C_i))$, $i = 1,\ldots,r$.

(iii) The induced map ϕ_ν: $(C^\nu)' \to (D^{\phi(\nu)})'$, is an equivalence.

This is an inductive definition of (a)-equivalence [**7**].

A deformation $f(x, y, t)$ of germs of plane curves at 0 is called *equisingular* if C_t is equivalent to C_0 where $C_t = \{(x, y) \in C^2;\ f(x, y, t) = 0\}$.

It is known that equivalent curves have the same Milnor number (see, for example, [**1**]). Though the converse is not true in general, we have

LEMMA 3.1 (LÊ [**4, 5, 7**]). *A μ-constant deformation is equisingular.*

4. Proof of Theorem 1.1. Let $f(x, y, t)$ be a μ-constant deformation as in §1. The following is the key step to the proof.

LEMMA 4.1. *Suppose that* $\Gamma(f_t) \cap \{x=0\} = \Gamma(f_0) \cap \{x=0\} = \{(0,b)\}$ *and* $\Gamma(f_t) \cap \{y=0\} = \Gamma(f_0) \cap \{y=0\} = \{(a,0)\}$ *for* $|t| \leq \delta$. *Then* $\Gamma(f_t) = \Gamma(f_0)$.

PROOF. The proof will be done by the induction on $\sigma^*(f_0)$ where $\sigma^*(f_0)$ is the maximum of the number of successive quadratic transformations to get a resolution of $f_0 = 0$ such that the total transform of $f_0 = 0$ has only ordinary double points and no two proper transforms of irreducible components intersect.

Case 1. Suppose that $\sigma^*(f_0) = 0$. Then either a or $b = 1$. $\Gamma(f_t)$ has only two possibilities and the assertion is obvious.

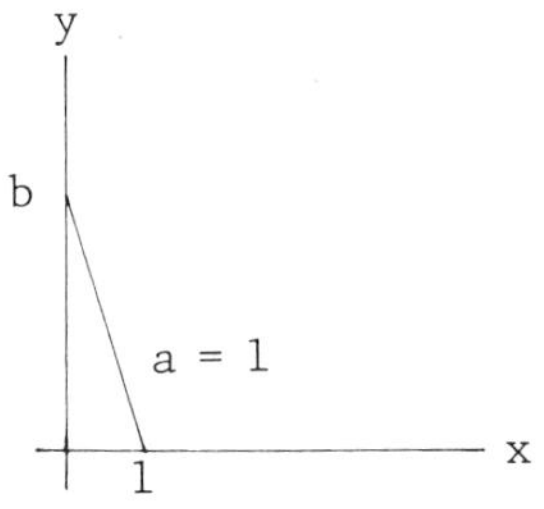

Figure I

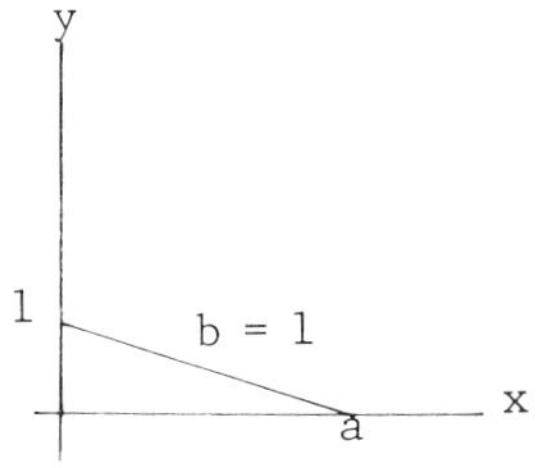

Figure I'

Case 2. Assume that $\sigma^*(f_0) > 0$. Let $m_1 = m(f_0)$ and let Δ_t be the tangential face of f_t. Let

$$f_{0,\Delta_0}(x, y) = x^{\gamma_1} y^{\beta_1} \prod_{\nu=1}^{d} (a_\nu x + b_\nu y)^{n_\nu}, \qquad a_\nu b_\nu \neq 0.$$

Assume that $\gamma_1 = \beta_1 = 0$. Then Δ_0 contains $(m_1, 0)$ and $(0, m_1)$ and this implies $\Gamma(f_0) = \Delta_0$ and $a = b = m_1$. Thus $\Delta_t = \Gamma(f_t) = \Gamma(f_0) = \Delta_0$. Assume that $\gamma_1 > 0$. We can write

$$f_{t,\Delta_t}(x, y) = (x + \varepsilon_1(t) y)^{\gamma_1} (y + \varepsilon_2(t) x)^{\beta_1} \prod_{\nu=1}^{d} (a_\nu(t) x + b_\nu(t) y)^{n_\nu}$$

because of Lemma 3.1 and the continuity $f_{t,\Delta_t} \to f_{0,\Delta_0}$. We can see that $\varepsilon_i(t) \to 0$, $a_\nu(t) \to a_\nu$, $b_\nu(t) \to b_\nu$ and $\varepsilon_i(t)$ $(i = 1, 2)$ are analytic in t for $|t|$ small enough. ($a_\nu(t)$, $b_\nu(t)$ are analytic in $t^{1/c}$ for some positive integer c.)

Assertion 1. $\varepsilon_i(t) \equiv 0$ and $\Delta_t = \Delta_0$.

PROOF. If $\varepsilon_1(t) \not\equiv 0$, Δ_t contains $(0, m_1)$ for t small enough which implies $(0, m_1) \in \Gamma(f_t) \cap \{x = 0\} = \{(0, b)\}$. This is impossible because $(0, m_1) \notin \Delta_0$. Similarly $\varepsilon_2(t) \equiv 0$ if $\beta_1 > 0$. Therefore $(\gamma_1, m_1 - \gamma_1)$ and $(m_1 - \beta_1, \beta_1)$ are the end points of Δ_t and $\Delta_t = \Delta_0$.

Now we consider the quadratic transformation of f_t. Let $(x, y) = (uv, v) = (u', u'v')$. Let $f_t^+(u, v)$ be the defining equation of the proper transform of the tangential component to $x = 0$. We have $f_t^+(u, v) = v^{-m_1} f_t(uv, v)$. Note that $\Gamma(f_t^+; (u, v)) \cap \{u = 0\} = \{(0, b - m_1)\}$ and $\Gamma(f_t^+; (u, v)) \cap \{u = 0\} = \{(\gamma_1, 0)\}$. $\Gamma(f_t^+)$ is the image of $\Gamma(f_t) \cap \{x \leqslant \gamma_1\}$ by the linear map $(i, j) \to (i, i + j + m_1)$. Note that f_t^+ depends on the choice of the coordinate (x, y) (see Figures II and III).

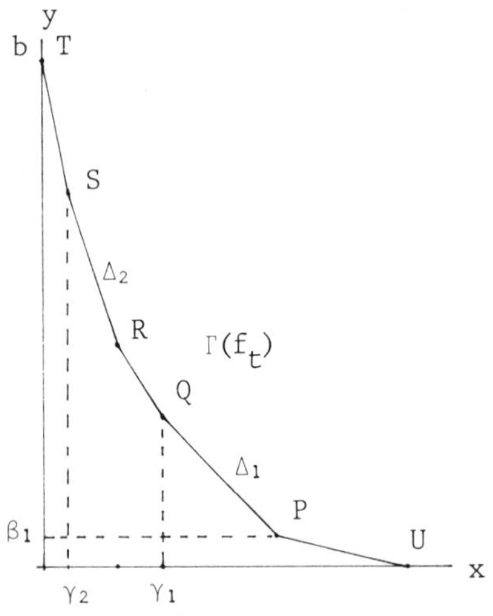

Figure II

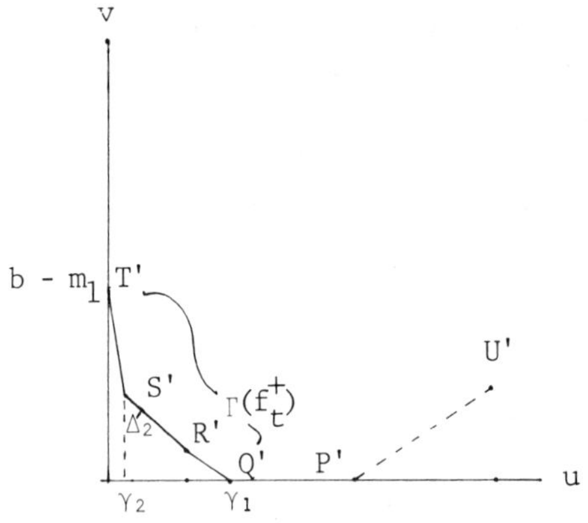

Figure III

Similarly the equation $f_0^-(u', v')$ of the proper transform of the tangential component to $y = 0$ is defined by $(u')^{-m_1} f_0(u', u'v')$ if $\beta_1 > 0$. We denote the equation of the proper transform of the tangential component to $a_\nu x + b_\nu y = 0$ by $h_\nu(u, v)$. Let $f_t^+(u, v)$, $f_t^-(u', v')$, $h_{\nu,t}(u, v)$ be the respective equation of the proper transform of the tangential components of $f_t = 0$ to $x = 0$, $y = 0$ and $a_\nu(t)x + b_\nu(t)y = 0$.

Assertion 2. f_t^+ and f_t^- are μ-constant deformations.

PROOF. By the semicontinuity of the Milnor number, we have $\mu(f_t^+) \leqslant \mu(f_0^+)$, $\mu(f_t^-) \leqslant \mu(f_0^-)$ and $\mu(h_{\nu,t}) \leqslant \mu(h_\nu)$. However the total sums of each side must be equal by Lemma 3.1. Thus we have $\mu(f_t^+) = \mu(f_0^+)$ and $\mu(f_t^-) = \mu(f_0^-)$.

Recall that $\Gamma(f_t^+) \cap \{u = 0\} = \{(0, b - m_1)\}$ and $\Gamma(f_t^+) \cap \{v = 0\} = \{(\gamma_1, 0)\}$. Applying the induction's hypothesis on f_t^+, we obtain $\Gamma(f_t^+) = \Gamma(f_0^+)$. This implies $\Gamma(f_t) \cap \{x \leqslant y_1\} = \Gamma(f_0) \cap \{x \leqslant \gamma_1\}$. Similarly we get $\Gamma(f_t^-) = \Gamma(f_0^-)$ which is equivalent to $\Gamma(f_t) \cap \{y \leqslant \beta_1\} = \Gamma(f_0) \cap \{y \leqslant \beta_1\}$. This completes the proof of Lemma 4.1.

PROOF OF THEOREM 1.1. Let $\{(a, 0)\} = \Gamma(f_0) \cap \{y = 0\}$ and $\{(0, b)\} = \Gamma(f_0) \cap \{x = 0\}$. Let Δ_1 be the tangential face of f_0, $m_1 = m(f_0)$ and let (γ_1, δ_1), (α_1, β_1), $(\gamma_1 \leqslant \alpha_1)$ be the end points of Δ_1. If $\gamma_1 \neq 0$, we define Δ_2 to be the face of $\Gamma(f_0)$ which corresponds to the tangential face Δ_2' of f_0^+ and let $m_2 = m(f_0^+)$ and $\gamma_2 = \min\{x; (x, y) \in \Delta_2 \text{ for some } y\}$. Δ_2 is defined by $\{(x, y) \in \Gamma(f_0); 2x + y = m_1 + m_2\}$ (see Figure II). Inductively we define $m_i = m(f_0^{(i-1)+})$ and

$\Delta_i = \{(x, y) \in \Gamma(f_0); ix + y = m_1 + \cdots + m_i\}$. Here f_0^{k+} is defined inductively by $f_0^{1+} = f_0^+$ and $f_0^{k+} = (f_0^{(k-1)+})^+$. Let $\gamma_i = \min\{x; (x, y) \in \Delta_i$ for some $y\}$. Δ_i corresponds to the tangential face of $f_0^{(i-1)+}$ by the correspondence $x^\alpha y^\beta \to u_{i-1}^\alpha v_{i-1}^{\beta+(i-1)\alpha-(m_1+\cdots+m_{i-1})}$.

First step. Let $f_{\Delta_1}(x, y) = x^{\gamma_1} y^{\beta_1} \prod_{\nu=1}^d (a_\nu x + b_\nu y)^{n_\nu}$ and $a_\nu b_\nu \neq 0$ for $\nu = 1, \ldots, d$. As $f_t = 0$ is equisingular, the tangential equation of f_t can be written as

$$(x + \varepsilon_1(t)y)^{\gamma_1}(y + \varepsilon_2(t)x)^{\beta_1} \prod_{\nu=1}^{d} (a_\nu(t)x + b_\nu(t)y)^{n_\nu}$$

where $\varepsilon_i(t) \to 0$ and $a_\nu(t) \to a$, $b_\nu(t) \to b$ and $\varepsilon_i(t)$ is analytic in t for $|t|$ sufficiently small. Let $\phi_1(x, y, t) = (x_1(t), y_1(t)) = (x + \varepsilon_1(t)y, y + \varepsilon_2(t)x)$. In this coordinate ϕ_1, the tangential face of f_t, coincides with Δ_1. The second step is the following assertion. We assume that $\gamma_r = 0 < \gamma_{r-1} \leqslant \gamma_{r-2} \leqslant \cdots \leqslant \gamma_1$.

Assertion 3. There exists an analytic family of coordinates $\phi_1(x, y, t) = (x_i(t), y_i(t))$ for $2 \leqslant i \leqslant r-1$ such that:

(i) $\phi_i(x, y, 0) = (x, y)$ and $\phi_i(0, 0, t) = 0$.

(ii) $\Gamma(f_t; \phi_i(t)) \cap \{\gamma_i \leqslant x \leqslant \gamma_1\} = \Gamma(f_0) \cap \{\gamma_i \leqslant x \leqslant \gamma_1\}$, and the tangential face of $f_t^{(i-1)+}$ is equal to that of $f_0^{(i-1)+}$.

(iii) $\Gamma(f_t; \phi_i(t)) \cap \{x \geqslant \alpha_1\} = \Gamma(f_t; \phi_1(t)) \cap \{x \geqslant \alpha_1\}$.

PROOF OF ASSERTION 3. Suppose that we have chosen ϕ_k satisfying (i), (ii) and (iii). We use the coordinate (u_k, v_k) for f_t^{k+} where $(x_k, y_k) = (u_k v_k^k, v_k)$. By the same argument as in the proof of Lemma 4.1 and (ii), (iii) for ϕ_k, $f_t^{k+}(u_k, v_k)$ is a μ-constant family and $f_t^{k+}(u_k, v_k) = v_k^{-(m_1+\cdots+m_k)} f_t(u_k v_k^k, v_k)$. Let $g(u_k, v_k)$ be the tangential equation of f_0^{k+} and write it as

$$u_k^{\gamma_{k+1}} v_k^{\beta_{k+1}} \prod_{\nu=1}^{s} (c_\nu u_k + d_\nu u_k)^{h_\nu}, \qquad c_\nu d_\nu \neq 0.$$

Let

$$(u_k + \varepsilon_1(t)v_k)^{\gamma_{k+1}} \cdot (v_k + \varepsilon_2(t)u_k)^{\beta_{k+1}} \prod_{\nu=1}^{s} (c_\nu(t)u_k + d_\nu(t)v_k)^{h_\nu}$$

be the tangential equation of f_t^{k+}. Note that $\varepsilon_i(t)$ are analytic in t and $\varepsilon_i(t) \to 0$ and $c_\nu(t)$ and $d_\nu(t)$ are nonzero complex numbers for t small. Moreover $\varepsilon_2(t) \equiv 0$ because $(\Gamma f_t^{k+}) \cap \{u = 0\} = \{(\gamma_k, 0)\}$ and $\gamma_k > m_{k+1}$ if $\beta_{k+1} > 0$.

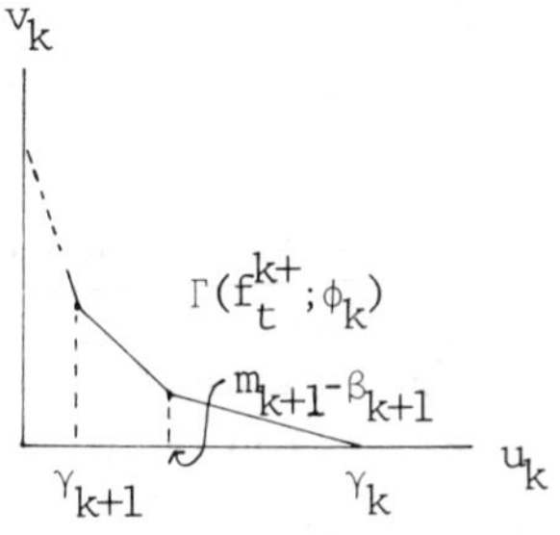

Figure IV

Let $\Delta_{k+1}(t)$ be the face of $\Gamma(f_t; \phi_k(t))$ which corresponds to the tangential face of f_t^{k+} by $(p, q) \mapsto (p, q + kp - m_1 - \cdots - m_k)$. We can see that $\Delta_{k+1}(t) = \{(p, q) \in \Gamma(f_t; \phi_k); (k+1)p + q = m_1 + \cdots + m_{k+1}\}$ and $f_{t,\Delta_{k+1}(t)}(x_k, y_k)$ is a weighted homogeneous polynomial of degree $m_1 \cdots + m_{k+1}$ under the weights $(k+1, 1)$. By the assumption, it is written as

$$y_k^a\left(x_k + \varepsilon_1(t) y_k^{k+1}\right)^{\gamma_{k+1}} \prod_{\nu=1}^{s} \left(c_\nu(t) x_k + d_\nu(t) y_k^{k+1}\right)^{h_\nu}$$

where $a = \sum_{i=1}^{k+1} m_i - (k+1)\delta$, $\delta = \gamma_{k+1} + h_1 + \cdots + h_s$ because the tangential equation of f_t^{k+} is equal to $v_k^{-(m_1 + \cdots + m_k)} f_{t,\Delta_{k+1}}(u_k v_k^k, v_k)$. (*Note.* $(x_k, y_k) = (u_k v_k^k, v_k)$.) Note also that $\beta_{k+1} = m_{k+1} - \delta$. Now we define $\phi_{k+1}(t) = (x_k + \varepsilon_1(t) y_k^{k+1}, x_k)$. As

$$\begin{aligned} x_k^p = y_k^q &= \left(x_{k+1} - \varepsilon_1(t) y_{k+1}^{k+1}\right)^p y_{k+1}^q \\ &= x_{k+1}^p y_{k+1}^q + \sum_{j=1}^{p} {}_pC_j \left(-\varepsilon_1(t)\right)^j x_{k+1}^{p-j} y_{k+1}^{q+(k+1)j}, \end{aligned}$$

one can see easily that the coefficients of $x_k^p y_k^q$ and $x_{k+1}^p y_{k+1}^q$ in the Taylor expansions of f_t in two coordinates (x_k, y_k) and (x_{k+1}, y_{k+1}) are equal if $(p, q) \in \Gamma(f_t; \phi_k) \cap \{p \geqslant \delta\}$ (see Figure IV′a and Figure IV′b).

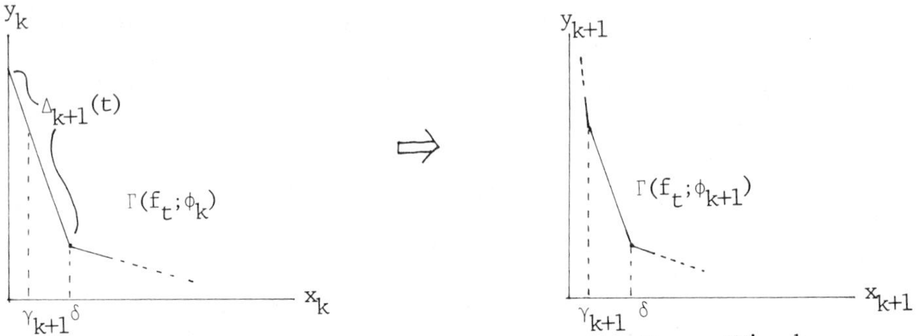

Figure IV' - a

Figure IV' - b

In particular, $\Gamma(f_t; \phi_k) \cap \{p \geqslant \delta\}$ is stable under the above change of the coordinates. We have

$$f_{t,\Delta_{k+1}(t)}(x_{k+1}, y_{k+1}) = y_{k+1}^a x_{k+1}^{\gamma_{k+1}} \prod_{\nu=1}^{s} \left(c'_\nu(t) x_{k+1} + d'_\nu(t) y_{k+1}^{k+1}\right)^{h_\nu}$$

where $c'_\nu(t) d'_\nu(t) \neq 0$ for any small t. Now the tangential equation of f_t^{k+} reduces to $\bar{u}_k^{\gamma_{k+1}} \bar{v}_k^{\beta_{k+1}} \prod_{\nu=1}^{s} d(c'_\nu(t)\bar{u}_k + d'_\nu(t)\bar{v}_k)^{h_\nu}$ where we put $(x_{k+1}, y_{k+1}) = (\bar{u}_k \bar{v}_k^k, \bar{v}_k)$. Now we apply Lemma 4.1 to the μ-constant family $(f_t^{k+})^-$ to see that (ii) is also true. This completes the proof of Assertion 3.

Now we consider the μ-constant family $f_t^{(r-1)+}$. Then $\Gamma(f_t^{(r-1)+}; (u_{r-1}, v_{r-1})) \cap \{v_{r-1} = 0\} = \{(\gamma_{r-1}, 0)\}$ and $m(f_t^{(r-1)+}) = m_r = b - (m_1 + \cdots + m_{r-1})$.

Thus $(0, m_r) \in \Gamma(f_0^{(r-1)+})$ and therefore $(0, m_r) \in \Gamma(f_t^{(r-1)+}; (u_{r-1}, v_{r-1}))$ for any sufficiently small t. Here $(x_{r-1}, y_{r-1}) = (u_{r-1}v_{r-1}^{r-1}, v_{r-1}^{r-1})$. Thus we apply Lemma 4.1 to obtain $\Gamma(f_t^{(r-1)+}) \cap \{x \leqslant \gamma_{r-1}\} = \Gamma(f_0^{(r-1)+}) \cap \{x \geqslant \gamma_{r-1}\}$. Therefore $\Gamma(f_t; \phi_{r-1}) = \Gamma(f_0)$ for $x \leqslant \gamma_1$. We do the same argument for f_t^-, $(f_t^-)^-$ etc. to complete the proof of Theorem 1.1.

REMARK. $(x(t), y(t))$ in Theorem 1.1 can be written as

$$x(t) = x_1(t) + \varepsilon_2(t)y_1(t)^2 + \cdots + \varepsilon_{r-1}(t)y_1(t)^{r-1}$$

and

$$y(t) = y_1(t) + \xi_2(t)x(t)^2 + \cdots + \xi_s(t)x(t)^s.$$

5. Newton boundaries of equivalent curves. Suppose that one is given two equivalent germs of plane curves C and D defined by reduced analytic functions f and g. Let (x, y) be a fixed coordinate. We will generalize Theorem 1.1 as follows.

THEOREM 5.1. *There exists a coordinate (z, w) such that* $\Gamma(f; (x, y)) \cap \{x \geqslant 1, y \geqslant 1\} = \Gamma(g; (z, w)) \cap \{z \geqslant 1, w \geqslant 1\}$.

The proof is completely parallel to the proof of Theorem 1.1.

LEMMA 5.2. *Suppose that C and D are equivalent and there exists two coordinates $(x, y), (z, w)$ such that*:

(i) $\tilde{C} = C \cup \{x = 0\} \cup \{y \neq 0\}$ *is equivalent to* $\tilde{D} = D \cup \{z = 0\} \cup \{w = 0\}$ *by the canonical bijection* $\tilde{\phi}$: $\tilde{C} \to \tilde{D}$ *where* $\tilde{\phi} \mid C$ *is the given equivalence and* $\phi(\{x = 0\}) = \{z = 0\}, \phi(\{y = 0\}) = \{w = 0\}$.

(ii) $\Gamma(f; (x, y)) \cap \{x = 0\} = \Gamma(g; (z, w)) \cap \{z = 0\} = \{(b, 0)\}$ *and* $\Gamma(f; (x, y)) \cap \{y = 0\} = \Gamma(g; (z, w)) \cap \{w = 0\} = \{(a, 0)\}$.

Then we have $\Gamma(f; (x, y)) = \Gamma(g; (z, w))$.

PROOF. We prove by the induction on $\sigma^*(C)$. The case $\sigma^*(C) = 0$ is trivial. Let Δ and Ξ be the tangential face of f and g in the respective coordinate. Let $f_\Delta(x, y) = x^\gamma y^\beta \prod_{\nu=1}^s (a_\nu x + b_\nu y)^{n_\nu}$ where $a_\nu b_\nu \neq 0$ for $\nu = 1, \ldots, d$. The integers γ and β are characterized by the multiplicities of the tangential components of C to $x = 0$ and $y = 0$ respectively. By the assumption, $g_2(z, w)$ can be expressed as

$$z^\gamma w^\beta \prod_{\nu=1}^{s} (c_\nu z + d_\nu w)^{n_\nu}, \qquad c_\nu d_\nu \neq 0.$$

Consider the quadratic transformation $(x, y) = (uv, v) = (u', u'v')$ and $(z, w) = (\zeta\xi, \xi) = (\zeta', \zeta'\xi')$. Then the exceptional divisors are defined by $v = 0$ (or $u' = 0$) and $\xi = 0$ (or $\zeta' = 0$) respectively and the lines $u = 0$ and $\zeta = 0$ are the proper transform of $x = 0$ and $z = 0$. Then $f^+(u, v)$ and $g^+(\zeta, \xi)$ are equivalent and satisfied the conditions in Lemma 5.2 by Lemma 1 of Zariski [7]. Thus we get $\Gamma(f^+; (u, v)) = \Gamma(g^+; (\zeta, \xi))$ by the induction hypothesis. Similarly we get $\Gamma(f^-; (u', v')) = \Gamma(g^-; (\zeta', \xi'))$. Thus we get $\Gamma(f; (x, y)) = \Gamma(g; (z, w))$. (Compare the proof with the proof of Lemma 4.1.)

PROOF OF THEOREM 5.1. We can assume that f is convenient by adding $x^N + y^N$ (N sufficiently large) if necessary. Let Δ_i, γ_i and m_i be as in the proof of Theorem 1.1 ($f_0 = f$). Let ϕ be the bijection of the irreducible components of C and D which gives an equivalence. Let $f_{\Delta_1}(x, y) = x^{\gamma_1} y^{\beta_1} \prod_{\nu=1}^{s} (a_\nu x + b_\nu y)^{n_\nu}$.

Case 1. $\gamma_1 = \beta_1 = 0$. Then $\Gamma(f; x, y) = \Delta_1$. Then we take any linear coordinates (z_1, w_1) whose axes are not tangent to any tangential components of D.

Case 2. $\gamma_1 > 0$ (or $\beta_1 > 0$). Let z_1 be the linear form which is tangent to the tangential component of D corresponding to the tangential component of C to $x = 0$. Define w_1 similarly. (If $\beta_1 = 0$, we choose w_1 so that $w_1 = 0$ is not tangent to any tangential components of D.) Then the tangential equation of $g(z_1, w_1)$ is of the form $z_1^{\gamma_1} w_1^{\beta_1} \prod_{\nu=1}^{s} (a'_\nu z_1 + b'_\nu w_1)^{n_\nu}$. Then ϕ maps the tangential component of C to $x = 0$ (respectively $y = 0$) to the tangential component of D to $z_1 = 0$ (respectively to $w_1 = 0$).

Let us carry out the quadratic transformation. Let $(x, y) = (u_1 v_1, v_1)$ and $(z_1, w_1) = (\zeta_1 \xi_1, \xi_1)$. $f^+(u_1, v_1)$ defines the proper transform of the tangential component of C to $x = 0$ and $v_1 = 0$ is the exceptional divisor. Similarly $g^+(\zeta_1, \xi_1)$ is the defining equation of the proper transform of the corresponding tangential component of D. $\xi_1 = 0$ is the exceptional divisor. $\{f^+ = 0\} \cup \{v_1 = 0\}$ is canonically equivalent to $\{g^+ = 0\} \cup \{\xi_1 = 0\}$ by Lemma 1 of [7]. Note that $\Gamma(f^+; (u_1, v_1)) \cap \{v_1 = 0\} = \Gamma(g^+; (\zeta_1, \xi_1)) \cap \{\xi_1 = 0\} = \{(\gamma_1, 0)\}$. Let $h(u_1, v_1)$ be the tangential equation of f^+. We can write

$$h(u_1, v_1) = u_1^{\gamma_2} v_1^{\beta_2} \prod_{\nu=1}^{e} (c_\nu u_1 + d_\nu v_1)^{p_\nu}.$$

β_2 is the multiplicity of the tangential component of $f^+ = 0$ to $v_1 = 0$. Thus we can write the tangential equation of $g^+ = 0$ as $(\zeta_1 + \varepsilon \xi_1) \cdot \xi_1^{\beta_2} \prod_{\nu=1}^{e} (c'_\nu \zeta_1 + d'_\nu \xi_1)^{p_\nu}$ where the tangential component of $\{g^+ = 0\}$ to $\zeta_1 + \varepsilon_1 = 0$ corresponds to the tangential component of $\{f^+ = 0\}$ to $u_1 = 0$.

Assertion 1. $\Gamma(f^+; (u_1, v_1)) \cap \{u_1 \geqslant m_2 - \beta_2\} = \Gamma(g^+; (\zeta_1, \xi_1)) \cap \{\zeta_1 \geqslant m_2 - \beta_2\}$.

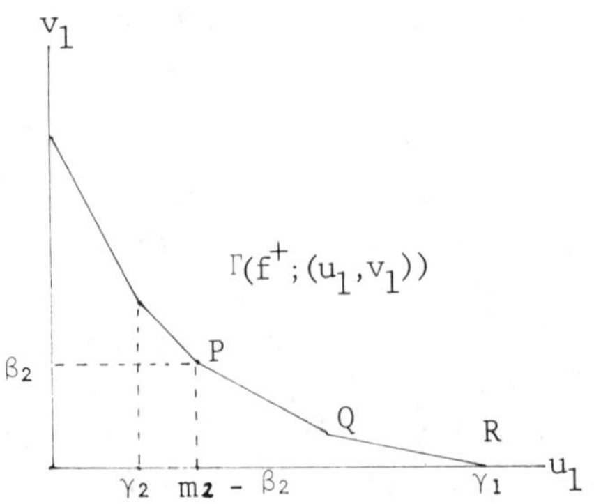

Figure V

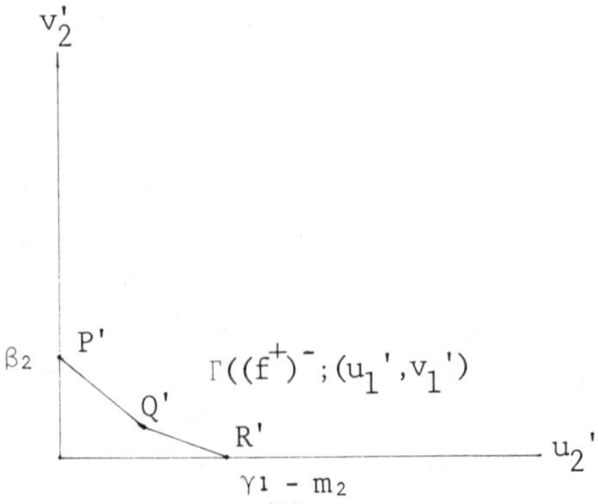

Figure VI

PROOF. We may assume that $\beta_2 \geqslant 0$. Consider the quadratic transformation $(u_1, v_1) = (u_2', u_2'v_2')$, $(\zeta_1, \xi_1) = (\zeta_2', \zeta_2'\xi_2')$. $(f^+)^-(u_2', v_2')$ and $(g^+)^-(\zeta_2', \xi_2')$ are the defining equations of the tangential component to $v_1 = 0$ and $\xi_1 = 0$ respectively. $u_2' = 0$ (respectivly $\zeta_2' = 0$) is the exceptional divisor and $v_2' = 0$ (or respectively $\zeta_2' = 0$) is the proper transform of $v_1 = 0$ (or $\xi_1 = 0$). Thus we can easily see that $(f^+)^-$, (u_2', v_2') and $(g^+)^-$, (ζ_2', ξ_2') satisfy the condition of Lemma 5.2, which implies the assertion (see Figures V and VI).

Take $z_2 = z_1 + \varepsilon w_1^2$ and $w_2 = w_1$. Then by the same discussion as in the proof of Assertion 3, we can see that

(i) $\Gamma(f; (x, y)) \cap \{\gamma_2 \leqslant x \leqslant \gamma_1\} = \Gamma(g; (z_2, w_2)) \cap \{\gamma_2 \leqslant z_2 \leqslant \gamma_1\}$,

(ii) $\Gamma(g; (z_2, w_2)) \cap \{w_2 \leqslant \beta_1\} = \Gamma(g; (z_1, w_1)) \cap \{w_1 \leqslant \beta_1\}$,

(iii) the tangential face of g^+ in $(\bar{\zeta}_1, \bar{\xi}_1)$ where $(z_2, w_2) = (\bar{\zeta}_1\bar{\xi}, \bar{\xi}_1)$ is equal to the tangential face f^+ of

(iv) $\{f^+ = 0\} \cup \{v_1 = 0\}$ is equivalent to $\{g^+ = 0\} \cup \{\bar{\xi}_1 = 0\}$ where

$$(x, y) = (u_1 v_1, v_1); \qquad (z_2, w_2) = (\bar{\zeta}_1\bar{\xi}_1, \bar{\xi}_1).$$

We proceed as in the proof of Theorem 1.1 to get a coordinate (z_k, w_k), $2 \leqslant k \leqslant r-1$, for g so that

(i) $\Gamma(f; (x, y)) \cap \{\gamma_k \leqslant x \leqslant \gamma_1\} = \Gamma(g; (z_k, w_k)) \cap \{\gamma_k \leqslant z_k \leqslant \gamma_1\}$,

(ii) $\Gamma(g; (z_k, w_k)) \cap \{z_k \geqslant \gamma_{k-1}\} = \Gamma(g; (z_{k-1}, w_{k-1})) \cap \{z_{k-1} \geqslant \gamma_{k-1}\}$,

(iii) the tangential face of $g^{(k-1)+}$ in (z_k, w_k) is equal to the tangential face of $f^{(k-1)+}$ and

(iv) $\{f^{(k-1)+} = 0\} \cup \{v_{k-1} = 0\}$ is equivalent to $\{g^{(k+1)+} = 0\} \cup \{\xi_{k-1} = 0\}$ where (u_{k-1}, v_{k-1}) (respectively (ζ_{k-1}, ξ_{k-1})) is the coordinate for $f^{(k-1)+}$ (respectively for $g^{(k-1)+}$) which is defined by $(x, y) = (u_{k-1}v_{k-1}^{k-1}, v_{k-1})$ and $(z_k, w_k) = (\zeta_{k-1}\xi_{k-1}^{k-1}, \xi_{k-1})$. (The equivalence is induced by the original equivalence.)

Now we consider $f^{(r-1)+}(u_{r-1}, v_{r-1})$ and $g^{(r-1)+}(\zeta_{k-1}, \xi_{k-1})$. By the assumption, $\Gamma(f^{(r-1)+}; (u_{r-1}, v_{r-1})) \cap \{v_{r-1} = 0\} = \Gamma(g^{(r-1)+}; (\zeta_{r-1}, \xi_{r-1})) \cap \{\xi_{r-1} = 0\} = \{(\gamma_{r-1}, 0)\}$ and $\Gamma(f^{(r-1)+}) \cap \{u_{r-1} = 0\} = \{(0, b - m_1 - \cdots - m_{r-1})\}$ and $m_r = m(f^{(r-1)+}) = b - m_1 - \cdots - m_{r-1}$. Let h be the tangential equation of $f^{(r-1)+}$ and write it as

$$v_{r-1}^{\beta_{r-1}} \prod_{\nu=1}^{d} (a_\nu u_{r-1} + b_\nu v_{r-1})^{n_\nu}, \qquad a_\nu b_\nu \neq 0.$$

Case 1. Assume that $d = 0$. Then the tangential equation of $g^{(r-1)+}$ must be $\xi_{r-1}^{\beta_{r-1}}$ because $m_r = \beta_{r-1} < \gamma_{r-1}$. Applying Lemma 5.2 to $(f^{(r-1)-}) = 0$ and $(g^{(r-1)-})^- = 0$, we conclude that

$$\Gamma(f; (x, y)) \cap \{x \leqslant \gamma_{r-1}\} = \Gamma(g; (z_{r-1}, w_{r-1})) \cap \{z_{r-1} \leqslant \gamma_{r-1}\}.$$

Case 2. Assume that $d > 0$. Then we have the possibility that the tangential equation of $g^{(r-1)+}$ is of the form $\zeta_{r-1}^{\alpha}\xi_{r-1}^{\beta_{r-1}}\prod_{\nu=1}^{d-1}(c_\mu\zeta_{r-1} + d_\mu\xi_{r-1})^{n_\nu'}$ where $c_\mu d_\mu \neq 0$ and $\{n_1, \ldots, n_d\} = \{\alpha, n_1', \ldots, n_{d-1}'\}$ (see Figure VII).

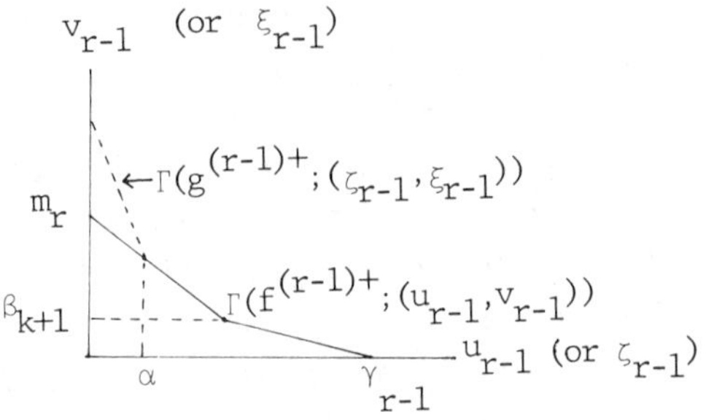

Figure VII

Case 2-1. Suppose that $\alpha = 0$. Then we have already $\Gamma(f;(x, y)) \cap \{x \leqslant \gamma_{r-1}\} = \Gamma(g;(z_{r-1}, w_{r-1})) \cap \{z_{r-1} \leqslant \gamma_{r-1}\}$.

Case 2.2. Suppose that $\alpha > 0$. Then we take $z_r = z_{r-1} + \varepsilon w_{r-1}^r$ and $w_r = w_{r-1}$ where $\varepsilon \neq 0$, $d\mu/c\mu$ for $\mu = 1,\ldots,d-1$. Then we can see that $\Gamma(f;(x, y)) \cap \{x \leqslant \gamma_{r-1}\} = \Gamma(g;(z_r, w_r)) \cap \{z_r \leqslant \gamma_{r-1}\}$.

For the lower half of $\Gamma(g;(z_r, w_r))$ we do the exact same argument for f^{k-} and g^{k-} to find a coordinate (z, w) for g such that $\Gamma(f;(x, y)) = \Gamma(g;(z, w))$. This completes the proof of Theorem 5.1.

Now we can sharpen Corollaries 1.2 and 1.3 as follows.

COROLLARY 5.3. *Let $f_t(x, y)$ be a μ-constant deformation and suppose that $\Gamma(f_0;(x, y))$ is nondegenerate. Then there is a coordinate (x_t, y_t) for f_t such that $\Gamma(f_t;(x_t, y_t))$ is nondegenerate.*

COROLLARY 5.4. *Let f and g be as in Theorem* 5.1. *Then $\alpha(f) = \alpha(g)$.*

The proof of Corollary 5.3 is obtained from the following observation.

Observation. Assume that f is nondegenerate in (x, y). Let Δ be the tangential face of f. Then we can write $f_\Delta(x, y) = x^\gamma y^\beta \prod_{\nu=1}^{d}(a_\nu x + b_\nu y)$ about $a_\nu b_\nu \neq 0$ and $a_\nu b_\mu - a_\mu b_\nu \neq 0$ for $\nu \neq \mu$. Moreover $f^+(u, v)$ and $f^-(u', v')$ is also nondegenerate in (u, v) and (u', v') respectively where $(x, y) = (uv, v) = (u', u'v')$.

The proof of Corollary 5.4 is derived from Theorem 5.1 and the fact that the Newton number of the triangle is zero if their vertices are $\{(0,0),(a,0),(a',1)\}$ or $\{(0,0),(0,b),(1,b')\}$.

REFERENCES

1. N. A'Campo, *La fonction zeta d'une monodromie*, Comment. Math. Helv. **50** (1975), 233–248.

2. A. G. Konchnirenko, *Polyèdres de Newton et nombres de Milnor*, Invent. Math. **32** (1976), 1–31.

3. A. N. Varchenko, *Zeta-function of monodromy and Newton's diagram*, Invent. Math. **37** (1976), 253–262.

4. D. T. Lê, *Sur un critère d'equisingularité*, C. R. Acad. Sci. Paris, Sér. A–B **272** (1971).

5. D. T. Lê and C. P. Ramanujam, *The invariance of Milnor's number implies the invariance of the topological type*.

6. M. Oka, *On the bifurcation of the multiplicity and topology of the Newton boundary*, J. Math. Soc. Japan **31** (1979), 435–450.

7. O. Zariski, *Studies in equisingularity*. I, Amer. J. Math. **87** (1965), 507–536; II, 972–1006.

TOKYO INSTITUTE OF TECHNOLOGY, JAPAN

Proceedings of Symposia in Pure Mathematics
Volume **40** (1983), Part 2

COXETER ARRANGEMENTS[1]

PETER ORLIK AND LOUIS SOLOMON

1. Introduction. An arrangement in $V = \mathbf{R}^l$ or $\mathbf{C}^l$ is a finite set $\mathcal{A}$ of hyperplanes, all containing the origin. Arrangements lead to interesting problems in topology and combinatorial geometry. For example: (i) If $H \in \mathcal{A}$ let α_H be a linear form with kernel H, so that $\bigcup_{H\in\mathcal{A}} H$ is the zero set of the polynomial $Q = \prod_{H\in\mathcal{A}}\alpha_H$. The singularity of Q at the origin may be studied using Saito's theory of vector fields with logarithmic poles [**15, 18**]. There is a paper by Terao on these singularities in this volume [**19**]. (ii) If $V = \mathbf{C}^l$ then $M = M(\mathcal{A}) = V - \bigcup_{H\in\mathcal{A}} H$ is an open connected submanifold of V. We may ask how various topological properties of M may be determined from $\mathcal{A}$. If $V = \mathbf{R}^l$ we may complexify and ask the same questions. This line of investigation began with work of Arnold [**2**], Brieskorn [**5**], and Deligne [**10**]. (iii) If $V = \mathbf{R}^l$ then each hyperplane H disconnects V. The number of connected components of $V - \bigcup_{H\in\mathcal{A}} H$ has been studied for one hundred fifty years and was determined by Zaslavsky [**20**] in 1975. It is remarkable that this number is equal to dim $H^*(M, \mathbf{C})$ where $M = M(\mathcal{A})$ is the corresponding complex manifold. Cartier [**9**] reviewed this interesting chapter of combinatorial geometry in the Bourbaki seminar.

In this paper we study certain arrangements $\mathcal{A}$ defined by Coxeter groups, and the Poincaré polynomials of the corresponding manifolds M. We begin with a summary of our results in [**12**] concerning the cohomology of M. Let $\mathcal{A}$ be any arrangement. Let $L = L(\mathcal{A})$ be the set of intersections of elements of $\mathcal{A}$. Partially order L by reverse inclusion so that L has V as its minimal element and $\mathcal{A}$ as its set of atoms. The poset L is a finite geometric lattice with rank function $r(X) = \dim(V/X)$, $X \in L$. The rank $r(L)$ of L is the rank of its maximal element $\bigcap_{H\in\mathcal{A}} H$. The Möbius function μ of L is an integer valued function defined recursively as follows: $\mu(X, X) = 1$, $\sum_{X\leq Z\leq Y}\mu(Z, Y) = 0$ if $X < Y$, and $\mu(X, Y) = 0$ otherwise. The characteristic polynomial $\chi(L, t)$ of L is defined by

$$\chi(L, t) = \sum_{X\in L} \mu(V, X)t^{r(L)-r(X)}. \tag{1.1}$$

1980 *Mathematics Subject Classification*. Primary 05B35, 14B05, 20C30, 57S25.

[1]This work was supported in part by the National Science Foundation.

0082-0717/81/0000-0805/$05.00

It is an important combinatorial invariant [**1**, p. 155]. We proved in [**12**] that the Poincaré polynomial $P(M, t) = \Sigma_{p \geq 0} \dim H^p(M, \mathbf{C}) t^p$ of the corresponding manifold M is given by the formula

$$P(M, t) = \sum_{X \in L} \mu(V, X)(-t)^{r(X)}. \tag{1.2}$$

Since L is a geometric lattice the sign of $\mu(V, X)$ is $(-1)^{r(X)}$ and thus (1.2) may be written as

$$P(M, t) = (-t)^{r(L)} \chi(L, -t^{-1}). \tag{1.3}$$

The cohomology ring $H^*(M, \mathbf{C})$ may be described as follows. If $H \in \mathcal{A}$ let α_H be a linear form with kernel H. Then $\omega_H = d\alpha_H / 2\pi i \alpha_H$ is a holomorphic 1-form on M. Let $[\omega_H]$ denote the corresponding deRham cohomology class. Let $\mathcal{R}$ be the graded $\mathbf{C}$-algebra of holomorphic differential forms on M generated by the $[\omega_H]$ and the identity. Brieskorn [**5**] showed that $\mathcal{R} \simeq H^*(M, \mathbf{C})$ as graded vector space. To describe the ring structure of $\mathcal{R}$ let $\mathcal{E} = \mathcal{E}(\mathcal{A})$ be the exterior algebra of a vector space with basis consisting of elements e_H in one to one correspondence with the hyperplanes $H \in \mathcal{A}$. Say that a subset S of $\mathcal{A}$ is independent if $\bigcap_{H \in S} H$ has codimension $|S|$, and is dependent otherwise. Thus S is independent when the hyperplanes of S are in general position. Define a $\mathbf{C}$-linear map $\partial: \mathcal{E} \to \mathcal{E}$ by $\partial 1 = 0$, $\partial e_H = 1$ and

$$\partial\left(e_{H_1} \cdots e_{H_p}\right) = \sum_{k=1}^{p} (-1)^{k-1} e_{H_1} \cdots \hat{e}_{H_k} \cdots e_{H_p}.$$

Let $\mathcal{I}$ be the ideal of $\mathcal{E}$ generated by all elements $\partial(e_{H_1} \cdots e_{H_p})$ where $\{H_1, \ldots, H_p\}$ is dependent. We proved in [**12**, Theorem 5.2] that the map $e_H \to [\omega_H]$ defines an isomorphism between the graded algebras $\mathcal{E}/\mathcal{I}$ and $H^*(M, \mathbf{C})$.

Let $G \subset \mathrm{GL}_l(\mathbf{R})$ be a finite reflection group and let $\mathcal{A} = \mathcal{A}(G)$ be the set of its reflecting hyperplanes. We call $\mathcal{A}$ a Coxeter arrangement and write $G = G(\mathcal{A})$. If $X \in L = L(\mathcal{A})$ let $\mathcal{A}^X$ be the arrangement in X consisting of all intersections $H \cap X$ where $H \in \mathcal{A}$ and H does not contain X. We may view G as a subgroup of $\mathrm{GL}_l(\mathbf{C})$ and, by abuse of notation, let $\mathcal{A}$ and $\mathcal{A}^X$ denote the complexified arrangements. Let $M^X = M(\mathcal{A}^X) = X - \bigcup_{H \in \mathcal{A}^X} H$ be the manifold corresponding to $\mathcal{A}^X$. Our main result in this paper is a factorization theorem for the Poincaré polynomial of M^X.

(1.4) THEOREM. *Let $\mathcal{A}$ be a Coxeter arrangement and let X be a p-dimensional subspace in the lattice $L(\mathcal{A})$. Let M^X be the manifold corresponding to the arrangement $\mathcal{A}^X$. Then there exist nonnegative integers $b_1^X, \ldots, b_p^X$ such that*

$$P(M^X, t) = \prod_{k=1}^{p} \left(1 + b_k^X t\right).$$

Since the lattice $L(\mathcal{A}^X)$ is isomorphic to the segment $L^X = \{Y \in L \mid Y \geq X\}$ of the lattice $L = L(\mathcal{A})$ formula (1.3) allows us to find the b_k^X by computing $\chi(L^X, t)$. Note that if $\bigcap_{H \in \mathcal{A}} H = 0$ then

$$\mu(X, 0) = b_1^X \cdots b_p^X. \tag{1.5}$$

We compute the b_k^X case by case. One might hope to apply Stanley's factorization theorem [**1**, p. 177]. This can be done for types A_{l-1}, B_l but not in general since there need not be a maximal chain of modular elements.

Suppose G is reducible and $V = V_1 \oplus V_2$ where V_i is stable under G. Then G defines a Coxeter arrangement $\mathcal{Q}_i$ in V_i. Let $L_i = L(\mathcal{Q}_i)$. Then $\chi(L, t) = \chi(L_1, t)\chi(L_2, t)$. The polynomials $\chi(L^X, t)$ factor in a similar way. Thus it is sufficient to compute the $\chi(L^X, t)$ for irreducible groups G. We assume G is irreducible. In the classical groups A_{l-1}, B_l, D_l we argue by induction on l. For types A_{l-1} and B_l, the lattices L^X depend only on $p = \dim X$ and are of type A_{p-1} and B_p. For type D_l the dimension of X does not determine the isomorphism type of L^X. In fact L^X never has type D_p when $p < l$. One must argue using a family of arrangements labeled D_l^k which are not Coxeter arrangements but interpolate between $B_l = D_l^l$ and $D_l = D_l^0$.

The exceptional groups require a very long computation. We say just enough to make the calculation reproducible and give few details. We use a square matrix $U = U(\mathcal{Q})$ whose rows and columns are indexed by the orbits of G on $L = L(\mathcal{Q})$. If $X \in L$ let $\mathcal{O}_X$ be the G-orbit of X. The entry of U in row X and column Y is the number of Z in L such that $Z \in \mathcal{O}_Y$ and $Z \geqslant X$. This entry depends only on the orbits $\mathcal{O}_X$ and $\mathcal{O}_Y$. Knowledge of the matrix U amounts to knowledge of the character values $\varphi_K(c_J)$ where φ_K is the character of G induced by the principal character of a parabolic subgroup G_K and c_J is a Coxeter element of a parabolic subgroup G_J. We use Carter's tables [**8**] of conjugacy classes in the exceptional Weyl groups. The group E_7 is particularly nasty; here we also use Frame's character table [**11**] and some numerical results of Beynon and Lusztig [**3**].

Brieskorn [**5**] has shown that

$$P(M, t) = (1 + m_1 t)\ldots(1 + m_l t) \tag{1.6}$$

where $m_1 \leqslant \cdots \leqslant m_l$ are the exponents of G. Thus when $X = V$ the integers b_k^X are the exponents of G. Since $\mathcal{Q}^X$ is not in general a Coxeter arrangement our theorem is not a consequence of this fact. Although we see no obvious connection between the b_k^X and the invariant theory of G, the numerical results show, when G is an irreducible group, that there exists for each $p = 0, \ldots, l$ a subspace $X = X_p$ of dimension p such that the b_k^X are $m_1, \ldots, m_p$. In particular, when $X = H$ is any hyperplane in $\mathcal{Q}$ the b_k^X are $m_1, \ldots, m_{l-1}$. In a sequel to this paper we will show that Theorem (1.4) has an analogue for unitary reflection groups. In case $X = V$ the integers b_k^X are the integers $n_1, \ldots, n_l$ we introduced in [**13**].

§2 of this paper contains the verification of Theorem (1.4) for the classical groups. In §3 we do the exceptional groups. §4 consists of miscellaneous remarks on the topology and combinatorics of Coxeter arrangements. We give the matrices $U(\mathcal{Q})$ and the integers b_k^X in an appendix.

We would like to thank Roger Carter for telling us the criterion "$W_1 = \overline{W_1}$" used in §3. We would also like to thank Anne Britt Orlik and Katrine Orlik for some help with the tables.

2. The classical groups. We choose a basis $x_1,\ldots,x_l$ for V^* so that the group of type A_{l-1} permutes $x_1,\ldots,x_l$ and the group of type B_l permutes $x_1,\ldots,x_l$ with sign changes. The group of type D_l permutes $x_1,\ldots,x_l$ with an even number of sign changes. Let $H_{ij} = \ker(x_i - x_j)$, let $\overline{H}_{ij} = \ker(x_i + x_j)$ and let $H_i = \ker(x_i)$. The arrangements $\mathcal{A}$ of classical type are $\mathcal{A}(A_{l-1}) = \{H_{ij}\}$, $\mathcal{A}(B_l) = \{H_{ij}\} \cup \{\overline{H}_{ij}\} \cup \{H_i\}$ and $\mathcal{A}(D_l) = \{H_{ij}\} \cup \{\overline{H}_{ij}\}$ where $1 \leqslant i < j \leqslant l$. If $X \in L(\mathcal{A})$ we let ρ: $V^* \to X^*$ be the restriction map; the dependence of ρ on X will be clear from the context. If $\mathcal{A}$ is the set of kernels of linear forms α, then $\mathcal{A}^X$ is the set of kernels of the nonzero linear forms $\rho(\alpha)$. If $H \in \mathcal{A}$ we let $y_i = \rho(x_i)$ for all i. We assume the known values of the exponents. Thus

$$\chi(L,t) = (t-1)(t-2)\cdots(t-(l-1)) \quad \text{in type } A_{l-1},$$
$$\chi(L,t) = (t-1)(t-3)\cdots(t-(2l-1)) \quad \text{in type } B_l.$$

If $\mathcal{A}$ has type A_{l-1} then L is isomorphic to the partition lattice and the formula for $\chi(L,t)$ is a known fact in combinatorics [**1**, p. 178].

(2.1) PROPOSITION. *Let $\mathcal{A}$ be an arrangement of type A_{l-1} and let $L = L(\mathcal{A})$. If $X \in L$ and* dim $X = p$ *then $\mathcal{A}^X$ has type A_{p-1} and thus* $\chi(L^X, t) = (t-1)(t-2)\cdots(t-(p-1))$.

PROOF. Arguing by induction it suffices to show that $\mathcal{A}^H$ has type A_{l-2} for all $H \in \mathcal{A}$. Since G acts transitively on $\mathcal{A}$ we may assume that $H = H_{l-1,l}$. Then $\rho(x_i - x_j) = y_i - y_j$ for $1 \leqslant i < j \leqslant l-1$, $\rho(x_i - x_l) = y_i - y_{l-1}$ for $1 \leqslant i < l-1$ and $\rho(x_{l-1} - x_l) = 0$. □

(2.2) PROPOSITION. *Let $\mathcal{A}$ be an arrangement of type B_l and let $L = L(\mathcal{A})$. If $X \in L$ and* dim $X = p$ *then $\mathcal{A}^X$ has type B_p and thus* $\chi(L^X, t) = (t-1)(t-3)\cdots(t-(2p-1))$.

PROOF. It suffices to show that $\mathcal{A}^H$ has type B_{l-1} for all $H \in \mathcal{A}$. Since G has two orbits on $\mathcal{A}$ we may assume that $H = H_l$ or $H = H_{l-1,l}$. If $H = H_l$ then the set of restrictions to H of the forms defining $\mathcal{A}$ consists of the forms 0 and $y_i - y_j$, $y_i + y_j$, y_i for $1 \leqslant i < j \leqslant l-1$ so the assertion is clear. If $H = H_{l-1,l}$ then $2y_{l-1} = \rho(x_{l-1} + x_l)$ also occurs as a restriction but it is superfluous since it defines the same hyperplane as y_{l-1}. □

We have just seen that for types A_{l-1}, B_l the lattice $L(\mathcal{A}^X) \simeq L^X$ depends only on dim X. This is not true for type D_l. For each $l \geqslant 2$ and $k = 0,\ldots,l$ let D_l^k be the arrangement consisting of the kernels of the linear forms $x_i \pm x_j$ for $1 \leqslant i < j \leqslant l$ and $x_1,\ldots,x_k$. These arrangements, which interpolate between $D_l^0 = D_l$ and $D_l^l = B_l$ were introduced by Zaslavsky [**21**]. They are not Coxeter arrangements if $0 < k < l$. In view of symmetry we may replace $x_1,\ldots,x_k$ by $x_{i_1},\ldots,x_{i_k}$ for any $1 \leqslant i_1 < \ldots < i_k \leqslant l$. We show in (2.3) that the restriction $\mathcal{A}^H$ of an arrangement $\mathcal{A}$ of type D_l^k to a hyperplane $H \in \mathcal{A}$ has type D_{l-1}^h for some h with $0 \leqslant h \leqslant l-1$. Thus the set of all arrangements D_l^k is closed under restriction to a hyperplane. Arguing by induction, we see that if $\mathcal{A}$ has type D_l and $X \in L(\mathcal{A})$ then the lattice $L^X \simeq L(\mathcal{A}^X)$ has type D_p^r where $p = \dim X$ and $0 \leqslant r \leqslant p$. In (2.6) we determine r in terms of X.

(2.3) PROPOSITION. *Let* $\mathcal{A}$ *be an arrangement of type* D_l^k *and let* $H \in \mathcal{A}$. *The type of* $\mathcal{A}^H$ *is given in the table*:

k	H		type $\mathcal{A}^H$
0	arbitrary		D_{l-1}^1
$1,\ldots,l-1$	H_{ij} or $\overline{H}_{ij}$	$1 \leqslant i < j \leqslant k < l$	D_{l-1}^{k-1}
$1,\ldots,l-1$	H_{ij} or $\overline{H}_{ij}$	$1 \leqslant i \leqslant k < j \leqslant l$	D_{l-1}^{k}
$1,\ldots,l-1$	H_{ij} or $\overline{H}_{ij}$	$1 < k < i < j \leqslant l$	D_{l-1}^{k+1}
$1,\ldots,l-1$	H_i	$1 \leqslant i \leqslant k$	$D_{l-1}^{l-1} = B_{l-1}$
l	arbitrary		$D_{l-1}^{l-1} = B_{l-1}$

PROOF. If $k = 0$ then $\mathcal{A}$ has type D_l. Since the group of type D_l acts transitively on $\mathcal{A}$ we may assume that $H = H_{l-1,l}$. Thus the set of restrictions to H of the linear forms defining $\mathcal{A}$ consists of the forms $y_i \pm y_j$ for $1 \leqslant i < j \leqslant l-1$ together with $2y_{l-1}$, so that $\mathcal{A}^H$ has type D_{l-1}^1. If $k = l$ then $\mathcal{A}$ has type B_l and it follows from (2.2) that $\mathcal{A}^H$ has type $B_{l-1} = D_{l-1}^{l-1}$. Suppose $1 \leqslant k \leqslant l-1$. All four cases are argued in similar fashion. We do the case where H is H_{ij} or $\overline{H}_{ij}$ and $1 < k < i < j \leqslant l$. We may assume without loss of generality that $H = H_{l-1,l}$. Then

$$\rho(x_i) = y_i \text{ for } 1 \leqslant i \leqslant k; \qquad \rho(x_i \pm x_j) = y_i \pm y_j \text{ for } 1 \leqslant i < j \leqslant l-1;$$

$$\rho(x_i \pm x_l) = y_i \pm y_{l-1} \text{ for } 1 \leqslant i < l-1; \qquad \rho(x_{l-1} + x_l) = 2y_{l-1};$$

$$\rho(x_{l-1} - x_l) = 0.$$

Thus $\mathcal{A}^H$ has type D_{l-1}^{k+1}. □

We compute the polynomial $\chi(L, t)$ for an arrangement of type D_l^k by induction. Let $\mathcal{A}$ be any arrangement in V and let $H \in \mathcal{A}$. Define $L' = L(\mathcal{A} - \{H\})$ and $L'' = L(\mathcal{A}^H)$. The method of deletion and contraction **[6]** yields the formula

$$\chi(L, t) = \chi(L', t) - \chi(L'', t). \tag{2.4}$$

(2.5) PROPOSITION. *Let* $\mathcal{A}$ *be an arrangement of type* D_l^k *and let* $L = L(\mathcal{A})$. *Then*

$$\chi(L, t) = (t-1)(t-3)\cdots(t-(2l-3))(t-(k+l-1)).$$

PROOF. We argue by induction on l and for fixed l by descending induction on k. The arrangement $\mathcal{A}$ of type D_2^k for $k = 0, 1, 2$ consists of $k+2$ lines and thus $\chi(L, t) = (t-1)(t-(k+1))$. This starts the induction. For $k = l$ the assertion follows from (2.2) since $D_l^l = B_l$. Suppose $k < l$. Write $P = (t-1)(t-3)\ldots(t-(2l-3))$. Let $\mathcal{A}$ be an arrangement of type D_l^{k+1} and let $H = H_{k+1}$. Then $\mathcal{A} - \{H\}$ has type D_l^k by definition and $\mathcal{A}^H$ has type B_{l-1} by (2.3), so that (2.4) and induction give $\chi(L, t) = (t-(k+l))P + P = (t-(k+l-1))P$. □

If $\mathcal{A}$ has type D_l and $X \in L(\mathcal{A})$ then it follows from (2.3) that $\mathcal{A}^X$ has type D_p^r for some $0 \leqslant r \leqslant p$. For given X we want to determine the value of r. Then the polynomial $\chi(L^X, t)$ is given by (2.5). Define a relation $\sim$ on $\{1,\ldots,l\}$ as follows: Write $i \sim j$ if either $i = j$ or $X \subseteq H_{ij}$ or $X \subseteq \overline{H}_{ij}$. Note that $i \sim j$ if and only if $y_i = y_j$ or $y_i = -y_j$ so $\sim$ is an equivalence relation. There are two kinds of equivalence classes:

(i) For all $i \neq j$ in the class either $X \subseteq H_{ij}$ or $X \subseteq \overline{H}_{ij}$ but not both. Label these classes $\Lambda_1, \ldots, \Lambda_p$.

(ii) There exists a pair of indices $i \neq j$ in the class such that $X \subseteq H_{ij} \cap \overline{H}_{ij}$. This implies $\rho(x_i - x_j) = 0 = \rho(x_i + x_j)$ and thus $y_i = 0 = y_j$. It follows that $y_k = 0$ for all k in the class and thus $X \subseteq H_{ij} \cap \overline{H}_{ij}$ for all $i \neq j$ in the class. There is at most one such class. If it exists call it Δ.

(2.6) PROPOSITION. *Let $\mathcal{Q}$ be an arrangement of type D_l. Suppose $X \in L$ has dimension p and the associated equivalence classes are $\Lambda_1, \ldots, \Lambda_p, \Delta$. Then $\mathcal{Q}^X$ has type $D_p^p = B_p$. If there is no class Δ then $\mathcal{Q}^X$ has type D_p^r where r is the number of Λ_i with $|\Lambda_i| > 1$. Thus*

$$\chi(L^X, t) = (t-1)(t-3) \cdots (t-(2p-3))(t-(p+r-1)).$$

PROOF. We may choose notation so that $i \in \Lambda_i$ for $i = 1, \ldots, p$. Then $\{y_1, \ldots, y_p\}$ is a basis for X^* and thus $\dim X = p$. The arrangement $\mathcal{Q}$ is defined by the set of linear forms $\mathcal{B} = \{x_i \pm x_j: 1 \leqslant i < j \leqslant l\}$. The arrangement $\mathcal{Q}^X$ consists of the kernels of the linear forms in $\rho(\mathcal{B})' = \{\rho(\alpha) \mid \alpha \in \mathcal{B}$ and $\rho(\alpha) \neq 0\}$. If $1 \leqslant i \leqslant l$ then either $y_i = 0$ or $y_i \in \{\pm y_1, \ldots, \pm y_p\}$. Thus $\rho(\mathcal{B})'$ consists of the $y_i \pm y_j$ with $1 \leqslant i < j \leqslant p$ and, possibly, some forms y_i or $2y_i$ with $1 \leqslant i \leqslant p$. Suppose first that there is a class Δ. Choose $m \in \Delta$. Then $\rho(x_i \pm x_m) = y_i$ for$1 \leqslant i \leqslant p$ so that $\rho(\mathcal{B})' \supseteq \{y_i \pm y_j, y_i: 1 \leqslant i < j \leqslant p\}$. Note that $\rho(\mathcal{B})'$ may also contain some forms $2y_i$ for $1 \leqslant i \leqslant p$ but these are superfluous. Thus $\mathcal{Q}^X$ has type B_p. If there is no class Δ then label the classes $\Lambda_1, \ldots, \Lambda_p$ so that $|\Lambda_k| > 1$ for $1 \leqslant k \leqslant r$ and $|\Lambda_k| = 1$ for $r+1 \leqslant k \leqslant p$. Recall that $k \in \Lambda_k$. If $1 \leqslant k \leqslant r$ we may choose $k' \in \Lambda_k$ with $k < k'$. Thus $X \subseteq H_{kk'}$ or $X \subseteq \overline{H}_{kk'}$ but not both. Say $X \subseteq H_{kk'}$. Then $\rho(x_k - x_{k'}) = 0$ and $\rho(x_k + x_{k'}) = 2y_k \neq 0$. Thus $\rho(\mathcal{B})' = \{y_i \pm y_j: 1 \leqslant i < j \leqslant p\} \cup \{y_k: 1 \leqslant k \leqslant r\}$, so that $\mathcal{Q}^X$ has type D_p^r. $\square$

3. The exceptional groups. We begin with some general remarks on Coxeter arrangements and then apply them to computing the characteristic polynomials $\chi(L^X, t)$ for the exceptional groups. We assume that $V = \mathbf{R}^l$ and for simplicity that $\bigcap_{H \in \mathcal{Q}} H = 0$. Thus the rank of L^X is $\dim X$. Since the Möbius function of L^X is the restriction to L^X of the Möbius function of L we have

$$\chi(L^X, t) = \sum_{\substack{Y \in L \\ Y \geqslant X}} \mu(X, Y) t^{\dim Y}. \tag{3.1}$$

Thus, using the definition of μ we get

$$\begin{aligned} \sum_{\substack{Y \in L \\ Y \geqslant X}} \chi(L^Y, t) &= \sum_{\substack{Y \in L \\ Y \geqslant X}} \left(\sum_{\substack{Z \in L \\ Z \geqslant Y}} \mu(Y, Z) t^{\dim Z} \right) \\ &= \sum_{Z \geqslant X} \left(\sum_{X \leqslant Y \leqslant Z} \mu(Y, Z) \right) t^{\dim Z} = t^{\dim X}. \end{aligned} \tag{3.2}$$

We will use (3.2) to compute $\chi(L^X, t)$ recursively starting with $X = 0$ where $\chi(L^X, t) = 1$. If $\dim X = 1$ then L^X consists of X and 0 so that $\chi(L^X, t) = t - 1$. If $\dim X = 2$ and X contains $n + 1$ lines of L then $\chi(L^X, t) = (t-1)(t-n)$. The integer n depends on X. This is the simplest nontrivial case, but even here the determination of n in type E_6, E_7, E_8 requires some effort.

The group G permutes L. We rewrite (3.2) as a sum over G-orbits. If $X \in L$ let $\mathcal{O}_X$ be the G-orbit of X. If $X, Y \in L$ let $u(X, Y)$ be the number of $Z \in L$ such that $Z \in \mathcal{O}_Y$ and $Z \geqslant X$. This number depends only on $\mathcal{O}_X$ and $\mathcal{O}_Y$. The polynomial $\chi(L^Y, t)$ depends only on $\mathcal{O}_Y$ and the sum on the left-hand side of (3.2) depends only on $\mathcal{O}_X$. Let $\Xi = \{X_1, \ldots, X_r\}$ be a set of representatives for the G-orbits on L numbered so that $\dim X_i \geqslant \dim X_j$ if $i \leqslant j$. Let $\mathcal{O}_i = \mathcal{O}_{X_i}$, let $u_{ij} = u(X_i, X_j)$ and let $\chi_i(t) = \chi(L^{X_i}, t)$. Then $U = U(\mathcal{A}) = (u_{ij})$ is an upper triangular $r \times r$ matrix with $u_{ii} = 1$. We may write (3.2) as

$$\sum_{j \geqslant i} u_{ij}\chi_j(t) = t^{\dim X_i}. \tag{3.3}$$

Thus to compute the characteristic polynomials $\chi(L^X, t)$ we must compute $U = U(\mathcal{A})$. To do this, the first step is to find the G-orbits. If $x \in V$ let $G_x = \{g \in G \mid gx = x\}$ be its fixer. If $X \subseteq V$ let $G_X = \bigcap_{x \in X} G_x$. If G_0 is a subgroup of G let $\mathrm{Fix}(G_0)$ be the subspace of V fixed by G_0 and let $N(G_0)$ be its normalizer in G.

(3.4) LEMMA. *Let $\mathcal{A}$ be a Coxeter arrangement and let $L = L(\mathcal{A})$. If $X \in L$ then* $\mathrm{Fix}(G_X) = X$. *Two elements $X, Y \in L$ lie in the same G-orbit if and only if G_X, G_Y are conjugate in G. Moreover $N(G_X) = \{g \in G \mid gX \subseteq X\}$ and* $|\mathcal{O}_X| = |G\colon N(G_X)|$.

PROOF. If $H \in \mathcal{A}$ then $H \subseteq \mathrm{Fix}(G_H)$ so $\mathrm{Fix}(G_H) = H$ or $\mathrm{Fix}(G_H) = V$. Since G_H contains a reflection we have $\mathrm{Fix}(G_H) = H$. If $X \in L$ and $X \subseteq H$ then $\mathrm{Fix}(G_X) \subseteq \mathrm{Fix}(G_H) = H$. Since $X \in L$ is the intersection of all $H \in \mathcal{A}$ containing it we have $\mathrm{Fix}(G_X) \subseteq X$ and thus $\mathrm{Fix}(G_X) = X$. Clearly if $Y = gX$ for some $g \in G$ then $G_Y = gG_Xg^{-1}$. Conversely if $G_Y = gG_Xg^{-1}$ then $Y = \mathrm{Fix}(G_Y) = \mathrm{Fix}(gG_Xg^{-1}) = \mathrm{Fix}(G_{gX}) = gX$. The last statement follows by taking $Y = X$. □

Since $V = \mathbf{R}^l$ the complement $V - \bigcup_{H \in \mathcal{A}} H$ of the union of the reflecting hyperplanes is a union of connected components, called chambers, which are congruent under G [**4**, V, §3, Théorème 2]. Choose a chamber C. The group G is generated by the set S of reflections in the hyperplanes which bound C and (G, S) is a Coxeter system [**4**, V, §3, Théorème 1]. For any $J \subseteq S$ let G_J be the subgroup generated by J. These are the parabolic subgroups. If $X \in L$ then G_X is generated by the reflections in G which fix X, and G_X is conjugate to G_J for some J [**4**, V, §3, Proposition 2]. Thus the orbits are in one-to-one correspondence with the conjugacy classes of parabolic subgroups. If $J \subseteq S$ let c_J be a Coxeter element of G_J, namely the product of elements of J. The conjugacy class of c_J in G_J does not depend on the order in the product and thus the conjugacy class in G depends neither on the chamber nor the order in the product. We call the conjugacy classes containing elements c_J parabolic classes.

(3.5) LEMMA. *Let (G, S) be a finite Coxeter system and let J, K be subsets of S. Then G_J and G_K are conjugate in G if and only if c_J and c_K are conjugate in G.*

PROOF. Lemma 7 of [**16**] shows that if c_J is conjugate to c_K then G_J is conjugate to G_K. For the converse we assume that G is a Weyl group so that we may use the results of [**7**]; the same arguments work in all finite Coxeter groups with obvious modifications, replacing the root system by a collection of unit vectors perpendicular to the hyperplanes $H \in \mathcal{Q}$. Let $\Phi \subset V$ be a root system for G and let $\Pi \subset \Phi$ be a fundamental system of roots corresponding to the chamber C. If $\alpha \in \Phi$ let $s_\alpha \in G$ be the corresponding reflection. Let $\Pi_J = \{\alpha \in \Pi \mid s_\alpha \in J\}$, let V_J be the subspace of V spanned by Π_J and let $\Phi_J = \Phi \cap V_J$.

Suppose $G_K = gG_Jg^{-1}$. Since [**7**, Proposition 2.5.1] $\alpha \in \Phi_J$ if and only if $s_\alpha \in G_J$ it follows that $\Phi_K = g\Phi_J$. Then both Π_K and $g\Pi_J$ are fundamental systems for the root system Φ_K and hence [**7**, Theorem 2.2.4] there exists $h \in G_K$ such that $hg\Pi_J = \Pi_K$. Thus $(hg)c_J(hg)^{-1}$ is a Coxeter element of G_K. □

Lemmas (3.4) and (3.5) show that there is a one-to-one correspondence between orbits of G on L and parabolic conjugacy classes. We determine the parabolic classes for the exceptional Weyl groups using Carter's tables [**8**, Tables 8–11]. The rows in these tables correspond to conjugacy classes in G. Carter [**8**, p. 6] associates to each conjugacy class a graph Γ. One sees directly from the definition that the graph Γ associated to the class of c_J is the Dynkin graph $\Gamma(\Phi_J)$ of the root system Φ_J. Thus the rows corresponding to parabolic classes are labeled by (full) subgraphs of $\Gamma(\Phi)$. One must first enumerate these subgraphs. It may happen that there are two rows in a table with the same graph. For example in E_7 we have pairs of rows $(A_1^3)'$, $(A_1^3)''$, $(A_1^4)'$, $(A_1^4)''$, $\cdots$. Some of these are not parabolic classes. Since $\Phi_J = \Phi \cap \mathbf{R}\Phi_J$ we see (using Carter's notation) that with $\Phi_1 = \Phi_J$ we have $W_1 = G_J$ and also $\overline{W}_1 = G_J$ [**8**, p. 39]. Thus the rows corresponding to parabolic classes must satisfy $W_1 = \overline{W}_1$. For example in E_7 the class labeled $(A_1^4)''$ has $W_1 = A_1^4$ and $\overline{W}_1 = D_4$ so it is not parabolic. If one of the classes in a pair is eliminated in this way then the other one must be parabolic. This picks out the parabolic classes in E_6 and E_8 and leaves only the following pairs in E_7:

$$(A_1^3)', (A_1^3)'', (A_1 \times A_3)', (A_1 \times A_3)'', A_5', A_5''. \tag{3.6}$$

These six classes are in fact parabolic.

In F_4 one must remember that there are long and short roots. Graphs of type A_i corresponding to long (short) roots are denoted $A_i(\tilde{A}_i)$. The criterion "$W_1 = \overline{W}_1$" picks out the parabolic classes here as in E_6, E_7, E_8. Note that column six in Carter's tables gives the cardinality $|\mathcal{O}_J| = |G: N(G_J)|$ of the orbit $\mathcal{O}_J$ corresponding to c_J.

It is easy to compute the parabolic classes and the cardinalities of the orbits for the remaining irreducible Coxeter groups of types H_3, H_4, $I_2(p)$. In types H_3, H_4 and $I_2(p)$ with p odd the group acts transitively on $\mathcal{Q}$ so there is just one orbit of type A_1. In type $I_2(p)$ with p even there are two orbits of type A_1, denoted A_1 and $\tilde{A}_1$. In types H_l ($l = 3$ or 4) the G_J with $|J| = l - 1$ have different orders and hence are not conjugate. In H_4 the same remark applies to the G_J with $|J| = 2$ except for the subgroups of type A_1^2. The Coxeter diagram of H_4 is

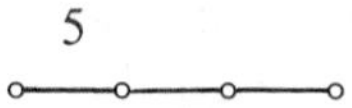

Number the nodes $1, 2, 3, 4$ from left to right. Then s_3 is conjugate to s_4 inside $\langle s_3, s_4 \rangle$ which centralizes s_1 and thus $G_{1,3}$ is conjugate to $G_{1,4}$. Similarly $G_{2,4}$ is conjugate to $G_{1,4}$. Thus if we label the class of the identity A_0, the list of parabolic classes is:

Type H_3:	$A_0, A_1, A_1^2, A_2, I_2(5), H_3$.	
Type H_4:	$A_0, A_1, A_1^2, A_2, I_2(5), A_1 \times A_2, A_1 \times I_2(5), A_3, H_3, H_4$	
Type $I_2(p)$:	$A_0, A_1, I_2(p)$	p odd
	$A_0, A_1, \tilde{A}_1, I_2(p)$	p even.

This completes the determination of the parabolic classes for all the exceptional groups.

Next we show how to compute the entries of U. Let $\mathfrak{T}$ be the set of all Coxeter graphs of finite Coxeter groups. The groups may be reducible and we allow the empty graph labeled A_0, corresponding to $G = 1$. We introduce a partial order on $\mathfrak{T}$. If $R, T \in \mathfrak{T}$ say $R \leqslant T$ if R is a subgraph of T. We mean a full subgraph consisting of a subset of the set of vertices of T and all edges of T connecting them. For example the graphs $\leqslant D_4$ are A_0, A_1, A_1^2, A_2, A_1^3, A_3, D_4. If $X \in L$, the fixer G_X is a Coxeter group so we may define the graph $\Gamma(X)$ as the graph of G_X. If $T \in \mathfrak{T}$ let $L(T)$ be the lattice of a Coxeter arrangement with graph T. Note that $R \leqslant T$ if and only if there exists $X \in L(T)$ such that $\Gamma(X) = R$. For $R, T \in \mathfrak{T}$ let $n(R, T)$ be the number of subspaces $X \in L(T)$ with $\Gamma(X) = R$. Note that $n(R, T) = 0$ unless $R \leqslant T$. We need the integers $n(R, T)$ to compute the matrix U. If T is irreducible, the integers $n(R, T)$ must be computed directly from the lattice $L(T)$. Since this requires some effort we exhibit the values we need in Table I. If $T = T_1 \times \cdots \times T_k$ where the T_i are irreducible then $n(R, T) = \Sigma n(R_1, T_1) \cdots n(R_k, T_k)$ summed over all k-tuples of graphs such that $R = R_1 \times \cdots \times R_k$ and $R_i \leqslant T_i$ for $1 \leqslant i \leqslant k$. Thus for example if $R = A_1 \times A_2$ and $T = A_2 \times D_5$ then

$$
\begin{aligned}
n(A_1 \times A_2, A_2 \times D_5) &= n(A_0, A_2)n(A_1 \times A_2, D_5) \\
&\quad + n(A_1, A_2)n(A_2, D_5) + n(A_2, A_2)n(A_1, D_5) \\
&= 1 \cdot 80 + 3 \cdot 40 + 1 \cdot 20 = 220.
\end{aligned}
$$

The values $n(\ ,\ )$ are in Table I.

(3.7) Lemma. *Let $\mathcal{A}$ be a Coxeter arrangement. Let $L = L(\mathcal{A})$ and let $X, Y \in L$. Let $X_1, \ldots, X_q$ represent the G-orbits for which $\Gamma(X_i) = \Gamma(X)$. Then*

$$\sum_{i=1}^{q} |\mathcal{O}_i| u(X_i, Y) = |\mathcal{O}_Y| n(\Gamma(X), \Gamma(Y)).$$

PROOF. We count the set

$$\Omega(X, Y) = \{(Z, W) \in L \times L \mid \Gamma(Z) = \Gamma(X), \mathcal{O}_W = \mathcal{O}_Y, Z \leqslant W\}$$

in two ways. Let $L_W = \{Z \in L \mid Z \leqslant W\}$. Then

$$\Omega(X, Y) = \bigcup_{W \in \mathcal{O}_Y} \{Z \in L_W \mid \Gamma(Z) = \Gamma(X)\}.$$

Since $L_W \simeq L(\Gamma(W)) = L(\Gamma(Y))$, each set in the union has cardinality $n(\Gamma(X), \Gamma(Y))$ and thus $|\Omega(X, Y)| = |\mathcal{O}_Y| \, n(\Gamma(X), \Gamma(Y))$. On the other hand,

$$\Omega(X, Y) = \bigcup_{\substack{Z \in L \\ \Gamma(Z) = \Gamma(X)}} \{W \in L \mid W \in \mathcal{O}_Y \text{ and } Z \leqslant W\}.$$

If $Z \in \mathcal{O}_i$ then $u(Z, Y) = u(X_i, Y)$ and thus $|\Omega(X, Y)| = \sum_{i=1}^{q} |\mathcal{O}_i| \, u(X_i, Y)$. □

(3.8) COROLLARY. *If $X \in L$ has the property that there is a unique G-orbit $X_i \in \Xi$ with $\Gamma(X_i) = \Gamma(X)$ then there is only one summand in* (3.7); *so setting $Y = X_j \in \Xi$ we have*

$$|\mathcal{O}_i| \, u_{ij} = |\mathcal{O}_j| \, n\big(\Gamma(X_i), \Gamma(X_j)\big).$$

Thus Table I *and the orbit sizes determine u_{ij}.* □

Since the set of G-orbits is indexed by the parabolic classes we may, when convenient, use the corresponding Dynkin graphs for the Weyl groups and Coxeter graphs for the rest, as index set for the rows and columns of U. If in E_6, E_7, E_8 only one class of a pair $(\ )'$, $(\ )''$ is parabolic then we omit the accent. Using the criterion $W_1 = \overline{W}_1$ in Carter's tables we see that the only orbits which do not have property (3.8) are: the orbits in E_7 given in (3.6); the orbits in F_4 labeled A_1, $\tilde{A}_1$, A_2, $\tilde{A}_2$, $A_1 \times \tilde{A}_2$, $\tilde{A}_1 \times A_2$, B_3, C_3; the orbits in $I_2(p)$ with p even, labeled A_1, $\tilde{A}_1$. It follows from these remarks that the matrices U for E_6, E_8, H_3, H_4 may be computed using (3.8). The case $I_2(p)$ is trivial and the case F_4 poses no real difficulty. In E_7 all the rows of U except those displayed in (3.6) may be computed using (3.8). The calculation of these six rows in the matrix for E_7 requires different methods.

First we use the Dynkin graph to show that certain $u(X, Y) = 0$. The classification of the orbits shows that the number q of summands in (3.7) is at most two so that if one of them is zero the other may be determined as before. If $u(X, Y) \neq 0$ then there exists $Z \in \mathcal{O}_Y$ such that $G_X \subseteq G_Z$ and thus $\Gamma(X) \leqslant \Gamma(Y)$ in the partial ordering of $\mathfrak{T}$. For $J \subseteq S$ let Γ_J be the full subgraph of the Dynkin graph of G which has vertex set J. Choose subsets J, K of S such that G_X, G_Y are conjugate to G_J, G_K. Then $\Gamma_J \leqslant \Gamma_K$. Thus if there is no pair $J \subseteq K$ with $G_J \sim G_X$ and $G_K \sim G_Y$ then $u(X, Y) = 0$. To use this method we must identify the subsets J of S such that c_J lies in a class (3.6). Label the Dynkin graph as in Bourbaki [**4**, p. 265]:

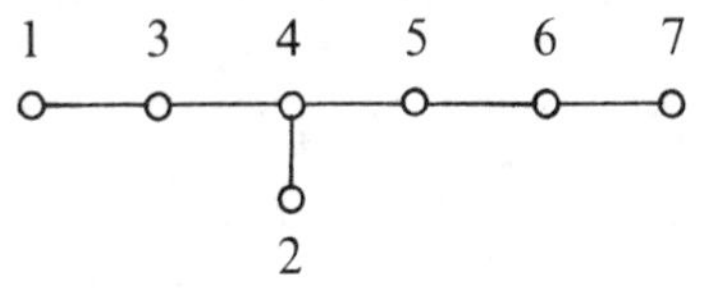

The classes labeled ( )′ correspond to subsets J in the following list:

Class	$(A_1^3)'$	$(A_1 \times A_3)'$	A_5'
J	$\{2,5,7\}$	$\{2,5,6,7\}$, $\{2,4,5,7\}$	$\{2,4,5,6,7\}$

All other sets J with these graphs correspond to classes labeled ( )″. This separation of classes may be done as follows. Compute the roots orthogonal to Π_J. This gives the subgroup W_2 of [**8**, p. 34] and now one can use the column $W_1 \times W_2$ of Carter's tables. This method of inspecting pairs $J \subseteq K$ provides enough entries $u(X, Y) = 0$ in these six rows to allow completion of all columns except

(3.9) $A_1^4,\ A_1^3 \times A_2,\ A_1^2 \times A_3,\ A_1 \times D_4,\ A_1 \times A_2 \times A_3,\ A_1 \times A_5,\ A_1 \times D_5,\ D_6.$

We may assume now that all $u(X, Y)$ have been computed for (X, Y) outside the intersection of the six rows of (3.6) and the eight columns of (3.9). Now we use Frame's character table [**11**] to complete the matrix U. We begin with some generalities about Coxeter groups and then apply them to E_7.

We proved in [**13**, Lemma 2] that $\mathrm{Fix}(g) \in L$ for any $g \in G$. Thus we may define an equivalence relation $\approx$ on G by $g \approx g'$ if and only if $\mathrm{Fix}(g)$ and $\mathrm{Fix}(g')$ are in the same G-orbit. Every $\approx$ class is a union of conjugacy classes.

(3.10) Lemma. *Let $\mathcal{A}$ be a Coxeter arrangement, $G = G(\mathcal{A})$ and $L = L(\mathcal{A})$. Every $\approx$ class contains a unique parabolic conjugacy class. Thus the map $g \to \mathrm{Fix}(g)$ induces a one to one correspondence between $\approx$ classes and orbits of G on L.*

Proof. Suppose $g \in G$. Then $X = \mathrm{Fix}(g)$ is in L. The group G_X is conjugate to some parabolic subgroup G_J [**4**, V, §3, Proposition 2]. By (3.4) $\mathrm{Fix}(G_X) = X$ and $\mathrm{Fix}(G_J)$ lie in the same orbit. The argument given in [**16**, Lemma 7] shows that $\mathrm{Fix}(G_J) = \mathrm{Fix}(c_J)$. Thus $\mathrm{Fix}(g)$, $\mathrm{Fix}(c_J)$ lie in the same orbit so $g \approx c_J$. To prove uniqueness suppose $c_J \approx c_K$. Then $\mathrm{Fix}(c_J)$ and $\mathrm{Fix}(c_K)$ lie in the same orbit. Thus $\mathrm{Fix}(G_J)$ and $\mathrm{Fix}(G_K)$ lie in the same orbit. By (3.4) G_J and G_K are conjugate. By (3.5) c_J and c_K are conjugate. □

(3.11) Lemma. *Let $\mathcal{A}$ be a Coxeter arrangement, $G = G(\mathcal{A})$ and $L = L(\mathcal{A})$. Let $X \in L$ and let φ_X be the character of G induced by the principal character of G_X. If $g \approx g'$ then $\varphi_X(g) = \varphi_X(g')$.*

Proof. Let $L_g = \{X \in L \mid X \geqslant \mathrm{Fix}(g)\}$. Since $g \approx g'$ we may choose $h \in G$ such that $\mathrm{Fix}(g') = h\,\mathrm{Fix}(g)$. Then $L_{g'} = hL_g$. It follows from the argument in [**14**, Proposition 3] that

$$(3.12) \qquad |\mathcal{O}_X|\,\varphi_X(g) = |L_g \cap \mathcal{O}_X|\,|G : G_X|.$$

The assertion is now clear since $|L_{g'} \cap \mathcal{O}_X| = |hL_g \cap \mathcal{O}_X| = |L_g \cap h^{-1}\mathcal{O}_X| = |L_g \cap \mathcal{O}_X|$. □

Recall that $\Xi = \{X_1, \ldots, X_r\}$ is a set of representatives for the orbits of G on L. Let $K_i = \{g \in G \mid \mathrm{Fix}(g) \in \mathcal{O}_i\}$ be the $\approx$ class corresponding to X_i. Choose $g_i \in K_i$ and let $\varphi_i = \varphi_{X_i}$. Let $X = X_j$, $G_j = G_{X_j}$ and $g = g_i$. Since $L_{g_i} = L^{X_i}$ and

$|L^{X_i} \cap \mathcal{O}_j| = u_{ij}$ we get from (3.12),

$$\varphi_j(g_i) = u_{ij}|N(G_j): G_j|. \tag{3.13}$$

Let ρ be an irreducible character of G and let (ρ, φ_j) be the multiplicity of ρ in φ_j. It follows from (3.11) and (3.13) that

$$|G|(\rho, \varphi_j) = |N(G_j): G_j| \sum_{i=1}^{r} u_{ij} \sum_{g \in K_i} \rho(g). \tag{3.14}$$

In most cases we can use the results of Beynon and Lusztig [3] to conclude that $(\rho, \varphi_j) = 0$ for certain ρ and φ_j as follows. Let $P_\rho(t) = \Sigma n_k(\rho)t^k$ be the associated polynomial in the indeterminate t defined in [3]. Let $P_j(t) = \Sigma(\rho, \varphi_j)P_\rho(t)$. We conclude from the definition of the $P_\rho(t)$ and Frobenius reciprocity that

$$P_j(t) = \prod_{i=1}^{l} (1 - t^{d_i}) / \prod_{i=1}^{l} (1 - t^{d_{ij}})$$

where the d_i are the degrees of the basic invariant polynomials of G and the d_{ij} are the corresponding degrees for G_j (acting on V). The $P_\rho(t)$ have nonnegative coefficients and the (ρ, φ_j) are nonnegative. Thus if $(\rho, \varphi_j) > 0$ then $\deg P_\rho(t) \leqslant \deg P_j(t)$ and we conclude that

$$\text{if } (\rho, \varphi_j) > 0 \text{ then } \deg P_\rho(t) \leqslant e - e_j \tag{3.15}$$

where $e = \Sigma\, d_i$ and $e_j = \Sigma d_{ij}$. When the multiplicity is not known to be zero one can still use the fact that the left-hand side of (3.14) is a nonnegative multiple of $|G|$ to derive congruences and inequalities for the u_{ij} which are sufficient to determine them in all cases. We give one example to illustrate the method, for the column labeled $A_1 \times D_4$. There are four unknown entries in this column because the matrix U is upper triangular. Let

$$x' = u((A_1^3)', A_1 \times D_4), \qquad x'' = u((A_1^3)'', A_1 \times D_4),$$
$$y' = u((A_1 \times A_3)', A_1 \times D_4), \qquad y'' = u((A_1 \times A_3)'', A_1 \times D_4).$$

From (3.7) we get the equations

$$315x' + 3780x'' = 945 \cdot 30, \qquad 1260y' + 7560y'' = 945 \cdot 12.$$

From these equations we get $x' \equiv 0 \bmod 6$ and $y' \equiv 0 \bmod 3$. Write $x' = 6\bar{x}$ and $y' = 3\bar{y}$. Then

$$\bar{x} + 2x'' = 15, \qquad \bar{y} + 2y'' = 3; \tag{3.16}$$

so we have two equations in four unknowns which are nonnegative integers. Table II in the appendix contains all the information necessary to determine them.

Table II is constructed as follows. Recall that all entries in the column X_j labeled $A_1 \times D_4$ are known except for x', x'', y', y''. The orbit column in Table II contains only those X_i for which $u_{ij} \neq 0$. The graph D_4 appears three times in the orbit column because the $\approx$ class of type D_4 is a union of three conjugacy classes which are labeled $(A_1^4)''$, D_4, $D_4(a_1)$. These are the conjugacy classes for which $\overline{W}_1 = D_4$ in Carter's Table 10 [8] and appear in column C of Table II. Similarly the $\approx$ class of type $A_1 \times D_4$ is the union of three conjugacy classes. The $\approx$ classes corresponding to the remaining orbits in this example are conjugacy classes. The second symbol in Column C is Frame's symbol **[11]** for the conjugacy class. We need this identification in order to use Frame's character table in (3.14). The identification is made by comparing characteristic polynomials and class sizes. The group G of type E_7 splits as $G = G^+ \times \langle z \rangle$ where z is the central involution. Frame's symbols refer to the classes in G^+ and we prefix his symbol by z if the class lies in zG^+. The columns ψ, θ in Table II are Frame's characters $56a$, $216a$ extended to G so that $\psi(zg) = \psi(g)$ and $\theta(zg) = \theta(g)$ for $g \in G^+$.

To solve (3.16) we use (3.14) and Table II. Since $|N(G_j): G_j| = 8$ we get

$$8 \cdot 15120(6\bar{y} - \bar{x} - 3) = |G|(\psi, \varphi_j),$$
$$8 \cdot 45360(\bar{x} - 2\bar{y} - 1) = |G|(\theta, \varphi_j).$$

Table 3 of **[3]** gives the polynomials for the characters labeled $56a'$ and $216a'$ where $\rho' = \rho \otimes$ sign. Since $P_{\rho'}(t) = t^{63}P_\rho(t^{-1})$ we find that $\deg P_\psi(t) = 60$ and $\deg P_\theta(t) = 54$. Since $e - e_j = 70 - 20 = 50$ we conclude from (3.15) that $(\psi, \varphi_j) = 0 = (\theta, \varphi_j)$. Thus $\bar{x} = 3$, $\bar{y} = 1$ so $x' = 18$, $x'' = 6$, $y' = 3$, $y'' = 1$. This completes the column $A_1 \times D_4$. The remaining columns listed in (3.9) are done in the same way.

4. Concluding remarks. *Iterated fibrations and* $K(\pi, 1)$. There are arrangements for which the manifold $M = M(\mathcal{Q})$ is a $K(\pi, 1)$ space. Brieskorn **[5]** proved this for the finite Coxeter groups of type A_l, B_l, D_l, G_2, F_4, $I_2(p)$. Deligne **[10]** then proved that if an arrangement $\mathcal{Q}$ in $V = \mathbf{R}^l$ is such that each component of $V - \bigcup_{H \in \mathcal{Q}} H$ is a simplicial cone then M is a $K(\pi, 1)$ space. All Coxeter arrangements $\mathcal{Q}$ satisfy this condition and so do the arrangements $\mathcal{Q}^X$ studied in this paper. Thus the corresponding manifolds M^X are $K(\pi, 1)$ spaces.

Deligne constructed the universal covering of M which is akin to a Tits building. On the other hand Brieskorn used an iterated fibration and his method may be used to prove that M is a $K(\pi, 1)$ space in cases where the arrangement is not real.

For example, let $r \geqslant 3$ be a fixed positive integer and let $\omega = \exp(2\pi i/r)$. For $l = 1, 2, 3, \ldots$ let $\mathcal{Q}_l$ be the arrangement in $\mathbf{C}^l$ consisting of all hyperplanes of the form

$$z_k = 0, \qquad 1 \leqslant k \leqslant l,$$
$$z_j - \omega^q z_k = 0, \qquad 1 \leqslant j < k \leqslant l, \quad 0 \leqslant q < r.$$

This is the arrangement of reflecting hyperplanes for the unitary reflection group consisting of all $l \times l$ monomial matrices whose nonzero entries are rth roots of unity. Let M_l be the manifold corresponding to $\mathcal{Q}_l$. The map $M_l \to M_{l-1}$ induced by $(z_1, \ldots, z_l) \to (z_1, \ldots, z_{l-1})$ is the projection map for a locally trivial fiber bundle. The fiber above the point $(z_1, \ldots, z_{l-1}) \in M_{l-1}$ is

$$F_l = \mathbf{C} - \{0, \omega^q z_k \mid 1 \leqslant k \leqslant l - 1, 0 \leqslant q < r\}.$$

Thus F_l is $\mathbf{C}$ with $r(l-1)+1$ points removed. This shows that M_l is the total space of an iterated fibration where the successive fibers are $\mathbf{C}$ with $r(l-1)+1$, $r(l-2)+1,\ldots,r+1$ points removed, and the last base space is $\mathbf{C}-\{0\}$. Thus M is a $K(\pi, 1)$ space.

In **[13]** we computed $P(M, t)$ for the manifold M_l and found that $P(M, t) = (1+n_1 t) \cdots (1+n_l t)$ where $n_k = r(k-1)+1$. Thus n_k has a geometric meaning; it is the number of points removed from $\mathbf{C}$ to get F_k. Since all the Poincaré polynomials $P(M^X, t)$ of the manifolds in this paper factor in this way it is natural to ask whether the M^X are total spaces of iterated fibrations corresponding to the factors.

In the opposite direction it is very easy to give an example of a real arrangement for which M is not a $K(\pi, 1)$ space and $P(M, t)$ does not factor in the above fashion. Let $V = \mathbf{R}^3$ and let $\mathcal{Q}$ consist of the four hyperplanes $x_1 = 0$, $x_2 = 0$, $x_3 = 0$, $x_1 + x_2 + x_3 = 0$. Then $P(M, t) = (1+t)(1+3t+3t^2)$. Furthermore M has the homotopy type of $S^1 \times N$ where N is a 3-dimensional torus with a point p removed. The boundary of a small ball with center p represents a nontrivial spherical homology class and hence a nontrivial element of $\pi_2(M)$.

Cell counts in the Coxeter complex. We use the notation of **[14]**. The Coxeter complex is Γ; there will be no confusion with our use of Γ in §3 for the Coxeter graph. If $X \in L$ let Θ_X be the set of all simplexes $\sigma \in \Gamma$ with linear span $\operatorname{lin}(\sigma) = X$. Let $\Gamma_X = \{\sigma \in \Gamma \mid \sigma \subseteq X\}$. Then Γ_X is a subcomplex of Γ and Θ_X is the set of top-dimensional simplexes of Γ_X. We showed in [**14**, (1.9) and (3.5)] that

$$|\Theta_X| = \sum_{Y \geqslant X} (-1)^{r(Y)-r(X)} \mu(X, Y)$$

and thus

$$|\Theta_X| = \prod_{k=1}^{\dim X} \left(1 + b_k^X\right). \tag{4.1}$$

On the other hand the numbers $|\Theta_X|$ may be computed by a different method. This serves as a check on the tables. Let $\Omega_X = \{\sigma \in \Gamma \mid \operatorname{lin}(\sigma) \in \mathcal{O}_X\}$. Since $\Omega_X = \bigcup_{Y \in \mathcal{O}_X} \Theta_Y$ we have $|\Omega_X| = |\Theta_X|\,|\mathcal{O}_X|$. Let $J_1,\ldots,J_\nu$ be the subsets of S such that G_X is conjugate to G_J for $J \in \{J_1,\ldots,J_\nu\}$. Sometimes we write $\nu = \nu_X$. Let $\sigma_1,\ldots,\sigma_\nu$ be the faces of $\sigma = C \cap S^{l-1}$ corresponding to $J_1,\ldots,J_\nu$ in the action of G on Γ. Then $\Omega_X = \mathcal{O}_{\sigma_1} \cup \cdots \cup \mathcal{O}_{\sigma_\nu}$. Since $|\mathcal{O}_{\sigma_1}| = \cdots = |\mathcal{O}_{\sigma_\nu}| = |G\colon G_X|$ we have $|\Omega_X| = |G\colon G_X|\,\nu_X$. Thus $|\Theta_X|\,|\mathcal{O}_X| = |G\colon G_X|\,\nu_X$ so that

$$|\Theta_X| = |N(G_X)\colon G_X|\,\nu_X. \tag{4.2}$$

Since there may be inconjugate G_J with isomorphic Dynkin graphs there is some subtlety in calculating the ν_X. Note that $N(G_X)/G_X$ permutes Θ_X without fixed point so that ν_X is the number of orbits in this action.

The matrix U may be used to compute the number of simplexes of Γ_X in any dimension. Let $P_j(t) = P(M^{X_j}, t)$. Since

$$\Gamma_X = \bigcup_{Y \geq X} \Theta_Y = \bigcup_{j=1}^{r} \bigcup_{\substack{Y \in \mathcal{O}_j \\ Y \geq X}} \Theta_Y$$

(4.1) shows that the number of q-simplexes in Γ_X is

$$|\Gamma_X^q| = \sum u_{ij} P_j(1), \qquad X \in \mathcal{O}_i, \tag{4.3}$$

where the sum is over all j such that $\dim X_j = q + 1$. For example, in type E_6, suppose $X \in L$ is a 3-dimensional subspace in the orbit labeled A_3. Then the 2-sphere $\Gamma_X = X \cap S^5$ has $|\Gamma_X^2| = |\Theta_X| = P_7(1) = 40$ two-simplexes. Similarly,

$$|\Gamma_X^1| = 2P_{10}(1) + 4P_{11}(1) + 2P_{12}(1) = 60,$$

$$|\Gamma_X^0| = 4P_{14}(1) + 2P_{15}(1) + 5P_{16}(1) = 22.$$

The inverse of U. Let L be a finite poset and let G be a group of automorphisms of L. Let $\mathcal{O}_1, \ldots, \mathcal{O}_r$ be the G-orbits. Choose $X_i \in \mathcal{O}_i$. Let $\mathcal{O}_j(X_i) = L^{X_i} \cap \mathcal{O}_j = \{Y \in \mathcal{O}_j \mid Y \geq X_i\}$. Let $u_{ij} = |\mathcal{O}_j(X_i)|$ and let $U = (u_{ij})$. Then

$$U^{-1} = \left(\sum_{Y \in \mathcal{O}_j(X_i)} \mu(X_i, Y) \right). \tag{4.4}$$

This is a special case of Lemma 1 of [**17**]. Suppose now that $\mathcal{A}$ is a Coxeter arrangement and $L = L(\mathcal{A})$. Then $U = U(\mathcal{A})$. The jth entry a_j in the first row of U^{-1} is thus $a_j = |\mathcal{O}_j| \mu(V, X_j)$. We proved in [**13**, Lemma 4.7] that

$$\mu(V, X) = \sum_{g \in F_X} \det(g)$$

where $F_X = \{g \in G \mid \mathrm{Fix}(g) = X\}$. Since, in this paper, G is an orthogonal group we have $\det(g) = (-1)^{r(X)}$. Thus writing $F_j = F_{X_j}$ we have $a_j = |\mathcal{O}_j| |F_j| = |K_j|$ where $K_j = \{g \in G \mid \mathrm{Fix}(g) \in \mathcal{O}_j\}$ is the $\approx$ class corresponding to X_j. Since F_j may be viewed as the set of fixed point free elements of G_j acting as reflection group in V/X_j the cardinality $|F_j|$ is the product of the exponents of G_j [**4**, Chapitre V]. Finally, note that (1.5) says that the entry in row i of the last column of U^{-1} is $b_1^i \cdots b_p^i$.

Appendix. Table I contains the values $n(R, T)$ for the irreducible T which occur in the exceptional groups. Table II contains the data needed to compute the entries in column 23 labeled $A_1 \times D_4$ in the E_7 table. Tables III–VIII contain the matrices $U(\mathcal{A})$ for the exceptional Coxeter groups. Each matrix U is bordered by two rows at the top indexed by $|N(G_j)\colon G_j|$ and by orbits $\mathcal{O}_j$ which are labeled by the graphs of their fixers. There are two columns at the left indexed by $|K_i|$ and $\mathcal{O}_i$. The entries in the first row of U, labeled A_0, are $|\mathcal{O}_j| = |G\colon N(G_j)|$. The columns on the right are the $b_1^i, b_2^i, \ldots$ corresponding to the orbit $\mathcal{O}_i$. The induced characters are $\varphi_j(g_i) = u_{ij} |N(G_j)\colon G_j|$. They are obtained from the matrix by multiplying the jth column by the number which appears above it.

TABLE I

	A_0	A_1	A_2	B_2	$I_2(5)$	A_3	B_3	H_3	A_4	D_4	A_5	D_5	A_6	D_6	E_6	A_7	D_7	E_7
A_0	1	1	1	1	1	1	1	1	1	1	1	1	1	1	1	1	1	1
A_1		1	3	4	5	6	9	15	10	12	15	20	21	30	36	28	42	63
A_1^2			0	0	0	3	6	15	15	18	45	70	105	195	270	210	441	945
A_2			1	0	0	4	4	10	10	16	20	40	35	80	120	56	140	336
B_2			0	1	0	0	3	0	0	0	0	0	0	0	0	0	0	0
$I_2(5)$			0	0	1	0	0	6	0	0	0	0	0	0	0	0	0	0
A_1^3						0	0	0	0	12	15	60	105	300	540	420	1260	4095
$A_1 \times A_2$						0	0	0	10	0	60	80	210	480	720	560	1680	5040
A_3						1	0	0	5	12	15	50	35	140	270	70	315	1260
B_3						0	1	0	0	0	0	0	0	0	0	0	0	0
H_3						0	0	1	0	0	0	0	0	0	0	0	0	0
A_1^4									0	0	0	0	0	180	0	105	1260	3780
$A_1^2 \times A_2$									0	0	0	40	105	240	1080	840	2520	15120
A_2^2									0	0	10	0	170	160	120	280	1120	3360
$A_1 \times A_3$									0	0	15	20	105	360	540	420	2100	8820
A_4									1	0	6	16	21	96	216	56	336	2016
D_4									0	1	0	5	0	15	45	0	35	315
$A_1^3 \times A_2$											0	0	0	0	0	0	1680	5040
$A_1 \times A_2^2$											0	0	0	0	360	280	0	10080
$A_1^2 \times A_3$											0	0	0	120	0	210	1260	7560
$A_2 \times A_3$											0	0	35	80	0	280	1680	5040
$A_1 \times A_4$											0	0	21	0	216	168	672	6048
A_5											1	0	7	32	36	28	224	1344
$A_1 \times D_4$											0	0	0	30	0	0	210	945
D_5											0	1	0	6	27	0	21	378
$A_1 \times A_2 \times A_3$													0	0	0	0	0	5040
$A_1^2 \times A_4$													0	0	0	0	336	0
A_3^2													0	0	0	35	280	0
$A_2 \times A_4$													0	0	0	56	0	2016
$A_1 \times A_5$													0	0	0	28	0	1008
A_6													1	0	0	8	64	288
$A_2 \times D_4$													0	0	0	0	140	0
$A_1 \times D_5$													0	0	0	0	42	378
D_6													0	1	0	0	7	63
E_6													0	0	1	0	0	28
A_7																1	0	0
D_7																0	1	0
E_7																0	0	1

TABLE II

C		$\|C\|$	orbit	u_{ij}	ψ	θ
A_0	1^7	1	A_0	945	56	216
A_1	$z1^{-5}2^6$	63	A_1	195	-24	-24
A_1^2	$1^3 2^2$	945	A_1^2	30	8	8
A_2	$1^4 3$	672	A_2	45	11	-9
$(A_1^3)'$	$z1^{-1}2^4(V)$	315	$(A_1^3)'$	$6\overline{x}$	-8	24
$(A_1^3)''$	$z1^{-1}2^4(-U)$	3780	$(A_1^3)''$	x''	0	0
$A_1 \times A_2$	$z1^{-2}2^3 3^{-1}6$	10080	$A_1 \times A_2$	3	-3	-3
A_3	$z1^{-3}2^3 4$	7560	A_3	9	-4	4
$(A_1^4)'$	$1^{-1}2^4(-U)$	3780	A_1^4	3	0	0
$(A_1 \times A_3)'$	124	7560	$(A_1 \times A_3)'$	$3\overline{y}$	4	-4
$(A_1 \times A_3)''$	124	45360	$(A_1 \times A_3)''$	y''	0	0
$(A_1^4)''$	$1^{-1}2^4(V)$	315	D_4	3	-8	24
D_4	$1^2 2 3^{-1}6$	10080	D_4	3	1	-3
$D_4(a_1)$	$1^3 2^{-2}4^2$	3780	D_4	3	0	0
A_1^5	$z1^3 2^2$	945	$A_1 \times D_4$	1	8	8
$A_1 \times D_4$	$z2^2 3$	30240	$A_1 \times D_4$	1	-1	-1
$A_1 \times D_4(a_1)$	$z1^{-1}4^2$	11340	$A_1 \times D_4$	1	0	0

TABLE III: H_3

	$\|N(G_j):G_j\|$	120	4	2	2	2	1			
$\|K_i\|$	O_i \ O_j	A_0	A_1	A_1^2	A_2	$I_2(5)$	H_3	b_1^i	b_2^i	b_3^i
1	A_0	1	15	15	10	6	1	1	5	9
15	A_1		1	2	2	2	1	1	5	
15	A_1^2			1	0	0	1	1		
20	A_2				1	0	1	1		
24	$I_2(5)$					1	1	1		
45	H_3						1			

TABLE IV: H_4

$\|K_i\|$	$\mathcal{O}_i$ \ $\mathcal{O}_j$	A_0	A_1	A_1^2	A_2	$I_2(5)$	$A_1\times A_2$	$A_1\times I_2(5)$	A_3	H_3	H_4	b_1^i	b_2^i	b_3^i	b_4^i
	$\|N(G_j):G_j\|$	14400	120	8	12	20	2	2	2	2	1				
1	A_0	1	60	450	200	72	600	360	300	60	1	1	11	19	29
60	A_1		1	15	10	6	40	36	30	15	1	1	11	19	
450	A_1^2			1	0	0	4	4	2	2	1	1	11		
400	A_2				1	0	3	0	6	3	1	1	11		
288	$I_2(5)$					1	0	5	0	5	1	1	9		
1200	$A_1\times A_2$						1	0	0	0	1	1			
1440	$A_1\times I_2(5)$							1	0	0	1	1			
1800	A_3								1	0	1	1			
2700	H_3									1	1	1			
6061	H_4										1				

TABLE V: F_4

$\|K_i\|$	$\mathcal{O}_i$ \ $\mathcal{O}_j$	A_0	A_1	$\tilde{A}_1$	$A_1\times\tilde{A}_1$	A_2	$\tilde{A}_2$	B_2	C_3	B_3	$A_1\times\tilde{A}_2$	$\tilde{A}_1\times A_2$	F_4	b_1^i	b_2^i	b_3^i	b_4^i
	$\|N(G_j):G_j\|$	1152	48	48	4	12	12	8	2	2	2	2	1				
1	A_0	1	12	12	72	16	16	18	12	12	48	48	1	1	5	7	11
12	A_1		1	0	6	4	0	3	3	6	4	12	1	1	5	7	
12	$\tilde{A}_1$			1	6	0	4	3	6	3	12	4	1	1	5	7	
72	$A_1\times\tilde{A}_1$				1	0	0	0	1	1	2	2	1	1	5		
32	A_2					1	0	0	0	3	0	3	1	1	5		
32	$\tilde{A}_2$						1	0	3	0	3	0	1	1	5		
54	B_2							1	2	2	0	0	1	1	3		
180	C_3								1	0	0	0	1	1			
180	B_3									1	0	0	1	1			
96	$A_1\times\tilde{A}_2$										1	0	1	1			
96	$\tilde{A}_1\times A_2$											1	1	1			
385	F_4												1				

TABLE VI: E_6

$\lvert K_i \rvert$	$\mathcal{O}_i$ \ $\mathcal{O}_j$	A_0	A_1	A_1^2	A_2	A_1^3	$A_1 \times A_2$	A_3	$A_1^2 \times A_2$	A_2^2	$A_1 \times A_3$	A_4	D_4	$A_1 \times A_2^2$	$A_1 \times A_4$	A_5	D_5	E_6	b_1^i	b_2^i	b_3^i	b_4^i	b_5^i	b_6^i
	$\lvert N(G_j):G_j \rvert$	51840	720	48	72	12	6	8	2	12	2	2	6	2	1	2	1	1						
1	A_0	1	36	270	120	540	720	270	1080	120	540	216	45	360	216	36	27	1	1	4	5	7	8	11
36	A_1		1	15	10	45	80	45	150	20	105	60	15	70	66	15	15	1	1	4	5	7	8	
270	A_1^2			1	0	6	8	3	28	4	18	12	3	20	20	6	7	1	1	4	5	7		
240	A_2				1	0	6	9	9	2	18	18	6	6	18	6	9	1	1	4	5	5		
540	A_1^3					1	0	0	6	0	3	0	1	6	6	1	3	1	1	4	5			
1440	$A_1 \times A_2$						1	0	3	1	3	3	0	4	6	3	3	1	1	4	5			
1620	A_3							1	0	0	2	4	2	0	4	2	5	1	1	3	4			
2160	$A_1^2 \times A_2$								1	0	0	0	0	2	2	0	1	1	1	4				
480	A_2^2									1	0	0	0	3	0	3	0	1	1	5				
3240	$A_1 \times A_3$										1	0	0	0	2	1	1	1	1	3				
5184	A_4											1	0	0	1	1	2	1	1	3				
2025	D_4												1	0	0	0	3	1	1	2				
1440	$A_1 \times A_2^2$													1	0	0	0	1	1					
5184	$A_1 \times A_4$														1	0	0	1	1					
4320	A_5															1	0	1	1					
11340	D_5																1	1	1					
12320	E_6																	1						

TABLE VIII: E_8

$\|K_i\|$	O_i \ O_j	A_0	A_1	A_1^2	A_2	A_1^3	$A_1 \times A_2$	A_3	A_1^4	$A_1^2 \times A_2$	A_2^2	$A_1 \times A_3$	A_4	D_4	$A_1^3 \times A_2$	$A_1 \times A_2^2$	$A_1^2 \times A_3$	$A_2 \times A_3$	$A_1 \times A_4$	A_5	$A_1 \times D_4$
	$\|N(G_j):G_j\|$	696729600	2903040	46080	103680	2304	1440	3840	384	96	288	96	240	1152	24	24	16	16	12	24	48
1	A_0	1	120	3780	1120	37800	40320	7560	113400	302400	67200	151200	24192	3150	604800	403200	453600	302400	241920	40320	37800
120	A_1		1	63	28	945	1344	378	3780	12600	3360	8820	2016	315	30240	23520	30240	22680	22176	5040	4095
3780	A_1^2			1	0	30	32	6	180	560	160	360	96	15	1920	1600	1920	1680	1600	480	300
2240	A_2				1	0	36	27	0	270	120	540	216	45	540	720	1620	1350	2160	720	540
37800	A_1^3					1	0	0	12	24	0	12	0	1	160	96	144	72	96	16	30
80640	$A_1 \times A_2$						1	0	0	15	10	15	6	0	45	80	90	135	120	60	15
45360	A_3							1	0	0	0	20	16	5	0	0	60	40	160	80	60
113400	A_1^4								1	0	0	0	0	0	16	0	12	0	0	0	4
604800	$A_1^2 \times A_2$									1	0	0	0	0	6	8	6	3	8	0	0
268800	A_2^2										1	0	0	0	0	6	0	18	0	6	0
907200	$A_1 \times A_3$											1	0	0	0	0	6	6	8	4	3
580608	A_4												1	0	0	0	0	0	10	10	0
141750	D_4													1	0	0	0	0	0	0	12
1209600	$A_1^3 \times A_2$														1	0	0	0	0	0	0
1612800	$A_1 \times A_2^2$															1	0	0	0	0	0
2721600	$A_1^2 \times A_3$																1	0	0	0	0
3628800	$A_2 \times A_3$																	1	0	0	0
5806080	$A_1 \times A_4$																		1	0	0
4838400	A_5																			1	0
1701000	$A_1 \times D_4$																				1
3175200	D_5																				
2419200	$A_1^2 \times A_2^2$																				
7257600	$A_1 \times A_2 \times A_3$																				
8709120	$A_1^2 \times A_4$																				
5443200	A_3^2																				
11612160	$A_2 \times A_4$																				
14515200	$A_1 \times A_5$																				
24883200	A_6																				
4536000	$A_2 \times D_4$																				
19051200	$A_1 \times D_5$																				
17860500	D_6																				
13798400	E_6																				
11612160	$A_1 \times A_2 \times A_4$																				
17418240	$A_3 \times A_4$																				
24883200	$A_1 \times A_6$																				
43545600	A_7																				
25401600	$A_2 \times D_5$																				
67359600	D_7																				
41395200	$A_1 \times E_6$																				
91891800	E_7																				
215656441	E_8																				

TABLE VIII: E_8 (continued)

$A_1^2\times A_2^2$	$A_1\times A_2\times A_3$	$A_1^2\times A_4$	A_3^2	$A_2\times A_4$	$A_1\times A_5$	A_6	$A_2\times D_4$	$A_1\times D_5$	D_6	E_6	$A_1\times A_2\times A_4$	$A_3\times A_4$	$A_1\times A_6$	A_7	$A_2\times D_5$	D_7	$A_1\times E_6$	E_7	E_8	b_1^i	b_2^i	b_3^i	b_4^i	b_5^i	b_6^i	b_7^i	b_8^i
8	4	4	8	4	4	4	12	4	8	12	2	2	2	2	2	2	2	2	1								
604800	604800	362880	151200	241920	120960	34560	50400	45360	3780	1120	241920	120960	34560	8640	30240	1080	3360	120	1	1	7	11	13	17	19	23	29
40320	50400	36288	15120	26208	16128	6048	6300	7938	945	336	28224	16128	6336	2016	5796	378	1036	63	1	1	7	11	13	17	19	23	
3520	4800	3456	1680	2880	1920	960	720	1080	195	80	3712	2496	1152	480	1040	126	272	30	1	1	7	11	13	17	19		
1080	2700	3240	1080	2376	2160	1080	765	1620	270	120	2376	1512	1080	432	1107	135	360	36	1	1	7	11	13	14	17		
384	480	384	144	288	192	96	88	156	30	16	576	384	192	96	216	36	72	13	1	1	7	11	13	17			
150	345	270	180	300	240	180	75	135	45	20	366	330	210	120	165	45	70	15	1	1	7	11	13	14			
0	80	240	40	160	240	160	80	300	70	40	160	96	160	80	200	45	120	20	1	1	7	9	11	13			
48	48	48	12	0	16	0	16	24	6	0	96	48	32	8	48	12	16	4	1	1	7	11	13				
28	42	36	12	36	24	12	3	18	3	4	76	60	36	24	35	9	20	6	1	1	7	11	13				
9	36	0	36	36	18	36	12	0	9	2	36	72	36	36	18	18	6	6	1	1	7	11	11				
0	16	24	12	24	24	24	12	21	9	4	32	32	32	24	34	15	18	7	1	1	7	9	11				
0	0	15	0	10	30	30	0	30	15	10	10	5	30	20	20	15	30	10	1	1	7	8	9				
0	0	0	0	0	0	0	16	72	18	16	0	0	0	0	48	12	48	12	1	1	5	7	11				
6	3	6	0	0	0	0	1	3	0	0	18	6	6	0	9	3	6	1	1	1	7	11					
3	6	0	0	6	3	0	0	0	0	1	12	12	6	6	6	0	4	3	1	1	7	11					
0	4	4	2	0	4	0	0	2	1	0	8	8	8	4	4	3	4	2	1	1	7	9					
0	2	0	4	4	0	4	2	0	1	0	4	12	4	8	5	6	0	2	1	1	7	9					
0	0	3	0	3	3	3	0	3	0	1	4	3	6	6	6	3	6	3	1	1	7	8					
0	0	0	0	0	3	6	0	0	3	1	0	0	6	6	0	6	3	4	1	1	5	7					
0	0	0	0	0	0	0	4	6	3	0	0	0	0	0	12	6	4	3	1	1	5	7					
0	0	0	0	0	0	0	0	6	3	4	0	0	0	0	4	3	12	6	1	1	5	7					
1	0	0	0	0	0	0	0	0	0	0	4	0	0	0	2	0	2	0	1	1	7						
	1	0	0	0	0	0	0	0	0	0	2	2	2	0	1	0	0	1	1	1	7						
		1	0	0	0	0	0	0	0	0	2	1	2	0	0	1	2	0	1	1	7						
			1	0	0	0	0	0	0	0	0	4	0	2	0	2	0	0	1	1	7						
				1	0	0	0	0	0	0	1	2	0	2	2	0	0	1	1	1	7						
					1	0	0	0	0	0	0	0	2	2	0	0	1	1	1	1	5						
						1	0	0	0	0	0	0	1	2	0	2	0	1	1	1	5						
							1	0	0	0	0	0	0	0	3	3	0	0	1	1	5						
								1	0	0	0	0	0	0	2	1	2	1	1	1	5						
									1	0	0	0	0	0	0	2	0	2	1	1	3						
										1	0	0	0	0	0	0	3	3	1	1	5						
											1	0	0	0	0	0	0	0	1	1							
												1	0	0	0	0	0	0	1	1							
													1	0	0	0	0	0	1	1							
														1	0	0	0	0	1	1							
															1	0	0	0	1	1							
																1	0	0	1	1							
																	1	0	1	1							
																		1	1	1							
																			1								

TABLE VII: E_7

$\|K_i\|$	O_i \ O_j	A_0	A_1	A_1^2	A_2	$(A_1^3)'$	$(A_1^3)''$	$A_1\times A_2$	A_3	A_1^4	$A_1^2\times A_2$	A_2^2	$(A_1\times A_3)'$	$(A_1\times A_3)''$	A_4	D_4	$A_1^3\times A_2$	$A_1\times A_2^2$	$A_1^2\times A_3$	$A_2\times A_3$	$A_1\times A_4$	A_5'	A_5''	$A_1\times D_4$	D_5	$A_1\times A_2\times A_3$	$A_2\times A_4$	$A_1\times A_5$	A_6	$A_1\times D_5$	D_6	E_6	E_7	b_1^i	b_2^i	b_3^i	b_4^i	b_5^i	b_6^i	b_7^i
	$\|N(G_j):G_j\|$	2903040	23040	768	1440	1152	96	48	96	48	8	24	48	8	12	48	12	4	4	4	2	12	4	8	4	2	2	2	2	2	2	2	1							
1	A_0	1	63	945	336	315	3780	5040	1260	3780	15120	3360	1260	7560	2016	315	5040	10080	7560	5040	6048	336	1008	945	378	5040	2016	1008	288	378	63	28	1	1	5	7	9	11	13	17
63	A_1		1	30	16	15	180	320	120	240	1200	320	140	840	320	60	480	1120	960	720	1056	80	240	195	120	800	416	256	96	126	30	16	1	1	5	7	9	11	13	
945	A_1^2			1	0	1	12	16	4	24	112	32	12	72	32	6	64	160	128	112	160	16	48	30	28	160	96	64	32	36	13	8	1	1	5	7	9	11		
672	A_2				1	0	0	15	15	0	45	20	15	90	60	15	15	60	90	75	180	20	60	45	45	75	66	60	30	45	15	10	1	1	5	7	8	9		
315	$(A_1^3)'$					1	0	0	0	12	0	0	12	0	0	0	16	0	72	0	0	16	0	18	0	48	0	48	0	12	12	0	1	1	5	7	11			
3780	$(A_1^3)''$						1	0	0	3	12	0	0	6	0	1	12	24	18	12	24	0	4	6	6	36	24	12	8	12	4	4	1	1	5	7	9			
10080	$A_1\times A_2$							1	0	0	6	4	1	6	4	0	3	16	12	18	24	4	12	3	6	23	20	16	12	9	6	4	1	1	5	7	8			
7560	A_3								1	0	0	0	1	6	8	3	0	0	6	4	24	4	12	9	15	4	8	12	8	15	7	6	1	1	5	5	7			
3780	A_1^4									1	0	0	0	0	0	0	4	0	6	0	0	0	0	3	0	12	0	4	0	6	3	0	1	1	5	7				
30240	$A_1^2\times A_2$										1	0	0	0	0	0	1	4	2	1	4	0	0	0	1	7	6	4	2	3	1	2	1	1	5	7				
13440	A_2^2											1	0	0	0	0	0	3	0	6	0	1	3	0	0	6	6	3	6	0	3	1	1	1	5	7				
7560	$(A_1\times A_3)'$												1	0	0	0	0	0	6	0	0	4	0	3	0	4	0	12	0	3	6	0	1	1	5	7				
45360	$(A_1\times A_3)''$													1	0	0	0	0	1	2	4	0	2	1	1	2	4	2	4	3	2	2	1	1	5	5				
48384	A_4														1	0	0	0	0	0	3	1	3	0	3	0	1	3	3	3	3	3	1	1	4	5				
14175	D_4															1	0	0	0	0	0	0	0	3	6	0	0	0	0	6	3	4	1	1	3	5				
10080	$A_1^3\times A_2$																1	0	0	0	0	0	0	0	0	3	0	0	0	3	0	0	1	1	5					
40320	$A_1\times A_2^2$																	1	0	0	0	0	0	0	0	2	2	1	0	0	0	1	1	1	5					
45360	$A_1^2\times A_3$																		1	0	0	0	0	0	0	2	0	2	0	1	1	0	1	1	5					
60480	$A_2\times A_3$																			1	0	0	0	0	0	1	2	0	2	0	1	0	1	1	5					
145152	$A_1\times A_4$																				1	0	0	0	0	0	1	1	1	1	0	1	1	1	4					
40320	A_5'																					1	0	0	0	0	0	3	0	0	3	0	1	1	5					
120960	A_5''																						1	0	0	0	0	0	2	0	1	1	1	1	3					
42525	$A_1\times D_4$																							1	0	0	0	0	0	2	2	0	1	1	3					
158760	D_5																								1	0	0	0	0	1	1	2	1	1	3					
60480	$A_1\times A_2\times A_3$																									1	0	0	0	0	0	0	1	1						
96768	$A_2\times A_4$																										1	0	0	0	0	0	1	1						
120960	$A_1\times A_5$																											1	0	0	0	0	1	1						
207360	A_6																												1	0	0	0	1	1						
158760	$A_1\times D_5$																													1	0	0	1	1						
297675	D_6																														1	0	1	1						
344960	E_6																															1	1	1						
765765	E_7																																1							

We give an example to show how the matrix U allows us to compute $b_1^i, b_2^i, \ldots$ using (3.3) and descending induction on i. Let $\mathcal{Q}$ be of type E_6 and suppose $X \in L$ is in the orbit $\mathcal{O}_7$ in row 7 labeled A_3. Then the values $u_{7,j}$ give

$$\chi_7(t) + 2\chi_{10}(t) + 4\chi_{11}(t) + 2\chi_{12}(t) + 4\chi_{14}(t) + 2\chi_{15}(t) + 5\chi_{16}(t) + \chi_{17}(t) = t^3.$$

By induction, using the values of $b_1^i, b_2^i, \ldots$ in the table for $i > 7$ we have $\chi_{17}(t) = 1$, $\chi_{16}(t) = \chi_{15}(t) = \chi_{14}(t) = t - 1$, $\chi_{12}(t) = (t-1)(t-2)$, $\chi_{11}(t) = \chi_{10}(t) = (t-1)(t-3)$ and thus $\chi_7(t) = t^3 - 8t^2 + 19t - 12 = (t-1)(t-3)(t-4)$. Thus b_1^7, b_2^7, b_3^7 are 1, 3, 4.

REFERENCES

1. M. Aigner, *Combinatorial theory*, Springer-Verlag, Berlin and New York, 1979.

2. V. I. Arnold, *The cohomology ring of the colored braid group*, Mat. Zametki **5** (1969), 227–231 = Math. Notes **5** (1969), 138–140.

3. W. M. Beynon and G. Lusztig, *Some numerical results on the characters of exceptional Weyl groups*, Math. Proc. Cambridge Philos. Soc. **84** (1978), 417–426.

4. N. Bourbaki, *Groupes et algèbres de Lie*, Hermann, Paris, 1968, Chapitres 4, 5, et 6.

5. E. Brieskorn, *Sur les groupes de tresses* (*d'après V. I. Arnold*), (Séminaire Bourbaki, 1971/72), Lecture Notes in Math., vol. 317, Springer-Verlag, Berlin and New York, 1973.

6. T. Brylawski, *A decomposition for combinatorial geometries*, Trans. Amer. Math. Soc. **171** (1972), 235–282.

7. R. W. Carter, *Simple groups of Lie type*, Wiley-Interscience, New York, 1972.

8. ______, *Conjugacy classes in the Weyl group*, Compositio Math. **25** (1972), 1–59.

9. P. Cartier, *Arrangements d'hyperplans: un chapitre de géométrie combinatoire*, Séminaire Bourbaki, 1980/81, Exposé 561.

10. P. Deligne, *Les immeubles des groupes de tresses généralisés*, Invent. Math. **17** (1972), 273–302.

11. J. S. Frame, *The classes and representations of the groups of* 27 *lines and* 28 *bitangents*, Ann. Mat. Pura Appl. (4) **32** (1951), 83–119.

12. P. Orlik and L. Solomon, *Combinatorics and topology of complements of hyperplanes*, Invent. Math. **56** (1980), 167–189.

13. ______, *Unitary reflection groups and cohomology*, Invent. Math. **59** (1980), 77–94.

14. ______, *Complexes for reflection groups*, Proc. Midwest Algebraic Geometry Conf., Lecture Notes in Math., vol. 862, Springer-Verlag, Berlin and New York, 1981, pp. 193–207.

15. K. Saito, *Theory of logarithmic differential forms and logarithmic vector fields*, J. Fac. Sci. Univ. Tokyo, IA, **27** (1980), 265–291.

16. L. Solomon, *A Mackey formula in the group ring of a Coxeter group*, J. Algebra **41** (1976), 255–264.

17. ______, *Partially ordered sets with colors*, Proc. Sympos. Pure Math., vol. 34, Amer. Math. Soc., Providence, R.I., 1979, pp. 309–329.

18. H. Terao, *Generalized exponents of a free arrangement of hyperplanes and Shephard-Todd-Brieskorn formula*, Invent. Math. **63** (1981), 159–179.

19. ______, *The exponents of a free hypersurface*, these PROCEEDINGS.

20. T. Zaslavsky, *Facing up to arrangements: face-count formulas for partition of space by hyperplanes*, Mem. Amer. Math. Soc. No. **154** (1975).

21. ______, *The geometry of root systems and signed graphs*, Amer. Math. Monthly **88** (1981), 88–105.

UNIVERSITY OF WISCONSIN, MADISON

Proceedings of Symposia in Pure Mathematics
Volume **40** (1983), Part 2

VANISHING FOLDS IN FAMILIES OF SINGULARITIES[1]

DONAL B. O'SHEA

1. Introduction. In this paper we summarize some results about topological equivalence in families of isolated hypersurfaces singularities along which the Milnor number is constant. Our approach has been strongly influenced by the paper of Lê and Ramanujam [7] where it is shown that, except for surfaces, "μ-constant implies constant topological type". Lê and Ramanujam use the h-cobordism theorem at a critical point in their argument. Our strategy is to try to determine the conditions under which the use of the h-cobordism theorem can be eliminated. In fact, it is a nontrivial question as to whether the situation which the h-cobordism theorem is used to cirumvent ever arises.

This paper is far from definitive. However, we do try to present a unified way of looking at questions of topological equivalence which we feel is quite instructive. Another approach to such questions is the use of controlled vector fields, which was pioneered by Mather [9] and, more recently, by Damon and Gaffney [3]. The latter technique works over the reals and is frequently more useful for computations. Nevertheless, our method—or, rather, the failure of our method—would indicate the existence of some rather bizarre geometrical behaviour in families of isolated singularities.

2. Definitions and conventions.

DEFINITION. Let f, $g \in \mathbf{C}[x_1,\ldots,x_n]$ have an isolated singularity at the origin $\mathbf{0} \in \mathbf{C}^n$ such that $f(\mathbf{0}) = g(\mathbf{0}) = 0$. Then f and g are said to have the *same topological type* (at the origin) if there exist balls $B_r = \{x \in \mathbf{C}^n\colon |x| \leq r\}$, $B_{r'} = \{x \in \mathbf{C}^n\colon |x| \leq r'\}$ and a homeomorphism $h\colon B_r \to B_{r'}$, $h(\mathbf{0}) = \mathbf{0}$, such that $h(f^{-1}(0) \cap B_r) = g^{-1}(0) \cap B_{r'}$. The polynomials f and g are said to be *topologically equivalent* (at the origin) if there exists a homeomorphism $h\colon B_r \to B_{r'}$, $h(\mathbf{0}) = \mathbf{0}$, such that $f \circ h^{-1} = g$ on $B_{r'}$.

1980 *Mathematics Subject Classification*. Primary 14D05; Secondary 14B07, 32C40, 32G11.

[1]This research was supported in part by a Faculty Grant from Mount Holyoke College.

0082-0717/81/0000-0652/$03.25

It is clear that if f and g are topologically equivalent they have the same topological type. A remarkable theorem, due to King [**5**], asserts that the converse is true provided that $n \neq 3$.

Questions concerning the variation (or constancy) of the topological type in families of isolated hypersurface singularities usually reduce to a consideration of families satisfying the following convention.

Convention. By a *family* $\{f_t\}$ of isolated hypersurface singularities, we mean a family of polynomials $f_t \in \mathbf{C}[x_1,\ldots,x_n]$ parameterized by the unit disk $\{t \in \mathbf{C}: |t| \leq 1\}$ such that if we set $F(x, t) \equiv f_t(x)$ then:

(1) $F(x, t)$ depends analytically on t (that is, $F(x, t)$ considered as a polynomial in $x_1,\ldots,x_n$ has coefficients which are analytic functions of t).

(2) $F(\mathbf{0}, t) = 0$.

(3) there exists $\varepsilon > 0$ such that for all t, $|t| \leq \varepsilon$, f_t has an isolated singularity at $\mathbf{0} \in \mathbf{C}^n$.

(4) for each t, $0 < |t| \leq \varepsilon$, the Milnor number $\mu(f_t)$ of the singular point $\mathbf{0}$ of f_t is a constant, $\mu(f_t) \equiv \mu$.

By a *μ-constant family* we mean a family $\{f_t\}$ which also satisfies:

(4′) $\mu(f_0) = \mu$, where μ is as in (4).

For convenience we shall often take t to be real, $t \in [0, \varepsilon]$.

Lê and Ramanujam [**7**] show that in a μ-constant family $\{f_t\}$, all the f_t have the same topological type for t sufficiently small, provided that $n \neq 3$. Timourian [**17**] has shown that a μ-constant family is topologically trivial (again, provided that $n \neq 3$). This latter result also follows directly from the Lê-Ramanujam theorem using King's results [**5**].

3. Three simple lemmas. Before reviewing the proof of the Lê-Ramanujam theorem, we shall need three simple lemmas.

LEMMA 1. *Let $f \in \mathbf{C}[x_1,\ldots,x_n]$ and $\Sigma = \{x \in \mathbf{C}^n: df(x) = \mathbf{0}\}$ be the singular set of f. Let g: $\mathbf{C}^n \to \mathbf{R}$ be any real-valued function which can be written as a polynomial in $x_1,\ldots,x_n$, $\bar{x}_1,\ldots,\bar{x}_n$. Then, for any $z \in \mathbf{C}$, the critical points of the restricted function $g|_{f^{-1}(z)-\Sigma}$ are precisely those points $x \in f^{-1}(z) - \Sigma$ such that*

$$\operatorname{rank}_{\mathbf{C}}\begin{pmatrix} \partial f/\partial x_1 & \cdots & \partial f/\partial x_n \\ \partial g/\partial x_1 & \cdots & \partial g/\partial x_n \end{pmatrix}_x = 1.$$

Thus, for example, if $g(x) = |x|^2$ is the squared distance function, the critical points of $g|_{f^{-1}(z)-\Sigma}$ are those points $x \in f^{-1}(z) - \Sigma$ for which $(\partial f/\partial x_1,\ldots,\partial f/\partial x_n) = c(\bar{x}_1,\ldots,\bar{x}_n)$ for some $c \in \mathbf{C} - 0$. A proof of Lemma 1 may easily be constructed along the lines of Lemma 2.7 in Milnor's book [**10**]. Alternatively, see O'Shea [**13**].

LEMMA 2. *Let $f \in \mathbf{C}[x_1,\ldots,x_n]$ have an isolated singular point at $\mathbf{0} \in \mathbf{C}^n$, $f(\mathbf{0}) = 0$. Then, for every sufficiently small real number $R > 0$, there exists a homeomorphism h: $B_R,\mathbf{0} \to B_R,\mathbf{0}$ which maps $f^{-1}(0) \cap B_R$ to* Cone$(f^{-1}(0) \cap S_R,\mathbf{0})$.

(Here, $B_R = \{x \in \mathbf{C}^n\colon |x| \leqslant R\}$, $S_R = \partial B_R$, and $\operatorname{Cone}(f^{-1}(0) \cap S_R, \mathbf{0}) = \{tx \in B_R\colon x \in f^{-1}(0) \cap S_R,\ 0 \leqslant t \leqslant 1\}$.) *Moreover, h can be chosen so that is the identity on $f^{-1}(0) \cap S_R$.*

The proof (which is not difficult) is given in Milnor [**10**, Theorem 2.10]. Lemma 2 means that the topological type of f is completely determined by the pair of topological spaces $(S_R, f^{-1}(0) \cap S_R)$.

One of the key ideas in the proof of the Lê-Ramanujam theorem is to characterize the topological type by a different pair of spaces. To describe this pair, we fix the following notation.

Let $D_r = \{z \in \mathbf{C}\colon |z| \leqslant r\}$ and $C_r = \partial D_r = \{z \in \mathbf{C}\colon |z| = r\}$. Put

$$T_{R,r}(f) = \left(f^{-1}(D_r) \cap S_R\right) \cup \left(f^{-1}(C_r) \cap B_R\right).$$

The latter set is indicated schematically below.

A better picture is the following.

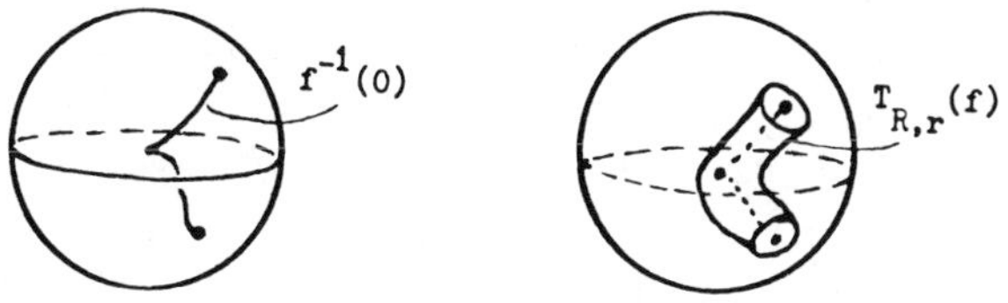

The set $T_{R,r}(f)$ is the boundary of a solid "tube" about $f^{-1}(0)$ with a "cap".

LEMMA 3. *Let $f \in \mathbf{C}[x_1,\ldots,x_n]$ have an isolated singularity at the origin, $f(\mathbf{0}) = 0$. Then, for every sufficiently small $R > 0$ and every sufficiently small r, $0 < r \ll R$, the pairs of spaces $(S_R, S_R \cap f^{-1}(0))$ and $(T_{R,r}(f), S_R \cap f^{-1}(0))$ are homeomorphic as pairs. The homeomorphism can be chosen to be the identity on $S_R \cap f^{-1}(0)$.*

For a proof, see Milnor [**10**, Chapter 5 or Chapter 10]. Lemmas 2 and 3 imply that the pair $(T_{R,r}(f), S_R \cap f^{-1}(0))$ completely characterizes the topological type of f.

In what follows, we shall let $T_{R,r,t} = T_{R,r}(f_t)$.

4. A sketch of the Lê-Ramanujam argument. In order to motivate some definitions which we introduce subsequently, it will be necessary to review the general idea behind Lê and Ramanujam's proof that "μ-constant families have the same topological type".

Let $\{f_t\}$ be a μ-constant family. We want to show that for all small t, f_t has the same topological type as f_0.

Choose $R > 0$ and r, $0 < r \ll R$, so small that:

1. $f_0^{-1}(0) \pitchfork S_{R'}$ for all R', $0 < R' \leqslant R$,
2. $\mathbf{0}$ is the only singular point of f_0 in B_R,
3. $f_0^{-1}(z) \pitchfork S_R$ for all $z \in \mathbf{C}$ such that $|z| \leqslant r$, and
4. $(T_{R,r,0}, S_R \cap f_0^{-1}(0))$ and $(S_R, S_R \cap f_0^{-1}(0))$ are homeomorphic as pairs.

(The above conditions are far from independent.) Then it is very easy to show that there exists $\tau > 0$ such that for all t, $|t| \leqslant \tau$:

1′. $f_t^{-1}(0) \pitchfork S_R$,
2′. $\mathbf{0}$ is the only singular point of f_t in B_R, and
3′. $f_t^{-1}(z) \pitchfork S_R$ for all z such that $|z| \leqslant r$.

Using standard fibration arguments (in particular, Ehresmann's theorem and the fact that $G(x, t) = (F(x, t), t)$ is the projection map of a smooth fibration when restricted to the set $(B_R \times D_\tau) - (G^{-1}(\mathbf{0}, D_\tau)))$, it easily follows that $(T_{R,r,0}, f_0^{-1}(0) \cap S_R)$ and $(T_{R,r,t}, f_t^{-1}(0) \cap S_R)$ are homeomorphic as pairs whenever $|t| < \tau$. (For a statement and proof of Ehresmann's theorem see Ehresmann [C. R. Acad. Sci. Paris **224** (1947), 1611–1612].)

It is very tempting to conclude that f_t and f_0 have the same topological type. However, reviewing the statement of Lemma 3, we see that the pair $(T_{R,r,t}, f_t^{-1}(0) \cap S_R)$ may not characterize the topological type of f_t. All that we know is that (for fixed t) for some $R' \leqslant R$ and $r' \leqslant r$ the pair $(T_{R',r',t}, f_t^{-1}(0) \cap S_{R'})$ characterizes the topological type. We know that $T_{R,r',0}$ is homeomorphic to $T_{R,r',t}$; but we do not know that $T_{R,r',t}$ is homeomorphic to $T_{R',r',t}$. If the latter two sets are homeomorphic, then we can conclude that f_t and f_0 have the same topological type.

On the other hand, suppose that $T_{R,r',t}$ is not homeomorphic to $T_{R',r',t}$. In fact, we may as well assume that $T_{R,r'',t}$ is not homeomorphic to $T_{R',r'',t}$ for all $r'' < r'$. If this is the case, then $g(x) = |x|^2$ restricted to $f_t^{-1}(C_{r'}) \cap (B_R - B_{R'})$ must have a critical point. Thus, g restricted to $f_t^{-1}(z') \cap (B_R - B_{R'})$ must have a critical point for some $z' \in C_{r'}$ and, hence, for every $z \in C_{r'}$ (the map f: $f_t^{-1}(C_{r'}) \cap (B_R - \operatorname{int}(B_{R'})) \to C_{r'}$ is the projection map of a smooth locally trivial fibration). Since this must be the case for all small $r' > 0$ and since the condition that $g|_{f_t^{-1}(z)}$ have a critical point is real algebraic (Lemma 1), $g|_{f_t^{-1}(0)}$ must have a critical point on $f_t^{-1}(0) \cap (B_R - \operatorname{int}(B_{R'}))$.

DEFINITION. A nonsingular point $x_0 \in f_t^{-1}(0)$ which is a critical point of the function $g(x) = |x|^2$ restricted to $f_t^{-1}(0) - \Sigma$ is called a *kink* of f_t.

Thus, if $T_{R,r',t}$ is not homeomorphic to $T_{R',r',t}$ for all sufficiently small r', f_t must have a kink at some point x such that $R' \leqslant |x| \leqslant R$. We might now go back and try to start with a smaller R. This we can always do, but we may have to take τ to be smaller. This will be the case if there exist sequences $t_i \to 0$ and $x_i \to \mathbf{0}$ such that x_i is a kink of f_{t_i}. By the curve selection lemma (Milnor [**10**, Chapter 3]), this happens if and only if there is a real analytic family of such kinks converging to the origin as $t \to 0$. To fix the above ideas we introduce the following definition.

DEFINITION. A *vanishing fold* (centered at $t = 0$) of the family $\{f_t\}$ is a real analytic curve $s \mapsto (x(s), t(s))$, $s \in [0, \varepsilon)$ such that:

(1) $0 \mapsto (\mathbf{0}, 0)$ and

(2) $x(s)$ is a kink of $f_{t(s)}^{-1}(0)$ if $s \in (0, \varepsilon)$.

REMARK. The above definition differs from another common in the literature in which a vanishing fold is taken to be an analytic curve along which the Whitney conditions fail. A vanishing fold in our sense is always a vanishing fold in the latter sense, but not conversely. We shall return to this point later.

The arguments above show that if $\{f_t\}$ is a μ-constant family and if there is no vanishing fold centered at $t = 0$ in $\{f_t\}$, then all the f_t have the same topological type for sufficiently small t. The condition that the family $\{f_t\}$ have no vanishing folds at $t = 0$ is, in Oka's language (Oka [**12**]), the same as saying that the family $\{f_t\}$ has a stable radius in a neighbourhood of $t = 0$.

Now let us rephrase the Lê-Ramanujam theorem so as to better reflect the structure of the arguments used to establish it.

THEOREM (LÊ-RAMANUJAM). *Let* $\{f_t\}$, $t \in [0, 1]$, *be a* μ*-constant family. Then*:

1. (*easy part*) *if* $\{f_t\}$ *does not have a vanishing fold centered at* $t = 0$, *all the* f_t *have the same topological type for small* t (*for any* $n \geqslant 2$); *and*

2. (*harder part*) *even if* $\{f_t\}$ *has a vanishing fold centered at* $t = 0$, *all the* f_t *have the same topological type for small* t, *provided that* $n \neq 3$.

We have just proved the first assertion above. To see that the second holds, note that (with the same notation employed previously) $f_t^{-1}(z) \cap B_R$ and $f_t^{-1}(z) \cap B_{R'}$ have the homotopy type of a bouquet of μ real spheres of dimension $n - 1$ for z sufficiently close to $0 \in C$ (the former because it is diffeomorphic to the Milnor fiber $f_0^{-1}(z) \cap B_R$, the latter because it is the Milnor fibre of f_t—see Milnor [**10**, Chapter 6]). If $n > 3$, $f_t^{-1}(z) \cap S_{R'}$ and $f_t^{-1}(z) \cap S_R$ are simply connected. A little fiddling (see [7]) allows us to apply the h-cobordism theorem to conclude that $f_t^{-1}(z) \cap B_{R'}$ is diffeomorphic to $f_t^{-1}(z) \cap B_R$ and thus that $T_{R',r',t}$ is diffeomorphic to $T_{R,r',t}$. The case $n = 2$ can be handled separately. the case $n = 3$ cannot, as yet, be dealt with in this manner.

The above analysis immediately suggests the question of whether there are vanishing folds in a μ-constant family. Surprisingly, this is not known. There is no known example of a μ-constant analytic family in which there is a vanishing fold. Pending the resolution of this question, we must face the embarrassing possibility that the use of the h-cobordism theorem in the Lê-Ramanujam argument (and the ensuing exclusion of the case $n = 3$) is merely to guard against a phenomenon that never occurs.

5. Families in which there are no vanishing folds. Before examining further the issue of the existence of vanishing folds, let us note that vanishing folds do not occur in many reasonable families of functions. We give a few examples, some of which are well known.

LEMMA 4. *If* f: $\mathbf{C}^n,\mathbf{0} \to \mathbf{C},0$ *is a quasihomogeneous polynomial with an isolated singular point at* $\mathbf{0} \in \mathbf{C}^n$, *then* f *has no kinks.*

PROOF. The proof is so simple that we give it. Let f have positive weights $w_1,\ldots,w_n$ and suppose that $x \in f^{-1}(0) - \mathbf{0}$ is a kink. Then, $(\partial f/\partial x_1,\ldots,\partial f/\partial x_n)_x = c(\bar{x}_1,\ldots,\bar{x}_n)$ for some $c \in C - 0$. Dot with $(x_1/w_1,\ldots,x_n/w_n)$ to get

$$\frac{x_1}{w_1}\frac{\partial f}{\partial x_1}(x) + \cdots + \frac{x_n}{w_n}\frac{\partial f}{\partial x_n}(x) = c\left(\frac{|x_1|^2}{w_1} + \cdots + \frac{|x_n|^2}{w_n}\right).$$

By the analogue of Euler's lemma for quasihomogeneous polynomials, the left-hand side equals $f(x)$ which is 0 by assumption. Thus $c = 0$, a contradiction.

From Lemma 4 and the "easy" part of the Lê-Ramanujam theorem we immediately obtain the following well-known proposition.

PROPOSITION 5. *Let* $\{f_t\}$ *be a* μ*-constant family of quasihomogeneous polynomials. Then all the* f_t *have the same topological type.*

The assumption that the family above be μ-constant can be replaced with almost no extra work by the assumption that the f_t have an isolated singularity for all t.

Another class of examples without vanishing folds may be obtained by making use of the following considerations. Let g be a polynomial with an isolated singularity at the origin, $g(\mathbf{0}) = 0$. Let h be a polynomial such that $h(\mathbf{0}) = 0$ which is "small", in some sense, compared to g. It seems clear intuitively that the family $\{f_t\} = \{g + th\}$ cannot have a vanishing fold centered at any $t_0 \in [0, 1]$. Making this idea precise yields the following two results (see O'Shea [**13**]).

THEOREM 6. *Let* h, g: $\mathbf{C}^n,\mathbf{0} \to \mathbf{C},0$ *be polynomials and suppose that* $\mathbf{0}$ *is an isolated singularity of* g *and that* $\lim_{x\to\mathbf{0}} |dh_x|/|dg_x| = 0$ *along any analytic curve. Then*:

(1) $\mathbf{0}$ *is an isolated singularity of* $f_t = g + th$ *for all* $t \in [0, 1]$ *and* $\mu(f_t) = \mu(g)$, *and*

(2) *the family* $\{f_t\}$ *has no vanishing folds centered at any* $t_0 \in [0, 1]$.

THEOREM 7. *Let* h, g: $\mathbf{C}^n,\mathbf{0} \to \mathbf{C},0$ *be polynomials and suppose that* $\mathbf{0}$ *is an isolated singularity of* g. *Set* $f_t = g + th$ *for* $t \in [0, 1]$. *If* $\mu(f_t) = \mu(g)$ *for all* $t \in [0, 1]$ *and if* $|dh_x|/|dg_x|$ *is bounded as* x *tends to the origin along any analytic curve lying in* $\bigcup_{t\in[0,1]} f_t^{-1}(0)$, *then there are no vanishing folds in* $\{f_t\}$ *centered at any* $t_0 \in [0, 1]$.

The proofs of Theorems 6 and 7 are by contradiction. One assumes that there is a vanishing fold and uses real analyticity to expand the gradients in a power series. Equating first order terms gives a contradiction.

Theorem 7 can be used to show that there are no vanishing folds in a family along which the Newton polygon is constant and the Newton principal parts are both commode and nondegenerate in the sense of Kouchnirenko [**4**]. Oka [**12**], however, has a nicer proof than mine. Damon and Gaffney [**3**] have given a proof along entirely different lines. I suspect, but cannot as yet prove, that there are no vanishing folds in a μ-constant family along which the Newton polygon is constant and the Newton principal parts define an isolated singularity.

6. Families in which the Milnor number jumps. It is instructive and reassuring to consider families $\{f_t\}$ in which $\mu(f_0) > \mu(f_t)$ for small $t \neq 0$. The following lemma is not difficult to establish (see [**14**]).

LEMMA 8. *If* $\{f_t\}$ *is a family satisfying conditions* (1)–(4) *of the convention in* §2 *and if* $\mu(f_0) > \mu(f_t) \equiv \mu$ *then there exists an analytic curve* p: $s \mapsto (x(s), t(s))$, $s \in [0, \varepsilon)$, $0 \mapsto (\mathbf{0}, 0)$, *such that* $|x(s)| > 0$ *if* $s > 0$ *and either*

(1) $x(s) \in f_{t(s)}^{-1}(0)$ *and* $x(s)$ *is a critical point of* $f_{t(s)}$ *for all* $s \in [0, \varepsilon)$ *or*

(2) p *is a vanishing fold.*

Both possibilities occur as may be seen by considering the examples $F(x, y, t) = xy(x + y - t)$ and $F(x, y, t) = y^2 - tx^2 + x^3$. In the former, the singularities $(0, t)$, $(t, 0) \in f_t^{-1}(0)$ of f_t tend to the origin as t tends to zero. In the latter example, $(t, 0)$ is a kink of f_t and thus the map $t \mapsto (t, 0, t)$ is a vanishing fold centered at $t = 0$.

7. Dependence on the choice of coordinates. It is very tempting to conjecture that there are no vanishing folds in a μ-constant family. One consideration that tempers this conjecture is that whether or not a polynomial has a kink obviously depends on the choice of coordinates. In fact, it is easy to show the following

LEMMA 9. *Let* f: $\mathbf{C}^n, \mathbf{0} \to C, 0$ *be a polynomial with an isolated singularity at* $\mathbf{0} \in \mathbf{C}^n$. *If* $x \in f^{-1}(0)$ *is such that* $\langle x, \overline{df_x} \rangle \neq 0$ (*here*, $\overline{df_x} = (\overline{\partial f/\partial x_1}, \ldots, \overline{\partial f/\partial x_n})_x$ *denotes the complex conjugate of* df_x *and* $\langle\ ,\ \rangle$ *the standard Hermitian inner product*), *then there exists a nonsingular linear map* A *such that* $A(x)$ *is a kink of* $f \circ A^{-1}$.

Another way of viewing this is to think in terms of changing the metric on $\mathbf{C}^n$ and working with the squared distance function with respect to this new metric. For simplicity, we consider the following class of metrics (which are linearly related to the standard metric). Let H: $\mathbf{C}^n \to \mathbf{C}^n$ be a positive definite Hermitian linear map. Consider the metric induced by the inner product $\langle\ ,\ \rangle_H$ defined by $\langle x, y \rangle_H = \langle x, H(y) \rangle$ for all $x, y \in \mathbf{C}^n$. Let $g_H(x) = \langle x, x \rangle_H$. Call a nonsingular point $x \in f^{-1}(0)$ an *H-kink* of f if x is a critical point of $g_H|_{f^{-1}(0)-\Sigma}$. By Lemma 1 a nonsingular point $x \in f^{-1}(0)$ is an H-kink if and only if $df_x = cH(\overline{x})$ for some nonzero $c \in \mathbf{C}$. An easy computation establishes the following result.

LEMMA 10. *Let* f *be as in Lemma* 9. *Then* $x \in f^{-1}(0)$ *is an H-kink of* f *if and only if* $A(x)$ *is a kink of* $f \circ A^{-1}$ *where* $A^*A = H$.

From Lemmas 9 and 10 we deduce the following

LEMMA 11. *Let f be as in Lemma 9. If $x \in f^{-1}(0)$ is a nonsingular point such that $\langle x, \overline{df_x}\rangle \neq 0$, then there exists a positive definite Hermitian linear map H such that x is an H-kink of f.*

The above considerations lend credence to (but, by no means establish that) the notion that a vanishing fold is an artifact of the coordinate choice. In an attempt to circumvent this shortcoming, define an *H-vanishing fold* by replacing the word "kink" in the definition of vanishing fold by the word "H-kink". By mimicing the "easy" part of the Lê-Ramanujam argument outlines above, we obtain the following

LEMMA 12. *Let $\{f_t\}$ be a μ-constant family. If there exists a positive definite Hermitian map H such that $\{f_t\}$ does not have an H-vanishing fold at $t = 0$, then all the f_t have the same topological type as f_0 for sufficiently small t.*

Conversely, if $\{f_t\}$ is as in Lemma 8, then it is not difficult to see that either assertion (1) of Lemma 8 holds or $\{f_t\}$ has an H-vanishing fold centered at $t = 0$ for every positive definite Hermitian linear map H. With this result and Lemma 12 in mind, we introduce the following definition.

DEFINITION. A family $\{f_t\}$ will be said to have a *linearly irremovable vanishing fold* (*lirvafold*) centered at $t = 0$ if $\{f_t\}$ has an H-vanishing fold centered at $t = 0$ for every positive definite Hermitian linear map $H: \mathbf{C}^n \to \mathbf{C}^n$.

We can now state our conjecture.

Conjecture. Let $\{f_t\}$ satisfy conditions (1)–(4) of the convention in §2. If $\{f_t\}$ has a lirvafold centered at $t = 0$, then $\mu(f_0) > \mu$.

It is, perhaps, worth remarking that if the above conjecture were true then the notion of a lirvafold (suitably modified to apply to families in which the singular stratum is not the t-axis) would be intrinsic. Given a family $\{f_t\}$ of isolated hypersurface singularities satisfying our convention, it would be very interesting to have necessary and sufficient geometric conditions on the varieties $f_t^{-1}(0)$ which guarantee that $\mu(f_0) > \mu(f_t)$ for small nonzero t. Lê and Saito [**8**] have given a necessary and sufficient condition for a family to be μ-constant. But their condition uses the notion of a "bonne stratification" which involves nonzero level sets of f_t. There ought to be some condition on the sets $f_t^{-1}(0)$ alone (because the Milnor number of f_t is determined by the embedding type of $f_t^{-1}(0)$—see, for example, Lê [**6**] or Teissier [**15**]).

8. An intrinsic notion? There is another, more speculative way to motivate the introduction of the notion of a lirvafold. One could begin with the final remark of the last section—namely, there ought to be a condition on the sets $f_t^{-1}(0)$ which implies that the Milnor number jumps. Moreover, since the Milnor number does not depend on the choice of coordinates, it seems reasonable to demand that the condition on the sets $f_t^{-1}(0)$ also be intrinsic.

In view of the lemmas in the previous section it is clear that one way to get an intrinsic condition would be to try to define an "irremovable vanishing fold" to be one that "persists" under any change of metric. It is easy to see that if $\mu(f_0) > \mu(f_t)$ there must be an irremovable vanishing fold (by the same argument used to establish Lemma 8—see **[14]**). Moreover, slight generalizations of Lemmas 9–12 show that the notion of an irremovable vanishing fold would be independent of the coordinate choice. The problem is, of course, that there are immense technical problems in dealing with "any change of metric". We can try to avoid these problems by restricting the types of metrics we consider. Hopefully, we could choose a class large enough to obtain an intrinsic condition, but small enough to be technically manageable.

A simple class of metrics on the sets $f_t^{-1}(0)$ are the restrictions of affine metrics on $\mathbf{C}^n$. The appropriate formulation of "irremovable" for this class of metrics leads directly to the notion of a lirvafold.

One of the problems with the notion of a vanishing fold is that if a given family of isolated hypersurface singularities is degenerate (in the sense of Kouchnirenko), then it is extremely difficult to determine whether or not there is a vanishing fold. Hopefully, this will turn out not to be the case for lirvafolds. In fact, the following conjecture appears to be true (and it easy to establish when $n = 2$).

Conjecture. Let $\{f_t\}$ be a family for which there exists a lirvafold centered at $t = 0$. Let $d_j f_t = \partial f_t/\partial x_j$ for $1 \leqslant j \leqslant n$ and set

$$V_i(t) = \{x \in \mathbf{C}^n : f_t(x) = d_1 f_t(x) = \cdots = d_{i-1} f_t(x) = d_{i+1} f_t(x) = \cdots = d_n f_t(x) = 0\}.$$

Then, for almost all coordinate choices (preserving conditions (1), (2) and (3) of our convention), there exist real analytic curves $\gamma_i\colon s \mapsto (x(s), t(s))$ defined for all $s \in [0, \varepsilon)$ and $1 \leqslant i \leqslant n$ such that $\gamma_i(0) = (\mathbf{0}, 0)$ and $|x(s)| > 0$ if $s > 0$ and $x(s) \in V_i(t(s))$ for all $s \in [0, \varepsilon)$.

If true, the conjecture above would give a useful criterion for detecting lirvafolds. The conclusion of the conjecture can be restated in terms of Teissier's polar curves **[16]**. It states that the intersection multiplicity (at the origin) of a polar curve of f_t and the set $f_t^{-1}(0)$ is strictly less than the intersection multiplicity (at the origin) of the polar curve of f_0, defined as above, and $f_0^{-1}(0)$.

9. Vanishing folds and Whitney conditions. In view of the above discussion it is interesting to investigate the connection between Whitney conditions, which are known to be intrinsic, and vanishing folds. Let $\{f_t\}$ be a family of isolated singularities. Set $X = \{(x, t) \in \mathbf{C}^n \times \mathbf{C}\colon F(x, t) = 0\}$ and $Y = \{(\mathbf{0}, t) \in \mathbf{C}^n \times \mathbf{C}\}$. Then if f_t has a vanishing fold centered at 0, the pair $(X - Y, Y)$ does not satisfy Whitney condition b at $(\mathbf{0}, 0)$. In fact, the latter fails rather badly along a vanishing fold. To see this, let

$$B_F(x, t) = \frac{x_1 F_1 + \cdots + x_n F_n}{|x|\,|(F_1, \ldots, F_n)|}$$

where $(x, t) = (x_1, \ldots, x_n, t) \in X - Y$ and $F_i = \partial F/\partial x_i$ for $1 \leq i \leq n$ is evaluated at (x, t). If $\alpha: s \mapsto (x(s), t(s))$ is a vanishing fold, then it follows immediately from Lemma 1 that the limit of $|B_F(\alpha(s))|$ as s tends to 0 is 1.

An example due to Briançon and Speder [**1**] shows that F may have no vanishing folds centered at $t = 0$, but the pair $(X - Y, Y)$ may fail to satisfy condition b. They let $F(x, y, z, t) = x^5 + txy^6 + zy^7 + z^{15}$ and show there is a path along which $B_F(x, y, z, t)$ does not tend to zero (as (x, y, z, t) tends to the origin). Since this choice of F gives a family of quasihomogeneous polynomials there are no vanishing folds (by Lemma 4).

There is, however, an intriguing connection between the failure of Whitney condition b in this example and vanishing folds. Consider the generic hyperplane section obtained by setting $z = ax + by$ and $\hat{F}(x, y, t) = F(x, y, ax + by, t)$. The Milnor number of the family of curves given by $\hat{F}$ jumps when $t = 0$. By Lemma 8, there is a vanishing fold $s \mapsto (x(s), y(s), t(s))$ of $\hat{F}$. An easy, but tedious, computation shows that the Whitney condition b fails along with the curve $s \mapsto (x(s), y(s), z(s), t(s))$ where $z(s) = ax(s) + by(s)$.

It is natural to ask whether this observation generalizes. That is, suppose that F is a family which is μ-constant, but which is such that the family $\hat{F}$ cut out by a generic hyperplane is not μ-constant (so F is not μ^*-constant). Lemma 8 provides a vanishing fold for $\hat{F}$. We ask whether it is always the case that the vanishing fold for $\hat{F}$ "lifts" to a curve along which Whitney condition b for F is violated. I suspect, but have not been able to prove, that this is the case. An affirmative answer would give an alternate proof of Briançon and Speder's result [**2**] that "Whitney conditions imply μ^*-constant".

We conclude with a final bit of speculation. It follows easily from Lemma 1 that the condition b also fails along any H-vanishing fold. Since a jump in the Milnor number gives an H-vanishing fold for every H, if H-vanishing folds were also to lift to curves along which condition b failed, then the nonconstancy of μ^* would imply the existence of many curves along which condition b fails. This suggests a connection with recent work of Navarro Aznar and Trotman [**11**] on measuring the extent of failure of the Whitney conditions. One could also speculate that any curve along which condition b fails projects to a vanishing fold (with respect to some metric) in a hyperspace section.

REFERENCES

1. J. Briançon and J.-P. Speder, *La trivialité topologique n'implique pas les conditions de Whitney*, C. R. Acad. Sci. Paris Sér. A–B **280** (1975), 365–367.

2. ______, *Les conditions de Whitney impliquent "$\mu^{(*)}$ constant"*, Ann. Inst. Fourier (Grenoble) **26** (1976), 153–163.

3. J. Damon and T. Gaffney, *Topological triviality of deformations and Newton filtrations*, preprint.

4. A. G. Kouchnirenko, *Polyhèdres de Newton et nombre de Milnor*, Invent. Math. **32** (1976), 1–31.

5. H. King, *Topological type of isolated critical points*, Ann. of Math. (2) **107** (1978), 385–397.

6. D. T. Lê, *Calcul du nombre de cycles évanouissants d'une hypersurface complexe*, Ann. Inst. Fourier (Grenoble), **23** (1973), 261–270.

7. D. T. Lê and C. P. Ramanujam, *The invariance of Milnor's number implies the invariance of the topological type*, Amer. J. Math. **98** (1976), 67–78.

8. D. T. Lê and K. Saito, *La constance du nombre de Milnor donne des bonnes stratifications*, C. R. Acad. Sci. Paris Sér. A–B **277** (1973), 793–795.

9. J. Mather, *Notes on topological stability*, Lecture notes, Harvard, 1970.

10. J. Milnor, *Singular points of complex hypersurfaces*, Ann. of Math. Studies, No. 61, Princeton Univ. Press, Princeton, N. J., 1968.

11. V. Navarro Aznar and D. A. Trotman, *Whitney regularity and generic wings*, Ann. Inst. Fourier (Grenoble) **31** (1981), 87–111.

12. M. Oka, *On the bifurcation of the multiplicity and the topology of the Newton boundary*, J. Math. Soc. Japan **31** (1979), 435–450.

13. D. B. O'Shea, *On μ-constant families of singularities*, Ph.D. thesis, Queen's Univ., 1980.

14. ______, *A criterion for the nonexistence of vanishing folds*, preprint.

15. B. Teissier, *Cycles évanescents, sections planes, et conditions de Whitney*, Asterisque **7–8** (1973), 285–362.

16. ______, *Introduction to equisingularity problems*, Amer. Math. Soc. Algebraic Geometry Sympos., Arcata 1974, Providence, R.I., 1975, pp. 593–632.

17. J. G. Timourian, *The invariance of Milnor's number implies topological triviality*, Amer. J. Math. **99** (1977), 437–446.

MOUNT HOLYOKE COLLEGE

Proceedings of Symposia in Pure Mathematics
Volume **40** (1983), Part 2

PROJECTIVE RESOLUTIONS OF HODGE ALGEBRAS: SOME EXAMPLES

JÜRGEN PESSELHOY AND OSWALD RIEMENSCHNEIDER[1]

De Concini, Eisenbud and Procesi developed in [**3**] a complete theory of Hodge algebras; in §7 of their paper they describe a method for constructing explicit free resolutions of a Hodge algebra A by stepwise lifting an explicit free resolution of the associated discrete Hodge algebra A_0. In this note, we want to apply this method to some examples and to emphasize the usefulness of replacing the stepwise procedure by constructing one single family depending on several variables with A as general and A_0 as special fibre. By this method we receive automatically interesting new complexes which are moreover grade-sensitive if A_0 is a Cohen-Macaulay algebra. So we get, for instance, by deforming the Eagon-Northcott complex for the 2×2 minors of a $2 \times s$ matrix a complex which leads to explicit minimal free resolutions of the local rings of many quotient surface singularities (which, in general, are not determinantal). This complex has been known to us for several years [**8**]; we would like to thank David Eisenbud for many helpful discussions on earlier versions of handling this complex and for introducing us to the new field of Hodge algebras.

For more background material we refer to [**3**] and the bibliography cited there (especially to the papers of Baclawski, Hochster, Reisner and Stanley).

1. Hodge algebras and their simplification. We restrict ourselves to Hodge algebras which are ordinal in the sense of [**3**]: Let R be a commutative ring and let A be a commutative R-algebra; suppose $H \subset A$ is a finite partially ordered set such that:

1. A is a free R-module admitting the set of *standard monomials* $\{x_1 \cdot \cdots \cdot x_s\colon s \in \mathbf{N}, x_\sigma \in H, x_1 \leqslant x_2 \leqslant \cdots \leqslant x_s\}$ as basis.

1980 *Mathematics Subject Classification*. Primary 13D25, 13D10, 14M12.

[1]Lecture on the present paper and a joint paper with D. Eisenbud and F.-O. Schreyer on *Projective resolutions of Cohen-Macaulay algebras*, Math. Ann. **257** (1981), 85–98, given by the second author.

2. If $y_1, y_2 \in H$ are incomparable and

$$(*) \qquad y_1 y_2 = \sum r_i x_{i,1} \cdot \cdots \cdot x_{i,s_i}, \qquad 0 \neq r_i \in R,$$

is the unique expression for $y_1 y_2 \in A$ as a linear combination of distinct standard monomials, then $x_{i,1} < y_1, y_2$ for all i;

then A is called a *Hodge algebra* on H (with straightening relations $(*)$). If the right-hand sides of all straightening relations are 0, then A is called the *discrete* Hodge algebra on H.

We suppose, moreover, that A is suitably graded (as is the case in all of our examples) such that the straightening relations $(*)$ give a presentation of A (see **[3**, Proposition 1.1**]**).

The simplification process from A to the discrete Hodge algebra A_0 on H works as follows:

Set $B_0 = A$, and let x_1 be a smallest element of H occurring on the right-hand sides of the straightening relations of A. Define the $R[t]$-algebra

$$\mathfrak{R}_1 = \mathfrak{R}(x_1, A) = \cdots \oplus At^k \oplus \cdots \oplus At \oplus A$$
$$\oplus (x_1 A)t^{-1} \oplus (x_1^2 A)t^{-2} \oplus \cdots.$$

Then, by Theorem 2.1 of **[3]**, $\mathfrak{R}_1$ is a Hodge algebra on H over $R[t]$ by sending $x \in H$ to $x \in A \subset \mathfrak{R}_1$ if $x \neq x_1$ and to xt^{-1} for $x = x_1$. The straightening relations for $\mathfrak{R}_1$ are formally received from $(*)$ by substituting tx_1 for x_1 and dividing a straightening formula by t if x_1 occurs on the left-hand side.

Setting $t = 1$, we get back the straightening relations for B_0, but setting $t = 0$ yields the straightening relations for the Hodge algebra

$$B_1 = \mathfrak{R}_1 / t\mathfrak{R}_1 = \mathrm{gr}_{\{x_1^i A\}} A = A/x_1 A \oplus x_1 A / x_1^2 A \oplus \cdots$$

on H over R which is simpler than B_0 since x_1 does not occur on any right-hand side of the straightening relations.

Repeating this process we get, after a finite number of steps, elements $x_1, \ldots, x_n \in H$ and Hodge algebras $B_0, \mathfrak{R}_1, B_1, \ldots, \mathfrak{R}_n, B_n$ such that $B_0 = A$, B_{i+1} is a normally flat deformation of B_i, and B_n is the discrete Hodge algebra A_0 on H. From now on we write $A = A_n, A_{n-1} = B_1, \ldots, A_0 = B_n$.

EXAMPLE 1. Consider the general 2×3 matrix

$$\begin{pmatrix} y_1 & y_2 & y_3 \\ x_1 & x_2 & x_3 \end{pmatrix}$$

and the determinantal variety on R defined by the vanishing of the 2×2 minors of this matrix. Then the affine coordinate ring A of this variety is a Hodge algebra on the partially ordered set

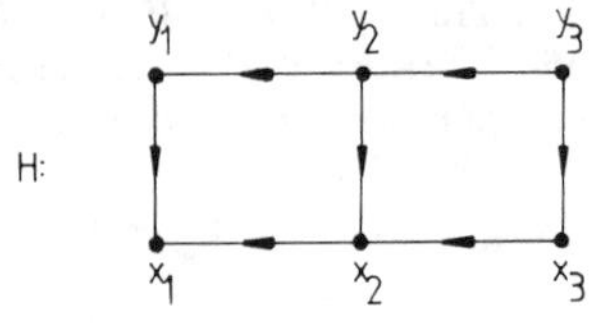

with straightening relations

$$x_2y_1 = x_1y_2, \quad x_3y_1 = x_1y_3, \quad x_3y_2 = x_2y_3.$$

In this case, $n = 2$, and the straightening formulas for $\mathcal{R}_1$, A_1, $\mathcal{R}_2$, A_0 are the following:

$\mathcal{R}_1 : x_2y_1 = tx_1y_2, x_3y_1 = tx_1y_3, x_3y_2 = x_2y_3,$
$A_1 : x_2y_1 = 0, x_3y_1 = 0, x_3y_2 = x_2y_3,$
$\mathcal{R}_2 : x_2y_1 = 0, x_3y_1 = 0, x_3y_2 = tx_2y_3,$
$A_0 : x_2y_1 = 0, x_3y_1 = 0, x_3y_2 = 0.$

Note. In the following, we always represent finite partially ordered sets as subsets of $\mathbf{Z}^k$ endowed with the induced partial ordering $(a_1,\dots,a_k) \leq (b_1,\dots,b_k) \Leftrightarrow a_j \leq b_j, j = 1,\dots,k$.

It is now very useful that the described stepwise deformations of A to A_0 can be embedded into one single family: Let $x_1,\dots,x_n$, A_0, $A_1,\dots,A_n = A$ be chosen as above.

Then there exists a Hodge algebra $\mathcal{A}$ on H over $R[t_1,\dots,t_n]$, such that

$$A_i \simeq \mathcal{A}/(t_1,\dots,t_{n-i}, t_{n-i+1} - 1,\dots,t_n - 1)\mathcal{A}.$$

We only sketch how to get inductively the straightening relations of $\mathcal{A}$: Each x_i has formally to be replaced by T_ix_i, where T_i is a product of certain elements t_j, $j \geq i$ (perhaps with multiplicities). Suppose that by our (descending) induction procedure we know already $T_{i+1},\dots,T_n$. During the process x_k was replaced by $T_k^{(i+1)}x_k$, $k \leq i$, where $T_k^{(i+1)}$ is a product of certain $t_j, j \geq i+1$. Then we define

$$T_k^{(i)} = t_i^{r_k}T_k^{(i+1)}, \qquad k \leq i, T_i = T_i^{(i)},$$

where the numbers $r_k \geq 0$ are again defined by descending induction on $k \leq i$ in the following way: Put $r_i = 1$, and if $r_i,\dots,r_{k+1}$ are constructed, replace in each straightening formula x_l by T_lx_l, $l \geq i$, resp. by $T_l^{(i)}x_l$ for $k+1 \leq l < i$ and defined for a fixed relation ρ_k to be the maximal difference of the t_i-degree of the left-hand side and the t_i-degrees of such monomials on the right-hand side which are divisible by x_k (if all these differences are negative, $\rho_k = 0$). r_k is then the maximal ρ_k (the maximum taken over all relations). Finally, after the substitutions $x_i \mapsto T_ix_i$, one has to divide each transformed straightening relation by the product of the t_j occuring on the left-hand side. (We are indebted to J. Behrmann who pointed out to us an error in the first version of the construction above.)

EXAMPLE 2. Take the example above. Then the procedure gives the substitutions $x_2 \mapsto t_2x_2$, $x_1 \mapsto t_2t_1x_1$, and the relations for $\mathcal{A}$ are

$$x_2y_1 = t_1x_1y_2, \quad x_3y_1 = t_1t_2x_1y_3, \quad x_3y_2 = t_2x_2y_3.$$

For the general $2 \times s$ matrix, $s \geq 2$,

$$\begin{pmatrix} y_1 & y_2 & \cdots & y_{s-1} & y_s \\ x_1 & x_2 & \cdots & x_{s-1} & x_s \end{pmatrix}$$

we get with $t_{i,i+1} := t_i$, $t_{i,j} := t_i \cdot \cdots \cdot t_{j-1}$, $i < j$, the relations

$$x_jy_i = t_{i,j}x_iy_j, \qquad 1 \leq i < j \leq s,$$

which we call *almost determinantal.* In **[9]** it is shown that many quotient surface singularities can be written in this form. For instance, the local rings of the cyclic quotients of $\mathbf{C}^2$ are of the form

$$\mathbf{C}\{z_1,\ldots,z_e\}/\left(z_{j+1}z_i - t_{i,j}z_{i+1}z_j;\ 1 \leqslant i < j \leqslant e-1\right)$$

with $t_{i,i+1} = z_{i+1}^{a_{i+1}-2}$, $a_{i+1} \geqslant 2$ for all i.

Generalizing this example to the case of the 2×2 minors of an arbitrary $r \times s$ matrix we have to regard the partially ordered set

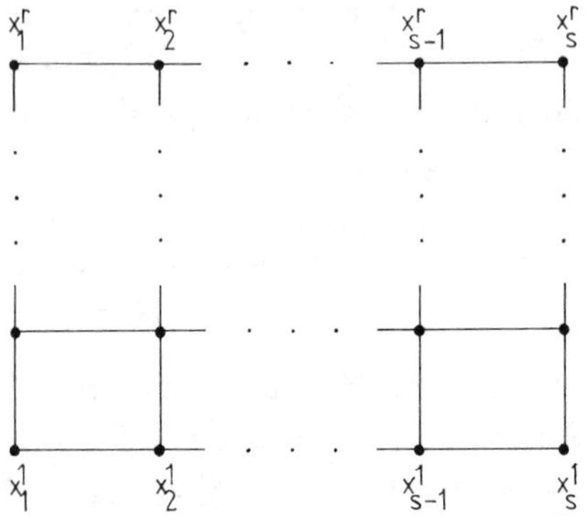

The set $\{x_1,\ldots,x_n\}$ is given by $\{x_j^i,\ i \leqslant r-1, j \leqslant s-1\}$, and the straightening formulas for the deformation $\mathcal{Q}$ consist of the equations

$$x_l^i x_k^j = \left(\prod_{\substack{i\leqslant\rho<j\\ k\leqslant\sigma<l}} t_\sigma^\rho\right) x_k^i x_l^j, \qquad i<j,\ k<l.$$

Since the Hodge property is preserved by base change, the algebra

$$\bar{A} = \mathcal{Q}/\left(t_1^{r-1}, t_j^i - 1, (i,j) \neq (r-1,1)\right)\mathcal{Q}$$

is a Hodge algebra on H over R with straightening relations

(a) $x_1^r x_j^i = 0$, $i<r, j>1$, and

(b) the vanishing of all 2×2 minors of the array

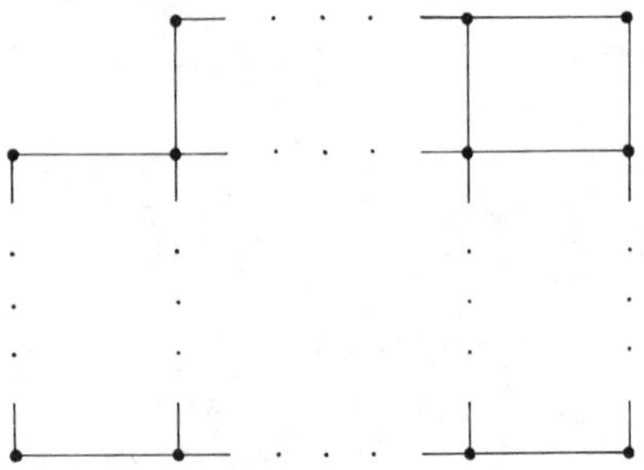

Obviously, the ideal $x_1^r\bar{A}$ is a free R-module admitting the standard monomials containing x_1^r as basis. Hence $B = \bar{A}/x_1^r\bar{A}$ is a Hodge algebra on $H \setminus \{x_1^r\}$.

Repeating this argument we conclude that the affine coordinate ring over R of the variety defined by the vanishing of all 2×2 minors of an array of the type

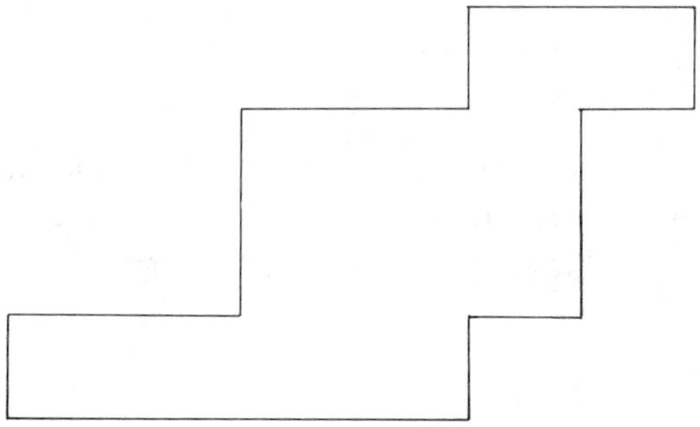

is a Hodge algebra. If R is Cohen-Macaulay then the coordinate ring is Cohen-Macaulay, too, since the underlying partially ordered set is wonderful (see [**3**, §8]).

This example may be used in the study of quotient surface singularities over **C**, since all quotients which are not almost determinantal may be represented by the vanishing of all 2×2 minors of an array of type

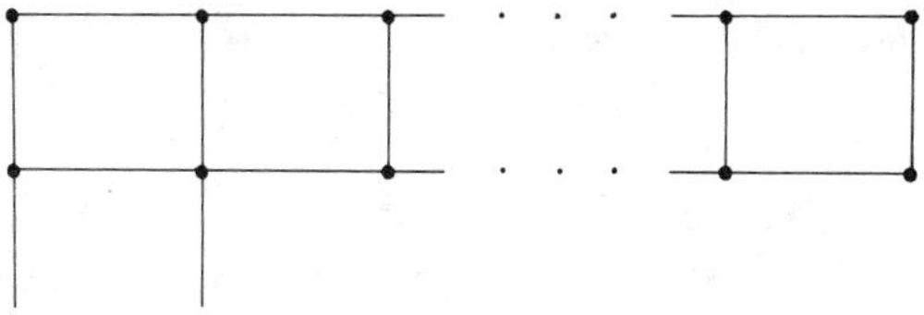

(cf. [**9**]). However, it should be noted that not all of the minors are independent and that henceforth the ideal generated by these minors has not maximal grade such that the general complex associated to the array above does not lead automatically to a free resolution of the local ring of the singularity. But by splitting off free summands lying in the kernels of the mappings of the complex it is possible to actually construct such free resolutions. We will not pursue this further in the present note.

EXAMPLE 3. The polynomial ring $A = R[x_1, x_2, x_3, y_1, y_2, y_3]$ is a Hodge algebra on the poset

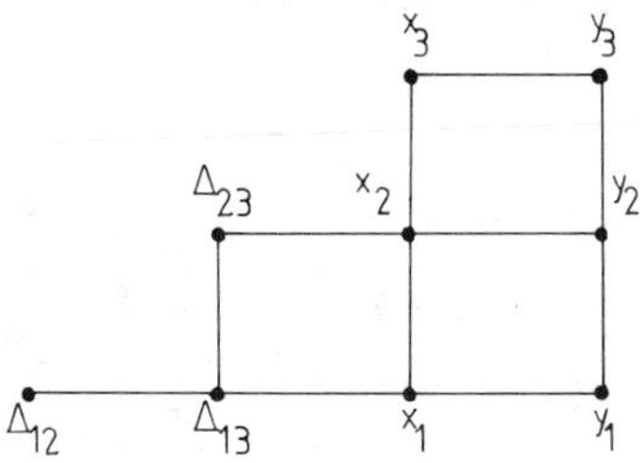

(see Doubilet, Rota and Stein [4] or De Concini, Eisenbud and Procesi [3, Chapter 12]). For the deformation to the discrete Hodge algebra, we get

$$\begin{aligned}
x_2 y_1 &= t_{12} t_{23} t_1 \Delta_{12} + t_1 x_1 y_2, \\
x_3 y_1 &= t_{13} t_{23} t_1 \Delta_{13} + t_1 t_2 x_1 y_3, \\
x_3 y_2 &= t_{23} \Delta_{23} + t_2 x_2 y_3, \\
\Delta_{23} x_1 &= t_{13} \Delta_{13} x_2 - t_{12} \Delta_{12} x_3, \\
\Delta_{23} y_1 &= t_{13} t_1 \Delta_{13} y_2 - t_{12} t_1 t_2 \Delta_{12} y_3.
\end{aligned}$$

EXAMPLE 4. The homogeneous coordinate rings of Grassmannians and Schubert varieties are Hodge algebras (see [3]). For instance, the partially ordered set for the Grassmannian of lines in projective 4-space is given by

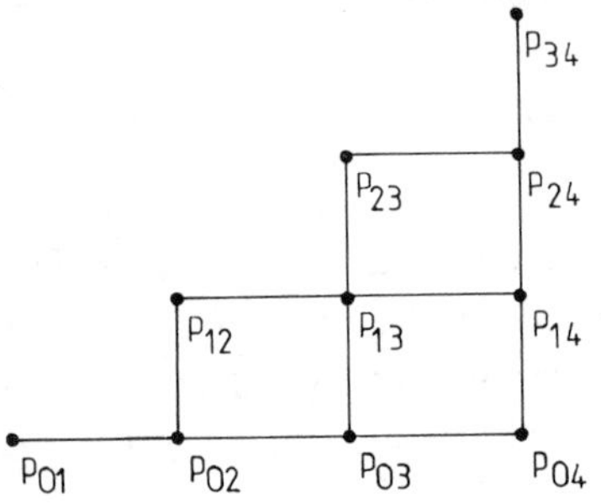

the straighening relations being the Plücker formulas

$$p_{il} p_{jk} = p_{ik} p_{jl} - p_{ij} p_{kl}, \qquad 0 \leqslant i < j < k < l \leqslant 4.$$

In our notation, $x_1 = p_{01}$, $x_2 = p_{02}$, $x_3 = p_{03}$, $x_4 = p_{12}$, $x_5 = p_{13}$. We write t_{01} instead of t_1 and so on, and get the substitutions

$$\begin{aligned}
&p_{13} \mapsto t_{13} p_{13}, \quad p_{12} \mapsto t_{12} p_{12}, \quad p_{03} \mapsto t_{03} t_{13} p_{03}, \\
&p_{02} \mapsto t_{02} t_{03} t_{12} p_{02}, \quad p_{01} \mapsto t_{01} t_{03} t_{12} t_{13} p_{01}
\end{aligned}$$

and the equations for the deformation:

$$\begin{aligned}
p_{03} p_{12} &= t_{02} p_{02} p_{13} - t_{01} p_{01} p_{23}, \\
p_{04} p_{12} &= t_{02} t_{03} p_{02} p_{14} - t_{01} t_{03} t_{13} p_{01} p_{24}, \\
p_{04} p_{13} &= t_{03} p_{03} p_{14} - t_{01} t_{03} t_{12} p_{01} p_{34}, \\
p_{04} p_{23} &= t_{13} t_{03} p_{03} p_{24} - t_{02} t_{03} t_{12} p_{02} p_{34}, \\
p_{14} p_{23} &= t_{13} p_{13} p_{24} - t_{12} p_{12} p_{34}.
\end{aligned}$$

Putting $p_{01} = p_{34} = 0$ yields the correct result for the 2×2-minors of an array of the form

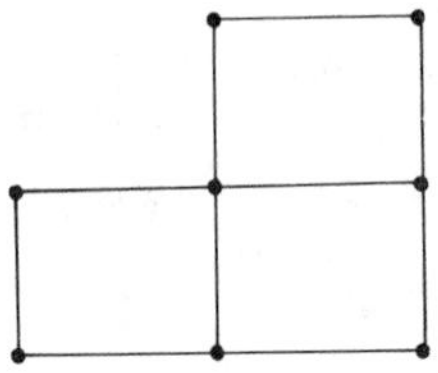

2. Some explicit projective resolutions. In all of the following examples, we represent finite free resolutions explicitly by concrete matrices. In each case, we first construct a free resolution of the corresponding discrete Hodge algebra (for instance by Diana Taylor's complex, cf. [**3**, Chapter 7]), then we extend this complex to the full deformation, and finally we specialize and (if necessary) minimalize.

EXAMPLE 5. We regard the discrete Hodge algebra belonging to the general 2×4 matrix

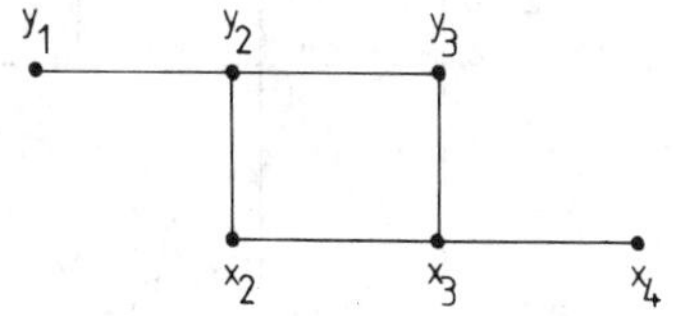

The ranks of the free modules in Taylor's complex are $\binom{6}{i}$, $i = 0, \ldots, 6$. We only write down the first two mappings:

1 2	1 3	1 4	2 3	2 4	3 4
x_2y_1	x_3y_1	x_4y_1	x_3y_2	x_4y_2	x_4y_3

	1 1 2 3	1 1 2 4	1 2 2 3	1 2 2 3	1 3 2 4	1 1 3 4	1 2 3 3
1 2	x_3	x_4	x_3y_2	x_4y_2	x_4y_3		
1 3	$-x_2$					x_4	y_2
1 4		$-x_2$				$-x_3$	
2 3			$-x_2y_1$				$-y_1$
2 4				$-x_2y_1$			
3 4					$-x_2y_1$		

	1 2 3 4	1 3 3 4	1 2 4 3	1 2 4 4	1 3 4 4	2 2 3 4	2 3 3 4	2 3 4 4
1 2								
1 3	x_4y_2	x_4y_3						
1 4			x_3y_2	y_2	y_3			
2 3			$-x_4y_1$			x_4	x_4y_3	
2 4	$-x_3y_1$			$-y_1$		$-x_3$		y_3
3 4		$-x_3y_1$			$-y_1$		$-x_3y_2$	$-y_2$

It is easy to see that the "quadratic" relations are all superfluous. The most effective but also tedious way to see this is to write down the next matrix. Hence, by minimalizing, we have as second matrix

x_3	x_4						
$-x_2$		x_4		y_2			
	$-x_2$	$-x_3$			y_2	y_3	
			x_4	$-y_1$			
			$-x_3$		$-y_1$		y_3
						$-y_1$	$-y_2$

For the third (and last) matrix, we receive in the same manner:

x_4		
$-x_3$		
x_2	$-y_2$	
	y_1	
	x_4	
	$-x_3$	y_3
		$-y_2$
		y_1

The extensions of these matrices for the deformation given by the matrix

x_2y_1 $-t_1x_1y_2$	x_3y_1 $-t_1t_2x_1y_3$	x_4y_1 $-t_1t_2t_3x_1y_4$	x_3y_2 $-t_2x_2y_3$	x_4y_2 $-t_2t_3x_2y_4$	x_4y_3 $-t_3x_3y_4$

are immediately found to be:

x_3	x_4			$-t_2y_3$	$-t_2t_3y_4$		
$-x_2$		x_4		y_2		$-t_3y_4$	
	$-x_2$	$-x_3$			y_2	y_3	
t_1x_1			x_4	$-y_1$			$-t_3y_4$
	t_1x_1		$-x_3$		$-y_1$		y_3
		$t_1t_2x_1$	t_2x_2			$-y_1$	$-y_2$

x_4	$-t_2t_3y_4$	
$-x_3$	t_2y_3	
x_2	$-y_2$	
$-t_1x_1$	y_1	
	x_4	$-t_3y_4$
	$-x_3$	y_3
	t_2x_2	$-y_2$
	$-t_1t_2x_1$	y_1

For $t_1 = t_2 = t_3 = 1$ these are precisely (up to signs) the matrices of the Eagon-Northcott complex of the general 2×4 matrix [**5**]. Hence our complex may be viewed as a deformation of that complex. We will present the general case in §3.

EXAMPLE 6. The Taylor complex for the discrete Hodge algebra on

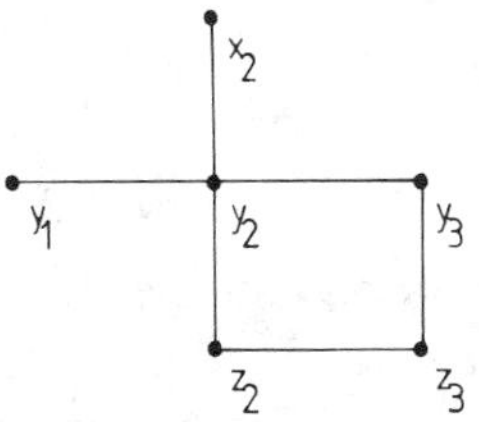

has ranks $\binom{5}{i}$, $i = 0, \ldots, 5$. After minimalizing we get a minimal free resolution with Betti numbers 1, 5, 6, 2, represented by the following matrices:

x_2y_3	y_2z_3	y_1z_2	y_1z_3	x_2z_3

z_3	y_1z_2				
		y_1	x_2		
	$-x_2y_3$			z_3	
		$-y_2$		$-z_2$	x_2
$-y_3$			$-y_2$		$-y_1$

		x_2	$-y_1$		y_2
$-y_1z_2$	z_3			x_2y_3	y_3z_2

EXAMPLE 7. Example 4 is a deformation of Example 6. Hence we get a free resolution by lifting the relations above to that deformation. This implies for instance for the second mapping

$-p_{04}$	$-p_{03}p_{14}+p_{02}p_{13}-p_{01}p_{23}$	p_{01}	p_{03}		p_{02}
$-p_{24}$		$-p_{12}$	p_{23}	p_{02}	
	$p_{14}p_{23}-p_{13}p_{24}+p_{12}p_{34}$	$-p_{14}$	$-p_{34}$	p_{04}	$-p_{24}$
p_{34}		p_{13}		$-p_{03}$	p_{23}
p_{14}			$-p_{13}$	$-p_{01}$	$-p_{12}$

It is immediately checked that the second column is dependent on the others. Hence by minimalizing and rearranging we receive

	p_{01}	$-p_{02}$	p_{03}	$-p_{04}$
$-p_{01}$		p_{12}	$-p_{13}$	p_{14}
p_{02}	$-p_{12}$		p_{23}	$-p_{24}$
$-p_{03}$	p_{13}	$-p_{23}$		p_{34}
p_{04}	$-p_{14}$	p_{24}	$-p_{34}$	

The third mapping is the transpose of the first one and the whole complex is evidently self-dual. That means, as is well known (see Stanley [**10**]), that the homogeneous coordinate ring of the Grassmannian of lines in projective 4-space is a Gorenstein ring, whereas the corresponding discrete Hodge algebra is not.

EXAMPLE 8. We leave it to the reader to lift the relations of Example 6 to the deformation given in Example 3. For $\Delta_{12} = \Delta_{13} = 0$ this complex minimalizes to a free resolution of $S = R[x_1, x_2, x_3, y_1, y_2, y_3]$ modulo $(x_1y_2 - x_2y_1, x_1y_3 - x_3y_1)$ as S-module, of course a Koszul complex since the ideal is a complete intersection.

EXAMPLE 9. For the discrete Hodge algebra belonging to the 2×2 minors of the general 3×3 matrix the Taylor resolution minimalizes to a complex with Betti numbers 1, 9, 17, 12 and 3 which lifts to a resolution of the deformed algebra with parameters t_1^1, t_2^1, t_1^2, t_2^2. It is interesting that this complex may be minimalized once more if $t_2^1 t_1^2 = 1$. In this case we get a self-dual complex with

Betti numbers 1, 9, 16, 9, 1, the first two mappings of which are represented by the transposes of the following matrices:

$x_2^1x_1^2 - t_1^1x_1^1x_2^2$
$x_2^1x_1^3 - t_1^1t_1^2x_1^1x_2^3$
$x_3^1x_1^2 - t_1^1t_2^1x_1^1x_3^2$
$x_3^1x_2^2 - t_2^1x_2^1x_3^2$
$x_3^1x_1^3 - t_1^1t_2^2x_1^1x_3^3$
$x_2^2x_1^3 - t_1^2x_1^2x_2^3$
$x_3^1x_2^3 - t_2^1t_2^2x_2^1x_3^3$
$x_3^2x_1^3 - t_1^1t_2^2x_1^1x_3^3$
$x_3^2x_2^3 - t_2^2x_2^2x_3^3$

		$-t_1^2t_2^2x_3^3$		x_3^2			$-x_3^1$	
	$-t_2^1t_2^2x_3^3$			x_2^3		$-x_1^3$		
					$-t_2^2x_3^3$		x_2^3	$-x_1^3$
$-t_1^2x_2^3$	x_2^2				$-x_2^1$			
$t_2^2x_3^3$			x_1^3	$-x_2^2$			$t_2^1x_2^1$	
$-t_2^1t_2^2x_3^3$		x_2^3				$-x_1^2$		$t_1^1t_2^1x_1^1$
		x_1^3		$-x_1^2$			$t_1^1t_2^1x_1^1$	
					x_3^2		$-x_2^2$	$t_1^2x_1^2$
			$-t_2^2x_3^3$			x_3^2		$-x_3^1$
$-t_2^1x_3^2$		x_2^2	$-x_1^2$					
$-t_2^2x_3^3$				x_2^2	$-x_3^1$	$-t_1^2x_1^2$		
$-t_1^2t_2^2x_3^3$	x_3^2						$-x_2^1$	$t_1^1t_1^2x_1^1$
x_3^1		$-x_2^1$	$t_1^1x_1^1$					
			x_2^3			$-x_2^2$		$t_2^1x_2^1$
	x_3^1			$-x_2^1$		$t_1^1t_1^2x_1^1$		
x_1^3	$-x_1^2$				$t_1^1x_1^1$			

Now compose the third matrix by taking as first row $(-t_1^2)$ times the fourth row of the second matrix, as second row t_1^2 times the 10th row, and then the 13th, $(-t_2^1) \times$ the first, $t_1^2 \times$ 6th, $t_1^2 \times$ 5th, $(-t_1^2) \times$ 14th, $t_2^1 \times$ 15th, the negative of the 16th, $t_2^1 \times$ the second, $t_2^1 \times$ 12th, $t_2^1 \times$ 11th, the third, $(-t_1^2) \times$ 7th, $t_2^1 \times$ 8th and the negative of the 9th row. Finally, the fourth matrix coincides with the first one above.

In particular, for $t_1^1 = t_2^1 = t_1^2 = t_2^2 = 1$, the complex above is a description of the Gulliksen-Negård complex [**6**] of the 2×2 minors of the general 3×3 matrix by concrete matrices.

3. A generalization of the Eagon-Northcott complex. In this section we describe a generalization of the Eagon-Northcott complex which can be obtained by a Hodge algebra deformation of the maximal minors of the general $r \times s$ matrix $(x_j^i)_{1 \leqslant i \leqslant r, 1 \leqslant j \leqslant s}$, $r \leqslant s$.

The theory of Hodge algebras sketched in §1 suggests the introduction of new variables t_j^i, $1 \leqslant i < r$, $1 \leqslant j < s$. We define inductively with

$$t^i_{j_1, j_2} = \prod_{j_1 \leqslant j < j_2} t^i_j$$

(the "empty" product being 1):

$$\Delta^2_{j_1, j_2} = x^2_{j_1} x^1_{j_2} - t^1_{j_1, j_2} x^1_{j_1} x^2_{j_2}, \qquad 1 \leqslant j_1 < j_2 \leqslant s,$$

$$\Delta^3_{j_1, j_2, j_3} = x^3_{j_1} \Delta^2_{j_2, j_3} - t^2_{j_1, j_2} x^3_{j_2} \Delta^2_{j_1, j_3} + t^2_{j_1, j_3} x^3_{j_3} \Delta^2_{j_1, j_2}, \qquad 1 \leqslant j_1 < j_2 < j_3 \leqslant s,$$

$$\cdots\cdots$$

$$\Delta^r_{j_1, \ldots, j_r} = x^r_{j_1} \Delta^{r-1}_{j_2, \ldots, j_r} - t^{r-1}_{j_1, j_2} x^r_{j_2} \Delta^{r-1}_{j_1, \hat{j_2}, \ldots, j_r} \pm \cdots, \qquad 1 \leqslant j_1 < \cdots < j_r \leqslant s,$$

and study the ideal I generated by the elements

$$\Delta^r_{j_1, \ldots, j_r}, \qquad 1 \leqslant j_1 < \cdots < j_r \leqslant s,$$

in $S = \mathbf{Z}[x_j^i\colon 1 \leqslant i \leqslant r, 1 \leqslant j \leqslant s;\ t_j^i\colon 1 \leqslant i < r, 1 \leqslant j < s]$. Obviously, if $t_j^i = 1$ for all i, j, the ideal I is generated by all maximal minors of the matrix (x_j^i). If we specialize to $t_j^i = 0$ for all i and j and to $x_1 := x_1^r = x_2^{r-1} = \cdots = x_r^1$, $x_2 := x_2^r = x_3^{r-1} = \cdots = x_{r+1}^1$, etc., all other $x_j^i = 0$, then the ideal I is generated by the products $x_{j_1} \cdot \cdots \cdot x_{j_r}$, $1 \leqslant j_1 \leqslant \cdots \leqslant j_r \leqslant s$; i.e. I is the rth power of the maximal ideal $(x_1, \ldots, x_{s-r+1})$ in $\mathbf{Z}[x_1, \ldots, x_{s-r+1}]$.

Now, denote by E the free S-module of rank s on a basis $e_1, \ldots, e_s$ and by F^* the free S-module of rank r with basis $f_1, \ldots, f_r$. We construct a free resolution

$$(+) \qquad 0 \to G_{s-r} \xrightarrow{d_{s-r}} G_{s-r-1} \to \cdots \to G_1 \xrightarrow{d_1} G_0 \xrightarrow{d_0} S \to S/I \to 0$$

with $G_l = \bigwedge^{l+r} E \otimes S_l F$, where $\bigwedge^{\cdot}$ denote the exterior and $S.$ the symmetric algebras, respectively. d_0 is the mapping sending

$$e_{j_1} \wedge \cdots \wedge e_{j_r} \in \textstyle\bigwedge^r E = G_0, \qquad 1 \leqslant j_1 < \cdots < j_r \leqslant s,$$

to $\Delta^r_{j_1, \ldots, j_r} \in S$, S being identified with $\bigwedge^r F$ via the dual basis $f_1^*, \ldots, f_r^*$. (For $t_j^i = 1$, all i and j, d_0 is naturally defined as the rth exterior power of the linear mapping $E \to F$ represented by the matrix (x_j^i).)

Each mapping d_l is decomposed into a sum $\Sigma_{i=1}^r d_l^i$, where d_l^i maps a basis element

$$e_{j_1, \ldots, j_{l+r}} \otimes f_{i_1, \ldots, i_l} := e_{j_1} \wedge \cdots \wedge e_{j_{l+r}} \otimes f_{i_1} \cdot \cdots \cdot f_{i_l} \in G_l,$$
$$1 \leqslant j_1 < \cdots < j_{l+r} \leqslant s,\ 1 \leqslant i_1 \leqslant \cdots \leqslant i_l \leqslant r,$$

to 0 if no i_λ is equal to i. If $i = i_\lambda$ (and then our definition does not depend on λ) define the sequence $k_1, \ldots, k_{r-1}$ as the ordered complement of

$$\{l + r, l + r - i_1, l + r - i_2 - 1, \ldots, l + r - i_{\lambda-1} - (\lambda - 2)\} \cup \{l + r - i_{\lambda+1} - (\lambda - 1), \ldots, l + r - i_l - (1 - 2), 1\}$$

in $\{1, \ldots, l + r\}$, and for $1 \leqslant \rho \leqslant l + r$:

$${}^i_\rho t^{i_1, \ldots, i_l}_{j_1, \ldots, j_{l+r}} := t^1_{j_{k_{r-1}}, j_\rho} \cdot \cdots \cdot t^{-1}_{j_{k_{r-i+1}}, j_\rho} t^i_{j_\rho, j_{k_{r-i}}} \cdot \cdots \cdot t^{r-1}_{j_\rho, j_{k_1}}.$$

Then

$$d^i_l\left(e_{j_1, \ldots, j_{l+r}} \otimes f_{i_1, \ldots, i_l}\right) := \sum_{\rho=1}^{l+r} (-1)^{\rho-1}\, {}^i_\rho t^{i_1, \ldots, i_l}_{j_1, \ldots, j_{l+r}} e_{j_1, \ldots, \hat{j}_\rho, \ldots, j_{l+r}} \otimes f_{i_1} \cdots \hat{f}_i \cdots f_{i_l}.$$

For $t^i_j = 1$, all i and j, this is precisely the Eagon-Northcott complex. For $t^i_j = 0$, all i and j, and the specialization of the x^i_j as indicated above, we obtain a complex which is isomorphic to a resolution of $\mathbf{Z}[x_1, \ldots, x_{s-r+1}]/(x_1, \ldots, x_{s-r+1})^r$ presented by Buchsbaum and Eisenbud [**2**] in an intrinsic form. Since one can check that $(+)$ is a complex (which obviously extends the Buchsbaum-Eisenbud resolution) we have constructed a finite free resolution of S/I. Since, moreover, S/I is a Hodge algebra on a wonderful partially ordered set it is Cohen-Macaulay and therefore, by Hochster's result [**7**], $(+)$ is a grade-sensitive complex. That means: if R is any commutative ring with distinguished elements a^i_j, $1 \leqslant i \leqslant r$, and b^i_j, $1 \leqslant i < r$, $1 \leqslant j < s$, and if $S \to R$ is the obvious mapping, then

$$g := \operatorname{grade}_R IR \leqslant s - r,$$

and the tensorized complex

$$(++) \qquad 0 \to G_{s-r} \otimes_S R \to \cdots \to G_1 \otimes_S R \to G_0 \otimes_S R \to R$$

is exact on $(g + 1)$ places from the left. In particular, if $g = s - r$, $(++)$ is a finite free resolution of R/IR as an R-module.

Bibliography

1. K. Baclawski, *Rings with lexicographic straightening law*, Advances in Math. **39** (1981), 185–213.

2. D. Buchsbaum and D. Eisenbud, *Generic free resolutions and a family of generically perfect ideals*, Advances in Math. **18** (1975), 245–301.

3. C. De Concini, D. Eisenbud and C. Procesi, *Hodge algebras*, preprint (1981).

4. P. Doubilet, G.-C. Rota and J. Stein, *On the foundations of combinatorial theory*. IX, Studies in Applied Math. **53** (1974), 185–216.

5. J. A. Eagon and D. G. Northcott, *Ideals defined by matrices and a certain complex associated to them*, Proc. Roy. Soc. London Ser. A **269** (1962), 188–204.

6. T. H. Gulliksen and O. G. Negård, *Un complexe résolvant pour certains idéaux déterminantiels*, C. R. Acad. Sci. Paris Sér. A-B **274** (1972), 16–18.

7. M. Hochster, *Grade-sensitive modules and perfect modules*, Proc. London Math. Soc. **29** (1974), 55–76.

8. J. Pesselhoy and O. Riemenschneider, *Eine Verallgemeinerung des Eagon-Northcott-Komplexes für zweireihige Matrizen* (unpublished).

9. O. Riemenschneider, *Zweidimensionale Quotientensingularitäten: Gleichungen und Syzygien*, Arch. Math. **37** (1981), 406–417.

10. R. Stanley, *Hilbert functions of graded algebras*, Advances in Math. **28** (1978), 57–83.

MATHEMATISCHES SEMINAR DER UNIVERSITÄT HAMBURG, FEDERAL REPUBLIC OF GERMANY

Proceedings of Symposia in Pure Mathematics
Volume **40** (1983), Part 2

VANISHING HOMOLOGIES AND THE n VARIABLE SADDLEPOINT METHOD[1]

FRÉDÉRIC PHAM[2]

Introduction. Our aim is to study the asymptotic behavior, as $\tau \to +\infty$, of an oscillatory integral

$$F(\tau) = \int_{\mathbf{R}^n} e^{-i\tau\varphi(x_1,\ldots,x_n)} a(x_1,\ldots,x_n)\, dx_1,\ldots,dx_n \tag{0}$$

where φ is a polynomial with real coefficients (the "phase function"), whereas a is a polynomial with possibly complex coefficients. Our starting point has been Malgrange's paper [**8**], but our method will be closer to the idea of the classical (one variable) *saddlepoint method* (*méthode du col*), which consists in pushing the integration contour in $\mathbf{C}^n$ towards the directions of "steepest descent" of $\operatorname{Im}\varphi$, thus achieving two goals:

Goal 1: give a meaning to the integral (which perhaps did not converge at infinity to start with);

Goal 2: replace the "Fourier-like" integral (0) by a rapidly convergent ("Laplace-like") integral, whose asymptotic behavior will be governed by the local behavior of the integrand near the—real or complex—critical points of the phase function.

Having achieved Goal 1 means that we have replaced the integral (0) by an integral

$$F_\Gamma(\tau) = \int_\Gamma e^{-i\tau\varphi(z_1,\ldots,z_n)} a(z_1,\ldots,z_n)\, dz_1 \wedge \cdots \wedge dz_n \tag{1}$$

of the "complexified" integrand over an "n-cycle" Γ chosen in such a way that the integrand is exponentially decreasing at infinity along Γ. We thus have to

---

1980 *Mathematics Subject Classification*. Primary 49F20; Secondary 81C35.

[1]Invited talk at the A.M.S. 1981 Summer Research Institute on Singularities at Arcata (California).

[2]The main part of this work was made when the author was the host of the Hanoi Institute of Mathematics, on a French-Vietnamese scientific cooperation contract during the academic year 1979/1980.

0082-0717/81/0000-0200/$04.25

study an integral of the form (1), where Γ is an n-cycle of $\mathbf{C}^n$ with support in the "*family of supports*"

$$\Phi = \{A \text{ closed} \subset \mathbf{C}^n \mid \operatorname{Im} \varphi \mid A \to -\infty \text{ quicker than } -\|z\|^q \text{ for some } q \in \mathbf{Q} \text{ as } \|z\| \to \infty\}.$$

It is obvious that Φ is a "family of supports" in the sense of homology theory, i.e.

$$\begin{cases} A \in \Phi, \quad B \text{ closed} \subset A \Rightarrow B \in \Phi, \\ A, B \in \Phi \Rightarrow A \cup B \in \Phi. \end{cases}$$

Now we would like to claim that the integral (1) *is well defined on cycle* Γ *with support in* Φ, *and only depends on the homology class of* Γ *in* $H_n^\Phi(\mathbf{C}^n)$, *the nth homology group of* $\mathbf{C}^n$ *with support in* Φ (*with integral coefficients*). Actually this is not quite true (as pointed out to me by B. Malgrange): in order to ensure the convergence at infinity, and to be able to apply Stokes' theorem, it is not enough to know that the integrand decreases exponentially (as ensured by the definition of Φ), we also have to make sure that the integration chain does not wiggle too much. Following a hint of Malgrange, we shall show in the Appendix A5 that this can be done without changing the homology group, *so that we are indeed entitled in considering integrals of the form* (1) *as integrals over homology classes in* $H_n^\Phi(\mathbf{C}^n)$.

From now on, *we can forget the assumption that* φ *has real coefficients* (this assumption will only appear in the example of §1.7).

The paper is divided into two sections and an appendix.

In §1 we construct, with the help of the appendix, a basis of $H_n^\Phi(\mathbf{C}^n)$ (see §1.5); the elements of that basis are the Lefschetz "*thimbles*" ("*onglets*" in Lefschetz's French original [**6**]) attached to the various critical points of φ; when an arbitrary homology clas in $H_n^\Phi(\mathbf{C}^n)$ is decomposed with respect to such a basis, the coefficients of that decomposition are given in §1.6 as suitable "intersection numbers" (in the sense of Kronecker), which also give an explicit way of achieving Goal 1 for an integration cycle whose support is not in Φ (§1.7).

When Γ is one of the "thimbles" constructed in §1, the integral (1) is a "Laplace-like" integral.

In §2 we show that the asymptotic behavior of such an integral is given by the singular behavior, near the corresponding critical value of φ, of an integral over the corresponding "vanishing cycle" (fiber of that thimble); when the critical point is *quadratic nondegenerate* this singular behavior can be computed explicitly by more or less elementary methods; for an arbitrary (but still isolated) critical point this behavior is analyzed by the theory of the "Gauss-Manin connection", which I sketch briefly, giving my personal version "*à la Sato*" of Malgrange's work [**7**], where the relation between local asymptotic integrals and the Gauss-Manin connection was first discovered.

The reduction of a *global* Fourier-like integral (0) to a sum of Laplace-like integrals is the topic of Malgrange's more recent paper [**8**], motivated by an idea of Balian-Parisi-Voros [**2**]. But Malgrange made a strong assumption on the behavior at infinity of the real polynomial $\varphi(x)$ (Hypothèse (p.1) of [**8**]): this was forced on him because he integrated along the fibers of φ *before* making the reduction to Laplace form. Another improvement of our approach on Malgrange's is our decomposition formula for homology, cf. §1.7 which gives a precise—although theoretical—answer to Malgrange's final "disillusioned remark" in [**8**] on the difficulty of determining the integration cycle of the Laplace-like integral. The key idea of these improvements can be wrapped in the following formulation: *the reduction from Fourier to Laplace integrals is not a problem of analysis but a problem of pure topology (homology).*

I like to dream that the same idea could help understanding the reduction of "Feynman path integrals" to "Wiener integrals", a very important challenge put on Mathematics by Theoretical Physics.

Note. When this work was completed I took notice of a paper by Fedoryuk [**4**], whose ideas bear some similarity to ours. The main difference is that he works with *relative homology* instead of homology in a family of supports (compare for instance his Lemma 3 with our result 1.5); this is because he replaces his integral over $\mathbf{R}^n$ by *truncated* integrals (over relative cycles). Of course this makes no difference so long as "asymptotic expansions" are understood in the usual sense of mathematicians, not in the stronger sense advocated by Malgrange in [**8**], and recalled in §2.2.1 of the present paper.

A fair view on the history of that subject should also include reference to the important ideas of V. Maslov, I. N. Bernstein and V. Arnold [**1**].

1. Homology of $\mathbf{C}^n$ with supports in Φ. Let φ: $\mathbf{C}^n \to \mathbf{C}$ be a nonconstant polynomial function.

Let us define the "*bifurcation points*" of φ as those points of $\mathbf{C}$ above which ϕ is not a $\mathcal{C}^\infty$ locally trivial fibration. It is well known that the set of bifurcation points of φ is a finite subset of $\mathbf{C}$, consisting of

(i) all critical values of φ;

(ii) perhaps some other points, which we shall call "*bifurcation points of the second type*": they can be understood as images of "critical points at infinity" of $\mathbf{C}^n$ (see Appendix A1 for a more precise statement).

Note that in the $n = 1$ case there can be no bifurcation points of the second type, because φ in that case is a *proper* map.

1.1. *Enlarging the family of supports* Φ. We replace Φ by the wider family of supports

$$\hat{\Phi} = \{A \text{ closed} \subset \mathbf{C}^n \mid \operatorname{Im}\varphi \mid A \to -\infty \text{ as } \|z\| \to \infty\}$$

(i.e. we forget the condition on the speed of growth of $\operatorname{Im}\varphi$). Then it is proved in the appendix that

$$H_*^{\Phi}(\mathbf{C}^n) = H_*^{\hat{\Phi}}(\mathbf{C}^n).$$

1.2. *Localization above a "basic system of cuts"* (T_β^-). To every bifurcation point t_β of φ let us attach a closed half-line T_β^- starting at t_β and going *downwards* to infinity in the complex plane, in such a way that no two such T_β^- intersect each other (Figure 1).

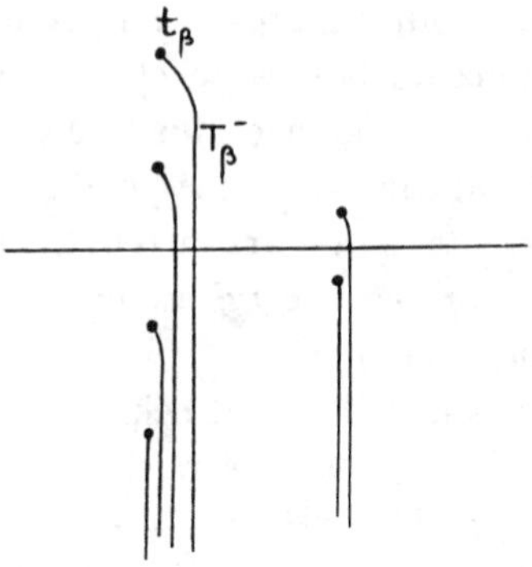

FIGURE 1. A possible choice of a "basic system of cuts"

PROPOSITION. *$H_*^{\hat\Phi}(\mathbf{C}^n) = \oplus_\beta H_*^{\mathrm{pr}}(\varphi^{-1}(T_\beta^-))$ where H_*^{pr} means homology with support in the family of proper sets (i.e. sets A such that $\varphi \mid A$ is proper).*

PROOF. It is easy to construct a deformation retraction of $\mathbf{C}$ on $\bigcup_\beta T_\beta^-$ such that $\operatorname{Im} t$ decreases monotonically all along the deformation. Since $\varphi: \mathbf{C}^n \to \mathbf{C}$ is a locally trivial fiber space above the complement of $\bigcup_\beta T_\beta^-$, this deformation of the base space can be lifted, defining a deformation retraction of $\mathbf{C}^n$ on $\bigcup_\beta \varphi^{-1}(T_\beta^-)$, obviously preserving the family $\hat\Phi$. The proposition then follows from the fact that $\hat\Phi \mid \varphi^{-1}(T_\beta^-)$ coincides with the family of proper subsets of $\varphi^{-1}(T_\beta^-)$.

1.3. *Further localization above neighborhoods of the bifurcation points.* Let U_β be a connected neighborhood of infinity in T_β^-, i.e. the complement of a closed segment $[t_\beta, u_\beta] \subset T_\beta^-$. Then

$$H_*^{\mathrm{pr}}\big(\varphi^{-1}(T_\beta^-)\big) = H_*\big(\varphi^{-1}(T_\beta^-), \varphi^{-1}(U_\beta)\big) = H_*\big(\varphi^{-1}\big([t_\beta, u_\beta]\big), \varphi^{-1}(u_\beta)\big)$$

(relative homology with compact support).

The second equality is just excision.

The first one follows from the fact that the complex of chains with proper supports $C_*^{\mathrm{pr}}(\varphi^{-1}(T_\beta^-))$ is the projective limit, as u_β goes to infinity, of the complex of relative chains (with compact supports) $C_*(\varphi^{-1}(T_\beta^-), \varphi^{-1}(U_\beta))$; since the fiber space $\varphi^{-1}(T_\beta^-)$ is trivial above the complement of t_β, the corresponding projective system of complexes is homologically trivial.

1.4. *Further localization around the critical points.*

Hypothesis. Above t_β there is just a finite number of isolated critical points, none of which are "at infinity".

This implies that for any small enough ε, every u_β close enough to t_β fulfills the following two conditions:

(i) every fiber $\varphi^{-1}(t)$, $t \in [t_\beta, u_\beta]$ is transverse to the boundary of B_ε, the closed ball of radius ε centered at the critical point (or the union of these disjoint closed balls if there are several critical points).

(ii) The complement of B_ε, restricted to $\varphi^{-1}([t_\beta, u_\beta])$, is a trivial fibration above $[t_\beta, u_\beta]$.

The latter condition is a consequence of the "*no critical points at infinity*" hypothesis, which can be understood *either* as meaning that the "strata at infinity" of Appendix A1 (relative to *some* resolution of singularities) have no critical points, *or* as meaning the existence of some minoration $\|\mathrm{grad}_z \varphi\| \geqslant \delta > 0$ as z goes to infinity in $\varphi^{-1}(t_\beta)$ (this is Fedoryuk's condition (8) in [**4**]). It is not clear to me whether these two possible meanings are equivalent (both can be proved true in the example of the final remark in Appendix A1).

It is an easy exercise in homology to show, using (ii), that

$$H_*\big(\varphi^{-1}\big([t_\beta, u_\beta]\big), \varphi^{-1}(u_\beta)\big) = H_*\big(B_\varepsilon \cap \varphi^{-1}\big([t_\beta, u_\beta]\big), B_\varepsilon \cap \varphi^{-1}(u_\beta)\big)$$

and using (i), that the latter homology group is trivial in all dimensions except n, where it is given by

$$H_n\big(B_\varepsilon \cap \varphi^{-1}\big([t_\beta, u_\beta]\big), B_\varepsilon \cap \varphi^{-1}(u_\beta)\big) \xrightarrow{\simeq} H_{n-1}\big(B_\varepsilon \cap \varphi^{-1}(u_\beta)\big)$$

(the boundary homomorphism is an *isomorphism*). Now $H_{n-1}(B_\varepsilon \cap \varphi^{-1}(u_\beta))$ is the well-known "*vanishing homology*" of the fiber $\varphi^{-1}(u_\beta)$, corresponding to the path $[t_\beta, u_\beta]$. Its rank is easily computed (cf. [**9**]): it is the so-called "*Milnor number*" μ of the critical point (or the sum of the Milnor numbers if there are several critical points).

In the special case of a *quadratic* (nondegenerate) critical point the Milnor number is 1, and $H_{n-1}(B_\varepsilon \cap \varphi^{-1}(u_\beta))$ is generated by the Lefschetz "*vanishing cycle*" e_β; the corresponding generator of $H_n(\varphi^{-1}([t_\beta, u_\beta]), \varphi^{-1}(u_\beta))$ is the Lefschetz "*thimble*" Δ_β^- ("*onglet de Lefschetz*").

1.5. *Conclusion about* $H_n^\Phi(\mathbf{C}^n)$. Let us make the following two hypotheses:

H1. *Every critical value of* φ *comes from isolated critical points, and there are no bifurcation points of the second type* ("no critical points at infinity").

H2. *The critical points are all quadratic* (*nondegenerate*), *and have different critical values*.

Conclusion. Then $H_n^\Phi(\mathbf{C}^n)$ *is the free group on the set of Lefschetz "thimbles"* Δ_β^- *associated to the basic set of cuts* T_β^-.

Concretely, the "thimble" Δ_β^- may be thought of as the n-dimensional cycle generated in $\mathbf{C}^n$ by the motion of a "vanishing" $(n-1)$-cycle $e_\beta(t)$ as t runs along the line T_β^-.

REMARK 1.5.1. The corresponding basis (Δ_β^-) of $H_n^\Phi(\mathbf{C}^n)$ only depends on the choice—up to homotopy—of the basic set of cuts (T_β^-). Any change of the basic set of cuts induces a change of basis in $H_n^\Phi(\mathbf{C}^n)$, which can be computed in terms of the "*monodromy*" of the vanishing homology.

REMARK 1.5.2. *Hypothesis* H2 *is not important*: if we forget about it, we just have to allow for *several* (exactly μ_β) "thimbles" above the cut T_β^- (μ_β = sum of the Milnor numbers of the critical points above t_β). Furthermore, any polynomial φ satisfying H1 *but not* H2 can be considered as a degenerate member of a continuous family of polynomials whose generic member satisfies H1 *and* H2.

REMARK 1.5.3. Discussion with Mark Goresky convinced me that hypothesis H1 should not be too difficult to get rid of, provided all critical points at infinity are isolated: in that case we can use some of the methods in M. Goresky and R. Mac Pherson (*Stratified Morse theory*, these Proceedings), which can be made quite explicit in the case of a Morse critical point on a divisor with normal crossings.

1.6. *The dual of* $H_n^\Phi(\mathbf{C}^n)$. Let $\overline{\Phi}$ be the family of supports in $\mathbf{C}^n$ deduced from Φ through complex conjugation $t \to \bar{t}$, i.e. $\overline{\Phi} = \{A \text{ closed} \subset \mathbf{C}^n \mid \operatorname{Im}\varphi \mid A \to +\infty$ quicker than $\|z\|^q$ for some $q \in \mathbf{Q}$ as $\|z\| \to \infty\}$.

Clearly $\Phi \cap \overline{\Phi}$ is the family of *compact* subsets of $\mathbf{C}^n$. Therefore the "*intersection number*" (or "Kronecker index") $\langle \Gamma', \Gamma \rangle$ of two n-cycles Γ', resp. Γ, with support in $\overline{\Phi}$, resp. Φ, can always be defined, and gives a bilinear map

$$H_n^{\overline{\Phi}}(\mathbf{C}^n) \otimes H_n^\Phi(\mathbf{C}^n) \to \mathbf{Z}.$$

PROPOSITION. *Under hypothesis* H1 *of* 1.5 *this bilinear map is nondegenerate, so that* $H_n^{\overline{\Phi}}(\mathbf{C}^n)$ *is the* **Z**-*dual of* $H_n^\Phi(\mathbf{C}^n)$.

PROOF. Let us assume that hypothesis H2 is fulfilled too (according to Remark 1.5.2 this does not restrict the generality). Let (T_β^-) be a basic set of cuts, and let (T_β^+) be a "dual" set of cuts, starting from the t_β's *upward* to infinity, in such a way that $T_\beta^+ \cap T_{\beta'}^- = \varnothing$ for $\beta \neq \beta'$. Let (Δ_β^-), resp. (Δ_β^+), be the bases of $H_n^\Phi(\mathbf{C}^n)$, resp. $H_n^{\overline{\Phi}}(\mathbf{C}^n)$, defined by the Lefschetz "thimbles" above (T_β^-), resp. (T_β^+).

The proposition then follows from the

LEMMA. *The bases* (Δ_β^+) *and* (Δ_β^-) *are dual to each other, i.e.* $\langle \Delta_\beta^+, \Delta_{\beta'}^- \rangle = \pm\delta_{\beta\beta'}$ (*the sign depends on orientations, which we have left imprecise*).

PROOF OF THE LEMMA. It is obvious that $\langle \Delta_\beta^+, \Delta_{\beta'}^- \rangle = 0$ if $\beta \neq \beta'$, since in that case $T_\beta^+ \cap T_{\beta'}^- = \varnothing$.

Now if $\beta = \beta'$, Δ_β^+ and Δ_β^- will intersect only at the critical point c_β, and can be realized respectively (at least locally near c_β) as the "outgoing" resp. "incoming" manifold of the gradient flow of $\operatorname{Im}\varphi$. That these are manifolds intersecting transversally, therefore having intersection number ± 1, follows from the quadratic nondegenerate hypothesis on c_β.

COROLLARY. *Let* Γ *be an arbitrary element of* $H_n^\Phi(\mathbf{C}^n)$. *Then with the above notations, assuming* H1 *and* H2 *to be fulfilled, we have an "explicit" decomposition of* Γ:

$$\Gamma = \sum_\beta N_\beta \Delta_\beta^-$$

with $N_\beta = \langle \Delta_\beta^+, \Gamma \rangle$ (*with suitable choices of orientations for* Δ_β^+ *and* Δ_β^-).

1.7. *Enlarging again the family of supports* Φ. Set $\Psi = \{A \text{ closed} \subset \mathbf{C}^n \mid \operatorname{Im}\varphi \mid A$ bounded above$\}$.

PROPOSITION. *Under hypothesis* H1,

$$H_*^\Phi(\mathbf{C}^n) = H_*^\Psi(\mathbf{C}^n).$$

PROOF. Using the same deformation retract argument as in 1.2, one proves that

$$H_*^\Psi(\mathbf{C}^n) \doteq \bigoplus_\beta H_*^F(\varphi^{-1}(T_\beta^-))$$

where $F = \Psi \mid \varphi^{-1}(T_\beta^-)$ is simply the family of all closed sets. Now, as in 1.3 one proves that

$$H_*^F(\varphi^{-1}(T_\beta^-)) = H_*(\varphi^{-1}([t_\beta, u_\beta]), \varphi^{-1}(u_\beta) \cup V)$$

where V is a "tubular neighborhood of infinity", whose boundary can be assumed to be transverse to the fibers of φ (under the assumption H1 or more specifically the hypothesis of 1.4); then V is a trivial fiber space above $[t_\beta, u_\beta]$, so that $\varphi^{-1}(u_\beta)$ is a deform-retract of $\varphi^{-1}(u_\beta) \cup V$. It follows that the pair inclusion

$$(\varphi^{-1}([t_\beta, u_\beta]), \varphi^{-1}(u_\beta)) \subset (\varphi^{-1}([t_\beta, u_\beta]), \varphi^{-1}(u_\beta) \cup V)$$

induces an isomorphism in homology, so that the corresponding homology group coincides with that which we computed in 1.3.

The proposition thus proved allows us to achieve Goal 1 of the Introduction: if the integration cycle Γ has support in Ψ but not in Φ what we have just shown means that Γ *is homologous* in $H_n^\Psi(\mathbf{C}^n)$ *to a* unique element of $H_n^\Phi(\mathbf{C}^n)$, *over which the integral converges*. According to 1.6 this element of $H_n^\Phi(\mathbf{C}^n)$ is given by the "explicit" formula $\Sigma N_\beta \Delta_\beta^-$, with $N_\beta = \langle \Delta_\beta^+, \Gamma \rangle$ (notice that $\bar{\Phi} \cap \Psi =$ the family of compact sets, so that the intersection number is well defined).

EXAMPLE. $\Gamma = \mathbf{R}^n$, $\varphi \in \mathbf{R}[z_1, \ldots, z_n]$.

Then

$$N_\beta = \langle \Delta_\beta^+, \Gamma \rangle = \begin{cases} 0 & \text{if } \operatorname{Im} t_\beta > 0, \\ \pm 1 & \text{if } \operatorname{Im} t_\beta = 0 \text{ (under assumptions H1, H2)}, \\ \langle e_\beta(t), \gamma(t) \rangle & \text{if } \operatorname{Im} t_\beta < 0, \end{cases}$$

where the last expression is an intersection number of $(n-1)$-cycles in $\varphi^{-1}(t)$, $t = T_\beta^+ \cap \mathbf{R}$; $e_\beta(t)$ is the "vanishing cycle" (fiber of Δ_β^+), whereas $\gamma(t) = \operatorname{Re} \varphi^{-1}(t)$ (fiber of Γ).

2. Local asymptotics of Laplace integrals.

2.0. The aim of this section is to study the asymptotic behavior of an integral

$$F_{\Delta^-}(\tau) = \int_{\Delta^-} e^{-i\tau\varphi(z_1, \ldots, z_n)} a(z_1, \ldots, z_n)\, dz_1 \wedge \cdots \wedge dz_n \tag{2}$$

where Δ^- is a Lefschetz thimble starting from some isolated critical point of φ, down to infinity in the negative $\operatorname{Im} \varphi$ direction. The relevance of this problem for studying asymptotic integrals (such as (1) of the Introduction) is obvious from the above Corollary 1.6: replacing $\int_\Gamma$ by $\Sigma_\beta N_\beta \int_{\Delta_\beta^-}$ will achieve Goal 2 of the

Introduction. Furthermore as it will turn out that the asymptotic behavior of (2) is governed by the *local* analytic behavior of the functions φ and a near the critical point, the present study is also relevant for oscillatory integrals which are only *locally* similar to that of the Introduction, for instance the local asymptotic integrals considered by Malgrange in [7].

From the description of the thimble Δ^- as an orbit of vanishing cycles $e(t)$ as t runs along the cut T^-, it follows that the integral (2) can be rewritten as

$$(3) \qquad F_{\Delta^-}(\tau) = \int_{T^-} e^{-i\tau t} f(t)\, dt,$$

with

$$(4) \qquad f(t) = \int_{e(t)} \frac{\omega}{d\varphi}\bigg|_{\varphi^{-1}(t)},$$

where the integrand of (4) is the $(n-1)$-differential form in $\varphi^{-1}(t)$ ($t \neq t_\beta$, the critical value), quotient by $d\varphi$ of the n-differential form $\omega = a(z)\, dz_1 \wedge \cdots \wedge dz_n$: explicitly, $\omega/d\varphi\,|_{\varphi^{-1}(t)}$ can be written locally as the restriction to $\varphi^{-1}(t)$ of any $(n-1)$-differential form χ in $\mathbf{C}^n$ such that $\omega = d\varphi \wedge \chi$.

Through the change of variable $t = t_\beta - is$, $s \in \mathbf{R}^+$, the integral (3) can be rewritten as

$$(3') \qquad F_{\Delta^-}(\tau) = -ie^{-i\tau t_\beta} \int_0^\infty e^{-\tau s} f(t_\beta - is)\, ds,$$

i.e. as a Laplace-Borel transform, whose asymptotic behavior as $\tau \to +\infty$ will be governed, as is well known, by the singular behavior of the function $f(t_\beta - is)$ near $s = 0$ (see §2.2 below).

2.1. *Singular behavior of an integral* (4). In order to simplify the notations we shall suppose that the critical point is the origin and has critical value $t_\beta = 0$; we shall also perform a rotation of $-\pi/2$ in the complex plane, so that what we now call t is actually the variable s of (3′).

The dependence of the integral on ω will be made explicit by writing

$$(4') \qquad f^\omega(t) = \int_{e(t)} \frac{\omega}{d\varphi}\bigg|_{\varphi^{-1}(t)}.$$

In the case of a *quadratic* nondegenerate critical point, it has been known for a long time that $f^\omega(t)$ can be written near the origin as $f^\omega(t) = t^{n/2-1} \cdot g^\omega(t)$, where $g^\omega(t)$ is a germ of holomorphic function whose Taylor expansion can in principle be computed explicitly from those of φ and ω: the idea is to choose local coordinates such that $\varphi = z_1^2 + \cdots + z_n^2$ (Morse lemma), thus getting an integral on an $(n-1)$-sphere of radius $\sqrt{t}$; in the simplest case where $\omega = dz_1 \wedge \cdots \wedge dz_n$ in these coordinates, one thus sees that

$$g^\omega(t) = \frac{n}{2} \frac{\pi^{n/2}}{\Gamma(n/2+1)},$$

half the surface of the unit $(n-1)$-sphere.

Although the quadratic case is the *generic* case (Remark 1.5.2), it is also important to study the *nonquadratic* cases, which will appear as special members of *generic families* depending on extra parameters.

For an arbitrary (isolated) critical point, general information on the singular behavior of integrals (4′) is given by the theory of the "*Gauss-Manin connection*" of φ, which I shall now sketch.

2.1.1. Noticing that $f^\omega(t)=0$ if $\omega = d\varphi \wedge d\theta$ (by Stokes' theorem), we can say that $f^\omega(t)$ only depends on the class $[\omega]$ of ω in $G^{(0)} = \Omega^n/d\varphi \wedge d\Omega^{n-2}$ (where Ω^p means the space of germs of holomorphic differential forms of degree p). Clearly $G^{(0)}$ is a $\mathbf{C}\{t\}$-module (multiplication by any function of t is given by $g(t)[\omega] = [g(\varphi)\omega]$). As a $\mathbf{C}\{t\}$-module it can be proved to be *free of rank* $\mu =$ the Milnor number of the critical point. I call $G^{(0)}$ the "Gauss-Manin lattice" of the germ of the analytic function φ.

2.1.2. Now if we try to differentiate (4′) with respect to t under the integral, the integrand yields

$$\frac{d}{dt}\frac{[\omega]}{d\varphi} = \frac{[d\chi]}{d\varphi} \quad \text{if } \omega = d\varphi \wedge \chi$$

(the resulting $[d\chi]$ does not depend on the choice of χ); the trouble is that to be able to write $\omega = d\varphi \wedge \chi$ one must allow for *polar* singularities of χ along $\varphi = 0$, i.e. take $\chi \in \Omega^{n-1}[1/\varphi]$. We are thus led to consider

$$G_{(t)} = G^{(0)}\left[\frac{1}{t}\right] = \mathbf{C}\{t\}\left[\frac{1}{t}\right] \bigotimes_{\mathbf{C}\{t\}} G^{(0)}.$$

With the above action of d/dt on $G_{(t)}$, the $\mathbf{C}\{t\}$-module structure of $G_{(t)}$ extends to a $\mathfrak{D}$-module structure, where $\mathfrak{D}$ is the ring of differential operators with coefficients in $\mathbf{C}\{t\}$. This $\mathfrak{D}$-module $G_{(t)}$ is the "*Gauss-Manin connection*" of φ. Notice that $G_{(t)}$ is also a μ-dimensional vector space on the field $\mathbf{C}\{t\}[1/t]$ of germs of meromorphic functions. In any basis $[\omega_1],\ldots,[\omega_\mu]$ of this vector space, the action of d/dt is given by a $\mu \times \mu$ matrix with meromorphic coefficients, i.e. an *ordinary differential system of the* 1*st order* in the sense of classical analysis.

THEOREM. *This ordinary differential system has regular singularity.*

Noticing that $(f^{\omega_1}(t),\ldots,f^{\omega_\mu}(t))$ is a solution of that differential system, we thus get the important

COROLLARY. *f^ω is given near $t=0$ by a finite sum*

$$f^\omega(t) = \sum_{\substack{p\in\mathbf{N}\\ \alpha\in\mathbf{C}}} g^\omega_{\alpha,p}(t)t^\alpha(\operatorname{Log} t)^p \quad \textit{where } g^\omega_{\alpha,p} \in \mathbf{C}\{t\}. \tag{5}$$

One further proves that the α's in that expansion are *rational numbers, greater than* -1.

Practically the exponents α and p are very difficult to compute (except in the case of "quasi-homogeneous" singularities, a "trivial" case). The proof of the above theorem and of the rationality of the α's is very sophisticated (cf. Brieskorn [**3**]).

REMARK. One must not forget that $f^\omega = f_e^\omega$ *actually depends on the integration cycle* e, which can be considered an element of the μ-dimensional space of "vanishing homology". The situation is most conveniently summarized using *Sato's* language: for every multivalued section e of the space of vanishing homology, the map $[\omega] \mapsto f_e^\omega$ defines a homomorphism of the $\mathcal{D}$-module $G_{(t)}$ into the $\mathcal{D}$-module of multivalued analytic functions (on arbitrarily small punctured neighborhoods of the origin). i.e. a (multivalued analytic) "*solution of the differential system* $G_{(t)}$"; more precisely, one thus gets an *isomorphism* between the space of vanishing homology and the space of multivalued analytic solutions of the Gauss-Manin connection (see the article of Maisonobe and Rombaldi in [**11**]).

2.1.3. Set

$$G^{(-1)} = \left(\frac{d}{dt}\right)^{-1} G^{(0)} = \left\{[\omega] \in G^{(0)} \,\middle|\, \frac{d}{dt}[\omega] \in G^{(0)}\right\}.$$

$G^{(-1)}$ is a sub-$\mathbf{C}\{t\}$-module of $G^{(0)}$, and one verifies that the derivation d/dt: $G^{(-1)} \to G^{(0)}$ has an inverse $(d/dt)^{-1}$, which can be considered as an operator in $G^{(0)}$.

THEOREM. *The action of* $(d/dt)^{-1}$ *on* $G^{(0)}$ *extends to an action of the ring* $\mathcal{E}^{(0)}$ *of microdifferential operators of order* $\leqslant 0$. *Moreover the* $\mathcal{E}^{(0)}$*-module* $G^{(0)}$ *is free of rank* μ *on the subring* $\mathbf{C}\{\{(d/dt)^{-1}\}\}$ *of microdifferential operators with constant coefficients.*

The idea of this theorem, which is proved in [**10**] (see also [**11**]), was suggested to me by Malgrange's paper [7]. I shall not recall here the definition of the ring $\mathcal{E}^{(0)}$ but only the definition of its subring of microdifferential operators *with constant coefficients*:

$$\mathbf{C}\left\{\left\{\frac{d}{dt}^{-1}\right\}\right\} = \left\{c_0 + c_1\frac{d}{dt}^{-1} + c_2\frac{d}{dt}^{-2} + \cdots \in \mathbf{C}\left[\left[\frac{d}{dt}^{-1}\right]\right] \,\middle|\, \sum_r c_r\frac{T^r}{r!} \in \mathbf{C}\{T\}\right\}.$$

2.1.4. REMARK. If we define $G = \mathcal{E} \otimes_{\mathcal{E}^{(0)}} G^{(0)}$, where $\mathcal{E}$ is the ring of microdifferential operators of arbitrarily (finite) order, we get a microdifferential system (i.e. a noetherian $\mathcal{E}$-module) which I call the "*Gauss-Manin system*" of φ. The inclusion of rings $\mathcal{D} \subset \mathcal{E}$ also makes it a differential system (i.e. a noetherian $\mathcal{D}$-module). The relation between this "Gauss-Manin system" G and the "Gauss-Manin connection" $G_{(t)}$ of 2.1.2 is as follows:

$$G_{(t)} = \mathbf{C}\{t\}\left[\frac{1}{t}\right] \bigotimes_{\mathbf{C}\{t\}} G$$

i.e. the Gauss-Manin connection is the differential system deduced from G by "localization with respect to (t)".

2.2. *Asymptotics of the Borel transform* (3′). With a slight change of notation, to study (3′) means to study the Laplace-Borel transform

$$'F^{\omega}_{\Delta^-}(\tau)=\int_0^{\infty}e^{-\tau s}f^{\omega}(s)\,ds$$

of a function f^{ω} given near $s=0$ by an expansion (5):

$$f^{\omega}(s)=\sum_{\alpha,p}g^{\omega}_{\alpha,p}(s)s^{\alpha}(\operatorname{Log}s)^{p}=\sum_{\alpha,p}c^{\omega}_{\alpha,p}s^{\alpha}(\operatorname{Log}s)^{p} \tag{5′}$$

where the latter summation is a convergent sum over an *infinite* number of α's, deduced from those in the first line by replacing each $g^{\omega}_{\alpha,p}$ by its Taylor expansion $c^{\omega}_{\alpha,p}+c^{\omega}_{\alpha+1,p}s+c^{\omega}_{\alpha+2,p}s^2+\cdots$. Taking the Borel transform, we thus get an asymptotic expansion

$$'F^{\omega}_{\Delta^-}(\tau)\propto\sum_{\alpha,p}c^{\omega}_{\alpha,p}\frac{\Gamma(\alpha+1)}{\tau^{\alpha+1}}(\operatorname{Log}\tau)^{p}. \tag{6}$$

2.2.1. Although this expansion does not converge (because of the $\Gamma(\alpha+1)$ in the numerator), it contains much more information on the function $'F^{\omega}_{\Delta^-}(\tau)$ than what mathematicians usually mean by an "asymptotic expansion": actually as emphasized by *Malgrange* in **[8]**, the expansion (6) *gives the exact value* of $'F^{\omega}_{\Delta^-}(\tau)$ if one "resums" it by a method similar to the *Nörlund* "factorial series" method; more naively, one may also hope to get good numerical accuracy, as in the examples worked out by **[2]**, by summing it in what Poincaré calls "the way of astronomers".

2.2.2. Let us now study the dependence on ω of the asymptotic expansion (6): since the operator d/ds corresponds through Borel transform to the multiplication operator by τ, Theorem 2.1.3 yields the following result:

Let $[\omega_1],\ldots,[\omega_\mu]$ *be a basis of the Gauss-Manin lattice* $G^{(0)}$; *then the asymptotic expansion of* $'F^{\omega}_{\Delta^-}(\tau)$, *for any* $[\omega]$ *in* $G^{(0)}$, *can be written in a unique way as a linear combination of the asymptotic expansions of* $'F^{\omega_1}_{\Delta^-}(\tau),\ldots,'F^{\omega_\mu}_{\Delta^-}(\tau)$, *with coefficients in the ring* $\mathbf{C}\{\{1/\tau\}\}$.

The interest of getting such a result for the ring $\mathbf{C}\{\{1/\tau\}\}$, not just the ring $\mathbf{C}[[1/\tau]]$ of formal power series as in **[7]**, is apparent from the above Remark 2.2.1.

Appendix. Geometry of polynomial maps φ: $\mathbf{C}^n\to\mathbf{C}$.

A1. *Construction of the bifurcation set.*

In order to replace the map φ: $\mathbf{C}^n\to\mathbf{C}$ by a *proper* one let us compactify the source space (setting $\overline{\mathbf{C}^n}=\mathbf{P}^n(\mathbf{C})$, for instance); the closure in $\overline{\mathbf{C}^n}\times\mathbf{C}$ of the graph of φ is an algebraic, possibly singular variety; by *resolving its singularities* (Hironaka **[5]**) we get a *smooth* algebraic manifold X together with a *proper* map π: $X\to\mathbf{C}$ such that the following diagram is commutative:

$$\begin{array}{ccc} X & \hookleftarrow & \mathbf{C}^n=X-D\\ \pi\downarrow & & \swarrow\varphi\\ \mathbf{C} & & \end{array}$$

$D =$ divisor with normal crossings. The pair (X, D) has an obvious stratification. By a *critical point* [resp. *critical value*] of π: $(X, D) \to \mathbf{C}$ we mean a critical point [resp. critical value] of π restricted to any of the strata of (X, D). Call $S \subset \mathbf{C}$ the set of critical values; it is a finite set (zero-dimensional algebraic variety). Restricted above $\mathbf{C} - S$, the projected pair (X, D) is a proper stratified submersion, therefore a $\mathcal{C}^\infty$ locally trivial fibration (the present "normal crossing" situation yields an easy generalization of the "Ehresmann fibration theorem", a trivial special case of the difficult "Thom-Mather isotopy theorem"). We conclude that $\varphi^{-1}(\mathbf{C} - S) = (X - D)|_{\mathbf{C}-S}$ is a locally trivial fiber space, so that *the finite set S contains the bifurcation set of φ*. More precisely, apart from the critical values of φ, all of which are bifurcation points (as follows from the work of Milnor [**9**]) the remaining bifurcation points (which we call "*bifurcation points of the second type*") consist of *some* of the critical values of the restriction of π to the "strata at infinity" (strata of D); I do not know of any simple way of telling *which* of those critical values are relevant.

REMARK. In some simple cases we can be sure that there are *no bifurcation points of the second type*: this is the case for instance if the leading homogeneous part φ_m of φ defines a smooth hypersurface in $\mathbf{P}^{n-1}(\mathbf{C})$, i.e. if $z = 0$ is the only point where $\operatorname{grad}_z \varphi_m$ (and therefore φ_m) vanishes (this is condition (3) of Fedoryuk's paper [**4**]).

A2. *Compactifying the target space*. In the above construction we might have replaced $\overline{\mathbf{C}^n} \times \mathbf{C}$ by $\overline{\mathbf{C}^n} \times \overline{\mathbf{C}}$, where $\overline{\mathbf{C}} = \mathbf{P}^1(\mathbf{C})$; using again Hironaka's resolution of singularities, we thus get a smooth compactification of the above construction, i.e. a commutative diagram

$$\begin{array}{ccccccc} \overline{X} & \hookleftarrow & X & = & \overline{X} - D' & \hookleftarrow & \mathbf{C}^n = X - D = \overline{X} - D \cup D' \\ \bar{\pi}\downarrow & & \downarrow\pi & & & & \\ \overline{\mathbf{C}} & \hookleftarrow & \mathbf{C} & = & \overline{\mathbf{C}} - \{\infty\} & & \end{array}$$

where $D' = \bar{\pi}^{-1}(\infty)$, and $D \cup D'$ is a divisor with normal crossings.

This compactification will be used in A4.

A3. *Approaching infinity in affine charts of algebraic manifolds*. Let $\overline{X}$ be a compact algebraic manifold, covered by a finite number of Zariski open sets (U_α) with affine charts λ_α, one of which (U_0, λ_0) will be distinguished.

Let $D_0 = \overline{X} - U_0$.

Calling $\operatorname{dist}_{\lambda_\alpha}$ the distance function in U_α defined by the "Euclidean distance" in the chart λ_α, let us define, for any $x \in U_0$,

$$d_0(x) = \inf_{\alpha \neq 0} \left\{ \operatorname{dist}_{\lambda_\alpha}(x, D_0 \cap U_\alpha) \mid x \in U_\alpha \right\}.$$

PROPOSITION. *Given any fixed point $x_0 \in U_0$, $\operatorname{dist}_{\lambda_0}(x, x_0)$ is bounded by an inverse power of $d_0(x)$ as x varies in U_0.*

PROOF. Let $(K_\alpha \subset U_\alpha)$ be a subcovering of $\overline{X}$ by compact sets. In order to prove the proposition, we just have to prove that for every α, $\operatorname{dist}_{\lambda_0}(x, x_0)$ is bounded by an inverse power of $\mathrm{dist}_{\lambda_\alpha}(x, D_\alpha \cap U_0)$ as x varies in $K_\alpha \cap U_0$. Thus it is enough to prove the

LEMMA. *Call g_α the Riemann metric on U_α induced by the Euclidean metric in the chart λ_α. For every $x \in U_0 \cap U_\alpha$, define*

$$\| g_0(x)/g_\alpha(x) \| \stackrel{\mathrm{def}}{=} \sup_{\xi \in T_x \bar{X}} \left(\langle \xi, \xi \rangle_{g_0(x)} / \langle \xi, \xi \rangle_{g_\alpha(x)} \right).$$

Then this function of x is bounded, locally near $D_0 \cap U_\alpha$, by an inverse power of $\mathrm{dist}_{\lambda_\alpha}(x, D_0 \cap U_\alpha)$.

PROOF OF THE LEMMA. $\| g_0(x)/g_\alpha(x) \|$ is just the Euclidean norm, with respect to the Euclidean metric of the chart λ_α, of the bilinear form $g_0(x)$. The Lemma follows from the fact that this bilinear form is represented in that chart by a matrix with rational coefficients, with polar locus D_0.

A.4. *Proof of* 1.1. $H_*^\Phi(\mathbf{C}^n) = H_*^{\hat{\Phi}}(\mathbf{C}^n)$. Consider $\varphi\colon \mathbf{C}^n \to \mathbf{C}$ as a "projected space". In the category of projected spaces we have an obvious notion of *homotopy* (between maps of projected spaces), *deformation-retraction* (of a projected space onto a "projected subspace") etc. We shall construct a deformation-retraction of the projected space $\varphi\colon \mathbf{C}^n \to \mathbf{C}$ onto a closed projected subspace $E \xrightarrow{\varphi|E} L$ such that

(i) this deformation-retraction preserves families Φ and $\hat{\Phi}$,

(ii) $\Phi \mid E = \hat{\Phi} \mid E = \mathrm{pr}(E)$, the family of subsets of E which are *proper* above L.

This will prove that

$$\left.\begin{aligned} H_*^\Phi(\mathbf{C}^n) &= H_*^{\Phi|E}(E) \\ H_*^{\hat{\Phi}}(\mathbf{C}^n) &= H_*^{\hat{\Phi}|E}(E) \end{aligned}\right\} = H_*^{\mathrm{pr}}(E).$$

For the construction, we shall need the following

DEFINITION. A closed subset $E \subset \mathbf{C}^n$ will be said to have *moderate growth at infinity*, relative to φ, if the distance between the origin of $\mathbf{C}^n$ and an arbitrary point x of E is bounded by some power of $|\varphi(x)|$ when $|\varphi(x)|$ is large enough.

This means that after removing the inverse image $\varphi^{-1}(K)$ of some compact set $K \subset \mathbf{C}$, what remains of E is proper, and the diameters of its fibers E_t as well as their distances to any given fixed point are bounded by some power of $|t|$.

Notice that property (ii) is implied by the property

(iii) E has moderate growth at infinity, and $\mathrm{Im}\, t < -c|t| + c'$ (with $c > 0$) for $t \in L = \varphi(E)$.

Construction of the deform-retract E of $\mathbf{C}^n$.

First step. Construct a deformation-retraction of $\mathbf{C}^n \xrightarrow{\varphi} \mathbf{C}$, *above the identity map of* $\mathbf{C}$, onto a closed subset $F = \varphi^{-1}(K) \cup F'$, where K is a compact subset of $\mathbf{C}$ containing all bifurcation points, whereas F' is a *proper* subfiber space of $\mathbf{C}^n - \varphi^{-1}(K)$ *with moderate growth at infinity*.

To perform that construction, consider situation A2. Near every point of $D' = \bar{\pi}^{-1}(\infty)$, $D \cup D'$ is a divisor with normal crossings and $\bar{\pi} | D$ is a locally trivial fibration. Then it is easy to construct in $\bar{\pi}^{-1}(U)$, with U some open neighborhood of $\infty \in \bar{\mathbf{C}}$, a "carpeting function" for the stratum $\bar{X} - D$ (see **[12]**), i.e. a nonnegative $\mathcal{C}^\infty$ function g such that

$$g^{-1}(0) = D \cap \bar{\pi}^{-1}(U);$$
$$dg \neq 0 \quad \text{outside } g^{-1}(0);$$

for any small ε, $V = g^{-1}(]0, \varepsilon])$ is retractible on $\partial V = g^{-1}(\varepsilon)$ by a deformation of the projected space $\bar{\pi}^{-1}(U)$ compatible with the identity map of the basis U.

Setting $U^* = U - \{\infty\}$, $K = \mathbf{C} - U^*$, $F' = \pi^{-1}(U^*) \cap (\bar{X} - V)$, one gets an $F = \varphi^{-1}(K) \cup F'$ having the desired properties (*the moderate growth property of F' follows from Proposition* A3).

Second step. Let L be the union of all vertical half lines in $\mathbf{C}$, starting in K and going downwards. Then it is easy to construct a deformation-retraction of $\mathbf{C}$ on L such that $\operatorname{Im} t$ is monotonically decreasing all along the deformation. Since this deformation-retraction is the identity on K, and F is a locally trivial fiber space above $\mathbf{C} - K$, the deformation can be "lifted" on F, defining a deformation retraction of the projected space $F \to \mathbf{C}$ onto the projected space $E = F \cap \varphi^{-1}(L) \to L$.

Composition of the above two steps gives the required deformation-retraction.

A5. *Construction of a subcomplex of $C_*^\Phi(\mathbf{C}^n)$, over which the integrals* (1) *converge*. In $C_*^\Phi(\mathbf{C}^n)$, the chain complex of $\mathbf{C}^n$ with support in Φ, let $\tilde{C}_*^\Phi(\mathbf{C}^n)$ be the subcomplex of those chains whose support is semianalytic in $\bar{X}$, some compactification of $\mathbf{C}^n$ chosen once and for all, of the type A2. We claim the following

LEMMA. (i) *Integrals of the form* (1) *are well defined for every* $\Gamma \in \tilde{C}_n^\Phi(\mathbf{C}^n)$, *and only depend on the homology class of* Γ *in* $H_n(\tilde{C}_*^\Phi(\mathbf{C}^n))$.

(ii) $H_*(\tilde{C}_*^\Phi(\mathbf{C}^n)) \approx H_*^\Phi(\mathbf{C}^n)$, *where the isomorphism is induced by the inclusion of complexes.*

PROOF OF (i). The proof follows from Herrera's results on integration over semianalytic sets (Bull. Soc. Math. France (2) **94** (1966)), according to which the integration current on a semianalytic set satisfies a boundedness condition C_p (part (c) of Theorem 2.1 in Herrera's Chapter II): this condition C_p will imply an *at most polynomial* divergence of the integral of a rational form in our compactified space $\bar{X}$, and this divergence will be killed by the exponentially decreasing factor.

PROOF OF (ii). Just read again construction A4, and get yourself convinced that the closed set E can be chosen to be semianalytic in $\bar{X}$. Since all chains—and all homologies—of $C_*^\Phi(\mathbf{C}^n)$ can be "pushed" into E, the claim then follows from the well-known result of Lojasiewicz on the existence of semianalytic triangulations of semianalytic sets.

REFERENCES

1. V. I. Arnold, *Integrals of rapidly oscillating functions*, Funkcional. Anal. i Priložen. (3) **6** (1972).

2. R. Balian, G. Parisi and A. Voros, *Quartic oscillator*, Math. Problems in Feynman Path Integrals, Lecture Notes in Math., Springer-Verlag, Berlin and New York, 1979.

3. E. Brieskorn, *Die Monodromie der isolierten singularitäten von Hyperflächen*, Manuscripta Math. **2** (1970).

4. M. V. Fedoryuk, *The asymptotics of the Fourier transform of the exponential function of a polynomial*, Dokl. Akad. Nauk SSSR **227** (3) (1976) = Soviet Math. Dokl. **17** (1976).

5. H. Hironaka, *Resolution of singularities of an algebraic variety*, Ann. of Math. (2) **79** (1964).

6. S. Lefschetz, *L'analysis situs et la géométrie algébrique*, Gauthier-Villars, Paris, 1924.

7. B. Malgrange, *Intégrales asymptotiques et monodromie*, Ann. Sci. École Norm. Sup. **4** (1974).

8. ______, *Méthode de la phase stationnaire et sommation de Borel*, Complex Analysis, Microlocal Calculus and Relativistic Quantum Theory, Lecture Notes in Physics, vol. 126, Springer, Berlin and New York, 1980.

9. J. Milnor, *Singular points of complex hypersurfaces*, Ann. of Math. Studies, No. 61, Princeton Univ. Press, Princeton, N.J., 1961.

10. F. Pham, *Caustiques, phase stationnaire et microfonctions*, Acta Math. Vietnam. **2** (1977).

11. ______, *Singularités des systèmes différentiels de Gauss-Manin* (*avec des contributions de Lo kam Chan, Ph. Maisonobe et J. E. Rombaldi*), Progress in Math., no. 2, Birkhaüser, Basel, 1980.

12. R. Thom, *Ensembles et morphismes stratifiés*, Bull. Amer. Math. Soc. **75** (1969).

UNIVERSITÉ DE NICE, FRANCE

Proceedings of Symposia in Pure Mathematics
Volume 40 (1983), Part 2

A NOTE ON HIGHER ORDER DUAL VARIETIES, WITH AN APPLICATION TO SCROLLS

RAGNI PIENE[1]

1. Introduction. The dual variety $X^{\vee} \subset \check{\mathbf{P}}^n$ of a variety $X \subset \mathbf{P}^n$ is the closure of the set of hyperplanes containing the tangent space to X at some smooth point. We define the *m-dual variety* $X_m^{\vee} \subset \check{\mathbf{P}}^n$ of X as the closure of the set of hyperplanes containing an mth osculating space to X, in particular, $X_1^{\vee} = X^{\vee}$. More generally, if $G = \mathrm{Grass}_{a+1}(V)$ denotes the Grassmann variety of a-spaces in $\mathbf{P}(V) = \mathbf{P}^n$, we can define mth osculating spaces of a variety $X \subset G$, using the sheaves of principal parts, and hence its m-dual variety $X_m^{\vee} \subset \mathrm{Grass}_{a+1}(V^{\vee})$ as the closure of the set of $(n-a-1)$-spaces containing an mth osculating space to X. This is of course closely related to Pohl's *associated varieties* [**Pohl**].

We show, in Proposition 1, a weak biduality result for m-duals: one always has $X \subset (X_m^{\vee})_m^{\vee}$, and equality holds under a dimension hypothesis, which is always satisfied in the classical case.

It is natural to ask for the degree of $X_m^{\vee}$ in terms of characters of X. Since we are working with "modified" bundles of principal parts, this can be done—at least in principle!—as in the classical case (see e.g. [**P2**, **U**]).

Here we only deal explicitly with the case of a scroll (i.e., a ruled, nondevelopable surface) $X \subset \mathbf{P}(V)$, or, equivalently, a curve $C \subset \mathrm{Grass}_2(V)$. In general, a scroll has 2nd osculating spaces of dimension 4. We give a formula for deg $X_2^{\vee}$. In the case $\mathbf{P}(V) = \mathbf{P}^5$, we call $X^* = X_2^{\vee}$ the *strict dual* of X. If dim $X^* = 2$, then X^* is also a scroll and $X^{**} = X$ holds. Moreover, the dual variety $X^{\vee}$ (the normal bundle of X) is equal to the osculating developable of X^*, and the dual plane of a tangent plane to X is the tangent plane to X^* at the corresponding point. Hence we get, for scrolls in $\mathbf{P}^5$, a complete parallel to the duality existing between a curve $C \subset \mathbf{P}^3$, its strict dual curve $C^* \subset \check{\mathbf{P}}^3$, and their developables (see e.g. [**P1**, §5; **P3**, Remark 1 on p. 111]).

1980 *Mathematics Subject Classification.* Primary 14N99; Secondary 14J25, 14M15, 14N10.

[1]Partially supported by the Norwegian Research Council for Science and the Humanities.

Scrolls are examples of surfaces with "too small" osculating spaces of higher order, hence are "of type Φ" in the terminology of Corrado Segre [S]. I am grateful to Gianni Sacchiero for bringing these varieties—in particular the scrolls and their strict duals—to my attention.

2. Higher order dual varieties. Fix the following notations: V is an $(n + 1)$-dimensional vector space over an algebraically closed field k of characteristic 0, G is the Grassmann variety $\mathrm{Grass}_{a+1}(V)$ consisting of $(a + 1)$-quotients of V (identified with a-dimensional linear subspaces of $\mathbf{P}(V)$), and $V_G \to Q$ is the universal $(a + 1)$-quotient on G.

For each integer m there is a natural homomorphism

$$\alpha^m\colon V_G = H^0(G, Q)_G \to P_G^m(Q),$$

where $P_G^m(Q)$ denotes the bundle of principal parts of order m of Q (see [**P1**, §6]).

Let $X \subset G$ be a subvariety, of dimension r, and set $E = Q|_X$. The restriction of α^m, composed with the natural map $P_G^m(Q)|_X \to P_X^m(E)$, gives a homomorphism

$$a^m\colon V_X \to P_X^m(E).$$

A point $x \in X$ is called *m-regular* if x is smooth and if $a^m(x)$ is surjective; if these points are dense in X, we say that X is generically m-regular. At each m-regular point $x \in X$ there is a well-defined *mth osculating space*, of dimension

$$(a + 1)\binom{r + m}{m} - 1,$$

defined by $a^m(x)$. Hence a generically m-regular X has an *mth associated variety* $X^{(m)} \subset \mathrm{Grass}_{(a+1)\binom{r+m}{m}}(V)$, defined as the closure of the set of the mth osculating spaces (see [**Pohl**, §IV]). We define the *m-dual variety* $X_m^{\vee} \subset G^{\vee} = \mathrm{Grass}_{a+1}(V^{\vee}) = \mathrm{Grass}_{n-a}(V)$ to be the closure of the set of $(n - a - 1)$-spaces containing an mth osculating space.

Even if X is nowhere m-regular, we can define mth osculating spaces: let $U \subset X$ be an open dense smooth subscheme such that $K_U = \mathrm{Ker}(a^m)|_U$ is a sub-bundle of V_U, or, equivalently, such that $\mathrm{Im}(a^m)|_U$ is a bundle. If

$$\mathrm{rk}\,\mathrm{Im}(a^m)|_U = s + 1,$$

then each point $x \in U$ has an mth osculating space, of dimension s, defined by $a^m(x)$. The *m-dual variety* $X_m^{\vee} \subset G^{\vee}$ of X is the closure of the set of $(n - a - 1)$-spaces containing the mth osculating spaces. Let $\tilde{X} \subset G \times G^{\vee}$ be the closure of $\mathrm{Grass}_{a+1}(K_U^{\vee}) \subset \mathrm{Grass}_{a+1}(V_U^{\vee}) = U \times G^{\vee}$; then $X_m^{\vee} = \mathrm{pr}_2(\tilde{X})$. Let $(X_m^{\vee})^{\sim} \subset G \times G^{\vee}$ denote the corresponding variety constructed for $X_m^{\vee}$, so that $(X_m^{\vee})_m^{\vee} = \mathrm{pr}_1((X_m^{\vee})^{\sim})$.

The following proposition gives a weak biduality for m-dual varieties, generalizing the classical biduality for projective varieties (see [**K**]).

PROPOSITION 1. *If $X_m^{\vee} \neq \varnothing$, then $\tilde{X} \subset (X_m^{\vee})^{\sim}$ and $X \subset (X_m^{\vee})_m^{\vee}$. In particular, if $\tilde{X} = (X_m^{\vee})^{\sim}$ (i.e., if $\dim \tilde{X} = \dim(X_m^{\vee})^{\sim}$), then $X = (X_m^{\vee})_m^{\vee}$ holds.*

REMARK. In the classical case ($a = 0$, $m = 1$), $\dim \tilde{X} = \dim(X^{\vee})^{\sim} = n - 1$ always. An example where the equality does not hold: $X \subset \mathbf{P}^6$ a generically 2-regular surface contained in a hyperplane H. Then $X_2^{\vee} = \{H\} \in \check{\mathbf{P}}^6$ and $(X_2^{\vee})_2^{\vee} = H$.

PROOF. It suffices to show the inclusion $\tilde{X} \subset (X_m^{\vee})_m^{\vee}$ on an open dense of $\tilde{X}$. Let $p\colon \tilde{X} \to X$ and $q\colon \tilde{X} \to X_m^{\vee}$ denote the projections. Consider a point $(x, y) \in \tilde{X} \subset G \times G^{\vee}$ such that $x \in U$, y is in the corresponding $V \subset X_m^{\vee}$, and q is smooth at (x, y). Let F denote the restriction of the universal $(a + 1)$-quotient of $V_{G^{\vee}}$ to $X_m^{\vee}$, and consider the following diagram (restricted to $p^{-1}U$):

$$\begin{array}{ccccccc}
0 & \to & p^*K_U & \to & V_{\tilde{X}} & \to & p^*P_X^m(E) \searrow \\
 & \nearrow & & & \| & & \downarrow \qquad\quad p^*E \\
q^*F^{\vee} & & & & V_{\tilde{X}} & \to & P_{\tilde{X}}^m(p^*E) \nearrow \\
 & \searrow & & & \| & & \\
 & & q^*P_{X_m^{\vee}}^m(F)^{\vee} & \to & V_{\tilde{X}} & &
\end{array}$$

To show that $(x, y) \in (X_m^{\vee})_m^{\vee}$ amounts to showing that the composition $q^*P_{X_m^{\vee}}^m(F)^{\vee} \to p^*E$ is zero (locally at (x, y)). The map $q^*F^{\vee} \to p^*P_X^m(E)$, and hence also $q^*F^{\vee} \to P_{\tilde{X}}^m(p^*E)$, is zero; since the composition $q^*F^{\vee} \to V_{\tilde{X}} \to p^*E$ is zero, we obtain, by "differentiating" (i.e. applying the differential operators of order $\leqslant m$, corresponding to $P_{\tilde{X}}^m$, to this composition), that $P_{\tilde{X}}^m(q^*F)^{\vee} \to p^*E$ is zero. Since $q^*P_{X_m^{\vee}}^m(F) \to P_{\tilde{X}}^m(q^*F)$ is locally split at (x, y), we obtain that $q^*P_{X_m^{\vee}}^m(F)^{\vee} \to p^*E$ is zero at (x, y). (This is the same as the argument used for curves in $\mathbf{P}^n$, as in [**P1**, §5].)

Suppose X is generically m-regular. Then $\operatorname{rk} K_U = n + 1 - (a + 1)\binom{r+m}{m}$. If $\operatorname{rk} K_U \geqslant a + 1$, then $\tilde{X}$ is defined and has dimension

$$r + (a + 1)\left(n - a - (a + 1)\binom{r + m}{m}\right).$$

Set $r^{\vee} = \dim X_m^{\vee}$. If $X_m^{\vee}$ is also generically m-regular, then biduality holds if and only if $r - (a + 1)^2\binom{r+m}{m} = r^{\vee} - (a + 1)^2\binom{r^{\vee}+m}{m}$. This is possible only if $a = 0$ and $m = 1$ (the classical case), or if $r = r^{\vee}$. In fact, when $\operatorname{rk} K_U > a + 1$, $X_m^{\vee}$ is ruled, and hence should not be generically m-regular. Note that the surjections $P_X^i(E) \to P_X^{i-1}(E)$ give a sequence of inclusions

$$X^{\vee} = X_1^{\vee} \supset X_2^{\vee} \supset \cdots \supset X_m^{\vee},$$

and that one could, instead of $\tilde{X}$, construct an $\tilde{X}_{1,2,\ldots,m}$ in the product of all the Grassmannians.

As in the case of classical duality, there is an invariance of m-duals under sections and projections: Suppose $W \subset V$ is a subspace, $\dim W \geqslant a + 1$. For $X \subset G = \mathrm{Grass}_{a+1}(V)$, consider the projection

$$X \dashrightarrow \mathrm{Grass}_{a+1}(W) \quad \text{corresponding to } W_G \to E.$$

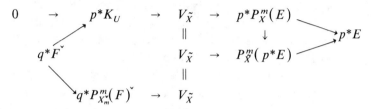

If the center of projection $\mathbf{P}(V/W)$ is reasonable (i.e., if most of the a-spaces corresponding to points of X are projected to a-spaces in $\mathbf{P}(W)$), this map is rational, and we denote by $\overline{X}$ the closure of its image. From the functorial properties of the sheaves of principal parts, we get

PROPOSITION 2. *The m-dual of a projection is the corresponding section of the m-dual, i.e.,*

$$\overline{X}^{\vee}_m = X^{\vee}_m \cap \mathrm{Grass}_{a+1}(W^{\vee})$$

holds.

(The proof is similar to the one in the classical ($a = 0$, $m = 1$) case; see [**P2**, p. 269], and observe that the genericity assumption made there is unnecessary.)

The *degree* of $X \subset G$ is its degree in $\mathbf{P}(\Lambda^{a+1}V)$ via the Plücker embedding. Thus we have $\deg X = c_1(E)^r \cap [X]$, and $\deg X^{\vee}_m = c_1(F)^{r^{\vee}} \cap [X^{\vee}_m]$. Whenever we can express F (or q^*F) in terms of known bundles, we get an expression for $\deg X^{\vee}_m$. When X is generically m-regular, F is determined by $P^m_X(E)$ and the singularities of a^m; hence we get, at least in principle, an expression for $\deg X^{\vee}_m$ in terms of the degree of X and its Chern classes (or rather, the Chern classes of a desingularization of X) and the various singularities of X and a^m. The very simplest case occurs when X is smooth and m-regular, $n - a = (a + 1)\binom{r+m}{m}$ and $r^{\vee} = r$. Then $\deg X^{\vee}_m = c_1(q^*F)^r = c_1(K^{\vee})^r = c_1(P^m_X(E))^r$. (For formulas in the classical case, see [**P2**, **U**]; see also [**Pohl**] for associated varieties.)

In the case of curves, formulas exist: Let $X \subset \mathbf{P}(V)$ be a curve spanning $\mathbf{P}(V)$. Then X is generically m-regular, for $m \leqslant n$, and we have associated curves $X^{(m)} \subset \mathrm{Grass}_{m+1}(V)$ and corresponding *osculating developables* $Y_m \subset \mathbf{P}(V)$. We also have m-dual varieties $X^{\vee}_m \subset \mathbf{P}(V^{\vee})$—these are nothing but the osculating developables Y^*_{n-m-1} of the *strict dual curve* $X^* = X^{(n-1)} \subset \mathbf{P}(V^{\vee})$, and they are also equal to the dual of the osculating developables of X. More precisely, for each m we have

$$X^{\vee}_m = Y^*_{n-m-1} = (Y_{m-1})^{\vee}.$$

The first equality follows from the duality of certain exact sequences on X and X^* (see [**P1**, 5.2]), the second holds because the tangent spaces to Y_{m-1} are the mth osculating spaces to X. Thus we have formulas

$$\deg X^{\vee}_m = (m + 1)(d + m(g - 1)) - \sum_{i=0}^{m-1} (m - i)k_i,$$

where $d = \deg X$, $g =$ (geometric) genus of X, and k_i is the ith stationary index of X [**P1**, 3.2].

3. Dual varieties of a scroll. Let $X \subset G$ be as in the preceding section. If m is such that $\tilde{X} \to X$ is birational, i.e., if there is a uniquely determined mth osculating $(n - a - 1)$-space to X at x for most points $x \in X$, we shall call $X^* = X^{\vee}_m$ the *strict dual variety* of X.

For example, if $C \subset \mathbf{P}^n$ is a curve spanning $\mathbf{P}^n$, then $C^* = C^{\vee}_{n-1}$. If $X \subset \mathbf{P}^6$ is a surface which is generically 2-regular, then $X^* = X^{\vee}_2 \subset \check{\mathbf{P}}^6$ is the strict dual.

An example of surfaces that are nowhere 2-regular (C. Segre [S] called them "of type Φ") are the ruled surfaces: scrolls, developables, and cones. The theory of duals of developables and cones reduces to that of curves in projective space; let us now look at the scrolls. By definition, a scroll $X \subset \mathbf{P}(V)$ is a ruled surface such that the tangent planes to X along a (general) generator are nonconstant. Suppose $l \subset X$ is a generator, $x \in l$. The 2nd osculating space to X at x, defined by a^2: $V_X \to P_X^2(1)$, is the space spanned by the tangent planes to X along l (this gives a $\mathbf{P}^3$) and the 2nd osculating space to a curve on X at x. If X is not contained in a $\mathbf{P}^3$, one expects this space to be of dimension 4; if X is not contained in a $\mathbf{P}^4$, one expects these 4-spaces to vary along l, so that X^* has dimension 2.

We shall now generalize to scrolls in $\mathbf{P}^5$ the duality results for curves in $\mathbf{P}^3$ [P1, P3]: Let $C \subset \mathbf{P}^3 = \mathbf{P}(V)$ be a (nonplanar) curve, and let $C^* \subset \check{\mathbf{P}}^3$ denote its strict dual. The dual $C^{\vee} \subset \check{\mathbf{P}}^3$ is the normal bundle to C and the tangent developable of C^*, and similarly for $(C^*)^{\vee}$. Moreover, the dual line of a tangent line to C is the tangent line to C^* at the corresponding point—in other words, the associated curves $C^{(1)} \subset \mathrm{Grass}_2(V)$ and $C^{*(1)} \subset \mathrm{Grass}_2(V^{\vee}) = \mathrm{Grass}_2(V)$ are equal.

PROPOSITION 3. *Let* $X \subset \mathbf{P}(V) = \mathbf{P}^5$ *be a scroll which admits a strict dual* $X^* = X_2^{\vee}$, *and assume* dim $X^* = 2$. *Then* X^* *is a scroll. The dual* $X^{\vee} \subset \mathbf{P}(V^{\vee})$, *the normal bundle of* X, *is equal to the tangent developable of* X^*, *and vice versa. Moreover, the dual plane of a tangent plane to* X *is the tangent plane to* X^* *at the corresponding point—in other words, the associated varieties* $X^{(1)} \subset \mathrm{Grass}_3(V)$ *and* $X^{*(1)} \subset \mathrm{Grass}_3(V^{\vee}) = \mathrm{Grass}_3(V)$ *are equal.*

PROOF. Let X' be a modification of X and of X^* such that, on X^1, $\mathrm{Im}(a^1)$ and $\mathrm{Im}(a_*^1)$ admit quotient bundles P_1 and P_1^* of rank 3. Then $K = \ker(V_{X'} \to P_1)$ and $K^* = \ker(V_{X'}^{\vee} \to P_1^*)$ are bundles of rank 3, and the sequences $0 \to K \to V_{X'} \to P_1 \to 0$ and $0 \leftarrow P_1^* \leftarrow V_{X'}^{\vee} \leftarrow K^* \leftarrow 0$ are dual to each other: as in the proof of Proposition 1, one shows that (generically on X') the composition of $(a_*^1)^{\vee}: P_{X^*}^1(1)^{\vee} \to V_{X^*}$ with $a^1: V_X \to P_X^1(1)$ is zero; since a_*^1 and a^1 both have rank 3, the result follows. In particular the existence of the exact sequence

$$0 \to (P_1^*)^{\vee} \to V_{X'} \to P_1 \to 0$$

shows that the tangent planes to X^* are the dual planes of the tangent planes to X; hence if X is a scroll, so is X^*. The other statements also follow directly from that exact sequence.

There is still another parallel to the curve case, namely to the fact that the strict dual curve $C^* \subset \check{\mathbf{P}}^3$ is a *cuspidal edge* of the dual variety $C^{\vee} \subset \check{\mathbf{P}}^3$ of a curve $C \subset \mathbf{P}^3$.

PROPOSITION 4. *If* $X \subset \mathbf{P}^5$ *is a scroll, then its strict dual* X^* *is a "cuspidal edge" of the dual variety* $X^{\vee} \subset \check{\mathbf{P}}^5$.

PROOF. Assume X is smooth, and $\bar{X} \subset \mathbf{P}^3$ a generic projection. Scrolls with ordinary singularities in $\mathbf{P}^3$ are numerically selfdual, so $\bar{X}^{\vee} \subset \check{\mathbf{P}}^3$ has a finite number of pinch points, corresponding to the pinch points of $\bar{X}$. If $L \subset \mathbf{P}^5$ is the centre of projection, a pinch point of $\overline{X}$ occurs when L intersects a tangent, i.e., when L intersects the $\mathbf{P}^3$ spanned by the tangent planes along a generator. But then L and this $\mathbf{P}^3$ span a $\mathbf{P}^4$ which is necessarily a point in X^* and also in $\check{\mathbf{P}}^3 = L^{\vee} \subset \check{\mathbf{P}}^5$. Since $\overline{X}^{\vee} = X^{\vee} \cap \check{\mathbf{P}}^3$, it follows that the "ramified singularities" of $X^{\vee}$ are precisely the points of X^*. The same argument works if X is not assumed smooth. Note that there might be other "cuspidal edges", as in the case of curves, where inflectionary tangents are cuspidal edges on the developable.

In order to compute the degree of X^*, it is convenient to consider X as a curve $C \subset G = \mathrm{Grass}_2(V)$. Consider a^1: $V_C \to P_C^1(E)$, where E is the restriction of the universal 2-quotient of V on G. The subspaces of V defined by a^1 can be interpreted by choosing, locally, a trivialization of E, corresponding to two curve sections of X. Hence a^1 defines, at a generator $l \in C$ of X, the space spanned by l and the tangent to the curves at the points of intersection with l; hence it is equal to the space spanned by the tangent planes to X along l. Since X is a scroll, this space has dimension 3, so C is generically 1-regular. It follows that $C^* = C_1^{\vee} \subset \mathrm{Grass}_4(V) = \mathrm{Grass}_2(V^{\vee})$ is the strict dual of C (and $C^* = C^{(1)}$, the 1st associated curve of C). If X^* is a scroll, then C^* is generically 1-regular, and $C^{**} = C$ (by Proposition 1). Moreover, the 2nd osculating spaces to X along a generator l are just the 4-spaces containing the 3-space spanned by the tangent planes. In other words, C^* is equal to X^* considered as a curve in $\mathrm{Grass}_2(V^{\vee})$. Thus we have proved

PROPOSITION 5. *If $X \subset \mathbf{P}(V) = \mathbf{P}^5$ is a scroll such that X^* is a scroll, then $X^{**} = X$ holds.*

The next proposition gives a formula for the degree of X^* (see also [**Pohl**, Theorem 4.5]).

PROPOSITION 6. *Let $X \subset \mathbf{P}(V) \cong \mathbf{P}^5$ be a scroll of degree d and genus g, and suppose $X^* \subset \mathbf{P}(V^{\vee})$ is a scroll. Then*

$$\deg X^* = 2(d + 2g - 2) - k,$$

where k is the stationary index of $C \subset \mathrm{Grass}_2(V)$.

REMARK. Let ν: $C' \to C$ denote the normalization, then, by definition, $k = \lg(\mathrm{Coker}\, V_{C'} \to P_{C'}^1(\nu^*E))$. By trivializing E one sees that an ordinary cusp of C counts *twice* in k (which checks with [**Edge**, §349]).

COROLLARY. *The stationary index k^* of C^* is given by*

$$k^* = 3(d + 2(2g - 2)) - 2k.$$

PROOF. On C', $P_C^1(E)$ admits a 4-quotient, namely $P_1 = \mathrm{Im}(V_{C'} \to P_{C'}^1(\nu^*E))$. Hence,

$$\deg X^* = \deg C^* = c_1(P_1) = c_1\left(P_{C'}^1(\nu^*E)\right) - k = 2(d + 2g - 2) - k.$$

The corollary follows from the duality $X^{**} = X$ of Proposition 5, $d = \deg X = 2(d^* + 2g - 2) - k^*$.

Note that if X has no singular generators ($k = 0$), then $k^* = 3(d + 2(2g-2))$, and hence X^* has no singular generators if and only if $d = d^* = 4$, $g = 0$. (Such scrolls are linearly normal in $\mathbf{P}^5$.)

We shall now look at some other approaches to the degree of X^*.

Because of the following (classical) proposition, the degree of X^* is equal to $\deg(X^*)^{\check{}}$, hence to the degree of the tangent developable of X (Proposition 3).

PROPOSITION 7. *Let $X \subset \mathbf{P}(V) \cong \mathbf{P}^n$ be a scroll, $X^{\check{}} \subset \mathbf{P}(V^{\check{}})$ its dual. Then*

$$\deg X^{\check{}} = \deg X.$$

PROOF. The classical proof goes like this: project X to a scroll $\overline{X} \subset \mathbf{P}(W) \cong \mathbf{P}^3$ with $d = \deg \overline{X} = \deg X$. Then $\overline{X}^{\check{}} = X^{\check{}} \cap \mathbf{P}(W^{\check{}})$, so $\deg \overline{X}^{\check{}} = \deg X^{\check{}}$ holds. If $L \subset \mathbf{P}(W)$ is a general line,

$$\begin{aligned} \deg \overline{X}^{\check{}} &= \#\{H \supset L,\ H \text{ tg. to } X\} = \#\{H \supset L \cup l \mid l \text{ generator of } X\} \\ &= \# L \cap X = \deg X. \end{aligned}$$

Note that $\overline{X}$ and $\overline{X}^{\check{}}$ are in fact *equal* considered as curves in $\mathrm{Grass}_2(W) = \mathrm{Grass}_2(W^{\check{}})$.

For a "modern" proof, one reduces to the case that $X \subset \mathbf{P}(V)$ is smooth, say $X = \mathbf{P}(E) \to C$. Then

$$\deg X^{\check{}} = c_2(P_X^1(1)) = c_2(\Omega_X^1(1)) + c_1(\Omega_X^1(1))c_1(O_X(1)),$$

which, by standard exact sequences, reduces to

$$\deg X^{\check{}} = 2c_1(E)c_1(O_X(1)) - c_1(O_X(1))^2 = 2d - d = d.$$

From the exact sequence given in the proof of Proposition 3 we obtain (using [**P2**, §2])

$$\begin{aligned} \deg X^* &= \deg(X^*)^{\check{}} = c_2(P_1^*) = c_1(P_1)^2 - c_2(P_1) \\ &= c_1(P_1)^2 - \deg X^{\check{}} = c_1(P_1)^2 - d. \end{aligned}$$

Suppose $X = \mathbf{P}(E) \to C$ is smooth. Then $X' = X$ and $P_1 = P_X^1(1)$, so we get

$$\deg X^* = c_1(P_X^\prime(1))^2 - d = 3d + 2(2g-2) - d = 2(d + 2g - 2).$$

In the general case, X is the image of a smooth $Y = \mathbf{P}(E) \to C$, and X' is a blow-up of Y. Then $c_1(P_1) = c_1(P_{X'}^1(1)) - [R]$, where R is the ramification divisor of $X' \to X$, and we obtain the earlier formula, but with k expressed "in terms of" R.

Two other approaches have been communicated to me by I. Vainsencher and F. Ronga, respectively.

1. (Vainsencher). Let $X \subset \mathbf{P}(V) \cong \mathbf{P}^5$ be a *smooth* scroll,

$$Y = \mathbf{P}(N(-1)) = \{(x, H) \mid H \text{ tg. to } X \text{ at } x\} \subset \mathbf{P}(V) \times \mathbf{P}(V^{\check{}}),$$

and set $Z = \{(x, H) \in Y \mid H \cap X = l_x \cup D \text{ with } D \text{ singular at } x\}$. Then $X^* = \mathrm{pr}_2(Z)$. One shows that Z is the zeros of a section of a certain rank 2 bundle on

Y; since the class of Y in $X \times \mathbf{P}(\check{V})$ is the 3rd Chern class of a rank 3 bundle, this gives the class of Z in $X \times \mathbf{P}(\check{V})$ as a 5th Chern class, and allows us to compute $\deg X^* = 2(d + 2g - 2)$, provided $\dim X^* = 2$.

2. (Ronga). Assume $X \subset \mathbf{P}(V)$ as above. Now one interprets Z as a modified $\Sigma^{2,2}$ (again by "forgetting" the generators of X) of the projection map $X \times \mathbf{P}(\check{V}) \to \mathbf{P}(\check{V})$. By computing all the normal bundles in sight, one gets an expression for the class of Z in $X \times \mathbf{P}(\check{V})$, which allows one to compute $\deg X^* = 2(d + 2g - 2)$.

Bibliography

[**Edge**] W. L. Edge, *The theory of ruled surfaces*, Cambridge Univ. Press, London, 1931.

[**K**] S. L. Kleiman, *Concerning the dual variety*, Proc. 18th Scandinavian Congr. Math., 1980 (E. Balslev, Editor), Birkhäuser, Boston, Basel and Stuttgart, 1981, pp. 386–396.

[**P1**] R. Piene, *Numerical characters of a curve in projective n-space*, Real and Complex Singularities, Oslo, 1976 (P. Holm, Editor), Sijthoff and Noordhoff, Groningen, 1978, pp. 475–496.

[**P2**] ______, *Polar classes of singular varieties*, Ann. Sci. Ecole Norm. Sup. **11** (1978), 247–276.

[**P3**] ______, *Cuspidal projections of space curves*, Math. Ann. **256** (1981), 95–119.

[**Pohl**] W. F. Pohl, *Differential geometry of higher order*, Topology **1** (1962), 169–211.

[**S**] C. Segre, *Su una classe di superficie degl'iperspazi legate colle equazioni lineari alle derivate parziale di 2° ordine*, Atti Accad. Sci. Torino Cl. Sci. Fis. Mat. Natur. **42** (1906/07), 559–591.

[**U**] T. Urabe, *Generalized Plücker formulas*, Publ. Res. Inst. Math. Sci., Kyoto University **17** (1981), 347–362.

Matematisk Institutt, P.B. 1053 Blindern, Oslo 3, Norway

Proceedings of Symposia in Pure Mathematics
Volume **40** (1983), Part 2

FACTORIZATION OF BIRATIONAL MAPS IN DIMENSION 3

HENRY C. PINKHAM[1]

This expository paper surveys some recent work on the problem of factorization of proper birational maps between smooth threefolds.

The problem, essentially, is to understand all proper birational maps between smooth threefolds by decomposing them into "elementary" morphisms which are well understood. Unlike dimension 2, what these elementary morphisms should be in dimension 3 is not completely clear. The standard approach is to allow blow ups of smooth subvarieties, and their inverses (cf. §2). Some results in this direction have been obtained by B. Crauder and M. Schaps (cf. §5), and using this approach V. Danilov has apparently settled the standard conjecture, which we formulate in §2, for toric varieties and toric maps (my source of information for this result is [**R2**]). However, recent work of S. Mori (§6) and V. Kulikov (§7) shows it is perhaps better to enlarge the category somewhat.

In keeping with the topic of the Summer Institute I will focus mainly on the aspects of the problem that deal with singularities, and indeed singularities of dimensions 1, 2 and 3 will appear, in the following ways.

The indeterminacy locus of a proper birational map has codimension at least 2, so for threefolds its components are either points or curves. Points or smooth curves can be readily blown up, but we cannot blow up singular curves without introducing singularities in the ambient space. Singular curves therefore play an important role in the factorization problem. The usual procedure has been to blow up the threefold at the singular points of the indeterminacy locus C until it becomes smooth, but this is often too crude a process. Kulikov does better by first blowing up curves meeting C at its singular points, in order to resolve C, and then blowing up C. However, this is still not completely satisfying since one would prefer to blow up only subvarieties of the indeterminacy locus. It seems there will be singular curves one will have to blow up without first resolving their singular locus, but it is not yet clear exactly which; see Examples 4, 5 and 6 of §4.

1980 *Mathematics Subject Classification*. Primary 14E05, 14E35, 14B05.
[1]Supported by NSF Grant MCS 8005802 at Columbia University, and by the Sloan Foundation.

0082-0717/83/0000-0654/$06.50

The surface singularities that occur in this paper are the rational double points. Because deformations of rational double points can be simultaneously resolved after base change [**B2**], one can construct from one parameter deformations birational maps of smooth threefolds which are isomorphisms in codimension 1; in other words there is no divisor either in the domain or the range of the map which is contracted to a lower-dimensional variety (see §8).

Finally threefold singularities arise by blowing up singular curves and also by Mori's contraction procedure (§6). Here again it is an interesting question to determine exactly which singularities should be considered: it seems it should be a subclass of M. Reid's canonical singularities [**R1**].

I have tried to give as many examples as possible; one example that occurs in various guises is the ordinary double point of dimension 3. We get it by all of the three procedures mentioned above: by blowing up a curve with an ordinary double point on the threefold (Example 5 of §4) together with its two resolutions with exceptional locus $\mathbf{P}^1$, by simultaneous resolution of the nontrivial deformation of a (surface) ordinary rational double point (Examples 1 and 2, §4 and Example 9, §8) and finally as the contraction of $\mathbf{P}^1 \times \mathbf{P}^1$ in Mori's theorem (§6).

I would like to thank B. Crauder, R. Miranda, B. Moishezon, S. Mori, D. Morrison, M. Reid and especially N. Shepherd-Barron for helpful conversations and correspondence on these matters, and S. Mori and V. Kulikov for sending me preprints of their papers [**M** and **Ku2**].

Notation and conventions. If X is a smooth variety of dimension 3, K_X is its canonical divisor (the zero set of a section of its canonical sheaf $\Lambda^n\Omega_{X/k}$, [**H**, p. 180]). If X is singular we use ω_X instead. If $X \hookrightarrow Y$, then $N_{X/Y}$, or N_X, or N will denote the normal bundle [**H**, p. 182]. If f: $X \to Y$ is a birational morphism, then E_f denotes the locus in X where f is not an isomorphism; if both X and Y are smooth, then E_f is a divisor, as one readily sees by computing the Jacobian. $S_{f^{-1}}$ denotes the indeterminacy locus of f^{-1}; $S_{f^{-1}}$ has codimension $\geqslant 2$ [**H**, Lemma V.5.1]. If D is a divisor on Y, f^*D denotes its total transform and $\bar{D}$ its proper transform. Occasionally when we blow up [**H**, V.3] we will identify, by abuse of language, a divisor with its proper transform. If we blow up the space X along the subvariety C, we will sometimes simply say, by abuse of language, when the space X is understood, that we blow up C.

If $\mathbf{P}^1$ is embedded in a smooth threefold we write its normal bundle $N = (a, b)$ when it is $\mathcal{O}(a) \oplus \mathcal{O}(b)$. (a, b) will in another context denote the invertible sheaf on $\mathbf{P}^1 \times \mathbf{P}^1$ whose intersection with fibres of the two rulings is a and b, respectively. Finally $\mathbf{F}_n$, $n \geqslant 0$, is the unique (minimal) ruled surface over $\mathbf{P}^1$ with a section S with $S^2 = -n$ [**H**, V.2.13].

1. The dimension two case. Let us first review the situation in dimension 2. Let X and Y be smooth complex surfaces and f: $X \to Y$ a proper birational map. Then it has been known for many years that f can be factored as

$$X \xrightarrow{f_1} X_1 \xrightarrow{f_2} X_2 \to \cdots \to X_{n-1} \xrightarrow{f_n} Y$$

where each f_i is the blow down of a projective line with self-intersection -1; or conversely each f_i^{-1} is the blow up of a smooth point. Each f_i is a projective morphism and therefore so is f, and the intermediate surfaces X_i are smooth.

There are at least two methods for proving such a result. The first is to choose a point $P \in S_{f^{-1}}$ (= locus where f^{-1} is not defined—recall that the indeterminacy locus of f^{-1} is of codimension $\geqslant 2$ and so consists at most of points). Now blow up the point P. One shows [**H**, Chapter 5, Proposition 5.3] that the birational map g:

$$\begin{array}{ccc} X & \overset{g}{\dashrightarrow} & \tilde{Y} \\ & {\scriptstyle f}\searrow \quad \swarrow{\scriptstyle f_P} & \\ & Y & \end{array}$$

where f_P: $\tilde{Y} \to Y$ is the blow up of P, is actually a morphism. By induction (for example on the difference $h^2(X, \mathbf{C}) - h^2(Y, \mathbf{C})$, since h^2 increases by 1 each time one blows up a point) we are done. This method could be called factorizing "from the bottom up."

The second method is to try to find in the exceptional locus of $X \to Y$ a rational curve E with self-intersection -1 (E is called an exceptional curve of the first kind). By Castelnuovo's theorem [**H**, Chapter V, Theorem 5.7] E can be contracted *projectively* to a smooth point P, this morphism is of course the inverse of the blow up of P; so once again we are done by induction, say on the number n of irreducible components of the exceptional locus $\bigcup E_i$. We can write

$$(1) \qquad K_X = f^*K_y + \sum_{i=1}^{n} r_i E_i, \qquad r_i > 0.$$

The intersection matrix (E_i, E_j) is negative definite [**Mu1**], so that we can find a component E_{i_0} such that $K_X \cdot E_{i_0} = \sum_{i=1}^{n} r_i E_i \cdot E_{i_0}$ is strictly negative. On the other hand, by adjunction [**H**, V.1.5] $(K_X + E_{i_0}) \cdot E_{i_0}$ is the arithmetic genus of E_{i_0}, and is thus $\geqslant 0$. These two inequalities yield the unique solution $K_X \cdot E_{i_0} = -1$ and $E_{i_0} \cdot E_{i_0} = -1$, which says precisely that E_{i_0} is exceptional of the first kind.

For higher-dimensional generalizations it is perhaps instructive to find E without using the negative definiteness of $(E_i \cdot E_j)$. Instead, let H be a curve through P. Then

$$(2) \qquad f^*H = \overline{H} + \sum_{i=1}^{n} s_i E_i$$

where the s_i are positive integers, and $\overline{H}$ is the proper transform of H. Now the dualizing sheaf ω_{E_j} of E_j is

$$\omega_{E_j} = (K_X + E_j)_{|E_j} = \underset{\substack{\| \\ 0}}{f^*K_Y \cdot E_j} + \sum_{\substack{i=1 \\ i \neq j}}^{n} r_i E_i \cdot E_j + (r_j + 1) E_j \cdot E_j$$

But dotting (2) with E_j

$$0 = \overline{H} \cdot E_j + \sum_{i=1}^{n} s_i E_i \cdot E_j.$$

So we use this equation to eliminate $E_j \cdot E_j$ from the previous one, obtaining

$$\omega_{E_j} = \sum_{i \neq j} \left(r_i - (r_j + 1)\frac{s_i}{s_j} \right) E_i \cdot E_j - \left(\frac{r_j + 1}{s_j} \right) \overline{H} \cdot E_j.$$

Now choose j so that $(r_j + 1)/s_j$ is maximum. Then the equation shows that ω_{E_j} is of strictly negative degree (so that $E_j \approx \mathbf{P}^1$ and ω_{E_j} is of degree -2) and that E_j meets at most two other components transversally. Further analysis of this situation leads to an exceptional curve of the first kind (unfortunately not necessarily one with $(r_j + 1)/s_j$ maximum). This trick of maximizing $(r_j + 1)/s_j$ is used by Crauder in dimension 3 [**C2**].

REMARK. To prove that $(E_j \cdot E_j)$ is negative definite one uses (2), so the two methods sketched above are closely related.

2. The general conjecture. We would like to generalize the result of the preceding section to dimension three. First we must decide what should replace the blow up of a point in dimension 2 as the elementary terms in the factorization. The obvious suggestion is to blow up points and smooth curves: we will call these blow ups *permissible*—and the inverse maps permissible blow downs. (Although the conjecture of this section will be phrased in terms of permissible blow ups and blow downs, we shall see in §5 that it is useful to consider a wider class of contractions.)

Blow up of a point. If P is a point on the smooth threefold Y, f_P: $\tilde{Y} \to Y$ will denote the blow up of P. The exceptional divisor $E = f_P^{-1}(P)$ is isomorphic to $\mathbf{P}^2$. A local computation shows that $K_{\tilde{Y}} = f_P^*(K_Y) + 2E$, so that by the adjunction formula $K_E = (K_{\tilde{Y}} + E)|_E = (f_P^*(K_Y) + 3E)|_E = 3E|_E$. But $K_E \approx \mathcal{O}_{\mathbf{P}^2}(-3)$, so $E|_{E'}$ which is the normal bundle N_E of E in $\tilde{Y}$, is $\mathcal{O}(-1)$.

Blow up of a smooth curve. If C is a smooth curve in Y and f_C: $\tilde{Y} \to Y$ the blow up of C, then the exceptional divisor $E = f_C^{-1}(C)$ is a (minimal) ruled surface over C. Here the local computation of differentials shows that $K_{\tilde{Y}} = f^*K_Y + E$, so $K_E = (K_{\tilde{Y}} + E)|_E \simeq (f^*K_Y + 2E)|_E$. Now if f is a fibre of E so that f is isomorphic to $\mathbf{P}^1$, we have $K_f = (K_E + f)|_f = 2N_E \cdot f + f \cdot_E f$. So deg $K_f = -2$. Since f is a fibre of the ruling of E, $f \cdot_E f = 0$, so $N_E \cdot f = -1$.

Conversely any time one has a divisor E in a smooth threefold of one of the above two types, and whose normal bundle fits the description above, then E can be contracted to a point or smooth curve lying on a smooth threefold (theorem of Nakano [**N**]). However, in the case of contraction to a curve, the contraction morphism may not be algebraic, i.e. the contracted variety may not be algebraic.

Although there is a numerical sufficient condition for algebraic contraction [**G**], there can be no numerical necessary condition on the normal bundle as we shall see later (see Examples 2 and 3). This gives rise to difficulties; cf. §§4 and 6.

When the varieties involved are projective, Mori has a global necessary condition for algebraic contraction.

As we shall see by examples in §§4 and 5, the following conjecture is the best one can hope for.

CONJECTURE. Given a proper birational morphism $f\colon X \to Y$, where X and Y are smooth threefolds, then f can be factored as a composition of permissible blow ups and blow downs, all of which are isomorphisms between $X - E_f$ and $Y - S_f$.

Schematically we have

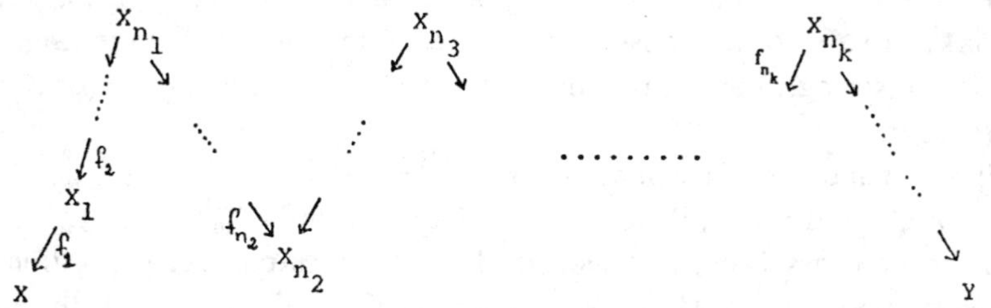

where each arrow is a permissible blow down and hence all the intermediate X_i are smooth. It does not seem realistic to expect that all the blow ups can be put first, followed by the blow downs, but to my knowledge there are no counterexamples to this slightly stronger conjecture.

The last condition in the conjecture, that the f_i are isomorphisms between $X - E_f$ and $Y - S_f$, is to ensure that we do not perturb the map on the open set where it is an isomorphism. (By the obvious abuse of language we are identifying $X - E_f$ and $Y - S_f$ with subspaces of the intermediate spaces X_i.)

3. Torus embeddings. The conjecture stated in §2 has been solved by Danilov [**R2**] in the special case where X and Y are torus embeddings, the map f is equivariant and the (permissible) factorizing maps are also required to be equivariant for the torus structure. In order to describe his result more precisely and also because many of the examples we will meet later in this paper are toric (or toroidal, that is, formally toric) and can be conveniently described using torus embeddings, we need to set up some terminology. For further material on torus embeddings, see Kempf *et al.* [**K**], Danilov [**D1**] or Oda [**O**]. We will follow Oda's presentation.

Let T be an r-dimensional torus: $T \approx (\mathbf{C}^*)^r$. A *torus embedding* $T \subset X$ is an algebraic variety X containing T as a Zariski open dense subset, and an action of T on X extending the group law of T:

$$\begin{array}{ccccc} T & \times & X & \to & X \\ & & \cup & & \cup \\ T & \times & T & \to & T \end{array}$$

A morphism between torus embeddings X and X' is a map f such that the following diagram commutes:

$$\begin{array}{ccc} & \nearrow & X \\ T & & \downarrow f \\ & \searrow & X' \end{array}$$

It is possible to describe all *normal* torus embeddings combinatorially. What we will do here is describe the combinatorial object, and then show how to pass from this object to a torus embedding.

DEFINITION. A finite rational partial polyhedral decomposition (f.r.p.p. decomposition) is a pair (N, Δ) consisting of a free $\mathbf{Z}$ module N of finite rank and a collection of strongly convex rational polyhedral cones with apex O (or *cones* for short) such that:

(i) If $\sigma \in \Delta$ and τ is a face of σ, then $\tau \in \Delta$.

(ii) If $\sigma \in \Delta$ and $\tau \in \Delta$, then $\sigma \cap \tau$ is a face of σ and of τ.

Here is how one associates a torus embedding to an f.r.p.p. decomposition. First some more notation. If M is the dual $\operatorname{Hom}(N, \mathbf{Z})$ of N, $\langle \ , \ \rangle$ the pairing between N and M and $N_{\mathbf{R}} = N \otimes_{\mathbf{Z}} \mathbf{R}$, then if σ is a cone in N, let

$$\check{\sigma} = \{x \in M_{\mathbf{R}} \colon \langle x, y \rangle \geqslant 0 \quad \text{for all } Y \in \sigma\}.$$

If σ is a cone, then $\check{\sigma} \cap M$ is a subsemigroup S_σ of M which contains O, is finitely generated as a semigroup, generates M as a group, and is *saturated*, i.e. if $v \in M$ and as a positive integer such that $av \in S$, then $v \in S$. To any such S_σ one can associate a normal affine torus embedding $X_\sigma = \operatorname{Spec} \mathbf{C}[S]$ as follows: identifying M with $\mathbf{Z}^r$, we can view S as a subsemigroup of $\mathbf{Z}^r$. Then let $\mathbf{C}[S] = \mathbf{C}[\ldots, t^s, \ldots]_{s \in S}$ with the usual multi-index notation. Note that $T = \operatorname{Spec} \mathbf{C}[M]$, and that T acts on X_σ. Given two elements $\sigma, \tau \in \Delta$ we can glue X_σ to X_τ along the common face $\sigma \cap \tau$ which is also a cone. It is easy to see this is compatible with the torus action. The resulting torus embedding is called X_Δ. It is also easy to see that X_σ is *smooth* if and only if the generators of σ form part of a basis of N. Given two f.r.p.p. decompositions Δ and Δ' of N, suppose that for every $\sigma \in \Delta$ there is a $\sigma' \in \Delta'$ such that $\sigma \subset \tau'$. Then there exists a unique equivariant morphism $f\colon X_\Delta \to X_{\Delta'}$ realizing the inclusion. The map f is proper if $\bigcup_{\sigma \in \Delta} \sigma = \bigcup_{\sigma' \in \Delta'} \sigma'$. All these statements are proved in [**K**, **D1** or **O**].

We will be concerned with torus embeddings exclusively in the dimension 3 case. In order to represent 3-dimensional f.r.p.p. decompositions conveniently we will intersect $\mathbf{R}^+ \Delta = \{\bigcup \sigma_{\mathbf{R}} \mid \sigma \in \Delta$, where $\sigma_{\mathbf{R}}$ is the cone generated by σ in $N_{\mathbf{R}}\}$, with a sphere centered around the origin and label each vertex of the intersection with coordinates in $\mathbf{Z}^3$ of the first rational point on the half line connecting the vertex to the origin. For example, $\mathbf{C}^3$ which is an obvious torus embedding will be represented as shown in the diagram.

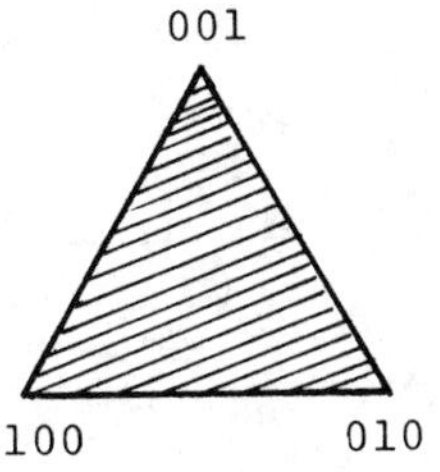

The vertices correspond to the coordinate hyperplanes,
the edges to the axes and the triangle to the origin in $\mathbf{C}^3$.
In the diagram, 100 means $(1,0,0)$, etc.

If we blow up the origin in $\mathbf{C}^3$, we get the torus embedding as shown in the diagram.

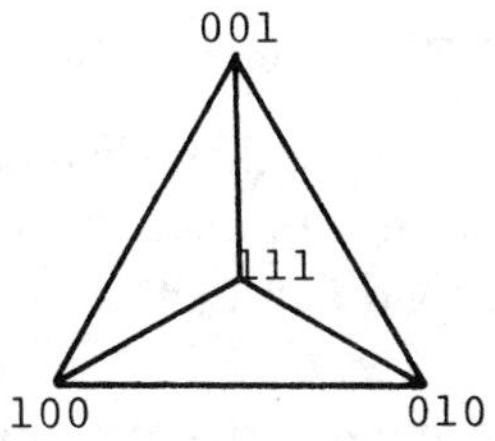

Notice that the condition given above for properness is of course satisfied.
From now on we will no longer shade in the 2-simplices, since it will always be obvious which ones we are considering.

In all such diagrams vertices correspond to divisors, edges to curves and 2-simplices to points. The study of proper birational toric maps $f\colon X \to Y$, where X and Y are themselves toric, is therefore reduced to the study of f.r.p.p. decompositions Δ and Δ' such that:

(1) if $\sigma \in \Delta$, then there exists a $\sigma' \in \Delta'$ such that $\sigma \subset \sigma'$, and

(2) $\bigcup_{\sigma\in\Delta}\sigma = \bigcup_{\sigma'\in\Delta'}\sigma'$.

We now describe permissible toric blow ups combinatorially.

Blow up of a point. The point must lie on the intersection of three invariant divisors, and hence corresponds in our combinatorial picture to a triangle. Since we are assuming the torus embedding is smooth, the vertices n_1, n_2, n_3 of the triangle form a basis of $\mathbf{Z}^3$. This is schematically shown in the diagram.

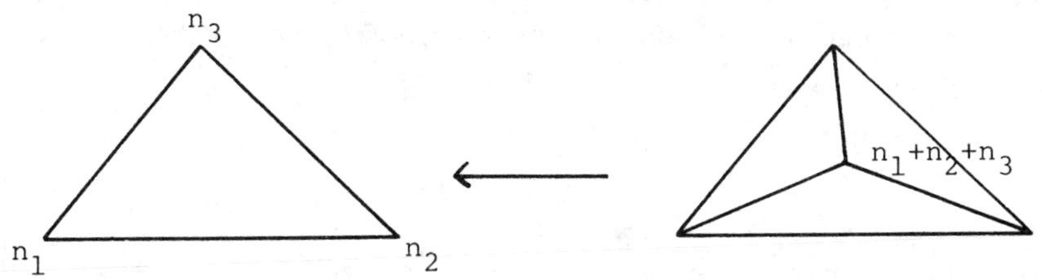

Blow up of a curve. The curve corresponds to an edge e of the f.r.p.p. decomposition. It lies on two 2-simplices if it is proper, one otherwise. The picture in the proper case is shown in the diagram.

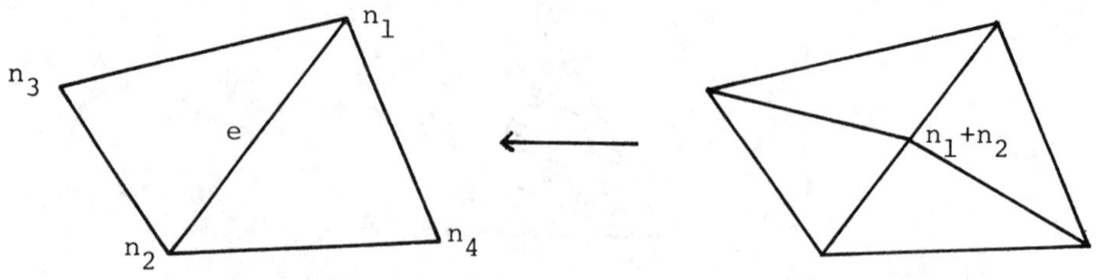

Otherwise:

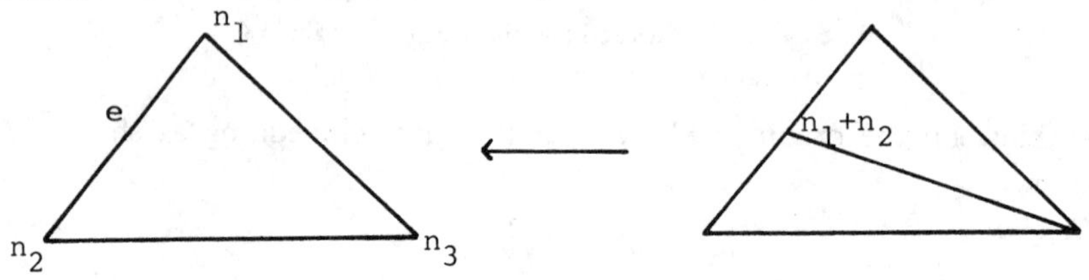

Therefore, what Danilov has shown is that when Δ and Δ' are as above, one can pass from Δ to Δ' by a series of moves, each one of which is like one of the two above or its inverse, with the condition that at each move, all the cones must be generated by part of a basis for $N \approx \mathbf{Z}^3$.

REMARKS. (1) From the combinatorial description $\Delta \to \Delta'$ one can readily recover the relative canonical divisor of $f\colon X_\Delta \to X_{\Delta'}$ [**O**, 6.6], the triple intersection of equivariant divisors [**O**, 6.7], and we can determine whether f is projective [**O**, 6.5].

(2) Most of the results on smooth torus embeddings are due to Demazure [**De**].

(3) Reid gives an interesting geometric interpretation of Danilov's proof in [**R2**].

EXAMPLE 1 (KEMPF ET AL. [**K**, p. 38]). Let σ be the cone with vertices $(1,0,0)$, $(0,1,0)$, $(0,0,1)$ and $(1,-1,1)$. The associated semigroup S_σ in the dual space is generated by $(1,0,0)$, $(1,1,0)$, $(0,1,1)$ and $(0,0,1)$ so that

$$\mathbf{C}[S] \approx \mathbf{C}[t_1, t_1t_2, t_2t_3, t_3].$$

Setting $x = t_1$, $y = t_1t_2$, $z = t_2t_3$ and $w = t_3$ we see that

$$\mathbf{C}[S] \approx \mathbf{C}[x, y, z, w]/(xz - yw)$$

so that the associated torus embedding is just the ordinary double point. We want to resolve this singularity torically. In our projection to the unit sphere, σ is given as shown in the diagram.

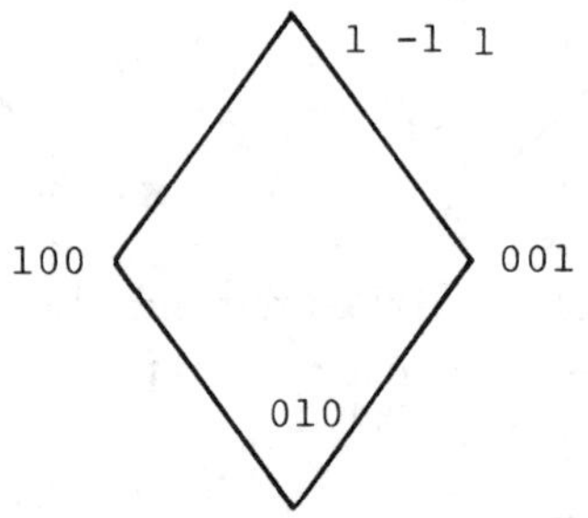

To resolve this we have three possibilities seen in the figure.

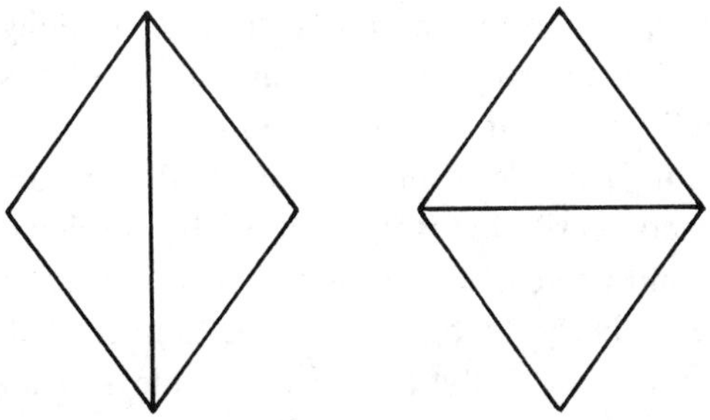

and

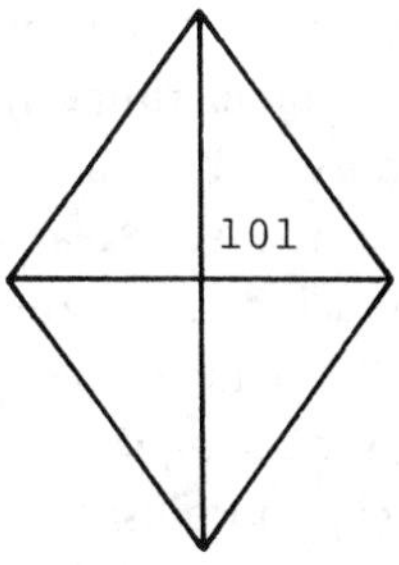

(One should check that in each case all the cones are generated by part of a basis of $\mathbf{Z}^3$.) The first two resolutions correspond to the blow up of the ideals (x, y) and (x, w) respectively (these are the ideals of the cones over lines in each of the two rulings of $xz - yw = 0$ considered as a projective variety), and the third is the blow up of the maximal ideal. The central vertex in the third resolution corresponds to a $\mathbf{P}^1 \times \mathbf{P}^1$ with normal bundle $(-1, -1)$, so that by the contraction criterion it can be blown down smoothly in either ruling, then yielding each of the first two resolutions. The central edge in both of the first two resolutions is of course a $\mathbf{P}^1$ with normal bundle $(-1, -1)$. This gives rise to an important "move" on f.r.p.p. decomposition of dimension 3: If one has

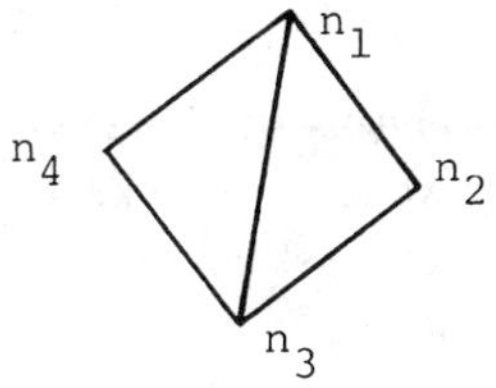

where $n_1 + n_3 = n_2 + n_4$, then we can replace it by:

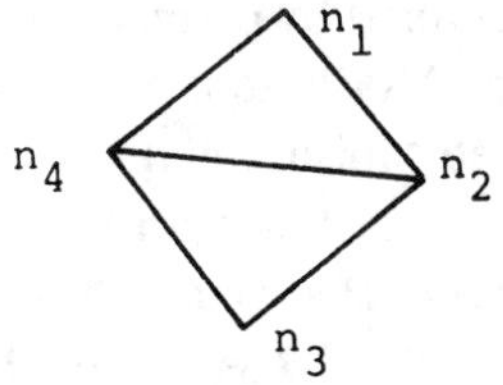

Note that the induced birational map from the first to the second f.r.p.p. decomposition is an isomorphism everywhere except at the $\mathbf{P}^1$'s corresponding to the central edge on either side. This type of phenomenon will be investigated further in §8.

The construction above is the toric version of a "modification of type II" (Kulikov [**Ku1**]).

EXAMPLE 2. Let $C \approx \mathbf{P}^1$ be embedded in a smooth threefold X with the normal bundle $(-1,-1)$. Blow up C: f_C: $\tilde{X} \to X$ with exceptional divisor E. E is isomorphic to $\mathbf{P}^1 \times \mathbf{P}^1$ with normal bundle $(-1,-1)$, so that E can be contracted along the other ruling to a curve C_1 in a smooth threefold X_1. Of course $C_1 \approx \mathbf{P}^1$ has normal bundle $(-1,-1)$. Therefore, at least as far as normal bundles are concerned we get a situation that is indistinguishable from Example 1. But in fact much more is true. By Grauert's contraction criterion [**Gr**] C can be blown down to a point P singular on a threefold Y and one can easily show that this singular point is *analytically* isomorphic to the ordinary double point of Example 1 [**L**]. The morphism g: $X \to Y$ is the blow up of an analytic ideal I, the ideal of an analytic Weil divisor of Y, i.e. an element of the analytic local Picard group of Y at P, which is $\approx \mathbf{Z}$. If this element is not in the *algebraic* local Picard group, then $X \to Y$ is not algebraic. As we shall see later (Example 3 and §6) it is possible for the algebraic local Picard group to be trivial (i.e. Y is algebraically factorial [**Sa**]), in which case the exceptional divisor C of $X \to Y$ meets all algebraic divisors D of X trivially: $C \cdot D = 0$. Indeed, let $\bar{D}$ be the image of D in Y. If the support of D meets C, then $\bar{D}$ goes through P. Since $\bar{D}$ is algebraic, by hypothesis $\bar{D}$ is a Cartier divisor. But then $D = g^{-1}(\bar{D})$, so $D \cdot C = 0$ as desired. In this case X is not an algebraic variety.

REMARK. This discussion should be compared to the theorem of Van der Waerden which says that if f: $X' \to X$ is a birational morphism with X locally factorial, then E_f is a divisor [**Mu2**, Proposition 1, p. 415]. The analogous theorem holds in the analytic category.

4. Examples of birational morphisms in dimension 3. We now give a series of examples, in part to explain why we did not make a stronger conjecture in §2.

The most naive question one can ask about factorization in dimension 3 is: Given a proper birational morphism f: $X \to Y$ of smooth threefolds, can one factor it by permissible blow downs, as in the dimension 2 case, without allowing blow ups? The answer is no, the first obstructions being:

(i) Not all proper birational morphisms between smooth threefolds are projective (but any one that can be factored by permissible blow downs is).

(ii) Not all smooth compact Moishezon threefolds are algebraic (a complex compact analytic variety is called Moishezon if its field of meromorphic functions has transcendence degree equal to its dimension).

Here is an example due to Hironaka which illustrates (ii). This example and a related one illustrating (i) are given in [**H**, Appendix B, 3.4.1 and 3.4.2].

EXAMPLE 3. Let C be an irreducible curve with only one ordinary double point P as singularity in a smooth threefold Y. Cover a neighborhood of C in Y by two analytic opens U_1 and U_2: U_1 is a small neighborhood of P, and U_2 any open not containing P. In U_2, C is smooth, so just blow it up. We get a proper map $\tilde{U}_2 \to U_2$.

On U_1, C consists of two smooth analytic branches C_1 and C_2. Blow up C_1 first, then C_2, and call the composition $\tilde{U}_1 \to U_1$. On $U_1 \cap U_2$ these two morphisms can be patched giving a birational morphism $X \to Y$, X smooth with irreducible exceptional divisor E which is neither $\mathbf{P}^2$ nor a minimal ruled surface, so that there is nothing permissible to contract.

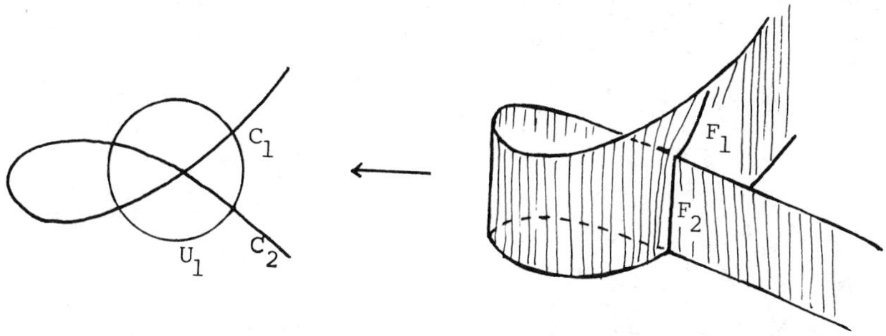

The fibre F_2 above the branch C_2 of C at P is irreducible, but the fibre above the branch C_1 at P has 2 irreducible components F_1 and F_2. Since the fibres of the nonminimal ruled surface E are algebraically equivalent we have $F_2 \sim F_1 + F_2$ so $F_1 \sim 0$ ($\sim =$ algebraic equivalence). Therefore X cannot be algebraic, in particular not projective. However, if we blow up F_1, getting a smooth $\tilde{X}$, the composite $\tilde{X} \to Y$ is projective, as one sees by first contracting the $\mathbf{P}^1 \times \mathbf{P}^1$ to a point. This example can be described toroidally.

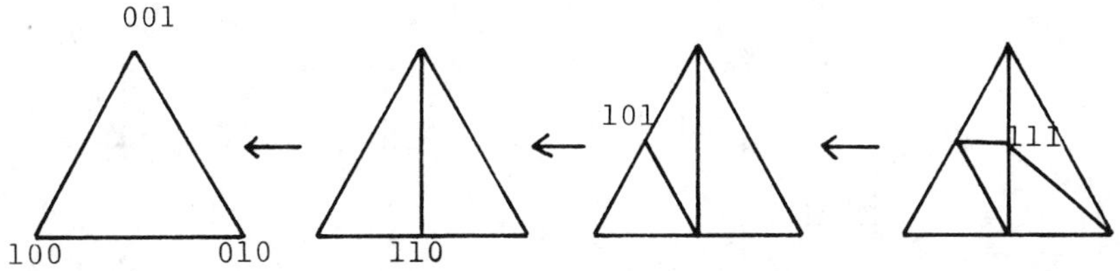

graph of $\mathbf{C}^3$ | blow up of one of the coordinate axes | blow up of a second axis | blow up of F_1

If the order of blow ups is interchanged one gets:

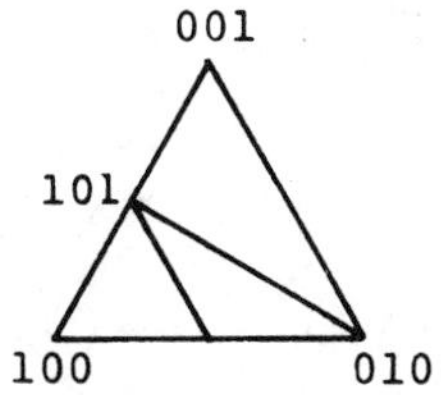

So we may ask: Suppose f is a morphism between projective varieties. Can f be factored by permissible blow downs? The answer is again no. Here is the first example of this, due to Hironaka [**Hi**].

EXAMPLE 4. Let C be a curve in a smooth threefold Y with a singular point at P consisting of 3 smooth branches C_1, C_2 and C_3 with distinct directions. Blow up

P, let $E \approx \mathbf{P}^2$ be the exceptional divisor, $\tilde{C}_i$ the proper transform of C_i, and L_{ij} the line on E joining $\tilde{C}_i \cap E$ and $\tilde{C}_j \cap E$, then blow up the proper transforms of the branches. Finally perform a modification of type II at each of the proper transforms of the L_{ij} (it is easy to see they have the right normal bundle, namely $(-1,-1)$), and call the resulting space X.

EXAMPLE 4

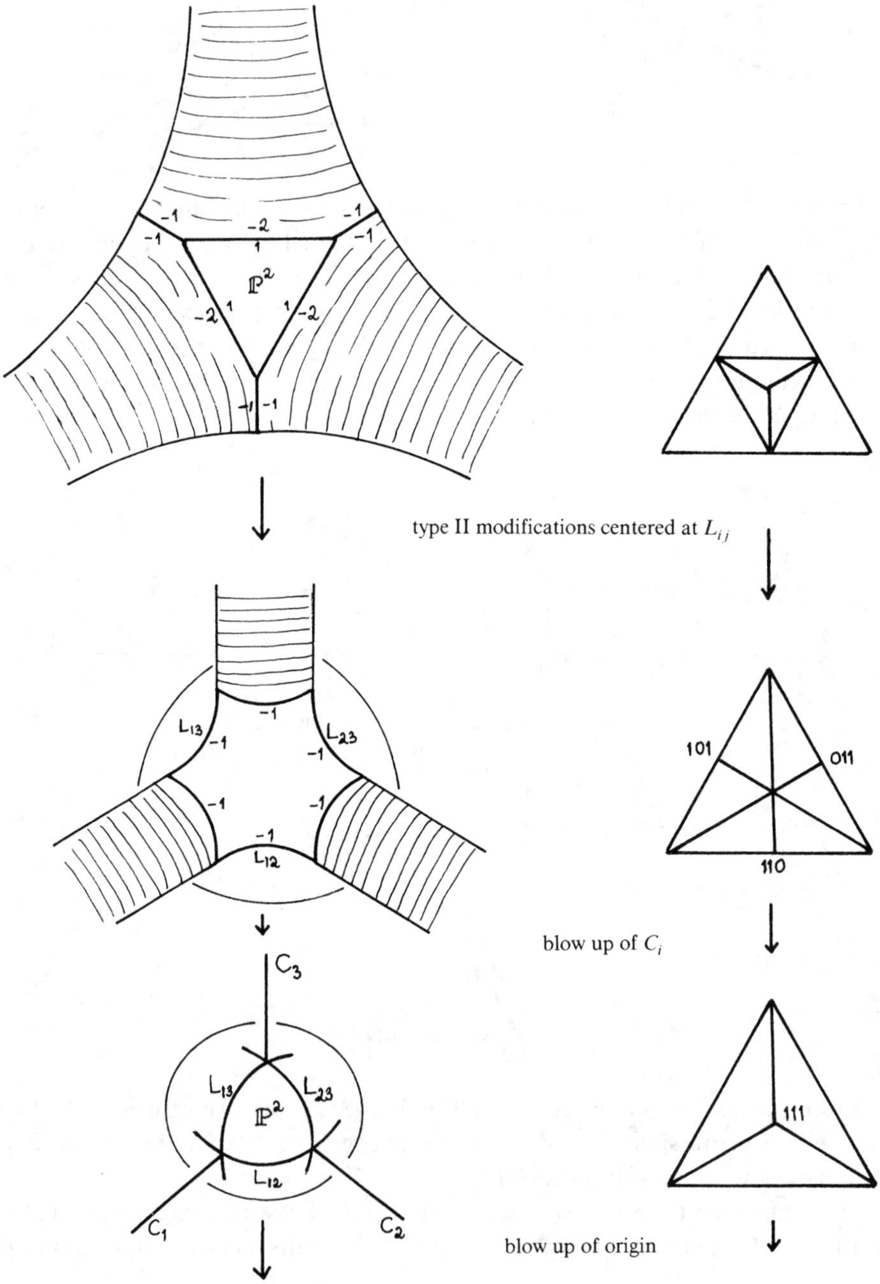

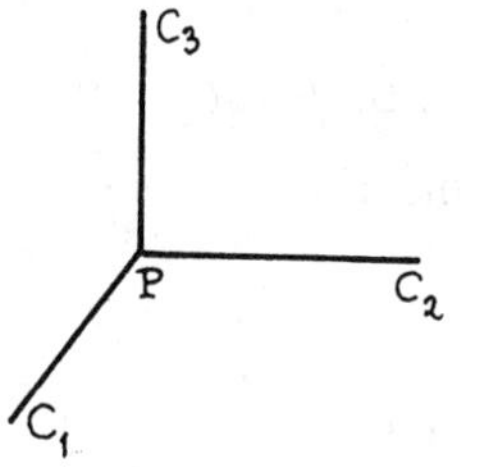

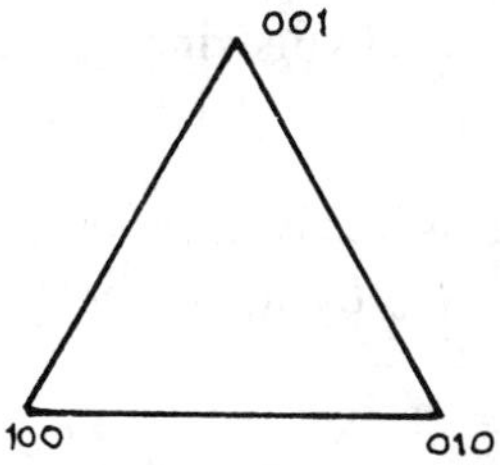

It remains to show that the morphism $X \to Y$ is projective. Hironaka did this by exhibiting an ideal sheaf on C whose blow up gave X. It can also be done by torus embedding theory. Here is a different proof. First blow up the ideal of C at P. In suitable analytic coordinates this is the ideal (xy, xz, yz). A computation (a similar computation is carried out in Example 6 below) shows that the exceptional divisor of the blow up consists of an $E \approx \mathbf{P}^2$ above the point P and three ruled surfaces E_i projecting to the branches C_i. The total space has 3 singular points on E, and E is not Cartier at these points.[2] So now blow up with center E. The resulting space is X, hence the map $X \to Y$ is projective, since it is a composition of blow ups.

For future reference note that the component of the exceptional divisor of $X \to Y$ above the origin is a $\mathbf{P}^2$ with normal bundle $\mathcal{O}_{\mathbf{P}^2}(-2)$.

It is amusing to note that the blow up of the relatively complicated ideal (xy, xz, yz), which is not even Gorenstein, gives a total space X with isolated singularity. One might ask: For what reduced curves C does the total space of the blow up of C have only isolated singularities?

EXAMPLE 5. If C is locally a complete intersection (in 3 space) it is easy to answer this question: only when C is locally a plane curve, i.e., its ideal can be written in suitable analytic coordinates $(f(x, y), z)$. Then the blow up of this ideal has a total space with one singular point with equation $zw + f(x, y) = 0$, as is seen by direct computation. Note that this singularity is *canonical* in the sense of Reid [**R1**]; indeed, the generic hyperplane section is a rational double point of the form A_{m-1}, where m is the multiplicity of $f(x, y) = 0$. It seems that all these singularities will be needed to continue Mori's program (§6).

EXAMPLE 6. Here is another example of a curve singularity in $\mathbf{C}^3$, not a complete intersection, such that the blow up of the ideal yields a total space with isolated singularities. The curve C is given parametrically by

$$\begin{Bmatrix} x = t^3 \\ y = t^4 \\ z = t^5 \end{Bmatrix},$$

so that its ideal is generated by 2×2 minors of the matrix

$$\begin{pmatrix} x & y & z \\ y & z & x^2 \end{pmatrix}.$$

[2] These points are ordinary double points (Example 1).

Therefore the blow up of the ideal has projective coordinates U_0, U_1, U_2 over $\mathbf{C}[x, y, z]$ satisfying

$$xU_0 + yU_1 + zU_2 = 0, \qquad yU_0 + zU_1 + x^2U_2 = 0.$$

This is easily seen to be smooth except in the chart where $U_2 \neq 0$. Setting $u_0 = U_0/U_2$, $u_1 = U_1/U_2$ we have

$$\begin{Bmatrix} xu_0 + yu_1 + z = 0 \\ yu_0 + zu_1 + z^2 = 0 \end{Bmatrix},$$

in other words, after eliminating z we obtain the hypersurface singularity P',

$$yu_0 - (xu_0 + yu_1)u_1 + x^2 = 0.$$

(This is analytically isomorphic to A_5: $x^2 + y^2 + z^2 + w^6 = 0$.) Note that the exceptional divisor has 2 components: a ruled surface over C and a $\mathbf{P}^2$ over the origin in $\mathbf{C}^3$. The ideal of the $\mathbf{P}^2$ at the singular point P' is (x, y), so blow it up (since the $\mathbf{P}^2$ is Cartier at all other points, blowing it up will have no effect except at P'). This replaces the singular point P' by a $\mathbf{P}^1$, and the new ambient space $\tilde{X}$ is smooth. The proper transform of the $\mathbf{P}^2$ is again $\mathbf{P}^2$, and its normal bundle is $\mathcal{O}(-2)$. As in Example 4 the only exceptional divisor of the morphism on Mori's list (§6) is the $\mathbf{P}^2$. Note that the factorization strategy we have adopted in these two examples is *not* the one given by Mori.

5. Morphisms with an exceptional divisor with a small number of components: results of Schaps and Crauder. In this section we will assume the morphism f: $X \to Y$, where as usual X and Y are smooth threefolds, and f is proper and birational, is in normal form; that is, the exceptional divisor of f has smooth irreducible components E_i, $1 \leqslant i \leqslant n$, meeting normally. By the results of Hironaka this can be achieved by performing permissible blow ups on X, and this of course does not affect the conjecture stated in §2.

Crauder and Schaps have recently classified morphisms of f in normal form such that $n \leqslant 3$, and the indeterminacy locus S of f^{-1} is either a smooth curve (Schaps [**S**]) or a point (Crauder [**C1**]). Very roughly the proofs go as follows. Write $K_X = f^*K_Y + \sum r_i E_i$ where the r_i are positive integers. A local study of K_X due to Schaps, together with the use of "test curves", permits the authors to determine all possible combinations of r_i that can arise. ("Test curves" are analytic branches Γ in Y meeting S in one point. The proper transform $\overline{\Gamma}$ of Γ in X intersects E in one point. As Γ varies, so does the intersection point, thus giving information about f.) A case by case study of the possible combinations of the r_i (and s_i, defined in §1) then allows them to classify all possible morphisms that occur. In Schaps' case this is a fairly easy consequence of a result of Danilov [**D2**]; Crauder goes through a long and delicate analysis of the E_i, using the type of "factorization from the top down" argument mentioned in connection with the 2-dimensional case in §1. This type of argument had already been used by Persson [**P**] and Kulikov [**Ku1**] to study families of surfaces with $\kappa = 0$; unfortunately in the dimension 3 birational morphism situation, unlike in dimension 2,

Crauder was not able to set up an inductive procedure that would have permitted him to handle a larger number of components. It seems clear that to push his method through for larger n, a new idea will be required.

Note that both authors were able to show that when $n \leqslant 3$ f factors through the blow up of the indeterminacy locus of f^{-1} (which in the two cases we are considering is smooth). Therefore these morphisms can be simplified by factorization "from the bottom up" (see §1). However, when $n = 4$ in neither case does f necessarily factor through the blow up of S, even when f is projective.

EXAMPLE 7 (CRAUDER [**C2**]). Blow up a point P on a smooth threefold Y, then blow up a point P_1 on the exceptional divisor E_1 of the first blow up. Let L be a line on the exceptional divisor E_2 of the second blow up, and let P_2 be the point of intersection of L with the intersection curve D_{12} of the two exceptional divisors.

$L \approx \mathbf{P}^1$ has normal bundle $(0, -1)$, so it blows up to a ruled surface $E_3 \approx \mathbf{F}_1$. There now is a curve L_1 on E_1, $L_1 \approx \mathbf{P}^1$ meeting E_2, with self-intersection -1 on E_1. Its normal bundle is $(-1, -2)$. Blow it up: the exceptional divisor E_4 is $\approx \mathbf{F}_1$. Blow the curve L_1 up again: it now has normal bundle $(-1, -1)$, so that by Example 2 one can contract it in the other direction obtaining a smooth threefold X which maps to Y: X factors *generically* through the blow up of P, the generic point of E_1 going to the generic point of the exceptional divisor, but obviously cannot factor completely since $\mathbf{F}_2$ does not dominate $\mathbf{P}^2$.

EXAMPLE 7

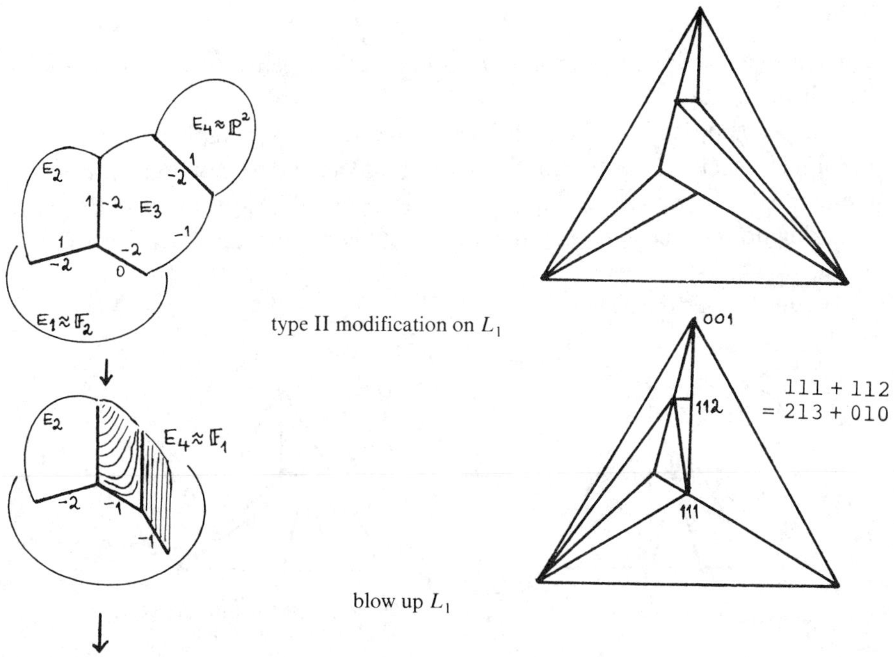

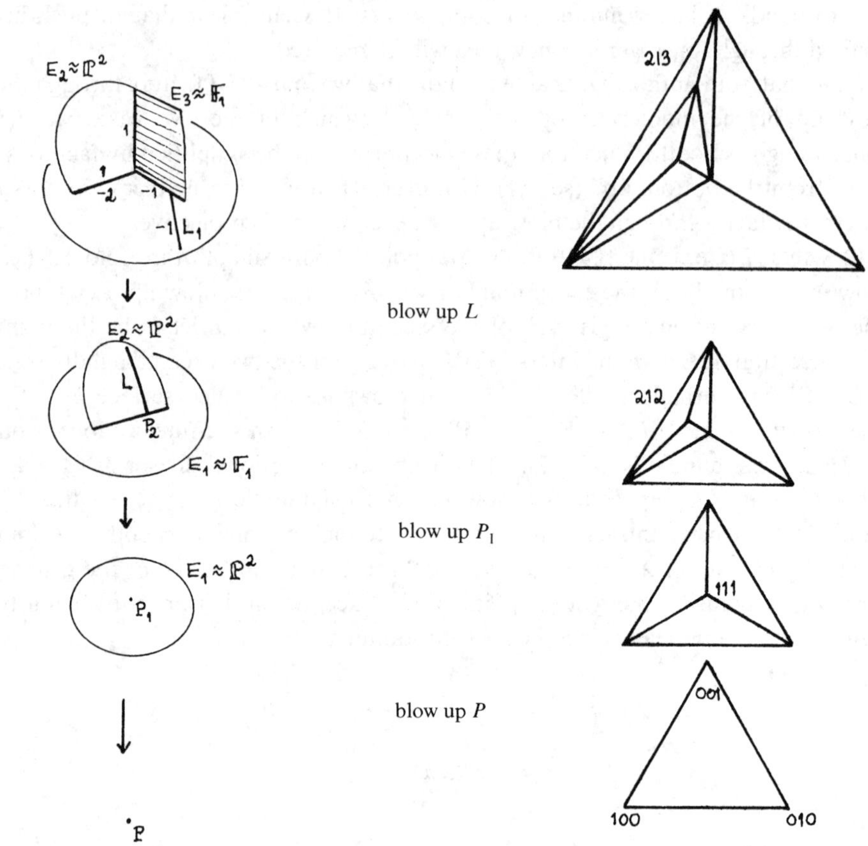

The toric description of f: $X \to Y$ shows it is projective (it is Example 8.10 of [**O**, Theorem 9.6]).

A slight modification of Crauder's example gives an example where $n = 4$ and the singular locus of f is a smooth curve C, and yet f does not factor through the blow up of C. Strangely enough, since this example is so similar to Example 7, according to [**O**, Theorem 9.6], this morphism is not projective; it is his Example 8.5.

EXAMPLE 8. See diagram.

EXAMPLE 8

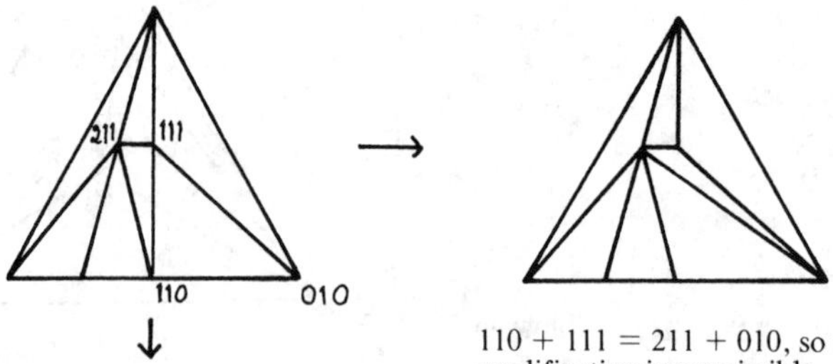

110 + 111 = 211 + 010, so type II modification is permissible

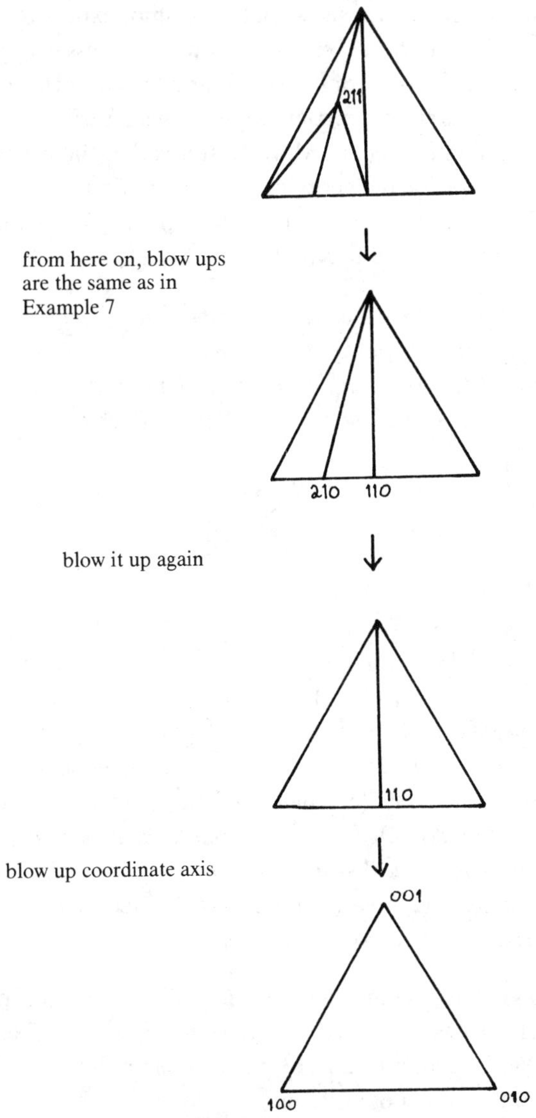

REMARKS. (1) Crauder [**C2**] has proved the main conjecture in the case f is in normal form and its exceptional divisor has no triple points.

(2) Crauder [**C2**] and Kulikov [**Ku2**] have independently reduced the main conjecture to the case where the indeterminacy locus of f^{-1} consists of points. Unfortunately this hypothesis is not very stable, so it is not clear how useful this result will be.

6. Results of Mori in the projective case. Mori only considers morphisms between projective varieties; his strategy is to factor "from the top down" (cf. the second method outlined in §1) without allowing blow ups, by finding an irreducible component of the exceptional divisor to contract. Many of the examples we have constructed in this paper show that we cannot expect to contract the divisor to a smooth threefold; in other words, we will not necessarily be able to find a $\mathbf{P}^2$ with normal bundle $\mathcal{O}_{\mathbf{P}^2}(-1)$ or a contractible ruled surface in the exceptional divisor, so that the first contraction will not necessarily be a permissible blow down. The astounding fact about Mori's result is that the list of divisors one must allow in order to contract algebraically without first blowing up is extremely short. Here is the result. The proof uses Mori's theory of extremal rays and is too involved to be even sketched here.

THEOREM (S. MORI [**M**]). *Let $f\colon X \to Y$ be a nontrivial birational morphism of nonsingular projective varieties of dimension* 3. *Then one can find a divisor E in the exceptional locus of f, which can be contracted by a birational morphism $f'\colon X \to X'$ to a projective variety X' and such that f factors through f'*:

$$\begin{array}{ccc} X & \xrightarrow{f'} & X' \\ & \searrow^{f} & \downarrow \\ & & Y \end{array}$$

The possibilities for E and X' are

(1) *f' is a permissible blow down.*

(2) *$E \approx \mathbf{P}^1 \times \mathbf{P}^1$, $N_E \approx (-1,-1)$, and $f'(E)$ is an isolated singular point on X', analytically isomorphic to $x^2 + y^2 + z^2 + w^2 = 0$.*

(3) *$E \approx$ quadratic cone in $\mathbf{P}^3$, $N_E \approx \mathcal{O}_E \otimes \mathcal{O}_{\mathbf{P}^3}(-1)$ and $f'(E)$ is an isolated singular point on X' analytically isomorphic to $x^2 + y^2 + z^2 + w^3 = 0$.*

(4) *$E \approx \mathbf{P}^2$, $N_E \approx \mathcal{O}_{\mathbf{P}^2}(-2)$, $f'(E)$ is an isolated singular point on X', analytically isomorphic to the cone of the Veronese surface.*

Furthermore in cases (2) *and* (3) *the algebraic local Picard group of the singular point of X' is trivial; in case* (4) *it is* $\mathbf{Z}/2\mathbf{Z}$.

REMARKS. (1) In case (4) the algebraic local Picard group cannot be trivial since the dualizing sheaf has order 2 in it. As we have seen in Examples 1 and 2 the analytic local Picard group of an ordinary double point is *not* 0, so the condition on the algebraic Picard is not vacuous.

(2) To decide whether a $\mathbf{P}^1 \times \mathbf{P}^1$ with $N \approx (-1,-1)$ should be contracted to a curve (see (1)) or a point (case (2)) one uses the criterion: If f_1 and f_2 are fibres for the two rulings of $\mathbf{P}^1 \times \mathbf{P}^1$, and if $f_1 \cdot D = f_2 \cdot D$ for all divisors D on X then we are in case (2); otherwise we are in case (1). Compare with Example 2 and the Remarks in §3.

(3) As X' is projective we see that the only hypothesis that is not preserved after the blow down of E is that of smoothness. However, X' is only mildly singular: the singular points are isolated canonical singularities, and there is only one case of a Weil divisor class which is not Cartier. So there is some hope of continuing the program to X'. The class of singularities that will have to be considered are probably the terminal singularities of Reid [**R3**].

(4) We will now construct birational morphisms where the only component that is on Mori's list is of type (2), (3) or (4). To get an example of (2), blow up an irreducible plane curve with an ordinary double point; by Example 5 we get a threefold with an ordinary double point. Resolving it we are in case (2). Note that we cannot be in case (1) since the contraction of $\mathbf{P}^1 \times \mathbf{P}^1$ to one of the rulings gives us Example 3, where Y is not projective. To get an example of (3) blow up a plane curve with a cusp (local analytic equation $x^2 + y^3 = 0$). We then get a threefold with an isolated singularity with analytic equation $z^2 + w^2 + x^2 + y^3 = 0$ (Example 5 again). In both cases by construction it is clear the singularities are algebraically factorial. These examples are due to Mori.

We have already seen examples of case (4): Examples 4, 6, 7 and 8.

7. Work of V. S. Kulikov.

THEOREM [**Ku2**]. *Let f: $X \dashrightarrow Y$ be a birational map of smooth proper algebraic threefolds. Then there exists a diagram*

$$\begin{array}{ccc} X_n & \overset{f'}{\dashrightarrow} & Y_m \\ \downarrow & & \downarrow \\ \vdots & & \vdots \\ \downarrow & & \downarrow \\ X_1 & & Y_1 \\ \downarrow & & \downarrow \\ X & \overset{f}{\dashrightarrow} & X \end{array}$$

where the vertical arrows are permissible blow downs, such that the induced map f': $X_n \dashrightarrow X_m$ is an isomorphism in codimension 1; *that is, there exist irreducible curves $C_i \subset X_n$ and $D_j \subset Y_m$ such that $X_n - \bigcup_i C_i$ is isomorphic to $Y_m - \bigcup D_j$. Furthermore the C_i and D_j are isomorphic to $\mathbf{P}^1$ and meet transversely.*

As already mentioned, Example 1 in §3 gives a birational map such as f'. More interesting examples are given in §8.

Let me give a grossly simplified idea of the proof. By results of Hironaka and Moishezon, after performing permissible blow ups on X and Y we may assume that X and Y are projective and f is a morphism. We can write $K_X = f^*K_y + \Sigma r_i E_i$, where E_i are components of the exceptional divisor. The proof will be by induction on the maximal value of the r_i and the number of components with the maximum value. Pick a component E_i such that r_i is maximal, and look at its image $f(E_i)$ in Y. It is either a point, a smooth curve or a singular curve. The difficult part of the proof is dealing with a singular curve; we will do that later. Suppose for the time being that $f(E_i) = C$ is smooth; blow it up: g: $\tilde{Y} \to Y$. As we have seen (Example 7, §5) there is no reason to expect f to factor through the blow up; but this only creates minor difficulties, so let us assume for simplicity it does. Then either E_i maps onto the exceptional divisor F of the blow up g, or not; if it does, then we have reduced the number of components of the exceptional divisor of f with maximal multiplicity r_i by one, so we are done by induction; if not, repeat the process. It is an easy exercise involving the local computation of the canonical divisor that this must stop in a finite number of steps, because the multiplicity of F in the canonical divisor of $\tilde{Y}$ (which at this first stage we are considering is one or two depending on whether $f(E_i)$ is a curve or a point) must be strictly smaller than r_i.

It remains to deal with the case where $f(E_i)$ is a singular curve C. Of course, we are not allowed to blow C up; on the other hand we do not want to first blow up the singular points of C since we could then lose control of the codimension one structure. Instead Kulikov looks for smooth curves L in Y going through one of the singular points of C and otherwise not intersecting $S_{f^{-1}}$. M, the proper image of L under f, is also smooth. Now blow up M on X and L on Y. The new exceptional divisor on X is not collapsed on Y, so we have not introduced any new codimension 1 phenomena, and yet by judicious choice of L we can eventually make it smooth.

So we have boiled down the proof to finding smooth curves L. To do this Kulikov needs the following key:

THEOREM. *Let f: $X \to Y$ be a proper birational morphism of smooth threefolds, with Y projective. Then for any x in the exceptional E_f of f, and any $\gamma \neq 0$ in the tangent space of X at x, there exists a smooth curve L in X passing through x with tangent direction γ and with no other intersection with E_f.*

To return to the first theorem of this section, notice that the method of proof systematically requires blow ups outside of singular locus of f^{-1}, and therefore the theorem is not a partial solution of the conjecture stated in §2. It would perhaps be worthwhile to explore in more detail the blowing up of singular curves: cf. Examples 3, 4, 5 and 6 in §4, in order to avoid blowing up curves outside of $S_{f^{-1}}$. This would require enlarging the class of permissible blow ups.

8. Birational maps that are isomorphisms outside of curves. In view of Kulikov's theorem stated in the previous section it is natural to try to classify birational maps f: $X \dashrightarrow Y$ between smooth threefolds such that f is an isomorphism on $X - \bigcup C_i$ to $Y - \bigcup D_j$ where the C_i and D_j are irreducible curves, which have rational normalization. I do not want to assume the C_i and D_j are smooth or meet transversally although this can be achieved by Kulikov's method, because to do so requires blowing up inside the domain of f, something I do not want to allow. We may as well assume that $\bigcup C_i$ and $\bigcup D_j$ are connected.

Once these maps are classified, one can attempt to factor them by permissible blow ups.

A complete description can be given in the following case:

(*) $\bigcup C_i$ can be contracted to an isolated singular point P on a Cohen-Macaulay threefold V.

The condition that the singular point be isolated is put in for simplicity of exposition only. The essential conditions are that the C_i's can be contracted, and that the resulting singular point is Cohen-Macaulay. It is an extremely interesting question to determine whether $(*)$ is implied by the other hypotheses or not.

At any rate $(*)$ implies that X and Y are simply two distinct resolutions of the singular variety V, such that the exceptional locus of both is one dimensional. Let us call any threefold singularity which admits a resolution with one-dimensional exceptional locus a *small* singularity and any such resolution a *small* resolution. Then $(*)$ amounts to classifying all isolated Cohen-Macaulay threefold singularities with at least two distinct small resolutions.

Let us first consider any isolated Cohen-Macaulay singularity V with small resolution $\pi: X \to V$. The following analysis is given in [**L**]. Let $\mathcal{I}$ be the ideal sheaf of any subvariety of Z of X supported on the exceptional locus. The exact sequence

$$(1) \qquad 0 \to \mathcal{I} \to \mathcal{O}_X \to \mathcal{O}_Z \to 0$$

and the vanishing of $R^1\pi_*\mathcal{O}_X$ (V is Cohen-Macaulay) and $R^2\pi_*\mathcal{I}$ (X is a small resolution) imply that $H^1(Z, \mathcal{O}_Z) = 0$. Therefore we have

PROPOSITION 1. *The irreducible components C_i of the exceptional locus are smooth rational curves, meeting transversally with no cycles.*

PROOF. Same as Artin's in the two-dimensional case [**A2**].

Now take for Z a reduced, irreducible component C of the exceptional locus. We have just seen that $C \approx \mathbf{P}^1$, so $\mathcal{I}/\mathcal{I}^2$ can be written $\mathcal{O}(a) \oplus \mathcal{O}(b)$, since all vector bundles on $\mathbf{P}^1$ split [**H**, V.2.14].

PROPOSITION 2. $a + b \geqslant 2$, *and* $a \geqslant -1$, $b \geqslant -1$.

PROOF. The exact sequence (1) shows that $R^1\pi_*\mathcal{I} = 0$. Now use the exact sequence

$$(2) \qquad 0 \to \mathcal{I}^2 \to \mathcal{I} \to \mathcal{I}/\mathcal{I}^2 \to 0.$$

We have just seen that $R^1\pi_*\mathcal{I} = 0$; $R^2\pi_*\mathcal{I}^2 = 0$ since X is a small resolution. Therefore $H^1(C, \mathcal{I}/\mathcal{I}^2) = 0$ so that both a and b are $\geqslant -1$. On the other hand since C is a component of the exceptional locus of a singularity, its Hilbert scheme in X is 0 dimensional. Thus $\chi(C, N_{X/C}) \leqslant 0$. Since $N_{X/C} \approx \mathcal{O}(-a) \oplus \mathcal{O}(b)$ we get, by Serre duality on $\mathbf{P}^1$, $a + b \geqslant 2$. Q.E.D.

Laufer [**L**] also exhibits small singularities that are not Cohen-Macaulay. Note that for small singularities, Cohen-Macaulay is equivalent to rational.

Let us now proceed with the classification of $(*)$.

PROPOSITION 3. *V is Gorenstein.*

The proof uses the following theorem, essentially due to Lipman [**Li2**]. The key ingredients are 12.1 and 7.4 of [**Li1**].

THEOREM 1. *Let V be a Cohen-Macaulay threefold with isolated singular point P and small resolution π: $X \to V$. Let $\overline{V}$ be the blow up of the dualizing sheaf ω_V. Then $\overline{V}$ is Gorenstein with isolated singularities, and π factors as $X \xrightarrow{\overline{\pi}} \overline{V} \to V$. The exceptional curves for $\overline{\pi}$: $X \to \overline{V}$ are precisely those for which $a + b = 2$.*

PROOF OF PROPOSITION 3. Applying Theorem 1 to both X and Y we obtain the diagram:

$$\begin{array}{ccc} X & \overset{f}{\dashrightarrow} & Y \\ & {\scriptstyle\overline{\pi}}\searrow \quad \swarrow & \\ & \overline{V} & \\ & \downarrow & \\ & V & \end{array}$$

Now if any curve C_i in X is not exceptional for $\overline{\pi}$, then this diagram shows that f is defined at its generic point, a contradiction to our original assumption. Therefore $\overline{V} \approx V$ and V is Gorenstein.

Questions. (1) What happens in Theorem 1 if we drop the hypothesis that V is Cohen-Macaulay? (Blow up $i_*\omega_u$, where $U = V - P$ and i: $U \hookrightarrow V$ is the inclusion.)

(2) In Theorem 1, is the generic hyperplane section of V rational?

Now that we know that V is Gorenstein, we can apply a result of Miles Reid [**R3**, 1.1 and 1.14].

THEOREM 2. *Let V be an isolated Gorenstein threefold singularity and π: $X \to V$ a small resolution. Then the generic hyperplane section H through the singular point is a rational double point, and its proper transform $\tilde{H}$ on X has isolated singularities, and hence is normal, and is dominated by the minimal resolution of H.*

Therefore we can view V as the total space of a one parameter family of deformations of its generic hyperplane H, and the small resolution X as the total space of a one parameter family of deformations of a partial resolution $\tilde{H}$ of H. We are thus in the presence of the phenomenon of simultaneous partial resolution of rational double points. We will now study this systematically, at the level of the versal deformation spaces. All the necessary ingredients for this study are found in Brieskorn's beautiful paper [**B1**, §3]. To do this universally we need some extra results from [**A2** and **BW**]. A similar set up has been studied by Wahl [**W1, 2**] but unfortunately his results do not quite apply to our situation.

So let H be a rational double point. Its minimal resolution $\overline{H}$ has as dual graph a Dynkin diagram, and therefore we can associate to H a Weyl group W_n (which is A_n, D_n or E_6, E_7 or E_8); n = number of irreducible curves in the exceptional divisor. Let $\tilde{H}$ be the partial resolution of H obtained from the minimal resolution $\overline{H}$ by contracting any set of r exceptional curves. Let W_r be the Weyl group associated to these exceptional curves, and W_{n-r} the Weyl group associated to the remaining, noncontracted curves. In general neither W_r nor W_{n-r} will be irreducible.

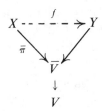

EXAMPLE. Take for H the rational double point E_8:

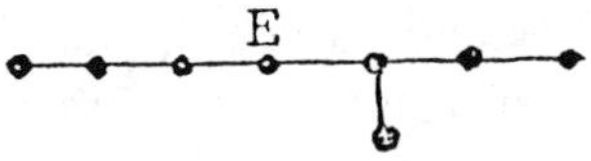

and let $\tilde{H}$ be the partial resolution obtained by contracting all curves except E. Then $n = 8$, $r = 7$, $W_r = A_3 \times A_4$, $W_{n-r} = A_1$.

Let $\mathfrak{X} \to R$ (resp. $\mathfrak{Y} \to S$, resp. $\mathfrak{Z} \to T$) be a versal deformation of H (resp. $\tilde{H}$, resp. $\overline{H}$). R, S, and T are complex analytic germs with closed point 0. We have maps $T \to S \to R$. The fundamental result of Brieskorn and Tjurina [**B1** and **T**] (see also [**Pi1**]) shows that $T \to R$ is Galois with group W_n. Similarly (here we must use a minor refinement of a result of [**BW**], given in [**Pi2**]), $T \to S$ is Galois with group W_r. In general $S \to R$ will *not* be Galois.

However, we can count the number of "simultaneous partial resolutions" (a precise definition is given later) of $\mathfrak{X}_S = \mathfrak{X} \times_R S$, by generalizing a result of Brieskorn [**B1**, Satz 0.3].

We need some results from [**B1**] on the (analytic) divisor class group $\text{Cl}(\mathfrak{X}_S)$ of $\mathfrak{X}_S$, in order to state the result.

Step 1. Since $f \colon \mathfrak{Y} \to \mathfrak{X}_S$ is a resolution which does not contract any divisors, $\text{Cl}(\mathfrak{X}_S) \approx \text{Pic}(\mathfrak{Y}) \approx R^1 f_* \mathcal{O}_{\mathfrak{Y}}^*$.

Step 2. Since $\mathfrak{X}_S$ is rational, $R^i f_* \mathcal{O}_{\mathfrak{Y}} = 0$, $i > 0$, so that $R^1 f_* \mathcal{O}_{\mathfrak{Y}}^* = R^2 f_* \mathbf{Z}$. Combining with Step 1, we get $\text{Cl}(\mathfrak{X}_S) \approx R^2 f_* \mathbf{Z}$.

Step 3. Any simultaneous partial resolution of $\mathfrak{X}_S$ is the blow up of an element $\mathfrak{p}$ of $\text{Cl}(\mathfrak{X}_S)$. This is proved in Brieskorn [**B1**, 3.1]. For more details see [**Li1**].

For $\mathfrak{p} \in \text{Cl}(\mathfrak{X}_S)$ let $C(\mathfrak{p})$ be the characteristic cone of $\mathfrak{p}$, i.e. the cone of elements $\mathfrak{p}' \in \text{Cl}(\mathfrak{X}_S)$ whose blow up give the same birational map as the blow up of $\mathfrak{p}$. The rest of the proof consists in studying the partition of $\text{Cl}(\mathfrak{X}_S)$ into characteristic cones.

Step 4. Consider the hyperplane sections

$$\overline{H} \xrightarrow{} \tilde{H} \xrightarrow{f_0} H$$
$$\underbrace{\phantom{\overline{H} \to \tilde{H} \to H}}_{g_0}$$

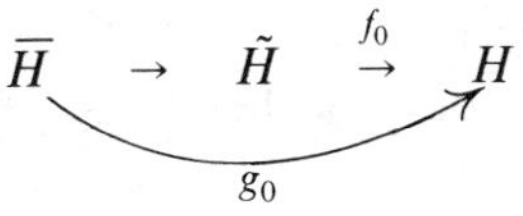

(f_0 is the restriction of $f \colon \mathfrak{Y} \to \mathfrak{X}_S$ to the fibre above the closed point; g_0 that of $g \colon \mathfrak{Z} \to \mathfrak{X}_T$).

Since the singularities of H and $\tilde{H}$ are rational, we have

$$R^1 f_{0*} \mathcal{O}_{\overline{H}}^* \approx R^2 f_{0*} \mathbf{Z} \approx \mathbf{Z}^{n-r}, \qquad R^1 g_{0*} \mathcal{O}_{\overline{H}}^* \approx R^2 g_{0*} \mathbf{Z} \approx \mathbf{Z}^n.$$

Let us call this second space M. Since $\overline{H}$ is smooth, M carries a negative definite bilinear symmetric form, the intersection form of $\overline{H}$. It is easy to see that the first space, call it N, is just the subspace of M orthogonal to the classes in M of the curves contracted by $\overline{H} \to \tilde{H}$. For convenience let us label the exceptional curves of $\overline{H}$ C_1 through C_n, so that the ones contracted by $\overline{H} \to \tilde{H}$ are $C_1, \ldots, C_r$.

Now M is clearly the group of weights for a root system of type W_n, and $C_1, \ldots, C_n$ (or rather the images of $\mathcal{O}_{\overline{H}}(C_i)$ in M, but we will abuse language) form a system of simple roots. N is the intersection, in M, of the walls corresponding to the roots $C_1, \ldots, C_r$.

The reflection hyperplanes of M which do not contain N divide $N \otimes_{\mathbf{Z}} \mathbf{R}$ into connected components. Let us call the intersection with N of each one of these connected components a *chamber* of N (even though N will not, in general, be a root system). The *fundamental chamber* for the set $C_1, \ldots, C_n$ of simple roots is that given by $L \cdot C_i < 0$, $i = r+1, \ldots, n$, $L \in N$ (by definition of N, $L \cdot C_i = 0$, $i = 1, \ldots, r$).

Step 5. An easy generalization of an argument of Brieskorn [**B1**] shows that

$$R^2 f_* \mathbf{Z} \approx R^2 f_{0*} \mathbf{Z}.$$

Combining Step 2 with Step 4, we can identify $\mathrm{Cl}(\mathcal{X}_S)$ with N. In the same way we can identify $\mathrm{Cl}(\mathcal{X}_T)$ with M.

By *simultaneous partial resolution* we mean a flat family $\mathcal{Y}' \to S$ such that there is a commutative diagram

$$\begin{array}{ccc} \mathcal{Y}' & \to & \mathcal{X}_S \\ \searrow & & \swarrow \\ & S & \end{array}$$

where, for all $s \in S$, the map on fibres $\mathcal{Y}'_s \to \mathcal{X}_{S,s}$ is a partial resolution of the rational double point $\mathcal{X}_{S,s}$ dominated by the minimal resolution. We do *not* require that $\mathcal{Y}'_0$ be $\tilde{H}$.

Of course $\mathcal{Y} \to S$ itself is one such simultaneous partial resolution.

We can now state the result.

THEOREM 3. *The number of simultaneous partial resolutions of* $\mathcal{X}_S = \mathcal{X} \times_R S$ *is* $\psi(\tilde{H})$, *the number of chambers of* N.

Recall that $\psi(\tilde{H})$ = number of connected components of $N \otimes \mathbf{R}$ − {reflection hyperplanes of $M \otimes \mathbf{R}$ which do not contain N}.

Step 6. The following claim, combined with Step 3, finishes the proof of Theorem 3.

Main claim. Each chamber of N is the characteristic cone of a simultaneous partial resolution of $\mathcal{X}_S$.

Note that Brieskorn proved this for $\mathcal{X}_T = \mathcal{X} \times_R T$. As in his case it is quite easy to see that the characteristic cone for any simultaneous partial resolution obtained by the blow up of a divisor class $\mathfrak{p}$ is the chamber containing $\mathfrak{p}$. To show that every chamber in N is the characteristic cone of a simultaneous partial resolution, use Brieskorn's theorem on $\mathcal{X}_T$: Take a chamber in M whose closure contains the given chamber in N, and take the simultaneous resolution $\mathcal{Z}' \to \mathcal{X}_T$ associated to the chamber by Brieskorn's result. Now just consider the effect of blowing up a divisor class in the given chamber in N (which we consider as an element in $\mathrm{Cl}(\mathcal{X}_T)$ by $N \hookrightarrow M$) on $\mathcal{X}_T$ to get the result. This concludes the proof of Theorem 3.

We can now give an answer to the classification problem (*). Let B be the germ of a smooth curve in S, not contained in the image of the singular locus of $\mathcal{X}_S$.

Let V be the inverse image of B under the map $\mathcal{X}_S \to S$. Finally assume that the inverse image X of B under $\mathcal{Y} \to S$ is smooth (this will be true for general B). By construction V is a Gorenstein threefold with small resolution X, and by Theorem 3, V has exactly $\psi(\tilde{H})$ distinct small resolutions. Moreover by M. Reid's results and our earlier analysis all singularities as in $(*)$ arise this way.

Although we have a construction for the singularities that arise, we do not really have a classification. For example, the construction does not explain how to get the generic hyperplane of V, which of course is a rational double point, from the original rational double point H. It would also be useful to have a method of factoring the birational map f into permissible blow ups and blow downs using "reflection group" data. D. Morrison, R. Miranda and I have worked this out in a few cases by *ad hoc* methods (see the examples below). Actually Miranda's example, which is too complicated to be given here, suggests that this may not be a worthwhile line of inquiry.

We conclude with a few additional remarks in the case where the exceptional locus C of the small resolution $\pi\colon X \to V$ is irreducible, and hence just $\mathbf{P}^1$. In this case Theorem 3 tells us that V has exactly two small resolutions which are either both algebraic or both nonalgebraic: the algebraic local Picard group of V injects into the analytic one, and the ideals $\mathfrak{p}$ and $\mathfrak{p}'$ corresponding to $\pi\colon X \to V$ and the other resolution $Y \to V$ are inverses of one another in these groups.

We now define the *sequence of normal bundles* associated to C. Since $C \approx \mathbf{P}^1$ its normal bundle splits and can be written $\mathcal{O}_{\mathbf{P}^1}(a_0) \oplus \mathcal{O}_{\mathbf{P}^1}(b_0)$, $a_0 \geqslant b_0$, which we write (a_0, b_0) for short. If $a_0 = b_0$, then we stop. Otherwise blow up C. The exceptional divisor is a natural ruled surface $\mathbf{F}_n$ with $n > 0$, and thus has a unique section C_1 with negative self-intersection with normal bundle (a_1, b_1), $a_1 \geqslant b_1$. If $a_1 = b_1$, stop. Otherwise repeat the process. The sequence $(a_0, b_0), (a_1, b_1), \ldots$ is called the sequence of normal bundles. It gives some information about the infinitesimal structure of X near C, much less than that required for factoring f, however. Our last result is

THEOREM 4. *The possibilities for the sequence of normal bundles are*

(1) $(-1,-1)$,

(2_t) $\underbrace{(0,-2),\ldots,(0,-2)}_{t \text{ times}},\ (-1,-1)$,

(3) $(1,-3),\ (-1,-2),\ (-1,-1)$,

(4) $(1,-3),\ (0,-3),\ (-1,-2),\ (-1,-1)$,

(5) $(1,-3),\ (0,-3),\ (0,-3),\ (-1,-2),\ (-1,-1)$,

and all these possibilities in fact occur.

This list should be compared to [**R3**, § 5; and **Wi**].

We now give examples illustrating cases (1), (2_t) and (3). Miranda's example falls in (4). Case (5) only arises from contracting on the double point E_8 as in the example following Theorem 2.

EXAMPLE 9. Take the ordinary double point $x^2 + y^2 + z^2 = 0$. Its versal deformation is $x^2 + y^2 + z^2 + s = 0$, so $S = \operatorname{Spec} \mathbf{C}[[s]]$. W is the group of order 2. If $T = \operatorname{Spec} \mathbf{C}[[t]]$ it acts on t by $t \to -t$, so that $s = t^2$. In this case the only choice for B is T itself, so that $\mathcal{X}_B = \{x^2 + y^2 + z^2 + t^2 = 0\}$. This is the ordinary (threefold) double point, and its two small resolutions obtained by the recipe above are those already computed in Example 1 of §3.

Now make the base change $\tilde{B} \to B$ given by $t = u^n$. Then $\mathcal{X}_{\tilde{B}} = \{x^2 + y^2 + z^2 + u^{2n}\}$, and again this has two small resolutions. It is easy to determine how to pass from one small resolution X to the other Y. Assume $n > 1$. The exceptional locus on either side is a $\mathbf{P}^1$ with normal bundle $(0, -2)$. Blow it up on X. The exceptional divisor is a $\mathbf{F}_2$; blow up its minimal section and so on. For $n - 1$ steps the exceptional divisor is an $\mathbf{F}_2$, as the normal bundle of the previous minimal section splits along the $\mathbf{F}_2$ and its normal direction and hence is $(0, -2)$. However, at the nth step the normal bundle is $(-1, -1)$, so that the exceptional divisor is $\mathbf{P}^1 \times \mathbf{P}^1$; it meets the previous $\mathbf{F}_2$ along its diagonal. So just blow down the $\mathbf{P}^1 \times \mathbf{P}^1$ along the other ruling, and then blow down the $\mathbf{F}_2$'s (an easy computation of normal bundles shows all these blow downs are permissible). For more details on this example see [**L** and **R3**].

EXAMPLE 10 (LAUFER [**L**], PINKHAM, D. MORRISON). Consider the hypersurface singularity Z

$$v_4^2 + v_2^3 - v_1 v_3^2 - v_1^3 v_2 + \lambda(v_1 v_2^2 - v_1^4) = 0$$

for any value of λ. This hypersurface admits the involution

$$i\colon \begin{Bmatrix} v_j \to v_j, \ i \leqslant 3 \\ v_4 \to -v_4 \end{Bmatrix}.$$

It is easy to see that its generic hyperplane section is the rational double point D_4. The blow up $g\colon X \to Z$ of the ideal I generated by the 2×2 minors of

$$\begin{pmatrix} v_1 & v_2 & v_3 & v_4 \\ -v_3 & -v_4 & v_1(v_2 + \lambda v_1) & v_2(v_2 + \lambda v_1) \end{pmatrix}$$

resolves the singularity of Z, so that X is smooth, and the exceptional locus of g is a $\mathbf{P}^1$ with normal bundle $(1, -3)$. If we blow up the ideal $i(I)$ instead we get a small resolution Y different from X. The exceptional locus in Y is again $\mathbf{P}^1$ with normal bundle $(1, -3)$. The lift of the generic hyperplane section of Z through the singular point to X or Y resolves the central curve of D_4, leaving three ordinary double points on it, so here we are in presence of *partial* simultaneous resolution, which explains why the normal bundle is $(1, -3)$ and not $(0, -2)$.

The description of the map $X \dashrightarrow Y$ is extremely complicated (but completely worked out) when $\lambda = 0$; it is somewhat simpler when $\lambda \neq 0$. This description was found by D. Morrison. First blow up the $\mathbf{P}^1$. The exceptional divisor is $E_1 \approx \mathbf{F}_4$ and the minimal section has normal bundle $(-1, -2)$. Blow it up. Get $E_2 \approx \mathbf{F}_1$; its minimal section intersects the $\mathbf{F}_4$ transversally in two distinct points (if $\lambda = 0$ it is tangent to the $\mathbf{F}_4$), and has normal bundle $(-1, -1)$. Blow it up; we get $E_3 \approx \mathbf{P}^1 \times \mathbf{P}^1$.

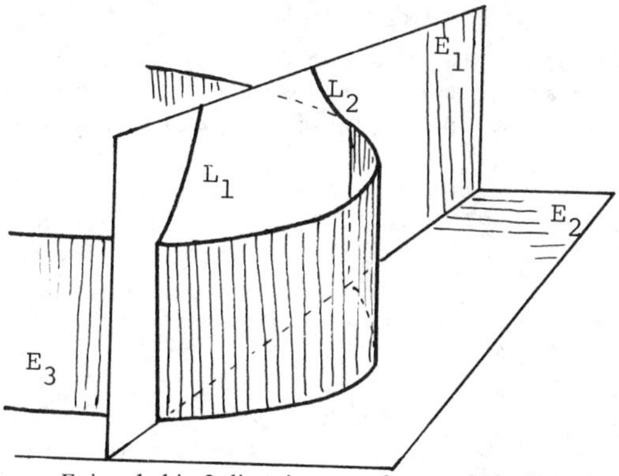

E_3 is ruled in 2 directions; we have only indicated one

E_1 now has 2 exceptional divisors L_1 and L_2 which are not double curves and which have normal bundle $(-1, -2)$. Blow them up once; they now have normal bundle $(-1, -1)$, so perform a modification of type II on each.

We now have the picture shown in the diagram.

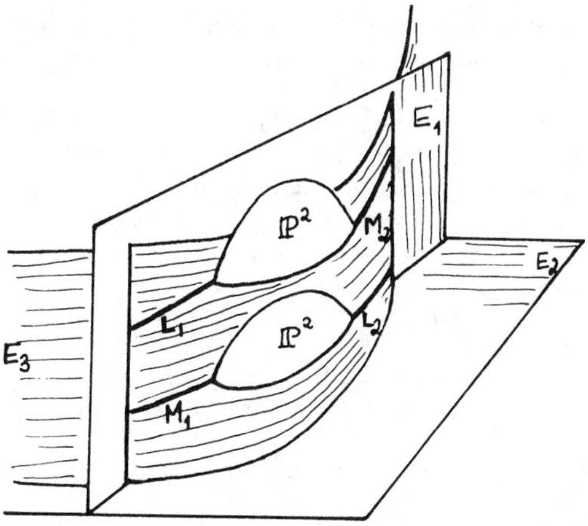

Here L_1 and L_2 are the images of the modifications of type II and therefore have normal bundles $(-1, -1)$. Note, however, that M_1 and M_2, which are proper transforms of sections of $\mathbf{P}^1 \times \mathbf{P}^1$ for the ruling given by the blow up, also have normal bundle $(-1, -1)$, so the picture is completely symmetric; so just interchange the L_i with the M_i to get back down to the second resolution.

Complete proofs of the results of this section, plus a few auxiliary results, will be given elsewhere.

REMARK. As Shepherd-Barron has pointed out, some of the assertions concerning the analytic divisor class group of Cohen-Macaulay singularities with a small resolution can be established using results of J. Bingener and U. Storch, *Zur Berechnung der Divisorenklassengruppen kompletter lokaler Ringe*, Nova Acta Leopoldina (N.F.) **240** (1981), 7–63. In particular they establish explicitly (Satz 8.8) a fact that is only implicit here: Given two small resolutions of a Cohen-Macaulay threefold singularity, the number of irreducible components of the exceptional locus is the same.

REFERENCES

[**A1**] M. Artin, *On isolated rational singularities of surfaces*, Amer. J. Math. **88** (1966), 129–137.

[**A2**] ______, *Algebraic construction of Brieskorn's resolutions*, J. Algebra **29** (1974), 330–348.

[**B1**] E. Brieskorn, *Die Auflösung der rationalen Singularitäten holomorpher Abbildungen*, Math. Ann. **178** (1968), 255–270.

[**B2**] ______, *Singular elements of semi-simple algebraic groups*, Proc. Internat. Congr. Math. (Nice, 1970), vol. 2, Gauthier-Villars, Paris, 1971, pp. 279–284.

[**BW**] D. Burns and J. Wahl, *Local contributions to global deformations of surfaces*, Invent. Math. **26** (1974), 67–88.

[**C1**] B. Crauder, *Birational morphisms of smooth algebraic threefolds collapsing three surfaces to a point*, Duke Math. J. **48** (1981), 589–632.

[**C2**] ______, Thesis, Columbia University, 1981.

[**D1**] V. I. Danilov, *The geometry of toric varieties*, Russian Math. Surveys (2) **33** (1978), 97–154.

[**D2**] ______, *The decomposition of certain birational morphisms*, Izv. Akad. Nauk SSSR Ser. Mat. **44** (1980), 465–477; English transl., Math. USSR-Izv. **44** (1980).

[**De**] M. Demazure, *Sous-groupes de rang maximum de groupe de Cremona*, Ann. Sci. École Norm. Sup. (4) **3** (1970), 507–588.

[**G**] D. Gieseker, *On two theorems of Griffiths about embeddings with ample normal bundle*, Amer. J. Math. **99** (1977), 1137–1150.

[**Gr**] H. Grauert, *Über Modifikationen und exzeptionelle analytische Mengen*, Math. Ann. **146** (1962), 331–368.

[**H**] R. Hartshorne, *Algebraic geometry*, Springer-Verlag, Berlin and New York, 1977.

[**Hi**] H. Hironaka, *An example of a non-Kaehlerian complex-analytic deformation of Kaehlerian complex structure*, Ann. of Math. (2) **75** (1962), 190–208.

[**K**] G. Kempf, F. Knudson, D. Mumford and B. Saint-Donat, *Toroidal embeddings*. I, Lecture Notes in Math., vol. 339, Springer-Verlag, Berlin and New York, 1973.

[**Ku1**] V. S. Kulikov, *Degenerations of K-3 surfaces and Enriques surfaces*, Math. USSR-Izv. **11** (1977), 957–989.

[**Ku2**] ______, *Decomposition of birational maps of smooth algebraic threefolds modulo codimension two*, Math. USSR-Izv. (to appear).

[**L**] H. Laufer, *On* $\mathbf{CP}^1$ *as an exceptional set*, Ann. of Math. Studies, No. 100, Princeton Univ. Press, Princeton, N. J., 1981, pp. 261–275.

[**Li1**] J. Lipman, *Rational singularities*, Inst. Hautes Études Sci. Publ. Math. **36** (1969), 195–279.

[**Li2**] ______, *Double point resolutions of deformations of rational singularities*, Compositio Math. **38** (1979), 37–43.

[**M**] S. Mori, *Threefolds whose canonical divisors are not numerically effective*, Ann. of Math. (2) **116** (1982), 133–176.

[**Mu1**] D. Mumford, *The topology of normal singularities on an algebraic surface*, Inst. Hautes Études Sci. Publ. Math. **9** (1961).

[**Mu2**] ______, *Introduction to algebraic geometry*, Preliminary version of first 3 chapters, distributed by Harvard University.

[**N**] S. Nakano, *On the inverse of monoidal transformations*, Publ. Res. Inst. Math. Sci. **6** (1970-71), 483–503.

[**O**] T. Oda, *Toris embeddings and applications*, Tata Institute Lecture Notes, vol. 57, Bombay, 1978.

[**P**] U. Persson, *On degenerations of algebraic surfaces*, Mem. Amer. Math. Soc., No. 189 (1977).

[**Pi1**] H. Pinkham, *Résolution simultanée de points doubles rationnels*, Sém. sur les Singularités des Surfaces, Lecture Notes in Math., vol. 777, Springer-Verlag, Berlin and New York, 1980.

[**Pi2**] ______, *Some local obstructions to deforming global surfaces*, Nova Acta Leopoldina (N.F.) **240** (1981), 173–178.

[**R1**] M. Reid, *Canonical 3-folds*, Journées de Géometrie Algébrique d'Angers (A. Beauville, ed.), Sijthoff and Noordhoff, Alphen, 1980, pp. 273–310.

[**R2**] ______, News from Moscow, February 11, 1981.

[**R3**] ______, *Minimal models of canonical 3-folds*, Proc. Sympos. Algebraic and Analytic Varieties (Tokyo, June 1981), Sympos. in Math., vol. 1, Kinokuniya, Tokyo and North-Holland, Amsterdam, (to appear).

[**Sa**] P. Samuel, *On unique factorization domains*, Tata Institute Lecture Notes, vol. 30, Bombay, 1946.

[**S**] M. Schaps, *Birational morphisms of smooth threefolds collapsing three surfaces to a curve*, Duke Math. J. **48** (1981), 401–420.

[**T**] G. N. Tjurina, *Resolution of singularities for flat deformations of rational double points*, Funkcional Anal. i Priložen **4** (1970), 77–83. (Russian)

[**W1**] J. Wahl, *Simultaneous resolution and discriminant loci*, Duke Math. J. **46** (1979), 341–375.

[**W2**] ———, *Elliptic deformations of minimally elliptic singularities*, Math. Ann. **253** (1980), 241–262.

[**Wi**] P. M. H. Wilson, *Base curves of multicanonical systems on threefolds*, preprint.

COLUMBIA UNIVERSITY

Proceedings of Symposia in Pure Mathematics
Volume **40** (1983), Part 2

SMOOTHINGS OF THE D_{pqr} SINGULARITIES, $p + q + r = 22$

HENRY C. PINKHAM[1]

This is an appendix to Looijenga's paper, *The smoothing components of a triangle singularity* (in these PROCEEDINGS). We use the notation of Looijenga's paper and refer to it as **[L]**. Other references are those given in his paper.

Looijenga states that using results of **[Nikulin]** the D_{pqr} with $p + q + r < 22$ can be smoothed, in fact negatively. Wahl **[2]** has shown that the ones with $p + q + r > 22$ cannot be smoothed at all. So it remains to consider the case $p + q + r = 22$. We show here that all but $D_{2,10,10}$ can be smoothed negatively, and that $D_{2,10,10}$ cannot be smoothed at all. Here are the steps of the proof.

PROPOSITION 1. *All the T_{pqr} lattices with $p + q + r = 22$ except $T_{2,6,14}$, $T_{2,10,10}$ and $T_{6,6,10}$ admit primitive embeddings into the K-3 surface lattice L.*

Therefore as in [L] the corresponding singularity can be smoothed negatively.

PROPOSITION 2. (a) *$T_{2,6,14}$, $T_{2,10,10}$ and $T_{6,6,10}$ have no primitive embeddings into L.*

(b) *$T_{2,10,10}$ has no imprimitive embeddings into L.*

(c) *$T_{2,6,14}$ has an overlattice S in which it has index 2, which admits a primitive embedding into L. Furthermore this embedding of $T_{2,6,14}$ is good (in the sense of* **[L**, §3]). *The same assertion holds for $T_{6,6,10}$.*

Therefore $D_{2,10,10}$ cannot be smoothed negatively, but $D_{2,6,14}$ and $D_{6,6,10}$ can. The computations show that it would be an easy matter to compute all imprimitive embeddings of these T_{pqr} if so desired.

PROPOSITION 3. *$D_{2,10,10}$ has no smoothings at all.*

The method of proof of Proposition 3 shows in principle how to determine if a D_{pqr}, $p + q + r = 22$, has a smoothing component which is not a negative

1980 *Mathematics Subject Classification*. Primary 14J17, 14B07.
[1]Supported in part by NSF Grant #MCS 8005802, Columbia University.

0082-0717/82/0000-0681/$01.75

smoothing component. It is generally believed that this will never happen (see [**L**, §1]), no matter what (p, q, r) is.

The proofs of Propositions 1 and 2 are straightforward computations using the results of [**Nikulin** and **L**] and are therefore only sketched. The proof of Proposition 3 is given in more detail. Some cases of Proposition 1 were settled by Lee McEwan, using a different method. Lemmas 1 and 2 are given in a form suggested by C. T. C. Wall and E. Looijenga, respectively.

I. Proof of Proposition 1. To apply [**Nikulin**, Theorem 1.12.2], we must first compute the finite abelian group $G = T^*_{pqr}/T_{pqr}$ and the quadratic form on it. G has order $pqr(1 - 1/p - 1/q - 1/r)$, as is well known since this number is the absolute value of the discriminant of T_{pqr}. The following computations holds for all (p, q, r). We label the vertices of T_{pqr} as in [**L**, §2]. Consider the following elements of T_{pqr}:

$$A = A_1 + 2A_2 + \cdots + (p-1)A_{p-1},$$
$$B = B_1 + 2B_2 + \cdots + (q-1)B_{q-1},$$
$$C = C_1 + 2C_2 + \cdots + (r-1)C_{r-1}.$$

Note that A is orthogonal to all the vertices of T_{pqr} except E and A_{p-1}, etc. Now consider the elements A^*, B^*, C^* and E^* in T_{pqr}, duals of A_{p-1}, B_{q-1}, C_{r-1} and E respectively (i.e., $A^*(A_{p-1}) = 1$ and A^* on any other vertex $= 0$, etc.). We can write A^*, B^*, C^* and E^* in terms of A, B, C and E. Set $d = 1 - 1/p - 1/q - 1/r$.

$$\begin{aligned}
E^* &= 1/dE & &+1/pdA & &+1/qdB & &+1/rdC,\\
A^* &= (1-1/p)/dE & &+(1/q+1/r)/pdA & &+(1-1/p)/qdB & &+(1-1/p)/rdC,\\
B^* &= (1-1/q)/dE & &+(1-1/q)/pdA & &+(1/p+1/r)/qdB & &+(1-1/q)/rdC,\\
C^* &= (1-1/r)/dE & &+(1-1/r)/pdA & &+(1-1/r)/qdB & &+(1/q+1/p)/rdC.
\end{aligned}$$

Finally let a, b and c denote the images in G of $E^* - A^*$, $E^* - B^*$ and $E^* - C^*$, respectively.

LEMMA 1. *a, b and c generate G with relations $a + b + c = pa = qb = rc$. Hence a and b alone always generate and if θ is the greatest common divisor of (p, q, r) then $G \approx \mathbf{Z}/\theta \oplus \mathbf{Z}/\varphi$, where θ divides φ.*

REMARK (C. T. C. WALL). Furthermore G is isomorphic to the group of "monomial" automorphisms $(x \to \lambda x,\ y \to \mu y,\ z \to \nu z)$ of the "cusp" singularity $x^p + y^q + z^r + xyz = 0$. I will show elsewhere that this is not a coincidence.

The proof of Lemma 1 is an easy computation which is left to the reader. By [**Nikulin**, 1.12.3], this shows that if $p + q + r \leqslant 21$, then T_{pqr} has a primitive embedding into L (indeed, this is the computation Looijenga made to get the result originally), and if $p + q + r = 22$ then T_{pqr} has a primitive embedding anytime G is cyclic. So we need only consider the case where the g.c.d. of p, q and r is greater than one.

In that case the 2-subgroup is not cyclic but of the form $\mathbf{Z}/2\mathbf{Z} \oplus \mathbf{Z}/2^k\mathbf{Z}$ for some $k \geqslant 1$. Here is the table (for 4, 6, 12 the 3-part is cyclic):

p	q	r	discriminant
2	4	16	$2^3 \cdot 3$
2	6	14	$2^2 \cdot 11$
2	8	12	$2^3 \cdot 7$
2	10	10	$2^2 \cdot 3 \cdot 5$
4	4	14	$2^5 \cdot 3$
4	6	12	$2^4 \cdot 3^2$
4	8	10	$2^3 \cdot 3 \cdot 7$
6	6	10	$2^2 \cdot 3 \cdot 17$
6	8	8	$2^5 \cdot 7$

If $k \geqslant 2$ then the form on G decomposes as $q_\theta^{(2)}(2) \oplus q_\theta^{(2)}(2^k)$ so there is nothing further to check in Nikulin's theorem, and T_{pqr} embeds. Proposition 1 is proved.

II. Proof of Proposition 2. We first need to compute the quadratic form $G \to \mathbf{Q}/2\mathbf{Z}$.

LEMMA 2. $(E^* - A^*)^2 = -1 + 1/p + 1/p^2 d$, $(E^* - A^*)\cdot(E^* - B^*) = 1/pqd$.

The other products can be obtained by the obvious permutation of (p, q, r).

We now treat the cases where $k = 1$. The following table shows that Nikulin's condition fails in all three cases, so there is no primitive embedding into L.

p	q	r	$\mathbf{Z}_2$ lattice	congruence to be satisfied
2	6	14	$U^{(2)}(2)$	$2^2 \cdot 11 \equiv \pm 2^2 \mod (\mathbf{Z}_2^*)^2$
2	10	10	$V^{(2)}(2)$	$2^2 \cdot 3 \cdot 5 \equiv \pm 2^2 \cdot 3 \mod (\mathbf{Z}_2^*)^2$
6	6	10	$U^{(2)}(2)$	$2^2 \cdot 3 \cdot 17 \equiv \pm 2^2 \mod (\mathbf{Z}_2^*)^2$

The lemma also shows that $T_{2,10,10}$ has no overlattices, so that it has no embeddings at all. So parts (a) and (b) of Proposition 2 are proved. For (c) notice that the other two lattices do have overlattices which embed into the K-3 surface lattice. Instead of attempting to check Looijenga's condition for a good embedding, we will construct directly a K-3 surface such that its Picard group contains $T_{2,6,14}$ (resp. $T_{6,6,10}$), and the vertices of the T correspond to $\mathbf{P}^1$'s. Its Picard group will of course be isomorphic to the overlattice constructed above.

Consider the lattice T_{pqrs} with the diagram (same conventions as in [L]):

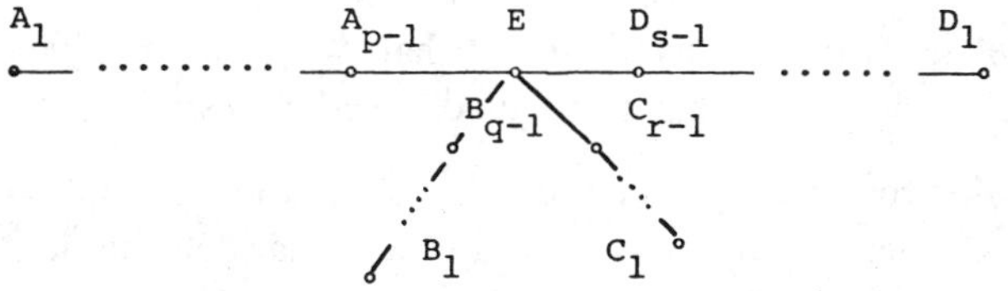

Just as in the three branch case we can show that $G = T^*_{pqrs}/T_{pqrs}$ is generated by A^*, B^*, C^*, D^* and E^* where these elements are defined as in the three branch case. Therefore, one can decide when T_{pqrs} admits a primitive embedding into L; in particular one has

LEMMA. $T_{3,5,6,6}$ *(which has rank* 17) *admits a primitive embedding into L.*

The order of G_{pqrs} is $pqrs(2 - 1/p - 1/q - 1/r - 1/s)$, so in this case is $2^2 \cdot 3^2 \cdot 17$, so G has at most 4 generators (it actually has 2), so the easy part of Nikulin's theorem applies.

Now take the involution on $T_{3,5,6,6}$ which permutes C_i and D_i and leaves the other vertices fixed. Using Theorem 4.3 of Nikulin's paper *Finite automorphism groups of Kahler K-3 surfaces*, Transl. Moscow Math. Soc. **38** (2) (1980), one can check there exists a K-3 surface X with Picard group isomorphic to $T_{3,5,6,6}$ together with an involution i acting on cohomology as above. (We need to know that i extends to an involution of the K-3 surface lattice, but this follows from [**Nikulin**, §14].) Next by using a generalization of Proposition 3 of [**L**], we see that the vertices of the $T_{3,5,6,6}$ can be realized by effective curves. Finally, it is easy to see that the desingularization of the quotient of X by i has a $T_{6,6,10}$ of effective $\mathbf{P}^1$'s.

To do the $T_{2,6,14}$ case consider the lattice $T_{6,6,7}$ and the involution permuting A_i and B_i, and fixing the other vertices. By Nikulin again we can find a K-3 surface X with Pic $\cong T_{6,6,7}$ with an involution acting on cohomology as above. The desingularization of the quotient of X by the involution has a $T_{2,6,14}$ of $\mathbf{P}^1$'s. So we are done.

REMARKS. (1) The double cover we have constructed is ramified above $A_1 + A_3 + A_5 + C_1 + C_3 + \cdots + C_9$ (for $T_{6,6,10}$) and $A_1 + C_1 + C_3 + \cdots + C_{13}$ (for $T_{2,6,14}$). This class is *even* in the K-3 by construction of the overlattice of the T_{pqr}.

(2) More generally, any T_{pqr} such that p and q are even and $p + q \equiv 0$ (8) has an overlattice M in which it has index 2, since

$$\tfrac{1}{2}(A_1 + A_3 + \cdots + A_{p-1} + B_1 + \cdots + B_{q-1})$$

is an element of the dual lattice with square in $2\mathbf{Z}$. However, if $M \hookrightarrow L$ is a good imbedding of M then the induced embedding of T_{pqr} is good iff $p + q = 16$. When $p + q = 8$ we can identify the overlattice M:

$$T_{2,6,d} \hookrightarrow T_{2,3,d+3}, \qquad T_{4,4,d} \hookrightarrow T_{2,4,d+2}.$$

These are the lattices that occur in cases (b) and (d) of Wahl [**1**, Theorem 5.6], giving "exotic" elliptic deformations. In the same vein, (e) comes from the index 3 inclusion:

$$T_{3,6,d} \hookrightarrow T_{2,3,d+4}.$$

Returning to the lattices at hand we see that $T_{2,6,14}$ has one "good" and one "bad" overlattice while $T_{6,6,10}$ has two "good" overlattices.

III. Proof of Proposition 3. Assume by contradiction that $D_{2,10,10}$ has a smoothing. Pick a germ of a curve $(B, 0)$ in the versal deformation S of $D_{2,10,10}$ over which it smooths; by Proposition 2 B cannot lie in S^-= the negative part of the versal deformation, nor can it lie in S^+, which is just the "equisingular" locus. Therefore B is transverse to the $\mathbf{C}^*$ orbits. Applying $\mathbf{C}^*$ to $B - 0$, we can move $B - 0$ to a punctured germ $B' - 0$ such that B' intersects S^- in a point other than the origin. Let V be the fiber in the versal deformation above $B' \cap S^-$; V can be compactified to $\bar{V}$ following the usual $\mathbf{C}^*$ procedure. By Proposition 2 it is easy to see that $\bar{V}$ must have one minimally elliptic singularity, which by construction must be smoothable. Let Y be the minimal good desingularization of $\bar{V}$. At infinity Y contains the $T_{2,10,10}$ lattice. Let Z be the fundamental cycle of the desingularization of the minimally elliptic singularity. Z is an anticanonical divisor for Y. $-Z \cdot Z$ is uppersemicontinuous here, since it is just the multiplicity of the corresponding minimally elliptic singularities (Laufer generalizes this to all Gorenstein singularities in this volume), so we have $-Z \cdot Z \leqslant 12$, the value for $D_{2,10,10}$.

Since K^2 decreases by one for each blow up, we see that the Picard group of Y has dimension $10 - Z \cdot Z$ (use the fact that $\operatorname{Pic} \mathbf{P}^2 = \mathbf{Z}$, $K_{\mathbf{P}^2}^2 = 9$ and $\operatorname{Pic} \mathbf{F}_n = \mathbf{Z} \oplus \mathbf{Z}$, $K_{\mathbf{F}_n}^2 = 8$). But Y contains $T_{2,10,10}$ at infinity, and this has rank 20 in $\operatorname{Pic} Y$. So $10 - Z \cdot Z \geqslant 21$, since we must have some contribution from the singularity itself.

If $-Z \cdot Z = 12$, then the deformation of $D_{2,10,10} \to V$ (using the notation of Wahl [**1**]) comes from a deformation of the resolution, and by his Theorem 5.4, we see that V must be a nonsmoothable singularity, a contradiction (we need Theorem 5.6 on a numerical condition for the smoothing of cusps, in Wahl's paper *Smoothing of normal surface singularities*, Topology **20** (1981)).

So we may assume $-Z \cdot Z \leqslant 11$. But then $\operatorname{Pic} Y$ has dimension $\leqslant 21$, so that the exceptional locus on Y of the elliptic singularity on $\bar{V}$ is irreducible. Hence Z is just the reduced exceptional locus. Since the lattice $\operatorname{Pic} Y$ is unimodular, the discriminant of the orthogonal complement $Z^\perp$ of Z in $\operatorname{Pic} Y$ has absolute value $|Z \cdot Z| = 11$ (Z is well known to be $T_{2,3,17}$, but we shall not need this fact). Now $T_{2,10,10}$ also is in $Z^\perp$ and has the same rank. So $|\operatorname{disc} T_{2,10,10}| = 2^2 \cdot 3 \cdot 5$ must be divisible by 11, which it is not. Contradiction.

COLUMBIA UNIVERSITY, NEW YORK

Proceedings of Symposia in Pure Mathematics
Volume **40** (1983), Part 2

THE NORMAL SINGULARITIES OF SURFACES IN $\mathbf{R}^3$

IAN R. PORTEOUS

This paper develops the approach to the geometry of smooth submanifolds of Euclidean space begun in [**9**]. That paper preceded the introduction by Arnol'd [**1**] of his by now standard notations for the simple singularity types of functions, so we begin by going over some old ground using his notations freely—at times too freely!

Curves in $\mathbf{R}^2$. The circle of points r in $\mathbf{R}^2$ with centre c and radius ρ has equation

$$(r-c)\cdot(r-c)=\rho^2$$

and if a parametric representation $t \mapsto r = r(t)$, with $r_1 \neq 0$ everywhere, is chosen for the circle, with subscripts denoting differentiation with respect to t, it follows that

$$\begin{aligned} r_1\cdot(c-r)&=0,\\ r_2\cdot(c-r)&=r_1\cdot r_1,\\ r_3\cdot(c-r)&=3r_1\cdot r_2,\\ r_4\cdot(c-r)&=3r_2\cdot r_2+4r_1\cdot r_3, \end{aligned}$$

and so on. For any immersed curve $t \mapsto r(t)$ the first of these equations defines the (parametrised) normal bundle of the curve, loosely thought of as the "A_1 set" of the distance-squared function V_c: $t \mapsto (r(t)-c)\cdot(r(t)-c)$, though strictly the A_1 set consists of those points of $\mathbf{R} \times \mathbf{R}^2$ where the first equation is satisfied, but the second is not. The first and second together define the normal focal set of the curve, this being loosely thought of as the "A_2 set" of the distance-squared function V_c. This projects to the evolute, or locus of centres of curvature of the curve, under the projection $\mathbf{R} \times \mathbf{R}^2 \to \mathbf{R}^2$.

1980 *Mathematics Subject Classification.* Primary 53A05.

0082-0717/81/0000-0201/$03.50

There will be a unique centre of curvature on each normal unless r_2 depends linearly on r_1, when the centre of curvature is "at infinity" and the curve has an inflection. For brevity, in what follows, we generally omit the discussion of singularities at infinity. Looijenga [**5**] and Wall [**12**] have shown how to extend our approach to handle them.

If the first three equations all hold then generically the curve has a vertex, where the evolute has a cusp and the radius of curvature is stationary. These vertex centres constitute the "A_3 set" of V_c. If the fourth equation also holds, but not the fifth, the curve has five point contact with its circle of curvature, the centre of curvature being an A_4 point of V_c, with $c_2 = 0$ as well as $c_1 = 0$ for the evolute $t \mapsto c(t)$ and with $\rho_2 = 0$ as well as $\rho_1 = 0$ for the radius of curvature ρ. Such points do not occur on a generic curve but may occur on individual members of generic one-parameter families of curves.

There are simple applications of this approach to curvature theory in the theory of mechanisms, for example to the derivation of the cubic of stationary curvature and the Burmester points of the instantaneous motion of a plane lamina.

Curves in $\mathbf{R}^3$. For a curve on a sphere in $\mathbf{R}^3$, with centre c and radius ρ, the same equations hold for the derivatives of r, and for any embedded curve in $\mathbf{R}^3$ the first equation defines the normal bundle, the "A_1 set" of V_c, and the first and second together the normal focal set or "A_2 set" of V_c, consisting of a focal line in each normal plane. For a generic curve r_2 will always be linearly independent of r_1, the focal line then being everywhere defined.

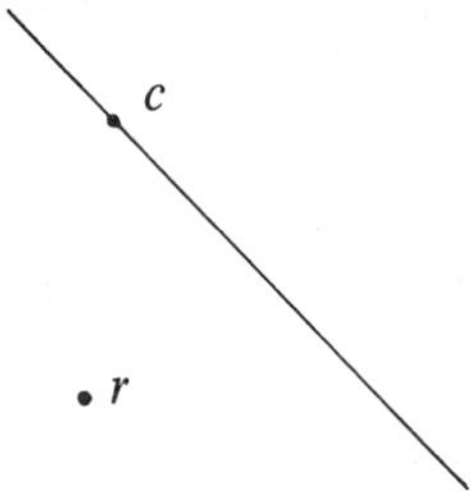

FIGURE 1

The A_3 locus consists of the centres of spherical curvature of the curve. If r_3 depends linearly on r_1 and r_2, which happens generically at isolated points, the centre of curvature is at infinity. Nongenerically it can happen in such a case that the entire focal line satisfies the A_3 equation. In that case there is one point of the focal line where the A_4 equation holds, namely the limit point of the centres of spherical curvature as the point is approached where r_3 depends on r_1 and r_2. In Thom's language this is a bec-à-bec singularity, a nongeneric A_4 singularity of V_c. For a generic curve in $\mathbf{R}^3$ an A_4 point occurs wherever the curve of centres of spherical curvature has a cusp. The radius of spherical curvature is stationary either at such points or at points where the vector $c - r$ is orthogonal to the focal line.

Surfaces in $\mathbf{R}^3$. Consider a smooth surface given parametrically locally by a smooth embedding $g: \mathbf{R}^2 \to \mathbf{R}^3$; $w \mapsto g(w)$, dgw therefore being injective everywhere. The domain of g need not, of course, be the whole of $\mathbf{R}^2$. The distance-squared function $V: \mathbf{R}^2 \times \mathbf{R}^3 \to \mathbf{R}$ is then defined by

$$V(w, x) = (g(w) - x) \cdot (g(w) - x).$$

Differentiation of V with respect to w will be indicated by subscripts. The various singularities of V or normal singularities of the surface g are then defined cumulatively as follows, using the criteria set out in detail in [**10**].

The A_1 points of V are characterised by $V_1 = 0$ and the nonexistence of $a \in \mathbf{R}^2 \backslash \{0\}$ such that $V_2 a = 0$. Now

$$V_1 = 0 \Leftrightarrow (g(w) - x) \cdot dgw = 0 \Leftrightarrow x - g(w) = \rho n(w),$$

where $n(w)$ is one of the two normal vectors of unit length at $g(w)$. This is, therefore, just the parametrised normal bundle of the surface.

The A_2 points are characterised by $V_1 = 0, V_2 a = 0$, for some $a \neq 0$, but $V_2 \neq 0$, with the nonexistence of b such that $V_3 a^2 + V_2 b = 0$, or equivalently $V_3 a^3 \neq 0$. Now

$$V_2 a = 0 \Leftrightarrow \rho n \cdot d^2 gwa = dgwa \cdot dgw \Leftrightarrow \rho \mathrm{II} a = \mathrm{I} a,$$

where $\mathrm{I} = dgw \cdot dgw$ is the first fundamental form and $\mathrm{II} = n \cdot d^2 gw$ is the second fundamental form. The vector a is then a principal vector and ρ is the corresponding principal radius of curvature. This eigenvalue problem has two real solutions which are distinct, since $V_2 \neq 0$. The eigenspaces are one-dimensional, are mutually orthogonal with respect to I and integrate to give lines of curvature on the surface. The subset of the normal bundle defined by $V_1 = 0$ and $V_2 a = 0$ for some nonzero $a \in \mathbf{R}^2$ is called the normal focal surface. Locally at least it is two-sheeted and it is sometimes helpful to think of the two sheets as being coloured with different colours, say red and blue. The normal focal surface projects under the projection $\mathbf{R}^2 \times \mathbf{R}^3 \to \mathbf{R}^3$ to the surface of centres of the original surface.

The A_3 points are characterized by $V_1 = 0, V_2 a = 0$, for some $a \neq 0$, but $V_2 \neq 0$, and $V_3 a^3 = 0$, with one further condition, namely that the equations $V_3 a^2 + V_2 b = 0$ and $V_4 a^4 + 6V_3 a^2 b + 3V_2 b^2 = 0$, or equivalently the equations $V_3 a^2 + V_2 b = 0$ and $V_4 a^4 + 3V_3 a^2 b = 0$, are not simultaneously solvable for b, this being equivalent to asserting that the quadratic equation for b is not parabolic. There are thus two types of A_3 points, the elliptic or sterile type and the hyperbolic or fertile type. Either type may lie on either sheet of the normal focal surface where generically they form curves which we call the ribs of the surface. These project under the projection $\mathbf{R}^2 \times \mathbf{R}^3 \to \mathbf{R}^3$ to cuspidal edges on the surface of centres and under the projection $\mathbf{R}^2 \times \mathbf{R}^3 \to \mathbf{R}^2$ (followed by g) to the rib-lines, or ridges in the terminology of [**9**], of the surface itself.

The A_4 points are just the parabolic points which we have just excluded, with the additional condition that $V_5 a^5 + 10V_4 a^3 b + 15V_3 ab^2 \neq 0$. We regard these as points of the ribs of the surface also. They form cusps on the ribs when these are projected to $\mathbf{R}^3$ but remain nonsingular points when projected to $\mathbf{R}^2$, when they become the turning points of the rib-lines with respect to the grid of lines of curvature, that is the places where the rib-line is tangential to the line of curvature with the relevant principal vector as tangent vector. See Figure 2.

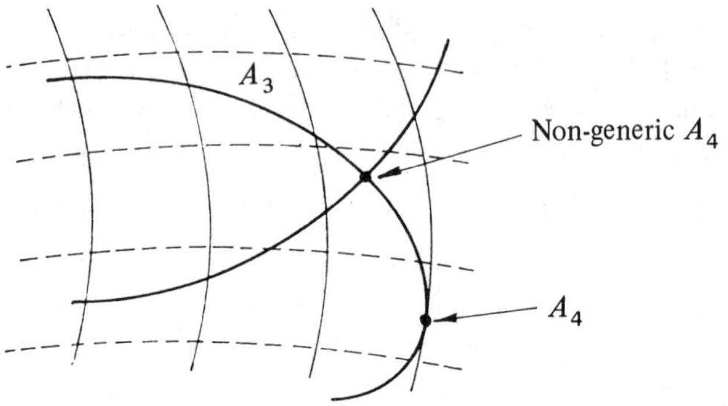

FIGURE 2

The A_3, A_4 and higher A_k points can be characterised in terms of the lines of curvature as follows. Consider a line of curvature as a space curve. Then on its focal line at any point we may mark the centre of spherical curvature and also the appropriate centre of curvature of the surface. These points coincide exactly at points of rib-lines of the surface and if the centre then happens to be A_k for the curve, for $k \leqslant 3$, then it is also an A_k point for the surface, and vice versa. Moreover the vectors $a, b, c, \ldots$ which enter into the A_k equations can be taken to be the successive derivatives of the line of curvature at the point in question. Crossings of rib-lines on a surface are of two types, flyovers, where the ribs belong to different sheets of the focal surface and so do not intersect, and level crossings, where the associated ribs belong to the same sheet of the focal surface. Only flyovers may occur on a generic surface. At a level crossing, which is a nongeneric A_4 point, the two ribs come together bec-à-bec. Amusingly the relevant line of curvature through such a point also exhibits a bec-à-bec, or nongeneric A_4, singularity, this occurring also at isolated nongeneric A_4 points which may be regarded as birth or death points for ribs. Though such events do not occur for a generic surface they may occur generically on individual members of a one-parameter family of surfaces.

Umbilics. So far we have ignored those points of our surface for which the two principal centres of curvature coincide. Then $V_2 = 0$, or equivalently $\rho\mathrm{II} = \mathrm{I}$, and the sphere more closely approximates a sphere than elsewhere. Such points are the umbilics of the surface. Umbilical centres generically are conic nodes of the normal focal surface.

An umbilical centre is a D_4 point for the distance-squared function V if $V_1 = 0, V_2 = 0$ and there exists no nonzero $a \in \mathbf{R}^2$ such that $V_3 a^2 = 0$. There are

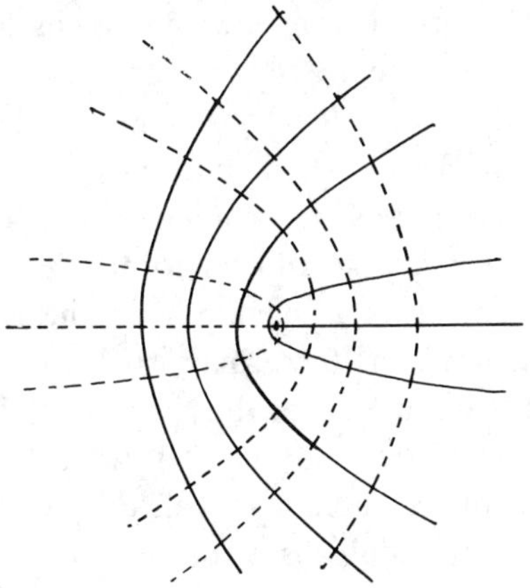

Lemon

Hyperbolic Umbilic – Index ½

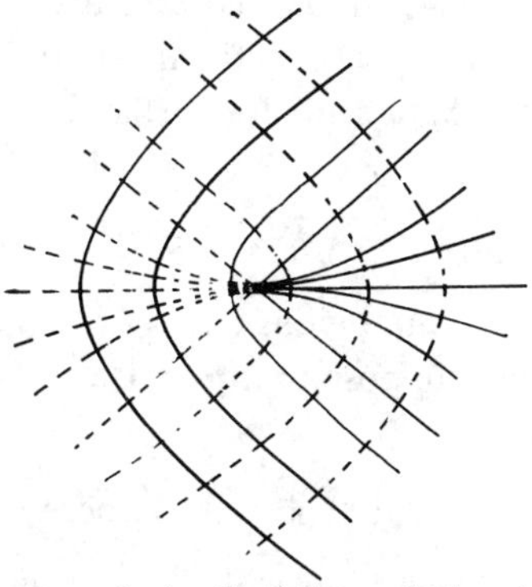

(Le) Monstar

Hyperbolic Umbilic – Index ½

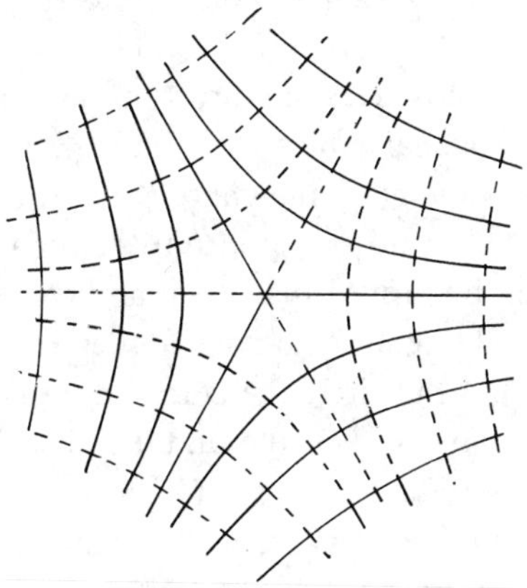

Star

Umbilic – Index −½

FIGURE 3

two types, the elliptic and the hyperbolic, which we discuss in detail in a moment. It is D_5 if $V_1 = 0$, $V_2 = 0$ and a nonzero $a \in \mathbf{R}^2$ does exist such that $V_3a^2 = 0$ but $V_3a \neq 0$ and $V_4a^4 \neq 0$. It is E_6 if $V_1 = 0$, $V_2 = 0$ and a nonzero $a \in \mathbf{R}^2$ exists such that $V_3a = 0$ but $V_4a^4 \neq 0$. Here V_3 is, of course, a symmetric trilinear form on $\mathbf{R}^2$.

The Hessian directions of a symmetric trilinear form C on $\mathbf{R}^2$ are defined as follows. Let $u \in \mathbf{R}^2\backslash\{0\}$. Then Cu is a symmetric bilinear form, in general nondegenerate. For certain u, however, there may exist $v \in \mathbf{R}^2\backslash\{0\}$ such that $Cuv = 0$, in which case u and v are interchangable. These two Hessian directions are explicitly given by $\det Cu = 0$. Both may be real, the hyperbolic case, when only one of the root directions of the cubic C is real; or they may be a complex conjugate pair, the elliptic case, when all three root directions of C are real; or they may coincide, $u = v = a$, say, when $Ca^2 = 0$ and two of the three root directions of C coincide with a. This is the case for surfaces at D_5 points, the parabolic umbilics. Yet another possibility is that $a \in \mathbf{R}^2$ is such that $Ca = 0$, in which case the Hessian directions are not defined. In this case all three root directions of C coincide, which is what occurs for surfaces at E_6 points.

A cubic form on $\mathbf{R}^2$ may be denoted by

$$\alpha z^3 + 3\bar{\beta}z^2\bar{z} + 3\beta z\bar{z}^2 + \bar{\alpha}\bar{z}^3,$$

where $z = x + iy$ and $\alpha, \beta \in \mathbf{C}$. By an obvious change of coordinates involving a choice of a cube root of α we may suppose that $\alpha = 1$, unless of course $\alpha = 0$ in which case the form is of the form

$$3(\bar{\beta}z + \beta\bar{z})z\bar{z} = (lx + my)(x^2 + y^2), \quad \text{where } l, m \in \mathbf{R}.$$

With $\alpha = 1$ a cubic form may then be represented in the β-plane by a point. It may be shown that the discriminant, or parabolic, locus is then the tricuspidal quartic with parametric equation $\beta = \frac{1}{3}(\gamma^2 + 2\bar{\gamma})$, where $|\gamma| = 1$, while the circle $|\beta| = 1$ consists of those cubics whose Hessian directions are mutually orthogonal and which we therefore call orthogonal cubics. See Figure 4.

In discussing a particular umbilic $g(w)$ the chart g may be supposed to be so chosen that the first fundamental form $\mathrm{I} = \cdot$ on $\mathbf{R}^2$ at the point w. For an umbilic on a generic surface it can then be proved that the umbilical centre is of type D_4 with V_3 not orthogonal; that is the representative complex number β does not lie on the unit circle. If β lies within the circle the corresponding umbilic is said to have index $-\frac{1}{2}$, while if it lies outside then the umbilic is said to have index $\frac{1}{2}$. These indices refer to the configuration of lines of curvature near such umbilics, as drawn by Darboux [3]. See Figure 3, where the nomenclature is due to Berry and Hannay [2]. The directions of lines of curvature through an umbilic are given by the cubic which is the Jacobian of V_3 and the first fundamental form I. Its discriminant is the larger of the two tricuspidal quartics in Figure 4.

There are three ribs or one through an umbilical centre and therefore three rib-lines or one through an umbilic according as the cubic form V_3 is elliptic or hyperbolic, the limiting values of the principal directions at points of the rib-line

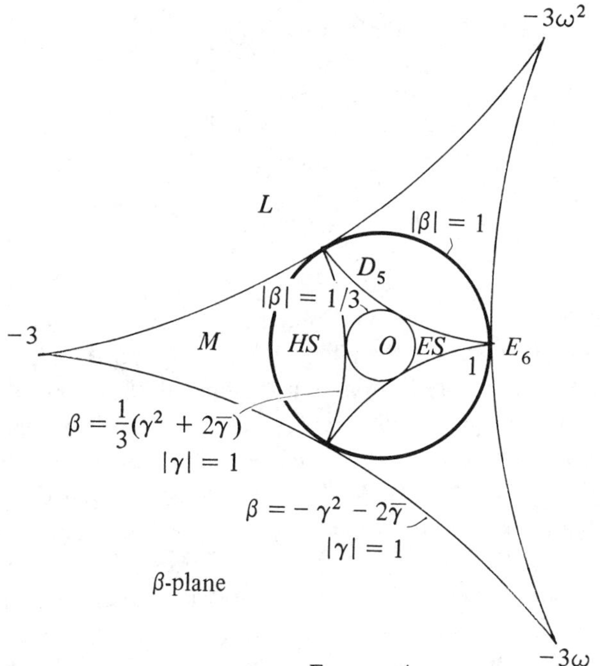

FIGURE 4

being the root-directions of V_3. A rib is necessarily of fertile type in the neighbourhood of an umbilical centre but it changes colour as it passes through the node, that is it passes there from one sheet of the focal surface to the other. The tangent direction t to a rib-line through an umbilic, with limiting principal direction u, is given by $V_3 tuv = 0$, where v is orthogonal to u. Two rib-lines pass through the umbilic in the same direction whenever β representing V_3 lies on the circle $|\beta| = \frac{1}{3}$. The vertical components of the tangents to the ribs will however differ there.

The sum of the indices of the umbilics of a compact smooth surface in $\mathbf{R}^3$ is equal to the Euler characteristic, from which it follows that when umbilics are born in a one-parameter family of surfaces they are born as twins at a spot on a fertile rib where V_3 is orthogonal, the umbilics born being subsequently one of index $-\frac{1}{2}$ and the other of index $\frac{1}{2}$.

An ellipsoid has four umbilics of lemon type, each of index $\frac{1}{2}$. They lie on the major-minor plane of symmetry of the ellipsoid. For any surface with a plane of symmetry the intersection of the surface and plane is automatically a rib-line. Moreover it is easy to prove that the vertices of such a rib-line are flyover points, where a rib-line of the opposite colour crosses it at right angles. There are six such flyover points on an ellipsoid. One of the rib-lines is wholly red, one is wholly blue and the third changes from red to blue and back to red again as it passes through the various umbilics. Monge [**8**] envisaged the lines of curvature on an ellipsoid as bounding tiles on the semiellipsoidal roof of a High Court chamber, with chandeliers hanging from the umbilics! See Figure 5.

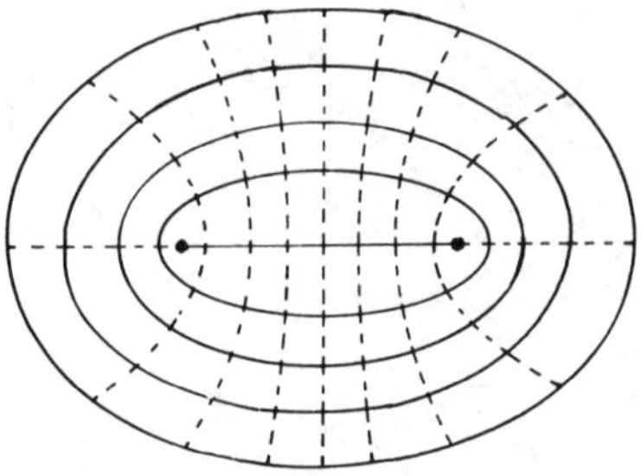

FIGURE 5

Symmetrical singularities. We have just remarked that for an ellipsoid in $\mathbf{R}^3$ there are reflections of $\mathbf{R}^3$ mapping the ellipsoid to itself, and that, in that case, the fixed point set for such a reflection is a rib-line of the ellipsoid. Several singularities of the distance-squared function for a surface in $\mathbf{R}^3$ can occur in such a way, and it is convenient to recode these, in accord with Slodowy [**11**]. Explicitly we recode A_3 with $\mathbf{Z}/2\mathbf{Z}$ symmetry as B_2, A_5 as B_3, D_4 as C_3, D_5 as C_4 and E_6 as F_4 (although we should have been happier to have had the labels B_k and C_k interchanged in this context).

We shall also encounter examples of surfaces invariant under actions of the symmetric group $\mathfrak{S}_3$ on $\mathbf{R}^3$. A fixed point of the surface under such an action will in general be an elliptic umbilic whose intrinsic third derivative V_3 is a harmonic cubic, represented in the β-plane by $\beta = 0$. Such an umbilic will be said to be of type G_2.

Some $\Sigma^{2,0}$ singularities. In [**4**] Les Lander discussed the geometry of the $\Sigma^{2,0}$ singularities of maps $f\colon \mathbf{R}^2 \to \mathbf{R}^2$, following the classification of such singularities given by Mather in [**7**].

There is particular interest in maps of the form

$$(x, y) \mapsto \left(B(x, y)^2, C(x, y)^3\right),$$

where B is symmetric bilinear and C is symmetric trilinear. Such maps are characterised by the quadratic form B and the part of the cubic form C that is harmonic with respect to B, the Hessian of this harmonic part then being a multiple of B. The harmonic part of C is the unique cubic form of the form $C + AB$, where A is linear, that has Hessian a multiple of B. A map of the given form is the Mather type IV_3 or of type $\mathrm{I}_{3,3}$ according as the harmonic part of C is nonzero and B, its Hessian, is elliptic or hyperbolic. In the parabolic case the singularity is of Thom-Boardman type $\Sigma^{2,1}$.

Part of what Lander showed was that when these singularities are presented transversally then in the elliptic case there are three branches and in the hyperbolic case one branch of Σ^{11111}, of codimension 5, through the $\Sigma^{2,0}$ points, of codimension 6, these branches being controlled as they pass through the $\Sigma^{2,0}$ points by the harmonic part of the cubic, though this last point is not made quite explicit in the published paper. Indeed a really nice elucidation of the geometry here is still to be given.

Bumpy circles. Recently, in lectures given in Liverpool, I have considered the geometry of the contours of a surface in $\mathbf{R}^3$ given in the "Monge form"

$$z = h(x, y).$$

Round a local maximum or minimum, where $dh(x, y) = 0$, the contours are slightly deformed ellipses, each with four vertices if one is near enough to the critical point, and it is easy to see that the locus of vertices of contours near the critical point consists of two smooth curves passing mutually orthogonally through the critical point in the directions of principal curvature.

Now, round an umbilical critical point, the ellipses will be nearly circular—indeed they are circles if the surface is an ellipsoid turned so that one of its umbilics is at its highest or lowest point. The question may then be asked as to how many vertices in general such a 'bumpy' circle has. It is natural to conjecture 2 or 6 according to the configuration of lines of curvature, but the four vertex theorem reminds us that there must be at least 4. However the whole geometry of the contours is most naturally expressed in terms of the singularities of the map

$$\begin{array}{ccc} \mathbf{R}^2 \times \mathbf{R}^2 & \to & \mathbf{R} \times \mathbf{R} \times \mathbf{R}^2, \\ (w, x) & \mapsto & \left(\tfrac{1}{2}|w - x|^2, h(w), x\right) \end{array}$$

(where I have replaced (x, y) above by the vector w and x also is a vector). The Σ^1 set of this map consists of the normal bundles to the contours and the Σ^{11} set consists of their evolutes, while the Σ^{111} set gives the centres of curvature at the vertices. When the map has a $\Sigma^{2,0}$ singularity $x = w$ and $dhw = 0$ and the intrinsic second derivative consists of the pair of quadratic forms $\cdot$ and d^2hw. These are multiples of each other in the case of an umbilical critical point when the singularity has the type IV_3 discussed above. The geometry is then controlled by the harmonic part of the cubic d^3hw, which as we have seen plays the role of intrinsic third derivative in this case. Indeed it turns out that the locus of the vertices consists locally of three curves passing harmonically through the critical point, in directions orthogonal to the root directions of the harmonic cubic. The locus of the vertex centres likewise consists locally of three curves passing through the critical point in the same three directions. Between adjacent branches the angle is $\pi/3$. So a generic bumpy circle has six vertices equally spaced around it. We sketch the proof.

For an umbilical point at O with radius of curvature $1/\kappa$ a contour of h near O is of the form

$$\frac{\kappa}{2}(w \cdot w) + \frac{1}{6}Cw^3 + O(w^4) = \text{const} = \frac{\kappa^3}{2}\varepsilon^2,$$

say, where $C = d^3hO$ and ε is small. Under the radial transformation $w = \kappa\varepsilon r$ this transforms to

$$r \cdot r + \frac{\varepsilon}{3}Cr^3 + \varepsilon^2 O(r^4) = 1,$$

where we can disregard the ε^2 term entirely, for ε small.

Our task therefore reduces to locating the vertices of

$$2F(r) = r \cdot r + \tfrac{1}{3}\varepsilon C r^3 = 1 \quad \text{for } \varepsilon \text{ small},$$

ignoring the far-away noncompact component of this cubic curve. Now, for any parametrisation $t \mapsto r(t)$ of this bumpy circle, the vertex centres c are given by

$$\begin{aligned} (c-r)\cdot r_1 &= 0, & && F_1 r_1 &= 0,\\ (c-r)\cdot r_2 &= r_1 \cdot r_1 & \text{where} && F_2 r_1^2 + F_1 r_2 &= 0,\\ (c-r)\cdot r_1 &= 3r_1 \cdot r_2 & && F_3 r_1^3 + 3F_2 r_1 r_2 + F_1 r_3 &= 0, \end{aligned}$$

where once again we denote differentiation by subscripts. Writing $F_1 = \lambda(c-r)\cdot$ and eliminating λ, noting that $F_2 = \ \cdot + \varepsilon Cr$ and that $F_3 = \varepsilon C$, we find that

$$(r_1 \cdot r_1)(\varepsilon C r_1^3 + 3r_1 \cdot r_2 + 3\varepsilon C r r_1 r_2) = 3(r_1 \cdot r_2)(r_1 \cdot r_1 + \varepsilon C r r_1^2),$$

that is

$$(r_1 \cdot r_1)(C r_1^3 + 3C r r_1 r_2) = 3(r_1 \cdot r_2)(C r r_1^2).$$

We may take the parametrisation of the bumpy circle to be $r = e^{it} +$ a small correction term, where we identify $\mathbf{R}^2$ with $\mathbf{C}$. Then $r_1 = ie^{it} + \cdots$ and $r_2 = -e^{it} + \cdots = -r + \cdots$. So to the first order $r_1 \cdot r_2 = 0$ with $r_2 \cdot r_2 = r_1 \cdot r_1$. Let u, v be the limits of r_1, r_2 as ε reaches 0. Then in this limit

$$Cu^3 - 3Cuv^2 = 0,$$

where $u \cdot v = 0$ and $u \cdot u = v \cdot v$. The expression on the left-hand side of this equation is just four times the harmonic part of $C = d^3hO$, evaluated at u, as is quickly deduced from the complex form for a cubic form on $\mathbf{R}^2$ that we have discussed before, namely

$$\alpha z^3 + 3\bar{\beta} z^2 \bar{z} + 3\beta z \bar{z}^2 + \bar{\alpha} \bar{z}^3.$$

The various assertions we have made follow at once from this on the assumption that the harmonic part of C is not zero, that is that $\alpha \neq 0$. This generically will be the case.

Now, by reference back to Lander's work, we might have expected not three branches of Σ^{111}'s but three branches of Σ^{11111}'s through an umbilical critical point. It seems likely that this is achievable by considering the contact with our surface not only of horizontal circles, as we have implicitly been doing here, but also of arbitrarily oriented circles.

A first result in this direction is a new characterisation of principal directions and of rib-lines, as follows.

Choose any point of a surface and any tangent vector there, and consider any plane through the corresponding tangent line. This will intersect the surface in a curve, and one can show that in general there is exactly one such plane such that the curve of intersection has an A_3 vertex at the point we first thought of. However it may be that the plane is not uniquely determined. This occurs if and only if our tangent direction is principal in which case our section never has a vertex at the point in question unless it lies on a rib-line in which case every section has a vertex there. It is to be expected that for a generic point there will be a finite number of tangent directions for which the vertex on the distinguished section is an A_4 vertex and that there will be curves on the surface of points at which A_5 contact with circles (or a Σ^{11111} singularity) is attainable. Then the general theory would predict that three such curves pass harmonically through each umbilical point as will the curves consisting of the corresponding centres of curvature. For obvious reasons it is not particularly helpful to take an ellipsoid as an example!

The answers to these conjectures should not be difficult to obtain, and are currently being sought by my student James Montaldi. [ADDED IN PROOF. The actual situation is slightly different from that predicted. It is the three limits of tangent lines to A_5 contact circles that lie harmonically around an umbilic, rather than the curves along which the umbilic is approached.]

Bumpy spheres. Bumpy circles have been the stimulus for the work of my student Stelios Markatis [6] on bumpy spheres. Instead of considering height functions $h\colon \mathbf{R}^2 \to \mathbf{R}$ we turn to height functions $h\colon \mathbf{R}^3 \to \mathbf{R}$ and consider the level surfaces round a local maximum or minimum. The most interesting case is that least likely to occur, when these level surfaces are bumpy spheres. So, by the sort of radial transformation we made before, we are led to study surfaces of the form

$$x^2 + y^2 + z^2 + \varepsilon C(x, y, z)^3 = 1,$$

where ε is small and the noncompact component far away is once again disregarded. Markatis has shown that there is in the family of cubic forms $C(x, y, z)^3 + A(x, y, z)(x^2 + y^2 + z^2)$, where A is linear, a unique harmonic cubic and also a unique cubic of the form LMN, where L, M and N are real linear forms. The geometry of the bumpy sphere, namely its configuration of rib-lines and umbilics, is uniquely determined by either of these. Technically the LMN form is easiest to handle.

Obvious examples to look at initially are

$$x^2 + y^2 + z^2 + \varepsilon xyz = 1 \quad \text{(the bumpy cube)},$$

$$x^2 + y^2 + z^2 + \varepsilon(x^3 - 3xy^2) = 1 \quad \text{(the bumpy orange)},$$

$$x^2 + y^2 + z^2 + \varepsilon xy^2 = 1 \quad \text{(the bumpy tennis ball)},$$

and

$$x^2 + y^2 + z^2 + \varepsilon x^3 = 1 \quad \text{(the bumpy sphere of revolution)},$$

these four types being the vertices of a three-dimensional tetrahedron, representing all possible types.

It is not difficult to figure out that in the case of the bumpy cube the sections of symmetry illustrated in Figure 6 are all rib-lines, with elliptic (or rather G_2) umbilics at the eight vertices of the cube and hyperbolic (or rather lemon B_3) umbilics at the mid-points of the twelve sides. At the mid-point of each face there is a flyover point where a red rib-line crosses a blue one. Moreover each of these rib-lines is a bumpy circle, with six vertices, each at a flyover, and so we can mark

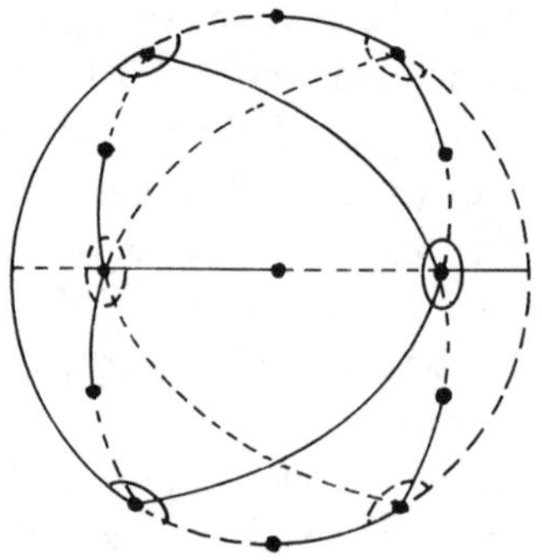

FIGURE 6

four additional flyovers on each of these circles, lying close to but not at the vertices of the cube. Thus we can detect the existence of eight small almost circular rib-lines each surrounding one of the G_2-points. (These were in fact first detected in a computer print-out!) Four of these are blue and four are red. Each crosses in turn the various rib-lines emanating from the vertex, alternately by a flyover and by a crossing on the level (a nongeneric A_4—in fact here a symmetric A_5 or C_3 point, the points of the straight ribs being C_2's).

For the bumpy orange it is clear that the equator and three equally spaced meridian circles through the poles are rib-lines. All are bumpy; there are two nearly circular rib-lines, one blue and one red, entwining around the north pole, and likewise in the southern hemisphere. The positions of two G_2 umbilics, each of index $-\frac{1}{2}$, at the poles, and of six lemon B_3 umbilics, each of index $\frac{1}{2}$, are indicated on Figure 7, which is an orthogonal projection of the northern hemisphere. There are in this case eight umbilics in all, as opposed to the twenty in the case of the bumpy cube.

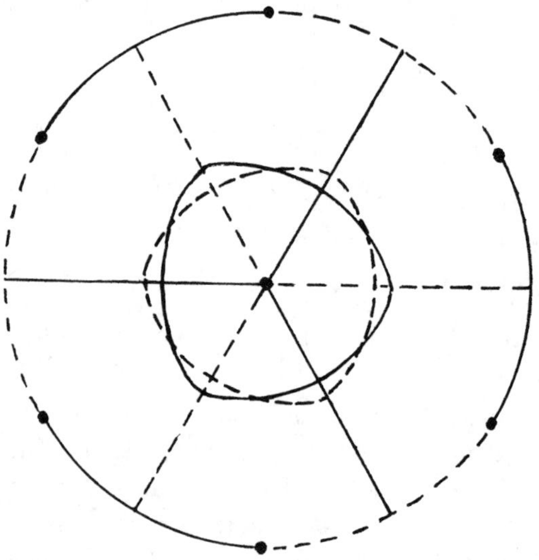

FIGURE 7

The bumpy tennis ball also has eight umbilics, while every point of the equator is a nongeneric umbilic, the poles also being nongeneric umbilics.

Markatis has drawn computer pictures of bumpy spheres of every type and has studied in particular how the twenty umbilics of the bumpy cube reduce to the eight of the bumpy orange or tennis ball. What he has shown is that no matter how the deformation is performed, all deaths occur at once when either six pairs annihilate each other or six triples become six singletons. The latter possibility occurs at an E_6 (or rather F_4) point in the sequence in Figure 8, which illustrates the unknotting of a 'G_2-circle'.

There is a great deal of interest in this three-parameter family of bumpy spheres. Markatis' pictures of one piece of the tetrahedron of all possible types is given in Figure 9. You will notice the bumpy cube at one vertex, the bumpy orange at another and the bumpy tennis ball at a third. A paper by Stelios Markatis, giving details, is in course of preparation.

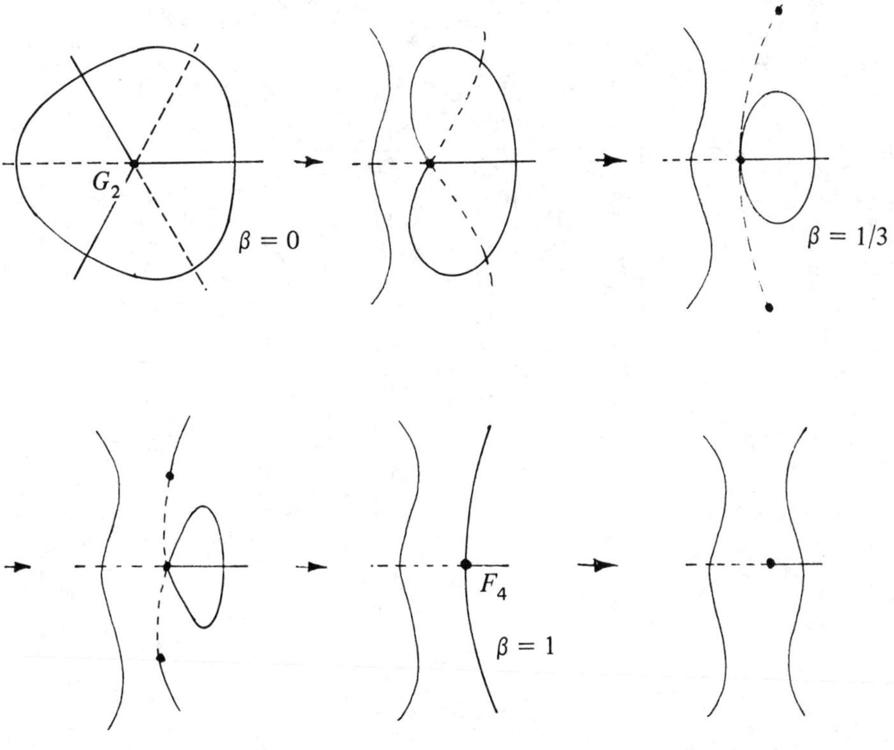

FIGURE 8

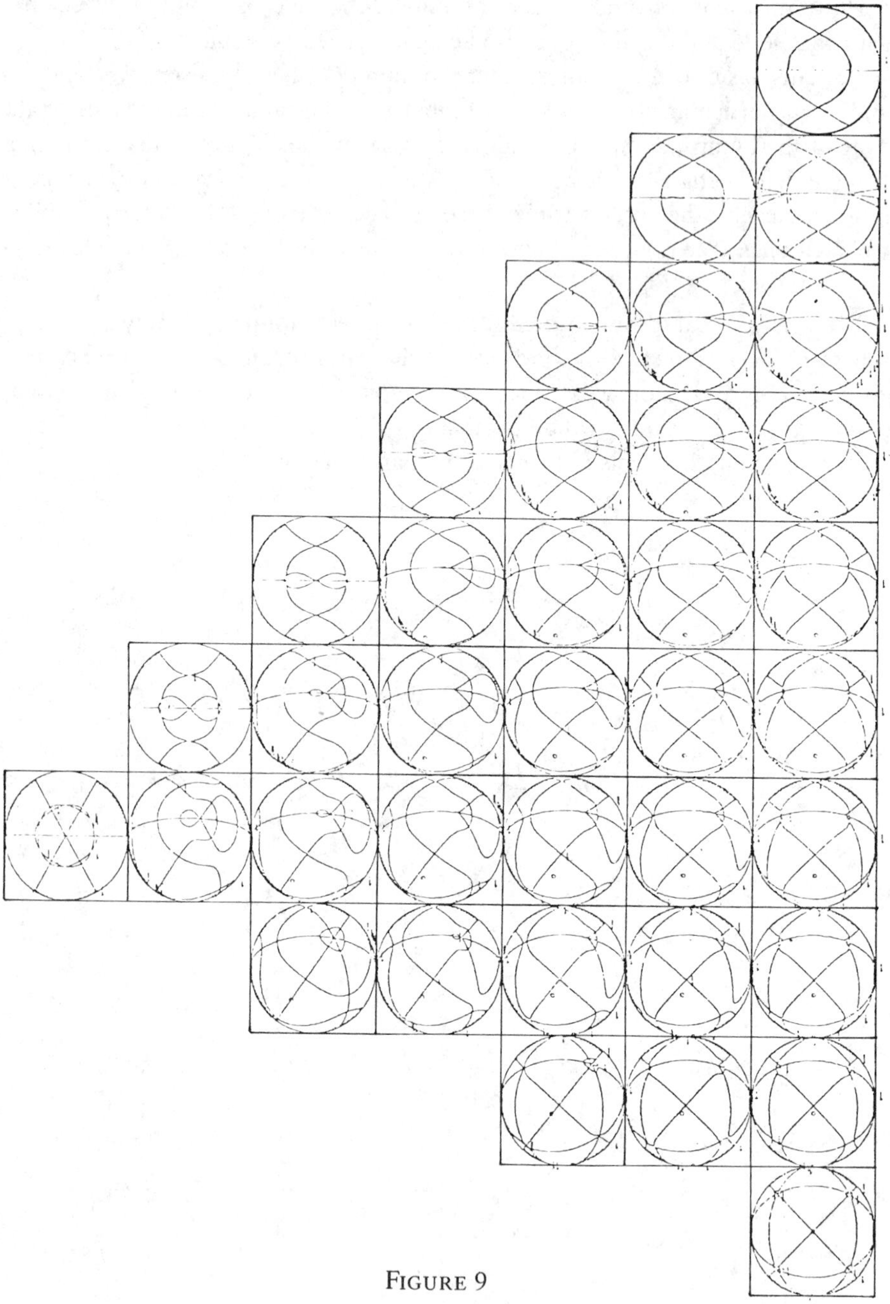

FIGURE 9

REFERENCES

1. V. I. Arnol'd, *Normal forms of functions near degenerate critical points, the Weyl groups A_k, D_k, E_k and Lagrangean singularities*, Funkcional Anal. i Priložen. (4) **6** (1972), 3 − 25 in Russian; Functional Anal. Appl. 254–272. (English)

2. M. V. Berry and J. H. Hannay, *Umbilic points on Gaussian random surfaces*, J. Phys. A **10** (1977), 1809–1821.

3. G. Darboux, *Leçons sur la théorie générale des surfaces*, Vol. 4, Gauthiers-Villars, Paris, 1896.

4. L. Lander, *The structure of the Thom-Boardman singularities of stable germs with type* $\Sigma^{2,0}$, Proc. London Math. Soc. (3) **33** (1976), 113–137.

5. E. Looijenga, *Structural stability of families of* C^∞*-functions and the canonical stratification of* $C^\infty(N)$, Inst. Hautes Etudes Sci. Publ. Math (1974).

6. S. Markatis, *Some generic phenomena in families of surfaces in* $\mathbf{R}^3$, Thesis, Liverpool, 1981.

7. J. Mather, *Stability of* C^∞*-mappings*. VI. *The nice dimensions*, Proc. Liverpool Singularities Sympos. I (C.T.C. Wall, editor), Lecture Notes in Math., vol. 192 Springer-Verlag, Berlin and New York, 1971, pp. 207–253.

8. G. Monge, *Application de l'algèbre à la géométrie*, 2nd ed., Paris, 1850.

9. I. R. Porteous, *The normal singularities of a submanifold*, J. Differential Geom. **5** (1971), 543–564.

10. ______, *Probing singularities*, these PROCEEDINGS.

11. P. Slodowy, *Simple singularities and simple algebraic groups*, Lecture Notes in Math., vol. 815, Springer-Verlag, Berlin and New York, 1981.

12. C.T.C. Wall, *Geometric properties of generic differentiable manifolds*, Lecture Notes in Math., vol. 597, Springer-Verlag, Berlin and New York, 1977, pp. 707–774.

UNIVERSITY OF LIVERPOOL, ENGLAND

Proceedings of Symposia in Pure Mathematics
Volume **40** (1983), Part 2

PROBING SINGULARITIES

IAN R. PORTEOUS

Two related topics are discussed in this paper. In the first part an elementary recognition tool is introduced that works well for singularities of germs of analytic functions $\mathbf{C}^2, 0 \to \mathbf{C}, 0$ or C^∞-functions $\mathbf{R}^2, 0 \to \mathbf{R}, 0$ of low codimension and is very convenient in applications to the differential geometry of surfaces (see [7]). It can be sharpened and put to more extended use. In particular it has prompted a fresh presentation of the Thom-Boardman singularity types for map-germs $\mathbf{R}^n, 0 \to \mathbf{R}^p, 0$. In the second part of the paper a sketch is given of a new "three-line" proof that these types are smooth submanifolds of appropriate jet spaces of appropriate codimension.

Stelios Markatis worked with the author on the initial exploration of Tables I and II.

1. Probing function-germs $\mathbf{R}^2, 0 \to \mathbf{R}, 0$. Suppose that we have a C^∞ function-germ $\mathbf{R}, 0 \overset{f}{\leftarrow} \mathbf{R}^2, 0$ and a C^∞ curve-germ $\mathbf{R}^2, 0 \overset{\phi}{\leftarrow} \mathbf{R}, 0$, the *probe*. The curve ϕ can then be composed with each of the derivatives of f as follows:

$$L(0^m\mathbf{R}^2, \mathbf{R}) \overset{d^mf}{\leftarrow} \mathbf{R}^2 \overset{\phi}{\leftarrow} \mathbf{R}$$

where $0^m\mathbf{R}^2$ denotes the m-fold symmetric power of $\mathbf{R}^2$.

In the discussion which follows it is convenient to use the capital letter alphabet to denote the successive derivatives of f at 0: thus $A = df0$, $B = d^2f0$, $C = d^3f0$, and so on, with Ω denoting the first infinite ordinal in the alphabet, and to use the lower case alphabet to denote the successive derivatives of ϕ at 0, it being permissible here to regard the latter as vectors in $\mathbf{R}^2$. The successive derivatives of the composite of two maps are given by the analog of the Leibnitz formula known as Faa' de Bruno's formula [**4**], the history of which is to be found in a paper by Felice Ronga in these proceedings [**8**]. The first few derivatives of $(d^mf) \circ \phi$ are

1980 *Mathematics Subject Classification.* Primary 58C27.

0082-0717/81/0000-0202/$03.25

exhibited in Table I for the first few values of m. It is to be noted that, for any nonnegative integers m, n,

$$d^{n+1}(d^m f)\circ\phi 0=\sum_{k=0}^{n}\binom{n}{k}d^{n-k}(d^{m+1}f\circ\phi)0(d^{k+1}\phi 0).$$

TABLE I

0	Aa	Ba^2+Ab	$Ca^3+3Bab+Ac$	$Da^4+6Ca^2b+3Bb^2$ $+4Bac+Ad$	$Ea^5+10Da^3b+15Cab^2+10Ca^2c$ $+10Bbc+5Bad+Ae$
	A	Ba	Ca^2+Bb	$Da^3+3Cab+Bc$	$Ea^4+6Da^2b+3Cb^2$ $+4Cac+Bd$
		B	Ca	Da^2+Cb	$Ea^3+3Dab+Cc$
			C	Da	Ea^2+Db
				D	Ea
					E

The next entry to the right in the first row is

80

$$Fa^6+15Ea^4b+45Da^2b^2+20Da^3c+15Cb^3+60Cabc+15Ca^2d+10Bc^2+15Bbd+6Bae+Af$$

For example,

$$\begin{array}{l} Ea^5+10Da^3b+15Cab^2+10Ca^2c+10Bbc+5Bad+Ae \\ = \begin{array}{l} (Ea^4+6Da^2b+3Cb^2+4Cac+Bd)a \\ +4(Da^3+3Cab+Bc)b \\ +6(Ca^2+Bb)c \\ +4(Ba)d \\ +Ae. \end{array} \end{array}$$

It follows that if the first few entries in any row are zero then each of the entries in the preceding row directly above these is also equal to zero.

The game, for a given function-germ f, is to find the curve probes ϕ which make the greatest number of entries zero in Table I. It is convenient to code such a probe ϕ by a word of nondecreasing letters. For example, the code CDE describes a probe ϕ for which the first entry in the fourth row is nonzero but for which the entries up to Ca in the third row, up to $Da^3+3Cab+Bc$ $(=Da^3$, when $B=0$, $Ca=0)$ in the second row, and up to $Ea^5+\cdots$ $(=Ea^5$, when $A=0$, $B=0$, $Ca=0$, $Da^3=0)$ in the first row, are all zero, but the next entry is zero. Likewise, the code $B\Omega$ describes a probe ϕ for which the first entry in the third row is nonzero, but for which the entries up to Ba in the second row and every entry in the first row '$ad\Omega$' is zero.

We shall consider only nonsingular curve probes ϕ; that is, probes for which $a=d\phi 0$ is nonzero. The higher derivatives of ϕ may well be zero, in particular cases, and may indeed be varied simply by reparametrising ϕ. Generally it will be convenient to assume that coordinates in $\mathbf{R}^2$ around 0 have been so chosen that $a=\binom{0}{1}$, and that ϕ is so parametrised that $b=\binom{\beta}{0}$, $c=\binom{\gamma}{0}$, and so on.

TABLE II

A_1: AA, 0 *or* $2A\Omega$				$\tilde{E}_7$: $CCCC$, 0 or 2 or $4CCC\Omega$	
A_2: BB				T_{245}: $CCDD$; 0 or $2CCC\Omega$	
A_3: CC, 0 or $2B\Omega$				T_{246}: $CCEE$, 0 or $2CCCD\Omega$; 0 or $SCCC\Omega$	
A_4: DD	D_4: 1 or $3BB\Omega$			T_{247}: $CCFF$; 0 or $2CCC\Omega$	
				⋮	
A_5: EE, 0 or $2C\Omega$	D_5: BCC; $BB\Omega$				
A_6: FF	D_6: BDD, 0 or $2BC\Omega$; $BB\Omega$	E_6: CCC		T_{255}: $2CCDD$	
A_7: GG, 0 or $2D\Omega$	D_7: BEE; $BB\Omega$	E_7: $CC\Omega$		T_{256}: $CCDD$; $CCEE$ 0 or $2CCD\Omega$	
A_8: HH	D_8: BFF, 0 or $2BD\Omega$; $BB\Omega$		E_8: DDD	⋮	
⋮	⋮			S_{245}: $CDDD$; $CCC\Omega$	
				S_{246}: $CDD\Omega$; $CCC\Omega$	S_{255}: $DDDD$
	$\tilde{E}_8$: DDE, 0 or $2CEE$, 1 or $3CD\Omega$			S_{247}: $CEEE$; $CCC\Omega$	S_{256}: $DDD\Omega$
	$\tilde{E}_8^{\infty}$: 1 or $3DD\Omega$, 0 or $2CEEE$		$\tilde{E}_8^0$: EEE, $CD\Omega$		
	T_{237}: DDE, CFF		S_{237}: EEF	S_{247}^0: FFF	
	T_{238}: DDE, CGG, 0 or $2CE\Omega$		S_{238}: EEG, $DE\Omega$		
	T_{239}: DDE, CHH		S_{238}^0: $EE\Omega$	S_{239}: FFG, 0 or $2DGG$	S_{239}^0: GGG
	⋮				

Table II shows the highest codes that are attainable for a range of singularity types, where the notations are those of Arnol'd [**1**] or Brieskorn [**3**]. Clearly within this range the probes are diagnostic of the various types.

Thus A_1 germs are characterised by $A = 0$ and the nonexistence of $a \in \mathbf{R}^2 \setminus \{0\}$ such that $Ba = 0$. There are then two cases according to whether the quadratic equation for a, $Ba^2 = 0$, is elliptic or hyperbolic. It cannot be parabolic, for in that case we should have $Ba = 0$, with $a \neq 0$, which we have assumed not to be the case. In the hyperbolic case for either of the root directions a the second derivative b of ϕ at 0 can then be chosen such that $Ca^3 + 3Bab + Ac = Ca^3 + 3Bab = 0$; then c can be chosen such that $Da^4 + 6Ca^2b + 3Bb^2 + 4Bac = 0$, and so on along the first row of Table I $ad\Omega$. Thus the best probe in the elliptic case has code AA, but in the hyperbolic case there are two probes with code $A\Omega$.

The A_2 germs are characterised by $A = 0$ and $Ba = 0$, for some $a \neq 0$, but $B \neq 0$, with the nonexistence of b such that $Ca^2 + Bb = 0$. It is easy to see that an equivalent condition is that $Ca^3 \neq 0$, for, when $Ba = 0$, $Ca^3 = 0$ is implied by $Ca^2 + Bb = 0$, and the converse is easily established also. Thus the best probe in this case has code BB.

The A_3 germs are characterised by $A = 0$, $Ba = 0$, for some $a \neq 0$, but $B \neq 0$, and $Ca^3 = 0$, with one further condition, namely that the equations $Ca^2 + Bb = 0$ and $Da^4 + 6Ca^2b + 3Bb^2 = 0$ are not simultaneously solvable for b, this being equivalent to asserting that the quadratic equation for b is not parabolic. In the elliptic or sterile case the best probe has code CC. In the hyperbolic or fertile case such a probe also exists but additionally there are two probes with code $B\Omega$.

The A_4 case is just the parabolic case that we have just excluded, with the additional condition that $Ea^5 + \cdots = Ea^5 + 10Da^3b + 15Cab^2 \neq 0$. However c can be found such that $Da^3 + 3Cab + Bc = 0$; so the best probe in this case has code DD.

The higher A_n's are handled analogously.

Now suppose that $A = 0$ and $B = 0$. The equation $Ca^3 = 0$ is always solvable for $a \neq 0$, but there are several cases, the *elliptic* case, when the Hessian of the cubic form C is elliptic and the cubic equation has three distinct real root directions, the *hyperbolic* case, when the Hessian of C is hyperbolic, but the cubic has only one real root direction, the *parabolic* case, when the Hessian is parabolic and two root directions of the cubic equation coincide, the *perfect* case, when the Hessian vanishes and all three root directions of the cubic coincide, this being the case that the cubic form is a perfect cube, and finally the case that C itself is zero. The perfect case is characterised by the existence of $a \neq 0$ such that $Ca = 0$, when a is then the repeated root, while the parabolic case is characterised by the existence of $a \neq 0$ such that $Ca^2 = 0$, but $Ca \neq 0$, when a again is the repeated root.

The D_4 singularities are characterised by $A = 0$ and $B = 0$ and the nonexistence of any $a \neq 0$ such that $Ca^2 = 0$. Thus there are just two cases, the elliptic and the hyperbolic. In the elliptic case there are three probes with code $BB\Omega$ and in the hyperbolic case just the one.

The D_5 singularity type, by contrast, is completely characterised by the existence of $a \neq 0$ such that $A = 0$, $B = 0$, $Ca^2 = 0$, but $Ca \neq 0$, and $Da^4 + \cdots = Da^4 \neq 0$, C in this case being parabolic but not perfect. The best probe in this case thus has code BCC.

If $Da^4 = 0$ we are faced with the equations $Da^3 + 3Cab = 0$ and $Ea^5 + 10Da^3b + 15Cab^2 = 0$ for b and the situation is analogous to the one encountered in the A_3 case. If there is no common root then there is always a probe with code BDD. Such a probe exists in the hyperbolic case also but there are then two additional probes with code $BC\Omega$. These are the D_6 singularity types. The D_7 case is just the parabolic case where the above equations do have a common root, with the further condition that $Fa^6 + \cdots = Fa^6 + 15Ea^4b + 45Da^2b^2 + 15Cb^3 \neq 0$. The best code is then BEE. The higher D_n singularities are handled analogously.

The singularity types E_6, E_7 and E_8 are briefly dealt with. For E_6 there exists $a \neq 0$ such that $A = 0$, $B = 0$ and $Ca = 0$, but $C \neq 0$ and $Da^4 \neq 0$, this clearly corresponding to the code CCC. As we have already remarked, the equation $Ca = 0$ is equivalent to the assertion that the cubic equation $Ca^3 = 0$ has three coincident roots; that is, that the cubic form is a perfect cube. For E_7 there exists $a \neq 0$ such that $A = 0$, $B = 0$ and $Ca = 0$, but $C \neq 0$, and $Da^4 = 0$ but $Da^3 \neq 0$. But then the equation $Ea^5 + \cdots = Ea^5 + 10Da^3b = 0$ is solvable for b, and so on along the top row of Table I $ad\Omega$. So the code in this case is $CC\Omega$. Finally, for E_8 there exists $a \neq 0$ such that $A = 0$, $B = 0$, $Ca = 0$, but $C \neq 0$, and $Da^3 = 0$, but $Ea^5 + \cdots = Ea^5 \neq 0$. Then there exists b such that $Da^2 + Cb = 0$. So the code is DDD.

The next stage is of greater interest, namely when $A = 0$, $B = 0$, $C \neq 0$, but there exists $a \neq 0$ such that $Ca = 0$, $Da^3 = 0$ and $Ea^5 = 0$. Our attention is then drawn simultaneously to the three equations

$$Da^2 + Cb = 0 \in L(0^2\mathbf{R}^2, \mathbf{R}),$$
$$Ea^4 + 6Da^2b + 3Cb^2 = 0 \in L(\mathbf{R}^2, \mathbf{R}),$$

and $Fa^6 + 15Ea^4b + 45Da^2b^2 + 15Cb^3 = 0 \in \mathbf{R}$ for $b \in \mathbf{R}^2$. Note that the left-hand sides of the second and first equations are up to numerical factors the first and second 'derivatives' of the left-hand side of the third equation 'with respect to b'. Despite appearances the first equation is essentially a single linear equation rather than three and the second is a single quadratic equation rather than two, since $Ca = 0$, $Da^3 = 0$ and $Ea^5 = 0$. With $a = \binom{0}{1}$ the 6-jet of f is then reducible to a cubic in x and y^2

$$x^3 + 3\lambda x^2y^2 + 3\mu xy^4 + \nu y^6,$$

where, after further changes of coordinates, either λ or μ may be chosen to be zero and ν to be 1, -1 or 0. The three equations for $b = \binom{\beta}{0}$, after cancellation of numerical factors, then become

$$8\nu + 3\mu(4\beta) + 3\lambda(2\beta^2) + \beta^3 = 0,$$

$$4\mu + 2\lambda(2\beta) + \beta^2 = 0 \quad \text{and} \quad 2\lambda + \beta = 0.$$

The general $\tilde{E}_8$ case is when no two of these three equations have a common root, in which case the cubic is either elliptic or hyperbolic as also is the quadratic. There will certainly be a probe with code DDE, with $\beta = -2\lambda$. There will also be two supplementary probes of type CEE when the quadratic is hyperbolic, which is certainly the case when the cubic is elliptic and possibly so when the cubic is hyperbolic. Finally, since the cubic has either one or three real roots there is a further probe of type $CD\Omega$ in the case that the cubic is hyperbolic and three further such probes when the cubic is elliptic. The canonical form for $f(x, y)$ is $x^3 + 3\mu xy^4 + y^6$, with $\mu \neq 0$ and $\mu^3 \neq -\frac{1}{4}$, the quadratic equation being elliptic if $\mu > 0$ and hyperbolic if $\mu < 0$ and the cubic equation being elliptic if $\mu^3 < -\frac{1}{4}$ and hyperbolic if $\mu^3 > -\frac{1}{4}$.

The case that the quadratic equation is parabolic is when the linear and quadratic equations have a common solution; that is, when there exists b such that $Da^2 + Cb = 0$ and $Ea^4 + 3Da^2b$ $(= Ea^4 + 6Da^2b + 3Cb^2) = 0$. There is then a probe of type EEE, for, since $Ca = 0$ and $Da^2 + Cb = 0$, c can be found such that $Ea^3 + 3Dab + Cc = 0$. In this case the 5-jet of f at 0 is completable to a perfect cube. In coordinate terms the 5-jet is completable to a perfect cube in x and y^2, by the addition of a suitable multiple of y^6, and so by an obvious coordinate change it is reducible to x^3. That is, coordinates may be so chosen around 0 in $\mathbf{R}^2$ that $D = 0$ and $E = 0$ for the map-germ f. Then $b = 0$ and $c = 0$ for the probe. Provided that the common root b does not satisfy the cubic equation there is no probe of type EEF and the canonical form reduces to $x^3 + y^6$. We denote this singularity type by $\tilde{E}_8^0$, the modulus μ being 0. There is a supplementary probe of type $CD\Omega$ in this case, the cubic for b being hyperbolic.

The next case to consider is where the linear and cubic equations have a common solution. This case turns out to be that in which the 6-jet is reducible to $x^3 \pm 3xy^4$, or, if one prefers it, to $x^3 \pm xy^4$, the sign being positive or negative according to whether the cubic is hyperbolic or elliptic or, equivalently, whether the quadratic is elliptic or hyperbolic. There are accordingly either one or three probes with code $DD\Omega$, with two supplementary probes of type CEE in the latter case. These are the singularities of type $\tilde{E}_8^\infty$.

The types $\tilde{E}_8^0$ and $\tilde{E}_8^\infty$ are known as the *equianharmonic* and *harmonic* $\tilde{E}_8$'s respectively, preferably regarded as distinct singularity types.

The third possibility is when the quadratic and cubic equations have a common solution. Then the cubic is parabolic, the infinite run along the top row of Table I does not in general occur, but $c \in \mathbf{R}^2$ can be found such that $Fa^5 + 10Ea^3b + 15Dab^2 + 10(Da^2 + Cb)c = 0$. The 6-jet of f at 0 can then be taken to be $x^3 + \lambda x^2y^2$, with $\lambda \neq 0$, provided that the solution of the linear equation remains distinct. In the case that $Ga^7 + \cdots \neq 0$ the canonical form is $x^3 + \lambda x^2y^2 + y^7$, the code CFF supplements the code DDE and the singularity is of type T_{237}. The further study of the T_{23r} singularities is analogous to our study of the D_k singularities.

There remains the possibility that all three equations share a common root. In that case we can suppose that $D = 0$, $E = 0$ and that the common root b is equal to zero. Then the first three rows of Table I from and to the right of the 'E' column reduce to:

0	0	Ga^7	$Ha^8 + 56Fa^5c$	$Ia^9 + \cdots$
0	Fa^5	$Ga^6 + 10Cc^2$	$Ha^7 + 35Fa^4c + 35Ccd$	$Ia^8 + \cdots$
Cc	$Fa^4 + Cd$	$Ga^5 + Ce$	$Ha^6 + 20Fa^3c + Cf$	$Ia^7 + \cdots$

Suppose first that $Fa^5 \neq 0$ and that $Ga^7 \neq 0$. This is the first 'triangular' singularity type, denoted by Brieskorn [3] by S_{237}, with canonical form $x^3 + \alpha xy^5 + y^7$, with $\alpha \neq 0$. The best code is EEF. If $\alpha = 0$ then $Fa^5 = 0$ and $d \in \mathbf{R}^2$ can be found such that $Fa^4 + Cd = 0$. The best code is now FFF. We denote this distinct type by S_{237}^0.

Next suppose that $Fa^5 \neq 0$, that $Ga^7 = 0$ and that $Cc = 0$ and $Ha^8 + 56Fa^5c = 0$ are not simultaneously solvable for c, that is that $Ha^8 \neq 0$. Then the best code is EEG, with supplementary code $DE\Omega$. This is the triangular singularity of type S_{238}, with canonical form $x^3 + xy^5 + \alpha y^8$, with $\alpha \neq 0$. When $\alpha = 0$ the two equations have the common solution $c = 0$, $Ha^8 = 0$ and there is a probe of type $EE\Omega$. We denote this distinct type by S_{238}^0.

Finally suppose that $Fa^5 = 0$ and $Ga^7 = 0$ but that $Ha^8 \neq 0$. Then $Cc = 0$ and $Ga^6 + 10Cc^2 = 0$ in general do not have a common solution c. This is the triangular singularity of type S_{239}, with canonical form $x^3 + \alpha xy^6 + y^8$, with $\alpha \neq 0$, the quadratic being elliptic or hyperbolic according as α is positive or negative. The best code is FFG, with two supplementary probes of type DGG in the hyperbolic case. In the parabolic case $c = 0$, $Ga^6 = 0$ and $e \in \mathbf{R}^2$ can be found such that $Ga^5 + Ce = 0$. The canonical form is then $x^3 + y^8$ and the best code is GGG. We denote this distinct type by S_{239}^0.

Until now we have assumed that $A = 0$ and $B = 0$ but that $C \neq 0$. We look very briefly at the first few options when $C = 0$ also, when the first four rows of the D, E and F columns of Table I reduce to:

Da^4	$Ea^5 + 10Da^3b$	$Fa^6 + 15Ea^4b + 45Da^2b^2 + 20Da^3c$
Da^3	$Ea^4 + 6Da^2b$	$Fa^5 + 10Ea^3b + 15Dab^2 + 10Da^2c$
Da^2	$Ea^3 + 3Dab$	$Fa^4 + 6Ea^2b + 3Db^2 + 4Dac$
Da	$Ea^2 + Db$	$Fa^3 + 3Eab + Dc$

We proceed just far enough to note the first cases of failure of the probe analysis in its simple form, but nevertheless to determine characteristic codes for the remaining five triangular singularity types which can occur for function germs $\mathbf{R}^2, 0 \to \mathbf{R}, 0$.

The $\tilde{E}_7$ singularity type is characterised by the nonexistence of $a \neq 0$, real or complex, such that $Da^3 = 0$. By mentioning the complex case here we are able to exclude the case that the quartic form D has repeated complex conjugate roots. Any probe satisfies $CCCC$ trivially, but there may be zero, two or four real probes of code $CCC\Omega$, according to the special nature of the quartic form D.

The case that $Da^3 = 0$ has one nonzero solution a but $Da^2 = 0$ has none leads to the T_{24r} series of singularity types, where $r > 4$, while if $Da^3 = 0$ has two distinct solutions, either both real or a complex conjugate pair, that is if the 4-jet is reducible either to x^2y^2 or to $(x^2 + y^2)^2$, then we have the T_{2qr} singularity types, with q and r both greater than 4. In the complex case they will be of type T_{2qq}. The characteristic probe codes are easily determined in these cases.

Next we suppose that $Da^2 = 0$ has a nonzero solution a but that $Da = 0$ has none. If $Ea^5 \neq 0$ we have the triangular singularity type S_{245}, with canonical form $x^3y + \alpha xy^4 + y^5$ and code $CDDD$. If $Ea^5 = 0$ but $Ea^4 \neq 0$ we have the triangular type S_{246} with canonical form $x^3y + \alpha x^2y^3 + xy^4$ and code $CDD\Omega$. If $Ea^4 = 0$ but $Fa^6 \neq 0$ we have the type S_{247} with canonical form $x^3y + \alpha xy^5 + y^6$ and code $CEEE$, for b can then be found such that $Ea^3 + 3Dab = 0$.

Finally we suppose that $Da = 0$ has a nonzero solution a but that $D \neq 0$. If $Ea^5 \neq 0$ we have the triangular type S_{255} with canonical form $x^4 + y^5 + \alpha x^2y^3$ and code $DDDD$, while if $Ea^5 = 0$ but $Ea^4 \neq 0$ we have the type S_{256} with canonical form $x^4 + xy^4 + \alpha y^6$ and code $DDD\Omega$.

In none of these five cases does any probe pick out any particular value of the modulus α as being significantly different from the others.

In conclusion we suppose still that $Da = 0$ has a nonzero solution a, with $D \neq 0$, but $Ea^4 = 0$. Here our probe analysis fails to analyse the quadratic forms in x^2 and y^3 that then turn up.

Despite this apparent failure of probe analysis, curve probes may be used to define all the Thom-Boardman singularities of function-germs $\mathbf{R}^2, 0 \to \mathbf{R}, 0$. The only probes used in this context are those whose code word, of finite length, consists in repetitions of a single letter. The best probes for the first few Thom-Boardman symbols are set out in Table III.

TABLE III

Thom-Boardman symbol	Best probe	Singularity types included
2, 0	AA	A_1
2, 1, 0	BB	A_2
2, 1, 1, 0	CC	A_3 and so on
2, 2, 0	BBB	D_k, $k \geq 4$
2, 2, 1, 0	CCC	E_6, E_7
2, 2, 1, 1, 0	DDD	E_8, $\tilde{E}_8$, $\tilde{E}_8^\infty$, T_{23r}, $r \geq 7$
2, 2, 1, 1, 1, 0	EEE	$\tilde{E}_8^0$, S_{237}, S_{238}, S_{238}^0
2, 2, 1, 1, 1, 1, 0	FFF	S_{237}^0, S_{239}
2, 2, 1, 1, 1, 1, 1, 0	GGG	S_{239}^0 and so on
2, 2, 2, 0	$CCCC$	$\tilde{E}_7$, T_{2qr}, $q \geq 4$, $r \geq 5$, S_{245}, S_{246}, S_{247}
2, 2, 2, 1, 0	$DDDD$	S_{255}, S_{256} and so on.

II. Probing Thom-Boardman singularities. We have just been using curve probes to characterise Thom-Boardman singularity types of map-germs $\mathbf{R}^2, 0 \to \mathbf{R}, 0$. In this section we generalise our concept of a probe to treat the general case of Thom-Boardman singularities of C^∞ map-germs $\mathbf{R}^n, 0 \to \mathbf{R}^p, 0$.

In what follows $L_r(p, n)$ will denote $L(0^r\mathbf{R}^n, \mathbf{R}^p)$ and $f_r \in L_r(p, n)$ will denote the rth derivative d^rf0 of a C^∞ map-germ $f\colon \mathbf{R}^n, 0 \to \mathbf{R}^p, 0$. The open subset of $L_1(p, n)$ consisting of all injective linear maps from $\mathbf{R}^n$ to $\mathbf{R}^p$ will be denoted by $\mathrm{GL}(p, n)$, with $\mathrm{GL}(k) = \mathrm{GL}(k, k)$. The set $\mathrm{GL}(p, n)$ is, of course, empty if $p < n$. A Boardman index is just a finite nonincreasing sequence of nonnegative integers, Boardman indices of the same length being ordered lexicographically.

Certain vector bundles are associated to a Boardman index. Consider, for example, the index (k, j, i). For each integer $n > k$ we define

$$X_{kji}^n = \mathrm{GL}(n, k) \times \mathrm{GL}(k, j) \times \mathrm{GL}(j, i).$$

For each positive integer p the trivial bundle over X_{kji}^n with fibre $L_3(p, k)$ then has a quotient bundle

$$\pi_{kji}^{p,n}\colon B_{kji}^{p,n} \to X_{kji}^n,$$

whose fibre at

$$(\phi_1, \chi_1, \psi_1) \in X_{kji}^n$$

consists of the subspace of $L(\mathbf{R}^i, L(\mathbf{R}^j, L(\mathbf{R}^k, \mathbf{R}^p)))$ and quotient space of $L_3(p, k)$ consisting of all $f_3(\phi_1, \chi_1, \psi_1)(\phi_1\chi_1)\phi_1$, where $f_3 \in L_3(p, n)$. Let λ_{kji} denote the rank of the bundle $\pi_{kji}^{1,n}$. Then $p\lambda_{kji}$ is the rank of the bundle $\pi_{kji}^{p,n}$; similarly for other Boardman indices.

We are now in a position to define the Thom-Boardman singularity type Σ^K, for map-germs f: $\mathbf{R}^p, 0 \leftarrow \mathbf{R}^n, 0$, for any Boardman index $K = (k, j, i, \ldots)$, with $n \geqslant k$. For simplicity we give the details for $K = (k, j, i)$.

We say that the map-germ f: $\mathbf{R}^p, 0 \leftarrow \mathbf{R}^n, 0$ is of type Σ^{kji} if there exists a probe consisting of map-germs

$$\mathbf{R}^n, 0 \overset{\phi}{\leftarrow} \mathbf{R}^k, 0 \overset{\chi}{\leftarrow} \mathbf{R}^j, 0 \overset{\psi}{\leftarrow} \mathbf{R}^i, 0$$

with first differentials ϕ_1, χ_1, ψ_1 injective, such that

$$d(f\phi)0 = 0, \qquad d(d(f\phi)\chi)0 = 0,$$
$$d(d(d(f\phi)\chi)\psi)0 = 0,$$

provided that such a probe does not exist for any larger Boardman index of the same length (k', j', i'), the singularity type Σ^{kji} itself being the set of 3-jets of germs of map-germs f: $\mathbf{R}^p, 0 \leftarrow \mathbf{R}^n, 0$ for which there exist

$$\phi_1 \in \mathrm{GL}(n, k), \quad \chi_1 \in \mathrm{GL}(k, j), \quad \psi_1 \in \mathrm{GL}(j, i),$$
$$\phi_2\chi_1 = d((d\phi)\chi) \in B^{n,k}_{kj}, \quad \chi_2\psi_1 = d((d\chi)\psi)0 \in B^{k,j}_{ji}$$

and

$$\phi_3(\chi_1\psi_1)\chi_1 + \phi_2\chi_2\psi_1 = d(d((d\phi)\chi)\psi)0 \in B^{n,k}_{ji},$$

for which the stated conditions all hold, namely

$$f_1\phi_1 = 0, \qquad f_1(\phi_1\chi_1)\phi_1 + f_1(\phi_2\chi_1) = 0,$$

and

$$f_3(\phi_1\chi_1\psi_1)(\phi_1\chi_1)\phi_1 + f_2(\phi_2\chi_1\psi_1\chi_1)\phi_1 + f_2(\phi_1\chi_2\psi_1)\phi_1 + f_2(\phi_1\chi_1)(\phi_2\chi_1\psi_1)$$
$$+ f_2(\phi_1\chi_1\psi_1)(\phi_2\chi_1) + f_1(\phi_3(\chi_1\psi_1)\chi_1 + \phi_2\chi_2\psi_1) = 0,$$

together with the maximality condition.

If such a probe (ϕ, χ, ψ) exists it will not be unique, for if γ: $\mathbf{R}^k, 0 \leftarrow \mathbf{R}^k, 0$, ζ: $\mathbf{R}^j, 0 \leftarrow \mathbf{R}^j, 0$ and η: $\mathbf{R}^i, 0 \leftarrow \mathbf{R}^i, 0$ are any germs of diffeomorphism at 0 then a new probe (ϕ', χ', ψ') is defined by the diagram

$$\begin{array}{ccccccc}
 & & \mathbf{R}^k & \overset{\chi}{\leftarrow} & \mathbf{R}^j & \overset{\psi}{\leftarrow} & \mathbf{R}^i \\
\mathbf{R}^n & \overset{\phi}{\swarrow} & \uparrow \gamma & & \uparrow \zeta & & \uparrow \eta \\
 & \nwarrow \phi' & & & & & \\
 & & \mathbf{R}^k & \overset{\chi'}{\leftarrow} & \mathbf{R}^j & \overset{\psi'}{\leftarrow} & \mathbf{R}^i
\end{array}$$

where the derivatives of γ, ζ, η are related to those of ϕ, χ, ψ, ϕ', χ', ψ' by the equations one gets by differentiating the commutativity relations

$$\phi\gamma = \phi', \quad \chi\zeta = \gamma\chi', \quad \psi\eta = \zeta\psi'.$$

It turns out that an adequate partial converse holds; namely, that if the probes (ϕ, χ, ψ) and (ϕ', χ', ψ') each satisfy the Σ^{kji} condition, then there exist unique

$$\gamma_1 \in \mathrm{GL}(k), \quad \zeta_1 \in \mathrm{GL}(j), \quad \eta_1 \in \mathrm{GL}(i),$$

$$\gamma_2\chi_1' = d((d\gamma)\chi')0 \in B_{kj}^{k;k}, \quad \zeta_2\eta_1' = d((d\zeta)\eta')0 \in B_{ji}^{j;j}$$

and

$$\gamma_3(\chi_1'\psi_1')\chi_1' + \gamma_2\chi_2'\psi_1' = d(d((d\gamma)\chi')\psi')0 \in B_{kji}^{k;k},$$

for which the relevant relations between the various derivatives all hold. (I am grateful to Terry Gaffney for assisting in the proof of this. Details will appear elsewhere.)

We now indicate the proof that the Thom-Boardman singularities are submanifolds of the jet space of the correct codimension, taking as our model the proof in the particular case Σ^k.

We require one elementary lemma.

LEMMA. *Let X be a smooth manifold, let π: $B \to X$ be a vector bundle, let Π be a smooth submanifold of B of codimension c and let Γ be a smooth bundle of d-dimensional Lie groups over X such that, for each $x \in X$, Γ_x acts effectively on B_x on the right, $\Pi \cap B_x$ being an orbit of the action, or null. Then $\Sigma = \pi(\Pi)$ is a smooth submanifold of X of codimension $c - \operatorname{rk}\pi + d$, where $\operatorname{rk}\pi$ denotes the dimension of the fibres of π.*

In the case of Σ^k, X will be $L_1(p, n)$ and Γ the trivial bundle over X with fibre $\mathrm{GL}(k)$, a maximal probe $\mathbf{R}^n, 0 \overset{\phi}{\leftarrow} \mathbf{R}^k, 0$ for a Σ^k map-germ $\mathbf{R}^p, 0 \overset{f}{\leftarrow} \mathbf{R}^n, 0$ being one for which the sequence $(f_1, \phi_1, 0)$ of linear maps

$$\mathbf{R}^p \overset{f_1}{\leftarrow} \mathbf{R}^n \overset{\phi_1}{\leftarrow} \mathbf{R}^k \overset{0}{\leftarrow} \{0\}$$

is exact.

Consider the bundle

$$\pi: L_1(p, n) \times \mathrm{GL}(n, k) = B \to L_1(p, n) = X$$

and let Π be the subset of B consisting of all pairs (f_1, ϕ_1) such that the sequence $(f_1, \phi_1, 0)$ is exact. Then Π is a smooth submanifold of B of codimension kp. For the map $B \to L(p, k)$; $(f_1, \phi_1) \to f_1\phi_1$ has differential matrix $(\circ\,\phi_1 \; f_1\,\circ)$ at (f_1, ϕ_1), and this is surjective, since $\circ\,\phi_1$: $L_1(p, n) \to L_1(p, k)$ is surjective, since ϕ_1 is injective. Moreover, the condition that $\dim\ker f_1 \leqslant k$ is open in $L_1(p, n)$.

Establishing that Π is a smooth submanifold of B of codimension kp is 'line one'. 'Line two' consists in the observation that Π is a principal $\mathrm{GL}(k)$ bundle. For suppose that $f_1\phi_1 = 0$ and $f_1\phi_1' = 0$, where the sequence $(f_1, \phi_1, 0)$ and $(f_1, \phi_1', 0)$ are exact. Then $\phi_1' = \phi_1\gamma_1$ for a unique $\gamma_1 \in \mathrm{GL}(k)$ and, conversely, if $(f_1, \phi_1, 0)$ is exact and $\gamma_1 \in \mathrm{GL}(k)$, then $(f_1, \phi_1\gamma_1, 0)$ is exact. Hence, by the trivial form of the lemma, $\Sigma^k = \Sigma = \Pi/\mathrm{GL}(k)$ is a smooth submanifold of $L_1(p, n)$ of codimension $c - \operatorname{rk}\pi + d = kp - kn + k^2 = k(p - n + k)$.

The proof for $\Sigma^{kji\cdot\cdot}$ goes similarly, 'line one' being almost as short, but 'line two' being replaced by two lines, 'line two' using the full force of the Lemma and 'line three' using only the trivial form of it.

In the case of Σ^{kj}, for example, Π is defined to be the open submanifold of the submanifold of zeros of the fibre-preserving map

$$L_1(p,n)\times L_2(p,n)\times B^{n,k}_{kj}\to L(p,k)\times B^{p,n}_{kj},$$

$$(f_1,f_2,(\phi_1,\chi_1,\phi_2\chi_1))\mapsto(f_1\phi_1,(\phi_1,\chi_1,f_2(\phi_1\chi_1)\phi_1+f_1\phi_2\chi_1))$$

with $\phi_1,\chi_1,\phi_2\chi_1$ maximal for f_1,f_2. From the injectivity of ϕ_1 and χ_1 it follows that $\circ\,\phi_1$ and $\circ\,(\phi_1\chi_1)\chi_1$ are surjective and hence that Π has codimension $pk+p\lambda_{kj}$.

Given another maximal system $\phi_1',\chi_1',\phi_2'\chi_1'$ there exist unique $\gamma_1,\zeta_1,\gamma_2\chi_1'$ such that

$$\phi_1'=\phi_1\gamma_1,\quad \chi_1'=\gamma_1^{-1}\chi_1\zeta_1$$

and

$$(*)\qquad \phi_1'\chi_1'=\phi_2\chi_1\zeta_1\gamma_1+\phi_1\gamma_1(\gamma_1^{-1}\gamma_2\chi_1').$$

We now apply the Lemma twice, first letting the group bundle $B^{k,k}_{kj}$ act on Π, with $(\phi_1',\chi_1',\gamma_1^{-1}\gamma_2\chi_1')$ sending $\phi_2'\chi_1'$ to $\phi_2\chi_1$ by $(*)$ to get a submanifold of $L_1(p,n)\times L_2(p,n)\times X^{n,k}_{kj}$, Θ say, of codimension $pk+p\lambda_{kj}-n\lambda_{kj}+k\lambda_{kj}$. Finally, using the simple form of the Lemma, we factor the submanifold Θ by the group $\mathrm{GL}(k)\times\mathrm{GL}(j)$ to obtain $\Sigma^{kj}=\Sigma$ as a smooth submanifold of $L_1(p,n)+L_2(p,n)$ of codimension

$$\begin{aligned}&pk+p\lambda_{kj}-n\lambda_{kj}+k\lambda_{kj}-kn-jk+k^2+j^2\\&\quad=(p-n+k)(\lambda_k+\lambda_{kj})-(k-j)\lambda_j,\end{aligned}$$

the codimension first computed by Thom and Levine [5] and generalised by Boardman [2]. The codimension of Σ^{kji} likewise is

$$(p-n+k)(\lambda_k+\lambda_{kj}+\lambda_{kji})-(k-j)(\lambda_j+\lambda_{ji})-(j-i)\lambda_i,$$

and so on, for any Boardman symbol $(k,j,i,\ldots)$.

The derivatives $d(f\phi)0$, $d(d(f\phi)\chi)0$, $d(d(d(f\phi)\chi)\psi)0,\ldots$ presented here give explicitly, in a form that involves kernels only and not cokernels, the intrinsic derivatives of f first encountered by the author in [6].

References

1. V. I. Arnol'd, *Normal forms of functions near degenerate critical points, the Weyl groups A_k, D_i, E_k and Lagrangean singularities*, Funkcional Anal. i Priložen **6** (4) (1972), 3–25 (Russian); Functional Anal. Appl. **6** (1972), 254–272.

2. J. M. Boardman, *Singularities of differentiable maps*, Inst. Hautes Études Sci. Publ. Math. **33** (1967), 21–57.

3. E. Brieskorn, *Die Hierarchie der 1-modularen Singularitäten*, Manuscripta Math. **27** (1979), 183–219.

4. Faa' de Bruno, *Note sur une nouvelle formule de calcul differentiel*, Quart. J. Math. **1** (1857), 359–360.

5. H. I. Levine, *Singularities of differentiable mappings*, Notes of lectures by R. Thom, Bonn, 1959; also in Proc. Liverpool Singularities Sympos. I, Lecture Notes in Math., vol. 192 (C. T. C. Wall, ed.), Springer-Verlag, Berlin and New York, 1971, pp. 1–89.

6. I. R. Porteous, *Simple singularities of maps*, Columbia Univ. Notes, 1962; also in Proc. Liverpool Singularities Sympos. I, Lecture Notes in Math., vol. 192 (C. T. C. Wall, ed.), Springer-Verlag, Berlin and New York, 1971, pp. 286–307.

7. ______, *The normal singularities of surfaces in* $\mathbf{R}^3$, these PROCEEDINGS.

8. F. Ronga, *A new look at Faa' de Bruno's formula for higher derivatives of composite functions and the expression of some intrinsic derivatives*, these PROCEEDINGS.

UNIVERSITY OF LIVERPOOL, UNITED KINGDOM

Proceedings of Symposia in Pure Mathematics
Volume **40** (1983), Part 2

$\mathbf{C}^*$-EQUIVARIANT DEFORMATIONS OF GERMS OF COHERENT SHEAVES

FERNANDO PUERTA

0. Let X be an algebraic variety with a G_m-action. In [**3**] it is shown that there exists a formal semiuniversal deformation of X which is G_m-equivariant. Such deformation is unique up to noncanonical isomorphism. More recently D. S. Rim [**5**] has obtained a similar result for the case of an algebraic variety with a G-action, where G is a linear reductive group, generalizing the case of G_m.

If $(X, 0)$ is a germ of analytic variety with a $\mathbf{C}^*$-action we have shown the existence of an analytic $\mathbf{C}^*$-equivariant semiuniversal deformation of $(X, 0)$ if $\dim T^1(X)$ is finite. The proof, which follows Donin's method of constructing the semiuniversal deformation, was sketched in [**4**]. More generally, the same method allows us to prove a similar result for the case of a $\mathbf{C}^*$-equivariant germ of a coherent sheaf over the germ $(X, 0)$.

We are going to make the definitions and the statement of the results precise. In what follows equivariant means $\mathbf{C}^*$-equivariant.

1. Let $(X, 0)$ be a germ of analytic space and $\mathcal{F}$ a germ of a coherent sheaf over $(X, 0)$. The fundamental definitions concerning the deformation of $\mathcal{F}$ are contained in [**6**] for the global case. The adaptation to the local case is obvious.

For the definition of a $\mathbf{C}^*$-equivariant germ of a coherent sheaf $\mathcal{F}$ over $(X, 0)$ we begin by recalling that we can suppose the given $\mathbf{C}^*$-action on $\mathbf{C}^n$ of the form

$$x = (x_1, \ldots, x_n) \to (\lambda^{q_1} x_1, \ldots, \lambda^{q_n} x_n) = \lambda \cdot x$$

where $\lambda \in \mathbf{C}^*$ and $q_i \in \mathbf{Z}$, $i = 1, \ldots, n$.

Then we get a $\mathbf{C}^*$-action on $\mathcal{O}_n$ by

$$(\lambda \cdot f)(x) = f(\lambda \cdot x), \qquad f \in \mathcal{O}_n.$$

If $\lambda \cdot f = \lambda^k f$, $k \in \mathbf{Z}$, we say that f is a quasihomogeneous element of $\mathcal{O}_n$ of degree k. We denote by $\mathcal{O}_n^{(k)}$ the set of quasihomogeneous elements of $\mathcal{O}_n$ of degree

1980 *Mathematics Subject Classification.* Primary 32G13, 32L10.

k. Then, each $f \in \mathcal{O}_n$ can be written in a unique way as $f = \Sigma f_k$ where $f_k \in \mathcal{O}_n^{(k)}$, $k \in \mathbf{Z}$, and the series converges for the analytic topology of $\mathcal{O}_n$ [**2**]. In general $\dim \mathcal{O}_n^{(k)}$ is infinite and f_k is not a polynomial.

Now, let $\mathcal{O}_n^p$ be a free $\mathcal{O}_n$-module. For each $\alpha = (\alpha_1, \ldots, \alpha_p) \in \mathbf{Z}^p$ and $k \in \mathbf{Z}$ we define

$$\mathcal{O}_{n,\alpha}^{p(k)} = \mathcal{O}_n^{(k+\alpha_1)} \times \cdots \times \mathcal{O}_n^{(k+\alpha_p)}.$$

Then, each $u \in \mathcal{O}_n^p$ can be written in a unique way as a convergent series $u = \Sigma u_k$ where $u_k \in \mathcal{O}_{n,\alpha}^{p(k)}$. We say, as in the former case, that u_k is the quasihomogeneous component of u of degree k and we write $\mathcal{O}_{n,\alpha}^p$ to indicate that each $u \in \mathcal{O}_n^p$ can be written in the above form. So, we have well-defined projections

$$\pi_k \colon \mathcal{O}_{n,\alpha}^p \to \mathcal{O}_{n,\alpha}^{p(k)}$$

with $\pi_k(u) = u_k$, $k \in \mathbf{Z}$.

A sub-$\mathcal{O}_n$-module M of $\mathcal{O}_{n,\alpha}^p$ is said to be equivariant if $\pi_k(M) \subset M$. This is equivalent to saying that M is generated by quasihomogeneous elements.

A continuous morphism of $\mathcal{O}_n$-modules

$$\varphi \colon \mathcal{O}_{n,\alpha}^p \to \mathcal{O}_{n,\beta}^q$$

is equivariant if $\phi(\mathcal{O}_{n,\alpha}^{p(k)}) \subset \mathcal{O}_{n,\beta}^{q(k)}$ for every $k \in \mathbf{Z}$. This is equivalent to saying that the columns of φ are quasihomogeneous elements of $\mathcal{O}_{n,\beta}^q$ of degrees $-\alpha_1, \ldots, -\alpha_p$.

Finally, let $\mathcal{F}$ be a germ of a coherent sheaf over $(X, 0)$. We suppose that X is a representative of the germ contained in some neighborhood U of $\mathbf{C}^n$ and that $0 \in U$ is the origin of $\mathbf{C}^n$. We say that $\mathcal{F}$ is equivariant if there exists a presentation of the fiber $\mathcal{F}_0$ of $\mathcal{F}$ over 0

$$\mathcal{O}_{n,\alpha}^p \xrightarrow{\varphi} \mathcal{O}_{n,\beta}^q \to \mathcal{F}_0 \to 0$$

such that φ is an equivariant morphism. The definition of equivariant morphism between equivariant germs of coherent sheaves is the natural one.

A germ of analytic space is said to be equivariant if the germ of the sheaf of ideals defining it is equivariant.

Here is, now, the definition of an equivariant deformation. Let $(X, 0)$ be a germ of an analytic space and $\mathcal{F}$ an equivariant germ of a coherent sheaf over $(X, 0)$. An equivariant deformation of $\mathcal{F}$ over the germ $(S, 0)$ is a quadruple $(\mathcal{G}, S, 0, \varphi)$ consisting of:

(a) An equivariant germ $(S, 0)$.

(b) An equivariant germ of an S-flat coherent sheaf on $X \times S$, $\mathcal{G}$.

(c) An equivariant isomorphism $\varphi \colon \mathcal{G}(0) \to \mathcal{F}$ where $\mathcal{G}(0) = j_0^* \mathcal{F}$ and

$$j_0 \colon X \to X \times S$$

is the canonical injection $x \to (x, 0)$.

2. We can now state the result generalizing that of [**4**].

We denote by $T^1(\mathfrak{F})$ the set of equivalence classes of deformations of $\mathfrak{F}$, not necessarily equivariant, over the double point.

THEOREM. *Let $(X, 0)$ be a germ of analytic space and $\mathfrak{F}$ a* **C*****-equivariant germ of a coherent sheaf over $(X, 0)$. If* dim $T^1(\mathfrak{F})$ *is finite there exists a semiuniversal deformation of $\mathfrak{F}$ which is equivariant. It is unique up to noncanonical* **C****-*equivariant isomorphism.*

3. Remarks. (1) Let $S(\mathfrak{F})$ be the singular locus of the coherent sheaf $\mathfrak{F}$, i.e., $S(\mathfrak{F}) = \{x \in X, \mathfrak{F}_x \text{ is not free and } \mathfrak{F}_x \neq 0\}$. If the support of $S(\mathfrak{F})$ consists of a finite number of points, dim $T^1(\mathfrak{F})$ is finite.

(2) Let $(X, 0)$ be an equivariant germ. It is easy to check that the germ of the sheaf of differentials Ω_X of X, is also equivariant. This provides a natural way to obtain equivariant germs of coherent sheaves.

REFERENCES

1. I. F. Donin, *Complete families of deformations of germs of complex spaces*, Math. USSR-Sbo. **18** (1972), 397–406.

2. H. Grauert and R. Remmert, *Analytische Stellenalgebren*, Grundl. Math. Wiss., band 176, Springer-Verlag, Heidelberg, 1971.

3. H. Pinkham, *Deformations of algebraic varieties with G_m-action*, Asterisque n° **20** (1974).

4. F. Puerta, *Deformations semiuniverselles et germs d'espaces analytiques* **C****-équivariantes*, Internat. Congr. Algebraic Geom. (La Rábida, Spain), 1981.

5. D. S. Rim, *Equivariant G-structure on versal deformations*, Trans. Amer. Math. Soc. **257** (1980), 217–226.

6. G. Trautman, *Deformations of coherent analytic sheaves with isolated singularities*, Proc. Sympos. Pure Math., vol. 30, Amer. Math. Soc., Providence, R.I., 1977, pp. 85–89.

UNIVERSIDAD POLITÉCNICA DE BARCELONA, ETSEIB, SPAIN

Proceedings of Symposia in Pure Mathematics
Volume 40 (1983), Part 2

THE REAL JACOBIAN PROBLEM

JOHN D. RANDALL[1]

ABSTRACT. Let $f\colon R^n \to R^n$ be a *polynomial mapping*, that is $f(x) = (f_1(x),\dots,f_n(x))$ where $f_i \in R[x_1,\dots,x_n]$. Let $Jf = \det df$ be the Jacobian determinant of f. The *real Jacobian problem* is to show that if Jf never vanishes, then f is a diffeomorphism. This is known to be the case for $n = 1$, but is undecided for $n \geqslant 2$. We shall give simple sufficient conditions for such mappings to be diffeomorphisms, and show that with respect to the vector space topology on monomials, these conditions are satisfied generically. The results are extended to polynomial local homeomorphisms.

1. Introduction. Let $P_{m,n}(d)$ denote the set of *polynomial mappings* $f\colon R^m \to R^n$ of *degree* $d = (d_1,\dots,d_n)$, that is, $f = (f_1,\dots,f_n)$ where $f_i \in R[x_1,\dots,x_n]$ and $\deg f_i = d_i$. We can write each f_i as

$$f_i = f(i,0) + \cdots + f(i, d_i),$$

where $f(i, j)$ is a homogeneous polynomial of degree j. Let $f^i = f(i, d_i)$. In order to simplify notation, let $P_n(d) = P_{n,n}(d)$, $P_{m,n} = \bigcup P_{m,n}(d)$, and $P_n = \bigcup P_n(d)$, where the unions are taken over all possible multi-indices. Let $J\colon P_n \to P_{n,1}\colon f \mapsto \det df$ be the Jacobian determinant mapping, so that

$$J(P_n(d)) \subset \bigcup \{P_{n,1}(r)\colon r \leqslant |d| - n\},$$

where $|d| = d_1 + \cdots + d_n$.

We shall fix a positive integer n throughout, and further simplify by setting $P(d) = P_n(d)$ and $P = P_n$. Let $D \subset P$ be the set of $f \in P$ such that for all $x \in R^n$, $Jf(x) \neq 0$. Then $f \in D$ if and only if f is a local diffeomorphism at each point of R^n. Let $D(d) = D \cap P(d)$.

CONJECTURE. *If $f \in D$, then f is a diffeomorphism.*

REMARK. By *diffeomorphism* we mean diffeomorphism *onto* R^n.

In order to show that the Conjecture is true, it would suffice to show that any $f \in D$ is proper. This follows from Palais [3, pp. 128–129] where it is shown that any proper local diffeomorphism of R^n is a diffeomorphism. If we were able to

1980 *Mathematics Subject Classification.* Primary 26B10; Secondary 57R50.
[1]Partially supported by an NSF grant.

0082-0717/82/0000-0655/$01.75

check directly that f were proper, then the solution of this problem would be easy. Unfortunately, no simple necessary and sufficient algebraic condition is known. We shall use the following sufficient condition.

PROPOSITION. *Let $f \in D$. If the f^i, $i = 1,\ldots,n$, have no common zeros, i.e. if $x \in R^n$, and $f^i(x) = 0$ for $i = 1,\ldots,n$, imply that $x = 0$, then f is a diffeomorphism.*

PROOF. We show that f is proper. Let $\bar{d} = d_1 \cdots d_n$. For $r > 0$, let $V_r = f^{-1}(W_r)$, where W_r is the algebraic set in R^n given by $\Sigma X_i^{2p(i)} = r^2$, where $p(i) = \bar{d}/d_i$. Thus if $(x_1,\ldots,x_n)$ are coordinates for R^n, then V_r is the zero-set of

$$\sum (f(i,0) + \cdots + f(i, d_i))^{2p(i)} - r^2.$$

Let $(x_0:\ldots:x_n)$ be homogeneous coordinates for RP^n, and let $i\colon R^n \to RP^n$ be the inclusion $(x_1,\ldots,x_n) \to (1:x_1:\ldots:x_n)$. Let $\bar{V}_r$ be the projective closure of V_r, obtained by homogenizing the defining polynomial for V_r, so that $\bar{V}_r$ is the zero-set of

$$\sum (f(i,0)x_0^{d_i} + \cdots + f(i, d_i - 1)x_0 + f(i, d_i))^{2p(i)} - r^2 x_0^{2\bar{d}}.$$

Then $\bar{V}_r - i(V_r)$ consists of those points $(x_0:\ldots:x_n)$ in $\bar{V}_r$ for which $x_0 = 0$, i.e. points for which $\Sigma f(i, d_i)^{2p(i)} = 0$. Since the $f(i, d_i) = f^i$, $i = 1,\ldots,n$, have no common zeros by hypothesis, we have that $\bar{V}_r - i(V_r) = \varnothing$. Thus V_r is compact for each r.

Let $\tilde{W}_r = \{X \mid \Sigma X_i^{2p(i)} \leqslant r^2\}$, the compact manifold whose boundary is W_r. Let $\tilde{Z}_r = \{X \mid \Sigma X_i^{2p(i)} \geqslant r^2\}$. In order to show that f is proper, it suffices to show that $f^{-1}(\tilde{W}_r)$ is compact for each r. Since f is a local diffeomorphism and W_r is a manifold, V_r is a manifold, which we have shown to be compact. It separates R^n into several components, exactly one of which has noncompact closure. Let the closure of this component be $\tilde{V}_r$. Then $f(\tilde{V}_r)$ is contained in either $\tilde{W}_r$ or $\tilde{Z}_r$. Since f is a polynomial mapping and $\deg f^i \geqslant 1$, $f(\tilde{V}_r)$ is not compact. Thus $f(\tilde{V}_r) \subset \tilde{Z}_r$, and $f^{-1}(\tilde{W}_r)$ is compact. Hence f is proper, and so is a diffeomorphism of R^n onto R^n.

REMARK. This condition is not necessary: for example, $f(x, y) = (x, y + x^2)$ is a diffeomorphism which does not satisfy it. The condition will be strong enough to prove genericity results.

The classical Jacobian problem, posed by Keller [**2**], asks whether a polynomial endomorphism of C^2 is an automorphism. This is so far undecided, although a number of mistaken proofs have been published. See Wang [**4**, §5.1, p. 475] for an account of the progress made with this problem, and its generalizations. The problem considered here is different. Over an algebraically closed field, if the Jacobian determinant never vanishes, then it is constant, and the inverse of a polynomial homeomorphism must be a polynomial mapping. In our case, we allow the Jacobian determinant to be any nonzero polynomial (and extend the results to certain nonnegative polynomials in §3) so that inverses need not be polynomial mappings.

2. Genericity. We now show that 'most' elements of D are diffeomorphisms. There is some difficulty in deciding upon an appropriate topology, as the Whitney C^∞-topology is too fine, since R^n is not compact, and if we topologize $P(d)$ as a vector space on monomials, then $D(d)$ is not an open subset. We will adopt this latter topology, and indicate its limitations.

THEOREM 2.1. *For each multi-index d, there is an open subset $G(d)$ of $P(d)$ such that if $f \in D \cap G(d)$, then f is a diffeomorphism.*

PROOF. Let $G(d)$ consist of those $f \in P(d)$ whose highest-degree homogeneous terms have no common zeros.

REMARK. Although $G \cap D$ is dense in D with the subspace topology, the fact that D is not open in P leads to problems. For example, if $f(x, y) = (x, y + x^2)$ then f cannot be approximated by elements of $G \cap D_2(1, 2)$.

We can adopt another viewpoint. Suppose $h \in P_{n,1}$ is *nonzero*, that is, $h(x) \neq 0$ for all $x \in R^n$. Such an h is a plausible Jacobian determinant for an element f of D, and we can try to decide whether the diffeomorphisms are dense in the set of *realizations* of h, i.e. the set of $f \in D$ satisfying $Jf = h$. Realizations of high degree will not satisfy the hypotheses of the Proposition, but if we restrict our attention to realizations of the 'right' degree, we obtain another genericity result.

Let $P^*(r)$ denote the set of nonzero polynomials in $P_{n,1}(r)$, and let $h \in P^*(r)$. We say that $f \in P(d)$ is an *exact realization* of h if $Jf = h$ and $r = |d| - n$.

THEOREM 2.2. *For each even integer $r > 0$ there is an open dense subset $G(r)$ of $P^*(r)$ such that if $h \in G(r)$, then every exact realization of h is a diffeomorphism.*

PROOF. Write $h \in P^*(r)$ as $h = h_0 + \cdots + h_r$, where h_i is a homogeneous polynomial of degree i, and define $G(r)$ to be the set of $h \in P^*(r)$ such that h_r has no nontrivial zeros, i.e. such that $h_r(x) = 0$ implies $x = 0$. This subset is clearly open, and is dense since any $k \in P^*(r)$ can be approximated by $h_t = k + t(x_1^r + \cdots + x_n^r)$, and for either $t > 0$ or $t < 0$, $h_t \in G(r)$.

Suppose that $f \in P(d)$ is an exact realization of $h \in G(r)$. Let $f_j^i = \partial f^i / \partial x_j$. Since f is an exact realization, we have that $h_r = \det(f_j^i)$. Suppose that the f^i have a common zero. Then they have a common linear factor, say x_1, so h_r has a factor of x_1^{n-1}. However, $h \in G(r)$, so h_r has no nontrivial zeros. Thus the f^i have no common zeros, and f satisfies the hypotheses of the Proposition. Thus f is a diffeomorphism.

3. Polynomial local homeomorphisms. The methods of this paper apply equally well to the set H of polynomial local homeomorphisms of R^n. The Proposition and Theorem 2.1 are true with H substituted for D, and with the same proofs. Theorem 2.2 does not extend so readily, since the Jacobian determinant alone cannot be used to characterize local homeomorphisms. Algebraic sufficient conditions exist: if H_1 is the set of $f \in P_n$ with $\det df_x \geqslant 0$ and rank $df_x \geqslant n - 1$ for all $x \in R^n$, then [**1**, 1.4] implies $H_1 \subset H$. If $\bar{P}^*(r)$ is the set of $f \in P(r)$ which are either nonnegative or nonpositive, we have

THEOREM 2.2′. *For each even integer r, there is an open dense subset $G(r)$ of $\bar{P}^*(r)$ such that if $h \in G(r)$, then any exact realization of h in H is a homeomorphism.*

ACKNOWLEDGEMENT. The author would like to thank P. T. Church for reading an earlier version of this paper and suggesting improvements which have been incorporated into the present work.

REFERENCES

1. P. T. Church, *Differentiable open maps on manifolds*, Trans. Amer. Math. Soc. **109** (1963), 87–100.
2. O. H. Keller, *Ganze Cremona-Transformationem*, Monatsh. Math. **47** (1939), 299–306.
3. R. S. Palais, *Natural operations on differential forms*, Trans. Amer. Math. Soc. **92** (1959), 125–141.
4. S. S.-S. Wang, *A Jacobian criterion for separability*, J. Algebra **65** (1980), 453–494.

RUTGERS UNIVERSITY

Proceedings of Symposia in Pure Mathematics
Volume **40** (1983), Part 2

MILNOR FIBERS AND ALEXANDER POLYNOMIALS OF PLANE CURVES

RICHARD RANDELL

Let C be an irreducible algebraic curve of degree n in the projective plane $\mathbf{P}^2$. Since C has real codimension two, it is natural to try to study the "knot theory" of the pair $(\mathbf{P}^2, C)$. From an algebraic point of view, an immediate difficulty is that the first homology of $\mathbf{P}^2 - C$ is cyclic of order n instead of infinite cyclic as in the case of classical knots. Thus, there is not immediately a theory of Alexander invariants. There are at least two possible ways to avoid this difficulty. In this note we introduce the second such way, and show that it generically yields the same answer as the first, more classical way, at least for the Alexander polynomial.

The first approach is used in several recent papers of A. Libgober [**1, 2, 3**]. One removes a line L to form $\mathbf{P}^2 - (C \cup L)$. It is easily seen that the first homology of the latter space is $\mathbf{Z}$, as desired. We will denote the Alexander polynomial resulting from this approach by $\Delta_C^L(t)$. This approach is inspired by work of O. Zariski, though he works with branched covers and not explicitly with the Alexander polynomial.

An alternative approach is to consider the Hopf bundle over $\mathbf{P}^2$. This bundle has total space S^5; if we restrict to $\mathbf{P}^2 - C$, we obtain a bundle with fiber S^1 and total space $S^5 - K$, where K is the link of the singularity of the homogeneous polynomial f which defines the curve C. Now K is a three-dimensional complex, and it is not difficult to show that $H_1(S^5 - K) \cong \mathbf{Z}$. Thus, there is an Alexander polynomial $\Delta_K(t)$. The following two results show that $\Delta_K(t)$ is of interest.

THEOREM 1. *If the line L is in general position with respect to the curve C, then $\Delta_K(t) = \Delta_C^L(t)$. For any L, $\Delta_K(t)$ divides $\Delta_C^L(t)$.*

REMARK. In the first case, we actually show that $\pi_1(\mathbf{P}^2 - (C \cup L)) \cong \pi_1(S^5 - K)$.

1980 *Mathematics Subject Classification*. Primary 14H30, 57Q45.

Now let $F = f^{-1}(1)$ be the Milnor fiber of the singularity of f at the origin. That is, there is a bundle $F \to S^5 - K \to S^1$. Furthermore, there is a monodromy map $h\colon F \to F$.

THEOREM 2. *$\Delta_K(t)$ is the characteristic polynomial of the monodromy*

$$h_*\colon \frac{H_1(F;\mathbf{Z})}{\textit{torsion}} \to \frac{H_1(F;\mathbf{Z})}{\textit{torsion}}.$$

We will write $\Delta(t)$ for $\Delta_K(t)$. The advantages of our approach basically stem from Theorem 2, which shows that the current theory is analogous to the theory of classical *fibered* knots. Later we will state several corollaries of these results.

EXAMPLE 1. Let C be defined by $\{x^2z = y^3\} \subset \mathbf{P}^2$, and take L to be the line $\{z = 0\}$. Then, $L \cap C$ is a single point, so L and C are not in general position. In [**1**] it is noted that $\mathbf{P}^2 - (C \cup L)$ deformation retracts to the complement of a trefoil knot in S^3. Thus, $\Delta_C^L(t) = t^2 - t + 1$. But it is easily seen that $\Delta_K(t) = 1$ in this case (using the usual techniques for computing fundamental groups, for example). So, in general, $\Delta_K(t) \neq \Delta_C^L(t)$.

EXAMPLE 2. Let C be the sextic with six cusps on a conic given by the equation $(x^2 + z^2)^3 + (y^3 + z^3)^2$. Then $\Delta(t) = t^2 - t + 1$. We defer until later a direct computation which uses Theorem 2 and avoids fundamental groups.

PROOFS. Theorem 2 is analogous to the folklore result about fibered knots in classical knot theory, and is proved in the same way. Note that we have an exact sequence

$$(1) \qquad \to H_1(F) \xrightarrow{h_* - I_*} H_1(F) \to H_1(S^5 - K) \to H_0(F) \cong \mathbf{Z} \to 0.$$

We prove Theorem 1 by considering presentations of the fundamental groups in question. The classical presentation for $\mathbf{P}^2 - C$ is

$$(2) \qquad \pi_1(\mathbf{P}^2 - C) \cong \langle a_1, \ldots, a_n \mid a_1 a_2 \cdots a_n, R_i \rangle.$$

Here we pick a pencil of lines in $\mathbf{P}^2$. A general line L' of the pencil intersects C in n points. The generators $a_1, \ldots, a_n$ are just loops on L' which circle these intersections once. Then on $L' \cong S^2$, we have the relation $a_1 a_2 \cdots a_n$. The relations R_i are obtained from lines of the pencil which intersect C in fewer than n points.

Next, Zariski has shown that if L is in general position with respect to C, one has

$$(3) \qquad \pi_1(\mathbf{P}^2 - (C \cup L)) \cong \langle a_1, \ldots, a_n, b \mid a_1 a_2 \cdots a_n b, R_i, [a_i, b] \rangle.$$

The notation here is as for (2). The generators a_j and relations R_i are identical to the above, while the generator b is represented by a loop in L' which circles $L \cap L' = \{x\}$ once.

Next we wish to obtain a presentation of $\pi_1(S^5 - K)$ which we can relate to (2) and (3). To do this, we pull the pencil of $\mathbf{P}^1$'s in $\mathbf{P}^2$ back to a "pencil" of S^3's in

S^5. That is, the inverse image of a $\mathbf{P}^1$ in $\mathbf{P}^2$ is an S^3 in S^5. In $\mathbf{P}^2$, the pencil consists of all $\mathbf{P}^1$'s containing a fixed point. In S^5, the inverse image consists of all S^3's containing a fixed circle.

To get the desired presentation we first consider $L' - (L' \cap C) \cong S^2 - \{n \text{ points}\}$. The inverse image of this set is the complement of n circles in S^3. These circles form an (n, n) torus link in S^3. (This link consists of n parallel $(1, 1)$ curves on a torus in S^3.) In terms of fundamental groups we have

$$\begin{aligned} \pi_1(S^3\text{-link}) &\cong \langle a_1', \ldots, a_n' \mid a_1' \cdots a_n' = a_2' \cdots a_n' a_1' = \cdots = a_n' a_1' \cdots a_{n-1}' \rangle \\ &\cong \langle a_1', \ldots, a_{n-1}' \rangle \times \langle C \rangle, \qquad C = a_1' a_2' \cdots a_n', \\ \pi_1(S^2\text{-points}) &\cong \langle a_1, \ldots, a_n \mid a_1 \cdots a_n = 1 \rangle. \end{aligned}$$

Furthermore, there is the commutative diagram

$$\begin{array}{ccccccccc} & & & & 1 & & 1 & & \\ & & & & \uparrow & & \uparrow & & \\ 1 & \to & \pi_1(S^1) \cong \mathbf{Z} & \to & \pi_1(S^5 - K) & \to & \pi_1(\mathbf{P}^2 - C) & \to & 1 \\ & & & & \uparrow i_* & & \uparrow i_* & & \\ 1 & \to & \pi_1(S^1) \cong \mathbf{Z} & \to & \pi_1(S^3\text{-link}) & \underset{\sigma}{\overset{\to}{\leftarrow}} & \pi_1(S^2\text{-pts}) & \to & 1 \\ & & & & \uparrow & & \uparrow & & \\ & & & & G = \sigma(G') & & G' & & \\ & & & & \uparrow & & \uparrow & & \\ & & & & 1 & & 1 & & \end{array}$$

where the rows and columns are exact, and i_* are inclusion-induced. The map σ takes a_i to a_i', $i = 1, \ldots, n-1$, $\sigma(a_n) = (a_1' \cdots a_{n-1}')^{-1}$. Here G' is the normal subgroup generated by the relations R_i. It then follows that G is the kernel of i_*: $\pi_1(S^3\text{-link}) \to \pi_1(S^5 - K)$.

Thus, a presentation of $\pi_1(S^5 - K)$ is

(4) $\pi_1(S^5 - K)$

$$\cong \langle a_1', a_2', \ldots, a_n' \mid a_1' \cdots a_n' = a_2' \cdots a_n' a_1' = \cdots = a_n' a_1' \cdots a_{n-1}', R_i' \rangle,$$

where the set R_i' is just the set R_i with all a_i's primed.

Then we define $\psi(a_i) = a_i'$, $\psi(b) = (a_1' \cdots a_n')^{-1}$ and obtain an isomorphism ψ: $\pi_1(\mathbf{P}^2 - (C \cup L)) \to \pi_1(S^5 - K)$.

This concludes the proof when L and C are in general position. If not, we can consider $L = L_0$ as the limit of a family L_u of lines, where for $u \neq 0$, L_u is in general position with C. The result follows since $\pi_1(\mathbf{P}^2 - (C \cup L))$ maps onto $\pi_1(\mathbf{P}^2 - (C \cup L_u))$, implying $\Delta_K(t) = \Delta_C^{L_u}(t)$ divides $\Delta_C^L(t)$, $u \neq 0$.

We gather several immediate consequences of the above theorems. These results were also proved by Libgober in [**1**].

COROLLARIES. $\Delta(t)$ *satisfies*

(i) $\Delta(1) = \pm 1$.

(ii) $\Delta(0) = \pm 1$.

(iii) $\deg \Delta(t) = 2q$, *where* q *is the irregularity of the cyclic n-fold cover of* $\mathbf{P}^2$ *branched along* C.

(iv) $\Delta(t)$ *is monic and cyclotomic.*

(v) $\Delta_C^L(t)$ *is independent of L as long as L is in general position with respect to C.*

Furthermore, $\Delta(t)$ can sometimes be calculated without reference to fundamental groups. Consider Example 2. Here C is defined by

$$C = \left\{ (x^2 + z^2)^3 + (y^3 + z^3)^2 = 0 \right\} \subset \mathbf{P}^2.$$

Thus, $F = \{(x^2 + z^2)^3 + (y^3 + z^3)^2 = 1\} \subset \mathbf{C}^3$.

Define $g\colon F \to G$ by $g(x, y, z) = (u, v)$, where $u = x^2 + z^2$, $v = y^3 + z^3$, and $G = \{(u, v) \in \mathbf{C}^2 \mid u^3 + v^2 = 1\}$. We want to use the Leray spectral sequence of g to identify the monodromy on $H^1(F; \mathbf{Z})$. This spectral sequence has

$$E_2^{p,q} \cong H^p(G; \mathcal{H}^q(g)),$$

where $\mathcal{H}^q(g)$ is the sheaf on G which associates $H^q(g^{-1}(U))$ to the open set U of G. Also, $E_2^{p,q} \Rightarrow H^{p+q}(F)$. We will sketch a proof that $H_1(F) \cong E_2^{1,0}$.

First, for dimensional reasons, there are no nonzero differentials d_r involving $E_r^{1,0}$, for $r \geqslant 2$. Thus $E_2^{1,0} \cong E_\infty^{1,0}$. Because of this, $H^1(F) \cong E_2^{1,0}$ follows from $E_2^{0,1} \cong 0$. We give a sketch of the proof of this latter fact.

We may identify $E_2^{0,1} \cong H^0(G; \mathcal{H}^1(g))$ with $\Gamma(\mathcal{H}^1(g))$, the set of sections of $\mathcal{H}^1(g)$. To show $\mathcal{H}^1(g)$ has no nontrivial sections requires some effort, since $\mathcal{H}^1(g)$ is not trivial.

Fix $(u_0, v_0) \in G$, and let $E_{(u_0,v_0)} = g^{-1}(u_0, v_0) \subset F$. Let p be the projection of $E_{(u_0,v_0)} \subset \mathbf{C}^3$ to the z-axis. Then $p^{-1}(z_0) = \{(x, y, z) \in E_{(u_0,v_0)} \mid z = z_0, x^2 = u_0 - z_0^2, y^3 = v_0 - z_0^3\}$. Thus, p represents $E_{(u_0,v_0)}$ as a 6-fold branched cover of $\mathbf{C}$, branched over the points $z_0^2 = u_0$, $z_0^3 = v_0$. From this it follows that $E_{(u_0,v_0)}$ is an open 2-manifold. For most points (u_0, v_0), the Euler characteristic of $E_{(u_0,v_0)}$ is -12. For special values $u_0 = 0$; $v_0 = 0$; or $v_0^2 = u_0^3 = \frac{1}{2}$ the first betti number of $E_{(u_0,v_0)}$ drops. Analysis of the monodromy around these special points shows that it is impossible to extend local sections to a nontrivial global section.

Thus, the spectral sequence of the map g yields $H^1(F; \mathbf{Z}) \cong E_2^{1,0}$ and the action of the monodromy h on $E_2^{1,0}$ is just the well-known monodromy for G as the fiber of the trefoil knot; $h_* = \left(\begin{smallmatrix} 0 & 1 \\ -1 & 1 \end{smallmatrix}\right)$. Thus,

$$\Delta(t) = \operatorname{Det}\begin{pmatrix} t & -1 \\ 1 & t-1 \end{pmatrix} = t^2 - t + 1.$$

REMARK. The invariants considered here are, of course, rational invariants. There are interesting torsion (or integral) invariants also. For example, one has the torsion in $H_1(F; \mathbf{Z})$, which is precisely the torsion of the abelianization of the commutator subgroup of $\pi_1(\mathbf{P}^2 - C)$. The possibilities for this torsion are limited as follows: For each singular point p_i of C, one has a link $L_i \subset S^3$ in the usual way. Let K_i be the n-fold cyclic cover of S^3 branched along L_i. Then we always have [**4**]

$$\operatorname{Tor}(H_1(F; \mathbf{Z})) \subset \bigoplus_i \operatorname{Tor} H_1(K_i; \mathbf{Z}).$$

Furthermore, for certain L_i and n, sharper results may be obtained [**4**]. Similar results have recently been obtained by Libgober [**2**].

REFERENCES

1. A. Libgober, *Alexander invariants of plane algebraic curves*, these PROCEEDINGS.

2. ______, *Alexander polynomials and cyclic multiple planes*, Duke Math. J. **49** (1982), to appear.

3. ______, *Alexander modules of plane algebraic curves*, Proc. 1981 AMS Special Session on Low Dimensional Topology (San Francisco), Contemporary Math., Amer. Math. Soc., Providence, R. I., to appear.

4. R. Randell, *Some topology of Zariski surfaces*, Proc. 1979 Siegen Topology Sympos. (U. Korschorke and W. D. Neumann, Ed.), Lecture Notes in Math., vol. 788, Springer-Verlag, Berlin and New York, pp. 145–164.

5. O. Zariski, *On the irregularity of cyclic multiple planes*, Ann. of Math. (2) **58** (1931), 485–511.

6. ______, *On the linear connection index of the algebraic surfaces* $z^n = f(x, y)$, Proc. Nat. Acad. Sci. U.S.A. **6** (1929), 494–501.

UNIVERSITY OF OKLAHOMA

Current address: Department of Mathematics, University of Iowa, Iowa City, Iowa 52242

Proceedings of Symposia in Pure Mathematics
Volume **40** (1983), Part 2

MULTIPLE POINTS OF REAL MAPPINGS

FELICE RONGA

Let $f\colon V^n \to W^{n+r}$ be a proper smooth map between smooth differentiable manifolds. We shall say that f is regular if the k-fold product $f^k\colon V^{(k)} \to W^{(k)}$ is transversal to $\delta_W(k) = \{(y,\ldots,y) \in W^{(k)}\}$ outside $\Delta_V(k) = \{(x_1,\ldots,x_k) \in V^{(k)} \mid \exists i \neq j, x_i = x_j\}$ and if the jet extension $j^h(f)\colon V \to J^h(V, W)$ is transversal to all Boardman singularities. We give here a straightforward generalisation of the formula for the dual class to the k-tuple points of f (mutatis mutandis) given in **[2]**, where the case of an immersion was considered.

I am grateful to C. McCrory for his skepticism towards the first version of this note.

Define as in **[2]**:

$N_k(f) = \{y \in W \mid \#(f^{-1}(y)) = k\}$, $M_k(f) = f^{-1}(N_k)$,

$\hat{M}_k(f) = \{(x_1,\ldots,x_k) \in V^{(k)} - \Delta_V(k) \mid f(x_i) = f(x_j)\}$.

The group of permutations of k objects S_k acts in the obvious way on $M_k(f)$. We set

$$\tilde{N}_k(f) = \hat{M}_k/S_k, \quad \tilde{M}_k(f) = \hat{M}_k/S_{k-1},$$

where S_{k-1} acts on the last $k-1$ coordinates. We write $[x_1,\ldots,x_k]$, resp. $(x_1,[x_2,\ldots,x_k])$ for the class of $(x_1,\ldots,x_k) \in \hat{M}_k$ in $\tilde{N}_k$, resp. $\tilde{M}_k$. We define f_k: $\tilde{M}_k \to V, f_k(x_1,[x_2,\ldots,x_k]) = x_1$ and $g_k\colon \tilde{N}_k \to W$, $g_k([x_1,\ldots,x_k]) = f(x_1)$.

Set now $\hat{M}_2' =$ union of components of $\hat{M}_2$ not closed in $V^{(2)}$ and $\hat{M}_2'' =$ union of components of $\hat{M}_2$ containing some (x_1, x_2) such that x_1 or $x_2 \in \Sigma(f)$, where $\Sigma(f) = \{x \in V \mid df_x \text{ is not injective}\}$. We set $\hat{M}_2^- = \hat{M}_2' \cup \hat{M}_2''$, $\hat{M}_2^+ = \hat{M}_2$, $\hat{M}_2^{++} = \hat{M}_2 - \hat{M}_2^-$ and $\hat{N}_2^{++} = \hat{M}_2^{++}/S_2$. We define now inductively $\hat{M}_k^+$, $\hat{M}_k^{++}$ and $\hat{N}_k^{++}$:

$\hat{M}_k^+ = \{(x_1,[x_2,\ldots,x_k]) \in \hat{M}_k \mid [x_2,\ldots,x_k] \in \hat{N}_{k-1}^{++}\}$ and $\hat{M}_k^{++} =$ union of components of $\hat{M}_k^+$ not containing $(x_1,[x_2,\ldots,x_k])$ such that $x_1 \in \Sigma(f)$. Set $\tilde{N}_k^{++} = \hat{M}_k^{++}/S_k$ and define in the obvious way $\tilde{M}_k^+$, $\tilde{M}_k^{++}$, $\tilde{N}_k^{++}$, M_k^+, M_k^{++}, N_k^{++}, f_k^+, f_k^{++} and g_k^{++}.

1980 *Mathematics Subject Classification*. Primary 57R45.

0082-0717/81/0000-0206/$01.50

THEOREM. *Let* $f\colon V \to W$ *be a regular map. Then* $\overline{M}_k^+$, $\overline{M}_k^{++}$ *and* $\overline{N}_k^{++}$ *carry fundamental classes over the integers* mod *two. Denoting by* m_k^+, m_k^{++} *and* n_k^{++} *respectively their Poincare duals and by* w_r *the rth Stiefel-Whitney class of the virtual normal bundle* $N_f = f^*TW - TV$ *of* f, *we have*

$$m_k^+ = f^*(n_{k-1}^{++}) + m_{k-1}^{++} \cdot w_r. \tag{1}$$

PROOF. One verifies as in [2] that f_k^+, f_k^{++} and g_k^{++} are proper and that $\operatorname{Im}(f_k^+) = \overline{M}_k^+$, $\operatorname{Im}(f_k^{++}) = \overline{M}_k^{++}$ and $\operatorname{Im}(g_k^{++}) = \overline{N}_k^{++}$. Since $\tilde{M}_k^+$, $\tilde{M}_k^{++}$ and $\tilde{N}_k^{++}$ are smooth they carry fundamental classes, which project on fundamental classes of $\overline{M}_k^+$, $\overline{M}_k^{++}$ and $\overline{N}_k^{++}$ respectively. Consider now the diagram:

$$\begin{array}{ccc} \tilde{M}_k^+ \cup \tilde{M}_{k-1}^{++} & \longrightarrow & \tilde{N}_{k-1}^{++} \\ {\scriptstyle f_k^{++} \cup f_{k-1}^+}\big\downarrow & & \big\downarrow{\scriptstyle g_{k-1}^{++}} \\ V & \xrightarrow{\ f\ } & W \end{array}$$

It is subcartesian, as defined in [2, §2], except that f is no longer an immersion. However $f \times g_{k-1}^{++}\colon V \times \tilde{N}_{k-1}^{++} \to W \times W$ is transversal to $\delta_W(2)$ on $\tilde{M}_k^+$ and it is an immersion near $\tilde{M}_{k-1}^{++}$. Hence the arguments of the proof of [2, Theorem 1.1] extend to the present situation and the stated formula follows.

Assume now that $\Sigma^{1,1}(f) = \varnothing$ (this is automatic if $n \geqslant 2(r+1)$). Then $\Sigma(f) = \Sigma^1(f)$ and $\overline{M}_2^- = \overline{M}_2 - \overline{M}_2^{++}$ is smooth near $\Sigma(f)$. Hence it carries a fundamental class whose dual m_2^- verifies $m_2^- = m_2 + m_2^+$ (recall that we work mod two). Assuming $W = \mathbf{R}^{n+r}$ and using that $m_2 = w_r$ (see [1]) we have

$$\begin{aligned} m_3^+ &= m_2^{++} \cdot w_r = (m_2 + m_2^-) \cdot m_2 \\ &= m_2^2 + (m_2^-)^2 = (m_2^{++})^2 = w_r^2 + (m_2^-)^2. \end{aligned} \tag{2}$$

If $\dim(M_k) = 0$, i.e. if $r(k-1) = n$, f is homotopic to a map for which $M_k^- = M_k - M_k^+$ is empty. Let us sketch an argument for this: if $x \in M_k^-$ it must be a regular point of f and there exists a smooth path $c\colon [0,1] \to N_k$ such that c is an embedding, $c([0,\frac{1}{2}[) \cap \Sigma(f) = \varnothing$, $c(\frac{1}{2}) \in f(\Sigma^{1,0}(f))$, $c'(\frac{1}{2})$ is not tangent to $f(\Sigma^{1,0}(f))$ and $c(]\frac{1}{2},1]) \cap f(V) = \varnothing$. Take a small disc D around x and push $f(\mathring{D})$ all the way through $c(1)$ in a small neighborhood of c. Thus we eliminate one point of M_k^-, and so on.

Hence if $\dim M_k^- = 0$, the points of M_k^- deserve to be called "disposable".

For example, if $f\colon V^2 \to R^3$, (2) says that the number of nondisposable triple points of f mod two equals w_1^2 plus the number of twisted components of M_2^-, namely those whose tubular neighborhood in V is a Möbius band. This fact is also a consequence of a more precise result of T. Banchoff (to appear).

REFERENCES

1. F. Ronga, *La classe duale aux points doubles d'une application*, Compositio. Math. **27** (1973), 223–232.

2. ______, *On multiple points of smooth immersions*, Comment. Math. Helv. **55** (1980), 521–527.

UNIVERSITY OF GENEVA, SWITZERLAND

Proceedings of Symposia in Pure Mathematics
Volume **40** (1983), Part 2

A NEW LOOK AT FAA' DE BRUNO'S FORMULA FOR HIGHER DERIVATIVES OF COMPOSITE FUNCTIONS AND THE EXPRESSION OF SOME INTRINSIC DERIVATIVES[1]

FELICE RONGA

Let $f(t)$ be a real function defined for t near t_0 and $g(u)$ a real function defined for u near $u_0 = f(t_0)$, both f and g of class C^m. The following expression of the mth derivative of the composite $g \circ f$ has been given by Faa' de Bruno in [**2**]:

$$(1) \quad (g \circ f)_{(t)}^{(m)} = \sum \frac{m!}{k_1! \cdots k_m! \cdot (1!)^{k_1} \cdots (m!)^{k_m}} \cdot g^{(n)} \cdot (f')^{k_1} \cdots (f^{(m)})^{k_m},$$

where the sum ranges over $n = 1, \ldots, m$ and all nonnegative integers $k_1, \ldots, k_m$ such that $k_1 + \cdots + k_m = n$ and $k_1 + 2k_2 + \cdots + mk_m = m$. All the derivatives of f are taken at t, those of g at $f(t)$.

Many proofs of this formula have appeared in the literature (see e.g. [**3**, **6**, **7**, **10**]). In [**4**], L. E. Fraenkel gives a generalisation of it to functions of several variables.

We give here our version of this formula, which we think is the most natural, and give a combinatorial proof of it. More precisely, let E, F and G be normed vector spaces (over **R** or **C**) and let $f\colon U \to F$ and $g\colon V \to G$ be mappings of class C^m, where $U \subset E$ and $V \subset F$ are open and $V \supset f(U)$. We shall prove the following formula:

$$(2) \qquad d^m(g \cdot f)_{(a)} = \sum \operatorname{Sym}\left(d^n g\left(df^{k_1}, \ldots, d^m f^{k_m}\right)\right),$$

---

[1]Presented on August 4, 1981, in a seminar, part of the A.M.S. Summer Institute on Singularities held at Humboldt State University (Arcata, California).

1980 *Mathematics Subject Classification*. Primary 57R45, 58C27.

where n and $k_1,\ldots,k_m$ range over the same values as for (1) and where the derivatives of f are taken at $a \in U$, those of g at $f(a)$. The hth derivative $d^h f_a$ is considered as a multilinear map

$$\underbrace{E \times \cdots \times E}_{h \text{ times}} \to F.$$

The expression $df^{k_1},\ldots,d^m f^{k_m}$: $E^{k_1+2k_2+\cdots+mk_m} \to F^{k_1+\cdots+k_m}$ means the product of the maps df k_1-times, d^2f k_2-times, and so forth; if $k_h = 0$, $d^h f$ is meant to be omitted. The meaning of Sym is explained in §2; roughly it is the minimal effort symmetrisation of the multilinear map in question and consists of a sum of $m!/(k_1! \cdots k_m! \cdot (1!)^{k_1} \cdots (m!)^{k_m})$ terms, whence Faa' de Bruno's formula follows.

Originally we obtained (2) as a byproduct of considerations leading to an expression for the intrinsic derivative of a map (a notion due to I. Porteous [**5**]). We discuss this in a special case in §3; a first attempt toward this was done in [**9**].

1. The isotropy group of a surjection. For a finite set X we denote by $B(X)$ the group of bijections of X.

1.1. DEFINITION. Let φ: $X \to Y$ be a surjection. The isotropy group $I(\varphi)$ of φ is the subgroup of $B(X)$ defined by $I(\varphi) = \{\sigma \in B(X) \mid \varphi(x) = \varphi(x') \Rightarrow \varphi(\sigma(x)) = \varphi(\sigma(x'))\}$.

Thus $I(\varphi)$ is the largest subgroup of $B(X)$ for which there is an induced homomorphism φ_*: $I(\varphi) \to B(Y)$. Also, $I(\varphi)$ is the projection on $B(X)$ of the group $I'(\varphi) = \{(\sigma, \sigma') \in B(X) \times B(Y) \mid \sigma'^{-1} \cdot \varphi \cdot \sigma = \varphi\}$; since φ is a surjection, the projection of $B(X) \times B(Y)$ on the first factor induces an isomorphism $I'(\varphi) \xrightarrow{\cong} I(\varphi)$.

Let $N_h(\varphi) = \{y \in Y \mid \#(\varphi^{-1}(y)) = h\}$ and $M_h(\varphi) = \varphi^{-1}(N_h)$. We have that $Y = \bigcup_{h \geq 1} N_h$ and $X = \bigcup_{h \geq 1} M_h$, the unions being disjoint.

1.2. LEMMA. *There is an exact sequence*

$$1 \to \prod_{y \in Y} B(\varphi^{-1}(y)) \xrightarrow{i} I(\varphi) \xrightarrow{\varphi_*} \prod_{h \geq 1} B(N_h(\varphi)) \to 1. \tag{3}$$

PROOF. It is clear how to define i, that it is injective and that $\mathrm{Im}(i) = \mathrm{Ker}(\varphi_*)$. If $\sigma \in B(N_h(\varphi))$, define $s(\sigma) \in B(M_h(\varphi))$ by choosing a bijection $s(\sigma) \mid \varphi^{-1}(y)$: $\varphi^{-1}(y) \to \varphi^{-1}(\sigma(y))$ for every $y \in N_h(\varphi)$. Then $\varphi_*(s(\sigma)) = \sigma$, hence φ_* surjects on $\prod_{h \geq 1} B(N_h(\varphi))$.

Let $m = \#(X)$, $n = \#(Y)$ and $k_h = \#(N_h)$, $h = 1,\ldots,m$ (we set $k_h = 0$ if $N_h = \varnothing$). Clearly $k_1 + \cdots + k_m = n$ and $k_1 + 2k_2 + \cdots + mk_m = m$ and it follows from Lemma 1.2 that $\#(I(\varphi)) = k_1! \cdots k_m! \cdot (1!)^{k_1} \cdots (m!)^{k_m}$.

Given integers a and b we denote by $[a, b]$ the set of integers k satisfying $a \leq k \leq b$. Take now a surjection φ: $[1, m] \to [1, n]$. We shall say that φ is well ordered if φ is nondecreasing and if $\#(\varphi^{-1}(y)) \leq \#(\varphi^{-1}(y'))$ whenever $y \leq y'$. For such a φ we have $N_h = [k_{h-1} + 1, k_h]$; setting $K_h = k_1 + 2k_2 + \cdots + hk_h$, M_h decomposes as a disjoint union $M_h = M_h^1 \cup \cdots \cup M_h^{k_h}$, where $M_h^j = [K_{h-1} + h(j-1) + 1, K_{h-1} + jh]$, $j = 1,\ldots,k_h$.

In fact φ is entirely determined by the infinite sequence $\underline{k} = (k_1, \ldots, k_h, \ldots)$ of nonnegative integers satisfying $k_1 + \cdots + hk_h + \cdots = m$ (whence $k_h = 0$ for $h > m$) and $k_1 + \cdots + k_m = n$. We shall denote by $K(m, n)$ the set of such sequences.

Given a surjection $\varphi\colon X \to Y$ it is easily verified that a numbering of X and Y making φ well ordered is well defined up to an element of $I(\varphi)$.

Let $M(\varphi)$ be the mapping cylinder of a well-ordered $\varphi\colon [1, m] \to [1, n]$, that is $M(\varphi) = \{(i, t) \mid i \in [1, m],\ 0 \leqslant t \leqslant 1,\ t \text{ a real number}\}/\!\sim$, where $\sim$ is the equivalence relation defined by $(i, t) \sim (i', t)$ if $t = 0$ and $\varphi(i) = \varphi(i')$. $M(\varphi)$ looks as follows:

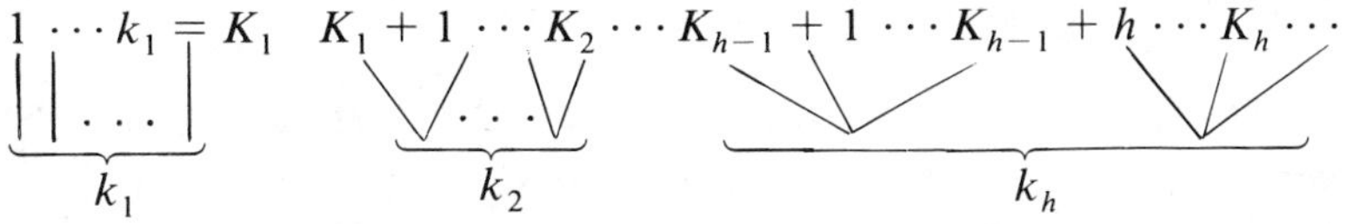

The above picture might be helpful to understand the sequel.

1.3. LEMMA. *Let* $S_m = B([1, m])$ *and let* $\varphi\colon [1, m] \to [1, n]$ *be well ordered. Every class* $[\sigma] \in I(\varphi) \setminus S_m$ *(where* σ *identifies with* $\rho \cdot \sigma$ *for* $\rho \in I(\varphi)$*) has a unique representative* σ *satisfying*:

(i) $\sigma \mid M_h^i$ *is increasing for any* i *and* h.

(ii) $\sigma(K_{h-1} + rh) < \sigma(K_{h-1} + (r+1)h)$, *for any* h *and* $1 \leqslant r \leqslant k_h - 1$.

The proof is quite obvious.

We deduce from the above lemma an embedding of $I(\varphi) \setminus S_m$ as a subset of S_m, that we shall denote by $A(\varphi)$, or $A(\underline{k})$. We shall also write $I(\underline{k})$ for $i(\varphi)$.

We can visualize conditions (i) and (ii) above for $\sigma \in S_m$ as follows. We label the top of the branches of the trees of $M(\varphi)$ with $\sigma(h)$, $h = 1, \ldots, m$, starting from the left. Then on every tree the labels must be increasing and for consecutive trees having the same number of branches the labels on the last branches must be increasing.

Let $\underline{k} \in K(m, n)$. We define $D^1(\underline{k}) \in K(m+1, n+1)$ by $D^1(\underline{k}) = (k_1 + 1, k_2, k_3, \ldots)$. For $i = 2, \ldots, m$, provided $k_{i-1} \geqslant 1$ we set $D^i(\underline{k}) = (k_1, \ldots, k_{i-2}, k_{i-1} - 1, k_i + 1, k_{i+1}, k_{i+2}, \ldots)$ an element of $K(m, n+1)$.

Take now $\sigma_m \in A(\underline{k})$; for any $i = 2, \ldots, m$ and $r = 1, \ldots, k_{i-1}$ we define $D_r^i(\sigma_m) \in S_{m+1}$ by:

$$D_r^i(\sigma_m)_{(j)} = \begin{cases} \sigma_m(j), & 1 \leqslant j \leqslant K_{i-2} + (r-1)(i-1), \\ \sigma_m(j + i - 1), & K_{i-2} + (r-1)(i-1) + 1 \\ & \qquad \leqslant j \leqslant K_i - (i-1), \\ \sigma_m(j - k_i i - (k_{i-1} - r)(i-1)), & K_i - (i-1) + 1 \leqslant j \leqslant K_i, \\ m+1, & j = K_i + 1, \\ \sigma_m(j-1), & K_i + 2 \leqslant j \leqslant m+1. \end{cases}$$

For $i = 1$, we define $D_0^1(\sigma_m) \in S_{m+1}$ by:

$$D_0^i(\sigma_m)_{(j)} = \begin{cases} \sigma_m(j), & 1 \leqslant j \leqslant k_1, \\ m+1, & j = k_1 + 1, \\ \sigma_m(j-1), & k_1 + 2 \leqslant j \leqslant m+1. \end{cases}$$

Here we set $K_i = 0$ whenever $i \leqslant 0$.

We can visualize $D_r^i(\sigma_m)$ as follows. For $i \geqslant 2$ we label the top of the branches of $M(\varphi)$ with σ_m starting from the left. Then we add a branch labelled $m+1$ to the right of the $(k_1 + \cdots + k_{i-1} + r)$th tree; finally we put the modified tree at its right place, namely as a $(k_1 + \cdots + k_i)$th tree. For $i = 1$, $D_0^1(\sigma_m)$ is obtained by adding a $(k_1 + 1)$th one-branch tree labelled $m + 1$.

1.4. LEMMA. *$D_r^i(\sigma_m) \in A(D^i(\underline{k}))$ and for any $\underline{k} \in K(m, n)$ and $\sigma_m \in A(\underline{k})$ there exist unique $\underline{k}'$, i, r and $\sigma_{m-1} \in A(\underline{k}')$ such that $\underline{k} = D^i(\underline{k}')$, $\sigma_m = D_r^i(\sigma_{m-1})$. Here $\underline{k}' \in K(m-1, n-1)$ if $i = 1$, $\underline{k}' \in K(m-1, n)$ if $i \geqslant 2$.*

PROOF. It is readily verified that $D_r^i(\sigma_m) \in A(D^i(\underline{k}))$. Also, if $\sigma_{m-1}, \sigma'_{m-1} \in A(\underline{k}')$ then $D_r^i(\sigma_{m-1}) \neq D_{r'}^i(\sigma'_{m-1})$ provided that $r \neq r'$ or if $r = r'$ but $\sigma_{m-1} \neq \sigma'_{m-1}$.

Take now $\sigma_m \in A(\underline{k})$. Let $j_0 = \sigma^{-1}(m)$; there is an i such that $j_0 \in M_i$ and because of (i) and (ii) in Lemma 1.3 we must have $j_0 = K_i$. If $i = 1$, we set $\underline{k}' = (k_1 - 1, k_2, \ldots) \in K(m-1, n-1)$ (note that since in this case $k_1 \geqslant 1$, $k_1 + 2k_2 + \cdots + mk_m = m$ implies $k_m = 0$) and define $\sigma_{m-1} \in S_{m-1}$ by:

$$\sigma_{m-1}(j) = \begin{cases} \sigma_m(j), & 1 \leqslant j \leqslant k_1 - 1, \\ \sigma_m(j+1), & k_1 \leqslant j \leqslant m-1. \end{cases}$$

Clearly $D^1(\underline{k}') = \underline{k}$ and $D_0^1(\sigma_{m-1}) = \sigma_m$. If $i \geqslant 2$, set $\underline{k}' = (k_1, \ldots, k_{i-2}, k_{i-1} + 1, k_i - 1, k_{i+1}, \ldots)$. If $\sigma_m(K_{i-2} + k_{i-1}(i-1)) < \sigma_m(j_0 - 1)$ we set $r = k_{i-1} + 1$. Otherwise there is a unique r, $1 \leqslant r \leqslant k_{i-1}$, such that $\sigma_m(K_{i-2} + (r-1)(i-1)) < \sigma_m(j_0 - 1) < \sigma_m(K_{i-2} + r(i-1))$. We define $\sigma_{m-1} \in S_{m-1}$ by:

$$\sigma_{m-1}(j) = \begin{cases} \sigma_m(j), & 1 \leqslant j \leqslant K_{i-2} + (r-1)(i-1), \\ \sigma_m(j + (k_{i-1} - r)(i-1) + (k_i - 1)i), & K_{i-2} + (r-1)(i-1) \\ & \quad + 1 \leqslant j \leqslant K_{i-2} + r(i-1), \\ \sigma_m(j - (i-1)), & K_{i-2} + r(i-1) + 1 \leqslant j \leqslant K_{i-1} + i(k_i - 1), \\ \sigma_m(j+1), & K_{i-1} + i(k_i - 1) + 1 \leqslant j \leqslant m-1. \end{cases}$$

It can be verified that $\sigma_{m-1} \in A(\underline{k}')$ and that $D_r^i(\sigma_{m-1}) = \sigma_m$.

We can visualize the construction of $\underline{k}'$ and σ_{m-1} in the above proof as follows. We take φ: $[1, m] \to [1, n]$ associated to $\underline{k}$ and label again the top of the trees of $M(\varphi)$ with σ_m starting from the left. There is some branch labelled m: since

$\sigma_m \in A(\underline{k})$, this must be the last branch of the last tree of N_i, for some i. We cut out this branch, obtaining a new tree with $i-1$ branches (or drop the tree if $i=1$), that we put then in its right place.

2. Derivatives of composite mappings. Give normed vector spaces E and F we denote by $L_s^n(E,F)$ the space of continuous n-linear symmetric maps. For $\alpha \in L_s^n(E,F)$ and $\sigma \in S_n$ we define ${}^{\sigma}\alpha(x_1,\ldots,x_n) = \alpha(x_{\sigma(1)},\ldots,x_{\sigma(n)})$. Given $\underline{k} \in K(m,n)$, $\alpha_h \in L_s^h(E,F)$, $h=1,\ldots,m$, and $\beta_n \in L_s^n(F\cdot G)$, where G is a normed vector space, we define

(4) $$\mathrm{Sym}\big(\beta_n\cdot\big(\alpha_1^{k_1},\ldots,\alpha_m^{k_m}\big)\big) = \sum_{\sigma\in A(\underline{k})} {}^{\sigma}\big(\beta_n\cdot\big(\alpha_1^{k_1},\ldots,\alpha_m^{k_m}\big)\big),$$

where $\alpha_h^{k_h}$ denotes the k_h-fold product of α_h. We shall also write $(\alpha_1,\ldots,\alpha_m)^{\underline{k}}$ for $(\alpha_1^{k_1},\ldots,\alpha_m^{k_m})$. Although not indicated, it is clear that the operation Sym depends on $\underline{k}$.

2.1. LEMMA. $\mathrm{Sym}(\beta_n\cdot(\alpha_1,\ldots,\alpha_m)^{\underline{k}}) \in L_s^m(E,G)$.

PROOF. Clearly $I(\underline{k})$ acts trivially on $\beta_n\cdot(\alpha_1,\ldots,\alpha_m)^{\underline{k}}$. Hence we can write

$$\begin{aligned}\mathrm{Sym}\big(\beta_n\cdot(\alpha_1,\ldots,\alpha_m)^{\underline{k}}\big) &= \sum_{[\sigma]\in I(\varphi)\backslash S_m} {}^{[\sigma]}\big(\beta_n\cdot(\alpha_1,\ldots,\alpha_m)^{\underline{k}}\big)\\ &= (1/\#(I(\underline{k})))\cdot \sum_{\sigma\in S_m} {}^{\sigma}\big(\beta_n\cdot(\alpha_1,\ldots,\alpha_m)^{\underline{k}}\big)\end{aligned}$$

and the last expression is clearly symmetric.

In fact, writing $\phi(\beta_n,\alpha_1,\ldots,\alpha_m) = \beta_n\cdot(\alpha_1,\ldots,\alpha_m)^{\underline{k}}$, considering $\phi\colon E^m \to L(L_s^n(F,G)\times_{h=1,\ldots,m} L_s^h(F,G),K)$ (where $K=\mathbf{R}$ or $\mathbf{C}$), it can be shown that $I(\underline{k})$ is the isotropy group of ϕ. This shows that Sym is the minimal effort symmetrisation.

2.2. PROPOSITION. *Let $f\colon U\to F$ and $g\colon V\to G$ be of class C^m, where $U\subset E$ and $V\subset F$ are open and $V\supset f(U)$. We have*:

$$d^m(g\cdot f)_{(a)} = \sum_{\substack{n=1,\ldots,m\\ k_1+\cdots+k_m=m\\ k_1+2k_2+\cdots+mk_m=m}} \mathrm{Sym}\big(d^n g\cdot\big(df,d^2f,\ldots,d^mf\big)^{\underline{k}}\big),$$

where the derivatives of f are taken at a, those of g at $f(a)$.

PROOF. By induction on m. Assuming the formula for $m-1$, elementary derivation rules show that for $\sigma\in A(\underline{k})$, where $\underline{k}\in K(m-1,n')$, we have:

$$d\big({}^{\sigma}\big(d^{n'}g\cdot\big(df,\ldots,d^{m-1}f\big)^{\underline{k}}\big)\big) = \sum_{r,i} {}^{D_r^i(\sigma)}\big(d^{n'+\delta(i,1)}g\cdot\big(df,\ldots,d^mf\big)^{D^i(\underline{k})}\big),$$

where r, i range over all integers verifying $i\geqslant 2$, $k_{i-1}\geqslant 1$, $1\leqslant r\leqslant k_{i-1}$ or $i=1$, $r=0$; $\delta(i,1)$ equals one for $i=1$, zero otherwise. We conclude then with Lemma 1.4.

2.3. REMARK. In the next section we will use the operation Sym in a slightly more general situation. Namely, let T be a vector space such that $E \subset T$ and let $L^h(E^{h-1} \times T, F)$ denote the space of continuous h-linear maps from $E^{h-1} \times T$ to F. If $\underline{k} \in K(m, n)$, $\alpha_h \in L^h(E^{h-1} \times T, F)$, $h = 1,\ldots,m$, $\beta_n \in L^n_s(F, G)$ and $\sigma \in A(\underline{k})$, for $(x_1,\ldots,x_m) \in E^{m-1} \times T$ the expression

$$^{\sigma}\left(\beta_n \cdot (\alpha_1,\ldots,\alpha_m)^{\underline{k}}\right)_{(x_1,\ldots,x_m)}$$

makes sense, since because of 1.3(i) x_m will go in the last entry of some α_h. Hence $\mathrm{Sym}(\beta_n \cdot (\alpha_1,\ldots,\alpha_m)^{\underline{k}}) \in L^m(E^{m-1} \times T, G)$ still makes sense; it will be symmetric provided the α_h are, in the sense of [**1**, §13].

3. The intrinsic derivative for Σ^{i_m}. We give here a formula expressing the mth intrinsic derivative of a germ of map $f\colon (E, 0) \to (F, 0)$, where E and F are finite dimensional vector spaces with $\dim(E) \leqslant \dim(F)$, assuming that $0 \in \Sigma^{i_m}(f)$, and that f is Σ^{i_m} transversal, where i_m denotes the Boardman symbol

$$\Big(\underbrace{i, i,\ldots,i}_{m \text{ times}}, 0, 0,\ldots\Big).$$

See [**1**] for the definition of Σ^{i_m} as well as for the intrinsic derivative, originally considered by I. Porteous (we will not use the results of [**1**], except for Lemma 16.4 which is self-contained). Let us recall that for f generic the $\Sigma^{i_m}(f)$ are smooth subvarieties, $\Sigma^i(f)$ is the germ at 0 of $\{x \in E \mid \dim(\ker(df_x)) = i\}$ and $\Sigma^{i_m}(f) = \Sigma^i(f \mid \Sigma^{i_{m-1}}(f))$.

We shall work with representatives $f\colon U \to F$ of f, U a varying open neighborhood of $0 \in E$. For $x \in \Sigma^i(f)$ let $K_x = \ker(df_x)$, an i-dimensional vector subspace of E, and let K denote the corresponding bundle over $\Sigma^i(f)$. If U is small enough we can find a fixed complementary subspace K_0' to K_x, $x \in \Sigma^i(f) \cap U$. Given $\alpha_1' \in L(K_0, K_0')$ and $\alpha_h \in L^h_s(K_0, K_0')$, $h = 2,\ldots,m$, we define

$$w_m(x, f, \alpha_1',\ldots,\alpha_m) = \sum_{\substack{n=1,\ldots,m\\ \underline{k}\in K(m,n)}} \mathrm{Sym}\, d^n f(\alpha_1,\ldots,\alpha_m)^{\underline{k}} \in L^m_s(K_0, F),$$

where $\alpha_1 \in L(K_0, E)$ is defined by $\alpha_1(v) = v + \alpha_1'(v)$ (hence the image of α_1 is the graph of α_1').

Let $\tilde{\Sigma}^{i_m}(f) = \{(x, \alpha_1', \alpha_2,\ldots,\alpha_m) \subset U \times L(K_0, K_0') \times \cdots \times L^m_s(K_0, K_0')\}$ such that $w_i(x, f, \alpha_1',\ldots,\alpha_i) = 0$ for $i \leqslant m$ and let $p_m\colon \tilde{\Sigma}^{i_m} \to U$ denote the natural projection.

3.1. PROPOSITION. *Assume that the set of equations $w_i(x, f, \alpha_1',\ldots,\alpha_i) = 0$, $i = 1,\ldots,h$, is of maximal rank for $h = 1,\ldots,m$. Then $\tilde{\Sigma}^{i_h}$ is a submanifold of $U \times L(K_0, K_0') \times \cdots \times L^h_s(K_0, K_0')$ for $h = 1,\ldots,m$ and we have*:

(i) *$p_m\colon \tilde{\Sigma}^{i_m} \to U$ is an embedding with image $\Sigma^{i_m}(f)$ (hence for $x \in \Sigma^{i_m}(f)$ we can consider $\alpha_h = \alpha_h(x)$, $h = 1,\ldots,m$).*

(ii) *Let $Q_x = F/\mathrm{Im}(df_x)$, $x \in \Sigma^i(f)$. For $x \in \Sigma^{i_m}(f)$ we have an exact sequence*

$$0 \to T_x\Sigma^{i_m}(f) \hookrightarrow T_x\Sigma^{i_{m-1}}(f) \xrightarrow{\tilde{d}^{m+1}f_x} L^m_s(K_0, Q_x) \to 0, \tag{5}$$

where T_x *denotes the tangent space at* x *and where* $\tilde{d}^{m+1}f_x$ *considered as an* $(m+1)$*-linear map*: $K_0^m \times T_x\Sigma^{i_{m-1}} \to Q_x$ *is defined as follows. Let* $\bar{\alpha}_1(x)$: $T_x\Sigma^{i_{m-1}} \to E$ *denote the inclusion and consider* $d_x\alpha_h$ *as an* $(h+1)$*-linear map*: $K_0^h \times T_x\Sigma^{i_m} \to K_0'$; *we set* $\tilde{d}^{m+1}f_x = w_{m+1}(x, f, \bar{\alpha}_1, d_x\alpha_1', d_x\alpha_2, \ldots, d_x\alpha_m)$ (*see Remark* 2.3).

(iii) *Let* $v_1, \ldots, v_m \in K_0$. *For* $x \in \Sigma^{i_m}(f)$ *we have*:

$$\alpha_2(x)_{(v_1,v_2)} = d_x\alpha'_{1(v_1,\,\alpha_1(v_2))} \quad \textit{and} \quad \alpha_h(x)_{(v_1,\ldots,v_h)} = d_x\alpha_{h-1(v_1,\ldots,v_{h-2},\alpha_1(v_h))}$$

(*note that since* $x \in \Sigma^{i_m}$, $T_x\Sigma^{i_{m-1}} \supset K_x \ni \alpha_1(v)$ *for* $v \in K_0$; *cf. the proof below*).

(iv) $\tilde{d}^{m+1}f_x$ *is the* $m+1$*th intrinsic derivative of* f *at* x.

PROOF. We proceed by induction on m. The case $m = 1$ is treated as in [**8**, Proposition 1.1]. Note that $\alpha_1'(x)$: $K_0 \to K_0'$ is defined by the condition $df_x \cdot \alpha_1(x) = 0$, $x \in \Sigma^i(f)$, and hence $\alpha_1'(x)$ is the linear map whose graph is K_x (this has been used in (iii) above).

$(\mathrm{i})_m$ Assume that $(0, 0, \ldots, 0, a_m) \in E \times L(K_0, K_0') \times \cdots \times L_s^m(K_0, K_0')$ is tangent to $\tilde{\Sigma}^{i_m}$ at some point $(x, \alpha_1', \alpha_2, \ldots, \alpha_m)$. Then, taking the derivative of w_m at that point in the direction $L_s^m(K_0, K_0')$ we see that this would imply that $df_x \cdot a_m = 0$. Since $df_x \mid K_0'$ is injective, we must have $a_m = 0$. Since p_{m-1} is an immersion, this implies that p_m is an immersion. By the induction hypothesis the equations $w_h(x, f, \alpha_1', \ldots, \alpha_h) = 0$, $h = 1, \ldots, m-1$, determine $\alpha_1', \ldots, \alpha_{m-1}$ as functions of x; since $df_x \mid K_0'$ is injective, the equation $w_m(x, \ldots) = 0$ determines α_m as a smooth function of x. Hence p_m is an embedding. From $(\mathrm{ii})_{m-1}$ we deduce that $\ker(df_x \mid T_x\Sigma^{i_{m-1}}) = \ker(d^m f_x)$. From $(\mathrm{iii})_{m-1}$ we deduce that $p_m(\tilde{\Sigma}^{i_m}) = \{x \in \Sigma^{i_{m-1}} \mid \ker(df \mid T_x\Sigma^{i_{m-1}}) = K_x\}$ and we can take the last defined set as a definition for $\Sigma^{i_m}(f)$.

$(\mathrm{ii})_m$ is immediate from the proof of $(\mathrm{i})_m$ and Lemma 1.4.

$(\mathrm{iii})_m$ For $x \in \Sigma^{i_m}(f)$ we have $K_x \subset T_x\Sigma^{i_{m-1}}(f)$ and hence $d^m f_x \mid K_x = 0$. But by $(\mathrm{iii})_{m-1}$ $\tilde{d}^m f_x \mid K_x = w_m(x, f, \alpha_1, \alpha_2, \ldots, \alpha_{m-1}, d_x\alpha_m(\ldots, \alpha_1)) = 0$. Since the equation $w_m(x, f, \alpha_1, \ldots, \alpha_m) = 0$ defines α_m uniquely, we must have $d_x\alpha_m(\ldots, \alpha_1) = \alpha_{m+1}$.

$(\mathrm{iv})_m$ By going through the constructions of [**1**] it can be seen that $\tilde{d}^{m+1}f$, if we identify K_0 with K_x via α_1, coincides with the bundle map constructed in [**1**, Theorem 16.15, item 8m]. But it can be seen more directly that at least $\tilde{d}^{m+1}f$ is intrinsic (i.e. depends only on the linear part of change of coordinates on source and target of f). Indeed, consider $\tilde{d}^m f_x \cdot \alpha_1(x)$: $K_0 \to L_s^{m-1}(K_0, Q_x)$ as a bundle map over Σ^{i_m}. Then $\tilde{d}^{m+1}f$ is the corresponding intrinsic derivative in the sense of [**1**, Lemma 16.4] (this lemma is self-contained).

3.2. REMARKS. (i) It can be proved with the usual transversality techniques that for almost every map f the transversality assumptions in 3.1 are satisfied. Also, it can be shwon that these assumptions are equivalent to saying that the jet extension $j^m(f)$: $U \to J^m(E, F)$ is transversal to Σ^{i_m}. These assumptions are somewhat clumsy, but they allow us to work with maps rather than on the jet space, and thus to have a simpler exposition.

(ii) If f does not satisfy the transversality assumptions of 3.1 and if we assume that its symbol (as defined in [**1**, 11.1]) at $0 \in E$ is i_m, we still have an intrinsic derivative $\tilde{d}^{m+1}f_0\colon T^{m-1} \to L_s^m(K_0, Q_0)$, where T^{m-1} is to be defined. To do this, take $g\colon U \times K^k \to F \times K^k$, $g(x, t_1, \dots, t_k) = (f(x) + \Sigma_{i=1,\dots,k} t_i \cdot g_i(x), t)$. For a suitable choice of the g_i's g has the required transversality properties. Also, $\ker(dg_0) = \ker(df_0)$, $\operatorname{coker}(df_0) = \operatorname{coker}(dg_0)$ and the symbol of g at 0 is i_m. We defined

$$\tilde{d}^{m+1}f_0 = \tilde{d}^{m+1}g_0 \mid T^{m-1}\colon T^{m-1} \to L_s^m(K_0, Q_0),$$
$$\text{where } T^{m-1} = T_0\Sigma^{i_m}(g) \cap E \times 0.$$

Let us consider some examples (actually, the genuine intrinsic derivative is a map: $T_x\Sigma^{i_{m-1}} \to L_s^m(K_x, Q_x)$ and is deduced from the one we considered via the isomorphism $\alpha_1\colon K_0 \overset{\cong}{\to} K_x$).

For $m = 2$, $\tilde{d}^2 f_x\colon E \to L(K_x, Q_x)$ is deduced from the ordinary second derivative.

For $m = 3$, we have:

$$\tilde{d}^3 f_0 = d^3 f|_{T^2\times K_0 \times K_0} + \operatorname{Sym}\big(d^2 f_0(d_0\alpha_1(\ ,\ ),\ )|_{T^2\times K_0\times K_0}\big) \quad \operatorname{mod}(\operatorname{Im} df_0).$$

Since $d^2 f|_{T^1\times K_0} + df_0(d_0\alpha_1) = 0$, we have $d_0\alpha_1 = -(df|_{K_0})^{-1}\cdot d^2 f|_{T^1\times K_0}$, where $T^m = T_0\Sigma^{i_m}$. Hence

$$\begin{aligned}\tilde{d}^3 f_0(v_1, v_2, v_3) = d^3 f_0(v_1, v_2, v_3) &- d^2 f_0\big((df|_{K_0})^{-1}\cdot d^2 f_0(v_1, v_2), v_3\big)\\ &- d^2 f_0\big((df|_{K_0})^{-1}\cdot d^2 f_0(v_1, v_3), v_2\big)\\ &- d^2 f_0\big((df|_{K_0})^{-1}\cdot d^2 f_0(v_2, v_3), v_1\big) \quad \operatorname{mod}(\operatorname{Im} df_0).\end{aligned}$$

For $m = 4$, omitting to indicate restrictions, the subscript 0 and mod Im df, we find

$$\begin{aligned}\tilde{d}^4 f(v_1, \dots, v_4) = d^4 f(v_1, \dots, v_4) &- d^3 f\big(df^{-1}d^2 f(v_1, v_2), v_3, v_4\big)\\ &- d^3 f\big(df^{-1}d^2 f(v_1, v_3), v_2, v_4\big)\\ &- d^3 f\big(df^{-1}d^2 f(v_1, v_4), v_2, v_3\big) - d^3 f\big(df^{-1}d^2 f(v_2, v_3), v_1, v_4\big)\\ &- d^3 f\big(df^{-1}d^2 f(v_2, v_4), v_1, v_3\big) - d^3 f\big(df^{-1}d^2 f(v_3, v_4), v_1, v_2\big)\\ &+ d^2 f\big(df^{-1}d^2 f(v_1, v_2), df^{-1}d^2 f(v_3, v_4)\big)\\ &+ d^2 f\big(df^{-1}d^2 f(v_1, v_3), df^{-1}d^2 f(v_2, v_4)\big)\\ &+ d^2 f\big(df^{-1}d^2 f(v_1, v_4), df^{-1}d^2 f(v_2, v_3)\big)\\ &- d^2 f\big(df^{-1}\tilde{d}^3 f(v_1, v_2, v_3), v_4\big) - d^2 f\big(df^{-1}\tilde{d}^3 f(v_1, v_2, v_4), v_3\big)\\ &- d^2 f\big(df^{-1}\tilde{d}^3 f(v_1, v_3, v_4), v_2\big) - d^2 f\big(df^{-1}\tilde{d}^3 f(v_2, v_3, v_4), v_1\big).\end{aligned}$$

If instead of i_m we take a general symbol $(i, j, \dots, k)$, for $m = 3$ the above formula is still valid, but for $m = 4$ we get extra terms.

ADDED IN PROOF. I have been informed that E. Stamm in *Formulae for higher F-derivatives and the solution of analytic equations in Banach spaces*, preprint, ETH Zurich (1979) has obtained similar and more general formulae than those considered here in §§2 and 3. Also, I. Porteous has informed me that he has elaborated a procedure to obtain expressions for the intrinsic derivatives similar to ours (see his paper *Probing singularities* in these Proceedings; see there also for more references).

BIBLIOGRAPHY

1. J. M. Boardman, *Singularities of differentiable maps*, Inst. Hautes Etudes Sci. Publ. Math. **33** (1967), 383–419.

2. Faa' de Bruno, *Note sur une nouvelle formule de calcul différentiel*, Quart. J. Math. **I** (1857), 359–360.

3. L. Comtet, *Advanced combinatorics*, Reidel, Boston, 1974, pp. 137–139.

4. L. E. Fraenkel, *Formulae for high derivatives of composite functions*, Proc. Cambridge Philos. Soc. **83** (1978), 159–165.

5. I. R. Porteous, *Simple singularities of maps*, preprint, Columbia University (1962); Reprinted in Proc. Liverpool Singularities Sympos., Lecture Notes in Math., vol. 192, Springer-Verlag, Berlin, 1971, pp. 217–236.

6. J. Riordan, *An introduction to combinatorial analysis*, Wiley, New York, 1958, pp. 36–37.

7. S. Roman, *Faa' di Bruno's formula*, Amer. Math. Monthly **87** (1980), 805–809.

8. F. Ronga, *Le calcul des classes duales aux singularitiés de Boardman d'ordre deux*, Comment. Math. Helv. **47** (1972), 15–35.

9. ______, *Dual classes to Boardman's singularities*, preprint, Institute for Advanced Study, Princeton, 1973.

10. P. G. Todorov, *New explicit formulas for the n-th derivative of composite functions*, Pacific J. Math. **92** (1981), 217–236.

UNIVERSITY OF GENEVA, SWITZERLAND

Proceedings of Symposia in Pure Mathematics
Volume **40** (1983), Part 2

LE TYPE TOPOLOGIQUE ÉCALTÉ D'UNE APPLICATION ANALYTIQUE

C. SABBAH

Introduction. Soient $f\colon X \to Y$ et $g\colon X' \to Y'$ deux applications analytiques entre espaces analytiques complexes réduits. On dit que f et g ont même type topologique si il existe des homéomorphismes φ et ψ qui rendent le diagramme suivant commutatif:

$$\begin{array}{ccc} X & \xrightarrow{\psi} & X' \\ f\downarrow & & \downarrow g \\ Y & \xrightarrow[\varphi]{} & Y' \end{array}$$

Fukuda a démontré [F] la conjecture suivante de R. Thom: Le nombre de types topologiques de polynômes $p\colon \mathbf{C}^n \to \mathbf{C}$ de degré inférieur ou égal à d (n et d fixés) est fini. On sait cependant [Th], [N] que, pour les applications, polynomiales p: $\mathbf{C}^n \to \mathbf{C}^m$ de degré inférieur ou égal à d, et pour n, m et d supérieurs ou égaux à 3, le type topologique présente des "modules". Ceci a été expliqué par R. Thom comme résultant de situations d'"éclatement" pour les applications p.

Si une telle application p présente de l'éclatement, on peut trouver une suite d'éclatements de $\mathbf{C}^m$ de sorte que l'application $\tilde{p}$ obtenue par changement de base n'en présente plus [S]. Ceci nous conduit à considérer une relation plus grossière
éclatés d'applications polynomiales p: $\mathbf{C}^n \to \mathbf{C}^m$ de degré inférieur ou égal à d.
certaines notations, et pour la démonstration de certains résultats. Les espaces

Quand $n = 2$, on peut alors montrer que le nombre de types topologiques est lui aussi fini, ce qui répond à une question posée par Nakai. Pour $n = m = 2$, le résultat était déjà démontré (cf. [A]).

Nous démontrerons ces résultats pour des applications entre espaces analytiques complexes compacts, le cas des polynômes en étant un corollaire.

1980 *Mathematics Subject Classification.* Primary 32C42, 14F45, 57N80.

Remarquons enfin qu'on pourrait de même définir le type topologique éclaté pour des germes d'applications analytiques. Il faudrait alors utiliser la notion d'éclatement local [$\mathbf{H}_3$]. D'une manière générale, nous renvoyons à [**S**] pour certaines notations, et pour la démonstration de certains résultats. Les espaces analytiques considérés seront toujours munis de leur structure réduite, et les stratifications seront analytiques complexes, les strates étant connexes.

1. Le type topologique éclaté. On considère deux applications analytiques $f_i\colon X_i \to Y_i$ $(i = 1, 2)$, où X_i et Y_i sont des espaces analytiques complexes réduits et compacts, et la relation suivante entre elles: Pour chaque i, il existe une suite finie d'éclatements au-dessus de Y_i, dont le composé est noté $\pi_i\colon \tilde{Y}_i \to Y_i$, telle que:

(a) Soit $\Sigma_i \subset Y_i$ la réunion des images des centres d'éclatement dans Y_i. Alors Σ_i est nulle part dense dans $f_i(X_i)$ $(i = 1, 2)$.

(b) On considère les produits fibrés:

$$\begin{array}{ccc} \tilde{X}_i & \rightarrow & X_i \\ \tilde{f}_i \downarrow & & \downarrow f_i \\ \tilde{Y}_i & \underset{\pi_i}{\rightarrow} & Y_i \end{array}$$

Alors les applications $\tilde{f}_1$ et $\tilde{f}_2$ sont topologiquement équivalentes, c'est-à-dire qu'il existe des homéomorphismes $\tilde{\varphi}$ et $\tilde{\psi}$ faisant commuter le diagramme:

$$\begin{array}{ccc} \tilde{X}_1 & \overset{\tilde{\psi}}{\rightarrow} & \tilde{X}_2 \\ \tilde{f}_1 \downarrow & & \downarrow \tilde{f}_2 \\ \tilde{Y}_1 & \underset{\tilde{\varphi}}{\rightarrow} & \tilde{Y}_2 \end{array}$$

(c) Soit $\widehat{f_i(X_i)}$ le transformé strict de $f_i(X_i)$ par π_i, et $\hat{X}_i$ celui de X_i.

Alors $\tilde{\varphi}$ induit un homéomorphisme $\hat{\varphi}\colon \widehat{f_1(X_1)} \to \widehat{f_2(X_2)}$, et $\tilde{\psi}$ un homéomorphisme $\hat{\psi}\colon \hat{X}_1 \to \hat{X}_2$, de sorte qu'on a un diagramme commutatif:

$$\begin{array}{ccc} \hat{X}_1 & \overset{\hat{\psi}}{\rightarrow} & \hat{X}_2 \\ \hat{f}_1 \downarrow & & \downarrow \hat{f}_2 \\ \widehat{f_1(X_1)} & \underset{\hat{\varphi}}{\rightarrow} & \widehat{f_2(X_2)} \end{array}$$

DÉFINITION. Soient $f\colon X \to Y$ et $g\colon X' \to Y'$ deux applications entre espaces analytiques compacts. On dira que f et g ont même type topologique éclaté si il existe une suite finie $f_0 = f, f_1, \ldots, f_n = g$ d'applications $f_i\colon X_i \to Y_i$, telle que pour chaque i, f_i et f_{i+1} vérifient la relation précédente.

REMARQUE. Si f et g ont même type topologique, elles ont même type topologique éclaté, la réciproque étant fausse. Par contre, si Y et Y' sont des courbes lisses, les deux notions coincident.

THÉORÈME 1. *Soit F: $X \to Y$ un morphisme propre entre espaces analytiques réduits de dimension finie, et p: $Y \to S$ un autre morphisme propre, S étant l'espace des paramètres. Le nombre de types topologiques éclatés apparaissant dans la famille F d'applications F_s: $X_s \to Y_s$, $s \in S$, est localement fini sur S.*

DÉMONSTRATION. On peut supposer que $S = p \circ F(X)$, puisque $p \circ F$ est propre.

Rappelons maintenant quelques résultats de [**S**]. Soit F: $X \to Y$ un morphisme propre entre espaces analytiques complexes réduits, et $\mathbb{S}$ une stratification **C**-analytique de F. On dit que $(F, \mathbb{S})$ est sans éclatement si la stratification de Y satisfait les conditions de Whitney, et celle de X la condition A_F de Thom et si la restriction de la stratification de X *au-dessus de chaque strate de* Y satisfait les conditions de Whitney.

Etant donné un morphisme propre F: $X \to Y$, il existe une modification propre (composée d'une suite d'éclatements) π: $\tilde{Y} \to Y$ telle que le morphisme $\tilde{F}$: $X \times_Y \tilde{Y} \to \tilde{Y}$, image inverse de F par π, soit stratifié sans éclatement [**S**, Théorème 1]. Appliquons ce théorème à F, et considérons maintenant la famille $\tilde{X} \overset{\tilde{F}}{\to} \tilde{Y} \overset{\tilde{p}}{\to} S$, où $\tilde{p} = p \circ \pi$. On note $\tilde{\mathbb{S}}$ la stratification de $\tilde{F}$ donnée par le théorème. Comme $\tilde{p}$ est propre, on peut trouver des stratifications $\mathcal{C}_{\tilde{Y}}$ et $\mathcal{C}_S$ de $\tilde{Y}$ et S, de sorte que $\tilde{p}$ soit stratifié, et on peut supposer que la stratification de $\tilde{Y}$ donnée par $\tilde{\mathbb{S}}$ n'est autre que $\mathcal{C}_{\tilde{Y}}$ (voir [**S**]). Soit Ω la strate de dimension maximum de S.

Soit $Z = F(X)$ et $\hat{Z}$ le transformé strict de Z par π. On peut supposer que $\hat{Z}$ est une union de strates de $\mathbb{S}_{\tilde{Y}}$. Par suite, c'est aussi une union de strates de $\mathcal{C}_{\tilde{Y}}$.

Soit $\hat{X}$ le transformé strict de X par π. Comme $\hat{X}$ est une union de composantes irréductibles de $\tilde{X}$, et que les strates de $\mathcal{C}_{\tilde{X}}$ sont connexes, $\hat{X}$ est une union de strates de $\mathcal{C}_{\tilde{X}}$.

Assertion. Il existe un ouvert de Zariski dense Ω' de S tel que, pour $s \in \Omega'$, on ait:

(1) $\tilde{Y}_s = \tilde{p}^{-1}(s)$ est le transformé strict de Y_s par π.

(2) $\hat{Z}_s = \hat{Z} \cap \tilde{Y}_s$ est le transformé strict de $Z_s = Z \cap p^{-1}(s)$ par π.

(3) $\hat{X}_s = \hat{X} \cap (\tilde{p} \circ \tilde{F})^{-1}(s)$ est le transformé strict de X_s par π.

Démontrons le (1). Le (2) et le (3) se démontrent de la même manière.

Soit $E \subset \tilde{Y}$ le diviseur exceptionnel de π, qui est partout de codimension 1 dans $\tilde{Y}$. Il suffit de trouver Ω' tel que, pour $s \in \Omega'$, $E_s = E \cap \tilde{p}^{-1}(s)$ soit nulle part dense dans $\tilde{Y}_s$, puisque $\hat{Y}_s$ est l'adhérence dans $\tilde{Y}_s$ de $\tilde{Y}_s \setminus E_s$. L'existence de Ω' est alors claire.

Soit maintenant $\Omega'' = \Omega \cap \Omega'$. On peut appliquer le deuxième lemme d'isotopie au morphisme stratifié $\tilde{F}$: $\tilde{X}|_{\Omega''} \to \tilde{Y}|_{\Omega''}$. Si s_1 et s_2 sont dans la même composante connexe de Ω'', $\tilde{F}_{s_1}$ et $\tilde{F}_{s_2}$ ont même type topologique. Comme $\hat{X}$ et $\hat{Z}$ sont union strates, et que les homéomorphismes donnés par le deuxième lemme d'isotopie sont stratifiés, on en déduit que F_{s_1} et F_{s_2} ont même type topologique éclaté. On en déduit le théorème par une récurrence sur dim S.

REMARQUES. (1) L'utilisation du 2ème lemme d'isotopie de Thom-Mather nécessite l'hypothèse que les espaces stratifiés sont contenus dans des variétés lisses C^∞, et que les morphismes sont restrictions de morphismes C^∞ entre ces variétés. Le théorème suivant permet de justifier l'utilisation faite plus haut:

THÉORÈME (VOIR [**A-H**, COROLLARY (3.2)] PAR EXEMPLE). *Soit X un espace analytique réel et K un compact de X. Il existe une application analytique réelle $f\colon X \to \mathbf{R}^n$ qui est injective et propre, et qui est un plongement analytique en tout point de K.*

(2) On peut montrer, en utilisant seulement le critère donné par Hironaka ([$\mathbf{H_2}$, Theorem 1, §5], voir aussi [**S**, Théorème (1.2.1)]) que si la dimension relative de Z sur s est partout $\leqslant 1$, le nombre de types topologiques d'applications F_s: $X_s \to Y_s$ est fini.

2. Finitude locale du nombre de types topologiques. On considère dans ce paragraphe une famille d'applications analytiques

$$F\colon X \times S \to Y \times S \qquad \begin{array}{c} \searrow \;\; \swarrow \\ S \end{array}$$

paramétrée par un espace analytique réduit S. On suppose que X et Y sont compacts, et que X est lisse, connexe, de dimension 2.

THÉORÈME 2. *Sous ces conditions, le nombre de types topologiques d'applications F_s: $X \to Y$, pour $s \in S$ est localement fini sur S.*

On va montrer qu'il existe un ouvert de Zariski dense et lisse Ω de S tel que le type topologique de F_s soit constant sur chaque composante connexe de Ω. Le théorème s'en déduit par récurrence sur $\dim S$. Soit $Z = F(X \times S)$. Si Z est génériquement de dimension relative $\leqslant 1$ sur S, l'existence de Ω provient de la remarque faite à la fin du §1. On supposera donc que Z est de dimension relative 2 sur S, et que S est irréductible.

Considérons d'abord la situation générale suivante: G: $T \to Z$ est un morphisme propre, T et Z étant des espaces analytiques réduits, et p: $Z \to S$ est un morphisme propre. On suppose $p \circ G$ surjectif, et S irréductible. Soit $\mathbb{S}$ une stratification de G.

LEMME. *Il existe une modification propre π: $\tilde{Z} \to Z$ (qui est une suite d'éclatements) telle que, pour tout sous-ensemble analytique fermé Σ de Z, génériquement de dimension relative* 1 *sur S, il existe des stratifications de Whitney $\mathcal{C}_Z$ et $\mathcal{C}_{\tilde{Z}}$ de Z et $\tilde{Z}$, et une stratification $\mathcal{C}_{\tilde{T}}$ de $\tilde{T} = T \times_Z \tilde{Z}$, compatibles avec $\mathbb{S}$, telles que*:

(1) *$(\mathcal{C}_{\tilde{Z}}, \mathcal{C}_Z)$ est une stratification de π.*

(2) *$\mathcal{C}_{\tilde{Z}}$ est compatible avec $E = \pi^{-1}(\Sigma)$, et $\mathcal{C}_Z$ avec Σ.*

(3) *$(\mathcal{C}_{\tilde{T}}, \mathcal{C}_{\tilde{Z}})$ est une stratification sans éclatement de $\tilde{F}$.*

(4) *Il existe un ouvert de Zariski dense et lisse Ω de S tel que la restriction au-dessus de Ω de la stratification de π: $E \to \Sigma$ induite par $\mathcal{C}_{\tilde{Z}}$ et $\mathcal{C}_Z$ soit sans éclatement.*

(5) *Si* Z_β *est une strate de* $\mathcal{C}_Z$, *telle que* $p(\overline{Z}_\beta) \cap \Omega \neq \varnothing$, *alors la restriction* $p\colon Z_\beta\,|_\Omega \to \Omega$ *est une submersion.*

DÉMONSTRATION DU THÉORÈME 2. On applique le lemme à $F\colon X \times S \to Y \times S$. Soit $\Sigma \subset Y \times S$ la réunion des images des centres d'éclatements définissant π. Alors $\Sigma \subset F(X \times S)$ et est nulle part dense dans $F(X \times S)$. Par suite, les fibres de Σ sur S sont génériquement sur S de dimension $\leqslant 1$. Σ vérifie donc la condition du lemme. Soit $\hat{F}\colon \widehat{X \times S} \to \widetilde{Y \times S}$ le transformé strict de F par π. Comme $\widehat{X \times S}$ est une composante irréductible de $\widetilde{X \times S}$, c'est aussi une union de strates de la stratification donnée par le lemme. On considère le diagramme:

$$\begin{array}{ccc} \widehat{X\times S} & \overset{\bar\omega}{\to} & X\times S \\ \hat F \downarrow & & \downarrow F \\ \widetilde{Y\times S} & \underset{\pi}{\to} & Y\times S \\ & \tilde p \searrow & \downarrow p \\ & & S \end{array}$$

Remarquons que, d'après les hypothèses faites, $\bar\omega$ est surjectif.

Le morphisme $\bar\omega$ est, par définition, une suite d'éclatements, les centres étant les images inverses de ceux définissant π. Soit $\Theta \subset X \times S$ le plus petit sous-ensemble analytique fermé de $X \times S$ tel que le morphisme

$$\bar\omega\colon \widehat{(X \times S)} \backslash \bar\omega^{-1}(\Theta) \to (X \times S) \backslash \Theta$$

soit un isomorphisme. Alors $\Theta \subset F^{-1}(\Sigma)$. On a vu au §1 qu'il existe un ouvert de Zariski dense et lisse de S tel que, pour tout s dans cet ouvert, $\widehat{(X \times S)}_s$ soit le transformé strict de $X \times \{s\}$ par $\bar\omega$.

Le morphisme $\bar\omega\colon \widehat{(X \times S)}_s \to X \times \{s\}$ est donc une suite d'éclatements, l'image des centres étant la fibre Θ_s. Comme X est lisse, connexe de dimension 2, Θ_s est un nombre fini de points. Il existe donc un ouvert de Zariski dense et lisse Ω' de S tel que le morphisme $p \circ G\colon \Theta\,|_{\Omega'} \to \Omega'$ soit un revêtement (non ramifié).

D'autre part, on sait que $\bar\omega^{-1}(F^{-1}(\Sigma)) = \hat{F}^{-1}(\pi^{-1}(\Sigma))$ est une union de strates de $\widehat{X \times S}$.

Il existe un ouvert de Zariski dense Ω'' de S tel que les fibres $F^{-1}(\Sigma)_s$, pour $s \in \Omega''$, soient de dimension $\leqslant 1$. En effet, Σ est nulle part dense dans $F(X \times S)$, donc $F^{-1}(\Sigma)$ est nulle part dense dans $X \times S$ (qui est irréductible). On peut alors supposer, quitte à diminuer Ω'', que toute composante Γ de $F^{-1}(\Sigma)$ de dimension relative générique nulle est un revêtement (non ramifié) de Ω'', quand on la restreint à Ω''. On remarque que sur $\Omega' \cap \Omega''$, Γ est aussi une composante de $\Theta\,|_{\Omega' \cap \Omega''}$.

Soit maintenant $\Omega_1 = \Omega \cap \Omega' \cap \Omega''$, où Ω est l'ouvert donné par le lemme. On restreint alors la situation à Ω, et on veut montrer que pour s_1 et s_2 dans Ω_1, F_{s_1} et F_{s_2} ont même type topologique.

Soit η un champ de vecteur C^∞ sur Ω_1 (on se place en fait sur un ouvert relativement compact de S, contenu dans Ω_1). On va remonter ce champ en plusieurs étapes.

(1) On peut relever η en un champ contrôlé ξ_Σ sur Σ (cf. [**M**]).

(2) On peut relever ξ_Σ en un champ ξ_E sur $E = \pi^{-1}(\Sigma)$, contrôlé au-dessus de ξ_Σ, d'après la propriété A_π.

(3) On peut étendre ξ_E en un champ $\xi_{\widetilde{Y\times S}}$ sur $\widetilde{Y\times S}$, qui vérifie les conditions de contrôle entre strates de $\widetilde{Y\times S}\setminus E$, ou entre strates de $\widetilde{Y\times S}\setminus E$ et strates de E.

(4) On peut relever $\xi_{\widetilde{Y\times S}}$ en un champ $\xi_{\widehat{X\times S}}$ sur $\widehat{X\times S}$, contrôlé au-dessus de $\xi_{\widetilde{Y\times S}}$.

Les résultats classiques de [**M**] montrent que si s_1 et s_2 sont sur une même ligne du champ η, le flot de $\xi_{\widetilde{Y\times S}}$ définit un homémorphisme $\tilde{\varphi}$: $(\widetilde{Y\times S})_{s_1} \to (\widetilde{Y\times S})_{s_2}$. Cet homéomorphisme induit un homéomorphisme $\tilde{\varphi}_E$: $E_{s_1} \to E_{s_2}$, puisque E est une union de strates, et, puisque ξ_E provient d'un champ ξ_Σ, il existe un homéomorphisme φ_Σ rendant le diagramme suivant commutatif:

$$\begin{array}{ccc} E_{s_1} & \xrightarrow{\tilde{\varphi}_E} & E_{s_2} \\ \pi\downarrow & & \downarrow\pi \\ \Sigma_{s_1} & \xrightarrow[\varphi_\Sigma]{} & \Sigma_{s_2} \end{array}$$

L'homéomorphisme $\tilde{\varphi}$ définit un homéomorphisme φ: $Y\times\{s_1\} \to Y\times\{s_2\}$. En effet, considérons le diagramme:

$$\begin{array}{ccc} (\widetilde{Y\times S})_{s_1} & \xrightarrow{\tilde{\varphi}} & (\widetilde{Y\times S})_{s_2} \\ \pi\downarrow & & \downarrow\pi \\ Y\times\{s_1\} & & Y\times\{s_2\} \end{array}$$

Il suffit alors de montrer que si deux points de $(\widetilde{Y\times S})_{s_1}$ ont même image par π, ils ont même image par $\pi\circ\tilde{\varphi}$. Comme π est un isomorphisme en dehors de E et Σ, et que $\tilde{\varphi}$ conserve E, il suffit de choisir ces deux points dans E_{s_1}. Mais si x_1, $x_1' \in E_{s_1}$ et si $\pi(x_1) = \pi(x_1')$, $\varphi_\Sigma\circ\pi(x_1) = \varphi_\Sigma\circ\pi(x_1')$, donc $\pi\circ\tilde{\varphi}_E(x_1) = \pi\circ\tilde{\varphi}_E(x_1')$ et $\pi\circ\tilde{\varphi}(x_1') = \pi\circ\tilde{\varphi}(x_1')$. Il existe donc un homéomorphisme φ qui fait commuter le diagramme ci-dessus.

Considérons maintenant l'homéomorphisme $\hat{\psi}$: $\widehat{(X\times S)}_{s_1} \to \widehat{(X\times S)}_{s_2}$, donné par l'intégration du champ $\xi_{\widehat{X\times S}}$. Il suffit de montrer qu'il définit un homéomorphisme ψ: $X\times\{s_1\} \to \mathrm{X}\times\{s_2\}$ faisant commuter le diagramme:

$$\begin{array}{ccc} \widehat{(X\times S)}_{s_1} & \xrightarrow{\hat{\psi}} & \widehat{(X\times S)}_{s_2} \\ \overline{\omega}\downarrow & & \downarrow\overline{\omega} \\ X\times\{s_1\} & \xrightarrow[\psi]{} & X\times\{s_2\} \end{array}$$

En effet, $\overline{\omega}$ et π étant des isomorphismes sur des ouverts denses, la relation $\varphi\circ F_{s_1} = F_{s_2}\circ\psi$ sera vérifiée sur un ouvert dense de $X\times\{s_1\}$, donc partout.

Soit $\Delta = \overline{\omega}^{-1}(\Theta)$. Δ est une union de composantes irréductibles de $\overline{\omega}^{-1}(F^{-1}(\Sigma))$, les autres étant celles du transformé strict de $F^{-1}(\Sigma)$ par $\overline{\omega}$. Δ est donc une union de strates. Pour montrer l'existence de ψ, il suffit de montrer que si deux points x_1 et x_1' de $\widehat{(X \times S)}_{s_1}$ ont même image par $\overline{\omega}$, ils ont même image par $\overline{\omega} \circ \hat{\psi}$. On peut supposer que ces deux points sont sur Δ, puisque $\overline{\omega}$ est un isomorphisme en dehors de Δ.

Comme Θ est un revêtement de Ω, le champ η se remonte de manière unique en un champ C^∞ sur Θ, noté η_Θ. Le champ $\xi_{\widehat{X \times S}}$ restreint à Δ est donc un relèvement de η_Θ. Par suite les flots commutent à $\overline{\omega}$, et $\hat{\psi}$ définit un homéomorphisme ψ. □

DÉMONSTRATION DU LEMME. Construisons une modification π: $\tilde{Z} \to Z$ comme dans [**S**, Théorème 1]. On dispose alors d'une stratification sans éclatement du morphisme $\tilde{G}$: $\tilde{T} \to \tilde{Z}$, notée $\tilde{\mathbb{S}} = (\mathbb{S}_{\tilde{T}}, \mathbb{S}_{\tilde{Z}})$, compatible avec $\mathbb{S}$. Tout raffinement (de Whitney) de $\mathbb{S}_{\tilde{Z}}$ permet de définir un raffinement de $\mathbb{S}_{\tilde{T}}$ qui reste sans éclatement. Ainsi, le (3) sera facilement vérifié. Faisons alors l'hypothèse de récurrence suivante:

(H_l) Il existe un raffinement $\mathbb{S}^l_{\tilde{Z}}$ de $\mathbb{S}_{\tilde{Z}}$ vérifiant les conditions de Whitney, compatible avec E, un raffinement $\mathbb{S}^l_Z$ de $\mathbb{S}_Z$, compatible avec Σ, et un ouvert de Zariski dense Ω_l de S, tels que, pour toute strate $\tilde{Z}_\beta$ de $\mathbb{S}^l_{\tilde{Z}}$ de dimension $\geqslant l$, et telle que $\pi(\tilde{Z}_\beta)$ soit une composante irréductible de Z, on ait:

(a) $\pi(\tilde{Z}_\beta)$ est une strate de $\mathbb{S}^l_Z$ et π: $\tilde{Z}_\beta \to \pi(\tilde{Z}_\beta)$ est une submersion.

(b) Si $Z_\beta \subset E$ et si $\tilde{Z}_\gamma \subset \tilde{Z}_\beta$, le couple $(\tilde{Z}_\beta, \tilde{Z}_\gamma)$ satisfait la condition A_π au-dessus de Ω_l.

(c) Le morphisme p: $Z_\beta|_{\Omega_l} \to \Omega_l$ est une submersion si $Z_\beta = \pi(\tilde{Z}_\beta)$ et si $p(\overline{Z}_\beta)$ est une composante irréductible de S.

Montrons que $(H_{l+1}) \to (H_l)$. C'est clair pour les propriétés (a) et (c), et la propriété (b) provient de la remarque (2) du §1.

Si $l = \dim X + 1$, (H_l) est trivialement satisfaite. On en déduit que (H_0) est vraie. On fait alors une récurrence descendante sur dim Z. Les propriétés (1), (2), (4), et (5) du lemme seront alors satisfaites. □

3. L'exemple des applications polynomiales. Soit P_d: $\mathbf{C}^n \times \mathbf{C}^N \to \mathbf{C}^m \times \mathbf{C}^N$ la famille des applications polynomiales de degré $\leqslant d$ de $\mathbf{C}$ dans $\mathbf{C}^m$, où $\mathbf{C}^N$ est l'espace des coefficients.

COROLLAIRE 1. *Le nombre de types topologiques éclatés apparaissant dans la famille P_d est fini.*

DÉMONSTRATION. On considère dans $\mathbf{P}^n \times \mathbf{P}^1 \times \cdots \times \mathbf{P}^1 \times \mathbf{P}^N$ l'adhérence G du graphe de P_d, considérée comme application de $\mathbf{C}^n \times \mathbf{C}^N$ dans $\mathbf{C}^m$. On stratifie le morphisme p: $G \to \mathbf{P}^1 \times \cdots \times \mathbf{P}^1 \times \mathbf{P}^N$ de manière compatible à $\mathbf{C}^n \times \mathbf{C}^N$ et $\mathbf{C}^m \times \mathbf{C}^N$. On peut alors lui appliquer le Théorème 1. Les homéomorphismes obtenus étant stratifiés, on en déduit le Corollaire 1.

COROLLAIRE 2. *Le nombre de types topologiques apparaissant dans la famille P_d: $\mathbf{C}^2 \times \mathbf{C}^N \to \mathbf{C}^m \times \mathbf{C}^N$ est fini.*

La démonstration se fait comme dans le Théorème 2: On peut appliquer le lemme au morphisme p: $G \to \mathbf{P}^1 \times \cdots \times \mathbf{P}^1 \times \mathbf{P}^N$. Seule l'existence du morphisme ψ impose la lissité des fibres de p. Les stratifications et les champs de vecteurs étant construits pour p, on se restreint à $\mathbf{C}^2 \times \mathbf{P}^N$ pour l'existence de ψ. $\square$

BIBLIOGRAPHIE

[A] K. Aoki, *On topological types of polynomial map germs of plane to plane*, Preprint.

[A-H] A. Andreotti and P. Holm, *Quasianalytic and parametric spaces*, Real and Complex Singularities, (Oslo, 1976), Noordhoff, Groningen, 1977.

[F] T. Fukuda, *Types topologiques des polynômes*, Inst. Hautes Etudes Sci. Publ. Math. **46** (1976), 87–106.

[$\mathbf{H}_1$] H. Hironaka, *Flattening theorem in complex analytic geometry*, Amer. J. Math. **97** (1975).

[$\mathbf{H}_2$] ______, *Stratifications and flatness*, Real and Complex Singularities (Oslo, 1976), (P. Holm, Editor), Noordhoff, Groningen, 1977.

[$\mathbf{H}_3$] ______, *La voûte étoilée*, Singularités à Cargèse, Astérisque no. **7/8** (1973).

[M] J. Mather, *Notes on topological stability*, Preprint.

[N] I. Nakai, *On topological types of polynomial map germs*, Preprint.

[S] C. Sabbah, *Morphismes stratifiés sans éclatement et cycles évanescents*, Preprint.

[T] B. Teissier, *Variétés Polaires locales et conditions de Whitney*, C. R. Acad. Sci. Paris Sér. A-B **290** (1980), 799–802.

[Th] R. Thom, *La stabilité topologique des applications polynomiales*, Enseign. Math. **8** (1962), 24–33.

CENTRE DE MATHEMATIQUES DE L'ECOLE POLYTECHNIQUE, PALAISEAU, FRANCE

Proceedings of Symposia in Pure Mathematics
Volume **40** (1983), Part 2

THE HIGHER RESIDUE PAIRINGS $K_F^{(k)}$ FOR A FAMILY OF HYPERSURFACE SINGULAR POINTS[1]

KYOJI SAITO

0. Introduction. In the study of the *period mapping* associated to a universal unfolding F of a function f with an isolated critical point, we introduced a concept of *higher residue pairings* $K_F^{(k)}$, $k = 0, 1, \ldots$, which are defined on the relative de Rham cohomology module $\mathcal{H}_F^{(0)}$ of the family F and take the values in the ring $\mathcal{O}_T$ of holomorphic functions on the parameter space T of the unfolding F (see [**3**, §§9, 10]).

The pairings $K^{(k)}$ enable us to compute the Poincaré-duality of the general fiber of the family F, and hence the pairings themselves seem to be of some interest. Since [**3**] is not yet published, in this note we want to expose a construction of the pairings $K^{(k)}$ and to give some properties which characterize them. Another exposition on the pairing will be given by Y. Namikawa in [**2**]. For a general survey of [**3**] consult [**4**].

In §1, we first describe an unfolding F of a function f over a parameter space T using the notion of a Hamiltonian system [**3**, §§1–8]. The unfolding F induces a family of n-dimensional (open) varieties $\varphi\colon X \to S$ and a projection $\pi\colon S \to T$ of codimension 1 whose fibers are integral curves of a vector field δ_1 on S. We study the $\mathcal{O}_S$-free module $\mathcal{H}_F^{(0)}$ of the vector bundle of the middle cohomology group of the fibers of the family $X \to S$, which was first studied by E. Brieskorn [**1**]. Then using certain local cohomology groups with support in the critical set of φ, one constructs the $\mathcal{O}_T$-dual module $\check{\mathcal{H}}_F^{(0)}$ of $\pi_*\mathcal{H}_F^{(0)}$.

The vector field δ_1 on S acts on $\check{\mathcal{H}}_F^{(0)}$ and $\pi_*\mathcal{H}_F^{(0)}$ through the Gauss-Manin connection so that it induces a filtration, $\pi_*\mathcal{H}_F^{(0)} \supset \pi_*\mathcal{H}_F^{(-1)} \supset \cdots$ and a co-filtration $\check{\mathcal{H}}_F^{(0)} \to \check{\mathcal{H}}_F^{(1)} \to \cdots$ such that $\pi_*\mathcal{H}^{(-k)}$ and $\check{\mathcal{H}}_F^{(k)}$ are each others'

1980 *Mathematics Subject Classification.* Primary 30E99.

[1] The contents of this paper is based on a part of a course at R.I.M.S. Kyoto University which was taught by the author in the academic year 1980/81.

0082-0717/81/0000-0657/$07.00

$\mathcal{O}_T$-duals. Put

$$\hat{\pi_*}\mathcal{H}^{(0)} = \varprojlim \pi_*\mathcal{H}_F^{(0)}/\pi_*\mathcal{H}_F^{(-k)}, \qquad \check{\mathcal{H}}_{F,\text{finite}}^{(0)} = \varinjlim \ker\left(\check{\mathcal{H}}_F^{(0)} \to \check{\mathcal{H}}_F^{(k)}\right).$$

In §2, we compute the module $\mathbf{R}^{n+1}q_*(\Omega^\cdot, \hat{d})$ where $q\colon X \to T$ is the composition $\pi \circ \varphi$, $\Omega^\cdot = \Omega^\cdot_{X/T}\delta_1^{-1}\delta_1$ and $\hat{d} = \delta_1^{-1}d - dF_1 \wedge$.

In §3, we construct exact sequences

$$0 \to \hat{\pi_*}\mathcal{H}_F^{(-k-1)} \overset{\alpha}{\to} \mathbf{R}^{n+1}q_*(\Omega^\cdot, \hat{d}) \overset{\nabla^{(k)}}{\to} \check{\mathcal{H}}_{F,\text{finite}}^{(0)} \to 0.$$

In particular we get an $\mathcal{O}_T$-isomorphism

$$(*) \qquad \pi_*\mathcal{H}_F^{(0)}/\pi_*\mathcal{H}_F^{(-k-1)} \cong \ker\left(\check{\mathcal{H}}_F^{(0)} \to \check{\mathcal{H}}_F^{(k+1)}\right).$$

On the other hand, due to the duality studied in §1, both sides of $(*)$ are the $\mathcal{O}_T$-duals of each other. Hence $\pi_*\mathcal{H}_F^{(0)}/\pi_*\mathcal{H}_F^{(-k-1)}$ becomes a self-dual $\mathcal{O}_T$-module. This bilinear form on the quotient module induces the pairing

$$K^{(k)}\colon \pi_*\mathcal{H}_F^{(0)} \times \pi_*\mathcal{H}_F^{(0)} \to \mathcal{O}_T.$$

In §4, we study several properties of $K^{(k)}$'s, which are summarized in Theorem (4.10).

Generally on this note, all the computations are done in the category of formal power series $\mathcal{O}_T[[\delta_1^{-1}]]$ instead of the convergent power series. This does not hurt the essence of our constructions of $K^{(k)}$. The parallel calculations can be carried out without much difficulty for the convergent series.

The author would like to express his thanks to Y. Namikawa for several valuable discussions.

1. Relative de Rham cohomology $\mathcal{H}_F^{(0)}$ and its dual $\check{\mathcal{H}}_F^{(0)}$. In this section we give a quick review of §§1–8 of [**3**]. First we define an unfolding of a function f using a Hamiltonian system over a parameter space T ((1.2), for more details see [**3**, §§1–6]). Then the aim is to introduce two modules, the relative de Rham cohomology group $\mathcal{H}_F^{(0)}$ and its $\mathcal{O}_T$-dual module $\check{\mathcal{H}}_F^{(0)}$, with certain differential module structure.

(1.1) Let us consider the following Cartesian diagram of smooth complex manifolds Z, X, S, T, with base points.

$$(1.1.1) \qquad \begin{array}{ccc} (Z,0) & \overset{\hat{\pi}}{\to} & (X,0) \\ \downarrow p & & \downarrow q \\ (S,0) & \overset{\pi}{\to} & (T,0) \end{array}$$

All the base points are denoted by the same 0, so far as no confusion occurs. We assume that q, p, π, $\hat{\pi}$ are smooth submersions which are Stein whose fibers are contractible. Furthermore, assume that there is a holomorphic vector field δ_1 on S such that

$$\pi^{-1}\mathcal{O}_T = \{g \in \mathcal{O}_S\colon \delta_1 g = 0\}.$$

Then there is a unique vector field on Z, which we shall denote again by δ_1, which is a lifting of δ_1 such that

$$\hat{\pi}^{-1}\mathcal{O}_X = \{g \in \mathcal{O}_Z : \delta_1 g = 0\}.$$

(Here we denote by $\mathcal{O}_Z$, $\mathcal{O}_X$, $\mathcal{O}_S$, $\mathcal{O}_T$ the structure sheaves of the manifolds respectively.)

Let us define

(1.1.2) $$\mathcal{G} := \{\delta \in \pi_* \operatorname{Der}_S : [\delta_1, \delta] = 0\}$$

where Der_S is the module of holomorphic vector fields on S and [,] means the bracket product of vector fields.

Let $\dim_{\mathbf{C}} S = m$ and $\dim_{\mathbf{C}} X = n + m$. Then $\dim Z = n + m + 1$, $\dim T = m - 1$. We introduce local coordinate systems $t' = (t_2, \ldots, t_m)$ for T, $t = (t_1, t_2, \ldots, t_m)$ for S, $(x, t') = (x_0, x_1, \ldots, x_n, t_2, \ldots, t_m)$ for X and $(x, t) = (x_0, \ldots, x_n, t_1, \ldots, t_m)$ for Z, where we assume $\delta_1 t_1 = 1$ so that in these local coordinates $\delta_1 = \partial/\partial t_1$.

(1.2) DEFINITION. A Hamiltonian system on the frame (1.1) is one of the following equivalent datas:

(i) A holomorphic function $F(x, t)$ on Z s.t. $\delta_1 F \equiv 1$, $F(0,0) = 0$.

(ii) A holomorphic mapping φ: $(X, 0) \to (S, 0)$, which commutes with the diagram (1.1.1).

(iii) A holomorphic section σ: $(X, 0) \to (Z, 0)$, i.e. $\hat{\pi} \circ \sigma = \mathrm{id}_X$, $\sigma(0) = 0$.

We shall call the Hamiltonian system F *an unfolding* of the function $f(x) := -F(x, 0) =$ first component of $\varphi(x, 0)$. As long as $f \not\equiv 0$, the mapping φ is flat.

Using local coordinates on (1.1), we decompose

(1.2.1) $$F(x, t) = t_1 - F_1(x, t').$$

Then the map φ is given by $t_1 = F_1(x, t')$.

(1.3) In this note we shall always assume that f has an isolated critical point at 0. Then $\partial F_1/\partial x_0, \ldots, \partial F_1/\partial x_n$ form a parameter system in $\mathcal{O}_{X,0}$. Hence the critical set $C := \{(x, t') \in X : \partial F_1/\partial x_0 = \cdots = \partial F_1/\partial x_n = 0\}$ is a complete intersection of dimension $m - 1$ and $q \mid C$: $C \to T$ is a flat ramified covering map so that $q_* \mathcal{O}_C$ is an $\mathcal{O}_T$-free module, whose rank will be denoted by μ and will be called the Milnor's number.

(1.4) Let us denote by $\Omega^i_{X/T} := \Omega^i_X / q^{-1}\Omega^1_T \wedge \Omega^{i-1}_X$ the sheaf of modules of relative holomorphic i-forms on X, $i = 0, \ldots, n + 1$.

Then, since q is a submersion, we have an exact sequence

(1.4.1) $$0 \to q^{-1}\mathcal{O}_T \to \mathcal{O}_X \xrightarrow{d_{X/T}} \Omega^1_{X/T} \to \cdots \xrightarrow{d_{X/T}} \Omega^{n+1}_{X/T} \to 0$$

where $d_{X/T}$ is the relative exterior differentiation.

Using a generalized de Rham lemma [5, theorem] and the fact that $\partial F_1/\partial x_0, \ldots, \partial F_1/\partial x_n$ is a parameter system, one obtains an exact sequence

(1.4.2) $$0 \to \mathcal{O}_X \xrightarrow{dF_1} \Omega^1_{X/T} \to \cdots \xrightarrow{dF_1} \Omega^{n+1}_{X/T} \to \Omega_F \to 0$$

where

$$\Omega_F := \Omega^{n+1}_{X/T}/dF_1 \wedge \Omega^n_{X/T} = \Omega^{n+1}_{X/S}$$

is a $\mathcal{O}_C$-free module of rank 1, so that supp $\Omega_F = C$ and $q_*\Omega_F$ is $\mathcal{O}_T$-free of rank μ.

(1.5) By choosing suitable neighbourhoods of 0's, we may assume from the beginning that, $X - \varphi^{-1}\varphi(C) \to S - \varphi(C)$ is a smooth fiber bundle, whose general fiber has a homotopy type of bouquet of μ n-spheres.

DEFINITION (CF. BRIESKORN'S $\mathcal{H}''$ AND $\mathcal{H}'$ IN [**1**]; SEE ALSO [**3** (3.8) Lemma]).

(1.5.1) $$\pi_*\mathcal{H}_F^{(0)} := q_*\Omega^{n+1}_{X/T}/dF_1 \wedge dq_*\Omega^{n-1}_{X/T},$$

(1.5.2) $$\pi_*\mathcal{H}_F^{(-1)} := q_*\Omega^n_{X/T}/\left(dF_1 \wedge q_*\Omega^{n-1}_{X/T} + dq_*\Omega^{n-1}_{X/T}\right).$$

We shall denote by $[\omega]$, $[\zeta]$ etc. the elements of $\pi_*\mathcal{H}^{(0)}$, $\pi_*\mathcal{H}^{(-1)}$, which are represented by the differential forms $\omega \in q_*\Omega^{n+1}_X$, $\zeta \in q_*\Omega^n_X$. From (1.4.2), one obtains an exact sequence

(1.5.3) $$0 \to \pi_*\mathcal{H}_F^{(-1)} \xrightarrow{dF_1} \pi_*\mathcal{H}_F^{(0)} \to q_*\Omega_F \to 0.$$

We regard $\pi_*\mathcal{H}_F^{(-1)}$ as a submodule of $\pi_*\mathcal{H}_F^{(0)}$ by this.

Actually $\pi_*\mathcal{H}_F^{(0)}$ and $\pi_*\mathcal{H}_F^{(-1)}$ are $\pi_*\mathcal{O}_S$-free modules of rank μ due to a result of Sebastiani [7].

(1.6) There exists an operation, the Gauss-Manin connection,

(1.6.1) $$\nabla : \mathcal{G} \times \pi_*\mathcal{H}_F^{(-1)} \to \pi_*\mathcal{H}_F^{(0)}$$

by

$$\nabla_{\partial/\partial t_i}[\zeta] := \left[(-1)^{i-1}(dt_2 \wedge \cdots \wedge dt_m)^{-1} dt_1 \wedge \cdots \wedge \widehat{dt_i} \wedge \cdots \wedge dt_m \wedge d\zeta\right].$$

In particular, the differentiation δ_1 induces an $\mathcal{O}_T$-isomorphism

(1.6.2) $$\nabla_{\delta_1} : \pi_*\mathcal{H}_F^{(-1)} \cong \pi_*\mathcal{H}_F^{(0)}, \quad [\zeta] \mapsto [d\zeta].$$

Hence one obtains a decreasing filtration of $\pi_*\mathcal{H}_F^{(0)}$ defined by

(1.6.3) $$\pi_*\mathcal{H}^{(-k-1)} := \left\{\omega \in \pi_*\mathcal{H}_F^{(-k)} : \nabla_{\delta_1}\omega \in \pi_*\mathcal{H}_F^{(-k)}\right\} \quad \text{for } k \geqslant 1.$$

By definition,

(1.6.4) $$0 \to \pi_*\mathcal{H}_F^{(-k-1)} \hookrightarrow \pi_*\mathcal{H}_F^{(-k)} \to q_*\Omega_F \to 0,$$

(1.6.5) $$\nabla : \mathcal{G} \times \pi_*\mathcal{H}_F^{(-k-1)} \to \pi_*\mathcal{H}_F^{(-k)}.$$

In §6 of [**3**] we give a geometric way of constructing the above filtration.

(1.7) The connection is regular singular. Hence the topology of $\pi_*\mathcal{H}^{(0)}$ induced by the filtration coincides with a certain adic topology.

Therefore, we conclude

(1.7.1) $$\bigcap_{k=0}^{\infty} \pi_*\mathcal{H}_F^{(-k)} = 0.$$

Let us denote by $\pi_*^\wedge\mathcal{H}^{(0)}$ the completion of $\pi_*\mathcal{H}^{(0)}$ with respect to the above topology. We have a natural inclusion homomorphism

(1.7.2) $$\pi_*\mathcal{H}_F^{(0)} \hookrightarrow \pi_*^\wedge\mathcal{H}_F^{(0)} = \varprojlim \pi_*\mathcal{H}_F^{(0)}/\pi_*\mathcal{H}_F^{(-k)}.$$

(1.8) *Notation.* Let us denote by $R^i{}_C q_*(\mathcal{F})$ ($\mathbf{R}^i{}_C q_*(\mathcal{F}^\cdot)$ resp.) for an abelian sheaf $\mathcal{F}$ on X (a complex $\mathcal{F}^\cdot$ on X respectively), the ith derived functor (the ith hypercohomology group) of the functor $q_*\Gamma_C$, which associates to $\mathcal{F}$ a sheaf on T given by $\mathcal{U} \subset T \mapsto \Gamma_{C\cap q^{-1}(\mathcal{U})}(q^{-1}(\mathcal{U}), \mathcal{F})$ (cf. [3, §7]).

(1.9) Using an extension theorem by Scheja [6] and Trautman [8], one sees that $R^i{}_C q_*\mathcal{F} = 0$ for $i \neq n+1$, and an $\mathcal{O}_X$-free module $\mathcal{F}$. Then (1.4.1) and (1.4.2) induce the following exact sequences.

(1.9.1) $$0 \to R_C^{n+1}q_*\mathcal{O}_X \overset{d_{X/T}}{\to} R_C^{n+1}q_*\Omega^1_{X/T} \to \cdots$$
$$\overset{d_{X/T}}{\to} R_C^{n+1}q_*\Omega^{n+1}_{X/T} \overset{\text{res.}}{\to} R_C^{2n+2}q_*q^{-1}\mathcal{O}_T \to 0.$$

(1.9.2) $$0 \to q_*\Omega_F \to R_C^{n+1}q_*\mathcal{O}_X \overset{dF_1}{\to} R_C^{n+1}q_*\Omega^1_{X/T} \overset{dF_1}{\to} \cdots$$
$$\to R_C^{n+1}q_*\Omega^{n+1}_{X/T} \to 0.$$

(1.10) DEFINITION.

(1.10.1) $$\check{\mathcal{H}}_F^{(0)} := \left\{\alpha \in R_C^{n+1}q_*\mathcal{O}_X : dF_1 \wedge d\alpha = 0 \text{ in } R_C^{n+1}q_*\Omega^2_{X/T}\right\},$$

(1.10.2) $$\check{\mathcal{H}}_F^{(1)} := \left\{\beta \in R_C^{n+1}q_*\Omega^1_{X/T} : dF_1 \wedge \beta = d\beta = 0 \text{ in } R_C^{n+1}q_*\Omega^2_{X/T}\right\}.$$

From (1.9.2), one obtains an exact sequence,

(1.10.3) $$0 \to q_*\Omega_F \to \check{\mathcal{H}}_F^{(0)} \overset{d\check{F}_1}{\to} \check{\mathcal{H}}_F^{(1)} \to 0.$$

By (1.10.3) we regard $\check{\mathcal{H}}_F^{(1)}$ as a quotient of $\check{\mathcal{H}}_F^{(0)}$.

(1.11) We also have the dual Gauss-Manin connection,

(1.11.1) $$\check{\nabla} : \mathcal{G} \times \check{\mathcal{H}}_F^{(0)} \to \check{\mathcal{H}}_F^{(1)}$$

by $\check{\nabla}_{\partial/\partial t_i}\alpha = (-1)^{i-1}(dt_2 \wedge \cdots \wedge dt_m)^{-1}\, dt_1 \wedge \cdots \wedge \widehat{dt_i} \wedge \cdots \wedge dt_m \wedge d\alpha$ in $R_C^{n+1}q_*\Omega^m_X$.

In particular, the differentiation δ_1 induces an $\mathcal{O}_T$-isomorphism

(1.11.2) $$\check{\nabla}_{\delta_1} : \check{\mathcal{H}}_F^{(0)} \cong \check{\mathcal{H}}_F^{(1)}, \qquad \alpha \mapsto d\alpha.$$

Hence we get an increasing sequence of $\mathcal{O}_T$-free submodules $\check{\Omega}_F^{(k)}$ of rank $(k+1)\mu$, $k = 0, 1, 2, \ldots$, by

(1.11.3) $$\check{\Omega}_F^{(k+1)} := (\vee_1)^{-1}\left(\check{\nabla}_{\delta_1}\check{\Omega}_F^{(k)}\right), \qquad k = 0, 1, 2, \ldots,$$
$$\check{\Omega}_F^{(0)} := q_*\Omega_F.$$

By putting

$$\check{\mathcal{H}}_F^{(k+1)} := \check{\mathcal{H}}_F^{(0)}/\check{\Omega}_F^{(k)} \tag{1.11.4}$$

we get

$$0 \to q_*\Omega_F \to \check{\mathcal{H}}_F^{(k)} \to \check{\mathcal{H}}_F^{(k+1)} \to 0, \tag{1.11.5}$$

$$\check{\nabla}: \mathcal{G} \times \check{\mathcal{H}}_F^{(k)} \to \check{\mathcal{H}}_F^{(k+1)}. \tag{1.11.6}$$

(1.12) DEFINITION. We put

$$\check{\mathcal{H}}^{(0)}_{F,\text{finite}} = \bigcup_{k=0}^{\infty} \check{\Omega}_F^{(k)}. \tag{1.12.1}$$

(1.13) Combining with the trace map (which is the Lefschetz dual),

$$\text{tr}: R_C^{2n+2} q_* q^{-1}\mathcal{O}_T \to \mathcal{O}_T \tag{1.13.1}$$

and the res. of (1.9.1), one defines the residue morphism,

$$\operatorname{Res}_{X/T}: R_C^{n+1} q_*\Omega_{X/T}^{n+1} \to \mathcal{O}_T. \tag{1.13.2}$$

Combining with the residue morphism, one obtains the usual local dualities with support in C,

$$q_*\Omega^i_{X/T} \times R_C^{n+1} q_*\Omega_{X/T}^{n-i+1} \to R_C^{n+1} q_*\Omega_{X/T}^{n+1} \overset{\operatorname{Res}_{X/T}}{\longrightarrow} \mathcal{O}_T. \tag{1.13.3}$$

(1.14) The above local duality induces $\mathcal{O}_T$-bilinear maps

$$\pi_*\mathcal{H}_F^{(0)} \times \check{\mathcal{H}}_F^{(0)} \to \mathcal{O}_T, \tag{1.14.1}$$

$$\pi_*\mathcal{H}_F^{(-1)} \times \check{\mathcal{H}}_F^{(1)} \to \mathcal{O}_T, \tag{1.14.2}$$

which are compatible with the operations of ∇ and $\check{\nabla}$,

$$\delta \operatorname{Res}_{X/T}[dF_1\zeta \cdot \alpha] = \operatorname{Res}_{X/T}[\nabla_\delta\zeta \cdot \alpha] + \operatorname{Res}_{X/T}[\zeta \cdot \check{\nabla}_\delta \alpha]$$

for $\zeta \in \pi_*\mathcal{H}_F^{(-1)}$, $\alpha \in \check{\mathcal{H}}_F^{(0)}$, $\delta \in \mathcal{G}$ (cf. (8.11), (8.12), (8.13) of [**3**]).

Then one can check that the above duality induces an infinite sequence of $\mathcal{O}_T$-dualities,

$$\pi_*\mathcal{H}_F^{(-k)} \times \check{\mathcal{H}}_F^{(k)} \to \mathcal{O}_T, \tag{1.14.3}$$

$$\pi_*\mathcal{H}_F^{(0)}/\pi_*\check{\mathcal{H}}_F^{(-k-1)} \times \Omega_F^{(k)} \to \mathcal{O}_T. \tag{1.14.4}$$

In particular (1.14.4) is a bilinear form on $\mathcal{O}_T$-finite ($= (k+1)\mu$) rank modules which is nondegenerate (cf. [**3**, (8.14)]).

When $k = 0$, using (1.5.3) and (1.10.3), one obtains an exact sequence,

$$0 \to \pi_*\mathcal{H}^{(-1)} \to \pi_*\mathcal{H}^{(0)} \overset{\chi}{\to} \check{\mathcal{H}}_F^{(0)} \to \check{\mathcal{H}}_F^{(1)} \to 0, \tag{1.14.5}$$

where χ is given by the correspondence $\varphi\, dx \mapsto \varphi/(\partial F/\partial x_0 \cdots \partial F/\partial x_n)$.

Then $q_*\Omega_F$ becomes a self-dual module, on which the bilinear form is given by

$$(1.14.6)\quad J\colon q_*\Omega_F \times q_*\Omega_F \to \mathcal{O}_T,\ \varphi\, dx \times \psi\, dx \mapsto \operatorname{Res}_{X/T}\begin{pmatrix} \varphi\psi\, dx \\ \dfrac{\partial F}{\partial x_0}, \ldots, \dfrac{\partial F}{\partial x_n} \end{pmatrix}.$$

Note. In §3, we shall construct an $\mathcal{O}_T$-bijection,

$$\nabla^{(k)}\colon \pi_*\mathcal{H}_F^{(0)}/\pi_*\mathcal{H}_F^{(-k-1)} \cong \check{\Omega}^{(k)}.$$

This together with (1.14.4) means that $\check{\Omega}^{(k)}$ is a self-dual $\mathcal{O}_T$-module.

2. $\mathbf{R}^{n+1}q_*(\Omega^{\cdot}_{X/T}[[\delta_1^{-1}]][\delta_1])$. In this section we compute $\mathbf{R}^{n+1}q^*(\Omega^{\cdot}_{X/T}[[\delta_1^{-1}]][\delta_1])$ with its natural filtration and $\mathcal{G}$-module structure.

(2.1) Let us consider the set of formal Laurent series in δ_1^{-1} with coefficients in $\Omega^{\cdot}_{X/T}$.

$$(2.1.1)\qquad \Omega^{\cdot}_{X/T}[[\delta_1^{-1}]][\delta_1] := \left\{ \sum_{k\leq k_0} \omega_k\delta_1^k \colon k_0 \in \mathbf{Z},\ \omega_k \in \Omega^{\cdot}_{X/T} \right\}.$$

Let us denote this module by $\Omega^{\cdot}$, i.e.

$$(2.1.2)\qquad \Omega^{\cdot} := \Omega^{\cdot}_{X/T}[[\delta_1^{-1}]][\delta_1]$$

which we regard as an $\mathcal{O}_X[[\delta_1^{-1}]][\delta_1]$-module.

$\Omega^{\cdot}$ has an increasing filtration

$$(2.1.3)\qquad F^k\Omega^{\cdot} := \left\{ \omega \in \Omega^{\cdot} : \omega = \sum_{m\leq k} \omega_m\delta_1^k,\ \omega_m \in \Omega^{\cdot}_{X/T} \right\}.$$

(2.2) The wedge product $\wedge$ and the exterior differentiation $d_{X/T}$ of the Poincaré complex $\Omega^{\cdot}_{X/T}$ naturally extend to $\Omega^{\cdot}$ by formally requiring that these operations commute with the multiplication δ_1^{-1}. We shall use the same notations $\wedge$ and $d_{X/T}$ for these extensions.

(2.3) Let us define an $\mathcal{O}_T[[\delta_1^{-1}]][\delta_1]$-homomorphism,

$$(2.3.1)\qquad \hat{d}\colon \Omega^{\cdot} \to \Omega^{\cdot+1} \quad \text{by } \hat{d} = \delta_1^{-1}d_{X/T} - dF_1 \wedge .$$

It is obvious by definition that one has:

(i) $\hat{d} \circ \hat{d} = 0$,

(ii) $\hat{d}F^k\Omega^{\cdot} \subset F^k\Omega^{\cdot+1}$.

Thus $(\Omega^{\cdot}, \hat{d})$, $(F^k\Omega^{\cdot}\hat{d})$ for $k \in \mathbf{Z}$ are complexes.

Since $\delta_1\hat{d} = \hat{d}\delta_1$, the multiplication by δ_1 induces an isomorphism

$$(2.3.2)\qquad \delta_1\colon (F^k\Omega^{\cdot}, \hat{d}) \cong (F^{k+1}\Omega^{\cdot}, \hat{d}).$$

(2.4) PROPOSITION. (i) *The complexes* $(\Omega^{\cdot}, \hat{d})$ *and* $(F^k\Omega^{\cdot}, \hat{d})$ *for* $k \in \mathbf{Z}$ *are pure by* $n+1$ *dimensional, i.e.*

$$H^p(F^k\Omega^{\cdot}, \hat{d}) = 0, \qquad p \neq n+1,\ k \in \mathbf{Z},$$

$$H^p(\Omega^{\cdot}, \hat{d}) = 0, \qquad p \neq n+1.$$

(ii) *The support* $(H^{n+1}(\Omega^{\cdot}, \hat{d}))$ *and the support* $(H^{n+1}(F^k\Omega^{\cdot}, \hat{d}))$ *for* $k \in \mathbf{Z}$ *are contained in the critical set* C.

(iii) *The inclusion* $F^k\Omega^{\cdot} \to \Omega^{\cdot}$ *induces an injective homomorphism*,

$$H^{n+1}(F^k\Omega^{\cdot}, \hat{d}) \to H^{n+1}(\Omega^{\cdot}\hat{d}). \tag{2.4.1}$$

So we can regard $H^{n+1}(F^k\Omega^{\cdot}, \hat{d})$ *for* $k \in \mathbf{Z}$ *as an increasing sequence of submodules of* $H^{n+1}(\Omega^{\cdot}, \hat{d})$ *such that*:

(1) $H^{n+1}(\Omega^{\cdot}, \hat{d}) = \bigcup_k H^{n+1}(F^k\Omega^{\cdot}, \hat{d})$.

(2) $\bigcap_k H^{n+1}(F^k\Omega^{\cdot}, \hat{d}) = 0$.

(3) *The natural inclusions induce the following exact sequences for* $k \in Z$,

$$0 \to H^{n+1}(F^{k-1}\Omega^{\cdot}, \hat{d}) \to H^{n+1}(F^k\Omega^{\cdot}, \hat{d}) \to \Omega_F \to 0.$$

(4) $H^{n+1}(\Omega^{\cdot}, \hat{d})$ *is complete with respect to the topology given by putting* $H^{n+1}(F^k\Omega^{\cdot}, \hat{d})$, $k \in \mathbf{Z}$, *to be a fundamental system of neighbourhoods of* 0 (*i.e. if there exists an infinite sequence* ω_k *of* $H^{n+1}(\Omega^{\cdot}, \hat{d})$ *such that* $\omega_k \equiv \omega_{k-1} \bmod H^{n+1}(F^k\Omega^{\cdot}, \hat{d})$ *for* $k \to -\infty$, *then there exists a unique element* $\omega \in H^{n+1}(\Omega^{\cdot}, \hat{d})$ *such that* $\omega \equiv \omega_k \bmod H^{n+1}(F^k\Omega^{\cdot}, \hat{d})$ *as* $k \to -\infty$).

PROOF. The statements about $H^p(\Omega^{\cdot}, \hat{d})$ in (i) and (ii) follow from the corresponding statements about $H^p(F^k\Omega^{\cdot}, \hat{d})$, since $H^p(\Omega^{\cdot}, \hat{d}) = \bigcup_{k\in\mathbf{Z}} \operatorname{Im}(H^p(F^k\Omega^{\cdot}, \hat{d}) \to H^p(\Omega^{\cdot}, \hat{d}))$.

In the proof of (i), (ii), (iii), it suffices to carry out the proof for the case $k = 0$, due to the isomorphism (2.3.2).

PROOF OF (i). Take an element $\omega = \Sigma_{k\leqslant 0}\, \omega_k\delta_1^k \in F^0\Omega^p$.

By definition

$$\begin{aligned} \hat{d}\omega = &-dF_1 \wedge \omega_0 + (d\omega_{-1} - dF_1 \wedge \omega_0)\delta_1^{-1} \\ &+ (d\omega_{-2} - dF_1 \wedge \omega_{-1})\delta_1^{-2} + \cdots. \end{aligned} \tag{2.4.2}$$

Thus if $d\omega = 0$, then each term of (2.4.2) vanishes, and then by a repeated use of the de Rham lemma (1.4.2) (notice that $p \leqslant n$), one finds an infinite sequence $\eta_i \in \Omega^p_{X/T}$, $i = 0, -1, -2, \ldots$, such that

$$\begin{aligned} \omega_0 &= -dF_1\eta_0, \\ \omega_{-1} &= d\eta_0 - dF_1\eta_1, \\ &\vdots \\ \omega_{-2} &= \cdots. \end{aligned}$$

Hence, putting $\eta := \Sigma_{i\leqslant 0}\, \eta_i\delta_1^{-i}$, one gets $\omega = d\eta$.

PROOF OF (iii). For an $\omega \in F^0\Omega^{n+1}$, suppose that there exists some $\eta = \Sigma_{m\leqslant k}\, \eta_m\delta_1^m \in F^k\Omega^n$ for some $k > 0$, such that $\omega = \hat{d}\eta$. Then as in (i),

$$\begin{aligned} -dF_1 \wedge \eta_k &= 0, \\ d\eta_k - dF_1 \wedge \eta_{k-1} &= 0, \\ &\vdots \\ d\eta_2 - dF_1 \wedge \eta_1 &= 0. \end{aligned}$$

Then one finds $\xi_k, \ldots, \xi_1 \in \Omega_{X/T}^{n-1}$ such that

$$\begin{aligned} \eta_k &= -dF_1 \wedge \xi_k, \\ \eta_{k-1} &= d\xi_k - dF_1 \wedge \xi_{k-1}, \\ &\vdots \\ \eta_1 &= d\xi_2 - dF_1 \wedge \xi_1. \end{aligned}$$

Hence $\eta - \hat{d}(\Sigma_{m=1}^k \xi_m \delta_1^m) \in F^0\Omega^n$ and $\omega = \hat{d}\eta = \hat{d}(\eta - \hat{d}\Sigma_{m=1}^k \xi_m \delta_1^m)$. Thus the injectivity of the homomorphism in (2.4.1) is proven.

From the injectivity, (1) and (2) follow directly.

We have a short exact sequence of complexes,

$$0 \to (F^{k-1}\Omega^{\cdot}, \hat{d}) \to (F^k\Omega^{\cdot}, \hat{d}) \to (\Omega^{\cdot}_{X/T}, -dF_1) \to 0.$$

Applying (1.4.2), one gets the exact sequence of (3).

The statement (4) follows from the completeness of the ring of formal Laurent series.

PROOF OF (ii). As we see from the above, for an $\omega \in F^0\Omega^{\cdot}$ with $\hat{d}\omega = 0$, the obstruction to construct $\eta \in F^0\Omega^{\cdot -1}$ with $\omega = \hat{d}\eta$ is described by the de Rham lemma. As we saw in (1.4.3) the obstruction module Ω_F has the support in C.

(2.5) COROLLARY. (i)

$$\mathbf{R}^p q_*(\Omega^{\cdot}, \hat{d}) = \begin{cases} 0 & \forall p \neq n+1, \\ H^{n+1}\big(q_*\Omega^{\cdot}\big), & p = n+1. \end{cases}$$

(ii) *For all* $k \in \mathbf{Z}$

$$\mathbf{R}^p q_*(F^k\Omega^{\cdot}, \hat{d}) = \begin{cases} 0, & p \neq n+1, \\ H^{n+1}\big(q_*F^k\Omega^{\cdot}\big), & p = n+1. \end{cases}$$

(iii) *The natural inclusion* $F^k\Omega^{\cdot} \hookrightarrow \Omega^{\cdot}$ *induces an injection*

$$0 \to \mathbf{R}^{n+1}q_*(F^k\Omega^{\cdot}, \hat{d}) \to \mathbf{R}^{n+1}q_*(\Omega^{\cdot}, \hat{d}). \tag{2.5.1}$$

So we can regard $\mathbf{R}^{n+1}q_*(F^k\Omega^{\cdot}, \hat{d})$ *for* $k \in \mathbf{Z}$ *as an increasing sequence of submodules of* $\mathbf{R}^{n+1}q_*(\Omega^{\cdot}, \hat{d})$ *such that*:

(1) $\mathbf{R}^{n+1}q_*(\Omega^{\cdot}, \hat{d}) = \bigcup_k \mathbf{R}^{n+1}q_*(F^k\Omega^{\cdot}, \hat{d})$.

(2) $\bigcap_k \mathbf{R}^{n+1}q_*(F^k\Omega^{\cdot}, \hat{d}) = \{0\}$.

(3) *The natural inclusions induce the following exact sequences*,

$$0 \to \mathbf{R}^{n+1}q_*(F^{k-1}\Omega^{\cdot}, \hat{d}) \to \mathbf{R}^{n+1}q_*(F^k\Omega^{\cdot}, \hat{d}) \to q_*\Omega_F \to 0$$

for $k \in \mathbf{Z}$.

(4) $\mathbf{R}^{n+1}q_*(\Omega^{\cdot}, \hat{d})$ *is complete with respect to the topology defined by putting* $\mathbf{R}^{n+1}q_*(F^k\Omega^{\cdot}, d)$, $k \in \mathbf{Z}$, *to be a fundamental system of neighbourhoods of* 0.

PROOF. Since (i) and (ii) are proven analogously, we prove (i) only.

The spectral sequence $H^p(R^q q_*\Omega^\cdot)$ converges to $\mathbf{R}^{p+q}q_*(\Omega^\cdot)$. Since the map q: $X \to T$ is Stein, $R^q q_*\Omega^\cdot = 0$ for $q > 0$, and we get the isomorphism $\mathbf{R}^p q_*(\Omega^\cdot) = H^p(q_*\Omega^\cdot)$. On the other hand, the spectral sequence $R^q q_*(H^p(\Omega^\cdot))$ converges to $\mathbf{R}^{p+q}q_*(\Omega^\cdot)$. Since $H^p(\Omega^\cdot) = 0$ for $p \neq n+1$, the supp $H^{n+1}(\Omega^\cdot)$ is contained in C (cf. (2.4)) and C is finite over T with the map $q \mid C$, the only terms which remain in the spectral sequence are $q_*(H^{n+1}(\Omega^\cdot))$. Hence,

$$\text{(2.5.2)} \qquad \mathbf{R}^p q_*\Omega^\cdot = \begin{cases} 0 & \text{for } p \neq n+1, \\ q_*(H^{n+1}(\Omega^\cdot)) = H^{n+1}(q_*\Omega^\cdot), & p = n+1. \end{cases}$$

(iii) Using the representation (2.5.2) and (2.4)(iii), we get the injectivity of the homomorphism in (2.5.1).

The other properties (1)–(4) follow from the representation (2.5.2) and the corresponding properties in (2.4)(iii).

(2.6) *Note*. The invertible multiplication by δ_1 in $\Omega^\cdot$ induces an invertible multiplication by δ_1 in $\mathbf{R}^{n+1}q_*(\Omega^\cdot, \hat{d})$, which induces an isomorphism

$$\text{(2.6.1)} \qquad \delta_1 \colon \mathbf{R}^{n+1}q_*(F^k\Omega^\cdot, \hat{d}) \simeq \mathbf{R}^{n+1}(F^{k+1}\Omega^\cdot, \hat{d}).$$

Then we have shown in (2.5)(iii) that

(i) $\mathbf{R}^{n+1}q_*(F^0\Omega^\cdot, \hat{d})$ is complete with respect to the δ_1^{-1}-adic topology, so that it may be regarded as a free $\mathcal{O}_T[[\delta_1]]$-module such that

$$q_*\Omega_F = \mathbf{R}^{n+1}q_*(F^0\Omega^\cdot, \hat{d})/\delta_1^{-1}\mathbf{R}^{n+1}q_*(F^0\Omega^\cdot, \hat{d}).$$

(ii) One has a natural identification

$$\text{(2.6.2)} \qquad \mathbf{R}^{n+1}q_*(\Omega^\cdot, \hat{d}) \simeq \mathbf{R}^{n+1}q_*(F^0\Omega^\cdot, \hat{d}) \underset{\mathcal{O}_T[[\delta_1^{-1}]]}{\otimes} \mathcal{O}_T[[\delta_1^{-1}]][\delta_1].$$

(2.7) DEFINITION. Let us define a $\mathcal{G} = \sum_{i=1}^m \mathcal{O}_T\partial/\partial t_i$-module on $\mathbf{R}^{n+1}q_*(\Omega^\cdot, \hat{d})$ as follows.

For the moment we fix a coordinate system $t_1, t_2, \ldots, t_m, x_0, \ldots, x_n$, and we denote by $\hat{\partial}/\partial t_i$, $i = 2, \ldots, m$, the vector fields on X to distinguish them from $\partial/\partial t_i$ on T. The element dF_1 is also defined in Ω_X^1.

For an element $[\omega] \in H^{n+1}(q_*\Omega^\cdot)$, let us choose a representative ω in $q_*\Omega_X^{n+1}[[\delta_1^{-1}]][\delta_1]$. Then one defines

$$\text{(2.7.1)} \qquad \frac{\partial}{\partial t_i}[\omega] \underset{\text{def}}{:=} \delta_1\left[\left\langle \frac{\hat{\partial}}{\partial t_j}, \hat{d}_X\omega \right\rangle\right] \in H^{n+1}(q_*\Omega^\cdot)$$

for $j = 2, \ldots, m$. Here $\hat{d}_X := \delta_1^{-1}d_X - dF_1$, where d_X is the exterior differentiation on X and $\langle\ ,\ \rangle$ means the inner product of vector fields with forms on X.

We put, also,

$$\text{(2.7.2)} \qquad \frac{\partial}{\partial t_1}[\omega] := \delta_1[\omega].$$

(i) Let us show that the right-hand side of (2.7.1) does not depend on the choice of ω. ($\because$ If ω and ω' represent the same class in $H^{n+1}(q_*\Omega^\cdot)$, then there exist ξ_i, $i = 1,\dots,m$, of $q_*\Omega^n_X[[\delta_1^{-1}]][\delta_1]$, such that $\omega - \omega' = -\hat{d}_X\xi_1 + \Sigma_{i=2}^m dt_i \wedge \xi_i$. Then $\hat{d}_X(\omega - \omega') = -\Sigma\, dt_i \wedge \hat{d}_X\xi_i$ and therefore

$$\left\langle \frac{\hat{\partial}}{\partial t_j}, \hat{d}_X(\omega - \omega') \right\rangle \equiv -\hat{d}_X\xi_j \bmod \Omega'_T \equiv 0$$

in $H^{n+1}(q_*\Omega^\cdot)$.)

(ii) The right-hand side of (2.7.1) does not depend on the choice of coordinates (proof omitted).

(iii) (2.7.1) and (2.7.2) together define an action of $\mathcal{G}$ on $\mathbf{R}^{n+1}q_*(\Omega^{n+1})$, by an $\mathcal{O}_T$-linear extension.

(iv) The action of $\mathcal{G}$ is integrable. One needs to check that

$$\frac{\partial}{\partial t_i}\frac{\partial}{\partial t_j}[\omega] = \frac{\partial}{\partial t_j}\frac{\partial}{\partial t_i}[\omega] \quad \text{for } i, j = 1,\dots,m$$

(proof omitted). As a consequence, the universal enveloping algebra $U(\mathcal{G})$ of $\mathcal{G}$ (= a noncommutative $\mathcal{O}_T$-algebra generated by $\mathcal{G}$ with relations $\delta\delta' - \delta'\delta = [\delta, \delta']$ for $\delta, \delta' \in \mathcal{G}$) operates on $\mathbf{R}^{n+1}q_*(\Omega^\cdot)$.

(v) Let us give a modification of the definition (2.7.1).

Let us represent an element $[\omega]$ of $H^{n+1}(q_*\Omega^\cdot)$ by $\omega = d_X\zeta$ for some $\zeta \in q_*\Omega^n_X[[\delta_1^{-1}]][\delta_1]$. (This is possible since q is a smooth contractible map and hence $q_*\Omega^{n+1}_X = d_X q_*\Omega^n_X + \Sigma_{i=2}^m dt_i \wedge q_*\Omega^n_X$.)

Then (2.7.1), (2.7.2) are equivalent to

(2.7.3)

$$\frac{\partial}{\partial t_i}[\omega] \underset{\text{def}}{:=} \delta_1\left[(-1)^{i-1}(dt_2 \wedge \cdots \wedge dt_m)\, dt_1 \wedge \cdots \wedge \widehat{dt_i} \wedge \cdots \wedge dt_m \wedge \omega\right]$$

for $i = 1,\dots,m$.

$$\because \quad \frac{\partial}{\partial t_i}[\omega] = \delta_1\left[\left\langle \frac{\hat{\partial}}{\partial t_i}, \hat{d}_X\omega \right\rangle\right] = \delta_1\left[\left\langle \frac{\hat{\partial}}{\partial t_i}, (\delta_1^{-1}d_X - dF_1)d_X\zeta \right\rangle\right]$$

$$= \delta_1\left[\left\langle \frac{\hat{\partial}}{\partial t_i}, -dF_1 \wedge d_X\zeta \right\rangle\right]$$

$$= \delta_1\big[(-1)^{m-i}(dt_2 \wedge \cdots \wedge dt_m)^{-1} dt_2 \wedge \cdots \wedge \widehat{dt_i} \wedge \cdots \wedge dt_m(-dF_1 \wedge d_X\zeta)\big]$$

$$= \delta_1\left[(-1)^{i-1}(dt_2 \wedge \cdots \wedge dt_m)^{-1} dt_1 \wedge \cdots \wedge \widehat{dt_i} \wedge \cdots \wedge dt_m \wedge d_X\zeta\right].$$

(vi) The right-hand side of (2.7.3) does not depend on the choice of $\omega = d_X\zeta$.

$\because$ Let $d_X\zeta$ and $d_X\zeta'$ represent the same class in $H^{n+1}(\Omega^\cdot, \hat{d})$. This means that there exist $\eta_1,\dots,\eta_m$ in $\Omega^n_X[[\delta_1^{-1}]][\delta_1]$, such that,

(2.7.4)
$$d_X\zeta - d_X\zeta' = -\hat{d}_X\eta_1 + \sum_{i=2}^m dt_2 \wedge \eta_i,$$

i.e. $d_X(\zeta - \zeta' + \delta_1^{-1}\eta_1) = \Sigma_{i=2}^m dt_2 \wedge \eta_i$. By taking the wedge product of $dt_2 \wedge \cdots \wedge \widehat{dt_i} \wedge \cdots \wedge dt_m$ with the exterior derivative of the equation, one gets the relation,

(2.7.5)

$$d\eta_i = (-1)^{i-1}(dt_2 \wedge \cdots \wedge dt_m)^{-1}(dt_1 \wedge \cdots \wedge \widehat{dt_i} \wedge \cdots \wedge dt_m)\, d\eta_1 \mod \Omega_T^1.$$

Then using (2.7.4), (2.7.5) one computes

$$\begin{aligned}(-1)^{i-1}&(dt_2 \wedge \cdots \wedge dt_m)^{-1}(dt_1 \wedge \cdots \wedge \widehat{dt_i} \wedge \cdots \wedge dt_m) \wedge (d_X\zeta - d_X\zeta') \\ &= -\delta_1^{-1} d\eta_i + dt_1 \wedge \eta_i \mod \Omega_T' \\ &= -\hat{d}\eta_i \equiv 0 \quad \text{in } H^{n+1}(q_*\Omega^\cdot) \quad \text{for } i = 2, \ldots, m.\end{aligned}$$

(2.8) We shall show in §3 (3.9), that $\mathbf{R}^{n+1}q_*(\Omega^\cdot, \hat{d})$ has a $\pi_*\mathcal{O}_S$-module structure.

3. The comparison of $\mathbf{R}^{n+1}q_*(\Omega^\cdot, \hat{d})$ with $\mathcal{H}_F^{(0)}$ and $\check{\mathcal{H}}_F^{(0)}$. Our aim in this section is to show that the sequence,

$$0 \to \pi_*^\wedge \mathcal{H}_F^{(-k-1)} \underset{\alpha}{\hookrightarrow} \mathbf{R}^{n+1}q_*(\Omega^\cdot, \hat{d}) \overset{\nabla^{(k)}}{\to} \check{\mathcal{H}}_{F,\text{finite}}^{(0)} \to 0$$

of (3.7) is exact, which together with the duality of (1.14.1), shows the self-duality of $\pi_*\mathcal{H}_F^{(0)}/\pi_*\mathcal{H}_F^{(-k-1)}$.

(3.1) Consider the natural inclusion,

$$\Omega_{X/T}^{n+1} \to \Omega_{X/T}^{n+1}[[\delta_1^{-1}]] = F^0\Omega^{n+1}, \qquad \omega \mapsto \omega,$$

which induces

$$q_*\Omega_{X/T}^{n+1} \to q_*F^0\Omega^{n+1} \to H^{n+1}(q_*F^0\Omega^\cdot, \hat{d}) \simeq \mathbf{R}^{n+1}q_*(F^0\Omega^\cdot, \hat{d}).$$

For an element ω of $dF_1 \wedge dq_*\Omega_{X/T}^{n-1}$, there exists some η such that $\omega = dF_1 \wedge d\eta = \hat{d}(-d\eta)$. Hence the image of ω in $\mathbf{R}^{n+1}q_*(F^0\Omega^\cdot, \hat{d})$ is zero. Thus by definition of $\pi_*\mathcal{H}_F^{(0)}$ of (1.5.1), we get a homomorphism

(3.1.1) $$\alpha : \pi_*\mathcal{H}_F^{(0)} \to \mathbf{R}^{n+1}q_*(F^0\Omega^\cdot, \hat{d}).$$

(3.2) PROPOSITION. (i) *The homomorphism α is equivariant with the operation δ_1^{-1} (i.e. $\alpha\nabla_{\delta_1}^{-1} = \delta_1^{-1}\alpha$), and it induces isomorphisms*

(3.2.1) $$\hat{\alpha} : \pi_*^\wedge\mathcal{H}_F^{(-k)} \simeq \mathbf{R}^{n+1}q_*(F^{-k}\Omega^\cdot, \hat{d}) \quad \textit{for } k \geqslant 0.$$

(ii) *The homomorphism α is equivariant with the operation $\mathcal{G}\delta_1^{-1}$ (i.e. $\alpha\nabla_\delta\nabla_{\delta_1}^{-1} = \delta\delta_1^{-1}\alpha$ for $\delta \in \mathcal{G}$).*

PROOF. (i) Let us compute $\alpha \mid \pi_*\mathcal{H}_F^{(-1)}$. Since the inclusion $\pi_*\mathcal{H}_F^{(-1)} \to \pi_*\mathcal{H}_F^{(0)}$ is defined by the wedge product with dF_1 (cf. (1.5.3)), for an $[\zeta] \in \pi_*\mathcal{H}_F^{(-1)}$, $\alpha([\zeta])$ is represented by $dF_1 \wedge \zeta \in q_*F^0\Omega^{n+1}$. Since

$$dF_1 \wedge \zeta - \delta_1^{-1} d\zeta = \hat{d}(-\zeta) \equiv 0 \mod \hat{d}(q_*F^0\Omega^n),$$

$\alpha([\zeta])$ is represented by $\delta_1^{-1} d\zeta \in q_* F^{-1}\Omega^{n+1}$. Hence $\alpha([\zeta]) = \delta_1^{-1}\alpha([d\zeta]) \in H^{n+1}(F^{-1}\Omega^{\cdot}, \hat{d})$, and $\delta_1\alpha([\zeta]) = \alpha([d\zeta]) = \alpha(\nabla_{\delta_1}[\zeta])$ (cf. (1.6.2)). This proves the equivariance of α with δ_1.

Since $\mathbf{R}_C^{n+1} q_*(F^0\Omega^{\cdot}, \hat{d})$ is complete with respect to the δ_1^{-1}-adic topology (cf. (2.6)), by completing α one get a homomorphism

$$\hat{\alpha}\colon \pi_*^{\wedge}\mathcal{H}_F^{(0)} \to \mathbf{R}^{n+1} q_*(F^0\Omega^{\cdot}, \hat{d}).$$

Let us show the injectivity of α. For an element $\omega \in q_*\Omega_{X/T}^{n+1}$, suppose that $\alpha([\omega]) = 0$. Then there exists $\zeta = \sum_{k=0}^{\infty} \zeta_k \delta_1^{-k} \in q_* F^0\Omega^n = q_*\Omega_{X/T}^n[[\delta_1^{-1}]]$, such that

$$\omega = \hat{d}\zeta = -dF_1 \wedge \zeta_0 + (d\zeta_0 - dF_1 \wedge \zeta_{-1})\delta_1^{-1} + (d\zeta_{-1} - dF_1 \wedge \zeta_{-2})\delta_1^{-2} + \cdots.$$

Thus

$$[\omega] = -[dF_1 \wedge \zeta_0] = -\nabla_{\delta_1}^{-1}[\zeta_{-1}] = -\nabla_{\delta_1}^{-2}[\zeta_{-2}] = \cdots.$$

Hence $[\omega] \in \bigcap_{k=0}^{\infty} \pi_*\mathcal{H}_F^{(-k)} = 0$ (cf. (1.7.1)). This proves the injectivity of α, which proves automatically the injectivity of $\hat{\alpha}$.

Let us show the surjectivity of $\hat{\alpha}$. By comparing (1.6.4) with (2.5)(iii)(3), one gets

$$\begin{array}{ccccccccc}
0 & \to & \pi_*\mathcal{H}_F^{(-k-1)} & \to & \pi_*\mathcal{H}^{(-k)} & \to & q_*\Omega_F & \to & 0 \\
 & & \downarrow\alpha & & \downarrow\alpha & & \| & & \\
0 & \to & H^{n+1}(F^{-k-1}\Omega^{\cdot}) & \to & H^{n+1}(F^{-k}\Omega^{\cdot}) & \to & q_*\Omega_F & \to & 0
\end{array}$$

for $k \geqslant 0$. This shows by induction on k that α induces $\mathcal{O}_T$-bijections for $k \geqslant 0$

$$\pi_*\mathcal{H}_F^{(0)}/\pi_*\mathcal{H}_F^{(-k-1)} \simeq H^{n+1}(F^0\Omega^{\cdot})/H^{n+1}(F^{-k-1}\Omega^{\cdot}).$$

By taking the projective limit as $k \to \infty$, we get an isomorphism

$$\hat{\alpha}\colon \widehat{\pi_*\mathcal{H}_F^{(0)}} \simeq H^{n+1}(F^0\Omega^{\cdot})^{\cdot} \simeq \mathbf{R}^{n+1} q_*(F^0\Omega, \hat{d}).$$

(ii) Take an element $[\zeta] \in \pi_*\mathcal{H}_F^{(-1)}$, so that $\alpha[\zeta] = [dF_1 \wedge \zeta] = \delta_1^{-1}[d\zeta]$ in $H^{n+1}(q_*\Omega^{\cdot})$.

Recall the definition of the Gauss-Manin connection in (1.6.1) and compare it with the formula (2.7.3). Then,

$$\begin{aligned}
\frac{\partial}{\partial t_i}\alpha[\zeta] &= \left[(-1)^{i-1}(dt_2 \wedge \cdots \wedge dt_m)^{-1} dt_1 \wedge \cdots \wedge dt_i \wedge \cdots \wedge dt_m \wedge d\zeta\right] \\
&= \alpha\left(\nabla_{\partial/\partial t_i}[\zeta]\right).
\end{aligned}$$

(3.3) Let us now construct an $\mathcal{O}_T$-homomorphism

$$\beta\colon \mathbf{R}^{n+1} q_*(\Omega^{\cdot}, \hat{d}) \to \left(R_C^{n+1} q_*\mathcal{O}_X\right) \underset{\mathcal{O}_T}{\otimes} \mathcal{O}_T[[\delta_1^{-1}]][\delta_1^{-1}] \tag{3.3.1}$$

as follows.

Let $\mathfrak{A}$ be a Stein open covering of $X - C$ and let us consider the double Čech complex $C^p(\mathfrak{A}, \Omega^q)$, $0 \leq p$, $0 \leq q \leq n+1$, where ∂: $C^p(\mathfrak{A}, \Omega^q) \to C^{p+1}(\mathfrak{A}, \Omega^q)$ is the Čech coboundary operator and $\hat{d}$: $C^p(\mathfrak{A}, \Omega^q) \to C^p(\mathfrak{A}, \Omega^{q+1})$ is a homomorphism induced from (2.3.1).

Due to the fact (2.4)(ii), for any fixed p, the complex $(C^p(\mathfrak{A}, \Omega^{\cdot}), \hat{d})$ is exact. Thus by the Leray's spectral map of the double complex $C^p(\mathfrak{A}, \Omega^q)$, one gets a homomorphism

$$H^{n+1}(\Gamma(X - C, \Omega^{\cdot})) \to H^n(C^{\cdot}(\mathfrak{A}, \Omega^{\cdot})) = H^n\left(X - C, \mathcal{O}_X[[\delta_1^{-1}]][\delta_1]\right).$$

By localizing the above construction for open sets of T, one gets a sheaf homomorphism,

$$H^{n+1}\left((q \mid X - C)_* \Omega^{\cdot}, \hat{d}\right) \to R_C^{n+1} q_*(\mathcal{O}_X)[[\delta_1^{-1}]][\delta_1].$$

Combining this homomorphism with the restriction homomorphism,

$$H^{n+1}\left(q_* \Omega^{\cdot}, \hat{d}\right) \to H^{n+1}\left((q \mid X - C)_* \Omega^{\cdot}, \hat{d}\right),$$

one obtains the homomorphism β of (3.3.1).

(3.4) Let us expand the image of β in the Laurent series in δ_1^{-1}, so that, one defines an infinite sequence of $\mathcal{O}_T$-homomorphisms,

$$\nabla^{(k)}: \mathbf{R}^{n+1} q_*(\Omega^{\cdot}, \hat{d}) \to R_C^{n+1} q_*(\mathcal{O}_X) \quad \text{for } k \in \mathbf{Z}, \tag{3.4.1}$$

such that

$$\beta(\omega) = \sum_k \nabla^{(k)}(\omega) \delta_1^{-k} \quad \text{for } \omega \in \mathbf{R}^{n+1} q_*(\Omega^{\cdot}, \hat{d}). \tag{3.4.2}$$

Since by definition β is equivariant to the multiplication by δ_1, one gets a relation,

$$\nabla^{(k)} \delta_1 = \nabla^{(k+1)}. \tag{3.4.3}$$

(3.5) PROPOSITION. *The image of the "iterated differentiations" $\nabla^{(k)}$ are contained in $\check{\mathcal{H}}^{(0)}_{F,\mathrm{finite}} \subset R_C^{n+1} q_* \mathcal{O}_X$. Furthermore we have a formula, for $\omega \in \mathbf{R}^{n+1} q_*(\Omega^{\cdot}, \hat{d})$,*

$$\nabla^{(k+1)}(\omega) \operatorname{mod} \check{\Omega}^{(0)}_F = \check{\nabla}_{\delta_1} \nabla^{(k)}(\omega) \quad \textit{in } \check{\mathcal{H}}^{(1)}_F \textit{ for } k \in \mathbf{Z} \tag{3.5.1}$$

(*cf.* (1.11.12) *for the definition of $\check{\nabla}_{\delta_1}$, and $\check{\mathcal{H}}^{(0)}_{F,finite}$*).

PROOF. Let us compute $\beta(\omega) = \sum \nabla^{(k)}(\omega) \delta_1^{-k}$ of (3.3) more explicitly. Suppose ω is represented by an element $\tilde{\omega} \in q_* \Omega^{n+1}_{X/T}[[\delta_1^{-1}]][\delta_1]$ by the isomorphism (2.5)(i). Denote by $\iota\tilde{\omega}$ its image in $C^0(\mathfrak{A}, \Omega^{n+1})$. The image of $\iota\tilde{\omega}$ in $C^n(\mathfrak{A}, \Omega^0) = C^n(\mathfrak{A}, \mathcal{O}_X[[\delta_1^{-1}]][\delta_1])$ by the Leray homomorphism is given by $\hat{d}^{-1}(\partial \hat{d}^{-1})^n \iota\tilde{\omega}$. (Here we choose representatives of $\hat{d}^{-1}(\partial \hat{d}^{-1})^j \iota\tilde{\omega}$, $0 \leq j \leq n$, explicitly in the Čech chain complex.) Let us expand it in a Laurent series,

$$\hat{d}^{-1}(\partial \hat{d}^{-1})^n \iota\tilde{\omega} = \sum_{k \geq k_0} A_k \delta_1^{-k}, \tag{3.5.2}$$

where A_k, $k \geqslant k_0$, is a cocycle in $C^n(\mathfrak{A}, \mathcal{O}_X)$ which represents the cohomology class $\nabla^{(k)}\omega$ in $R_C^{n+1}q_*(\mathcal{O}_X)$.

Applying $\hat{d}$ to both sides of (3.5.2), we get

$$(3.5.3)\quad (\partial\hat{d}^{-1})^n \iota\tilde{\omega} = \hat{d} \sum_{k\geqslant k_0} A_k \delta_1^{-k}$$

$$= -dF_1 \wedge A_{k_0}\delta_1^{-k_0} + \sum_{l=0}^{\infty} \left(d_{X/T}A_{k_0+l} - dF_1 \wedge A_{k_0+l+1}\right)\delta_1^{-k_0-l-1}.$$

Since the left side of (3.5.3) is a coboundary in $\partial C^{n-1}(\mathfrak{A}, \Omega^1)$, the cohomology class on the right side of (3.5.3) is zero in $R_C^{n+1}q_*(\Omega^1)$. Hence one gets a sequence of relations in $R_C^{n+1}q_*(\Omega^1_{X/T})$, $-dF_1 \wedge \nabla^{(k_0)}(\omega) = 0$, $d\nabla^{(k_0+l)}(\omega) - dF_1 \wedge \nabla^{(k_0+l+1)}(\omega) = 0$, $l = 0, 1, \ldots$. Thus,

$$\nabla^{(k_0)}(\omega) \in \ker\left(dF_1 \wedge : R_C^{n+1}q_*\mathcal{O}_X \to R_C^{n+1}q_*\Omega^1_{X/T}\right) = q_*\Omega_F = \check{\Omega}^{(0)}.$$

Then by induction on l, one checks that

$$(3.5.4)\qquad \nabla^{(k_0+l)}(\omega) \in \check{\Omega}^{(l)} \subset \check{\mathcal{H}}^{(0)}_{F,\mathrm{finite}},$$

(cf. (1.11.3) for the definition of $\check{\Omega}^{(l)}$) and $\nabla^{(k_0+l+1)}\omega \bmod \check{\Omega}^{(0)} = \check{\nabla}_{\delta_1}\nabla^{(k_0+l)}\omega$ in $\check{\mathcal{H}}^{(1)}_F$. Q.E.D.

(3.6) Proposition. *The following sequence is exact for $k + l \geqslant -1$,*

$$(3.6.1)\qquad 0 \to \mathbf{R}^{n+1}q_*(F^{-k-1}\Omega^\cdot) \to \mathbf{R}^{n+1}q_*(F^l\Omega^\cdot) \xrightarrow{\nabla^{(k)}} \check{\Omega}^{(k+l)} \to 0.$$

Proof. In view of (2.6.1) and (3.4.3), it suffices to show the case $l = 0$.

Replacing the double complex $C^p(\mathfrak{A}, \Omega^q)$ of (3.3) by the double complex $C^p(\mathfrak{A}, F^0\Omega^q)$, one constructs an $\mathcal{O}_T$-homomorphism,

$$(3.6.2)\qquad \beta' \colon \mathbf{R}^{n+1}q_*(F^0\Omega^\cdot, \hat{d}) \to \left(\mathbf{R}_C^{n+1}q_*\mathcal{O}_X\right) \underset{\mathcal{O}_T}{\otimes} \mathcal{O}_T[[\delta_1^{-1}]],$$

such that $\beta' = \beta \mid \mathbf{R}^{n+1}q_*(F^0\Omega^\cdot, \hat{d})$.

Since the right-hand side of (3.6.2) consists of formal power series in δ_1^{-1}, one gets an expansion,

$$(3.6.3)\qquad \beta' = \sum_{k=0}^{\infty} \nabla^{(k)} \otimes \delta_1^{-k}.$$

In particular $\nabla^{(k)}R^{n+1}q_*(F^0\Omega^\cdot, \hat{d}) = 0$ for $k < 0$. This implies the semiexactness of (3.6.1) and the argument of (3.5.4) shows that $\nabla^{(k)}R^{n+1}q_*(F^0\Omega^\cdot, \hat{d}) \subset \check{\Omega}^{(k)}$.

Due to the semiexactness, $\nabla^{(k)}$ induces an $\mathcal{O}_T$-homomorphism,

$$(3.6.4)\qquad \mathbf{R}^{n+1}q_*(F^0\Omega^\cdot)/\mathbf{R}^{n+1}q_*(F^{-k-1}\Omega^\cdot) \to \check{\Omega}^{(k)}.$$

Since both sides of (3.6.4) are $\mathcal{O}_T$-free modules of the same rank $(k+1)\mu$ (cf. (2.5)(iii)(3) and (1.11.3)), to show the bijectivity of (3.6.4), it is enough to show the surjectivity of it.

We show the surjectivity by induction on k. For $k = 0$, the equality (3.6.3) says that $\nabla^{(0)} = \beta' \bmod(\delta_1^{-1})$. Hence, taking $\Omega^{\cdot}_{X/T} = F^0\Omega^{\cdot}/\delta_1^{-1}F^0\Omega^{\cdot}$, $\nabla^{(0)}$ is given by replacing the double complex $C^p(\mathfrak{A}, \Omega^q)$ in (3.3) by the double complex $C^p(\mathfrak{A}, \Omega^q_{X/T})$, so that $\nabla^{(0)}$ is equal to the morphism

$$\pi_*\mathcal{H}_F^{(0)} \simeq \mathbf{R}^{n+1}q_*F^0\Omega^{\cdot} \overset{\chi}{\to} \check{\mathcal{H}}_F^{(0)}$$

which was given in (1.14.5). Due to the exactness of (1.14.5), the image of $\nabla^{(0)} = q_*\Omega_F = \check{\Omega}^{(0)}$.

For $k \geqslant 0$ assume $\nabla^{(k)}\mathbf{R}^{n+1}q_*(F^0\Omega^{\cdot}) = \check{\Omega}^{(k)}$. Then

$$\begin{aligned}\nabla^{(k+1)}\mathbf{R}^{n+1}q_*(F^0\Omega^{\cdot}) &\supset \nabla^{(k+1)}\mathbf{R}^{n+1}q_*(F^{-1}\Omega^{\cdot})\\ &= \nabla^{(k)}\mathbf{R}^{n+1}q_*(F^0\Omega^{\cdot}) \quad (\because (2.6.1), (3.4.3))\\ &= \check{\Omega}^{(k)} \supset \check{\Omega}^{(0)}.\end{aligned}$$

Then applying (3.5.1), $\nabla^{(k+1)}\mathbf{R}^{n+1}q_*(F^0\Omega^{\cdot})/\check{\Omega}^{(0)} = \check{\nabla}_{\delta_1}\nabla^{(k)}\mathbf{R}^{n+1}q_*(F^0\Omega^{\cdot}) = \check{\nabla}_{\delta_1}\check{\Omega}^{(k)}$. Hence by definition of $\check{\Omega}^{(k+1)}$ in (1.11.3), one gets $\nabla^{(k+1)}\mathbf{R}^{n+1}q_*(F^0\Omega^{\cdot}) = \check{\Omega}^{(k+1)}$. This completes the proof of (3.6).

(3.7) COROLLARY. *The following is an exact sequence for $k \in \mathbf{Z}$,*

$$0 \to \widehat{\pi_*\mathcal{H}_F^{(-k-1)}} \overset{\hat{\alpha}}{\to} \mathbf{R}^{n+1}q_*(\Omega^{\cdot}) \overset{\nabla^{(k)}}{\to} \check{\mathcal{H}}^{(0)}_{F,finite} \to 0.$$

($\because$ *Take the union of* (3.6.1) *for* $l \geqslant -(k+1)$. *Then combine it with the isomorphism of* (3.2.1).)

(3.8) PROPOSITION (CF. [3, (9.6)]). *The homomorphism $\nabla^{(k)}$: $\mathbf{R}^{n+1}q_*(\Omega^{\cdot}, \hat{d}) \to \check{\mathcal{H}}_F^{(0)}$ is equivariant with $\mathcal{G}$-module structures, in the sense that the following diagram commutes for any $\delta \in \mathcal{G}$:*

$$\begin{array}{ccccc} \mathbf{R}^{n+1}q_*(\Omega^{\cdot}, \hat{d}) & \overset{\nabla^{(k)}}{\to} & \check{\mathcal{H}}_F^{(0)} & & \\ & & & \searrow^{\nabla_\delta} & \\ \downarrow \delta & & & & \\ \mathbf{R}^{n+1}q_*(\Omega^{\cdot}, \hat{d}) & \overset{\nabla^{(k)}}{\to} & \check{\mathcal{H}}_F^{(0)} & \overset{dF_1}{\to} & \check{\mathcal{H}}_F^{(1)} \end{array}$$

PROOF. Let us fix a coordinate system $(t_1, t_2, \ldots, t_m, x_0, \ldots, x_n)$, so that any vector field δ on T is uniquely lifted to a vector field $\hat{\delta}$ on X such that $\hat{\delta}x_j = 0$, $j = 0, \ldots, n$. Then by a direct calculation one computes the following [3, (9.7.6)]:

(3.8.1) $$\langle \hat{\delta}, d_X M \rangle \equiv -d_X \langle \hat{\delta}, M \rangle \quad \bmod \Omega_T^1 \wedge \Omega_X^p$$

for $M \in \Omega_T^1 \wedge \Omega_X^p$ and $\delta \in \mathrm{Der}_T$.

(3.8.2) $$\langle \hat{\delta}, dF_1 \wedge M \rangle \equiv -dF_1 \wedge \langle \hat{\delta}, M \rangle \quad \bmod \Omega_T^1 \wedge \Omega_X^p$$

for $M \in \Omega_T^1 \wedge \Omega_X^p$. Here $\langle\, ,\rangle$ is the inner product of vector fields with forms.

Getting back to the proof of the proposition, let us represent an element $[\omega] \in \mathbf{R}^{n+1}q_*(\Omega^\cdot, \hat{d}) \simeq H^{n+1}(q_*\Omega^\cdot)$ by $\omega \in q_*\Omega^{n+1}$. Let

$$L^i := \hat{d}^{-1}(\partial\hat{d}^{-1})^i \iota\omega \in C^i\left(\mathfrak{A}, \Omega^{n-i}_{X/T}[[\delta_1^{-1}]][\delta_1]\right), \qquad i = -1, \dots, n$$

(where $L^{-1} = \omega$ by convention), be a sequence defining the Leray homomorphism $\beta\colon H^{n+1}(q_*\Omega^\cdot) \to R_C^{n+1}q_*\mathcal{O}_X[[\delta_1^{-1}]][\delta_1]$.

Let $\hat{L}^i$ be a lifting of L^i to the module $C^i(\mathfrak{A}, \Omega_X^{n-i}[[\delta_1^{-1}]][\delta_1])$ for $i = -1, \dots, n$.

For any fixed $\delta \in \mathrm{Der}_T$, let us define a sequence

$$\text{(3.8.3)} \qquad N^i := \left\langle \hat{\delta}, \hat{d}_X\hat{L}^i - \partial\hat{L}^{i-1} \right\rangle \mod \Omega_T^1 \wedge \Omega_X^{n-i-1}$$
$$\in C^i\left(\mathfrak{A}, \Omega^{n-i}_{X/T}[[\delta_1^{-1}]][\delta_1]\right), \qquad i = -1, \dots, n.$$

Here by convention, $N^{-1} = \langle \hat{\delta}, \hat{d}_X\omega \rangle$.

Then by using (3.8.1) and (3.8.2), one computes

$$\begin{aligned}\hat{d}N^i &= \left(\delta_1^{-1}d - dF_1\right)\left\langle \hat{\delta}, \hat{d}_X\hat{L}^i - \partial\hat{L}^{i-1} \right\rangle \\ &= \left\langle \hat{\delta}, -\left(\delta_1^{-1}d - dF_1\right)\left(\hat{d}_X\hat{L}^i - \partial\hat{L}^{i-1}\right)\right\rangle \\ &= \left\langle \hat{\delta}, \hat{d}_X\partial\hat{L}^{i-1} \right\rangle = \left\langle \hat{\delta}, \partial\left(\hat{d}_X\hat{L}^{i-1} - \partial\hat{L}^{i-2}\right)\right\rangle \\ &= \partial\left\langle \hat{\delta}, \hat{d}_X\hat{L}^{i-1} - \partial\hat{L}^{i-2}\right\rangle = \partial N^{i-1}.\end{aligned}$$

This implies that the sequence $N^i \in C^i(\mathfrak{A}, \Omega^{n-i})$, $i = -1, \dots, n$, defines the Leray's homomorphism β for N^{-1}.

On the one hand by definition of N^n in (3.8.3),

$$\begin{aligned}dF_1 \wedge N^n &= dF_1 \wedge \left\langle \hat{\delta}, \delta_1^{-1}\hat{d}_X\hat{L}^n - \partial\hat{L}^{n-1}\right\rangle \\ &= -\delta_1^{-1}\left\langle \hat{\delta}, dF_1 \wedge \hat{d}_X\hat{L}^n\right\rangle - \partial\left(dF_1 \wedge \langle \hat{\delta}, \hat{L}^{n-1}\rangle\right) \\ &\equiv -\delta_1^{-1}\left\langle \hat{\delta}, dF_1 \wedge d_X\hat{L}^n\right\rangle \mod \partial C^{n-1}(\mathfrak{A}, \Omega^1) \\ &\equiv \delta_1^{-1}(dt_2 \wedge \cdots \wedge dt_m)^{-1} v \wedge d_X\hat{L}^n\end{aligned}$$

where $v = \sum_{i=2}^m (-1)^{i-1} g_i\, dt_1 \wedge \cdots \wedge \widehat{dt_i} \wedge \cdots \wedge dt_m$ and $\delta = \sum_{v=1}^m g_i \partial/\partial t_i$.

By definition of $\check{\nabla}$ in (1.11), the last line represents the class $\delta_1^{-1}\check{\nabla}_\delta[L^n]$ in $\check{\mathcal{H}}_F^{(1)} \otimes [[\delta_1^{-1}]][\delta_1]$. Thus one has

$$\text{(3.8.4)} \qquad \left[dF_1 \wedge N^n\right] = \delta_1^{-1}\check{\nabla}_\delta\left[L^n\right] = \delta_1^{-1}\check{\nabla}_\delta \sum_k \nabla^{(k)}(\omega)\delta_1^{-k}.$$

On the other hand by definition of the action of δ on $\mathbf{R}^{n+1}q_*(\Omega^\cdot, \hat{d})$ in (2.7), $N^{-1} = \langle \hat{\delta}, \hat{d}_X\omega \rangle$ represents the cohomology class $\delta_1^{-1}\delta[\omega]$ in $\mathbf{R}^{n+1}q_*(\Omega^\cdot, \hat{d})$. Hence

$$\begin{aligned}\text{(3.8.5)} \qquad \beta\left(\left[N^{-1}\right]\right) &= \beta\left(\delta_1^{-1}\delta[\omega]\right) \\ &= \sum_k \nabla^{(k)}\left(\delta_1^{-1}\delta[\omega]\right)\delta_1^{-k} = \sum_k \nabla^{(k-1)}\left(\delta[\omega]\right)\delta_1^{-k}.\end{aligned}$$

Since $\beta([N^{-1}]) = [N^n]$, (3.8.4) and (3.8.5) together imply the equality

$$\check{\nabla}_\delta\nabla^{(k)}(\omega) = dF_1 \wedge \nabla^{(k)}(\delta[\omega]). \quad \text{Q.E.D.}$$

(3.9) Let us define a $\pi_*\mathcal{O}_S$-module structure on $\mathbf{R}^{n+1}q_*(\Omega^\cdot \hat{d})$.

$$\pi_*\mathcal{O}_S \times \mathbf{R}^{n+1}q_*(\Omega^\cdot, \hat{d}) \to \mathbf{R}^{n+1}q_*(\Omega^\cdot, \hat{d}). \tag{3.9.1}$$

To define that, let us remember that $\mathbf{R}^{n+1}q_*(F^0\Omega^\cdot, \hat{d}) \cong \pi_*\mathcal{H}_F^{(0)}$ already has a natural $\pi_*\mathcal{O}_S$-module structure, $g[\omega] = [g\omega]$ for $g \in \pi_*\mathcal{O}_S$ and $\omega \in \pi_*\mathcal{H}_F^{(0)}$.

Then (3.9.1) is gotten by an extension of this structure as follows.

Any element $\eta \in \mathbf{R}^{n+1}q_*(\Omega^\cdot, \hat{d})$ can be expressed as $\delta_1^m\omega$ for some $m \in \mathbf{N}$ and $\omega \in \mathbf{R}^{n+1}q_*(F^0\Omega^\cdot, \hat{d})$. Then for $g \in \pi_*\mathcal{O}_S$

$$g\eta(= g\delta_1^m\omega) := \sum_{r=0}^{m} (-1)^r\binom{m}{r}\delta_1^{m-r}((\delta_1^r g)\omega). \tag{3.9.2}$$

Using the Leibnitz rule for δ_1, one checks easily that the definition (3.9.2) does not depend on the choice of m and ω, and that commutativity, and associativity hold for this multiplication. Also one checks that the usual Leibnitz rule holds, i.e.

$$\delta(g\eta) = (\delta g)\eta + g(\delta\eta) \tag{3.9.3}$$

for $\delta \in \mathcal{G}$, $g \in \pi_*\mathcal{O}_S$ and $\eta \in \mathbf{R}^{n+1}q_*(\Omega^\cdot, \hat{d})$.

(3.10) Remember that $\pi_*\mathcal{H}_F^{(0)}$, $\mathcal{H}_F^{(0)}$ and $\mathbf{R}^{n+1}q_*(\Omega^\cdot, \hat{d})$ have natural $\pi_*\mathcal{O}_S$-module structures and that α is a $\pi_*\mathcal{O}_S$-homomorphism. Whereas β is not a $\pi_*\mathcal{O}_S$-homomorphism as we see below (3.10.4).

PROPOSITION (CF. [3, (9.5)]). *$\nabla^{(k)}$'s satisfy the following "troisted" Leibnitz rule.*

$$\nabla^{(k)}(g\omega) = \sum_{r\geq 0}\binom{n+k}{r}\delta_1^r g\nabla^{(k-r)}\omega \tag{3.10.1}$$

for $g \in \pi_*\mathcal{O}_S$, $\omega \in \mathbf{R}^{n+1}q_*(\Omega^\cdot)$, $k \in \mathbf{Z}$. In particular, by putting $g = t_1$ one gets

$$\nabla^{(k)}(t_1\omega) - t_1\nabla^{(k)}(\omega) = (n+k)\nabla^{(k-1)}(\omega). \tag{3.10.2}$$

PROOF. Let us write (k, l), if the formula (3.10.1) is true for any $\omega \in \mathbf{R}^{n+1}q_*(F^l\Omega^\cdot)$. Then using (3.4.3), one checks easily (k, l) and $(k+1, l)$ implies $(k, l+1)$. Hence one needs only to prove $(k, 0)$ for $k \in \mathbf{Z}$. $(k, 0)$ for $k < 0$ is trivial, since both sides of (3.10.1) are zero.

Now we may assume that $\omega \in \mathbf{R}^{n+1}q_*(F^0\Omega^\cdot)$ is presented by $\hat{\omega} \in q_*\Omega_{X/T}^{n+1}$. As in the proof of (3.5), Leray homomorphism $\beta(\omega)$ is presented by a sequence, $L^i = \sum_{k\geq 0} L_k^i\delta_1^{-k} \in C^i(\mathfrak{A}, \Omega^{n-i})$, $i = 0, \ldots, n$, such that $\hat{d}L^i = \partial L^{i-1}$, $i = 0, \ldots, n$.

Put

$$M^i = \sum_{k\geq 0}\sum_{r=0}^{k}\binom{i+k}{r}\delta_1^r L_{k-r}^i\delta_1^{-k} \in C^i(\mathfrak{A}, \Omega^{n-i}), \qquad i = 0, \ldots, n.$$

Then one checks relations, $\hat{d}M^i = \partial M^{i=1}$, $i = 1, \ldots, n$, $\hat{d}M^0 = \iota(g\hat{\omega})$. Thus the sequence M^i gives the Leray homomorphism for $g\hat{\omega}$:

$$\beta(g\omega) = \sum_{k=0}^{\infty}\sum_{r=0}^{k}\binom{n+k}{r}\delta_1^r g\nabla^{(k-r)}(\omega)\delta_1^{-k}.$$

This proves the formula (3.10.1) for $\omega \in \mathbf{R}^{n+1}q_*(F^0\Omega^\cdot)$. Q.E.D.

COROLLARY. *For a holomorphic function g on S, let us define*

$$X(g) := \sum_{p=0}^{\infty} \binom{n+p}{p} (\delta_1^p g) \delta_1^{-p}. \tag{3.10.3}$$

Then one has a formula,

$$\beta(g\omega) = \left(\sum_{m=0}^{\infty} (-1)^m \frac{1}{m!} X(\delta_1^m g) \left(\frac{\partial}{\partial \delta_1} \right)^m \right) \beta(\omega). \tag{3.10.4}$$

In particular by putting $g = t_1$,

$$\beta(t_1\omega) = t_1\beta(\omega) = (n+1)\delta_1^{-1}\beta(\omega) - \frac{\partial}{\partial \delta_1}\beta(\omega). \tag{3.10.5}$$

4. Higher residues K_F and $K_F^{(k)}$, $k = 0, 1, 2, \ldots$. In this section we define a bilinear map (cf. (4.3))

$$K_F \colon \mathbf{R}^{n+1} q_*(\Omega^\cdot, \hat{d}) \times \mathbf{R}^{n+1} q_*(\Omega^\cdot, \hat{d}) \to \mathcal{O}_T[[\delta_1^{-1}]][\delta_1]$$

which induces the higher residue pairings, $K_F^{(k)} \colon \pi_* \mathcal{H}_F^{(0)} \times \pi_* \mathcal{H}_F^{(0)} \to \mathcal{O}_T$ for $k \in \mathbf{Z}$.

(4.1) *Notation*. For a Laurent series $P = \Sigma_k P_k \delta_1^k$, let us denote by $\overline{P}$ the Laurent series $\Sigma_k (-1)^k P_k \delta_1^k$.

It is obvious by the definition that:

(i) $\overline{\overline{P}} = P$.

(ii) $\overline{P \cdot Q} = \overline{P} \cdot \overline{Q}$ or $\overline{P \wedge Q} = \overline{P} \wedge \overline{Q}$ whenever these products are defined.

(4.2) First let us define an $\mathcal{O}_T$-bilinear map,

$$K_1 \colon H^{n+1}\left(q_*\Omega^\cdot, \hat{d}\right) \times q_*\Omega^{n+1} \to \mathcal{O}_T[[\delta_1^{-1}]][\delta_1], \tag{4.2.1}$$

$$K_1(\omega, \xi) = \mathrm{Res}_{X/T}\left[\beta(\omega) \cdot \overline{\xi}\right] = \sum_k \sum_l (-1)^l \delta_1^{l-k} \mathrm{Res}_{X/T}\left[(\nabla^{(k)}\omega)\xi_l\right]$$

for $\omega \in H^{n+1}(q_*\Omega^\cdot, \hat{d})$ and $\xi = \Sigma_l \xi_l \delta_1^l \in q_*\Omega_{X/Y}^{n+1}[[\delta_1^{-1}]][\delta_1]$.

Assertion. The restriction of K_1 to $\mathbf{R}^{n+1} q_*(\Omega^\cdot, \hat{d}) \times \hat{d}(q_*\Omega^n)$ is the zero map.

PROOF. Take $\zeta \in q_*\Omega_{X/T}^n$ and compute

$$\begin{aligned} K_1(\omega, \hat{d}\zeta) &:= \mathrm{Res}_{X/T}\left[\beta(\omega)\left(\overline{\hat{d}\zeta}\right)\right] \\ &= \mathrm{Res}_{X/T}\left[\sum_k \delta_1^{-k} \nabla^{(k)}(\omega)\left(-\delta_1^{-1} d\zeta - dF_1 \wedge \zeta\right)\right] \\ &= -\sum_k \delta_1^{-k-1} \mathrm{Res}_{X/T}\left[(\nabla^{(k)}\omega)\, d\zeta + (\nabla^{(k+1)}\omega)\, dF_1 \wedge \zeta\right]. \end{aligned}$$

Then recalling (3.5.1) and (1.14.2), the last line above equals $-\Sigma_k \delta_1^{-k-1} \mathrm{Res}_{X/T}[d((\nabla^{(k)}\omega)\zeta)]$, which turns out to be zero by (1.14.2). Q.E.D.

As a consequence of this, one gets a bilinear map

$$\mathbf{R}^{n+1} q_*(\Omega^\cdot, \hat{d}) \times \left(q_*\Omega^{n+1}/\hat{d}\left(q_*\Omega^n\right)\right) \to \mathcal{O}_T[[\delta_1^{-1}]][\delta_1].$$

Using the identification $\mathbf{R}^{n+1}q_*(\Omega^{\cdot}, \hat{d}) \simeq q_*\Omega^{n+1}/\hat{d}(q_*\Omega^n)$ of (2.5)(i), one gets a definition.

(4.3) DEFINITION. Let us denote by K_F the $\mathcal{O}_T$-bilinear map

$$K_F\colon \mathbf{R}^{n+1}q_*(\Omega^{\cdot}, \hat{d}) \times \mathbf{R}^{n+1}q_*(\Omega^{\cdot}, \hat{d}) \to \mathcal{O}_T[[\delta_1^{-1}]][\delta_1]$$

which is induced by K_1 of (4.2.1).

(4.4) PROPOSITION. *K_F has the following properties*:

(i) $\varphi K_F(\omega_1, \omega_2) = K_F(\varphi\omega_1, \omega_2) = K_F(\omega_1, \bar{\varphi}\omega_2)$ *for* $\omega_i \in \mathbf{R}^{n+1}q_*(\Omega^{\cdot}, \hat{d})$, $\varphi \in \mathcal{O}_T[[\delta_1^{-1}]][\delta_1]$.

(ii) $K_F(\omega_1, \omega_2) = K_F(\omega_2, \omega_1)$ *for* $\omega_i \in \mathbf{R}^{n+1}q_*(\Omega^{\cdot}, \hat{d})$.

PROOF OF (i). Since β is equivariant with the multiplication by δ_1, the equality $\varphi K(\omega_1, \omega_2) = K(\varphi\omega_1, \omega_2)$ is trivial. If we assume (ii), the second part of (i) follows immediately from the first part. Thus it suffices to prove (ii). The proof given here is essentially the same as in the proof of Theorem (10.17) of [**3**].

First we introduce some notions on the Čech cochain complexes with coefficients in $\Omega^{\cdot} = \Omega_{X/T}[[\delta_1^{-1}]][\delta_1]$.

(4.5) Let us consider the double complex $C^p(\mathfrak{A}, \Omega^q)$ as in (3.3), where $\mathfrak{A}$ is an open Stein covering of $X - C$. Since C is a complete intersection of codimension $n+1$ in X, one may assume that there are no nerves of dimension $n+1$ and hence $C^{n+1}(\mathfrak{A}, \Omega^q) = 0$ and $C^n(\mathfrak{A}, \Omega^q) = Z^n(\mathfrak{A}, \Omega^q)$.

Let us define two wedge products of cochains,

(4.5.1) $$\wedge : C^l(\mathfrak{A}, \Omega^i) \times C^m(\mathfrak{A}, \Omega^j) \to C^{l+m+1}(\mathfrak{A}, \Omega^{i+j})$$

for $l, m \geqslant -1$,

$$(L \wedge M)_{i_1,\ldots,i_{l+m+1}} = \frac{1}{(l+1)!\,(m+1)!} \sum_{I,J} \operatorname{sgn}\begin{pmatrix} i_0, \ldots, i_{l+m+1} \\ I, J \end{pmatrix} L_I \wedge M_J$$

for $L \in C^l(\mathfrak{A}, \Omega^{\cdot})$ and $M \in C^m(\mathfrak{A}, \Omega^{\cdot})$. Here the summation on the right-hand side runs over all pairs of ordered subsets I, J of $i_0, \ldots, i_{l+m+1}$ with $\#\{I\} = l+1$, $\#\{J\} = m+1$.

(4.5.2) $$\wedge{*}\colon C(\mathfrak{A}, \Omega^i) \times C^m(\mathfrak{A}, \Omega^j) \to C^{l+m}(\mathfrak{A}, \Omega^{i+j}) \quad \text{for } l, m \geqslant 0,$$

$$(L \wedge M)_{i_0,\ldots,i_{l+m}} = \frac{1}{l!m!} \sum_{i_p, I, J} \operatorname{sgn}\begin{pmatrix} i_0, \ldots, i_{l+m} \\ i_p, I, J \end{pmatrix} L_{i_p, I} \wedge M_{i_p, J}$$

for $L \in C^l(\mathfrak{A}, \Omega^{\cdot})$ and $M \in C^m(\mathfrak{A}, \Omega^{\cdot})$. Here the summation on the right-hand side runs over all triples (i_p, I, J) where $0 \leqslant p \leqslant k + l$ and I and J are ordered subsets of $i_0, \ldots, i_{l+m}$ with $\#\{I\} = l$ and $\#\{J\} = m$.

Those wedge products $\wedge$ and $\wedge{*}$ behave as usual with respect to the exterior differentiation $d = d_{X/T}$ and satisfy the usual commutation relations. The following needs some computations. (For the proof see [**3**, (10.17)].)

Formulae ((10.17) Assertion 3, [**3**]). Let the notation be as above.

(4.5.3) $$\partial L \wedge {*}M = (-1)^{l+1}(m+1) L \wedge M + \partial(L \wedge {*}M),$$

$$L \wedge {*}\partial M = (l+1) L \wedge M + (-1)^l \partial(L \wedge {*}M).$$

In particular

$$\iota L \wedge {}^*M = (m+1)L \wedge M \quad \text{for } l = -1, \tag{4.5.4}$$
$$L \wedge {}^*\iota M = (l+1)L \wedge M \quad \text{for } m = -1,$$

where $\iota\colon \Gamma(X - C, \Omega^i) \to C^0(\mathfrak{A}, \Omega^i)$ is the canonical inclusion homomorphism.

(4.6) PROOF OF (4.4)(ii). For $\omega_1, \omega_2 \in H^{n+1}(q_*\Omega^\cdot)$, let us take representatives $\tilde{\omega}_1, \tilde{\omega}_2 \in q_*\Omega^{n+1}$. Put $L^{-1} = \tilde{\omega}_1$ and $M^{-1} = \tilde{\omega}_2$ and as in (3.5) let us choose sequences of cochains,

$$L^i = \hat{d}^{-1}(\partial\hat{d}^{-1})^i_\iota\tilde{\omega}_1, \qquad M^i = \hat{d}^{-1}(\partial\hat{d}^{-1})^i\iota\tilde{\omega}_2 \in C^i(\mathfrak{A}, \Omega^{n-i}) \tag{4.6.1}$$

for $i = 0, 1, \ldots, n$. By definition of these sequences, they satisfy relations

$$(\delta_1^{-1}d - dF_1)L^q = \partial L^{q-1}, \qquad (\delta_1^{-1}d - dF_1)M^q = \partial M^{q-1} \tag{4.6.2}$$

for $k = 0, \ldots, n$.

Now using the relations (4.6.2)and (4.5.3), let us compute

$$\begin{aligned}
\delta_1^{-1}d\left(L^{n-q} \wedge {}^*\overline{M^q}\right) &= \left(\delta_1^{-1}dL^{n-q}\right) \wedge {}^*\overline{M^q} + (-1)^q L^{n-q} \wedge {}^*\left(\delta_1^{-1}d\overline{M^q}\right) \\
&= \left(dF_1 \wedge L^{n-1} + \partial L^{n-q-1}\right) \wedge {}^*\overline{M^q} \\
&\quad + (-1)^q L^{n-q} \wedge {}^*\left(-\overline{\left(dF_1 \wedge M^q + \partial M^{q-1}\right)}\right) \\
&= dF_1 \wedge \left(L^{n-q} \wedge {}^*\overline{M^q} - L^{n-q} \wedge {}^*\overline{M^q}\right) \\
&\quad + (-1)^{n-q}(q+1)L^{n-q-1} \wedge \overline{M^q} - (-1)^q(n-q+1)L^{n-q} \wedge \overline{M^{q-1}} \\
&\quad + \partial\left(L^{n-q-1} \wedge {}^*\overline{M^q} - (-1)^{q+n-q}L^{n-q} \wedge {}^*\overline{M^{q-1}}\right).
\end{aligned}$$

Hence in $\mathbf{R}_C^{n+1}q_*(\Omega_{X/T}^{n+1}[[\delta_1^{-1}]][\delta_1])$,

(4.6.3)

$$\begin{aligned}
(q+1)\mathrm{Res}_{X/T}&\left[L^{n-q-1} \wedge \overline{M^q}\right] - (-1)^n(n-q+1)\mathrm{Res}_{S/T}\left[L^{n-q} \wedge \overline{M^{q-1}}\right] \\
&= (-1)^{n-q}\delta_1^{-1}\,\mathrm{Res}_{X/T}\left[d\left(L^{n-q} \wedge {}^*\overline{M^q}\right)\right] = 0.
\end{aligned}$$

Hence

$$\mathrm{Res}_{X/T}\left[L^{n-q-1} \wedge \overline{M^q}\right] = (-1)^{n(q+1)}\binom{n+1}{q+1}\mathrm{Res}_{X/T}\left[L^n \wedge \overline{M^{-1}}\right]. \tag{4.6.4}$$

In particular

$$\mathrm{Res}_{X/T}\left[L^{-1} \wedge \overline{M^n}\right] = \mathrm{Res}_{X/T}\left[L^n \wedge \overline{M^{-1}}\right]. \tag{4.6.5}$$

Thus

$$\begin{aligned}
K_F(\omega_1, \omega_2) &= \mathrm{Res}_{X/T}\left[L^n \wedge \overline{M^{-1}}\right] = \mathrm{Res}_{X/T}\left[L^{-1} \wedge \overline{M^n}\right] \\
&= \overline{\mathrm{Res}_{X/T}\left[\overline{L^{-1}} \wedge M^n\right]} = K_F\overline{(\omega_2, \omega_1)}. \quad \text{Q.E.D.}
\end{aligned}$$

(4.7) PROPOSITION. *For* $\delta \in \mathcal{G}$ *and* $\omega_1, \omega_2 \in \mathbf{R}^{n+1}q_*(\Omega^{\cdot}, \hat{d})$,

$$\delta K_F(\omega_1, \omega_2) = K_F(\delta\omega_1, \omega_2) + K_F(\omega_1, \delta\omega_2).$$

Here δ *operates on* $\mathcal{O}_T$, *by the natural projection*

$$0 \to \mathcal{O}_T\delta_1 \to \mathcal{G} \to \mathrm{Der}_T \to 0.$$

PROOF. Put

$$\mathcal{Q}_F := \mathbf{R}_C^{n+1}q_*\left(\Omega_{X/T}^{n+1}\right)/dF_1 \wedge d\,\mathbf{R}_C^{n+1}q_*\left(\Omega_{X/T}^{n-1}\right).$$

(i) [**3**, (8.4)] Then $\mathcal{Q}_F$ has a natural structure of an integrable $\mathcal{G}$-module

$$\hat{\nabla} : \mathcal{G} \times \mathcal{Q}_F \to \mathcal{Q}_F.$$

(ii) [**3**, (7.14), (8.7)] The structure $\hat{\nabla}$ is compatible with the residue map:

$$\begin{array}{ccc}
\hat{\nabla} : \mathcal{G} \times \mathcal{Q}_F & \rightarrow & \mathcal{Q}_F \\
\downarrow(\mathrm{id}, \mathrm{res}) & & \downarrow \mathrm{res} \\
\hat{\nabla} : \mathcal{G} \times R_C^{2n+2}q_*\left(q^{-1}\mathcal{O}_T\right) & \rightarrow & R_C^{2n+2}q_*\left(q^{-1}\mathcal{O}_T\right) \\
\downarrow(\mathrm{id}, \mathrm{tr}) & & \downarrow \mathrm{tr} \\
\mathcal{G} \times \mathcal{O}_T & \rightarrow & \mathcal{O}_T
\end{array}$$

(iii) The local duality (1.13.3) induces a commutative diagram:

$$\begin{array}{ccccc}
 & \swarrow^{(\mathrm{id}, dF_1)} & \check{\mathcal{H}}_F^{(0)} \times \pi_*\mathcal{H}_F^{(-1)} & \searrow^{(dF_1, \mathrm{id})} & \\
 & & \downarrow dF_1 & & \\
\check{\mathcal{H}}_F^{(0)} \times \pi_*\mathcal{H}_F^{(0)} & \rightarrow & \mathcal{Q}_F & \leftarrow & \check{\mathcal{H}}_F^{(1)} \times \pi_*\mathcal{H}_F^{(-1)}
\end{array}$$

(iv) [**3**, (8.12)] In the above situation, one has an equality $\hat{\nabla}_\delta(\omega \cdot \zeta) = \check{\nabla}_\delta\omega \cdot \zeta + \omega \cdot \nabla_\delta\zeta$ for $\omega \in \check{\mathcal{H}}_F^{(0)}$, $\omega \in \pi_*\mathcal{H}_F^{(-1)}$, and $\delta \in \pi_*\,\mathrm{Der}_S$.

For the proof of this equality, shifting by a power of δ_1 one may assume that $\omega_2 = \alpha(\zeta)$ for a $\zeta \in \pi_*\mathcal{H}_F^{(-1)}$ ($\because$ (4.4)(ii)). Then using above (i)–(iv)

$$\begin{aligned}
\delta K_F(\omega_1, \omega_2) &= \delta\,\mathrm{Res}_{X/T}[\beta(\omega_1)\cdot\zeta] \\
&= \mathrm{Res}_{X/T}\left[\hat{\nabla}_\delta(\beta(\omega_1)\cdot\zeta)\right] \\
&= \mathrm{Res}_{X/T}\left[(\check{\nabla}_\delta\beta(\omega_1))\cdot\zeta + \beta(\omega_1)\cdot\nabla_\delta\zeta\right] \\
&= \mathrm{Res}_{X/T}\left[\beta(\delta(\omega_1))\cdot\zeta + \beta(\omega_1)\cdot\delta(\omega_2)\right] \\
&= K_F(\delta\omega_1, \omega_2) + K_F(\omega_1, \delta\omega_2). \quad \text{Q.E.D.}
\end{aligned}$$

(4.8) PROPOSITION. *Let us fix a coordinate* $t_1 = F_1$. *Then one has an equality, for* $\omega_1, \omega_2 \in \mathbf{R}_{q*}^{n+1}(\Omega^{\cdot}, \hat{d})$,

(4.8.1)
$$K_F(t_1\omega_1, \omega_2) - K_F(\omega_1, t_1\omega_2) = (n+1)\delta_1^{-1}K_F(\omega_1, \omega_2) - \frac{\partial}{\partial\delta_1}K_F(\omega_1, \omega_2).$$

PROOF. Using the formula (4.4)(i) and the commutation relations $[\delta_1, t_1] = 1$ and $[\partial/\partial\delta_1, \delta_1] = 1$, one checks easily that the formula (4.8.1) is invariant under the multiplication by δ_1 on ω_1 or on ω_2. Hence one may assume that $\omega_2 \in \alpha(\pi_* \mathcal{H}_F^{(0)})$ so that ω_2 is represented by $\hat{\omega}_2 \in q_* \Omega_{X/T}^{n+1}$.

Then applying the formula (3.10.5), one obtains the formula (4.8.1).

(4.9) DEFINITION. Let us expand K_F in a Laurent series in δ_1^{-1}

$$K_F(\omega_1, \omega_2) = \sum_k K^{(k)}(\omega_1, \omega_2)\delta_1^{-k} \tag{4.9.1}$$

so that we get an infinite sequence of $\mathcal{O}_T$-bilinear forms,

$$K^{(k)}: \mathbf{R}^{n+1}q_*(\Omega^{\cdot}) \times \mathbf{R}^{n+1}q_*(\Omega^{\cdot}) \to \mathcal{O}_T, \qquad k \in \mathbf{Z}. \tag{4.9.2}$$

(4.10) THEOREM. *$K^{(k)}$'s have the following properties*:

(i) *$K^{(k)}$ is symmetric for even k and skew-symmetric for odd k.*

(ii) $K^{(k+1)}(\omega_1, \omega_2) = K^{(k)}(\delta_1\omega_1, \omega_2) = -K^{(k)}(\omega_1, \delta_1\omega_2)$.

(iii) $\delta K^{(k)}(\omega_1, \omega_2) = K^{(k)}(\delta\omega_1, \omega_2) + K^{(k)}(\omega_1, \delta\omega_2)$ *for* $\delta \in \mathcal{G}$.

(iv) $K^{(k)}(t_1\omega_1, \omega_2) - K^{(k)}(\omega_1, t_1\omega_2) = (n+k)K^{(k-1)}(\omega_1, \omega_2)$.

(v) *$K^{(0)}$ induces the zero map on $\mathbf{R}^{n+1}q_*(F^{-1}\Omega^{\cdot}) \times \mathbf{R}^{n+1}q_*(F^0\Omega^{\cdot})$ (and hence on $\mathbf{R}^{n+1}q_*(F^0\Omega^{\cdot}) \times \mathbf{R}^{n+1}q_*(F^{-1}\Omega^{\cdot})$) so that the induced bilinear map on the module*

$$q_*\Omega_F \times q_*\Omega_F \left(\cong \mathrm{gr}^0 \mathbf{R}^{n+1}q_*(\Omega^{\cdot}) \times \mathrm{gr}^0 \mathbf{R}^{n+1}q_*(\Omega^{\cdot})\right)$$

coincides with J of (1.14.6).

REFERENCES

1. E. Brieskorn, *Die Monodromie der isolierten Singularitäten von Hyperflächen*, Manuscripta Math. **2** (1970), 103–160.

2. Y. Namikawa, *Higher residues associated with an isolated hypersurface singularity*, Symposia in Math., Kinokuniya, Tokyo and North-Holland, Amsterdam, 1981 (to appear).

3. K. Saito, *On the periods of primitive integrals*. I, Res. Inst. Math. Sci. Kyoto Univ. **412** (1982), preprint.

4. ______, *Primitive forms for a universal unfolding of a function with an isolated critical point*, J. Fac. Sci. Univ. Tokyo Sect. IA Math. **28** (1982), 775–792.

5. ______, *On a generalization of de-Rham lemma*, Ann. Inst. Fourier (Grenoble) **26** (1976), 165–170.

6. G. Scheja, *Fortsetzungssätze der Komplex analytischen Cohomologie und ihre algebraische Charakterisierung*, Math. Ann. **157** (1964), 75–94.

7. M. Sebastiani, *Preuve d'une conjecture de Brieskorn*, Manuscripta Math. **2** (1970), 301–308.

8. G. Trautman, *Ein Endlichkeitssatz in der analytischen Geometrie*, Invent. Math. **8** (1969), 143–174.

9. B. Malgrange, *Intégrales asymtotiques et monodromy*, Ann. Sci. École. Norm. Sup. **7** (1974).

KYOTO UNIVERSITY, JAPAN

Proceedings of Symposia in Pure Mathematics
Volume 40 (1983), Part 2

ON THE EXPONENTS AND THE GEOMETRIC GENUS OF AN ISOLATED HYPERSURFACE SINGULARITY

MORIHIKO SAITO

1. Introduction. Let $f \in \mathcal{O}_{\mathbf{C}^{n+1},0}$ be a germ of a holomorphic function such that $f(0) = 0$ and f has an isolated singularity at 0. Using the limit mixed Hodge structure J. H. M. Steenbrink defined some rational numbers associated with f (cf. [**St2**]), which we call the *exponents* of f (cf. Definition (3.2)).

The *geometric genus* p_g of the hypersurface $V := \{f = 0\}$ is defined via a resolution of singularity ρ: $\tilde{V} \to V$:

$$p_g := \dim_{\mathbf{C}}\left(R^{n-1}\rho_*\mathcal{O}_{\tilde{V}}\right)_0 \quad \text{for } n \geqslant 2,$$

$$p_g := \delta = \dim_{\mathbf{C}}\left(\rho_*\mathcal{O}_{\tilde{V}}/\mathcal{O}_V\right)_0 \quad \text{for } n = 1.$$

The main result of this paper is the following.

THEOREM 1. *The geometric genus p_g equals the number of the exponents not greater than* 1.

REMARKS. (1) By the definition of exponents this theorem is equivalent to the formula $p_g = \dim_{\mathbf{C}} \mathrm{Gr}_F^n H^n(X_\infty, \mathbf{C})$ (cf. §4).

(2) For the quasihomogeneous case, this theorem was proved by Kimio Watanabe (cf. [**W**, Theorem 1.13, p. 71]).

Combining the theorem with the result of Steenbrink [**St2**, Theorem (4.11)], we obtain the following.

COROLLARY 1. $2p_g = \mu_+ + \mu_0$ *for* $n = 2$, $2\delta = \mu + \mu_0$ *for* $n = 1$.

Here μ is the Milnor number of f and μ_+ (resp. μ_0) is the number of positive (resp. zero) eigenvalues of the intersection form (cf. [**St2**]).

1980 *Mathematics Subject Classification*. Primary 57R45.

The formula for $n = 1$ is the so-called Milnor relation because $\mu_0 + 1$ equals the number of the irreducible components of $(V, 0)$. The case $n = 2$ was proved by A. Durfee, I. Dolgachev and M. Reid (cf. [**Dur**]).

As an application of our argument, we obtain the following result.

Let V be a hypersurface of degree d in $\mathbf{P}^{n+1}$. We assume that $\Sigma(V)$, the singular locus of V, is discrete. Let $\rho: \tilde{V} \to V$ be a resolution. Put $h^{pq}(\tilde{V}) := \dim_{\mathbf{C}} H^q(\tilde{V}, \Omega^p_{\tilde{V}})$ and $p_g(\tilde{V}) := h^{n0}(\tilde{V})$. Let $p_g(V, x)$ be the geometric genus of $x \in \Sigma(V)$. Then we have

$$\sum_{x \in \Sigma(V)} p_g(V, x) = h^{n-1,0}(\tilde{V}) + \binom{d-1}{n+1} - p_g(\tilde{V}) \quad \text{for } n \geqslant 2.$$

Suppose V is rational. In particular, this is the case when there exists $x \in \Sigma(v)$ such that $\operatorname{mult}_x(V) = d - 1$. Then the above formula turns out to be

$$\sum p_g(V, x) = \binom{d-1}{n+1} \quad \text{for } n \geqslant 2.$$

We remark that the number on the right-hand side is the geometric genus of a nonsingular hypersurface of degree d in $\mathbf{P}^{n+1}$.

I thank K. Saito, I. Naruki, A. Fujiki, K. Watanabe and T. Yano for helpful discussions.

2. Limit mixed Hodge structure. We review the theory of the limit mixed Hodge structure of Deligne-Steenbrink (cf. [**St2**]).

(2.1) *Compactification.* Let $f \in \mathcal{O}_{\mathbf{C}^{n+1},0}$ be a holomorphic function with $f(0) = 0$. We assume that f has an isolated critical point at 0. Then we may assume that f is a polynomial and $Y_0 := \overline{\{x \in \mathbf{C}^{n+1}: f(x) = 0\}} \subset \mathbf{P}^{n+1}$ is nonsingular away from 0. Define

$$Y := \overline{\{(x, t) \in \mathbf{C}^{n+1} \times S: f(x) = t\}} \subset \mathbf{P}^{n+1} \times S,$$

and

$$Y_t := Y \cap (\mathbf{P}^{n+1} \times \{t\}) \quad \text{for } t \in S,$$

where $S := \{t \in \mathbf{C}: |t| < \eta\}$ and the number η is sufficiently small such that the projection $Y \to S$ is smooth away from 0.

(2.2) *Milnor's fibration.* Let $B := \{x \in \mathbf{C}^{n+1}: \|x\| < \varepsilon\}$ be an open ball with radius ε. If ε and η are sufficiently small, the projection

$$X := Y \cap (B \times S) \to S$$

is smooth away from 0 and its restriction over $S^* := S - \{0\}$ is a topological fibration and the topological type of a general fiber X_t does not depend on ε. If $i \neq 0, n$, $H_i(X_t, \mathbf{C}) = 0$ and $\dim_{\mathbf{C}} H_n(X_t, \mathbf{C})$ is called the Milnor number and is denoted by μ. Since $\bigcup_{t \in S^*} H^n(X_t, \mathbf{C})$ is a flat vector bundle, the monodromy transformation γ acts on $H^n(X_t, \mathbf{C})$, and we call it the local monodromy.

In the same way γ acts on $H^i(Y_t, \mathbf{C})$ and the restriction morphism $H^i(Y_t, \mathbf{C}) \to H^i(X_t, \mathbf{C})$ is equivariant with respect to γ.

(2.3) *Resolution*. Let $\rho\colon \mathcal{Y} \to Y$ be an embedded resolution of Y_0, i.e., ρ is a proper holomorphic map of a manifold $\mathcal{Y}$ to Y such that $\rho|_{\mathcal{Y}-\rho^{-1}(0)}\colon \mathcal{Y} - \rho^{-1}(0) \to Y - \{0\}$ is biholomorphic and $E := \rho^{-1}(Y_0)$ is a divisor with normal crossings in $\mathcal{Y}$. Let $E = \bigcup_{i\geq 0} E_i$ be the decomposition into irreducible components. We may assume that E_i is nonsingular and E_0 is the proper transform of Y_0. We put $m_i = \operatorname{ord}_{E_i}(f \circ \rho)$ and $m = \operatorname{LCM}(m_i)$.

(2.4) *Base changes*. Let S_∞ be a universal covering space of S^* and set $X_\infty := X \times_S S_\infty$ and $Y_\infty := Y \times_S S_\infty$.

Let $\pi'\colon S \ni t \to t^m \in S$ be an m-fold covering and $\tilde{\mathcal{Y}}$ be the normalization of $\mathcal{Y} \times_S S$. We remark that $\tilde{\mathcal{Y}}$ is a V-manifold and $D := \pi^{-1}(E)$ is a divisor with V-normal crossings in $\tilde{\mathcal{Y}}$, where π is the natural map $\tilde{\mathcal{Y}} \to \mathcal{Y}$ (cf. [**St2**]). We put $D_i := \pi^{-1}(E_i)$ and $C_i := D_0 \cap D_i$. We have $\pi|_{D_0}\colon D_0 \tilde{\to} E_0$ and $\pi|_{C_i}\colon C_i \tilde{\to} E_0 \cap E_i$, since $m_0 = 1$.

(2.5) *The exact sequence*. According to Deligne-Steenbrink [**St2**], we can put mixed Hodge structures on $\tilde{H}^i(Y_0)$, $H^i(Y_\infty)$ and $H^i(X_\infty)$ such that

$$(2.5.1)\qquad \cdots \to \tilde{H}^i(Y_0) \to H^i(Y_\infty) \to H^i(X_\infty) \to \tilde{H}^{i+1}(Y_0) \to \cdots$$

is an exact sequence of mixed Hodge structures, where $\tilde{H}^i$ is the reduced cohomology and the coefficient field is assumed to be $\mathbf{C}$ unless explicitly specified.

(2.6) *Spectral sequences*. We have the following spectral sequences with respect to the weight and Hodge filtrations (cf. [**St1, St2**]):

$$({}_W\mathrm{I})\qquad {}_WE_1^{pq} = H^q(\tilde{C}^{(p)}) \Rightarrow \tilde{H}^{p+q}(Y_0), \quad \text{s.t. } {}_WE_2^{pq} \cong \operatorname{Gr}_q^W \tilde{H}^{p+q}(Y_0),$$

$$({}_W\mathrm{II})\qquad {}_WE_1^{-r,q+r} = \bigoplus_{k\geq 0,-r} H^{q-r-2k}(\tilde{D}^{(2k+r+1)})(-r-k) \Rightarrow H^q(Y_\infty),$$
$$\text{s.t. } {}_WE_2^{-r,q+r} \cong \operatorname{Gr}_{q+r}^W H^q(Y_\infty),$$

$$({}_F\mathrm{II})\qquad {}_FE_1^{pq} = H^q\big(D, \tilde{\Omega}^p_{\tilde{\mathcal{Y}}/S}(\log D) \otimes_{\mathcal{O}_{\tilde{\mathcal{Y}}}} \mathcal{O}_D\big) \Rightarrow H^{p+q}(Y_\infty),$$
$$\text{s.t. } {}_FE_1^{pq} \cong \operatorname{Gr}_F^p H^{p+q}(Y_\infty),$$

$$({}_W\mathrm{II}')\qquad {}_WE_2^{-r,q+r} = \bigoplus_{k\geq 0,-r} H^{q-r-2k}(\tilde{E}^{(2k+r+1)})(-r-k) \Rightarrow H^q(Y_\infty)_1,$$
$$\text{s.t. } {}_WE_2^{-r,q+r} \cong \operatorname{Gr}_{q+r}^W H^q(Y_\infty)_1,$$

where we define $\tilde{C}^{(p)} := \coprod_{i_1<\cdots<i_p} C_{i_1} \cap \cdots \cap C_{i_p}$ (and $\tilde{D}^{(p)}$, $\tilde{E}^{(p)}$ similarly) and $H^q(Y_\infty)_1 := \{u \in H^q(Y_\infty)\colon (\gamma - 1)^k u = 0, \exists k \in \mathbf{Z}_+\}$.

We remark that $({}_W\mathrm{II}')$ is a direct factor of $({}_W\mathrm{II})$ on which γ acts unipotently, that $\operatorname{Im}(\tilde{H}^i(Y_0) \to H^i(Y_\infty))$ is contained in $H^i(Y_\infty)_1$ and it induces a morphism of $({}_W\mathrm{I})$ to $({}_W\mathrm{II}')$.

3. Exponents.

(3.1) For $\lambda \in \mathbf{C}$ we put

$$H^i(X_\infty)_\lambda := \{u \in H^i(X_\infty)\colon (\gamma - \lambda)^k u = 0, \exists k \in \mathbf{Z}_+\},$$
$$H_\lambda^{pq} := \operatorname{Gr}_F^p \operatorname{Gr}_{p+q}^W H^n(X_\infty)_\lambda, \quad h_\lambda^{pq} := \dim_{\mathbf{C}} H_\lambda^{pq}.$$

Since the direct decomposition $H^n(X_\infty) = \oplus_\lambda H^n(X_\infty)_\lambda$ is compatible with both filtrations, we have $\Sigma_{\lambda,p,q} h_\lambda^{pq} = \mu$ $(= \dim_{\mathbf{C}} H^n(X_\infty))$. Due to the monodromy theorem, if $h_\lambda^{pq} \neq 0$, λ is a root of unity.

(3.2) DEFINITION. We define μ rational numbers $\{\alpha_1,\ldots,\alpha_\mu\}$ as follows, and we call them the *exponents* of f.

(a) $\alpha_1 \leqslant \alpha_2 \leqslant \cdots \leqslant \alpha_\mu$,

(b)

$$\forall\lambda \in \mathbf{C}, \forall p \in \mathbf{Z}, \lambda \neq 1 \Rightarrow \#\left\{j\colon \exp 2\pi\sqrt{-1}\,\alpha_j = \lambda^{-1}, [\alpha_j] = n-p\right\} = \sum_q h_\lambda^{pq},$$

$$\lambda = 1 \Rightarrow \#\{j\colon \alpha_j = n-p+1\} = \sum_q h_1^{pq},$$

where $[\alpha_j]$ is the Gauss symbol, i.e., $[\alpha_j] = \max\{k \in \mathbf{Z}\colon k \leqslant \alpha_j\}$. This is well defined because of $\Sigma_{\lambda,p,q} h_\lambda^{pq} = \mu$.

(3.3) PROPOSITION (DUALITY). $\alpha_i + \alpha_{\mu+1-i} = n+1$.

PROOF. Set $N := \log \gamma_u$, where γ_u is the unipotent part of the local monodromy γ. Due to Steenbrink [**St2**], we have

$$N^{p+q-n}\colon H_\lambda^{pq} \xrightarrow{\sim} H_\lambda^{n-q,n-p} \quad \text{for } \lambda \neq 1,$$

and

$$N^{p+q-n-1}\colon H_1^{pq} \xrightarrow{\sim} H_1^{n+1-q,n+1-p}.$$

Since γ is defined over $\mathbf{Z}$, we have $\overline{H_\lambda^{pq}} = H_{\bar\lambda}^{qp}$, which gives the desired result. Q.E.D.

4. Proof of Theorem 1. By the definition of the exponents, it suffices to prove the following.

(4.1) PROPOSITION. $p_g = \dim_{\mathbf{C}} \mathrm{Gr}_F^n H^n(X_\infty)$, *for* $n \geqslant 1$.

First we reduce (4.1) to the following proposition.

(4.2) PROPOSITION.

$$\dim_{\mathbf{C}} \mathrm{Im}\left(H^{n-1}(D_0, \mathcal{O}_{D_0}) \to \bigoplus_i H^{n-1}(C_i, \mathcal{O}_{C_i})\right)$$
$$= \dim_{\mathbf{C}} \mathrm{Coker}(\mathrm{Gr}_F^n H^n(Y_\infty) \to \mathrm{Gr}_F^n H^n(X_\infty)) + \delta_{1,n} \quad \textit{for } n \geqslant 1,$$

where $\delta_{1,n}$ *is the Kronecker* δ.

We further reduce (4.2) to the following and prove (4.3).

(4.3) PROPOSITION. *Let* $\psi_{22}\colon \mathrm{Gr}_F^{n-1} H^{n-1}(\tilde E^{(2)} - \tilde C^{(1)}) \to \mathrm{Gr}_F^n H^{n+1}(\tilde E^{(1)} - E_0)$ *be the Gysin morphism* (*cf. Lemma* (4.5.4)). *Then we have*

$$\dim_{\mathbf{C}} \mathrm{Coker}\, \psi_{22} = \delta_{1,n} \quad \textit{for } n \geqslant 1.$$

(4.4) Proposition (4.2) $\Rightarrow$ Proposition (4.1).

First we describe the geometric genus p_g as the difference of the Euler characteristics, i.e.,

(4.4.1) $$\chi(D_0, \mathcal{O}_{D_0}) = \chi(D, \mathcal{O}_D) + (-1)^{n-1} p_g.$$

PROOF. By the invariance of Euler characteristics under a flat deformation, we have

$$\chi(Y_0, \mathcal{O}_{Y_0}) = \chi(Y_t, \mathcal{O}_{Y_t}) = \chi(D, \mathcal{O}_D).$$

Therefore it remains to show that

$$\chi(D_0, \mathcal{O}_{D_0}) = \chi(Y_0, \mathcal{O}_{Y_0}) + (-1)^{n-1} p_g.$$

When $n \geqslant 2$, $\rho_* \mathcal{O}_{D_0} = \mathcal{O}_{Y_0}$ holds since Y_0 is normal. Using the Leray spectral sequence we obtain: $\chi(D_0, \mathcal{O}_{D_0}) = \bar{\chi}(Y_0, \mathcal{O}_{Y_0}) + \Sigma_{i \geqslant 1}(-1)^i \dim_{\mathbf{C}}(R^i\rho_* \mathcal{O}_{D_0})_0$, and $R^i\rho_*\mathcal{O}_{D_0} = 0$ for $0 < i < n-1$ since Y_0 is Cohen-Macaulay.

The case $n = 1$ follows from

$$\begin{aligned} \chi(D_0, \mathcal{O}_{D_0}) &= \chi(Y_0, \rho_*\mathcal{O}_{D_0}) = \chi(Y_0, \mathcal{O}_{Y_0}) + \chi(Y_0, \rho_*\mathcal{O}_{D_0}/\mathcal{O}_{Y_0}) \\ &= \chi(Y_0, \mathcal{O}_{Y_0}) + \delta. \quad \text{Q.E.D.} \end{aligned}$$

From the exact sequence (2.5.1) and the vanishing of $H^i(X_\infty)$ for $i \neq 0, n$, we obtain

(4.4.2) $$\dim_{\mathbf{C}} \mathrm{Gr}_F^p \tilde{H}^p(Y_0) = \dim_{\mathbf{C}} \mathrm{Gr}_F^p H^p(Y_\infty) \quad \text{for } 0 < p < n.$$

We calculate both sides of (4.4.2) using the spectral sequences in (2.6). First we have the following for $\forall p \in \mathbf{Z}$ by $({}_W\mathrm{I})$.

(4.4.3) $$\begin{aligned} \dim_{\mathbf{C}} \mathrm{Gr}_F^p \tilde{H}^p(Y_0) &= \dim_{\mathbf{C}} \mathrm{Gr}_F^p \mathrm{Gr}_p^W \tilde{H}^p(Y_0) \\ &= \dim_{\mathbf{C}} \mathrm{Ker}\big(\mathrm{Gr}_F^p H^p(\tilde{C}^{(0)}) \to \mathrm{Gr}_F^p H^p(\tilde{C}^{(1)})\big) \\ &= \dim_{\mathbf{C}} \mathrm{Ker}\Big(H^p(D_0, \mathcal{O}_{D_0}) \to \bigoplus_i H^p(C_i, \mathcal{O}_{C_i})\Big). \end{aligned}$$

Secondly, we have by $({}_F\mathrm{II})$

(4.4.4) $$\dim_{\mathbf{C}} \mathrm{Gr}_F^p H^p(Y_\infty) = \dim_{\mathbf{C}} \mathrm{Gr}_F^0 H^p(Y_\infty) = \dim_{\mathbf{C}} H^p(D, \mathcal{O}_D) \quad \forall p.$$

As $R^p\rho_*\mathcal{O}_{D_0} = 0$ for $0 < p < n-1$, we obtain

$$\rho^*\colon H^p(Y_0, \mathcal{O}_{Y_0}) \xrightarrow{\sim} H^p(D_0, \mathcal{O}_{D_0}) \quad \text{for } p \leqslant n-2.$$

Hence the restriction morphism $H^p(D_0, \mathcal{O}_{D_0}) \to \oplus_i H_p(C_i, \mathcal{O}_{C_i})$ is the zero map, and (4.4.3) turns out to be the following for $p \neq 0, n-1$.

(4.4.5) $$\dim_{\mathbf{C}} \mathrm{Gr}_F^p \tilde{H}^p(Y_0) = \dim_{\mathbf{C}} H^p(D_0, \mathcal{O}_{D_0}) \quad \text{for } p \neq 0, n-1.$$

Substituting (4.4.3)–(4.4.5) into (4.4.2) we have

(4.4.6) $$\dim_{\mathbf{C}} H^p(D, \mathcal{O}_D) = \dim_{\mathbf{C}} H^p(D_0, \mathcal{O}_{D_0}) \quad \text{for } p < n-1,$$

(4.4.7)

$$\dim_{\mathbf{C}} H^{n-1}(D, \mathcal{O}_D) = \dim_{\mathbf{C}} \operatorname{Ker}\Big(H^{n-1}(D_0, \mathcal{O}_{D_0}) \to \bigoplus_i H^{n-1}(C_i, \mathcal{O}_{C_i})\Big) + \delta_{1,n}.$$

Substituting these into (4.4.1), we obtain

$$\begin{aligned} p_g &= \left(\dim_{\mathbf{C}} H^{n-1}(D_0, \mathcal{O}_{D_0}) - \dim_{\mathbf{C}} H^{n-1}(D, \mathcal{O}_D)\right) \\ &\quad + \left(\dim_{\mathbf{C}} H^{n}(D, \mathcal{O}_D) - \dim_{\mathbf{C}} H^{n}(D_0, \mathcal{O}_{D_0})\right) \\ &= \dim_{\mathbf{C}} \operatorname{Im}\left(H^{n-1}(D_0, \mathcal{O}_{D_0}) \to \oplus H^{n-1}(C_i, \mathcal{O}_{C_i})\right) - \delta_{1,n} \\ &\quad + \dim_{\mathbf{C}} \operatorname{Coker}\left(\operatorname{Gr}_F^n \tilde{H}^n(Y_0) \hookrightarrow \operatorname{Gr}_F^n H^n(Y_\infty)\right) \\ &= \dim_{\mathbf{C}} \operatorname{Gr}_F^n H^n(X_\infty). \end{aligned}$$

The last equality follows from Proposition (4.2) and (2.5.1). Q.E.D.

(4.5) Proposition (4.3) $\Rightarrow$ Proposition (4.2).

From the exact sequence (2.5.1), we have

(4.5.1)
$$\begin{aligned} &\dim_{\mathbf{C}} \operatorname{Coker}(\operatorname{Gr}_F^n H^n(Y_\infty) \to \operatorname{Gr}_F^n H^n(X_\infty)) \\ &\qquad = \dim_{\mathbf{C}} \operatorname{Ker}\left(\operatorname{Gr}_F^n \tilde{H}^{n+1}(Y_0) \to \operatorname{Gr}_F^n H^{n+1}(Y_\infty)_1\right). \end{aligned}$$

Here we used the remark in (2.6).

Using the spectral sequences $({}_W\mathrm{I})$ and $({}_W\mathrm{II}')$ in (2.6), we obtain

(4.5.2)
$$\operatorname{Gr}_F^n \tilde{H}^{n+1}(Y_0) \cong \operatorname{Gr}_F^n H^{n+1}(D_0),$$

(4.5.3)
$$\begin{aligned} \operatorname{Gr}_F^n H^{n+1}(Y_\infty)_1 &= \operatorname{Gr}_F^n \operatorname{Gr}_{n+1}^W H^{n+1}(Y_\infty)_1 \\ &\cong \operatorname{Coker}\left(\operatorname{Gr}_F^n\left(H^{n-1}(\tilde{E}^{(2)})(-1)\right) \to \operatorname{Gr}_F^n H^{n+1}(\tilde{E}^{(1)})\right). \end{aligned}$$

We can verify that the morphism in the last expression is the Gysin morphism. We set

$$\begin{aligned} A_1 &:= \operatorname{Gr}_F^{n-1} H^{n-1}(\tilde{C}^{(1)}), \quad B_1 := \operatorname{Gr}_F^n H^{n+1}(\tilde{C}^{(0)}), \\ A_2 &:= \operatorname{Gr}_F^{n-1} H^{n-1}(\tilde{E}^{(2)} - \tilde{C}^{(1)}), \quad B_2 := \operatorname{Gr}_F^n H^{n+1}(\tilde{E}^{(1)} - \tilde{C}^{(0)}), \end{aligned}$$

and apply the following lemma to obtain (4.2).

(4.5.4) LEMMA. *Let*

$$\psi = \begin{pmatrix} \psi_{11} & \psi_{12} \\ \psi_{21} & \psi_{22} \end{pmatrix} : A_1 \oplus A_2 \to B_1 \oplus B_2$$

be a linear map of finite dimensional linear spaces, satisfying

(a) $\psi(A_2) \subset B_2$, *i.e.*, $\psi_{12} = 0$.

(b) $(\psi_{21}, \psi_{22})\colon A_1 \oplus A_2 \to B_2$ *is surjective.*

(c) $\operatorname{Ker}(\psi_{11}\colon A_1 \to B_1) \subset \operatorname{Ker}(\psi_{21}\colon A_1 \to \operatorname{Coker} \psi_{22})$.

Then we have

$$\dim \operatorname{Ker}(B_1 \to \operatorname{Coker} \psi) = \dim \operatorname{Im} \psi_{11} - \dim \operatorname{Coker} \psi_{22}.$$

We remark that the left-hand side of the last expression equals the right-hand side of (4.5.1) because of (4.5.2) and (4.5.3).

PROOF OF LEMMA (4.5.4). Since we have

$$\begin{aligned}\operatorname{Ker}(B_1 \to \operatorname{Coker}\psi) &= \psi_{11}(\{a_1 \in A_1 : \exists a_2 \in A_2, \text{ s.t., } \psi_{21}(a_1) = \psi_{22}(a_2)\}) \\ &= \psi_{11}(\psi_{21}^{-1}(\psi_{22}(A_2))) \\ &= \psi_{11}(\operatorname{Ker}(\psi_{21}: A_1 \to \operatorname{Coker}\psi_{22})),\end{aligned}$$

we can replace A_1, A_2 and B_2 by $\operatorname{Coim}\psi_{11}$, 0 and $\operatorname{Coker}\psi_{22}$ respectively. Then we have that $\psi_{11}: A_1 \to B_1$ is injective, and $\psi_{21}: A_1 \to B_2$ is surjective, and

$$\begin{aligned}\dim \operatorname{Ker}(B_1 \to \operatorname{Coker}\psi) &= \dim \operatorname{Ker}\psi_{21} = \dim A_1 - \dim B_2 \\ &= \dim \operatorname{Im}\psi_{11} - \dim \operatorname{Coker}\psi_{11}. \quad \text{Q.E.D.}\end{aligned}$$

First we consider the case $n \geqslant 2$. Due to Proposition (4.3), ψ_{22} is surjective. Hence all the hypotheses of Lemma (4.5.4) are satisfied, and we have

$$\dim_{\mathbf{C}} \operatorname{Ker}(B_1 \to \operatorname{Coker}\psi) = \dim_{\mathbf{C}} \operatorname{Im}\psi_{11}.$$

Since the Gysin morphism and the restriction morphism are dual to each other by the Serre duality, we have

$$\dim_{\mathbf{C}} \operatorname{Im}\psi_{11} = \dim_{\mathbf{C}} \operatorname{Im}(\operatorname{Gr}_F^0 H^{n-1}(\tilde{C}^{(0)}) \to \operatorname{Gr}_F^0 H^{n-1}(\tilde{C}^{(1)})),$$

which gives the desired result for $n \geqslant 2$.

The case $n = 1$ can be proved similarly. Q.E.D.

(4.6) PROOF OF PROPOSITION (4.3). We can calculate the mixed Hodge structure of $\mathbf{P}^{n+1} - \{0\}$ by using $\{E_i\}_{i>0}$, since $\bigcup_{i>0} E_i$ blows down to the nonsingular point 0 in Y (cf. [**D**]). In particular, we have the spectral sequence,

$$\begin{aligned}{}_W E_1^{-r,q+r} = H^{q-r}(\tilde{E}^{(r)} - \tilde{C}^{(r-1)})(-r) &\Rightarrow H^q(\mathbf{P}^{n+1} - \{0\}), \\ \text{s.t. } {}_W E_2^{-r,q+r} &= \operatorname{Gr}^W_{q+r} H^q(\mathbf{P}^{n+1} - \{0\}).\end{aligned}$$

Here we define the projective manifold $\tilde{E}^{(0)}$ to be $(\mathbf{P}^{n+1} - \{0\}) \cup (\bigcup_{i>0} E_i)$. Then we have

$$\begin{aligned}H(\operatorname{Gr}_F^{n+1}(H^{n-1}(\tilde{E}^{(2)} - \tilde{C}^{(1)})(-2)) &\to \operatorname{Gr}_F^{n+1}(H^{n+1}(\tilde{E}^{(1)} - E_0)(-1)) \\ &\to \operatorname{Gr}_F^{n+1} H^{n+3}(\tilde{E}^{(0)})) \\ &= \operatorname{Gr}_F^{n+1} \operatorname{Gr}^W_{n+3} H^{n+2}(\mathbf{P}^{n+1} - \{0\}) = 0.\end{aligned}$$

The last equality follows from the pureness of the Hodge structure on $H^{n+2}(\mathbf{P}^{n+1} - \{0\})$.

Since $\tilde{E}^{(0)}$ is rational, we have

$$\dim_{\mathbf{C}} \operatorname{Gr}_F^{n+1} H^{n+3}(\tilde{E}^{(0)}) = \dim_{\mathbf{C}} \operatorname{Gr}_F^0 H^{n-1}(\tilde{E}^{(0)}) = 0, \quad \text{if } n \geqslant 2.$$

Here the first equality comes from the Serre duality. Then we have the desired result for $n \geqslant 2$.

The case $n = 1$ can be proved similarly. Q.E.D.

REFERENCES

[D] P. Deligne, *Théorie de Hodge*. II, Inst. Hautes Études Sci. Publ. Math. **40** (1971), 5–58.

[St1] J. H. M. Steenbrink, *Limits of Hodge structure*, Invent. Math. **31** (1976), 229–257.

[St2] ______, *Mixed Hodge structure on the vanishing cohomology*, Real and Complex Singularities, Oslo, 1976, Proc. Nordic Summer School.

[Dur] A. H. Durfee, *The signature of smoothings of complex surface singularities*, Math. Ann. **232** (1978), 85–98.

[W] K. Watanabe, *On plurigenera of normal isolated singularities*. I, Math. Ann. **250** (1980), 65–94.

KYOTO UNIVERSITY, JAPAN

Current address: Institute Fourier, Université de Grenoble I, B. P. 116, 384 Saint-Martin-d'Hères, Cedex, France

Proceedings of Symposia in Pure Mathematics
Volume **40** (1983), Part 2

A NOTE ON TWO LOCAL HODGE FILTRATIONS

JOHN SCHERK

In [**2**], J. H. M. Steenbrink constructed a mixed Hodge structure on the cohomology of the generic fibre of the Milnor fibration of an isolated hypersurface singularity. Recently, A. N. Varchenko introduced the "asymptotic Hodge filtration" which, together with Steenbrink's weight filtration, also gives a mixed Hodge structure on $H^n(X_\infty)$. In [**3**] Varchenko proved that the two agree for curve singularities and quasihomogeneous singularities. In the general case, he showed that the two Hodge filtrations agree on $\mathrm{Gr}^W \cdot H^n(X_\infty, C)$ (cf. [**4**]). The purpose of this note is to give a simple example of a function for which the two filtrations are not the same.

Suppose $f = f(x_0,\ldots,x_n)$ is an analytic function defined in some open ball B about 0 in C^{n+1}. Assume that $f(0) = 0$, and that 0 is the only critical point of f. As is well known, one can find an open neighbourhood X of 0 in C^{n+1}, and a disc T about 0 in C such that $f(X) \subset T$ and

$$f\colon X - f^{-1}(0) \to T - \{0\}$$

is a C^∞ fibre bundle. Such neighbourhoods X form a neighbourhood basis of 0. Let $T' = T - \{0\}$, $X' = X - f^{-1}(0)$.

Pick a path in T' which travels once about 0 in a counter-clockwise direction as a generator of $\pi_1(T', t)$ $(t \neq 0)$. Then this determines an automorphism σ of $H^n(X_t, C)$, the monodromy of f. For different values of t, the corresponding automorphisms will be conjugate. Write $\sigma = \sigma_s\sigma_u$ where σ_s is semisimple and σ_u is unipotent.

$H^n = \bigcup_{t\in T'} H^n(X_t, C)$ is a flat vector bundle with a canonical connection ∂_t. Let $\mathcal{O}_{T'}(H^n)$ denote the sheaf of germs of holomorphic sections of H^n. $\mathcal{O}_{T'}(H^n)$

1980 *Mathematics Subject Classification*. Primary 55R55.

has a unique extension $\mathcal{S}$ over T such that

(i) $\mathcal{S}$ is a free $\mathcal{O}_T$ module,

(ii) ∂_t has a simple pole on $\mathcal{S}$,

(iii) the eigenvalues of $\mathrm{res}_0(\partial_t)$ lie in $(-1,0]$.

As in [**2**, 2.12], one can extend σ to an automorphism of $\mathcal{S}_0/t\mathcal{S}_0$. Then

$$\sigma = \exp(-2\pi i\, \mathrm{res}_0(\partial_t)).$$

We now briefly recall the definitions of Steenbrink's Hodge filtration $F^{\cdot}$ as explained in [**1** and **6**], and the asymptotic Hodge filtration $F_a^{\cdot}$. For details, we refer the reader to [**1**, **4** and **5**].

Let $\mathcal{H}'' = f_*\Omega_X^{n+1}/dt \wedge df_*\Omega_X^{n-1}$. $\mathcal{H}''$ is also a free extension of $\mathcal{O}_{T'}(H^n)$. Define a filtration $F_h^{\cdot}$ on $\mathcal{S}_0/t\mathcal{S}_0$ by

$$F_h^k(\mathcal{S}_0/t\mathcal{S}_0) = \left[\left(\mathcal{S}_0 \cap \partial_t^{n-k}\mathcal{H}_0''\right) + t\mathcal{S}_0\right]/t\mathcal{S}_0, \qquad 0 \leqslant k \leqslant n.$$

The subspace $F_h^h(\mathcal{S}_0/t\mathcal{S}_0)$ may not be invariant under σ_s (cf [**7**]). So let $\alpha_1 \leqslant \cdots \leqslant \alpha_r \in (-1,0]$ be the eigenvalues of $\mathrm{res}_0(\partial_t)$. Set $\lambda_j = \exp(-2\pi i\alpha_j)$ and $(\mathcal{S}_0/t\mathcal{S}_0)_j = \ker(\sigma_s - \lambda_j)$, $1 \leqslant j \leqslant r$. Define a filtration $V^{\cdot}$ on $\mathcal{S}_0/t\mathcal{S}_0$ by

$$V^k(\mathcal{S}_0/t\mathcal{S}_0) = \bigoplus_{j \geqslant k} (\mathcal{S}_0/t\mathcal{S}_0)_j.$$

Thus $\mathrm{Gr}_v(\mathcal{S}_0/t\mathcal{S}_0)$ is naturally isomorphic to $\mathcal{S}_0/t\mathcal{S}_0$. Now $F_h^{\cdot}$ induces a filtration $\mathrm{Gr}_v F_h^{\cdot}$ on $\mathrm{Gr}_v(\mathcal{S}_0/t\mathcal{S}_0)$. The corresponding filtration $F^{\cdot}$ on $\mathcal{S}_0/t\mathcal{S}_0$ is Steenbrink's Hodge filtration. If $F_h^{\cdot}$ is invariant under σ_s (as will be the case in the example below), then $F^{\cdot} = F_h^{\cdot}$.

$F_a^{\cdot}$ is a priori defined on H^n (or $\mathcal{O}_{T'}(H^n)$). For any $\omega \in \mathcal{S}_0[t^{-1}]$, we have an expansion

$$\omega = \sum_{\alpha,q} A_{\alpha,q}^{\omega} t^{\alpha}(\log t)^q/q!$$

where $A_{\alpha,q}^{\omega}$ is a locally constant multivalued section of H^n, for $A_{\alpha,q}^{\omega} \neq 0 \cdot e^{-2\pi i\alpha}$ is an eigenvalue of σ and $\alpha \in Q$, $0 \leqslant q \leqslant n$, and $\{\alpha \mid A_{\alpha,q}^{\omega} \neq 0$ for some $q\}$ is bounded below. The series converges absolutely and uniformly on any sufficiently small angular sector of T'. Set

$$\nu(\omega) = \min\{\alpha \mid A_{\alpha,q}^{\omega} \neq 0 \text{ for some } q\}.$$

Define the leading term ω_l of any such ω by

$$\omega_l = \sum_q A_{\nu(\omega),q}^{\omega} t^{\nu(\omega)}(\log t)^q/q!.$$

Then $\omega_l \in \mathcal{S}_0[t^{-1}]$. $\{\omega_l \mid \omega \in \mathcal{H}_0''\}$ generates $\mathcal{S}_0[t^{-1}]$ over $\mathcal{O}_{T,0}[t^{-1}]$. Varchenko defines subbundles $F_a^kH^n$, $0 \leqslant k \leqslant n$, by $F_a^kH^n =$ subbundle generated by sections ω_l, $\omega \in H^0(T, \mathcal{H}'')$, with $\nu(\omega) \leqslant n-k$. Since $\sigma_s\omega_l = e^{-2\pi i\nu(\omega)}\omega_l$, F_a^k is invariant under σ_s.

Let U denote the upper half-plane, and let $\rho = U \to T'$ be a universal covering of T'. Set $X_\infty = X'x_{T'}U$. Then there is a canonical isomorphism $\rho^*H^n \cong U \times H^n(X_\infty, C)$. Thus $\rho^*F_a^{\cdot}$ defines a holomorphic mapping from U into a flag manifold of $H^n(X_\infty, C)$. The limit of this mapping (in the sense of Schmid: cf. [**2**, §2, Appendix]) as $\operatorname{im} z \to \infty$, $z \in U$, exists. The limit filtration $F_a^{\cdot}$ on $H^n(X_\infty, C)$ is the Hodge filtration of a mixed Hodge structure.

There is a natural isomorphism

$$\Phi: \mathcal{S}_0/t\mathcal{S}_0 \overset{\cong}{\to} H^n(X_\infty, C).$$

By definition $F^{\cdot}H^n(X_\infty, C) = F^{\cdot}(\mathcal{S}_0/t\mathcal{S}_0)$.

Notice that $H^n(X_\infty, C)$ is naturally isomorphic to the space of locally constant multivalued sections of H^n.

Our example is $f(w, x, y, z) = w^2 + x^5 + y^5 + z^5 + xyz$. A basis of $C\{w, x, y, z\}/(f_w, f_x, f_y, f_z)$ is

$$\{1, x, x^2, x^3, x^4, y, y^2, y^3, y^4, z, z^2, z^3, z^4, xyz\}.$$

Thus $\mu = 14$. A basis of $\tilde{\mathcal{H}}_0''$ (the smallest submodule of $\mathcal{S}_0$ containing $\mathcal{H}_0''$, on which ∂_t has a simple pole) consists of the classes of

$$\{\omega, x\omega, x^2\omega, x^3\omega, x^4\omega, y\omega, y^2\omega, y^3\omega, y^4\omega, z\omega, z^2\omega, z^3\omega, z^4\omega, \partial_t xyz\omega\}.$$

Here $\omega = dw \wedge dx \wedge dy \wedge dz$. That $\partial_t xyz\omega \in \tilde{\mathcal{H}}_0''$ follows from the equation

$$t\partial_t\omega = \frac{\omega}{10} + \frac{2}{5}\partial_t xyz\omega.$$

The matrix of the residue of ∂_t with respect to this basis is

	ω	$\partial_t xyz\omega$	$x\omega$	$x^2\omega$	$x^3\omega$	$x^4\omega$	$y\omega$	$y^2\omega$	$y^3\omega$	$y^4\omega$	$z\omega$	$z^2\omega$	$z^3\omega$	$z^4\omega$
ω	$\frac{1}{10}$	$\frac{-2}{5}$												
$\partial_t xyz\omega$	$\frac{2}{5}$	$\frac{9}{10}$												
$x\omega$			$\frac{7}{10}$											
$x^2\omega$				$\frac{9}{10}$										
$x^3\omega$					$\frac{11}{10}$									
$x^4\omega$						$\frac{13}{10}$								
$y\omega$							$\frac{7}{10}$							
$y^2\omega$								$\frac{9}{10}$						
$y^3\omega$									$\frac{11}{10}$					
$y^4\omega$										$\frac{13}{10}$				
$z\omega$											$\frac{7}{10}$			
$z^2\omega$												$\frac{9}{10}$		
$z^3\omega$													$\frac{11}{10}$	
$z^4\omega$														$\frac{13}{10}$

The characteristic polynomial of the monodromy σ is

$$(s^{10} - 1)^3(s - 1)/(s^5 - 1)^3(s^2 - 1).$$

The Jordan normal form of σ has one 2×2 block, with eigenvalue -1. In $\tilde{\mathcal{H}}_0''/t\tilde{\mathcal{H}}_0''$, the invariant subspace belonging to the eigenvalue $\frac{1}{2}$ of $\operatorname{res}(\partial_t)$ is spanned by the classes of ω and $\partial_t xyz\omega$. Put

(1) $$v = \omega - \partial_t xyz\omega, \qquad u = \omega - \tfrac{7}{2}\partial_t xyz\omega.$$

Then

(2) $$t\partial_t v = v/2, \qquad t\partial_t u = v + u/2$$

in $\tilde{\mathcal{H}}_0''/t\tilde{\mathcal{H}}_0''$. Thus

(3) $$v_l = Bt^{1/2}, \qquad u_l = At^{1/2} + Bt^{1/2}\log t$$

where A and B are locally constant multivalued sections of H^n. So u_l and v_l are linearly independent sections of H^n which span the subbundle $H^n_{-1} = \ker(\sigma_s + 1)$. From (1) and (3), it follows that

(4) $$\begin{aligned} \omega_l &= \tfrac{7}{5}v_l - \tfrac{2}{5}u_l = (-\tfrac{2}{5}A + \tfrac{7}{5}B)t^{1/2} - \tfrac{2}{5}Bt^{1/2}\log t, \\ (\partial_t xyz\omega)_l &= \tfrac{2}{5}v_l - \tfrac{2}{5}u_l = (-\tfrac{2}{5}A + \tfrac{2}{5}B)t^{1/2} - \tfrac{2}{5}Bt^{1/2}\log t. \end{aligned}$$

Since 1 is not an eigenvalue of σ, (4) implies that

$$(xyz\omega)_l = (-\tfrac{4}{15}A + \tfrac{4}{9}B)t^{3/2} - \tfrac{4}{15}Bt^{3/2}\log t.$$

Thus ω_l and $(xyz\omega)_l$ also generate the subbundle H^n_{-1}. It follows then that $F_a^1H^n_{-1} = H^n_{-1}$, and that $F_a^2H^n_{-1}$ is generated by ω_l.

If we use (3) and "take limits", $u_l \to A$, $v_l \to B$. Equivalently

$$t^{-1/2}\exp(N\log t/2\pi i)u_l = A, \qquad t^{-1/2}\exp(N\log t/2\pi i)v_l = B,$$

where $N = \log\sigma_u$ (cf. [**2**, §2, Appendix]). Thus A and B span $H^n(X_\infty, C)_{-1} = \ker(\sigma_s + 1) \subset H^n(X_\infty, C)$. So $F_a^1H^n(X_\infty, C)_{-1} = H^n(X_\infty, C)_{-1}$, and $-\tfrac{2}{5}A + \tfrac{7}{5}B$ generates $F_a^2H^n(X_\infty, C)_{-1}$ (cf. (4)).

From (2) it follows that $\partial_t u, \partial_t v \in \mathcal{S}_0$, and therefore $\partial_t\omega, \partial_t^2 xyz\omega \in \mathcal{S}_0$. Furthermore, they represent a basis of $(\mathcal{S}_0/t\mathcal{S}_0)_{-1} = \ker(\sigma_s + 1) \subset \mathcal{S}_0/t\mathcal{S}_0$. By [**1**] then, $F^1(\mathcal{S}_0/t\mathcal{S}_0)_{-1} = (\mathcal{S}_0/t\mathcal{S}_0)_{-1}$ and the class of $\partial_t\omega$ generates $F^2(\mathcal{S}_0/t\mathcal{S}_0)_{-1}$. From (4) we have

(5) $$\begin{aligned} \partial_t\omega &= \tfrac{1}{2}(-\tfrac{2}{5}A + \tfrac{7}{5}B)t^{-1/2} - \tfrac{1}{5}Bt^{-1/2}\log t - \tfrac{2}{5}Bt^{-1/2} \pmod{t\mathcal{S}_0} \\ &= \tfrac{1}{2}t^{-1}\omega_l - \tfrac{2}{5}Bt^{-1/2} \pmod{t\mathcal{S}_0}. \end{aligned}$$

Therefore

$$\Phi(\partial_t\omega) = t^{1/2}\exp(N\log t/2\pi i)\partial_t\omega = -\tfrac{1}{5}A + \tfrac{3}{10}B.$$

So $-\tfrac{1}{5}A + \tfrac{3}{10}B$ generates $F^2H^n(X_\infty, C)_{-1}$, and $F_a^2 \neq F^2$.

In $H^n(X_\infty, C)_{-1}$, we have $N^2 = 0, (-N/2\pi i)A = B$. Therefore $W_4H^n(X_\infty)_{-1} = H^n(X_\infty)_{-1}$ and $W_2H^n(X_\infty)_{-1}$ is generated by B. Now

$$-\tfrac{1}{5}A + \tfrac{3}{10}B = \tfrac{1}{2}(-\tfrac{2}{5}A + \tfrac{7}{5}B) - \tfrac{2}{5}B \quad (\text{cf. } (5)).$$

So $F_a^2(W_4/W_3) = F^2(W_4/W_3)$, as predicted by Varchenko's theorem [**4**].

REFERENCES

1. J. Scherk and J.H.M. Steenbrink, *On the mixed Hodge structure on the Milnor fibre*, Math. Ann. (to appear).

2. J.H.M. Steenbrink, *Mixed Hodge structure on the vanishing cohomology*, Real and Complex Singularities, Proc. Nordic Summer School (Oslo,1976).

3. A. N. Varchenko, *Hodge properties of the Gauss-Manin connection*, Functional Anal. and Appl. **14** (1980), 1.

4. ______, *Asymptotics of holomorphic forms define a mixed Hodge structure*, Dokl. Akad. Nauk. SSSR **255** (5) (1980), 1035–1038.

5. ______, *Asymptotic mixed Hodge structure on vanishing cohomology*, Izv. Akad. Nauk. SSSR Ser. Mat. (3) **45** (1981), 540–591.

6. M. Saito, *Hodge filtrations on Gauss-Manin systems* (preprint).

7. ______, *Gauss-Manin system and mixed Hodge structure*, Proc. Japan Acad. Ser. A (1) **58** (1982), 29–32.

UNIVERSITY OF ALBERTA, CANADA

Proceedings of Symposia in Pure Mathematics
Volume **40** (1983), Part 2

A COBORDISM INVARIANT FOR SURFACE SINGULARITIES

JOSÉ A. SÉADE

In this work we describe and use the results of [**15**] to define a cobordism invariant $e(\mathcal{V}, P)$ for a noraml Gorenstein singularity P in a complex analytic surface $\mathcal{V}$, and we give an extension of these results (Theorem 3.1) as well as some applications. The idea is as follows: If $(\mathcal{V}, P)$ is a singularity as above, then $(\mathcal{V}, P)$ defines a canonical element (M, ρ) in the cobordism group Ω_3^{fr} of stably framed 3-manifolds (see §§1 and 2). Since this group is isomorphic to $\mathbf{Z}_{24}$, we may write the class (M, ρ) in Ω_3^{fr} as $E(M, \rho) \cdot \nu$, where ν is the standard generator of Ω_3^{fr} and $E(M, \rho)$ is a well-defined integer mod 24. The number $E(M, \rho)/24 \in \mathbf{Q}/\mathbf{Z}$ is the real Adams e-invariant $e(M, \rho)$ of (M, ρ) and our invariant $e(\mathcal{V}, P)$ is defined by $e(\mathcal{V}, P) = e(M, \rho) \in \mathbf{Q}/\mathbf{Z}$.

The number $e(\mathcal{V}, P)$ can be evaluated in terms of topological invariants of a resolution of P and if P is smoothable, $e(\mathcal{V}, P)$ can also be evaluated in terms of the Milnor number of a smoothing of P. By comparing these two expressions for $e(\mathcal{V}, P)$ we get some relations among invariants of the singualrity.

The arrangement of the paper is as follows: In §1 we review briefly some well-known results on framed manifolds and the Adams e-invariant. In §2 we define the invariant $e(\mathcal{V}, P)$ and we give a formula for evaluating this invariant in terms of a good resolution of P. In §3 we show that if P is Gorenstein and smoothable, then

$$e(\mathcal{V}, P) = \frac{\mu + 1}{24} \quad \mathrm{mod}\, \mathbf{Z},$$

where μ is the Milnor number of a smoothing of P, which generalizes previous results of [**14**, **15**] for complete intersections. Finally, in §4, we give some applications of these results, the main one being Corollary 4.1, that if P is Gorenstein and smoothable, then

$$\mu + 1 \equiv \chi(\tilde{V}) + K^2 + 12\,\mathrm{Arf}\,K \quad \mathrm{mod}\, 24$$

1980 *Mathematics Subject Classification*. Primary 57R45.

0082-0717/81/0000-0058/$02.50

where μ is the Milnor number of a smoothing of P and $\tilde{V}$ is a good resolution of P (see §2 for the definition of the terms of the RHS of this formula), which improves a result of Durfee [**4**, 1.7].

I would like to thank A. Durfee, X. Gomez-Mont, M. V. Nori and E. Rees for very helpful conversations.

1. Framed manifolds and the *e*-invariant. If M is a closed n-manifold with tangent bundle TM, then its stable tangent bundle $T_{\text{st}}M$ is the direct sum of TM with a large trivial bundle $M \times R^t$. A *framing* on M means a trivialization α of $T_{\text{st}}M$, or in other words, a framing is a set $\alpha = (\alpha_1, \ldots, \alpha_{n+t})$ of $(n + t)$-sections of $T_{\text{st}}M$, linearly independent everywhere. It is clear that a parallelism on M defines, canonically, a framing on M.

A framed manifold (M, α) is *null-cobordant*, or cobordant to zero, if M is the boundary of a compact manifold X^{n+1} endowed with a framing $\tilde{\alpha}$ that restricts on M to $\tau \oplus \alpha$, where τ is the unit, outward-pointing, normal field of M in X. The *inverse* $-(M, \alpha)$ of a framed manifold (M, α) is defined by taking M with the "opposite" framing, that is, with the framing obtained by reversing one of the sections of α. Two framed n-manifolds (M_1, α_1), (M_2, α_2) are *framed cobordant* if their disjoint union $(M_1, \alpha_1) \cup -(M_2, \alpha_2)$ is null-cobordant. This is an equivalence relation for framed manifolds, and the set of equivalence classes of framed n-manifolds forms an abelian group Ω_n^{fr} under disjoint union of manifolds. We refer to [**3**, **9**] for details on framed manifolds and cobordism.

Pontryagin (1951) established an isomorphism between Ω_n^{fr} and the stable homotopy group of the spheres $\pi_n^s \cong \pi_{n+q}(S^q)$, q large, and Adams [**1**] defined a homomorphism

$$e\colon \pi_n^s \to \mathbf{Q}/\mathbf{Z},$$

the (*real*) *e-invariant*. When $n \equiv 3 \bmod 4$, e can be defined in terms of spin-cobordism, and it is injection over the image of the classical J-homomorphism. Since for $n = 3$, J is into [**1**], it follows that two framed e-manifolds are framed cobordant if and only if they have the same e-invariant.

We refer to [**2**, **3**] for a definition of the e-invariant, and to [**15**] for a formula for evaluating e that does not require spin-structures (only for $n = 3$).

2. The invariant $e(\mathcal{V}, P)$. If $\mathcal{V}$ is a complex analytic surface with a normal singularity at P, then the link M of P is an oriented 3-manifold, so M has trivial tangent bundle. If α is a parallelism on M and if τ is the unit, outward-pointing, normal field of M in $\mathcal{V}$, then $\tau \oplus \alpha$ defines a trivialization of the bundle $T\mathcal{V}|M$ tangent to $\mathcal{V}$ over M, as a real vector bundle. If $\tau \oplus \alpha$ also defines a trivialization of $T\mathcal{V}|M$ as a complex bundle, then we say that α *is compatible with the complex structure* on $\mathcal{V}$. In [**15**] we prove

1. THEOREM. *If α, α' are parallelisms on M compatible with the complex structure on $\mathcal{V}$, then (M, α) and (M, α') are framed cobordant. That is, (M, α) and (M, α') represent the same class in Ω_3^{fr}.*

Therefore, the association,

$$P \to e(M, \alpha) \in \mathbf{Q}/\mathbf{Z},$$

defines an invariant $e(\mathcal{V}, P)$ of P which, clearly, depends only on its isomorphism class.

In [**15**] there is given a formula for evaluating $e(\mathcal{V}, P)$ in terms of topological invariants of a good resolution $\pi\colon \tilde{\mathcal{V}} \to \mathcal{V}$ of P: Assume $\mathcal{V}$ is embedded in some $\mathbf{C}^m$, and let $\tilde{V} = \pi^{-1}(V)$, where $V = \mathcal{V} \cap D_\varepsilon$ is the intersection of $\mathcal{V}$ with a small closed disc $D_\varepsilon \subset \mathbf{C}^m$ about P. Then

2. THEOREM. *With the above hypothesis and notation,*

$$e(\mathcal{V}, P) = \tfrac{1}{24}\big(\chi(\tilde{V}) + K^2 + 12 \operatorname{Arf} K\big) \mod \mathbf{Z},$$

where $\chi(\tilde{V})$ is the topological Euler characteristic of $\tilde{V}$, $K \in H_2(\tilde{V};\ \mathbf{Z})$ is a nonsingular representative of the canonical divisor of $\tilde{V}$, and Arf $K \in \{0, 1\}$ *is the* Arf *invariant of a certain quadratic form on $H_1(K;\ \mathbf{Z}_2)$.*

The above Arf invariant is precisely that defined by Rochlin in [**12**] (see also [**5**]). We refer to [**13**] for a clear account of Arf invariants.

Note that Theorems 1 and 2 above were stated, and proved, in [**15**] together as Theorem 3. Also, the formula given in [**15**] is somehow more general than the one given here. This extra generality is useful for applications, as for example to the quotient singularities of Dolgachev. (This shall be given elsewhere.) The formula that we give here is deduced from that in [**15**] by choosing a nonsingular representative of the canonical class K and letting $W = K$. (Such a representative of K always exists, see [**15**, p. 5].)

3. On Gorenstein singularities. We recall [7, V.§9] that an analytic space X of dimension r is *Gorenstein* at a point P if $\operatorname{Ext}^i_{\mathcal{O}_{X,P}}(\mathbf{C}, \mathcal{O}_{X,P})$ vanishes for $i \neq r$ and $\operatorname{Ext}^r_{\mathcal{O}_{X,P}}(\mathbf{C}, \mathcal{O}_{X,P})$ is $\mathbf{C}$. For a normal singularity P in an analytic surface $\mathcal{V}$ this is equivalent to demanding the existence of a nowhere vanishing holomorphic 2-form ω on a punctured neighbourhood of P in $\mathcal{V}$. Such a form ω defines around P a reduction to $SU(2) \cong Sp(1)$ of the structure group of the bundle $T\mathcal{V}$ tangent to $\mathcal{V} - \{P\}$, and so defines a multiplication by the quaternions i, j, k at each fibre of $T\mathcal{V}$ in a neighborhood of P. If we let τ denote the unit, outward-pointing, normal field of the link M of P in $\mathcal{V}$, then multiplying τ by i, j, k at each point of M according to the Sp(1)-structure, we get three linearly independent vector fields on M, which endow this manifold with a parallelism ρ compatible with the complex structure on $\mathcal{V}$. We note that the homotopy class of the framing ρ on M possibly depends on the choice of the form ω. However, by Theorem 2.1 above, the class in Ω_3^{fr} so represented is independent of it.

1. THEOREM (CF. [**14**, 1.2] AND [**15**, THEOREM 1]). *If P is Gorenstein and smoothable, then*

$$e(\mathcal{V}, P) = \frac{\mu + 1}{24} \mod \mathbf{Z}$$

where μ is the Milnor number of a smoothing of P.

We recall that P is smoothable if there exist (the germ of) an analytic space $\mathcal{W}$, and a flat map

$$\mathcal{F}: \mathcal{W} \to \Delta \subset \mathbf{C}$$

onto a small open disc Δ in $\mathbf{C}$, centered at 0, with $\mathcal{F}^{-1}(t)$ nonsingular for t in $\Delta - \{0\}$ and $\mathcal{F}^{-1}(0) = \mathcal{V}$. If we embed $\mathcal{W}$ in some complex space $\mathbf{C}^m$, and if we choose $\varepsilon > 0$ sufficiently small [**10**], then the intersection F of each nonsingular fibre $\mathcal{F}^{-1}(t)$, $|t|$ small, with the closed ε-disc D_ε in $\mathbf{C}^m$ centered at P, is an oriented 4-manifold with boundary diffeomorphic to M, and the diffeomorphism type of F is independent of the embedding of $\mathcal{W}$ and of the choices of ε, t. The manifold F is called a *smoothing* of P. It has the homotopy type of a CW-complex of dimension 2, and the number

$$\mu = \operatorname{Rank} H_2(F; \mathbf{Z})$$

is its *Milnor number*. In [**16**] Wahl conjectured that the 1st-Betti number b_1 of F always vanishes, and this was recently proved in [**6**]. Therefore, the Euler characteristic of F is $\chi(F) = \mu + 1$, and Theorem 1 will be proved if we show that the Adams e-invariant of the pair (M, ρ) is $\chi(F)/24 \bmod \mathbf{Z}$, where ρ is the framing of M defined above by the holomorphic 2-form ω. The proof of this is, essentially, the same as that of Theorem 1 in [**15**] (see also [**14**]). The only difference is that if F is the Milnor fibre of a complete intersection singularity, then its canonical bundle $\Lambda^2 T^*F$ is automatically holomorphically trivial. In the general case of a smoothing of a normal Gorenstein surface singularity, this is not automatic, but still true.

2. Lemma. *The canonical bundle* $\Lambda^2 T^*(\operatorname{Int} F)$ *of the interior of* F *is holomorphically trivial.*

Proof. If $\mathcal{F}: \mathcal{W} \to \Delta \subset \mathbf{C}$ is a smoothing of P, then each fibre $\mathcal{F}^{-1}(t)$ is nonsingular, for t in $\Delta - \{0\}$. Since $\mathcal{V}$ is Gorenstein at P, it follows that $\mathcal{W}$ is also Gorenstein at P (see [7, V.9.6]). Hence, there exists a nowhere vanishing holomorphic 3-form $\tilde{\omega}$ on $\mathcal{W} - \{P\}$. Since $\mathcal{F}$ is flat, we have an exact sequence of vector bundles over $\mathcal{W} - \{P\}$,

$$0 \to T_{\mathcal{F}} \to T(\mathcal{W} - \{P\}) \to \mathcal{F}^*(T\Delta) \to 0$$

where $T_{\mathcal{F}}$ denotes the tangent bundle along the fibers of $\mathcal{F}$. Hence,

$$\Lambda^3 T^*(\mathcal{W} - \{P\}) \cong \Lambda^2 T^*_{\mathcal{F}} \otimes \mathcal{F}^* T^* \Delta$$

and the above holomorphic 3-form on $\mathcal{W} - \{P\}$ defines a holomorphic trivialization of $\Lambda^2 T^*_{\mathcal{F}}$, i.e. a nowhere-vanishing holomorphic 2-form on each nonsingular fibre of the smoothing (and on the regular points of $\mathcal{V}$).

3. Remark. We note that the above Lemma 2 combines with Durfee's lemma [**4**, 1.1] to show that the tangent bundle TF of F is trivial, as conjectured in [**4**, 1.6]. Therefore, Durfee's theorem [**4**, 1.7] applies to all normal Gorenstein singularities.

4. Application. Let P denote a normal, Gorenstein singularity in an analytic surface $\mathcal{V}$. The following is an immediate consequence of Theorems 2.2 and 3.1.

1. COROLLARY. *If P is smoothable, then*

$$\mu + 1 \equiv \chi(\tilde{V}) + K^2 + 12\,\mathrm{Arf}\,K \mod 24$$

where μ is the Milnor number of a smoothing of P, and K is a nonsingular representative of the canonical divisor of a good resolution $\tilde{V}$ of P.

This formula is similar to those of Laufer [**8**, Theorem 1] and Durfee [**4**, 1.7]. Indeed, it combines with Laufer's Theorem [**8**, Theorem 1] to show that if P is complete intersection, then $\mathrm{Arf}\,K \equiv \rho_g \bmod 2$, where ρ_g is the geometric genus of P. That is, the mod(2)-reduction of ρ_g is Rochlin's Arf invariant [**12**] of the canonical class K. We do not know if a similar formula holds for all normal, Gorenstein singularities. (For P smoothable, this is equivalent to Laufer's conjecture mod 24 [**8**, c.1].)

2. COROLLARY. *If F, F' are smoothings of P, and if μ, μ' are their Milnor numbers, then*

$$\mu \equiv \mu' \mod 24.$$

The proof of this result is immediate from Corollary 1 above.

3. COROLLARY. *If $(\mathcal{V}_1, P_1)$ and $(\mathcal{V}_2, P_2)$ are singularities as above, whose links M_1 and M_2 are (orientation preserving) diffeomorphic, then*

$$e(\mathcal{F}_1, P_1) - e(\mathcal{V}_2, P_2) \equiv \tfrac{1}{2}(\mathrm{Arf}\,K_1 - \mathrm{Arf}\,K_2) \mod \mathbf{Z}.$$

PROOF. Since $M_1 \cong M_2$, it follows that the minimal good resolutions $\tilde{V}_1$, $\tilde{V}_2$ of P_1, P_2 are diffeomorphic [**17**]. Therefore $\chi(\tilde{V}_1) = \chi(\tilde{V}_2)$ and $K_1 = K_2$, so the result is a consequence of Theorem 2.2.

It is worth noting that the above corollary shows that there are Gorenstein singularities $(\mathcal{V}_1, P_1)$ and $(\mathcal{V}_2, P_2)$ with diffeomorphic links M_1, M_2, such that there exists no parallelism on M_1 compatible with the complex structure on $\mathcal{V}_1$ which is framed cobordant to a parallelism on m_2 compatible with the complex structure on $\mathcal{V}_2$. In other words, this shows that the invariant $e(\mathcal{V}, P)$ does not depend only on the C^∞-structure of the singularity. (However, $2e(\mathcal{V}, P)$ depends only on the oriented diffeomorphism type of the link!)

4. EXAMPLE. Consider the singularities at 0 defined by the complex polynomimals,

$$f_1(z) = z_1^2 + z_2^7 + z_3^{14}; \quad f_2(z) = z_1^3 + z_2^4 + z_3^{12}.$$

By [**11**], both singularities have resolutions which are line bundles with Chern number -1 over Riemann surfaces of genus 3; hence their links M_1, M_2 are diffeomorphic. However $\mu_1 = 78$ and $\mu_2 = 66$, therefore $e(\mathcal{V}_1, 0) \neq e(\mathcal{V}_2, 0)$.

5. COROLLARY. *With the hypothesis of Corollary* 3 *above, if* P_1, P_2 *are smoothable, then*

(i) $\mu_1 - \mu_2 \equiv 12(\mathrm{Arf}\, K_1 - \mathrm{Arf}\, K_2) \bmod 24$,

(ii) $\sigma(F_1) - \sigma(F_2) \equiv 8(\mathrm{Arf}\, K_1 - \mathrm{Arf}\, K_2) \bmod 16$,

where F_i, $i = 1, 2$, *is a smoothing of* P_i, μ_i *is the Milnor number of* F_i, *and* $\sigma(F_i)$ *is its signature.*

The first part of this result follows from Corollary 3 above and Theorem 3.1. For the second part one also applies Durfee's theorem [**4**, 1.7].

REFERENCES

1. F. Adams, *On the group* $J(X)$. IV, Topology **5** (1966), 21–71.

2. M. F. Atiyah and L. Smith, *Compact Lie groups and the stable homotopy of spheres*, Topology **13** (1974), 135–142.

3. P. E. Conner and E. E. Floyd, *The relation of cobordism to K-theories*, Lecture Notes in Math., vol. 28, Springer-Verlag, Berlin and New York, 1966.

4. A. Durfee, *The signature of smoothings of complex surface singularities*, Math. Ann. **232** (1978), 85–98.

5. M. Freedham and R. Kirby, *A geometric proof of Rochlin's Theorem*, Proc. Sympos. Pure Math., vol. 32, part 2, Amer. Math. Soc., Providence, R.I., pp. 85–97.

6. G.-M. Greuel and J. Steenbrink, *On the topology of smoothable singularities*, preprint.

7. R. Hartshorne, *Residues and duality*, Lecture Notes in Math., vol. 20, Springer-Verlag, Berlin and New York, 1966.

8. H. B. Laufer, *On* μ *for surface singularities*, Proc. Sympos. Pure Math., vol. 30, part 1, Amer. Math. Soc., Providence, R.I., 1977, pp. 45–49.

9. J. Milnor, *Topology from the differentiable viewpoint*, University Press of Virginia, 1965.

10. ______, *Singular points of complex hypersurfaces*, Ann. of Math. Studies, No. 61, Princeton Univ. Press, Princeton, N.J.

11. P. Orlik and P. Wagreich, *Isolated singularities of algebraic surfaces with* $\mathbf{C}^*$-*action*, Ann. of Math. (2) **93** (1971), 205–228.

12. V. A. Rochlin, *Proof of Gudkov's hypothesis*, Functional Anal. Appl. **6** (1972), 136–138.

13. C. P. Rourke and D. P. Sullivan, *On the Kervaire obstruction*, Ann. of Math. (2) **94** (1971), 397–413.

14. J. A. Seado and B. F. Steer, *The elements of* π_3^s *represented by invariant framings of quotients of* $SL_2(\mathbf{R})$ *by certain discrete subgroups*, Adv. in Math. (to appear).

15. J. A. Seade, *Singular points of complex surfaces and homotopy*, Toploogy **21** (1982), 1–8.

16. J. Wahl, *Smoothings of normal surface singularities*, Topology **20** (1981), 219–246.

17. W. D. Neumann, *A calculus for plumbing applied to the topology of complex surface singularities and degenerating complex curves*, Trans. Amer. Math. Soc. **268** (1981), 299–344.

UNIVERSIDAD NACIONAL AUTÓNOMA DE MÉXICO, MÉXICO

Proceedings of Symposia in Pure Mathematics
Volume **40** (1983), Part 2

ISOLATED LINE SINGULARITIES

DIRK SIERSMA

Introduction. In this paper we study germs of functions f: $(\mathbf{C}^{n+1}, 0) \to \mathbf{C}$ with a smooth 1-dimensional critical set Σ. After a change of coordinates we can suppose

$$\Sigma = L = \{(x, y) \in \mathbf{C} \times \mathbf{C}^n \mid y = 0\}.$$

We call those singularities *line singularities.*

Isolated line singularities are defined by the condition that for every $x \neq 0$ the germ of f at $(x, 0) \in \mathbf{C} \times \mathbf{C}^n$ is equivalent to $y_1^2 + \cdots + y_n^2$ and so is a Morse singularity in the transversal direction. In a certain sense isolated line singularities are the first generalizations of isolated (point) singularities. For isolated line singularities we prove that the Milnor fibre of f is homotopy equivalent to a bouquet of spheres.

For the study of nonisolated singularities in general we refer to the work of Lê Dũng Tráng [**Lê-1**, **2**] and Randell [**Ra**]. The special case of singularities with a 1-dimensional critical locus is especially studied by Iomdin [**Io-1–4**], who gave formulas for the Euler characteristic of the Milnor fibre; see also Lê Dũng Tráng [**Lê-3**]. Kato and Matsumoto [**K-M**] proved that in this case the Milnor fibre is $(n-2)$-connected. Moreover it is proved in [**Lê-Sa**] that the Milnor fibre is simply connected when $n = 2$ and f is irreducible.

The paper is organized as follows.

In §1 we treat line singularities from the point of view of Thom-Mather theory. Let $(y) = (y_1, \ldots, y_n)$ and $\mathfrak{m} = (x, y_1, \ldots, y_n)$ be ideals in $\mathcal{E} = \mathcal{E}_{n+1}$ = the ring of germs of holomorphic functions at $0 \in \mathbf{C}^{n+1}$. The action of germs of diffeomorphisms of $\mathbf{C}^{n+1}$ preserving L define an orbit structure in $\mathcal{E}$. For a singular germ $f \in (y)^2$ we define the codimension of f as $\dim_{\mathbf{C}}(y)^2/\tau(f)$ where $\tau(f)$ is the tangent space to the orbit of f in $(y)^2$.

1980 *Mathematics Subject Classification.* Primary 32B30, 32C40, 58C27.

0082-0717/81/0000-0614/$04.00

Just as in the case of isolated singularities we define k-determinacy and obtain algebraic conditions for being finitely determined. There is also a list of singularities of low codimension. The beginning of this list is as follows:

$$\begin{array}{llll} \text{codimension} & 0 & (A_\infty) & y_1^2 + y_2^2 + \cdots + y_n^2, \\ \text{codimension} & 1 & (D_\infty) & xy_1^2 + y_2^2 + \cdots + y_n^2, \\ \text{codimension} & 2 & (J_\infty) & x^2y_1^2 + y_1^3 + y_2^2 + \cdots + y_n^2. \end{array}$$

In §2 we give some other characterizations of finite codimension. Among others we show that equivalent are:

(a) $\operatorname{cod}(f) < \infty$,

(b) f has an isolated line singularity.

In §3 we study the topology of the Milnor fibre. We mimic a construction of Lê (cf. [**Br**]) and construct a nice approximation of f having only a finite number of A_1-points and a finite number of D_∞-points. With use of hyperplane sections $x = c$ we show

THEOREM. *Let f be an isolated line singularity (not of type A_∞), then the Milnor fibre of f is homotopy equivalent to a bouquet of μ spheres S^n,*

$$\mu = \sigma + 2\tau - 1$$

where σ is the number of A_1-points and τ is the number of D_∞-points in a generic approximation of f.

§4 contains remarks and questions.

I thank Lê for his remark, which simplified the proof of (2.5).

1. Line singularities.

(1.1) We consider the ring $\mathcal{E}_{n+1}$ of germs at 0 of holomorphic functions $f\colon \mathbf{C}^{n+1} \to \mathbf{C}$. We write $(x, y) = (x, y_1, \ldots, y_n)$ for coordinates in $\mathbf{C}^{n+1}$ and $L = \{(x, y) \mid y = 0\}$. We set $\mathcal{E} = \mathcal{E}_{n+1}$ and define ideals:

$$\mathfrak{m} = \mathcal{E}(x, y_1, \ldots, y_n) = \{f \in \mathcal{E} \mid f(0) = 0\},$$

$$(y) = \mathcal{E}(y_1, \ldots, y_n) = \{f \in \mathcal{E} \mid f(x, 0) = 0 \text{ for all } x\}.$$

The objects of our study are elements of $(y)^2$.

Let $\mathcal{D} = \mathcal{D}_{n+1}$ be the group of germs at 0 of local diffeomorphisms of the source space and let $\mathcal{D}_L$ be the subgroup of $\mathcal{D}$, consisting of $\phi \in \mathcal{D}$ with $\phi(L) = L$. There is a right action of $\mathcal{D}_L$ on $\mathcal{E}$. The orbit of f in $\mathcal{E}$ under $\mathcal{D}_L$ is denoted by $\operatorname{Orb}(f)$.

(1.2) We next define the *tangentspace* $\tau(f)$ to $\operatorname{Orb}(f)$ at f. Let ϕ_t be a curve in $\mathcal{D}_L$ with $\phi_0 = \mathrm{Id}$.

The chain rule gives:

$$\left.\frac{df\phi_t(p)}{dt}\right|_{t=0} = \left.\frac{\partial f}{\partial x}\frac{d\phi_t^0}{dt}(p)\right|_{t=0} + \sum_{j=1}^{n} \left.\frac{\partial f}{\partial y_j}\frac{d\phi_t^j}{dt}(p)\right|_{t=0}$$

$$= \xi(x, y)\frac{\partial f}{\partial x}(x, y) + \sum_{j=1}^{n} \eta_j(x, y)\frac{\partial f}{\partial y_j}(x, y)$$

with

$$\xi(0,0) = \frac{d\phi_t^0}{dt}(0,0)\bigg|_{t=0} = 0 \quad \text{and} \quad \eta_j(x,0) = \frac{d\phi_t^j}{dt}(x,0)\bigg|_{t=0} = 0.$$

So $\xi \in \mathfrak{m}$ and $\eta_j \in (y)$. For this reason we define

$$\tau(f) = \mathfrak{m}\frac{\partial f}{\partial x} + (y)\left(\frac{\partial f}{\partial y}\right) \quad \text{where} \quad \left(\frac{\partial f}{\partial y}\right) = \left(\frac{\partial f}{\partial y_1}, \ldots, \frac{\partial f}{\partial y_n}\right).$$

(1.3) DEFINITION. For $f \in (y)^2$ we define the *codimension*

$$c(f) = \operatorname{codim}(f) = \dim \frac{(y)^2}{\tau(f)}.$$

(1.4) DEFINITION. $f \in (y)^2$ is called *k-determined* if

$$f + \mathfrak{m}^{k-1}(y)^2 \subset \operatorname{Orb}(f)$$

(so every $g \in (y)^2$ with the same k-jet as f is right-equivalent with f).

(1.5) PROPOSITION. *Let* $f \in (y)^2$.
(a) *If f is k-determined then* $(y)^2\mathfrak{m}^{k-1} \subset \tau(f) + (y)^2\mathfrak{m}^k$.
(b) *If* $(y)^2\mathfrak{m}^{k-1} \subset \mathfrak{m}\tau(f) + (y)^2\mathfrak{m}^k$ *then f is k-determined.*

PROOF. (a) We work modulo $(y)^2\mathfrak{m}^k$ in a finite dimensional subspace $j^{k+1}((y)^2)$ of $J^{k+1}(n,1)$. Since $j^{k+1}(f + \mathfrak{m}^{k-1}(y)^2)$ is an affine subspace of $J^{k+1}(n,1)$ its tangent space at $j^{k+1}f$ is $j^{k+1}(\mathfrak{m}^{k-1}(y)^2)$. The tangent space to $j^{k+1}(\operatorname{Orb}(f))$ at $j^{k+1}f$ is $j^{k+1}(\tau(f))$. Since f is k-determined, we have

$$(y)^2\mathfrak{m}^{k-1} \subset \tau(f) + (y)^2\mathfrak{m}^k.$$

(b) Let $f \in (y)^2$ and suppose for $g \in (y)^2$ we have $j^k f = j^k g$, so $g - f \in (y)^2\mathfrak{m}^{k-1}$

$$F(x, y, t) = f(x, y) + t(g(x, y) - f(x, y)).$$

We consider F as an element of $\mathcal{E}_{n+2}$, the ring of germs at $(0, t_0)$. We denote its maximal ideal by $\mathfrak{m}_{n+2}$.

We have inclusions: $\mathcal{E} = \mathcal{E}_{n+1} \subset \mathcal{E}_{n+2}$ and $\mathfrak{m} = \mathfrak{m}_{n+1} \subset \mathfrak{m}_{n+2}$. In the rest of the proof the notations $\mathfrak{m}$, (y), $(y)^2$, etc. are actually used for $\mathcal{E}_{n+3}\mathfrak{m}$, $\mathcal{E}_{n+2}(y)$, $\mathcal{E}_{n+2}(y)^2$, etc. Let

$$\tau^*(F) = \left\{\xi\frac{\partial F}{\partial x} + \sum \eta_j \frac{\partial F}{\partial y_j} \,\middle|\, \xi \in \mathfrak{m} \text{ and } \eta_j \in (y)\right\} \subset \mathcal{E}_{n+2}.$$

Remark that $\tau(f) \subset \tau^*(F) + (y)^2\mathfrak{m}^{k-1}$. So $(y)^2\mathfrak{m}^{k-1} \subset \mathfrak{m}\tau(f) + (y)^2\mathfrak{m}^k \subset \mathfrak{m}\tau^*(F) + (y)^2\mathfrak{m}^k \subset \mathfrak{m}\tau^*(F) + \mathfrak{m}_{n+2}(y)^2\mathfrak{m}^{k-1}$. By Nakayama's lemma:

$$(y)^2\mathfrak{m}^{k-1} \subset \mathfrak{m}\tau^*(F) \subset \tau^*(F).$$

So there exist time dependent vector fields $(\xi, \eta_1, \ldots, \eta_n)$ defined in a neighborhood U of $(0, 0, t_0)$ such that

$$\frac{\partial F}{\partial x}(x, y, t)\xi(x, y, t) + \sum_{j=1}^{n} \frac{\partial F}{\partial y_j}(x, y, t)\eta_j(x, y, t) + g(x, y) - f(x, y) = 0$$

for all $(x, y, t) \in U$.

Moreover $\xi \in \mathfrak{m}$ and $\eta_j \in (y)$. So $\xi(0,0,t) = 0$ for all $(0,0,t) \in U$ and $\vec{\eta}(x, 0, t) = 0$ for all $(x, 0, t) \in U$. The differential equation

$$\begin{cases} \dfrac{\partial h^x}{\partial t}(x, y, t) = \xi(h(x, y, t), t), \\ \dfrac{\partial h^y}{\partial t}(x, y, t) = \vec{\eta}(h(x, y, t), t), \\ h(x, y, t_0) = (x, y), \end{cases}$$

has a unique solution, generating a family of local diffeomorphisms h_t (for all t near t_0) satisfying

$$\begin{cases} F_{t_0} = F_t h_t, \\ h_t \in \mathfrak{D}_L. \end{cases}$$

By "continuous induction" over the interval [0, 1] we find that $g = F_1$ and $f = F_0$ are right equivalent.

(1.6) COROLLARY. *Let* $f \in (y)^2$. *Then* $\operatorname{codim}(f) < \infty \Leftrightarrow f$ *is k-determined for some* $k \in \mathbf{N}$.

PROOF. f is k-determined for some $k \in \mathbf{N} \Leftrightarrow$

$$\exists k\ (y)^2 \mathfrak{m}^{k-1} \subseteq \tau(f) \Leftrightarrow \operatorname{cod}(f) < \infty.$$

CLASSIFICATION OF LINE SINGULARITIES

The same computational methods as in the case of ordinary singularities can be applied. We find the following beginning of the list:

Type	Residual singularity	codim(f)	determined jet
A_∞	0	0	2
D_∞	xy^2	1	3
$J_{k,\infty}$ $(k \geqslant 2)$	$y^3 + x^k y^2$	$k \geqslant 2$	$k + 2$
$T_{\infty,k,2}$ $(k \geqslant 4)$	$x^2y^2 + y^k$	$k - 1 \geqslant 3$	k
$Z_{k,\infty}$ $(k \geqslant 1)$	$xy^3 + x^{k+2}y^2$	$k + 3 \geqslant 4$	$k + 4$
$W_{1,\infty}$	$y^4 + y^2x^3$	5	5
$T_{\infty,q,r}$ $(q \geqslant 3, r \geqslant 3)$	$xyz + y^q + z^r$	$q + r - 3 \geqslant 3$	$\max(q, r)$
$Q_{k,\infty}$ $(k \geqslant 2)$	$xz^2 + y^3 + x^k y^2$	$k + 2 \geqslant 4$	$k + 2$
$S_{1,\infty}$	$y^2z + xz^2 + x^2y^2$	5	4

The list contains all simple singularities and all line-singularities of codimension $\leqslant 6$. All nonsimple line singularities are adjacent to one of the following three families of one modular singularities.

$y^4 + Ay^3x^2 + y^2x^4$	7	6	$(A^2 \neq 4)$
$x^3y^2 + Axy^4 + \varepsilon y^5$	7	5	$(\varepsilon^2 = \varepsilon)$
$y^2z + xz^2 + Axy^3 + x^3y^2$	7	5	

2. Isolated line singularities.

(2.1) We shall give some other characterizations for finite codimension. Let $f \in (y)^2$. Write $f(x, y) = \Sigma_{i,j=1}^n g_{ij}(x, y) y_i y_j$ with $g_{ij} = g_{ji}$ e.g. take

$$g_{ij}(x, y) = \int_0^1 \int_0^1 \frac{\partial^2 f(x, sty)}{\partial y_i \partial y_j} ds\, dt.$$

We define the *Hessian of f* (relative x) by

$$h_f(x, y) = \det\big(g_{ij}(x, y)\big).$$

The 2-jet of f in $(c, 0)$ is equal to $\Sigma\, g_{ij}(c, 0) y_i y_j$. So we see $h_f(c, 0) \neq 0 \Leftrightarrow f$ has type A_∞ at $(c, 0)$.

(2.2) DEFINITION. $f \in (y)^2$ has a *line singularity* if its singular locus is

$$\Sigma(f) = L = \{(x, y) \in \mathbf{C} \times \mathbf{C}^n \mid y = 0\}.$$

The *line singularity* is called *isolated* if for $c \neq 0$ the germ of f at $(c, 0)$ has only A_∞-singularities.

(2.3) EXAMPLES. (a) A_∞: $f(x, y) = y^2$,

(b) D_∞: $f(x, y) = xy^2$,

(c) $f(x, y) = x^2 y^2$ is not a line singularity,

(d) $f(x, y) = y^3$ is not an isolated line singularity.

(2.4) We consider the following ideals in $\mathcal{E}$:

$$\tau(f) = \mathfrak{m}\frac{\partial f}{\partial x} + (y)\left(\frac{\partial f}{\partial y}\right) \subset (y)^2 \quad \text{(tangent space)},$$

$$J(f) = \left(\frac{\partial f}{\partial x}, \frac{\partial f}{\partial y_1}, \dots, \frac{\partial f}{\partial y_n}\right) \subset (y), \quad \text{(Jacobian ideal)},$$

$$h(f) = \left(h_f, \frac{\partial f}{\partial x}, \frac{\partial f}{\partial y_1}, \dots, \frac{\partial f}{\partial y_n}\right) \quad \text{(Hessian ideal)}.$$

(2.5) THEOREM. *Let $f \in (y)^2$. Equivalent are*:

(A) $c(f) := \dim_{\mathbf{C}} (y)^2/\tau(f) < \infty$,

(B) $j(f) := \dim_{\mathbf{C}} (y)/J(f) < \infty$,

(C) $\lambda(f) := \dim_{\mathbf{C}} \mathcal{E}/h(f) < \infty$ *and* $\Sigma(f) = L$.

(D) *f has an isolated line singularity at* $\underline{0}$.

PROOF. We take a representative f of the germ. Define sheafs of $\mathcal{E}_{n+1}$-modules as follows:

$$\mathcal{F}^1(U) = \frac{(y)}{J(f)} \quad \text{and} \quad \mathcal{F}^2(U) = \frac{(y)^2}{\tau(f)}$$

where (y), $(y)^2$, $J(f)$ and $\tau(f)$ are considered as modules over the holomorphic functions on U. It is clear that $\mathcal{F}^1$ and $\mathcal{F}^2$ are coherent. We intend to use the fact that $\mathcal{F}^i$ is concentrated in a point $\Leftrightarrow \dim \Gamma(\mathcal{F}^i) < \infty$.

(i) (D) $\Rightarrow$ (A) and (B). For $y \neq 0$ f is regular at (x, y) and we have $\dim \mathcal{F}^1_{(x,y)} = \dim \mathcal{F}^2_{(x,y)} = 0$ since $(y) \cong \mathcal{E}_{n+1}$, $J(f) \cong \mathcal{E}_{n+1}$, $(y)^2 \cong \mathcal{E}_{n+1}$ and $J(f) \cong \mathcal{E}_{n+1}$ at

(x, y). If $y = 0$ and $x \neq 0$ then f is of type A_∞ at $(x, 0)$ and we have $\dim \mathfrak{F}^1_{(x,0)} = \dim \mathfrak{F}^2_{(x,0)} = 0$, since $(y) \cong J(f)$ and $(y)^2 \cong \tau(f)$ at $(x, 0)$. So both $\mathfrak{F}_1$ and $\mathfrak{F}_2$ are concentrated at 0 and this implies $c(f) < \infty$ and $j(f) < \infty$.

(ii) (B) $\Rightarrow$ (D). Since $j(f) < \infty$ we have $\dim \mathfrak{F}^1_{(x,y)} = 0$ for $(x, y) \neq (0, 0)$. Since $\dim \mathcal{E}/J(f) = 0$ implies f is regular, and $\dim (y)/J(f) = 0$ implies f is of type A_∞, we have (D).

(iii) (A) $\Rightarrow$ (D) is similar.

(iv) (C) $\Leftrightarrow$ (D) is trivial.

3. The topology of the Milnor fibre.

(3.1) We recall that we consider an *isolated line singularity*, that is an analytic germ f with singular locus the line $L = \{(x, y) \in \mathbf{C} \times \mathbf{C}^n \mid y = 0\}$ such that for every $x \neq 0$ the germ of f at $(x, 0)$ is of type A_∞, i.e. equivalent to $y_1^2 + \cdots + y_n^2$. Let B_ε be the closed ε-ball in $\mathbf{C}^{n+1}$ and D_η be the closed 1-disc in $\mathbf{C}$. We select $\varepsilon > 0$ and $\eta > 0$ such that the restriction

$$f\colon B_\varepsilon \cap f^{-1}(D_\eta) \to D_\eta$$

satisfies the conditions for the Milnor construction, and so f is a C^∞-locally trivial fibre bundle above $D_\eta - 0$.

(3.2) In the case of an ordinary isolated singularity it is useful to consider a generic approximation g of f with only ordinary Morse points (cf. Brieskorn [**Br**]). At every Morse point one can study its local Milnor fibration, with Milnor fibre homotopy equivalent to one n-sphere S^n ("the vanishing cycle"). The Milnor fibre of the original f then has the homotopy type of the wedge of those μ spheres.

We like to mimic the construction in the case of an isolated line singularity. First we prove the existence of a nice approximation.

(3.3) Lemma. *Let f have an isolated line singularity. There exist a deformation g of f such that g has*:

(i) *only D_∞ and A_∞ singularities on L,*

(ii) *only A_1 (Morse) singularities outside L.*

[*Recall D_∞ singularity is locally given by $xy_1^2 + y_2^2 + \cdots + y_n^2$.*]

Proof. Define $F\colon \mathbf{C} \times \mathbf{C}^n \times S \times T \to \mathbf{C}$ by

$$F(x, y, (a_{ij}), (b_{ij})) = f(x, y) + \sum_{i,j} (a_{ij} + b_{ij}x) y_i y_j.$$

A computation shows that the 2-jet extension

$$j^2F\colon \mathbf{C} \times \mathbf{C}^n \times S \times T \to J^2(n+1, 1)$$

is transversal to the A_1-stratum outside L and transversal to the D_∞ stratum on L. The assertion follows now as an application of Sard's theorem.

Example.

$$f(x, y) = y^2(x^2 - y^2),$$

$$g(x, y) = y^2(x^2 - (y - t)^2).$$

$f^{-1}(0)$ $\qquad$ $g^{-1}(0)$

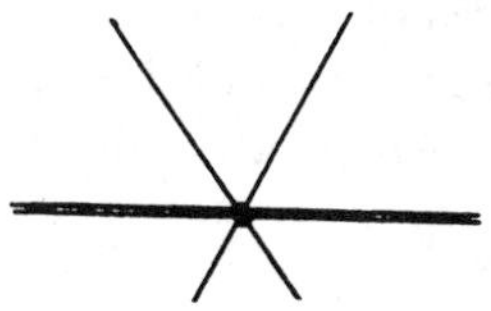

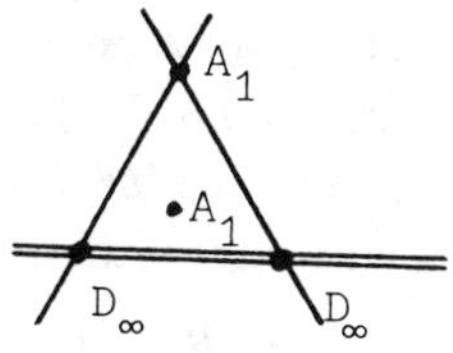

(3.4) PROPOSITION. *There exists an approximation g of f (as in Lemma* (3.3)) *with the additional property: The Milnor fibrations of g and f above the boundary* ∂D_η *are equivalent.*

PROOF. Let $\varepsilon > 0$ such that $(f^{-1}(0)\setminus L) \pitchfork S_{\varepsilon'}$ for all $0 < \varepsilon' \leqslant \varepsilon$. Let $f_\lambda(x, y)$ be a 1-parameter deformation of f, which satisfies (3.3) for $\lambda \neq 0$. We claim now, that there exist $\delta > 0$ and $\eta > 0$ such that $f_\lambda^{-1}(t) \pitchfork S_\varepsilon$ for all $0 < \lambda \leqslant \delta$ and $0 < |t| \leqslant \eta$. This follows from:

(1) $f^{-1}(0) \cap S_\varepsilon$ is compact,

(2) $f | S_\varepsilon$ is a submersion in points of $f^{-1}(0)\setminus L$,

(3) on $L \cap S_\varepsilon$ we have only A_∞-points, so near points of $L \cap S_\varepsilon$ we can change coordinates $(y_1, \ldots, y_n)$, smoothly depending on (x, λ) such that

$$f_\lambda(x, y) \sim y_1^2 + \cdots + y_n^2$$

(parameter version of the Morse lemma).

For $t \neq 0$ the tangent space in (x, y) to $f_\lambda^{-1}(t)$ contains L, since it is given by $\xi_1 y_1 + \cdots + \xi_n y_n = 0$. So we find locally

$$f_\lambda^{-1}(t) \pitchfork S_\varepsilon \quad \text{for } t \neq 0.$$

Let $F(x, y, \lambda) = f_\lambda(x, y), \lambda)$. The proposition follows now from the fact that

$$F\colon F^{-1}(\partial D_\eta \times [0, \delta]) \cap (B_\varepsilon \times [0, \delta]) \to \partial D_\eta \times [0, \delta] \to [0, \delta]$$

and the restriction to $S_\varepsilon \times [0, \delta]$ are submersions.

(3.5) REMARK. The equivalence of f and an approximation g is generally nonvalid for nonisolated singularities. Here we have the equivalence because of special properties of isolated line singularities and of the approximation g.

(3.6) We now take a generic approximation g of f as in the above lemma and suppose moreover that the approximation is so close that the Milnor fibrations of g and f above the boundary ∂D_η are the same. We can also suppose that all critical values of g are different (this is mostly for notational convenience). The critical value 0 corresponds to the nonisolated singularities on the line L.

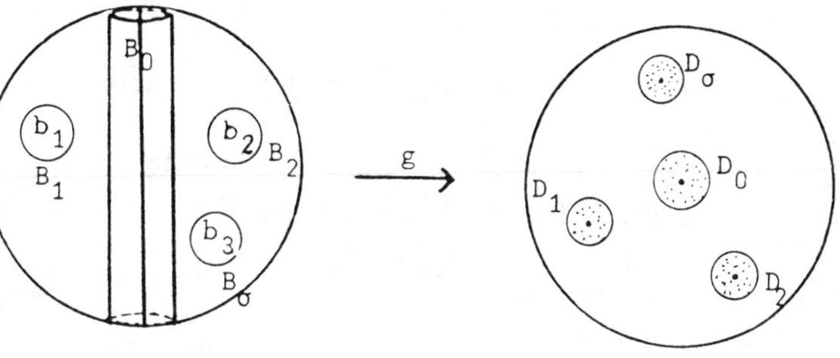

Let $b_1, \ldots, b_\sigma$ be the Morse points of g with critical values $g(b_1), \ldots, g(b_\sigma)$. Define $B_1, \ldots, B_\sigma$ disjoint $(2n+2)$ balls around $b_1, \ldots, b_\sigma$ and inside $B = B_\varepsilon$. Let $D_1, \ldots, D_\sigma$ be disjoint 2-discs around $g(b_1), \ldots, g(b_\sigma)$ and inside $D = D_\eta$, chosen in such a way that we get local fibrations

$$g\colon B_i \cap g^{-1}(D_i) \to D_i \qquad (i = 1, \ldots, \sigma)$$

satisfying the usual transversality condition

$$\partial B_i \pitchfork g^{-1}(t) \qquad \text{if } t \in D_i.$$

We also define a small cylinder B_0 around L and a 2-disc D_0 around 0 such that

$$\partial B_0 \pitchfork g^{-1}(t) \qquad \text{if } t \in D_0.$$

Of course we can take all $B_0, \ldots, B_\sigma$ and $D_0, \ldots, D_\sigma$ to be disjoint. We first study the fibres of

$$g\colon B_0 \cap g^{-1}(D_0) \to D_0.$$

We take hyperplane sections $x = c$. A fibre $g^{-1}(t) \cap B_0$ is now fibered by the projection π on L. This projection can have singularities. It is convenient to consider the map

$$\Phi_g\colon g^{-1}(D_0) \cap B_0 \to \mathbf{C} \times \mathbf{C}$$

defined by $\Phi_g(x, y) = (g(x, y), x)$. The singular locus of Φ_g consists of the line L and the so called *polar curve* Γ. The projection

$$\pi\colon g^{-1}(t) \cap B_0 \to L$$

is smooth outside points of Γ_g $(t \neq 0)$.

(3.7) LEMMA. *The polar curve Γ_g can cut L only in the D_∞ points of g.*

PROOF. We have to show, that if g is of type A_∞, then L is locally the discriminant locus of Φ_g.

So let $g(x, y) = \Sigma\, g_{ij}(x, y) y_i y_j$ with $\det(g_{ij}(0,0)) \neq 0$

$$\begin{cases} \dfrac{\partial g}{\partial y_1} = \Sigma \dfrac{\partial g_{ij}}{\partial y_1} y_i y_j + \Sigma\, g_{1j} y_j \\ \vdots \\ \dfrac{\partial g}{\partial y_n} = \Sigma \dfrac{\partial g_{ij}}{\partial y_n} y_i y_j + \Sigma\, g_{nj} y_j. \end{cases}$$

$\det(g_{ij}(0,0)) \neq 0$ now implies that we can modulo $(y)^2$ solve for $y_1, \ldots, y_n$. So $(y) \subset (\partial g/\partial y) + (y)^2$.

Nakayama's lemma now gives $(y) = (\partial g/\partial y)$. So the variety defined by $\partial g/\partial y_1 = \cdots = \partial g/\partial y_n = 0$ is just L and it is clear that this is the critical set of Φ_g.

Next we study the D_∞-points.

(3.8) PROPOSITION. *Let g be of type* D_∞ *and let* $\Phi_g(x, y) = (g(x, y), x)$.

(a) *The diffeomorphism type of the pair of Milnor fibres of* Φ_g *and g is independent of the choice of g within the type* D_∞.

(b) *The pair of Milnor fibres of g and* Φ_g *is homotopy equivalent to the pair of standard spheres* (S^n, S^{n-1}).

PROOF. (a) Let g_0 and g_1 be of type D_∞. Select representatives such that the transversality conditions for the Milnor fibrations of g_0, g_1, Φ_{g_0} and Φ_{g_1} are satisfied for certain $(s_0, t_0) \in L \times \mathbf{C}$. Consider the complex family

$$\overline{\Psi}(x, y, \tau) = (x, \tau g_1(x, y) + (1-\tau) g_0(x, y)) = (x, \Psi'(x, y, \tau)).$$

The variety $\Psi^{-1}(s_0, t_0)$ intersects

$$\left\{(x, y, \tau) | 0 = \frac{\partial \Psi'}{\partial y_1} = \cdots = \frac{\partial \Psi'}{\partial y_n}\right\}$$

only in a finite number of points. Choose a path $\lambda(t)$ from 0 to 1 in the τ-plane missing τ-coordinates of those points. The real homotopy $g_{\lambda(t)}$ between g_0 and g_1 induces a diffeomorphism between the pairs of Milnor fibres.

(b) It is sufficient to study $g = xy_1^2 + y_2^2 + \cdots + y_n^2$. Take first $n = 1$, $g(x, y) = xy^2 = \delta$, $|x| \cdot |y|^2 = \delta$, $\arg x + 2 \arg y = 0 \bmod 2\pi$.

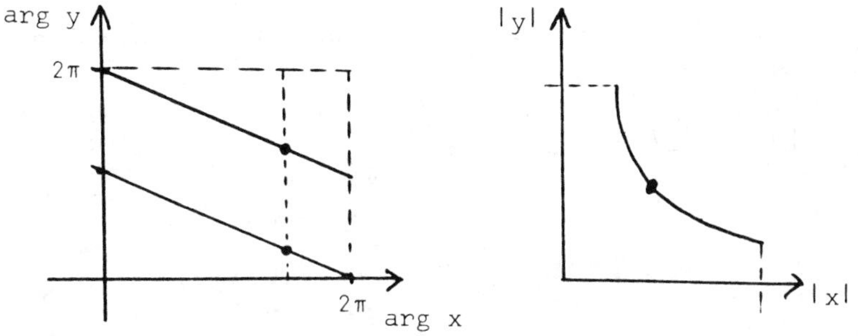

In (torus) $\times$ $\mathbf{R}^2$ it is clear that the Milnor fibre of g is homotopy equivalent to S^1, the hyperplane section $x = s_0$ is homotopy equivalent to S^0. In the general case we have to take double suspensions of (S^1, S^0).

(3.9) LEMMA. *Let f be not of type* A_∞. *Let* $s_0 \in \partial(B_0 \cap L)$. *The fibre*

$$X_t' = g^{-1}(t) \cap B_0$$

is homotopy equivalent to 2τ *n-balls glued together along their common boundary* $S^{n-1} \overset{h}{\simeq} Y_t' = \Phi_g^{-1}(t, s_0) \cap B_0$. *So* X_t *is homotopy equivalent to a bouquet of* $(2\tau - 1)$ *n-spheres, where* τ *is the number of* D_∞*-points in a generic approximation g.*

PROOF. Set $S = B_0 \cap L$. Let $S_1, \ldots, S_\tau$ be small disjoint discs inside S around the D_∞ points $c_1, \ldots, c_\tau$. Choose B_0 so small that above $S \setminus \bigcup_{i=1}^{\tau} S_i$ the projection

$$\pi \colon g^{-1}(t) \cap B_0 \to L \qquad (t \neq 0)$$

is locally trivial.

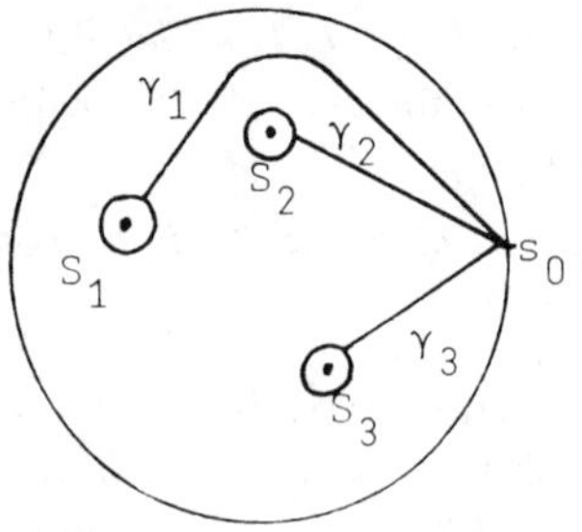

Choose a system of paths $\gamma_1,\dots,\gamma_\tau$ from s_0 to $S_1,\dots,S_\tau$ (in the usual way; see the diagram). Set $Z = S_1 \cup \cdots \cup S_\tau \cup \gamma_1 \cup \cdots \cup \gamma_\tau$, $W = \gamma_1 \cup \cdots \cup \gamma_\tau$. Z is a deformation retract of S; $\{s_0\}$ is a deformation retract of W. Since π is locally trivial over the complement of $S_1 \cup \cdots \cup S_\tau$ it follows from the homotopy lifting property that $\pi^{-1}(Z)$ is a deformation retract of X_t' and Y_t' is a deformation retract of $\pi^{-1}(W)$. Moreover Y_t' is the Milnor fibre of the hyperplane section of an A_∞-singularity and so homotopy equivalent to an S^{n-1} (since the hyperplane section has an A_1-singularity). It follows that $(\pi^{-1}(Z), \pi^{-1}(W))$ is relatively homotopy equivalent to

$$\left(\pi^{-1}(W) \cup e_1^+ \cup e_1^- \cup \cdots \cup e_\tau^+ \cup e_\tau^-, \pi^{-1}(W)\right),$$

where for each D_∞ point two n-cells e^+ and e^- are attached to the vanishing cycle S^{n-1} in the standard way. So

$$(X_t', Y_t') \overset{h}{\simeq} \left(S^{n-1} \cup_{\gamma_1} e_1^+ \cup e_1^- \cup \cdots \cup_{\gamma_\tau} e_\gamma^+ \cup_{\gamma_\tau} e_\tau^-, S^{n-1}\right).$$

So

$$X_t' \overset{h}{\simeq} S^n \vee \cdots \vee S^n \quad (2\tau - 1 \text{ copies}).$$

(3.10) THEOREM. *Let f be an isolated line singularity (not of type A_∞); then the Milnor fibre of f is homotopy equivalent to a bouquet of μ spheres S^n, $\mu = \sigma + 2\tau - 1$, where σ is the number A_1-points and τ is the number D_∞-points in a generic approximation of f.*

PROOF. Take D, D_0, $D_1,\dots,D_\sigma$ and B, B_0, $B_1,\dots,B_\sigma$ as before. Let $t \in \partial D_0$. Choose a system of paths $\psi_1,\dots,\psi_\sigma$ from t to $D_1,\dots,D_\sigma$. For $T \subset D$ set $X_T = g^{-1}(T) \cap B$. As in the preceding lemma there is a homotopy equivalence

$$(X_D, X_t) \overset{h}{\simeq} \left(X_{D_0} \cup_{\psi_1} e_1^{n+1} \cup \cdots \cup_{\psi_\sigma} e_\sigma^{n+1}, X_t\right).$$

Moreover,

$$(X_{D_0}, X_t) \overset{h}{\simeq} (X_{D_0} \cap B_0 \cup X_t, X_t).$$

Let $\phi_1,\dots,\phi_{2\tau-1}$: $S^n \to X_t' = X_t \cap B_0$ represent the $2\tau - 1$ generators of $\pi_n(X_t \cap B_0)$. Use $\phi_1,\dots,\phi_{2\tau-1}$ to attach $(n+1)$-cells $f_1^{n+1},\dots,f_{2\tau-1}^{n+1}$ to $X_t \cap B_0$.

The inclusion mapping

$$X_t \cap B_0 \hookrightarrow X_{D_0} \cap B_0$$

extends to a homotopy equivalence

$$X_t \cup_{\phi_1} f_1^{n+1} \cup \cdots \cup_{\phi_{2\tau-1}} f_{2\tau-1}^{n+1} \to X_{D_0} \cap B_0,$$

since both spaces are contractible. So we get a homotopy equivalence

$$\left(X_{D_0}, X_t\right) \overset{h}{\simeq} \left(X_t \cup_{\phi_1} f_1^{n+1} \cup \cdots \cup_{\phi_{2\tau-1}} f_{2\tau-1}^{n+1}, X_t\right).$$

X_D is obtained from X_t by attaching $\sigma + 2\tau - 1$ $(n+1)$-cells. So X_t is $(n-1)$-connected, since X_D is contractible. Since X_t has the homotopy type of an n-dimensional finite CW-complex it follows that X_t has the homotopy type of a bouquet of $\mu = \sigma + 2\tau - 1$ n-spheres.

4. Remarks and questions.

(4.1) In the case of isolated (point) singularities, there is the algebraic description of the Milnor number

$$\mu = \dim \mathcal{E}_{n+1} / \left(\frac{\partial f}{\partial x_0}, \ldots, \frac{\partial f}{\partial x_n} \right).$$

For isolated line singularities we have

$$\tau = {}^{\#}(D_\infty \text{ points}) = \dim \mathcal{E}_1 / \left(h_f(x,0)\right).$$

A question to prove is

$$c(f) = {}^{\#}(A_1 \text{ points}) + 1 = \sigma + 1,$$
$$j(f) = {}^{\#}(A_1 \text{ points}) + {}^{\#}(D_\infty \text{ points}) = \sigma + \tau,$$

which is true in all known examples.

(4.2) Find the intersection forms for isolated line singularities. For $n = 2$ one can use the method of A'Campo and Guzein-Zade. Here follows an example:

$$f(x, y) = x^2y^2 + y^4.$$

Nice approximation: $g(x, y) = y^2(x^2 - (y - t)^2)$. Level curves:

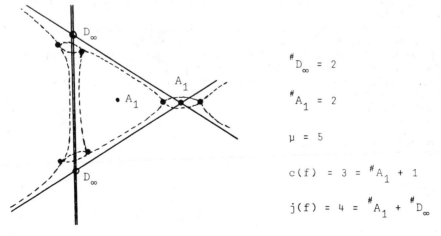

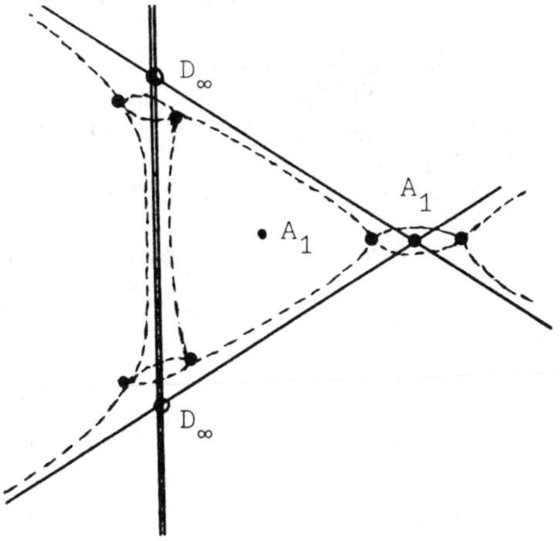

$^{\#}D_\infty = 2$

$^{\#}A_1 = 2$

$\mu = 5$

$c(f) = 3 = {}^{\#}A_1 + 1$

$j(f) = 4 = {}^{\#}A_1 + {}^{\#}D_\infty$

In the diagram you can easily find the 5 cycles.
There is some freedom in choice.
A diagram for the intersection matrix is:

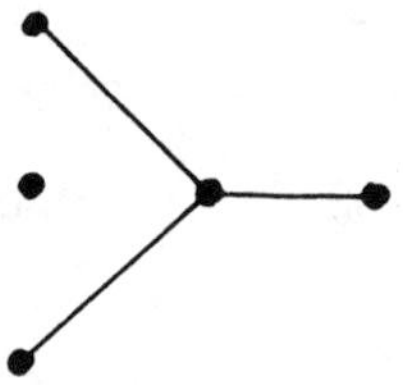

(c) Find the relation between isolated line singularities and certain series of isolated singularities, especially relate the topology of their Milnor fibres. From Iomdin [**Io-4**] it already follows that for k sufficiently large $\chi(F) = \chi(F_k') - k$ where χ is the Euler characteristic, F is the Milnor fibre of $f(x, y)$ and F_k' is the Milnor fibre of $f(x, y) + x^k$.

(d) Study line singularities which have other transversal singularities than A_1 (outside 0).

(e) Study in general singularities with 1-dimensional critical locus, which have transversally A_1-singularities (outside 0).

REFERENCES

[**Br**] E. Brieskorn, *Die Monodromie der isolierten Singularitäten von Hyperflächen*, Manuscripta Math. **2** (1970), 103–161.

[**Io-1**] I. N. Iomdin, *Some properties of isolated mappings of real polynomial singularities*, Mat. Zametki **13** (4), 565–572.

[**Io-2**] ______, *The Euler characteristic of the intersection of a complex surface with a disc*, Sibirsk. Mat. Ž. **14** (2) (1973), 322–336.

[**Io-3**] ______, *Local topological properties of complex algebraic sets*, Sibirsk. Mat. Ž. **15** (4) (1974), 784–805.

[**Io-4**] ______, *Complex surfaces with a one dimensional set of singularities*, Sibirsk. Mat. Ž. **15** (5) (1974), 1061–1082.

[**K-M**] M. Kato and Y. Matsumoto, *On the connectivity of the Milnor fibre of a holomorphic function at a critical point*, Proc. 1973 Tokyo Manifolds Conf., pp. 131–136.

[**Lê-1**] Lê Dũng Tráng, *Calcul du nombre de cycles évanouissants d'une hypersurface complexe*, Ann. Inst. Fourier (Grenoble) **23** (1973), 261–270.

[**Lê-2**] ______, *La monodromie n'a pas de point fixes*, J. Fac. Sci. Univ. Tokyo Sect. IA Math. **22** (1975), 409–427.

[**Lê-3**] ______, *Ensembles analytiques complexes avec lieu singulier de dimension un (d'après I. N. Iomdin)*, Sém. sur les Singularitiés, Publ. Math. de l'Univ. Paris VII, pp. 87–95.

[**Lê-Sa**] Lê Dũng Tráng and K. Saito, *The local π_1 of the complement of a hypersurface with normal crossings in codimension 1 is abelian*, Preprint, RIMS-350, Kyoto, Japan.

[**Mi**] J. Milnor, *Singular points of complex hypersurfaces*, Ann. of Math. Studies, Princeton Univ. Press, Princeton, N.J., 1968.

[**Ra**] R. Randell, *On the topology of non isolated singularities*, Proc. 1977 Georgia Topology Conference, pp. 445–473.

RIJKSUNIVERSITEIT UTRECHT, THE NETHERLANDS

Proceedings of Symposia in Pure Mathematics
Volume **40** (1983), Part 2

A CONVEXITY THEOREM

ANDREW JOHN SOMMESE

In this article I continue the investigation started in [**So**$_2$] of the convexity of the complement of a complex submanifold of a homogeneous complex manifold. The following theorem gives the flavour of my results (the full result allows arbitrary products of Abelian varieties, Grassmannians, and quadrics).

(2.1) COROLLARY. *Let A be a connected complex submanifold of an Abelian variety X. If k is the maximal dimension of a proper Abelian subvariety of X, the $X - A$ is* $\operatorname{cod}_{\mathbf{C}} A + k$ *convex in the sense of Andreotti-Grauert* [*cf.* (0.4.5)].

In [**So**$_2$], I could only prove this result for $k = 0$. It has a number of immediate consequences:

(a) finiteness theorems [**A** + **Gr**] for coherent sheaf cohomology on $X - A$,

(b) transplanting theorems [**So**$_3$], e.g. given any coherent analytic sheaf $\mathcal{S}$ on X as above,

$$H^j(X - A, \mathcal{S}) = 0 \quad \text{for } j \geqslant 2\operatorname{cod}_{\mathbf{C}} A + k - 1,$$

(c) generalized Lefschetz theorems [**So**$_4$], e.g. for (X, A) as above, and any $a \in A$,

$$\pi_j(X, A, a) = 0 \quad \text{for } j \leqslant \dim_{\mathbf{C}} A - \operatorname{cod}_{\mathbf{C}} A - k + 1.$$

Of course, the applications in (b) and (c) are modeled on the results for $\mathbf{P}_{\mathbf{C}}$ of Barth and Larsen ([**F** + **L**] gives a good survey of these and related results).

In §0, I review some background material, in particular the concepts of k ampleness and $k + 1$ convexity for holomorphic vector bundles. The former is an algebro-geometric notion [**So**$_1$] which coincides with ampleness in the sense of Grothendieck (see [**Ha**$_1$]) for $k = 0$. The latter is a differential geometric notion which coincides with Grauert negativity for $k = 1$. For $k = 0$, E is 0 ample if and only if E^* is 1 convex, but this is not true for $k > 0$ [see (0.4.4)].

1980 *Mathematics Subject Classification*. Primary 32F10.

In §1, I show that k ampleness of a vector bundle E implies the $k+1$ convexity of its dual E^*. This is important because k ampleness is much more easily checked than $k+1$ convexity, but it is the latter concept that is used in [**So$_2$**] to construct exhaustion functions on complements of submanifolds. I also obtain other results as simple corollaries of the proof, e.g. that if $\phi: X \to Y$ is a holomorphic map with finite fibres of a compact complex manifold X to a Kaehler manifold Y, then X is Kaehler.

§2, I use the results of §1 to give criteria for convexity of complements of submanifolds of products of Abelian varieties and Grassmannians. These criteria extend the results of [**So$_2$**].

In §3 I discuss the sharpness of my results. I also mention some recent work where N. Goldstein computes the k ampleness of the holomorphic tangent bundle of a homogeneous rational projective manifold, X, for all such X.

I would like to thank R. Hamilton for a helpful suggestion towards the proof of Lemma (1.1). I would also like to thank the referee for his suggestions.

This work was partially supported by the National Science Foundation and the Sonderforschungsbereich "Theoretische Mathematik" in Bonn.

0. Notation and background material.

(0.1) Complex spaces are reduced unless otherwise stated.

(0.2) I denote the holomorphic tangent bundle of a complex manifold X, by T_X and its dual by T_X^*. I denote the conjugate of T_X^* by $\overline{T}_X^*$. Given a C^∞ function f on a complex manifold, X, I denote the Levi form of f by $\mathfrak{L}(f)$. Recall $\mathfrak{L}(f)$ is a C^∞ section of $T_X^* \otimes \overline{T}_X^*$ given by

$$\sum_{i,j} \frac{\partial^2 f}{\partial z_i \partial \bar{z}_j} dz_i \otimes d\bar{z}_j$$

in local coordinates. The following lemma is well known; I have heard it attributed to A. Andreotti.

(0.2.1) LEMMA. *Let A be a complex subspace of a complex manifold X. There exists a C^∞ nonnegative function f_A on X with the properties*:

(a) *$\mathfrak{L}(f_A)$ is positive semidefinite on $T_X|_x$ for all $x \in A$,*

(b) *given any manifold point $x \in A$, $\mathfrak{L}(f_A)$ is positive definite on a linear subspace of $T_X|_x$ that surjects onto the fibre at x of the normal bundle of the smooth points of A.*

PROOF. Let $\{U_\alpha \mid \alpha \in \mathcal{Q}\}$ be a locally finite cover of X with $\{g_{1,\alpha}, \ldots, g_{n_\alpha,\alpha}\}$ holomorphic functions on U_α whose vanishing defines $A \cap U_\alpha$. Let $\{\rho_\alpha \mid \alpha \in \mathcal{Q}\}$ be a C^∞ partition of unity subordinate to $\{U_\alpha \mid \alpha \in \mathcal{Q}\}$. Then

$$f_A = \sum_\alpha \rho_\alpha \left(\sum_{i=1}^{n_\alpha} |g_{i,\alpha}|^2 \right)$$

is easily checked to have the desired properties. $\square$

The following is easy (cf. [**So**$_2$, Lemma (1.3.1)] for a proof).

(0.2.2) LEMMA. *Let $\mathfrak{Q}_1$ and $\mathfrak{Q}_2$ be two Hermitian forms on a complex vector space V. Assume $\mathfrak{Q}_1$ is positive semidefinite on V and positive definite on a complex subspace W of V. Assume $\mathfrak{Q}_2$ is positive definite on a complex subspace W' of V. Then $\lambda\mathfrak{Q}_2 + \mathfrak{Q}_1$ is positive definite on W' for all $\lambda > 0$ and, there exists an $\varepsilon > 0$ such that $\lambda\mathfrak{Q}_2 + \mathfrak{Q}_1$ is positive definite on W for $0 < \lambda < \varepsilon$.*

(0.3) Let ψ: $X \to Y$ be a holomorphic map from a complex space X to a complex manifold Y. Let $\mathbb{S}(X, \psi)$, the *singular set of the map* ψ, be the analytic subset of X consisting of points $x \in X$ such that either

(a) x is a singular point of X, or

(b) there is a neighborhood U of x consisting only of smooth points and the rank of $d\psi$ on $T_U|_x$ is less than the rank of $d\psi$ on $T_U|_y$ for a dense set of $y \in U$.

The fact that $\mathbb{S}(X, \psi)$ is a complex analytic set is most simply seen by using Hironaka's desingularization theorem [**Hi**]. Let π: $\tilde{X} \to X$ be a desingularization of X, i.e. π is proper and π: $\tilde{X} - \pi^{-1}(\mathbb{S}(X)) \to X - \mathbb{S}(X)$ is a biholomorphism where $\mathbb{S}(X)$ is the singular set of X. Then it is easily checked that $\mathbb{S}(X, \psi)$ is the image under π of the analytic set where $\psi \circ \pi$ is not of maximal rank. For a more elementary proof of this fact see [**Siu**, Lemma 2.A.4, p. 97].

(0.3.1) DEFINITION. *Let ψ: $X \to Y$ be a holomorphic map from a compact complex manifold to a complex manifold Y. The singular filtration of ψ,*

$$A_{n+1} \subseteq A_n \subseteq \cdots \subseteq A_0,$$

is defined by $A_0 = X$, $A_{i+1} = \mathbb{S}(A_i, \psi \mid A_i)$, and n being the least integer for which A_{n+1} is the empty set.

The important properties of this filtration for us are given by the following lemma:

(0.3.2.) LEMMA. *Let ψ, X, Y and $A_{n+1} \subseteq \cdots \subseteq A_0$ be as above. Then $n \leqslant \dim_{\mathbf{C}} X$ and $A_i - A_{i+1}$ is smooth. Further,*

(a) *the rank of $d\psi$ on $T_{A_i - A_{i+1}}|_x$ is constant for x in any connected component of $A_i - A_{i+1}$, and*

(b) *if g is a Hermitian metric on Y, then the restriction of ψ^*g to $T_{A_i - A_{i+1}}|_x$ is semipositive and has at most k nonpositive eigenvalues for all $x \in A_i - A_{i+1}$ where $k = \max_{y \in Y} \dim_{\mathbf{C}} \eta^{-1}(y)$.*

PROOF. All but (b) are immediate from Definition (0.3.1). To see (b) simply note that by (a), $\psi|_{A_i - A_{i+1}}$ is locally a product projection and thus the kernel of $d(\psi|_{A_i - A_{i+1}})$ on $T_{A_i - A_{i+1}}|_x$ is at most

$$\dim_{\mathbf{C}}(\psi^{-1}\psi(x)) \cap (A_i - A_{i+1}) \leqslant \max_{y \in Y} \dim_{\mathbf{C}} \psi^{-1}(y). \quad \square$$

For technical reasons I need X, a space, in the next definition.

(0.4) If E is a holomorphic vector bundle on a complex space X, then $P(E)$ denotes $(E^* - X)/_{\mathbf{C}^*}$, the space of lines of E^*. Let π_E: $P(E) \to X$ denote the obvious projection and ζ_E^* the tautological subbundle of $\pi_E^* E^*$. For more details see [**Ha**$_1$, **So**$_1$].

(0.4.1) DEFINITIONS. *Let X be compact complex manifold. A holomorphic vector bundle E on X is called k ample if ζ_E^r is spanned by global sections for some $r > 0$ and the map to $\mathbf{P_C}$ associated to $\Gamma(P(E), \zeta_E^r)$ has at most k dimensional fibres.*

I have need for the simplest criteria of k ampleness (cf. [**So₁**] for a detailed discussion of k ampleness).

(0.4.2) LEMMA. *Let E be a holomorphic vector bundle on a compact complex space X. Assume ζ_E^r is spanned by global sections over $P(E)$ for some $r > 0$. The following conditions are equivalent:*

(a) *E is k ample,*

(b) *given any finite-to-one holomorphic map ϕ: $Z \to X$ from a compact irreducible complex space, Z, to X and a nowhere vanishing holomorphic section of ϕ^*E^*, it follows that* $\dim_{\mathbf{C}} Z \leq k$.

PROOF. The equivalence of (a) and (b) is (1.7.3) of [**So$_1$**]. □

(0.4.3) DEFINITION. *Let E be a holomorphic vector bundle on a complex manifold X. E^* is said to be $k + 1$ convex if ζ_E possesses a Hermitian norm, $\| \ \|$, on $P(E)$, such that the Hermitian form associated to the curvature form of $\| \ \|$ has at most k nonpositive eigenvalues at each point of $P(E)$.*

Thus E^* is $k + 1$ convex if and only if ζ_E is metrically $k + 1$ convex [cf. [**Sc**]]. I would like to mention a confusing misprint which comes up in the discussion of $k + 1$ convexity in [**So$_2$**]; the phrase "E is equivalent to ζ_E" on line 8 of p. 39 should read "E^* is equivalent to ζ_E^*".

It follows from the Kodaira embedding theorem that $k + 1$ convexity of E^* implies k ampleness of E when $k = 0$. This is false for $k > 0$, even if there are no negative eigenvalues, as the next example illustrates (see (1.3) also).

(0.4.4) EXAMPLE. Let $\mathfrak{E}$ be an elliptic curve. There exist holomorphic line bundles L_1 on $\mathfrak{E}$ with the property that L_1^n has no sections for $n \neq 0$ and L_1 has a norm with 0 curvature form. Let L_2 be any very ample line bundle on $\mathfrak{E}$. Let $\mathfrak{L} = \pi_1^* L_1 \otimes \pi_2^* L_2$ on $\mathfrak{E} \times \mathfrak{E}$ where π_1 and π_2 are the product projections. Using any metric on L_2 with positive curvature form and the 0 curvature metric on L_1 we induce a metric on $\mathfrak{L}$ which makes $\mathfrak{L}^*$ 2 convex and which has the property that the Hermitian form associated to the curvature form is positive semidefinite. By construction $\mathfrak{L}^n$ has no section for any $n \neq 0$ and is thus not k ample for any integer k.

(0.4.5) DEFINITION. *A complex manifold G is q convex in the sense of Andreotti and Grauert* [**A + Gr**] *if there is a $C_0 \in \mathbf{R}$ and a proper C^∞ funciton φ: $G \to [0, \infty)$ whose Levi form $\mathfrak{L}(\varphi)$ has at most $q - 1$ nonpositive eigenvalues on $T_G|_x$ for all* $x \in G - \varphi^{-1}([0, C_0])$.

(0.4.6) PROPOSITION. *Let A be a connected complex submanifold of a compact complex manifold X. If N_A, the normal bundle of A, is spanned by global sections and N_A^* is $1 + q$ convex, then $X - A$ is $\operatorname{cod}_{\mathbf{C}} A + q$ convex in the sense of Andreotti and Grauert* (0.4.5).

This result is simply a special case of the main result, Proposition (1.3) of [$\mathbf{So_2}$], combined with Fritzsche's Proposition (1.5) of [$\mathbf{So_2}$].

(0.5) LEMMA. *Let X denote the Grassmannian* $\mathrm{Gr}(r, n)$ *of* $\mathbf{P}^r_{\mathbf{C}}$*'s in* $\mathbf{P}^n_{\mathbf{C}}$*. The tangent bundle,* T_X*, is* $r(n-r-1)$ *ample.*

PROOF. Assume (0.5) is false and use (0.4.2) to conclude that there is a holomorphic finite-to-one map ϕ: $Z \to X$ of a compact analytic space Z to X where $\dim_{\mathbf{C}} Z > r(n-r-1)$ and $\phi^* T_X^*$ has a nowhere zero section s.

In [$\mathbf{So_2}$] I constructed a function f on T_X^* (the square of the norm of the metric μ of (2.4.1) of [$\mathbf{So_2}$]) with the property that $\mathfrak{L}(f)$ is positive semidefinite and has at most $(n-r-1)\,r$ nonpositive eigenvalues at any point of $T_X^* - X$. Since the graph of s in $\phi^* T_X^*$ misses Z, I can pull f back to get $\tilde{f}$, a plurisubharmonic function on Z with $\mathfrak{L}(\tilde{f})$ having at least one positive eigenvalue at each smooth point of Z. By the maximum principle $\tilde{f}$ must be constant and thus this is impossible. This absurdity proves the result. □

(0.6) LEMMA. *Let* $\mathfrak{Q} \subseteq \mathbf{P}^n_{\mathbf{C}}$ *be a smooth quadric. Then* $T_{\mathfrak{Q}}$ *is one-ample.*

PROOF. By (0.4.2) I must simply bound the dimension of a compact irreducible analytic set Z where ϕ: $Z \to \mathfrak{Q}$ is a finite-to-one holomorphic map and $\phi^* T^*_{\mathfrak{Q}}$ possesses a nontrivial section s.

I can assume Z is normal by replacing it by its normalization, and also that $\dim_{\mathbf{C}} Z \geqslant 2$ since otherwise there is nothing to prove.

I have the exact sequence

$$0 \to N^*_{\mathfrak{Q}} \to T^*_{\mathbf{P}^n_{\mathbf{C}}}\big|_{\mathfrak{Q}} \to T^*_{\mathfrak{Q}} \to 0.$$

Pull back this sequence by ϕ and note that $\phi^* T^*_{\mathbf{P}^n_{\mathbf{C}}}$ can have no nontrivial section since $T_{\mathbf{P}^n_{\mathbf{C}}}$ is 0 ample and $\dim_{\mathbf{C}} Z \geqslant 1$. We thus see that $H^1(Z, \phi^* N^*_{\mathfrak{Q}}) \neq 0$ since it contains the image of s. Since $N_{\mathfrak{Q}}$ is very ample, $\dim_{\mathbf{C}} Z \geqslant 2$, and Z is normal, this is impossible. To see this use [**Mu**]. □

1. A perturbation lemma and its consequences.

(1.1) LEMMA. *Let* ψ: $X \to Y$ *be a holomorphic map from a compact complex manifold* X *to a complex manifold* Y*. Let* g *be a Hermitian metric on* Y*, and let* $k = \max_{y \in Y} \dim_{\mathbf{C}} \psi^{-1}(y)$*. Then there exists a nonnegative* C^∞ *function* f *on* X *such that* $\mathfrak{L}(f) + \psi^* g$ *has at most* k *nonpositive eigenvalues on* $T_X|_x$ *for each* $x \in X$*.*

PROOF. Let $\phi = A_{n+1} \subseteq \cdots \subseteq A_0 = X$ be the singular filtration (0.3.1) of ψ. Consider the property $*_j$:

$$*_j \quad \begin{cases} \text{There exists a nonnegative } C^\infty \text{ function } f_j \text{ on } X \text{ with the} \\ \text{property that for each } x \in A_j,\ \mathfrak{L}(f_j) + \psi^* g \text{ is positive definite} \\ \text{on a vector subspace of } T_X|_x \text{ of codimension at most } k. \end{cases}$$

The lemma is the statement $*_0$. By induction I am reduced to the following:

(1.1.1) *Claim.* $*_n$ *is true and* $*_{j+1}$ *implies* $*_j$ *for* $j < n$.

PROOF OF CLAIM. Let f_{A_j} be a function that exists for A_j by (0.2.1). $\mathfrak{L}(f_{A_j}) + \psi^*g$ is positive semidefinite on $T_X|_x$ for each $x \in A_j$, and is positive definite on a vector subspace of $T_X|_x$ of codimension at most k for all $x \in A_j - A_{j+1}$. This follows from (0.2.1) and (0.3.2), and proves $*_n$ by taking $j = n$. Thus we can assume without loss of generality that $n \neq 0$.

Assume $*_{j+1}$. Clearly $\mathfrak{L}(f_{j+1}) + \psi^*g$ is positive definite on a vector subspace of $T_X|_x$ of codimension at most k for each x in some neighborhood $\mathfrak{U}$ of A_{j+1}. Let $\mathfrak{V}$ be a relatively compact open set of $X - A_{j+1}$ with the properties that $\mathfrak{U} \cup \mathfrak{V} = X$.

Consider $\mathfrak{Q}(\lambda) = \lambda[\mathfrak{L}(f_{j+1}) + \psi^*g] + [\mathfrak{L}(f_{A_j}) + \psi^*g]$ for $\lambda > 0$. By the first paragraph of the proof of (1.1.1) above and Lemma (0.2.2) it follows that given $x \in A_j - A_{j+1}$ there is an $\varepsilon(x) > 0$, such that for $0 < \lambda \leq \varepsilon(x)$, $\mathfrak{Q}(\lambda)$ is positive definite on a vector subspace of $T_X|_x$ of codimension at most k. By openness of this property and compactness of $\overline{\mathfrak{V}}$, I can choose an ε independent of $x \in A_j \cap \overline{\mathfrak{V}}$ for which this is true.

Again by (0.2.2) and the defining property of $\mathfrak{U}$, $\mathfrak{Q}(\lambda)$ is positive definite on a vector subspace of $T_X|_x$ of codimension at most k for all $x \in \mathfrak{U}$ and for all $\lambda > 0$.

Putting these last two paragraphs together we see that $*_j$ is true with

$$f_j = \left(\frac{1}{\varepsilon + 1}\right) \cdot \left(\varepsilon f_{j+1} + f_{A_j}\right). \quad \square$$

(1.2) COROLLARY. *If* $\psi\colon X \to Y$ *is a holomorphic finite-to-one map from a compact complex manifold* X *to a Kaehler manifold* Y, *then* X *is Kaehler.*

PROOF. Let g be the Kaehler metric on Y. Let f be the function whose existence (1.1) asserts. Then $\mathfrak{L}(f) + \psi^*g$ is the Kaehler metric on X. $\square$

(1.3) COROLLARY. *Let* E *be a* k *ample vector bundle on a compact complex manifold* X. *Then* E^* *is* $k + 1$ *convex.*

PROOF. Since both definitions are in terms of ζ_E on $P(E)$, I can assume without loss of generality that E is a line bundle.

I have a holomorphic map $\psi\colon X \to \mathbf{P}_{\mathbf{C}}$ associated to $\Gamma(X, E^r)$ where I have chosen $r > 0$ so that E^r is spanned by global sections. Let $\|\ \|$ be the usual norm on $O(1)$, the hyperplane section bundle on $\mathbf{P}_{\mathbf{C}}$, with the property that the Hermitian form g, associated to its curvature form is the Fubini- Study metric on $\mathbf{P}_{\mathbf{C}}$. Let f be the function whose existence is guaranteed by (1.1). Note that $e^{-f}\psi^*\|\ \|$ is a norm on E with $\mathfrak{L}(f) + \psi^*g$ the Hermitian form associated to its curvature form. $\square$

(1.4) COROLLARY. *Let* A *be a submanifold of a compact complex manifold. Assume* N_A, *the normal bundle of* A, *is spanned by global sections and is* k *ample. The* $X - A$ *is* $\operatorname{cod}_{\mathbf{C}} A + k$ *convex in the sense of Andreotti and Grauert* (*cf.* (0.4.5)).

PROOF. Apply (1.3) and (0.4.6). □

The following theorem is proven in greater generality using stratified Morse theory by Goresky and Mcpherson (see [**F** + **L**]); a version using integer homology was given in [**So**$_1$, Proposition (1.16)]. It seems worth putting here since it is an immediate consequence of the preceding considerations.

(1.5) COROLLARY. *Let $\psi: X \to \mathbf{P}_{\mathbf{C}}^N$ be a holomorphic map from a compact complex manifold X to $\mathbf{P}_{\mathbf{C}}^N$. If the maximal fibre dimension of ψ is k then given any linear $\mathbf{P}_{\mathbf{C}}^{N-r}$,*

$$\pi_j\left(X, \psi^{-1}\left(\mathbf{P}_{\mathbf{C}}^{N-r}\right), x\right) = 0, \qquad j \leqslant \dim_{\mathbf{C}} X - r - k,$$

for any $x \in \psi^{-1}(\mathbf{P}_{\mathbf{C}}^{N-r})$.

PROOF. $\psi^{-1}(\mathbf{P}_{\mathbf{C}}^{N-r})$ is defined by the vanishing of a holomorphic section s of $\bigoplus_{r\,\text{copies}} \psi^*O(1) = E$. Putting the norm $\|\ \|$ on E, which is gotten from the square root of the sum of squares of norms on the $\psi^*O(1)$ by (1.3), we can apply Morse theory to $\|s\|$ (cf. [**Bo** and **So**$_4$]). □

2. Criteria for convexity. The following is the main theorem of this paper (recall (0.5) and (0.6)):

(2.1) PROPOSITION. *Let X_0 be an Abelian variety with k_0, the maximal dimension of any proper Abelian subvariety. Let X_i for $i = 1,\ldots,r$ be a compact homogeneous complex manifold whose tangent bundle T_{X_i} is k_i-ample (cf. (0.5) and (0.6) for X_i a Grassmannian or a quadric). Let A be a submanifold of $X = \prod_{i=0}^r X_i$ and let π_i: $X \to X_i$ for $i = 0,\ldots,r$ be the product projections. Then N_A is k ample and $X - A$ is $\operatorname{cod}_{\mathbf{C}} A + k$ convex in the sense of Andreotti and Grauert where*

$$k = \max_i \left\{k_i + \max_{y\in X_i}\left[\dim_{\mathbf{C}} \pi_i^{-1}(y) \cap A\right]\right\}. \tag{2.1.1}$$

This k is bounded by $\dim_{\mathbf{C}} X - \min_i\{\dim_{\mathbf{C}} X_i - k_i\}$.

PROOF. First note that the last assertion is true. Indeed

$$\dim_{\mathbf{C}} \pi_i^{-1}(y) \cap A \leqslant \dim_{\mathbf{C}} \pi_i^{-1}(y) = \dim_{\mathbf{C}} \prod_{j\neq i} X_j = \dim_{\mathbf{C}} X - \dim_{\mathbf{C}} X_i.$$

To prove the theorem it suffices by (0.4.2) to bound the dimension of a compact irreducible analytic space Z with the properties:

(2.1.2) there is a finite-to-one holomorphic map $\phi: Z \to A$, and

(2.1.3) there is a nowhere vanishing holomorphic section s of $\phi^*N_A^*$.

Noting that

$$0 \to N_A^* \to \bigoplus_i \pi_i^* T_{X_i}^*\big|_A \to T_A^* \to 0 \tag{2.1.4}$$

we see that we can write $s = \Sigma_i s_i$ where s_i is a holomorphic section of $\phi^*T_{X_i}^*$.

If s_i is not identically zero for each $i > 0$ then (2.1) is proven. To see this, note that $(\pi_i \circ \phi)^*T_{X_i}^*$ is $k_i + \max_{y\in X_i} \dim_{\mathbf{C}} \phi^{-1}(\pi_i^{-1}(y))$ ample by [**So**$_1$, Corollary (1.9)] for $i > 0$. Thus by (0.4.2) and the fact that $(\pi_i \circ \phi)^* T_{X_i}$ is spanned, it follows that s_i is identically zero unless

$$\dim_{\mathbf{C}} Z \leqslant k_i + \max_{y \in X_i} \dim_{\mathbf{C}} \phi^{-1}(\pi_i^{-1}(y)).$$

I can now assume without loss of generality that each s_i is identically zero for $i > 0$. Therefore s gives rise to a nonzero section of $(\pi \circ \phi)^*(T^*_{X_0})$ gotten by pulling back a nonzero holomorphic one form ω first from X_0 to X by π_0^* and then to Z by ϕ. By (2.1.4) $\pi_0^* \omega \mid_A$ is zero as a section of T^*_A at all points of $A = \phi(Z)$. Therefore ω is zero when restricted to any neighborhood on $\mathfrak{Z} = \pi_0(\phi(Z))$ of a smooth point of $\mathfrak{Z}$.

Let $g\colon \tilde{\mathfrak{Z}} \to \mathfrak{Z}$ be a projective desingularization of $\mathfrak{Z}$. (Note that $\mathfrak{Z}$ is projective since it is an analytic subset of X_0, an Abelian variety, and use **[Hi]**.)

Since $g^* \omega$ is zero on $\tilde{\mathfrak{Z}}$ the Albanese variety of $\tilde{\mathfrak{Z}}$ maps onto a proper Abelian subvariety of X_0 that contains $\mathfrak{Z} = \pi_0(\phi(Z))$. This proves the proposition. □

(2.2) REMARKS. The trick used above with the Abelian variety goes back to Hartshorne **[Ha$_2$]**; it was extensively exploited in **[So$_1$]**. In the case of a nonsingular A, Proposition (2.1) considerably improves (2.3) and (2.4) of **[So$_2$]**.

(2.3) COROLLARY. *If $A, X_0, \ldots, X_r$ are as above and $\pi_0(A)$ contains no Abelian variety then k_0 can be replaced by $k_0 - 1$ in* (2.1.1).

PROOF. I use the notation of (2.2) above. Under the hypotheses of (2.3), $\mathfrak{Z}$ cannot equal the image of the Albanese variety of $\tilde{\mathfrak{Z}}$. □

(2.4) COROLLARY. *Let A be a submanifold of $X = \prod_{i=0}^r X_i$ where X_0 is a simple Abelian variety and X_i for $i > 0$ is $\mathbf{P}_{\mathbf{C}}^{n_i}$. Assume $\pi_i\colon A \to X_i$ is finite-to-one for all i where $\pi_i\colon X \to X_i$ is the product projection. Then N_A is* 0 *ample* (*i.e. ample in the sense of Grothendieck*).

(2.5) REMARK. Using **[So$_4$]**, (2.1) and (0.5) I get corollaries such as the following: *Assume that the A of* (2.1) *is of codimension* 2 *and X_i for $i > 0$ is the Grassmannian of $\mathbf{P}_{\mathbf{C}}^{r_i}$'s in $\mathbf{P}_{\mathbf{C}}^{n_i}$. Let $n_0 = \dim_{\mathbf{C}} X_0 - k_0$. Then the restriction*

$$\operatorname{Pic}(X) \to \operatorname{Pic}(A)$$

is an isomorphism if $\min_i\{n_i\} \geqslant 6$ *and an injection with torsion free cokernel if* $\min_i\{n_i\} \geqslant 5$.

3. Concluding remarks. The results of this paper raise two questions. The first is concerned with the exact relationship of k ampleness and q convexity in the sense of Andreotti and Grauert. The following conjecture holds for all examples I know:

Conjecture. Let A be a smooth connected projective submanifold of a homogeneous projective manifold X. Then N_A is k-ample if and only if $X - A$ is $\operatorname{cod}_{\mathbf{C}} A + k$ *convex in the sense of Andreotti and Grauert.*

The 'only if' part of the above is of course (1.4) of this paper.

The second question concerns what is the smallest k such that the holomorphic tangent bundle of a rational homogeneous projective manifold, X, is k ample; it is not hard to check that the results in (0.5) and (0.6) are sharp. N. Goldstein has recently given a complete and very elegant solution to the above (see **[Go]**).

REFERENCES

[A + Gr] A. Andreotti and H. Grauert, *Théorèmes de finitude pour la cohomologie des espaces complexes*, Bull. Soc. Math. France **90** (1962), 193–259.

[Bo] R. Bott, *On a theorem of Lefschetz*, Michigan Math. J. **6** (1959), 211–216.

[F + L] W. Fulton and R. Lazarsfeld, *Connectivity and its applications in algebraic geometry*, Algebraic Geometry: Proceedings of the 1980 Chicago Circle Conference (A. Liebgober and P. Wagreich, Editors), Lecture Notes in Math., vol. 862, Springer-Verlag, New York, 1981, pp. 26–92.

[Go] N. Goldstein, *Ampleness and connectedness in complex G/P*, Trans. Amer. Math. Soc. (to appear).

[Ha_1] R. Hartshorne, *Ample vector bundles*, Inst. Hautes Études Sci. Publ. Math. **29** (1966), 63–94.

[Ha_2] ______, *Ample vector bundles on curves*, Nagoya Math. J. **43** (1971), 73–89.

[Hi] H. Hironaka, *Bimeromorphic smoothing of a complex-analytic space* (Summary), Math. Inst. Warwick University, England (1971).

[Mu] D. Mumford, *Pathologies*. III, Amer. J. Math. **89** (1967), 94–104.

[Sc] M. Schneider, *Über eine Vermutung von Hartshorne*, Math. Ann **201** (1973), 221–229.

[Siu] Yum-Tong Siu, *Techniques of extension of analytic objects*, Lecture Notes in Pure and Appl. Math., vol. 8, Marcel Dekker, New York, 1974.

[So_1] A. J. Sommese, *Submanifolds of Abelian varieties*, Math. Ann. **233** (1978), 229–256.

[So_2] ______, *Concavity theorems*, Math. Ann. **235** (1979), 37–53.

[So_3] ______, *Complex subspaces of homogeneous complex manifolds*. I. *Transplanting theorems*, Duke Math. J. **46** (1979), 527–548.

[So_4] ______, *Complex subspaces of homogeneous complex manifolds*. II. *Homotopy results*, Nagoya Math J. **86** (1982), 101–129.

UNIVERSITY OF NOTRE DAME, INDIANA

Proceedings of Symposia in Pure Mathematics
Volume **40** (1983), Part 2

TRIPLE CONTACT OF PLANE CURVES: SCHUBERT'S ENUMERATIVE THEORY

ROBERT SPEISER

Among other questions, Schubert [**1**] wanted, and found, answers to these:

(1) Let $C = \{C_t\}$ be a 2-parameter family of curves in $\mathbf{P}^2$, general enough so that the numbers

$M = \sharp C_t$ tangent to a general line at a general point,
$K = \sharp C_t$ with given general point as cusp and
$K' = \sharp C_t$ with given general line as inflectional tangent

make sense. Let D be another curve, with

$$n = \deg(D), \quad n' = \text{class}(D) = \deg(\check{D}), \quad e = \sharp \text{ flexes of } D.$$

Then how many points of D have triple contact with members of $C = \{C_t\}$? If D is in general position, the answer is

$$(3n + e)M + nK' + n'K.$$

(2) For a one-parameter family $C = \{C_t\}$ of projective plane curves, assume

$\mu = \sharp C_t$ through a general point,
$\mu' = \sharp$ tangent to a given general line,
$k =$ degree of $\{\text{cusps of } C_t\} \subset \mathbf{P}^2$,
$k' =$ degree of $\{\text{infl. tangents of } C_t\} \subset \check{\mathbf{P}}^2$

make sense, where $\check{\mathbf{P}}^2$ is the dual plane. Given 2 such families (in general position), how many triple contacts are there, between members of one family and members of the other? The answer is

$$\mu_1 k_2' + \mu_1' k_2 + k_1' \mu_2 + k_1 \mu_2' + 3(\mu_1 \mu_2 + \mu_1' \mu_2'),$$

where each subscript indicates the family whose invariant is to be computed.

Triple contact means: a common point, a common tangent at that point, and, essentially, the same second-degree Taylor term, in local coordinates. But, to solve enumerative problems, we need to see also how this *second-order data* fits together

---

1980 *Mathematics Subject Classification*. Primary 14B05.

0082-0717/81/0000-0659/$02.25

globally over $\mathbf{P}^2$, and, more precisely, over the first-order data on $\mathbf{P}^2$. We thus have a basic construction problem, and we shall solve it in 2 stages, starting with a construction for the first-order information.

First-order data. At a point p on a plane curve C, first-order data of p is represented by a line L through p, tangent to C at p.

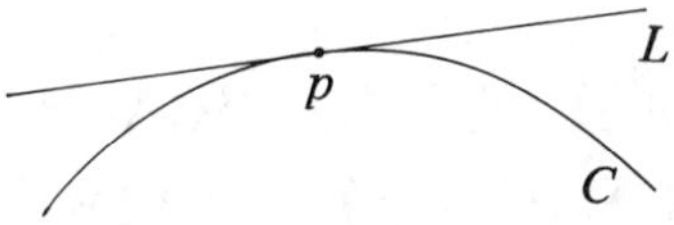

Thus the incidence correspondence

$$X = \{(p, L) \mid p \in L\} \subset \mathbf{P}^2 \times \check{\mathbf{P}}^2$$

parametrizes first-order data on $\mathbf{P}^2$. Let α_1 and α_2 be the pullbacks to X of the hyperplane sheaves on the 2 factors of $\mathbf{P}^2 \times \check{\mathbf{P}}^2$. Then the intersection ring is

$$A^{\cdot}(X) = \mathbf{Z}[\alpha_1, \alpha_2]/\left(\alpha_1^3, \alpha_2^3, \alpha_1^2 - \alpha_1\alpha_2 + \alpha_2^2\right)$$

as one sees easily, since $X = \mathbf{P}(\Omega^1_{\mathbf{P}^2})$.

Second-order data. Here we are after a $\mathbf{P}'$-bundle $X^* \to X$ whose fibres measure, in effect, curvature. Schubert's basic idea is to consider 3 points on a curve,

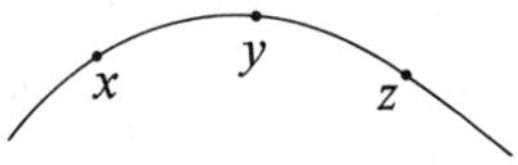

that is, a triangle, and let the vertices approach $p \in C$. Throwing in the sides gives first-order data in the limit; hence we shall use

$$W = \left\{((x, y, z), (L, M, N)) \in (\mathbf{P}^2)^3 \times (\check{\mathbf{P}}^2)^3 \;\middle|\; \begin{matrix} x, y \in L \\ x, z \in M \\ y, z \in N \end{matrix}\right\}$$

as base space. For second-order data, let w be a general point of W (x, y, z distinct, say). Let

$$\Sigma(w) = \{\text{all conics through the vertices of } w\}.$$

Thus, Σ gives a rational map from W to the $G(2,5)$ parametrizing 2-parameter linear families of conics in $\mathbf{P}^2$. Define

$$W^* = \text{closure of } \Gamma_\Sigma \text{ in } W \times G(2,5).$$

This is a concrete realization of the collection of *complete triangles* studied by Schubert.

Identify X with the subvariety {points of W with $x = y = z$, $L = M = N$} of W, and let X^* = pullback of X to W^* under the projection $W^* \to W$. Then X^* gives the second-order data on $\mathbf{P}^2$! In fact, one can show that each smooth point (or cusp) of C gives just one point of X^*, namely {all conics meeting C triply at P}, each point of X^* comes up in this way, and if 2 curves have the same second-order data at p, then they give the same point on X^*.

(The interpretation given here of second-order data via conics is due originally to Study (1901); the present version follows Semple [**2**].)

To compute $A^{\cdot}(X^*)$, we have the following difficult central result, proved jointly with Joel Roberts.

THEOREM. *We have* 2 *sections*, s_0 *and* s_∞, *of the projection* $X^* \to X$, *corresponding to curvature* $= 0$ (*at flexes*) *and curvature* $= \infty$ (*at cusps*). *Hence*,

$$X^* = \mathbf{P}(L \oplus M)$$

for L, $M \in \operatorname{Pic} X$, *when* L *is the quotient giving* s_0, M *giving* s_∞. *Moreover*,

$$c_1(L) = 3\alpha_2$$

and

$$c_1(M) = 3\alpha_1.$$

SKETCH OF PROOF. The sections s_0 and s_∞ are constructed directly. Since $s_0(X) \cap s_\infty(X) = \varnothing$, we have the direct sum as claimed. For the Chern classes: let $\overline{W} =$ blowup of W along X, $B =$ scheme gotten by first blowing up the triple diagonal $\Delta \subset (\mathbf{P}^2)^3$, then the 3 double diagonals Δ_{ij}. Then $\overline{W}$ and B are smooth, and we have a commutative diagram

$$\begin{array}{ccccc} & & \overline{W} & & \\ & \swarrow & & \searrow & \\ B & & & & W^* \\ & \searrow & & \swarrow & \\ & & W & & \end{array}$$

of birational surjections, where both upper arrows are blowups, along the pullbacks, $X_B \subset B$ and $X^* \subset W^*$, of $X \subset W$. Looking deeper, one sees that the diagram of pullbacks

$$\begin{array}{ccccc} & & \overline{X} & = & \text{exceptional locus on } \overline{W} \\ & \swarrow & & \searrow & \\ X_B & & & & X^* \\ & \searrow & & \swarrow & \\ & & X & & \end{array}$$

is Cartesian, all maps $\mathbf{P}^1$-bundles. But $\overline{X} \to X_B$ is $\mathbf{P}$ (conormal sheaf of X_B in B), and

$$X_B = C \underset{\text{complete}}{\bigcap} D,$$

where $C =$ {triangles on B with collinear vertices} and $D =$ {triangles on B with concurrent vertices}.

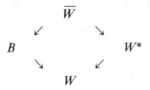

Thus $\overline{X} \to X_B$ is a $\mathbf{P}(E)$, E decomposable of rank 2, unique up to a twist, with known summands (modulo Pic(X), we can take the C-summand with $c_1 = 3\alpha_1$, and the D-summand with $c_1 = 3\alpha_2$). But the lifts of s_0 and s_∞ show that $X \to X_B$ is also of the form $\mathbf{P}(E')$, with $E' = L' \oplus M'$, where L', M' are the pullbacks of L and M to X_B. Further, modulo Pic(X), only *one* twist of E has summands which descend to X, hence must be, after a twist, perhaps, E'. By Krull-Schmidt, the indecomposable summands are unique up to isomorphism, so the summands of E are, essentially, L' and M'. But, *in what order*?

To settle this crucial point one can use a divisorial formula (proved readily on B and applied on W^*, whose Pic is isomorphic):

$$[C] + h_1 + h_2 + h_3 = [D] + \check{h}_1 + \check{h}_2 + \check{h}_3,$$

where the h_i are pullbacks of the hyperplane sheaves on the factors of $(\mathbf{P}^2)^3$, and the $\check{h}_i$ are similar, but on $(\check{\mathbf{P}}^2)^3$.

To distinguish the summands, one reads this formula on W^*, and restricts to X^*. Then one has

$$[C \mid X^*] + 3\alpha_1 = [D \mid X^*] + 3\alpha_2,$$

which forces the c_1's to be as claimed, and, as a bonus, implies the well-known Plücker formula for cusps and flexes of a plane curve.

COROLLARY. *We have*

$$\begin{aligned} A^{\cdot}(X^*) &= A^{\cdot}(X)[\eta]/\eta^2 + 3(\alpha_1 - \alpha_2)\eta \\ &= A^{\cdot}(X)[\zeta]/\zeta^2 + 3(\alpha_2 - \alpha_1)\zeta \end{aligned}$$

where

$$\begin{aligned} \eta &\Leftrightarrow s_0(X) \quad (\textit{flexes}), \\ \zeta &\Leftrightarrow s_\infty(X) \quad (\textit{cusps}). \end{aligned}$$

Duality. To work out the coefficients of arbitrary classes in $A^{\cdot}(X)$ relative to the bases suggested by the last Corollary, we need to know the duality structure on $A^{\cdot}(X)$, with respect to the intersection pairing, and this can be done by elementary calculation.

(1) The class of a point in $A^4(X^*)$ is

$$\alpha_1\alpha_2^2\eta = \alpha_1^2\alpha_2\eta = \alpha_1\alpha_2^2\zeta = \alpha_1^2\alpha_2\zeta.$$

(2) For $A^3(X^*) \otimes A^1(X^*) \to A^4(X^*) = \mathbf{Z}$, the basis $\{\alpha_1^2\alpha_2, \alpha_2^2\eta, \alpha_1^2\zeta\}$ is dual to $\{\eta + 3\alpha_1, \alpha_1, \alpha_2\}$.

(3) Similarly, for $A^2(X^*) \otimes A^2(X^*) \to A^4(X^*) = \mathbf{Z}$, the bases $\{\alpha_1^2, \alpha_2^2, \alpha_1, \eta, \alpha_1\zeta\}$ and $\{\alpha_2\eta + 3\alpha_1^2, \alpha_1\eta - 3\alpha_2^2, \alpha_2^2, \alpha_1^2\}$ are dual. The pairing on $A^2(X^*)$ cannot be diagonalized over $\mathbf{Z}$.

Deriving the formulas. First, one lifts the total spaces of our curves or families to subvarieties of X^*, to obtain varieties of 2nd order data, whose intersections we can count. For smooth points the lifts to X^* are clear, so for nonsmooth points one passes to the closure. To each curve $C \subset \mathbf{P}^2$ one obtains, thus, a *proper transform* $\tilde{C} \subset X^*$; similarly for families. For a family $C = \{C_t\}$ the proper transform $\tilde{C}$ may be bigger than the union of the proper transforms $\tilde{C}_t$: smooth points of nearby curves could give limits in the neighborhood of a singularity.

From here, one uses duality to express the classes, in $A^{\cdot}(X^*)$ of the proper transforms in terms of the given bases; then the evaluation of the products is easy.

That the formulae also make enumerative sense, i.e., count actual solutions, is also true, but we shall skip the details here.

REFERENCES

1. H. Schubert, *Anzahlgeometrische Behandlung des Dreiecks*, Math. Ann. **17** (1880), 153–212.

2. J. G. Semple, *The triangle as a geometric variable*, Mathematica **1** (1954), 80–88.

3. J. Roberts and R. Speiser, *Schubert's enumerative geometry of triangles from a modern viewpoint*, Proc. Midwest Alg. Geom. Conf., Chicago Circle (Libgoter and Wagreich, Eds.), Lecture Notes in Math., Springer-Verlag, Berlin and New York, 1981.

4. ______, *Enumerative geometry of triangles*. I, II, III (in preparation).

[The next to the last item is a brief announcement, giving some extra indications about the constructions of B, $\overline{W}$, etc., but without the main results on $A^{\cdot}(X^*)$ given here. Full details will be given in the 3 papers in the last item.]

ILLINOIS STATE UNIVERSITY, NORMAL, ILLINOIS

Proceedings of Symposia in Pure Mathematics
Volume **40** (1983), Part 2

MIXED HODGE STRUCTURES ASSOCIATED WITH ISOLATED SINGULARITIES

J. H. M. STEENBRINK

Introduction. This paper deals with applications of the theory of mixed Hodge structures to the study of isolated singularities. In contrast with most of the work that has been done in mixed Hodge structures, we try to avoid arguments which refer to the singularities as lying on a compact algebraic variety. Instead, our work depends heavily on resolution of singularities and on the (global) geometric properties of the exceptional divisor. We first study the mixed Hodge structure on the local cohomology groups of an isolated singularity of a complex space. The main result is Corollary (1.12), which is a consequence of the decomposition theorem in intersection cohomology, proved by P. Deligne, O. Gabber, A. Beilinson and I. Bernstein. It provides a bound on the weight that may occur on the local cohomology groups. It has been observed first by Deligne on the Metz Colloquium 1974 "sur la monodromie" that the vanishing (co)homology groups of a one-parameter smoothing of an isolated singularity carry a mixed Hodge structure. The theory has been developed in [**26**] for isolated hypersurface singularities and in §2 we describe its properties for the general case. The main properties are listed in (2.6); an important role is played by the "variation of the monodromy". We find that certain invariants of the mixed Hodge structure are in fact independent of the smoothing (Proposition (2.13) and Theorem (2.24)). For the surface case we find proofs of results, conjectured by J. Wahl [**30**] and proved before under the assumption that the smoothing can be globalized. In the last section we introduce the concept of a Du Bois singularity, and relate this to the problem of cohomological insignificance of smoothings. We classify two-dimensional isolated Gorenstein singularities with the Du Bois property and study smoothings of isolated Du Bois singularities. This work has been stimulated very much by conversations and correspondence with many people, among whom I like to mention R.-O. Buchweitz, A. Durfee, J. Carlson, I. N. Shepherd-Barron

---

1980 *Mathematics Subject Classification*. Primary 14C30, 14D05, 14H20.

0082-0717/81/0000-0660/$06.50

and G.-M. Greuel; I also would like to thank those at the Arcata conference who have given their comments on the earlier draft of this paper.

1. Local cohomology.

(1.1) A mixed Hodge structure is a triple $(H_{\mathbf{Z}}, W, F)$ where

(1) $H_{\mathbf{Z}}$ is a finitely generated abelian group;

(2) W is an increasing filtration of $H_{\mathbf{Q}} = H_{\mathbf{Z}} \otimes \mathbf{Q}$;

(3) F is a decreasing filtration of $H_{\mathbf{C}} = H_{\mathbf{Z}} \otimes \mathbf{C}$.

These data must satisfy the conditions, that for all integers p,

$$F^{p+1} \cap W_k \otimes \mathbf{C} + \overline{F^{k-p}} \cap W_k \otimes \mathbf{C} + W_{k-1} \otimes \mathbf{C} = W_k \otimes \mathbf{C};$$

$$\left(F^{p+1} + W_{k-1} \otimes \mathbf{C}\right) \cap \left(\overline{F^{k-p}} + W_{k-1} \otimes \mathbf{C}\right) = W_{k-1} \otimes \mathbf{C}.$$

W is called the weight filtration and F is called the Hodge filtration. Write F_k the filtration, induced by F on $\mathrm{Gr}_k^W H_{\mathbf{C}} = W_k \otimes \mathbf{C}/W_{k-1} \otimes \mathbf{C}$, and let

$$H^{pq} = F^p_{p+q} \cap \overline{F^q_{p+q}} \subset \mathrm{Gr}^W_{p+q} H_{\mathbf{C}}.$$

Then

$$\mathrm{Gr}_k^W H_{\mathbf{C}} = \bigoplus_{p+q=k} H^{pq} \quad \text{with } H^{qp} = \overline{H^{pq}}.$$

This is the Hodge decomposition of $\mathrm{Gr}_k^W H_{\mathbf{C}}$.

(1.2) In many important cases, the space $H_{\mathbf{Z}}$ has the form $H^k(X, \mathbf{Z})$ where X is some algebraic variety over $\mathbf{C}$. In this case P. Deligne **[6]** has constructed a canonical and functional mixed Hodge structure on $H_{\mathbf{Z}}$. The main idea is: to lift the filtrations F and W to a complex $K^{\cdot}$ of sheaves on some space X', which, has the property that its hypercohomology groups satisfy

$$\mathbf{H}^k(X', K^{\cdot}) = H^k(X, \mathbf{C})$$

and that F induces a pure Hodge structure of weight $r + k$ on the hypercohomology groups

$$\mathbf{H}^k\left(X', \mathrm{Gr}_r^W(K^{\cdot})\right).$$

(1.3) Our main interest lies in cohomology groups, associated with a smoothing of an isolated singularity of a complex space. Among these one finds of course those associated with the singularity as such. We review them first and study some of their properties.

(1.4) Let (X, x) be a germ of an n-dimensional complex space with an isolated singular point $x \in X$. We choose the representative X such that it is a contractible Stein space, e.g. a closed subset of a sufficiently small open ball in $\mathbf{C}^N$ with center x.

Let $\pi\colon \tilde{X} \to X$ be a resolution of the singularity x; π is a proper, bimeromorphic map, which maps $\tilde{X} \setminus \pi^{-1}(x)$ isomorphically onto $X \setminus \{x\}$; and $D = \pi^{-1}(x)$ is a union of smooth divisors with normal crossings on $\tilde{X}$. The cohomology groups which interest us are $H^*(D)$, $H^*_D(\tilde{X})$ and $H^*_{\{x\}}(X)$. They all carry canonical mixed Hodge structures which are closely related.

(1.5) Let us start with $H^*(D)$. Its mixed Hodge structure has been described in detail in [**16** and **25**]. It is obtained by the isomorphism

$$H^k(D, \mathbf{C}) \cong \mathbf{H}^k(D, A^{\cdot}_D)$$

where $A^{\cdot}_D$ is the single complex of sheaves, associated with the double complex $A^{\cdot\cdot}_D$ defined as follows. Choose an ordering of the components of D,

$$D = D_1 \cup \cdots \cup D_m.$$

Let $\tilde{D}^{(q)}$, $q \geqslant 1$, denote the disjoint union of all q-fold intersections of different components of D,

$$\tilde{D}^{(q)} = \coprod_{i_1 < \cdots < i_q} D_{i_1} \cap \cdots \cap D_{i_q}.$$

We have natural maps $\delta_j\colon \tilde{D}^{(q)} \to \tilde{D}^{(q-1)}, j = 1, \ldots, q$, given by the inclusions

$$D_{i_1} \cap \cdots \cap D_{i_q} \subset D_{i_1} \cap \cdots \cap D_{i_{j-1}} \cap D_{i_{j+1}} \cap \cdots \cap D_{i_q},$$

and projections $a_q\colon \tilde{D}^{(q)} \to D$.

Set $A^{pq}_D = (a_{q+1})_* \Omega^p_{\tilde{D}^{(q+1)}}$ $(p, q \geqslant 0)$. Let $d'\colon A^{pq}_D \to A^{p+1,q}_D$, be differential in $(a_{q+1})_* \Omega^{\cdot}_{\tilde{D}^{(q+1)}}$ and $d''\colon A^{pq}_D \to A^{p,q+1}_D$ be given by $d'' = \sum_{j=1}^{q+1} (-1)^{q+j} \delta^*_j$. Then $(A^{\cdot\cdot}_D, d', d'')$ is a double complex of sheaves; its total single complex is a resolution of the sheaf $\mathbf{C}_D$. It carries filtrations W and F, which are given by

$$F^p A^{\cdot\cdot}_D = \bigoplus_{r \geqslant p} A^{r\cdot}_D \quad \text{and} \quad W_q A^{\cdot\cdot}_D = \bigoplus_{s \geqslant -q} A^{\cdot s}_D$$

and which induce Hodge and weight filtrations on $H^*(D, \mathbf{C})$; the weight filtration is already defined over $\mathbf{Q}$.

However, for W and F we have a different convention:

$F^p H^k(D, \mathbf{C})$ is the image of $\mathbf{H}^k(D, F^p A^{\cdot}_D)$ in $H^k(D, \mathbf{C})$, but
$W_r H^k(D, \mathbf{C})$ is the image of $\mathbf{H}^k(D, W_{r-k} A^{\cdot}_D)$ in $H^k(D, \mathbf{C})$.

Remark that $\mathrm{Gr}^W_{-p} A^{\cdot}_D$ is a resolution of $(a_{p+1})_* \mathbf{C}_{\tilde{D}^{(p+1)}}$ so the weight spectral sequence reads

$${}_W E_1^{pq} \cong H^q(\tilde{D}^{(p+1)}, \mathbf{C}) \Rightarrow H^{p+q}(D, \mathbf{C}).$$

It degenerates at E_2 and d_1^{pq} is induced by d''. Thus

$$\mathrm{Gr}^W_r H^k(D) = 0 \quad \text{for } r > k.$$

Moreover it is an easy local computation, that $A^{p\cdot}_D$ is a resolution of

$$\Omega^p_{\tilde{X}} / I_D \Omega^p_{\tilde{X}}(\log D),$$

where $\Omega^p_{\tilde{X}}(\log D)$ is the sheaf of germs of meromorphic p-forms on $\tilde{X}$ which are holomorphic outside D and have at most logarithmic poles among D. Because the spectral sequence

$$E_1^{pq} = \mathbf{H}^{p+q}(D, \mathrm{Gr}^p_F A^{\cdot}_D) \Rightarrow H^{p+q}(D, \mathbf{C})$$

degenerates even at E_1, one obtains

$$\mathrm{Gr}_F^p H^{p+q}(D,\mathbf{C}) \cong H^q(D, \Omega_{\tilde{X}}^p / I_D \Omega_{\tilde{X}}^p(\log D)).$$

In particular $\mathrm{Gr}_F^0 H^k(D,\mathbf{C}) \cong H^k(D, \mathcal{O}_D)$.

(1.6) The mixed Hodge structure on $H_D^*(\tilde{X})$ has been studied by Fujiki [**13**]. It has been shown that one has an isomorphism of mixed Hodge structures (over **Q**):

$$H_D^k(\tilde{X}) \xrightarrow{\sim} \mathrm{Hom}(H^{2n-k}(D), \mathbf{Q}(-n)).$$

Here $\mathbf{Q}(-n)$ is the mixed Hodge structure on **C** with rational lattice $(2\pi i)^{-n}\mathbf{Q}$ which is purely of type (n, n).

A direct description of $H_D^*(\tilde{X})$ can be given as follows. The complex $\Omega_{\tilde{X}}^{\cdot}(\log D)$ has filtrations W and F, given by

$$W_k \Omega_{\tilde{X}}^p(\log D) = \begin{cases} \Omega_{\tilde{X}}^{p-k} \wedge \Omega_{\tilde{X}}^k(\log D) & \text{if } 0 \leqslant k \leqslant p, \\ 0 & \text{if } k < 0, \\ \Omega_{\tilde{X}}^p(\log D) & \text{if } k > p. \end{cases}$$

$$F^r \Omega_{\tilde{X}}^p(\log D) = \begin{cases} \Omega_{\tilde{X}}^p(\log D) & \text{if } p \geqslant r, \\ 0 & \text{if } p < r. \end{cases}$$

Then let

$$A_D^p(\tilde{X}) = \Omega_{\tilde{X}}^{p-1}(\log D)/\Omega_{\tilde{X}}^{p-1},$$
$$W_k A_D^p(\tilde{X}) = W_{k+1}\Omega_{\tilde{X}}^{p+1}(\log D)/\Omega_{\tilde{X}}^{p+1},$$
$$F^r A_D^p(\tilde{X}) = \begin{cases} A_D^p(\tilde{X}) & \text{if } p > r+1, \\ 0 & \text{if } p < r+1. \end{cases}$$

Then $(A_D^{\cdot}(\tilde{X}), W, F)$ determine the mixed Hodge structure on $H_D^*(\tilde{X}, \mathbf{C})$. One obtains

$$\mathrm{Gr}_r^W H_D^k(\tilde{X}) = 0 \quad \text{for } r < k.$$

(1.7) A way to describe the mixed Hodge structure on $H_{\{x\}}^*(X)$ is to use mapping cones. For details and motivation for the definitions we refer to [**11**]. If $\tau \colon B^{\cdot} \to A^{\cdot}$ is a map of complexes of sheaves, with $d\tau(b) = \varepsilon\tau(db)$, $\varepsilon = \pm 1$, we define the mapping cone $C_\tau^{\cdot}$ to be the complex with

$$C_\tau^m = B^m \oplus A^{m-1}, \qquad d(b, a) = (-\varepsilon\, db, da + \tau(b)).$$

One has an exact sequence of complexes

$$0 \to A^{\cdot -1} \to C_\tau^{\cdot} \to B^{\cdot} \to 0.$$

If $A^{\cdot}$ and $B^{\cdot}$ carry filtrations F and W such that τ is compatible with these filtrations, $C_\tau^{\cdot}$ gets two filtrations W and F in a "mixed" way:

$$\begin{cases} F^k C_\tau^m = F^k B^m \oplus F^k A^{m-1}, \\ W_r C_\tau^m = W_r B^m \oplus W_{r+1} A^{m-1}. \end{cases}$$

Sometimes one calls $(C_\tau^\cdot, W, F)$ the mixed cone of τ (cf. [**12**]). If τ is injective, one has a quasi-isomorphism of complexes

$$\lambda\colon C_\tau \to A^{\cdot-1}/\tau(B^{\cdot-1}), \qquad \lambda(b, a) = a \bmod \tau B^{\cdot-1}.$$

Honestly speaking, one should do this construction on the level of "cohomological mixed Hodge complexes" (cf. [**6**]) instead of bifiltered complexes of sheaves. For the sake of simplicity we omit the parts which take care of the rational level.

(1.8) Consider the morphisms of complexes

$$\phi\colon \Omega_{\tilde{X}}^\cdot \to A_D^\cdot, \qquad \phi(\omega) = a_1^*(\omega)$$

and

$$i\colon \Omega_{\tilde{X}}^\cdot \to \Omega_{\tilde{X}}^\cdot(\log D).$$

Then i is injective, and $\operatorname{Coker}(i)[-1]$ is isomorphic to $A_D^\cdot(\tilde{X})$, hence

$$\mathbf{H}^*(C_i) \cong H_D^n(\tilde{X}).$$

However, remark that $\Omega_{\tilde{X}}^\cdot(\log Y)$, $\Omega_{\tilde{X}}^\cdot$ and C_i are *not* part of cohomological Hodge complexes, because $\tilde{X}$ is not complete algebraic; this is only true for $A_D^\cdot(\tilde{X})$, which has support on D.

It follows from [**11**] that $H_{\{x\}}^k(X) \cong \mathbf{H}^k(\tilde{X}, C_{(i,\phi)})$ where $(i, \phi)\colon \Omega_{\tilde{X}}^\cdot \to \Omega_{\tilde{X}}^\cdot(\log D) \oplus A_D^\cdot$. Again $C_{(i,\phi)}$ is not part of a cohomological mixed Hodge complex, but it is quasi-isomorphic to $\operatorname{Coker}(i, \phi)[-1]$, which has support on D and is part of a cohomological mixed Hodge complex.

To make things more precise, choose a complete complex algebraic variety V which contains X as an open subset, and has x as its only singular point. Let $\tilde{V} \to V$ be the resolution of V, obtained by replacing X by $\tilde{X}$. Then $H^*(V, V \setminus \{x\})$ carries a canonical mixed Hodge structure (cf. [**6**]), which does not depend on the choice of the resolution. By excision, $H^*(V, V \setminus \{x\}) \cong H^*(X, X \setminus \{x\})$, and the reasoning above just shows that the resulting mixed Hodge structure on $H_{\{x\}}^*(X)$ does not depend on the choice of V, but then it clearly is also independent of the choice of the resolution.

Summarizing, we get

(1.9) THEOREM. *If (X, x) is an isolated singularity of a complex space, the local cohomology groups $H_{\{x\}}^k(X)$ carry a canonical mixed Hodge structure; if $(\tilde{X}, D) \to (X, x)$ is a resolution, and*

$$A_{\{x\}}^\cdot(X) := \left(\Omega_{\tilde{X}}^{\cdot-1}(\log D) \oplus A_D^{\cdot-1}\right)/(i, \phi)\Omega_{\tilde{X}}^{\cdot-1},$$

with filtrations W, F given by

$$F^pA_{\{x\}}^m(X) = \text{Image of } F^p\Omega_{\tilde{X}}^{m-1}(\log D) \oplus F^pA_D^{m-1} \quad \text{in } A_{\{x\}}^m(X),$$

$$W_kA_{\{x\}}^m(X) = \text{Image of } W_{k+1}\Omega_{\tilde{X}}^{m-1}(\log D) \oplus W_{k+1}\Omega_D^{m-1} \quad \text{in } A_{\{x\}}^m(X)$$

then

$$H_{\{x\}}^m(X, \mathbf{C}) \cong \mathbf{H}^m\left(D, A_{\{x\}}^\cdot(X)\right)$$

and the Hodge and weight filtrations on it are induced by the filtrations F and W on $A_{\{x\}}^\cdot(X)$.

(1.10) Remark that one has a short exact sequence of complexes

$$0 \to A_D^{\cdot -1} \to A^{\cdot}_{\{x\}}(X) \to A^{\cdot}_D(\tilde{X}) \to 0$$

giving rise to a long exact sequence of mixed Hodge structures

$$\to H^{k-1}(D) \xrightarrow{u} H^k_{\{x\}}(X) \to H^k_D(\tilde{X}) \to H^k(D) \to \cdots .$$

Here the map u fits in a commutative diagram

$$\begin{array}{ccc} H^{k-1}(D) & \xleftarrow{\sim} & H^{k-1}(\tilde{X}) \\ \downarrow u & & \downarrow \bar{u} \\ H^k_{\{x\}}(X) & \xrightarrow{\sim} \tilde{H}^{k-1}(X \setminus \{x\}) = & \tilde{H}^{k-1}(\tilde{X} \setminus D) \end{array}$$

with $\bar{u}$ the restriction map. The following generalizes [**25**, Lemma 4.7], and is a corollary of the decomposition theorem in intersection cohomology.

(1.11) THEOREM (GORESKI-MACPHERSON). *The restriction map*

$$H^{k-1}(\tilde{X}, \mathbf{Q}) \to H^{k-1}(\tilde{X} \setminus D, \mathbf{Q})$$

is surjective for $k \leqslant n$ and the zero map for $k > n$.

PROOF. See [**14**].

(1.12) COROLLARY. *One has short exact sequences of mixed Hodge structures*

$$0 \to H^k_D(\tilde{X}) \to H^k(D) \to H^{k+1}_{\{x\}}(X) \to 0 \qquad (k < n)$$

$$0 \to H^n_D(\tilde{X}) \to H^n(D) \to 0$$

$$0 \to H^k_{\{x\}}(X) \to H^k_D(\tilde{X}) \to H^k(D) \to 0 \qquad (k > n).$$

Hence

$$\begin{array}{ll} \mathrm{Gr}_r^W H^k_D(\tilde{X}) = 0 & \textit{for } r \neq k \ \textit{ if } k \leqslant n, \\ \mathrm{Gr}_r^W H^k(D) = 0 & \textit{for } r \neq k \ \textit{ if } k \geqslant n, \\ \mathrm{Gr}_r^W H^k_{\{x\}}(X) = 0 & \textit{for } r \geqslant k \ \textit{ if } k \leqslant n, \\ \mathrm{Gr}_r^W H^k_{\{x\}}(X) = 0 & \textit{for } r < k \ \textit{ if } k > n. \end{array}$$

As an important consequence we quote

(1.13) THEOREM. *Let Z be an n-dimensional complete variety over* $\mathbf{C}$ *with only isolated singularities. Then $H^k(Z)$ carries a pure Hodge structure of weight k for $k > n$.*

PROOF. Let $\Sigma = \operatorname{Sing} Z$. One has a long exact sequence of mixed Hodge structures

$$\to H^k_\Sigma(Z) \to H^k(Z) \to H^k(Z \setminus \Sigma) \to \cdots .$$

If $k > n$, then $\mathrm{Gr}_r^W H_\Sigma^k(Z) = 0$ for $r < k$, and also $\mathrm{Gr}_r^W H^k(Z \setminus \Sigma) = 0$ for $r < k$ because $Z \setminus \Sigma$ is smooth. Hence also $\mathrm{Gr}_r^W H^k(Z) = 0$ for such r, k. Because Z is complete, $\mathrm{Gr}_r^W H^k(Z) = 0$ for $r > k$. Hence $H^k(Z)$ is purely of weight k for $k > n$.

(1.14) COROLLARY. *The intersection homology groups $IH^k(Z)$ of Z carry a pure Hodge structure of weight k for all k.*

PROOF. For $k > n$, one has $IH^k(Z) \cong H^k(Z)$. For $k < n$, $IH^k(Z) \cong H^k(Z - \Sigma)$ which is pure by duality. Finally

$$IH^n(Z) = \text{Image of } H^n(Z) \text{ in } H^n(Z - \Sigma)$$

which is pure of weight n because $H^n(Z)$ is of weight $\leqslant n$ and $H^n(Z - \Sigma)$ of weight $\geqslant n$.

This fact has been proved by A. Dubson (unpublished).

(1.15) COROLLARY. *Under Poincaré duality on $M = \partial X = \partial \tilde{X}$, the mixed Hodge structures on $H^{k+1}_{\{x\}}(X) \cong H^k(M)$ and $H^{2n-k}_{\{x\}}(X) \cong H^{2n-k-1}(M)$ are dual to each other for $k = 1, \ldots, 2n-2$.*

PROOF. It follows from Fujiki's work [**13**] that the maps

$$\alpha_k \colon H_D^k(\tilde{X}) \to H^k(D)$$

satisfy $\alpha_{2n-k} = {}^t\alpha_k$ under the duality between $H_D^k(\tilde{X})$ and $H^{2n-k}(D)$. If $k < n$, then

$$H^{k+1}_{\{x\}}(X) = \mathrm{Coker}(\alpha_k)$$

and

$$H^{2n-k}_{\{x\}}(X) = \mathrm{Ker}(\alpha_{2n-k})$$

which is dual to $\mathrm{Coker}({}^t\alpha_{2n-k}) = \mathrm{Coker}(\alpha_k)$.

2. Vanishing cohomology. In this section we will study the vanishing cohomology groups, associated with a one-parameter smoothing of an isolated singularity. We start with the geometric setting.

(2.1) Let Y be a complex space of dimension $n+1$ with at most one singular point x. Let f be a flat holomorphc map from Y to the unit disk S, which is of maximal rank outside x and such that $f(x) = 0$. Let $X = f^{-1}(0)$. We assume that X and Y are contractible Stein spaces and that f induces a differentiable fibration above the punctured disk. Let Y_∞ denote a smooth fibre of this fibration. The space X has at most an isolated singularity at x.

By the semistable reduction theorem [**17**] we can find a natural number e and a complex manifold $\tilde{Y}$ with the following properties. Let $\tilde{S}$ be another copy of the unit disk and map $\tilde{S}$ onto S by sending t to t^e. The space $\tilde{Y}$ admits holomorphic mappings π to Y and $\tilde{f}$ to $\tilde{S}$, which make it a resolution of singularities of the fibre product $Y \times_S \tilde{S}$. Moreover, $\tilde{f}^{-1}(0)$ is a reduced divisor with normal crossings on Y without self-intersection.

DEFINITION. The smoothing Y of X is called *semistable* if we can take $e = 1$ in the above construction.

Write $\tilde{f}^{-1}(0) = \tilde{X} \cup E$ where $E = \pi^{-1}(x)$ and $\tilde{X}$ is the closure of $\pi^{-1}(X \setminus \{x\})$ in $\tilde{Y}$. Then $\tilde{X}$ is a resolution of singularities for X with exceptional divisor $D = \tilde{X} \cap E$.

For the rest of this section we will fix a semistable smoothing $f: Y \to S$ of X and keep the notations $\tilde{Y}$, $\tilde{X}$, E, D, π.

(2.2) We will construct mixed Hodge structures on the groups $H_c^*(Y_\infty)$ and $H^*(Y_\infty)$, and relate them with the mixed Hodge structures associated with (X, x) which have been studied in §1.

Recall that local triviality of f over the punctured disk gives rise to a geometric monodromy $h: Y_\infty \to Y_\infty$ which can be constructed in such a way that h is the identity outside a compact subset of Y. The map $\omega \mapsto h^*(\omega)$ induces monodromy transformations

$$T: H^*(Y_\infty) \to H^*(Y_\infty) \quad \text{and} \quad T_c: H_c^*(Y_\infty) \to H_c^*(Y_\infty).$$

Because Y_∞ is a Stein manifold of dimension n, $H^k(Y_\infty) = 0$ for $k > n$ and $H_c^k(Y_\infty) = 0$ for $k < n$. One has an exact sequence

$$\cdots \to H^k_{\{x\}}(X) \to H_c^k(Y_\infty) \xrightarrow{j} H^k(Y_\infty) \to H^{k+1}_{\{x\}}(X) \to \cdots \tag{2.3}$$

which under the identification of $H^k_{\{x\}}(X)$ with $H^{k-1}(\partial X) \cong H^{k-1}(\partial Y_\infty)$ corresponds with the long exact cohomology sequence of the pair $(Y_\infty, \partial Y_\infty)$. Hence the natural map $\tilde{H}^k(Y_\infty) \to H^{k+1}_{\{x\}}(X)$ is an isomorphism for $k < n - 1$ and injective for $k = n - 1$, so T acts as the identity on $H^k(Y_\infty)$ for $k \neq n$.

(2.4) If ω is an n-form on Y_∞, $h^*(\omega) - \omega$ has compact support on Y_∞. This gives one a variation map

$$\text{Var}: H^n(Y_\infty) \to H_c^n(Y_\infty)$$

with $\text{Var}([\omega]) = [h^*(\omega) - \omega]$. Then the following relations hold:

$$T = I + j \circ \text{Var}, \qquad j \circ T_c = T \circ j,$$
$$T_c = I + \text{Var} \circ j, \qquad T_c \circ \text{Var} = \text{Var} \circ T.$$

Because our smoothing is semistable, T and T_c are unipotent, and moreover $(T - I)^{n+1} = 0$. Put

$$N = \log T; \qquad N_c = \log T_c.$$

Then $j \circ N_c = N \circ j$ and $N_c \circ \text{Var} = \text{Var} \circ N$. One has a natural pairing

$$\langle\, , \rangle: H_c^n(Y_\infty) \otimes H^n(Y_\infty) \to \mathbf{C}, \qquad \langle \omega, \eta \rangle = \int_{Y_\infty} \omega \wedge \eta.$$

It satisfies: $\langle T_c\omega, T\eta \rangle = \langle \omega, \eta \rangle$ hence $T_c = {}^tT^{-1}$, and ${}^tN = -N_c$. Moreover ${}^t\text{Var} = -{}^tT \circ \text{Var}$.

(2.5) We define a map $V: H^n(Y_\infty) \to H_c^n(Y_\infty)$ which is related to Var just like N is related to $T - I$, as

$$V = \mathrm{Var} + \sum_{k=1}^{n-1} (-1)^k \,\mathrm{Var} \circ (T - I)^k / (k+1).$$

Then $j \circ V = N$ and $V \circ j = N_c$. Finally the intersection form on $H_c^n(F)$ is given by

$$S(x, y) = \langle x, j(y) \rangle .$$

Before passing to the construction of our mixed Hodge structures we list a number of properties:

(2.6) (a) the sequence (2.3) is an exact sequence of mixed Hodge structures;

(b) the map V fits in an exact sequence of mixed Hodge structures

$$0 \to H_{\{x\}}^{n+1}(Y) \to H^n(Y_\infty) \xrightarrow{V} H_c^n(Y_\infty) \xrightarrow{\partial} H_{\{x\}}^{n+1}(Y) \to 0$$

where V is a morphism of type $(-1,-1)$, i.e. $V(W_k) \subset W_{k-2}$ and $V(F^p) \subset F^{p-1}$, and ∂ is a morphism of type $(1, 1)$;

(c) for $k > 0$, let ϕ_k denote the map from $\mathrm{Gr}_{n+k}^W H^n(Y_\infty)$ to $\mathrm{Gr}_{n-k}^W H_c^n(Y_\infty)$, induced by $V \circ N^{k-1}$. Then ϕ_k is an isomorphism of Hodge structures of type $(-k, -k)$;

(d) for $k \geqslant 0$, let ψ_k denote the map from $\mathrm{Gr}_{n+k}^W H_c^n(Y_\infty)$ to $\mathrm{Gr}_{n-k}^W H^n(Y_\infty)$, induced by $N^k \circ j$. Then ψ_k is an isomorphism of Hodge structures of type $(-k, -k)$. Moreover for $k > 0$ one has $\psi_k = j \circ \phi_k \circ j$;

(e) the mixed Hodge structures on $H_c^n(Y_\infty)$ and $H^n(Y_\infty)$ are dual to each other with respect to $\mathbf{Q}(-n)$: this means that

$$W_k H_c^n(Y_\infty) = [W_{2n-1-k} H^n(Y_\infty)]^\perp ,$$
$$F^p H_c^n(Y_\infty) = [F^{n-p+1} H^n(Y_\infty)]^\perp ;$$

(f) if $f\colon Y \to S$ arises by restriction to Y from a projective family $\bar{f}\colon \bar{Y} \to S$ for which x is the only critical point, one has a long exact sequence of mixed Hodge structures

$$\cdots \to \tilde{H}^k(\bar{Y}_0) \to H^k(\bar{Y}_\infty) \to H^k(Y_\infty) \to H^{k+1}(\bar{Y}_0) \to \cdots$$

where $\bar{Y}_0 = \bar{f}^{-1}(0)$ and $H^k(\bar{Y}_\infty)$ is the cohomology of a general fibre of $\bar{f}$, equipped with the limit mixed Hodge structure (cf. [**22** and **25**]).

(2.7) We now construct bifiltered complexes of sheaves $K_c^\cdot$ and $K^\cdot$ on E which by taking hypercohomology will give $H_c^*(Y_\infty)$ and $H^*(Y_\infty)$ with their mixed Hodge structure.

We consider the complex $A^{\cdot\cdot}$ on $\tilde{X} \cup E$ as defined in [**25**],

$$A^{pq} = \Omega_{\tilde{Y}}^{p+q+1} (\log \tilde{X} + E)/W_q ,$$

where $d'\colon A^{pq} \to A^{p+1,q}$ is induced by the differential in $\Omega_{\tilde{Y}}^\cdot (\log \tilde{X} + E)$ and $d''\colon A^{pq} \to A^{p,q+1}$ is induced by cup product with the global section dt/t of

$\Omega^1_{\tilde{Y}}(\log \tilde{X} + E)$, where t is a parameter on the disk S. One has filtrations F and W on $A^{\cdot\cdot}$,

$$F^p A^{\cdot\cdot} = \bigoplus_{s \geq p} \bigoplus_q A^{sq},$$

$$W_r A^{\cdot\cdot} = \bigoplus_{p,q} W_r A^{pq},$$

where $W_r A^{pq}$ is the image of $W_{2q+r+1}\Omega^{p+q+1}_{\tilde{Y}}(\log \tilde{X} + E)$ in A^{pq}. Define $K_c^{\cdot\cdot}$ to be the subcomplex of $A^{\cdot\cdot}$ given by

$$K_c^{p0} = \Omega^{p+1}_{\tilde{Y}}(\log E)/W_0,$$
$$K_c^{pq} = A^{pq} \text{ for } q > 0$$

with induced filtrations F and W.

Define $K^{\cdot\cdot}$ to be the quotient complex of $A^{\cdot\cdot}$ given by

$$K^{pq} = \Omega^{p+q+1}_{\tilde{Y}}(\log \tilde{X} + E)/\Omega^1_{\tilde{Y}}(\log \tilde{X}) \wedge W_q \Omega^{p+q}_{\tilde{Y}}(\log E),$$

with the induced filtrations F and W.

To understand these complexes, put $\Omega^{\cdot} = \Omega^{\cdot}_{\tilde{Y}/S}(\log \tilde{X} + E)$. Then one has a filtered quasi-isomorphism

$$(\Omega^{\cdot}/I_{\tilde{X}\cup E}\Omega^{\cdot}, F) \underset{\wedge dt/t}{\rightarrow} (A^{\cdot}, F)$$

(cf. [**25**]). In a similar way one obtains filtered quasi-isomorphisms

$$(I_{\tilde{X}}\Omega^{\cdot}/I_{\tilde{X}\cup E}\Omega^{\cdot}, F) \underset{\wedge dt/t}{\rightarrow} (K_c^{\cdot}, F)$$

and

$$(\Omega^{\cdot}/I_E\Omega^{\cdot}, F) \underset{\wedge dt/t}{\rightarrow} (K^{\cdot}, F).$$

(2.8) THEOREM. *Fix a coordinate function t on S. Then there are natural isomorphisms*

$$H^k(Y_\infty) \tilde{\rightarrow} \mathbf{H}^k(K^{\cdot}) \quad \textit{and} \quad H^k_c(Y_\infty) \tilde{\rightarrow} \mathbf{H}^k(K_c^{\cdot})$$

such that the map j is induced by the morphism of complexes

$$K_c^{\cdot} \hookrightarrow A^{\cdot} \twoheadrightarrow K^{\cdot}.$$

The filtrations F and W induce mixed Hodge structures on $H^*_c(Y_\infty)$ and $H^*(Y_\infty)$.

The proof of this theorem is essentially contained in [**26**]. The details of it will appear elsewhere.

(2.9) We sketch a proof of some of the properties (b)–(f). As for (b), we define

$$\nu\colon K^{pq} \rightarrow K_c^{p-1,q+1} \qquad (p, q \geq 0)$$

by

$$\nu\left(x \bmod \Omega^1_{\tilde{Y}}(\log \tilde{X}) \wedge W_q\right) = (-1)^{p+q+1} x \bmod W_{q+1}.$$

One checks the following facts:

$$\nu(F^pK^\cdot) \subset F^{p-1}K_c^\cdot, \qquad \nu(W_rK^\cdot) \subset W_{r-2}K_c^\cdot;$$

hence ν induces a morphism of mixed Hodge structures

$$V\colon H^n(Y_\infty) \to H_c^n(Y_\infty)$$

which is of type $(-1, 1)$. That this V is the same map as the one constructed before from the variation morphism can be shown in a similar way as in [**25**, (4.22)].

The exact sequence (b) is now constructed by giving a suitable quasi-isomorphism between $A_{\{x\}}^{\cdot+1}(Y)$ (see §1) and the mixed cone C_ν.

It can be checked easily now that (c) and (d) are true for the E_1-terms of the spectral sequences for the filtered complexes $(K^\cdot, W)$ and $(K_c^\cdot, W)$, because the corresponding components of the graded complexes are the same. A reasoning as in [**25**, §5] then enables one to conclude for the E_2-terms, hence for E_∞ because these spectral sequences degenerate at E_2.

To prove (e), one considers first the Hodge filtrations. These come from the trivial ("bête") filtrations on the complexes

$$I_{\tilde{X}}\Omega^\cdot / I_{\tilde{X}\cup E}\Omega^\cdot \text{ resp. } \Omega^\cdot/I_E\Omega^\cdot$$

where

$$\Omega^p = \Omega^p_{\tilde{Y}/S}(\log \tilde{X} + E).$$

Remark that the dualing sheaf ω_E on E satisfies

$$\omega_E \cong \Omega_{\tilde{Y}}^{n+1}(\log E)/I_E\Omega_{\tilde{Y}}^{n+1}(\log E)$$

and

$$I_{\tilde{X}}\Omega^n \cong I_{\tilde{X}}\Omega_{\tilde{Y}}^{n+1}(\log \tilde{X} + E) \cong \Omega_{\tilde{Y}}^{n+1}(\log E)$$

hence

$$\omega_E \cong I_{\tilde{X}}\Omega^n / I_{\tilde{X}\cup E}\Omega^n.$$

Clearly under cup product $F^k\mathbf{H}^n(I_{\tilde{X}}\Omega^\cdot/I_{\tilde{X}\cup E}\Omega^\cdot)$ is orthogonal to $F^{n-k+1}\mathbf{H}^n(\Omega^\cdot/I_E\Omega^\cdot)$, and Serre-duality provides us with a perfect pairing between

$$\mathrm{Gr}_F^k \mathbf{H}^n(I_{\tilde{X}}\Omega^\cdot/I_{\tilde{X}\cup E}\Omega^\cdot) \cong H^{n-k}(E, I_{\tilde{X}}\Omega^k/I_{\tilde{X}\cup E}\Omega^k)$$

and

$$\mathrm{Gr}_F^{n-k}\mathbf{H}^n(\Omega^\cdot/I_E\Omega^\cdot) \cong H^k(E, \Omega^{n-k}/I_E\Omega^{n-k})$$

because

$$\Omega^{n-k}/I_E\Omega^{n-k} \cong \mathrm{Hom}_{\mathcal{O}_E}(I_{\tilde{X}}\Omega^k/I_{\tilde{X}\cup E}\Omega^k, \omega_E).$$

As for the weight filtrations, consider again the sequence (2.3):

$$\cdots \to H_x^n(X) \xrightarrow{\beta} H_c^n(Y_\infty) \xrightarrow{j} H^n(Y_\infty) \xrightarrow{{}^t\beta} H_x^{n+1}(X) \to \cdots.$$

Clearly j induces an isomorphism of mixed Hodge structures between Coker(β) and Ker(${}^t\beta$). Therefore it follows from properties (c) and (d) that W on Coker(β) is the monodromy weight filtration determined by N_c. As N_c is an infinitesimal isometry, this filtration is dual to the corresponding one on Ker(${}^t\beta$). One may conclude (e) using the duality between $H^n_{\{x\}}(X)$ and $H^{n+1}_{\{x\}}(X)$ (see §1) and properties (c) and (d) again. Property (f) has been the starting point of the construction above in [**26**].

(2.10) EXAMPLE. We consider two different one-parameter smoothings of the curve singularity $(X, 0)$: $x^3 + y^4 = 0$. The first one is the map germ $(\mathbf{C}^2, 0) \to (\mathbf{C}, 0)$ given by $(x, y) \mapsto x^3 + y^4$. For this, $H^1(Y_\infty)$ is purely of weight 1, and in fact is isomorphic to $H^1(C)$ where $C \subset \mathbf{P}^2$ is given by $x^3z + y^4 - z^4 = 0$.

Let S denote the surface in $\mathbf{C}^3$ with equation $x^3 + y^4 + t^3 + txy = 0$, which has an isolated singularity of type $T_{3,3,4}$ at 0. As a second smoothing we consider the map germ $(S, 0) \to (\mathbf{C}, 0)$, $(x, y, t) \mapsto t$. Here $\mathrm{Gr}_0^W H^1(Y_\infty)$ and $\mathrm{Gr}_2^W H^1(Y_\infty)$ both are one-dimensional and $\mathrm{Gr}_1^W H^1(Y_\infty) \cong H^1(C')$ where C' is the 5-fold covering of $\mathbf{P}^1$, totally ramified at 0, 1 and ∞. Hence even if the Milnor fibres of two smoothings are diffeomorphic, the mixed Hodge structures can be different.

(2.11) It can be shown without much difficulty that the mixed Hodge structures on $H_c^*(Y_\infty)$ and $H^*(Y_\infty)$ do not depend on the resolution map π chosen. Moreover, the fact that we assumed the smoothing to be semistable is not too much a restriction. If one starts with a smoothing which is not semistable, one can make a finite base change to get one, and this one carries an action of a finite cyclic group, which can be lifted to a suitable resolution. One then obtains an action of this cyclic group on $H_c^*(Y_\infty)$, $H^*(Y_\infty)$, $H_x^*(Y)$ and can extract information about the original smoothing, e.g. about the local cohomology of its total space, by taking invariants.

(2.12) We define numerical invariants of the smoothing by

$$h^{pq} = \dim \mathrm{Gr}_F^p \mathrm{Gr}_{p+q}^W H^n(Y_\infty)$$

and

$$h_c^{pq} = \dim \mathrm{Gr}_F^p \mathrm{Gr}_{p+q}^W H_c^n(Y_\infty).$$

Properties (c) and (d) imply

$$h^{pq} = h_c^{n-p,n-q} = h_c^{n-q,n-p} = h^{qp}.$$

Clearly

$$\dim \mathrm{Gr}_F^p H^n(Y_\infty) = \sum_q h^{pq}.$$

We will relate $\dim \mathrm{Gr}_F^0 H^n(Y_\infty)$ and $\dim \mathrm{Gr}_F^n H^n(Y_\infty)$ with invariants of the singularity (X, x).

Recall the definition of the geometric genus $p_g(X)$,

$$p_g(X) = \sum_{i=1}^{n-1} (-1)^{n-i+1} \dim_{\mathbf{C}} H^i(\tilde{X}, \mathcal{O}_{\tilde{X}}) + (-1)^{n-1}\delta$$

where $\delta = \dim_{\mathbf{C}} H^0(\mathcal{O}_{\tilde{X}})/H^0(\mathcal{O}_X)$.

If (X, x) is Cohen-Macaulay, then $p_g(X) = \dim H^{n-1}(\mathcal{O}_{\tilde{X}})$.

(2.13) PROPOSITION. $\dim \mathrm{Gr}_F^n H^n(Y_\infty) = p_g(X) = \dim \mathrm{Gr}_F^0 H_c^n(Y_\infty)$.

PROOF. As has been remarked above,

$$\mathrm{Gr}_F^n H^n(Y_\infty) \cong H^0(E, \omega_E(D))$$

which is dual to

$$\mathrm{Gr}_F^0 H_c^n(Y_\infty) \cong H^n(E, \mathcal{O}_E(-D)).$$

Consider the exact sequences

$$0 \to \mathcal{O}_{\tilde{Y}}(-\tilde{X}) \to \mathcal{O}_{\tilde{Y}} \to \mathcal{O}_{\tilde{X}} \to 0$$

and

$$0 \to \mathcal{O}_{\tilde{Y}} \xrightarrow{t} \mathcal{O}_{\tilde{Y}}(-\tilde{X}) \to \mathcal{O}_E(-D) \to 0.$$

The start of their long exact sequences can be transformed into

$$0 \to H^0(\mathcal{O}_{\tilde{X}})/H^0(\mathcal{O}_X) \to H^1(\mathcal{O}_{\tilde{Y}}(-\tilde{X})) \to H^1(\mathcal{O}_{\tilde{Y}}) \to \cdots$$

and

$$0 \to H^1(\mathcal{O}_{\tilde{Y}}) \to H^1(\mathcal{O}_{\tilde{Y}}(-X)) \to \cdots.$$

Now the fact that $H^k(\mathcal{O}_E(-D)) \cong \mathrm{Gr}_F^0 H_c^k(Y_\infty)$ implies

$$H^k(\mathcal{O}_E(-D)) = 0 \quad \text{for } k \neq n$$

because

$$H_c^k(Y_\infty) = 0 \quad \text{for } k < n.$$

One obtains the theorem by comparison of dimensions in both long exact sequences.

REMARK. As mentioned to me by J. Wahl, this proposition implies Mumford's necessary condition for smoothability of a nonnormal isolated surface singularity, $\delta \leqslant h^1(\mathcal{O}_{\tilde{X}})$.

REMARK. This proposition is due to K. Watanabe **[31]** in the case of quasi-homogeneous hypersurfaces and to M. Saito **[20]** for the general hypersurface case.

The following lemma ought to be well known:

(2.14) LEMMA. *Let (X, x) be an isolated singularity where X is a contractible Stein space; let p: $\tilde{X} \to X$ be a resolution of X such that $D = p^{-1}(x)^{\mathrm{red}}$ is a divisor with normal crossings on $\tilde{X}$. Then for all $i \geqslant 0$ the natural map*

$$H^i(\tilde{X}, \mathcal{O}_{\tilde{X}}) \to H^i(D, \mathcal{O}_D)$$

is surjective.

PROOF. Because X is contractible, D is a deformation retract of $\tilde{X}$. Hence $H^i(\tilde{X}, \mathbf{C})$ is isomorphic to $H^i(D, \mathbf{C})$ for all $i \geqslant 0$. Because D has normal crossings the natural maps $H^i(D,\mathbf{C}) \to H^i(D,\mathcal{O}_D)$ are surjective. One has a commutative diagram

$$\begin{array}{ccc} H^i(\tilde{X},\mathbf{C}) & \xrightarrow{u} & H^i(\tilde{X},\mathcal{O}_{\tilde{X}}) \\ \downarrow v' & & \downarrow v \\ H^i(D,\mathbf{C}) & \xrightarrow{u'} & H^i(D,\mathcal{O}_D) \end{array}$$

Because $u'v' = vu$ is surjective, v must be surjective too. □

As is well known, for $i > 0$, $\dim H^i(\mathcal{O}_{\tilde{X}})$ is an invariant of the singularity for every resolution $\tilde{X}$. The preceding lemma shows, that for a good resolution, $\dim H^i(\mathcal{O}_D)$ is an invariant of (X, x) for any i.

(2.15) PROPOSITION. $\dim \mathrm{Gr}_F^0 H^n(Y_\infty) - \dim \mathrm{Gr}_F^0 H^{n-1}(Y_\infty) = p_g(X) - \dim_{\mathbf{C}} H^{n-1}(\mathcal{O}_D)$.

PROOF. Because $\mathrm{Gr}_F^0 K^{\cdot\cdot}$ is a resolution of $\mathcal{O}_E$,

$$\mathrm{Gr}_F^0 H^k(Y_\infty) \cong H^k(\mathcal{O}_E) \quad \text{for } k = 0,\ldots,n.$$

Because $H^k(\mathcal{O}_E(-D)) = 0$ for $k \neq n$, one obtains a short exact sequence

$$0 \to H^{n-1}(\mathcal{O}_E) \to H^{n-1}(\mathcal{O}_D) \to H^n(\mathcal{O}_E(-D)) \to H^n(\mathcal{O}_E) \to 0.$$

The claim follows now easily from Lemma (2.14).

(2.16) COROLLARY. *Suppose (X, x) is an isolated Cohen-Macaulay singularity. Then* $\dim \mathrm{Gr}_F^0 H^n(Y_\infty)$ *does not depend on the smoothing, but only on (X, x). For the Cohen-Macaulay property implies $H^i(\mathcal{O}_E) = 0$ for $i \neq 0, n$.*

(2.17) We will treat some special cases. First suppose that $\dim X = 1$, X reduced. Then $H^0(Y_\infty) \cong H^{0,0}(Y_\infty) \cong \mathbf{C}$, $\dim H^{n-1}(\mathcal{O}_D) = \dim H^0(\mathcal{O}_D) = r$: the number of branches of X at x, and $p_g(X) =$ the number $\delta = \dim_{\mathbf{C}}(\mathcal{O}_{\tilde{X}}/\mathcal{O}_X)$. Thus

$$\dim \mathrm{Gr}_F^0 H^1(Y_\infty) = \delta - r + 1, \qquad \dim \mathrm{Gr}_F^1 H^1(Y_\infty) = \delta,$$

adding up to $\mu = 2\delta - r + 1$, which does not depend on the smoothing (see [**2**]).

(2.18) Recall from [**15**] that for an isolated normal singularity (X, x) one has $b_1(Y_\infty) = 0$ for every smoothing. This can also be concluded in an easy way from (2.6)(a) and (1.12): because $H^1(Y_\infty) \subset H^2_{\{x\}}(X)$, it carries a mixed Hodge structure of type $(0, 0) + (1, 0) + (0, 1)$, and $\dim \mathrm{Gr}_F^0 H^1(Y_\infty) = h^{00} + h^{0,1} = \dim H^1(\mathcal{O}_E)$. For a normal singularity, $H^1(\mathcal{O}_E) = 0$, so $h^{00} = h^{01} = 0$, hence also $h^{10} = 0$, so $b_1(Y_\infty) = 0$. The same reasoning shows that for (X, x) a 3-dimensional Cohen-Macaulay singularity $H^2(Y_\infty)$ is purely of type $(1, 1)$.

(2.19) We consider the intersection form on $H_c^2(Y_\infty, \mathbf{R})$ for a smoothing of an isolated surface singularity, and will express its invariants μ_0, μ_+ and μ_- in terms of the Hodge numbers h^{pq} of $H^2(Y_\infty, \mathbf{C})$.

Consider again the exact sequence

$$H^2_{\{x\}}(X) \xrightarrow{\beta} H_c^2(Y_\infty) \xrightarrow{j} H^2(Y_\infty) \xrightarrow{{}^t\beta} H^3_{\{x\}}(X).$$

Then $\mu_0 = \dim \mathrm{Ker}(j) = \Sigma \dim \mathrm{Gr}_r^W \mathrm{Ker}(j)$.

Because $\operatorname{Ker}(j) \subset W_1 H_c^2(Y_\infty)$ and $W_1 \operatorname{Coker}(j) = 0$, we obtain

$$\mu_0 = \dim \operatorname{Ker}(j) = \sum_{p+q \leqslant 1} (h_c^{pq} - h^{pq}) = h^{22} + 2h^{21} - h^{00} - 2h^{01}.$$

Clearly, if $b_1(Y_\infty) = 0$, e.g. if X is normal, then β is injective and

$$\mu_0 = \dim \operatorname{Im}(\beta) = \dim H_x^2(X) = \dim H^1(D).$$

(2.20) To determine u_+, we can use the weight spectral sequence abutting to $H_c^2(Y_\infty)$ resp. $H^2(Y_\infty)$ and properties (2.6)(c), (d). These result in a decomposition of $\operatorname{Gr}_2^W H_c^2(Y_\infty)$, and we can determine the intersection form on the factors, using results of Clemens [**3**].

We first consider the bilinear form on $\operatorname{Gr}_4^W H^2(Y_\infty)$ given by

$$Q_2(x, y) = \langle \phi_2(x), y \rangle.$$

If P is a triple point of $\tilde{X} \cup E$, let V_P be the intersection of a small ball around P with Y_∞, such that in local coordinates

$$V_P \cong \{z \in \mathbf{C}^3 \mid |z| < 1 \text{ and } z_0 z_1 z_2 = \varepsilon\}$$

for ε sufficiently small. Let δ_P be a generator of $H_c^2(V_P)$ and let δ'_P be the dual generator of $H^2(V_P)$. Then $W_0 H^2(Y_\infty)$ is just the subspace generated by all δ_P for P a triple point of $\tilde{X} \cup E$. Clemens has shown that one can construct the monodromy h on Y_∞ in such a way that $h(V_P) = V_P$ and

$$(h^* - 1)^2(\delta'_P) = -2\delta_P$$

for suitable representatives δ_P, δ'_P. Indeed, V_P is diffeomorphic to $U_1 \times U_2$, where $U_i = \{z \in \mathbf{C} \mid 0 < |z| < 2\}$. Let ρ_i: $U_i \to [0, 1]$ be a C^∞-function with $\rho_i \equiv 0$ for $|z| < \frac{1}{2}$, $\rho_i \equiv 1$ for $|z| > \frac{3}{2}$, ρ_i only depending on $|z|$. Let θ_i be the argument on U_i. Then

$$\int_{V_P} \frac{d\theta_1}{2\pi} \wedge d\rho_1 \wedge \frac{d\theta_2}{2\pi} \wedge d\rho_2 = 1.$$

One can take $\delta_P = [d\rho_1 \wedge d\rho_2]$, $\delta'_P = -[(d\theta_1 \wedge d\theta_2)/4\pi^2]$, and h such that

$$h^*(d\rho_i) = d\rho_i, \qquad h^*\left(\frac{d\theta_i}{2\pi}\right) = \frac{d\theta_i}{2\pi} + d\rho_i.$$

Then $(h^* - 1)^2 \delta'_P = -2\delta_P$.

Moreover there is a natural map

$$\operatorname{Gr}_4^W H^2(Y_\infty) \to \bigoplus_P C \cdot \delta'_P$$

which is injective and has as its image the the set of those $\Sigma \lambda_P \delta'_P$, for which

$$\sum_{P \in C} \lambda_P = 0$$

for every component C of the double curve of $\tilde{X} \cup E$. As a consequence,

(2.21) LEMMA. *The form Q_2 is symmetric negative definite on* $\operatorname{Gr}_4^W H^2(Y_\infty, \mathbf{R})$.

It follows from properties (2.6)(c), (d) that we have an orthogonal decomposition

$$\mathrm{Gr}_2^W H_c^2(Y_\infty) = V\,\mathrm{Gr}_4^W H^2(Y_\infty) \oplus P_2$$

where

$$P_2 = \mathrm{Ker}\big(N_c\colon \mathrm{Gr}_2^W H_c^2(Y_\infty) \to \mathrm{Gr}_0^W H_c^2(Y_\infty)\big).$$

Write T for the triple point set of E and T' for the double point set of D. In the sequence

$$H^0(\tilde{E}^{(2)}) \xrightarrow{(\gamma,\rho)} H^2(\tilde{E}) \oplus H^0(T \cup T') \xrightarrow{\rho'-\gamma'} H^2(\tilde{E}^{(2)} \cup \tilde{D})$$

with γ, γ' Gysin maps and ρ, ρ' restriction maps, one has

$$\mathrm{Gr}_2^W H_c^2(Y_\infty) \cong \mathrm{Ker}(\rho' - \gamma')/\mathrm{Im}(\gamma, \rho),$$

and

$$P_2 \cong \mathrm{Ker}(\rho')/\mathrm{Ker}(\rho') \cap \mathrm{Im}(\gamma).$$

We will show that P_2 is a polarized sub Hodge-structure of $\mathrm{Gr}_2^W H_c^2(Y_\infty)$, i.e.

$$P_2 = P_2^{2,0} \oplus P_2^{1,1} \oplus P_2^{02}$$

and the cup product is positive definite on $(P_2^{20} \oplus P_2^{02}) \cap P_2(\mathbf{R})$ and negative definite on $P_2^{1,1} \cap P_2(\mathbf{R})$.

Write ε_i, $i = 1,\ldots,m$ the fundamental class of E_i in $H^2(\tilde{E})$. Suppose that $E_i \cap \tilde{X} = \varnothing$ for $i = k+1,\ldots,m$, $E_i \cap \tilde{X} \neq \varnothing$ for $i = 1,\ldots,k$. Then $\varepsilon_i \in \mathrm{Im}(\gamma)$ for $i = k+1,\ldots,m$ and $\mathrm{Ker}(\rho') \subset [\varepsilon_1,\ldots,\varepsilon_m]^\perp$. One easily checks that $\varepsilon_i \cdot \varepsilon_j = -D_i \cdot D_j$, $D_i = E_i \cap \tilde{X}$ hence

$$[\varepsilon_{k+1},\ldots,\varepsilon_m] = [\varepsilon_1,\ldots,\varepsilon_m] \cap [\varepsilon_1,\ldots,\varepsilon_m]^\perp.$$

One concludes that $[\varepsilon_1,\ldots,\varepsilon_m]^\perp/[\varepsilon_{k+1},\ldots,\varepsilon_m]$ is a polarized Hodge structure of weight 2, containing P_2 as a sub Hodge-structure. Hence P_2 is also polarized.

We conclude from Lemma (2.21) that S is positive definite on $V\,\mathrm{Gr}_4^W H^2(Y_\infty)$. Indeed

$$S(Vx, Vy) = \langle Vx, jVy\rangle = \langle Vx, Ny\rangle = -\langle N_c Vx, y\rangle = -Q_2(x, y)$$

for $x, y \in H^2(Y_\infty)$.

The proof of the following lemma is left to the reader:

(2.22) LEMMA. *Let Q be a symmetric bilinear form on a real finite-dimensional vector space V. Let $E \subset V$ be a linear subspace with $E^\perp \subset E$, and let Q' be the induced form on $E/E^\perp$. Then Q and Q' have the same signature.*

Here the signature is the difference between the number of positive and negative eigenvalues of the matrix of the bilinear form on some basis.

We apply this lemma to $V = H_c^2(Y_\infty, \mathbf{R})$, $E = W_2 H_c^2(Y_\infty, \mathbf{R})$. Because S is nondegenerate on $\mathrm{Gr}_2^W H_c^2(Y_\infty, \mathbf{R})$, $E^\perp = W_1$, so we obtain

(2.23) THEOREM. *Let Y_∞ be the Milnor fibre of a smoothing of an isolated surface singularity, and let h^{pq} denote the Hodge numbers of $H^2(Y_\infty)$. Then the signature σ of the intersection form is given by*

$$\sigma = h^{20} + h^{02} + 2h^{22} - h^{11}.$$

PROOF. $V\,\mathrm{Gr}_4^W H^4(Y_\infty)$ has Hodge numbers $(0, h^{22}, 0)$ and P_2 has Hodge numbers $(h^{20}, h^{11} - h^{22}, h^{02})$.

Summarizing, we get

$$\mu_0 = h^{22} + h^{12} + h^{21} - h^{10} - h^{01} - h^{00},$$

$$\mu_+ = h^{22} + h^{20} + h^{02} + h^{10} + h^{01} + h^{00},$$

$$\mu_- = -h^{22} + h^{11} + h^{10} + h^{01} + h^{00}.$$

By using $h^{pq} = h^{qp}$ we obtain

(2.24) THEOREM. $\mu_0 + \mu_+ = 2p_g(X)$.

PROOF. $\mu_0 + \mu_+ = 2(h^{22} + h^{21} + h^{20}) = 2\dim_{\mathbf{C}} \mathrm{Gr}_F^2 H^2(Y_\infty, \mathbf{C})$. Now use Proposition (2.13).

For globalisable smoothings of normal surface singularities this is due to J. Wahl [**30**] and for complete intersections to A. Durfee [**10**].

(2.25) COROLLARY. *If (X, x) is a normal surface singularity, then the numbers μ_0 and μ_+ are independent of the smoothing.*

It would be interesting to know if for an n-dimensional isolated singularity there can exist two smoothings such that $b_{n-1}(Y_\infty)$ takes two different values.

(2.26) We will prove a formula of Laufer for the Milnor number of a normal Gorenstein surface singularity:

$$\mu = 12p_g(X) - b_1(D) + b_2(D) + K_{\tilde{X}}^2.$$

Laufer has proved this formula for hypersurfaces, and Wahl gave a proof for the case that the smoothing can be globalized [**30**, (3.15.1)]. The following proof shows that in many cases arguments from a globalization can be replaced by a careful study of the mixed Hodge structure. It would be very interesting to know if one can also prove Wahl's Theorem (3.13)(a), (c) in this way. In view of (2.13) and (2.16),

$$\mu = \sum_{i=0}^{2} \dim \mathrm{Gr}_F^i H^2(Y_\infty) = 2p_g(X) - h^1(\mathcal{O}_D) + \dim \mathrm{Gr}_F^1 H^2(Y_\infty).$$

As $H^i(E, K^1) = 0$ for $i \neq 1$, one has

$$\dim \mathrm{Gr}_F^1 H^2(Y_\infty) = -\chi(K^1) = \chi(K^{11}) - \chi(K^{10}).$$

One has exact sequences (from the weight filtration),

$$0 \to \Omega^1_{\tilde{E}} \to K^{10} \to \mathcal{O}_{\tilde{E}(2)} \oplus \mathcal{O}_{\tilde{D}} \to 0$$

and

$$0 \to \Omega^1_{\tilde{E}(2)} \to K^{11} \to \mathbf{C}_T \oplus \mathbf{C}_{T'} \to 0$$

where T and T' are as in (2.21). Let $t = \sharp T$, $t' = \sharp T'$.Then one obtains

$$\dim \operatorname{Gr}^1_F H^2(Y_\infty) = 2\chi(\Omega^1_{\tilde{E}(2)}) - \chi(\Omega^1_{\tilde{E}}) - \chi(\mathcal{O}_{\tilde{D}}) + t + t'.$$

As (X, x) is Gorenstein, (Y, x) is also Gorenstein and hence $\omega_{\tilde{Y}} \cong \mathcal{O}_{\tilde{Y}}(-Z)$, $|Z| \subset E$.

Using the adjunction formula one obtains

$$2\chi(\Omega^1_{\tilde{E}(2)}) = \sum_{i<j} E_i \cdot E_j \cdot (E_i + E_j - Z),$$

$$t = \sum_{i<j<k} E_i E_j E_k,$$

$$t' = -\sum_{i<j} E \cdot E_i \cdot E_j \qquad (\text{for } \tilde{X} + E \sim 0).$$

Moreover

$$\chi(\Omega^1_{\tilde{E}}) = 2\chi(\mathcal{O}_{\tilde{E}}) - \chi(\mathbf{C}_{\tilde{E}}) = K^2_{\tilde{E}} - 10\chi(\mathcal{O}_{\tilde{E}})$$

and

$$K^2_{\tilde{E}} = \sum_j E_j(E_j - Z)^2,$$

$$\chi(\mathcal{O}_{\tilde{E}}) = \chi(\mathcal{O}_E) + \chi(\mathcal{O}_{\tilde{E}(2)}) - t,$$

$$\chi(\mathcal{O}_E) = p_g(X) + \chi(\mathcal{O}_D),$$

$$\chi(\mathcal{O}_{\tilde{E}(2)}) = -\chi(\Omega^1_{\tilde{X}(2)}).$$

Finally $K^2_{\tilde{X}} = -E(E + Z)^2$, hence all terms involved can be expressed in triple intersection products of Z and components of E. After substitution one obtains a polynomial of degree 2 in Z, for which all coefficients but the constant term vanish identically. The constant term appears to be

$$-\sum_{i<j} E_i E_j(E_i + E_j) + \sum_{i<j} E \cdot E_i \cdot E_j - 3 \sum_{i<j<k} E_i \cdot E_j \cdot E_k$$

which also vanishes identically: for $E_i \cdot E_j \cdot (E_i + E_j)$ equals the sum of the self-intersections of $E_i \cap E_j$ as a divisor on E_i and on E_j, and this is minus the number of triple points on $E_i \cap E_j$. Hence

$$-\sum_{i<j} E_i \cdot E_j(E_i + E_j) = 3t + t'$$

while

$$\sum_{i<j} E \cdot E_i \cdot E_j = -\tilde{X} \cdot \sum_{i<j} E_i \cdot E_j = -t'$$

and

$$-3 \sum_{i<j<k} E_i \cdot E_j \cdot E_k = -3t.$$

Remark that in this formula $b_2(D)$ and $K_{\tilde{X}}^2$ are not invariants of (X, x), but their sum is an invariant of (X, x).

(2.27) COROLLARY. *If Y is a smoothing of a Gorenstein normal surface singularity, then the signature of the intersection form on $H_c^2(Y_\infty)$ is*

$$\sigma = -8p_g(X) - b_2(D) - K_{\tilde{X}}^2.$$

This generalizes formulas of Durfee [**10**] and Wahl [**30**].

(2.28) Much work has been done in the last few years on the special case of hypersurfaces, where f arises from a germ $(\mathbf{C}^{n+1}, 0) \to (\mathbf{C}, 0)$ with isolated singularity by a base change $t \to t^e$. Then the cyclic group $\mathbf{Z}/e\mathbf{Z}$ acts on Y_∞ and one defines

$$h_\lambda^{pq} = \dim\left(\operatorname{Ker}\left(\gamma^* - e^{2\pi i\lambda}\right) \cap H^{pq}(Y_\infty)\right),$$

where γ is a generator of $\mathbf{Z}/e\mathbf{Z}$.

V. I. Danilov [**4**] has found a way to express the numbers h_λ^{pq} in terms of the Newton diagram in case the function f is nondegenerate. He uses the theory of toric varieties.

A. N. Varchenko [**28, 29**] has constructed a mixed Hodge structure on $H^n(Y_\infty)$ which has the same associated graded mixed Hodge structure as above, and the Hodge filtration of which is defined using asymptotic developments of integrals of relative n-forms on Y over horizontal cycles. This can be considered as a local variant of W. Schmid's construction. A similar approach, in terms of Gauss-Manin differential systems, is used in joint work of J. Scherk and the author [**21**]. These definitions of the Hodge filtrations which do not use resolution of singularities can be used to solve some problems formulated in [**26**]. One can show that the numbers h_λ^{pq} are constant in a deformation with constant Milnor number and prove the Thom-Sebastiani conjecture from [**26**].

V. I. Arnol'd [**1**] has formulated a nice conjecture concerning the numbers h_λ^{pq}. Define the singularity spectrum of an isolated hypersurface singularity as the nondecreasing sequence $n_1 \leqslant \cdots \leqslant n_\mu$ of rational numbers such that for all q, λ the number of those j for which $e^{2\pi i n_j} = \lambda$ and $[n_j] = q$ is equal to $\dim_{\mathbf{C}} \operatorname{Gr}_F^q H^n(Y_\infty)_\lambda = \Sigma_p h_\lambda^{pq}$. In particular $n_1 \geqslant 1$ if and only if $H^n(\mathcal{O}_E) = 0$, and $n_1 > 1$ iff $p_g = 0$.

Conjecture. The singularity spectrum is semicontinuous in the following sense: let a singularity S be adjacent to a singularity S' (with $\mu' < \mu$), then $n_k \leqslant n'_k$ for $k = 1, \ldots, \mu'$.

3. Cohomological insignificance and Du Bois singularities.

(3.1) In [**23**] J. Shah studies a class of singularities, introduced by D. Mumford and baptized insignificant limit singularities. He gives a list of surface singularities in $\mathbf{C}^3$ with this property. For all singularities of this list he shows that for all smoothings as in §2 one has $\mathrm{Gr}_F^0 H^k(Y_\infty) = 0$. This was the reason for I. Dolgachev to introduce the concept of a cohomologically insignificant degeneration (= smoothing, but from a different point of view).

(3.2) One considers a projective flat family $\bar{f}\colon \bar{Y} \to S$ which has only 0 as a critical value. One has specialization maps

$$\mathrm{sp}_k\colon H^k(\bar{X}) \to H^k(\bar{Y}_\infty)$$

where $\bar{X}$ is the special fibre and $\bar{Y}_\infty$ is the general fibre. If $H^k(\bar{X})$ carries the canonical mixed Hodge structure of Deligne and $H^k(\bar{Y}_\infty)$ the limit Hodge structure, then sp_k is a morphism of mixed Hodge structures. One says that $\bar{f}$ is a cohomologically in significant smoothing of $\bar{X}$ if sp_k induces an isomorphism between $\mathrm{Gr}_F^0 H^k(\bar{X})$ and $\mathrm{Gr}_F^0 H^k(\bar{Y}_\infty)$ for all k. In [7] Dolgachev proves the following results:

(3.3) THEOREM. *Assume $\bar{f}$ is a family of curves such that $\bar{Y}$ is smooth and $\bar{X}$ does not contain exceptional curves of the first kind. Then $\bar{f}$ is a cohomologically insignificant smoothing if and only if $\bar{X}^{\mathrm{red}}$ has only ordinary double points as singularities and either $\bar{X} = mC$ with C reduced and $H^1(C, \mathcal{O}_C) = \mathbf{C}$, $m > 1$ or $\bar{X}$ is reduced.*

(3.4) THEOREM. *Assume $\bar{f}$ is a family of surfaces with $\bar{Y}$ smooth and such that $\bar{X}$ has only isolated singularities. Then $\bar{f}$ is cohomologically insignificant if and only if each of its critical points is either a rational double point or a simple elliptic point or a cusp point.*

(3.5) It appears that cohomological insignificance of smoothings is connected with local Hodge theoretic properties of the special fibre, which have been introduced by Ph. Du Bois [8]. He has shown that Deligne's construction in fact defines for every complex algebraic variety V a pair $(\underline{\Omega}_V^\cdot, F)$ consisting of a complex $\underline{\Omega}_V^{\cdot}$ of quasi-coherent analytic sheaves on V with a decreasing filtration F, which is unique in the filtered derived category and has the following properties:

(1) $d\colon \underline{\Omega}_V^i \to \underline{\Omega}_V^{i+1}$ is a first order differential operator and the induced map on $\mathrm{Gr}_F \underline{\Omega}_V^\cdot$ is $\mathcal{O}_V$-linear;

(2) $\underline{\Omega}_V^\cdot$ is a resolution of the constant sheaf $\mathbf{C}$ on V; the cohomology sheaves of $\mathrm{Gr}_F \underline{\Omega}_V^\cdot$ are coherent;

(3) there is a natural morphism of filtered complexes

$$(\Omega_V^\cdot, \sigma) \to (\underline{\Omega}_V^\cdot, F)$$

which is a filtered quasi-isomorphism if V is smooth. Here $\Omega_V^\cdot$ is the ordinary De Rham complex of V and σ is its "filtration bête";

(4) if V is complete, the spectral sequence of hypercohomology of the filtered complex $(\underline{\Omega}_V^\cdot, F)$ degenerates at E_1; the resulting filtration on $H^k(V, \mathbf{C})$ is Deligne's Hodge filtration.

We define (X, x) to be a *Du Bois singularity* if the natural map from $\mathcal{O}_{X,x}$ to $\mathrm{Gr}_F^0 \underline{\Omega}^{\cdot}_{X,x}$ is a quasi-isomorphism.

(3.6) Among the singularities which have the Du Bois property we find normal crossings singularities, singularities of generic projections, and certain other classes like quotient singularities. A normal isolated singularity is a Du Bois singularity if and only if for a resolution $p\colon \tilde{X} \to X$ with $p^{-1}(x)$ a divisor with normal crossings one has the natural maps $(R^i p_* \mathcal{O}_{\tilde{X}})_x \to H^i(D, \mathcal{O}_D)$ where $D = p^{-1}(x)^{\mathrm{red}}$ are isomorphisms for all $i > 0$. A reduced curve singularity is a Du Bois singularity if and only if it is an ordinary multiple point, i.e. isomorphic to the germ at the origin of the union of some coordinate lines in $\mathbf{C}^n$.

(3.7) PROPOSITION. *If (X, x) is a rational isolated singularity, then X is Du Bois.*

PROOF. Rational singularities are normal and from Lemma (2.14) it follows that the maps $H^i(\tilde{X}, \mathcal{O}_{\tilde{X}}) \to H^i(D, \mathcal{O}_D)$ are isomorphisms as soon as $H^i(\tilde{X}, \mathcal{O}_{\tilde{X}}) = 0$.

(3.8) THEOREM. *Let (X, x) be a normal Gorenstein surface singularity. Then (X, x) is a Du Bois singularity if and only if it is a rational double point, a simple elliptic or a cusp singularity.*

PROOF. Let $p\colon \tilde{X} \to X$ be the minimal good resolution of X, which is assumed to be a contractible Stein space. Then according to Reid [**19**] (see also Dolgachev [**7**]), the dualizing sheaf of X has the form

$$\omega_{\tilde{X}} = p^*\omega_X \otimes \mathcal{O}_{\tilde{X}}(-Z)$$

with $Z = 0$ or Z is effective and has support equal to $D = p^{-1}(x)$. If $Z \neq 0$, $\mathrm{supp}(Z) = D$. Moreover Z is also minimal with respect to the property that $H^1(X, \mathcal{O}_{\tilde{X}}) \xrightarrow{\sim} H^1(X, \mathcal{O}_Z)$ (see Dolgachev, loc. cit. Proposition 4.8). Thus by the adjunction formula $\omega_Z = \mathcal{O}_Z$, hence $h^1(\mathcal{O}_Z) = h^1(\omega_Z) = h^0(\mathcal{O}_Z) = 1$, so (X, x) is a minimally elliptic singularity. It follows from Laufer [**18**, Proposition 3.5] that the only minimally elliptic singularities with reduced fundamental cycle are the simple elliptic and the cusp singularities. See H. Flenner, J. Reine Angew. Math. **328** (1981), 128–160 for the graded case.

(3.9) THEOREM. *Let $\bar{f}$ be a projective family as above and assume that the special fibre has only Du Bois singularities. Then $\bar{f}$ is cohomologically insignificant.*

PROOF. See [**27**].

(3.10) Let us restrict our attention now to families with only isolated critical points. Then, according to property (2.6)(f) such a smoothing is cohomologically insignificant if and only if for every Milnor fibre of a critical point one has $\mathrm{Gr}_F^0 H^k(Y_\infty) = 0$, so it only depends on the local smoothings associated to the critical points whether or not $\bar{f}$ is cohomologically insignificant. Hence let us define cohomological insignificance of a local smoothing $f\colon Y \to S$ to mean that $\mathrm{Gr}_F^0 H^k(Y_\infty) = 0$ for all $k > 0$. Recall that this means that $H^k(\mathcal{O}_E) = 0$ for all $k > 0$.

Our aim is to compare this property of the smoothing with the Du Bois property.

(3.11) We can as well restrict our attention to semistable smoothings: passing to a finite ramified covering of the disk S does not change the special fibre or the Milnor fibre, nor the mixed Hodge structure on its cohomology. For the analysis of the situation two exact sequences are important:

$$(1) \qquad 0 \to \mathcal{O}_{\tilde{Y}}(-E) \to \mathcal{O}_{\tilde{Y}} \to \mathcal{O}_E \to 0$$

and

$$(2) \qquad 0 \to \mathcal{O}_{\tilde{Y}} \xrightarrow{u} \mathcal{O}_{\tilde{Y}}(-E) \to \mathcal{O}_{\tilde{X}}(-D) \to 0$$

where the notations are as in §1 and u is given by multiplication with the parameter t on S. In virtue of Lemma (2.14) we obtain from (1) short exact sequences

$$(1)_i \qquad 0 \to H^i(\mathcal{O}_{\tilde{Y}}(-E)) \to H^i(\mathcal{O}_{\tilde{Y}}) \to H^i(\mathcal{O}_E) \to 0.$$

(3.11) THEOREM *Let (X, x) be a normal surface singularity and $f\colon Y \to S$ a semistable smoothing of X. Then the following conditions are equivalent*:

(i) *(X, x) is a Du Bois singularity*;

(ii) *(Y, x) is a rational singularity*;

(iii) *Y is a cohomologically insignificant smoothing of X.*

PROOF. Assume that (X, x) is a Du Bois singularity. Then $H^i(\mathcal{O}_{\tilde{X}}(-D)) = 0$ for all $i > 0$ (apply $(1)_i$ for $(\tilde{X}, D)$). Therefore the maps $H^i(\mathcal{O}_{\tilde{Y}}(-E)) \to H^i(\mathcal{O}_{\tilde{Y}})$ induced by u are injective for $i = 1, 2$. By $(1)_i$ for $i = 1, 2$ this implies that multiplication with t is injective on $H^i(\mathcal{O}_{\tilde{Y}})$ for $i = 1, 2$. Nakayama's Lemma then implies that $H^i(\mathcal{O}_{\tilde{Y}}) = 0$ for $i = 1, 2$ so Y has a rational singularity at x. Assume that (Y, x) is a rational singularity. Then $(1)_i$ implies that $H^i(\mathcal{O}_E) = 0$ for $i = 1, 2$. This means that Y is a cohomologically insignificant smoothing of X.

Finally assume that Y is a cohomologically insignificant smoothing of X. So from $(1)_i$ we get $\dim_{\mathbf{C}} H^i(\mathcal{O}_{\tilde{Y}}(-E)) = \dim_{\mathbf{C}} H^i(\mathcal{O}_{\tilde{Y}})$ for $i = 1, 2$. By the long exact sequence from (2) we obtain $H^1(\mathcal{O}_{\tilde{X}}(-D)) = 0$, so X is a Du Bois singularity.

Remark that only the last argument in the proof is not valid in higher dimensions. However with the same methods one can prove the following:

(3.12) THEOREM. *Let (X, x) be a normal isolated singularity and $f\colon Y \to S$ a semistable smoothing of X. Then the following conditions are equivalent*:

(i) *(X, x) is a Du Bois singularity*;

(ii) *(Y, x) is a rational singularity*;

(iii) *(Y, x) is a Du Bois singularity and Y is a cohomologically insignificant smoothing of X.*

Some questions concerning Du Bois singularities remain open:

1. Is every insignificant limit singularity a Du Bois singularity?
2. Classify all nonnormal Gorenstein Du Bois singularities of surfaces.

3. Are all rational singularities Du Bois singularities? There seems no difficulty to prove this for the toroidal singularities.

4. Is every (small) deformation of a Du Bois singularity again Du Bois?

Note (*added in proof*). The discussion in §1 is too careless regarding the difference between $H^q_{\{x\}}(X)$ and $H^{q-1}(X\setminus\{x\})$ where X is a contractible Stein space. The complex $A^{\cdot}_{\{x\}}(X)$ in fact satisfies

$$\mathbf{H}^q\big(A^{\cdot}_{\{x\}}(X)\big) \cong H^{q-1}(X\backslash\{x\}),$$

instead of $H^q_{\{x\}}(X)$. The reader will have no difficulty in making the substitution where necessary (e.g. Theorem (1.9), Corollary (1.12) and sequence (2.3)).

REFERENCES

1. V. I. Arnol'd, *On some problems in singularity theory*, Geometry and Analysis, Papers Dedicated to the Memory of V. K. Patodi, Tata Institute, Bombay, 1980.

2. R. O. Buchweitz and G-M. Greuel, *The Milnor number and deformations of complex curve singularities*, Invent. Math. **58** (1980), 241–281.

3. C. H. Clemens, *Degeneration of Kähler manifolds*, Duke. Math. J. **44** (1977), 215–290.

4. V. I. Danilov, *Newton polyhedra and vanishing cohomology*, Functional Anal. Appl **13** (1979), 103–114.

5. P. Deligne, *Equations différentielles à points singuliers réguliers*, Lecture Notes in Math., vol. 163, Springer-Verlag, Berlin and New York, 1970.

6. ______, *Théorie de Hodge*. III, Inst. Hautes Études Sci Publ. Math. **44** (1972), 5–77.

7. I. Dolgachev, *Cohomologically insignificant degenerations of algebraic varieties*, Compositio Math. **42** (1981), 279–313.

8. Ph. Du Bois, *Complexe de De Rham filtré d'une variété singulière*, Bull. Soc. Math. France **109** (1981), 41–81.

9. A. Durfee, *Fifteen characterizations of rational double points and simple critical points*, l'Enseignement Math. **25** (1979), 131–163.

10. ______, *The signature of smoothings of complex surface singularities*, Math. Ann. **232** (1978), 85–98.

11. A. Durfee, *Mixed Hodge structure on punctured neighborhoods*, Preprint, 1981.

12. F. Elzein, *Structures de Hodge mixtes*, C. R. Acad. Sci. Paris Sér. A–B **292** (1981), 409–412.

13. A. Fujiki, *Duality of mixed Hodge structures of algebraic varieties*, Publ. Res. Inst. Math. Sci. **16** (1980), 635–667.

14. M. Goresky and R. MacPherson, *On the topology of complex algebraic maps*, these PROCEEDINGS.

15. G-M. Greuel and J. H. M. Steenbrink, *On the topology of smoothable singularities*, these PROCEEDINGS.

16. Ph. Griffiths and W. Schmid, *Recent developments in Hodge theory*, Discrete Subgroups of Lie Groups, Bombay, 1973.

17. G. Kempf et al., *Toroidal embeddings*. I, Lecture Notes in Math., vol. 339, Springer-Verlag, Berlin and New York, 1973.

18. H. B. Laufer, *Minimally elliptic singularities*, Amer. J. Math. **99** (1977), 1257–1295.

19. M. Reid, *Elliptic Gorenstein singularities*, Preprint, 1973.

20. M. Saito, *On the exponents and the geometric genus of an isolated hypersurface singularity*, these PROCEEDINGS.

21. J. Scherk and J. H. M. Steenbrink, *On the mixed Hodge structure on the cohomology of the Milnor fibre*, Math. Ann. (submitted).

22. W. Schmid, *Variations of Hodge structure: the singularities of the period mapping*, Invent. Math. **22** (1973), 211–330.

23. J. Shah, *Insignificant limit singularities and their mixed Hodge structure*, Ann. of Math. **109** (1979), 497–536.

24. N. I. Shepherd-Barron, *Some questions on singularities in* 2 *and* 3 *dimensions*, Preprint, University of Warwick, 1980.

25. J. H. M. Steenbrink, *Limits of Hodge structures*, Invent. Math. **31** (1976), 229–257.

26. ______, *Mixed Hodge structure on the vanishing cohomology*, Real and Complex Singularities, Oslo, 1976.

27. ______, *Cohomologically insignificant degenerations*, Compositio Math. **42** (1981), 315–320.

28. A. N. Varchenko, *Asymptotics of holomorphic forms define mixed Hodge structure*, Dokl. Akad. Nauk SSSR **255** (1980), 1035–1038.

29. ______, *Asymptotic mixed Hodge structure on vanishing cohomology*, Izv. Akad. Nauk SSSR (1981).

30. J. Wahl, *Smoothings of normal surface singularities*, Topology **20** (1981), 219–246.

31. K. Watanabe, *On plurigenera of normal isolated singularities*. I, Math. Ann. **250** (1980), 65–94.

UNIVERSITY OF LEIDEN, THE NETHERLANDS

Proceedings of Symposia in Pure Mathematics
Volume **40** (1983), Part 2

THE TANGENT SPACE AND EXPONENTIAL MAP AT AN ISOLATED SINGULAR POINT[1]

DAVID A. STONE

1. Introduction and heuristic examples. The work to be described arose from a project to adapt variational methods to search for geodesics in Riemannian manifolds with singularities, such as complex analytic varieties. In seeking an analogue of the first variational formula in such a space M it seemed helpful to linearize the problem locally by using an "exponential map" $\exp_P$ at a point P (assumed singular) to shift the problem into some sort of "tangent space" $T_P M$ at P. My present purpose is to define these entities in case P is an isolated singular point of M, and to state geometric conditions on M that ensure their existence. Proofs can be found in [**9**]. Only the local situation will be discussed, so I shall usually assume that $M^0 = M - \{P\}$ is an n-manifold with a fixed Riemannian metric.

One can now define *piecewise* C^1 paths in M and the *length* and *energy* of such paths as in the smooth case. The *intrinsic metric* on M is defined by the rule

$$d^M(X, Y) = \inf\{\text{length}(\beta)\},$$

where β ranges over all piecewise C^1 paths from X to Y. β is a *geodesic from X to Y* if it has least energy among all such paths having the same domain as β; then β also has shortest length.

Caveat. Under mild hypotheses on the metric on M^0 any geodesic β satisfies the usual differential equation $D^M_{\beta'}\beta' \equiv 0$ on that part of β which lies in M^0. The converse need not be true, however, even in this local setting: a solution of the boundary-value problem, which will be called a *weak geodesic*, can fail to be a *shortest* path from X to Y.

1980 *Mathematics Subject Classification.* Primary 53B20; Secondary 32C40, 32B15.

[1] Research supported in part by grants from the National Science Foundation (MCS 7802147) and from the PSC-BHE Research Award Program of CUNY (13079).

0082-0717/81/0000-0057/$04.50

In the following heuristically presented examples it will be enough to assume that when $T_P M$ and $\exp_P$ exist, they satisfy (A)–(D):

(A) $T_P M$ reduces to the usual tangent space if M is smooth at P, and $\exp_P$ to the usual exponential map.

(B) $T_P M$ is homeomorphic to M near P, and $\exp_P$ a special homeomorphism between neighbourhoods of P in these spaces.

(C) $T_P M$ parametrizes in linear fashion the geodesics from P, so it is a cone with one generator for each geodesic (parametrized by arc length) from P, and $\exp_P$ carries the generators into the geodesics.

(D) $T_P M$ has a natural metric, and $\exp_P$ is approximately an isometry near P.

(Unlike the smooth case we cannot expect $\exp_P$ to be defined, even as a continuous function, globally on $T_P M$, even when M is metrically complete. For example let M be the infinitely-extended 1-dimensional space indicated in Figure 1. Then on the ray $P\hat{Q}$ in $T_P M$, $\exp_P$ cannot be defined beyond $\hat{Q}$.)

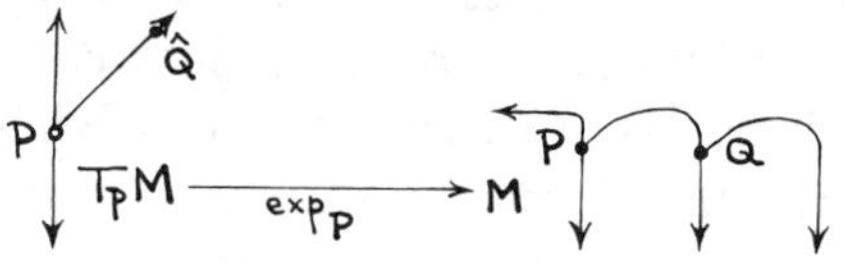

FIGURE 1

EXAMPLE 1. Let $M \subseteq \mathbf{C}^2$ have equation $x^2 - x^3 - y^2 = 0$, and let P be the origin. Near P, M is the one-point union of two nonsingular manifolds, $M = M_1 \cup_P M_2$. Then $T_P M = (T_P M_1) \cup_P (T_P M_2)$ and $\exp_P$ is defined by $\exp_P \restriction T_P M_i = \exp_P^{(M_i)}$. Thus $T_P M$ is simply the Whitney tangent cone (C_3 in Whitney [11]) T of M at P, whose equation is $x^2 - y^2 = 0$. Note that $T_P M$ is no longer a vector space nor even a topological disk.

EXAMPLE 2. The difficulties caused by nonisolated singularities can arise even with normal crossings of nonsingular hypersurfaces in higher dimensions. In $\mathbf{C}^3$ let M have equation $x(y^2 + 2y + z^2) = 0$ and let P be the origin. Then $M = M_1 \cup_C M_2$, where $M_1 = \{x = 0\}$, $M_2 = \{y^2 + 2y + z^2\} = 0$ and $C = M_1 \cap M_2$. Now $T_P M = (T_P M_1) \cup_{T_P C} (T_P M_2)$, which is indeed homeomorphic to M near P. But $\exp_P$ does not exist, for it would have to satisfy $\exp_p\colon T_P C \to C$ and $\exp_P \restriction T_P M_i = \exp_P^{(M_i)}$; yet C is not geodesically closed in M_1.

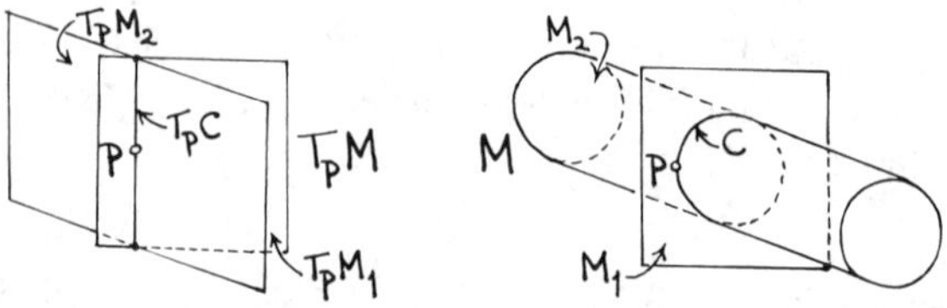

FIGURE 2

EXAMPLE 3. Let $M \subseteq \mathbf{C}^2$ have equation $f(x, y) = x^3 - y^2 = 0$ and let P be the origin.

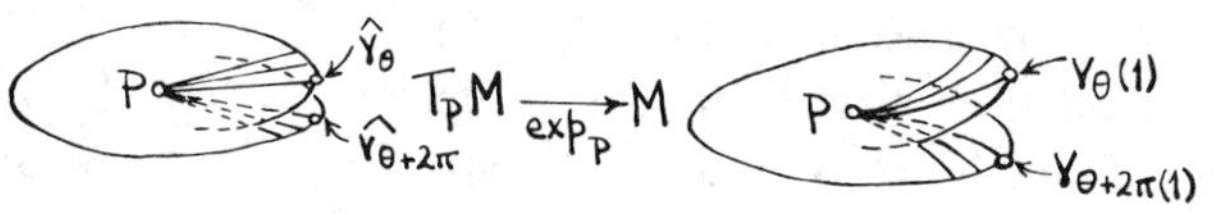

FIGURE 3

For each θ, taken mod 4π, let α_θ be the path from P defined by $\alpha_\theta(t) = (te^{i\theta}, t^{3/2}e^{i(3/2)\theta})$ for $t \geqslant 0$; and let γ_θ be the reparametrization of α_θ by arc length. It can be proved that:

(A) $\gamma_\theta \restriction [0, t^*]$ is a geodesic for every $t^* > 0$.

Let L be a circle parametrized as $\{\hat{\gamma}_\theta, \theta$ taken mod $4\pi\}$; then $T_P M$ is the cone on L and is thus topologically $\mathbf{R}^2$. The exponential map is defined by $\exp_P(t\hat{\gamma}_\theta) = \gamma_\theta(t)$. The Whitney tangent cone T of M at P is the x-axis $\mathbf{C}_x$. Algebraically T arises with multiplicity 2. The corresponding geometric phenomenon is the following. There is a natural map $\hat{\pi}$: $T_P M \to T$ defined by

$$\hat{\pi}(t\hat{\gamma}_\theta) = \gamma'_\theta(0) = (te^{i\theta}, 0).$$

It turns out that $\hat{\pi}$ is a 2-sheeted covering map, branched at P ($\hat{\pi}$ identifies $t\hat{\gamma}_\theta$ and $t\widehat{\gamma_{\theta+2\pi}}$); and that $\hat{\pi}$ is a local isometry except at P. The metric structure of $T_P M$ is thus that it has total angle 4π at P and is flat elsewhere.

EXAMPLE 4. Let M be any cone, defined by homogeneous polynomials in some $\mathbf{C}^N$ and let P be the origin. Then, whether P is an isolated singularity or not, surely $T_P M = M$ and $\exp_P$ is the identity map. More generally M can be any "metric cone" as defined in §2.

2. Definitions of tangent space and exponential map. Let L be a compact space. The *cone* cL with *vertex* c and *base* L is $L \times [0, \infty)/L \times 0$ in the identification topology, with $L \times 0$ named c and $L \times 1$ identified with L. The *truncated cone* is defined for any $\underline{t} > 0$ by

$$c^{\underline{t}}L = \{(Z, t) \in cL \text{ such that } 0 \leqslant t \leqslant \underline{t}\}.$$

When L has a metric there is a natural metric to put on cL, which is then called a *metric cone*. If the metric on L is the intrinsic metric associated to a Riemannian metric ds^L, then the metric on cL is also intrinsic, determined by the Riemannian metric ds^M on M^0, where

$$(ds^M)^2(Z, t) = t^2(ds^L)^2(Z) + dt^2.$$

For any $r > 0$ the *dilation with factor* r is defined by $r(Z, t) = (Z, rt)$; when cL is a metric cone, this multiplies the metric by r.

In defining the "tangent space" at a singular point P, I want to think of a "tangent vector" as an equivalence class of paths from P which are "differentiable" at P; but there is no *a priori* criterion for differentiability at P. So we must assume that certain germs of paths β: $[0, 1] \to M$ have been designated to form the *class* $C^1(P)$ *of paths* C^1 *at* P. I do not know how to characterize such a class axiomatically except in case M is a metric cone; but the following, at least, must surely be true.

($C^1$1) The constant path $\beta(s) \equiv P$ is in $C^1(P)$.

($C^1$2) Every other germ in $C^1(P)$ has a representative β: $[0, \delta] \to M$ such that $\beta(0) = P$, β: $(0, \delta] \to M^0$ and β is C^1 on $(0, \delta]$. We shall also require that $\lim_{s \to 0+} |\beta'(s)|$ exists.

($C^1$3) $C^1(P)$ is closed under linear reparametrization: given $\beta \in C^1(P)$ and $r > 0$, define $r_*(\beta)$ by $r_*\beta(s) = \beta(rs)$; then $r_*(\beta) \in C^1(P)$.

In case M is a metric cone, write $\beta(s) = (Z(s), t(s))$; then the remaining axioms are:

($C^1$4) $\lim_{s \to 0+} Z(s)$ exists; and

($C^1$5) $\lim_{s \to 0+} t(s)/s$ exists.

(In $\mathbf{R}^2$ with polar coordinates—so $Z \equiv \theta$, $t \equiv r$—these axioms characterize paths which are C^1 at the origin. In the absence of ($C^1$4) and ($C^1$5) one can have paths such as an equiangular spiral parametrized by arc length, which surely has no "tangent vector" at the origin.)

Returning to general M and P, assume that a suitable class $C^1(P)$ has been fixed. Define α and β in $C^1(P)$ to be *tangent* if $d^M(\alpha s, \beta s) = o(s)$.

EXAMPLE. In Example 3 of §1, the paths γ_θ and $\gamma_{\theta+2\pi}$ are tangent at P in $\mathbf{C}^2$ but not in M in the sense just defined. The shortest path *in* M from $\gamma_\theta(s)$ to $\gamma_{\theta+2\pi}(s)$ consists of the entire segments $\gamma_\theta([0, s]) \cup \gamma_{\theta+2\pi}([0, s])$; so $d^M(\gamma_\theta s, \gamma_{\theta+2\pi} s) = 2s$ (see Figure 3).

A *tangent vector* at P is an equivalence class of mutually tangent paths of $C^1(P)$. The class of β will be denoted $\hat{\beta}$, except that the class of the constant path will usually be written as P. $T_P M$ is defined to be the set of tangent vectors. It has the natural metric

$$d^T(\hat{\alpha}, \hat{\beta}) = \limsup_{s \to 0+} d^M(\alpha s, \beta s)/s.$$

This makes $T_P M$ a metric cone with vertex P; that is,

$$d^T(\widehat{r_*\alpha}, \widehat{r_*\beta}) = r d^T(\hat{\alpha}, \hat{\beta}).$$

To be of use in studying geodesics from P, $C^1(P)$ and $T_P M$ must satisfy, at least:

(T1) every geodesic from P is in $C^1(P)$; and

(T2) $T_P M$ is homeomorphic to M near P.

When (T1) and (T2) hold I shall say that M has a *tangent space at* P, namely $T_P M$. M will be said to *have an exponential map at* P, $\exp_P$, if also (E1)–(E6) below hold.

(E1) There exists a neighbourhood U of P in M such that for every $X \in U$ there is a geodesic γ_X from P to X with domain $[0, 1]$, which lies in U.

(E2) For each $X \in U$ (chosen sufficiently small), γ_X is unique.

There is now an *inverse-exponential map* e_P^{-1}: $U \to T_P M$ defined by $e_P^{-1}(X) = \widehat{\gamma_X}$.

Caveat. The nomenclature is not intended to imply that e_P^{-1} is the inverse of any map.

We further require:

(E3) e_P^{-1} is continuous and one-to-one (when U is sufficiently small); and

(E4) $\operatorname{im}(e_P^{-1})$ is a neighbourhood V of P in T_PM.

Now e_P^{-1} has an inverse function $\exp_P\colon V \to U$. To ensure this map is a homeomorphism near P it remains to require:

(E5) $\exp_P$ is continuous on V.

One way to express the requirement that $\exp_P$ be approximately an isometry at P is:

(E6) $|d^T(\widehat{\gamma_X}, \widehat{\gamma_Y}) - d^M(X, Y)|/d^M(X, Y) \to 0$ as $X, Y \to P$ with $X \neq Y$, uniformly in X and Y.

These axioms imply the principles (A)–(D) used in §1.

3. The main theorems. The two main theorems give geometric conditions sufficient to ensure that M has a tangent space and an exponential map, respectively, in the following context. Suppose we have a candidate for T_PM, namely a cone cL, and a candidate for $\exp_P$, namely a homeomorphism $h\colon cL \to M$ defined near P. Assume that h carries generating rays of cL into "approximate geodesics" from P, that is, into curves which do not bend very badly. If their second-order bending is not too bad, then M has a tangent space at P (Theorem 1). If the third-order bending is also under control, then M has an exponential map at P (Theorem 2).

Notation. Let N be a Riemannian manifold, $X \in N$ and $v, w \in T_XN$. Then $\langle v, w \rangle$ denotes the inner product of v and w, and $v \vee w$ their Grassmann product. If $v \neq 0$, then $v^{\parallel}(w)$ and $v^{\perp}(w)$ are the components of w respectively parallel and perpendicular to v. When N is a Hermitian manifold, $\underline{\langle v, w \rangle}$, $v_{\underline{\parallel}}(w)$ and $v_{\underline{\perp}}(w)$ will denote the Hermitian analogues of the concepts above.

D_v^N denotes the operation of covariant differentiation in the direction v. R^N is the curvature tensor on N, so $R^N(v, w)\colon T_XN \to T_XN$. Π will denote a 2-plane tangent to N, and $K^N(\Pi)$ its sectional curvature; if v and w span Π, then $K^N(\Pi) = \langle R^N(v, w)v, w \rangle / |v \vee w|^2$.

DEFINITION. Let L be a closed, compact, smooth manifold and let $\underline{t} > 0$. A *chart for M at P* is a homeomorphism, for some choices of L and $\underline{t}$, $h\colon c^{\underline{t}}L \to M$ satisfying (C1)–(C3) below. Let $\underline{\partial}_t$ denote the unit radial vector field on $(cL)^0 = cL - \{c\}$:

$$\underline{\partial}_t(Z, t) = \frac{\partial}{\partial t}(Z, t).$$

It is required that:

(C1) $h(c) = P$, $h\colon (c^{\underline{t}}L)^0 \to M^0$ is a diffeomorphism.

Thus h induces a vector field $\underline{\rho} = Dh \cdot \underline{\partial}_t$ on M^0. We also require:

(C2) for each $Z \in L$, $\lim_{t \to 0+} |\underline{\rho}(h(Z, t))| = 1$ uniformly in Z; and

(C3) whenever $t \in (0, \underline{t}]$ and $u \in T_{(Z,t)}(L \times t)$, then $\langle Dh \cdot u, \underline{\rho} \rangle = 0$.

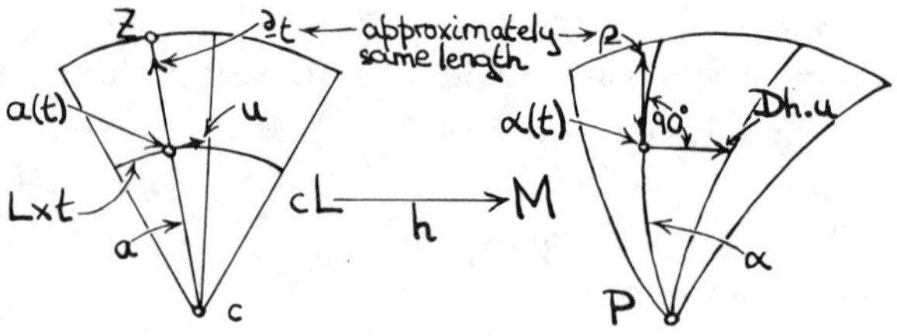

FIGURE 4

THEOREM 1. *Assume that M has a chart at P and that there exists $k > 0$ such that whenever $t \in (0, \underline{t}]$ and $\psi \in T_{h(Z,t)}M^0$, then*

$$(D^M)\ |D_\psi^M \underline{\rho} - t^{-1}\underline{\rho}^{\perp}(\psi)| = |\psi|\, O(t^{-1+2k})$$

uniformly in Z and ψ.

Let $C^1(c)$ be determined by (C^11)–(C^15) for cL, and define $C^1(P)$ to be the class of germs of paths β from P such that $h^{-1} \circ \beta \in C^1(c)$. Then $C^1(P)$ is independent of the choice of chart h satisfying (D^M), and the resulting invariant T_PM of M satisfies (T1) *and* (T2).

REMARK 1. In case M is a metric cone and h the identity, the right-hand side of (D^M) is identically 0. In polar coordinates on $\mathbf{R}^2$, for instance, (D^M) boils down to the fact that $D_\psi(\partial/\partial r) - r^{-1}(\partial/\partial\theta\text{-component of } \psi) = 0$.

REMARK 2. In (D^M) take $\psi = \underline{\rho}(h(Z, t))$; then in view of (C2) we obtain the following. Fix $Z \in L$ and let $a(t) = (Z, t)$ describe the corresponding generator of cL. Set $\alpha = h \circ a$, so $\alpha'(t) = \underline{\rho}(\alpha t)$ (see Figure 4). Then the geodesic curvature of α is controlled by $|D_{\alpha'}^M\alpha'(t)| = O(t^{-1+2k})$.

THEOREM 2. *Assume that M has a chart at P satisfying (D^M) and further that:*

(K^M) there exists an upper bound on the sectional curvature of all 2-planes tangent to M^0 (in some neighbourhood of P); and

(R^M) let $\phi, \psi \in T_{h(Z,t)}M^0$; then with $k > 0$ as in (D^M),

(i) $\|R^M(\phi, \psi)\| = |\phi \vee \psi|\, O(t^{-2})$,

(ii) $\|R^M(\underline{\rho}, \psi)\| = |\psi|\, O(t^{-2+k})$, and

(iii) $|R^M(\underline{\rho}, \psi)\underline{\rho}| = |\psi|\, O(t^{-2+2k})$,

all uniformly in Z, ϕ and ψ.

Then M has an exponential map at P satisfying (E1)–(E6).

REMARK 1. Assume (E1) and (E2) hold; then (R^M) can be interpreted intrinsically to M. Set $s = d^M(P, X)$, so that $X = \gamma_X(s)$. Define the unit radial vector field $\underline{\partial}_s$ on M^0 near P by $\underline{\partial}_s(X) = \gamma_X'(s)$. Then (R^M) implies the following.

$(R^M{}_*)$ Let $X \in M^0$ and let $\Pi \subseteq T_XM$ be a 2-plane. Then:

(i*) $K^M(\Pi) = O(s^{-2})$, and

(ii*) if $\underline{\partial}_s(X) \in \Pi$, then $K^M(\Pi) = O(s^{-2+2k})$,

both uniformly in X and Π.

This form of (R^M), though conceptually simpler—see the next remark—is not practicable because one can hardly ever calculate s or $\underline{\partial}_s(X)$.

REMARK 2. Let us examine (K^M) and (R^{M_*}) in case M is a metric cone. The right-hand side of (ii*) is 0, since Π is developable. As for (i*), say $X = (Z, s)$ and suppose $\Pi \subseteq T_X(L \times s)$. Then Π corresponds to a 2-plane $\Pi^* \subseteq T_Z L$, and in fact

$$K^M(\Pi) = s^{-2}\big(K^L(\Pi^*) - 1\big).$$

Since L is compact, K^L is bounded above, and it is now not hard to prove (i*). However (K^M) holds if and only if $K^L \leqslant 1$ everywhere. Now M has an exponential map whether or not K^L is $\leqslant 1$ (see Example 4); and this suggests

Conjecture. Theorem 2 is valid without the hypothesis (K^M).

An example due to Aleksandroff [**1**, Chapter I, §10] shows that some control from above on sectional curvature is needed in order to have an exponential map; but perhaps (R^M), in view of (R^{M_*}), is enough. The conjecture would be very useful in applications to complex hypersurfaces (see the Conjecture to Theorem 4 in §4).

EXAMPLE 5. This example shows the extreme of negative curvature allowed by (R^M) in Theorem 2. It is used in the proof of Theorem 2. Let L be any closed, compact Riemannian $(n-1)$-manifold with metric ds^L. Let a positive number $\mathcal{Q}$ and a positive integer q be given. Consider the o.d.e. on $(0, \infty)$

(A) $$r''(t) = \mathcal{Q}t^{-2+1/q}r(t); \qquad \lim_{t\to 0+} r(t) = 0; \qquad \lim_{t \to 0+} r'(t) = 1.$$

The substitution $x = t^{1/q}$ transforms (A) into an equation with a regular singularity at 0; it follows that (A) has a solution on some $(0, \delta)$ of the form

$$r(t) = t\left(1 + \sum_1^\infty a_m t^{m/q}\right).$$

Let $\mathfrak{M}$ be the cone cL, let $\mathcal{P}$ be its vertex, and let $\mathfrak{M}^0$ have a metric which near $\mathcal{P}$ is defined by

$$(ds^{\mathfrak{M}})^2(Z, t) = r(t)^2(ds^L)^2(Z) + dt^2.$$

Then $\mathfrak{M}$ has an exponential map at $\mathcal{P}$, namely the identity map id: $cL \to \mathfrak{M}$. In this case $\exp_P$ is a chart for $\mathfrak{M}$ at $\mathcal{P}$, and (D^M) and (R^M) hold. In (ii*) of (R^{M_*}) for example, if $\Pi \subseteq T_{(Z,s)}$ is a 2-plane containing $\underline{\partial}_s(Z, s)$, then

$$K^{\mathfrak{M}}(\Pi) = -\mathcal{Q}s^{-2+1/q}.$$

(K^M) is irrelevant here; compare the Conjecture in Remark 2 above. But any attempt to weaken (R^M) in Theorem 2 will probably encounter the problem of solving a more difficult o.d.e. along the lines of (A).

EXAMPLE 6. In $\mathbf{R}^3$ let M have equation $f(x, y, z) = x^2 + y^2 - z^3 = 0$ and let P be the origin. Then (K^M) holds but (R^{M_*}) fails. There can be no exponential map at P because (E3) fails. In fact $T_P M$ reduces to the nonnegative z-axis, so (T2) fails; and therefore by Theorem 1, (D^M) fails in any chart.

PROOF. Let β be the curve $\beta(t) = (t^{3/2}, 0, t)$ for $t \geqslant 0$. M is of course the surface obtained by rotating β around the z-axis. Let γ be the reparametrization of β by arc length s. It can be proved that γ is a geodesic. Moreover $s(t) = t(1 + O(t))$; and it follows that

$$\text{(A)} \qquad t(s) = s(1 + O(s)).$$

For $t > 0$ the principal directions of curvature at $\beta(t)$ are $\beta'(t)$ and $v(t) = (0, 1, 0)$. The Gaussian curvature of M^0 at $\beta(t)$ is given by

$$\text{(B)} \qquad K^M(\beta t) = \frac{\langle D_{\beta' t} \operatorname{grad} f, \beta' t \rangle \langle D_{vt} \operatorname{grad} f, vt \rangle}{|\operatorname{grad} f|^2 \, |\beta' t|^2 \, |v|^2}.$$

(Here D_u denotes the usual derivative in $\mathbf{R}^3$ applied in the direction u, and $\operatorname{grad} f = (\partial f/\partial x, \partial f/\partial y, \partial f/\partial z)$.) (A) and (B) together give

$$K^M(\gamma s) = -(3/2)s^{-2}(1 + O(s)).$$

In view of the rotational symmetry of M, this proves (K^M) but contradicts (ii*) of (R^{M_*}).

To see that (E3) fails, let γ_0 and γ_1 be distinct geodesics parametrized by arc length. In cylindrical coordinates,

$$\gamma_i(s) = (r(s), \theta_i, z(s)), \qquad i = 1, 2, \ \theta_0 \neq \theta_1.$$

The meridian circle $\{z = z(s)\}$ provides two arcs joining γ_0 to $\gamma_1(s)$; one such arc must be no longer than $\pi r(s)$. Thus

$$\begin{aligned} d^M(\gamma_0 s, \gamma_1 s) &\leq \pi r(s) = \pi[z(s)]^{3/2} \\ &= \pi[s(1 + O(s))]^{3/2}, \quad \text{by (A)}, \\ &= O(s^{3/2}) = o(s). \end{aligned}$$

Thus γ_0 and γ_1 are equivalent in $T_P M$, so (E3) fails. More generally any geodesic is reduced in $T_P M$ simply to its speed, which may conveniently be expressed as the tangent vector (in $\mathbf{R}^3$) $\gamma'(0)$ on the nonnegative z-axis. This supports, though it does not prove, the final assertion about $T_P M$.

4. Affine complex analytic hypersurfaces. Let $M \subseteq \mathbf{C}^{n+1}$ have equation $f(x_1, \ldots, x_{n+1}) = 0$, where f is an analytic function defined in some neighbourhood of the origin, P, and $f(P) = 0$. Following Milnor [7], define a vector field on $\mathbf{C}^{n+1}$ by

$$\underline{\operatorname{grad}} \, f = \left(\overline{\frac{\partial f}{\partial x_1}}, \ldots, \overline{\frac{\partial f}{\partial x_{n+1}}} \right).$$

Assume henceforth that P is an *isolated singularity of* M, that is, that near P, $\underline{\operatorname{grad}}(f)$ is zero exactly at P. At a point $X \in M^0$, $\underline{\operatorname{grad}}(f)$ generates (over $\mathbf{C}$) the normal bundle to M: a vector $v \in T_X \mathbf{C}^{n+1}$ lies in $T_X M^0$ if and only if the Hermitian product $\langle v, \underline{\operatorname{grad}} \, f \rangle = 0$.

The simplest way to try to construct a chart for M is to change as little as necessary the natural conical structure of $\mathbf{C}^{n+1}$. Let $\underline{R}$ be the unit radial vector field on $\mathbf{C}^{n+1} - \{0\}$, $\underline{R}(X) = X/|X|$. Assume:

(A) $\underline{R}$ and $\underline{\operatorname{grad}}(f)$ are linearly independent over $\mathbf{C}$ at every $X \in M^0$.

Let S be a sphere of small radius t about P. Under assumption (A), S is transverse to M^0, so $L = S \cap M$ is a closed, compact, smooth manifold. There is a unique map h: $cL \to M$ defined by the following rules, in which X denotes $h(Z, t)$:

(B) $|X| = t$, and $\partial h(Z, t)/\partial t$ is a positive, real-scalar multiple of $(\text{grad } f)_\perp \underline{R}(X)$, the orthogonal projection into $T_X M$ of $\underline{R}(X)$ (see Figure 5). Then (C1) and (C3) hold, so to check that h is a chart at P reduces to verifying (A) and (C2).

Let us first consider the case of curves in $\mathbf{C}^2$, when $n = 1$.

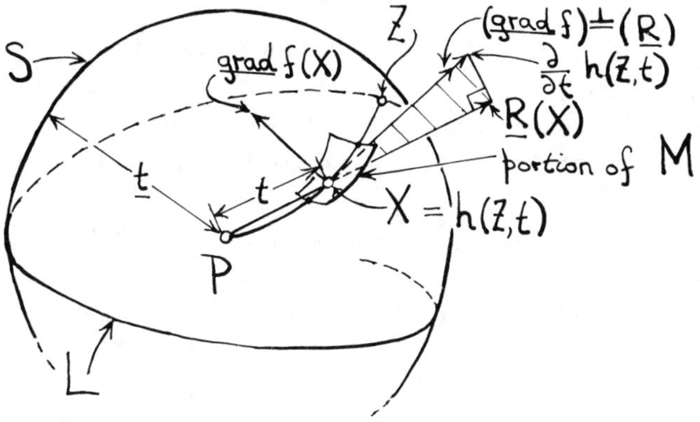

FIGURE 5

THEOREM 3. *Let $M \subseteq \mathbf{C}^2$ be a complex analytic curve and let P be a singular point of M. Then M has a tangent space and an exponential map at P.*

OUTLINE OF PROOF. First consider the case that, under a congruence of $\mathbf{C}^2$, coordinates can be chosen so that P is the origin and M has equation

$$f(x, y) = ax^q + \sum_{q+1}^{\infty} a_m x^m - y^p = 0, \quad \text{where } q > p \geqslant 2$$

(when $p \leqslant 1$, M is nonsingular). Then (A) and (C2) hold, so h is a chart. M^0 has Gaussian curvature everywhere negative, so (K^M) holds. It is tedious, rather than difficult, to show that (D^M) and (R^M) hold. It turns out that one may choose $k = (1/2)\min\{1, (q/p) - 1\}$. So Theorems 1 and 2 establish the result in this case; one might note that they are stretched to their utmost when $p \gg 0$ and $q \approx p$. $T_P M$ can be described exactly, as in Example 3 of §2. In the coordinate system above, there is a natural map $\hat{\pi}$: $T_P M \to \mathbf{C}_x$ defined by $\hat{\pi}(\hat{\alpha}) = \alpha'(0)$. Then $\hat{\pi}$ is a p-fold covering (p is the multiplicity of M at P) branched at P; and $\hat{\pi}$ is a local isometry except at P, where $T_P M$ has total angle $2\pi p$.

In the general case M can be expressed near P as a one-point union of curves $M_1, \ldots, M_r$, each of which can be cast into the form just described by using suitable coordinates. Then each M_i has a tangent space and exponential map and as in Example 1 of §2, $T_P M$ is the one-point union at P of the $T_P M_i$ and $\exp_P$ is determined by $\exp_P \restriction T_P M_i = \exp_P^{(M_i)}$. Q.E.D.

I now turn to the case $n > 1$. Let g be the initial homogeneous polynomial of f and say g has degree p; so $f = g +$ (terms of degree $> p$). The *Whitney tangent cone of* M *at* P is $T = \{g = 0\}$ ($C_3(M, P)$ in the notation of Whitney [11]). The *quasisingular locus of* M *at* P is $B = \{X \in M$ such that $\underline{\text{grad}}\, g(X) = 0\}$. P is called a *strictly isolated singularity of* M if it is an isolated singularity of T, so that, near P, $\underline{\text{grad}}(g)$ vanishes exactly at P on $\mathbf{C}^{n+1}$. This implies that g is irreducible and that P is in fact an isolated singularity of M.

THEOREM 4. *Let P be a strictly isolated singularity of a complex analytic hypersurface $M \subseteq \mathbf{C}^{n+1}$, with $n > 1$. Then M has a tangent metric space at P.*

The proof is similar to that of Theorem 3. One checks that (A) and (C2) hold, so the procedure described above determines a chart at P. To check (D^M) is easier than in the previous proof: one need only use $k = 1/2$. The theorem now follows from Theorem 1. In fact (R^M) also holds with $k = 1/2$, so Theorem 2 gives

Scholium. Let M and P be as in Theorem 4. If also (K^M) holds, then M has an exponential map at P.

Conjecture. Let M and P be as in Theorem 4. Then M has an exponential map at P.

This is a special case of the Conjecture to Theorem 2 in §3. There are elementary examples—even of cones—of M having a strictly isolated singularity at which (K^M) fails and yet an exponential map exists. On the other hand, results of Bochner [2] on Kähler manifolds show that the holomorphic sectional curvature and the Ricci curvature of M^0 are everywhere negative. So the present conjecture might be true even if the more general one is false.

EXAMPLE 7. Let $M \subseteq \mathbf{C}^3$ have equation $f(x, y, z) = y(x^3 - y^2) - z^2 = 0$ and let P be the origin; it is an isolated singularity but not strictly so. There can be no exponential map at P: either (E2) or (E3) fails. Incidentally, f is a weighted-homogeneous polynomial.

PROOF. The quasi-singular locus $B = \{y^2 = x^3, z = 0\} \cup \{y = 0 = z\}$. B is the intersection of M with the (x, y)-plane H; so under reflection h in H, M is invariant and B is the fixed-point set of $H \upharpoonright M$. Let $X \in B$ and suppose there is a geodesic γ from P to X which does not lie in B. Then $h \circ \gamma$ is a second geodesic from P to X; so (E2) fails.

Now assume (E2) holds. Let $\beta(t) = (t, t^{3/2}, 0)$ for $t \geqslant 0$, and let γ be the reparametrization of β by arc length s. By (A) of Example 3 in §1 for any $s^* > 0$, $\gamma \upharpoonright [0, s^*]$ is a geodesic in B from P to $\gamma(s^*)$ and so, by the preceding argument, is a geodesic in M. The path $\gamma^*(t) = (t, 0, 0)$ is also a geodesic in M. I claim that:

(A) β and γ^* are tangent in $T_P M$.

The proof of (A) in Example 6 of §3 applies here to show that β and γ are tangent in $T_P M$; so (A) shows that γ and γ^* are tangent, contradicting (E3).

For each t let α_t be the path in M from $\gamma^*(t)$ to $\beta(t)$ defined by

$$\alpha_t(u) = \left(t, ut^{3/2}, z_t(u)\right) \quad \text{for } u \in [0, 1],$$

where $z_t(u) = [ut^{3/2}(t^3 - u^2t^3)]^{1/2}$. Regarding α_t as a curve in the real (y, z)-plane, it is concave downwards. For implicit differentiation yields

$$\frac{\partial z}{\partial y} = \frac{x^3 - 3y^2}{2y(x^3 - y^2)} = \frac{1}{2y} - \frac{y}{x^3 - y^2};$$

hence

$$\frac{\partial^2 z}{\partial y^2} = -\frac{1}{2y^2} - \frac{1}{x^3 - y^2} - \frac{2y^2}{(x^3 - y^2)^2},$$

which shows that $\alpha_t''(u) < 0$ on $(0, 1)$. It follows (see Figure 6) that

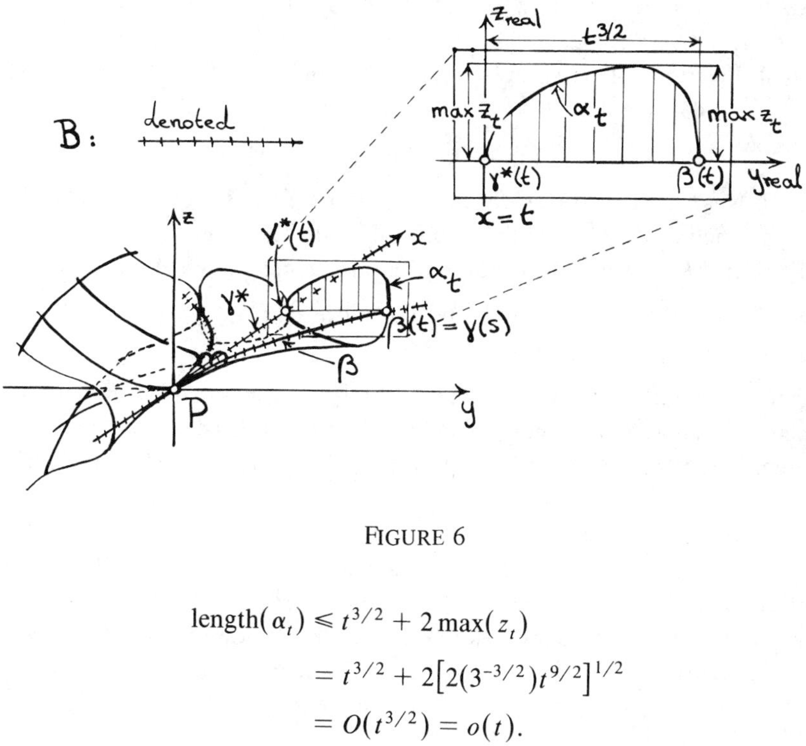

FIGURE 6

$$\begin{aligned}\text{length}(\alpha_t) &\leqslant t^{3/2} + 2\max(z_t)\\ &= t^{3/2} + 2\left[2(3^{-3/2})t^{9/2}\right]^{1/2}\\ &= O(t^{3/2}) = o(t).\end{aligned}$$

Therefore $d^M(\gamma^*t, \beta t) \leqslant \text{length}(\alpha_t) = o(t)$, which proves (A).

EXAMPLE 8. Let $M \subseteq \mathbf{C}^3$ we have equation $f(x, y, z) = xy - z^3 = 0$, and let P be the origin. As in Example 7, P is an isolated singularity but not strictly so. This time, however, the quasisingular locus is the cone $B = \{xy = z = 0\}$, so the previous argument cannot be applied. Here (K^M) fails, and in any chart at least one of (D^M), (E2) and (R^M) fails. I do not know whether M has a tangent space or an exponential map at P (see the next Conjecture). Incidentally, M is a toric variety (see Danilov [**12**]).

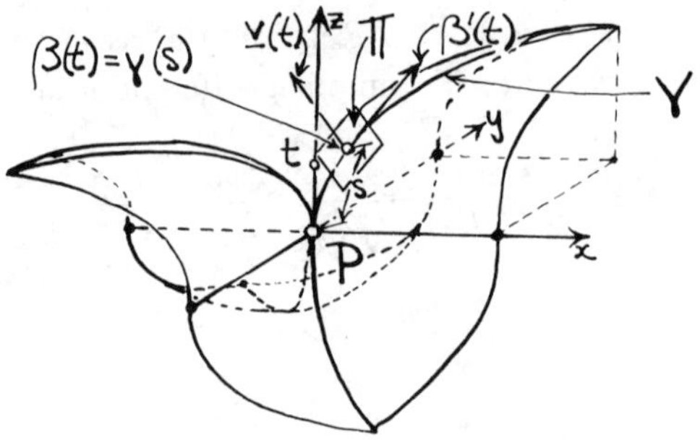

FIGURE 7

PROOF. Set $\beta(t) = (t^{3/2}, t^{3/2}, t)$ for $t \geqslant 0$, and let γ be the reparametrization of β by arc length s from P. One can show that γ is a weak geodesic. Let $\underline{v}$ be the (constant) vector field along β defined by $\underline{v}(t) = (1, -1, 0)$. Let M^* be the real part of M. Then $\beta'(t)$ and $\underline{v}(t)$ span the plane $\Pi = T_{\beta t}M^*$, and in fact are principal directions of curvature for the hypersurface $(M^*)^0 \subseteq \mathbf{R}^3$. From (B) in Example 6 of §3 we obtain

$$K^{M^*}(\Pi) = (3/4)t^{-2}(1 + O(t)).$$

Since $\underline{v}$ is a constant vector field, it is *a fortiori* parallel along the weak geodesic γ in the geometry of M^0. By the lemma of Synge (see Hicks [**5**]), $K^M(\Pi t) \geqslant K^{M^*}(\Pi t)$; so (K^M) fails. The proof of (A) in Example 6 of §3 applies, so

$$K^M(\Pi) \geqslant (3/4)s^{-2}(1 + O(s)).$$

I claim that

(A) if (E2) holds, then for all small s, $\gamma \restriction [0, s]$ is a geodesic from P to $\gamma(s)$.

Then the last estimate shows that (ii*) of (R^{M^*}) fails; so by Remark 2 to Theorem 2 in §3, there can be no chart at P in which (D^M), (E2) and (R^M) all hold.

To prove (A) we first show that there does exist a geodesic from P to $X = \gamma(s)$. Let γ^* be any such geodesic parametrized by arc length. Now M is invariant under complex conjugation, and P and X are fixed points. Hence, by (E2), γ^* must also be invariant and therefore must lie in M^*. Similarly M^* is invariant, and P and X fixed, under reflection of $\mathbf{R}^3$ in the hyperplane $H = \{x = y\}$; hence γ^* must lie in $M^* \cap H$, which is the real curve $\{x = y,\ z^3 = x^2\}$. It is now clear that γ^* can only be γ. This implies (A).

Conjecture. Let M be a complex analytic hypersurface in $\mathbf{C}^{n+1}$ with an isolated singular point. Assume that the quasisingular locus B is a cone (as in Example 8). Then M has a tangent space and an exponential map at P.

Further, I believe that T_PM can be described as follows, based on Example 3 and the discussion following Theorem 3. Let T be the Whitney tangent cone of M at P; then $B \subseteq T$ and $\exp_P\colon T_PB \to B$ is an isometry (see Example 4 of §1). Define a map $\hat{\pi}\colon T_PM \to T$ by $\hat{\pi}(\hat{\alpha}) = \alpha'(0)$. Then trivially $\hat{\pi} \restriction T_PB = \exp_P \restriction T_PB$. I conjecture that $\hat{\pi}\colon (T_PM - T_PB) \to (T - B)$ is a covering map and a local isometry, so that T_PM is metrically a covering of T branched along B.

In particular, when P is unbranched, $\hat{\pi}$: $T_P M \to T$ should be a global isometry. In this case I can prove, using (T2) and an equisingularity theorem due to Teissier [**10**], that $T_P M$ and T are homeomorphic, but without reference to $\hat{\pi}$.

A positive answer to the final problem would have several desirable consequences: the immediate generalization of Theorems 3 and 4 to hypersurfaces in arbitrary complex manifolds; a more powerful proof of Theorem 3 itself; and perhaps the conjecture just formulated.

Problem. Let f: $N \to N^*$ be a diffeomorphism between Riemannian manifolds. Let $M \subseteq N$ be a submanifold with an isolated singular point P, and give M^0 the Riemannian metric induced from that on N. Set $M^* = f(M)$ and $P^* = f(P)$; then M^* has an isolated singularity at P^*. If M has a tangent space at P, is the same true for M^* at P^*, and does f induce a map $T_P f$: $T_P M \to T_{P^*} M^*$? If Df: $T_P N \to T_{P^*} N^*$ is an isometry, is $T_P f$ also one? If M has an exponential map at P, is the same true for M^* at P^*?

The case $N^* = T_P N$, $f = (\exp_p^N)^{-1}$ is particularly interesting. We are then asking whether, in studying the local behaviour of geodesics from P in M by linearizing with an exponential map, we can separate the local intrinsic geometry of M from its extrinsic geometry in an ambient manifold N.

BIBLIOGRAPHY

1. A. D. Aleksandroff, *Die innere Geometrie der konvexe Flächen*, Akademie-Verlag, Berlin, 1958.

2. S. Bochner, *Curvature in Hermitian metric*, Bull. Amer. Math. Soc. **53** (1947), 179–195.

3. J. Cheeger, *On the Hodge theory of Riemannian pseudomanifolds*, Proc. Sympos. Pure Math., vol. 36, Amer. Math. Soc., Providence, R. I., 1980.

4. D. Gromoll, W. Klingenberg and W. Meyer, *Riemannische Geometrie im Grossen*, Lecture Notes in Math., vol. 55, Springer-Verlag, Berlin, 1968.

5. N. J. Hicks, *Notes on differential geometry*, Van Nostrand-Reinhold, London, 1971.

6. S. Lang, *Introduction to differentiable manifolds*, Wiley, New York, 1962.

7. J. Milnor, *Singular points of complex hypersurfaces*, Ann. of Math. Studies, No. 61, Princeton Univ. Press, Princeton, N. J., 1968.

8. L. Ness, *Curvature on algebraic plane curves*. I, Compositio Math. **35** (1977), 57–63.

9. D. Stone, *The exponential map at an isolated singular point*, Mem. Amer. Math. Soc. No. 256 (1982).

10. B. Teissier, *Cycles évanescents, sections planes et conditions de Whitney*, Astérisque (Soc. Math. France) **7** & **8** (1973), 285–362.

11. H. Whitney, *Tangents to an analytic variety*, Ann. of Math. (2) **81** (1965), 496–549.

12. V. I. Danilov, *The geometry of toric varieties*, Uspehi Mat. Nauk **33** (1978), 85–134; translated as Russian Math. Surveys **33** (1978), 97–154.

BROOKLYN COLLEGE OF THE CITY UNIVERSITY OF NEW YORK

Proceedings of Symposia in Pure Mathematics
Volume **40** (1983), Part 2

SINGULARITIES OF COMPLEX ANALYTIC FOLIATIONS

TATSUO SUWA[1]

This is a summary of the papers [**12** – **15**], which mainly study unfoldings of complex analytic foliation singularities. Certain aspects of foliation singularities can be treated and explained systematically from the unfolding theoretical viewpoint, for example, some extension theorems and stability theorems, solving differential equations with singularities by unfolding, etc.

§§1 and 2 contain terminologies and definitions. In §3, we classify the equivalence classes of first order unfoldings of a (codim 1) local foliation ((3.3) Theorem). In §4, we explain how Malgrange's theorem on Frobenius with singularity is related to the unfolding theory. In §5, we give a theorem of versality ((5.2) Theorem), which says that if an unfolding of a codim 1 local foliation is infinitesimally versal, then it is versal, and we list some of its consequences. The theorem also enables us to construct universal unfoldings of some foliation singularities. In §6, we show how some of the residues of Baum and Bott can be expressed in terms of the local Chern classes of a certain sheaf which came up in the unfolding theory of global foliations. This proves the rationality of these residues. Also in many cases the residues can be computed using the Riemann-Roch theorem for analytic embeddings of Atiyah and Hirzebruch.

1. Complex analytic foliations with singularities. Let M be an n-dimensional complex manifold. We denote by $\mathcal{O}_M$, Θ_M and Ω_M, respectively, the sheaves of germs of holomorphic functions, holomorphic vector fields and holomorphic 1-forms. For a coherent sub-$\mathcal{O}_M$-module F of Ω_M, the quotient sheaf Ω_M/F is denoted by Ω_F and the singular set $S(F)$ of F is defined to be the singular set of the coherent sheaf Ω_F;

$$S(F) = \{z \in M \mid \Omega_{F,z} \text{ is not a free } \mathcal{O}_{M,z}\text{-module}\},$$

1980 *Mathematics Subject Classification.* Primary 32C40, 32G11, 58F18; Secondary 14B07, 14F12, 57R20, 57R30, 58C27, 58F14.

[1]Partially supported by the National Science Foundation.

0082-0717/81/0000-0005/$03.00

where for a sheaf $\mathcal{S}$ on M, $\mathcal{S}_z$ denotes the stalk of $\mathcal{S}$ over z. Also the annihilator F^a of F is defined by

$$F^a = \{\theta \in \Theta_M \mid \omega(\theta) = 0, \omega \in F\}.$$

Note that there is a canonical isomorphism $F^a \cong \mathcal{H}om_{\mathcal{O}_M}(\Omega_F, \mathcal{O}_M)$. The annihilator $(F^a)^a$ of F^a is defined dually and is a sub-$\mathcal{O}_M$-module of Ω_M. Note that $F \subset (F^a)^a$ in general. We say that F is full in Ω_M if $F = (F^a)^a$.

(1.1) DEFINITION. A foliation on M is a coherent sub-$\mathcal{O}_M$-module F of Ω_M which is integrable in the sense that

$$dF_z \subset (\Omega_M \wedge F)_z \quad \text{for all } z \text{ in } M - S(F).$$

A foliation F is said to be reduced if it is full in Ω_M.

If F is a foliation on M, it defines an ordinary foliation (without singularity) on $M - S(F)$. Its codimension is called the codimension of F. We call the quotient sheaf $\Omega_F = \Omega_M/F$ the sheaf of germs of holomorphic 1-forms along the leaves of F. We record the exact sequence for later use:

$$0 \to F \to \Omega_M \to \Omega_F \to 0. \tag{1.2}$$

We say that a foliation F on M is a Haefliger foliation if each point of M has a neighborhood on which F is generated by exact forms. Also we say that F is of complete intersection type if F is a locally free $\mathcal{O}_M$-module.

In what follows, when we say simply a foliation, we usually mean a reduced foliation of complete intersection type.

2. Unfoldings of foliations. Let F be a codim q foliation (of complete intersection type) on a complex manifold M.

(2.1) DEFINITION. An unfolding of F consists of:

(I) A deformation $\pi: \mathfrak{M} \to T$ of M, i.e., π is a holomorphic submersion of a complex manifold $\mathfrak{M}$ onto another T such that $\pi^{-1}(o) = M$ for some point o of T. We assume that π is proper when M is compact.

(II) A codim q foliation $\mathfrak{F}$ on $\mathfrak{M}$ such that the restriction (as forms) of $\mathfrak{F}$ to M is equal to F.

(2.2) DEFINITION. An unfolding $(\mathfrak{M}' \xrightarrow{\pi'} T', \mathfrak{F}')$ of (M, F) is said to be induced from another $(\mathfrak{M} \xrightarrow{\pi} T, \mathfrak{F})$ if there exist holomorphic maps $\Phi: \mathfrak{M}' \to \mathfrak{M}$ and $\varphi: T' \to T$ such that (i) φ maps the specific point o' in T' to the specific point o in T, (ii) $\varphi \circ \pi' = \pi \circ \Phi$, (iii) for each t' in T', Φ maps $\pi'^{-1}(t')$ biholomorphically onto $\pi^{-1}(\varphi(t'))$, and (iv) $\mathfrak{F}'$ is equal to the pull-back (as forms) of $\mathfrak{F}$ by Φ.

An unfolding of (M, F) is versal if every unfolding of (M, F) is induced from it.

For an unfolding $(\mathfrak{M} \xrightarrow{\pi} T, \mathfrak{F})$ of (M, F), we have the infinitesimal unfolding map [**12**, (2.8)]

$$\rho: \mathbf{T}_{T,o} \to \operatorname{Ext}^1_{\mathcal{O}_M}(\Omega_F, \mathcal{O}_M), \tag{2.3}$$

where $\mathbf{T}_{T,o}$ is the holomorphic tangent space to T at o. The map ρ is $\mathbf{C}$-linear and, roughly, picks up the first order terms in $\mathfrak{M}$ and $\mathfrak{F}$ in the given tangential direction. For the explicit description of (2.3), see [**12**, §3]. It is especially simple if $\mathcal{F}$ is an unfolding of a local foliation F (see the next section). As we see later, the set of first order unfoldings of (M, F) is, in general, smaller than $\mathrm{Ext}^1_{\mathcal{O}_M}(\Omega_F, \mathcal{O}_M)$.

If we consider the spectral sequence for Ext, we have the following commutative diagram with exact rows:

(2.4)

$$\begin{array}{ccccccc} 0 \to H^1(M, F^a) & \to & \mathrm{Ext}^1_{\mathcal{O}_M}(\Omega_F, \mathcal{O}_M) & \xrightarrow{\varepsilon} & H^0(M, \mathcal{E}xt^1_{\mathcal{O}_M}(\Omega_F, \mathcal{O}_M)) & \to & H^2(M, F^a) \\ \downarrow & & \downarrow & & & & \\ H^1(M, \Theta_M) & \xrightarrow{\sim} & \mathrm{Ext}^1_{\mathcal{O}_M}(\Omega_M, \mathcal{O}_M). & & & & \end{array}$$

The composition of (2.3) and the second vertical map in (2.4) is identical with the usual infinitesimal deformation map of $\mathcal{M}$ at o. The sheaf $\mathcal{E}xt^1_{\mathcal{O}_M}(\Omega_F, \mathcal{O}_M)$ is supported on the singular set $S(F)$ of F and contains some information about the singularity (see §6).

3. First order unfoldings of local foliations. We denote by ${}_n\mathcal{O}$ or simply by $\mathcal{O}$ the local ring $\mathcal{O}_{\mathbf{C}^n,O}$ of germs of holomorphic functions at the origin O in $\mathbf{C}^n = \{(z_1,\ldots,z_n)\}$ and by ${}_n\Omega$ or simply by Ω the ${}_n\mathcal{O}$-module $\Omega_{\mathbf{C}^n,O}$. For simplicity, here we consider only foliations of codim 1. Thus a codim 1 local foliation at O in $\mathbf{C}^n$ is a rank 1 free sub-$\mathcal{O}$-module $F = (\omega)$ of Ω satisfying the integrability condition $d\omega \wedge \omega = 0$, where ω is a generator of F. The singular set $S(F)$ of F is the germ at O of the set of zeros of a 1-form representing ω. We assume that F is reduced. Thus codim $S(F) \geqslant 2$ [**12**, (5.1) Lemma]. For a local foliation F we denote the quotient module Ω/F by Ω_F;

$$0 \to F \to \Omega \to \Omega_F \to 0.$$

From this we have a presentation of $\mathrm{Ext}^1_{\mathcal{O}}(\Omega_F, \mathcal{O})$;

$$\mathrm{Hom}_{\mathcal{O}}(\Omega, \mathcal{O}) \to \mathrm{Hom}_{\mathcal{O}}(F, \mathcal{O}) \to \mathrm{Ext}^1_{\mathcal{O}}(\Omega_F, \mathcal{O}) \to 0.$$

Hence if we write $\omega = \sum_{i=1}^n f_i(z)\,dz_i$ with $f_i \in \mathcal{O}$, we have

$$\mathrm{Ext}^1_{\mathcal{O}}(\Omega_F, \mathcal{O}) = \mathcal{O}/(f_1,\ldots,f_n), \tag{3.1}$$

where $(f_1,\ldots,f_n)$ is the ideal generated by $f_1,\ldots,f_n$. For any element h in $\mathcal{O}$, we denote by $[h]$ its class in $\mathrm{Ext}^1_{\mathcal{O}}(\Omega_F, \mathcal{O})$. Now we consider an unfolding $\mathcal{F} = (\tilde{\omega})$ of F, i.e., $\mathcal{F}$ is a codim 1 local foliation at the origin in $\mathbf{C}^n \times \mathbf{C}^m$ with coordinates $(z, t) = (z_1,\ldots,z_n;\, t_1,\ldots,t_m)$, such that there is a generator $\tilde{\omega}$ whose restriction to $t = 0$ is equal to ω. We express $\tilde{\omega}$ as

$$\tilde{\omega} = \sum_{i=1}^n f_i(z, t)\,dz_i + \sum_{k=1}^m h_k(z, t)\,dt_k, \qquad f_i(z, t), h_k(z, t) \in {}_{n+m}\mathcal{O}.$$

Note that $f_i(z,0) = f_i(z)$. We set $h_k(z) = h_k(z,0)$. Then the infinitesimal unfolding map

$$\rho\colon \mathbf{T}_{\mathbf{C}^m,O} \to \mathrm{Ext}^1_{\mathcal{O}}(\Omega_F, \mathcal{O})$$

of $\mathfrak{F}$ sends the tangent vector $\partial/\partial t_k$ to the element $[h_k]$ in $\operatorname{Ext}^1_{\mathcal{O}}(\Omega_F, \mathcal{O})$. Consider ${}_{n+1}\Omega = \Omega_{\mathbf{C}^n\times\mathbf{C},O}$ and let $(z_1,\ldots,z_n;\, t)$ be a coordinate system on $\mathbf{C}^n \times \mathbf{C}$.

(3.2) DEFINITION. A first order unfolding $\mathfrak{F}^{(1)} = (\tilde\omega)$ of $F = (\omega)$ is a rank 1 free sub-${}_{n+1}\mathcal{O}$-module of ${}_{n+1}\Omega$ having a generator $\tilde\omega$ such that the restriction of $\tilde\omega$ to $t = 0$ is equal to ω and that $\tilde\omega$ satisfies the first order integrability condition

$$d\tilde\omega \wedge \tilde\omega \equiv 0, \quad \operatorname{mod} t^2,\, t\,dt.$$

If $\mathfrak{F}^{(1)} = (\tilde\omega)$ is a first order unfolding of $F = (\omega)$, then we may write

$$\tilde\omega \equiv \omega + \omega^{(1)}t + h\,dt, \quad \operatorname{mod} t^2,\, t\,dt,$$

with $\omega^{(1)} \in {}_n\Omega$ and $h \in {}_n\mathcal{O}$. It is not difficult to show that the germ h is uniquely determined by $\mathfrak{F}^{(1)}$ (and ω). Moreover, if $\mathfrak{F}^{(1)\prime} = (\tilde\omega')$, $\tilde\omega' \equiv \omega + \omega^{(1)\prime}t + h'\,dt$, is another first order unfolding of $F = (\omega)$, then $\mathfrak{F}^{(1)}$ and $\mathfrak{F}^{(1)\prime}$ are equivalent if and only if $[h] = [h']$ in $\operatorname{Ext}^1_{\mathcal{O}}(\Omega_F, \mathcal{O})$. Thus if we denote by $U(F)$ the set of equivalence classes of first order unfoldings of F, the assignment $\mathfrak{F}^{(1)} \mapsto [h]$ induces an embedding of $U(F)$ into $\operatorname{Ext}^1_{\mathcal{O}}(\Omega_F, \mathcal{O})$. The proof of the following is straightforward.

(3.3) THEOREM. *If $F = (\omega)$ is a* codim 1 *local foliation, then*

$$U(F) = \left\{[h] \in \operatorname{Ext}^1_{\mathcal{O}}(\Omega_F, \mathcal{O}) \;\middle|\; h\,d\omega = \eta \wedge \omega \text{ for some } \eta \in \Omega\right\}.$$

In the higher codimensional case, we need some local cohomology to describe $U(F)$. Also, for a global foliation, the description of the first order unfoldings involves the map ε in (2.4).

Note that if $\mathfrak{F}$ is an unfolding of F, then the image of the infinitesimal unfolding map of $\mathfrak{F}$ is contained in $U(F)$. We say that an element in $U(F)$ is unobstructed if it is in the image of the infinitesimal unfolding map of some unfolding of F.

4. Malgrange's theorem. Let $F = (\omega)$ be a codim 1 local foliation at O in $\mathbf{C}^n$. We denote by Ω^r the $\mathcal{O}$-module of germs of holomorphic r-forms at O and define $\delta\colon \Omega^r \to \Omega^{r+1}$ by $\delta(\xi) = \xi \wedge \omega$. Let H^r be the rth cohomology group of the complex $(\Omega^\cdot, \delta)$. If we denote by $[\xi]$ the cohomology class of a cocycle ξ, then we may express $U(F)$ as ((3.3) Theorem)

$$U(F) = \left\{[h] \in \operatorname{Ext}^1_{\mathcal{O}}(\Omega_F, \mathcal{O}) \;\middle|\; [h\,d\omega] = 0 \text{ in } H^2\right\}.$$

If F is a Haefliger foliation, then $U(F) = \operatorname{Ext}^1_{\mathcal{O}}(\Omega_F, \mathcal{O})$ and moreover, none of the elements in $U(F)$ is obstructed. The unfolding theory for $F = (df)$ is equivalent to the unfolding theory for the function germ f with respect to right-morphisms ((5.10) Corollary). Also, if $H^2 = 0$, then $U(F) = \operatorname{Ext}^1_{\mathcal{O}}(\Omega_F, \mathcal{O})$. This is the case if codim $S(F) \geqslant 3$. In fact we have the following strong result:

(4.1) THEOREM (MALGRANGE [**8**]). *Let $F = (\omega)$ be a local foliation. If* codim $S(F) \geqslant 3$, *then F is a Haefliger foliation.*

The higher codim (of F) case is proved in Malgrange [**9**]. The proof of this theorem can be interpreted as construction of an unfolding corresponding to the element [1] in $U(F)$. From the condition $\operatorname{codim} S(F) \geqslant 3$, it is possible to find $\omega^{(p)}$ in Ω for each nonnegative integer p such that $\omega^{(0)} = \omega$ and that the formal series

$$\tilde{\omega} = \sum_{p \geqslant 0} \omega^{(p)} t^p + dt$$

satisfies the integrability condition. Malgrange proved the convergence of the series using his theorem on privileged neighborhoods. Thus $\tilde{\omega}$ is an integrable holomorphic 1-form on (a neighborhood of the origin in) $\mathbf{C}^n \times \mathbf{C} = \{(z, t)\}$. Since $\tilde{\omega}$ is nonsingular, by the Frobenius theorem, $\tilde{\omega}$ has an integrating factor. If we restrict it to $t = 0$, we get an integrating factor for ω.

It is not difficult to show the following:

(4.2) PROPOSITION. *For a local foliation F the following are equivalent*:
(i) *F is a Haefliger foliation.*
(ii) $U(F) = \operatorname{Ext}^1_{\mathcal{O}}(\Omega_F, \mathcal{O})$ *and none of the elements in $U(F)$ is obstructed.*
(iii) [1] *is in $U(F)$ and is not obstructed.*

(4.3) EXAMPLE. If $F = (\omega)$ is a local foliation at O in $\mathbf{C}^2$ generated by

$$\omega = z_2\, dz_1 + \left(z_1 + z_1^2 z_2\right) dz_2,$$

then we have $U(F) = \operatorname{Ext}^1_{\mathcal{O}}(\Omega_F, \mathcal{O}) = \mathbf{C}$. However, every nonzero element in $U(F)$ is obstructed.

5. Theorem of versality. Let $F = (\omega)$ be a local foliation.

(5.1) DEFINITION. An unfolding $\mathfrak{F}$ of F is infinitesimally versal if the image of the infinitesimal unfolding map of $\mathfrak{F}$ is equal to $U(F)$.

(5.2) THEOREM [**13**]. *Let F be a* codim 1 *local reduced foliation and let $\mathfrak{F}$ be an unfolding of F. If $\mathfrak{F}$ is infinitesimally versal, then it is versal.*

Analogous theorems of versality have been proved in various cases (see, for example, [**5**, **6**, **7**, **10**, **16**, **17**]). When $F = (df)$ is a Haefliger foliation, the theorem reduces to the versality theorem for the function germ f with respect to right-morphisms ((5.10) Corollary). A detailed proof of the theorem is given in [**13**]. It consists of: (I) construction of a map, as a formal series, inducing an arbitrarily given unfolding of F from the unfolding $\mathfrak{F}$, and (II) proof of the convergence of the series constructed in (I). Part (I) is technically complicated and requires extensive use of the integrability condition. Part (II) is done basically by comparing the formal series with the convergent series used in Kodaira and Spencer [**7**]. In our case, we have to use Malgrange's estimate for derivatives and his theorem on privileged neighborhoods [**8**]. Thus we have to modify Kodaira and Spencer's series accordingly.

In what follows we give some of the direct consequences of the theorem.

(5.3) DEFINITION. Let $F = (\omega)$ be a local foliation at O in $\mathbf{C}^n$. An unfolding $\mathfrak{F}$ of F is trivial if it is generated by the pull-back of ω by a local holomorphic submersion $(\mathbf{C}^n \times \mathbf{C}^m, O) \to (\mathbf{C}^n, O)$ where $\mathbf{C}^m$ is the parameter space of $\mathfrak{F}$.

(5.4) DEFINITION. An unfolding $\mathcal{F}$ of F is universal if it is versal and if the infinitesimal unfolding map of $\mathcal{F}$ is injective.

It is not difficult to show that if $\mathcal{F}$ is a universal unfolding of some other foliation, then every unfolding of $\mathcal{F}$ is trivial.

(5.5) COROLLARY. *If $U(F) = 0$, then F is a universal unfolding of F itself and thus every unfolding of F is trivial.*

(5.6) COROLLARY (CERVEAU ET NETO [4]). *Let $\bar{F} = (\bar{\omega})$ be a local foliation at O in $\mathbf{C}^n$ generated by*

$$\bar{\omega} = z_1 \cdots z_n \sum_{i=1}^{n} a_i \frac{dz_i}{z_i}, \qquad a_i \in \mathbf{C}.$$

If $a_i \neq a_j \neq 0$, then every unfolding of $\bar{F}$ is trivial.

In fact, it is not difficult to show that $U(\bar{F}) = 0$. In [4], it is shown that every unfolding of a form whose $n-1$ jet is $\bar{\omega}$ is trivial. The case where two or more of the a_i's are the same is discussed in [14].

If $F = (df)$ is a Haefliger foliation, then we have $\mathrm{Ext}^1_{\mathcal{O}}(\Omega_F, \mathcal{O}) = \mathcal{O}/(\partial f)$, where (∂f) denotes the ideal generated by $\partial f/\partial z_1, \ldots, \partial f/\partial z_n$. In what follows we assume that $n \geqslant 2$. Thus the finiteness of $\dim_{\mathbf{C}} \mathcal{O}/(\partial f)$ implies that $F = (df)$ is reduced.

(5.7) COROLLARY. *If $\dim_{\mathbf{C}} \mathcal{O}/(\partial f)$ is finite, then $F = (df)$ has a universal unfolding which is also Haefliger.*

In fact, if $[h_1], \ldots, [h_m]$ is a $\mathbf{C}$-basis of $\mathcal{O}/(\partial f)$, then $\mathcal{F} = (\tilde{\omega})$,

$$\tilde{\omega} = d\left(f + \sum_{k=1}^{m} h_k t_k\right) = df + \sum_{k=1}^{m} dh_k \cdot t_k + \sum_{k=1}^{m} h_k \, dt_k,$$

is a universal unfolding of F.

(5.8) COROLLARY. *If $\dim_{\mathbf{C}} \mathcal{O}/(\partial f)$ is finite, then every unfolding of a Haefliger foliation $F = (df)$ is Haefliger.*

(5.9) COROLLARY (MATTEI ET MOUSSU [11]). *Let $\mathcal{F} = (\tilde{\omega})$ be a local foliation at O in $\mathbf{C}^N$ and let ι be an embedding of $(\mathbf{C}^2, O)$ into $(\mathbf{C}^N, O)$ such that $S(F) = \{O\}$, where F is the local foliation at O in $\mathbf{C}^2$ generated by $\omega = \iota^* \tilde{\omega}$. If F is a Haefliger foliation genrated by df, then $\mathcal{F}$ is also Haefliger and is generated by $d\tilde{f}$ with $\tilde{f}$ an unfolding of f.*

This is the extension theorem for strong premier integrals [11]. The proof in [11] works also for (weak) premier integrals. Also, if we use their theorem of transversality (the existence of ι with the property in (5.9) Corollary), we see that (5.8) Corollary holds without the finiteness of $\dim_{\mathbf{C}} \mathcal{O}/(\partial f)$.

Since an unfolding $\tilde{f}$ of a function germ f is infinitesimally right-versal if and only if $\mathcal{F} = (d\tilde{f})$ is an infinitesimally versal unfolding of $F = (df)$, we have

(5.10) COROLLARY (CF. MATHER **[10]** AND WASSERMANN **[17]**). *If $\tilde{f}$ is an infinitesimally right-versal unfolding of f, then it is right-versal.*

(5.11) EXAMPLE. If $F = (\omega)$ is the local foliation at O in $\mathbf{C}^2$ generated by

$$\omega = z_2^2\, dz_1 - z_1^2\, dz_2,$$

then we have $\mathrm{Ext}^1_{\mathcal{O}}(\Omega_F, \mathcal{O}) = \mathbf{C}^4$ and we may take $[1]$, $[z_1]$, $[z_2]$ and $[z_1z_2]$ as basis elements. Also we have $U(F) = \mathbf{C}^2$ and we may take $[z_1 - z_2]$ and $[z_1z_2]$ as basis elements. By the versality theorem, $\mathcal{F} = (\tilde{\omega})$,

$$\begin{aligned}\tilde{\omega} = {} & \left(z_2^2 - z_2t_1 - t_2\right) dz_1 - \left(z_1^2 + z_1t_1 - t_2\right) dz_2 + \left(z_1z_2 - t_2\right) dt_1 \\ & + \left(z_1 - z_2 + t_1\right) dt_2,\end{aligned}$$

is a universal unfolding of F with parameter space $\mathbf{C}^2 = \{(t_1, t_2)\}$.

For more examples, see **[14]**.

6. Residues of Baum and Bott. Now we go back to the global situation. Let M be an n-dimensional complex manifold. In Baum and Bott **[3]**, a foliation on M is defined to be a full integrable coherent subsheaf ξ of Θ_M. Let Q denote the quotient sheaf Θ_M/ξ;

(6.1) $$0 \to \xi \to \Theta_M \to Q \to 0.$$

This is equivalent to our definition of a reduced foliation if we set $\xi = F^a$ or $F = \xi^a$. By taking the duals of (1.2), we obtain the exact sequence:

(6.2)

$$\begin{array}{ccccccccc} 0 \to & \mathcal{H}om_{\mathcal{O}_M}(\Omega_F, \mathcal{O}_M) & \to & \mathcal{H}om_{\mathcal{O}_M}(\Omega_M, \mathcal{O}_M) & \to & \mathcal{H}om_{\mathcal{O}_M}(F, \mathcal{O}_M) & \to & \mathcal{E}xt^1_{\mathcal{O}_M}(\Omega_F, \mathcal{O}_M) & \to 0 \\ & \| & & \| & & \| & & & \\ & F^a & & \Theta_M & & F^* & & & \end{array}$$

Comparing (6.1) and (6.2), we get the exact sequence:

(6.3) $$0 \to Q \to F^* \to \mathcal{E}xt^1_{\mathcal{O}_M}(\Omega_F, \mathcal{O}_M) \to 0.$$

Let Z be a connected component of the singular set $S(F)$ of F and assume that Z is compact. If φ is a symmetric and homogeneous polynomial of degree d in n variables and if $d > q$ ($=$ codim of F), then we have the residue of Baum and Bott:

$$\mathrm{Res}_{\varphi}(F, Z) \in H_{2n-2d}(Z; \mathbf{C}).$$

From now on we assume that $F = \xi^a$ is of complete intersection type, i.e., F is a locally free $\mathcal{O}_M$-module (of rank q). We take an open neighborhood U of Z in M such that Z is a deformation retract of U. Since the restriction of $\mathcal{E}xt^1_{\mathcal{O}_M}(\Omega_F, \mathcal{O}_M)$ to U is a coherent sheaf on U with support in Z, there is the associated Grothendieck element $\gamma_Z(\mathcal{E}xt^1_{\mathcal{O}_M}(\Omega_F, \mathcal{O}_M))$, which we simply denote by $\mathcal{E}$, in

$K^0(U, U - Z)$ [1]. The Chern character gives the map

$$\mathrm{ch}\colon K^0(U, U - Z) \to H^*(U, U - Z; \mathbf{Q}).$$

Since Z is a deformation retract of U, we have $H^*(U, U - Z; \mathbf{Q}) = H_c^*(U; \mathbf{Q})$ (cohomology with compact support). If we denote by κ the canonical homomorphism $H_c^*(U; \mathbf{Q}) \to H^*(U; \mathbf{Q})$, then $\mathrm{ch}(\mathcal{E})$ determines the local Chern classes $c_1(\mathcal{E}), \ldots, c_n(\mathcal{E})$ in $H^*(U, U - Z; \mathbf{Q}) = H_c^*(U; \mathbf{Q})$ such that $1 + \kappa(c_1(\mathcal{E})) + \cdots + \kappa(c_n(\mathcal{E}))$ is the total Chern class of the coherent sheaf $\mathcal{E}xt^1_{\mathcal{O}_M}(\Omega_F, \mathcal{O}_M)$ on U. For each integer k with $1 \leqslant k \leqslant n$, we set

$$d_k(\mathcal{E}) = \sum_{r=1}^{k} (-1)^r \sum_{j_1 + \cdots + j_r = k,\, j_\nu > 0} c_{j_1}(\mathcal{E}) \cdots c_{j_r}(\mathcal{E}).$$

Also let $c(F^*) = 1 + c_1(F^*) + \cdots + c_n(F^*)$ be the total Chern class in $H^*(U; \mathbf{Q})$ of F^*. Using the canonical pairing

$$H^*(U; \mathbf{Q}) \times H_c^*(U; \mathbf{Q}) \to H_c^*(U; \mathbf{Q}),$$

we define the elements $c_1(F^* - \mathcal{E}), \ldots, c_n(F^* - \mathcal{E})$ as follows: If $q < j \leqslant n$, then $c_j(F^* - \mathcal{E})$ denotes the element

$$c_q(F^*)\, d_{j-q}(\mathcal{E}) + \cdots + c_1(F^*)\, d_{j-1}(\mathcal{E}) + d_j(\mathcal{E})$$

in $H_c^{2j}(U; \mathbf{Q})$, and if $1 \leqslant j \leqslant q$, then $c_j(F^* - \mathcal{E})$ denotes the element $c_j(F^*) + c_{j-1}(F^*)\kappa(d_1(\mathcal{E})) + \cdots + c_1(F^*)\kappa(d_{j-1}(\mathcal{E})) + \kappa(d_j(\mathcal{E}))$ in $H^{2j}(U; \mathbf{Q})$. We denote the elementary symmetric functions in n variables by $\sigma_1, \ldots, \sigma_n$ and we let L be the composition of the two isomorphisms

$$H_c^{2d}(U; \mathbf{Q}) \overset{D_U}{\underset{\sim}{\to}} H_{2n-2d}(U; \mathbf{Q}) \overset{i_*^{-1}}{\underset{\sim}{\to}} H_{2n-2d}(Z; \mathbf{Q}),$$

where D_U denotes the Poincaré duality map and i is the embedding $Z \hookrightarrow U$. The following essentially comes from the exact sequence (6.3).

(6.4) THEOREM [15]. *Let F be a foliation of* codim q *on M and let U and Z be as above. If $\varphi = \sigma_{j_1} \cdots \sigma_{j_r}$ with $j_\nu > q$ for some ν, then*

$$\mathrm{Res}_\varphi(F, Z) = L\big(c_{j_1}(F^* - \mathcal{E}) \cdots c_{j_r}(F^* - \mathcal{E})\big).$$

Note that the condition $j_\nu > q$ guarantees that the quantity in $L(\ )$ above is in $H_c^{2d}(U; \mathbf{Q})$, $d = j_1 + \cdots + j_r$.

(6.5) COROLLARY. *Let F and Z be as before and let φ be a symmetric and homogeneous polynomial of degree d in n variables. If each monomial in the expression of φ as a polynomial of $\sigma_1, \ldots, \sigma_n$ contains σ_j with $j > q$, then* $\mathrm{Res}_\varphi(F, Z)$ *is rational, i.e., it is in $H_{2n-2d}(Z; \mathbf{Q})$* (*cf.* [3, *p.* 287, *Rationality Conjecture*]).

Now we suppose that Z is nonsingular and that there exists a holomorphic vector bundle E on Z such that $\mathcal{E}xt^1_{\mathcal{O}_M}(\Omega_F, \mathcal{O}_M) = i_!\mathcal{O}_Z(E)$ (= the sheaf $\mathcal{O}_Z(E)$ of germs of holomorphic sections of E extended by zero on $U - Z$), where i is the embedding $Z \hookrightarrow U$. Then the (stronger version of) Riemann-Roch theorem for analytic embeddings (Atiyah and Hirzebruch [**2**]) gives the local Chern classes of $\mathcal{E}$,

$$\mathrm{ch}(\mathcal{E}) = i_*\left(td(N)^{-1}\,\mathrm{ch}(E)\right),$$

where i_* is the Thom-Gysin homomorphism

$$i_*\colon H^*(Z;\mathbf{Q}) \to H^*(U, U - Z;\mathbf{Q}) = H^*_c(U;\mathbf{Q}).$$

This can be used to compute the residue in (6.4) Theorem. For example, if the singularity is isolated, we have [**15**].

(6.6) PROPOSITION. *Let U be a polydisk about the origin O in $\mathbf{C}^n$ and let $F = (\omega)$ be a* codim 1 *foliation on U with an isolated singularity at O. If we denote the stalks $\mathcal{O}_{\mathbf{C}^n,O}$ and $\Omega_{F,O}$ simply by $\mathcal{O}$ and Ω_F, respectively, we have*

$$\mathrm{Res}_{\sigma_n}(F,\{O\}) = (-1)^n(n-1)!\dim_{\mathbf{C}}\mathrm{Ext}^1_{\mathcal{O}}(\Omega_F,\mathcal{O}) \text{ in } H_0(\{O\};\mathbf{Q}) = \mathbf{Q}.$$

For an example with nonisolated singularity, see [**15**].

REFERENCES

1. M. F. Atiyah and F. Hirzebruch, *Analytic cycles on complex manifolds*, Topology **1** (1962), 25–45.

2. ______, *The Riemann-Roch theorem for analytic embeddings*, Topology **1** (1962), 151–166.

3. P. Baum and R. Bott, *Singularities of holomorphic foliations*, J. Differential Geom. **7** (1972), 279–342.

4. D. Cerveau et A. L. Neto, *Formes intégrables tangentes à des actions commutatives*, Université de Dijon, 1979.

5. S. Izumiya, *On versality for unfoldings of smooth section germs*, Publ. Res. Inst. Math. Sci. Kyoto Univ. **18** (1982), 135–157.

6. A. Kas and M. Schlessinger, *On the versal deformation of a complex space with an isolated singularity*, Math. Ann. **196** (1972), 23–29.

7. K. Kodaira and D. C. Spencer, *A theorem of completeness for complex analytic fibre spaces*, Acta Math. **100** (1958), 281–294.

8. B. Malgrange, *Frobenius avec singularités*. 1. *Codimension un*, Inst. Haute Études Sci. Publ. Math. **46** (1976), 163–173.

9. ______, *Frobenius avec singularités*. 2. *Le cas général*, Invent. Math. **39** (1977), 67–89.

10. J. Mather, unpublished notes on right equivalence.

11. J. F. Mattei et R. Moussu, *Holonomie et intégrales premières*, Ann. Sci. École Norm. Sup. **13** (1980), 469–523.

12. T. Suwa, *Unfoldings of complex analytic foliations with singularities*, preprint.

13. ______, *A theorem of versality for unfoldings of complex analytic foliation singularities*, Invent. Math. **65** (1981), 29–48.

14. ______, *Kupka-Reeb phenomena and universal unfoldings of certain foliation singularities*, Osaka J. Math. (to appear).

15. ______, *Residues of complex analytic foliation singularities and the Riemann-Roch theorem for embeddings*, preprint.

16. G. Tjurina, *Locally semi-universal flat deformations of isolated singularities of complex spaces*, Izv. Akad. Nauk SSSR **33** (1970), 967–999.

17. G. Wassermann, *Stability of unfoldings*, Lecture Notes in Math., vol. 393, Springer-Verlag, Berlin, 1974.

HOKKAIDO UNIVERSITY, SAPPORO, JAPAN

Proceedings of Symposia in Pure Mathematics
Volume **40** (1983), Part 2

THE EXPONENTS OF A FREE HYPERSURFACE

HIROAKI TERAO

0. Introduction. In this article we briefly survey recent results on free hypersurfaces and their exponents. Although the exponents are defined for any homogeneous hypersurface, we are most interested in arrangements of hyperplanes. Our main result here, which generalizes a result of Orlik-Solomon [3], concerns the relation between the exponents and some combinatorial data (Möbius functions). We have a much simpler proof than the original one in [**10**] and shall give its outline here. The two problems which seem to be very important for our theory will be posed at the end of this article.

1. The exponents of a free hypersurface. Throughout this article let $(D, 0)$ be a germ of a hypersurface in $(\mathbf{C}^{n+1}, 0)$ and let $Q \in \mathcal{O}_{\mathbf{C}^{n+1},0}$ be a defining equation of D.

DEFINITION 1. Define

$$\Omega^q(\log D)_0 := \{\text{germs at 0 of meromorphic differential } q\text{-forms } \omega;\ Q\omega \text{ and } Q(d\omega) \text{ are both holomorphic at } 0\}.$$

Then $\Omega^q(\log D)_0$ has a natural $\mathcal{O} = \mathcal{O}_{\mathbf{C}^{n+1},0}$-module structure. We call an element of $\Omega^q(\log D)_0$ the *germ of a logarithmic q-form along D*.

The various properties of $\Omega^q(\log D)_0$ are studied in [**6**]. When D has a normal crossing, $\Omega^q(\log D)_0$ is a free $\mathcal{O}$-module. For a general D, $\Omega^q(\log D)_0$ is not necessarily a free $\mathcal{O}$-module.

DEFINITION 2. We call D *free* if $\Omega^1(\log D)_0$ is a free $\mathcal{O}$-module. (In this case all $\Omega^q(\log D)_0$ $(q \geq 1)$ are free.)

EXAMPLE. Let $X \to S$ be a universal unfolding of a holomorphic function with an isolated critical point. Then a germ of the discriminant is free [7] (cf. [**13**]).

The following proposition (see [**8**, 2.4]) clarifies the ring-theoretic meaning of freeness.

1980 *Mathematics Subject Classification.* Primary 32B30, 05A15, 52A99.

Key words and phrases. Logarithmic form, logarithmic vector field, exponent, Macaulay ring, arrangement of hyperplanes, Betti number, Möbius function, reflection group.

0082-0717/82/0000-0661/$02.25

PROPOSITION 3. *A germ D of a divisor is free if and only if*

$$\mathcal{O}/(Q, \partial Q/\partial z_0, \ldots, \partial Q/\partial z_n)$$

is Macaulay of dimension $(n-1)$, *where* $(z_0, \ldots, z_n)$ *is a coordinate system of* $(\mathbf{C}^{n+1}, 0)$.

DEFINITION 4. Define

$$\mathrm{Der}(\log D)_0 := \{\text{germs at 0 of holomorphic vector fields } \theta;\ \theta \cdot Q \in Q\mathcal{O}\}.$$

Then $\mathrm{Der}(\log D)_0$ is an $\mathcal{O}$-module and a dual module of $\Omega^1(\log D)_0$ **[6]**, i.e.,

$$\mathrm{Hom}_{\mathcal{O}}(\Omega^1(\log D)_0, \mathcal{O}) \simeq \mathrm{Der}(\log D)_0.$$

Thus D is free if and only if $\mathrm{Der}(\log D)_0$ is free. We call an element of $\mathrm{Der}(\log D)_0$ a *germ of a logarithmic vector field along D*.

An element $\theta \in \mathrm{Der}(\log D)_0$ is said to be homogeneous of degree d, denoted by $\deg \theta = d$, if θ has a local expression

$$\theta = \sum_{i=0}^{n} f_i \frac{\partial}{\partial z_i}$$

such that the f_i are homogeneous polynomials and all nonzero f_i have the same degree d. Assume that D is free and defined by a homogeneous polynomial Q. Then one can find a homogeneous free basis $\theta_0, \ldots, \theta_n$ for $\mathrm{Der}(\log D)_0$. It is easy to see that the set $(\deg \theta_0, \ldots, \deg \theta_n)$ depends only on D.

DEFINITION 5. We call $(\deg \theta_0, \ldots, \deg \theta_n)$ the *exponents* of D.

2. Free arrangement of hyperplanes. We define an n-arrangement as a finite family of hyperplanes through the origin in $\mathbf{C}^{n+1}$. Let X be an n-arrangement. Denote $\bigcup_{H \in X} H$ by $|X|$.

DEFINITION 6. Let $L(X) = \{\bigcap_{H \in A} H;\ A \subset X\}$, and agree that $\mathbf{C}^{n+1} = \bigcap_{H \in \varnothing} H$. Define the join and meet operations in $L(X)$ by

$$s \vee t = s \cap t,$$
$$s \wedge t = \bigcap H \quad (\text{where } \{H \in X;\ H \supset s \cup t\})$$

for $s, t \in L(X)$. Then $L(X)$ becomes a lattice which is called *the lattice associated with X*. Write $s < t$ if $s \vee t = t$ $(s, t \in L(X))$.

DEFINITION 7. Let ν be an indeterminate. A map

$$F\colon L(X) \to \mathbf{Q}[\nu]$$

is called *cumulative* if there exists another map

$$f\colon L(X) \to \mathbf{Q}[\nu]$$

satisfying

I. $\nu^{r(s)} \mid f(s)$ for any $s \in L(X)$, where $r(s) = \mathrm{codim}_{\mathbf{C}^{n+1}} s$ is the rank function of $L(X)$,

II. $F(s) = \sum_{t<s} f(t)$ for any $s \in L(X)$.

Denote the set of all cumulative maps of $L(X)$ by $\mathrm{Cum}(X)$. Then $\mathrm{Cum}(X)$ is, of course, a subset of the set $\mathrm{Map}(L(X), \mathbf{Q}[\nu])$ of all maps of $L(X)$ to $\mathbf{Q}[\nu]$. Since

$\mathbf{Q}[\nu]$ is naturally a graded ring, $\mathrm{Map}(L(X), \mathbf{Q}[\nu])$ has a graded $\mathbf{Q}[\nu]$-algebra structure. Then we have

PROPOSITION 8. *$\mathrm{Cum}(X)$ is a graded $\mathbf{Q}[\nu]$-subalgebra of $\mathrm{Map}(L(X), \mathbf{Q}[\nu])$, in other words, $\mathrm{Cum}(X)$ is a $\mathbf{Q}[\nu]$-subalgebra and $F_i\nu^i \in \mathrm{Cum}(X)$ for any $\Sigma_{i\geqslant 0} F_i\nu^i \in \mathrm{Cum}(X)$ with $F_i \in \mathrm{Map}(L(X), \mathbf{Q})$ $(i = 0, 1, 2, \ldots)$.*

PROOF. Only the closedness under the multiplication is not obvious. Let F, $G \in \mathrm{Cum}(X)$. Choose f and $g \in \mathrm{Map}(L(X), \mathbf{Q}[\nu])$ satisfying the conditions in Definition 7 for F and G respectively. Define $h \in \mathrm{Map}(L(X), \mathbf{Q}[\nu])$ by

$$h(t) = \sum_{u \vee v = t} f(u)g(v) \qquad (t \in L(X)).$$

Then $\nu^{r(t)} \mid h(t)$ for any $t \in L(X)$ because $r(u \vee v) \leqslant r(u) + r(v)$ $(u, v \in L(X))$. Moreover

$$(FG)(s) = F(s)G(s) = \Big(\sum_{u<s} f(u)\Big)\Big(\sum_{v<s} g(v)\Big)$$

$$= \sum_{t<s}\Big(\sum_{u\vee v=t} f(u)g(v)\Big) = \sum_{t<s} h(t).$$

This implies that $FG \in \mathrm{Cum}(X)$. Q.E.D.

There is a function μ called the Möbius function on $L(X)$ (see [**10**]). Define a map

$$M\colon L(X) \to \mathbf{Q}[\nu]$$

by $M(s) = \Sigma_{t<s} |\mu(t)|\,\nu^{r(t)}$. Then we have

THEOREM 9 [**10**]. *The map $M\colon L(X) \to \mathbf{Q}[\nu]$ satisfies conditions* I–IV *below. Conversely* I–IV *characterize the map M.*

I. $M(\mathbf{C}^{n+1}) = 1$,

II. $(\nu + 1) \mid M(s)$ *for* $s \in L(X) \setminus \{\mathbf{C}^{n+1}\}$,

III. $\deg M(s) \leqslant r(s)$ *for any* $s \in L(X)$,

IV. *M is cumulative.*

The map M has a topological interpretation. Define $P\colon L(X) \to \mathbf{Q}[\nu]$ by $P(s) :=$ the Poincaré polynomial of $\mathbf{C}^{n+1} \setminus |X_s|$ $(s \in L(X))$, where $X_s := \{H \in X;\ H \supset s\}$. Then $P(\mathbf{C}^{n+1}) =$ the Poincaré polynomial of $\mathbf{C}^{n+1} = 1$. Since $\mathbf{C}^{n+1} \setminus |X_s|$ is a $\mathbf{C}^*$-bundle, $(\nu + 1) \mid P(s)$ $(s \in L(X) \setminus \{\mathbf{C}^{n+1}\})$. Moreover $\mathbf{C}^{n+1} \setminus |X_s|$ is obviously homotopy equivalent to $A \cap (\mathbf{C}^{n+1} \setminus |X_s|)$ if A is an $r(s)$-dimensional plane orthogonal to s through the origin. These imply that P satisfies I–III in Theorem 9. Define $p\colon L(X) \to \mathbf{Q}[\nu]$ by $p(t) = (\dim H^{r(t)}(\mathbf{C}^{n+1} \setminus |X_t|; \mathbf{C}))\nu^{r(t)}$ for $t \in L(X)$, and $P(s) = \Sigma_{t<s}\, p(t)$ due to Brieskorn [**1**, Lemma 3]. Thus condition IV yields for P and we have

COROLLARY 10 (ORLIK-SOLOMON [**2**]). $M = P$.

DEFINITION 11. An n-arrangement X is said to be *free* if $(|X|, 0)$ is free (see Definition 2). The exponents of $(|X|, 0)$ are called *the exponents* of X.

Let X be a free n-arrangement. Define a map

$$E\colon L(X) \to \mathbf{Q}[\nu]$$

by $E(s) = \prod_{i=0}^{n}(1 + d_i(s)\nu)$ $(s \in L(X))$, where $(d_0(s), \ldots, d_n(s))$ are the exponents of X_s.

Then it is easy to see that E satisfies I–III in Theorem 9. To show the cumulativeness of E (IV), we need the exact sequence

$$0 \to D(X) \to \overline{D}(\varnothing) \to \bigoplus_{r(s)=1} \overline{D}(X_s) \to \bigoplus_{r(s)=2} \overline{D}(X_s) \to \cdots,$$

where $D(X) := \mathrm{Der}_{\mathbf{C}^{n+1}}(\log|X|)_0$ and $\overline{D}(X_s)$ is an $\mathcal{O}$-module with its support on s. ($\overline{D}(\varnothing)$ has its support on $\mathbf{C}^{n+1}$.) From the exact sequence we can deduce that a map

$$H\colon s \in L(X) \rightsquigarrow \nu^n H(D(X_s); 1/\nu) \in \mathbf{Q}[\nu]$$

is cumulative, where $H(D(X_s); \nu)$ stands for the Hilbert polynomial of $D(X_s)$. By using the exponents $(d_0(s), \ldots, d_n(s))$ of X_s, we have

$$\nu^n H(D(X_s); 1/\nu) = \sum_{k=0}^{n} \prod_{j=1}^{n} \{1 - (d_k(s) - j)\nu\}.$$

Denote the homogeneous component of degree i of H by $H_i\nu^i$ with $H_i \in \mathrm{Map}(L(X), \mathbf{Q})$ $(i = 0, \ldots, n)$:

$$H(s) = \sum_{i=0}^{n} H_i(s)\nu^i.$$

Explicitly we have

$$H_i(s) = (-1)^i \sum_{k=0}^{n} \sum_{j_1 < \cdots < j_i} (d_k(s) - j_1) \cdots (d_k(s) - j_i).$$

Therefore the graded $\mathbf{Q}[\nu]$-subalgebra generated by $\{H_i\nu^i;\ i = 1, \ldots, n\}$, which is a subset of $\mathrm{Cum}(X)$ thanks to Proposition 8, equals

$\{\sum F_i\nu^i \in \mathrm{Map}(L(X), \mathbf{Q}[\nu]);\ \ ^{\exists}S_i(\nu_0, \ldots, \nu_n) \in \mathbf{Q}[\nu_0, \ldots, \nu_n]$ such that S_i is a symmetric polynomial of degree $\leqslant i$ and $F_i(s) = S_i(d_0(s), \ldots, d_n(s))$ $(i = 0, 1, 2, \ldots)$ for any $s \in L(X)\}$.

This implies that the map

$$E\colon s \rightsquigarrow \prod_{i=0}^{n} (1 + d_i(s)\nu)$$

is cumulative. These arguments above give a much simplified proof of

THEOREM 12 [**10**]. $E = M = P$.

REMARK A. This can be generalized for any finite set of hyperplanes in $\mathbf{C}^{n+1}$ or $\mathbf{P}^{n+1}\mathbf{C}$ [**11**].

REMARK B. Let $G \subset \mathrm{GL}(n+1; \mathbf{C})$ be a finite reflection group acting on $\mathbf{C}^{n+1}$. Then the set of the reflecting hyperplanes of the reflections in G makes an n-arrangement X. In this case we know

$$D(X) = (\mathrm{Der}_{\mathbf{C}^{n+1},0})^G \otimes_{\mathcal{O}^G} \mathcal{O}$$

is a free $\mathcal{O}$-module [**9**]. Here $(\mathrm{Der}_{\mathbf{C}^{n+1}})^G$ and $\mathcal{O}^G$ stand for the G-invariant parts. In this case Theorem 12 was obtained by Orlik and Solomon [**3**]. But these free arrangements form a small subset of the set of all free arrangements. In fact many examples show that the freeness of an arrangement may be a combinatorial property [**8**]. Here we pose

Problem 1. Can the freeness of an arrangement X be determined from the lattice $L(X)$?

The following criterion [**8**] may suggest the plausibility of an affirmative answer to Problem 1.

THEOREM 13. *Let X be a free n-arrangement. Let $H \in X$ and $\dot{X} = X \setminus \{H\}$. Denote an $(n-1)$-arrangement $\{H \cap L;\ L \in \dot{X}\}$ by X^H. Then any two among the following three conditions imply the third*:

1. *X is free with exponents $(d_0, d_1, \ldots, d_n)$,*
2. *$\dot{X}$ is free with exponents $(d_0, d_1, \ldots, d_i - 1, \ldots, d_n)$,*
3. *X^H is free with exponents $(d_0, d_1, \ldots, \hat{d}_i, \ldots, d_n)$.*

The following question was raised by P. Orlik [**4**]:

Problem 2. If X is free, is X^H free for any $H \in X$?

Let $s \in L(X)$. Then s is identified with $\mathbf{C}^{n+1-r(s)}$. Define an $(n+1-r(s))$-arrangement

$$X^s = \{H \cap s;\ H \in X,\ H \not\supset s\}$$

for an n-arrangement X. If the answer to Problem 2 is affirmative and X is free, then X^s is of course also free for any $s \in L(X)$. Thus Problem 2 is equivalent to

Problem 2′. If X is free, is X^s free for any $s \in L(X)$?

Define a map

$$M^*\colon L(X) \to \mathbf{Q}[\nu]$$

by $M^*(s) = M(\bigcap_{H \in X^s} H)$, $s \in L(X)$, where the map $M\colon L(X^s) \to \mathbf{Q}[\nu]$ was already defined by using the Möbius function on $L(X^s)$. Orlik and Solomon [**5**, **12**] computed M^* for each free arrangement X attached to a finite reflection group (see Remark B). Their computations show that there are nonnegative integers $d_0^s, \ldots, d_{n-r(s)}^s$ for each $s \in L(X)$ such that

$$M^*(s) = \prod_{i=0}^{n-r(s)} (1 + d_i^s \nu).$$

This strongly suggests that X^s is free with exponents $(d_0^s, \ldots, d_{n-r(s)}^s)$.

References

1. E. Brieskorn, *Sur les groupes de tresses (d'après V. I. Arnold)*, Sém. Bourbaki 24ᵉ année 1971/72, Lecture Notes in Math., vol. 317, Springer-Verlag, Berlin, Heidelberg and New York, 1973.
2. P. Orlik and L. Solomon, *Combinatorics and topology of complements of hyperplanes*, Invent. Math. **56** (1980), 167–189.
3. ______, *Unitary reflection groups and cohomology*, Invent. Math. **59** (1980), 77–94.
4. P. Orlik, Personal correspondence, April 8, 1981.
5. P. Orlik and L. Solomon, *Coxeter arrangements*, these PROCEEDINGS.
6. K. Saito, *Theory of logarithmic differential forms and logarithmic vector fields*, J. Fac. Sci. Univ. Tokyo Sect. IA Math. **27** (1980), 265–291.
7. ______, *Primitive forms for a universal unfolding of a function with an isolated critical point*, J. Fac. Sci. Univ. Tokyo Sect. IA Math. **28** (1982), 775–792.
8. H. Terao, *Arrangements of hyperplanes and their freeness*. I, J. Fac. Sci. Univ. Tokyo Sect. IA Math. **27** (1980), 293–312.
9. ______, *Free arrangements and unitary reflection groups*, Proc. Japan Acad. Ser. A Math. Sci. **56** (1980), 389–392.
10. ______, *Generalized exponents of a free arrangement of hyperplanes and Shephard-Todd-Brieskorn formula*, Invent. Math. **63** (1981), 159–179.
11. ______, *On Betti numbers of complement of hyperplanes*, Publ. Res. Inst. Math. Sci. **17** (1981), 657–663.
12. P. Orlik and L. Solomon, *Arrangements defined by unitary reflection groups* (to appear).
13. H. Terao, *Discriminant of a holomorphic map and logarithmic vector field* (to appear).

INTERNATIONAL CHRISTIAN UNIVERSITY, TOKYO, JAPAN

Proceedings of Symposia in Pure Mathematics
Volume **40** (1983), Part 2

LOCAL TRIVIALITY OF FAMILIES OF MAPS

J. G. TIMOURIAN

1. In [**11**] M. E. Hamstron surveyed the work that had been done up to that time on regular mappings.

1.1. DEFINITION. A map $f: X \to Y$ is completely regular if, for each $y \in Y$ and $\varepsilon > 0$, there is a $\delta > 0$ such that if the distance from y to y' is less than δ, then there is an ε-homeomorphism from $f^{-1}(y)$ onto $f^{-1}(y')$.

In [**9**] E. Dyer and M. E. Hamstrom prove that under certain conditions completely regular maps are locally trivial fiber maps. A discussion of how this is done appears in [**11**].

The results surveyed in [**11**] are purely topological in nature. In [**15**] a stronger version of completely regular was used to demonstrate that a parameterized family of complex n-dimensional hypersurfaces with constant Milnor number is (for $n \neq 2$) topologically a product family. The techniques of Lê and Ramanujam [**12**] were used to apply the topological results needed.

In [**15**] the methods are presented in the generality needed to solve the problem addressed. However the techniques are much more general and apply to topological and differentiable maps, as well as to algebraic ones. In [**14**] the full generality of the lemmas required to prove topological triviality of a product family is discussed. In this talk I will state what those lemmas are and show how they can be applied.

The following theorem was proved by P. T. Church and the author in [**8**]. The original proof did not use the techniques described in [**14** or **15**]. We will show how they could have been used. The illustrations are particularly easy to understand, yet the ideas are similar to those used in the proof of the main theorem of [**15**].

1.2. DEFINITION. The branch set B_f of a map $f: M \to N$ is defined by: $x \in M^n - B_f$ if and only if f is topologically equivalent to the projection map $\rho: R^p \times R^{n-p} \to R^p$ in a neighborhood of x.

1980 *Mathematics Subject Classification.* Primary 57R35.

1.3. THEOREM (CHURCH AND TIMOURIAN). *Let* $f\colon M^{p+1} \to N^p$ *be a* C^3 *open map with* $p \geqslant 1$ *and let* $\dim(B_f \cap f(y)) \leqslant 0$ *for each* $y \in N^p$. *Then there is a closed set* $X \subset M^{p+1}$ *such that* $\dim f(X) \leqslant p-2$ *and, for every* $x \in M^{p+1} - X$, *there is a natural number* $d(x)$ *with* f *in a neighborhood of* x *topologically equivalent to*

$$\phi_{d(x)}\colon C \times R^{p-1} \to R \times R^{p-1}$$

defined by

$$\phi_{d(x)}(z, t_1, \dots, t_{p-1}) = \left(\operatorname{Re} z^{d(x)}, t_1, \dots, t_{p-1}\right).$$

Thus this theorem says that off of an exceptional set the map f is locally topologically equivalent to a product map.

1.4. EXAMPLES. (1) The branch set B_f is contained in the critical set of f, by the Rank Theorem. The map $f\colon R^2 \to R^1$ defined by $f(x, y) = x^3$ illustrates the fact that a map can have a nonempty critical set (in this case the y axis), but have $B_f = \varnothing$.

(2) The map $f\colon C \to R$ defined by $f(z) = \operatorname{Re} z^n$ has zero set consisting of n lines through the origin. Other level curves (for $n = 4$) are sketched in Figure A. The plus or minus sign in the figure indicates whether the region is mapped to the right or left of zero.

The origin is the branch set and the critical set. The map is clearly open.

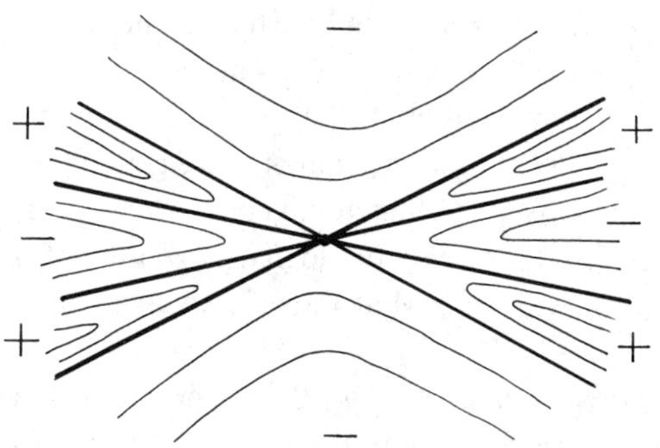

FIGURE A

(3) A condition such as $\dim(B_f \cap f^{-1}(y)) \leqslant 0$ is needed to avoid examples ;of the type illustrated in Figure B.

A map $f\colon R^2 \to R^1$ is defined with $f^{-1}(1/n)$, n an integer > 0, the curves sketched in the figure. Between these curves the map is defined to be equivalent to a product map. The interval forming the limit of the ripples is the branch set.

By a lemma of Church (see [5, p. 151]) this map can be made into a differentiable one. Note that f is an open map.

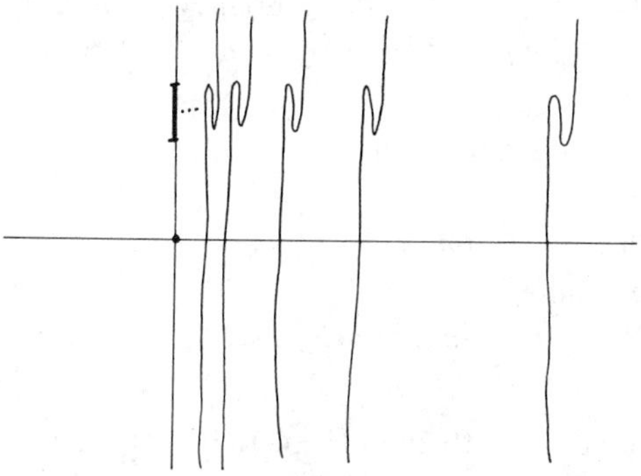

FIGURE B

(4) The openness condition is used to eliminate examples which would have sequences of local extrema converging to a point, so that local finiteness (for maps from $R^2 \to R^1$) is lost.

(5) The exceptional set is necessary. Let $f: C \times R^1 \to R^2$ be defined by $f(z, t) = (\mathrm{Re}(z^3 - 3t^2 z), t)$. For any fixed $t \neq 0$ the critical set (and the branch set) of f consists of two points. They coalesce when t goes to 0. The exceptional set is $(0, 0)$.

A more interesting nonanlytic (but C^3) example is $f(z, t) = (\mathrm{Re}(z^3 - 3(e^{-1/t^2} \sin 1/t)^2 z), t)$ which has the zero set illustrated in Figure C. The exceptional set consists of a sequence of points converging to the origin.

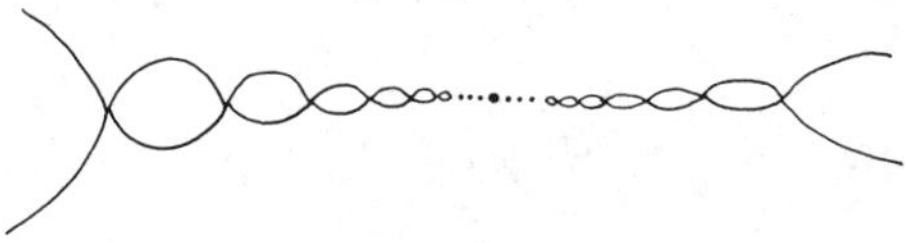

FIGURE C

2. Outline of proof and lemmas requiring differentiability. There are three basic parts to the proof of a theorem such as this one. First the problem is reduced to one in which we have a parameterized family of maps (in this case from R^2 to R^1). Next the topological structure of each of the maps in the family is analyzed, and third it is shown that the family is, in fact, topologically equivalent to a product family.

The main technical lemma used to reduce the problem to a parameterized version is this one:

2.1. LEMMA. *Let $f: M^n \to N^p$ be a C^r proper map, let $q = 1, 2, \ldots,$ let $r \geqslant \max(n - q + 1, 1)$ and let $Y \subset N^p$ be closed. Suppose that either*

(a) $\dim Y \geqslant q$ *or*

(b) $q \leqslant p - 1$ *and for some* $j, j = 0, 1, \ldots, p - q - 1,$

W open in N^p, and $x \in W - Y$, the homeomorphism $i_\#: \pi_j(W - Y, x) \to \pi_j(W, x)$ induced by inclusion is not a monomorphism.

Then there are an open $U \subset W$, C^r *diffeomorphisms* α *and* β, *a closed manifold* K^{m+n-p} (*possibly empty*) *and a* C^r *layer map* $h: K^{m+n-p} \times R^{p-m} \to S^m \times R^{p-m}$ *such that*

(i) $\rho \circ h \circ \alpha: f^{-1}(U) \to U$ *is restriction*,

(ii) $m = 0, 1, \ldots, p - q$,

(iii) $(X^m \times t) \cap \beta^{-1}(Y) \neq \varnothing$ *for each* t.

Instead of (iii) *we could require*

(iii)′ $(S^m \times 0) \cap \beta^{-1}(Y) \neq \varnothing$ *and* $\dim(S^m \times t \cap \beta^{-1}(Y)) \leqslant 0$ *for each* t.

This version of Lemma 2.1 is stated in [**5**, p. 166].

Lemma 2.1 is applied in the case of Theroem 1.3 by letting $X \subset B_f$ be the complement of the set on which f has the desired structure. We suppose $\dim f(x) \geqslant p - 1$. Since f is C^3, $\dim f(R_{p-2}(f)) \leqslant p - 2$ [**5**, p. 156], where $R_{p-2}(f)$ is the set of points at which f has rank at most $p - 2$. Thus we can assume that there is an $x \in M^{p+1} - f^{-1}f(R_{p-2}(f))$ such that for every open neighborhood $U \subset M^{p+1} - f^{-1}f(R_{p-2}(f))$ of x, $\dim f(U \cap X) \geqslant p - 1$.

Apply Lemma 2.1 to conclude that in some neighborhood f is diffeomorphically equivalent to

$$h: K^2 \times R^{p-1} \to S^1 \times R^{p-1}$$

and (using (iii)′) $(S^1 \times 0) \cap \beta^{-1}f(X) \neq \varnothing$ and

$$\dim(S^1 \times 0) \cap \beta^{-1}f(X) \leqslant 0 \quad \text{for each } t.$$

Note that $r \geqslant p + 1 - (p - 1) + 1 = 3$. (The lemma requires f to be proper, but this can be handled in this case.)

The differentiability hypothesis in Lemma 2.1 comes from using Sard's Theorem in Thom's Transversality Theorem. In the proof in [**8**], the differentiability is used so that Lemma 2.1 can be applied. Otherwise as much as possible is done in the topological category.

Onece it is demonstrated that there is a layered or parameterized structure we show that, in fact, the map h described above does have the local topological structure described in the theorem.

In general, without any differentiability hypothesis, it is possible if $n - p \geqslant 0$, $n - p \neq 4, 5$ and $\dim B_h \cap h^{-1}(y) \leqslant 0$, to define the analog of the boundary of a Milnor ball for complex hypersurfaces [7, (2.6)]. Of course, what goes on inside such a ball can be peculiar.

In the situation at hand it is easy to construct the analog of the Milnor ball desired. If $f^{-1}(y)$ looks as in Figure D, a cell is constructed whose boundary is disjoint from the branch set. Call the set so constructed N, let N_t be $N \cap K^2 \times t$.

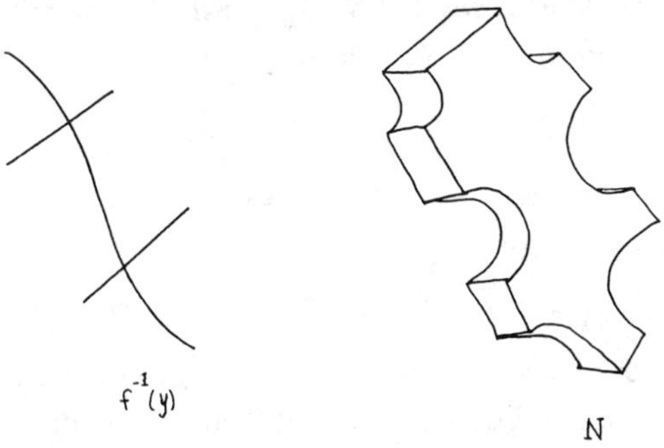

FIGURE D

The topological details of why the branch set and branch set image of h must be discrete will be omitted. The end result of these arguments is that each h has level curves that look like what they are supposed to look like for Theorem 1.3 to be true, if N is chosen small enough. In particular, N can be chosen small enough to insure that each N_t intersects B_h in exactly one point in the interior of N_t.

3. The product structure. To show a family of differentiable maps is topologically a product family the standard method is to construct vector fields which can be integrated to yield a product structure.

If $\{f_t\}$ is a parameterized family of complex hypersurfaces having an isolated singularity at the origin with constant Milnor number, then there are examples [**2**] to show that the stratified set $\bigcup_t(f_t^{-1}(0), 0 \times t)$ does not satisfy Whitney condition b. Thus it is not clear how to go about defining a vector field to integrate to show that $\{f_t\}$ is a product family.

For the problem presented by Theorem 1.3 there is a direct method that was originally used to demonstrate the product structure [**8**, (3.1)]. Here the techniques from [**15**] will be used, as an illustration of how they are applied.

The first step is to state a purely topological lemma that uses the original idea of complete regularity and the results in [**9**] to obtain a product structure. (A proof of a general version of this lemma, with details, appears in [**14**].) The next step is to show how the lemma can be applied in this case.

3.1. LEMMA. *Suppose $t \in R^p$ and N_t is a compact metric space with $N = \bigcup N_t$ complete. Let Γ_t be any family of closed subsets of N_t. Consider θ: $N \to R^p$ defined by $\theta(N_t) = t$. Suppose that there exists a homeomorphism β_t from (N_t, Γ_t) to (N_0, Γ_0) and that, for each $s \in R^p$ and $\varepsilon > 0$, there exists a $\delta > 0$ such that if $|s - t| < \delta$, then there exists an ε-homeomorphism from (N_s, Γ_s) to (N_t, Γ_t) which maps $\beta_s^{-1}(K)$ to $\beta_t^{-1}(K)$ for each $K \in \Gamma_0$.*

If the space of homeomorphisms of (N_0, Γ_0) *to itself is locally contractible, then there exists a homeomorphism* α *so that the diagram*

$$\begin{array}{ccc} (N, \Gamma) & \xrightarrow{\alpha} & (N_0, \Gamma_0) \times R^p \\ \downarrow \theta & \swarrow \pi & \\ R^p & & \end{array}$$

commutes, where π *is product projection,* $\Gamma = \bigcup_t \Gamma_t$.

Consider the elementary situation which arises in the proof of Theorem 1.3. The locally contractible condition is satisfied if any homeomorphism in a neighborhood of the identity can be connected by a small arc to the identity.

The N and N_t in Lemma 3.1 are as in our situation. Let Γ_t be the components of fibers of $h_t \mid N_t$ and the branch point of h_t in N_t. Suppose it was already known that there existed a homeomorphism β_t from (N_t, Γ_t) to (N_0, Γ_0) so that we already know h_t is topologically equivalent to h_0. We would like to show that for any $\varepsilon > 0$ there exists a $\delta > 0$ such that if $|s - t| < \delta$, then we have a homeomorphism from (N_s, Γ_s) to (N_t, Γ_t) that moves points less than ε.

The homeomorphism can be constructed in the following manner: Consider N_s and N_t, and take smaller sets N_s', N_t' as illustrated in Figure E.

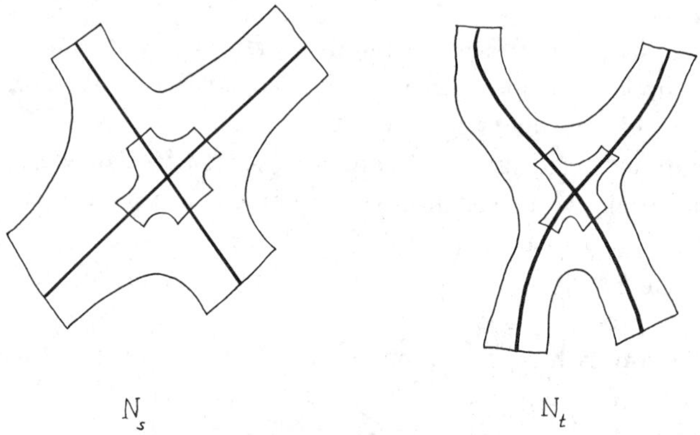

FIGURE E

There are no branch points in the regions described by $N_s - N_s'$ and $N_t - N_t'$, if s and t are close enough. As a result the regions inside of N_s or N_t and outside of N_s' or N_t' are a union of products and there will be obvious homeomorphisms which can be used. (According to [**4**] if a map has empty branch set it is a topological submersion. From this one can conclude that if s and t are taken close enough there is an ε-homeomorphism from $N_s - N_s'$ to $N_t - N_t'$.) Inside N_s' and N_t' extend this ε-homeomorphism to any homeomorphism that preserves the sets Γ_s and Γ_t. If N_s' and N_t' are chosen small enough, no point will be moved by the homeomorphism more than a distance ε.

Now we have to examine why the homeomorphisms of (N_0, Γ_0) to itself are locally contractible.

4. Local contractibility of homeomorphisms. Homeomorphisms which preserve the map h, that is, only move points around within the fibers of h, form a locally contractible group. This is true in this and many other cases, essentially because of the cone structure of the map h.

I will not go through the details of how one can show that the homeomorphism group indicated is locally contractible. Instead I will review the method due to Alexander [**1**] for showing that the homeomorphisms of the disk D in the complex plane are locally contractible. The essential ingredient in showing the local contractibility in our case is present in the methods of [**1**]. Of course for arbitrary compact manifolds the result is due to Černavskii [**3**] and Edwards and Kirby [**10**]. For stratified sets the results are in Siebenmann's paper [**13**]. For the required result in the case of complex polynomial maps with isolated singularity see [**15**].

Let $\alpha\colon D \to D$ be a homeomorphism, and suppose $\alpha =$ identity on ∂D. Let $r\colon D \times [0,1] \to D$ be defined by $r(z,t) = tz$. Define $\alpha_t\colon D \to D$ by $\alpha_t(z) = t\alpha(z/t)$ for $z \in$ range r_t, $t \neq 0$, and $\alpha_t(z) =$ identity for $z \notin$ range r_t or $t = 0$.

The above arc concentrates all of the turmoil of the homeomorphism α inside a smaller and smaller neighborhood of 0, until it is finally squeezed off to a point and disappears.

Now we have the conclusion of Lemma 3.1, and we observe that α is a homeomorphism of the family Γ to the family $\Gamma_0 \times R^{p-1}$, so that we have the commutative diagram

$$\begin{array}{ccc} (N,\Gamma) & \xrightarrow{\alpha} & (N_0,\Gamma_0)\times R^{p-1} \\ \uparrow h & & \downarrow h_0 \times \mathrm{id} \\ R^1 \times R^{p-1} & \xleftarrow{\lambda} & R^1 \times R^{p-1} \end{array}$$

where λ is a homeomorphism induced by β_t.

5. Concluding remarks. Any open real analytic map satisfies the hypotheses, and hence the conclusions, of Theorem 1.3.

5.1. REMARK. If $\alpha = (\alpha_1, \alpha_2, \ldots, \alpha_{2n})$ is any sequence with range $\{+,-1\}$, then it can be shown by a direct construction that there is a polynomial map $p\colon C \to R$, such that in a neighborhood of 0, p is topologically equivalent to a map $\phi\colon C \to R$, where

(i) $|\phi(z)| = |\operatorname{Re} z^n|$ and

(ii) $\alpha = (\operatorname{sgn}\phi(z_1), \operatorname{sgn}\phi(z_2), \ldots, \operatorname{sgn}\phi(z_{2n}))$,

where $z_1, z_2, \ldots, z_{2n}$ are chosen from distinct components of $C - \phi^{-1}(0)$.

Thus every real analytic germ $g\colon R^2 \to R$ is topologically equivalent to a polynomial germ.

5.2. COROLLARY. *For an arbitrary real analytic map it is possible to prove* Theorem 1.3 *without the hypothesis that* $\dim(B_f \cap f^{-1}(y)) \leq 0$, *if the first component function* $\phi_1 = \operatorname{Re} z^{d(x)}$ *of* $\phi_{d(x)}$ *is replaced by a map* ϕ *as in Remark* 5.1 (*see* [**6**] *for a proof*).

REFERENCES

1. J. W. Alexander, *On the deformation of an n-cell*, Proc. Nat. Acad. Sci. U.S.A. **9** (1923), 406–407.

2. J. Briançon and J.-P. Speder, *La trivialité topologique n'implique pas les conditions de Whitney*, C. R. Acad. Sci. Paris Sér. A-B **280** (1975), 365–367.

3. A. V. Černavskii, *Local contractibility of the homeomorphism group of a manifold*, Soviet Math. Dokl. **9** (1968), 1171–1174 = Dokl. Akad. Nauk SSSR **182** (3) (1968).

4. J. Cheeger and J. M. Kister, *Counting topological manifolds*, Topology **9** (1970), 149–151.

5. P. T. Church, *Differentiable monotone mappings and open mappings*, Proc. Conf. Monotone Mappings and Open Mappings (SUNY Binghamton, 1970), SUNY, Binghamton, pp. 145–183.

6. ______, *Real analytic maps of manifolds into manifolds of one lower dimension*, Math. Ann. **238** (1978), 1–14.

7. P. T. Church and J. G. Timourian, *Differentiable maps with* 0-*dimensional critical set*. I, Pacific J. Math. **41** (1972), 615–630.

8. ______, *Differentiable open maps of* $(p+1)$-*manifold to p-manifold*, Pacific J. Math. **48** (1973), 35–45.

9. E. Dyer and M. E. Hamstrom, *Completely regular mappings*, Fund. Math. **45** (1957), 103–118.

10. R. D. Edwards and R. C. Kirby, *Deformations of spaces of imbeddings*, Ann. of Math. (2) **93** (1971), 63–88.

11. M. E. Hamstrom, *Regular mappings: a survey* (Proc. Conf. Monotone Mappings and Open Mappings, SUNY Binghamton, 1970), pp. 238–254.

12. Lê Dũng Tráng and C. P. Ramanujam, *The invariance of Milnor's number implies invariance of the topological type*, Amer. J. Math. **98** (1976), 67–78.

13. L. C. Siebenmann, *Deformation of homeomorphisms on stratified sets*, Comment. Math. Helv. **47** (1972), 123–163.

14. J. G. Timourian, *Completely regular mappings and parametrized families of maps* (to appear).

15. ______, *The invariance of Milnor's number implies topological triviality*, Amer. J. Math. **99** (1977), 437–466.

UNIVERSITY OF ALBERTA, CANADA

Proceedings of Symposia in Pure Mathematics
Volume **40** (1983), Part 2

COMPARING REGULARITY CONDITIONS ON STRATIFICATIONS

DAVID TROTMAN

We give a list of the regularity conditions most often studied on differentiable stratified sets, together with all known implications between these conditions. We give examples whenever a given condition is known not to be a consequence of another given condition.

This survey was coaxed into existence by Bob MacPherson, Larry Siebenmann, and René Thom, each of whom I thank.

Introduction. Throughout X and Y will be disjoint C^1 submanifolds of $\mathbf{R}^n$ (all our conditions are defined locally) with $0 \in Y \subset \overline{X} - X$. The reader may care to fix ideas by taking $X \cup Y$ to be a topological manifold-with-boundary M with $\operatorname{Int} M = X$ and $\partial M = Y$. When one of the regularity (or equisingularity) conditions, E say, holds for (X, Y) at 0, then X is *E-regular* over Y at 0. When an equisingularity condition E fails to hold for (X, Y) at 0, we call 0 an *E-fault* of (X, Y). We shall indicate the status of implications between equisingularity conditions in a diagram following the list of the conditions studied. Then we shall discuss each implication giving references to published proofs and presenting counterexamples when appropriate.

There is now an extensive literature concerning the equisingularity of complex analytic varieties beginning with the exploratory papers of Hironaka [**10**] and Zariski [**51**]. After the provocative list of questions posed by Zariski [**52**] the hypersurface case has become well understood. For details see the papers by Teissier [**29**–**32**] and the work of Briançon and Speder in their 1976 Nice theses. The case of families of complex space curves has been thoroughly treated by Briançon, Galligo and Granger [**2**] and Buchweiss and Greuel [**5**]. For varieties of arbitrary dimension and codimension see the work of Navarro Aznar [**25**], and Teissier [**33**, **34**]. Three more papers have reached me in spring 1982 (Navarro

1980 *Mathematics Subject Classification.* Primary 58A35, 32C42, 32B20.

0082-0717/81/0000-0663/$03.25

Aznar [26], Henry and Merle [9] and Teissier [35]). These correct and extend the announcement of Teissier [33].

Equisingularity conditions on smooth stratified sets. In this list X and Y are disjoint C^1 submanifolds of $\mathbf{R}^n$ with $0 \in Y \cap \bar{X}$. In parentheses after the statement of each condition is given the paper in which the condition first appeared.

(a) Given a sequence x_i in X which tends to 0 such that $T_{x_i}X$ tends to τ in $G^n_{\dim X}$, then $T_0 Y \subset \tau$ (Whitney [49]).

(a^k) Given any C^k chart for Y as a submanifold of $\mathbf{R}^n$ at 0,

$$(U, U \cap Y, 0) \overset{\phi}{\rightarrow} (\mathbf{R}^n, \mathbf{R}^m \times 0^{n-m}, 0),$$

there is a neighbourhood of 0 in which $\pi_\phi|_X$ is a submersion, where $\pi_\phi = \phi^{-1} \circ \pi_m \circ \phi$ and $\pi_m(u_1, \ldots, u_n) = (u_1, \ldots, u_m, 0, \ldots, 0)$ (Wall [48] (k imprecise), Trotman [42]).

(b) Given sequences x_i in X and y_i in Y which tend to 0 such that $T_{x_i}X$ tends to τ in $G^n_{\dim X}$ and $x_i y_i / |x_i y_i|$ tends to λ in $\mathbf{P}^{n-1}$, then $\lambda \subset \tau$ (Whitney [50]).

(b^k) Given any C^k chart for Y as a submanifold of $\mathbf{R}^n$ at 0,

$$(U, U \cap Y, 0) \overset{\phi}{\rightarrow} (\mathbf{R}^n, \mathbf{R}^m \times 0^{n-m}, 0),$$

there is a neighbourhood of 0 in which $(\pi_\phi, \rho_\phi)|_X$ is a submersion, where π_ϕ is as in (a^k) and $\rho_\phi = \rho \circ \phi$ where $\rho(u_1, \ldots, u_n) = \sum_{j=m+1}^n u_j^2$ (Wall [48] (k imprecise), Trotman [42]).

(fin^k) There are finitely many homeomorphism types among germs at 0 of $S \cap X$ where S is a C^k submanifold of a given dimension transverse to Y at 0 (this paper, but see hints in Kuo-Lu [20]).

(h^k) For every pair of C^k submanifolds S_1, S_2 transverse to Y at 0 and of a given dimension, the germs at 0 of $S_1 \cap X$ and $S_2 \cap X$ are homeomorphic (Trotman [42]; when $k = \infty$, Kuo [19]).

(r) For all $\varepsilon > 0$ there is a neighbourhood U of 0 in $\mathbf{R}^n$ such that

$$\frac{d(T_x X, T_y Y)\,|x|}{|x - y|} < \varepsilon$$

for all $x \in X \cap U$, and all $y \in Y \cap U$ (Kuo [18], in an equivalent form).

(t^k) Every C^k submanifold transverse to Y at 0 is transverse to X in some neighbourhood of 0 (Thom [37] (k imprecise), Trotman [42]).

(w) There is $C > 0$ and a neighbourhood U of 0 in $\mathbf{R}^n$ such that

$$d(T_x X, T_y Y) < C|x - y|$$

for all $x \in X \cap U$, and all $y \in Y \cap U$ (Verdier [47], but see Hironaka [11]).

($\cdots *$) There is an open dense set of i-planes P containing $T_0 Y$, $n > i > \dim Y$, such that if M is an i-manifold containing Y such that $T_0 M = P$, then $(X \cap M, Y)$ verifies ($\cdots$) at 0 (Teissier [30]).

Implications

(w^*)	$\overset{\omega}{\leftarrow}$	(w)								
$\Downarrow 1^*$		$\Downarrow 1$								
(r^*)	$\overset{\rho}{\leftarrow}$	(r)								
$\downarrow\dashuparrow 2^*$		$\downarrow\dashuparrow 2$								
(b^*)	$\overset{\beta}{\dashleftarrow}$	(b)	$\overset{\beta^1}{\Leftrightarrow}$	(b^1)	$\overset{\beta^2}{\Rightarrow}$	(b^2)	$\overset{\beta^3}{\Rightarrow}$	$\cdots$	$\Rightarrow$	(b^∞)
$\Downarrow 3^*$		$\Downarrow 3$		$\Downarrow 3^1$		$\Downarrow 3^2$				$\Downarrow 3^\infty$
(a^*)	$\overset{\alpha}{\dashleftarrow}$	(a)	$\overset{\alpha^1}{\Leftrightarrow}$	(a^1)	$\overset{\alpha^2}{\Rightarrow}$	(a^2)	$\overset{\alpha^3}{\Rightarrow}$	$\cdots$	$\Rightarrow$	(a^∞)
$\Downarrow\!\Uparrow 4^*$		$\Downarrow\!\Uparrow 4$		$\Downarrow\!\Uparrow 4^1$		$\Downarrow 4^2$				$\Downarrow 4^\infty$
(t^*)	$\overset{\tau}{\dashleftarrow}$	(t)	$\equiv$	(t^1)	$\overset{\tau^2}{\Rightarrow}$	(t^2)	$\overset{\tau^3}{\Rightarrow}$	$\cdots$	$\Rightarrow$	(t^∞)
				$\Updownarrow 5^1$		$\Uparrow 5^2$				$\Uparrow 5^\infty$
$\Updownarrow 5^*$		$\Updownarrow 5$		(fin^1)	$\overset{\phi^2}{\Rightarrow}$	(fin^2)	$\overset{\phi^3}{\Rightarrow}$	$\cdots$	$\Rightarrow$	(fin^∞)
				$\Updownarrow 6^1$		$\Uparrow 6^2$				$\Uparrow 6^\infty$
(h^*)	$\overset{\eta}{\dashleftarrow}$	(h)	$\equiv$	(h^1)	$\overset{\eta^2}{\Rightarrow}$	(h^2)	$\overset{\eta^3}{\Rightarrow}$	$\cdots$	$\Rightarrow$	(h^∞)

Key

$\Rightarrow$: Implication always valid.
$\rightarrow$: Implication valid in the subanalytic case.
$\dashrightarrow$: Implication valid in the subanalytic case when $\dim Y = 1$.
$\equiv$: Two notations for the same condition.

Details of the implications. 1. It is obvious that (w) implies (r) by looking at the definitions. See Brodersen and Trotman **[4]** for real algebraic examples of (w)-faults for which (r) holds. The simplest example is as follows: let $V = \{(x, y, z) \in \mathbf{R}^3 \mid y^3 = z^2x^3 + x^5\}$, then let Y be the z-axis and $X = V - Y$. Because $(z^2x^3 + x^5)^{1/3}$ is a C^1 function of x and z, it follows that V is the graph of a C^1 map, and in particular V is a C^1 submanifold. It follows (for example by **[44]**) that (X, Y) is a (b)-regular pair. Kuo **[18]** showed that (r) and (b) are equivalent when Y is one-dimensional and the strata are semianalytic, so that (r) holds in our example. The reader may verify that (w) fails along the curve $(t^3, 2^{1/3}t^5, t^3)$.

It is a consequence of Teissier **[33]**, as corrected by **[26, 9** and **35]**, that (r) and (w) are equivalent for complex analytic strata.

2. Kuo **[18]** showed for subanalytic strata that (r) implies (b) and when Y is one-dimensional, then (b) implies (r). It is shown by Brodersen and Trotman **[4]** that for any subanalytic (w)-fault (X, Y) (r) fails for $(X \times \mathbf{R}, Y \times \mathbf{R})$ at each point of $0 \times \mathbf{R}$. However if (X, Y) is (b)-regular, $(X \times \mathbf{R}, Y \times \mathbf{R})$ is (b)-regular.

Thus the real algebraic variety V defining the (r)-regular (w)-fault described above (in 1) gives a (b)-regular (r)-fault in $\mathbf{R}^4$ by taking products with $\mathbf{R}$.

For complex analytic (X, Y), (b) and (r) are equivalent: use the same references as for (r) and (w) in 1.

3. It is easy to show that (b) implies (a) (see [**22** or **44**]). The standard example of an (a)-regular (b)-fault was given in the original paper of Whitney [**50**]. Let V be $\{y^2 = z^2x^2 + x^3\}$ in $\mathbf{R}^3$, let Y be the z-axis, and let $X = V - Y$.

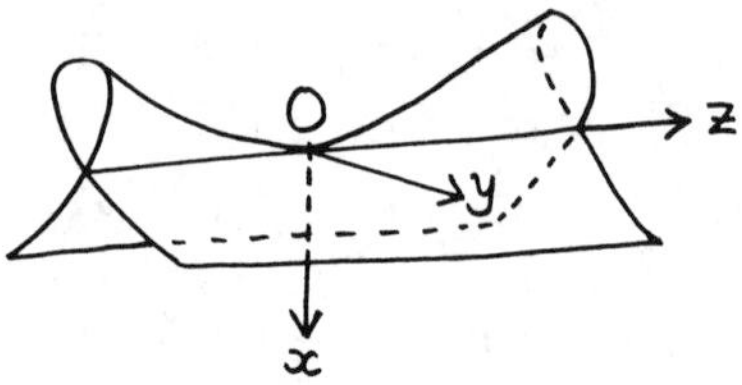

Then (b) fails along the curve $\{y = 0, x + z^2 = 0\}$. Choose for $\{x_i\}$ any sequence tending to 0 in this curve, and for y_i the z-coordinate of x_i for all i.

4. It is easy to show that (a) implies (t); this was observed by Whitney [**49**]. The converse in the case of subanalytic X follows from the proof for semianalytic X given in [**40**], either by quoting the curve selection lemma for subanalytic sets of [**12**], or by using the definition of a subanalytic set as in [**28**]. The simplest example of a (t)-regular (a)-fault was given in [**41**]: let X be obtained from a half-plane by inserting an infinite sequence of bumps (the round barrows defined below) contained in a horn tangent to the line $Y = \partial \overline{X}$.

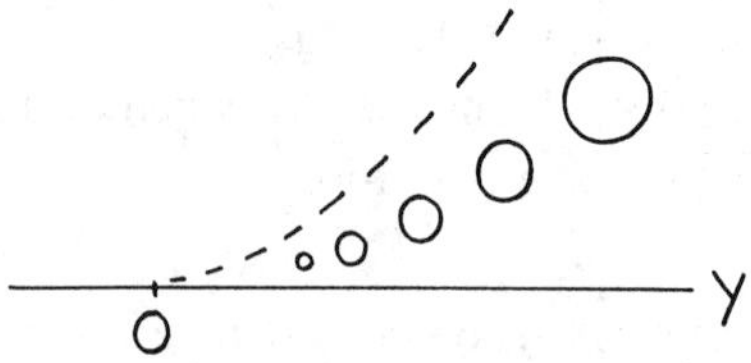

This example is (t)-regular because a C^1 transversal to Y at 0 in $\mathbf{R}^3$ will not intersect the horn. A more subtle example is given by Kambouchner and Trotman [**15**]: define a sequence of long barrows on a line L_1 in a half-plane X such that the limiting tangent planes all contain a line L_2 in X perpendicular to L_1. No transversals to Y can both meet infinitely many barrows (which entails containing L_1) and be nontransverse at 0 to a limiting tangent plane (which entails containing L_2), for Y is contained in the span of L_1 and L_2.

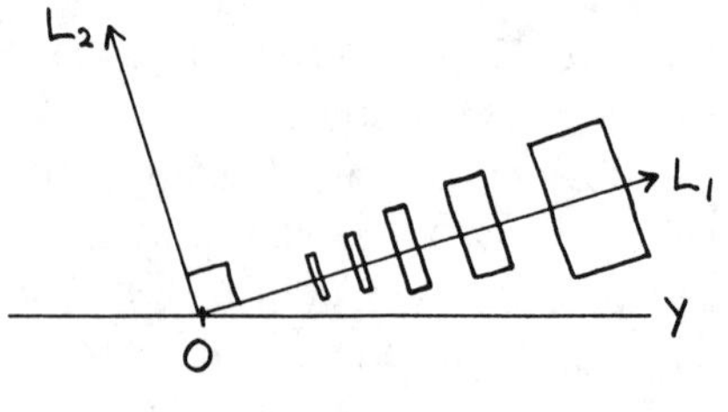

DEFINITIONS. A *round barrow* is a semialgebraic surface of the form

$$\{(x, y, z) \in \mathbf{R}^3 \mid m^3 z = (x^2 + y^2 - m^2)^2, x^2 + y^2 \leqslant m^2\}$$

$(0 < m < \infty)$ which inserts into the plane $z = 0$ with the disc $\{z = 0, x^2 + y^2 \leqslant m^2\}$ removed to give a semialgebraic C^1 submanifold of $\mathbf{R}^3$. Note that the height and radius of the base of the barrow are equal.

A *long barrow* is a semialgebraic surface of the form $\{(x, y, z) \in \mathbf{R}^3 \mid m^7 r^3 z = (m^2 - x^2)^2 (m^2 r^2 - y^2)^2,\ x^2 \leqslant m^2,\ y^2 \leqslant m^2 r^2\}$ which inserts into the plane $z = 0$ with the rectangle $\{z = 0, x^2 \leqslant m^2, y^2 \leqslant m^2 r^2\}$ removed to give a semialgebraic C^1 submanifold of $\mathbf{R}^3$. The height is mr and the sides of the base are $2mr$ and $2m$. As m varies the barrow varies in size but the set of normals (or tangents) remains the same. As r tends to 0 the normals tend to lie in the arc $\{(1 : 8\lambda/3\sqrt{3} : 0) \mid \lambda \in [-1, 1]\}$.

5. That (h) implies (t) is shown in [**42**], using Morlet's theorem that a weakly closed subspace of $C^1(N, P)$ is strictly Baire [**24**, **14**]. This and the converse ((t) implies (h)) will appear in [**46**]: it is assumed that $Y = \overline{X} - X$, as in the proof by Kuo [**19**] that (a) implies (h^∞).

Note. 1* (resp. 2*, etc.) is contained in 1 (resp. 2, etc.).

$\{3^i\}$ $(i \geqslant 1)$. These are obvious. The converses are false: use the example showing (a) does not imply (b).

$\{4^i\}$ $(i \geqslant 2)$. These are obvious. The converses are false in general as shown by the (t)-regular (a)-faults described above (4). It is not known if (t^i) implies (a^i) for subanalytic sets when $i \geqslant 2$.

$\{5^i\}$ $(i \geqslant 1)$. These follow in principle from the proof that (h^i) implies (t^i) given in [**42**] by giving a more precise perturbation argument for nontransverse submanifolds as used by Bochnak and Kuo [**1**]. I conjecture that (t^i) implies (fin^i) when $i \geqslant 2$ and the strata are subanalytic. For $i = 1$ this is known without subanalyticity because (t^1) implies (h^1), although we suppose $Y = \overline{X} - X$ (see 5 above).

$\{6^i\}$ $(i \geqslant 2)$. That (h^i) implies (fin^i), $i \geqslant 2$, is immediate. The converse is false as is the converse to (h^i) implies (t^i). There are real algebraic counterexamples. Let V be $\{(x, y, z) \mid x^3 - 3xz^5 + yz^6 = 0\} \subset \mathbf{R}^3$; let Y be the y-axis and $X = V - Y$. There are precisely 2 topological types of germs at 0 of intersections of X with a transverse submanifold to Y of class C^k, $k \geqslant 2$. Note that (t^2) holds, but (t^1) fails—there is a C^1 submanifold transverse to Y at 0 and *tangent* to X along a curve in the critical locus in V of the projection of V onto the (y, z)-plane. For more details and discussion see [**45**, **46**]. This example arises in the theory of sufficiency of jets [**16**]: see [**20**] for some extraordinary results translating jet-sufficiency among representatives of class C^{r+k} of an r-jet z in terms of (t^k)-regularity of the deformation by monomials of order r of the variety of zeros over the space of such monomials.

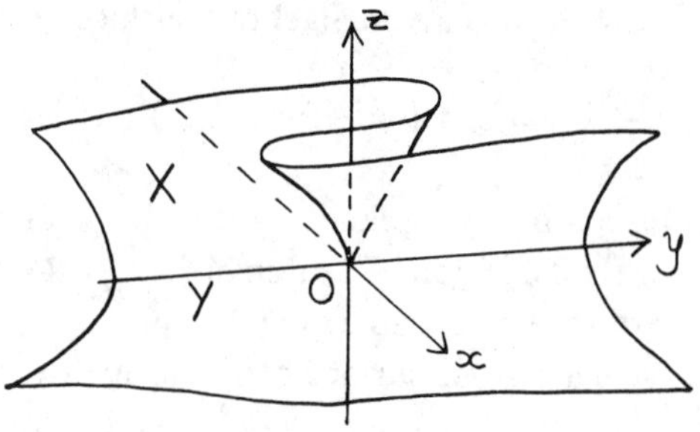

Notes. (i) A recent paper of Koike [**17**] suggests considering a refined form of (h^i): demanding a homeomorphism between germs of the triples $(S_1, S_1 \cap X, 0)$ and $(S_2, S_2 \cap X, 0)$.

(ii) Although the proof of $(t) \Rightarrow (h)$ is rather difficult and not yet available, one can use the result (conjectured by Thom and proved by Kuo [**19**]) that (a) implies (h^∞) to show that (a) implies (h^1) by a lemma smoothing everything in sight. See [**45** and **46**] for details.

We have completed our discussion of the vertical arrows and pass to the horizontal.

ω, ρ: These are proved for subanalytic strata in [**27**]. The key technical result is as follows.

TRANSVERSALITY LEMMA. *Let X be a subanalytic C^1 submanifold of $\mathbf{R}^m \times \mathbf{R}^s$ such that $Y = \mathbf{R}^m \times 0 \subset \overline{X} - X$. Suppose that either* (I) $m = 1$, *or* (II) (X, Y) *verifies (r)-regularity at* 0. *Then for each i, $1 \leqslant i \leqslant s$, there is an open dense subset U^i of the grassmannian of $(m + s - i)$-planes containing Y such that if $P \in U^i$ and if $a(t) \in \mathbf{R}^m \times \mathbf{R}^s$ and $P(t) \in U^i$ are analytic curves with $a(0) = 0$, $P(0) = P$, and $a(t) \in X \cap P(t)$ when $t \neq 0$, then P is transverse to* $\lim_{t \to 0} T_{a(t)} X$.

(The case (I) $m = 1$ was proved also by Henry and Merle [**8**].) We apply the Transversality Lemma in the case (II), using that (w) implies (r), to obtain the implications ω and ρ.

β: By case (I) of the Transversality Lemma it is easily shown that (b) implies (b^*) when $\dim Y = 1$ (see Navarro Aznar and Trotman [**27**]). For complex analytic strata with Y of arbitrary dimension (b), (r) and (w) are equivalent, as noted above, so that β may be deduced from ω and ρ.

For semialgebraic strata with Y of dimension two it is false that (b) implies (b^*). An example was given in [**27**] based on an observation of [**8**]. Let Y be the (u, v)-plane in $\mathbf{R}^4$ with coordinates (x, y, u, v), let X_1 be $\{x^2 + y^2 + ux + vy = 0\} - Y$, and finally let X be X_1 intersected with $\{x^2 + y^2 < (u^2 + v^2)^2\}$. It may be checked that (b)-regularity holds for (X, Y) at 0 but that (r) fails (along (t^2, t^2, t, t) for example), and if we intersect $X \cup Y$ with a generic 3-plane P which contains Y then (b) fails to hold for $(X \cap P, Y)$ at 0 for reasons of dimension. However Y and X are not strata of a Whitney stratification of a closed subset of $\mathbf{R}^4$ so this example is pathological: a better example should be looked for.

Applications of the fact that (b) implies (b^*) for complex analytic strata are given by [**8**, **25**, **26**, **27**, and **33**].

α: Again when Y is one-dimensional and strata are subanalytic, this is an easy consequence of case (I) of the Transversality Lemma. The semialgebraic example of a (b)-regular pair not satisfying (b^*) above also serves here.

τ, η: Since in the subanalytic case (t) and (h) are equivalent to (a), the implications τ and η are equivalent to α.

α^1, β^1: The implications (a) implies (a^1) and (b) implies (b^1) were first proved in [**37**]. The converses appear in my thesis [**42**] and in [**44**].

α^i, β^i $(i \geqslant 2)$: The implications given are obvious. Counterexamples to the converses were given in [**15**]. Let Y be the y-axis in $\mathbf{R}^3$ and let X_1 be the open half-plane $\{z > 0, x = 0\}$. Now perturb X_1 by an infinite sequence $\{B_n\}$ of long barrows tending to 0 with centres on $\{x = y - z = 0\}$ and with axes at an angle θ_n to the z-axis so that θ_n as a function of the distance δ_n of B_n to Y is "continuous but not C^1" at 0. This may be done so that any C^1 foliation nontransverse to infinitely many barrows and transverse to Y at 0 (the leaf through 0 transverse to Y at 0) is automatically not a C^2 foliation. Then (b^2) holds but (b^1) fails. The same example gives an (a^2)-regular (a^1)-fault. Choosing θ_n to be a "C^k but not a C^{k+1}" function of δ_n $(k \geqslant 2)$ will give (b^{k+1})-regular (a^k)-faults (hence (a^{k+1})-regular (b^k)-faults). See [**15**] for details.

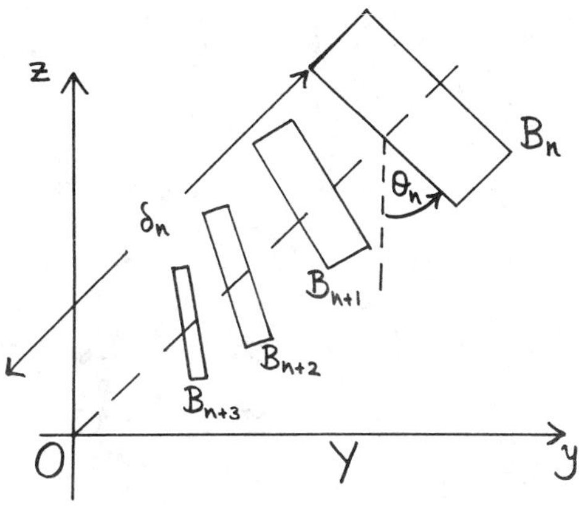

τ^i, η^i, ϕ^i: The implications $(t^k) \Rightarrow (t^{k+1})$, $(\text{fin}^k) \Rightarrow (\text{fin}^{k+1})$, and $(h^k) \Rightarrow (h^{k+1})$, $k \geqslant 1$, are all obvious. All the converses are false at least for semialgebraic strata, as is clear for $k = 1$, from the real algebraic examples given above (5^i) to show that (fin^2) does not imply (h^2). Semialgebraic examples for all $k \geqslant 1$ are straightforward to construct. They look as follows (this construction was discovered by

Kuo and the author at Inst. Hautes Etudes Scientifiques in January 1979):

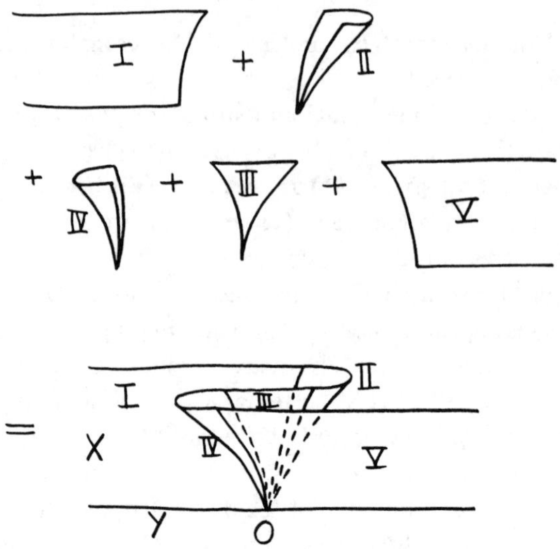

The curves along which the semialgebraic pieces (I–V) are glued together will be C^k-diffeomorphic with a half-line but not C^{k+1}-diffeomorphic. The infinitely many topological types of C^k intersections that arise (contradicting (t^k), (h^k) and (fin^k)) begin with (omitting the last eight obvious arcs):

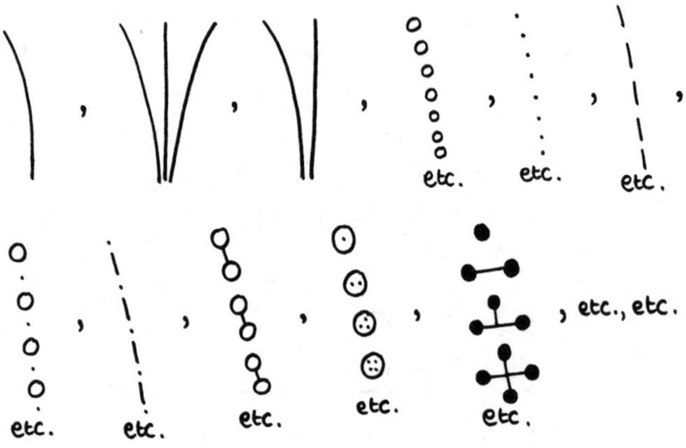

I call these "Comic Cuts". Note that we can obtain any sequence $\{C_i\}_{i=1}^{\infty}$ of compact subsets C_i of the plane. There are uncountably many topologically distinct cuts (we take the topological type of the germ at 0 of $S \cap X$ where S is transverse to Y at 0).

This completes our discussion of the implications between the conditions considered.

Further remarks. I have not as yet mentioned the two most quoted properties of Whitney stratifications. Firstly, a (b)-regular stratification is locally topologically trivial—the famous Thom-Mather Isotopy Theorem [**38**, **39**, **22**, **23**]. It is a conjecture of Sullivan [**53**] that any analytic variety admits a locally Lipschitz trivial stratification: the trivialisations are Lipschitz homeomorphisms. How useful (b)-regularity is for this question remains unclear. Certainly by [**22**], (b)-regular stratifications have *rugose* trivialisations since those constructed by Mather are controlled. Rugosity is a Lipschitz condition relative to strata: a map $f\colon \mathbf{R}^n \to \mathbf{R}^k$ is rugose with respect to a stratification $\mathbb{S}$ of a closed subset of $\mathbf{R}^n$ if for all $X, Y \in \mathbb{S}$ and for all $y_0 \in Y$, there exists $C > 0$ and a neighbourhood U of y_0 such that $|f(x) - f(y)| < C|x - y|$ for all $x \in X \cap U$, $y \in Y \cap U$. C may not be uniform as one approaches the frontier of X. Verdier [**47**] showed that (w)-regular stratifications admit rugose vector fields tangent to strata which are integrable.

The result that (a) (and indeed (t)) implies (h) may be considered as a localised version of the Thom-Mather Isotopy Theorem. The second famous result about Whitney stratifications, which has too many applications to mention, is that transverse maps to an (a)-regular stratification form an open subset of the space of smooth maps. In fact this is *equivalent* to (a)-regularity by [**42**, **43**]. It is striking that transversals to a stratification with just 2 subanalytic strata have intersections which are topologically stable under perturbations of the transversals if and only if the stratification is (a)-regular and locally topologically trivial (use $(a) \Leftrightarrow (h)$).

Stratified maps. Almost all of the conditions described here have *relative* versions, for example (a_f) (also known as the Thom condition on maps, it was introduced by Thom [**36**] and studied by Mather [**22**], Gibson [**6**] and Hironaka [**13**]) is a relative version of (a) imposed on a map f; these provide notions of stratified maps between stratified sets. Goresky [**7**] considers a similar (equivalent?) notion of stratified map (his condition D). Shortly after the symposium I received a paper by Brodersen [**3**] which considers relative versions (w_f) and (t_f) (of (w) and (t)) in the jet space context studied by Kuo and Lu [**20**]. Some relative versions of our results were studied in Chapter 4 of my thesis [**42**]; see also [**26** and **34**]. Relative results have much interest, as there is as yet no satisfactory theory of stratified maps.

Analyticity of regularity conditions. All the conditions studied here are generic: given a subanalytic set V there exists a stratification of V which is (w)-regular (proved by Verdier [**47**]) and hence verifies all the other conditions considered. For complex analytic varieties the equivalence of (b)-regularity and equimultiplicity of polar varieties [**9**, **26** and **33**] implies that (b) (and thus (r) and (w)) is an *analytic* condition: the set of points of an analytic subvariety Y (of a variety $\overline{X}$ containing Y) at which (X, Y) is not (b)-regular is an analytic subvariety of Y. However Zariski [**51**] gave an example showing that (a) is not an analytic condition: let V be $\{(x, y, u, v) \mid x^2 = uvy^2 + y^3\} \subset \mathbf{C}^4$ or $\mathbf{R}^4$, let Y be $\{x = y = 0\}$, the singular set of V, and let $X = V - Y$. Then X is (a)-regular over Y at all points of Y except on $\{uv = 0\} - \{0\}$. It is (a)-regular at $(0, 0, 0, 0) = \{0\}$. So the bad set for (a)-regularity is not analytic. It follows that (t) and (h) are not analytic either.

Local topological triviality. Following the referee's suggestion we add a description of the place in our diagram of the condition of local topological triviality:

(TT) $(\overline{X}, Y)$ is locally homeomorphic near 0 to a product $(Y \times \overline{X}_0, Y \times 0)$.

Now the usual application of Thom's isotopy lemmas gives that (b^∞) implies (TT); however the semialgebraic topologically trivial (a)-faults given for the implications τ^i, η^i, ϕ^i preclude any other implication being verified. Probably real algebraic counterexamples exist generalising that given for the implication 6^2 which shows that (TT) does not imply (h^2) or (t^1). In the complex analytic case it is unknown if (a)-regularity follows from (TT), but it is true for hypersurfaces by [**29** or **21**] together with the topological invariance of the Milnor number.

References

1. J. Bochnak and T.-C. Kuo, *Different realizations of a non-sufficient jet*, Indag. Math. **34** (1972), 24–31.

2. J. Briançon, A. Galligo and M. Granger, *Déformations équisingulières des germes de courbes gauches réduites*, Mem. Soc. Math. France (Nouv. série) **1** (1980).

3. H. Brodersen, *A note on sufficient and non-sufficient jets*, University of Oslo preprint (1981), 17 pages.

4. H. Brodersen and D. Trotman, *Whitney (b)-regularity is strictly weaker than Kuo's ratio test for real algebraic stratifications*, Math. Scand. **45** (1979), 27–34.

5. R. Buchweiss and G.-M. Greuel, *The Milnor number and deformations of complex curve singularities*, Invent. Math. **58** (1980), 241–281.

6. C. G. Gibson, *Construction of canonical stratifications*, Topological Stability of Smooth Mappings, Lecture Notes in Math., vol. 552, Springer-Verlag, Berlin and New York, 1976, pp. 8–34.

7. M. Goresky, *Whitney stratified chains and cochains*, Trans. Amer. Math. Soc. **267** (1981), 175–196.

8. J.-P. Henry and M. Merle, *Sections planes, limites d'espaces tangents et transversalité de variétés polaires*, C. R. Acad. Sci. Paris Sér. A **291** (1980), 291–294.

9. ______, *Limites de normales, conditions de Whitney et éclatement d'Hironaka*, these PROCEEDINGS.

10. H. Hironaka, *Equivalences and deformations of isolated singularities*, Woodshole Seminar in Algebraic Geometry (1964).

11. ______, *Normal cones in analytic Whitney stratifications*, Inst. Hautes Etudes Sci. Publ. Math. **36** (1969), 27–138.

12. ______, *Subanalytic sets*, Number Theory, Algebraic Geometry and Commutative Algebra, volume in honour of Y. Akizuki, Kinokuniya, Tokyo, 1973, pp. 453–493.

13. ______, *Stratification and flatness*, Real and Complex Singularities, (Oslo, 1976) (P. Holm, Editor), Sijthoff & Noordhoff, Alphen aan den Rijn, 1977, pp. 199–265.

14. M. Hirsch, *Differential topology*, Springer-Verlag, Berlin, 1976.

15. A. Kambouchner and D. J. A. Trotman, *Whitney (a)-faults which are hard to detect*, Ann. Sci. École Norm. Sup. (4) **12** (1979), 465–471.

16. S. Koike and W. Kucharz, *Sur les réalisations de jets non-suffisants*, C. R. Acad. Sci. Paris Sér. A **288** (1979), 457–459.

17. S. Koike, *On v-sufficiency and $(\bar{h})$-regularity*, Publ. Res. Inst. Math. Sci. Kyoto Univ. **17** (1981), 565–575.

18. T.-C. Kuo, *The ratio test for analytic Whitney stratifications*, Liverpool Singularities Sympos. I, Lecture Notes in Math., vol. 192, Springer-Verlag, Berlin and New York, 1971, pp. 141–149.

19. ______, *On Thom-Whitney stratification theory*, Math. Ann. **234** (1978), 97–107.

20. T.-C. Kuo and Y.-C. Lu, *Sufficiency of jets via stratification theory*, Invent. Math. **57** (1980), 219–226.

21. Lê Dũng Tráng and K. Saito, *La constance du nombre de Milnor donne des bonnes stratifications*, C. R. Acad. Sci. Paris Sér. A-B **277** (1973), 793–795.

22. J. Mather, *Notes on topological stability*, Harvard University notes, 1970.

23. ______, *Stratifications and mappings*. Dynamical Systems (M. Peixoto, Editor), Academic Press, New York, 1971, pp. 195–223.

24. C. Morlet, *Le lemme de Thom et les théorèmes de prolongement de Whitney*. I. *Les Topologies des espaces d'applications*, Sém. Henri Cartan (1961/62).

25. V. Navarro Aznar, *Conditions de Whitney et sections planes*, Invent. Math. **61** (1980), 199–225.

26. ______, *Stratifications régulières et variétés polaires locales*, Manuscript (1981).

27. V. Navarro Aznar and D. J. A. Trotman, *Whitney regularity and generic wings*, Ann. Inst. Fourier (Grenoble) **31** (1981), 87–111.

28. J. Stasica, *A transversality property weaker than the Whitney* (*a*)*-regularity*, Zezyty Nauk. Uniw. Jagielloń. Prace. Mat. **21** (1979), 81–83.

29. B. Teissier, *Cycles évanescents, sections planes et conditions de Whitney*, Singularités à Cargèse, Soc. Math. France Astérisque **7–8** (1973), 285–362.

30. ______, *Introduction to equisingularity problems*, AMS Algebraic Geometry Sympos., Arcata 1974, Amer. Math. Soc., Providence, Rhode Island, 1975, pp. 593–632.

31. ______, *The hunting of invariants in the geometry of discriminants*, Real and Complex Singularities, Oslo 1976 (P. Holm, Editor), Sijthoff & Noordhoff, Alphen aan den Rijn, 1977, pp. 565–677.

32. ______, *Résolution simultanée et cycles évanescents*, Sém. sur les Singularités des Surfaces, Lecture Notes in Math., vol. 777, Springer-Verlag, Berlin and New York, 1980, pp. 82–146.

33. ______, *Variétés polaires locales et conditions de Whitney*, C. R. Acad. Sci. Paris Sér. A **290** (1980), 799–802.

34. ______, *Variétés polaires locales: quelques resultats*, Nancy, May 1980, Publication de l'Institut Élie Cartan, 1980.

35. ______, *Variétés polaires*. II. *Multiplicités polaires, sections planes, et conditions de Whitney*, École Polytechnique préprint (1982).

36. R. Thom, *La stabilité topologique des applications polynomiales*, Enseign. Math. **8** (1962), 24–33.

37. ______, *Local topological properties of differentiable mappings*, Differential Analysis, Bombay Colloquium, Oxford Univ. Press, London, 1964, pp. 191–202.

38. ______, *Propriétés différentielles locales des ensembles analytiques*, Séminaire Bourbaki no. 281 (1964/65).

39. ______, *Ensembles et morphismes stratifiés*, Bull. Amer. Math. Soc. **75** (1969), 240–284.

40. D. J. A. Trotman, *A transversality property weaker than Whitney* (*a*)*-regularity*, Bull. London Math. Soc. **8** (1976), 225–228.

41. ______, *Counterexamples in stratification theory: two discordant horns*, Real and Complex Singularities, Oslo 1976 (P. Holm, Editor), Sijthoff & Noordhoff, Alphen aan den Rijn, 1977, pp. 679–686.

42. ______, *Whitney stratifications: faults and detectors*, Doctoral thesis, Warwick (1977), 93 pages.

43. ______, *Stability of transversality to a stratification implies Whitney* (*a*)*-regularity*, Invent. Math. **50** (1979), 273–277.

44. ______, *Geometric versions of Whitney regularity for smooth stratifications*, Ann. Sci. Ecole Norm. Sup (4) **12** (1979), 453–463.

45. ______, *Regular stratifications and sufficiency of jets*, Proc. of the Algebraic Geometry Conf. at La Rabida, January 1981, Lecture Notes in Math., Springer-Verlag, Berlin and New York (to appear).

46. ______, *Transverse transversals and homeomorphic transversals* (to appear).

47. J.-L. Verdier, *Stratifications de Whitney et théorème de Bertini-Sard*, Invent. Math. **36** (1976), 295–312.

48. C. T. C. Wall, *Regular stratifications*, Dynamical Systems—Warwick 1974, Lecture Notes in Math., vol. 468, Springer-Verlag, Berlin and New York, 1974, pp. 332–344.

49. H. Whitney, *Local properties of analytic varieties*, Differential and Combinational Topology (S. Cairns, Editor), Princeton Univ. Press, Princeton, N. J., 1965, pp. 205–244.

50. ______, *Tangents to an analytic variety*, Ann. of Math. **81** (1965), 496–549.

51. O. Zariski, *Contributions to the problem of equisingularity*, C.I.M.E. Varenna 1969, Edizioni Cremonese, Rome, 1969.

52. ______, *Some open questions in the theory of singularities*, Bull. Amer. Math. Soc. **77** (1971), 481–491.

53. L. Siebenmann and D. Sullivan, *On complexes that are Lipschitz manifolds*, Geometric Topology (J. C. Cantrell, Ed.), Academic Press, New York, 1979, pp. 503–525.

UNIVERSITÉ DE PARIS-SUD, FRANCE

UNIVERSITÉ D'ANGERS, FRANCE

Proceedings of Symposia in Pure Mathematics
Volume 40 (1983), Part 2

ON SINGULARITIES ON DEGENERATE DEL PEZZO SURFACES OF DEGREE 1, 2

TOHSUKE URABE

This is the summary of joint work with I. Naruki.

Let X be a degenerate Del Pezzo surface of degree k with $k = 1$ or 2. Let $\tilde{X} \to \mathbf{P}^2$ be the composition of $9 - k$ blowing-ups of a projective space of dimension 2 such that any irreducible curve with a selfintersection number less than -2 does not appear on $\tilde{X}$. By definition the variety obtained from $\tilde{X}$ by contracting the configuration of curves with the selfintersection number -2 is X.

We say that two degenerate Del Pezzo surfaces X, X' have the same singularity type if we can deform X to X' by finite steps of deformations all of whose fibres have the same number of isolated singularities of each type.

We note that, for example, there are two Del Pezzo surfaces of degree 2 with only $4A_1$ singularities such that we cannot deform one to the other by deformations whose fibres have only $4A_1$ singularities.

MAIN THEOREM. *Singularity types of degenerate Del Pezzo surfaces of degree $k = 2$ (resp. $k = 1$) and isomorphism classes of subsystems of the root system E_7 (resp. E_8) except for the subsystem of type $7A_1$ (resp. subsystems of type $7A_1$, $8A_1$, and $D_4 + 4A_1$) are in one-to-one correspondence such that the configuration of singularities on the surface coincides with the type of the corresponding root system.*

As for the classification of root subsystems, there is the complete result by Dynkin, which is as in the next table.

Why are there exceptions? We can give two independent reasons. One is geometrical: For $7A_1$, $8A_1$ and $D_4 + 4A_1$, the rational elliptic surface obtained by blowing up certain k points on $\tilde{X}$ has the Euler-Poincaré characteristic greater than the sum 12 of $\chi(\mathbf{P}^2) = 3$ and the total number 9 of blowing-ups, which should be prohibited. The other is graph-theoretical. For a given set of Dynkin diagrams, we call the following procedure an elementary transformation: Add one

1980 *Mathematics Subject Classification*. Primary 55R55.

0082-0717/82/0000-0210/$02.00

vertex and several edges to an arbitrary chosen component and make it the extended Dynkin diagram of corresponding type. Then, remove one vertex or more and connecting edges from the extended component. The diagrams $7A_1$, $8A_1$, and $D_4 + 4A_1$ are exceptional ones in the next table which cannot be obtained from E_7 or E_8 by repeating elementary transformation only *twice*.

The origin of our problem goes back to deformation theory of simple elliptic singularities. A simple elliptic singularity $\tilde{E}_7$ (resp. $\tilde{E}_8$) are defined by the equality

$$z^2 - xy(x-y)(x-ay) = 0, \qquad a \neq 0, 1$$

(resp. $z^2 - x(x-y^2)(x-ay^2) = 0$, $a \neq 0, 1$). Each fibre in the versal deformation of $\tilde{E}_r$ $(r = 7, 8)$ has a compactification, which is either a cone over an elliptic curve or a degenerate Del Pezzo surface of degree $9 - r$. Thus, our theory also gives the answer to the next question: What singularities and how many of each type can appear in the fibres of the versal deformation of a simple elliptic singularity $\tilde{E}_7$ or $\tilde{E}_8$?

THEOREM. *The only configuration of singularities in fibres of the versal deformation of simple elliptic singularities* $\tilde{E}_r$, $r = 7, 8$, *is*

(i) *simple elliptic singularities of the same type*,

(ii) *configuration of rational double points for which the dual graph of the minimal resolution is* $r = 7$ (*resp.* 8); *any subgraph obtained from the Dynkin diagram* E_7 (*resp.* E_8) *by elementary transformation twice.*

Moreover, the compactification $\overline{F}$ cited above of any fibre of the deformation of $\tilde{E}_7$ has a canonical structure as a double covering over $\mathbf{P}^2$ branching along a quartic curve B. Note that the structure of $\overline{F}$ is uniquely determined by B and that singular points on $\overline{F}$ and those on B are in one-to-one correspondence. For $\tilde{E}_8$, the compactification of a fibre is a double-covering over Λ, where Λ is a conic in $\mathbf{P}^3$ with one singular point S such that the branching locus B is the union of S and a sextic curve not passing through the singular point S of Λ. Thus, our theory also gives the classification of singularities of plane quartic curves and that of space sextics on Λ not passing through S.

THEOREM. *The ambient isotopy classes of plane quartic curves without multiple components are described as follows.*

(i) *Four lines passing through one point.*

(ii) *Ones except for the one cited in* (i) *and isomorphism classes of root subsystems of the root system* E_7 *except for the one of type* $7A_1$ *are in one-to-one correspondence such that if the configuration of singular points in the curve is* $\sum a_k A_k + \sum b_l D_l + \sum c_m E_m$, *then the corresponding root subsystem is of type* $\sum a_k A_k + \sum b_l D_l + \sum c_m E_m$.

THEOREM. *Let* Λ *be a conic in* $\mathbf{P}^3$ *with one singular point* S. *The ambient isotopy classes of sextic space curves lying on* Λ *not passing through* S *and without multiple components are described as follows.*

(a) *Three quadratic curves passing through one point and with a common tangent line at that point.*

(b) *Ones except for the one cited in* (a) *and isomorphism classes of subsystems of the root system* E_8 *except for the ones of type* $7A_1, 8A_1$ *and* $D_4 + 4A_1$ *are in one-to-one correspondence such that if the configuration of singular points on the curve is* $\Sigma a_k A_k + \Sigma b_l D_l + \Sigma c_m E_m$, *then the corresponding root subsystem is of type* $\Sigma a_k A_k + \Sigma b_l D_l + \Sigma c_m E_m$.

The most important step in the proof is to construct the moduli space of the so-called overmarked Del Pezzo surfaces. Let E be an elliptic curve with a group law.

DEFINITION. We call the following object $\tilde{X} = (\tilde{X}, \alpha, C, \iota)$, consisting of 4 items, an *overmarked Del Pezzo surface over* E of degree k. We assume $1 \leqslant k \leqslant 6$.

(1) First, $\tilde{X}$ is the variety obtained by successive $r = 9 - k$ blowing-ups of the projective space of dimension 2 such that any irreducible curve with the selfintersection number less than -2 does not appear on $\tilde{X}$.

(2) The second item $\alpha\colon P \to \operatorname{Pic}(\tilde{X})$ is an isomorphism, where P is a standard model of $\operatorname{Pic}(\tilde{X})$ equipped with the intersection form and the canonical class $\kappa \in P$. The map α is an isomorphism which preserves the intersection form and such that $\alpha(\kappa)$ coincides with the canonical line bundle K of $\tilde{X}$.

(3) C is a smooth member of the linear system $|-K|$ which has the following isomorphism ι.

(4) ι is a group isomorphism, $\iota\colon \operatorname{Pic}(C)^0 \to E$.

Let Γ be the orthogonal complement of $\mathbf{Z}\kappa$ in P and $L = \alpha(\Gamma)$. The composition

$$\phi_{\tilde{X}}\colon \Gamma \xrightarrow{\alpha} L \to \operatorname{Pic}(C)^0 \xrightarrow{\iota} E$$

is called the *characteristic homomorphism* of $\tilde{X}$. Here the middle arrow is the restriction map.

THEOREM. *The set of isomorphism classes of overmarked Del Pezzo surfaces over* E *of degree* k *with* $1 \leqslant k \leqslant 6$ *and the group* $\operatorname{Hom}(\Gamma, E)$ *have one-to-one correspondence by means of associating* $\tilde{X} = (\tilde{X}, \alpha, C, \iota)$ *to its characteristic homomorphism* $\phi_{\tilde{X}} \in \operatorname{Hom}(\Gamma, E)$.

We call an element η with $\eta^2 = -2$ which is orthogonal to the canonical class a *root*.

PROPOSITION. *Let* $R \subset \operatorname{Pic}(\tilde{X})$ *be the root system generated by effective roots on* $\tilde{X}$. *Then*, $\alpha^{-1}(R)$ *coincides with the maximal root system contained in* $\operatorname{Ker} \phi_{\tilde{X}}$.

It is known that Γ is isomorphic to the root lattice of type E_r with $r = 9 - k$ if $6 \leqslant r \leqslant 8$. Let W be the Weyl group acting on Γ. We can deduce our problem to the classification of the fixed points on $\operatorname{Hom}(\Gamma, E)$ by the action of W. Going up to the universal covering $\operatorname{Hom}(\Gamma, \mathbf{C}) \cong \mathbf{C}^r$, with the aid of theory of affine Weyl groups, we get our result.

The origin of our problem goes back to the deformation theory of singularities. What singularities can appear in the fibres of the versal deformation of simple elliptic surface singularities? In [6] Saito answered this question for simple elliptic

TABLE

E_6		E_7			E_8				
A_5+A_1	A_2+2A_1	D_6+A_1	$3A_2$	D_4	A_8	$A_4+A_2+A_1$	A_4+A_3	A_4+2A_1	A_2+3A_1
$3A_2$	$4A_1$	A_5+A_2	$2A_3$	A_4	D_8	A_5+A_2	A_5+2A_1	A_6	$2A_2+A_1$
A_5	D_4	$2A_3+A_1$	A_6	$[A_3+A_1]'$	A_7+A_1	$3A_2+A_1$	$[A_7]'$	$A_3+A_2+A_1$	D_4
$2A_2+A_1$	A_3	A_7	$6A_1$	$[A_3+A_1]''$	$A_5+A_2+A_1$	E_6+A_1	$[A_7]''$	$[A_5+A_1]'$	$[4A_1]'$
A_4+A_1	A_2+A_1	D_4+3A_1	D_5	$2A_2$	$2A_4$	E_7	$3A_2$	$[A_5+A_1]''$	$[4A_1]''$
D_5	$3A_1$	$7A_1$	A_4+A_1	A_2+2A_1	$4A_2$	D_7	E_6	A_4+A_2	A_2+2A_1
A_3+2A_1	A_2	E_6	$2A_2+A_1$	$[4A_1]'$	E_6+A_2	D_5+2A_1	D_6	$2A_2+2A_1$	$2A_2$
A_4	$2A_1$	D_5+A_1	$[A_5]'$	$[4A_1]''$	E_7+A_1	D_4+3A_1	D_4+2A_1	D_5	A_3+A_1
A_3+A_1	A_1	A_4+A_2	$[A_5]''$	A_3	D_6+2A_1	$2A_3+A_1$	$[2A_3]'$	$[A_3+2A_1]'$	A_4
$2A_2$		$A_3+A_2+A_1$	D_4+A_1	A_2+A_1	D_5+A_3	$7A_1$	$[2A_3]''$	$[A_3+2A_1]''$	A_3
		$[A_5+A_1]'$	A_3+A_2	$[3A_1]'$	$2D_4$	D_6+A_1	D_5+A_1	A_3+A_2	A_2+A_1
		$[A_5+A_1]''$	$5A_1$	$[3A_1]''$	D_4+4A_1	D_5+A_2	A_3+3A_1	A_5	$3A_1$
		D_6	A_2+3A_1	A_2	$2A_3+2A_1$	$A_3+A_2+2A_1$	D_4+A_2	$5A_1$	A_2
		D_4+2A_1	$[A_3+2A_1]'$	$2A_1$	$8A_1$	D_4+A_3	$6A_1$	A_4+A_1	$2A_1$
		A_3+3A_1	$[A_3+2A_1]''$	A_1	A_6+A_1	A_3+4A_1	A_2+4A_1	D_4+A_1	A_1

singularities of degree 1, 2 and 3. However he did not discuss the number of singularities of each type in the fibres. Pinkham gave the list of the configurations of singularities appearing in the fibres of the versal deformation of a simple elliptic singularity of degree $k \geqslant 3$ in [**5**] by using Del Pezzo's classification of surfaces of degree k in $\mathbf{P}^k$. Studying primitive integrals, Looijenga claimed in [**2**] that a similar result to Pinkham's one holds for degree 1 and 2. (It was noted in [**5**].) But we can easily construct a counterexample to Looijenga's claim. Although he essentially solved the problem, misunderstanding something at the final step of the proof, he got a wrong result.

Thus our result is the correction of Looijenga's one.

Recently we noticed that the enumeration itself can be already found in Timms [**7**] and Du Val [**1**] and that J. Mérindol treated the same question in greater generality in [**4**]. Moreover we hear that the correction by Looijenga himself will be published [**3**].

However our elementary and systematic methods are different from Looijenga's one. Thus, we believe it helps to understand simple elliptic singularities clearly.

The details of the proof will appear elsewhere.

REFERENCES

1. P. Du Val, *On isolated singularities which do not affect the conditions of adjunction*. I, II, III, Proc. Cambridge Philos. Soc. **30** (1934), 453–465, 483–491.

2. E. Looijenga, *On the semi-universal deformation of a simple elliptic hypersurface singularity*. II, Topology **17** (1978), 23–40.

3. ______, *Rational surfaces with an anti-canonical cycle*, Ann. of Math. (2) **114** (1981), 267–322.

4. J.-Y. Mérindol, *Déformations des surfaces de Del Pezzo, de points rationnels et des cones sur une courbe élliptique*, Thèse 3-ème cycle, Université Paris VII, 1980.

5. H. C. Pinkham, *Simple elliptic singularities, Del Pezzo surfaces and Cremona transformations*, Proc. Sympos. Pure Math., vol. 30, Amer. Math. Soc., Providence, R. I., 1979, pp. 69–71.

6. K. Saito, *Einfach elliptsiche Singularitäten*, Invent. Math. **23** (1974), 289–325.

7. G. Timms, *The nodal cubic surfaces and the surfaces from which they are derived by projection*, Proc. Roy. Soc. Ser. A **119** (1928), 213–248.

TOKYO METROPOLITAN UNIVERSITY, JAPAN

Proceedings of Symposia in Pure Mathematics
Volume **40** (1983), Part 2

THE STRUCTURE OF QUASIHOMOGENEOUS SINGULARITIES

PHILIP WAGREICH[1]

Abstract. In this paper the close relationship between singularities with **C***-action, graded algebras of finite type and automorphy factors is discussed. It is shown how tools arising in the study of graded algebras (such as the Poincaré power series) can be used to compute invariants of singularities with **C***-action. Results of Van Dyke and Wagreich on the structure of the coordinate ring of an affine surface with **C***-action are described.

Once upon a time three blind mathematicans were led to examine a singularity. The first examined its geometry. 'I do believe this is a complex surface with **C***-action', he declared. The second examined its algebra. 'No, this is a graded algebra of finite type', she stated. The third examined it analytically and insisted that it was an automorphy factor.

The purpose of this expository paper is to describe results which show that each of the mathematicians was correct. The three types of objects mentioned, singularities with **C***-action, graded algebras of finite type and automorphy factors are closely related. In fact, under suitable restrictions the respective categories are equivalent. Thus results in any one of these categories can be applied in the other —often to great effect.

The problem of classifying all isolated singularities of complex surfaces is quite difficult. Therefore, in order to get interesting structure theorems, one restricts attention to singularities having certain properties or additional structure. A very strong assumption that can be made is that $V \subset \mathbf{C}^n$ and is a cone, i.e. V is defined by homogeneous polynomials. In this case V is determined by the corresponding projective curve $X \subset \mathbf{CP}^{n-1}$ and the line bundle induced by the restriction of the hyperplane bundle to X. Since curves and line bundles are well understood, we can completely classify cones. A larger class of singularities that appears in many

1980 *Mathematics Subject Classification.* Primary 14J17, 14L30, 16A03, 32N10.
[1]Research partially supported by grants from the National Science Foundation and the University of Illinois at Chicago Circle Research Board.

places is the class of singularities defined by quasihomogeneous polynomials. Such singularities are invariant under a natural action of the multiplicative group of complex numbers, **C***. Surfaces with **C***-action were studied in special cases by Hirzebruch [**Hi**], and Brieskorn [**Br**] and others. One of the first systematic treatments was [**O-W1**]. The topology of these surfaces is easy to understand. The essential results were already contained in Seifert's work in the thirties [**S**]. A powerful tool for studying these singularities is the observation that the category of affine varieties with **C***-action is equivalent to the category of graded algebras of finite type over **C**. Thus elegant tools arising in the study of graded algebras (such as the Poincaré power series) can be applied in the study of these singularities. Conversely, the theory of singularities can be applied to tell us about such rings. One interesting type of graded ring is a ring of automorphic forms (relative to a Fuchsian group). The theory of singularities of surfaces has been applied to get results on the structure of such rings [**Do2, Do3, Sh, W2, W3**]. Moreover, this led to the observation that almost all singularities with **C***-action can be constructed from automorphy factors. That is, for almost all affine surfaces with good **C***-action, there is a (Fuchsian) group of fractional linear transformations of the upper half-plane, Γ, and an extension of the action to $H_+ \times \mathbf{C}^*$ so that $V - \{0\} = (H_+ \times \mathbf{C}^*)/\Gamma$.

This paper is organized as follows. In the first section we define singularities with **C***-action and their geometric invariants, such as the genus of the orbit space, orbit invariants and the virtual degree. The second section is devoted to graded algebras and the calculation of the genus and virtual degree from the Poincaré power series. In §3 we define an automorphy factor and summarize some results of Dolgachev, Pinkham, Neumann, etc. In the final section we summarize some applications of the theory to the structure of quasihomogeneous singularities. The general problem is to determine the structure of the coordinate ring of a surface with **C***-action, given the topological type of the singularity. We describe results of Wagreich and VanDyke to the effect that under suitable conditions on the topological type (the virtual degree is sufficiently negative), the degrees of the generators and relations for the coordinate ring can be determined.

I would like to thank the referee for carefully reading the manuscript and making many valuable suggestions.

1. Singularities with C*-action.

(1.1) Suppose V is an algebraic variety defined over the complex numbers **C**.

DEFINITION (1.1.1). A **C***-*action* on V is an algebraic map

$$\begin{array}{ccc} \mathbf{C}^* \times V & \to & V, \\ t, v & \mapsto & t \cdot v \end{array}$$

so that

(i) $1 \cdot v = v$,

(ii) $(t_1 t_2) \cdot v = t_1 \cdot (t_2 \cdot v)$.

The action is said to be *effective* if $t \cdot v = v$ for all v implies $t = 1$.

EXAMPLE (1.1.2). Let $f(X, Y, Z) = X^{12} + Y^8 - YZ^5$ and let

$$V = \{(x, y, z) \in \mathbf{C}^3 \mid f(x, y, z) = 0\}.$$

Then V is a surface with an isolated singularity at 0 and V is invariant under the action on $\mathbf{C}^3$ defined by $t \cdot (x, y, z) = (t^{10}x, t^{15}y, t^{21}z)$.

DEFINITION (1.1.3). Assume V is affine. We say a $\mathbf{C}^*$-action is *good* if there is a point $v \in V$ which is in the closure of every orbit.

If V is an affine variety with good $\mathbf{C}^*$-action, we shall see that there is an action of $\mathbf{C}^*$ on affine space $\mathbf{C}^n$ of the form

$$t(z_1, \ldots, z_n) = (t^{q_1}z_1, \ldots, t^{q_n}z_n)$$

and an equivariant embedding of V in $\mathbf{C}^n$ with all the q_i having the same sign. Moreover, the defining equations of V in $\mathbf{C}^n$ are quasihomogeneous relative to the weights $q_1, \ldots, q_n$. The action is *effective* if and only if g.c.d.$(q_1, \ldots, q_n) = 1$.

(1.2) *Geometric invariants*: *the orbit space*. Henceforth V is an affine variety with $\mathbf{C}^*$-action. Let V_0 denote $V - \{0\}$. Then $\mathbf{C}^*$ acts on V_0 without fixed points. We shall see that if the action is good, the orbit space $X = V_0/\mathbf{C}^*$ is a projective variety and the only singularities of X are cyclic quotient singularities, i.e., singularities locally of the form $\mathbf{C}^{d-1}/\mu_\alpha$ where $d =$ dimension of V and μ_α is a finite cyclic group of order α acting linearly on $\mathbf{C}^{d-1}$. In particular, if $\dim_{\mathbf{C}} V = 2$ then X is a nonsingular curve.

Now let $S^1 = \{z \in \mathbf{C}^* \mid |z| = 1\}$. This is a closed topological subgroup of $\mathbf{C}^*$ and, in fact, $\mathbf{C}^* = S^1 \times \mathbf{R}^*_+$ where $\mathbf{R}^*_+$ denotes the multiplicative group of positive real numbers. The link of $0 \in V$ is defined to be

$$K = V \cap S_\varepsilon^{2n-1}$$

where S_ε^{2n-1} is the sphere of radius ε around 0. Now K is independent of ε (up to diffeomorphism) so we may choose $\varepsilon = 1$. The action of S^1 on V leaves K invariant (even when the action is not good). For a good action one can easily prove that V_0 is equivariantly diffeomorphic to $K \times \mathbf{R}^*_+$ under the map

$$\begin{aligned} \psi: K \times \mathbf{R}^*_+ &\to V_0, \\ (k, t) &\mapsto t \cdot k. \end{aligned}$$

The action of $\mathbf{C}^* = S^1 \times \mathbf{R}^*_+$ on $K \times \mathbf{R}^*_+$ is given by S^1 on the first factor and $\mathbf{R}^*_+$ on the second, hence $K_0 = V_0/\mathbf{R}^*_+$ for any good action.

EXAMPLE (1.2.1). Let $V = \mathbf{C}^2$ and $t \cdot (x, y) = (tx, ty)$. The orbits in V_0 are straight lines through 0 and $X = P^1_{\mathbf{C}}$.

(1.2.2) Let $V = \mathbf{C}^2$ and $t \cdot (x, y) = (tx, t^{-1}y)$. The orbits are the curves $xy = c$, $c \neq 0$, and the coordinate axes. X is not Hausdorff (the coordinate axes cannot be separated). Note that S^1 acts freely on $K = S^3$; hence K is an S^1 bundle over K/S^1. One can easily see that K/S^1 must be S^2 (since it is a simply connected 2-manifold). Thus K/S^1 is not homeomorphic to $V_0/\mathbf{C}^*$ in this case.

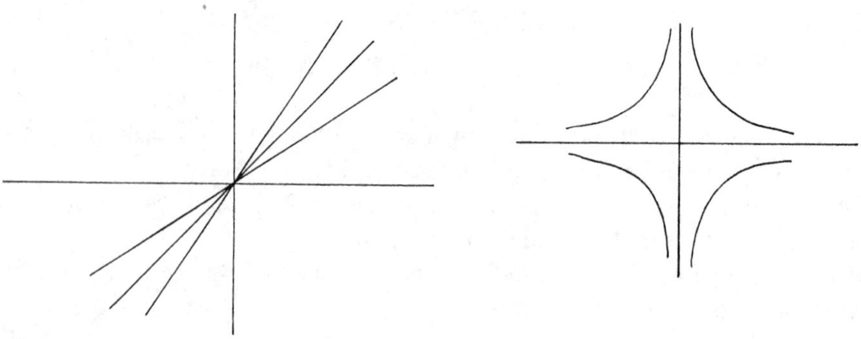

FIGURE 1

and $X = P^1_{\mathbf{C}}$.

(1.2.3) Let $V = \mathbf{C}^2$ and $t \cdot (x, y) = (tx, y)$. Points on the y axis are fixed. The y axis must be removed to get a Hausdorff orbit space.

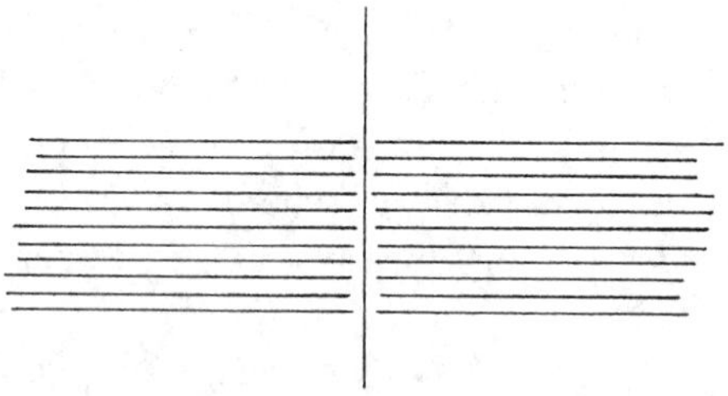

(1.2.4) Let $V = \mathbf{C}^2$ and $t \cdot (x, y) = (t^p x, t^q y)$, where $p, q > 0$. The orbits in V_0 are the coordinate axes and cuspidal curves $x^q = cy^p$, $c \neq 0$. Using the fact that X is an algebraic curve and $\pi_1(X) = 0$ we see that X is $\mathbf{C}P^1$.

In the case that V is a surface with good $\mathbf{C}^*$-action and isolated singularity at 0, X is a nonsingular projective curve. One can compute the genus of X geometrically, as was done in **[O-W1]**. In this paper we shall compute the genus using an algebraic method (2.2.4).

(1.3) *Geometric invariants*: *orbit invariants.* Assume V has a good $\mathbf{C}^*$-action. The orbit map $\pi : V_0 \to X$ has fibers homeomorphic to $\mathbf{C}^*$, however it is not a locally trivial bundle. The local structure of π is determined by certain numerical orbit invariants, as follows.

THE SLICE THEOREM (1.3.1) **[Ho]**. *Suppose G is a complex Lie group of dimension n and V_0 is an analytic d-manifold with a proper G action* (*this is the case above*). *If $v \in V_0$ and $G(v)$ is the orbit of v then there exist*

(i) *a representation σ of the isotropy group G_v in* $\mathrm{GL}(d - n, \mathbf{C})$,

(ii) *a neighborhood U of* 0 *in $\mathbf{C}^{d-n}$ invariant under σ and*

(iii) *an invariant tubular neighborhood W of $G(v)$ such that W is equivariantly biholomorphic to $(G \times U)/G_v$ where G_v acts on $G \times U$ via $g \cdot (g_1, u) = (g_1 g^{-1}, \sigma(g)u)$.*

This is an analogous theorem for actions of compact transformation groups (see for instance [**B**]).

In the case $G = \mathbf{C}^*$ and $d = 2$ we have a one-dimensional representation, so the geometry is particularly simple. The isotropy group is a closed proper subgroup of $\mathbf{C}^*$, hence finite, cyclic. Let $G_v = \mu_\alpha$, where μ_α denotes the group of αth roots of unity. If $\xi \in \mu_\alpha$ then the slice representation must be of the form $\sigma(\xi)(z) = \xi^\nu z$ for some ν, $0 < \nu < \alpha$. Thus, we can attach two integers α and ν to the orbit which determine the local structure in a neighborhood of the orbit. The effectiveness of the action implies g.c.d.$(\alpha, \nu) = 1$. It will turn out that the inverse of $-\nu$ modulo α is a nicer invariant than ν and hence we make the following definition.

DEFINITION (1.3.2). The orbit invariants of the orbit $G(v)$ are the integers (α, β) where β is defined by $\nu\beta \equiv -1$, $0 < \beta < \alpha$.

It should be noted that this is the definition of orbit invariants used in [**O-W1, 2, 3**]. If one chooses the opposite orientation, as is done in [**Ne** and **Or**], then one gets orbit invariants $(\bar{\alpha}, \bar{\beta})$, where $\bar{\alpha} = \alpha$ and $\bar{\beta} = \alpha - \beta$.

The orbits with nontrivial isotropy form a set of codimension $\geqslant 1$. These orbits are called *exceptional.*

EXAMPLE (1.3.3). We return to Example (1.1.2). Recall $V = \{x^{12} + y^8 - yz^5 = 0\} \subset \mathbf{C}^3$ with the action $t \cdot (x, y, z) = (t^{10}x, t^{15}y, t^{21}z)$. The exceptional orbits and their isotropy groups are

orbit	defining equations	isotropy group
O_1	$X = Y^7 - Z^5 = 0$	μ_3
O_2	$X = Y = 0$	μ_{21}
$O_{3+i}, 0 \leqslant i \leqslant 3$	$Z = X^3 - \eta_i Y^2 = 0$	μ_5

where the η_i are the primitive fourth roots of 1.

The slice representation can be read off the action of μ_α on the tangent line to a slice.

O_1: Let $p = (0, 1, 1) \in O_1$. Then the equation of the tangent line to a slice is $y = z = 0$. Then the slice representation is $\xi(x, 0, 0) = (\xi^{10}x, 0, 0)$, $\xi \in \mu_3$. Therefore $\nu_1 \equiv 10 \equiv 1 \pmod 3$.

O_2: Let $p = (0, 0, 1) \in O_2$. The tangent plane to V at p is $y = 0$. The tangent line to the orbit is $x = y = 0$. Then the slice representation is $\xi(x, 0, 0) = (\xi^{10}, 0, 0)$, $\xi \in \mu_{21}$. Therefore $\nu_2 \equiv 10 \pmod{21}$.

Finally, it is not hard to see that the orbits $O_3, \ldots, O_6$ all have the same orbit invariants. Let $p \in O_3$. Then the tangent line to a slice is given by $x = y = 0$. Hence the slice representation is $\xi(0, 0, z) = (0, 0, \xi^{21}z)$. Thus $\nu \equiv 21 \equiv 1 \pmod 5$. If we let (α_i, β_i) be the orbit invariants of O_i then we have

$$(\alpha_1, \beta_1) = (3, 2), \quad (\alpha_2, \beta_2) = (21, 2), \quad (\alpha_i, \beta_i) = (5, 4), \qquad i = 3, \ldots, 6.$$

COROLLARY OF THE SLICE THEOREM. *X has only quotient singularities. In particular, if $d = 2$ then X is a nonsingular curve.*

PROOF. The question is local so we may assume $V_0 = (\mathbf{C}^* \times U)/G_v$. Then

$$V_0/\mathbf{C}^* = (\mathbf{C}^* \times U)/G_v/\mathbf{C}^* = U/G_v.$$

(1.4) *Geometric invariants: the virtual degree (Euler class).* Assume for simplicity that $\dim V = 2$. If π: $V_0 \to X$ is a bundle it is classified topologically by its degree (Chern class, Euler number) in $H^2(X, \mathbf{Z}) \approx \mathbf{Z}$. One can define a rational number which generalizes this notion to arbitrary π. Its significance appears to have first been observed by Dolgachev (letter to Wagreich, April 1976), Bailey **[Ba]** and Neumann and Raymond **[N-R]** (see also **[Pi]**). To define this invariant we shall first need the following proposition.

PROPOSITION (1.4.1) (DOLGACHEV **[Do3]**, PINKHAM **[Pi]**). *There exists a finite Galois branched cover $Y \to X$ with covering group H, an honest $\mathbf{C}^*$-bundle $L \to Y$ and an action of H on L (commuting with the action of $\mathbf{C}^*$) so that $L/H \approx V_0$ and the following diagram commutes:*

$$\begin{array}{ccc} L & \overset{H}{\to} & V_0 \\ \downarrow & & \downarrow \\ Y & \overset{H}{\to} & X \end{array}$$

DEFINITION AND PROPOSITION (1.4.2). *virtual degree π = degree σ/order$(H) \in$* $\mathbf{Q}$.

The fact that this is independent of the choice of Y follows easily from the fact that for any two such Y there exists a third which covers them both.

SKETCH OF THE PROOF OF (1.4.1). We first claim that the proposition is true locally over X. In that case we can assume $V_0 = (\mathbf{C}^* \times U)/\mu_\alpha$ where U is a disk in $\mathbf{C}$ and U/μ_α is also a disk. Now let $Y = U$ and $Y \to X = U/\mu_\alpha$ be the quotient map. If we choose local coordinates in Y and X this is just the map $z \to z^\alpha$. The normalization of the pull-back of V_0 to Y is merely $\mathbf{C}^* \times U$, which is trivial over Y.

One proves the proposition globally by constructing a branched cover $Y \to X$ which is branched at the points $x_1, \ldots, x_n \in X$ corresponding to the exceptional orbits, and has order of ramification α_i at x_i. Such a branched cover exists by **[B-N, Fo]**.

Seifert defined a generalization of the Chern class in a different way. His invariant, denoted by b, was defined as an obstruction to extending a certain partial section of π. One can recover b as follows.

DEFINITION AND THEOREM (1.4.3) **[O-W1, N-R, Do3]**. *b = –virtual degree $\pi + \Sigma_{i=1}^r \beta_i/\alpha_i$ is an integer.*

As mentioned in (1.3.2), some authors use the opposite orientation. In this case the corresponding invariant is defined by

$$\bar{b} = -\text{virtual degree } \pi - \sum_{i=1}^{r} \bar{\beta}_i/\bar{\alpha}_i = b + r.$$

EXAMPLE (1.4.4). Returning to Example (1.1.2), let $\phi\colon \mathbf{C}^3 \to \mathbf{C}^3$ be defined by $\phi(x, y, z) = (x^{10}, y^{15}, z^{21})$. Let $W = \phi^{-1}(V)$. Then W is defined by

$$x^{120} + y^{120} - y^{15}z^{105} = 0,$$

a homogeneous polynomial of degree 120. Let $Z = W - \{0\}/\mathbf{C}^*$ (where the action is the diagonal action $t\cdot(x, y, z) = (tx, ty, tz)$). Then we have an induced map $\phi\colon Z \to X$ and W_0 is the pull-back of V_0 to Z. Now Z has a singular point. We let Y be the normalization of Z and let L be the normalization of the pull-back of W_0 to Y. Then L is a $\mathbf{C}^*$-bundle over Y of degree -120. On the other hand, degree of the map $Y \to X$ is $10\cdot 15\cdot 21$. Therefore vdegree$(V_0 \to X)$ = $-120/(10\cdot 15\cdot 21) = -4/105$. Now

$$b = 4/105 + 2/3 + 2/21 + 4\cdot 4/5 = 4.$$

(1.5) TOPOLOGICAL CLASSIFICATION THEOREM (SEIFERT [S]). *Let g = genus X. The invariants $\{g; b; (\alpha_1, \beta_1),\ldots,(\alpha_r, \beta_r)\}$ determine the link K up to orientation preserving equivariant homeomorphism. The set of invariants is called the numerical invariants of K.*

2. Graded algebras.

(2.1) We shall see, in this section, that positively graded algebras of dimension n are equivalent to affine varieties of dimension n with $\mathbf{C}^*$-action. When $V = 2$ this can be used to efficiently compute the invariants b and g of a surface with $\mathbf{C}^*$-action. Conversely, the classification of isolated surface singularities with $\mathbf{C}^*$-action gives us a classification theorem for the corresponding graded rings.

DEFINITION (2.1.1). A *graded ring* R is a ring together with a direct sum decomposition $R = \bigoplus_{i=-\infty}^{\infty} R_i$ so that if $x \in R_i$ and $y \in R_j$ then $xy \in R_{i+j}$. Suppose F is a subring of R_0. Then the R_i are F submodules of R and we say that R is a graded F-algebra.

In what follows we will always assume that $F = R_0 = \mathbf{C}$. The results described in this section can be easily generalized to varieties with G_m-action over an arbitrary algebraically closed field F.

EXAMPLE (2.1.2). Let $R = F[X_1,\ldots,X_n]$. We can define many gradings on R. Choose $q_1,\ldots,q_n \in \mathbf{Z}$. Define degree $X_i = q_i$ for all i and let R_i be the F-subspace of R spanned by the monomials of degree i. Equivalently

$$R_i = \{f \in R \mid f(t^{q_1}z_1,\ldots,t^{q_n}z_n) = t^i f(z_1,\ldots,z_n)\}$$

where t is an indeterminate. An element $f \in R_i$ as above is called (quasi)*homogeneous of degree i relative to the weights* $q_1,\ldots,q_n$. A grading as above is called a *diagonal grading* of R.

DEFINITION (2.1.3). An F-algebra R is said to be of finite type if there exist $x_1,\ldots,x_r \in R$ so that every element of R is a polynomial in the x_i with coefficients in F. Equivalently, R is isomorphic to $F[X_1,\ldots,X_r]/I$ where the grading on the polynomial ring is a diagonal grading, and I is generated by homogeneous polynomials.

DEFINITION (2.1.4). A homomorphism $f\colon M \to N$ of graded modules is said to be of degree d if $f(M_i) \subset N_{i+d}$ for all i.

PROPOSITION (2.1.5). *There is a (contravariant) equivalence of categories between*

$$\left(\left(\begin{matrix}\textit{affine varieties with } \mathbf{C}^*\textit{-action}\\ \textit{equivariant algebraic morphisms}\end{matrix}\right)\right)$$

$$\Psi \downarrow \qquad \uparrow \Phi$$

$$\left(\left(\begin{matrix}\textit{graded integral domains of finite type over } \mathbf{C}\\ \textit{ring homomorphisms of degree } 0\end{matrix}\right)\right)$$

defined as follows:

If A is a graded algebra of finite type and $A = \mathbf{C}[X_1,\ldots,X_n]/I$ where I is a homogeneous ideal as in (2.1.3) *then $\Phi(A) = Z(I)$, the zero set of I in $\mathbf{C}^n$ and the action of $\mathbf{C}^*$ on $Z(I)$ is given by*

$$t(x_1,\ldots,x_n) = (t^{q_1}x_1,\ldots,t^{q_n}x_n)$$

(where $q_i =$ degree X_i in A).

If V is an affine variety with $\mathbf{C}^*$-action let $\Psi(V) = \mathbf{C}[V]$, the ring of polynomial functions on V, and define a grading by

$$\mathbf{C}[V]_i = \{f \in \mathbf{C}[V] \mid f(t\cdot v) = t^i f(v) \textit{ for all } t \in \mathbf{C}^*\}.$$

SKETCH OF PROOF. Once the correspondence above is shown to be well defined, the fact that it is an equivalence of categories follows immediately from the usual proof that the categories of affine algebraic varieties and integral $\mathbf{C}$-algebras of finite type are equivalent categories (see **[Ha]**, for example). The fact that the action defined above leaves $Z(I)$ invariant follows immediately from the fact that I is generated by homogeneous polynomials. It only remains to show that

$$\mathbf{C}[V] = \bigoplus_{i=-\infty}^{\infty} \mathbf{C}[V]_i.$$

Clearly the constant functions $\mathbf{C} \subset \mathbf{C}[V]_0$, and $\mathbf{C}[V]_i \cap \mathbf{C}[V]_j = \{0\}$ for $i \neq j$. Now the action is equivalent to an algebraic map $\sigma : \mathbf{C}^* \times V \to V$. This gives us a map $\sigma^* : \mathbf{C}[V] \to \mathbf{C}[t, t^{-1}] \otimes \mathbf{C}[V]$. If $f \in \mathbf{C}[V]$ then $\sigma^*(f) = \Sigma_{i=-\infty}^{\infty} f_i t^i$ for some $f_i \in \mathbf{C}[V]$ (note this is a finite sum). The fact that $f = \Sigma_{i=-\infty}^{\infty} f_i$ follows from the commutativity of the diagram

$$\begin{array}{ccc} V & \xrightarrow{i} & \mathbf{C}^* \times V \\ & \searrow \quad \swarrow & \\ & V & \end{array}$$

where $i(v) = (1, v)$. The fact that $(t_1t_2)\cdot v = t_1 \cdot (t_2 \cdot v)$ translates into a commutative diagram

$$\begin{array}{ccc} \mathbf{C}^* \times \mathbf{C}^* \times V & \to & \mathbf{C}^* \times V \\ \downarrow & & \downarrow \\ \mathbf{C}^* \times V & \to & V \end{array}$$

which gives rise to a corresponding commutative diagram of rings. This last commutative diagram easily implies that $f_i \in \mathbf{C}[V]_i$, for all i, which shows that the definition we gave does indeed give a grading on $\mathbf{C}[V]$.

DEFINITION (2.1.6). We say a graded ring A is *positively graded* if $A_i = 0$, for all $i < 0$.

It is not hard to see that under the correspondence of (2.1.5) positively graded algebras correspond to varieties with good $\mathbf{C}^*$-action.

EXAMPLE (2.1.7). Let V be the surface in Example (1.1.2). Then the coordinate ring of V is

$$A = \mathbf{C}[X, Y, Z]/(X^{12} + Y^8 - YZ^5)$$

and A is graded by letting degree $X = 10$, degree $Y = 15$ and degree $Z = 21$. The grading on A is well defined since $f(X, Y, Z) = X^{12} + Y^8 - YZ^5$ is homogeneous (relative to this grading) of degree 120.

(2.2) *Calculating the genus and virtual degree*: *The Poincaré power series*. In this section we show how to define the arithmetic genus and virtual degree in the category of graded rings. These definitions have two virtues. They are valid for graded algebras of finite type over an arbitrary field F and can be effectively computed. When $F = \mathbf{C}$ the definitions agree with the corresponding geometric notion.

Suppose A is a positively graded algebra of finite type over field F.

DEFINITION (2.2.1). Let $a_i = \dim_F A_i$ and define $P_A(t) = \Sigma_{i=0}^{\infty} a_i t^i$ as the *Poincaré power series* of A.

The Poincaré power series is a rational function and the order of its pole at $t = 1$ is equal to the dimension of A (see Atiyah and MacDonald [**A-M**, Chapter 11]). If A is generated by forms of degree 1 then the function $f(n) = a_n$ is a polynomial function for $n \gg 0$. This polynomial is called the Hilbert polynomial of A. The fact that $P_A(t)$ is a rational function is a generalization of the fact that f is a polynomial for $n \gg 0$.

We summarize a few facts about the Poincaré power series (for proofs see [**A-M**]).

PROPOSITION (2.2.2). (1) *If* $0 \to M' \xrightarrow{f} M \xrightarrow{g} M'' \to 0$ *is an exact sequence of graded A-modules of finite type and if f and g are degree* 0 *homomorphisms then* $P_{M'}(t) - P_M(t) + P_{M''}(t) = 0$.

(2) *If* $C = A \otimes B$ *then* $P_C(t) = P_A(t) \cdot P_B(t)$.

(3) *If* $A = \mathbf{C}[X_1, \ldots, X_n]$ *with degree* $X_i = q_i$ *then*

$$P_A(t) = 1/(1 - t^{q_1}) \cdots (1 - t^{q_n}).$$

(4) *If A is as in* (3) *and I is an ideal* $I = (f_1, \ldots, f_r)$ *generated by an A-sequence* (*i.e.*, $V(I)$ *is a complete intersection*) *and degree* $f_i = d_i$ *then*

$$P_A(t) = (1 - t^{d_1}) \cdots (1 - d^{d_r})/(1 - t^{q_1}) \cdots (1 - t^{q_n}).$$

SKETCH OF PROOF. (1) and (2) are easy. For (3), suppose $n = 1$ and let $q = q_1$. Then

$$P_A = 1 + t^q + t^{2q} + \cdots = 1/(1 - t^q).$$

The general case follows from (2) by induction. For (4) it is sufficient to show that if B is a graded algebra of finite type, f is a nonzero divisor of degree d and $C = B/f$ then $P_C(t) = (1 - t^d)P_B(t)$. This follows from the exact sequence

$$0 \to B \xrightarrow{\phi} B \to C \to 0$$

where $\phi(b) = f \cdot b$. Noting that ϕ is a homomorphism of degree d, this gives us $t^d p_B(t) - p_B(t) + p_C(t) = 0$ and the desired result follows.

THEOREM (2.2.3). *Let A be a graded algebra of finite type generated by forms of degree $q_1, \ldots, q_n$ and $X = \operatorname{Proj}(A)$ the corresponding projective variety. (Recall that if $F = \mathbf{C}$ and $V = \operatorname{Spec}(A)$ then $X = V_0/\mathbf{C}^*$.) Let q be a common multiple of $q_1, \ldots, q_n$. Define $f(i) = \dim A_{qi}$. Then there exists a polynomial g so that $g(i) = f(i)$ for $i \gg 0$. The constant term of g is $\chi(X, O_X)$. If X is a nonsingular curve this is just $1 - \operatorname{genus}(X)$.*

PROOF. Let $A^{(q)} = A_0 \oplus A_q \oplus A_{2q} \oplus \cdots$. Then $A^{(q)}$ is generated by A_q and $\operatorname{Proj}(A) = \operatorname{Proj}(A^{(q)})$ [**EGA**, II]. Now we grade $A^{(q)}$ so that $A_i^{(q)} = A_{qi}$ and so that $A^{(q)}$ is generated by forms of degree 1. But then $f(i)$ is just the Hilbert function of $A^{(q)}$ and g is the Hilbert polynomial. But Serre [**Se**] showed that $g(i) = \chi(X, O_X(i))$ and hence $g(0) = \chi(X, O_X)$.

COROLLARY (2.2.4) [**O-W3**]. *Suppose $V \subset \mathbf{C}^3$ is a surface with an isolated singularity and V is defined by a polynomial of degree d relative to the grading with degree $Z_i = q_i$. If $\text{g.c.d.}(q_0, q_1, q_2) = 1$ then the genus of the algebraic curve $X = V_0/\mathbf{C}^*$ is given by*

$$g = 1/2\big[d^2/q_0q_1q_2 - d(q_0, q_1)/q_0q_1 - d(q_1, q_2)/q_1q_2 - d(q_0, q_2)/q_0q_2 + (d, q_0)/q_0 + (d, q_1)/q_1 + (d, q_2)/q_2 - 1\big].$$

DEFINITION (2.2.5). Suppose A is a positively graded algebra of finite type, the dimension of A is n, and A is generated by elements of degree $q_0, q_1, \ldots, q_r$. Suppose that $\text{g.c.d.}(q_0, q_1, \ldots, q_r) = 1$. (This is the case when the corresponding action is effective.) Then the *virtual degree of A* is defined to be the coefficient of $1/(1 - t)^n$ in the Laurent expansion of $p_A(t)$ at $t = 1$.

THEOREM (2.2.6) [**Do3**]. *If V is a surface with good $\mathbf{C}^*$-action and A is the coordinate ring of V then the virtual degree of $\pi: V_0 \to X$ is equal to –virtual degree A.*

COROLLARY (2.2.7). *If V is as in (2.2.4) then the virtual degree of $\pi: V_0 \to X$ is $-d/(q_0q_1q_2)$.*

EXAMPLE (2.2.8). If $f(X, Y, Z) = X^{12} + Y^8 - YZ^5$ and $A = \mathbf{C}[X, Y, Z]/(f)$ then genus $X = 0$ and virtual degree $A = 4/105$ (c.f. (1.4.4)).

(2.3) *Equivariant resolution*. Every surface with an isolated singularity and good $\mathbf{C}^*$-action has a canonical equivariant resolution. The graph of the resolution can be determined from the Seifert invariants.

THEOREM (2.3.1) (ORLIK-WAGREICH [**O-W1, O-W3**]). *If V is an affine surface with an isolated singularity and a good $\mathbf{C}^*$-action then there is a unique equivariant resolution $\pi : \tilde{V} \to V$ so that:*

(1) $\pi^{-1}(0) = \bigcup_{i=0}^{s} X_i$, *where exactly one of the X_i, say X_0, is fixed (pointwise) by the action.*

(2) $(X_i \cdot X_i) \leqslant -2$ *for* $i > 1$.

(3) *The X_i meet transversely.*

(4) *For $i \geqslant 1$, X_i is a nonsingular rational curve.*

(5) *The graph of the resolution is star-shaped, with the center vertex corresponding to the fixed curve.*

(6) *X_0 is isomorphic to $V_0/\mathbf{C}^*$.*

(7) *The 'arms' of the graph are in one-one correspondence with the exceptional orbits of V_0. The self-intersections of the curves on an arm are determined by*

$$\alpha_i/\beta_i = n_{i_1} - 1/\left(n_{i_2} - 1/\left(\cdots - n_{i_{ri}}\right)\cdots\right)$$

where (α_i, β_i) are the orbit invariants of the corresponding orbit.

(8) $b = -$*virtual degree*$(V_0 \to X) + \Sigma(\beta_i/\alpha_i)$.

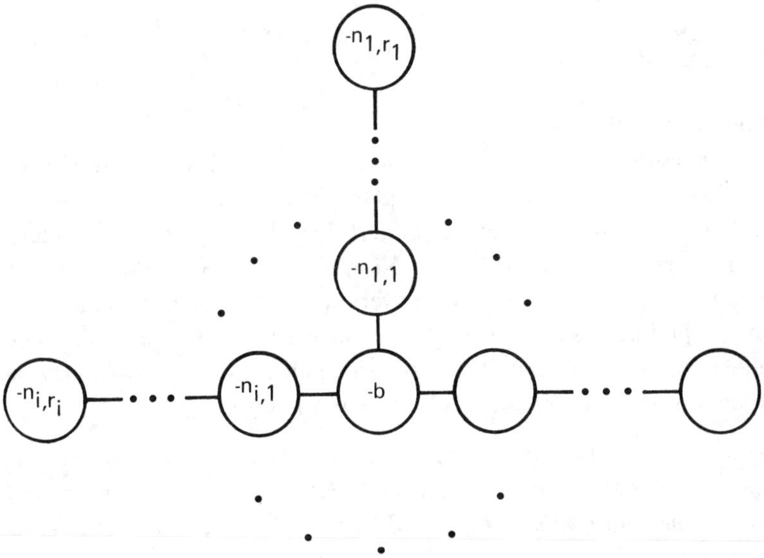

EXAMPLE (2.3.2). If $V \subset \mathbf{C}^3$ is defined by the polynomial $f(X, Y, Z) = X^{12} + Y^8 - YZ^5$ then we have computed all invariants of V and the graph of the resolution is shown in the diagram.

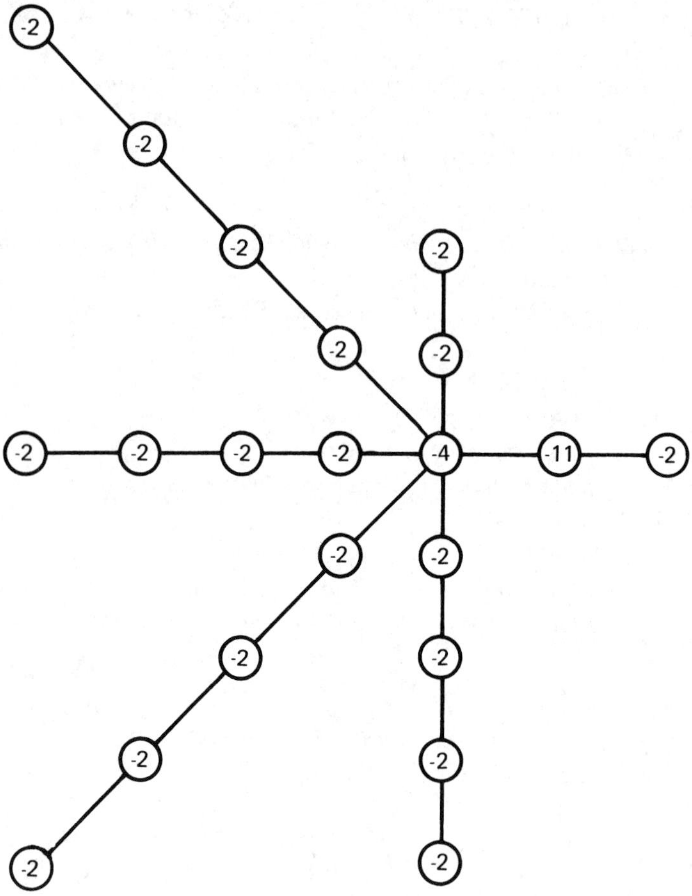

3. Automorphic forms.

(3.1) The results in this section are due to Conner and Raymond [**C-R3**], Carrell [**Ca**], Dolgachev [**Do2, Do3**], Milnor [**Mi**], Neumann [**Ne**] and Pinkham [**Pi**]. Recall that in (1.4) we pulled a Seifert bundle $V_0 \to X$ back to an honest $\mathbf{C}^*$-bundle over a finite cover Y of X. Now there is no reason to stop at Y, so we can "pull back" to the universal cover of Y. The operation is not, strictly speaking, a pull-back since we must normalize after pulling back. This construction allows one to show that all surface singularities with good $\mathbf{C}^*$-action can be constructed from generalized rings of automorphic forms.

THEOREM (3.1.1). *If V is a complex surface with isolated singularity and good $\mathbf{C}^*$-action there exist a simply connected Riemann Surface U, a line bundle L on U, a discrete properly discontinuous group of automorphisms Γ of U and an extension of Γ to L so that*:

(1) *The action of Γ on L commutes with the action of* $\mathbf{C}^*$.
(2) $V_0 = L/\Gamma$.
(3) $X = U/\Gamma$.

(4) *The following diagram commutes*:

$$\begin{array}{ccc} L & \xrightarrow{\Gamma} & V_0 \\ \downarrow & & \downarrow \\ U & \xrightarrow{\Gamma} & X \end{array}$$

DEFINITION (3.1.2). The triple (L, U, Γ) is called an automorphy factor.

The theorem above was initially proven by Conner and Raymond [**C-R1**] for a compact smooth manifold M with injective S^1 action, i.e., an action for which the map $\pi_1(S^1) \to \pi_1(M)$ (induced by the map of S^1 into a general orbit) is injective. They later extended their results to the holomorphic case [**C-R3**].

(3.2) Suppose the numerical orbit invariants of V are

$$\{g; b; (\alpha_1, \beta_1), \ldots, (\alpha_r, \beta_r)\}.$$

The surface singularities with $\mathbf{C}^*$-action fall into three classes.

(1) $\pi_1(V_0)$ finite. This holds if and only if $g = 0$ and $\Sigma_{i=1}^r 1/\alpha_i > r - 2$. The condition on the α_i holds if and only if $r \leqslant 2$ or $r = 3$ and $(\alpha_1, \alpha_2, \alpha_3)$ is one of the Platonic triples (up to a permutation of the subscripts)

$$(2, 2, n), \quad (2, 3, 3), \quad (2, 3, 4), \quad (2, 3, 5).$$

(2) $\pi_1(V_0)$ is infinite nilpotent. This holds only if either

(i) $g = 1$ and $r = 0$,

(ii) $g = 0$ and $\Sigma_{i=1}^r 1/\alpha_i = r - 2$. The condition on the α_i holds only for

$$(3, 3, 3), \quad (2, 3, 6), \quad (2, 4, 4), \quad (2, 2, 2, 2).$$

(3) $\pi(V_0)$ is infinite and not nilpotent.

Returning to the discussion above, in case (1) we can take $U = \mathbf{P}^1$; in case (2) we can take $U = \mathbf{C}$ and in case (3) $U = H_+$, the upper half-plane. The singularities covered by case (1) were studied by Brieskorn [**Br**].

(3.3) We shall now assume we are in case (3), i.e., $U = H_+$. Any $\mathbf{C}^*$-bundle L on H_+ is trivial so we can write $L = H_+ \times \mathbf{C}^*$.

There is a very natural automorphy factor which arises in the study of automorphic forms.

DEFINITION (3.3.1). Suppose Γ is a discrete, discontinuous group of automorphisms of H_+. The canonical automorphy factor is $(H_+ \times \mathbf{C}^*, H_+, \Gamma)$ where Γ acts on $H_+ \times \mathbf{C}^*$ via $\gamma(z, t) = (\gamma(z), d\gamma/dz \cdot t)$.

DEFINITION (3.3.2). If Γ is as above, a holomorphic function on H_+ is said to be an (entire) *automorphic form* of weight k if $f(\gamma(z)) = (d\gamma/dz)^{-k} f(z)$ for all $z \in H_+, \gamma \in \Gamma$.

If Γ is a finitely generated Fuchsian group of the first kind (i.e., H_+/Γ has finite volume and only a finite number of punctures) then A_k is a finite-dimensional vector space and we let $A = \bigoplus_{k=0}^{\infty} A_k$. This has a natural graded ring structure since $f \in A_k$ and $f' \in A_j$ implies $ff' \in A_{j+k}$.

DEFINITION (3.3.3). A, as above, is called the (graded) *algebra of automorphic forms relative to* Γ.

PROPOSITION (3.3.4) (CF. [**Do2, Do3**]). *If* H_+/Γ *is compact then* A *is a graded algebra of finite type. Let* $0 \in \operatorname{Spec}(A)$ *be the point corresponding to the maximal ideal* $m = \bigoplus_{k=1}^{\infty} A_k$. *Then* $(H_+ \times \mathbf{C}^*)/\Gamma = \operatorname{Spec}(A) - \{0\}$.

REMARKS. (1) If H_+/Γ is finite volume then A is still finitely generated. In this case $\operatorname{Spec}(A) - \{0\}$ is a 'partial compactification' of $(H_+ \times \mathbf{C}^*)/\Gamma$ [**Wa3**].

(2) Dolgachev [**Do3**] has generalized the proposition to higher dimensions assuming (among other things) the quotient U/Γ is compact.

(3) Dolgachev [**Do3**], Milnor [**Mi**] and Neumann [**Ne**] have studied automorphic forms with fractional weights and have shown that rings of complete intersections of Brieskorn polynomials arise naturally as rings of automorphic functions with fractional weights.

4. Applications to structure theorems.

(4.1) The ideas of §3 can be used to prove a structure theorem for normal positively graded rings of dimension 2. Suppose X is an algebraic curve. A divisor D with rational coefficients is a finite formal sum $D = \sum_{i=1}^{r} a_i p_i$ where the a_i are rational numbers and $p_i \in X$. Now, generalizing the usual definition, one can write $L(D) = \{f \mid (f) \geqslant -D\}$ where f is a rational function on X and (f) denotes the divisor of f.

THEOREM (4.1.1). *Suppose* V *is a normal surface with good* $\mathbf{C}^*$*-action. Then there exists a divisor with rational coefficients* D *on* X *so that*

$$\mathbf{C}[V] = \bigoplus_{k=0}^{\infty} L(kD).$$

REMARKS (4.2). (1) This theorem has been generalized by Demazure to normal varieties with good $\mathbf{C}^*$-action [**De**]. Thus all normal positively graded algebras of finite type can be written in the form above. The theorem was proven for surfaces by Dolgachev [**Do3**] and Pinkham [**Pi**]. A special case was proven earlier by Gunning [**Gu**, Chapter II, Theorem 1].

(2) We can write D uniquely in the form

$$D = D_0 + \sum_{i=1}^{r} (f_i/e_i) p_i$$

where D_0 has integer coefficients, $p_i \in X$, e_i, f_i are positive integers, g.c.d.$(e_i, f_i) = 1$ and $0 < f_i < e_i$, for all i. Then the p_i are the points corresponding to orbits with nontrivial isotropy. The orbit corresponding to p_i has orbit invariants $(\alpha_i, \beta_i) = (e_i, e_i - f_i)$ (notation of (1.3)). Thus

$$D = (D_0 + p_1 + \cdots + p_r) - \sum (\beta_i/\alpha_i) p_i$$

and

$$\text{Degree } D = -\text{virtual degree } \pi = b - \sum (\beta_i/\alpha_i).$$

(3) $D_0 + p_1 + \cdots + p_r$ is the divisor of the conormal bundle to X in V (see [**Pi**]).

(4) If V arises from the canonical automorphy factor then $D_0 = K$, where K is a canonical divisor on X and $\beta_i = 1$, for all i [**Gu, Do3**].

EXAMPLE (4.3). Let $X = \mathbf{P}^1$, $p_0 = 0$, $p_1 = 1$, $p_2 = \infty \in \mathbf{P}^1$. Let $D = -p_0 + \frac{2}{3}p_1 + \frac{2}{3}p_2$. Then if t is the coordinate in $\mathbf{P}^1$ we have

$$L(D) = \{0\}, \quad L(2D) = \langle X \rangle, \quad L(3D) = \langle Y, Z \rangle,$$

where $X = t^2/(t-1)$, $Y = t^3/(t-1)$ and $Z = t^3/(t-1)^2$. Now one can easily see that $A = \bigoplus_{n=0}^{\infty} L(nD)$ is generated by X, Y and Z. Moreover there is a relation $X^3 = YZ$ of degree 6. Hence A is isomorphic to $\mathbf{C}[X, Y, Z]/(X^3 - YZ)$. If, in this example, we choose the p_i to be arbitrary points on $\mathbf{P}^1$ the resulting ring is isomorphic to the above, since we can find an automorphism of $\mathbf{P}^1$ mapping any three points to 0, 1 and ∞.

(4.4) COROLLARY TO THEOREM (4.1.1). *The analytic Seifert invariants*

$$\{X;\ D_0;\ (p_1, \alpha_1, \beta_1), (p_2, \alpha_2, \beta_2), \ldots, (p_r, \alpha_r, \beta_r)\}$$

are a complete set of analytic invariants for V. To be precise, suppose V' is a second surface with invariants

$$\{X',\ D_0';\ (p_1', \alpha_1', \beta_1'), \ldots, (p_{r'}', \alpha_{r'}', \beta_{r'}')\}.$$

Then there is a $\mathbf{C}^*$ equivariant isomorphism $\phi: V \to V'$ if and only if $r = r'$ and there exists an isomorphism $\psi: X \to X'$ and a permutation σ so that:

(i) *$\psi^*(D_0')$ is linearly equivalent to D_0.*

(ii) *$\psi(p_i) = p'_{\sigma(i)}$.*

(iii) *$(\alpha_i, \beta_i) = (\alpha'_{\sigma(i)}, \beta'_{\sigma(i)})$.*

DEFINITION (4.4.1). We call $\{X;\ D_0;\ (p_1, \alpha_1, \beta_1), \ldots, (p_r, \alpha_r, \beta_r)\}$ the *analytic Seifert invariants* of V.

One can ask whether it is possible to determine the structure of A from the analytic or numerical Seifert invariants. In particular one can ask for

(1) generators,

(2) relations,

(3) syzygies.

The analytic Seifert invariants have moduli, in general. The only case in which there are no moduli, i.e., in which the topological Seifert invariants determine the analytic structure, is $g = 0$ and $r \leqslant 3$. Behnke, Riemenschneider and Schreyer [**R**] have explicitly calculated the coordinate rings of many surfaces in this case (for example quotient singularities).

In the case that there are moduli one can ask for the *numbers and degrees* of generators, relations and syzygies.

EXAMPLE (4.4.2). Suppose V is a cone, i.e., $r = 0$. Then $A = \bigoplus_{n=0}^{\infty} L(nD)$ where D is a divisor with integer coefficients. In the case that degree D is sufficiently large, we are reduced to the problem of finding generators and relations for the coordinate ring of a curve. There are two classical theorems.

THEOREM [**Mu, S-D**]. *If degree $D \geqslant 2g+1$ then A is generated by elements of degree* 1. *The ideal of relations is generated by elements of degree* 2 *and* 3. *If degree $D \geqslant 2g+2$ then the ideal of relations is generated by elements of degree* 2.

THEOREM (MAX NOETHER, ENRIQUES, PETRI, SAINT-DONAT) [**S-D2**]. *If $D = K$ and X is nonhyperelliptic then A is generated by elements of degree* 1 *and the ideal of relations is generated by elements of degree* 2 *and* 3. *The ideal of relations is generated by elements of degree* 2 *except in the following cases.*

(i) *C is a nonsingular plane quintic* ($g = 6$).

(ii) *C is trigonal* (*i.e., a triple covering of* $\mathbf{P}^1$).

(4.5) Now we would like to describe how the theorems above generalize to the case where D has rational coefficients. Since the general case is somewhat more complicated we will need a handy way to keep track of the degrees of the generators of A. We do this as follows. Note that $m = \bigoplus_{n=1}^{\infty} A_i$ is a homogeneous, maximal ideal. The vector space m/m^2 is finite dimensional and graded. Any basis for m/m^2 lifts to a minimal set of algebra generators for A (and conversely). The Poincaré power series of m/m^2, denoted by $p_{m/m^2}(t)$ is a polynomial. The coefficient of t^i in this polynomial is precisely the number of independent generators of A having degree i.

Now we must introduce some notation. Let $[u_1, \ldots, u_n]$ denote the continued fraction $u_1 - 1/(u_2 - 1/(\cdots - 1/u_n)\cdots)$.

DEFINITION (4.5.1). Suppose α, β are relatively prime positive integers, $0 < \beta < \alpha$. Then there exist unique $u_i \geqslant 2$ so that $\alpha/(\alpha - \beta) = [u_1, \ldots, u_n]$. Define s_i, t_i for $i = 1, \ldots, n$ by $s_i/t_i = [u_1, \ldots, u_i]$ and g.c.d.$(s_i, t_i) = 1$. Now define $f_{\alpha,\beta} = t^{s_1} + \cdots + t^{s_n}$.

EXAMPLE (4.5.2). Suppose $\beta = 1$. Then $n = \alpha - 1$ and $u_i = 2$, for all i and $s_i = i + 1$. Thus $f_{\alpha,1} = t^2 + t^3 + \cdots + t^{\alpha}$.

THEOREM (4.5.3) (F. VANDYKE [**Va**]). *If V has numerical invariants $\{g; b; (\alpha_1, \beta_1), \ldots, (\alpha_r, \beta_r)\}$, $A = \mathbf{C}[V]$ and $b \geqslant 2g + 1 + r$ then the degrees of generators for A are given by*

$$p_{m/m^2}(t) = (b - r + 1 - g)t + \sum_{i=1}^{r} f_{\alpha_i,\beta_i}(t).$$

THEOREM (4.5.4) [**W3**]. *Suppose A is a ring of automorphic forms, i.e., $D_0 = K$ and $\beta_i = 1$ for all i. Then*

$$p_{m/m^2}(t) = h(t) + \sum_{i=1}^{r} f_{\alpha_i,1}(t)$$

where $h(t)$ is given in the table below. Note that $p_{m/m^2}(1)$ is the embedding dimension of A. Let $\alpha = \sum_{i=1}^{r} \alpha_i$.

Seifert Invariants	$f(t)$
$g \geqslant 3$, X *nonhyperelliptic*	gt
$g \geqslant 2, g + r \geqslant 3$, X *hyperelliptic*	$gt + (g-2)t^2$
$g = 1, \alpha - r \geqslant 3$	$t - t^2$
$g = 0, r \geqslant 4, \alpha \geqslant 11$	$-3t^2 + (r-5)t^3$
$g = 0, r = 3, \alpha_i \geqslant 3$, *for all* i, $\alpha \geqslant 12$	$-3t^2 - 2t^3 - t^4$
$g = 0, r = 3, \alpha_1 = 2, \alpha_2, \alpha_3 \geqslant 4, \alpha \geqslant 13$	$-3t^2 - 2t^3 - t^4 - t^5$
$g = 0, r = 3, \alpha_1 = 2, \alpha_2 = 3, \alpha_3 \geqslant 9$	$-3t^2 - 2t^3 - t^4 - t^5 - t^7$

The finite number of cases that do not appear above are all hypersurfaces. They have been classified by several people [**Do1, Sh, W2**]. The theorem proven in [**W3**] is more general than the above in that it applies to rings of automorphic forms relative to a Fuchsian group Γ where H_+/Γ is not necessarily compact.

(4.6) In [**W2**] all rings of automorphic forms generated by four (or fewer) elements were classified. One remaining problem is to describe the relations (and syzygies) under the hypotheses above.

We give an example that illustrates the theorems that have been proven and the kinds of more general results one could expect.

EXAMPLE. Suppose V is an affine surface with $\mathbf{C}^*$-action and an isolated singularity and (with the notation of (4.4.1)) assume $g = \text{genus}(X) = 1$, $r = 1$, $p_0 \in X$ is an arbitrary point, $D = 3p_0 + p_1$ and $(\alpha, \beta) = (2, 1)$. Let f denote the Weierstrass $\wp$-function with pole (of order 2) at p_0 (and no other pole). The derivative of f, f', has a pole of order 3 at p_0. Thus

$$A_1 \approx L(3p_0) = \langle 1, f, f' \rangle,$$

$$A_2 \approx L(6p_0 + p_1) = \left\langle 1, f, f', f^2, ff', (f')^2, g \right\rangle,$$

where $g \in L(6p_0 + p_1) - L(6p_0)$. Let $X = 1 \in A_1$, $Y = f \in A_1$, $Z = f' \in A_1$, $W = g \in A_2$. Then it follows from (4.5.3) that X, Y, Z, W generate A. It is not hard to verify this directly using the following theorem (cf. [**S-D2, Mu**]).

THEOREM. *If D_1 and D_2 are divisors on X and degree $D_i \geqslant 2g + 1$ then the map*

$$\begin{array}{ccc} L(D_1) \otimes L(D_2) & \to & L(D_1 + D_2), \\ f \otimes g & \mapsto & fg \end{array}$$

is onto.

Thus one can see that $A_1 \otimes A_2$ maps into A_3, etc. Now to find relations we first compute the Poincaré power series of A:

$$\begin{aligned} p_A(t) &= 1 + \sum_{k=1}^{\infty} (3k + [k/2])t^k \\ &= 1 + 3t/(1-t)^2 + t^2/(1-t)^2(1+t) \\ &= (1 - 3t^3 + t^4 + t^5)/(1-t)^3(1-t^2). \end{aligned}$$

Let R be the polynomial ring $\mathbf{C}[X, Y, Z, W]$ with the grading defined by letting degree $X =$ degree $Y =$ degree $Z = 1$ and degree $W = 2$. The Poincaré power series above leads us to conjecture the existence of a free graded resolution of the form

$$0 \to R(4) \oplus R(5) \to R(3) \oplus R(3) \oplus R(3) \to R \to A \to 0.$$

Now by noting that $\dim A_3 = 10$ and $\dim R_3 = 13$ we see that there are three relations of degree 3. Let I be the ideal generated by these relations and $J = \text{kernel } \phi$. Then one can show that $\dim(R/I)_i \leqslant \dim(R/J)_i$ for all i. But $I \subset J$ and hence $I = J$. Finally, since A is Cohen-MacCauley it has a free graded resolution of length 2 and hence the kernel of Φ is free (of rank 2). The grading on the kernel can be read off the Poincaré power series.

Recent work of Frances VanDyke indicates that the results of Mumford and Saint-Donat [**Mu, S-D1**] on relations can be generalized to normal graded rings of dimension 2 with $b \geqslant 2g + 1 + r$.

Bibliography

[**A-M**] M. Atiyah and I. MacDonald, *Introduction to commutative algebra*, Addison-Wesley, Reading, Mass., 1969.

[**Ba**] Bailey, Ph. D. Thesis, Birmingham, England, 1977.

[**B**] G. Bredon, *Introduction to compact transformation groups*, Academic Press, New York, 1972.

[**Br**] E. Brieskorn, *Rationale Singularitaten komplexer Flachen*, Invent. Math. **4** (1968), 336–358.

[**B-N**] S. Bungard and J. Nielsen, *On normal subgroups with finite index in F-groups*, Matematisk Tidsskr. B (1951), 56–58.

[**Ca**] J. Carrel, *Holomorphically injective complex toral actions*, Proc. Second Conf. on Compact Transformation Groups. II, Lecture Notes in Math., vol. 299, Springer-Verlag, Berlin, pp. 205–236.

[**C-R1**] P. Conner and F. Raymond, *Injective operations of the toral groups*, Topology **10** (1971), 283–296.

[**C-R2**] ______, *Injective operations of the toral groups*. II, Proc. Second Conf. on Compact Transformation Groups. II, Lecture Notes in Math., vol. 299, Springer-Verlag, Berlin.

[**C-R3**] ______, *Holomorphic Seifert fiberings*, Proc. Second Conf. on Compact Transformation Groups. II, Lecture Notes in Math., vol. 299, Springer-Verlag, Berlin, pp. 124–204.

[**De**] M. Demazure, *Anneaux gradues normaux*, Sém. sur les Singularités des Surfaces (Demazure-Giraud-Teissier), École Polytechnique, Paris, 1979.

[**Do1**] I. Dolgachev, *Quotient conical singularities on complex surfaces*, Functional Anal. Appl. **8** (1974), 160–161.

[**Do2**] ______, *Automorphic forms and quasi-homogeneous singularities*, Functional Anal. Appl. **9** (1975), 149–151.

[**Do3**] ______, *Automorphic forms and quasi-homogeneous singularities* (preprint).

[**EGA**] A. Grothendieck, *Éléments de géométrie algébrique*, Inst. Hautes Études Sci. Publ. Math., nos. 4, 8, 11, ..., (1961).

[**Fo**] R. H. Fox, *On Fenchel's conjecture about F-groups*, Matematisk Tidsskr. B (1952), 61–65.

[**Gu**] R. Gunning, *Lectures on modular forms*, Princeton Univ. Press, Princeton, N.J., 1962.

[**Ha**] R. Hartshorne, *Algebraic geometry*, Springer-Verlag, Berlin, 1977.

[**Hi**] F. Hirzebruch, *Differentiable manifolds and quadratic forms*, revised by W. D. Neumann, Dekker, New York, 1972.

[**Ho**] H. Holmann, *Seifertsche Faserraume*, Math. Ann. **157** (1964), 138–166.

[**Mi**] J. Milnor, *On the 3-dimensional Brieskorn manifolds $M(p, q, r)$*, Knots, Groups and 3-Manifolds, Ann. of Math. Studies, No. 84, Princeton Univ. Press, Princeton, N.J., 1975.

[**Mu**] D. Mumford, *Varieties defined by quadratic equations*, C.I.M.E., 1969.

[Ne] W. Neumann, *Brieskorn complete intersections and automorphic forms*, Invent. Math. **42** (1977), 285–293.

[N-R] W. Neumann and F. Raymond, *Seifert manifolds, plumbing, μ-invariant and orientation reversing maps*, Algebraic Geometry and Topology, Lecture Notes in Math., vol. 664, Springer-Verlag, Berlin, 1978, pp. 162–195.

[Or] P. Orlik, *Seifert manifolds*, Lecture Notes in Math., vol. 291, Springer-Verlag, Berlin, 1972.

[O-W1] P. Orlik and P. Wagreich, *Isolated singularities of algebraic surfaces with* $\mathbf{C}^*$*-action*, Ann. of Math. (2) **93** (1971), 205–228.

[O-W2] ______, *Singularities of algebraic surfaces with* $\mathbf{C}^*$*-action*, Math. Ann. **93** (1971), 205–226.

[O-W3] ______, *Equivariant resolution of singularities with* $\mathbf{C}^*$*-action*, Proc. Second Conf. on Compact Transformation Groups. II, Lecture Notes in Math., vol. 299, Springer-Verlag, Berlin.

[Pi] H. Pinkham, *Normal surface singularities with* $\mathbf{C}^*$*-action*, Math. Ann. **227** (1977), 183–193.

[R] O. Riemenschneider, *Dihedral singularities: Invariants, equations and infinitesimal deformations*, Bull. Amer. Math. Soc. **82** (1976), 745–747.

[S-D1] B. Saint-Donat, *Sur les équations définissant une courbe algebrique*, C. R. Acad. Sci. Paris Ser. A **274** (1972), 324–327.

[S-D2] ______, *On Petri's analysis of the linear system of quadrics through a canonical curve*, Math. Ann. **206** (1975), 157–175.

[S] H. Seifert, *Topologie dreidimensionaler gefaserter Raume*, Acta Math. **60** (1933), 147–238.

[Se] J.-P. Serre, *Faisceaux algébriques coherents*, Ann. of Math. (2) **61** (1955), 197–278.

[Sh] I. G. Sherbak, *Algebras of automorphic forms with* 3 *generators*, Functional Anal. Appl. **12** (1978), 156–158.

[Va] F. VanDyke, Thesis, University of Illinois at Chicago (in preparation).

[W1] P. Wagreich, *Singularities of complex surfaces with solvable local fundamental group*, Topology **11** (1972), 51–72.

[W2] ______, *Algebras of automorphic forms with few generators*, Trans. Amer. Math. Soc. **262** (1980), 367–389.

[W3] ______, *Automorphic forms and singularities with* $\mathbf{C}^*$*-action*, Illinois J. Math. **25** (1981), 359–382.

UNIVERSITY OF ILLINOIS AT CHICAGO

Proceedings of Symposia in Pure Mathematics
Volume 40 (1983), Part 2

DERIVATIONS, AUTOMORPHISMS AND DEFORMATIONS OF QUASI-HOMOGENEOUS SINGULARITIES

JONATHAN M. WAHL[1]

0. Introduction. Let $R = \bigoplus_{m=0}^{\infty} R_m$ be a finitely generated algebra over $R_0 = \mathbf{C}$. Choosing homogeneous generators, we write $R = P/I$, where $P = \mathbf{C}[z_1, \ldots, z_n]$ is a graded (or weighted) polynomial ring. We may as well assume the $\{w_i = \text{weight } z_i\}$ are relatively prime; and $I = (f_1, \ldots, f_t)$, where f_j is weighted homogeneous, with weight d_j.

A complete local $\mathbf{C}$-algebra (= singularity) is *quasi-homogeneous* (q.-h.) if it is the completion $\hat{R}$ of a graded R. If the singularity is isolated, quasi-homogeneity is equivalent to having a good $\mathbf{C}^*$-action on the completion ("good" means the weights are > 0). The same singularity may have essentially different graded structures. For example, $R = \mathbf{C}[x, y, z]/xz - y^2$ is bigraded, so has many (single) gradings (e.g., use weights $\{1, a+1, 2a+1\}$). A general problem is, for q.-h. $\hat{R}$, to choose in a natural way a graded R which is unique up to graded isomorphism.

For instance, K. Saito has proved **[11]** that an isolated q.-h. hypersurface singularity has unique *normalized* weights. After analytic change of coordinates, one may write the defining equation as a q.-h. polynomial with (possibly new) weights

$$(0.1) \qquad f = \sum_{1}^{r} z_i^2 + h(z_{r+1}, \ldots, z_n), \qquad \text{mult } h \geqslant 3,$$
$$2w_i = \text{wt } f \quad (i \leqslant r), \qquad 2w_j < \text{wt } f \quad (j > r).$$

Saito has also written down for us a direct proof (in the spirit of **[11]**) of the graded uniqueness of the representation (0.1).

1980 *Mathematics Subject Classification*. Primary 14B05.

[1]Partially supported by NSF.

Saito's choice of weights gives a graded R ($= \mathbf{C}[z_1, \ldots, z_n]/(f)$) for which there are no derivations of negative weight (see (1.1)); in fact, if $r \neq 2$, only normalized weights have this property. So, we conjecture (1.4) that every isolated normal q.-h. singularity admits a grading for which there are no derivations of negative weight. There are many reasons why we expect and want such a result to be true; these are discussed (along with partial results) in §§1 and 2. For instance, the nonexistence of derivations of negative weight on a normal cone is equivalent to a theorem of ours giving a cohomological characterization of $\mathbf{P}^n$ (Theorem 2.5).

We asked in our talk whether a complete intersection q.-h. isolated singularity $\hat{R}$ uniquely determines a graded R (assuming $\dim R > 0$, and excluding the case of multiplicity 2 hypersurfaces, covered by (0.1)). This had already been proved by C. T. C. Wall [**21**], as a direct consequence of a theorem of G. Scheja and H. Wiebe on derivations [**13**]. We still do not know (except in dimension $\leqslant 2$) that there are no derivations of negative weight in this case. A letter to us from Scheja describes a proof by M. Kersken of a result (Theorem 2.8), stated as a conjecture in our talk, which describes explicit generators for the module of derivations.

Motivated by Wall's point of view, we include a new section (§3) on automorphisms of $\hat{R}$, especially the "maximal reductive automorphism group G" of $\hat{R}$. Any two $\mathbf{C}^*$-actions can be conjugated into G; so, $\hat{R}$ admits a unique $\mathbf{C}^*$-action if $\dim G = 1$. With help from Peter Slodowy, we prove Proposition 3.10 that if $\hat{R}$ is q.-h., then there is a good $\mathbf{C}^*$-action in the center of G. In the hypersurface case, the only $\mathbf{C}^*$-action in the center of G is the one giving Saito's normalized representation Corollary 3.11; this gives another proof of the graded uniqueness of the representation (0.1).

In particular, one might optimistically conjecture that every isolated normal q.-h. singularity is the completion of a graded R for which all automorphisms of $\hat{R}$ preserve the weight filtration Conjecture 3.14; unfortunately, such R is not unique (Example 3.6.2).

The last section is motivated by a result of Jim Damon [**3**] on graded T^1 of certain codimension 2 complete intersections. Recall that the modules θ_R of derivations and T^1_R of first-order deformations sit in an exact sequence

$$0 \to \theta_R \to \theta_P \otimes R \to \operatorname{Hom}_R(I/I^2, R) \to T^1_R \to 0. \tag{0.2}$$

(The map in the middle sends a derivation of P to the induced map of I into P/I.) Our Theorem 4.5 is a generalization of local duality for a q.-h. hypersurface to a q.-h. Gorenstein surface singularity; the proof will be published elsewhere. The duality gives an explicit description of the "high weight" graded pieces of T^1 and yields, for example, that "modality" $\geqslant$ geometric genus.

The final version of this paper benefitted from conversations at the Conference with C. T. C. Wall, E. Looijenga, and H. Hironaka, and correspondence with P. Slodowy and G. Scheja.

1. Graded derivations.

(1.1) With $R = P/I$ as above, derivations of R are induced by derivations of P sending I into I. As R is graded, we may write

$$\theta_R = \bigoplus_{j \in \mathbf{Z}} \theta_R(j)$$

where

$$\theta_R(j) = \{D \in \theta_R \mid D(R_m) \subset R_{m+j}, \text{ all } m\}.$$

There is a natural *Euler derivation* $E \in \theta_R(0)$, where $E \mid R_m$ = multiplication by m. In general, a derivation $D = \Sigma a_i(z)\partial/\partial z_i$ has weight j if wt $a_i = w_i + j$, all i. The following result is well known.

PROPOSITION 1.2. *Suppose $R = P/(f)$ is a graded isolated hypersurface singularity. Then θ_R is generated as an R-module by E and the "trivial derivations"*

$$\frac{\partial f}{\partial z_j}\frac{\partial}{\partial z_j} - \frac{\partial f}{\partial z_i}\frac{\partial}{\partial z_j} \in \theta_R(d - w_i - w_j).$$

PROOF. Note $Ef = df$. So, if $Df = \lambda f$, subtracting a multiple of E gives a derivation D' with $D'f = 0$. Writing $D' = \Sigma a_i\partial/\partial z_i$, one obtains a relation $\Sigma a_i \partial f/\partial z_i = 0$. But $\{\partial f/\partial z_1, \ldots, \partial f/\partial z_n\}$ forms a regular sequence (since the singularity is isolated), whence the only relations are those generated by the obvious ones.

(1.3) It follows easily from Proposition 1.2 that for normalized hypersurface weights (0.1), since $w_i + w_j \leqslant d$, there are no derivations of negative weight. Of course, the hypersurface $xz = y^2$, with weights $\{1, 2, 3\}$, has a (trivial) derivation of negative weight. It is not hard to show that if r (the rank of the quadratic part of f) is $\neq 2$, then a $\mathbf{C}^*$-action gives no derivations of negative weight iff it gives normalized weights. We offer the

Conjecture 1.4. Let R be a normal graded ring, with isolated singularity. Then there is a normal graded $\bar{R}$, with $\hat{R} \simeq \hat{\bar{R}}$, so that $\bar{R}$ has no derivations of negative weight.

(1.5) In the next section we shall discuss some cases in which the above conjecture (and some related ones) are true. However, we first give some motivation for the conjecture as well as reasons for its interest.

(1.6) If all weights $w_i = 1$, then a derivation $D \neq 0$ of negative weight, being determined by its action on R_1, sends R_1 onto $R_0 = \mathbf{C}$; so, $Dt = 1$, some $t \in R_1$. By Zariski's lemma [**23**], $R = S[t]$ for a subring S of R, contradicting that R has an isolated singularity.

(1.7) If R has only E as a derivation of weight $\leqslant 0$, then $\hat{R}$ uniquely determines R up to graded isomorphism (see Proposition 3.9(d)).

(1.8) Letting $F_j = \oplus_{i \geqslant j} R_i$ be the weight filtration, consider π: $Y = \operatorname{Proj} \oplus_{j=0}^{\infty} F_j \to \operatorname{Spec} R$, the natural "partial resolution." Then π is equivariant (i.e., $\pi_*\theta_Y \xrightarrow{\sim} \theta_R$) iff R has no derivations of negative weight [**19**].

(1.9) If R is graded with isolated singularity, let Def^+ be the functor (or space) of flat deformations of R plus flat deformations of the filtration $\{F_j\}$ above. According to an idea of Dock Rim, $\mathrm{Def}^+ =$ equisingular deformations (in one sense) = deformations of weight $\geqslant 0$. But to have a sensible functorial notion, one has initially to carry the extra baggage of a section along with the deformation. Our result **[19]** says that R has no derivations of negative weight iff the natural forgetful functor $\mathrm{Def}^+ \to \mathrm{Def}$ (into the moduli space of R) is injective, i.e., iff one has "uniqueness of equisingular sections."

(1.10) In the work of Pinkham **[9]**, the existence of coarse moduli spaces for certain problems depends on a graded R having no derivations of negative weight (or only E of weight $\leqslant 0$). The point is that the derivations of weight $\leqslant 0$ yield exactly the derivations of the natural compactification of $\operatorname{Spec} R$ which are logarithmic at ∞ (see also **[18]**).

2. Special cases of graded derivations.

(2.1) Assume from now on R is graded, with isolated singularity. It is not hard to show that a q.-h. curve singularity has E as the only derivation of weight $\leqslant 0$ **[6]**. The following result uses the detailed description of Orlik-Wagreich of the compactification $\operatorname{Proj} R[t]$.

THEOREM 2.2 **[18]**. *Let R be a graded normal domain of dimension* 2. *If R is not a cyclic quotient singularity* ($= \mathbf{C}^2/\mathbf{Z}_n$), *then E is the only derivation of weight* $\leqslant 0$.

(2.3) If X is a projective nonsingular variety, L an ample line bundle, one can form the *cone* $R = C(X, L) = \bigoplus_0^\infty \Gamma(X, L^{\otimes n})$, a normal domain with isolated singularity. R is singular unless $(X, L) = (\mathbf{P}^n, \mathcal{O}(1))$.

PROPOSITION 2.4 **[17]**. *There is an exact sequence*

$$0 \to R_i E \to \theta_R(i) \to H^0(\theta_X \otimes L^i), \quad \text{all } i \in \mathbf{Z};$$

if $-i < 0$ *and* $\dim X > 1$, *then*

$$\theta_R(-i) = H^0(\theta_X \otimes L^{-i}).$$

THEOREM 2.5 **[20]**. (a) *Let X be a nonsingular projective variety, L an ample line bundle, so that $H^0(\theta_X \otimes L^{-1}) \neq 0$. Then $(X, L) = (\mathbf{P}^n, \mathcal{O}(1))$ or $(\mathbf{P}^1, \mathcal{O}(2))$.*

(b) *A singular cone $C(X, L)$ has no derivations of negative weight.*

(2.6) Part (a) of Theorem 2.5 was conjectured in **[17]**, and proved in case $H^0(L) \neq 0$ by Mori and Sumihiro **[8]**; in fact, a lemma of theirs is crucial to our proof. That $(\mathbf{P}^1, \mathcal{O}(2))$ has $H^0(\theta_X \otimes L^{-1}) \neq 0$ shows up not as a derivation of negative weight, but in the special contribution rational -2 curves make to the deformation theory of a projective surface **[2]**.

(2.7) Now let $R = \mathbf{C}[z_1, \ldots, z_n]/(f_1, \ldots, f_t)$ be a graded complete intersection. Analogously to Proposition 1.2, a derivation D of P induces a "trivial derivation" of R if $Df_i = 0$, all i; this notion depends on the choice of generators. Such D can be described explicitly as follows: if $1 \leqslant i_1 < \cdots < i_{t+1} \leqslant n$, consider the cofactor expansion along the first row:

$$(2.7.1) \qquad D = \det \begin{pmatrix} \partial/\partial z_{i_1} & \cdots & \partial/\partial z_{i_{t+1}} \\ \partial f_1/\partial z_{i_1} & & \partial f_1/\partial z_{i_{t+1}} \\ \vdots & & \\ \partial f_r/\partial z_{i_1} & \cdots & \partial f_r/\partial z_{i_{t+1}} \end{pmatrix}.$$

Clearly, $Df_j = 0$, $\forall_j$. That these span all trivial derivations follows from some homological algebra considerations due to Saito and Eisenbud-Buchsbaum. The following theorem, analogous to Proposition 1.2, has been proved by M. Kersken, a recent student of U. Storch.

THEOREM 2.8. *Let R be a graded normal isolated complete intersection. Then θ_R is generated as an R-module by E and the trivial derivations* (2.7.1).

(2.9) The idea of Kersken's proof is as follows: Let $\Omega^i = \Lambda^i\Omega^1_R$, $\omega^i = (\Omega^i)^{**} = H^0(\operatorname{Spec} R - \{m\}, \Omega^i)$. If $q = n - t = \dim R$, then ω^q is free of rank 1, with an explicit generator written in terms of residue symbols (once we choose explicit generators for P and I, with $R = P/I$). There is a natural map $\Omega^i \to \omega^i$, an isomorphism for $i \leqslant q - 2$ (as follows from the standard presentation of Ω^1). Kersken shows ω^{q-1} is generated by the image of Ω^{q-1} and the image of the Euler contraction $\omega^q \to \omega^{q-1}$. But the natural isomorphism $\theta_R \otimes \omega^q \xrightarrow{\sim} \omega^{q-1}$ gives an isomorphism $\omega^{q-1} \xrightarrow{\sim} \theta_R$; making the map explicit (via residue symbols) shows Ω^{q-1} gives trivial derivations in θ_R (of the form (2.7.1)), and the Euler contraction gives E. This method also shows $H^1_{\{m\}}(\Omega^{q-1})$ is a principal R-module, generated by the Euler contraction class in $H^0(\operatorname{Spec} R - \{m\}, \Omega^{q-1}) = \omega^{q-1}$.

REMARKS. (2.10.1) We mentioned in the Introduction that a q.-h. isolated complete intersection $\hat{R}$ has a unique graded structure. So, Conjecture 1.4 in this case, combined with Theorem 2.8, reduces to showing that if $w_1, \ldots, w_n$ are the weights of the variables, $d_1, \ldots, d_t$ the weights of the defining equations, then $\Sigma_1^t d_i \geqslant$ sum of any $(t + 1)$ of the w_j. Perhaps this can be proved analogously to [**13**, §2].

(2.10.2) In our lecture, we indicated a proof of Theorem 2.8 in the homogeneous case, when R is the cone over a smooth complete intersection $X \subset \mathbf{P}^{n-1}$. This was accomplished by computing $H^0(\theta_X(i))$ (compare Proposition 2.4), which is less easy than it looks. This method (or Kersken's) yields the following result, perhaps well known, which we could not locate in the literature.

PROPOSITION 2.11. *Let $X \subset \mathbf{P}^m$ be a nonsingular complete intersection. Then $H^0(\theta_X) = 0$, unless X is a quadric hypersurface, or X is a cubic in $\mathbf{P}^2$ or intersection of 2 quadrics in $\mathbf{P}^3$.*

3. The reductive automorphism group.

(3.1) Let $(\hat{R}, m)$ be a complete local $\mathbf{C}$-algebra, reduced and equi-dimensional, with isolated singularity. Consider the groups of $\mathbf{C}$-algebra automorphisms. $\hat{G} = \operatorname{Aut} \hat{R}$ and $\hat{G}_n = \operatorname{Aut}(\hat{R}/m^{n+1})$, and the natural homomorphisms $\hat{G} \to \hat{G}_n$. An action of the affine algebraic group G on $\hat{R}$ is a group homomorphism $G \to \hat{G}$ such that the induced maps $G \to \hat{G}_n$ are algebraic group homomorphisms. G is reductive if G^0 (= connected component of the identity) is reductive, i.e., if every finite-dimensional representation is completely reducible (in characteristic zero, reductive = linearly reductive). Necessarily, G/G^0 is finite. A general reference for algebraic groups is [**1**]; see also [**16**, pp. 1–15].

THEOREM 3.2 (WALL [**21**], JÄNICH [**5**]). *There is a maximal reductive subgroup $G \subset \hat{G} = \operatorname{Aut} \hat{R}$, such that every reductive subgroup of $\hat{G}$ is conjugate to a subgroup of G. The action of G can be linearized so that $G \subset \mathrm{GL}(n)$, where $n = \dim m/m^2$. If G contains a $\mathbf{C}^*$ which acts on the tangent space with positive weights, then there is a G-invariant ideal $I \subset P = \mathbf{C}[z_1, \ldots, z_n]$ so that $\hat{R} = (P/I)\hat{}$ G-equivariantly.*

(3.3) The aforementioned theorem is slightly different from the results of Jänich and Wall, but the basic proof is the same. It is useful to use Hironaka's theorem on (t, r)-indices [**4**]. That is, there exist integers t, $r > 0$ so that if $m \geqslant t$, then $\operatorname{Image}(\hat{G}_m \to \hat{G}_{m-r}) = \operatorname{Image}(\hat{G} \to \hat{G}_{m-r})$; define this (algebraic) group to be $\overline{G}_{m-r}$. Using conjugacy of Levi complements in the $\overline{G}_n$ (see, e.g., [**24**, p. 117] for a proof in the disconnected case), one gets an eventually stable sequence of reductive subgroups of the $\overline{G}_n$. The action on each $\hat{R}/m^{n+1}$ can be linearized, hence eventually on $\hat{R}$. Writing $\hat{R} = \mathbf{C}[[z_1, \ldots, z_n]]/I$, where $G \subset \mathrm{GL}(n)$, a good $\mathbf{C}^*$-action means I is generated by the ideal of leading forms (with respect to the weights); thus, one may use polynomials.

(3.4) With this point of view (explained to us by Terry Wall), two positive gradings on some $\hat{R}$ correspond to two good $\mathbf{C}^*$-actions. If the $\mathbf{C}^*$-actions are conjugate by some $g \in \hat{G}$, then the corresponding weight filtrations are isomorphic via g, and the associated graded rings are then isomorphic via g. Thus, the original graded rings are isomorphic. If the two $\mathbf{C}^*$-actions are not conjugate, then the reductive rank of G (= dimension of a maximal torus) is > 1, whence R has a $\mathbf{C}^* \times \mathbf{C}^*$ action.

(3.5) How do we compute G? The finite group G/G^0 can be quite complicated, even for R a homogeneous hypersurface (one must know the projective automorphism group of a hypersurface). But, Proposition 3.9(c) says that if R is graded, with no derivations of negative weight, then $\dim G \leqslant \dim \theta_R(0)$ (the inequality may be strict, e.g. Remark (3.12.2)). So, in principle one should be able to compute G^0. This result, plus Proposition 1.2 and Theorem 2.2, yield the first two examples below (see also [**14**]).

EXAMPLES. (3.6.1) If $f = \Sigma_1^r z_i^2 + h(z_{r+1}, \ldots, z_n)$ defines a q.-h. hypersurface as in (0.1), then G^0 is the connected component of the identity of

$$\mathbf{C}^* \times O(r)/(-1, -I).$$

(3.6.2) If $\hat{R}$ is a q.-h. normal surface singularity, then $G^0 = \mathbf{C}^*$, except for a cyclic quotient singularity $\mathbf{C}^2/H$, where H is generated by $\left(\begin{smallmatrix}\zeta & 0\\ 0 & \zeta^q\end{smallmatrix}\right)$, $\zeta^n = 1$, $(q, n) = 1$. In these cases, one has

$$\begin{array}{lll} q = 1: & G = \mathrm{GL}(2, \mathbf{C})/H, & \\ q^2 \equiv 1(n), q \neq 1, & G^0 = \mathbf{C}^* \times \mathbf{C}^*, & G/G^0 = \mathbf{Z}/2, \\ q^2 \not\equiv 1(n), & G = \mathbf{C}^* \times \mathbf{C}^*. & \end{array}$$

(3.6.3) For any quotient singularity $\mathbf{C}^n/H$, where $H \subset \mathrm{GL}(n)$ is a finite group acting freely on $\mathbf{C}^n - \{0\}$, let $N(H) =$ normalizer of H in $\mathrm{GL}(n)$. Then the reductive group G of the quotient singularity is $N(H)/H$.

(3.7) The verification of the above examples is straightforward, except the proof in (3.6.3) that $N(H)$ is reductive; equivalently, one must show the centralizer $Z(H)$ is reductive. Peter Slodowy's argument is as follows: Decompose $\mathbf{C}^n$ into H-isotypic components V_i. Thus $V_i \cong$ (irreducible H-module) $\otimes$ (trivial H-module $\mathbf{C}^{m_i}$). Then $Z(H)$ is the direct product of the corresponding groups on the spaces V_i; but by Schur's lemma, the group of H-automorphisms of V_i is isomorphic to $\mathrm{GL}(m_i)$.

(3.8) Suppose we are given a good $\mathbf{C}^*$-action on $\hat{R}$, hence a graded R, with $\mathbf{C}^* \subset G$ (as in Theorem 3.2). Then the Lie algebra $\mathcal{G}$ of G is graded (since $\mathbf{C}^*$ acts on it), and is contained in the (graded) module θ_R. Since G is reductive, one can prove fairly easily

PROPOSITION 3.9. (a) *G preserves the weight filtration of R iff G preserves the grading iff* $\mathbf{C}^* \subset Z(G)$.

(b) $\mathbf{C}^* \subset Z(G^0)$ *iff $\mathcal{G}$ has weight $\geqslant 0$ iff $\mathcal{G}$ has weight* 0.

(c) *If R has no derivations of negative weight, then* $\mathbf{C}^* \subset Z(G^0)$ *and* $\mathcal{G} \subset \theta_R(0)$.

(d) *If $\{E\}$ is the only derivation of weight $\leqslant 0$, then* $\dim G = 1$, *and $\hat{R}$ admits a unique* $\mathbf{C}^*$*-action.*

(e) Aut $\hat{R}$ *preserves the weight filtration on $\hat{R}$ iff* $\mathbf{C}^* \subset Z(G)$ *and R has no derivations of negative weight.*

PROPOSITION 3.10. *Suppose $\hat{R}$ admits a good* $\mathbf{C}^*$*-action, then there is a good* $\mathbf{C}^*$*-action in the center of G.*

PROOF (DUE TO PETER SLODOWY). We show that if $G \subset \mathrm{GL}(V)$ is reductive and $\mathbf{C}^* \subset G$ gives only positive weights on V, then there is a $\mathbf{C}^* \subset Z(G)$ with positive weights.

Assume first that G is connected. Let $T \subset G$ be a maximal torus containing the given $\mathbf{C}^*$, and $X_*(T)$ the lattice of one-parameter subgroups; denote the given $\mathbf{C}^*$ by a (primitive) $\lambda' \in X_*(T)$. Let $P \subset X^*(T) = \mathrm{Hom}_{\mathbf{Z}}(X_*(T), \mathbf{Z})$ be the set of weights (characters) of T for the given representation; so, $p \in P$ iff for some $v \neq 0$, $t(v) = p(t)v$, all $t \in T$. There is a natural pairing

$$X_*(T) \times P \to \mathbf{Z},$$

denoted by $\langle\,,\rangle$. The Weyl group $W = N_G(T)/Z_G(T)$ acts on each set,

$$\lambda^w(t) \equiv w^{-1}\lambda(t)w,$$

$$\chi^w(t) \equiv \chi(wtw^{-1}).$$

(Note that v is an eigenvector for p iff $w^{-1}(v)$ is an eigenvector for p^w.) Note $\langle\lambda^w, p^w\rangle = \langle\lambda, p\rangle$.

Since λ' has positive weights, we have $\langle\lambda', p\rangle > 0$, all $p \in P$. Thus, $\langle\lambda', p^{w^{-1}}\rangle = \langle(\lambda')^w, p\rangle > 0$, all p. Let $\lambda_1 = \Sigma_{w\in W}(\lambda')^w$. Then $\langle\lambda_1, p\rangle > 0$, all p, so λ_1 has positive weights, and it is W-invariant. If necessary, divide λ_1 by an integer to make it primitive in the lattice $X_*(T)$.

We claim λ_1 is central in G. For, let $\alpha \in X^*(T)$ be any root; then for some $w \in W$, $\alpha^w = \alpha^{-1}$. But

$$\langle\lambda^w, \alpha^w\rangle = \langle\lambda, \alpha^{-1}\rangle = \langle\lambda, \alpha\rangle,$$

whence $\langle\lambda, \alpha\rangle = 1$, so $\lambda(t) \in \operatorname{Ker}\alpha$. Since $Z(G) = \bigcap \operatorname{Ker}\alpha$, we have $\lambda \in Z(G)$.

If G is not necessarily connected, we have just seen that $Z(G^0)^0$ is a torus T of positive dimension s, containing a $\mathbf{C}^*$ with positive weights. Now, T is invariant under all automorphisms of G^0 (being a center), so T is a normal subgroup of G. The conjugation map $G \to \operatorname{Aut} T$ factors via G^0, giving a representation of the finite group $H = G/G^0 \to \operatorname{Aut} T = \mathrm{GL}(s, \mathbf{Z})$. As above, H acts on $X_*(T)$ and P (the weights with respect to this torus T); averaging as above gives a one-parameter subgroup on which G/G^0 acts trivially, hence is in the center of G, with positive weights on V.

COROLLARY 3.11. *Let $\hat{R}$ be an isolated complete intersection singularity of positive dimension, and suppose* $\dim G \neq 0$. *Then* $\dim Z(G) = 1$, *hence there is a unique $\mathbf{C}^*$-action in the center of G, and the weights are positive. In the hypersurface case, the representation* (0.1) *is unique up to graded isomorphism.*

PROOF. In the nonhypersurface case, this is the result of Scheja-Wiebe and Wall mentioned in the Introduction; since $G^0 = \mathbf{C}^*$, $\mathbf{C}^*$ is central in G by "rigidity of tori" (or, by Proposition 3.10).

In the hypersurface case, suppose a $\mathbf{C}^*$-action gives a representation (0.1) with good weights. By Propositions 1.2 and 3.9(d), $\mathbf{C}^* \subset Z(G^0)$. But Example 3.6.1 implies $\dim Z(G^0) = 1$, unless $r = 2$; excluding this case, then $\dim Z(G) = 1$ (by Proposition 3.10), so $\mathbf{C}^* = Z(G)^0$. If $r = 2$, a graded coordinate change gives a representation of the defining equation as $z_1z_2 + h(z_3,\ldots,z_n)$. Then G^0 is isogenous to $\mathbf{C}^* \times \mathbf{C}^*$, the second factor corresponding to $z_1 \mapsto sz_1$, $z_2 \mapsto s^{-1}z_2$, $z_i \mapsto z_i$ $(i > 2)$. But G contains also the involution switching z_1 and z_2, and the only $\mathbf{C}^*$ in G^0 commuting with this is our original $\mathbf{C}^*$-action. So, $Z(G)^0 \subset \mathbf{C}^*$, hence (Proposition 3.10) $Z(G)^0 = \mathbf{C}^*$. Therefore, representations as in (0.1) give the $\mathbf{C}^*$ in $Z(G)$, hence they are graded isomorphic (Remark (3.4)).

REMARKS. (3.12.1) We do not know in what generality $\dim Z(G) = 1$; but, recall Example 3.6.2, third case. Proposition 3.10 implies G is never semisimple if $\hat{R}$ has a good $\mathbf{C}^*$-action. Most seriously, one does not know (except in dimension

2 or for a complete intersection) if an isolated singularity with $\mathbf{C}^*$-action admits a good $\mathbf{C}^*$-action.

(3.12.2) By blowing up appropriately the fixed locus of a translation on $\mathbf{P}^2$ one obtains a smooth rational surface X with $(\operatorname{Aut} X)^0 = \mathbf{G}_a$. The cone (with respect to an appropriate ample line bundle) will have $\dim G = 1$ and $\dim \theta(0) = 2$.

(3.12.3) According to Slodowy [**15**] and Rim [**10**], one should consider the action of the reductive group G on the semiuniversal deformation of R, as well as on its tangent space T_R^1.

(3.12.4) If $\hat{R}$ is a q.-h. singularity, H a finite group of automorphisms of $\hat{R}$, then the ring of invariants $\hat{R}^H$ is also q.-h. (In case $\hat{R}$ is two-dimensional normal, this is Corollary 3.2 of [**25**].) For, H is conjugate to a subgroup of G (by maximality), and there is a good $\mathbf{C}^*$ in the center of G (Proposition 3.10); so, H commutes with a good $\mathbf{C}^*$-action on $\hat{R}$.

(3.13) We close this section with a conjecture, stronger than Conjecture 1.4, and verified for hypersurfaces and in dimension 2.

Conjecture 3.14. Every isolated normal q.-h. singularity admits a good $\mathbf{C}^*$-action with $\mathbf{C}^* \subset Z(G)$ and with no derivations of negative weight; equivalently, there is a good $\mathbf{C}^*$-action for which the corresponding weight filtration is preserved by every formal automorphism.

4. Graded T^1.

(4.1) If R is graded, then T_R^1 is graded from (0.2); say $\phi \in \operatorname{Hom}(I/I^2, R)$ has weight j if $\operatorname{wt}\phi(f_i) = \operatorname{wt} f_i + j$, all i. The deformations of weight $\geqslant 0$ should be thought of as equisingular (1.9).

(4.2) In case $R = P/f$, then since f is q.-h.,

$$T_R^1 = \mathbf{C}[z_1, \ldots, z_n] \Big/ \left(\frac{\partial f}{\partial z_1}, \ldots, \frac{\partial f}{\partial z_n} \right).$$

The element "1" on the right has weight $-d$. There is a multiplication $T_R^1 \otimes_{\mathbf{C}} T_R^1 \to T_R^1$, since the space on the right is an algebra. Let $H = \det(\partial^2 f/\partial z_i \partial z_j)$ be the class of the Hessian in T_R^1. As $h = nd - 2\Sigma w_i$ is the weight of H in R, we have $H \in T_R^1(h-d)$. According to local duality (e.g., [**2**] or Griffiths-Harris),

(4.2.1) $T_R^1(h-d)$ has dimension 1, with H as basis,

(4.2.2) $T_R^1(h-d+i) = 0$, $i > 0$,

(4.2.3) $T_R^1 \otimes_{\mathbf{C}} T_R^1 \to T_R^1 \overset{\alpha}{\to} T_R^1(h-d) \simeq \mathbf{C}$ is a symmetric perfect pairing (α is projection to the top graded piece).

(4.3) According to (4.2) the one-parameter deformation defined by $f + tH = 0$ is "most equisingular"; in fact, Looijenga and Wirthmüller [**22**] have proved that the whole versal deformation of f, except for f a simple singularity, is topologically trivial along the "modulus parameter" (given by H). Note also that (4.2) implies the symmetry result

(4.3.1) $$T_R^1(i) = T_R^1(h-2d-i), \qquad i \in \mathbf{Z}.$$

(4.4) J. Damon proved [**3**] that for certain two-dimensional q.-h. complete intersections in $\mathbf{C}^4$, one has topological triviality of the versal deformation along a certain modulus parameter. One of the ingredients of his proof is the use of the pairing

$$R \otimes_{\mathbf{C}} T_R^1 \to T_R^1 \to \mathbf{C},$$

where the first map is module multiplication, and the second is the projection of T_R^1 into its top graded piece (which turns out to be one-dimensional). He also remarked that the top weight piece of T_R^1 need not be one-dimensional for a complete intersection of dimension $\neq 2$. His use of this pairing, and the question he asked us about a more general result, motivated our proof of the following:

THEOREM 4.5 [**19**]. *Let R be a graded Gorenstein surface singularity, not a rational double point. Then*

(a) *there is an integer $k \geqslant 0$, computable from the resolution graph or weights of R, so that*

$$\dim T^1(2k) = 1,$$
$$\dim T^1(2k+i) = 0, \qquad i > 0;$$

(b) *the multiplication*

$$R \otimes_{\mathbf{C}} T_R^1 \to T_R^1 \to T_R^1(2k) \simeq \mathbf{C}$$

induces a perfect pairing of vector spaces

$$(R/\operatorname{Ann} T^1) \otimes_{\mathbf{C}} T^1 \to \mathbf{C}.$$

(4.6) So, in dimension 2, one recovers part of local duality as in (4.2), but not the symmetry of (4.3.1). The proof of Theorem 4.5 in the case of a cone $R = C(X, L)$ follows fairly easily from [**17**, §3]; R Gorenstein means $L^{\otimes k} = K_X$, some k, and one proves

$$T_R^1(i) \subset H^1(X, \theta_X \otimes L^i) = H^1(X, L^{i-k}),$$

with $=$ if $i > k$. The perfectness of the pairing in Theorem 4.5(b) then follows from Serre duality on X. For the general graded case, the idea is the same, but the proof is much more technical. Our method gives some results as well for q.-h. normal surfaces.

COROLLARY 4.7. *Let R be as in Theorem* 4.5. *Then*

$$\dim \bigoplus_{k}^{2k} T^1(i) = h^1(\mathcal{O}_M) \qquad (M \to \operatorname{Spec} R \textit{ a resolution}).$$

COROLLARY 4.8. *Let R be a q.-h. normal surface singularity. Then $T_R^1 \neq 0$; in fact, $\oplus_{i\geqslant 0} T_R^1(i) \neq 0$ unless R is taut* [7].

(4.9) Corollary 4.8 says that in the graded Gorenstein case, "modality" $\geqslant$ geometric genus. It also implies $\dim T_R^1 \geqslant h^1(\mathcal{O}_M)$; but stronger inequalities are available by the same method, or by other methods (see Stephen Yau's talk in these PROCEEDINGS). For a graded complete intersection, one has $\dim T_R^1 = \mu \geqslant 2h^1(\mathcal{O}_M) + 1$.

(4.10) Preliminary calculations with cones suggest that, for q.-h. Gorenstein singularities in dimension 3, one has a symmetric perfect pairing

$$T^1 \otimes_{\mathbf{C}} T^1 \to \mathbf{C};$$

but, the top graded piece of T^1 need not be one-dimensional.

(4.11) As Damon points out, Theorem 4.5 means that the R-module $\mathrm{Hom}_{\mathbf{C}}(T^1_R, \mathbf{C})$ is *principal*; but by local duality, this module is R-isomorphic (if R Gorenstein, of dimension 2) to $H^1_{\{m\}}(\Omega^1_R)$. Also, for a q.-h. complete intersection of dimension $q \geqslant 2$, the discussion in (2.9) shows $H^1_{\{m\}}(\Omega^{q-1}_R)$ is principal, generated by the Euler contraction class. However, as H. Flenner has kindly informed us, one must be careful in conjecturing a generalization: if $R = \mathbf{C}[z_1, z_2, z_3]^G$, where G is the diagonal cyclic group of order 3, then R is graded Gorenstein of dimension 3, but $H^1_{\{m\}}(\Omega^2_R)$ requires 9 generators.

REFERENCES

1. A. Borel, *Linear algebraic groups*, Benjamin, New York, 1969.

2. D. Burns and J. Wahl, *Local contributions to global deformations of surfaces*, Invent. Math. **26** (1974), 67–88.

3. J. Damon, *Topological triviality of versal unfoldings of complete intersections*, preprint.

4. H. Hironaka, *On the equivalence of singularities*. I, *Arithmetical Algebraic Geometry*, (O. Schilling, editor), Harper and Row, New York, 1965.

5. K. Jänich, *Symmetry properties of singularities of* C^∞*-functions*, Math. Ann. **238** (1979), 147–156.

6. J-M. Kantor, *Dérivations sur les singularités quasi-homogènes: cas des courbes*, C. R. Acad. Sci. Paris Sér. A-B **287** (1978), 1117–1119.

7. H. Laufer, *Taut two-differential singularities*, Math. Ann. **205** (1973), 131–164.

8. S. Mori and H. Sumihiro, *On Hartshorne's conjecture*, J. Math. Kyoto Univ. **18**, (1978), 523–533.

9. H. Pinkham, *Groupe de monodromie des singularités unimodularies exceptionelles*, C. R. Acad. Sci. Paris Sér. A-B **284** (1977), 1515–1518.

10. D. S. Rim, *Equivariant G-structure on versal deformations*, Trans. Amer. Math. Soc. **257** (1980), 217–266.

11. K. Saito, *Quasihomogene isolierte Singularitäten von Hyperflächen*, Invent. Math. **14** (1971), 123–142.

12. ———, *Einfach-elliptische Singularitäten*, Invent. Math. **23** (1974), 289–325.

13. G. Scheja and H. Wiebe, *Über Derivationen in isolierten Singularitäten auf vollständigen Durchschnitten*, Math. Ann. **225** (1977), 161–171.

14. ———, *Derivationen in zweidimensionalen normalen Singularitäten*, Abh. Math. Sem. Univ. Hamburg (to appear).

15. P. Slodowy, *Einige Bemerkungen zur Entfaltung symmetrischer Funktionen*, Math. Z. **158** (1978), 157–170.

16. ———, *Simple singularities and simple algebraic groups*, Lecture Notes in Math., vol. 815, Springer-Verlag, Berlin and New York, 1980.

17. J. Wahl, *Equisingular deformations of normal surface singularities*. I, Ann. of Math. (2) **104** (1976), 325–356.

18. ———, *Derivatives of negative weight and non-smoothability of certain singularities*, Math. Ann. **258** (1982), 383–398..

19. ———, T^1*-duality for graded Gorenstein surface singularities*, informal research announcement (1980).

20. ———, *A cohomological characterization of* $\mathbf{P}^n$, Invent. Math. (to appear).

21. C. T. C. Wall, *A second note on symmetry of singularities*, Bull. London Math. Soc. **12** (1980), 347–354.

22. K. Wirthmüller, *Universell Topologische Triviale Deformationen*, Thesis (Regensburg).

23. O. Zariski, *Studies in equisingularity*. I, Amer. J. Math. **87** (1965), 507–536.

24. G. Hochschild, *Basic theory of algebraic groups and Lie algebras*, Graduate Texts in Math., vol. 75, Springer-Verlag, Berlin and New York, 1981.

25. G. Scheja and H. Wiebe, *Zur Chevalley-Zerlegung von Derivationen*, Manus. Math. **33** (1980), 159–176.

UNIVERSITY OF NORTH CAROLINA, CHAPEL HILL

Proceedings of Symposia in Pure Mathematics
Volume **40** (1983), Part 2

CLASSIFICATION OF UNIMODAL ISOLATED SINGULARITIES OF COMPLETE INTERSECTIONS

C. T. C. WALL

Despite the title, I shall think of singularities in terms of map-germs $(\mathbf{C}^n, \mathbf{0}) \to (\mathbf{C}^p, \mathbf{0})$ with $n > p$ (the case $n = p = 2$ was considered by Gibson [**7**]). The reasons for undertaking this classification include (a) the attempt to give explicit descriptions of the structure of C^0-stable maps in a wider range than the nice dimensions (e.g. improve $p \leqslant 7$ to $p \leqslant 10$), (b) confirmation of the calculation of the codimension of the bimodal singularities needed by du Plessis [**15**] to determine in which dimensions finite $\mathcal{A}$-determinacy holds in general, and (c) a hope that systematic phenomena would emerge, as with Arnold's list of function-germs unimodal for right-equivalence. The project is partly successful in all three aspects.

The concept of modality is somewhat delicate, and I will describe it in two alternative ways. In general, given an algebraic group action of G on a variety V, a theorem of Rosenlicht [**16**] implies that V has a finite stratification $\mathbb{S}$ such that the action of G on each stratum S defines a (differentiably) locally trivial fibration $S \to S/G$. The *r-modal set* of the action, $M_r(V)$, is the *closure* of $\bigcup\{S\colon \dim S/G \geqslant r\}$. This is independent of the stratification (it would not be if you did not take closures). We recover Arnold's [**1**] notion of modality by taking V as the set of k-jets $(\mathbf{C}^n, \mathbf{0}) \to (\mathbf{C}, \mathbf{0})$ which are sufficient for right equivalence, and G as the group $\mathcal{R}$ of right-equivalences (induced by diffeomorphisms of $(\mathbf{C}^n, \mathbf{0})$). For target dimension $p > 1$, this group is too small, and we enlarge to the group $\mathcal{K}$ of contact-equivalence, by including n-parameter families of diffeomorphisms of $(\mathbf{C}^p, \mathbf{0})$.

Alternatively, we consider a versal unfolding $\{f_t\}$ of a map-germ f, and seek to classify the germs of the f_t (t small) at points $\mathbf{x}$ near $\mathbf{0}$ up to $\mathcal{K}$- (or $\mathcal{R}$-) equivalence. If these fall into finitely many families of equivalence classes, each depending on r parameters (at most) then f is r-modal (at most).

1980 *Mathematics Subject Classification*. Primary 57R45.

0082-0717/81/0000-0211/$04.25

In this picture we recognize that an f_t may have several singular points bifurcating from $\mathbf{0}$. If (for example) $f_t^{-1}(\mathbf{0})$ has two singular points, one r-modal and one s-modal, we shall say that the *strict $\mathcal{K}$-modality* of the germ of f at $\mathbf{0}$ is $\geqslant (r+s)$. At the end of this paper we shall examine our lists to see which germs are strictly unimodal.

Even when $p = 1$, the modality for $\mathcal{K}$ differs from the (more familiar) modality for $\mathcal{R}$: for example, the theorem of Gabrielov [**6**] that an r-modal family depends on exactly r parameters no longer applies. We begin by reexamining this case.

PART I. HYPERSURFACE SINGULARITIES

Given a singular function-germ $f(z_1,\ldots,z_n)$, Thom's splitting theorem [**10**, **18**] shows that f is right-equivalent to a function of the form $g(z_1,\ldots,z_r) + \Sigma_{r+1}^n z_i^2$, with g of order $\geqslant 3$, and that g is unique up to equivalence. The number r is called the *corank*. From now on we omit the surplus variables z_i $(i > r)$, and consider f and g to be equivalent in an extended sense.

Each of the parabolic (alias simple-elliptic) singularities

$$\tilde{E}_6 = T_{333},\quad x^3 + y^3 + z^3 + \lambda xyz,$$
$$\tilde{E}_7 = T_{244},\quad x^4 + y^4 + \lambda x^2y^2,$$
$$\tilde{E}_8 = T_{236},\quad x^3 + y^6 + \lambda x^2y^2$$

involves a modulus λ for classification up to $\mathcal{K}$-equivalence (as well as for the $\mathcal{R}$-classification). For a $\mathbf{0}$-modal germ f, none of these may occur nearby (in jet space or in a versal unfolding). Since $\tilde{E}_6$ does not appear, f has corank $\leqslant 2$. If the corank $\leqslant 1$, we can reduce f to

$$A_k,\quad f(x) = x^{k+1}\quad (k \geqslant 1).$$

If it is 2, since $\tilde{E}_7$ does not appear f has order 3. And since $\tilde{E}_8$ does not, its Newton polygon lies strictly below that for T_{236}. The terms giving the polygon are now easily shown to determine f up to $\mathcal{R}$-equivalence: we have

$$D_k,\quad x^{k-1} + xy^2\quad (k \geqslant 4),\qquad E_6,\quad x^3 + y^4,$$
$$E_7,\quad x^3 + xy^3,\qquad E_8,\quad x^3 + y^5$$

giving Arnold's list of simple singularities.

We can obtain the $\mathcal{K}$-unimodals of corank 2 similarly. The families

$$X_{2,0},\quad x^4 + \lambda x^2y^4 + \mu xy^6 + y^8,$$
$$N_{16},\quad x^5 + \lambda x^3y^2 + \mu x^2y^3 + y^5$$

each involve 2 moduli for $\mathcal{K}$-classification. It suffices to consider the terms of least weight: for N_{16}, the 5 roots of the homogeneous quintic determine two cross-ratios; for $X_{2,0}$ similarly we have the roots of the quartic for y^2/x, with 0 as a further preferred value. If f is $\mathcal{K}$-unimodal, as N_{16} does not occur nearby f has order 3 or 4, and in the latter case as $X_{2,0}$ does not occur, the Newton polygon for f has at least one vertex below the polygon for $X_{2,0}$.

Now Arnold **[2]** contains complete lists of normal forms for all such functions. In each case, we can write the normal form as

$$f(x, y) = f_0(x, y) + x^a y^b \phi(y)$$

where f_0 is weighted homogeneous, $x^a y^b$ has strictly higher weight, and ϕ is polynomial.

PROPOSITION 1. *In this situation, the* $\mathcal{K}$*-equivalence class of* f *is determined by that of* f_0 *and* ord ϕ.

For, holding f_0 fixed, we have a subspace V of $J^k(2, 1)$, where k is chosen large enough so that all the f of the above form are k-determined. To determine the (orbits of $\mathcal{K}$) $\cap$ V, it is enough to find their tangent spaces. But if f_0 has weights w_1, w_2, and we write $\phi(y) = \Sigma\, c_r y^r$, the tangent space contains

$$f - w_1 x \frac{\partial f}{\partial x} - w_2 y \frac{\partial f}{\partial y} = x^a y^b \sum \lambda_r c_r y^r$$

where $\lambda_r = 1 - aw_1 - (b + r)w_2 < 0$. Since $T\mathcal{K}(f)$ is an $\mathcal{E}_{x,y}$-module, if ϕ has order r it contains $x^a y^{b+r}$. Hence the set of all f with ord $\phi = r$ is contained in a single $\mathcal{K}$-orbit in V.

Since the classification of f_0 depends on one parameter at most, these cases are all $\mathcal{K}$-unimodal. We list the corresponding f_0 at the end of this paper (they are labelled E, Z, W or T). In a few cases, ϕ is restricted to have order O; we obtain a single $\mathcal{K}$-orbit: here we list $f_0 + x^a y^b$ instead.

The above technique of considering the intersection of the tangent space $T\mathcal{K}(f)$ to the $\mathcal{K}$-orbit of f with a suitably chosen linear space V of 'prenormal forms' will be used in all the other cases also. In the hypersurface case ($p = 1$), $T\mathcal{K}(f)$ is the ideal in $\mathcal{E}_{x_1,\ldots,x_n}$ generated by f and the $x_i \partial f/\partial x_j$.

If the corank $n \geqslant 4$, the classification of $j^3 f$ is equivalent to that of cubic surfaces in $P_3\mathbf{C}$ (or hypersurfaces in P_{n-1}) and so involves 4 moduli. We can thus restrict to $n = 3$. Then the 3-jet determines a plane cubic curve Γ. Assume this curve has a simple point P. If this is $(1, 0, 0)$ with tangent $y = 0$, then $j^3 f(x, y, z) = yx^2 + q(y, z)x + r(y, z)$. Since $\partial f/\partial y$ is regular in x of order 2 (in the sense of Weierstrass), it is easy to show that f can be reduced to the prenormal form

$$f(x, y, z) = yx^2 + 2b(y, z)x + c(y, z).$$

We define the *reduction* $\Delta_x(f) = \phi$ to be

$$\phi(y, z) = b(y, z)^2 - yc(y, z),$$

the discriminant of f as quadratic in x. Now compare the classification of f with that of ϕ by comparing $T\mathcal{K}(f)$ and $T\mathcal{K}(\phi)$. For a deformation of f which is still quadratic in x,

$$f_t = f + t(\alpha x^2 + \beta x + \gamma) + O(t^2),$$

we find

$$\phi_t = \phi + t(b\beta - c\alpha - y\gamma) + O(t^2),$$

so define

$$R(\alpha x^2 + \beta x + \gamma) = b\beta - c\alpha - y\gamma.$$

It is not hard to show that the intersection of $T\mathcal{K}(f)$ with the space of functions quadratic in x is the $\mathcal{E}_{y,z}$-module spanned by: f, $y\partial f/\partial y$, $z\partial f/\partial y$, $y\partial f/\partial z$, $z\partial f/\partial z$, $y\partial f/\partial x$, $z\partial f/\partial x$, $x\partial f/\partial x$, $x\partial f/\partial z$ and $x(f - y\partial f/\partial y)$. The images of these under R are, respectively, 2ϕ, $y\partial\phi/\partial y$, $z\partial\phi/\partial y$, $y\partial\phi/\partial z$, $z\partial\phi/\partial z$, 0, 0, 2ϕ, $b\partial c/\partial z - 2c\partial b/\partial z$, and $2(b\partial\phi/\partial y - \phi\partial b/\partial y)$. Thus if

$$(\dagger) \qquad (b\partial c/\partial z - 2c\partial b/\partial z) \in T\mathcal{K}(\phi),$$

we have $R(T\mathcal{K}f \cap V) = T\mathcal{K}(\phi)$. So a 1-parameter family of functions f is $\mathcal{K}$-trivial if and only if the corresponding family of ϕ's is $\mathcal{K}$ trivial; in this case we have a bijection of classifications.

It is not always the case that (†) holds. When it fails, the classification of f will involve *at most* as many parameters as for ϕ: so if ϕ is $\mathcal{K}$-unimodal, at least those deformations of f in normal form will also be. In fact the results below can again be verified by reference to Arnold's lists, and use of Proposition 1, but the technique can be used to extend these lists. There are various cases according to the nature of the curve Γ.

If Γ has nodes (at worst), then $j^4\phi$ has no root of multiplicity > 2, so ϕ is hyperbolic (or parabolic); f is likewise (type T).

If Γ is a cuspidal cubic (Q-series) we can take $j^3f = yx^2 - z^3$, $j^4\phi = yz^3$ in the Z-series. Here one can choose $b = 0$: (†) always holds and we obtain a bijection. In this case, ϕ and f are always $\mathcal{K}$-unimodal.

If Γ is a conic-plus-tangent (S-series) it is somewhat more convenient to choose P on the conic ($j^3f = x^2y + 2xz^2$, $j^4\phi = z^4$). Here if ϕ has type $X_{2,0}$ we see that classifying the terms of lowest weight in f involves 2 moduli. So f is $\mathcal{K}$-unimodal if (and only if) ϕ is $\mathcal{K}$-unimodal in the W-series.

If Γ consists of 3 concurrent lines (U-series), we again find ϕ in the W-series but do not obtain a bijection, since not all ϕ's can occur. However, the correspondence again predicts correctly which cases are unimodal.

Finally if Γ has a repeated line (V-series), ϕ has order 5, hence type N and is $\mathcal{K}$-bimodal. So all functions f of this type (or of types even more degenerate) have $\mathcal{K}$-modality $\geqslant 2$.

The list obtained is tabulated later. We defer detailed discussion till then. It is of interest to note however that when $\Delta_x f = \phi$, the relation of f and ϕ is very close. Both have the same $\mathcal{R}$-modality, and are of the same type (in terms of resolution diagrams for example); $\mu(\phi) = 1 + \mu(f)$, and if we write $\mu = \mu_- + \mu_0 + \mu_+$ to describe the signature of the quadratic form on the Milnor lattice (for surfaces), $\mu_0(\phi) = \mu_0(f)$ and $\mu_+(\phi) = \mu_+(f)$. The discriminants of the quadratic forms are also closely related. (These are tabulated in Ebeling's thesis **[5]**).

The method can be used to obtain complete lists of the μ-constant strata in the S-series, the U-series and even (in principle) the V-series. It seems likely that in the S-series one can also deduce the $\mathcal{K}$-classification.

PART II. COMPLETE INTERSECTIONS

We turn to map-germs $(\mathbf{C}^n, \mathbf{0}) \xrightarrow{f} (\mathbf{C}^p, \mathbf{0})$ with $n > p$; we may suppose $df\mathbf{0} = 0$ (otherwise $f^{-1}\mathbf{0}$ embeds in lower dimension than n). Already in the case $p = 3$, $n = 4$ the classification of the 2-jet $j^2 f$ (a net of quadrics in $P_3\mathbf{C}$) involves 6 moduli, so unimodal germs can only occur if $p = 2$, which we assume from now on. In this case the 2-jets A_1, A_2 of the components f_1, f_2 of f define a pencil of quadric surfaces, the structure of which is the key to proceeding.

A pencil is *regular* if some linear combination $\lambda_1 A_1 + \lambda_2 A_2$ is a nonsingular quadratic form. We may suppose A_1 nonsingular. Then the Jordan normal form of $A_1^{-1}A_2$ is an invariant of the pencil. The usual notation (Segre) is to list the sizes of the Jordan blocks, enclosing those corresponding to a given eigenvalue in parentheses. There are in general n eigenvalues, whose cross-ratios are invariants. Thus if $n \geqslant 5$, the lowest stratum has 2 moduli: for $\mathcal{K}$-unimodals we must have $n = 3$ or $n = 4$.

My first classification technique for this case is a splitting theorem [**19**] which has some analogy with Thom's splitting theorem for the case $p = 1$. If f: $(\mathbf{C}^s, \mathbf{0}) \to (\mathbf{C}^p, \mathbf{0})$ and g: $(\mathbf{C}^t, \mathbf{0}) \to (\mathbf{C}^p, \mathbf{0})$ are two map-germs, we define the sum $f \oplus g$: $(\mathbf{C}^{s+t}, \mathbf{0}) \to (\mathbf{C}^p, \mathbf{0})$ by $f \oplus g(\mathbf{x}, \mathbf{y}) = f(\mathbf{x}) + g(\mathbf{y})$. This operation does not quite behave well for $\mathcal{K}$-equivalence: we define $\mathcal{K}_1$ to be the subgroup of $\mathcal{K}$ of diffeomorphisms of $\mathbf{C}^n \times \mathbf{C}^p$ whose 1-jet at $(\mathbf{0}, \mathbf{0})$ is of the form $h_n \times I_p$ ($h_n \in GL_n\mathbf{C}$, I the identity). Call f *good* if $j^2 f$ is regular and f has finite singularity type (i.e. $f^{-1}\mathbf{0}$ has an isolated singularity at $\mathbf{0}$).

THEOREM. *Suppose F good. Given any partition into 2 sets of the set of eigenvalues of F, there exists a corresponding $\mathcal{K}_1$-equivalence $F \sim f \oplus g$. Moreover, f and g are good, and the $\mathcal{K}_1$-equivalence class of F determines those of f and g. Conversely, if f and g are good and have no common eigenvalue, then $f \oplus g$ is good; and the $\mathcal{K}_1$-equivalence classes of f and g determine that of $f \oplus g$.*

We can keep track of the $\mathcal{K}_1$-equivalence class by giving the $\mathcal{K}$-class and specifying the eigenvalues.

There is not time or space to describe the proof of this result. The basic idea is a simple piece of linear algebra. Since the eigenvalues of F are split, we can choose coordinates to put the 2-jet in the form

$$\Phi(\mathbf{x}, \mathbf{y}) = \left(\mathbf{x}^t A_1 \mathbf{x} + \mathbf{y}^t B_1 \mathbf{y}, \mathbf{x}^t A_2 \mathbf{x} + \mathbf{y}^t B_2 \mathbf{y}\right)$$

with A_1, B_1 nonsingular and $A_1^{-1}A_2$, $B_1^{-1}B_2$ having no common eigenvalue. (Here, $\mathbf{x}$ is a column vector, $\mathbf{x}^t$ its transpose, A_1 a square matrix, etc.) Now show that this implies that the $2st$ bilinear maps $\mathbf{C}^s \times \mathbf{C}^t \to \mathbf{C}^2$ given by

$$x_i \frac{\partial \Phi}{\partial y_j}, \quad y_j \frac{\partial \Phi}{\partial x_i}, \qquad (1 \leqslant i \leqslant s,\ 1 \leqslant j \leqslant t)$$

are linearly independent, so span the space of such maps. It now follows that $T\mathcal{K}(\Phi)$ contains all terms $(m, 0)$ and $(0, m)$ where m is a monomial divisible by at least one of the x_i and by at least one of the y_j. The same holds for $T\mathcal{K}(F)$, but only modulo terms of higher order. Now deform F inductively to eliminate these cross-terms.

The splitting theorem reduces the problem of classifying f to the case when there is just one eigenvalue. Here we proceed by the partition given by Jordan blocks.

Case (1). $f \sim (x^2, 0)$.

Case (2). The 2-jet has normal form (xy, x^2); we can reduce f to $(xy, x^2 + y^k)$ $(k \geqslant 3)$.

Case (1,1). The 2-jet has normal form $(xy, 0)$; we can reduce f to $(xy, x^k + y^l)$ $(3 \leqslant k \leqslant l)$.

These singularities are of type $\Sigma^{2,0}$, and the assertions are due to Mather [**14**]. We now have a classification for the case when no eigenvalue has multiplicity greater than 2.

If $n = 2$, we set $f(x, y) = (xy, x^k + y^l)$ $(2 \leqslant k \leqslant l)$: the case $k = l = 2$ is that when the eigenvalues are distinct.

If $n = 3$, we have the cases $f(x, y) = (xy, x^k + y^l + z^2)$ $(2 \leqslant k \leqslant l)$. Here the cases with $k = 2$ are 0-modal. Not all the others are, since the deformation $(xy + tz, x^k + y^l + z^2)$ is of type $x^k + \lambda x^2 y^2 + y^l$ if $t \neq 0$, so is parabolic if $k^{-1} + l^{-1} = \frac{1}{2}$ (and hyperbolic if $< \frac{1}{2}$), so only $(k, l) = (3, 3)$, $(3, 4)$ and $(3, 5)$ are 0-modal (cf. Giusti [**8**]). In the parabolic cases $(k, l) = (4, 4)$ or $(6, 3)$ this deformation is even weighted homogeneous, and so has μ constant.

If $n = 4$, we can write all cases in the form

$$f(x_1, x_2, y_1, y_2) = \left(x_1 x_2 + y_1^{l_1} + y_2^{l_2}, x_1^{k_1} + x_2^{k_2} + y_1 y_2\right).$$

These are the hyperbolic (alias cusp) singularities. The case $k_1 = k_2 = l_1 = l_2 = 2$ is exceptional: all eigenvalues are distinct, and their cross-ratio gives a parameter (so the above normal form must be replaced by e.g. $(x_1^2 + x_2^2 + x_3^2, x_2^2 + \lambda x_3^2 + x_4^2)$). This case is parabolic (alias simple elliptic).

Before we leave the splitting theorm, we note that in all good cases, $\mu(f \oplus g) = \mu(f) + \mu(g) + 1$, where $\mu(x^2, 0)$ is interpreted as 1 and in general, $\mu(xy, x^k + y^l) = k + l - 1$. Here μ denotes the extension of the Milnor number to complete intersections (see Hamm [**11**]).

Case (3). The 2-jet has normal form $(2xz - y^2, 2xy)$, and we can reduce f to

$$f(x, y, z) = \left(2xz - y^2, 2xy + yu(z) + v(z)\right),$$

which we regard as a prenormal form. Our method for dealing with this is to eliminate x to obtain

$$\phi(y, z) = y^3 + yzu(z) + zv(z),$$

and then compare the classifications of f and ϕ, in the same style as for $\Delta_x f$ in Part I. We shall use the same method in other cases too.

In general, consider maps with prenormal form

$$f(x, \mathbf{y}) = \left(xy_1 + a(\mathbf{y}), xy_2 + b(\mathbf{y})\right),$$

where $\mathbf{y} = (y_1, y_2)$ or (y_1, y_2, y_3). Using $\partial f/\partial y_1$ and $\partial f/\partial y_2$ we see that deformations of f can be reduced to the same form. Define *the linear-x-reduction* $L_x f = \phi = y_1 b - y_2 a$. Again this induces a map of tangent spaces

$$R(x\alpha + \beta, x\gamma + \delta) = b\alpha - y_2\beta - a\gamma + y_1\delta.$$

The intersection of these vectors linear in x with $T\mathcal{K}f$ forms the $\mathcal{E}_y$-module spanned by $x\partial f/\partial x$, $y_i\partial f/\partial x$, $y_i\partial f/\partial y_j$, $f_i\varepsilon_j$ $(i, j = 1, 2)$—here $\varepsilon_1 = (1, 0)$ and $\varepsilon_2 = (0, 1)$—and, if y_3 is present, $x\partial f/\partial y_3$. The images of these under R are: ϕ, 0, $y_i\partial\phi/\partial y_j$, $\delta_{ij}\phi$ and (perhaps) $b\partial a/\partial y_3 - a\partial b/\partial y_3$. All but the last span $T\mathcal{K}(\phi)$.

Thus if $\mathbf{y} = (y_1, y_2)$ we obtain a bijection between classifications of f and ϕ. If $\mathbf{y} = (y_1, y_2, y_3)$ the situation is similar to that for $\Delta_x f$ in Part I (save that we do not have Arnold's work to fall back on): by listing ϕ we can obtain a list for f, but we have no à priori guarantee that the cases so obtained are distinct. In our examples, in fact, they are.

Returning to Case (3) we now obtain a bijection between our new list and Arnold's *EJ* series. To economise notation, we denote Arnold's series by E and the new one by J. All are unimodal for $\mathcal{K}$ except for J_7, J_8, J_9 (corresponding to E_6, E_7, E_8) which are 0-modal.

Curve singularities: *remaining cases*. There are two further regular pencils.

Case (2, 1). This can be dealt with directly by eliminating x, but we obtain this series (denoted K) in a more systematic manner below.

Case (1, 1, 1). Here f has 2-jet $(x_1^2 + x_2^2 + x_3^2, 0)$. The plane cubic $j^3 f_2 = 0$ meets the conic $x_1^2 + x_2^2 + x_3^2 = 0$ in 6 points whose cross-ratios give invariants. These are thus (at least) 3-modal cases.

Next, we have three types of singular pencil.

Case (xy, xz). Here we can apply linear x-reduction (as above) to show that the list of f corresponds bijectively to the list of ϕ, where ϕ is a function of order $\geqslant 4$. Thus this series, say the F-series, breaks into cases corresponding to those in Part I: the unimodal ones must be of one of the types *FT*, *FW*, *FZ*.

Case (x^2, y^2). If neither f_1 nor f_2 contains a term in z^3, then examining the lowest degree terms $(x^2 + ayz^2 + bz^4, y^2 + cxz^2 + dz^4)$ yields 2 moduli for $\mathcal{K}$-equivalence. In the contrary case, it is not hard to reduce f explicitly to one of the forms

$$G_{2n+3} \quad (x^2 + z^3, y^2 + z^n), \qquad n \geqslant 3,$$
$$G_{2n+6} \quad (x^2 + z^3, y^2 + xz^n), \qquad n \geqslant 2.$$

The Milnor numbers here are calculated from the Greuel-Hamm formula **[9]** (weighted homogeneous case). The singularities G_9 and G_{10} are 0-modal.

Case (x^2, xy). The results in this case have been obtained by direct calculation (series H).

Finally there are cases when the 2-jet is yet more degenerate. However, a direct calculation shows that the generic case with 2-jet $(xy, 0)$ is already 2-modal (at least), so we do not need to look further.

In the cases when linear x-reduction applies we can show using the Buchweitz-Greuel formula [3] for curves that the Milnor number $\mu(L_x f) = \mu(f) - 1$.

SURFACE SINGULARITIES: ORGANISING THE LIST

It is useful to begin by analysing linear reductions of the 2-jet $j^2 f$ into $j^3\phi$, as there are so many types of pencil. We find (at least for regular pencils):

(1) The pencil can be written in prenormal form $\Leftrightarrow$ the base curve B contains a regular point $P \Leftrightarrow$ the pencil contains no repeated plane.

If the base curve has several simple components, there will be several choices of reduction.

(2) If $\Delta(\lambda, \mu)$ is the discriminant of $\lambda f_1 + \mu f_2$, then $\Delta(y_2, -y_1) = 0$ is the equation of the set of 4 tangents from $Z(0, 0, 1)$ to $\Gamma(\phi(y_1, y_2, y_3) = 0)$. Note that Z always lies on Γ. Moreover, the roots of Δ corresponding to plane-pairs in the pencil give just those lines through Z which are components of Γ.

(3) Z is singular on $\Gamma \Leftrightarrow$ the component of B containing P is a straight line.

From these simple observations we can sort out which types of pencil reduce to which types of cubic Γ. There are a score of cases, which we classify into three groups.

I. f is hyperbolic $\Leftrightarrow$ Δ has no root of multiplicity $> 2 \Leftrightarrow \Gamma$ has nodes at worst $\Leftrightarrow \phi$ is hyperbolic.

II. We have one of the cases in the table:

Segre symbol of f	3, 1	4	(2, 1) 1	(3, 1)	(3, 1)	4	1, 1, 1
Series for ϕ	Q	S	S	U	S	Q	U

Here the symbol 1, 1, 1 represents a singular pencil. The base loci for 4 and (3, 1) are reducible (as cubic + line resp. conic + 2 lines), so there are 2 choices for P. We return to these cases below.

III. If f is a regular pencil with symbol (2, 2), or a singular one of type $(xy, xz + w^2)$ or of type 2, 1, then ϕ is of type V (repeated line plus line). If we return now to mappings rather than jets, these ϕ's have $\mathcal{K}$-modality $\geqslant 2$; the corresponding f's likewise. As in earlier cases, this is seen on the bottom stratum by considering terms of lowest weight. Any other singular pencil is a specialisation of 2, 1 so none of these can give rise to $\mathcal{K}$-unimodal germs. Similarly we have specialisations $(2, 2) \to (2, 1, 1) \to (1, 1, 1, 1)$, so we need not consider these cases further. And the pencil of type (1, 1, 1), 1 is easily eliminated by using the splitting theorem. Thus it only remains to check the cases in group II.

Series I: symbol 1, 1, 1. A closer examination here shows that there is a single infinite series (there are ambiguities in reduction, as B has 4 components, but the symmetry sorts these out).

Series J: symbol 3, 1. Here we recover the classification we could have obtained using the splitting theorem.

Series K: symbol (2, 1),1. Linear reduction shows that the cases arising correspond to those in the S-series. Here we use the splitting theorem in reverse to obtain the list for (2, 1).

Series L: symbol 4. Here linear reduction gives a correspondence to cases in the S-series.

Note that correspondence with the Q-series does not work well. Taking normal form

$$f(w, x, y, z) = \left(wz + xy,\ y^2 + xz + x^2a(w) + xb(w) + c(w)\right)$$

gives

$$L_z f(w, x, y) = -x^2y + wy^2 + w\left(x^2a(w) + xb(w) + c(w)\right),$$

in normal form for the S series. If in f we take an alternative normal form, replacing $x^2a(w)$ by $yd(w)$, we can form

$$L_x f(w, y, z) = y^3 + y^2d(w) + yc(w) - wz^2 - wzb(w)$$

whose position in the Q-series is not clear.

Series M: symbol (3, 1). Here linear reduction gives a good correspondence with the U-series.

We now present a table of our results. To save space we shall only indicate the strata and not the complete $\mathcal{K}$-classification in each case.

If $\phi = L_x f$, and f is weighted homogeneous, so is ϕ. Using the Greuel-Hamm formula [**9**] for Milnor numbers, we find $\mu(\phi) = \mu(f) + 1$. It seems certain that this holds in general, and likely that an argument using resolutions will also give equality of μ_0 and μ_+.

Notation in the tables. Most strata contain weighted homogeneous singularities. In these cases, these examples are shown: the weights can easily be inferred. λ denotes the modulus when there is one.

The other strata are (a) all those with $i > 0$ in the notation, (b) $T_{p,q,r}$, $T_{p,q,r,s}$, $FT_{k,l}$ (apart from $T_{6,3,2}$, $T_{4,4,2}$, $T_{3,3,3}$, $T_{2,2,2,2}$ and $FT_{4,4}$) and (c) series HA, HB and HC. Each of these cases consists of a single $\mathcal{K}$-orbit.

I have set

$$l_i(x, y) = \begin{Bmatrix} xy^q & i = 2q, \\ y^{q+2} & i = 2q+1 \end{Bmatrix}$$

for brevity.

In (5), (12), (19), where weights are listed, the notation is $w_1, w_2, \ldots / d_1, \ldots,$ where w_i = weight x_i and d_i = weight $f_i(\mathbf{x})$.

TABLES OF $\mathcal{K}$-UNIMODALS. I. HYPERSURFACES

(1) A_n $\quad x^{n+1} \quad (n \geqslant 1)$,

D_n $\quad x^{n-1} + xy^2 \quad (n \geqslant 4)$,

E_6, $\quad E_7$, E_8 (as below).

(2) $T_{p,q,r}$ $\quad x^p + y^q + z^r + \lambda xyz \quad (p \geqslant q \geqslant r \geqslant 3)$,

$x^p + y^q + \lambda x^2y^2 \quad (r = 2;\ p \geqslant q \geqslant 4$ or $p \geqslant 6,\ q = 3)$,

$\mu = p + q + r - 1$. If $p^{-1} + q^{-1} + r^{-1} < 1$, take $\lambda = 1$.

(3)

E_{6m+6}	$x^3 + y^{3m+4}$,
E_{6m+7}	$x^3 + xy^{2m+3}$,
E_{6m+8}	$x^3 + y^{3m+5}$,
$E_{m+1,0}$	$x^3 + x^2y^{m+1} + \lambda y^{3m+3}$ $(\lambda \neq 0, -\frac{4}{27})$,
$E_{m+1,i}$	$x^3 + x^2y^{m+1} + y^{3m+3+i}$,
Z_{6m+5}	$y(x^3 + y^{3m+1})$,
Z_{6m+6}	$y(x^3 + xy^{2m+1})$,
Z_{6m+7}	$y(x^3 + y^{3m+2})$,
$Z_{m-1,0}$	$y(x^3 + x^2y^m + \lambda y^{3m})$ $(\lambda \neq 0, -\frac{4}{27})$,
$Z_{m-1,i}$	$y(x^3 + x^2y^m + y^{3m+i})$,
Q_{6m+4}	$yz^2 + x^3 + y^{3m+1}$,
Q_{6m+5}	$yz^2 + x^3 + xy^{2m+1}$,
Q_{6m+6}	$yz^2 + x^3 + y^{3m+2}$,
$Q_{m,0}$	$yz^2 + x^3 + x^2y^m + \lambda y^{3m}$ $(\lambda \neq 0, -\frac{4}{27})$,
$Q_{m,i}$	$yz^2 + x^3 + x^2y^m + y^{3m+i}$.

Take $m \geqslant 1$ throughout, except for E_6, E_7, E_8. However,

$$E_{2,i} = T_{6+i,3,2},\ Z_{0,i} = T_{4+i,4,2},\ Q_{1,i} = T_{3+i,3,3} \qquad (\text{for } i \geqslant 0).$$

$$\mu(E_{m+1,i}) = 6m + 4 + i,\ \mu(Z_{m-1,i}) = 6m + 3 + i,\ \mu(Q_{m,i}) = 6m + 2 + i.$$

In each case, m is the $\mathcal{R}$-modality.

(4)

W_{12}	$x^4 + y^5$,
W_{13}	$x^4 + xy^4$,
$W_{1,0}$	$x^4 + x^2y^3 + \lambda y^6$ $(\lambda \neq 0, \frac{1}{4})$,
$W_{1,i}$	$x^4 + x^2y^3 + y^{6+i}$,
$W^{\#}_{1,i}$	$(x^2 + y^3)^2 + xy^3 l_i(x, y)$,
W_{17}	$x^4 + xy^5$,
W_{18}	$x^4 + y^7$,
S_{11}	$z^2y + zx^2 + y^4$,
S_{12}	$z^2y + zx^2 + xy^3$,
$S_{1,0}$	$z^2y + zx^2 + x^2y^2 + \lambda y^5$ $(\lambda \neq 0, -1)$,
$S_{1,i}$	$z^2y + zx^2 + x^2y^2 + y^{5+i}$,
$S^{\#}_{1,i}$	$z^2y + zx^2 + x^2y^2 - y^5 + xy^2 l_i(x, y)$,
S_{16}	$z^2y + zx^2 + xy^4$,
S_{17}	$z^2y + zx^2 + y^6$,
U_{12}	$x^2z + z^3 + y^4$,
$U_{1,0}$	$x^2z + z^3 + zy^3 + \lambda xy^3$ $(\lambda \neq 0, \pm i)$,
$U_{1,i}$	$x^2z + z^3 + zy^3 + xy^2 l_{i-1}(x, y)$,
U_{16}	$x^2z + z^3 + y^5$,

$\mu(W_{1,i}) = \mu(W^{\#}_{1,i}) = 15 + i$,

$\mu(S_{1,i}) = \mu(S^{\#}_{1,i}) = \mu(U_{1,i}) = 14 + i$.

(5) Cases which are *not* $\mathcal{K}$-unimodal contain one (at least) of the following nearby:

Series	*Corank*	μ	*Weights*	*Codimension*
X	2	21	2, 1/8	16
S	3	20	3, 1, 2/7	15
U	3	20	2, 1, 2/6	14
N	2	16	1, 1/5	12
V	3	15	3, 2, 2/8	11
O	4	16	1, 1, 1, 1/3	10

The codimension c here is that of the stratum in the space of singular jets. For a stable map $f\colon N^n \to P^p$, the corresponding set $A(f)$ has dimension $p - 1 - c$, so $f(A(f))$ has codimension $c + 1$ in P.

TABLE II. SPACE CURVES (MAPS $\mathbf{C}^3 \to \mathbf{C}^2$)

(6) $P_{k,l}$ $\quad (xy, x^k + y^l + z^2) \quad (k \geqslant l \geqslant 2)$,
$\mu(P_{k,l}) = k + l + 1$.

(7) $FT_{4,4}$ $\quad (xy + z^3, xz + y^3 + \lambda yz^2)$,
$FT_{k,l}$ $\quad (xy + z^{l-1}, xz + y^{k-1} + yz^2) \quad (k \geqslant l \geqslant 4, k > 4)$,
FW_{13} $\quad (xy + z^3, xz + y^4)$,
FW_{14} $\quad (xy + z^3, xz + zy^3)$,
$FW_{1,0}$ $\quad (xy + z^3, xz + z^2y^2 + \lambda y^5) \quad (\lambda \neq 0, -\frac{1}{4})$,
$FW_{1,i}$ $\quad (xy + z^3, xz + z^2y^2 + y^{5+i})$,
$FW^{\#}_{1,i}$ $\quad (xy + z^3, xz + 2z^2y^2 - y^5 + zy^2 l_i(z, y))$,
FW_{18} $\quad (xy + z^3, xz + zy^4)$,
FW_{19} $\quad (xy + z^3, xz + y^6)$,
FZ_{6m+6} $\quad (xy, xz + z^3 + y^{3m+1})$,
FZ_{6m+7} $\quad (xy, xz + z^3 + zy^{2m+1})$,
FZ_{6m+8} $\quad (xy, xz + z^3 + y^{3m+2})$,
$FZ_{m-1,0}$ $\quad (xy, xz + z^3 + z^2y^m + \lambda y^{3m}) \quad (\lambda \neq 0, -\frac{4}{27})$,
$FZ_{m-1,i}$ $\quad (xy, xz + z^3 + z^2y^m + y^{3m+i})$,
$\mu(FT_{k,l}) = k + l + 2$, $\mu(FW_{1,i}) = \mu(FW^{\#}_{1,i}) = 16 + i$,
$\mu(FZ_{m-1,i}) = 6m + 4 + i$.

(8) G_{n+8} $\quad (x^2 + z^3, y^2 + zl_n(x, z)) \quad (n \geqslant 1)$,
$G_7 = P_{3,3}$, $G_8 = K_8$.

(9) HA_{r+11} $\quad (xy + z^3, x^2 + z^3 + yz^2 + y^{3+r}) \quad (r \geqslant 0)$,
HB_{r+13} $\quad (xy + z^3, x^2 + yz^2 + y^{4+r}) \quad (r \geqslant 0)$,
HC_{13} $\quad (xy + z^3, x^2 + z^3 + y^4)$,
HC_{14} $\quad (xy + z^3, x^2 + z^3 + zy^3)$,
HC_{15} $\quad (xy + z^3, x^2 + z^3 + y^5)$,
HD_{13} $\quad (xy + z^3, x^2 + zy^2)$,
HD_{14} $\quad (xy + z^3, x^2 + y^3)$.

(10) J_{6m+7} $(xy + z^2, xz + y^{3m+3})$,
J_{6m+8} $(xy + z^2, xz + zy^{2m+2})$,
J_{6m+9} $(xy + z^2, xz + y^{3m+4})$,
$J_{m+1,0}$ $(xy + z^2, xz + z^2y^m + \lambda y^{3m+2})$ $(\lambda \neq 0, -\frac{4}{27})$,
$J_{m+1,i}$ $(xy + z^2, xz + z^2y^m + y^{3m+2+i})$.
Here $m \geqslant 0$, but $J_{1,i} = P_{i+2,2}$. $\mu(J_{m+1,i}) = 6m + i + 5$.

(11) K_8 $(xy + z^2, x^2 + y^3)$,
K_9 $(xy + z^2, x^2 + zy^2)$,
$K_{1,0}$ $(xy + z^2, x^2 + z^2y + \lambda y^4)$ $(\lambda \neq 0, \frac{1}{4})$,
$K_{1,i}$ $(xy + z^2, x^2 + z^2y + y^{4+i})$,
$K^{\#}_{1,i}$ $(xy + z^2, x^2 + 2z^2y + y^4 + zyl_i(z, y))$,
K_{13} $(xy + z^2, x^2 + zy^3)$,
K_{14} $(xy + z^2, x^2 + y^5)$,
$\mu(K_{1,i}) = \mu(K^{\#}_{1,i}) = 11 + i$.
The 0-modal cases in Table II are [**8**] $P_{k,2}$ $(k \geqslant 2)$, $P_{3,3}$, $P_{4,3}$, $P_{5,3}$, J_7, J_8, J_9, K_8, K_9, G_9 and G_{10}.
(12) Cases which are *not* $\mathcal{K}$-unimodal contain one (at least) of the following nearby:

Series	μ	*Weights*	*Codimension*
F	22	2, 1, 5/7, 6	14
F	17	1, 1, 3/4, 4	9
G	13	2, 2, 1/4, 4	6
H	17	none*	11
H	19	2, 1, 1/3, 4	9
K	17	3, 1, 2/4, 6	9
(1, 1, 1)	13	1, 1, 1/2, 3	5

*$(xy + z^3, x^2 + z^3 + az^2y^2 + bzy^4 + dy^6)$.

Here c is as in (5); for f stable, $A(f)$ has codimension $c + 6$ and $f(A(f))$ has codimension $c + 5$.

TABLE III. SURFACES (MAPS $\mathbf{C}^4 \to \mathbf{C}^2$)

(13) $T_{2,2,2,2}$ $(x_1^2 + x_2^2 + x_3^2, x_2^2 + \lambda x_3^2 + x_4^2)$ $(\lambda \neq 0, 1)$,
$T_{p,q,r,s}$ $(x_1x_2 + y_1^q + y_2^s, y_1y_2 + x_1^p + x_2^r)$ $(p, q, r, s \geqslant 2)$,
$\mu(T_{p,q,r,s}) = p + q + r + s - 1$.
(14) $I_{1,0}$ $(x(y - z) + w^3, y(z - x) + \lambda w^3)$ $(\lambda \neq 0, 1)$,
$I_{1,i}$ $(x(y - z) + w^3, y(z - x) + w^2l_{i-1}(x, w))$,
$\mu(I_{1,i}) = 13 + i$.

(15) J'_{6m+9} $(xy + z^2, w^2 + xz + y^{3m+3})$,
J'_{6m+10} $(xy + z^2, w^2 + xz + zy^{2m+2})$,
J'_{6m+11} $(xy + z^2, w^2 + xz + y^{3m+4})$,
$J'_{m+1,0}$ $(xy + z^2, w^2 + xz + z^2y^m + \lambda y^{3m+2})$ $(\lambda \neq 0, -\frac{4}{27})$,
$J'_{m+1,i}$ $(xy + z^2, w^2 + xz + z^2y^m + y^{3m+2+i})$,
$\mu(J'_{m+1,i}) = 6m + i + 7$,
$m \geqslant 0$, but $J'_{1,i} = T_{i+2,2,2,2}$.

(16) K'_{10} $(xy + z^2, w^2 + x^2 + y^3)$,
K'_{11} $(xy + z^2, w^2 + x^2 + zy^3)$,
$K'_{1,0}$ $(xy + z^2, w^2 + x^2 + z^2y + \lambda y^4)$ $(\lambda \neq 0, \frac{1}{4})$,
$K'_{1,i}$ $(xy + z^2, w^2 + x^2 + z^2y + y^{4+i})$,
$K^{\flat}_{1,i}$ $(xy + z^2, w^2 + x^2 + 2z^2y + y^4 + zyl_i(z, y))$,
K'_{15} $(xy + z^2, w^2 + x^2 + zy^3)$,
K'_{16} $(xy + z^2, w^2 + x^2 + y^5)$,
$\mu(K'_{1,i}) = \mu(K^{\flat}_{1,i}) = 13 + i$.

(17) L_{10} $(wz + xy, y^2 + xz + w^3)$,
L_{11} $(wz + xy, y^2 + xz + xw^2)$,
$L_{1,0}$ $(wz + xy, y^2 + xz + x^2w + \lambda w^4)$ $(\lambda \neq 0, -1)$,
$L_{1,i}$ $(wz + xy, y^2 + xz + x^2w + w^{4+i})$,
$L^{\#}_{1,i}$ $(wz + xy, y^2 + xz + x^2w - w^4 + wxl_i(x, w))$,
L_{15} $(wz + xy, y^2 + xz + xw^3)$,
L_{16} $(wz + xy, y^2 + xz + w^5)$,
$\mu(L_{1,i}) = \mu(L^{\#}_{1,i}) = 13 + i$.

(18) M_{11} $(2wy + x^2 - z^2, 2wx + y^3)$,
$M_{1,0}$ $(2wy + x^2 - z^2, 2wx + y^2x + \lambda y^2z)$ $(\lambda \neq \pm 1)$,
$M_{1,i}$ $(2wy + x^2 - z^2, 2wx + y^2x + zl_{i+1}(z, y))$,
$M^{\#}_{1,i}$ $(2wy + x^2 - z^2, 2wx + y^2(x - z) + zl_{i+1}(z, y))$,
M_{15} $(2wy + x^2 - z^2, 2wx + y^4)$,
$\mu(M_{1,i}) = \mu(M^{\#}_{1,i}) = 13 + i$.

There are no 0-modals in Table III.

(19) Cases which are *not* $\mathcal{K}$-unimodal have one (at least) of the following nearby:

Series	μ	*Weights*	*Codimension*
I	19	1, 2, 2, 2/4, 4	7
K	19	1, 2, 3, 3/4, 6	9
L	19	1, 2, 3, 4/5, 6	9
M	19	1, 2, 2, 3/4, 5	8
(2, 2)	14	2, 2, 3, 3/5, 6	5
1 ; 1	14	2, 2, 3, 4/6, 6	5
(1, 1, 1) 1	15	2, 2, 2, 3/4, 6	5

Here again, c is as in (5). If f is stable, the stratum $A(f)$ has codimension $c + 8$, and $f(A(f))$ has codimension $c + 6$.

(20) The codimensions ${}^2\sigma(n, p)$ of the set of bimodal map-germs (see [**15**]) are as follows, for $n > p$: the critical cases are as shown below.

With the exceptions of ${}^2\sigma(2,1) = {}^2\sigma(3,1) = 14$, the numbers ${}^2\sigma(n, p) = \tau(n-p)$ depend only on $n - p$:

$\tau(1) = 11$		final case in (12),
$\tau(2) = 13$		final cases in (19),
$\tau(k) = 2k + 4$	for $3 \leqslant k \leqslant 7$	maps of rank $(p-2)$,
$\tau(k) = k + 11$	for $k \geqslant 7$	final case in (5).

Recognition of the list

The significance of Arnold's classification of $\mathcal{R}$-unimodal singularities lay in the fact that the lists had alternative characterisations: we thus seek the same here. I have not yet succeeded in interpreting Table II. For the others, it seems that the nice characterisations hold for strict unimodals.

We have the deformations

$E_{4,0} \to T_{2,3,6} + T_{2,3,6}$

$$y^3 + ay(x^2 - t^2)^2 + b(x^2 - t^2)^3 = f_3(y, x^2 - t^2),$$

$Z_{2,0} \to T_{2,3,6} + T_{2,4,4}$

$$\{x + (y - 2t)\}f_3(x, (y - 2t)(y + t)^2),$$

$Q_{3,0} \to T_{2,3,6} + T_{3,3,3}$

$$(y - 2t)z^2 + f_3(x, (y - 2t)(y + t)^2),$$

$J'_{3,0} \to T_{2,3,6} + T_{2,2,2,2}$

$$\{x(y - 2t) + z^2, w^2 + xz + az^2(y + t)^2 + b(y - 2t)^2(y + t)^6\};$$

$FZ_{2,0} \to T_{2,3,6} + FT_{4,4}$

$$(xy, xz + f_3(z, y(y - t)^2)),$$

$J_{4,0} \to J_{2,0} + J_{2,0}$

$$(xy + f_3(z, (y^2 - t^2)^2), xz).$$

Thus the strict $\mathcal{K}$-unimodals in Table I are those in (1), (2) and (4) together with those in (3) with $m \leqslant 2$. This coincides exactly with Arnold's list of singularities of $\mathcal{R}$-modality $\leqslant 2$. In Table II we have to exclude the FZ and the J with $m \geqslant 3$; and in Table III the J' with $m \geqslant 2$.

We now compare with Laufer's list [**13**] of minimally elliptic singularities. However, some of these are more minimal than others, and I am grateful to Jonathan Wahl for explaining this to me in detail. Call such a singularity *strictly minimal* if its resolution is either one of the exceptional cases of [**13**, (3.5)] (the exceptional set isomorphic to a plane cubic curve) or the exceptional set consists of embedded rational curves whose dual graph is an extended Dynkin diagram

($\tilde{A}_n$, $\tilde{D}_n$ or $\tilde{E}_n$). In the hypersurface case, this allows $z^2 + f(x, y)$ with f of order 3 or 4, but excludes f or order 5; for $g(x, y, z)$ it allows just the cases when the terms of order 3 in g define a reduced cubic curve in $P_2(\mathbf{C})$. Then we have

THEOREM. *The strictly $\mathcal{K}$-unimodal isolated singularities of complete intersection surfaces are precisely the strictly minimally elliptic surface singularities.*

With this insight, we can recognise the various classes, and also verify (using the fact that the genus increases on specialisation) that singularities in our list are indeed unimodal. Parabolic (or simple-elliptic) singularities ($T_{2,3,6} = \tilde{E}_8$, $T_{2,4,4} = \tilde{E}_7$, $T_{3,3,3} = \tilde{E}_6$, $T_{2,2;2,2} = \tilde{D}_5$) are well known [**17**]: the resolution graph is a nonsingular elliptic curve. Each type is classified by a modulus (that of the curve).

Hyperbolic (or cusp) singularities (the other $T_{p,q,r}$ and $T_{p,q;r,s}$) are also well known [**12**]. The resolution graph has type $\tilde{A}_n$ ($n \geqslant 0$: interpret $n = 0$ as a nodal plane cubic, $n = 1$ as two rational curves meeting transversely twice). Each type has a unique equivalence class.

Triangle singularities include Arnold's 14 exceptional unimodals, also J'_9, J'_{10}, J'_{11}, K'_{10}, K'_{11}, L_{10}, L_{11}, M_{11}. These were studied by Dolgachev [**4**]. The resolution graphs have nonordinary singularities (cuspidal cubic, conic + tangent, concurrent lines). In each case there are 2 $\mathcal{K}$-classes: one weighted homogeneous and one not.

'Quadrilateral' singularities (uniformised by a hyperbolic quadrilateral) correspond to cases with a modulus λ ($E_{3,0}$, $Z_{1,0}$, $Q_{2,0}$, $W_{1,0}$, $S_{1,0}$, $U_{1,0}$ also $I_{1,0}$, $J'_{2,0}$, $K'_{1,0}$, $L_{1,0}$ and $M_{1,0}$). Each has resolution graph of type $\tilde{D}_4$ and starts one or two series, with resolution graph $\tilde{D}_{4+i}$. For each value of λ, there are 2 equivalence classes (just one weighted homogeneous): but each $\tilde{D}_{4+i}$ case has a unique equivalence class.

Finally, resolution graphs of type $\tilde{E}_n$ correspond to Arnold's 14 exceptional bimodals, also J'_{15}, J'_{16}, J'_{17}, K'_{15}, K'_{16}, L_{15}, L_{16} and M_{15}. Here each case contains 3 classes for $\mathcal{K}$-equivalence.

I have not yet achieved a corresponding insight into the curve case. However, in case H it is not hard to show that $\mathcal{K}$-unimodals must have the form $(xy + z^3, x^2 + u(y, z))$. Eliminating x gives $\phi(y, z) = z^6 + y^2 u(y, z)$ (where u has order $\geqslant 3$). The list corresponds to a subset of Laufer's list of (nonstrictly) minimally elliptic $x^2 + \phi(y, z)$.

REFERENCES

1. V. I. Arnold, *Classification of unimodular critical points of functions*, Functional Anal. Appl. **7** (1973), 75–76.
2. ______, *Local normal forms of functions*, Invent. Math. **35** (1976), 87–109.
3. R. O. Buchweitz and G.-M. Greuel, *The Milnor number and deformations of complex curve singularities*, Invent. Math. **58** (1980), 241–281.
4. I. V. Dolgachev, *Factor-conical singularities of complex hypersurfaces*, Functional Anal. Appl. **8** (1974), 160–161.
5. W. Ebeling, *Quadratische Formen und Monodromiegruppen von Singularitäten*, Math. Ann. **255** (1981), 463–498.

6. A. M. Gabrielov, *Bifurcations, Dynkin diagrams and the moduli number for isolated singularities*, Functional Anal. Appl. **8** (1974), 7–13.

7. A. Dimca and C. G. Gibson, *Contact germs from the plane to the plane*, these PROCEEDINGS.[1]

8. M. Giusti, *Classification des singularités isolées d'intersections complètes simples*, C. R. Acad. Sci. Paris Sér. A-B **284** (1977), 167–169.

9. G.-M. Greuel and H. A. Hamm, *Invarianten quasihomogener vollständiger Durchschnitte*, Invent. Math. **49** (1978), 67–86.

10. D. Gromoll and W. Meyer, *On differentiable functions with isolated critical points*, Topology **8** (1969), 361–369.

11. H. A. Hamm, *Lokale topologische Eigenschaften Komplexer Räume*, Math. Ann. **191** (1971), 235–252.

12. U. Karras, *Deformations of cusp singularities*, Proc. Sympos. Pure Math., vol. 30(1), Amer. Math Soc., Providence, R.I., 1977, pp. 37–44.

13. H. B. Laufer, *On minimally elliptic singularities*, Amer. J. Math. **99** (1977), 1257–1295.

14. J. N. Mather, *Stability of C^∞-mappings*. VI. *The nice dimensions*, Lecture Notes in Math., vol. 192, Springer-Verlag, Berlin and New York, 1971, pp. 207–255.

15. A. A. du Plessis, *Genericity and smooth finite determinancy*, these PROCEEDINGS.

16. M. Rosenlicht, *A remark on quotient spaces*, An. Acad. Brasil Ci. **35** (1963), 487–489.

17. K. Saito, *Einfach-elliptische Singularitäten*, Invent. Math. **23** (1974), 289–325.

18. R. Thom, *Structural stability and morphogenesis*, Benjamin, New York, 1973.

19. C. T. C. Wall, *A splitting theorem for maps into* $\mathbf{R}^2$, Math. Ann. **259** (1982), 443–453.

UNIVERSITY OF LIVERPOOL, ENGLAND

[1] See also A. Dimca and C. G. Gibson, *Contact unimodular germs of the plane*, preprint no. 78/1981, Inst. Math. of Increst, Budapest.

Proceedings of Symposia in Pure Mathematics
Volume **40** (1983), Part 2

b-FUNCTIONS AND EXPONENTS OF HYPERSURFACE ISOLATED SINGULARITIES

TAMAKI YANO

1. Introduction.

(1.1) In 1961, M. Sato introduced a-, b- and c-functions associated with prehomogeneous vector spaces [**S1**, **2**, **4**]. The existence of b-functions for polynomials and holomorphic functions was established by [**Be3** and **Bj1**], respectively. As first clarified by [**M2, 4**], there is an intimate connection between the b-function of a function f and the singularity structure of $f^{-1}(0)$. The b-function is also called "Bernstein polynomial" or "Bernstein-Sato polynomial".

(1.2) The prototype of a b-function is the identity for a quadratic form $Q(x) = \sum_{i=0}^{n} x_i^2$:

$$\left(\sum_{i=0}^{n} (\partial/\partial x_i)^2 \right) Q(x)^{s+1} = 4(s+1)(s+(n+1)/2)Q(x)^s.$$

Here, the b-function $b(s) = (s+1)(s+(n+1)/2)$. This type of equality can be used for analytic continuation of the complex power of a function and it was in that situation where I. N. Bernstein proved the existence (cf. [**GS**] also).

(1.3) When f has an isolated singularity, the polynomial $b(s)/(s+1)$ equals the minimal polynomial of the minus of the residue of the Gauss-Manin connection on the standard saturated lattice [**M4**]. We can naturally construct the saturated lattice $\tilde{\mathcal{H}}$ of $\mathcal{H} = \Omega^{n+1}/df \wedge d\Omega^{n-1}$ in an algebro-analytic way. There are some results in this direction when f has a nonisolated singularity [**H2**, **Ks4**, **SM1**].

(1.4) Roots of $b(s) = 0$ are negative rational and they relate when so-called resolution exponents, Newton polyhedra and polar invariants [**Kt**, **L1**, **2**, **V1**, **Y10**, **15**].

1980 *Mathematics Subject Classification*. Primary 57R45.

0082-0717/81/0000-0665/$03.50

2. b-function.

(2.1) Let X be a complex manifold of dimension n with the origin 0, $\mathcal{O}_X$, Ω_X^p, $\mathcal{D}_X$ the sheaf of holomorphic functions, holomorphic p-forms, holomorphic linear differential operators of finite order, respectively. Let s be an indeterminate commuting with elements of $\mathcal{D}_X$ and set $\mathcal{D}_X[s] = \mathcal{D}_X \otimes_{\mathbf{C}} \mathbf{C}[s]$. Let $(x_1, \ldots, x_n)$ be local coordinates around 0. We use the notation $D_i = \partial/\partial x_i$. For an element $P \in \mathcal{D}$, $\operatorname{ord}(P)$ denotes its order.

(2.2) DEFINITION OF b-FUNCTION. The b-function $b(s) = b_{f,0}(s)$ of f at $0 \in X$ is the monic generator of the ideal $\{b'(s) \in \mathbf{C}[s];$ there exists $P(s) \in \mathcal{D}[s]_0$ such that $P(s, x, D)f^{s+1} = b'(s)f^s\}$.

(2.3) Fix an $f(x) \in \mathcal{O}_X$, $f(0) = 0$. Define an ideal of $\mathcal{D}[s]$ by the following:

$$\mathcal{J}(s) = \left\{ P(s, x, D) \in \mathcal{D}[s]; \right.$$

$$\left. \exists m, f^m P(s, x, D + s \cdot \operatorname{grad} f/f) \in \mathbf{C}[s] \otimes \sum_{i=1}^{n} \mathcal{D} D_i \right\}.$$

Here the notation $D + s \cdot \operatorname{grad} f/f$ means the substitution $D_i \mapsto D_i + s \cdot (\partial f/\partial x_i)/f$.

We employ formal powers f^{s+k}, $k \in \mathbf{Z}$, by the fundamental relation $f \cdot f^{s+k} = f^{s+k+1}$ and $D_i f^{s+k} = (s+k)\partial f/\partial x_i \cdot f^{s+k-1}$. Set $\mathcal{N}[f^{-1}] := \bigcup_{k \geqslant 0} \mathcal{D}[s] f^{s-k}$, the $\mathcal{D}[s]$-module generated by those formal powers.

(2.4) THEOREM-DEFINITION. *The ideal* $\mathcal{J}(s)$ *is* $\mathcal{D}$*-coherent* **[Ks1]**. *Set* $\mathcal{N} = \mathcal{D}[s]/\mathcal{J}(s)$ *and* $\mathcal{N}^{(0)} = \mathcal{D}/(\mathcal{D} \cap \mathcal{J}(s))$. *Then there are isomorphisms* $\mathcal{N} \xrightarrow{\sim} \mathcal{D}[s]f^s$ *and* $\mathcal{N}^{(0)} \simeq \mathcal{D} f^s$.

(2.5) DEFINITION. Define $\mathcal{O}$-linear maps s, t and ∇ on $\mathcal{N}[f^{-1}]$ by the following rule.

$$s\colon P(s)f^{s-k} \mapsto sP(s)f^{s-k},$$
$$t\colon P(s)f^{s-k} \mapsto P(s+1)f^{s-k+1},$$
$$\nabla\colon P(s)f^{s-k} \mapsto -sP(s-1)f^{s-k-1}.$$

They satisfy relations $ts - st = 1$, $s = -\nabla t$ (or $s + 1 = -t\nabla$).

(2.6) DEFINITION-PROPOSITION. *Set* $\mathcal{M} = \mathcal{N}/t\mathcal{N}$ *and* $\tilde{\mathcal{M}} = (s+1)\mathcal{M}$. *Then there are isomorphisms*

$$\mathcal{M} \simeq \mathcal{D}[s]/(\mathcal{J}(s) + \mathcal{D}[s]f),$$

$$\tilde{\mathcal{M}} \simeq \mathcal{D}[s]/(\mathcal{J}(s) + \mathcal{D}[s](\mathfrak{A} + \mathcal{O}f)), \qquad \mathfrak{A} = \sum_{i=1}^{n} \mathcal{O}\partial f/\partial x_i.$$

By the definition, the $b(s)$ (or $\tilde{b}(s) = b(s)/(s+1)$) is the minimal polynomial of $s \in \operatorname{End}_{\mathcal{D}}(\mathcal{M})_0$ (or $\operatorname{End}_{\mathcal{D}}(\tilde{\mathcal{M}})_0$).

(2.7) THEOREM (EXISTENCE). *The ideal in* (2.2) *does not reduce to* 0. (*Hence there exists nonzero* $b(s)$.)

This theorem was proved in [**Be2**] for polynomials and [**Bj1**] for holomorphic functions (see [**Bj2**]). In fact it is a corollary of the following

(2.8) THEOREM. *The $\mathfrak{D}$-Module $\mathfrak{N}$ is subholonomic and $\mathfrak{M}$ is holonomic* [**Ks1**].

A $\mathfrak{D}$-Module $\mathfrak{L}$ is called holonomic (resp., subholonomic) if its singular support $S\check{S}(\mathfrak{L})$ in the cotangent bundle T^*X is n ($= \dim X$) (resp., $n+1$) dimensional (see [**Me3**] and reference cited there).

Once (2.8) is established, we apply a theorem of M. Kashiwara which reads "Let $\mathfrak{L}_1$ and $\mathfrak{L}_2$ be holonomic. Then $\mathrm{Ext}^i_{\mathfrak{D}}(\mathfrak{L}_1, \mathfrak{L}_2)_x$ is a finite dimensional vector space for i." Then $\mathrm{End}_{\mathfrak{D}}(\mathfrak{M})_0$ is finite-dimensional and hence s has a nonzero minimal polynomial $b(s)$. (We refer the reader to [**Me3**] for the theory of $\mathfrak{D}$-modules.)

(2.9) THEOREM (RATIONALITY). *Let α satisfy $b(-\alpha) = 0$. Then α is positive rational.*

This theorem was proved in [**M4**] for isolated singularities and in [**Ks1** and **M5**] for the general case.

(2.10) EXAMPLE. The b-function of a cusp.

$$\left(D_y^3/27 + yD_x^2D_y/6 + D_x^3x/8\right)\left(x^2+y^3\right)^{s+1} = (s+1)(s+5/6)(s+7/6)\left(x^2+y^3\right)^s.$$

(2.11) *Total order.* Let $P(s, x, D) = \sum P_i(x, D)s^i$ be an element of $\mathfrak{D}[s]$. We define the total order of P by counting the order of s as one, that is,

$$\mathrm{ord}^T(P) = \max\{\mathrm{ord}(P_i) + i\}.$$

One may notice that in Examples (1.2) and (2.10), an equality $\mathrm{ord}^T(P) = \deg b(s)$ holds for P and b in the formula $P(s, x, D)f^{s+1} = b(s)f^s$. This is always true.

(2.12) THEOREM. *There exists $P(s) \in \mathfrak{D}[s]$ satisfying*

$$P(s)f^{s+1} = b(s)f^s \quad \textit{and} \quad \mathrm{ord}^T(P) = \deg b(s)$$

for b-function $b(s)$.

This theorem was proved in [**Y7**] for several types of functions and in [**Ks4**] for the general cases.

(2.13) DEFINITION. Fix a complex number c. Define ideals

$$\mathfrak{J}(c) = \{P \in \mathfrak{D};\ P = Q(c) \text{ for some } Q(s) \in \mathfrak{J}(S)\},$$

$$\mathfrak{J}_c = \left\{P \in \mathfrak{D};\ f^m P(x, D + c \cdot \mathrm{grad}\, f/f) \in \sum_{i=1}^n \mathfrak{D}D_i \text{ for some } m\right\}.$$

The module $\mathfrak{N}/(s-c)\mathfrak{N} = \mathfrak{N}_c$ is isomorphic to $\mathfrak{D}/\mathfrak{J}(c)$. The class 1 mod $\mathfrak{J}(c)$ is denoted by $(f^s)_c$. Consider the module $\mathfrak{D}/\mathfrak{J}_c$ and denote 1 mod $\mathfrak{J}_c$ by f^c. There is a canonical surjection $\mathfrak{N}_c \to \mathfrak{D}f^c$.

(2.14) THEOREM. *$\mathfrak{N}_c$ and $\mathfrak{D}f^c$ are holonomic* [**Ks1**].

In general, those two modules are not isomorphic. Set $R = \{r \in \mathbf{Q};\ b(r) = 0\}$, the set of roots of $b(s)$.

(2.15) THEOREM. $\mathfrak{N}_s \overset{\sim}{\to} \mathfrak{D}f^c$ *if and only if* $C \notin R + N$, *where* N *is the set of natural numbers* 1, 2,….

This was proved in **[Y9]** (also in **[Ks1]** for one direction). (2.14) has an interesting corollary **[Me3]**.

(2.16) COROLLARY. *Let* Y *be the hypersurface defined by* $f = 0$, $U = X \setminus Y$, j: $U \hookrightarrow X$. *The sheaf* $\mathcal{O}[*Y]$ *of meromorphic functions along* Y *is holonomic. There is an isomorphism* $\mathfrak{D}^\infty \otimes_{\mathfrak{D}} \mathcal{O}[*Y] \overset{\sim}{\to} j_*\mathcal{O}_U$, *where* $\mathfrak{D}^\infty$ *is the sheaf of differential operators of infinite order.*

This is due to **[Me1]**. Note that the existence of $b(s)$ implies that there is N such that $\mathcal{O}[*Y] \simeq \mathfrak{D}(f^{-N})$. Therefore, $\mathfrak{D}^\infty(f^{-N}) \simeq j_*\mathcal{O}_U$ by (2.16). Moreover, one can represent an element g of $j_*\mathcal{O}_U$ as $P \cdot f^{-N}$, $P \in \mathfrak{D}^\infty$, in an explicit manner using (2.12). This constructible proof of (2.16) was also noted in **[Me1]**.

3. Isolated singularities.

(3.1) In this section, we assume that f has an isolated critical point at 0. Then the ideal $\mathfrak{A}$ (cf. (2.6)) contains a power of the maximal ideal and hence the singular support of $\tilde{\mathfrak{M}}$ is concentrated in the conormal bundle of the origin, $T^*_{\{0\}}X$. Therefore, $\tilde{\mathfrak{M}}$ is a finite direct sum of $\mathcal{B}_{\{0\}} = \mathfrak{D}\delta(x)$, the $\mathfrak{D}$-module generated by the delta function. Thus we reduce the determination of $\tilde{b}(s)$ to the following **[Ks1]**

(3.2) PROPOSITION. *The* $\tilde{b}(s)(= b(s)/(s+1))$ *is the minimal polynomial of* s *in* $F = (\Omega^n \otimes_{\mathfrak{D}} \tilde{\mathfrak{M}})_0$ (*or in* $\mathrm{Hom}_{\mathfrak{D}}(\tilde{\mathfrak{M}}, \mathcal{B}_{\{0\}})_0$).

Concrete calculation of $b(s)$ can be found in **[Y7, 9]**.

(3.3) EXAMPLE. Let f be a quasihomogeneous isolated singularity with weight $(a_1, \ldots, a_n)$. Then $\tilde{\mathfrak{M}} \simeq \mathfrak{D}/\mathfrak{D}\mathfrak{A}$ and hence $F = \Omega^n/df \wedge \Omega^{n-1} \simeq \mathcal{O}/\mathfrak{A}$. The operation s is nothing but $\Sigma a_i x_i D_i =: X_0$ because of $X_0 f^s = sf^s$. Now expand the following $P(t)$ in a fractional polynomial:

$$P(t) = \prod_{i=1}^{n} \frac{t^{a_i} - t}{1 - t^{a_i}} = \Sigma q_\alpha t^\alpha.$$

Then $b(s) = \Pi_{q_\alpha \neq 0}(s + \alpha)$ (see **[Y7]**).

(3.4) DEFINITION. $r(f) :=$ the minimum of r such that $f^r \in \mathfrak{A}$.

\$l(f) :=$ the degree of integral dependence of f over $\mathfrak{A}$. $=$ the minimum of l such that $f^l \in \mathfrak{A}(\mathfrak{A} + \mathcal{O}f)^{l-1}$.

$L(f) :=$ the minimum of L such that there exists $P(s)$ in $\mathfrak{J}(s)$ of the form $P(s) = s^L + P_1 s^{L-1} + \cdots + P_L$, ord $P_j \leqslant j$.

Those three numbers are finite and $r(f) \leqslant l(f) \leqslant L(f)$. Briancon-Skoda's theorem tells us $r(f) \leqslant n$.

Other numerical invariants of the function f are $\mu(f) = \dim_C \Omega^n/df \wedge \Omega^{n-1} = \dim_C \mathcal{O}/\mathfrak{A}$, called the Milnor number, and $m(f) =$ modulus number or the modality (V. I. Arnold).

(3.5) THEOREM. *Suppose* $\mu(f) \leqslant 16$ *or* $m(f) \leqslant 2$. *Then* $L(f) \leqslant 2$.

This theorem is proved in **[Y7, 9, 13]** using Arnold's classification.

Examples of f with $L(f) = 3$ having least Milnor number are $x^2y + z^4 + yz^3 + y^5$ and $x^2y + y^2 - z^2 + y^3z + xz^3$. Both satisfy $\mu(f) = 17$, $m(f) = 3$, $r(f) = l(f) = 2$, and belong to V of Arnold's classification.

The structure of $\mathfrak{J}(s)$ when $l(f) = L(f) = 2$ and $l(f) = 2$, $L(f) = 3$ is extensively studied in **[Y7]**.

(3.6) *Exponents*. As we will see later in (4.9), $\dim F = \mu(f)$. The b-exponents of f or of $f^{-1}(0)$ are the totality of $\mu(f)$ eigenvalues $\alpha_1, \ldots, \alpha_\mu$ of $-s$ on F.

There are other exponents called MH-exponents (MH = Mixed Hodge structure) defined by **[St, SM2]**. These exponents $\beta_1, \ldots, \beta_\mu$ are symmetric in the sense that $\forall i\ \beta_i + \beta_{\mu+1-i} = n$. Moreover, they are invariant under μ-cte deformation **[SS]**.

The number $\alpha(f) = \min \alpha_i = \min \beta_i$ is called the minimal exponent.

(3.7) THEOREM. *Let f be nondegenerate with respect to its Newton polyhedron* $\Gamma(f)$. *Let t_0 be the diagonal distance to* $\Gamma(f)$, *that is*, $t_0 = \min(t; (t, \ldots, t) \in \Gamma(f))$. *Then* $\alpha(f) = t_0^{-1}$.

This theorem is proved in **[E-L]**, and also in **[Y10]** on the condition $t_0 > 1$. Mitsuo Kato also has a partial proof in case $t_0 \leqslant 1$.

(3.8) Let $\pi\colon Y \to X$ be proper projective with $Y \backslash \pi^{-1}(\mathrm{Sing}(f^{-1}(0)) \simeq X \backslash \mathrm{Sing}(f^{-1}(0))$, $\pi^{-1}(f^{-1}(0))$ is normal crossings. Define

> $\rho_Y(f) = \min\{(k+1)/m$: m is the multiplicity of a component of the normal crossing divisor $\pi^{-1}(f^{-1}(0))$, k is the multiplicity of the jacobian of π on that component$\}$.

In general $\rho_Y(f)$ depends on the choice of $\pi\colon Y \to X$ **[V1]**.

(3.9) THEOREM. $\alpha(f) \geqslant \rho_Y(f)$. *If one of them is smaller than one, they coincide* (**[Y10]**, *see also* **[Lo]**); *and hence $\rho_Y(f)$ is independent of the choice of* $\pi\colon Y \to X$.

(3.10) COROLLARY. *Let f be an irreducible plane curve with characteristic* $(n_0, b_1, \ldots, b_g)$. *Then* $\alpha(f) = 1/n_0 + 1/b_1$.

It is known that $\rho_Y(f) = 1/n_0 + 1/b_1$. For details see **[L2, I]**.

(3.11) THEOREM (3.9) holds for nonisolated singularities if we define $\alpha(f) = \min\{\alpha; s + \alpha \text{ divides } b(s)/(s+1)\}$.

4. Local monodromy, asymptotic expansion.

(4.1) Let $f\colon (\mathbf{C}^n, 0) \to (\mathbf{C}, 0)$ be a germ of holomorphic functions. For positive numbers ε and η, let $B = \{z \in \mathbf{C}^n; |z| < \varepsilon\}$, $T = \{t \in \mathbf{C}; |t| < \eta\}$, $T' = T \backslash 0$ and set $X = f^{-1}(T) \cap B$, $X' = f^{-1}(T') \cap B$. If ε is small enough and $0 < \eta \ll \varepsilon$, the mapping $f' = f|_{X'}$ defines a $\mathbf{C}^\infty$-fibre bundle (Milnor **[H1]**). Changing f if necessary, we assume $1 \in T$. For each i the generator of $\pi_1(T', 1)$ induces an automorphism h_f^i of $\tilde{H}^i(X_1, \mathbf{C})$ where $X_t = f^{-1}(t)$. This h_f^i is called the local Picard-Lefschetz monodromy in dimension i. The analogous automorphism of $\tilde{H}_i(X_1)$ is also called by the same name. We denote it by h.

(4.2) Let λ be an eigenvalue of h. Take a root vector $[\gamma_1]$ of height p in $\tilde{H}_i(X_1)$. That is,

$$(h - \lambda\,\mathrm{id})^p[\gamma_1] = 0, \quad (h - \lambda\,\mathrm{id})^{p-1}[\gamma_1] \neq 0.$$

Let $[\Gamma]$ be an element of $H_{i+1}(X, X_1)$ corresponding to $[\gamma_1]$ via the isomorphism $H_{i+1}(X, X_1) \simeq \tilde{H}_i(X_1)$. Let Ω'_X be the de Rham complex and $\mathcal{S} = \mathrm{Ker}(df\colon (\Omega'_X, d) \to (\Omega'^{+1}_X, d))$, $[\omega] \in H^{i+1}(\Gamma(X, \mathcal{S}'))$.

(4.3) THEOREM. *For $\tau \to \infty$, the integral $\int_\Gamma \exp(\tau f)\omega$ has the following asymptotic expansion*:

$$\sum_{\substack{\exp(2\pi i\mu)=\lambda \\ 0 \leqslant q \leqslant p-1}} c_{\mu,q}(\omega)\tau^{-\mu}(\log \tau)^q$$

where $c_{\mu,q}(\omega) \in \mathbf{C}$ vanishes for $\mu \leqslant 0$ **[H2]**.

(4.4) Let $\gamma(t)$ be a horizontal family in $\tilde{H}_i(X_t, C)$, $t \in T'$, such that $[\gamma(1)] = [\gamma_1]$. Then the regularity theorem of the Gauss-Manin connection entails

$$\int_{\gamma(t)}^{\omega/df} \sim \sum_{\substack{\exp(2\pi i\mu)=\lambda \\ 0 \leqslant q \leqslant p-1}} d_{\mu,q}(\omega)t^{\mu-1}(\log t)^q$$

where $d_{\mu,q}(\omega)$ vanishes for $\mu \leqslant 0$.

The expansion (4.3) follows because of $\int_p \exp(\tau f)\omega = \int_0^1 dt \exp(\tau t)\int_{\gamma(t)} \omega/df$.

(4.5) We *roughly* explain why an expansion of type (4.4) determines a factor of $b(s)$.

$$\begin{aligned}\int_p P(s)f^{s+1}\omega &= \int_0^1 dt\, t^{s+1}\int_{\gamma(t)} (P^*(s)\omega)/df \\ &= \int_0^1 dt \sum d_{\mu,q}(P^*(s)\omega)t^{s+\mu}(\log t)^q \\ &= \sum d'_{\mu,q}(d/ds)^q(1/(s+\mu+1))\end{aligned}$$

$$\int_p b(s)f^s\omega = \sum d_{\mu,q}(\omega)(d/ds)^q(1/(s+\mu))\cdot b(s).$$

Comparing two expressions, we conclude $(s+\mu)^{q+1}$ divides $b(s)$ where μ is the minimum for all numbers μ' for which $d_{\mu',q}(\omega) \neq 0$. (First noticed by **[M2]**.)

(4.6) From now on, we return to the case of isolated singularity. See **[H1]** for the nonisolated case.

Let $\Omega^p_{X/T}$ be relative p-forms defined by $\Omega^p_X/df \wedge \Omega^{p-1}_X$. Define three $\mathcal{O}_T$-modules $\mathcal{H}^{(i)}$, $i = -2, -1, 0$, by

$$\mathcal{H}^{(-2)} = \mathcal{H}^{n-1}\left(f^*\Omega_{\dot{X}/T}\right),$$
$$\mathcal{H}^{(-1)} = f_*\Omega^{n-1}_{X/T}/d\left(f_*\Omega^{n-2}_{X/T}\right),$$
$$\mathcal{H}^{(0)} = f_*\Omega^n_X/dt \wedge d\left(f_*\Omega^{n-2}_X\right).$$

They are coherent and extensions of the sheaf of analytic sections of $\bigcup_t H^{n-1}(X_t, C) \to T'$ [**Br1**]. The stalks at 0 are

$$H^{n-1}(\Omega_{\dot{X}/T,0}),\ \Omega^{n-1}_{X,0}/\{d\Omega^{n-2}_{X,0} + df \wedge {}^{n-2}_{X,0}\},$$

and $\Omega^n_{X,0}/df \wedge d\Omega^{n-2}_{X,0}$, respectively. They are free $\mathcal{O}_T$ modules of rank μ.

The mapping $dt \wedge : \mathcal{H}^{(-1)} \to \mathcal{H}^{(0)}$ is injective.

(4.7) The Gauss-Manin connection D: $dt \wedge \mathcal{H}^{(-1)} \to \mathcal{H}^{(0)}$ is defined by $D[df \wedge \phi] = [d\varphi]$.

The saturated lattice $\tilde{\mathcal{H}}$ of $\mathcal{H}^{(0)}$ means $\bigcup_{k\geq 0}(tD)^k\mathcal{H}^{(0)}$.

The residue of D on $\tilde{\mathcal{H}}$ is the linear mapping $\overline{Dt}$ on $\tilde{\mathcal{H}}/t\tilde{\mathcal{H}}$ induced by Dt.

(4.8) THEOREM. *D is regular singular.*

As a consequence there exists N such that $\tilde{\mathcal{H}} = \bigcup_{k=0}^{N}(tD)^k\mathcal{H}^{(0)}$.

(4.9) THEOREM. *There is an isomorphism $\Omega^n_X \otimes_{\mathfrak{D}} \mathfrak{M} \overset{\sim}{\to} \tilde{\mathcal{H}}/t\tilde{\mathcal{H}}$. Under the isomorphism, linear operator s is identified with $-\overline{Dt}$. Therefore, $b(s)$ is the minimal polynomial of $-\overline{Dt}$ and* $\exp(2\pi i s)$ *is equivalent to the local monodromy.*

This is the theorem of Malgrange [**M4**]. The isomorphism is based on the following lemma.

(4.10) LEMMA. *If f has an isolated singularity, $\mathcal{J}(s) \cap \mathfrak{D}$ is generated by $\partial f/\partial x_i D_j - \partial f/\partial x_j D_i$, $1 \leq i, j \leq n$.*

From this lemma, $\dot{\Omega}^n_X \otimes \mathfrak{N}^{(0)} \simeq \mathcal{H}^{(0)}$. Since $\mathfrak{N} = \bigcup_{k\geq 0} s^k\mathfrak{N}^{(0)}$ and $0 \to \mathfrak{N} \overset{t}{\to} \mathfrak{N} \to \mathfrak{M} \to 0$ is exact, the result follows.

(4.11) $\tilde{\mathcal{H}} = \bigcup_{k=0}^{L(f)-1}(tD)^k\mathcal{H}^{(0)}$.

5. Examples. In this section, we exhibit some typical examples of b-functions or exponents. As in [**Y7**], b-exponents are given by the generating function $P(t) = \sum_{i=1}^{\mu} t^{\alpha_i}$ and also MH-exponents by $Q(t) = \sum_{i=1}^{\mu} t^{\beta_i}$.

There is a conjecture of the author which gives MH-exponents from b-exponents and some data on the structure of $\Omega^{n+1}_X \otimes_D \tilde{\mathfrak{M}}$. In examples (5.4), (5.5), (5.6) $Q(t)$ are first determined by that conjecture. Moreover in (5.5) we determine $Q(t)$ directly from the data in [**St**]. Of course the two formulae coincide. See also [**Sch1**].

(5.1) $x^n + y^n + c(xy)^m$, $4 \leq m + 1 < n < 2m$.

Here c is a parameter. When $m = n - 2$, the above function is a so-called "Hessian deformation" of $x^n + y^n$.

When $c = 0$, the function is quasihomogeneous.

$$P_0(t) = Q_0(t) = \left(\frac{t^{1/n} - t}{1 - t^{1/n}}\right)^2.$$

When $c \neq 0$, $Q_c(t)$ remains the same because of the μ-constancy, whereas $P_c(t)$ changes.

$$P_c(t) = P_0(t) + t^{2m/n-1}(1-t)\left(\frac{t^{1/n} - t^{(n-m)/n}}{1 - t^{1/n}}\right)^2,$$

$$Q_c(t) = Q_0(t).$$

The above $P_c(t)$ is called (EEF) type 1 in **[Y7]**.

(5.2) $(xy)^m + x^n + y^n$, $4 \leqslant 2m < n$.

$$P(t) = t + 2\frac{t^{1/m}(1-t)(1-t^{(m-1)(n-m)/nm})}{(1-t^{1/n})(1-t^{(n-m)/nm})}.$$

Double roots are $2(t^{1/m} - t)/(1 - t^{1/m})$.

$$Q(t) = \frac{t^{1/m} - t^2}{1 - t^{1/m}} + 2\frac{(t^{1/m} - t)(t^{1/n} - t)}{(1 - t^{1/m})(1 - t^{1/n})}.$$

This $P(t)$ is called (EEF) type 2 in **[Y7]**. Note that double roots $1/m$, $2/m, \ldots, (m-1)/m$ split into $1/m$, $2/m, \ldots, (m-1)/m$, $(m+1)/m, \ldots$, $(2m-1)/m$ in $Q(t)$.

(5.3) $xyz + x^p + y^q + z^r$. (Γ_{pqr}),

$$P(t) = -t + t(1-t)\left(\frac{1}{1-t^{1/p}} + \frac{1}{1-t^{1/q}} + \frac{1}{1-t^{1/r}}\right),$$

$$Q(t) = P(t) + t(t-1).$$

In this case the double root 1 splits into 1 and 2.

(5.4) $(x^2 + y^3)^2 + xy^n$ ($W^{\#}_{1,2n-9}$), $n \geqslant 5$.

This is an irreducible plane curve with characteristic (4, 6, $2n - 3$). The b-exponents are determined in **[Y9]** (see also **[Y15]**.)

$$P(t) = t^{1/2}\frac{t^{1/(2n+3)} - t}{1 - t^{1/(2n+3)}} + t^{5/12}\frac{(1 - t^{1/3})(1 - t)}{(1 - t^{1/6})(1 - t^{1/2})}$$
$$+ t^{(2n+1)/2(2n+3)}(1 - t),$$

$$Q(t) = P(t) + (t-1)(t^{7/12} + t^{(2n+1)/2(2n+3)})$$
$$= t^{1/2}\frac{t^{1/(2n+3)-t}}{1 - t^{1/(2n+3)}} + t^{5/12}(1 + t^{1/2})(t + t^{2/3}).$$

(5.5) $(xyz)^2 + x^8 + y^8 + z^8$.

Set fractional polynomials $P_i(t)$, $Q_i(t)$, $i = 0, 1, 2$, as follows.

$$P_2(t) = 3t^{1/2},$$

$$P_1(t) = 6t^{5/8} + 9t^{3/4} + 15t^{7/8} + 14t + 15t^{9/8} + 9t^{5/4} + 6t^{11/8},$$

$$P_0(t) = 6t + 12t^{9/8} + 18t^{5/4} + 21t^{11/8}$$

$$+24t^{3/2} + 21t^{13/8} + 18t^{7/4} + 12t^{9/8} + 6t^2,$$

$$Q_2(t) = P_2(t) + t^{1/2}(t-1) + t^{1/2}(t^2-1) = t^{1/2} + t^{3/2} + t^{5/2},$$

$$Q_1(t) = P_1(t) + (t-1)(3t^{5/8} + 3t^{3/4} + 6t^{7/8} + 7t + 9t^{9/8} + 6t^{5/4} + 3t^{11/8})$$

$$= 3t^{5/8} + 6t^{3/4} + 9t^{7/8} + 7t + 6t^{9/8} + 3t^{5/4} + 3t^{11/8}$$

$$+3t^{13/8} + 3t^{7/4} + 6t^{15/8} + 7t^2 + 9t^{17/8} + 6t^{9/4} + 3t^{19/8},$$

$$Q_0(t) = P_0(t).$$

Then $P(t) = P_2(t) + P_1(t) + P_0(t)$ and $Q(t) = Q_2(t) + Q_1(t) + Q_0(t)$. We remark that (triple roots) $= P_2(t)$ and

$$\text{(double roots)} = 2t + 6t^{1/2}\frac{t^{1/8} - t}{1 - t^{1/8}}.$$

The polynomial $Q(t)$ has the duality $t^3Q(t^{-1}) = Q(t)$ but $P(t)$ does not. The theory of b-function, however, gives the hierarchy in $P(t)$ and each $P_i(t)$ has the duality $t^{3-i}P_i(t^{-1}) = P_i(t)$. We remark that superscripts p and q of $H^{p,q}$ should be interchanged in table of [**St**].

(5.6) $x^3 + yz^2 + x^2y^k + \mathbf{b}y^{3k+i}$, $k > 1$, $i > 0$, $(Q_{k,i})$.

Here b includes moduli parameters: $\mathbf{b} = b_0 = b_1y + \cdots + b_{k-1}y^{k-1}$, $b_0 \neq 0$. When k is odd, there is one pair of double roots 1.

$$P(t) = -t + (1-t)\left(\frac{t}{1-t^{1/3}} + \frac{t^{(5k+2i+1)/2(3k+i)}}{1-t^{1/(3k+i)}} + \frac{t^{(5k+1)/6k}}{1-t^{1/3k}}\right),$$

$$= t^{4/3} + t^{5/3} + \sum_{j=0}^{3k+i-1} t^{a(j)} + \sum_{j=0}^{3k-1} t^{b(j)},$$

$$Q(t) = P(t) + (t-1)\left(t^{(5k+2i+1)/2(3k+i)}\frac{1 - t^{[(k+1)/2]/(3k+i)}}{1-t^{1/(3k+i)}}\right.$$

$$\left. + t^{(5k+2[(k+1)/2]+1)/6k}\frac{1 - t^{[k/2]/3k}}{1-t^{1/3k}}\right)$$

$$= t^{4/3} + t^{5/3} + \left(\sum_{j=0}^{[(k-2)/2]} + \sum_{j=k}^{3k-1} + \sum_{j=[(7k+1)/2]}^{4k-1}\right) t^{b(j)}$$

$$+ \sum_{j=[(k+1)/2]}^{3k+i+[(k-2)/2]} t^{a(j)} \qquad (+t + t^2 \text{ if } k \text{ is odd}).$$

$$a(j) = (5k + 2i + 2j + 1)/2(3k+i),\ b(j) = (5k + 2j + 1)/6k.$$

(5.7) $x^3y^2 + x^4y + y^6$.

This is an example of a function of 2-variables with $2 = l(f) < L(f) = 3$ and $\mu(f) = 18$.

$$P(t) = t + \frac{t^{2/5}(1-t)}{1-t^{1/5}} + t^{7/18}(1+t^{1/8})\frac{1-t}{1-t^{1/6}},$$

$$\begin{aligned} Q(t) &= P(t) + (t-1)(t^{4/9} + t^{2/5} + t^{11/18}) \\ &= t + t^{3/5}\frac{1-t}{1-t^{1/5}} + t^{7/18}(1+t^{7/18})\frac{1-t}{1-t^{1/6}}. \end{aligned}$$

BIBLIOGRAPHY

[Ar1] V. I. Arnold, *The index of a singular point of a vector field, the Petrovskii-Oleinki inequalities, and mixed Hodge structures*, Functional Anal. Appl. **12** (1978), 1–14. MR **58** #16685.

[Be1] I. N. Bernstein, *Feasibility of the analytic continuation of f_+^λ for certain polynomials f*, Functional Anal. Appl. **2** (1968), 85–87.

[Be2] ______, *Modules over a ring of differential operators. An investigation of the fundamental solutions of equations with constant coefficient*, Functional Anal. Appl. **5** (1971), 89–101. MR **44** #7282.

[Be3] ______, *Analytic continuation of generalized functions with respect to a parameter*, Functional Anal. Appl. **6** (1972), 273–285. MR **47** #9269.

[Bj1] J. E. Björk, *Dimensions over algebras of differential operators* (Xerox copies were propagated but not published).

[Bj2] ______, *Rings of differential operators*, North-Holland, Amsterdam, 1979.

[Bo] J.-M. Bony, *Polynômes de Bernstein et monodromie (d'après B. Malgrange)*, Sém. Bourbaki (1974/1975), Exposé No. 459, pp. 97–110; Lecture Notes in Math., vol. 514, Springer-Verlag, Berlin and New York, 1976. MR **58** #5664.

[Br1] E. V. Brieskorn, *Die Monodromie der isolierten Singularitäten von Hyperflächen*, Manuscripta Math. **2** (1970), 103–161. MR **42** #2509.

[D] V. I. Danilov, *Newton polyhedra and vanishing cohomology*, Functional Anal. Appl. **13** (1979), 103–115. MR **80h** #14001.

[E-L] F. Ehlers and K.-C. Lo, *Minimal characteristic exponent of the Gauss-Manin connection of isolated singular point and Newton polyhedron* (to appear).

[GS] I. M. Gel'fand and G. E. Shilov, *Generalized functions. vol.* 1. *Properties and operations*, Academic Press, New York and London, 1964 (original 1958, in Russian).

[H1] H. A. Hamm, *Zur analytischen und algebraischen Beschriebung der Picard-Lefschetz-Monodromie*, Habilitationsschrift Göttingen, 1974.

[H2] ______, *Remarks on asymptotic integrals, the polynomial of I. N. Bernstein and the Picard-Lefschetz monodromy*, *Several complex variables*, Proc. Sympos. Pure Math., vol. 30, part 1, Williams College, Williamstown, Mass., 1975, Amer. Math. Soc., Providence, R. I., 1977, pp. 31–35. MR **58** #28653.

[I] J. Igusa, *On the first terms of certain asymptotic expansions*, Internat. Colloq. on Algebraic Geometry and Complex Analysis, 1977.

[Kn] J. M. Kantor, *Prolongement méromorphe de f^λ, et division des distributions, d'après I. N. Bernstein*, Sém. Goulaouic-Schwartz 1973–1974: Équations aux dérivées partielles et analyse fonctionelle, Exposé 9, Centre de Math., École Polytech., Paris, 1974, 16pp. MR **56** #12308.

[Ks1] M. Kashiwara, *B-functions and holonomic systems. Rationality of roots of b-functions*, Invent. Math. (1976/1977), 33–53.

[Ks2] ______, *On the holonomic systems of linear differential equations.* ii, Invent. Math. **49** (1978), 121–135.

[Ks3] M. Kashiwara ad T. Kawai, *On the characteristic variety of a holonomic system with regular singularities.*

[Ks4] M. Kaswara, *b-function and the singularity of a hypersurface*, Publ. Res. Inst. Math. Sci. Kokyuroku **225**, 16–53. (Japanese)

[Kt] M. Kato, *An estimate of minimal exponents by Newton polyhedra* (to appear).

[Ko] F. Kochman, *Bernstein polynomials and Milnor algebras*, Proc. Nat. Acad. Sci. U.S.A. **73** (1976), 2546. MR **56** #5932.

[L1] B. L. Lichtin, *A numerical study of the array of multiplicities for an irreducible plane curve* (to appear).

[L2] ______, *A connection between polar invariants and roots of the Bernstein-Sato polynomial*, these PROCEEDINGS.

[Lo] K.-C. Lo, *Exposants de Gauss-Manin, Singularité des systèmes différentiels de Gauss-Manin*, Progress in Math. **2**, Birkhauser, 1979, pp. 171–212.

[M1] B. Malgrange, *Intégrales asymptotiques et monodromie*, Ann. Sci. École Norm. Sup. (4) **7** (1974), 405–430 (1975). MR **51** #8459.

[M2] ______, *Sur les polynômes de I. N. Bernstein*, Sém. Goulaouic-Schwartz 1973–1974: Équations aux dérivees partielles et analyse fonctionnelle, Exposé No. 20, Centre de Math., École Polytech., Paris 1974, 10pp. MR **53** #13635.

[M3] ______, *On I. N. Bernštĭen's polynomials, Collection of articles dedicated to the memory of I. G. Petrovskiĭ*. II, Uspehi Math. Nauk **29** (1974), no. 4 (178), 81–88. (Translation of [**M2**]) MR **50** #3342.

[M4] ______, *Le polynome de Bernstein d'une singularité isolée*, *Fourier integral operators and partial differential equations* (Colloq. Internat., Univ. Nice, Nice 1974), Lecture Notes in Math., vol. 459, Springer-Verlag, Berlin and New York, 1975, pp. 98–119. MR **54** #7845.

[M5] ______, *Rationality of the roots of Bernstein polynomial* (in preparation).

[Me1] Z. Mebkhout, *Cohomologie locale d'une hypersurface*, Seminaire François Norguet, 22 Janvier 1976.

[Me2] ______, Thesis.

[Me3] ______, *Introduction to D-Modules*, These PROCEEDINGS.

[OS] P. Orlik and L. Solomon, *Unitary reflection groups and cohomology*, Invent. Math. **59** (1980), 77–94.

[Pa1] V. P. Palamodov, *Moduli and versal deformations of complex spaces*, Soviet Math. Dokl. **17** (1976), no. 5, 1251–1255.

[Pa2] ______, *The asymptotics of oscillating integrals with an isolated stationary phase point*, Soviet Math. Dokl. **21** (1980), no. 1, 323–327.

[Ph] F. Pham, *Caustiques, phase stationnaire et microfunctions*, Acta Math. Vietnam. **2** (1977), 35–101. MR **58** #22648.

[SK1] K. Saito, *Primitive forms for a universal unfoldings of a function with an isolated critical point*, preprint RIMS-368, Kyoto Univ. 1981.

[SK2] ______, *On the periods of primitive integrals* (in preparation).

[SM1] M. Saito, *Gauss-Manin connections and b-functions of non-isolated hypersurface singularities*, Thesis presented to Tokyo Univ., 1980.

[SM2] ______, *Exponents and the geometric genus of an isolated hypersurface singularity*, These PROCEEDINGS.

[S1] M. Sato, *Theory of prehomogeneous vector spaces*, Sugaku Ayumi **15** (1970), 85–157. (noted by T. Shintani in Japanese)

[S2] M. Sato and T. Shintani, *On zeta-functions associated with prehomogeneous vector spaces*, Ann. of Math. (2) **100** (1974), 131–170.

[S3] M. Sato, *Reduced b-functions*, Sûrikaisekikenkyusho Kôkyûroku **225** (1975), 1–15. (noted by T. Yano in Japanese) MR **58** #11506.

[S4] M. Sato, M. Kashiwara, T. Kimura and T. Oshima, *Micro-local analysis of prehomogeneous vector spaces*, Invent. Math. (1980).

[Sch1] J. Scherk, *Isolated singular points and the Gauss-Manin connection*, D. Phil. Thesis, Exeter College, Oxford, 1977.

[Sch2] ______, *On the Gauss-Manin connection of an isolated hypersurface singularity*, Math. Ann. **283** (1978), 23–32.

[Sch3] ______, *On the monodromy theorem for isolated hypersurface singularities*, Invent. Math. **58** (1980), 289–301.

[Se1] M. Sebastiani, *Preuve d'une cinfecture de Brieskorn*, Manuscripta Math. **2** (1970), 301–307.

[Se2], ______, *Monodromie et polynôme de Bernstein, d'après Malgrange. Fonctions de plusieurs variables complexes*. III, Sém. François Norguet, 1975–1977, pp. 370–381; Lecture Notes in Math., vol. 670, Springer-Verlag, Berlin and New York, 1978.

[St] J. H. M. Steenbrink, *Mixed Hodge structure on the vanishing cohomology*, Nordic Summer School, Symposium in Math., Oslo, 1976, pp. 525–562. MR **58** #5670.

[SS] J. H. M. Steenbrink and J. Scherk, *On the mixed Hodge structure on the cohomology of the Milnor fibre* (to appear).

[T1] H. Terao, *Generalized exponents of a free arrangement of hypersurfaces and Shepherd-Todd-Brieskorn formula*, Invent. Math. **69** (1981), 159–170.

[V1] A. N. Varchenko, *Newton polyhedra and estimation of oscillating integrals*, Functional Anal. Appl. **10** (1976), 175–196.

[V2] ______, *Gauss-Manin connection of isolated singular point and Bernstein polynomial*, Bull. Sci. Math. (2) **104** (1980), 205–223.

[V3] ______, *Hodge properties of Gauss-Manin connections*, Functional Anal. Appl. **14** (1980), 36–37.

[Y1] T. Yano, *Theory of b-functions*, Master thesis presented to Kyoto Univ. 1974 (almost the same with [**Y2**]).

[Y2] ______, *Theory of b-functions: examples*, Publ. Res. Inst. Math. Sci. Kokyuroku **225** (1975), 72–234. (Japanese) MR **58** #11508.

[Y3] ______, *Recent developments in b-functions*, Publ. Res. Inst. Math. Sci. Kokyuroku **248** (1975), 293–311. (Japanese) MR **58** #11509.

[Y4] ______, *Some remarks on the theory of b-functions*, Publ. Res. Inst. Math. Sci. Kokyuroku **266** (1976), 319–329. (Japanese) MR **58** #11531.

[Y5] ______, *On the invariance of $L(f)$ of the holomorphic functions $f(x)$ and the properties of the singular points of $f^{-1}(0)$*, Publ. Res. Inst. Math. Sci. Kokyuroku **295** (1977), 38–45. (Japanese) MR **58** #11510.

[Y6] ______, *On the holonomic system of f^s and b-function*, Publ. Res. Inst. Math. Sci. **12** (1976/77), supplement, 469–480. MR **58** #22650.

[Y7] ______, *On the theory of b-functions*, Publ. Res. Inst. Math. Sci. **14** (1978), 111–202. MR **80h** #32026.

[Y8] ______, *Certain filtration on Gauss-Manin connections*, 1978 (to appear). (Japanese)

[Y9] ______, *On the theory of b-functions.* II (to appear).

[Y10] ______, *The minimal exponent of a hypersurface singularity* (to appear).

[Y11] T. Yano and J. Sekiguchi, *The microlocel structure of weighted homogeneous polynomials associated with Coxeter systems.* 1, Tokyo J. Math. (1980),

[Y12] ______, *The microlocel structure of weighted homogeneous polynomials associated with Coxeter systems*, 2, Tokyo J. Math. (in print).

[Y13] T. Yano, *On the structure of differential equations associated with some isolated singularities*, Sci. Rep. Saitama Univ. Ser. A **9** (1977), 9–20.

[Y14] ______, *Open problems*, *Open problems in theory of singularities*, Proc. Seventh Internat. Sympos. held at Katata, Sept. 3–9, 1980, pp. 22–25.

[Y15] ______, *Exponents of singularities of plane irreducible curves*, Sci. Rep. Saitama Univ. Ser. A **10** (1982).

SAITAMA UNIVERSITY, SAITAMA, JAPAN

Proceedings of Symposia in Pure Mathematics
Volume 40 (1983), Part 2

ON IRREGULARITY AND GEOMETRIC GENUS OF ISOLATED SINGULARITIES

STEPHEN S.-T. YAU[1]

1. Introduction. Let $(V, 0)$ be a normal surface singularity. Wagreich first defined an invariant geometric genus p_g for the singularity $(V, 0)$. It turns out that this is an important invariant for the theory of normal surface singularities. In this paper we shall introduce another invariant called irregularity q of the singularity $(V, 0)$. This invariant is interesting for the following reason. It is a long-term conjecture that Gorenstein surface singularities are not rigid, i.e. $\dim T_V^1 \geqslant 1$ (cf. §3 for the definition of T_V^1). In the case of Gorenstein surface singularities this irregularity actually gives a lower bound for $\dim T_V^1$ (cf. Remark 3.2). Therefore it is of great interest to understand this irregularity more closely. In §2 we give formulae for the irregularity in case $(V, 0)$ is a hypersurface singularity (cf. Theorem 2.1) or a singularity with $\mathbf{C}^*$-action (cf. Theorem 2.2). An example will be computed explicitly by using these formulae.

In §3 we give lower estimate for $\dim T_V^1$ in terms of geometric genus (cf. Corollary 3.5). We also have lower estimate for irregularity in terms of geometric genus (cf. Theorem 3.7). It is well known that Gorenstein surface singularities with geometric genus equal to zero are rational double points A_k, D_k, E_6, E_7, E_8. These singularities admit a $\mathbf{C}^*$-action. The natural question is to classify Gorenstein singularities with $\mathbf{C}^*$-action and irregularity equal to zero. Theorem 3.8 gives a very simple answer to this question. In most of the above results, the fact that $(V, 0)$ admits a $\mathbf{C}^*$-action plays an important role.

The purpose of this paper is two-fold. On the one hand we describe some of our previous work [**11, 12, 13**]. On the other hand, we study irregularity for those singularities which do not admit $\mathbf{C}^*$-action (cf. Theorem 3.9). This is a beginning step towards classifying hypersurface singularities with irregularity equal to zero.

1980 *Mathematics Subject Classification.* Primary 57R45.
[1]Research supported in part by NSF grant MCS 77-18723 A04 and a Sloan fellowship.

2. Irregularity and geometric genus. Let $(V,0)$ be a normal Stein analytic space of dimension n $(n \geq 2)$ with 0 as its only isolated singularity. Let $\pi\colon M \to V$ be a resolution of the singularity of V with exceptional set $A = \pi^{-1}(0)$. We define invariants $s^{(i)}$, $0 \leq i \leq n$, of the singularity 0 to be $\dim \Gamma(M - A, \Omega^i)/\Gamma(M, \Omega^i)$.

Let $\overline{\Omega}^i_V$ be the 0th direct image sheaf $\pi_* \Omega^i_M$ of Ω^i_M. By Grauert's direct image theorem, $\overline{\Omega}^i_V$ is a coherent sheaf. Let $\theta\colon V - \{0\} \to V$ be the inclusion map. Then the 0th direct image sheaf $\overline{\overline{\Omega}} := \theta_* \Omega^i_{V-\{0\}}$ is coherent (cf. [7]). Hence the quotient sheaf $\overline{\overline{\Omega}}^i_V/\overline{\Omega}^i_V$ is coherent and supported on $\{0\}$. $s^{(i)}$ is exactly $\dim \overline{\overline{\Omega}}^i_V/\overline{\Omega}^i_V$. Therefore the invariants $s^{(i)}$, $1 \leq i \leq n$, are indeed invariants of isolated singularities. Wagreich defined geometric genus p_g of the singularity $(V,0)$ to be $\dim H^{n-1}(M, \mathcal{O}) = \dim R^{n-1}\pi_* \mathcal{O}_M$. It is proved in [**2**, **10**] that $p_g = s^{(n)}$. Hence it is quite natural to define irregularity q of the singularity $(V,0)$ to be $s^{(n-1)}$. In this section we shall assume that $(V,0)$ is a surface singularity. The following theorem which relates the irregularity q and $\dim H^1(M, \Omega^1)$ was proved in [**11**].

THEOREM 2.1. *Let $f(x,y,z)$ be holomorphic in N, a Stein neighborhood of $(0,0,0)$ with $f(0,0,0) = 0$. Let $V = \{(x,y,z) \in N\colon f(x,y,z) = 0\}$ have $(0,0,0)$ as its only singular point. Let*

$$\mu = \dim \mathbf{C}[[x,y,z]]/\left(\frac{\partial f}{\partial x}, \frac{\partial f}{\partial y}, \frac{\partial f}{\partial z}\right)$$

and

$$\tau = \dim \mathbf{C}[[x,y,z]]/\left(f, \frac{\partial f}{\partial x}, \frac{\partial f}{\partial y}, \frac{\partial f}{\partial z}\right).$$

Let $\pi\colon M \to V$ be a resolution of V and $A = \pi^{-1}(0,0,0)$. Then

$$(2.0) \quad q + \dim H^1(M, \Omega^1) = \tau - (\mu + 1) + \chi_T(A) + 2 \dim H^1(M, \mathcal{O})$$

where q is irregularity of the singularity and $\chi_T(A)$ is the topological Euler characteristic of A.

The following theorem gives us an explicit way to compute the irregularity and geometric genus of any isolated singularity $(V,0)$ with $\mathbf{C}^*$-action.

THEOREM 2.2. *Suppose $V \subseteq \mathbf{C}^m$ is an analytic variety of dimension two with the origin as its only isolated singularity. Suppose σ is a $\mathbf{C}^*$-action leaving V invariant, defined by*

$$\sigma(t, (z_1, \ldots, z_m)) = (t^{q_1} z_1, \ldots, t^{q_m} z_m),$$

q_i's are positive integers. Let $\varphi\colon \mathbf{C}^m \to \mathbf{C}^m$ be defined by $\varphi(z_1, \ldots, z_m) = (z_1^{q_1}, \ldots, z_m^{q_m})$ and let $V' = \varphi^{-1}(V)$ be the cone over V. Then V' has a natural $\mathbf{C}^*$-action defined by $\sigma'(t,(z_1, \ldots, z_m)) = (tz_1, \ldots, tz_m)$ and the induced map φ: $V' \to V$ commutes with the $\mathbf{C}^*$-action. Let $A' = (V' - \{0\})/\mathbf{C}^* \subseteq \mathbf{P}^{m-1}$. Let N' be the universal subbundle (i.e. dual of the hyperplane bundle) of $\mathbf{P}^{m-1}$ restricted to A'. Identify $\mathbf{Z}_{q_i}$ with the group of q_ith roots of 1. $G = \mathbf{Z}_{q_i} \oplus \cdots \oplus \mathbf{Z}_{q_m}$ acts on V' by coordinatewise multiplication. G also acts on A' and N'. Let $\pi\colon A'' \to A'$ be the

normalization and $N'' = \pi^*(N')$, *the pull back of* N' *by* π. *Then the irregularity* q *and the geometry genus* p_g *of the singularity* $(V, 0)$ *can be computed by the following formulae.*

$$q = \begin{cases} 0 & \text{if } g'' \leqslant 1, \\ \sum_{n=1}^{\infty} \dim \Gamma(A'', K_{A''}N''^n)^G & \text{if } g'' \geqslant 1; \end{cases} \tag{2.1}$$

$$p_g = \begin{cases} 0 & \text{if } g'' = 0, \\ \dim \Gamma(A'', K_{A''})G & \text{if } g'' = 1, \\ \sum_{n=1}^{\infty} \dim \Gamma(A'', K_{A''}N''^{n+1})^G & \text{if } g'' \geqslant 2, \end{cases} \tag{2.2}$$

where g'' *is the genus of* A'', $K_{A''}$ *is the canonical line bundle of* A'' *and* $\Gamma(A'', K_{A''}N''^n)^G$ *denotes the G-invariant sections.*

For the proof of the above theorem we refer the reader to [**11**].

COROLLARY 2.3. *Suppose* $V \subseteq \mathbf{C}^m$ *is an analytic variety of dimension two which admits a* $\mathbf{C}^*$-*action. Let* G, A'' *and* $K_{A''}$ *be defined as in Theorem* 2.2. *Then*

$$P_g = q + \dim \Gamma(A'', K_{A''})G. \tag{2.3}$$

In particular

$$P_g \geqslant q. \tag{2.4}$$

COROLLARY 2.4. q *is equal to zero for any 2-dimensional rational singularity with* $\mathbf{C}^*$-*action.*

In fact, for any 2-dimensional rational singularity q is equal to zero. This was proved by Pinkham [**4**] and Wahl [**9**]. In view of formula (2.0), if we know $\dim H^1(M, \Omega^1)$ then the irregularity q of the singularity can be computed explicitly. Recall that in [**3**] Narasimhan proved that given a finitely generated abelian group G and integers $k \geqslant 1$, $n \geqslant k + 3$, there is a Runge domain D in $\mathbf{C}^n$ with $H_k(D, \mathbf{Z}) \simeq G$. However, for a strongly pseudoconvex 2-dimensional manifold we have the following [**11**].

THEOREM 2.5. *Let* M *be a two-dimensional strongly pseudoconvex manifold in which the exceptional set may admit arbitrary singularities. Then* $\dim H^1(M, \Omega^1) \geqslant b_2$ *where* b_2 *is the second betti number of the tubular neighborhood of the exceptional set* A *of* M.

THEOREM 2.6. *Let* M *be a two-dimensional strongly pseudoconvex manifold with a nonsingular Riemann surface* A *of genus* g *as its maximal compact analytic set. Then* $\dim H^1(M, \Omega^1) = \dim \Gamma(M, \Omega^1_M \otimes \mathcal{O}_{n_0A}(n_0A))$ *where* $n_0 + 1 = \max\{[(2 - 2g)/(A \cdot A)] + 1, 2\}$. *If* $(2 - 2g)/(A \cdot A) \leqslant 1$, *then* $\dim H^1(M, \Omega^1) = 1$. *In particular, if* A *is a rational curve or an elliptic curve, then* $\dim H^1(M, \Omega^1) = 1$. *Hence irregularity for any simple elliptic singularity is equal to zero.*

The proof of Theorem 2.6 can be found in [**11**].

If M is a resolution of 2-dimensional rational singularity, then J. Wahl [**9**] has proved the following theorem.

THEOREM 2.7. *Let M be a two-dimensional strongly pseudoconvex manifold M with A as its maximally compact analytic set. Let b_2 be the second betti number of A. If $H^1(M, \mathcal{O}) = 0$, then $b_2 = \dim H^1(M, \Omega^1)$.*

EXAMPLE 2.8. Let $V = \{z^2 = y(x^4 + y^6)\} \subseteq \mathbf{C}^3$ as in Theorem 2.2. V admits a $\mathbf{C}^*$-action σ

$$\sigma: \mathbf{C}^* \times V \to V,$$
$$(t, (x, y, z)) \to (t^3x, t^2y, t^7z).$$

Then $V' = \{(x', y', z'): x'^{12}y'^2 + y'^{14} - z'^{14} = 0\}$. Identify $\mathbf{Z}_3$, $\mathbf{Z}_2$ and $\mathbf{Z}_7$ with the groups of 3rd roots, 2nd roots and 7th roots of 1 respectively. $G = \mathbf{Z}_3 \oplus \mathbf{Z}_2 \oplus \mathbf{Z}_7$ acts on V' by coordinate multiplication. Let A' be the curve defined by $x'^{12}y'^2 + y'^{14} - z'^{14} = 0$ in $\mathbf{CP}^2$. In $(\mathcal{U}x, y_1, z_1)$ coordinate patch, equation of A' is

$$y_1^2 + y_1^{14} - z_1^{14} = 0.$$

Clearly, any holomorphic one forms on A'' are of the forms $P(y_1, z_1)\, dy_1/14z_1^{13}$ where P is a polynomial of degree $\leqslant 11$. Let $g_1(\xi, 1, 1)$, $g_2 = (1, \eta, 1)$ and $g_3 = (1, 1, \zeta)$ be the three generators in G, where $\xi = 2\pi i/3$, $\eta = -1$ and $\zeta = 2\pi i/7$. Then the actions by g_1, g_2 and g_3 look like

$$(x, y, z) \overset{g_1}{\mapsto} (\xi x, y, z)\ (x, y, z) \overset{g_2}{\mapsto} (x, \eta y, z)\ (x, y, z) \overset{g_3}{\mapsto} (x, y, \zeta y).$$

In $(\mathcal{U}_x, y_1, z_1)$ coordinate patch

$$(1, y_1, z_1) \overset{g_1}{\mapsto} \left(1, \frac{1}{\xi} y_1, \frac{1}{\xi} z_1\right)\ (1, y_1, z_1) \overset{g_2}{\mapsto} (1, \eta y_1, z_1)(1, y_1, z_1) \overset{g_3}{\mapsto} (1, y_1, \zeta y_1),$$

$$\omega = \frac{\sum_{j+k \leqslant 11} P(y_1, z_1)\, dy_1}{14z_1^{13}} = \frac{\sum_{j+k \leqslant 11} a_{jk} y_1^j z_1^k\, dy_1}{14z_1^{13}}$$

$$\overset{g_1}{\mapsto} g_1 * \omega = \frac{\sum_{j+k \leqslant 11} a_{jk} \xi^{-j-k-1} y_1^j z_1^k\, dy_1}{14\xi^{-13} z_1^{13}}$$

$$= \frac{\sum_{j+k \leqslant 11} a_{jk} \xi^{12-j-k} y_1^j z_1^k\, dy_1}{14z_1^{13}}.$$

Hence ω is invariant under g_1 if and only if $a_{jk} = 0$ for $12 - j - k \neq 3, 6, 9, 12$, i.e. ω is invariant under g_1 if and only if ω is one of the following forms.

$$\omega_1 = \frac{a_{00}\, dy_1}{14z_1^{13}}, \qquad \omega_2 = \frac{\sum_{j+k=3} a_{jk} y_1^j z_1^k\, dy_1}{14z_1^{13}},$$

$$\omega_3 = \frac{\sum_{j+k=6} a_{jk} y_1^j z_1^k \, dy_1}{14 z_1^{13}}, \qquad \omega_4 = \frac{\sum_{j+k=9} a_{jk} y_1^j z_1^k \, dy_1}{14 z_1^{13}},$$

$$\omega_1 \overset{g_2}{\mapsto} g_2^* \omega_2 = \frac{a_{00} \eta \, dy_1}{14 z_1^{13}},$$

$$\omega_2 \overset{g_2}{\mapsto} g_2^* \omega_2 = \frac{\sum_{j+k=3} a_{jk} \eta^{j+1} y_1^j z_1^k \, dy_1}{14 z_1^{13}},$$

$$\omega_3 \overset{g_2}{\mapsto} g_2^* \omega_3 = \frac{\sum_{j+k=6} a_{jk} \eta^{j+1} y_1^j z_1^k \, dy_1}{14 z_1^{13}},$$

$$\omega_4 \overset{g_2}{\mapsto} g_2^* \omega_4 = \frac{\sum_{j+k=9} a_{jk} \eta^{j+1} y_1^j z_1^k \, dy_1}{14 z_1^{13}}.$$

Hence (i) ω_1 is invariant under g_2,

(ii) ω_2 is invariant under g_2 if and only if ω_2 is one of the forms

$$\omega_{21} = \frac{a_{12} y_1 z_1^3 \, dy_1}{14 z_1^{13}}, \qquad \omega_{22} = \frac{a_{30} y_1^3 \, dy_1}{14 z_1^{13}},$$

(iii) ω_3 is invariant under g_2 if and only if ω_3 is one of the forms

$$\omega_{31} = \frac{a_{15} y_1 z_1^5 \, dy_1}{14 z_1^{13}}, \qquad \omega_{32} = \frac{a_{33} y_1^3 z_1^3 \, dy_1}{14 z_1^{13}},$$

(iv) ω_4 is invariant under g_2 if and only if ω_4 is one of the forms

$$\omega_{41} = \frac{a_{18} y_1 z_1^8 \, dy_1}{14 z_1^{13}}, \qquad \omega_{42} = \frac{a_{36} y_1^3 z_1^6 \, dy_1}{14 z_1^{13}},$$

$$\omega_{43} = \frac{a_{54} y_1^5 z_1^4 \, dy_1}{14 z_1^{13}}, \quad \omega_{44} = \frac{a_{72} y_1^7 z_1^2 \, dy_1}{14 z_1^{13}}, \quad \omega_{45} = \frac{a_{90} y_1^9 \, dy_1}{14 z_1^{13}}.$$

It is easy to see that among ω_1, ω_{21}, ω_{22}, ω_{23}, ω_{24}, ω_{31}, ω_{32}, ω_{33}, ω_{34}, ω_{41}, ω_{42}, ω_{43}, ω_{44} and ω_{45}, only

$$\omega_{42} = \frac{a_{36} y_1^3 \, dy_1}{14 z_1^7}$$

is invariant under the action of g_3. Hence

$$\Gamma(A'', K_{A''})^G = \left\{ a_{36} \frac{y_1^3 \, dy_1}{14 z_1^7} : a_{36} \in \mathbf{C} \right\}.$$

By Theorems 2.1 and 2.2 we have

$$\begin{aligned} \dim H^1(M, \Omega^1) &= \chi_T(A) - 1 + \dim H^1(M, \mathcal{O}) + \dim \Gamma(A'', K_{A''})^G \\ &= (3 + 1 - 2) - 1 + 2 + 1 = 4 \end{aligned}$$

and

$$\dim \Gamma(M \setminus A, \Omega^1)/\Gamma(M, \Omega^1) = \dim H^1(M, \mathcal{O}) - \dim \Gamma(A'', K_{A''})^G = 2 - 1 = 1.$$

3. Lower estimate of $\dim T_V^1$. Let $V \subseteq \mathbf{C}^m$ be a local complex space. A deformation of V is a flat map $\pi\colon X \to T$ of local complex spaces, together with an isomorphism $\pi^{-1}(0) \cong V$. Let $I = (f_1, \ldots, f_r)$ be the ideal in $\mathcal{O}_{\mathbf{C}^m}$ defining the local complex space $V \subset \mathbf{C}^m$. Thus $\mathcal{O}_V = \mathcal{O}_{\mathbf{C}^m}/I$. If we take as parameter the (one point) space $T = \operatorname{spec} \mathbf{C}\{\varepsilon\}/(\varepsilon^2)$, then a deformation $X \to T$ of V is a first-order infinitesimal deformation of V. X will be given by equations

$$f_i(x) + \varepsilon g_i(x) = 0$$

and condition for flatness is simply that g_i determine an element of the normal sheaf $N_V = \operatorname{Hom}_{\mathcal{O}_{\mathbf{C}^m}}(I, \mathcal{O}_V)$. The deformation $X \to T$ is trivial if and only if there is an automorphism $x_j \to x_j + \varepsilon\delta_j(x)$ of $\mathbf{C}^m \times T$ over T such that $(f_i(x + \varepsilon\delta(x)))$ and $(f_i + \varepsilon g_i)$ determine the same ideal in $\mathcal{O}_{\mathbf{C}^m \times T}$; in other words

$$\sum_j \frac{\partial f_i}{\partial x_j}\delta_j(x) = g_i(x) \pmod{I}.$$

Now there is a homomorphism $\rho\colon \Theta_{\mathbf{C}^m} \to N_V$ (Θ denoting tangent sheaf) defined by mapping the vector field $\Theta = \Sigma_j \delta_j(x)\partial/\partial x_j$ to the homomorphism $f_i \mapsto \Theta(f_i)$ in N_V, and the element g in N_V induces a trivial deformation of V if and only if it lies in the image of ρ.

Following Schlessinger [6], we define an $\mathcal{O}_V$-module T_V^1 by the exact sequence

$$0 \to \Theta_V \to \Theta_{\mathbf{C}^m/V} \to N_V \to T_V^1 \to 0.$$

Then T_V^1 is the set of isomorphism classes of first order infinitesimal deformations of V, analogous to $H^1(Y, \Theta_Y)$ for a manifold Y. In [8] Tyurina shows that the T_V^1 may be replaced by $\operatorname{Ext}^1_{\mathcal{O}_V}(\Omega_V^1, \mathcal{O}_V)$ (Ω_V^1 denoting Kähler differentials) when V has positive depth along singular locus, e.g. when V is reduced of positive dimension. In [1] Grauert constructs a versal deformation $X \to S$ of V from which every other deformation $W \to T$ may be induced, up to isomorphism, by a map φ: $T \to S$, with $\varphi^*(X) \cong W$. Moreover, the map $t_\varphi\colon t_T \to t_S$ between Zariski tangent spaces is uniquely determined by the isomorphism class of W. As Grauert shows, the Zariski tangent space of S is isomorphic to T_V^1.

V is rigid when every deformation is trivial, or S is reduced to a point. Thus, $T_V^1 = 0$ is the necessary and sufficient condition for rigidity.

In [6] Schlessinger proves that quotient singularities of dimension ≥ 3 are rigid. It is a long-standing conjecture that there is no rigid normal surface singularity. The normality condition is important because the singularity obtained by taking two planes in $\mathbf{C}^4$ which meet at a point is rigid.

DEFINITION 3.1. Let $(V, 0)$ be a normal isolated singularity of dimension $n \geq 2$. Let $\overline{\Omega}_V^p$ and $\overline{\overline{\Omega}}_V^p$ be as defined in §2. Clearly, there are natural maps $\varphi_p\colon \Omega_V^p \to \overline{\Omega}_V^p$ and $\tau_p\colon \Omega_V^p \to \overline{\overline{\Omega}}_V^p$. The cokernels of these maps are finite-dimensional vector spaces over 0. In [**11** and **13**] the invariants $g^{(p)}$ and $\delta^{(p)}$ are defined to be the dimensions of coker φ_p and coker τ_p respectively.

REMARK 3.2. In case $(V, 0)$ is a Gorenstein surface singularity, then

$$\begin{aligned}\dim T_V^1 &= \dim \operatorname{Ext}^1_{\mathcal{O}_V}\left(\Omega_V^1, \mathcal{O}_V\right)\\ &= \dim H^1_{\{0\}}\left(V, \Omega_V^1\right) \quad \text{(Gorenstein duality)}\\ &= \delta^{(1)}\begin{cases} \geqslant g^{(1)},\\ \geqslant q,\end{cases}\end{aligned}$$

where q is the irregularity of the singularity $(V, 0)$.

In view of the above remark, it is important to know the lower estimates for $g^{(1)}$ and q. In [**13**] we have proved the following theorems.

THEOREM 3.3. *Suppose that $(V, 0)$ is a normal isolated singularity of dimension n which admits a* **C****-action. Then $g^{(n-1)} \geqslant g^{(n)}$ and $\delta^{(n-1)} \geqslant \delta^{(n)}$.*

THEOREM 3.4. *Let π: $M \to V$ be any resolution of normal isolated singularity $(V, 0)$ of dimension n. Then*:

(a) *For $n = 2$, $g^{(2)} \geqslant 1$ and $\delta^{(2)} \geqslant 1 + \dim H^1(M, \mathcal{O})$.*

(b) *For $n \geqslant 3$, $g^{(n)} \geqslant n - 1$ and $\delta^{(n)} \geqslant n - 1 + \dim H^{n-1}(M, \mathcal{O})$ if $\dim H^{n-1}(M, \mathcal{O}) > 0$.*

COROLLARY 3.5. *Let $(V, 0)$ be a normal surface singularity with* **C****-action. Then*

$$\delta^{(1)} \geqslant 1 + \dim H^1(M, \mathcal{O}). \tag{3.1}$$

In particular, if $(V, 0)$ is Gorenstein, then

$$\dim T_V^1 \geqslant 1 + \dim H^1(M, \mathcal{O}). \tag{3.2}$$

REMARK 3.6. In case $(V, 0)$ is a Gorenstein isolated surface singularity with **C***-action, Wahl has informed us that he has proved that $\dim T_V^1 \geqslant \dim H^1(M, \mathcal{O})$. Ours in formula (3.2) gives the fact that there exists a holomorphic 2-form ω defined on $V - \{0\}$ which is L^2-integrable in a neighborhood of 0; but ω is not holomorphic on V.

The second approach to the rigidity problem for Gorenstein surface singularities is to get lower estimate for irregularity. In [**12**] we have proved the following.

THEOREM 3.7. *Let $(V, 0)$ be a normal isolated singularity of dimension n with* **C****-action, $M \to V$ an equivariant resolution whose exceptional divisor A has normal crossings. Then*:

(a) $s^{(n-1)} > \dim H^{n-1}(M, \mathcal{O}_M) - \dim H^{n-1}(A, \mathcal{O}_A)$.

(b) *If V is Gorenstein and* $\dim H^{n-1}(M, \mathcal{O}_M) \geqslant 2$, *then* $s^{(n-1)} > 0$.

As an immediate application of the above theorem, we have the following classification theorem.

THEOREM 3.8. *Suppose that* $(V, 0)$ *is a Gorenstein surface singularity with* $\mathbf{C}^*$*-action. Then the irregularity is equal to zero if and only if* $(V, 0)$ *is either a rational double point or a simple elliptic singularity.*

For singularities without $\mathbf{C}^*$-action, it is quite difficult to get lower estimate of irregularity. However, the technique of the proof of the following theorem should be useful.

THEOREM 3.9. *Suppose that* $V = \{(x, y, z) \in \mathbf{C}^3\colon z^m = g(x, y)\}$ *has isolated singularity at origin. If* $(V, 0)$ *does not admit any* $\mathbf{C}^*$*-action and* $m \geqslant$ *multiplicity* $g(x, y)$*, then the irregularity* q *of the singularity is strictly greater than zero.*

PROOF. Since $(V, 0)$ does not admit any $\mathbf{C}^*$-action, it is clear that $\{(x, y) \in \mathbf{C}^2\colon g(x, y) = 0\}$ defines a reduced place curve singularity at origin without any $\mathbf{C}^*$-action. $\{(x, y) \in \mathbf{C}^2\colon g = \partial g/\partial x = \partial g/\partial y = 0\}$ defines a codimension 2 subvariety in $\mathbf{C}^2$. Consequently, it is determinantal **[5]**. There exists an exact sequence of the form

$$(3.3)\qquad 0 \to \mathbf{C}\{x, y\}^2 \xrightarrow{\begin{pmatrix} a_1 & b_1 \\ a_2 & b_2 \\ a & b \end{pmatrix}} \mathbf{C}\{x, y\}^3 \xrightarrow{(g_x,\, g_y,\, g)} \mathbf{C}\{x, y\}$$

$$\to \mathbf{C}\{x, y\}/(g_x, g_y, g) \to 0$$

with the following properties:

$$(3.4)\qquad a_2 b - b_2 a = g_x,$$

$$(3.5)\qquad ab_1 - a_1 b = g_y,$$

$$(3.6)\qquad a_1 b_2 - a_2 b_1 = g.$$

Exactness of (3.3) implies

$$(3.7)\qquad a_1 g_x + a_2 g_y + ag = 0,$$

$$(3.8)\qquad b_1 g_x + b_2 g_y + bg = 0.$$

Set $f(x, y, z) = z^m - g(x, y)$. Since $g = z^m - f = \frac{1}{m} z\partial f/\partial z - f$; we have

$$(3.9)\qquad a_1 \frac{\partial f}{\partial x} + a_2 \frac{\partial f}{\partial y} + \frac{az}{m}\frac{\partial f}{\partial z} = af,$$

$$(3.10)\qquad b_1 \frac{\partial f}{\partial x} + b_2 \frac{\partial f}{\partial y} + \frac{bz}{m}\frac{\partial f}{\partial z} = bf.$$

By a biholomorphic change of coordinate, we may assume that the least order terms of $g(x, y)$ involves the term x^r where $r =$ multiplicity of $g(x, y)$.

$$\omega = \frac{dy \wedge dz}{f_x} = \frac{dy \wedge dz}{g_x} = \frac{dz \wedge dx}{f_y} = \frac{dz \wedge dx}{g_y}$$

is a holomorphic 2-form on $V - \{0\}$ which cannot be extended across 0 holomorphically.

$$a_1\frac{\partial}{\partial x}+a_2\frac{\partial}{\partial y}+\frac{az}{m}\frac{\partial}{\partial z}\quad\text{and}\quad b_1\frac{\partial}{\partial x}+b_2\frac{\partial}{\partial y}+\frac{bz}{m}\frac{\partial}{\partial z}$$

are holomorphic vector fields on $(V,0)$. By contracting ω with the above two holomorphic vector fields, we get two holomorphic 1-forms on $V-\{0\}$

$$\gamma=\frac{az\,dy/m-a_2\,dz}{g_x}\quad\text{and}\quad\delta=\frac{bz\,dy/m-b_2\,dz}{g_x}.$$

Let $\sigma(t)=(\sigma_1(t),0,\sigma_3(t))$ be a normalization of a curve in V. We can choose $\sigma(t)$ such that $O(\sigma_3(t))\leqslant O(\sigma_1(t))$ where $O(\sigma_i(t))$ denotes the vanishing order of $\sigma_i(t)$ at origin. In fact, if $m\geqslant r=$ multiplicity of $g(x,y)$, we can choose $O(\sigma_1(t))=m\geqslant O(\sigma_3(t))=r$.

$$\sigma^*\gamma=\frac{-(a_2\circ\sigma)d\sigma_3/dt}{g_x\circ\sigma}dt,\qquad\sigma^*\delta=\frac{-(b_2\circ\sigma)\,d\sigma_3/dt}{g_x\circ\sigma}dt.$$

By (3.4) we have

$$(a_2\circ\sigma)(b\circ\sigma)-(b_2\circ\sigma)(a\circ\sigma)=g_x\circ\sigma.$$

Case 1. $O((a_2\circ\sigma)(b\circ\sigma))\leqslant O((b_2\circ\sigma)(a\circ\sigma))$.

Then

$$\begin{aligned}O(g_x\circ\sigma)-O\left((a_2\circ\sigma)\left(\frac{d\sigma_3}{dt}\right)\right)&=O(a_2\circ\sigma)+O(b\circ\sigma)\\&\quad-O(a_1\circ\sigma)-O(\sigma_3)+1\\&=O(b\circ\sigma)-O(\sigma_3)+1.\end{aligned}$$

Observe that since the singularity does not admit any $\mathbf{C}^*$-action, b vanishes at the origin. It follows that

$$O(b\circ\sigma)-O(\sigma_3)+1>0.$$

This implies that γ is not in $\Gamma(M,\Omega^1)$.

Case 2. $O((a_2\circ\sigma)(b\circ\sigma))\geqslant O((b_2\circ\sigma)(a\circ\sigma))$.

Similar arguments as above will show that δ is not in $\Gamma(M,\Omega^1)$. Q.E.D.

REFERENCES

1. H. Grauert, *Ube die Deformationen isolierter singularitaten analytischer Mengen*, Invent. Math. **15** (1972), 171–198.
2. H. Laufer, *On rational singularities*, Amer. J. Math. **94** (1972), 597–608.
3. R. Narasimhan, *On the homology groups of Stein spaces*, Invent. Math. **2** (1967), 377–385.
4. H. Pinkham, *Singularités rationnelles de surfaces*, *Appendice*, Sém. sur les Singularités des Surfaces, Lecture Notes in Math. vol. 777, Springer-Verlag, Berlin and New York, pp. 147–178.
5. M. Schaps, *Deformations of Cohen-Macaulay schemes of codimension two*, Tel Aviv University.
6. M. Schlessinger, *On rigid singularities*, Rice University Studies, Proc. Conf. on Complex Analysis, vol. 1, 1972, pp. 147–162.
7. Y.-T. Siu, *Analytic sheaves of local cohomology*, Trans. Amer. Math. Soc. **148** (1970), 347–366.
8. G. N. Tyurina, *Locally semiuniversal flat deformations of isolated singularities of complex spaces*, Math. USSR-Izv. **3** (1969), 967–999.
9. J. Wahl, Private communication.

10. Stephen S.-T. Yau, *Two theorems on higher dimensional singularities*, Math. Ann. **231** (1977), 55–59.

11. ______, *Various numerical invariants for isolated singularities*, Amer. J. Math. (to appear).

12. ______, $s^{(n-1)}$ *invariant for isolated n-dimensional singularities and its applications to module problems*, Amer. J. Math. **104** (1982), 829–841.

13. ______, *Existence of L^2-integrable holomorphic forms and lower estimates for* $\dim T_V^1$, Duke Math. J. **48** (1981), 537–547.

THE INSTITUTE FOR ADVANCED STUDY, PRINCETON

Current address: University of Illinois at Chicago Circle

Proceedings of Symposia in Pure Mathematics
Volume **40** (1983), Part 2

THE STRUCTURE OF STRATA μ = CONST IN A CRITICAL SET OF A COMPLETE INTERSECTION SINGULARITY

YOSEF YOMDIN

This report is a summary of results of [**9**, **10**, **11**].

Let f: $M^n \to N^k$, $n \geqslant k$, be a flat mapping of regular complex manifolds, such that $f/\Sigma(f)$ is finite (where $\Sigma(f)$ is the critical set of f). Then for any $z \in M$ the germ of f at z defines a complete intersection singularity (SCI). Setting $\mu_f(z)$ to be equal to the Milnor number of this singularity, we define an integral function μ_f on M. The function μ_f is upper semicontinuous and $\mu_f(z) > 0$ if and only if $z \in \Sigma(f)$ (see e.g. [**2**, **3**]).

Let $W_\nu(f) = \{z \in M \mid \mu_f(z) \geqslant \nu\}$, $V_\nu(f) = \{z \in M \mid \mu_f(z) = \nu\} = W_\nu - W_{\nu+1}$. Then each $W_r(f)$ is an analytic subset of M, $W_0(f) = M$, $W_1(f) = \Sigma(f)$ and if M is compact, $W_\nu(f) = \varnothing$ for ν sufficiently big.

The symbol f will be omitted below in the notation for $\Sigma(f)$, μ_f, $V_\nu(f)$ and $W_\nu(f)$.

In the local case of SCI, f: $(C^n, 0) \to (C^k, 0)$, μ, W_ν, V_ν are germs at $0 \in C^n$. Note that $V_{\mu(0)} = W_{\mu(0)}$ is an analytic germ.

Let f: $M^n \to N^k$, μ, V_ν, W_ν be as above, with M and N compact.

THEOREM 1 [**9**]. *Let F be a generic (nonsingular) fiber of f and χ denote topological Euler characteristics. Then*

$$\sum_{\nu\geqslant 1} \nu\cdot\chi(V_\nu) = \sum_{\nu\geqslant 1} \chi(W_\nu) = (-1)^{n-k}[\chi(F)\cdot\chi(N) - \chi(M)].$$

In the local case of the germ f: $(C^n, 0) \to (C^k, 0)$ of SCI we have the following corollaries.

1980 *Mathematics Subject Classification*. Primary 32B30; Secondary 14B05.

0082-0717/81/0000-0213/$02.00

Let L be a generic l-dimensional subspace of C^k, L' its generic sufficiently small parallel translation. Denote by $\mu^{k-l}(0)$ the Milnor number of the singularity of $f^{-1}(L)$ at 0, and let $V_\nu' = V_\nu \cap f^{-1}(L')$.

COROLLARY 2.

$$\sum_{\nu=1}^{\mu(0)} \nu \cdot \chi(V_\nu') = \mu(0) + (-1)^{l+1}\mu^{k-l}(0).$$

Finally, if we consider some small deformation f' of f as the section of an appropriate family, then for corresponding strata V_ν' of f' we obtain

COROLLARY 3.

$$\sum_{\nu=1}^{\mu(0)} \nu \cdot \chi(V_\nu') = \mu(0).$$

This formula can be considered as the generalization to complete intersections of the fact that the sum of Milnor numbers of singularities in a deformation of an isolated hypersurface singularity is equal to $\mu(0)$.

The proof of these results is based on a construction similar to that used in [**5**, Proposition 1]. In particular, in terms of the functor introduced in this proposition, the following relation for $f: M^n \to N^k$ is valid: for the constructible functions 1_M and μ_f on M and 1_N on N,

$$f_*\left(1_M + (-1)^{n-k}\mu_f\right) = \chi(F) \cdot 1_N.$$

Some examples can be found in [**10**]. In particular let $f = (f_1, f_2)$: $(c^2, 0) \to (C^2, 0)$. Here the critical set of f is defined by the equation $J = \det(\partial f_i/\partial z_j) = 0$. If J has an isolated critical point at 0 with the Milnor number μ_1, then, for any generic deformation f' of f, $W_1 = \Sigma^1$ is a smooth curve, and $\chi(W_1) = 1 - \mu_1$. The set $V_2 = W_2$ is a set of cusps $\Sigma^{1,1}$ on Σ^1, and by Corollary 3, the number of cusps is equal to $\chi(\Sigma^{1,1}) = \mu - \chi(\Sigma^1) = \mu + \mu_1 - 1$. For example, for $f = (z_1^p + z_2^q, z_1 \cdot z_2)$, $\mu = p + q + 1$, $\mu_1 = (p-1)(q-1)$ and the number of cusps for any generic deformation f' of f is $pq - 1$.

Actually, in [**10**] the special case is considered, where V_ν coincides with the Thom-Boardman strata $\Sigma_\nu = \Sigma^{n-k+1,1,\ldots,1}$ ($\nu - 1$ times). Theorem 1 in this case follows from the study of Morse functions on $\Sigma^\mu \subset \Sigma^{\mu-1} \subset \cdots \subset \Sigma^1$.

In [**11**] the following inequality is obtained. Let $f: (C^n, 0) \to (C^k, 0)$ be SCI. Let Σ_i, $i = 1, \ldots, l$, be irreducible components of Σ_{red} and let k_i be the multiplicity of Σ_i at 0. For each i the function μ is constant on a Zariski open subset of Σ_i and its value on this subset we denote by μ_i.

THEOREM 4. *Let* rank $df(0) \geqslant k - 2$. *Then* $\mu(0) \geqslant \sum_{i=1}^{l} k_i\mu_i$. *If* rank $df(0) = k - 2$, *then the strict inequality holds.*

COROLLARY 5. *If* rank $df(0) = k - 2$, *then* $k_i < \mu(0)/\mu_i$. *In particular, if* $\mu_i \geqslant \mu(0)/2$, *then* $(\Sigma_i)_{\text{red}}$ *is regular.*

Then we combine Theorem 4, Corollary 2 and the inequality of Corollary 1.6 [**1**] to prove the following theorem.

THEOREM 6. *Let* $f\colon (C^n, 0) \to (C^k, 0)$ *be SCI, and let* $\dim_0 V_{\mu(0)} = k - 1$. *Then f is a family of hypersurface singularities with* $\mu = \mu(0)$. *In particular for* $n - k \neq 2$ *this family is topologically trivial by* [**4, 8**].

REMARK. Theorem 6 shows that f is topologically equisingular along strata μ = const of maximal dimension. If the dimension of stratum μ = const is strictly less than $k - 1$, this is not true, even for rank $df = k - 1$ (see [**6**]). Rank df can also change in such deformations (see [**7**, Example 4.6.3]).

REFERENCES

1. M. Guisti and J. P. Henry, *Minorations de nombres de Milnor*, Bull. Soc. Math. France **108** (1980), 17–45.

2. G. M. Greuel, *Der Gauss-Manin-Zusammenhang isolierter Singularitaten von vollstandigen Durchschnitten*, Math. Ann. **214** (1975), 235–266.

3. Lê Dung Trãng, *Calculation of the Milnor number of isolated singularity of complete intersection*, Functional Anal. Appl. **8** (1974), 127–131.

4. Lê Dung Trãng and C. P. Ramanujam, *The invariance of Milnor's number implies the invariance of topological type*, Amer. J. Math. **98** (1976), 67–78.

5. R. D. MacPherson, *Chern classes for singular algebraic varieties*, Ann. of Math. (2) **100** (1974), 423–432.

6. F. Pham, *Remarques sur l'equisingularité universelle*, preprint, Université de Nice (1971).

7. B. Teissier, *The hunting of invariants in the geometry of discriminants*, Real and Complex Singularities, Proc. Nordic Summer School/NAVF, Oslo, 1976.

8. J. G. Timourian, *The invariance of Milnor's number implies topological triviality*, Amer. J. Math. **99** (1977), 437–446.

9. Y. Yomdin, *Euler characteristics of strata* μ = const *for isolated singularities of complete intersections*, preprint, Ben Gurion University of the Negev (1979).

10. I. N. Iomdin, *Euler characteristics of Boardman strata and the Milnor number of an isolated singularity of a complete intersection*, Functional Anal. Appl. **10** (1976), 231–232.

11. Y. Yomdin, *The structure of a critical set of a complete intersection singularity*, Proc. Amer. Math. Soc. **84** (1982), 383–388.

Current address: Max-Planck Institut für Mathematik, Bonn, Federal Republic of Germany

Proceedings of Symposia in Pure Mathematics
Volume 40 (1983), Part 2

SOME RESULTS ON FINITE DETERMINACY AND STABILITY NOT REQUIRING THE EXPLICIT USE OF SMOOTHNESS

YOSEF YOMDIN

Definitions of structural or topological equivalence of mappings ($f, g: M \to N$ are equivalent if $f = \psi^{-1} \circ g \circ \varphi$ for some diffeomorphisms (homeomorphisms) $\varphi: M \to M$, $\psi: N \to N$) do not require smoothness of f and g. However in order to obtain a sufficiently good structure of equivalence classes, usually smooth mappings are considered.

In §1 we give examples, showing that some results on (topological) finite determinacy of singularities of real functions remain valid under much weaker assumptions. The nonsmooth functions to which these results can be applied appear usually as a result of taking sup (inf) of a family of smooth functions or an interpolation of smooth functions. The approach here is to replace the assumptions on derivatives of high order by some assumptions on the behavior of the first derivative, which exists almost everywhere for functions considered.

The result reported in §2 is a joint work with J. Sanzot. This result concerns a method for studying equivalence of mappings, not requiring the use of derivatives. The idea is to minimize the distance between f and $\psi^{-1} \circ g \circ \varphi$ (in a suitable functional space), choosing φ and ψ. Then f and g are equivalent if and only if the minimum of the distance is zero. There are evident obstacles: possibility of improper solutions and nonconvexity of the minimization problem considered, even near the proper solution. We consider here the following simplest situation:

1. f and g are real functions of one variable;
2. the weighted mean square distance is considered;
3. we consider the case of right equivalence, i.e. only transformations φ of the domain are allowed;
4. we assume a priori that $f = g(\omega)$, where ω is a transformation, close to identity.

1980 *Mathematics Subject Classification.* Primary 58C27, 49D99.

It turns out that even in this simplest case both the difficulties, described above, are present. As a result, too precise a minimization procedure starting at an identical transformation inevitably leads to an improper solution.

However, some minimization procedure, starting at Id and converging to ω can be found (for g sufficiently good and ω sufficiently close to Id) because of the following fact.

If all the critical points of g are nondegenerate and if ω is sufficiently close to Id, although the initial minimization problem is not convex, its suitable (sufficiently rough) finite-dimensional approximation turns out to be convex. The best preciseness of such an approximation, which we can achieve, not destroying convexity, depends on the deviation of the replacement ω from the identity.

Thus our procedure consists of solving successive finite-dimensional convex minimization problems, which approximate more and more precisely the initial one. The starting point of each problem is the minimum point of the previous one.

In §2 we describe precisely this procedure and state the theorem, giving conditions under which it converges.

Note that the restriction to the case of right equivalence is not essential. The result for the case of right-left equivalence is completely similar. The only difference is that we do not need to assume the existence of the solution. It is sufficient to require all the critical points of g to be nondegenerate with distinct critical values and f to be sufficiently close to g.

Although smoothness of functions considered is used both in statement and in proof of the result, the procedure itself operates with values of f and g and not of their derivatives (only approximate values of g' are used in weights). Theorem 2.3 can be reformulated for nonsmooth functions.

These results as well as the detailed proof of results of this report will appear separately.

1. Sufficiency of jets for Lipschitzian functions. The local structure of a C^k-smooth function $f\colon R^n \to R$ in many cases is completely determined by its partial derivatives at x_0 up to some order $l \leqslant k$ (in a sense that by an appropriate change of coordinates near x_0, f can be reduced to its lth Taylor polynomial).

There are important classes of functions which are not smooth but have higher derivatives in some generalized sense. In particular, in many cases the kth Taylor polynomial (or the kth differential) of $f\colon R^n \to R$ can be defined at almost every point $x_0 \in R^n$ as the polynomial P_{x_0} of degree $\leqslant k$ such that

$$(*) \qquad |f(x) - P_{x_0}(x)| = o(|x - x_o|^k).$$

(See e.g. [**2, 3**].)

Then the natural question is: In what cases is the local structure of such a function defined by its Taylor polynomial?

Clearly, without additional restrictions no Taylor polynomial of f in the sense of $(*)$ does define the local topology of f because of possible oscillation of f near x_0.

We show that the situation becomes almost completely parallel to that of smooth functions, if we restrict ourselves to Lipschitzian functions and in definition of the kth differential require (*) not only for the function itself, but also for its first derivative (which exists almost everywhere, see [3]). The kth differential, defined in this way, is called below the kth L-differential.

In this paper we study functions, highly L-differentiable at a given point. However, the condition of k times L-differentiability is not much more restrictive than that of the existence of the kth differential (*), and thus highly L-differentiable almost everywhere Lipschitzian functions appear in various situations. For example, functions belonging to the Sobolev space $L_k^p(R^n)$ with $p(k-1)>n$ are C^1-smooth in the usual sense, but have the kth L-differential almost everywhere. $f=\max(g_1,\dots,g_l)$, where $g_1,\dots,g_l$ are C^k-smooth, is k times L-differentiable almost everywhere. The pointwise maximum of a bounded (in a C^2-norm) family of C^2-smooth functions is almost everywhere twice L-differentiable. Particular examples are the distance function to any closed subset in R^n, and any convex function. Results and methods of this paper can be used to study the local and global structure of singularities of such functions. For example in [5] the simplest variant of sufficiency results—the implicit function theorem—is applied to the (nonsmooth) distance function to obtain the topological description of the set of nondifferentiability of this function (which in the case of a distance to the submanifold appears as the cut locus of this submanifold).

We shall use the following notations. For the germ h at $x_0 \in R^n$, $h=o(r)$, $0 \leqslant r \leqslant \infty$, if

$$\lim_{x \to x_0} \frac{|h(x)|}{|x-x_0|^r} = 0$$

(for $r=\infty$, $\lim_{x\to x_0}(|h(x)|/|x-x_0|^N)=0$ for any positive integer N).

$h=O(r)$, $-\infty<r<\infty$, if for some constant $C>0$ and x sufficiently close to x_0, $|h(x)| \leqslant C|x-x_0|^r$.

Now let $f\colon (R^n, x_0) \to (R^m, y_0)$ be the Lipschitzian germ, i.e. for some constant K and for all x_1, x_2 near x_0 we have

$$|f(x_1)-f(x_2)| \leqslant K|x_1-x_2|.$$

Here and below $|\ |$ denotes the Euclidean norm in R^q and corresponding norms of linear operators.

DEFINITION 1.1. The germ $f\colon (R^n, x_0) \to (R^m, y_0)$ is said to be k times L-differentiable at x_0, $1 \leqslant k \leqslant \infty$, if the following condition is satisfied:

There exists a C^k-smooth germ $g\colon (R^n, x_0) \to (R^m, y_0)$, such that for any x outside of some subset of measure zero if $df(x)$ exists, then

$$|df(x)-dg(x)| \leqslant K(x), \qquad K=o(k-1).$$

The set of Lipschitzian germs $f\colon (R^n, x_0) \to R^m$, k times L-differentiable at x_0, will be denoted by $CL^k_{n,m}(x_0)$ or briefly $CL^k(x_0)$.

DEFINITION 1.2. For $f \in CL^k(x_0)$ the k-jet of f at x_0, $J^k_{x_0}(f)$, is the k-jet $J^k_{x_0}(g)$ of any C^k-smooth g, satisfying Definition 1.1. The coordinates of $J^k_{x_0}(f)$ will be called the partial L-derivatives of f at x_0.

The partial L-derivatives and L-differentials will be denoted below by the subscript L. Denote by $J^k_{x_0,y_0}(n, m)$ the space of k-jets at x_0 of $f:(R^n, x_0) \to (R^m, y_0)$.

Let $j_1 \in J^k_{x_0,y_0}(n,m)$, $j_2 \in J^k_{y_0,z_0}(m, p)$, $j_3 \in J^k_{a,b}(q, q)$, j_3-nondegenerate. Then $j_2 \circ j_1 \in J^k_{x_0,z_0}(n, p)$ and $j_3^{-1} \in J^k_{b,a}(q, q)$ are naturally defined.

PROPOSITION 1.3. *Let* f_1: $(R^n, x_0) \to (R^m, y_0)$, $f_1 \in CL^k(x_0)$, f_2: $(R^m, y_0) \to (R^p, z_0)$, $f_2 \in CL^k(y_0)$, $1 \leq k \leq \infty$. *Then* $f_2 \circ f_1 \in CL^k(x_0)$ *and*

$$J^k_{x_0}(f_2 \circ f_1) = J^k_{y_0}(f_2) \circ J^k_{x_0}(f_1).$$

PROPOSITION 1.4. *Let* f: $(R^n, x_0) \to (R^n, y_0)$ *be a Lipschitzian germ,* $f \in CL^k(x_0)$, $1 \leq k \leq \infty$, *with* $d_L f(x_0)$ *nondegenerate. Then* f *is a* CL^k*-diffeomorphism, and* $J^k_{y_0}(f^{-1}) = [J^k_{x_0}(f)]^{-1}$. *(*$f$ *is a* CL^k*-diffeomorphism, if* $f^{-1} \in CL^k(y_0)$.*)*

Now we shall consider some situations where highly L-differentiable functions naturally arise.

THEOREM 1.5. *Let* $g_1,\ldots,g_r$, g_i: $(R^n, x_0) \to (R, c)$, *be* C^k*-smooth germs,* $1 \leq k \leq \infty$, *with* $J^k_{x_0}(g_1) = \cdots = J^k_{x_0}(g_r) = j$. *Let* $f(x) = \max(g_1(x),\ldots,g_r(x))$. *Then* $f \in CL^k(x_0)$ *and* $J^k_{x_0}(f) = j$.

Examples of L-differentiable functions appear also in approximation of smooth functions. We shall consider the simplest case of a piecewise linear approximation.

Let Σ be a simplicial decomposition of a domain $U \subset R^n$. The Σ-approximation f_Σ of f: $U \to R^s$ is defined as follows: any $x \in U$ belongs to the interior of one and only one simplex Δ^q in Σ, $q \leq n$. Let $x_0,\ldots,x_q$ be the vertices of Δ^q and let $t_0,\ldots,t_q$ be baricentric coordinates of x in Δ^q. Then $f_\Sigma(x) = \Sigma^q_{i=0} t_i f(x_i)$. f_Σ is Lipschitzian if f is Lipschitzian (and f_Σ is always Lipschitzian if Σ is locally finite).

Now let U be a neighborhood of $x_0 \in R^n$, and let Σ be the simplicial decomposition of U. For $x \in U$ denote by $\Delta(x)$ the simplex in Σ for which $x \in \mathring{\Delta}(x)$. Define $r(\Sigma)$ as the biggest integer for which diam $\Delta(x) = o(r(\Sigma))$. If $r(\Sigma) \geq 0$, then the simplices of Σ are condensed at x_0.

THEOREM 1.6. *Let* g: $(R^n, x_0) \to (R^s, y_0)$ *be a* C^m*-smooth germ,* $m \geq 2$. *Let* q *be the biggest integer such that* $J^q_{x_0}(g - dg(x_0)) = 0$. *Let* Σ *be the simplicial decomposition of a neighborhood of* x_0, $r = r(\Sigma) \geq 0$. *Then if* $r = 0$, g_Σ *is once* L*-differentiable at* x_0 *and* $J^1_{x_0}(g_\Sigma) = J^1_{x_0}(g)$.

If $r > 0$, *then* $g_\Sigma \in CL^{r+l}(x_0)$, *where* $l = \min(m - r, q)$, *and* $J^{r+l}_{x_0}(g_\Sigma) = J^{r+l}_{x_0}(g)$.

Two germs f_1: $(R^n, 0) \to (R, 0)$ and f_2: $(R^n, 0) \to (R, 0)$ are said to be C^k-equivalent (respectively, CL^k-equivalent) if there exists a germ of a C^k-diffeomorphism (resp. CL^k-diffeomorphism) W: $(R^n, 0) \to (R^n, 0)$ such that $f_2 = f_1 \circ W$.

The jet $j \in J^l_{0,0}(n, 1)$ is said to be C^k-sufficient, if any two $C^{\max(l,k)}$-smooth germs g_1 and g_2 with $J^l_0(g_1) = J^l_0(g_2) = j$ are C^k-equivalent.

j is said to be CL^k-sufficient, if any two germs f_1, $f_2 \in CL^{\max(l,k)}(0)$ with $J^l_0(f_1) = J^l_0(f_2) = j$ are CL^k-equivalent.

Let g: $(R^n, 0) \to (R, 0)$ be a C^∞-smooth germ. Define, following [**4**], the numbers $r(g)$ and $r_0(g)$ as follows.

$r(g)$ is the smallest positive integer, such that there is a positive constant C with $|\operatorname{grad} g(x)| \geqslant C|x|^{r(g)-1}$ for all x near the origin, or, if no such integer exists, is ∞.

$r_0(g)$ is the biggest positive integer, such that $g = O(r_0(g))$, or, if no such biggest integer exists, is ∞.

Clearly, $r_0(g)$ is the biggest integer for which $J_0^{r_0(g)-1}(g) = 0$, and $r_0(g) \leqslant r(g)$.

THEOREM 1.7. *Let* g: $(R^n, 0) \to (R, 0)$ *be a* C^∞*-smooth germ with* $r(g) < \infty$. *Then for any* k, $1 \leqslant k \leqslant \infty$, *and* $l \geqslant k + 2r(g) - r_0(g) - 1$ *the* l*-jet of* g *at* 0 *is* CL^k*-sufficient.*

REMARK. It is interesting to compare Theorem 1.7 with the result of F. Takens [**4**]; in the same notations the l-jet of g is C^k-sufficient if $l \geqslant k + 2r(g) - r_0(g) - 1 + (k-1)[r(g) - r_0(g)]$. Thus for $k = 1$ or $r(g) = r_0(g)$ this condition coincides with the one of Theorem 1.7 and the l-jet of g will be both C^k-sufficient and CL^k-sufficient. (Our proof actually gives a C^1-smooth diffeomorphism W, if f is C^1-smooth.)

If we consider infinitely L-smooth germs only, then the following is true:

COROLLARY 1.8. *If the jet* $j \in J^l_{0,0}(n, 1)$ *is* C^k*-sufficient*, $0 \leqslant k \leqslant \infty$, *then it is* CL^k*-sufficient.*

PROOF. Let f_1 and $f_2 \in CL^\infty(0)$ and $J^l_0(f_1) = J^l_0(f_2) = j$. Since j is C^k-sufficient, then for any smooth g with $J^l_0(g) = j$, $r(g) < \infty$. Hence by Theorem 1.7, the jet $J^\infty_0(g)$ is CL^∞-sufficient. Now let g_1 and g_2 be C^∞-smooth germs, such that $J^\infty_0(g_i) = J^\infty_0(f_i)$, $i = 1, 2$. Then f_i is CL^∞-equivalent to g_i, $i = 1, 2$. But $J^l_0(g_1) = J^l_0(g_2) = j$, hence g_1 and g_2 are C^k-equivalent.

COROLLARY 1.9. *The jet* $j \in J^l_{0,0}(m, 1)$ *is* C^∞*-sufficient if and only if it is* CL^∞*-sufficient.*

PROOF. If j is C^∞-sufficient, it is CL^∞-sufficient by Corollary 1.8. Let j be CL^∞-sufficient. Passing to jets we obtain that the set $U \subset J^{l+1}_{0,0}(n, 1)$, consisting of jets j' with the projection on $J^l_{0,0}(n, 1)$ equal to j, is contained in the orbit of the action of the group of $(l+1)$-jets of diffeomorphisms. But in this case j is C^∞-sufficient (see e.g. [**1**, Corollary 11.9]).

2. Minimization procedure for testing equivalence of functions. Before we give a precise description of the procedure and state a theorem, we explain why the initial problem is not convex. Let $f\colon [0,1] \to R$, $g\colon [0,1] \to R$. Consider the problem of minimization of

$$\sigma_{f,g}(\varphi) = \int_0^1 \eta(x)(f(x) - g(\varphi(x)))^2\,dx \tag{2.1}$$

by choosing a one-to-one transformation $\varphi\colon [0,1] \to [0,1]$. Computing the second variation of $\sigma_{f,g}$ (at $\varphi \equiv \mathrm{Id}$) we have for the vector field ξ on $[0,1]$

$$\left(d^2\sigma_{f,g}\right)(\xi,\xi) = \frac{d^2}{dt^2}\sigma_{f,g}(\mathrm{Id} + t\xi)\Big|_{t=0}$$
$$= 2\int_0^1 \eta(x)\left[(g'(x))^2 - (f(x) - g(x))g''(x)\right]\xi^2(x)\,dx = \int_0^1 I(x)\xi^2(x)\,dx.$$

Since the weight function $\eta(x)$ is always positive, the sign of $I(x)$ is determined by the sign of $(g'(x))^2 - (f(x) - g(x))g''(x)$. We see that if g has critical points, then $I(x)$ can be negative in a neighborhood of these critical points unless $f \equiv g$. Moreover, if g has only nondegenerate critical points, and $f = g(\omega)$, where the replacement $\beta = \omega - \mathrm{Id}$ is sufficiently small, then necessarily $I(x) < 0$ in a neighborhood of each critical point of g. Indeed, at a critical point x_0 of g we have $f(x_0) = g(\omega(x_0)) \approx g(x_0) + \frac{1}{2}g''(x_0)\beta^2(x_0)$ ($g'(x_0) = 0$ since x_0 is a critical point of g, $g''(x_0) \neq 0$ since this point is nondegenerate). Hence $I(x_0) = \eta(x_0)(-(f(x_0) - g(x_0)\cdot g''(x_0))) \approx -\frac{1}{2}\eta(x_0)(g''(x_0))^2\beta^2(x_0) < 0$.

Thus, for $f = g(\omega)$, if ω is arbitrarily close, but not exactly equal to Id, the second variation $d^2\sigma_{f,g}$ is negative on vector fields ξ, concentrated near critical points of g. For $f \equiv g$, i.e., for $\omega \equiv \mathrm{Id}$, $d^2\sigma_{g,g}$ is positive, but has arbitrarily small eigenvalues.

Description of the procedure. In a finite-dimensional approximation of a minimization problem (2.1) we take φ in a form of a piecewise linear transformation on an equidistant network on $[0,1]$.

Let $N \geqslant 2$ be an integer,

$$\delta = \frac{1}{N+1}, \quad x_i = i\delta,\ i = 0,\dots,N+1, \quad x_0 = 0, \quad x_{N+1} = 1.$$

Since all the transformations considered preserve the ends of $[0,1]$, any such transformation is defined by its values $\alpha_1,\dots,\alpha_N$ at $x_1,\dots,x_N$, with $\alpha_0 = 0$, $\alpha_{N+1} = 1$. Thus for $\alpha = (\alpha_1,\dots,\alpha_N) \in R^N$, $\varphi(\alpha,\cdot)$ will denote the piecewise-linear transformation $\varphi(\alpha,\cdot)\colon [0,1] \to [0,1]$, $\varphi(\alpha, x) = \alpha_i + t(\alpha_{i+1} - \alpha_i)$ for $x = x_i + t(x_{i+1} - x_i)$, $0 \leqslant t \leqslant 1$, $i = 0,\dots,N$. We define the functional to be minimized as follows:

$$\sigma(\alpha) = \frac{1}{\delta}\int_0^1 \eta(x)(f(x) - g(\varphi(\alpha,x)))^2\,dx.$$

Here the weight function η is assumed to be constant on each interval $[x_i, x_{i+1}]$. It will be precisely described below.

Thus $\sigma(\alpha)$ can be written in a form

$$\sigma(\alpha) = \frac{1}{\delta} \sum_{i=0}^{N} \int_{x_i}^{x_{i+1}} \eta_i (f(x) - g(\varphi(\alpha, x)))^2 \, dx.$$

Let $N_1, N_2, \ldots, N_k$ be positive integers. We will define inductively procedures

$$A_k(f, g, N_1, \ldots, N_k), \qquad k = 0, 1, \ldots.$$

Define the transformation φ_0: $[0, 1] \to [0, 1]$ to be the identity, $\varphi_0(x) \equiv x$.

The procedure $A_0(f, g, -)$ is always defined and its result is the transformation φ_0. Now assume that the procedure $A_k(f, g, N_1, \ldots, N_k)$ is defined and its result is the transformation φ_k: $[0, 1] \to [0, 1]$.

Let N_{k+1} be a positive integer. Denote $\delta_{k+1} = 1/(N_{k+1} + 1)$, $x_i = i\delta_{k+1}$, and let $\alpha_i^0 = \varphi_k(x_i)$, $i = 1, \ldots, N_{k+1}$, $\alpha^0 = (\alpha_1^0, \ldots, \alpha_{N_{k+1}}^0) \in R^{N_{k+1}}$.

For $\alpha \in R^{N_{k+1}}$ let

$$\sigma_{k+1}(\alpha) = \frac{1}{\delta_{k+1}} \sum_{i=0}^{N_{k+1}} \frac{1}{\mu_i^2} \int_{x_i}^{x_{i+1}} (f(x) - g(\varphi(\alpha, x)))^2 \, dx,$$

where the weights $1/\mu_i^2$ are defined by $\mu_i = \max_{x \in [x_i, x_{i+1}]} |g'(\varphi(\alpha^0, x))|$, $i = 0, \ldots, N_{k+1}$. (Actually, we need only μ_i to be of the same order as this maximum, e.g., we can take

$$\mu_i = \frac{1}{\delta_{k+1}} \int_{x_i}^{x_{i+1}} |g'| \, dx.)$$

DEFINITION 2.1. The procedure $A_{k+1}(f, g, N_1, \ldots, N_k, N_{k+1})$ is said to be defined if there exists an open domain $U_{k+1} \subset R^{N_{k+1}}$ such that

1. $\alpha^0 \in U_{k+1}$,
2. σ_{k+1} is convex on U_{k+1},
3. σ_{k+1} has a minimum α^1 (necessarily unique) in U_{k+1}.

The result of the procedure $A_{k+1}(f, g, N_1, \ldots, N_{k+1})$ is defined as the transformation φ_{k+1}: $[0, 1] \to [0, 1]$, $\varphi_{k+1}(x) = \varphi(\alpha^1, x)$.

Now let $P = (N_1, N_2, \ldots, N_k, \ldots)$ be an infinite sequence of positive integers.

DEFINITION 2.2. The procedure $A(f, g, P)$ is said to be defined, if for each $k = 1, \ldots,$ the procedure $A_k(f, g, N_1, \ldots, N_k)$ is defined. The result of $A(f, g, P)$ is a sequence of results $\varphi_1, \varphi_2, \ldots, \varphi_k, \ldots,$ of procedures $A_k(f, g, N_1, \ldots, N_k)$.

The main theorem. First of all we will state the precise assumptions on functions g and f.

(*) We assume that g is twice continuously differentiable on $[0, 1]$. Then, we assume that $g(0) = g(1) = 0$, $g'(0) \cdot g'(1) \neq 0$, and all the critical points of g on $[0, 1]$ are nondegenerate, i.e., if for $z \in [0, 1]$, $g'(z) = 0$, then $g''(z) \neq 0$. It follows that there is only a finite number of critical points $z_1, \ldots, z_l \in (0, 1)$ of g.

Let

$$M_1 = \max_{x \in [0,1]} |g''(x)|, \qquad 2M_2 = \min_{p=1,\ldots,l} |g''(z_p)|.$$

We have $M_1 \geqslant 2M_2 > 0$. By continuity of g'' we can choose for each z_p the interval $V_p = [z_p - \xi_p, z_p + \xi_p]$, on which $|g''| \geqslant \frac{1}{2}|g''(z_p)| \geqslant M_2$. Let $D = \min_{p=1,\ldots,l} \xi_p$.

Denote by V the union $V = \bigcup_{p=1}^{l} V_p$. Let $M_3 = \min_{x \in [0,1] \setminus \operatorname{int} V} |g'(x)|$. Since g' does not vanish on $[0,1] \setminus \operatorname{int} V$, $M_3 > 0$.

Now we turn to f.

We assume that $f(x) = g(\omega(x))$, where $\omega\colon [0,1] \to [0,1]$ is a twice differentiable one-to-one transformation, $\omega(0) = 0$, $\omega(1) = 1$. Since ω is a monotone function, we assume that there exists a constant L_1, $0 < L_1 \leqslant 1$, such that for each $x \in [0,1]$, $\omega'(x) \geqslant L_1$. We also assume that for some constant L_2, $L_2 \geqslant L_1$, always $\omega' \leqslant L_2$, and for some constant $L_3 \geqslant 0$, always $|\omega''| \leqslant L_3$.

Finally, denote by $\beta(\omega) = \max_{x \in [0,1]} |\omega(x) - x|$ the deviation of ω from the identical transformation.

The constants M_1, M_2, M_3, D, L_1, L_2, L_3 will be considered as the absolute constants in what follows, while the condition of the theorem actually requires $\beta(\omega)$ to be sufficiently small.

THEOREM 2.3. *Let g and $f = g(\omega)$ satisfy conditions $(*)$. There exists the constant $Q > 0$, depending only on M_1, M_2, M_3, D, L_1, L_2, L_3, such that if $\beta(\omega) \leqslant Q$, then the sequence $P = (N_1, N_2, \ldots, N_k, \ldots)$ can be built, for which*

1. *The procedure $A(f, g, P)$ is defined.*
2. *The resulting sequence $\varphi_1, \ldots, \varphi_k, \ldots$, of the procedure $A(f, g, P)$ converges uniformly to the transformation ω.*

REFERENCES

1. Th. Brocker and L. Lander, *Differential germs and catastrophes*, London Math. Soc. Lecture Notes Series, vol. 17, Cambridge Univ. Press, Cambridge, 1975.
2. A. P. Calderón and A. Zygmund, *Local properties of solutions of elliptic partial differential equations*, Studia Math. **20** (1961), 171–225.
3. E. M. Stein, *Singular integrals and differentiability properties of functions*, Princeton Univ. Press, Princeton, N.J., 1970.
4. F. Takens, *A note on sufficiency of jets*, Invent. Math. **13** (1971), 225–231.
5. Y. Yomdin, *On the local structure of a generic central set*, Compositio Math. **43** (1981), 225–238.

BEN-GURION UNIVERSITY OF THE NEGEV, BEER-SHEVA, ISRAEL

Proceedings of Symposia in Pure Mathematics
Volume 40 (1983), Part 2

L_2-COHOMOLOGY AND INTERSECTION HOMOLOGY OF LOCALLY SYMMETRIC VARIETIES

STEVEN ZUCKER[1]

1. Introduction. From the perspective of the theme of the Summer Institute, L_2-cohomology is of interest because it offers the prospect of placing a natural Hodge structure on the middle intersection homology groups of singular projective varieties.[2] In fact, the use of L_2-cohomology (and the associated L_2 harmonic forms) is the only known way of producing "new" Hodge decompositions, i.e., ones whose existence cannot be deduced by standard constructions on known ones. The chief drawback of this method is that one must choose the right Kähler metric on the regular part of the variety, so that the corresponding L_2-cohomology is indeed isomorphic to the intersection homology. In two known instances, viz. [**C** and **Z2**], metrics of quite opposite nature are used, so they do not even suggest a unified theory. Moreover, L_2-cohomology is highly nonfunctorial, and it behaves poorly with respect to cup-products. If this is the best of all possible worlds, it will be possible to construct ultimately a maximally functorial cohomological Hodge complex (in the sense of [**D**, (8.1.2)]) that behaves well with respect to products (cf. [**Z1**]; also [**CGM**, §6, **Z4**] for the relation between [**Z1**] and intersection homology). This would clarify the incomplete, and not fully understood, functorial properties of intersection homology.

On the other hand, L_2-cohomology groups are of interest in their own right, as algebraic quasi-isometry invariants of Riemannian manifolds. One can then ask whether these groups provide (at least in cases) topological invariants of suitable, possibly singular, compactifications.

1980 *Mathematics Subject Classification*. Primary 22E40, 55N99, 58A14; Secondary 32M15, 57T99.

[1]Supported in part by the National Science Foundation, through Grant MCS81-01650 and a grant to the Institute for Advanced Study.

[2]One wants to have, in addition, that the Hodge structure be compatible with the existing mixed Hodge theory of the variety and its smooth locus.

0082-0717/81/0000-0214/$02.50

The results described in this article arose from this second point of view, and are part of the content of [**Z2**]. Specifically, the L_2-cohomology of quotients of symmetric spaces $X = G/K$ by arithmetic groups Γ, with respect to natural metrics, with its relation to the representation theory of $L_2(\Gamma\backslash G)$, has been studied for some time (see [**B1, B2, G, MM**], etc.). We have proved in certain cases that the L_2-cohomology of $\Gamma\backslash X$ is the intersection homology of its Baily-Borel-Satake compactification: for the case where X is the ball in $\mathbf{C}^n$, and for Hilbert modular varieties of any dimension. The common feature of these two cases is that the compactification is achieved by adjoining isolated singular points to $\Gamma\backslash X$. We have shown directly the vanishing of local cohomology which characterizes intersection homology for such spaces. The validity of the aforementioned isomorphism for general locally symmetric varieties (i.e., whenever X is Hermitian)—and also with twisted coefficients—is conjectured in [**Z2**].

This article is an expanded version of the author's seminar talk of August 6, 1981.

2. L_2-cohomology. Let M be a C^∞ Riemannian manifold. If ϕ is any tensor field on M, one may speak of its length $|\phi|$ as a function on M. Thus, the metric determines L_2 seminorms

$$\|\phi\|^2_{(2)} = \int_M |\phi|^2\, dV_M,$$

where dV_M is the volume density of M; one says that ϕ is in L_2 if this integral is finite.

Put $L^i_{(2)}(M, \mathbf{C}) = \{\mathbf{C}\text{-valued global } i\text{-forms } \phi \text{ on } M \text{ such that } \phi \text{ and } d\phi \text{ are in } L_2\}$. In the above, we are taking forms with only measurable coefficients—though it makes no difference for cohomological purposes if one uses C^∞ forms (see [**C**, §8])—so that $L^*_{(2)}(M, \mathbf{C})$ can be identified as the domain of the maximal closed extension of the exterior derivative on M. We see that $L^*_{(2)}(M, \mathbf{C})$ is a subcomplex of the full deRham complex of M, and one defines the L_2-cohomology, $H^*_{(2)}(M, \mathbf{C})$, as the cohomology of this complex. Specifically, if we let

$$Z^i_{(2)}(M, \mathbf{C}) = \{L_2\ i\text{-forms } \phi\colon d\phi = 0\},$$

then

$$H^i_{(2)}(M, \mathbf{C}) = Z^i_{(2)}(M, \mathbf{C})/dL^{i-1}_{(2)}(M, \mathbf{C}).$$

The following are the basic properties of L_2-cohomology:

(1) There are homomorphisms $H^i_{(2)}(M, \mathbf{C}) \to H^i(M, \mathbf{C})$, which need be neither injective nor surjective. (In some of the earlier literature, the *image* of this mapping was called L_2-cohomology. We wish to discourage this attitude.)

(2) $H^*_{(2)}(M, \mathbf{C})$ depends on the metric only to the extent that the L_2 spaces do. Quasi-isometric metrics (metrics g_1, g_2 for which there is a uniform estimate $Cg_1 \leqslant g_2 \leqslant C'g_1$ for some positive constants C, C') determine the same L_2-cohomology. (In practice, this says that for local calculations one may use the metric with the simplest formula in the quasi-isometry class.)

(3) If M is compact (with boundary,[3] even) all metrics are quasi-isometric, and the mappings in (1) are isomorphisms.

(4) [Hodge theorem] Let $\mathfrak{h}^i_{(2)}(M, \mathbf{C})$ be the orthogonal complement of $dL^{i-1}_{(2)}(M, \mathbf{C})$ in $Z^i_{(2)}(M, \mathbf{C})$. One can show that $\mathfrak{h}^*_2(M, \mathbf{C})$ is the space of solutions to the Laplace equation $(dd^* + d^*d)\phi = 0$ that satisfy (implicit) Neumann boundary conditions. (There are no boundary conditions if M is complete, without boundary.) Then

$$H^i_{(2)}(M, \mathbf{C}) \simeq \mathfrak{h}^i_{(2)}(M, \mathbf{C}) \oplus \left[\overline{dL^{i-1}_{(2)}(M, \mathbf{C})} / dL^{i-1}_{(2)}(M, \mathbf{C})\right].$$

The second factor on the right-hand side is either the zero space (if $dL^{i-1}_{(2)}(M, \mathbf{C})$ is closed) or is of infinite algebraic dimension. The former is the case if M is compact.

One can define the *reduced* L_2-cohomology

$$\overline{H}^i_{(2)}(M, \mathbf{C}) = Z^i_{(2)}(M, \mathbf{C}) / \overline{dL^{i-1}_{(2)}(M, \mathbf{C})},$$

which is always isomorphic to $\mathfrak{h}^i_{(2)}(M, \mathbf{C})$. A decomposition theorem for harmonic forms imparts a decomposition to the reduced L_2-cohomology.

The notion of L_2 localizes, and we can define L_2 complexes of sheaves as follows. Let $\overline{M}$ be *any* compactification of M as a Hausdorff topological space. Then $\mathfrak{L}^i_{(2)}(\overline{M}, \mathbf{C})$ is the sheaf on $\overline{M}$ associated to the presheaf that assigns to each open subset U of $\overline{M}$ the space $L^i_{(2)}(U \cap M, \mathbf{C})$. We observe

(5) $H^0(\overline{M}, \mathfrak{L}^i_{(2)}(\overline{M}, \mathbf{C})) \simeq L^i_{(2)}(M, \mathbf{C})$.

(6) $\mathfrak{L}^i_{(2)}(\overline{M}, \mathbf{C})$ is not necessarily a fine sheaf! To see what the problem is, let f be a cut-off function (part of a partition of unity on $\overline{M}$). We want $\phi \in L^*_{(2)}(M, \mathbf{C})$ to imply that $f\phi \in L^*_{(2)}(M, \mathbf{C})$. For the latter, we must have that

$$d(f\phi) = fd\phi + df \wedge \phi$$

is in L_2, and this requires $|df|$ to be a bounded function on M. Whether such f exists at the boundary points of M depends on $\overline{M}$.

(7) If $\mathfrak{L}^*_{(2)}(\overline{M}, \mathbf{C})$ is fine, then

$$\mathbf{H}^i\left(\overline{M}, \mathfrak{L}^*_{(2)}(\overline{M}, \mathbf{C})\right) \simeq H^i_{(2)}(M, \mathbf{C}).$$

More generally, one can replace $\mathbf{C}$ in the above by the locally constant sheaf of horizontal sections of a flat complex vector bundle equipped with a (not necessarily flat) Hermitian metric. If the bundle is a trivial line bundle, flatly though not metrically, one gets L_2-cohomology *with weights*.

3. Locally symmetric varieties. Let X be the symmetric space associated to the semisimple real algebraic group G; i.e., $X = G/K$, where K is a maximal compact subgroup of G. If G is defined over $\mathbf{Q}$, let Γ be a torsion-free arithmetic subgroup of G. The space X has G-invariant metrics, all of which are clearly in the same

[3] M is meant to be here a *Riemannian* manifold with boundary: the metric must extend to the boundary.

quasi-isometry class. We will consider the locally symmetric spaces $M = \Gamma \backslash X$, with the induced metric. For some groups G, X has an invariant complex structure, and then M becomes a complete Kähler manifold (in general noncompact), in fact a quasiprojective variety; in this case, one calls M a *locally symmetric variety*.

For the purposes of this article, we will not consider the general such M, but rather only the following examples:

A. $X =$ ball in $\mathbf{C}^n$ ($G = \mathrm{SU}(n, 1)$),

B. $M =$ Hilbert modular variety of dimension n ($G = \mathrm{SL}(2, \mathbf{R})^n$, with a strange structure over $\mathbf{Q}$).

We add a third example, which is not a complex manifold

C. $X = (2n)$-dimensional real hyperbolic space ($G = \mathrm{SO}(2n, 1)$).

A common feature of the three classes of examples is that the group G is of rank one over $\mathbf{Q}$.

Any locally symmetric space M has a compactification $\overline{M}$, which is a manifold with corners (if Γ is sufficiently small) [**BS**]. When the $\mathbf{Q}$-rank of G is one, there are no corners, so $\overline{M}$ is a compact manifold with boundary. We let M^* denote the space obtained by collapsing each connected component of the boundary of $\overline{M}$ to a point.

THEOREM. *In examples* A, B *and* C, *there is a natural isomorphism between* $H^i_{(2)}(M, \mathbf{C})$ *and the middle intersection homology* $IH^i(M^*, \mathbf{C})$.

By the local description of intersection homology [**GM**, §3], we must show two things. First, we must know that $\mathcal{L}^*_{(2)}(M^*, \mathbf{C})$ is a complex of fine sheaves. Since the singular points are isolated, this is evident. (However, $\mathcal{L}^*_{(2)}(\overline{M}, \mathbf{C})$ is *not* fine; this phenomenon is plainly visible in the case of the upper half-plane with the Poincaré metric.)

Secondly, we must show that for a fundamental system of deleted neighborhoods U' of a singular point

$$H^i_{(2)}(U', \mathbf{C}) \simeq \begin{cases} H^i(U', \mathbf{C}) & \text{if } i \leqslant n-1, \\ 0 & \text{if } i \geqslant n. \end{cases} \tag{8}$$

We may take U' to be a deleted collar of a boundary component N, shown in the figure. Of course, we have differentiably $U' \simeq (0, 1) \times N$, but the metric is *not* a product.

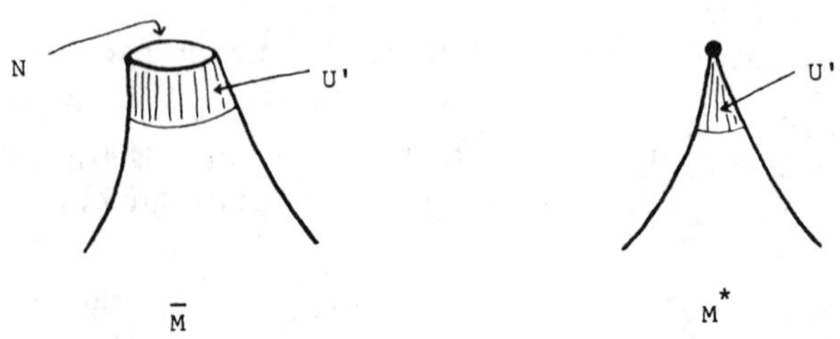

Let r denote the variable on $(0, 1)$. We have included example C especially because the metric has a simple formula, namely

$$(dr/r)^2 + r^2 g_N, \tag{9}$$

where g_N is a metric on N. We will sketch the proof of our theorem in this case only. For the other cases, the argument is somewhat more complicated, though similar in spirit, and we refer the reader to [**Z2**, §§3–6].

The formula (9) expresses U' as a so-called *warped product*, that is, a product of smooth manifolds $Y \times N$ equipped with a metric of the form

$$g_Y \oplus w^2 g_N, \tag{10}$$

where w is a positive smooth function on Y. We write $Y \times_w N$ for the Riemannian manifold. We consider the general problem of determining the L_2-cohomology of a warped product.

Let ϕ be a j-form on Y, and ψ a k-form on N, and consider the form $\phi \otimes \psi$ on $Y \times_w N$. One has by (10)

$$|\phi \otimes \psi| = w^{-k} |\phi| |\psi| .$$

Since $dV_{Y \times_w N} = w^\nu \, dV_Y dV_N$ ($\nu = \dim N$), we see that

$$\|\phi \otimes \psi\|^2_{(2)} = \int_Y |\phi|^2 w^{\nu - 2k} \, dV_Y \int_N |\psi|^2 \, dV_N .$$

Thus we obtain a decomposition of graded Hilbert spaces

$$L_2^*(Y \times_w N, \mathbf{C}) \simeq \bigoplus_k \left(L_2^*(Y, \mathbf{C};\, w^{\nu-2k}) \,\hat{\otimes}\, L_2^k(N, \mathbf{C})[-k] \right), \tag{11}$$

where L_2^i (subtly distinguishable from $L^i_{(2)}$) denotes the L_2 space of i-forms, $L_2^j(Y, \mathbf{C};\, w^{\nu-2k})$ denotes the appropriate weighted L_2 space, and $\hat{\otimes}$ denotes completed tensor product (Hilbert space completion of the algebraic tensor product). The following is contained in [**Z2**, §2].

PROPOSITION. *Suppose that* $Y \times_w N$ *is the interior of a complete Riemannian manifold with boundary, d has closed range on N, and w is a bounded function on Y. Then* (11) *induces an isomorphism*

$$H^*_{(2)}(Y \times_w N, \mathbf{C}) \simeq \bigoplus \left(H^*_{(2)}(Y, \mathbf{C};\, w^{\nu-2k}) \,\hat{\otimes}\, H^k_{(2)}(N, \mathbf{C})[-k] \right). \tag{12}$$

We apply the Proposition to U', where $Y = (0, 1)$, $g_Y = (dr/r)^2$ and $w = r$. It is a good idea to reparametrize the interval by arc-length; let $u = -\log r$. In terms of the variable u, $Y = (0, \infty)$, $g_Y = (du)^2$ and $w = e^{-u}$. To compute (12), we must therefore determine the L_2-cohomology of the half-line $(0, \infty)$ with weights e^{-lu}. One computes (see [**Z2**, (4.51)])

$$H^0_{(2)}((0, \infty), \mathbf{C};\, e^{-lu}) = \begin{cases} \mathbf{C} & \text{if } l > 0, \\ 0 & \text{if } l \leq 0; \end{cases}$$

$$H^1_{(2)}((0, \infty), \mathbf{C};\, e^{-lu}) \begin{cases} = 0 & \text{if } l \neq 0, \\ \text{is infinite dimensional} & \text{if } l = 0. \end{cases}$$

It is also true that $\overline{H}^1_{(2)}((0,\infty),\mathbf{C}) = 0$. Since ν is odd ($\nu = 2n-1$), the infinite dimensional $H^1_{(2)}$ above does not enter into (12), and we see that (8) holds.

In case B of the Theorem, the symmetric space X is the product of n copies of the upper half-plane. For such M, the L_2-cohomology has been, in effect, computed in [**MS**, §7] (cf. [**Z3**, (5.22)]). We obtain the following corollary of our Theorem.

COROLLARY. *Let M be a Hilbert modular variety (not compactified) of dimension n. Let $\omega_1,\dots,\omega_n$ be the Kähler classes (volume forms) of the factors of X, W the* **C** *vector space spanned by the ω_j's, and Λ^* the truncated polynomial algebra generated by W (with $\Lambda^1 = W$, and $\omega_j^2 = 0$). Then for $i < n$,*

(a) *$H^i(M,\mathbf{C}) \simeq \Lambda^m$ if $i = 2m$,*

(b) *$H^i(M,\mathbf{C}) = 0$ if i is odd.*

The dual assertion is: for $i > n$,

(c) *$H_i(M^*,\mathbf{C}) \simeq \Lambda^{n-m}$ if $i = 2m$,*

(d) *$H_i(M^*,\mathbf{C}) = 0$ if i is odd.*

A locally symmetric variety has a Baily-Borel-Satake compactification that is a projective variety [**BB**]. In examples A and B, it is homeomorphic to the space M^*. We have conjectured that our Theorem remains valid on any locally symmetric variety, if we take M^* to be this compactification of M.

REFERENCES

[**BB**] W. Baily and A. Borel, *Compactification of arithmetic quotients of bounded symmetric domains*, Ann. of Math. (2) **84** (1966), 442–528.

[**B1**] A. Borel, *Stable real cohomology of arithmetic groups*, Ann. Sci. Ecole Norm. Sup. **7** (1974), 235–272.

[**B2**] ______, *Stable real cohomology of arithmetic groups*. II (to appear).

[**BS**] A. Borel and J.-P. Serre, *Corners and arithmetic groups*, Comment. Math. Helv. **48** (1973), 436–491.

[**C**] J. Cheeger, *On the Hodge theory of Riemannian pseudomanifolds*, Proc. Sympos. Pure Math., vol. 36, Amer. Math. Soc., Providence, R. I., 1980, pp. 91–146.

[**CGM**] J. Cheeger, M. Goresky and R. MacPherson, *The L^2-cohomology and intersection homology of singular algebraic varieties*, Seminar on Differential Geometry (Yau (ed.)), Princeton Univ. Press, Princeton, N. J., 1982, pp. 303–340.

[**D**] P. Deligne, *Théorie de Hodge*. III, Inst. Hautes Études Sci. Publ. Math. **44** (1973), 5–77.

[**G**] H. Garland, *A finiteness theorem for K_2 of a number field*, Ann. of Math. (2) **94** (1971), 534–548.

[**GM**] M. Goresky and R. MacPherson, *Intersection homology*. II (to appear).

[**MM**] Y. Matsushima and S. Murakami, *On vector bundle valued harmonic forms and automorphic forms on symmetric Riemannian manifolds*, Ann. of Math. (2) **78** (1963), 365–416.

[**MS**] Y. Matsushima and G. Shimura, *On the cohomology groups attached to certain vector valued differential forms on the product of upper half planes*, Ann. of Math. (2) **78** (1963), 417–449.

[**Z1**] S. Zucker, *Hodge theory with degenerating coefficients: L_2 cohomology in the Poincaré metric*, Ann. of Math. (2) **109** (1979), 415–476.

[**Z2**] ______, *L_2 cohomology of warped products and arithmetic groups*, Invent. Math. (to appear).

[**Z3**] ______, *Locally homogeneous variations of Hodge structure*, Enseign. Math. (2) **27** (1981), 243–276.

[**Z4**] ______, *Hodge theory and arithmetic groups* (to appear).

INDIANA UNIVERSITY, BLOOMINGTON

INSTITUTE FOR ADVANCED STUDY, PRINCETON

ABCDEFGHIJ–CM–89876543